Advances in Cryogenic Engineering

VOLUME 41, PART B

A Continuation Order Plan is available for this series. A continuation order will bring delivery of each new volume immediately upon publication. Volumes are billed only upon actual shipment. For further information please contact the publisher.

A Cryogenic Engineering Conference Publication

Advances in Cryogenic Engineering

VOLUME 41, PART B

Edited by

Peter Kittel

NASA–Ames Research Center
Moffett Field, California

PLENUM PRESS · NEW YORK and LONDON

The Library of Congress cataloged the first volume of this title as follows:

Advances in cryogenic engineering. v. 1–
New York, Cryogenic Engineering Conference; distributed by Plenum Press, 1960–
v. illus., diagrs. 26 cm.
Vols. 1– are reprints of the Proceedings of the Cryogenic Engineering Conference, 1954–
Editor: 1960– K. D. Timmerhaus.

1. Low temperature engineering—Congresses. I. Timmerhaus, K. D., ed. II. Cryogenic Engineering Conference

TP490.A3 660.29368 57-35598

Proceedings of the 1995 Cryogenic Engineering Conference,
held July 17–21, 1995, in Columbus, Ohio

ISBN 0-306-45300-2

A Division of Plenum Publishing Corporation
233 Spring Street, New York, N.Y. 10013

10 9 8 7 6 5 4 3 2 1

Printed in the United States of America

CRYOFUELS, NOW AND IN THE FUTURE

J.A. Barclay[1]

[1]Cryofuel Systems, University of Victoria
Victoria, B.C. V8W 3P6, Canada

ABSTRACT

Our desire to use domestically available, cleaner, and less expensive vehicular fuels is creating a market demand for new business related to the supply of the alterative fuels, propane (LPG), natural gas (CNG/LNG), and hydrogen (CH_2/LH_2). These three fuels have the common attribute of equilibrium boiling points that are less than ambient temperature, i.e., -42 °C, -161 °C, and -253 °C, respectively, and in that sense, they are "cryofuels". Each of these fuels is either readily or potentially readily available in North America and many other countries of the world. Although properly designed engines that combust these fuels with substantially less harmful emissions than from gasoline and diesel are being integrated into several sizes of vehicles and substantial domestic supplies of cheap natural gas and propane are available, the present use of cryofuels in the transportation sector is negligible. To change this situation and enable incremental conversion to gaseous cryofuels, a distribution of fleet-size refueling systems that provide LPG, CNG, LNG, hythane, CH_2, and LH_2 in an integrated operation with the existing supply network are required. Our analysis shows that the creation of improved conversion economics is essential to initiate rapid fleet conversion. The price of the liquid or compressed form of cryofuels to fleet owners/operators or individuals depends on several factors including the capital cost and operating cost of the refueling stations. Such stations include many types of cryogenic materials and equipment such as 9% Ni steel, high strength Al, new composites, highly specific adsorbents, phase separators, liquefiers, cryogen pumps, vaporizers, cryogen and high-pressure storage tanks, pressure vessels, and dispensers. With existing stable natural gas feedstock prices, inexpensive dedicated alternate fuel engines/vehicles, and less expensive refueling stations, dispensed cryofuels are or can be significantly less expensive than gasoline and diesel for fleet owners/operators. Such price differentials will pay for fleet conversion, staff retraining, refueling systems, etc., and provide substantial returns on fuel conversion investments. This paper will summarize the present status of the transition to cryofuels, identify the technical and market barriers, and discuss some business opportunities for new cryogenic materials, equipment, and systems.

INTRODUCTION

Perspective for Opportunities

New opportunities in the fields of cryogenic engineering equipment and materials are constantly in review as managers in companies search for new revenue streams and the means to compete globally with differentiated products. Academics are constantly trying to find new ideas that land research contracts and provide great thesis projects to the best students. Federal and state staff are searching for better methods to implement policy

Advances in Cryogenic Engineering, Vol. 41
Edited by P. Kittel, Plenum Press, New York, 1996

decisions of government agencies. There is an interesting mix of visions in these views into the future; some are near term, i.e., a few years, some are far term with plans extending a decade or two; some must fit into the corporate bottom line within a few quarters; others are driven by the average tenure of a graduate student; and some are driven by the discovery of new knowledge. These visions into the future to search for new opportunities are a dynamic balance of technology-push versus market-pull influences.

Technology-push predictions are done by experts from time to time for several reasons.[1] General trends in new technology such as the influence of the computerized materials databases, faster more powerful, and less expensive computers, and flexible manufacturing systems are reasonably obvious but innovations in many areas such as advanced energy systems are harder to predict. It is interesting to note that one reviewer just preceding the discovery of high temperature superconductors mentioned that he thought some small increase in superconducting transition temperature would occur in the 90's. Perhaps the best lesson to learn from many years of exciting new technological developments is that if an environment for innovation is established, provided with sufficient capital, plant, and human resources, innovative new technology will be forthcoming! Creativity is more often shown in little steps rather than in giant leaps so identification of areas where progress will likely occur can start by identifying areas where R&D resources are being allocated to talented people with good facilities.

Market-driven predictions are difficult in any area of high technology. Predictions in cryogenic applications are more complicated because the products tend to have a higher level of sophistication than many other product areas. The cryogenic applications, ranging from commodities to materials to biological and medical products include:

- production of industrial gases;
- food preservation, especially freezing and storage;
- biology and medicine such as cryopreservation and cryosurgery;
- cryopumping;
- industrial applications such as metal processing and scrap reduction;
- applied superconductivity in areas such as magnets, medical diagnostics, electronics, etc; and
- cryogenic fuels for storage, marine transportation, peakshaving, and of course, fuel.

The predictors of the sizes of cryogenic equipment markets for these applications have traditionally had a difficult time establishing expected revenues. I suggest that some published predictions must have caused a great deal of laughter in certain executive circles and board rooms, although perhaps excitement in some lesser knowledgeable in cryogenics. For example, at a recent cryogenic engineering conference, a review of the prospects for commercialization of high and low temperature superconductors (LTSC/HTSC) was presented.[2] The revenues from LTSC/HTSC products were stated as a few million dollars in 1994 and projected to grow to US$ 1.5 billion annual revenues by the year 2000. Such an increase requires an average compounded annual growth rate of over 300%. Even accepting the difficulty of accurately estimating future markets in high technology products, it is safe to say that most cryogenic equipment and materials market projections are extremely optimistic. In my experience, excluding commodities such as liquefied industrial gases, food preservation, and cryogenic fuels, the present, annual global cryogenic materials and equipment market is realistically in the range of US$ 200 - 300 million and perhaps growing at a few per cent per year.

My goal in this paper is to suggest new growth opportunities for one cryogenics application with enough market-driven near-term perspective and yet with enough "blue sky" promise to make it exciting to government laboratory staff, corporate engineers and managers, and academic faculty and students. That area is cryogenic fuels or "cryofuels" for transportation applications.

Fuels and their Use

In 1993 the United States used ≈ 85 quads [quadrillion Btu's or exajoules] of energy.[3] The primary energy sources [fuels] for this energy included coal, natural gas, petroleum, hydro, nuclear, and other lesser sources. The users of this energy are conveniently segmented into four or five economic sectors including industrial, residential, commercial, electricity generation [sometimes called utilities], and transportation. Figure 1 schematically shows the mix of energy sources and users and the totals for each fuel source and use sector. Several interesting facts are apparent from this figure. For instance, the demand for energy is

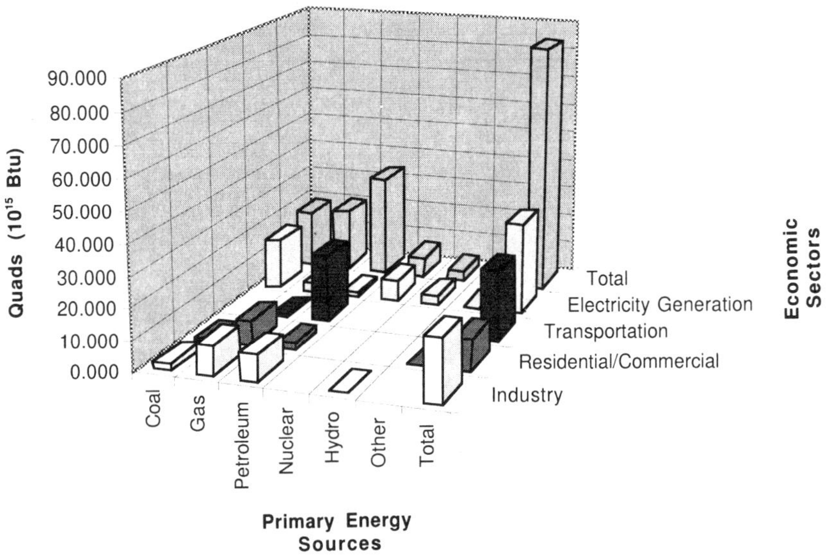

Figure 1. The United States energy use in 1993 segmented into primary energy sources and economics sectors (users).

relatively balanced among the user sectors with electricity generation largest and the combined residential/commercial sector the smallest. Coal, natural gas, and petroleum are the three dominant energy sources with natural gas supplying about 25% and petroleum supplying 42% of the total energy. Further, Figure 1 shows the total reliance [99.8%] of the transportation sector on petroleum. Indeed, the transportation sector consumes 63% of the petroleum and only 0.02% of the natural gas used in the U.S. [4,5]

Transportation Energy Users

From 1990 a survey initiated by the Natural Gas Vehicle/American Gas Association Coalition one can surmise that there are ≈ 210 million vehicles registered in North America (almost 190 million in the U.S., 15 million in Canada, and an estimated 5 million in Mexico).[6] This number is increasing at ≈ 4% per year in North America. These vehicles are at some 110 million different locations and annually consume ≈ 150 billion US gallons of fuel.[6] Several distinct segments of vehicles were defined in this report. One, called 'identified business fleets', includes government, business, and trade vehicles in fleets of 2-9 vehicles. Another segment includes the larger fleets with ≥ 10 vehicles. These two segments comprises ≈ 16% of total registered vehicles at ≈ 11 million locations.[6] Most 'heavy duty vehicles' are in the 10^+ fleets and they consume over 22% of the total fuel used. The largest segment of registered vehicles, some 176 million vehicles, is composed primarily of cars and light trucks called 'light duty vehicles'. These data on fleet distribution and fuel usage are summarized in Figure 2.

The Winds of Change

Although gasoline and diesel are firmly entrenched as the fuels of choice in North America, strong market forces are driving consideration of alternative fuels in the transportation fuel arena. These forces include:

- Favourable economics; users want a less expensive fuel than gasoline or diesel. Some of the alternative fuel choices are cheaper than gasoline or diesel. For example, in North America the price of natural gas is about 75% of the price of crude oil at the well head.

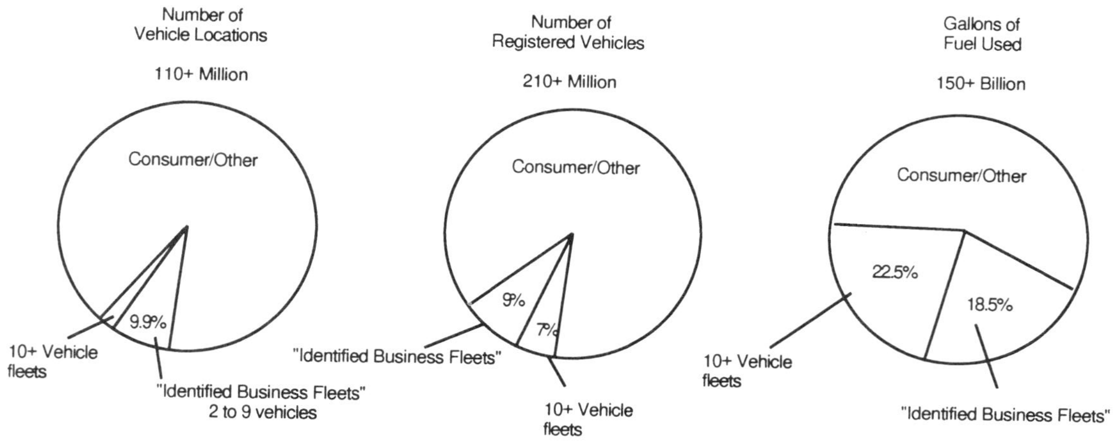

Figure 2. Pie charts indicating distribution of registered vehicles and fuel used in North America.

- Plentiful domestic supply; over 50% of the U.S.'s crude oil is imported from the middle east and concern over this dependence on foreign supply inspired the National Energy Security Act of 1992. This act mandates federal, state, and municipal fleets plus alternative fuel suppliers to convert their fleets to alternative fuels according to a near term schedule. It also specifies that reformulated gasoline and clean diesel are not alternative fuels! Canada presently supplies most of its own petroleum but it will be importing larger fractions of its oil in the next decade. Both Canada and the U.S. have enormous reserves of natural gas.
- Clean air regulations; the environmentally harmful emissions from the 150 billion gallons of high-carbon fuel combusted in vehicles per year are an important issue with the public. Legislation aimed at cleaning up emissions from vehicles has been enacted at the federal, state and municipal levels. Examples include the Clean Air Act Amendments of 1991 that targets fleets of 10 or more vehicles in certain highly polluted cites; the Energy Policy Act of 1992 mentioned above; and the Presidential Executive Order in 1993 to increase alternative fuel use in federal fleets. California and Texas had early plans to reduce emissions. Many cities such as Denver also have emission reductions plans.[7]

Selection and use of gasoline and diesel as vehicular fuels are no longer the "no-risk" choices they once were.

Alternative Fuels

The choices of alternative fuels includes a variety of alcohols, natural gas in two forms, hydrogen and its mixture with methane, and propane. Depending upon the document and/or proponent, other substitutes for traditional fuels include biodiesel, gasohol, gasoline with MTBE additives, superclean diesel, and reformulated gasoline. The U.S. Department of Energy recently published a report that lists the following as alternative fuels:[8]

- M-85 - a mixture of 85% methanol and 15% gasoline;
- M-100 - a fuel consisting of 100% methanol;
- E-85 - a mixture of 85% ethanol and 15% gasoline;
- E-95 - a mixture of 95% ethanol and 5% gasoline;
- LPG - liquefied petroleum gas or propane;
- CNG - compressed natural gas;
- LNG - liquefied natural gas; and
- Hydrogen.

This list could include hythane, a mixture of approximately 10% hydrogen in natural gas. Liquid ethane has also been suggested as a vehicular fuel.[9] Further, the list could be subdivided into compressed, adsorbed, and liquid hydrogen forms and have variations in LNG from over 99^{+}% methane concentration to approximately 90% methane concentration with the remainder ethane, other heavier hydrocarbons and nitrogen.

CRYOFUELS, THE PRESENT SITUATION

Definition of Cryofuels

The fuels in the list above separate nicely into those fuels that are liquids and those that are gaseous at room temperature and atmospheric pressure. It is these latter fuels that are of interest here because they must be cooled below room temperature and insulated and/or pressurized to be maintained as a liquid. In this sense, they are cryogenic fuels or "cryofuels" although the strict definition of "cryogenic" arbitrarily refers to temperatures below about -150 °C. These gaseous fuels are also of special interest because among all the alternative fuels, only LPG, CNG, and LNG are less expensive than gasoline or diesel fuel [to my knowledge, the price of liquid ethane as a fuel has not been established but it is probably priced close to LPG where both are available].

Characteristics of Cryofuels

The physical, thermal, combustive, safety, chemical, magnetic, mechanical, and associated properties of all the cryofuels are well known.[10] Typical values of the some of the key properties of interest for this paper are collected in Table 1.

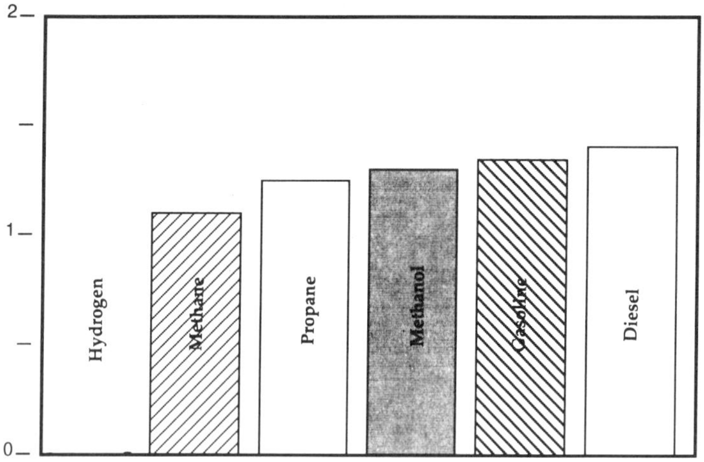

Figure 3. The amounts of CO_2 generated (kmol CO_2/GJ) for different fuels combusted stoichiometrically. {Ref. 12}

Note that the boiling point of propane at 101 kPa is -42 °C, that of ethane is -88 °C, that of methane is -161 °C, and that of hydrogen is -253 °C. The freezing point of the hydrocarbon cryofuels is almost the same at ≈ 90 K which means that minor levels of one of these in the others will not freeze out in the heat exchangers during liquefaction of any of the light hydrocarbon cryofuels. One property not shown in Table 1 is the density of gaseous fuel at room temperature and 101 kPa pressure; gasoline and propane are heavier than air while ethane, methane, and hydrogen are all lighter than air. Cold methane gas has a positive buoyancy once it warms to about -110 °C. This high buoyancy makes the natural gas and hydrogen safer if an accidental spill or vehicular crash occurs. It is well known that in the Hindenburg crash, most of the people who died were killed from the fall or from burning diesel fuel rather than from burning hydrogen. The flammability limits for the light hydrocarbons are remarkably narrow and add to their safety factor along with the higher ignition temperatures.

Properly designed engines for any of the cryofuels produce significantly less emissions such as CO, NO_x, particulates, volatile organic compounds, etc., than combustion of

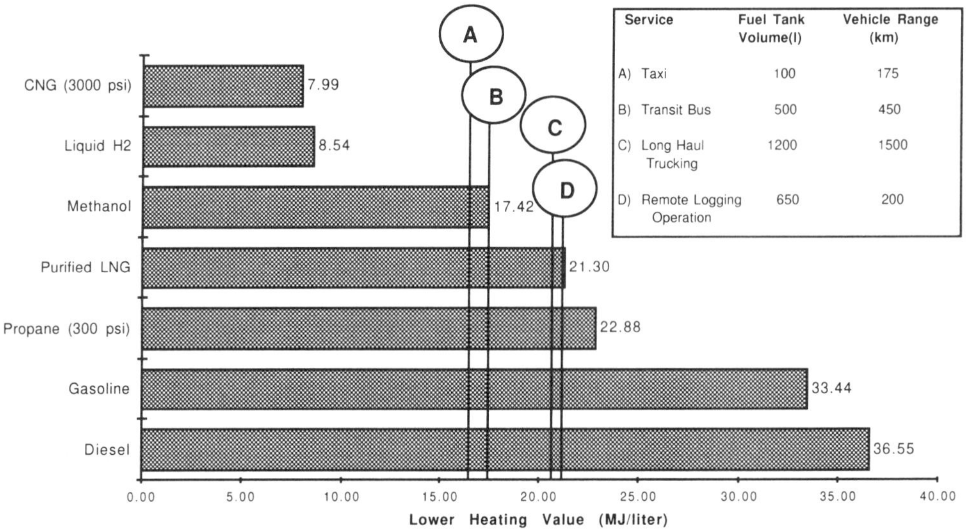

Figure 4. A comparison of the volumetric energy densities of cryofuels with gasoline and diesel. Four examples of vehicle applications, their estimated range and fuel tank requirements for LNG are included.

Table 1. Selected Properties of Cryofuels and Gasoline

Property	Propane	Ethane	Methane	Hydrogen	Gasoline
Molecular Wt	44.097	30.07	16.043	2.016	107
Boiling Pt. (K) at 101.3 kPa	231.07	184.55	111.63	20.268	310-478
Freezing Pt (K) at equil. pressure	85.52	90.35	90.69	13.80	180-220
Density of liquid at NBP (kg/m3)	507.3	357.8	422.3	70.8	700
Density ratio - NBP liquid to NTP gas	272	280	649	845	156
Heat of vaporization (kJ/kg) @ NBP	425	491	510	446	309
Heat of combustion (kJ/kg) and BTU/gal	50,368 91,500	51,950 94,375	55,530-50k 88,400	120,000 30,000	48,000 120,000
Hot surface ignition temperature (C)	399	------	632	427	360
Flammability limit 310 k & 101 kPa	2 - 9.5%	2.9 - 13%	5.0 - 15%	higher	higher+

gasoline or diesel.[11] NO_x reduction does require lean burn engines in the case of natural gas because of its higher ignition temperature at stoichiometric conditions but if this is done, the reductions in NO_x are ten-fold or more compared to combustion of gasoline or diesel. On an energy equivalent basis, the lower the carbon content of the cryofuel, the less carbon dioxide it emits as shown in Figure 3.[12]

The heats of combustion are quoted with higher or lower heating value; usually the lower heating value is taken in engine use because it is not easy to take advantage of the latent heat associated with the condensation of steam produced by combustion in this case. The lower volumetric energy density relative to gasoline or diesel is the Achilles' heel of the cryofuels. To obtain an equivalent range in vehicles with engines of equal thermal efficiency,

Table 2. A Comparison of the On-board Storage Properties of Cryofuels with Diesel

	Typical Diesel[a]	Typical Gaso-line	Methanol (M100)	Ethanol (E100)	LPG (HD-5)	CNG[b,c] (Methane)	LNG[d,e] (Methane)
Density (lb/gal)	7.0	6.2	6.63	6.59	4.2	1.4	3.4
Energy content, LHV (BTU/lb) (BTU/gal)	18,300 128,100	18,900 117,200	8.600 57,000	11,600 76,300	19,900 83,000	21,500 30,000	21,500 73,000
Relative fuel volume[f]	1	1.1	2.2	1.7	1.5	4.3	1.7
Relative fuel tank external volume[f,g]	1	1.1	2.2	1.7	1.5	4.7	2.0
Relative full fuel tank weight[f,g]	1	1.0	2.1	1.6	1.2	3.7	1.4

[a] Typical No. 2 diesel fuel.
[b] CNG (methane) at 3,000 psig.
[c] Possible non-isothermal tank filling neglected in tank volume and mass estimates.
[d] LNG (methane) saturated liquid at 20 psig and -238°F.
[e] Need for vapor space in "full" tank neglected in tank volume and mass estimates.
[f] For equivalent range; all engines assumed to have equal thermal efficiency.
[g] For typical 30-gal medium-duty truck tank.

larger volumes of the cryofuels are required, e.g., about 1.7 gallons of LNG has the same energy content as 1 gallon of diesel. [Note - all comparisons in this paper are on an energy equivalent basis, i.e., a diesel equivalent gallon or DEG.] Figure 4 shows the relative volumetric energy densities for most of the cryofuels in comparison to gasoline and diesel.

The on-board vehicular storage tanks for cryofuels other than LPG must be insulated and usually double-walled. All the fuel tanks have a pressure vessel rating capable of at least 1.67 MPa [250 psi]. These requirements give a volume and weight penalty for the cryofuels on-board a vehicle. Table 2 shows a comparison to diesel for several of the cryofuels [hydrogen needs a tank about five times larger than the equivalent diesel tank but in this case, fuel cells or gas turbine engines may increase the engine efficiency to help compensate for this large size penalty].

Enabling Conditions for a Transition to Cryofuels

In spite of the attractive features of the cryofuels, there is not a stampede of fleet owners to convert their vehicles. The big question is, 'why not'? Several reasons have been identified as barriers that must be overcome to shift from gasoline and diesel to alternatives among the cryofuels, especially to natural gas. Several key factors are being addressed by:

- Making vehicle conversion technology and OEM vehicles available and introducing mono-fuel engines with high efficiency;
- Insuring that pollution is much less by proper certification and testing procedures of engines and vehicles - natural gas is recognized as one of the least polluting fuels, in both regulated and unregulated emissions, if properly combusted. LPG is not far behind natural gas, hydrogen is better than natural gas;
- Educating users about alternative fuels and sharing in the fleet owner's risk of conversion;
- Insuring that favourable economics are proven even when feedstock costs,vehicle conversion or surcharge costs, refueling system costs, training, and taxes are taken into account; and
- Establishing refueling infrastructure!

FAVOURABLE ECONOMICS FOR USERS IS THE KEY

Although a combination of many essential factors are required to change from gasoline and diesel, attractive overall economics compared to other fuels will ultimately cause fleet owners to make the transition to cryofuels as transportation fuels. This is supported by reports that show fleet managers usually select their fuel on the basis of three generic sets of factors:[6]

- **Crucial costs:** fuel cost, maintenance, assurance of future fuel supply and vehicle cost premium;
- **Operations costs:** range of autonomy, on-site refueling, public refueling, warranties and emission reductions; and
- **Vehicle specific factors:** time to refuel, fuel tank size/weight and vehicle resale value.

Diesel/gasoline are formidable competition

The prices of gasoline and diesel fuels at public pumps across the U.S. is about \$1.35/DEG and about US\$ 0.05-0.10/DEG more expensive in Canada. Fleet owners/operators who use enough fuel to keep a 1000 DEG/day refueling system operating over 80% of the time also use enough diesel fuel to get a price reduction of at least \$0.10-0.20/DEG for a net price of about \$1.15/DEG in bulk. If large fleet owners qualify for non-taxed fuel, they can purchase bulk fuel at prices as low as ~US \$.60/DEG. The familiarity with the fuel and completely established refueling infrastructure makes it especially attractive to stay with gasoline or diesel. Fleets owners are not in the fuel business but realize that a reliable, affordable fuel supply that meets emissions regulations is essential to their successful business operations.

Table 3. CNG/LNG Fuel Cost Breakdown for Industrial Fleets

Cost Component	Percentage of Fuel Cost	Cost, $/DEG
System Cost	13.1%	0.138
Site Preparation Cost	1.0%	0.01
Vehicle Conversion Cost	11.8%	0.124
NG Feedstock Cost	35.6%	0.374
Liquefaction Cost	8.4%	0.088
Fixed Running Cost	4.4%	0.046
Taxes	25.7%	0.27
TOTAL	100%	1.05

Notes:
1. 90% of the system's capacity is used to deliver LNG at 60 psia;
2. Feedstock price is fixed at $2.50 (US) per Mcf;
3. Annual operating cost is $15,000 (US) per annum;
4. Interest rate and inflation rate are 8% and 2% respectively;
5. Refueling system cost of $ 350,000 with a Payback period of 10 years;
6. Annual operation of 325 days per year;
7. 16 HDV's converted to LNG at US$ 18,000 per vehicle;
8. Heavy duty vehicle operation of 300 miles per day at 5 mpg; and
9. Road taxes of $0.27 (US) per DEG.

LPG and LNG Prices can Compete

Fleet owners have consistently identified fuel costs as the most important factor influencing their decision to convert to an alternate fuel. Among the numerous ways to calculate return on investments made to obtain a supply of alternate fuels, the simplest comparison among the fuel choices is to calculate the cost of an energy equivalent gallon of CNG or LNG from the refueling system for different fleet types with all costs relevant to fleet owner operations included. Fleet owners with different types of fleets analyze all of the cost factors specific to their own perspective, e.g., some fleets may not have to pay taxes and others may be eligible for financial assistance from the state or federal level. Factors considered by essentially all fleet owners include:

- Refueling system capital cost, utilization, financing, and payback period;
- Site preparation costs;
- Operations and maintenance (O&M) costs; and
- Feedstock costs.

To illustrate the contributions of the various cost components to the final fuel cost, two situations have been analyzed for natural gas. They are (1) the basic refueling system without taxes or vehicle conversion costs, and (2) the industrial fleet where vehicle conversion costs and taxes are included in the cost analysis.

• **Basic System Perspective.** In this case the baseline fuel price includes only the system capital cost of a stand-alone refueling system, the costs of O&M, and average feedstock costs.[14] The LNG fuel price obtained in this case [the quantitative assumptions are shown in the footnotes of Table 3] is about US$ 0.66/DEG and the CNG fuel price is about 10% less than LNG per equivalent energy content.

• **Industrial fleet owner/operator.** Vehicle retrofit costs for LPG and NG are well known as are the surcharges that the automobile manufacturers charge for mono-fuel vehicles. These costs must be recoverable by the fleet owner from the fuel price differential and/or savings in maintenance. In this worst case scenario for natural gas, the vehicle conversion costs and taxes are also included in the cost of fuel to give a comparison to diesel fuel in states and provinces where CNG/LNG are or will be taxed.[13] No incentive packages or bulk discounts are included. A breakdown of the fuel cost elements for the industrial fleet owner who has to bear all these costs for CNG or LNG is given in Table 3.

As shown in Table 3, with an advanced CNG/LNG refueling system such as illustrated in Figure 5, reasonable feedstock prices, and good yet inexpensive vehicle conversion kits [some estimates for conversion costs may be double the number we have assumed], the comparable price of LNG is $1.05/DEG. This price does include taxes but does not include

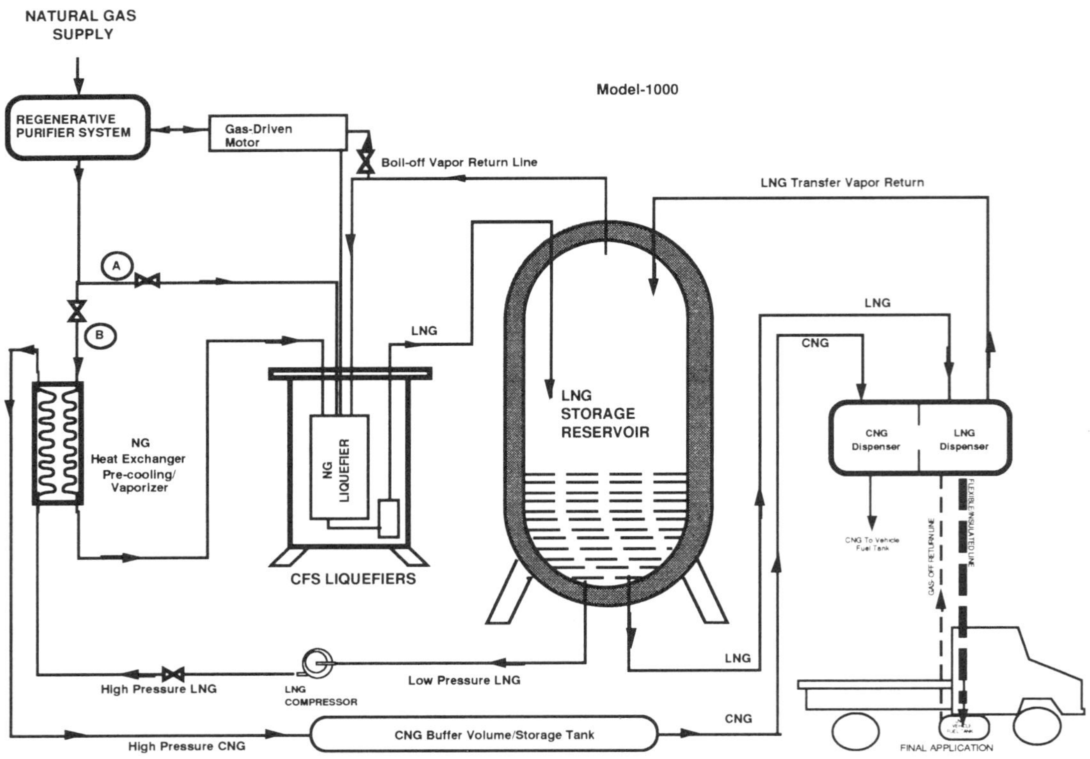

Note: All LNG and cold gas lines are insulated

Figure 5. A schematic of Cryofuel System's integrated, fleet-size CNG/LNG Refueling System.

incentives so it is a conservative estimate. Even with all these cost elements, the small industrial or commercial owner has attractive fuel economics if he/she uses sufficient CNG/LNG to fully utilize his/her fleet of vehicles and the refueling system. If taxes are reduced as they are in some states and all provinces in Canada and some of the existing incentives are taken, the price differential in favor of LNG or CNG is even greater. LPG prices are approximately the same as LNG in many places in Canada and the U.S. and the vehicle conversion kits are less expensive than for NG. So in places where LPG is available, it too offers an economic choice over gasoline or diesel.

Unfortunately, the largest fleet owner group in North America is the single user with a short commute each day. This user will not find switching to LPG, CNG, or LNG attractive with the present surcharge on LPG, CNG, and/or LNG vehicles. The LPG conversion is inexpensive enough that it may make sense even for the smaller user. The "single-vehicle" fleet owner simply does not use enough fuel in the life of the vehicle to pay back the initial costs of the vehicle conversion from the cryofuel savings. This situation will improve as the demand for mono-fuel vehicles increases and more refueling stations are established nation-wide. If there is a major disturbance in the middle east where most of the U.S. petroleum is produced, this situation will change dramatically because the price of gasoline and diesel will go up sharply!

Hydrogen fuel prices present a major barrier

According to a paper given at the 5th Canadian National Hydrogen Association meeting[19], the cost of an energy equivalent gallon of liquid hydrogen is approximately US$ 6.00/DEG. This cost is dominated by the steam reformer for methane to obtain the feedstock gaseous hydrogen and especially by the cost of liquefaction by existing gas-cycle technology. Clearly, a breakthrough is needed in this area to reduce the cost of the fuel.

REFUELING INFRASTRUCTURE IS ESSENTIAL

Although making the price of the alternative fuel less than gasoline or diesel is a key step, a remaining key barrier impeding the use of LPG and natural gas as a transportation fuel is the lack of cost-effective refueling infrastructure that interfaces between the existing propane or natural gas distribution systems and fuel tanks in fleets of vehicles.[8,15,16] For LPG, the refueling infrastructure integrates reasonably well with existing gasoline stations and is available in about 3300 sites.[17] For natural gas, about 1,040 CNG compressor-based stations exist in the U.S., and about 25 LNG stations exist.[17] Clearly, this is a small fraction of the ~200,000 gasoline and diesel refueling systems across North America. Liquid hydrogen supply infrastructure exists via tanker trucks from the existing merchant plants. Although no public distributed LH_2 refueling stations exist in the U.S., a liquid hydrogen refueling system has been recently built and demonstrated in Germany[18] and work is being done at Los Alamos National Laboratory on design/analysis of LH_2 refueling stations. [See proceedings of the 1995 CEC.]

New Distributed CNG/LNG Refueling Stations

To meet the CNG/LNG cost target in Table 3, it is essential to have an inexpensive, fleet-size refueling system with a local liquefier that has an efficiency of about 22% of ideal. Integrated CNG/LNG refueling systems schematically shown in Figure 5 that meet these specifications are being developed. Such a system takes natural gas from a pipeline network, purifies and enriches the gas as necessary, liquefies the gas, stores the liquid, and delivers CNG or LNG to vehicular fuel tanks under automated control.

This refueling system is capable of operating in several different modes with various feed gas pressures and fleet refueling needs. Two operating modes are:

- **Low pressure NG feed /LNG-only product** - The NG is purified and flows directly to the liquefier, where it is cooled to form a liquid at 111-129 K ($\approx$ -150 °C, depending upon operating pressure). The LNG is stored in an insulated storage tank and then transferred to the vehicle by the LNG dispenser.
- **Low pressure NG feed /LNG and CNG products** - In this case, the purified NG is precooled in vaporizer/heat exchanger and then liquefied and stored. CNG is

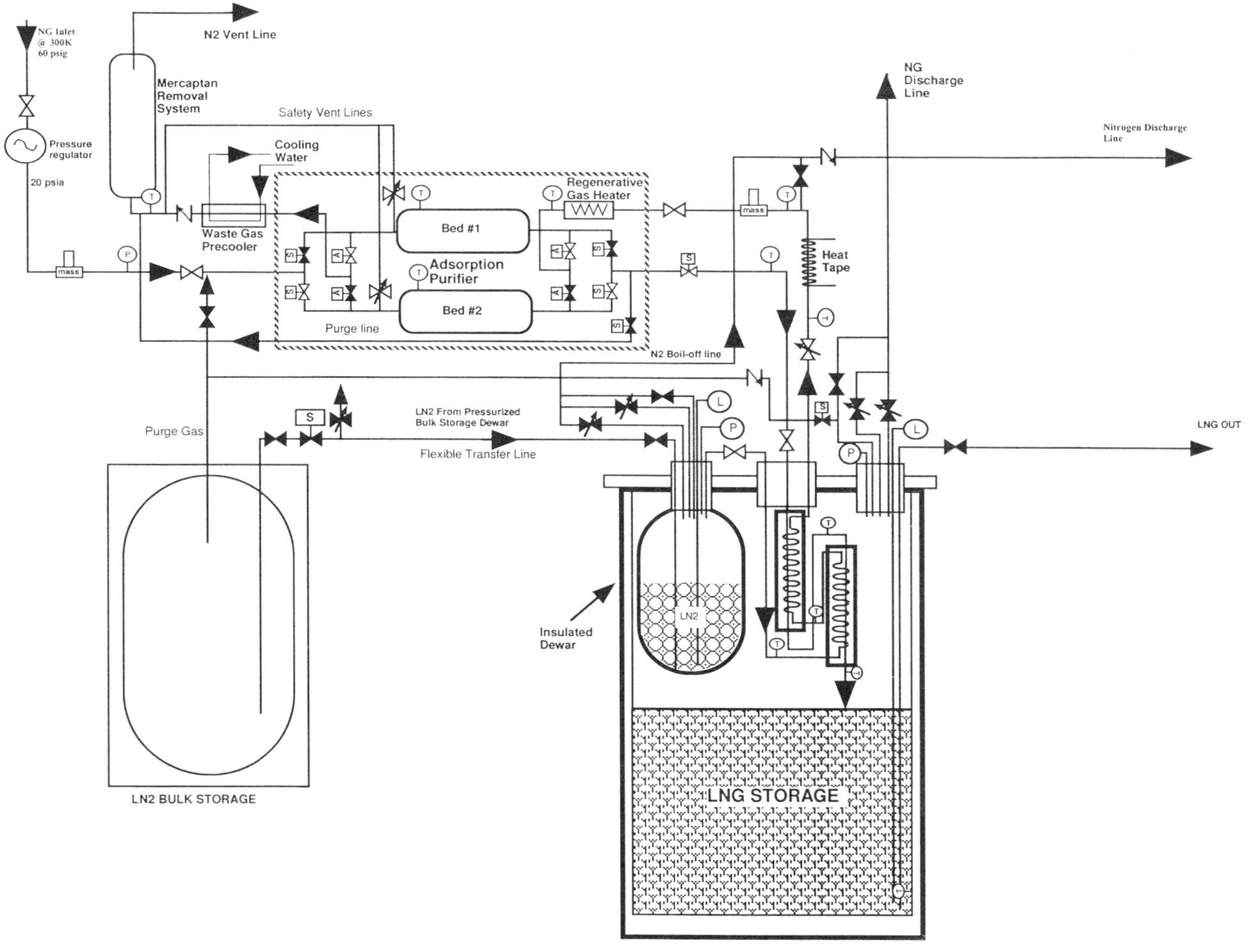

Figure 6. A schematic of an LNG liquefier that uses pressurized liquid nitrogen as its refrigerant.

produced by a flow through the LNG compressor from the storage tank, in counterflow through vaporizer/heat exchanger and into the CNG buffer tank/dispenser.

Obviously, both CNG and LNG could be dispensed simultaneously if desired by the fleet operator. Additional features of this refueling system with its local liquefaction capability are:

• **Recovery of some of the latent/sensible heat** from pressurized LNG to provide CNG and precool the incoming natural gas. This improves efficiency and reduces the cost of CNG.

• **Quick-fill CNG and/or LNG** - this system can dispense CNG from the vaporizer into the vehicle fuel tank at a temperature cold enough to allow for a quick and complete fill. LNG can be transferred as quickly as gasoline or diesel.

• **Capacity variation and flexibility** - Fleet operators can meet a change in usage and demand patterns without incurring the significant disturbances and capital outlays caused by changes from non-modular systems. On-site liquefaction requires smaller, less expensive storage tanks compared to refueling systems supplied by off-site liquefaction.

• **Reliquefaction of boil-off** - Gaseous natural gas coming from storage tank boiloff, pressure adjustments during transfer, and cool-down of transfer lines can be readily fed back to the liquefier in CFS' systems. Such versatile vapor control is very easy with an on-site liquefier. Weathering is no longer a problem is because the boiloff vapors are returned to the storage tank.

• **Acceptance of wide range of natural gas supplies** - Pipeline natural gas has a range of compositions and pressures depending upon location; this system is largely immune to these variations because the natural gas is cooled in a process stream separate from the liquefier's internal working fluid. High purity methane [99+%] can readily be provided as well as a more consistent grade of CNG due to mixing in the LNG storage tank and drawing from the bottom of the tank.

Methods of Liqucfying Cryofuels[20]

LPG is normally extracted by a Linde or vapor-compression cycle in natural gas processing plants or by compression of propane from the light-fraction stream in a oil refinery. Hydrogen is commercially liquefied via nitrogen pre-cooled Claude cycle units with a variety of different types of compressors and expanders.[21] Natural gas has been liquefied by many different methods depending upon the scale of plant. Steady improvement in large baseload [1000's of metric tons/day] technology has been reported.[22] A key element of distributed fleet-size [1-10 metric tons/day or ≈1000 DEG/day] refueling systems for CNG/LNG is the method of liquefaction. Eight different methods of liquefaction of natural gas including magnetic refrigeration were identified and analyzed for application in an economical refueling system.[23] These include:

- Cascade cycle systems with three different refrigerants;
- Claude cycle systems with mechanical expanders (these include Linde cycle systems and those that use high pressure pipeline gas;
- Mixed refrigerant cycle systems (including precooled MRCs);
- Stirling cycle systems;
- Gifford McMahon cycle systems;
- Orifice Pulse Tube cycle systems;
- Liquid nitrogen based systems; and
- Magnetic cycle systems.

Analyses of each of these techniques have shown that some novel gas-cycle liquefiers are much less expensive to manufacture than conventional Claude cycle or high-pressure, mixed-refrigerant cycle liquefiers used in large baseload or peakshaving plants and may have some promise in refueling systems where operating costs are not dominant. These novel gas-cycle liquefiers are also less expensive to manufacture than magnetic liquefiers but they have only modest efficiency compared to a magnetic cycle device. The best potential for the highest thermodynamic efficiency, hence lowest operating costs, is with magnetic cycles. The exact capital costs of magnetic cycles were uncertain because no complete prototype has been designed/built.

The simplest LNG liquefier is one that uses liquid nitrogen for its refrigeration source. A small system has been has been demonstrated[24] and is shown in Figure 6. If LN_2 is available for less than about 5-6 cents/liter, this method produces LNG less expensively than any other. Such a niche market may exist in some areas where there is a high demand for liquid oxygen from an air separation plant.

The literature for all types of regenerative magnetic liquefiers was reviewed to determine which conceptual design best matched the requirements for a natural gas liquefier operating from near 300 K to near 110 K. An active magnetic regenerative rotary design was selected as the best option. The heart of the magnetic liquefier is a well-designed magnetic regenerator that enables the liquefier to efficiently span large temperature ranges with the relatively small adiabatic temperature changes (≈10 K) produced by magnetization/demagnetization of certain ferromagnets in ≈8 Tesla magnetic fields. The thorough design and analysis of a laboratory-scale, active magnetic regenerative liquefier [an AMRL with ≈400 W of cooling power at 110 K] for natural gas has been completed and is schematically shown in Figure 7.

CRYOFUELS, THE FUTURE

The Old System Fights Back

The costs of diesel fuel mentioned earlier in this paper are well documented. Even though most large oil companies have holdings in natural gas as well as oil, there has been major reaction to the serious consideration of alternative fuels, especially natural gas and LPG. Substantial lobbying efforts are being made to make reformulated gasoline and superclean diesel remain the fuels of choice for the future. Some of the full-page advertisements in major newspapers by Mobil are cleverly done to "shade the truth." Similar television ads by Texaco and others convey different degrees of misinformation on emissions reductions, environmental compatibility, etc. The large engine and vehicle manufacturers are only making a modest attempt to provide monofuel engines and properly integrated vehicles; no business leadership in the form of serious capital investments is being provided. Even many natural gas distribution companies have adopted a "wait and see" attitude about the use of natural gas as a transportation fuel. The realities of big business, i.e., $150 billion/year, are that there is a major resistance to proposed changes that will require some new winners and some losers. There is a quiet war going on!

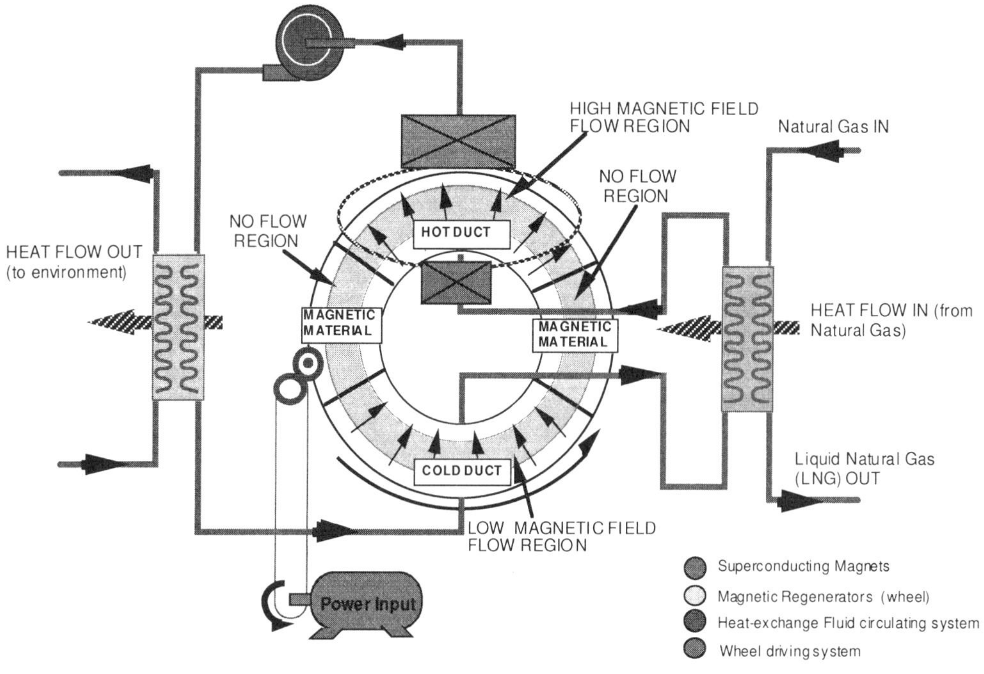

Figure 7. A schematic diagram of an active magnetic regenerative liquefier for natural gas.

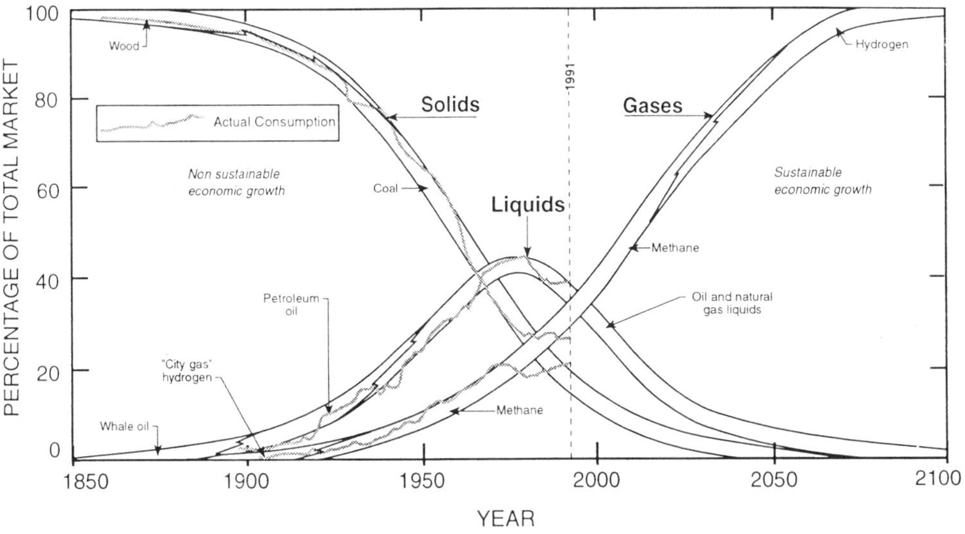

Figure 8. A diagram showing the evolution of energy fuel sources as a function of time. Taken from Ref. 27.

Crystal Ball vision of Future Transportation Fuels

There is hope for cryofuels in the transportation sector. Several well documented studies on the future of gaseous fuels indicate that over the next two to three decades, there will be a transition from liquid fuels to gaseous fuels.[25,26,27] This transition is based on energy substitution models where the gradual evolution from one fuel form to another is driven by technological innovation and societal needs rather than by depletion. Figure 8 shows the projections of these models. These curves indicate the transition to gaseous fuels is underway and will continue over the next several decades.

OPPORTUNITIES IN MATERIALS SCIENCE AND CRYOGENIC ENGINEERING

The transitions indicated in figure 8 are driven by technical innovations that cannot be programmed, only encouraged by providing adequate direction, resources and facilities. The innovation environment is also dynamic. Some general trends that influence new materials and equipment are:

- Developments have the tendency to go from simple to complex as technology becomes more sophisticated;
- Shorter product lifetimes are driving shorter development times;
- Information management systems are improving rapidly;
- Workstations capabilities are doubling every 18 months at present time;
- Products are becoming "smarter"; and
- Products are becoming more customized and of higher quality.

Within these constraints, many materials and equipment opportunities can be identified that will help the transition to cryofuels for transportation. In the near term, LPG and CNG/LNG create the following opportunities:

- Develop more affordable refueling systems for LNG/CNG which may occur if
  - * Much less expensive, more efficient, fleet-size liquefiers are found;
  - * Much less expensive LNG storage dewars are made;
  - * Inexpensive turbo-expanders are designed;
  - * Longer life, variable pressure, less expensive cryogenic pumps are manufactured;
  - * Automatic (card lock), smart, and fail-safe dispenser systems are made; and
  - * Sensors and related instrumentation improve because safety is a key item.

- Develop improved, vehicle-conversion technologies including
  - * Conformal, inexpensive vehicular tanks for LNG; e.g. $8000 as a typical price for a HDV tank for LNG is too high;
  - * Less expensive CNG storage tanks;
  - * Smart fuel injection and control systems with more sensors; and
  - * Adsorption materials for effective CNG storage at higher pressures.
- Develop advanced OEM engines and vehicles with much smaller surcharges which requires
  - * Continued R&D on fuel tanks, engines, new fuel injection systems, reduction of manufacturing costs; and
  - * Provision of OEM warranties on complete alternate fuel vehicles.

It is not difficult to see many opportunities for companies, national labs, and universities in this short list. In the longer term, the use of LNG as a fuel will be a pathway to LH2 but many technical and cost improvements are essential. Some of these items are identified in the following list

- Develop a more cost-effective, fleet-size steam reformer which may occur if a lower-temperature catalyst is developed.
- Develop other less expensive sources of hydrogen;
- Develop affordable refueling systems for CH_2/LH_2 that require much less expensive, more efficient, fleet-size liquefiers. This key development is more important for LH_2 than for LNG than for LPG.
- Develop sensors, transducers, and related instrumentation because safety is a key item.
- Develop a breakthrough in hydrogen storage! Two ideas that may make sense are:
  - * Conformal high-performance vehicular tanks for LH_2, and
  - * Adsorption materials for effective CH_2 storage

In summary, the cryofuels are excellent fuels that produce less emissions than gasoline or diesel. They are domestically available and in most cases, can be delivered into vehicles at a price that is less than gasoline or diesel. To enable a more rapid transition to these desirable fuels less expensive and higher performance technology is required in engines, fuel cells, gas turbines, and in vehicles. Affordable refueling infrastructure is an absolute must to sustain the use of cryofuels. A revolution is in progress, join it!

ACKNOWLEDGEMENTS

It is a pleasure to thank the Natural Science and Engineering Research Council of Canada, Centra Gas B.C., and the University of Victoria for their support of this project. Contributions from members of the Cryofuel Systems Group are also deeply appreciated.

REFERENCES

1. See, for example, several review articles published in journals such as *Cryogenics, American Scientist, American Academies of Science and Engineering*, etc. over the last decade in which the authors were asked to speculate on the technical advances likely to occur before the year 2000.
2. Plenary talk on Applied Superconductivity; Proceedings of the International Cryogenic Engineering Conference, Genoa, Italy, June (1994).
3. Energy Information Administration, "Annual Energy Outlook 1994 With Projections for 2010", U.S. Department of Energy, DOE/EIA-0383(94), January (1994).
4. D.G. Howell, editor. "The Future of Energy Gases", U. S. Geological Survey Professional Paper 1570; . S. Government Printing Office, Washington, D. C. (1993).
5. "Energy Statistics of OECD Countries; 1989-1990" International Energy Agency, 75775 Paris Cedex 16, France (1992).
6. Easton Consultants, Inc., "AGA/NGV Coalition Fleet Market Study", Volume 1, December, (1991).
7. J.E. Signor Consultants, Inc., Clean Fuels Legislation Directory in "The Clean Fuels Report," Vol. 6, No. 5, November (1994).

8. Energy Information Administration; Office of Coal, Nuclear, Electric, and Alternate Fuels; "Alternatives to Traditional Transportation Fuels; An Overview," U.S. Department of Energy, DOE/EIA-0585/O, June (1994).
9. Private communication to Dr. C. Simpson, Consumers' Gas, Scarborough, Ontario, April, (1995).
10. A wide range of data are available from the LPG Association, the Natural Gas Processors Association, numerous books on Hydrocarbons including methane, ethane, and propane; Hydrogen data are available from McCarthy's publication from the National Institute on Standards and Technology and now in a commercial software package called "Gaspak" from Cryodata, Inc. in Boulder, CO.
11. For example, see J. M. Beck, "Natural Gas - A Rational Approach to Clean Air," SAE Technical Paper Series: Truck and Bus Meeting and Exposition, SAE-902228, Oct. 27- Nov 1, (1990).
12. J. Stephenson, ed., "A Position Paper on Natural Gas Vehicles - 1990," pg. 65, Intl. Assoc. for Natural Gas Vehicles, Auckland, NZ, (1990).
13. B.C. Ministry of Environment, Lands, and Parks, "State of the Environment Report for British Columbia," (1993).
14. C. Freedenthal, "Natural Gas Pricing," Jofree Corporation, Houston, Texas, January 26, (1994).
15. J.E.Signor, "Rational for the use of LNG as a Transportation Fuel," Liquefied Natural Gas TOPTEC, pp 1-16, SAE, March 10-11, (1992).
16. J. S. Cannon, "Paving the Way to Natural Gas Vehicles," pg. 89, Inform, Inc., New York, (1993).
17. P. McArdel, U.S. Dept. of Energy, Invited talk at the Windsor Workshop on Alternative Fuels; Toronto, Ontario, June 12-14, (1995).
18. J. Tachtler and A. Szyszka, Car fueling with liquid hydrogen Solar Hydrogen Project: Experience and Results of the First Project Phase, *Int. J. Hydrogen Energy*, Vol. 19, 377-385, (1994).
19. J. A. Barclay, J. B. deGrasse, and H.H Rogner, Hydrogen for Transportation, What can be Learned from Natural Gas?, Proceedings of 5th Canadian Hydrogen Assoc.Conf., Victoria, February, (1994).
20. Timmerhaus, K. D., Flynn, T. M., "Cryogenic Process Engineering", Plenum Press, New York (1989).
21. M.Wanner, R.Gross, W. Otto, and A.Patzelt., Concept and Operation of a 4.4 ton/d Liquid Hydrogen Facility, Proc. of CEC - ICMC '93, (1994).
22. C. Collins et. al., Liquefaction plant design in the 1990s, Hydrocarbon Processing, April, pg 67-76, (1995).
23. J.A. Barclay, D.W. Goudie and C.E.J. Reid, Comparison of Natural Gas Liquefiers for Fleet-size LNG/CNG Refuelling Systems, Proc. of the 11th Windsor Workshop on Alternate Fuels, Toronto, ON, June, (1995).
24. A.J.Corless, S. Sarangi, J.L.Hall, and J.A.Barclay, Development of a Simple 5-15 litre per hour LNG Refuelling System, "Advances in Cryogenic Engineering," Vol. 39, pp. 55-62 (1994).
25. H. H. Rogner, Natural Gas as the Fuel for the Future, Annu. Rev. Energy, 14: 47-83 (1989).
26. C. Marchetti and N. Nakicenovic, The Dynamics of Energy Systems and the Logistic Substitution Model, RR-19-13, International Institute of Applied Systems Analysis, Laxenburg, Austria (1979).
27. R. Heffner, in "The Future of Energy Gases," U.S. Geological Survey Professional Paper 1570, U.S. Government Printing Office, D.G. Howell, ed., Washington, D.C. (1993).

1995 CEC/ICMC Conference: The Conference Chairs for the 1997 CEC/ICMC Conference.: (from left to right:) John A. Barclay, CEC and David B. Smathers, ICMC.

THE EFFECTS OF REFUELING SYSTEM OPERATING PRESSURE ON LNG AND CNG ECONOMICS

Adrian J. Corless and John A. Barclay

Cryofuel Systems Group, Dept. of Mechanical Engineering
University of Victoria
Victoria, B.C., Canada V8W 2Y2

ABSTRACT

Natural gas (NG) liquefaction and compression are energy intensive processes which make up a significant portion of the overall delivered price of liquefied NG (LNG) and compressed NG (CNG). Increases in system efficiency and/or process changes which reduce the required amount of work will improve the overall economics of NG as a vehicle fuel. This paper describes a method of reducing the delivered cost of LNG by liquefying the gas above ambient pressures. Higher pressure LNG is desirable because OEM NG engine manufacturers would like NG delivered to the engine intake manifold at elevated pressures to avoid compromising engine performance. Producing LNG at higher pressures reduces the amount of work required for liquefaction but it is only practical when the LNG is liquefied on-site. Using a thermo-economic approach, it is shown that NG fuel costs can be reduced by as much as 10% when producing LNG at higher pressures. A reduction in the delivered cost is also demonstrated for CNG produced on-site from high pressure LNG.

INTRODUCTION

Natural Gas (NG) is gaining acceptance as a viable alternate vehicular fuel. The two major modes of on-board NG storage are compressed NG (CNG) at ~22.5 MPa (3300) psig and liquified NG (LNG) at pressures under 1.8 MPa (250 psig). CNG is the appropriate choice for light duty and/or short range vehicles whereas LNG is much better suited for heavy duty and/or long range applications. While environmental and geopolitical factors clearly favor NG as an alternate fuel to diesel or gasoline, its widespread use has been limited by marginally favorable economics. As an energy feed-stock, NG is generally about 20% cheaper than crude oil[1]. However, most of this advantage is lost using current technologies to convert this feedstock into a usable vehicle fuel. The lack of a cost-effective refueling infrastructure for liquified and compressed natural gas is one of the major remaining barriers to its widespread adoption.

Much progress has been made in recent years in on-board vehicle LNG storage and delivery technology. This progress has been driven by the demands of OEM NG engine manufacturers who, to avoid compromising engine performance, would like fuel systems to deliver NG at elevated pressures of approximately 0.8 MPa (~100 psig) to the engine manifold. This may be done using on-board LNG pumps to boost the pressure before the liquid is vaporized or by storing the LNG on-board at elevated pressures. The latter of these methods is the simplist and is used in many of the new LNG fuel delivery systems. These

Advances in Cryogenic Engineering, Vol. 41
Edited by P. Kittel, Plenum Press, New York, 1996

engine requirements have implications as to how LNG must be liquefied and stored prior to being dispensed into the vehicle. This paper explores how fleet-size LNG/CNG liquefaction and refueling systems operating on-site may gain a competitive cost advantage by providing high pressure LNG.

THERMODYNAMICS OF NATURAL GAS LIQUEFACTION

Natural gas liquefaction and compression are very energy intensive processes that make up a significant portion of the overall delivered cost of LNG and CNG to the customer. Increases in system efficiency and/or process changes that reduce the required amount of work are translated into savings that improve the over-all economics of NG as a vehicular fuel. Pipeline natural gas consists primarily of methane with lesser quantities of ethane, propane, nitrogen and trace amounts of heavier hydrocarbons, CO_2, and H_2O. For the purpose of this analysis, methane thermodynamic properties are used as an estimate for NG.

At 0.1 MPa, the saturation temperature, density, and latent heat of vaporization of methane are 112 K, 423 kg/m^3, and 0.5 MJ/kg respectively. As the saturation pressure is increased, these saturation properties are affected as shown in Figure 1. The increased saturation temperature and the decreased latent heat act together to reduce the ideal amount of work required to liquefy a unit mass of NG. The ideal work is given by:

$$w = (h_f - h_i) - T_o(s_f - s_i)$$

and is shown in Figure 2 as a function of saturation pressure and available initial pipeline pressure. This figure illustrates that the work reduction is most pronounced between 0.1 and 1 MPa and is very dependent on the original line pressure of the NG. In most industrial settings, it is reasonable to expect pipeline pressures of 2.0-2.5 MPa. If this line pressure is not available, then the minimum work of liquefaction includes the work required to pre-compress as well as the work required to liquefy it at a given pressure.

Figure 1 shows that LNG density is reduced as a function of saturation pressure. This does not affect the minimum work of liquefaction but is still of practical importance because it reduces the energy storage capacity for a given volume. From 0.1 MPa to 0.8 MPa there is a 16% decrease in LNG energy density meaning a vehicle will either experience a decrease in

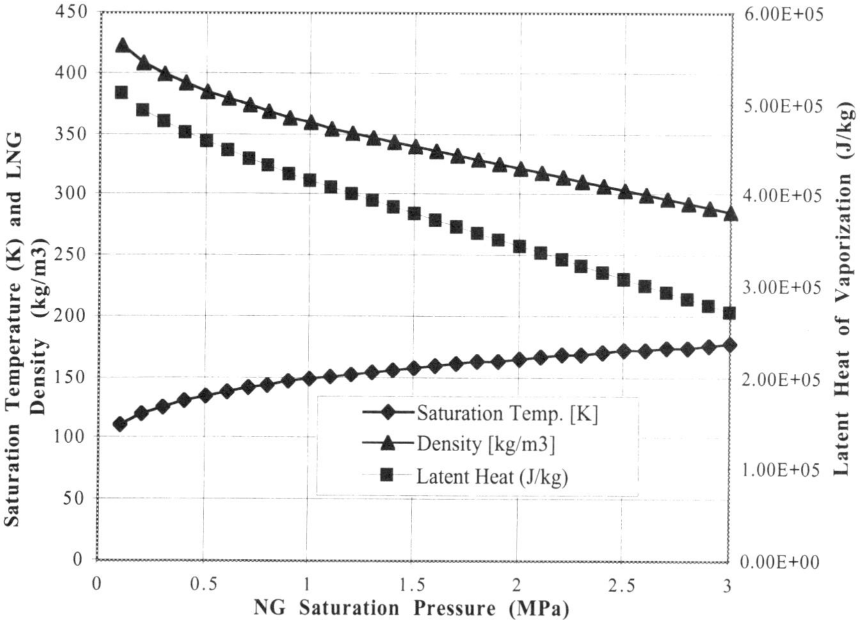

Figure 1. Saturation temperature, latent heat of vaporization, and saturated liquid density of NG as a function of saturation pressure.

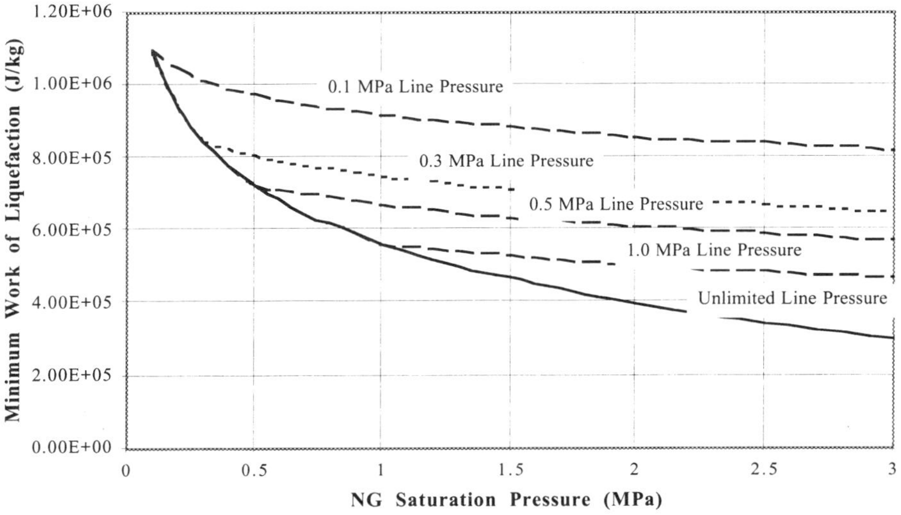

Figure 2. Minimum work of NG liquefaction as a function of saturation pressure and initial pipeline pressure.

range or will require a larger fuel tank. This effect must be considered when determining the over-all benefits of producing LNG at higher pressures.

THERMODYNAMICS OF CNG PRODUCED FROM LNG

To date, CNG has had the most important role to play in the NG vehicle marketplace. An alternate method of supplying CNG has been identified by Beale[2] and others that utilizes existing LNG infrastructure.

Multi-stage compressors are typically used to compress pipeline gas to 20-25 Mpa to provide CNG. These compressors are high maintenance items and generally have a low operating duty. An alternate method of providing CNG is to use LNG already available on-

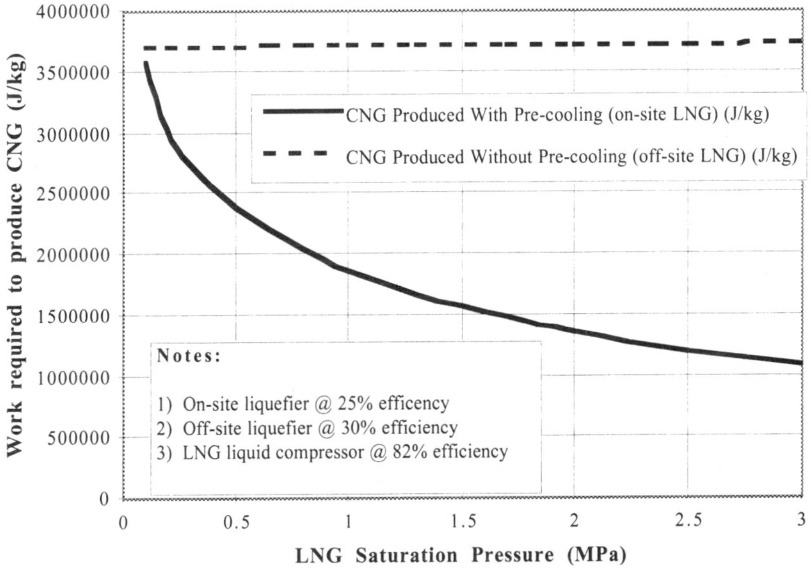

Figure 3. Work required to produce CNG from LNG as a function of LNG pressure both with and without NG pre-cooling.

site. LNG may be compressed as a liquid and then vaporized to form CNG. This method has several potential advantages over conventional compressor systems including:

- use of low maintenance, high pressure cryogenic liquid compressors which are an established technology and require low input power to raise the pressure of the LNG from 0.1 MPa to ~25 MPa;
- availability of both LNG and CNG without significant additional capital costs; and
- "fast filling" of CNG by charging the tanks with cold high pressure gas. (eliminating the need for large cascade storage systems)

When the CNG is produced simutaneously with the LNG, a portion of the vaporization may be done by pre-cooling the NG before it is liquefied. Figure 3 shows the increase in overall efficiency of the CNG production process when NG pre-cooling is applied. Furthermore, as the saturation pressure of the LNG is increased, the efficiency of this pre-cooling step improves as the enthalpy curves of the CNG and LNG become more closely matched and the work required to produce the LNG is decreased.

SUPPLY SCENARIOS FOR LNG/CNG

Most of the scenarios for supplying LNG to fleet operators have been identified by Powars et al.[3] Of these, the viable alternatives can be grouped into two broad categories:

- NG liquefied at a large central facility then transported to a refueling site; and
- NG supplied to the refueling site via existing pipelines and liquefied on-site by smaller liquefaction facilities.

LNG From Centralized Liquefaction Facilities

Most of the early LNG transportation projects have utilized the central liquefaction scenario. LNG is transferred from large storage vessels to cryogenic transport trucks for delivery to the refueling station where it is off-loaded into local storage tanks then dispensed directly into vehicles as LNG or compressed, vaporized, and dispensed as CNG. Large liquefaction plants may be dedicated to the production of LNG as a vehicle fuel, such as the Liquid Carbonics facility in Willis, Texas that supplies the Houston Metro bus fleet, or they may be LNG peak-shaving plants with excess capacity. In situations where large numbers of refueling stations are located near the central facility, the economy of scale of a large liquefaction plant allows NG fuel to be provided very economically.

While this scenario does not preclude LNG from being delivered to the vehicle's tanks at elevated pressures, it is simply not feasible to produce LNG at elevated pressures in centralized facilities for the following reasons:

- It is not practical to manufacture extremely large high pressure storage vessels because of the required capital expenditures; and
- The cryogenic liquid must retain enough cold exergy to absorb the heat leaks associated with several cryogen transfers and relatively long storage intervals without exceeding the final desired delivery pressure/temperature.

Also, when producing CNG on-site from LNG that is trucked in, it is not possible to recover any of the cold exergy contained in the cryogen by pre-cooling the NG supply stream before liquefaction.

Distributed Fleet-sized Refueling Systems

The operational limitations of the centralized LNG supply scenario may be overcome in a refueling system that utilizes on-site liquefaction technology. An extensive pipeline infrastructure exists to transport NG to most North American locations. An alternative LNG delivery mode is to tap into the NG distribution grid at a fleet operator's site and then to purify, liquefy, store and dispense the fuel locally. Several suppliers, including CryoFuel Systems Inc., have recently entered the NG fuel market place with small-scale distributed refueling systems capable of delivering LNG and/or CNG at prices competitive to diesel and

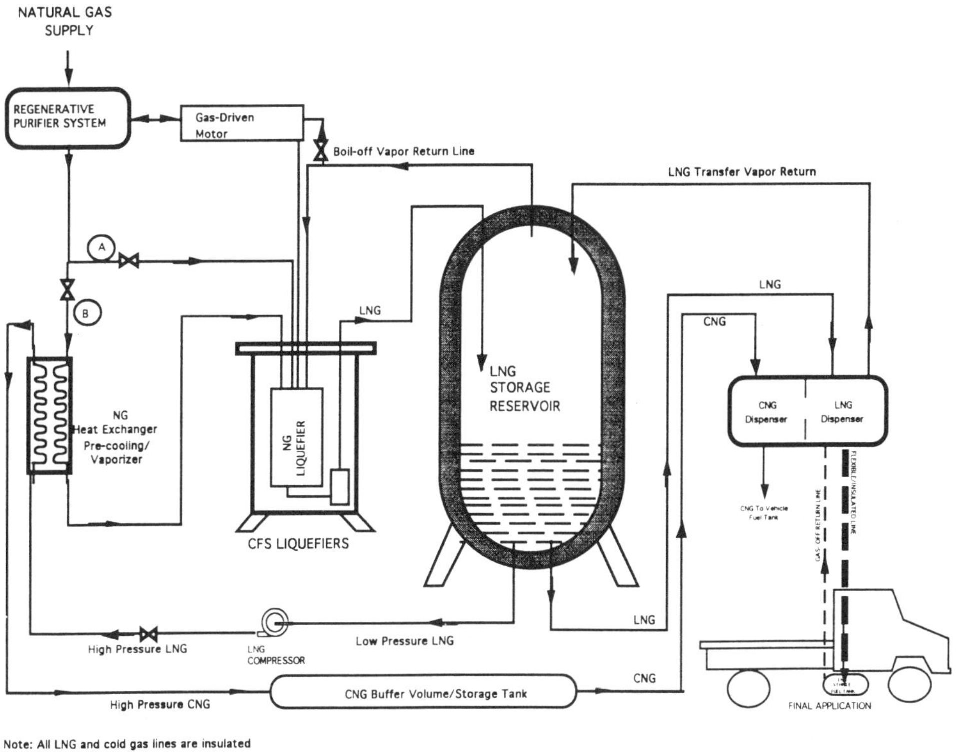

Figure 4. A schematic of a fleet-sized refueling system capable of converting pipeline NG to LNG and/or CNG for use a vehicular fuel.

gasoline. A distributed, fleet-sized LNG/CNG refueling system as shown schematically in Figure 4 offers several potential advantages to the fleet operator:

- Larger fleets may be converted incrementally. A large central liquefaction facility and LNG distribution network requires the conversion of a large number of fleets to be economically viable.

- Fleet operators have better control over fuel quality. Customized on-site purification systems can tailor fuel purity to the customers needs. Quick turn around of fuel results in less LNG degradation from aging.

- LNG and CNG can be supplied economically at sites which are distant from a large central liquefaction facility.

- NG boil-off vapors are easily handled because of the available on-site refrigeration. NG boil-off is also inherently lower since the cryogen is transferred less times and is held in storage for a much shorter time from liquefaction to use.

- Fleet owners may recover co-generation heat when an NG powered drive-motor is used for compression, further improving the economics of on-site liquefaction.

- CNG can be produced more efficiently on-site from LNG because a portion of the liquefaction work can be recovered by the CNG stream.

- LNG may be produced and maintained at or near its end-use pressure in smaller vessels and does not have to absorb the heat leaks associated with storage and transfers. This results in significant reductions in operating and capital costs as analyzed in the following section.

Thermo-economic Optimization of On-site Liquefaction Pressure

The purpose of a thermo-economic optimization is to minimize the overall life-cycle costs of a process in the presence of both thermodynamic and capital cost considerations. For this analysis the goal is to minimize the cost of fuel delivered to the customer by altering the operating pressure of the liquefier in a distributed refueling system. All capital investments are converted to the equivalent cost per diesel equivalent gallon (DEG - an energy equivalent comparison) of LNG using a capital recovery formula:

$$\$/\mathrm{DEG} = \frac{\text{Capital cost} \times \dfrac{i(1+i)^n}{(1+i)^n - 1}}{\text{Daily capacity} \times 365 \times \text{duty}}$$

All other costs are assessed per DEG. The overall cost of the fuel is made up of six major components:

Feedstock costs and road taxes

These two items are not affected by altering the systems operating pressure. The feedstock cost reflects the negotiated NG pipeline rate typically in $/MMBTU (or $/GJ, $/MCF). The road taxes vary from state to state and are usually assessed on an energy equivalence basis to gasoline or diesel.

Operating and maintenance (O&M) expenses

O&M are the costs incurred for a system operator if necessary, plus routine scheduled maintenance. It is assumed that the operational expenses will be independent of operating pressure but maintenance costs are a function of the system's normal operating pressure.

Energy costs

The work input required to produce the LNG is based upon the calculated minimum work of liquefaction and reasonable liquefier efficiencies[4]. The cost to provide this work is based on the cost of electricity or NG depending on what type of drive motor is used.

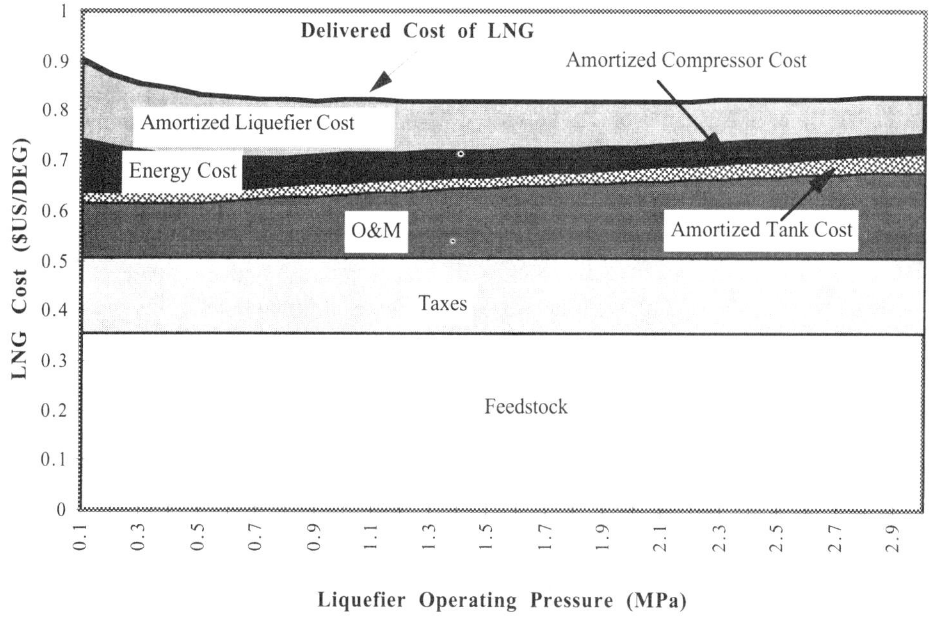

Figure 5. Delivered cost of LNG to a fleet-owner as a function of liquefier operating pressure.

Table 1. Model Parameters/Assumptions

Parameter	Value
NG feedstock price	\$2.50/MMBTU (GJ)
NG pipeline pressure	1.0 MPa
Liquefier capacity	1000 DEG/day (~ 1700 US US gallons)
Liquefier efficiency	25%
Liquefier operating duty	85%
NG engine efficiency	28%
Capital amortization period	10 years
Cost of capital	10%
Taxes	\$0.15/DEG

Amortized storage tank costs

The capital cost of the LNG storage tank is a function of tank capacity and operating pressure. As stated previously, the density of the LNG decreases as a saturation pressure increases. Additional capital cost is assessed based upon the need to increase storage capacity to maintain a constant energy content in the tank. LNG storage tanks under 56,400 litres (15,000 US gallons) are typically rated to 1.8 MPA (250 psig). An additional cost is assessed to account for higher operating pressures.

Amortized liquefier and dispenser costs

Based upon extensive analysis performed by CryoFuel Systems over the last 3 years and accepted costing standards[5], a realistic estimate of liquefier capital cost vs capacity has been developed. As the operating pressure of the system is increased and the work requirement is decreased, the size and capital cost of the liquefaction equipment required to produce an energy equivalent quantity of LNG are likewise reduced.

Amortized pre-compressor costs (if necessary)

In the event that there is insufficient NG pipeline pressure available on-site it would become necessary to install an NG compressor to boost the inlet pressure to the liquefaction system. The capital cost of the compressor is determined as a function of the required pressure ratio.

RESULTS

Figure 5 shows the overall delivered cost of LNG to the fleet-owner as a function of operating pressure. These results are based on the parameters listed in Table 1. which are representative of what a typical fleet owner might expect. Figure 5 demonstrates the relative contribution of each of the cost components to the overall delivered cost of LNG. From 0.1 MPa to 1.0 MPa the delivered cost of LNG is substantially reduced as operating pressure increases. The price levels off at around 1.0 MPa and reaches a minimum at approximately 1.5 MPa. Considering the current practice of storing LNG on-board a vehicle at approximately 0.8 MPa, there is a potential 7-8¢ cost reduction (~9%) per DEG over producing the fuel at 0.1 MPa. This is over and above the savings LNG provides compared to existing gasoline and diesel prices (ie. current gasoline and diesel prices are ~ US\$ 1.30/DEG) and could provide a fleet owner with a very large competitive advantage.

Referring again to Figure 2 it can be seen that when pre-compression is considered to boost the inlet NG pressure, the minimum work required for liquefaction is not reduced significantly at higher pressures. When the added capital cost of this compressor is factored in, it is shown in Figure 5 that little or no savings are realized by pre-compressing the pipeline NG to a higher pressure prior to liquefaction.

CONCLUSIONS

LNG and CNG are emerging as alternate fuels to gasoline and diesel. The largest barrier to their widespread adoption is a current lack of cost-effective NG refueling infrastructure. The delivered cost of these fuels are made up of several components. A model created to evaluate the effects of operating pressure on each of these components shows that the delivered cost of LNG may be reduced by as much as 10% when it liquefied on-site at higher pressures. This is primarily because the work required to liquefy NG is reduced as the saturation pressure increases. This high pressure LNG is also desirable on-board a vehicle because it improves delivery of NG to the engine manifold. These potential cost savings are only available when the LNG is produced on-site and in smaller quantities. CNG may be supplied in addition to LNG for only modest added capital costs by pumping the LNG to high pressure and then vaporizing it. When CNG is produced in this way from the LNG produced on-site significant reductions in delivered CNG cost may also be realized.

NOMENCLATURE

s_i = entropy of room temp gas (J/kg•K)
s_f = entropy of saturated liquid (J/kg•K)
h_i = enthalpy at room temp gas (J/kg)
h_f = enthalpy of saturated liquid ((J/kg)
T_o = ambient temperature (K)
i = investment interest rate (dimensionless)
n = capital amortization period (years)
w = work input per unit mass (J/kg)

ACKNOWLEDGEMENTS

The authors would like to thank the Natural Science and Engineering Research Council of Canada, Centra Gas Inc., and the University of Victoria for their on-going support.

REFERENCES

1. Rogner, H.H., "Natural Gas as the Fuel of the Future",*Anual Review of Energy*, Vol. 14, pp 47-93, 1989.
2. Powars, C., Luscher, D., Moyer, C., Browning, L., "A White Paper: Assessment of LNG Vehicle Technology, Economic, and Safety Issues" *Topical Report for GRI Contract # 5091-292-2153*, 1992.
3. Beale, J.P, "LNG's Role in Expanding the CNG Infrastructure" *Proceeding of NGV '94 Conference,* pp 879-888, Toronto, Ontario, 1994.
4. Barclay, J. A, Goudie, D.W., Reid, C.E.J., "Comparison of Natural Gas Liquefiers for Fleet-sized LNG/CNG Refuelling Systems", *Windsor Workshop on Alternate fuels*, Toronto, Ontario,1995.
5. Humphreys, K.K., Katell, S., Basic Cost Engineering, Marcell Dekker Inc., NY, 1981.

HYDROGEN VEHICLE FUELING STATION

D. E. Daney, F. J. Edeskuty, M. A. Daugherty, F.C. Prenger, and D. D. Hill

Los Alamos National Laboratory
Los Alamos, New Mexico, 87545, USA

ABSTRACT

We describe a hydrogen vehicle fueling station that receives and stores hydrogen in liquid form and dispenses it either as a liquid or compressed gas. The economics of distribution that accrue from the favorable weight and volume advantages of liquid hydrogen support this concept both now and for some time to come. Our model for liquid transfer to a 120 L vehicle tank shows that tank filling times under five minutes are feasible with pump-assisted transfer, or for pressure transfer with subcooling greater than 1 K. Our model for compressed gas transfer shows that vehicle tank underfilling of nearly 30 percent can occur during rapid refueling. Cooling the fill gas to 214 K completely eliminates the underfilling problem.

INTRODUCTION

Because of the need to improve urban air quality, hydrogen-fueled vehicles have become a subject of increasing interest in recent years; in fact, a number of demonstration projects are planned or in progress around the world. Hydrogen fueling stations are essential to a practical demonstration of these vehicles. Furthermore, only a practical demonstration can accurately and convincingly address a number of issues such as safety, efficiency, design, and operating procedures. Of particular importance is safety which encompasses both technical risk and public perception of that risk. Regardless of whether the vehicle is powered by an internal combustion engine or fuel cell, or how hydrogen is stored on-board, the fueling station is the critical technology that links the local hydrogen storage facility and the vehicle.

A fundamental assumption guiding the Los Alamos National Laboratory (LANL) program is that the station should receive hydrogen in liquid form because of near term economics, which result from the favorable weight and volume advantages that accrue from shipping liquid rather than gas. This view is confirmed by our conversations with US. industry (which delivered 92 percent of non-pipeline merchant hydrogen as liquid in 1993 [1]) and by systems studies (which favor LH2 supplied fueling stations for capacities up to about 300 vehicles/day [2]). At least three methods of onboard fuel storage (liquid, compressed gas and metal hydride) are under consideration. Because it is unclear what the preferred method will be, or if there will be only one method, we are designing our fueling station to deliver hydrogen as either a liquid, high pressure gas, or low pressure gas so that it can accommodate vehicles with any type of fuel tank. Figure 1. illustrates the LANL fueling station concept.

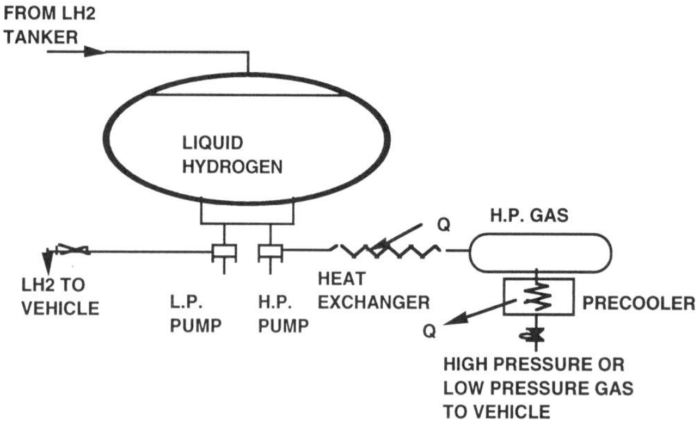

Figure 1. Los Alamos hydrogen fueling station concept.

Vehicle tank fill rates are important since refueling times must be short (less than five minutes) if the motoring public is to accept hydrogen fuel. In the case of liquid vehicle tanks, properly selected transfer and vent line sizes will give the required fill rates if venting is allowed. Because venting wastes hydrogen and is a minor safety hazard, no-vent filling is preferred. With no-vent filling, however, both the degree of subcooling of the liquid in the supply tank and the amount of assist from a transfer pump strongly influence the fill rate. In the case of compressed gas vehicle tanks, fast filling causes significant adiabatic warming in the tank which results in underfilling. With compressed natural gas tanks, underfilling by as much as 20 percent is possible.[3] With hydrogen, because of its higher ratio of specific heats, underfilling by as much as 30 percent can be expected. We have developed models of both the liquid and gas fill processes to study fill rates and the tank heating phenomenon.

FUELING STATION DESIGN

Figure 2 shows a detailed schematic of the LANL fueling station that will receive hydrogen as a liquid and dispense it as either liquid, high pressure gas or low pressure gas.

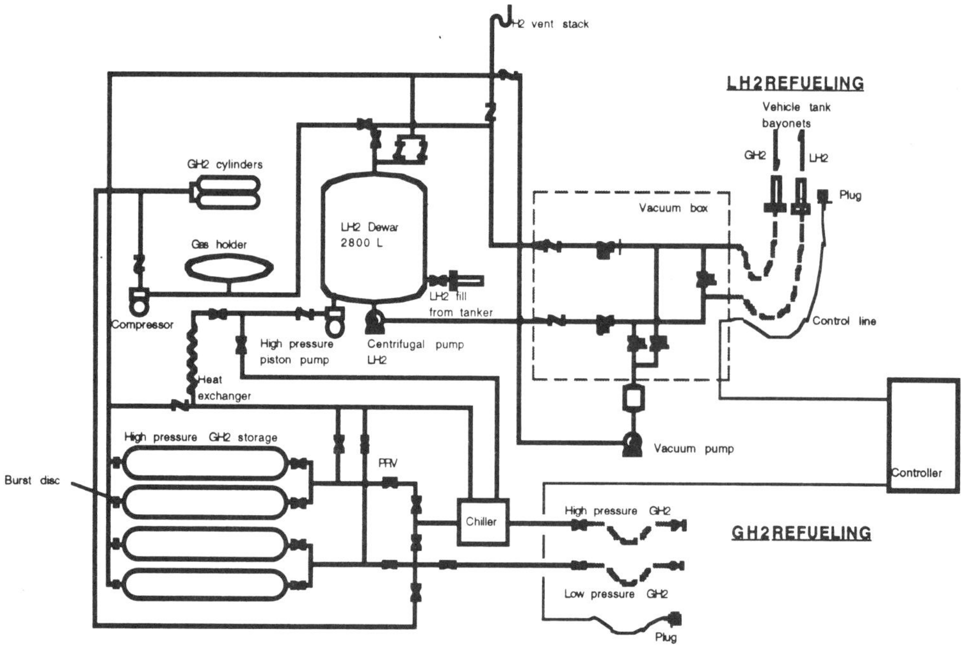

Figure 2. Detailed Schematic of Los Alamos hydrogen fueling station.

Near the center of the schematic is the liquid hydrogen (LH2) dewar which receives liquid from a commercial LH2 tanker truck. We anticipate that this dewar will be a leased Customer Service Station, as it is known in the trade.

For LH2 refueling, a centrifugal pump assists the transfer. From the pump the LH2 flows to the vacuum-insulated valve box where the flow is automatically regulated by sequencing solenoid valves with a microprocessor controller. Prior to the start of liquid transfer, the connecting lines are vacuum purged and checked for leaks by the pressure rise technique. The microprocessor, via a control line, sequences the vehicle tank valves and shuts off the vehicle ignition through an interlock. Under normal transfer conditions there will be no venting from the vehicle tank once the lines are cooled down. To prevent venting to the atmosphere of any gas generated in line cooldown, a gas holder, compressor, and gas storage cylinders are provided. This collected gas maintains the dewar pressure and supplies hydrogen for gaseous hydrogen (GH2) refueling.

For GH2 refueling, a pump-vaporizer unit (a high pressure piston pump in combination with a heat exchanger) generates high pressure hydrogen gas. To reduce the size of the pump-vaporizer required, high pressure gas storage cylinders are provided. These are interconnected for cascade-type discharge, which maximizes their effective storage capacity. As with liquid discharge, valve sequencing is automatic; and a vehicle ignition interlock prevents the vehicle from starting during refueling. A chiller, located between the GH2 storage tanks and vehicle coupling, mixes cold and ambient temperature GH2 and thus reduces vehicle tank heating during fast filling.

HYDROGEN TRANSFER MODELS

Liquid Transfer

Our liquid hydrogen transfer model is based on a computer code developed by Daney and co-workers at NIST [4] which simulates both vented and non-vented transfer between the fueling station tank and the vehicle tank. It is a thermodynamic model developed from the differential mass and energy balances applied to both the fueling station tank and vehicle tank, and it uses actual hydrogen properties. We consider both single-phase conditions in the vehicle tank (that occur during the initial cooldown of an empty tank) and two-phase conditions (that occur after liquid begins to collect in the tank). Either pump-assisted or pressure induced H2 transfer can be analyzed. The program accepts a wide range of input variables, including heat leaks, tank temperatures, and piping and equipment specifications.

The model begins with the first law of thermodynamics for this open system

$$dU = \delta Q + h_1 dm_1 + h_2 dm_2 \tag{1}$$

which for single-phase conditions in the vehicle tank gives

$$dT = \frac{\phi T dm + \frac{(h_1 - h)}{C_v} dm_1 + \frac{\delta Q}{C_v}}{\rho V + \frac{m_w C_w}{C_v}} \tag{2}$$

as the differential temperature equation and

$$dp = \frac{\phi\left[\theta dm + (h_1 - h)dm + \delta Q - m_w C_w dt\right]}{V} \tag{3}$$

as the differential pressure equation prior to venting.

For two-phase conditions in the vehicle tank the change in mass of the liquid phase is given by

$$dm_\lambda = \frac{-\frac{\rho_\lambda}{\rho_v}(dm_1 - dm_2) - \rho_\lambda \left[m_\lambda \left(\frac{\partial v}{\partial p}\right)_\lambda + m_v \left(\frac{\partial v}{\partial p}\right)_v \right] dp}{1 - \frac{\rho_\lambda}{\rho_v}} \tag{4}$$

and the differential pressure equation prior to venting is

$$dp = \frac{\left[h_l - u_v + \frac{\rho_\lambda}{\rho_v}(u_\lambda - h_l) \right] dm_l + \delta Q \left(1 - \frac{\rho_\lambda}{\rho_v}\right)}{F(p)} \tag{5}$$

where

$$F(p) = m_\lambda \left[\left(\frac{\partial u}{\partial p}\right)_\lambda \left(1 - \frac{\rho_\lambda}{\rho_v}\right) + \rho_\lambda \left(\frac{\partial v}{\partial p}\right)_\lambda (u_v - u_\lambda) \right]$$

$$+ m_v \left[\left(\frac{\partial u}{\partial p}\right)_v \left(1 - \frac{\rho_\lambda}{\rho_v}\right) + \rho_v \left(\frac{\partial v}{\partial p}\right)_v (u_v - u_\lambda) \right] - m_w C_w \left(\frac{\partial T}{\partial p}\right)_{sat} \left(1 - \frac{\rho_\lambda}{\rho_v}\right) \tag{6}$$

Gas Transfer

Because of the interest in relatively short fill times, our analysis of the temperature rise that occurs during the filling of compressed gas vehicle tanks assumes perfect mixing in the vehicle tank and no heat transfer with the tank wall. For the adiabatic case, combining equations (2) and (3) to eliminate m gives

$$dT = \left\{ \frac{v \left[\phi T + \frac{(h_i - h)}{C_v} \right]}{\phi \left[\theta + (h_i - h) \right]} \right\} dp \tag{7}$$

as the differential temperature equation for the tank. If we assume an ideal gas equation of state, then integration of equation (7) gives

$$T_2 = T_1 \left\{ \frac{\left(\frac{p_2}{p_1}\right)}{1 + \left(\frac{p_2}{p_1} - 1\right)\left(\frac{T_1}{\gamma T_i}\right)} \right\} \tag{8}$$

Equation (8) has the interesting property that as the pressure ratio, p_2/p_1 increases without limit

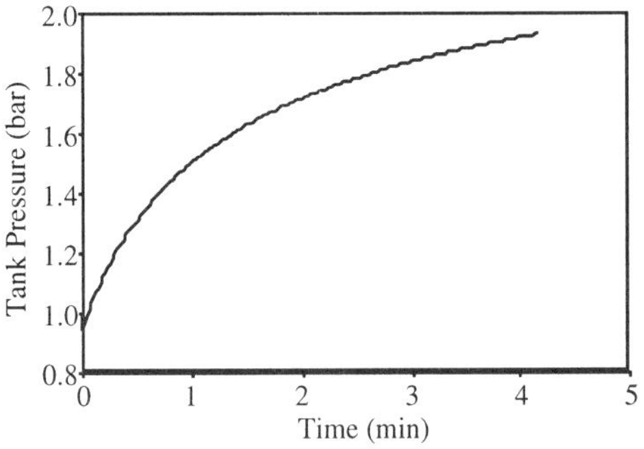

Figure 3. Typical vehicle tank calculated pressure history during pump-assisted, no-vent LH2 transfer with 1.0 K of subcooling. Initial vehicle tank pressure is 1.0 bar. Initial pressure difference is 1.0 bar. Pump power is 20 W and line OD is 12.7 mm.

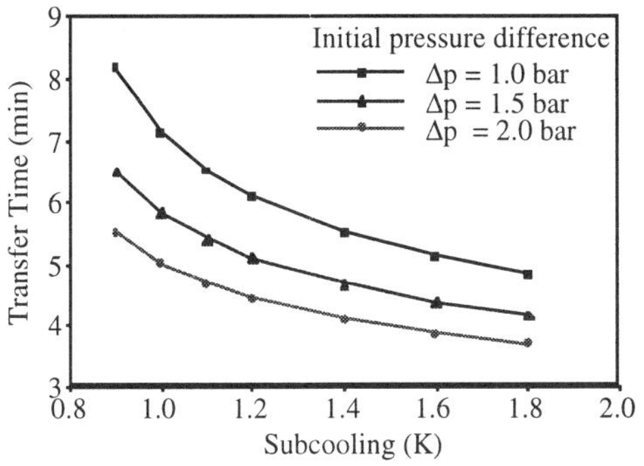

Figure 4. Effect of supply tank subcooling on calculated LH2 transfer time for no-vent pressure transfer. Initial vehicle tank pressure is 1.0 bar and line OD is 12.7 mm.

$$\frac{T_2}{T_1} \rightarrow \gamma \tag{9}$$

for the case of an ambient gas supply temperature ($T_i = T_1$). This approach to a fixed temperature ratio, in contrast to an increase without limit for a closed, adiabatic system, results from the cooling effect of the incoming gas.

The inlet gas temperature, T_i, required to give a final temperature, T_2, equal to the initial gas temperature, T_1, is found from equation (8) to be simply

$$T_i = \frac{T_1}{\gamma} \tag{10}$$

Thus, there is a single inlet temperature -- 214 K for a hydrogen tank initially at 300 K -- that will maintain a constant tank gas temperature for all fill pressure ratios.

MODELING RESULTS

Liquid Transfer

We are using our liquid transfer model to investigate the effects on transfer time of the initial conditions in the vehicle tank, the pressure and degree of subcooling in the supply tank, transfer-pipe size and configuration, and transfer-pump power. Figure 3, which shows the rise in vehicle tank pressure during a pump assisted, no-vent transfer, illustrates the detail inherent in the model. Figure 4, which gives the transfer time as a function of supply tank subcooling and initial pressure difference, illustrates the importance of subcooling. Close to

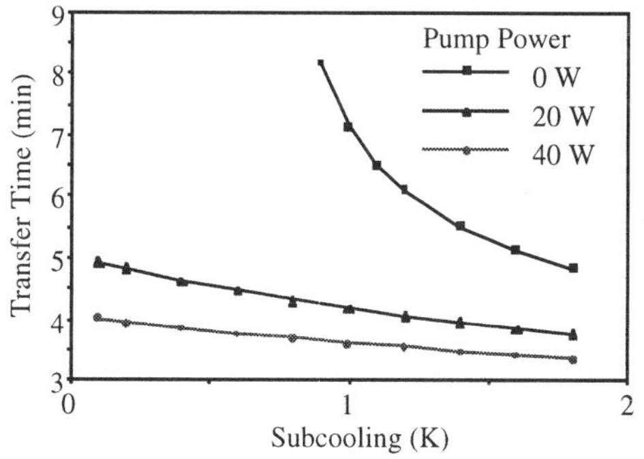

Figure 5. Effect of transfer-pump power on calculated LH2 transfer time for no-vent transfer. Initial pressure difference is 1.0 bar. Initial vehicle tank pressure is 1.0 bar and line OD is 12.7 mm.

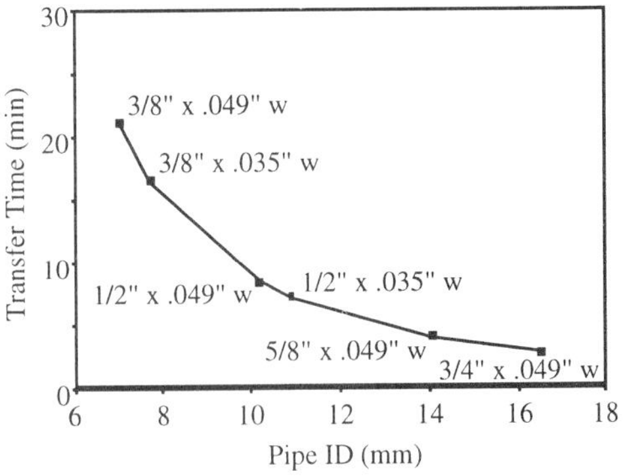

Figure 6. Calculated dependence of LH2 transfer time on pipe diameter for no-vent pressure transfer. The degree of subcooling is 1.0 K. The geometry is: 9.2 m of straight pipe, 2.6 m of flexible pipe, eight elbows and five globe valves.

0.9 K of subcooling is required, and more is desirable if there is no pump to assist the transfer . For the 1/2-inch line in these simulations, rapid (below 5 minutes) transfer is feasible with good subcooling. Figure 5 shows that use of a centrifugal transfer pump can further reduce the transfer time as well as extend the supply tank operating range. For vented transfer, of course, no subcooling is required in the supply tank. Figure 6, which gives transfer time as a function of transfer pipe size, illustrates how the model aids in the design of individual components. A 16 mm (5/8-inch) diameter transfer line size appears to be a good

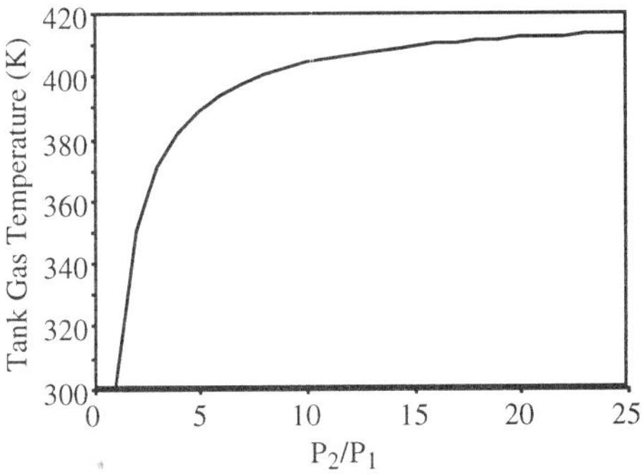

Figure 7. Warming in a GH2 fuel tank during fast fill. These calculations are for the limiting adiabatic case.

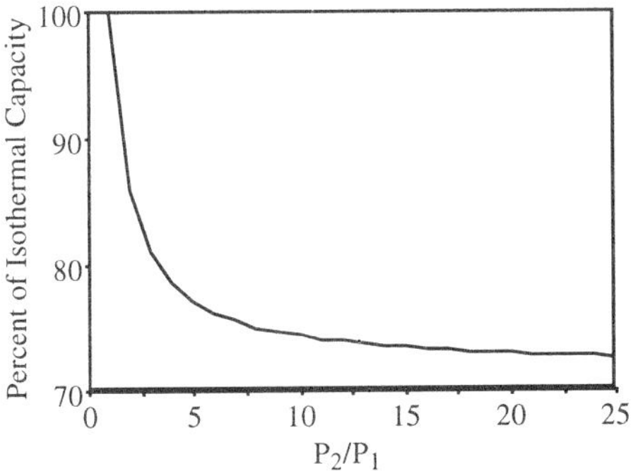

Figure 8. Calculated percent of isothermal GH2 tank capacity that results from an adiabatic fill. The asymptote at high pressure ratios is 71.4 percent full.

compromise between fast transfer times and easily handled flexible lines with low mass, and thus, low heat capacity.

Gas Transfer

The warming that occurs in a compressed gas hydrogen vehicle tank during a fast fill with ambient temperature (300 K) gas is illustrated in Figure 7. The curve, generated from equation (8), represents the limiting case of negligible heat transfer from the gas to the tank wall, such as would occur during rapid filling. The associated underfilling of the tank is illustrated in the Figure 8. Because the weight and volume penalties associated with compressed gas vehicle tanks degrade vehicle performance [5], reducing the loss of capacity due to fast filling becomes particularly important. As discussed above, precooling the fill gas to 214 K holds the tank gas at ambient temperature thus eliminating the problem of underfilling.

CONCLUSIONS

Refueling hydrogen powered vehicles in less than five minutes and without venting is feasible with a properly designed fueling station. In the case of LH2 refueling, this goal requires either pump-assisted transfer or pressure transfer with subcooling greater than 1 K. In the case of compressed gas refueling, the fill gas should be precooled to reduce or eliminate fuel tank underfilling that results from compression warming due to rapid fill. Precooling the fill gas to 214 K completely eliminates this problem for hydrogen.

NOMENCLATURE

C_p	specific heat capacity at constant pressure
C_V	specific heat capacity at constant volume
C_w	specific heat capacity of tank wall
h	specific enthalpy; unsubscripted, it refers to single-phase fluid in tank.
m	mass; unsubscripted, it refers to single-phase fluid in tank.
p	tank pressure
δQ	heat transferred to tank fluid
s	specific entropy
t	time
T	temperature
u	specific internal energy
U	total internal energy of fluid within tank
v	specific volume of fluid within tank; unsubscripted, it refers to single-phase fluid.
V	tank volume

Greek symbols

γ	specific heat ratio, C_p/C_v; $\dot{\gamma} = 1.4$ for hydrogen at ambient temperature.
θ	heat of expulsion, $-\rho\left(\frac{\partial h}{\partial \rho}\right)_p$
ϕ	Grüneisen parameter [6], $\frac{1}{\rho}\left(\frac{\partial p}{\partial u}\right)_\rho = \frac{\rho}{T}\left(\frac{\partial T}{\partial \rho}\right)_s$
ρ	density

Subscripts

i	inlet gas

v vapor phase
λ liquid phase
1 fill stream in liquid fill model; initial gas temperature in gas fill model
2 vent stream in liquid fill model; final gas temperature in gas fill model
w tank wall

ACKNOWLEDGMENT

This work was supported by Los Alamos National Laboratory, Directed Research and Development Funding.

REFERENCES

1. B. Heydorn, Hydrogen, in: "Chemical Economics Handbook," SRI International, Menlo Park, CA (1994).

2. J.M. Ogden, E. Dennis, M. Steinbugler, J.W. Strohbehn, "Hydrogen Energy Systems Studies," Technical Report, Center for Environmental Studies, Princeton University, Princeton, NJ (1995).

3. K.J. Kountz, Modeling the fast fill process in natural gas vehicle storage cylinders, presented at: 207th ASC National Meeting-Division of Fuel Chemistry, San Diego (1994)

4. B. Louie, N.J. Kemp, and D.E. Daney, "Cryogenic Propellant Scavenging", NBS Technical Report NBSIR 85-3023, National Bureau of Standards, Boulder, CO (1985).

5. M.A. Daugherty, F.C. Prenger, D.E. Daney, F.J. Edeskuty, and D.D. Hill, A comparison of the overall energy efficiency of on-board hydrogen storage systems using the EPA urban driving schedule, Paper FR-S1-5, CEC/ICMC 1995.

6. V. Arp , Thermodynamics of single-phase one-dimensional fluid flow, *Cryogenics* 15:285 (1975).

A COMPARISON OF HYDROGEN VEHICLE STORAGE OPTIONS USING THE EPA URBAN DRIVING SCHEDULE

M.A. Daugherty, F.C. Prenger, D.E. Daney, D.D. Hill, and F.J. Edeskuty

Los Alamos National Laboratory
Los Alamos, New Mexico, 87545, USA

ABSTRACT

The three standard options for the storage of hydrogen fuel on passenger vehicles are compressed gas, metal hydride and cryogenic liquid storage. The weight of the hydrogen storage system affects the performance of the vehicle. We examine vehicle performance as a function of hydrogen storage system type and capacity. The impact of storage system volume on vehicle performance is not addressed in this paper. Three vehicles are modeled, a metro commuter, a mid size sedan and a full size van. All vehicles are powered by a fuel cell and an electric drive train. The impact of auxiliary power requirements for air conditioning is also examined. In making these comparisons it is necessary to assume a driving cycle. We use the United States Environmental Protection Agency (EPA) urban dynamometer driving schedule in all simulations to represent typical urban driving conditions.

INTRODUCTION

Historically, on-board storage of hydrogen has been recognized as a major obstacle to the use of hydrogen as a transportation fuel.[1,2] Cost, weight and/or volume constraints severely limited the storage of adequate quantities of hydrogen for vehicular applications. Earlier comparisons of the relative merits of different fuel storage systems for hydrogen powered vehicles only considered storage in metal hydrides and cryogenic storage of liquid hydrogen.[3,4] These studies, which examine the use of hydrogen as a fuel for vehicles with internal combustion engines, both concluded that cryogenic storage of liquid hydrogen was the preferred option. The large weight of the metal hydride storage systems resulted in poor performance and increased primary energy consumption for vehicles with ranges in excess of 200 km. The energy penalty incurred by carrying the massive hydride storage system was found to be more than 100 percent for vehicles with a 400-500 km range.[3] This result included the energy penalty incurred to liquefy the hydrogen.

Two significant advances led us to reexamine on-board hydrogen storage options. The first is the development of highly efficient and non-polluting fuel cells for transportation applications.[5] Because of the high overall efficiency of fuel cell electric powered vehicles, less hydrogen storage is necessary for a given driving cycle and range. The second advance is the development of light weight high-pressure compressed gas tanks based on composite technology.[6,7] These new high performance tanks were originally developed for the aerospace industry and are currently being used for on-board storage of fuel for natural gas vehicles. They are made from high strength graphite composites filament wound over thin seamless liners.

Advances in Cryogenic Engineering, Vol. 41
Edited by P. Kittel, Plenum Press, New York, 1996

Table 1. Vehicle parameters

	Metro Commuter	Mid-Size Sedan	Full Size Van
Rolling Resistance Coef.	0.0135	0.0135	0.0135
Drag Coef.	0.28	0.28	0.35
Frontal Area	1.07 m^2	2.14 m^2	3.25 m^2
Vehicle Mass (Excluding Drive & H2 Storage)	573 kg	1064 kg	1671 kg
Mass Passengers	136 kg	136 kg	409 kg
Total Mass (Excluding Drive & H2 Storage)	709 kg	1200 kg	2080 kg
Air Conditioning	1.5 kW	3 kW	5 kW
Controller Efficiency	0.93	0.93	0.93
Motor Efficiency	0.94	0.94	0.94

In this paper we examine the impact of the hydrogen storage system weight on the performance of fuel cell electric powered vehicles. We have not modeled the effect of storage system volume on vehicle performance because it is strongly influenced by choices made by the vehicle designer. It should be noted that the differences in volume between different storage system types can be significant. Three different hydrogen storage systems are investigated; metal hydride, compressed gas and liquid hydrogen. Three different vehicles are modeled; a small metro commuter, a mid-size sedan, and a full size van. The impact of the mass of the various hydrogen storage systems on vehicle range and required drive train size are estimated. Issues associated with the fueling of the different types of hydrogen storage systems are examined in a related paper by Daney et al.[8]

VEHICLE PARAMETERS

The parameters used to model the three vehicles in this study are shown in table 1. These values are based on information contained in a report on the development of automotive fuel cell power systems prepared by Directed Technologies, Inc. (DTI).[5] The rolling resistance coefficient is assumed to be the same for all three vehicles. The drag coefficient is larger for the full size van than for the smaller vehicles and the frontal area increases uniformly with vehicle size. The vehicle masses shown do not include the drive train and hydrogen storage system masses. Two passengers were assumed for the metro commuter and mid-size sedan and six passengers were assumed for the full size van. Air conditioning, usually ignored in studies of this type, can have a significant impact on the vehicle range and performance. As table 1 shows, the larger vehicles require larger air conditioning systems. We modeled each vehicle with and without air conditioning. Motor and controller efficiencies are based on estimates by DTI.[5]

The fuel cell efficiencies are based on data taken by Swan et al. on a fuel cell system manufactured by Ballard Power Systems of Vancouver, Canada.[9] They measured efficiencies ranging from 45 percent at full power to 68 percent at low output. When these values are adjusted for the power necessary to run the fuel cell intake air compressor, net efficiencies ranging from 41 percent at full power to 65 percent at low loads were obtained.

Table 2. Drive train component masses for 40 kW fuel cell

Component	Mass (kg)
Gas Pressurization and Humidification	34
Fuel Cell Stack	45
Motor Controller	41
Motor	50
Step Down Gear & Differential	27
Radiator & Coolant	11
Total Drive Train	**209**
Total Drive Train (kg/kW)	**5.23**

We developed a curve fit to Swan's data using the percent of full power as the independent variable. A maximum efficiency of 65 percent was used when fuel cell output dropped below 20 percent of maximum. Based on the modularity of fuel cell systems we used the percent of full power to determine the efficiency of all fuel cells modeled regardless of size.

Drive train component masses for a 40 kW net output fuel cell system are shown in table 2. These values are based upon information gathered by DTI.[5] As table 2 shows, the fuel cell stack accounts for slightly less than one quarter of the total drive train mass. The values shown in table 2 were used to determine the mass of the total drive train per net fuel cell output. This value (5.23 kg/kW) was used to determine the drive train mass as a function of maximum fuel cell output.

HYDROGEN STORAGE

The mass fraction of hydrogen each storage system is capable of achieving is the important parameter for the purposes of this study. Mass fractions for metal hydride storage systems found in the literature vary from 0.5 to 4 percent.[10,11,12] A value of 1.5 percent hydrogen by mass was selected for this study after reviewing the literature. This represents a feasible hydride system such as FeTiMn with a reasonable disassociation temperature.

Literature values for compressed gas storage range from 1 to 6 percent for composite cylinders.[10,11,12] After discussions with Structural Composites, Inc. and EDO Canada, a value of 4 percent hydrogen by mass was chosen for this study. This represents a safety factor of 2.5 for a 33 MPa (4,800) psi composite tank. This is a conservative assumption which exceeds the DOT required safety factor of 2.25.

Published data on liquid hydrogen storage systems range from 7 to 25 percent.[1,10,11,12,13] Differences in boil off rates and size of the storage tank account affect the mass fraction of hydrogen. For this study we based our liquid hydrogen tank on a 65 kg tank developed by BMW-DFVLR with a 1.8 percent per day evaporation rate and a 9.2 kg liquid hydrogen capacity.[12] We did not scale the liquid hydrogen tank linearly. We assumed the tank would be roughly spherical, and that the mass versus storage capacity would scale as the surface to volume ratio of a sphere. Using this method, hydrogen mass percentages vary from 6 percent for 1 kg tanks to 15 percent for 8 kg tanks.

VEHICLE MODEL

The vehicle model computes aerodynamic drag, rolling resistance and kinetic energy requirements at each time step based on the vehicle parameters and the driving schedule velocity profile.[14] The vehicle parameters and the hydrogen storage capacity are read from one input file and the driving schedule velocity profile is read from a second input file. This allows different vehicles to be easily modeled using different driving cycles.

During acceleration the aerodynamic drag, rolling resistance and kinetic energy requirements are all met by the fuel cell. During deceleration some kinetic energy is available to help meet aerodynamic drag and rolling resistance requirements. If enough kinetic energy is available to completely satisfy aerodynamic drag and rolling resistance no power is required from the fuel cell for that time step. Any excess kinetic energy is dissipated by braking. If the kinetic energy available cannot meet aerodynamic drag and rolling resistance requirements the fuel cell contributes the necessary energy. Losses in the drive train are calculated using the efficiencies discussed above. Air conditioning energy requirements are drawn directly from the fuel cell. Figure 1 is a schematic of the energy flows used in the vehicle model.

The mass of the hydrogen storage system is calculated based on the type of storage and the specified fuel capacity. This mass is then increased by 15 percent to account for the mass of the structural supports for the fuel tank. The mass of the drive train is initially estimated based on the size of the fuel cell specified in the vehicle input file. The total vehicle mass is obtained by summing the masses of the drive train, fuel, hydrogen storage system, passengers, and base vehicle. The model then calculates the maximum fuel cell power required to carry this mass over the driving schedule.

Table 3. Energy dissipated (kJ) in one 11.9 km driving cycle with 7 kg compressed gas storage

	Metro Commuter	Mid-Size Sedan	Full Size Van
Rolling Resistance	1565	2476	4102
Aerodynamic Drag	461	920	1750
Braking	1419	2211	3626
Air Conditioning	2056	4113	6855
Motor & Controller Losses	495	807	1364
Fuel Cell Losses	3691	6487	10931
Total	9687	17014	28628

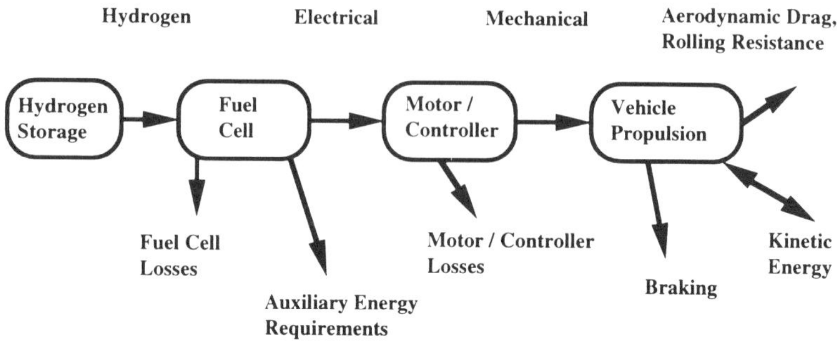

Figure 1. Schematic of energy flows used to model the hydrogen vehicle.

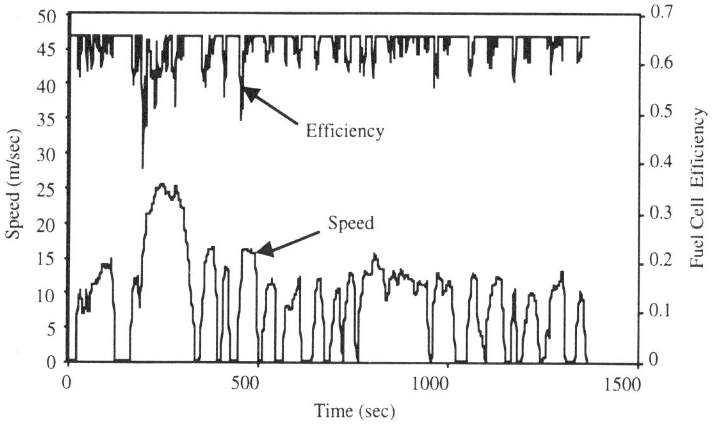

Figure 2. Urban driving schedule and corresponding efficiency.

Table 4. Hydrogen storage capacity and fuel cell size required for a 500 km range (Values are without / with air conditioning.)

	Liquid	Compressed Gas	Hydride
Metro Commuter			
Hydrogen Capacity (kg)	1.96 / 3.14	2.02 / 3.26	2.30 / 3.70
Fuel Cell Size (kW)	21.0 / 23.0	21.8 / 24.5	25.1 / 29.8
Mid-Size Sedan			
Hydrogen Capacity (kg)	3.37 / 5.73	3.51 / 5.99	3.98 / 6.79
Fuel Cell Size (kW)	35.5 / 39.4	37.2 / 42.6	42.9 / 52.3
Full Size Van			
Hydrogen Capacity (kg)	5.91 / 9.83	6.17 / 10.3	7.01 / 11.68
Fuel Cell Size (kW)	61.6 / 68.0	64.8 / 73.7	74.8 / 90.5

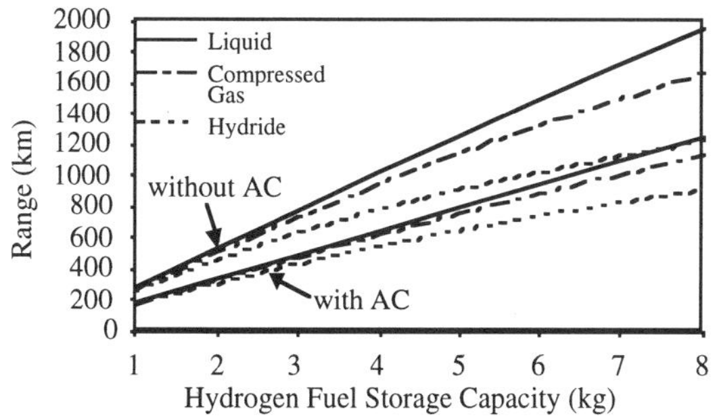

Figure 3. Metro commuter range as a function of hydrogen storage capacity. (Top three curves without air conditioning, bottom three with air conditioning.)

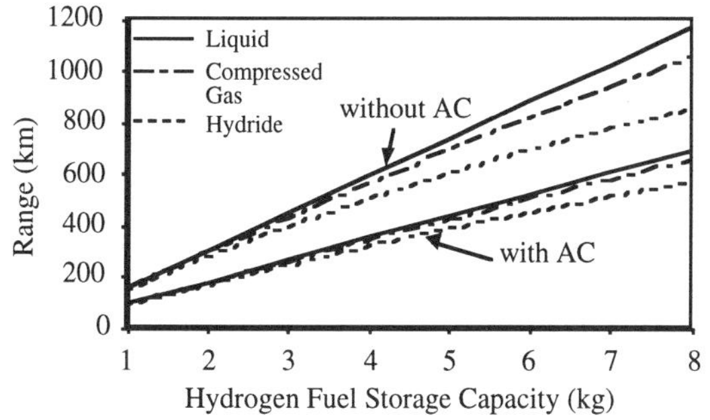

Figure 4. Mid-size sedan range as a function of hydrogen storage capacity. (Top three curves without air conditioning, bottom three with air conditioning.)

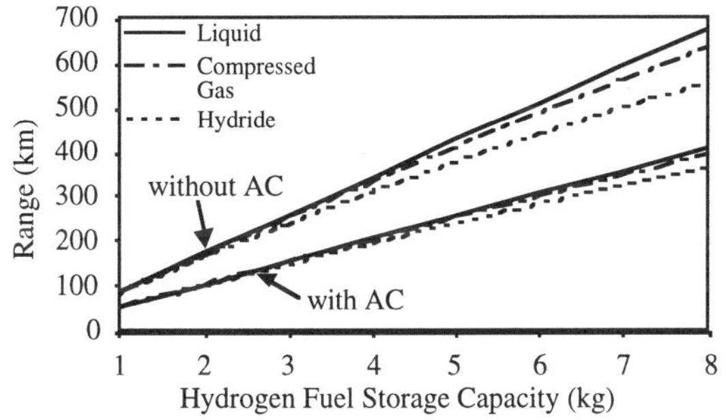

Figure 5. Full size van range as a function of hydrogen storage capacity. (Top three curves without air conditioning, bottom three with air conditioning.)

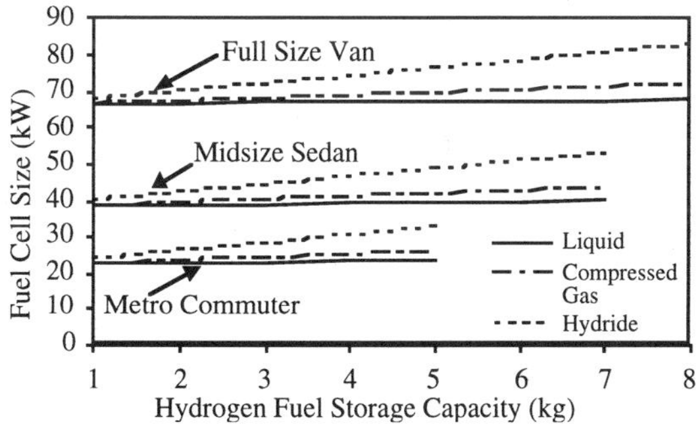

Figure 6. Fuel cell size as a function of hydrogen storage capacity.

Based on the maximum fuel cell power required a new drive train mass is calculated. Then the program iterates until the fuel cell size is consistent with the drive train mass and the driving schedule.

The vehicle range is calculated by dividing the distance traveled in one driving cycle by the mass of fuel consumed over the driving cycle and multiplying by the fuel system capacity. Energy balances are calculated by the model to verify energy conservation. In addition, the model was run using input parameters from Carpetis[3] Donnelly et al.[4] and Wicks and Marchionne[14] to verify the accuracy of our results.

RESULTS

The model was run with each of the three vehicles to determine the range and the maximum fuel cell power required when traveling over the EPA urban driving schedule. The three vehicles have considerably different energy requirements. Table 3 shows the energy dissipated in one 11.9 km driving cycle for each of the three vehicles equipped with a 7 kg compressed gas storage system. Air conditioning consumes more energy than rolling resistance or aerodynamic drag when traveling over this driving cycle. The full size van consumes almost three times the energy of the metro commuter.

Fuel cell losses are a strong function of the instantaneous fuel cell output. Figure 2 shows the instantaneous efficiency of the fuel cell for a mid-size sedan with 7 kg compressed gas storage. Spikes in the upper plot of efficiency correspond to periods of acceleration as shown by the lower plot of vehicle speed. The minimum fuel cell efficiency corresponds to the point of maximum fuel cell power. This point is used to determine the required size of the fuel cell and drive train for each combination of vehicle and storage system. It is interesting to note that, for the most part, the fuel cell is operating at relatively low power output with a correspondingly high efficiency. This behavior contrasts with that of an internal combustion engine, which operates with relatively poor efficiencies when traveling over this driving schedule.

Figures 3-5 show range as a function of hydrogen fuel storage capacity for each of the three vehicles with and without air conditioning. Air conditioning has a significant effect on vehicle range decreasing it by, on average, 40 percent. In all cases vehicles with liquid hydrogen storage systems have the longest range followed by compressed gas and then metal hydride systems. As vehicle range is increased the impact of storage system weight becomes much more significant. This effect can be seen most clearly in the upper set of plots in figure 3. The difference in range between storage systems is not as pronounced in the cases with air conditioning. This is because the air conditioning power consumption is independent of the weight of the hydrogen storage systems. Figure 6 shows the fuel cell size required for each of the vehicles as a function of hydrogen storage capacity. As hydrogen fuel storage capacity is increased larger drive trains are required to carry the more massive compressed gas and hydride storage systems.

Table 4 shows the hydrogen storage capacity and fuel cell size required for a range of 500 km. For all cases hydride storage requires about 18 percent more hydrogen than liquid storage to achieve the same vehicle range. The difference in required fuel cell size is even more dramatic. For hydride storage systems without air conditioning a 20 percent larger fuel cell and drive train is required and a 30 percent larger drive train is required for the cases with air conditioning compared to liquid storage. Compressed gas storage systems require about 4 percent more hydrogen than liquid systems for a 500 km range. The drive train for the compressed gas storage varies from 4 to 8 percent larger than that required for liquid storage. Hydride systems thus require significantly larger hydrogen capacities and drive trains to provide the same range and performance. Compressed gas storage systems, however, only require slightly larger hydrogen capacities and drive trains.

CONCLUSIONS

We have modeled the performance of three hydrogen vehicles; a metro commuter, a mid-size sedan and a full size van traveling over the EPA urban dynamometer driving schedule. Each vehicle was modeled with three different hydrogen storage systems; liquid hydrogen, compressed gas, and metal hydride. Vehicle performance is adversely affected by the weight of the hydrogen storage system. Vehicles with liquid hydrogen storage require the least amount of fuel and the smallest drive train to travel a given range over a fixed driving schedule, followed by compressed gas storage vehicles and then hydride storage vehicles.

In general, vehicles with hydride storage systems require approximately 18 percent more fuel and vehicles with compressed gas storage systems require approximately 4 percent more fuel than vehicles with liquid hydrogen storage systems to travel the same distance. Drive trains on vehicles with hydride storage are 20 percent larger for cases without air conditioning and 30 percent larger for cases with air conditioning than those equipped with liquid hydrogen storage. Vehicles with compressed gas storage average 5 percent larger drive trains for cases without air conditioning and 8 percent larger drive trains for cases with air conditioning than vehicles with liquid hydrogen storage.

A brief note regarding storage system volume is appropriate at this point. For a mid-sized sedan with air conditioning and a range of 500 km the volume of a liquid hydrogen storage tank is estimated to be 0.11 m^3 (3.85 ft^3). For the hydride storage system the volume is estimated to be 0.20 m^3 (7.05 ft^3) and for compressed gas storage 0.27 m^3 (9.56 ft^3). While the compressed gas storage volume requirement is about 2.5 times that of the liquid storage, the compressed gas can be relatively easily divided into multiple tanks which give the vehicle designer greater flexibility to distribute the storage volume in the vehicle. It is therefore difficult to estimate the effect of storage system volume on overall vehicle performance.

These results show that the high efficiencies achievable with fuel cell electric drive vehicles and the recent advances in light weight composite compressed gas tanks allow vehicles equipped with compressed gas storage to approach the performance of vehicles equipped with liquid hydrogen storage. Even with highly efficient fuel cell drives the vehicles with hydride storage systems incur significant performance penalties requiring substantially more fuel and larger drive trains. On vehicles with less efficient mechanical drives the performance penalties would be much larger.

ACKNOWLEDGMENTS

This work was supported by Los Alamos National Laboratory Directed Research and Development funding.

REFERENCES

1. W.F. Stewart, Operating Experience with a Liquid-Hydrogen Fueled Buick and Refueling System, *J. Hydrogen Energy*, vol. 9, no. 6, p. 525 (1984).
2. W.J.D. Escher, Survey of Liquid Hydrogen Container Techniques for Highway Vehicle Fuel System Applications, U.S. Department of Energy Report, no. HCP/M2752-01 (1979).
3. C. Carpetis, Comparison of the Expenses Required for the On-Board Fuel Storage Systems of Hydrogen Powered Vehicles, *Int. J. Hydrogen Energy*, vol. 7, no. 1, p. 61 (1982).
4. J.J. Donnelly, W.C. Greayer, R.J. Nichols, W.J.D. Escher, and E.E. Ecklund, Hydrogen-Powered vs Battery-Powered Automobiles, *Int. J. Hydrogen Energy*, vol. 4, p. 411 (1979).
5. B.D. James, G.N. Baum, and I.F. Kuhn, Jr., Technology Development Goals for Automotive Fuel Cell Power Systems, Argonne National Laboratory Report, no. ANL-94/44 (1994).
6. S.N. Sirosh, Application of Advanced Composites for Efficient On-Board Storage of Fuel in Natural Gas Vehicles, *Proc. Int. Conference on Composite Materials and Energy*, Montreal, Quebec, Canada, p. 793 (1995).
7. F.J. Darms, Space Age Pressure Vessels, *36th Int. SAMPE Symposium*, p. 818 (1991).
8. D.E. Daney, F.J. Edeskuty, M.A. Daugherty, F.C. Prenger, and D.D. Hill, Hydrogen Vehicle Fueling Station, Paper FR-S1-3, CEC/ICMC 1995.
9. D.H. Swan, B.E. Dickinson, and M.P. Arikara, Proton Exchange Membrane Fuel Cell Characterization for Electric Vehicle Applications, *Advancements in Electric and Hybrid Electric Vehicle Technology*, SAE Special Report-1023, p. 19 (1994).

10. M.A. DeLuchi, Hydrogen Vehicles: An Evaluation of Fuel Storage Performance, Safety, Environmental Impacts, and Cost, *Int. J. Hydrogen Energy*, vol. 14, no. 2, p. 81 (1989).
11. A.J. Appleby, Fuel Cells and Hydrogen Fuel, *Int. J. Hydrogen Energy*, vol. 19, no. 2, p. 175 (1994).
12. National Hydrogen Association, The Technology Assessment of Hydrogen Vehicles, NREL Report (1992).
13. W. Peschka, The Status of Handling and Storage Techniques for Liquid Hydrogen in Motor Vehicles, *Int. J. Hydrogen Energy*, vol. 12, no. 11, p. 753 (1987).
14. F.E. Wicks, D. Marchionne, Development of a Model to Predict Electric Vehicle Performance Over a Variety of Driving Conditions, (1992).

THE FIRST MINE COUNTERMEASURE DEVICES WITH SUPERCONDUCTING MAGNETS

V. R. Karasik,[1] A. A. Konjukhov,[2] G. A. Agapov,[3] A. I. Rusinov,[3]
G. S. Kurljandtsev,[3] V. S. Vysotsky,[4] V. N.Tsikhon,[3] V. A. Malginov,[3]
V. V. Matokhin,[3] and M. V. Sidorov [4]

[1]Babcock and Wilcox, Accelerator and Magnet Systems Dept.
P.O. Box 785, Lynchburg, VA 24505, USA
[2]Russian Research Center Kurchatov Institute
Kurchatov Sq., Moscow 123182, Russia
[3]P. N. Lebedev Physical Institute
Leninskiy Prospect 53, Moscow 117333, Russia
[4]MIT, Plasma Fusion Center
Albany St.185, NW-22115, Cambridge MA 02139, USA

ABSTRACT

Superconducting magnets, capable of operating with high current densities in a persistent mode within small overall dimensions, provide a good base for creating a new generation of autonomous, explosion stable, and effective mine countermeasure (MCM) devices. The first maritime and land mine countermeasure devices using superconducting magnets were developed in the Soviet Union during the period between 1976 and 1988. This large scale and long term multi-disciplinary program was tasked to develop new types of MCM devices to provide ships, tanks, and troop carriers with protection against modern mines, which use microprocessors and other elements of artificial intelligence. A joint team with contributors from the USSR Academy of Sciences, Soviet Industry and Navy executed the program. This program was chronologically divided into two parts: MCM Research and MCM Development. MCM Research undertook the creation of high performance superconducting magnets, and the design, manufacture, and land and sea testing of MCM prototypes. MCM Development was tasked with creating a mine sweeping craft towed by a helicopter that used a superconducting magnet, and a self propelled, high speed, radio controlled, sea-going MCM vehicle with a rotating superconducting magnet. It was shown that one rotating magnetic dipole or a combination of dipoles could precisely simulate the typical magnetic signatures of degaussed ships, tanks and troop carriers. The application of high performance, superconducting magnets to mine-sweeping was a break-through in anti-mine defense: it could provide an individual ship with protection, or it could clear mines from channels, shallow water, the surf zone of a beach, or the battlefield.

INTRODUCTION

Marine ground mines (mines that are deployed on the sea bed and that are activated by influence, not contact) are strategic weapons. They can be used to blockade commercial ports and naval bases, disrupt over sea transportation, destroy heavy warships and merchant fleets, thus changing the strategy of military operations. The most widely distributed ground mines are those with magnetic sensors. Ground magnetic mines were designed and produced by Germany after World War I (WWI). The results produced by the application of magnetic mines during World War II (WWII), and in many local wars, are very impressive. Great Britain's ship losses to magnetic mines were the same as to German submarines. During operation "Desert Storm," the mine fields laid by Iraq blockaded Kuwait, and changed the operational plans of the U.S. Marine Corp. Two heavy warships were damaged by mine explosions in the Persian Gulf. Magnetic mines were also used by Egypt, Argentina, the U.S.A., and other countries during local wars. In the future, mines in the hands of terrorists could be the tools of threat and blackmail.

The first form of MCM was the degaussing of ships. Degaussing was independently introduced by Great Britain and the USSR. During WWII, degaussing effectively protected the Russian Navy against German ground mines laid by aircraft. After WWII, the sensitivity of magnetic sensors was increased. An astatic magnetic suspension was replaced by magnetic film, and by multi-turn coils with permalloy cores connected to amplifiers. Modern magnetic sensors can detect a time rate of change in magnetic field (dB/dt) on the order of 10^{-9} T/s, which permits the detection of degaussed ships at great distances, thus rendering degaussing ineffective. Other MCM were required.

It is possible to subdivide the existing maritime MCM into two categories: mine-sweepers and mine hunters. Mine-sweepers simulate a ship's magnetic signature and detonate magnetic influence mines with an intense magnetic field produced by passing a current through a long resistive cable, between electrodes using the sea water to close the circuit, or with a huge resistive magnet. These devices are towed by a ship or helicopter. However, there are shortcomings to these approaches to mine-sweeping; they all demand large power supplies, and hydraulic resistance retards the vessel's speed making it liable to damage when a mine detonates. Often the cable or electrodes are destroyed when a mine detonates and must be replaced. Mine-sweeping is slow and ultimately non-productive with respect to the volume of mines that can be laid in a short period of time. Mine hunters discover mines and destroy them with explosive charges. They are made from non-magnetic materials and are equipped with a very powerful sonar. Additionally, they are equipped with small, remotely pilotable submarines for searching out and detonating ground mines. Mine-hunters are very expensive and their productivity is poor. With a draft of 12 feet, they could not operate in shallow water or in the surf zone.

After World War II, most anti-tank and anti-landing mines were supplied with magnetic sensors, drastically increasing their efficiency. A tank or an armored troop-carrier does not have enough electric power or carrying capacity to handle heavy electromagnets during assault or landing operations.

To defeat modern, sophisticated magnetic mines it was necessary to develop new, effective anti-mine weapons.

USSR PROGRAM TO APPLY SUPERCONDUCTING MAGNETS TO ANTI-MINE DEFENSE

A national research and development program for MCM, headed by the Applied Superconductivity Group of the P.N. Lebedev Physical Institute of the USSR Academy of Sciences (later the Applied Superconductivity Laboratory), was carried out during the period 1976 - 1988. A joint team from the USSR Academy of Sciences, the shipbuilding industry, the electronic and electro-engineering industry, the cryogenic industry and the navy participated in the program. MCM testing was carried out on the Black Sea, the Baltic Sea, and at the Army's Weapons Testing Grounds.

The Development of Superconducting Magnets and Cryostats for On-Board Operations [1-18]

Cooling Superconducting Magnets. Superconducting magnets intended for MCM duties at sea must be compact, light, explosion stable, operate in persistent mode while pitching and rolling, and withstand vibrations and pulse overloads of up to 30 gs. Two cooling schemes for these magnets were explored: helium pool boiling, and conductively cooled with a cryocooler. The use of pool boiling helium allows for long term, autonomous magnet operations in the persistent mode, eliminating the need for a power supply during operations. This approach allows the system to be quickly prepared for operations, and deployed on a high speed carrier, on land, sea, or in the air. Drawbacks include the need for periodic replenishment of the liquid helium boiled away by the heat flow through current leads, supports and radiation. Additionally, the need for liquid helium can create significant logistic problems. Systems with cryocoolers operate without liquid helium. However, to maintain the cold head temperature, the cryocooler compressor must be continuously supplied with power. Magnet cooldown with a cryocooler takes much longer than a pool boiling system, and cryocoolers are also sensitive to vibrations and explosions. Compressor weight, in some applications, can be a critical concern. Trade studies concluded that pool boiling was simpler and more reliable than cryocooler cooling, and was the appropriate choice for a MCM system.

Stress and Strain Distribution in the Superconducting Magnets. To provide the maximum magnetic moment within the smallest envelope, a superconducting magnet should be designed to maximize the current density. The magnets for MCM were manufactured with a Nb-Ti-Cu composite, wound on a stainless steel housing, and reinforced with bands of stainless steel wire and impregnated with epoxy. Analytical and computer methods were developed to predict the stress and strain distribution in the winding and housings during all stages of manufacture and operations: reeling, removal from the winding machine, cooldown, and charging. Based on these calculations, the optimum reeling tension was chosen to provide stable magnet operation. To measure the influence of tension on the magnet's quench current, a unique device was designed and successfully tested: very high tension in a thin solenoid wound from the desired conductor was developed with a magnetic force generated by the interaction between a uniform background magnetic field created by a second magnet and the solenoid's current. This tension could be measured with traditional techniques and the relationship between stress, strain and current established. Dozens of superconducting magnets were designed and tested to develop reproducible and effective manufacturing processes for explosive stable magnet production.

Superconducting Magnet Quench Protection. Superconducting magnets for MCM operate as autonomous devices in the persistent mode. Quench protection must be 100% reliable. Any commutating element could fail, therefore protection circuits without switches are preferable. Developed and successfully applied was the concept of protection by subdivision originally proposed by P.F. Smith [23]. This concept placed small resistors in parallel with sections of the magnet. Upon the occurrence of a quench, current is shared between the section and its associated resistor, accelerating quench propagation and allowing rapid dissipation of magnetic energy uniformly as heat. This approach to quench protection solved the problem of resistive zone propagation in magnets with high current densities, and the problem of quenching in magnets with inductively coupled sections (electromagnetic avalanche). An elegant method of measuring coil temperature distribution during quench was developed. Dozens of thermocouples and heaters were imbedded in the coil windings. By pulsing individual heaters, the conductor was driven normal and the time rate of change of coil temperature measured. Hundreds of quenches in multi-sectional magnets with stored energies in the megajoule range were performed, with the analyzed results compared with theoretical predictions. On the bases of these tests, Smith's concept was verified and reliable protection system for ship-board MCM superconducting magnets was obtained.

Selection of the Current Leads and Operating Current [19]. MCM efficiency is increased by reducing the loss of liquid helium. Considerable losses are introduce by the current leads: not less than 1.04 W per 1 kA of operating current. By operating the magnet in the persistent mode, liquid helium boiloff can be reduced by a factor of two. For mine-sweeping operations, the typical magnet charge time is much less than the time of operation. Therefore, current leads should either be removable, or should be forced cooled during magnet charging. Removable leads suffer from reliability concerns: drive gears can fail and contacts can oxidize. Experiments simulating forced cooled current lead behavior [19] demonstrated that the current density in copper could be between 60-150 A/mm^2 during charging in leads that are effectively cooled . Therefore, lead cross-section and the heat flow to the helium vessel from the environment during operations in the persistent mode can be drastically reduced. Operating current selection is a typical optimization problem: reducing I_{op} reduces LHe losses, but charging takes longer and magnet manufacturing is more complicate and expensive. Our magnet for MCM had forced cooled current leads and operated at currents of 100-300A.

Cryostat for MCM Prototypes [20]. The cryostat for a MCM device must be enough strong to withstand pressure overloads caused by explosions, but be light and portable while minimizing liquid helium losses. To meet these requirements, the helium vessel was suspended inside of the vacuum vessel with a stainless steel tube and surrounded by two radiation shields covered with multi-layer insulation. Both shields were attached to the tube in predetermined positions, providing an optimum temperature gradient inside of the cryostat. To protect the cryostat neck from damage from radial loads during transportation and operation, radial supports made from a high strength and low heat conductivity polymer were introduced. These supports were arranged as six (6) "bumpers" around the midplane of the cryostat. All cryostats were equipped with forced-flow current leads, safety valves for over-pressurization during quench, rupture discs, instrument lead feed-throughs for magnetic field, liquid level and temperature sensors. The MCM cryostats were mounted on a cradle and isolated with shock absorbers. Shock absorbers were selected on the basis of calculations that accounted for cryostat weight, resonance frequencies, projected overload to the system from explosions, and permissible load on the cryostat. Sea tests confirmed that the shock absorbers selected were appropriate.

Power Supply and Control System. All magnets were supplied with a superconducting switch to place the magnet in the persistent mode by shorting the cold end of the leads together. The switch was manufactured from Nb-Ti filaments imbedded in a high resistivity Cu-Ni matrix . The temperature of the switch was controlled with a Cu / Au + 0.07% Fe thermocouple. The switch was used in the following manner. After cooldown and in preparation for magnet charging the heater is activated and the switch opened by warming to above T_c. After charging started, the heater is switched off, its function being replaced by the heat generated in the conductor matrix. Once charged, the current in the magnet is stabilized and the switch cooled by lowering the power supply voltage, reducing the voltage across the switch and thus reducing switch current. With switch current reduction, the switch cools and places the magnet in the persistent mode. The heater is activated to open the switch and allow the magnet to discharge through diodes mounted on the cryostat head. A back-up battery to activate the switch heater was provided. All control systems operated automatically, and were stable and reliable. MCM devices with rotating magnets had control systems which managed the magnet rotation. The primary power supplies were either batteries or AC generators, which operated at either 50 or 400 Hz.

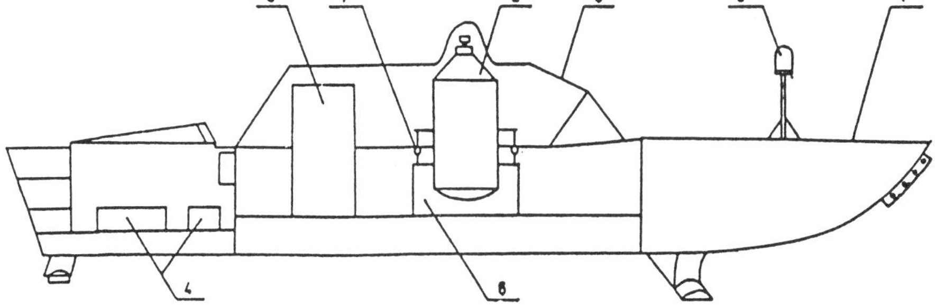

Figure 1.Prototype of mine counter measure system for helicopter tow. 1-hydrofoil cutter; 2-cryostat; 3-control system; 4-power supply; 5-flashing light; 6-cryostat cradle; 7-shock absorbers; 8-awning.

Performance of Prototype On-Board MCM Devices Using Superconducting Magnets

Helicopter Towed MCM Devices. Figure 1 is a sketch of the Russian prototype, helicopter towed, MCM device using a superconducting magnet. The magnet, superconducting switch, cryostat, and control system of this device were designed, manufactured and laboratory tested by the Applied Superconductivity Group at the P.N. Lebedev Institute in Moscow during the period 1976-1977. During the spring of 1977, all MCM and cryogenic equipment was moved from Moscow to Sevastopol, the base of the Soviet Black Sea Fleet. The MCM components were mounted on an aluminum hulled hydrofoil cutter. The magnet had the following parameters: ID - 370 mm, OD - 440 mm, height - 290 mm, inductance - 16 H, Iop - 100 A, maximum dipole magnetic moment - 0.85 10^5 A m^2. The magnet's magnetic axis was directed perpendicular to the sea surface. The cryostat was a modified SG100, an industrial design for liquid helium storage and transport, installed on 36 shock absorbers. For stable operation, the control system was mounted at a distance such that the fringing magnetic field was essentially background. All equipment was placed inside of water-proofed compartments. The first sea tests were to demonstrate the MCM vitality under rugged sea conditions and correct any teething problems. It was important to know the stability of magnet operations in the persistent mode when exposed to vibration and shock. Sea testing, which commenced in the fall of 1977, included the following: (1) Testing on the moored cutter, which evaluated the cutter's static stability with the MCM equipment on-board, cryostat filling and refilling with liquid helium on a pier and on-board the cutter, and magnet operation in the persistent mode. (2) Towing the MCM device in the harbor and at sea with a mine-sweeping ship at different speeds, sea conditions, and at different angles between sea waves and cutter heading. (At a sea roughness of 3 and a speed of 14 knots the pulse overloads from cutter hull collisions with waves recorded with tensiometers were 3 - 4 gs, and the maximum list was 40 degrees. Under these conditions the magnet operation was stable in the persistent mode.) (3) The MCM devices magnetic signature was recording by the Sevastopol Navy Magnetic Control Station. (4) The cutter mounted MCM device was towed with a helicopter. For the helicopter towing tests the MCM laden cutter was moved to the test area on the quarter-deck of a mine-sweeper and lowered to the sea surface with a crane. The free end of the tow-line was lifted by the hovering helicopter and the towing started. The tension on the tow-line was 0.3 metric tons. It was very difficult for the pilot to tow the cutter due to low tension and poor weather conditions. Due to collisions of the cutter's hull with sea waves, the tow-line tension changed spasmodically and it was necessary to constantly change the engine traction. The hydrofoil's angle of attack was designed for a speed of 25 knots. During trials on October 17, 1977, the pilot increased speed to 50 knots and more. The cutter was lifted above the sea surface and began to rotate around the tow-line. Within a few seconds, the cutter struck the water surface at a speed of 70 knots. The estimated overload was 30 gs. The tow-line was automatically disconnected under the excessive shock loads and the cutter flipped head-over-heels. It took more than three hours for the mine-sweeper to get the cutter on-board. The cutter was heavily damaged, but the cryostat survived and the magnet continued to operate in the persistent mode. This accident demonstrated the magnet's striking ruggedness. Servicing of the MCM system was demonstrated to be simple and acceptable for navy personal. During the sea-tests, the average loss of liquid helium was 0.5 L/hr.

Prototype of a Magneto-Acoustic, Self-Propelled, MCM (SPMCM) Device with a Superconducting Magnet. The main objectives of this part of the Program were: (1) Evaluating the possibility of engine operation in fringe magnetic fields (a few dozen gauss). (2) Mine-sweeping of magneto-acoustic ground marine mines. This device was equipped with a magnet with a dipole magnetic moment of 0.85 10^5 A m^2, a remote-control system, a 100 HP engine, and two different acoustic emitters. The first test of SPMCM took place on October 6, 1978 in the Sevastopol area. The magnet was charged and placed in the persistent current mode by radio. The engine functioned properly, propelling the cutter at 23 knots. The acoustic emitters and all other systems performed satisfactory. An influence of fringe magnetic fields on jellyfish behavior was discovered: during magnet operation, hundreds of meduses clustered around the anchored cutter. For the next test, a training mine with acoustic and magnetic sensors was laid on the sea floor and connected by cable to an on-shore control station to record the operation of the influence fuse. It became apparent that the acoustic emitters were not required since the mine's acoustic sensor recorded the screw propeller noise of the cutter. Fig. 2 shows the results of mine sweeping. The mine was located at the point X=Y=0. The cutter was moved along in the X direction at different distances from the X-axis. Continuous lines enclose calculated regions where the influence fuse should be detonated (when dB/dt = 1.5 X 10^{-4} gauss/s (15γ/s), or more). Crosses show the cutter position when the influence fuses were detonated during sea trials. Twelve (12) passes were performed with distances on the beam of the cutter from 5 to 45 m at a speed of 21 ks with a dipole magnetic moment 0.6 10^5 A m^2. The width of the sweep zone was seventy meters.

The probability of an influence mine firing inside of the sweep zone was 100%. It was confirmed that the SPMCM device using a superconducting magnet was highly efficient. These tests marked the first time that a marine ground magneto-acoustic mine was fired with a superconducting magnet.

Prototype SPMCM Device with a Rotating Superconducting Magnet. The magnetic signature of a typical degaussed ship has few maximums and minimums. A single moving magnetic dipole can not simulate this signature precisely. However, magnetic signature simulation can be accomplished with a rotating solenoid with the magnetic axis directed parallel to the sea surface. Additionally, with a rotating magnetic dipole the sweep zone is increased. With forward motion dB/dt is proportional 3v / R^4, where v is the cutter speed. With magnet rotation, dB/dt decreases with distance more slowly: dB/dt in this case is proportional to ω / R^3. This concept was demonstrated during sea trials of the SPMCM with an on-board rotating superconducting magnet. The dipole magnetic moment was 0.94 10^5 A m^2. The cryostat was rotated around the magnet's vertical axis with an electric drive. Also developed was a new power supply unit with a superconducting inductive filter, which was connected in series with the magnet and used to dampen the ripple of the half wave rectifier mounted on the head of the dewar. Application of the filter simplified drastically the power supply design.

On May 12, 1981, at the Navy Magnetic Control Station near Sevastopol, the magnetic signature of the SPMCM with a rotating superconducting magnet was recorded. Three magnetic field components, H_x, H_y and H_z, were recorded with magnetic sensors placed on the sea bed. Varying forward speed and rotation frequency, 99 magnetic signatures were recorded. Fig. 3. shows two types of magnetic signature: with forward motion only, and the combination of forward motion with magnet rotation. It is clearly seen that the latter case simulates a ship's multipole magnetic field much better than the former, and that the sweep zone is wider. The SPMCM with a rotating magnet has three independently adjustable parameters: forward speed, rotation frequency and dipole magnetic moment. Different combinations of these parameters allow the simulation of the magnetic signatures of different vessels with high precision.

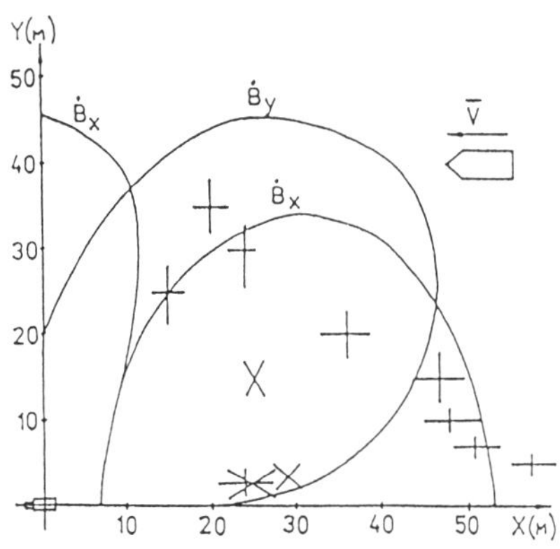

Figure 2. Firing of ground mine with acoustic and magnetic fuses. Upright crosses indicate MCM location at the moment of fuses firing at cutter speed of 15m/s, sloping crosses-at 5m/s.

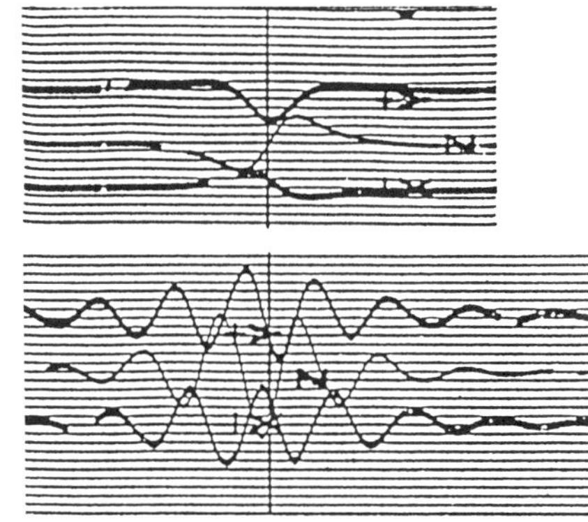

Figure 3. MCM magnetic signature.
1.The upper record-forward motion.
2.The lower record-forward motion and rotation with frequency 1/6 hertz.

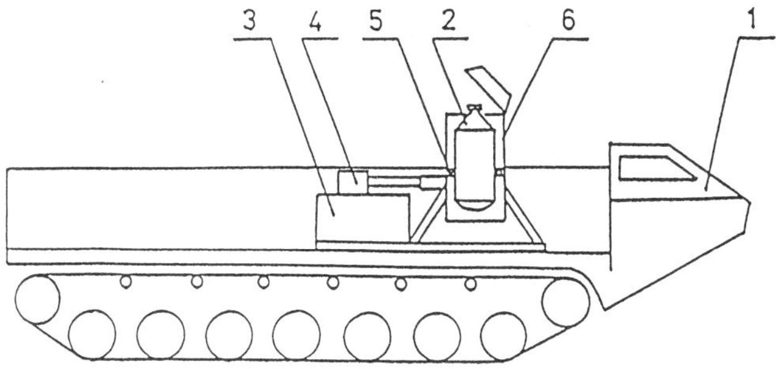

Figure 4. An prototype a mine counter measure for defeating anti-tank and anti-landing mines.
1. The armoured troop carrier. 2. A rotating cryostat with superconducting magnet. 3. A power supply and control system. 4. The cryostat electric drive. 5. A supporting bearing. 6 The armour-plated box-cryostat protection against damage from shell-splinters.

Prototype of the MCM Device for Anti-Tank and Anti-Landing Mines. The first MCM device with a superconducting magnet for defeating anti-tank and anti-landing mines was created and tested in 1984 by a joint effort between P.N. Lebedev Institute in Moscow and the MCM Division of the Soviet Army. A rotating superconducting magnet and radio-controlled power supply were mounted on an amphibious armored troop-carrier (Fig. 4). To protect the cryostat against damage from shell-splinters, it was placed inside of an armor-plated box. Since the magnet would operate while surrounded with ferromagnetic materials, the interactions of the magnet with the troop-carrier hull and with near-by tanks and other vehicles (magnetic field shielding, attractive force, breaking moment acting on the rotating magnet from its image in the surrounding material, additional load on cryostat electric drive, etc.) were calculated and measured during testing. The magnet had a dipole magnetic moment of 1.15 10^5 A m^2. To equalize all three magnetic field components on the land surface, the magnetic axis was tilted at 45° to the horizon. The magnet was operated in the persistent mode. During the period December 5-18, 1984, field tests were conducted at the Army's Firing Ground. A hard frost, heavy snowfall, temperatures of -15^oC, and the jolting of the carrier had no influence on the stable and reliable operation of the magnet. During the testing on December 18th, ignition of magnetic sensor equipped anti-tank mines was realized. Magnetic sensor sensitivity was B = 0.1 gauss or dB/dt = 0.1 gauss/s. Mines were laid at different traverse distances from troop-carrier track and along the direction of its forward motion. At a forward speed 10 km/hr and a magnet rotating frequency of 0.5 hertz the sweep zone had an elliptical shape with a major axis of 9 m and a minor axis of 8 m. This was enough for the reliable protection of the carrier against anti-tank mines. However, anti-landing mines can have a larger explosive charge. Therefore, to provide safe operations during landings, it is useful to increase the dipole magnetic moment.

Design, Manufacturing, Sea Trials and Commissioning of the MCM Device with an On-Board Superconducting Magnet Towed by a Helicopter

With the successful completion of the prototype tests in 1981, the program focused on transferring the technology to industry and the development of operational mine sweepers. This included the integration of the lessons learned from the various tests, and improved parameters such as an increased dipole magnetic moment to broaden the sweep zone. The goals for the first operational naval MCM device with a superconducting magnet included sweeping for marine ground mines with magnetic and acoustic sensors of high sensitivity, laid at depths of up to 60 m. The operational MCM device would have the following basic components: a superconducting magnet operating in the persistent mode with a horizontal magnetic axis, an explosive stable cryostat mounted on a hydrofoil cutter; a control system and power supply; a reinforced electrical cable-tow-line; and an acoustic emitter. The device was to be towed by a helicopter at speeds of 22-27 knots.

Superconducting Magnet. The new magnet had the following characteristics. The magnet had an operating current of 160 A, and a dipole magnetic moment of 0.6 10^6 A m^2. The conductor was a 1 mm diameter Nb-Ti wire, SC/Cu=3. The magnet envelope was : ID 420mm, OD 630mm, a height of 540 mm, and a weight of 350 kg. The wire was wound on a stainless steel housing with a tension of 21 kg and then impregnated with epoxy. The magnet was protected from overheating and electric breakdown during quench by winding subdivisions, and with shunts placed in the gaseous helium. The superconducting switch was mounted below the magnet in an area with low fringe field. Two magnet prototypes were manufactured and tested at the P.N. Lebedev Institute. Later, a full set of technical documents and drawings was issued and the technology was transferred to industry. Two of these magnets were manufactured by the Joint Stock Company "Electrosila".

Explosion Stable Cryostat. The cryostat was totally redesigned and manufactured by the Company "Heliymash" in cooperation with the P.N. Lebedev Physical Institute. The stainless steel helium vessel was suspended by the neck and fixed radially with six G-10 supports inside of the vacuum vessel, which was manufactured from an aluminum alloy. The cryostat was also supplied with one neck mounted aluminum shield, and with forced-cooled current leads. The cryostat envelope was an ID of 920 mm, an OD of 1,200 mm, a height of 1,850 mm, and a gross weight (with magnet) of 810 kg. When filled, the cryostat (with magnet) held 500 liters of helium. The cryostat was installed inside of a water-proofed compartment on shock absorbers, which allowed for stable operations during pulse overloads of up to 30 gs. Access to the cryostat head was through a hatchway from the cutter deck. The magnet could operate in the persistent mode for 3 days without refilling the cryostat.

Control System. The automatic control system handled magnet charge, operation in the persistent mode, magnet discharge, magnetic and acoustic sensor readings, and liquid helium level control. The magnet was charged from a 400 hertz AC power source, which was mounted on the cutter's mother ship or on the towing helicopter (through the cable-tow-line). The power consumed during magnet charge was 10 kW. Power consumption during operations in the persistent mode was 50 W. An on-board transformer and rectifier was installed which transformed 380 V AC current to 26 V DC current. Gross weight of the control system components installed inside of a water-proofed housing on the cutter was 296 kg. The control system was designed and manufactured by the shipbuilding industry.

MCM Cutter. A new MCM hydrofoil cutter was designed by the Russian Central (Hydrofoil) Design Office. The cutter had three water-proofed compartments which respectively

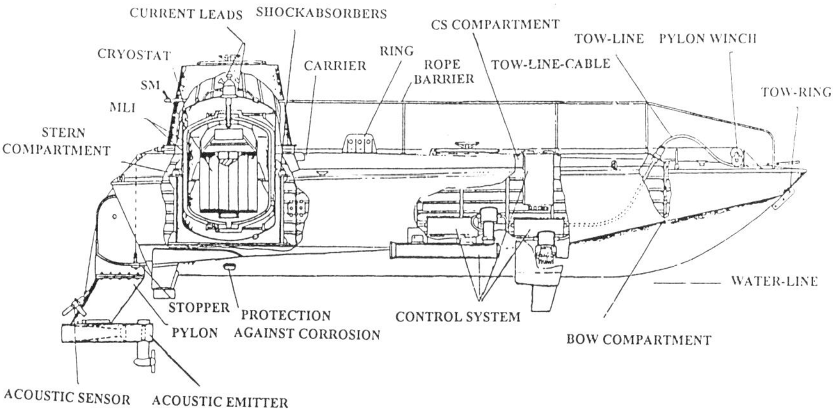

Figure 5. The first operational mine countermeasure device with an on-board superconducting magnet. Dipole magnetic moment 0.6 10^6 A m^2. Speed 22-27 knots. Seaworthiness 3. This new MCM device was intended for defeating ground marine mines equipped with high sensitivity acoustic and magnetic sensors.

housed the following components: the cryostat, which was mounted below the quarter deck; the control system, which was mounted amidship; and the auxiliary equipment, which was mounted in the bow. The acoustic emitter was mounted on a rotating pylon. The cutter had the following basic parameters: a length of 8,650 mm, a height of 2,500 mm, a beam of 2,500 mm, and a gross weight of 2,300 kg. The cutter was designed for a speed of 22 -27 knots. The tow-line tension at a speed of 22 knots was not more than 1,600 kg. The tow-line length was 250 m. MCM device is shown on Fig 5. During 1981-1982, the first industry manufactured MCM device with on-board superconducting magnet was built. Testing on this system was conducted in the Black Sea during 1982-1984.

MCM Preliminary Test. The program for preliminary testing included: (1) Towing the MCM device with a ship or a helicopter at speeds up to 27 knots in different sea conditions (more than 40 towing tests were conducted, in all more than 200 hours under good and bad weather conditions). (2) Long term, on-board, environmental test: the MCM device was transported over long distances to confirm cryostat ruggedness and helium losses under the influence of vibration, rolling and pitching. Results of preliminary tests were positive and the MCM device was admitted to the final testing for certification.

Final Test and Certification. During the final proof test, the following program was successfully realized: (1) Mooring Tests. (2) Towing Tests - 13 by ship and 17 by helicopter at speeds from 8 to 27 knots at seaworthiness of up to 3. (3) Verifying of liquid helium expenditures for magnet cooldown and operation. (4) Verifying the time to prepare the MCM device for mine-sweeping. (5) Control and Power System cyclic test. The magnet was repeatedly fully charged and discharged. (6) The magnetic and acoustic signatures of the MCM device were recorded. (7) The most interesting was the explosion stability test. A mine charged with 435 kg of TNT was laid on the sea floor at a depth of 19.5 m. A helicopter towed the MCM device with a superconducting magnet operating in the persistent mode, and with an acoustic emitter. The mine was detonated at a distance of 25 m from the moving hydrofoil cutter. Immediately after the explosion the sea surface above the mine had a protuberant shape and the carrier crossed at the center of the protuberance. A few seconds later the protuberance exploded in a giant water mushroom 50 m in height close behind the carrier. It was verified by radio that the magnet and acoustic emitter continued to operate: the MCM device passed the explosion stability test without any damage. The cutter survived since the hydrofoil was transparent to the shock wave. A shock wave is reflected by the sea surface and therefore could not damage the cutter hull moving above the sea surface. Outstanding explosive stability is a great advantage of the new MCM device. (8) Minefield sweeping. A defined area of water was mined in a pattern unknown to the MCM device operators with mines provided with magnetic and acoustic sensors. A helicopter towed a MCM device with a superconducting magnet operated in the persistent mode, and with an acoustic emitter. The mine-sweeping was carried out by equidistant parallel sweeping tacks. All mines were successful detonated. (9) The MCM device transport by air. The cutter with the superconducting magnet on-board was transported by air from a heliport to the open sea and placed on the sea surface. It was also demonstrated that mine-sweeping in the littoral zone by helicopter towing is possible without the need for accompanying ships. After successful completing final testing, the first MCM device with an on-board superconducting magnet was certified and added to the USSR Naval Armory by Order of the Commander in Chief of the Navy.

Design, Manufacturing and Test of a Radio Controlled SPMCM Device with a Rotating Superconducting Magnet

During 1977-1978, the USSR Shipbuilding Industry manufactured a SPMCM device with an on-board rotating superconducting magnet. The magnet parameters were an ID of 715 mm, an OD of 905 mm, a height of 180 mm, an inductance of 192 H, a maximum current of 164 A, a maximum magnetic field of 7.42 T, and a stored magnetic energy of 2.6 MJ. (To increase the dipole magnetic moment while maintaining the same physical dimensions, the magnet was wound with a wire with different SC to Cu ratio and different diameter) [21]. The maximum dipole magnetic moment was 1.27 10^6 A m^2. The following equipment was mounted on the cutter: a

1,000 HP diesel engine, a cryostat with magnet, a cryostat rotation subsystem, the main and auxiliary electric generators, remote controlled systems for steering and engine control, and for magnet operation and rotation, radio receiver and emitter, and radar emission reflector. Preliminary sea tests of the SPMCM with a rotating magnet were carried out in 1988 on the Baltic Sea. The SPMCM speed was 30 knots, and it had a displacement of 15 metric tons.

Other Superconducting Magnets for MCM Devices

For mine-sweeping at great depths, the Applied Superconductivity Group of P.N. Lebedev Institute designed a superconducting magnet with a dipole magnetic moment of 10^7 A m^2. The calculated maximum magnetic field on the winding was 6.75T, the magnetic field at the magnet center was 0.77T. Other magnet parameters were an ID of 2,700 mm, an OD of 2,900 mm, a height of 96 mm, an inductance of 5.6 H, a weight of 960 kg, and a maximum stored magnetic energy of 9.1 MJ. The conductor cross-section was 2.5 X 3.5 mm, with a Sc/Cu ratio of 0.66, and an operating current of 1,800 A. The magnet winding was placed in a stainless steel case, prestressed by reeling, and impregnated with epoxy. The magnet was transportable by helicopter while operating in the persistent mode. The magnet was manufactured by Industry [22].

Degaussed Ship Magnetic Signature Simulation

Simulating the magnetic signature of a degaussed ship with a rotating superconducting magnet was described above. The second independent method of degaussed ship magnetic signature simulation is by towing two or more magnetic dipoles connected in series. For better ship signature simulation, it is possible to change the dipole magnetic moments, their orientation in space, and the distance between dipoles. Fig. 6 shows the dH_x/dt component of the magnetic signature of two dipoles connected in series. The speed is 15 m/s, the distance between dipoles is 80 m, and the mine depth is 40 m. Magnetic signatures were calculated for three beam distances, 0, 25 and 50 m, respectively. A combination of two or more dipoles has more free parameters for magnetic signature simulation than one rotating dipole. This method of magnetic signature simulation could be used to clear a path through a minefield with influence fuses set to detonate at the recognition of a particular ship's signature.

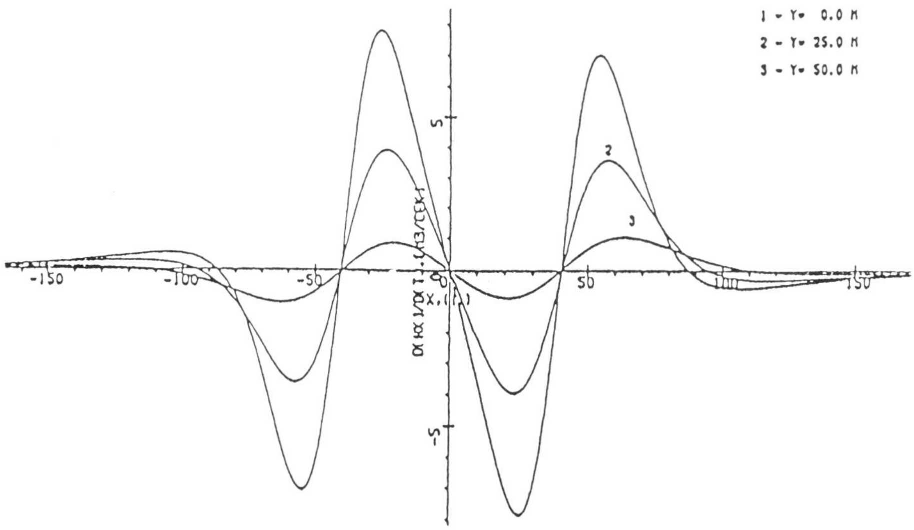

Figure 6. Magnetic signature derivative, dB/dt, of two point magnetic dipoles connected in series. The dipole magnetic moment of each dipole is 0.65 10^9 A m^2. Distance between dipoles is 80 m, depth is 40m, and speed is 15m/s. Beam distance: 0 m, 25 m and 50 m. X - travel direction, Z - depth, Y - beam. dB/dt [milligauss/second].

Conclusion

On the basis of many years of R&D, explosion and vibration stable superconducting magnets with high current density, operated in the persistent mode were designed, manufactured and tested. For the first time operational mine countermeasure devices with on-board superconducting magnets were added to a Navy armory. A new generation of MCM devices allows the defeat of the most sophisticated modern sea ground, anti-tank and anti-landing mines, and provides the possibility of mine-sweeping in shallow water and the surf zone.

Acknowledgment

The authors are very grateful to B.M. Vul, N.G. Basov, P.G. Kotov, A.A. Kondratovich, V.B. Zarin, Yu.I Usov, Yu.I Garanov, V.M. Konjakhin, A.V. Krugovenko, K.K. Staroseltsev, E.A. Andryushin, V.E. Ivanov and A.I. Vinnikov for support and important personal contribution to the first MCM with Superconducting Magnet R&D. The authors would also like to acknowledge the many enterprises, institutes, naval units, warships and helicopter crews that contributed their extensive efforts to the successful completion of the MCM Program.

The authors would like to thank J. Patrick Kelley for editing this article.

References

1. V.R. Karasik, N.V. Krivoluzkaya, A.I. Rusinov, Analysis of electromagnetic processes in superconducting magnet with winding subdivision, *Proc.of P.N. Lebedev Institute* 121:52 (1980),

2. A.I. Rusinov, N.V. Krivoluzkaya, Stress distribution calculation in prestressed superconducting magnets, taking into account wire tension by reeling, *Proc.of P.N.Lebedev Institute*, 121: 14 (1980),

3. V.R. Karasik, N.V. Krivoluzkaya, To the problem of critical mechanical and thermal loads in supeconducting magnets, *Proc. of P.N. Lebedev Institute* 150:22 (1984),

4. V.N. Tsikhon, S.S. Prgevsky, Power supply for superconducting magnets with extended operation mode, *Proc. of P.N. Lebedev Institute,* 121:101 (1980),

5. V.M. Zakosarenko, O.A. Kleshnina, Device for liquid helium level measurement, *Proc of P.N. Lebedev Institute,* 121:109 (1980).

6. V.S. Vysotsky, A.A. Konjukhov, Thermal properties and critical currents of superconducting switches, *Proc. of P.N.Lebedev Institute*, 150: 57 (1984),

7. A.I. Rusinov, N.V. Krivolutskaya, Velocity of resistive zone propagation and quench in thin superconducting solenoid, *Proc. of P.N. Lebedev Institute,* 150:70 (1984),

8. A.I. Rusinov, About some peculiarities of resistive zone propagation in thermo-isolated superconducting wire, *Proc. of P.N.Lebedev Institute,* 150:91 (1984),

9. V.S. Vysotsky, V.R. Karasik. A.A. Konjukhov, Voltage distribution by quench in superconducting magnet operated in persistent mode, in: "Applied Superconductivity in Electrical Power Engineering", CEA, Moscow, (1986), p. 104,

10. A.A. Konjukhov, V.A. Malginov, V.V. Matokhin, V.R. Karasik, Quench protection of superconducting magnet by subdivision with internal shunt resistors, *Proc. of P.N. Lebedev Institute,* 205:73 (1991),

11. A.A. Malginov, V.V. Matokhin, A.A. Konjukhov, V.R. Karasik, Heating by quench of superconducting winding with reinforced electrical insulation between layers, *Proc. of P.N. Lebedev Institute*, 205:63 (1991),

12. A.A. Malginov, V.V. Matokhin, V.R. Karasik, A.A. Konjukhov, Heat transfer in superconducting magnet by quench, *Proc of P.N. Lebedev Institute*, 150:48 (1984),

13. V.R. Karasik, A.I. Rusinov, V.S. Vysotsky, A.A. Konjukhov, Protection of superconducting magnets with high current density, *IEEE Trans.Magn. (USA)* 25:1541 (1989),

14. A.A. Konjukhov, V.A. Malginov, V.V. Matokhin, V.R. Karasik, Quenching of multisection superconducting magnets with internal and external shunt resistors, *IEEE Trans. Magn.(USA)* 25:1538 (1989),

15. B.M. Vul, V. R. Karasik, B.D. Kopylovskii, G.B. Kurganov, V.S. Vysotsky, D.V. Pronkin, Yu.A. Efimov, G.I. Agapov, Superconducting inductive energy storage unit for laser pumping system, *Sov. J. Quantum Electron. (USA)* 4:1099 (1975),
16. V.S. Vysotsky, V.R. Karasik, A.A. Konjukhov, A quench protection study of superconducting magnets, *Proc. of P.N. Lebedev Institute* 150:35 (1984),
17. V.S. Vysotsky, V.R. Karasik, A.A. Konjukhov, V.A. Malginov, Superconducting switches study for operation at the persistent mode, *Proc.of P.N. Lebedev Institute*, 121:82 (1980),
18. M.V. Sidorov, V.R. Karasik, Equipment for test of superconductors and superconducting magnets mechanical properties, *Proc of P.N.Lebedev Institute*, 205:140 (1991),
19. V.S. Vysotsky, V.R. Karasik, A.A. Konjukhov, Current leads with forced cooling for superconducting magnets operated in persitent current mode, *Proc. of P.N. Lebedev Institute,* 121:83 (1980),
20. E.A. Andryushin, V.E. Ivanov, Heat flow calculation in cryostat with two thermoshields, *Proc. of P.N. Lebedev Institute,* 121:89 (1980),
21. V.R. Karasik, O.A. Kleshnina, A.A. Konjukhov, V.A. Malgonov, V.V. Matokhin, V.S. Vysotsky and G.I. Agapov, Superconducting magnets for energy devices, in:"11th International Conference on Magnet Technology", Elsevier Applied Science, London and New York (1989), p.1312,
22. V.R. Karasik, O.A. Kleshnina, A.A. Konjukhov, V.A. Malginov, V.V. Matokhin, M.V. Sidorov, V.S. Vysotsky, A high current density superconducting magnet with 2.58m bore, Adv. Cryo. Eng. 37a:409 (1992).
23. P.F. Smith, Protection of superconducting coils, Rev. Sci. Instrum. 34:368 (1963).

DESIGN AND COIL PROTECTION OF THE HIGH CURRENT DENSITY AUTONOMOUS SUPERCONDUCTING MAGNETS FOR MINE SWEEPING

V.S.Vysotsky[1], A.A.Konjukhov[2], A.I.Rusinov, and V.R.Karasik[3]

P.N.Lebedev Physical Institute, 117924, Moscow, Russia

[1]Present address: PFC MIT, Cambridge, MA 02139 USA
[2]Present address: RSC Kurchatov Institute Moscow, Russia
[3]Present address: Babcock and Wilcox. Lynchburg, VA, USA

ABSTRACT

In the framework of the Russian R&D Program on the application of superconductivity to mine-sweeping, a series of high current density, compact superconducting magnets was developed, produced and tested. The diameters of the magnets varied from 0.4 m to 1 m and stored energy varied from 30 kJ up to 2.7 MJ. A low operating current was chosen to reduce helium losses.

The design and test results of the magnets are presented. The main focus was the development of the reliable coil protection because an overall current density in the magnets' windings exceeded 200 A/mm^2. A sub-divided winding with low resistance shunts was chosen as a passive quench protection scheme. Analytical and numerical methods of quench calculation in sectioned magnet windings were developed and are in good agreement with experimental results.

Magnets with 2.7 MJ stored energy and ~200 A operating current were quenched several dozen times without any problems. Most magnets were tested in severe outdoor conditions and further proving the reliability of the coils' design and protection scheme.

INTRODUCTION

Since 1976, more then a decade of work on the application of superconductivity for mine countermeasures (MCM) was performed and completed[1] for the P.N. Lebedev Physical Institute of Russian Academy of Science R&D Program. Several types of light, vibro--stable, high current density, autonomous superconducting magnet systems were developed, produced and tested in severe outdoor conditions. At the time there was no experience in the application of superconductivity for Navy purposes. Hence extensive research was done in optimization, design and quench protection development of superconducting magnets for MCM. The results of the research were published mainly in Russian [2,3,4] or in classified reports and remained unknown to the wide scientific society.

Nevertheless the results are important for general application of superconducting magnets in different military and civil purposes such as magnetically levitated trains, magnetic separation, space investigations, etc. Everywhere high magnetic fields in a large volume are necessary, our compact, self-protected and reliable magnets may be used. Since the results of the MCM Program were recently declassified we now present them to a general audience.

In the present paper we briefly survey the results of R&D for superconducting magnets with high current density.

GENERAL APPROACHES

In the mid 1970's there was no experience in producing superconducting magnets for outdoor conditions. Hence our main approach was to perform full cycle of magnet production: development, design, winding, test in laboratory, assembling with cryostat and other devices, outdoor-field test and design improvement for the next series.

Feedback from full-scale tests permitted us to complete the work rather quickly. Development time for each prototype was about one year. Starting from the small prototypes in 1976, we developed a full--size magnet for Navy in 1982 and some bigger prototypes in 1984 and 1987. During the years 1977 to 1988 we tested in outdoor conditions at least 8 full superconducting magnet systems including cryostats, power supplies, carrying platforms and cryogenic systems.

We found that for MCM the best solution is an Autonomous Superconducting Magnet System (ASMS). The ASMS consists of:

- A transportable superconducting solenoid with high current density and good vibration and shock stability. The magnet construction should include fully passive quench protection. A low operating current is very desirable to minimize helium consumption by ASMS.
- An optimized persistent current switch (PCS) with low helium losses adjustable to automatic power supply[5].
- A transportable and vibro--stable cryostat with low helium losses and non-detachable low loss current leads.
- A fully automatic power supply with "two button" remote control[5].

MAGNET DESIGN

Optimization of the Magnet and Material Choice

We performed the analysis of the optimal shape of the magnet taking into account critical current dependencies on magnetic field for real superconductors. Theoretically, it is possible to achieve any Dipole Magnetic Moment (DMM) with a solenoid with of wire length if any magnet diameter and crossection ("donut-shape") are allowed. In real life the outer diameter of the magnet is limited by the available cryostat and by the size of the platform carrying the cryostat.

In our studies, we solved the optimization problem for the maximum DMM for a given cryostat volume (diameter) and superconductor mass. It was shown that for solenoids optimal thickness of the winding is about 2.5--3 cm. If more thickness is used, the mass of the magnet increases faster than its DMM (see figure 1). The thin solenoid means low operating fields.

We rejected the use of NbSn wires in spite of their higher critical temperature and field than NbTi for two reasons. The first reason is the much lower overall current density in NbSn windings in low magnetic fields. The advantage of NbSn appear in high fields (more then 9 T) while to achieve high DMM we need not high field but high current (as was shown above).

The second reason is the low mechanical stability of the NbSn wires. NbSn superconductor is very brittle and allows much less strain than NbTi. Advanced techniques like "react after winding" do not improve mechanical properties and are very difficult in production. Thus, it is difficult to achieve high seismic stability of magnets wound from NbSn wire.

The only advantage of NbSn is the opportunity to use higher operational temperatures and to cool the magnet by cryocoolers ("dry magnets" [4]) to exclude liquid helium from the system. **But in more than ten years experience, we found the use of liquid helium in field conditions to be no more difficult than filling a car with gasoline.**

Magnets Design

We found it more reasonable to use the modular magnet design. The modules were each short solenoids wound on their own coil former. Each of the modules could be tested separately and the

Table 1. Parameters of modules

Series ##, modules and magnets tested Years of development	a_1, a_2, 2b, m	Wire, diameter, mm; Cu:Sc, num. of filaments	number of turns (average)	I_q after training (average) A	Fraction of minimum guaranteed I_c, % (average)	Notes
M1, 10 mod., 1 magnet with 8 modules 1976-1977	0.19, 0.21 0.023	0.7 2.33:1 42	884	175	75	no impreg. , strong training (5 -- 7 quenches)
M2, 24 mod., 7 magnets with 2--18 modules 1977-1978	0.179 0.212 0.0314	1.0 1.38:1 1045	843	380	90	partially impreg., strong training (4 -- 10 quenches)
M3, 16 mod., 8 magnets, with 2--10mod. 1979	0.179 0.212 0.0314	1.0 2.33:1 1132	810	370	~100	fully impregnated, no training
T, 26 modules, 4 magnets with 3 --7 modules 1980-1982	0.28 0.315 ~ 0.08	1.0 2.33:1 132	1500 2460 2340	300	90	fully impreg., non sufficient training (1 -- 3 quenches)
I1, one single module magnet 1983	0.19 0.215 0.119	0.85 1:1 210	2966	312	87	fully impreg., non sufficient train. (~3 que.)
K, two single mod. magnets 1986-1987	0.715 092 0.18	0.85-1.0 1:1-2.33:1 132-210	~18000	164	85	fully impreg., sufficient train. (~25 que.)
V, two single mod. magnets different wires 1986-1987	1.35 1.45 0.096	2x3.5-1.0 1.5 6290	960 - ~6700	~1800 - ~190 (calcul.)	??	fully impreg., both magnets produced but not tested

best ones were selected for the highest field area in the magnet assembly. The coil formers for our modules were made from stainless steel. The winding sides were insulated from the flanges of the coil former by textolite hoops.

During winding, we tensioned the wires to 190 MPa, close to the linear stress--strain limit for our superconductors. After removing the magnet from the winding machine and following cooling, the stress was obviously reduced. During energizing of the magnet, the stress never exceed our winding tension. This idea of preliminary pre--stressing during winding was supported by calculations done by software developed by our team[2,3,4]. This high tension allows a very good quality of winding and prevents the separation of the winding from the bobbin which might lead to preliminary quench. Our software permits us to calculate all mechanical stresses and strains in a winding at any stage of the magnets' production process. If it were necessary, we could calculate winding tension to provide completely uniform stresses inside of a magnet's winding.

As mentioned above, we developed and tested several types of modules and magnets, as listed in the Table 1. Detailed design and test results of the magnets labeled K are described in[7], design of one of the magnets "V" was published in[8].

The first two series of magnets were not impregnated as we wanted to simplify production. But absence of impregnation leads to higher degradation of the quench current of our magnets. So, after series M3, all other magnets were impregnated with epoxy resin. We used a simple method of "wet" winding when the epoxy compound with extender is painted on the each layer of winding.

One can see from Table 1 that use of impregnation eliminates training and reduced degradation for the magnets less than 0.6 m in diameter. Magnets of series "K" demonstrated a sufficient training and design operating current was only achieved after 15+ training events. We found that the training starts if **strain** in winding exceeds 0.05-0.07%. For our biggest magnets "V" we used strong banding to reduce strain at operation current [8].

To determine the operating magnet current, we used a method developed in[4]. We found that the most probable quench current (if the magnet is properly wound) is the minimum critical current guaranteed by the wire providers, but not the measured critical current of the wire shown in the producer's specification. This may be due to nonuniformity of the critical current along the wire.

The production of part of series "T" magnets and both "V" magnets was performed by Electrosila Co. in St.Petersburg in 1981 and 1987. The experience of work with the industry showed that our technological solutions were simple enough to be handled by the conventional electromachinery producer.

Several magnets were assembled from the modules of the different series. They varied in length, number of modules and stored energy. Some of them are listed in Table 2 where magnets used for outdoor-sea tests are also mentioned.

A magnet completed from 8 modules of M1 series was the first magnet in the world tested as the superconducting component of MCM in fall 1977. During the test, the platform with the magnet and cryostat was pulled off the sea at the speed more than 100 km/h due to pilot's mistake and then dropped to the sea surface. The platform remained upside--down in the sea with the cryostat's outlet underwater several hours before it was rescued. During this accident shock test the magnet remain superconducting and in persistent mode! The magnets from six "T" modules underwent explosion tests with perfect results[1] in 1984. Tests show that our main approaches to magnet design were correct.

MAGNET QUENCH PROTECTION

The demand for low system weight lead us to reject the use of cryostable superconductors for our magnets. We used intrinsically stable superconducting wires with high Cu:Sc ratio from 2.3 to 1

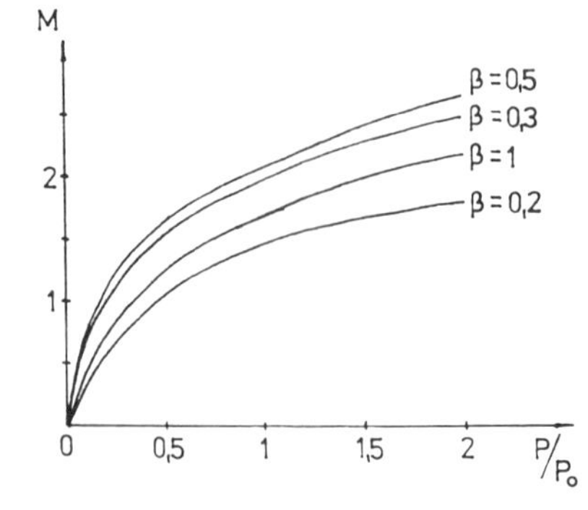

Figure 1. DMM of the magnet vs. scaled **magnet** mass for different β that is ratio of height to diameter. DMM is shown in units of 10^6 A/m^2.

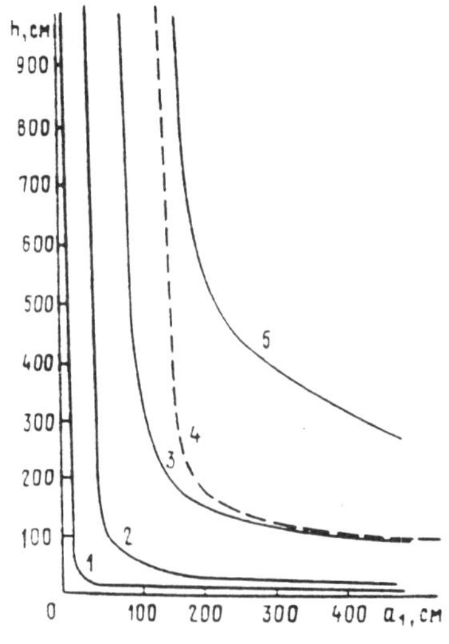

Figure 2. Phase diagrams show the limit for the size of the solenoid for stress not more than 300 MPa (solid lines) and heating not more than 300 K at uniform energy deposition (dashed line) at different characteristic parameters for superconductor (current density at zero field, and critical magnetic field).
1 -- J_0= 5 10^9 A/m^2, B_0=20 T (NbSn);
2 -- J_0= 2.2 10^9 A/m^2, B_0=11 T (NbTi);
3 -- J_0= 1.1 10^9 A/m^2, B_0=11 T (NbTi);
5 -- J_0= 0.7 10^9 A/m^2, B_0=11 T (NbTi);
4 -- uniform heating up to 300 K

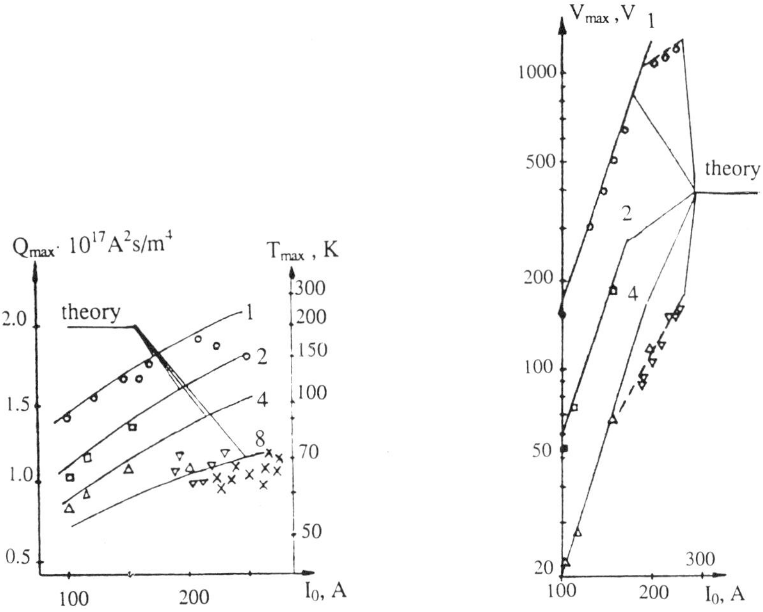

Figure 3. Maximum voltage and quench load (temperature) vs. initial current for the model solenoid with different number of sections. Digits in figure represent the number of sections in the same magnet.

for large magnets. High overall current density coupled with the demand for low operating currents completely excluded the possibility of the extraction of stored energy as the coil protection method. Any active devices to perform removal of energy such as quench detectors or commutators were also excluded. So, we met the completely new challenge: the protection of large high current density magnets operating at persistent mode at low current.

If extraction of stored energy from the magnet is unfeasible, the **only choice is to distribute it uniformly** in the magnet's winding. We found[2] that in this case no overheating will happen even for very large sizes of superconducting magnets. This conclusion is illustrated in figure 2. One can see that the stress limit on magnet sizes is stronger than the overheat limit.

Table 2. Parameters of magnets tested

Magnet ##, series	Energy stored, kJ	$I_{oper.}$,A; $J_{overall.}$ A/m^2	Magnetic moment, A·m^2	Quenches in tests	Notes about tests and pecul.
M1, 8 mod.	80	100	$0.85\cdot 10^5$	~ 4	at sea, '77--'84
M2, 18 mod.	~ 280	175; 1.4 10^8	3.2 10^5	4	at sea
17 M2 mod.	~ 300	190; 1.6 10^8	3.3 10^5	3	in laboratory
9 split M2 modules	~ 270	330; 2.8 10^8	3.0 10^5	5	in laboratory
M3, 3 modules	~ 300	345; 2.7 10^8	0.94 10^5	3	at sea, '79; cold diodes, choke, rotation
10 M3 modules	~ 1000	274; 2.1 10^8	2.5 10^5	2	in laboratory
I	200	312; 3.3 10^8	1.15 10^5	3	at snow and frost; sc choke, rotation
SPS-T0, two units	~1400	185; 1.4 10^8	7.3 10^5	~ 5 each	at sea ,'82--'87, explosion test
7 big T mod.	~1900	182; 1.4 10^8	12.3 10^5	2	in laboratory
K1+K2	~2700	~165; 1.2 10^8	~10.0 10^5	~ 25 + ~15	at sea, '87-'88, rotation

Therefore, to protect the magnet without energy extraction one should force the normal zone to propagate as fast as possible. There are different methods, both active and passive ones to achieve this. We chose the method of subdividing the magnet's winding in sections with shunting of all sections by shunts with low resistance. This method was proposed a long time ago[9] but was never before studied in detail. During our R&D activity, we performed this investigation after our second magnet assembled from 18 M2 modules accidentally burned. The results of this study were published in our collected papers[2,3,4] and partially in English. In this part, we present a brief review of those results.

We developed several experimental methods for quench study. Most of our magnets were heavily instrumented. That means a large number of potential taps to monitor voltages in winding, specially developed techniques of using small thermocouples to measure temperature in winding, miniature heaters to stimulate quench in different locations, etc. During our R&D activity we performed 345 quenches of single modules (from "M1" to "T") and about 170 quenches of assembled magnets, plus 39 quenches of "K" magnets. The stored energy ranged from 30 kJ (single "M1" module) up to 2.7 MJ ("K" magnets). We thus collected a huge experimental data base and a great deal of experience in quenching of superconducting magnets. Some of the magnets tested are listed in Table 2 with notes about their outdoor-sea or laboratory tests. Overall current density and stored energy are listed also.

Along with experimental research a lot of theory was developed. The analytical theory of quench processes in tight impregnated windings permitted us to develop the method of quick estimation of quench parameters of sectioned magnets. Several computer programs were developed to perform more exact numerical calculations of a quench.

During a quench there is a competition between the volume of normal zone and the energy dissipated in it. It was shown[4] that the ratio between the normal zone volume and the volume of the whole winding is the most important parameter for protection. The smaller this ratio is, the smaller the voltage and temperature of hot spot of the winding will be. So, if one is limited by the natural 3-D normal zone propagation velocity, the only way to protect the winding is to reduce its size. But if a large magnet is required the only way to overcome this constraint is to subdivide the winding with low resistance shunts.

The shunts allow each section of the magnets to behave like independent magnets with a smaller part of stored energy (if shunts resistances are low enough). Only a fraction of the total stored energy will be realized in a normal zone appearing in one section. That means of course reduced heating and voltage. In figure 3 the current load and voltage for one of the magnets that was subdivided in different number of sections are shown One can see that sectioning sharply reduces heating and voltage. The calculations done by analytical theory[4] are shown by the solid and dashed lines in figure 3. One can see the good coincidence of theory and the experiment that permits us to use this analytical theory for quick estimation of quench parameters.

Actually, in the sectioned magnets, quench development is more complicated. As the current decays in the first quenched section the current rises in an adjacent section. At some moment, this next section may also quench leading to current decay in it and current increase in another section. Under some conditions, quench will propagate quickly through the whole winding not by the usual heat propagation but by electromagnetic interaction only. One such example is shown in figure 4. This process, called "electromagnetic avalanche" was first modeled by numerical calculations in 1977 [2] and then was observed experimentally. In figure 4 the results of numerical simulation of quench by the software is shown. One can see that for complicated processes the software gives good coincidence with the experiment.

We found that generally the process of electromagnetic avalanche has a trend to uniform energy deposition in each section if the size of sections is small enough. The size criterion for small enough section is: the volume of the section should be equal to the volume that the normal zone would fill out during some characteristic time τ_h [4]. This time for our magnets is ~1 s if overall current density is ~ 0.8 10^8 A/m^2 and ~ 0.1 s if current density is ~2.5 10^8 A/m^2. This conclusion is illustrated in figure 5 where measured temperatures after quench of a several magnets are shown vs. stored energy. One can see that while stored energy changes about two orders of magnitude, the average temperature of windings after quench remains low and follows the heating corresponding to uniform energy deposition. In unsubdivided magnets the temperature would rise very quickly. The results show how effective coil protection by subdividing with low resistance shunts can be.

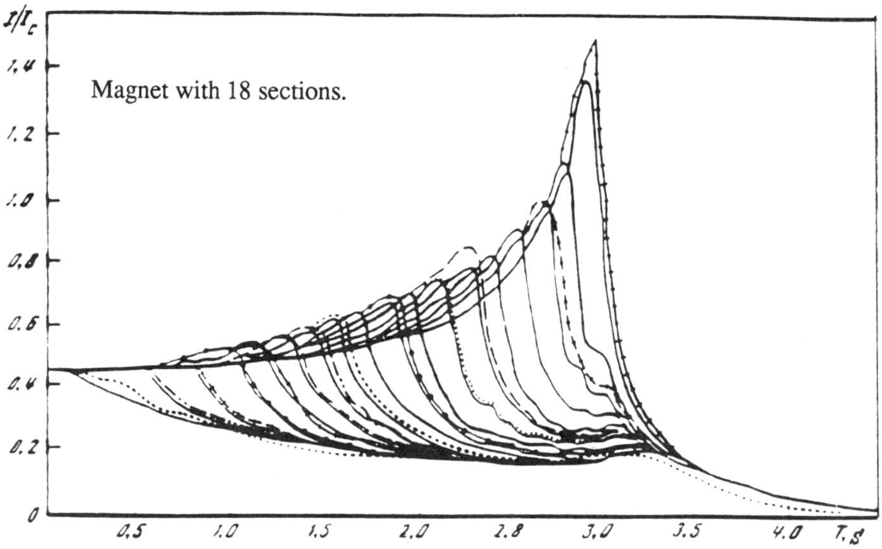

Figure 4. The example of the specific quench development process - electromagnetic avalanche. Solid lines - calculations; lines with symbols - experiment.

The role of shunts in this coil protection method is to catalyze the electromagnetic avalanche and to separate winding sections. Shunts completely change quench development in the magnet by permitting currents in sections to change non--synchronously. This leads to acceleration of the normal zone propagation and to almost uniform energy deposition in the winding and therefore to full magnet protection. If electromagnetic avalanche happens during quench only ~2-5% of the stored energy is released in the shunts, while the rest is released inside the winding. Thus shunts may be made small and light.

We found that for reliable coil protection shunts should have the same resistance as one magnet section has at 10 K. This rather low resistance of shunts may lead to some helium losses during charging-discharging magnets. To reduce helium loss one may wish to place the shunts above the helium level in the cryostat and use the enthalpy of helium gas for shunts' cooling. We found that this measure reduced helium loss by almost a factor of ten.

Another opportunity is to use cold diodes instead of shunts. The diodes at helium temperature have voltage--current characteristics ideal for coil protection[4]. We used them in M3 magnet (see Table 2) and its appears they were very reliable and worked well during all outdoor tests.

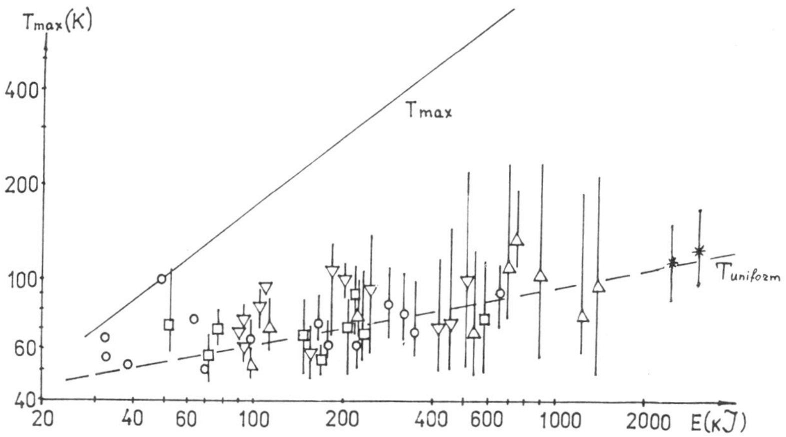

Figure 5. Peak temperature of sections vs. stored energy for almost all magnets tested. Solid line - heating in non protected magnet, dashed line -heating at uniform energy deposition.

The safety criteria for large magnets was proposed in[10]. The parameter Ej_{cu}^2 must be less than 10^{23} JA^2/m^4, where E is stored energy and j_{cu} is the current density in a matrix. Most of our magnets successfully operated at Ej_{cu}^2 ~2.2 10^{23} JA^2/m^4 while the "K" magnets operated at as much as 4.6 10^{23} JA^2/m^4 ! The reliability of our coil protection method for large superconducting magnets with high current density is thus proven.

Sometimes the size of sections cannot be made small enough as described above. In this case, to accelerate normal zone velocity in the sections, the protection shunt from one section may be imbedded inside of the winding of another section[11]. This method was successfully tested for small magnets with strong insulation which reduced normal zone propagation in the transversal direction. Later it was used in the design of our biggest produced magnet "V"[8].

SUMMARY

During R&D activity on superconducting magnets for MCM, the design, technology and protection method of vibro--stable, light superconducting magnets with high current density was developed and tested. The design and the protection method is described. The outdoor-sea tests prove the soundness and reliability of technical solutions and designs. The results of R&D are of general importance to superconducting magnet technology for different civilian and military purposes.

ACKNOWLEDGMENTS

This work was completed by a big team and we are very sorry that here is no room to list all of them. The authors are very grateful to everyone who participated and assure you that we remember your great personal contributions in our common efforts.

REFERENCES

1. V.R Karasik et al. "The first mine countermeasures with superconducting magnet", this conference
2. "The issues on the Applied Superconductivity", *Proc. of the P.N. Lebedev Physical Institute,* 121, Moscow, NAUKA, (1980), (in Russian)
3. "The Transient Processes in the Superconducting Magnet Systems", *Proc. of the P.N. Lebedev Physical Institute,* 150, Moscow, NAUKA, (1984), (in Russian)
4. "The Physical Processes in the Superconducting Devices", *Proc. of the P.N. Lebedev Physical Institute,* 205, Moscow, NAUKA, (1991), (in Russian)
5. V.S. Vysotsky, A.A.Konjukhov, V.N.Tsikhon and V.R.Karasik, this conference
6. E.M. Golda, J.D.Walters, G.F.Green, *Naval Engineering Journal,* May 1992, 53
7. V.R.Karasik et al. *Proc. MT-11,* Tsukuba, Japan, (1989), 1312
8. V.R.Karasik et al. *Advances in Cryog. Engin.,* 37A:409, 1992
9. P.F.Smith, *Rev.Sci.Instr.,* 34:368. (1963)
10. Eberhardt P.H. et al. *Proc. MT-6,* Bratislava, ALFA, (1977), 654
11. A.A.Konjukhov et al. *IEEE Trans.Magn.,* 25:1538, (1989)

PERSISTENT MODE SWITCHES AND AUTOMATIC POWER SUPPLIES FOR AUTONOMOUS SUPERCONDUCTING MAGNETS

V.S.Vysotsky[1], A.A.Konjukhov[2], V.N.Tsikhon, and V.R.Karasik[3]

P.N.Lebedev Physical Institute, 117924, Moscow, Russia

[1]Present address: PFC MIT, Cambridge, MA 02139 USA
[2]Present address: RSC Kurchatov Institute, Moscow, Russia
[3]Present address: Babcock and Wilcox, Lynchburg, VA, USA

ABSTRACT

In the framework of the Russian R&D Program on the application of superconductivity to mine-sweeping, autonomous superconducting magnets operating in persistent mode were developed, produced and tested. Special efforts were taken to develop optimal, low loss and reliable persistent current switches (PCS). The operational cycle of PCS working with superconducting magnets was considered. The methods of calculation and optimal design were developed. Seven dozen PCSs were produced and tested. Automatic two-button power supplies (APS) were developed simultaneously with PCS. Different approaches to electrical scheme design of APS were considered and tested. An operating algorithm for an APS with PCS and superconducting magnet was developed and realized in several real devices with remote control. Outdoor tests in sea conditions showed very high reliability for the newly developed APS and PCS.

INTRODUCTION

The demand for persistent mode operation of superconducting magnet systems for mine countermeasures (MCM) was one of the main priorities in the Navy's task code for R&D of superconducting MCM. That permits use of the main advantage of superconducting systems: long time operation without power supply. At the time there was no published source for design of reliable persistent mode switches nor automatic power supplies for superconducting magnets. Only recently some publi ations have appeared on the subject.[1,2]

During our work several dozen PCSs with different parameters were designed and tested. Since a superconducting magnet using PCS places special demands on the power supply, the design of the APS and PCS was done in parallel with adjusting one device to another.

In the end several reliable, optimal PCS switches that are good for rather large magnets and several automatic, two button power supplies were developed and tested [3,4,5]. In the present paper we offer the brief review of the results.

Advances in Cryogenic Engineering, Vol. 41
Edited by P. Kittel, Plenum Press, New York, 1996

PERSISTENT CURRENT SWITCHES

There are two main kinds of superconducting switches used with superconducting magnets: the interrupter and the persistent switch. They have different operating cycles and design criteria.

The main function of the switch-interrupter is to provide the transfer of energy stored in the superconducting magnet to a given load with small discharge time. Therefore, the winding of the switch-interrupter is heated adiabatically. The mass of the switch-interrupter has to be large enough to accept this heating and the resistance of switch-interrupter has to be high enough to provide the necessary commutation power. It may be placed in a separate cryostat from a magnet.

PCS is necessary to provide the persistent mode of a superconducting magnet. PCS must be placed in the same cryostat with the magnet and wired in parallel with the magnet winding. Thus, it is in the leakage magnetic field. The PCS winding is heated when charging--discharging the magnet over long time scales (usually more than hundreds of seconds). Therefore, the heat released in the switch winding can be transferred to helium. Thus, the switch winding is operated at heat balance conditions and heat release in PCS is an isothermal process.

The goal of the designer of the PCS is to choose the parameters of the switch winding so that they would be optimum from standpoint of minimal helium loss during energizing of the magnet. The PCS must have enough current carrying capacity and stability at given magnetic field. Also, it is necessary to take into consideration the problem of contacts between the conductors of the magnet and the switch, because the total resistance of the circuit influences the decay time of persistent mode operation.

Operating Cycle of PCS

To formulate the demands on a well--designed PCS, let us consider the operating cycle of a superconducting magnet with a persistent current switch. It has the following stages:

1. Transforming the persistent current switch to a normal state is carried out by its control element. The choice of control method depends on switching speed. As a fast switch is not needed, the simplest and most reliable heat control was chosen, i.e. contact of switch winding with heater.

2. The magnet energizing occurs after switch quenching when the power supply voltage is applied to the magnet. While the magnet is charging, the switch winding is in its normal state. Since the voltage is applied simultaneously to the switch, heat is generated in the switch winding leading to liquid helium loss. The amount of loss depends on the charge voltage, normal resistance of the switch and charging time. The last parameter depends on the power supply voltage, operating current and self-inductance of the magnet.

During magnet charging the resistance of the switch depends on the heat generated in the winding and on cooling conditions. The optimum case occurs if the power released in the winding is enough to maintain it at a temperature just higher than critical temperature of superconducting wire. In this case the whole winding of the PCS is in the normal state and the heater can be turned off.

3. Stabilization of the magnet current is necessary for cooling of the switch and transition of its winding to a superconducting state. The power supply must stabilize the operating current. In this step, voltage on the magnet becomes zero and heat generation in the PCS winding stops. The duration of cooling time depends on the switch design and its previous heating before the stabilization of the current. Voltage ripples in the power supply may induce alternating current in the switch winding which increases the cooling time or prevents the switch from cooling down. Reducing ripples is a special challenge in power supply design as the ripples must not exceed a certain level.

4. Decreasing power supply current to zero can be carried out after the switch winding becomes completely superconducting. During this process the PCS current increases from zero up to operating value. Since fast change of a current affects the stability of superconductor, the rate of current rise must not exceed a permissible value.

5. Persistent mode starts after turning off the power supply. The current of the magnet is closed through the switch. The power supply may be disconnected and current leads removed if necessary. For safer operation it is desirable to be able to connect the load to the magnet. This limits the maximum voltage on the switch under spontaneous quench.

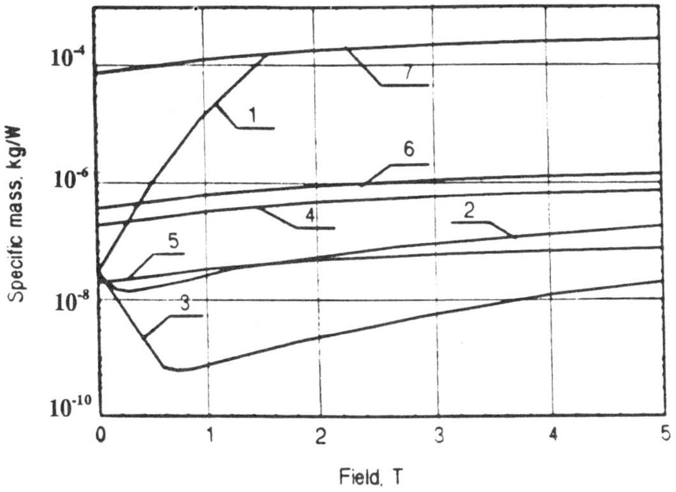

Figure 1. The specific mass of superconducting commutator vs. operating magnetic field
1 -- NbTi tape, field is perpendicular to the tape surface and current; 2 -- the same, but field is parallel to the tape surface; 3 -- NbTi tape, field is parallel to the tape surface and current; 4 -- NbTi wire in the CuNi matrix; 5 -- NbTi wire in the CuMn matrix; 6 -- NbTi wire in the brass matrix; 7 -- NbTi wire in Cu matrix

The rate of current decay depends on total contact resistance of closed superconducting circuit and the magnet inductance. If requirements on current decay time is not very strong and inductance of magnet is high enough (low operating current as in our case), the use of soldered contacts is possible. The total resistance of several tens of soldered contacts doesn't exceeded 0.1μOhm for wires with diameter of 1 mm and operating current about 200A.

6. Discharging of the magnet to the load occurs once the persistent current switch is quenched by dint of the heater. Helium loss issues are the same as during the charging process.

From the considered operation cycle of PCS one can see that the main problem is the stability in the switch at fast changing current in the switch (see item 4). This process mostly affects the maximum operating current of the PCS. To minimize helium loss the thermal properties of the switch should be considered.

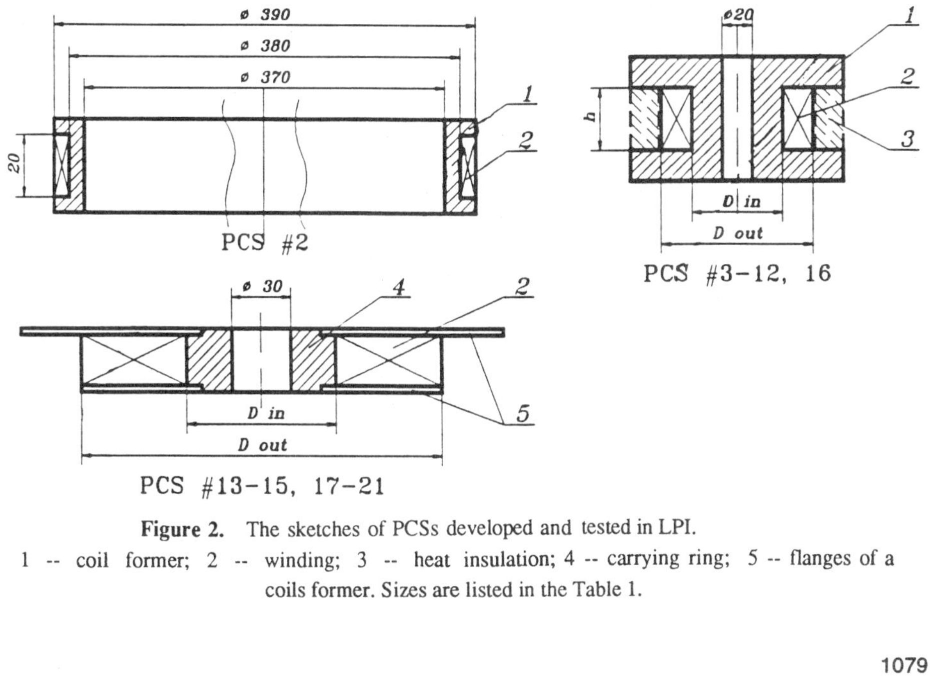

Figure 2. The sketches of PCSs developed and tested in LPI.
1 -- coil former; 2 -- winding; 3 -- heat insulation; 4 -- carrying ring; 5 -- flanges of a coils former. Sizes are listed in the Table 1.

Table 1. Parameters of the PCS

PC S ##	Superconductor, diameter matrix	d_{in},d_{out}, h mm (see fig.3)	Coil former material	Outer heat insulation	R_0 , at 10 K, Ohm	I_c at B=0, A	V_0, V, I_m, A	Notes, works with magnet: energy, current
1	NbTi tape	15,20, 10	brass	no	4.4	520	2.5 0.58	yes, 5kJ, 200 A
2	NbTi tape	380 390 20	textolite, 5mm	2 layer, fibercloth	10	1000	25 1.8	Heater: brass tape
3-9	Multifil. NbTi 1.5mm brass	40,60, 25	stearfoam plastic, 20mm	stearfoam glued or not	0.5-0.55	980	0.2-2.2 0.4-4.4	
10	Mf. NbTi, 1mm brass 2 parallel.	60,190 25	the same	the same with holes	6.2	880	3.5 1	yes, 300 kJ, 100 A
11	Mf., NbTi .5 mm CuNi 3 parallel.	80,130 20	the same	stearfoam plastic, 20 mm	22	750	10.5 0.47	yes, 0.5MJ, 150 A
12, 13	Mf., NbTi, 1mm, Cu	80,170 20	12- the same, 13 Al ring, textolite 5 mm	2 layers fibercloth	4-6	940	0.3-0.6 7.6-7.8	epoxy impregnated
14 15	Mf., NbTi, 1mm, CuNi	80,170 20	Al ring, textolite	the same	9.5	750	0.48	
16	the same, 1.5mm	80,200 20	stearfoam plastic	the same	6.25	1240	5 0.8	inductive winding
17	Mf. NbTi, 1 mm, CuNi	80,110 25	Al ring, textolite	the same	6	750	4.9 0.8	yes, 140kJ, 325 A
18-21	Mf. NbTi, 1 mm, CuNi	80,110 25	Al ring, textolite	the same	24-30	750	8-12 0.4-0.5	yes, see below*
22	the same	240 275 110	the same	the same	59	600	60 0.89	yes, 2.7 MJ, 180 A
23	Mf. NbTi, 1mm, CuMn	240 245 110	the same, extra heat insulation	the same	41	550	40 0.9	yes, 2.8 MJ, 190 A
24	the same	80, 113,25	Al ring, textolite	2 layers fibercloth	24	700	10.5 0.42	

* 18 - 120 kJ, 300 A; 20 - 1.4 MJ, 200 A; 21 - 1.4 kJ, 200 A

Selection of Materials

We have considered different superconducting materials for use in persistent current switches. We calculated the dependence of specific mass of the switches on magnetic field for different superconducting current carrying elements that are shown in figure 1.

The specific mass is the mass of the switch from the certain material at a certain field divided by required operating power[3]. At parallel magnetic field, the NbTi tape has the highest specific characteristics, superconducting multifilament wire with copper matrix has the lowest. The use of NbTi tape for PCS is limited by influence of leakage field on its current carrying capacity and by necessity for shielding.

In choosing the current carrying element for PCS, it is important to take into account its stability with respect to spontaneous disturbances. The stability is connected with heating conditions of the switch -- the lower the available heat transfer to helium, the more stable the switch has to be. The stability improves with decreasing wire matrix resistivity and current density. At the same time the mass of the switch rises with the decrease of resistivity and current density. An optimum design balances these two issues.

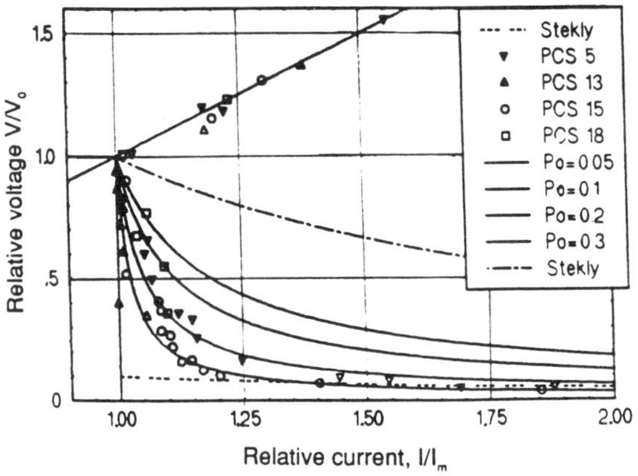

Figure 3. The scaled voltage--current characteristics for different PCS.

One solution could be the choice of composite material with matrix resistivity less than NbTi alloy but more than copper, i.e. brass or copper-nickel or copper-manganese alloys. These materials are most useful for persistent current switches operated in leakage fields.

In summary it is necessary to note that real current density in the switch is limited not only by current carrying capacity at certain magnetic field, but also by the maximum current during rapid current changes (see item 4 in operating cycle).

Experimental Investigations

During research and development activity for autonomous superconducting magnets, the goal to create reliable and optimum persistent current switches was accomplished. More than twenty PCS were produced and tested in LPI from 1975 to 1987 with the use of different superconducting materials such as niobium-titanium foil and different superconducting multifilament wires. Copper, brass, copper-nickel and copper-manganese were used as the matrix material. The typical view of some tested switches is shown in figure 2. Their main parameters are listed in Table 1.

To exclude the influence of self-field on switch parameters, all PCS, with the exception of PCS16 had bifillar winding. During the winding process two superconducting wires or tapes (four wires for PCS10 and six wires for PCS11) were jointly wound around the former. The heater was simultaneously mounted with the main winding. It was usually wound by copper-nickel alloy wire with diameter of about 0.4mm. For PCS2 the brass foil was used as a heater. Sometimes, a heater was subdivided in sections for parallel joining.

AuFe--Cu thermocouples were imbedded in the windings of all switches to check the temperature of switches. The windings of switches 12--24 were impregnated by epoxy resin.

Thermal Properties of PCS

To design an optimum persistent current switch for the charge-discharge of a superconducting magnet; it is necessary to study the thermal properties of the switch. The thermal properties of the switch are determined by its voltage-current characteristic (VCC) and cooling time.

We found that VCC has a common shape for all PCS that is shown in figure 3. One can see that VCC has two regions.

If voltage exceeds V_o, which is characteristic for individual switch, VCC is linear, which corresponds to constant switch resistance. In this region the whole winding of switch is at a normal state.

When voltage is less than V_o, VCC has vertical portion and there after the curve deflects to a range of large currents. Vertical portion and abscissa intersect at point I_m which can be named as minimum current for existence of normal zone, by analogy with work[6]. If current exceeds I_m the switch will never cool after turning off a heater. Basing on the minimum current for existence of a

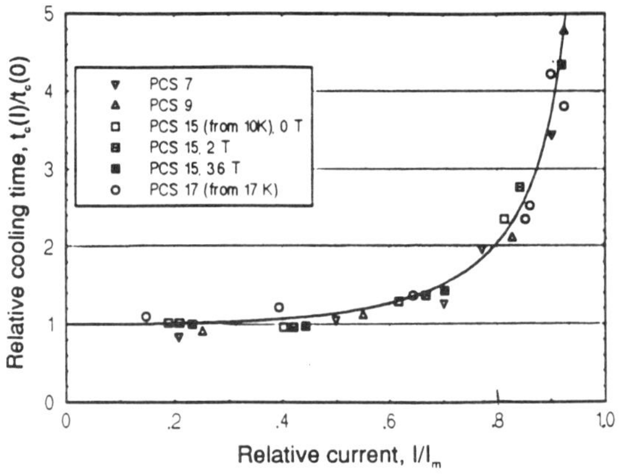

Figure 4. The scaled cooling time vs. relative switch current for different PCS.

normal zone one can state the requirement on the power supply: **during stabilization regime the current in the switch must not exceed Im.** If $I< I_m$, switch becomes superconducting and persistent mode is possible.

Voltage V_o, current I_m and corresponding power $P_o= I_m V_o$ are the parameters for **optimum charging-discharging regime.**

We found that VCC of PCS may be described by the expression below which reasonably matches the measured VCC (see figure 3). Here p_o is the parameter. The values of V_o and I_m may be calculated for PCS design to create optimal switch.

$$v = ip_0 / (i^2 - (1 - p_0)); \quad i = I / I_m \tag{1}$$

The cooling time of PCS -- t_c determines the stabilization time of the current. Excessive cooling times can lead to extra helium losses [1,3,4].

The cooling time depends on the current in the switch and initial temperature of the winding. We found common dimensionless dependence of cooling time on current in the switch as it is shown in figure 4. In general $t_c \sim 1/j_m$ where j_m is the minimum current density for existence of a normal zone.

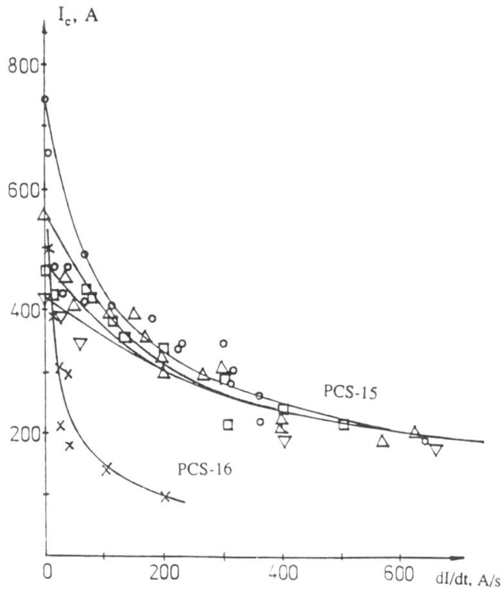

Figure 5. Current carrying capacity of PCS 15 and PCS 16 vs. dI/dt. Solid lines - theory[7].

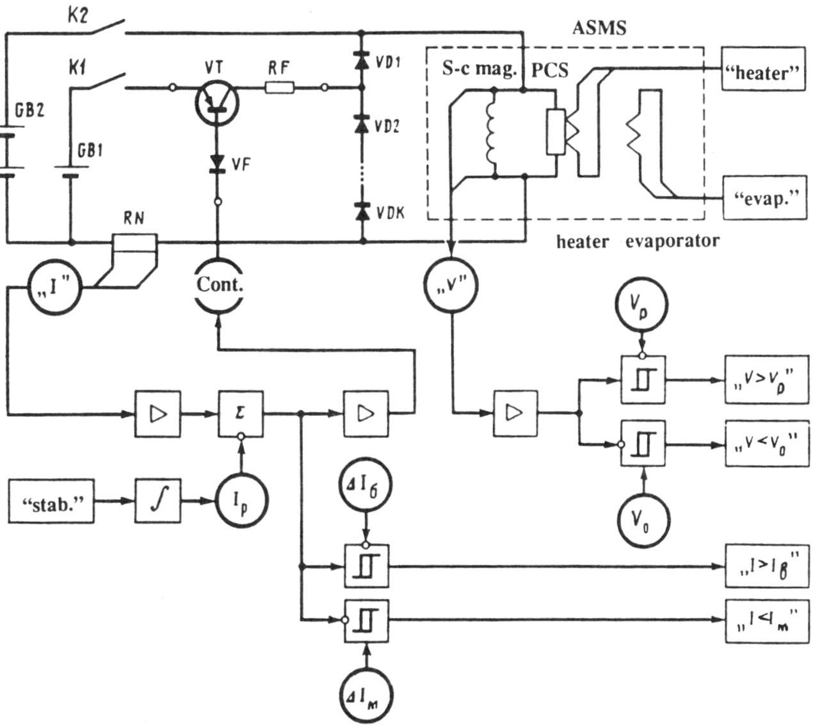

Figure 6. Automatic power supply with a transistor regulator. Logical scheme is shown.

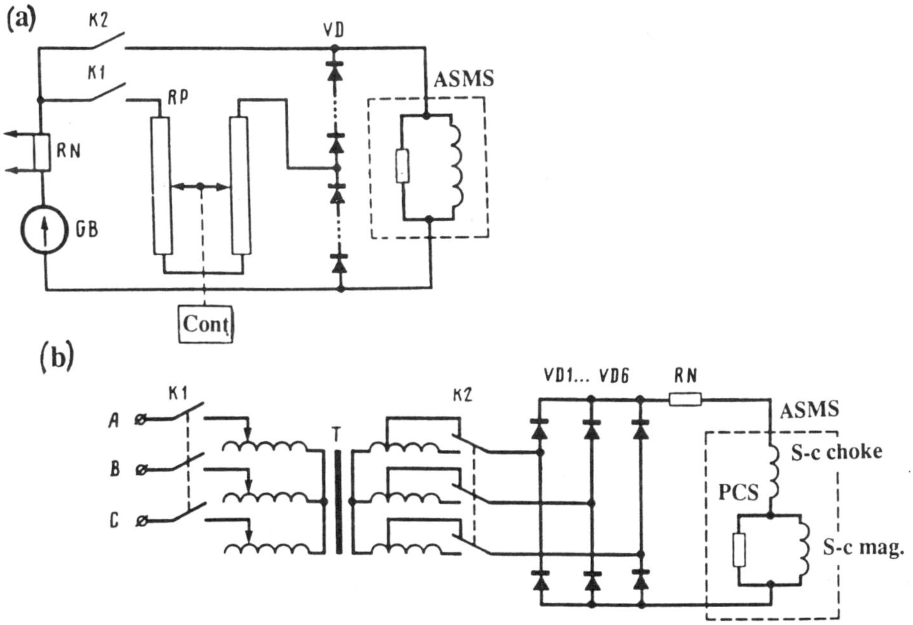

Figure 7. Automatic power supplies with rheostat (a) and superconducting choke (b).

Critical Current and Stability

The critical current of a switch is the parameter that determines its mass. It depends not only on superconductor used and the background magnetic field, but also on the time derivative of current during ramp--down of APS (see item 4 of Operation cycle). We studied this problem and found that superconducting multifilament wires in highly resistive matrix have good enough stability to work with the autonomous superconducting magnets.

The typical dependence of critical currents on dI/dt for PCS 15 and PCS 16 are shown in figure 5, for different DC background magnetic fields. One can see that reduction of quench current is more pronounced in the inductively wound PCS16. This confirms the choice of bifillar winding for PCS.

The solid line shown in figure 5 is a calculation using the theory of thermomagnetic instability[7]. One can see good agreement between theory and experiment thus permitting us to use this theory for design of PCS. From this calculations and/or measurements of current carrying capacity of PCS we may determine the demands on the PS for reducing its current after stabilization.

Design of PCS

Based on our investigations, we developed a method of engineering design for PCS. Having the initial data: operating current I_o, inductance of magnet L, magnetic field on the switch B, charging voltage V_o and allowable helium losses W, one can design the switch with the resistance R_0 and current I_m, using as an adjusting parameter the cooling surface A. The expressions for calculation are as follows:

$$R_0 = \frac{V_0}{I_m} = \frac{LI_0 V_0}{W}; \quad A = \frac{I_m V_0}{h_o \Delta T_o} = \frac{WV_0}{LI_0 h_o \Delta T_o}; \quad I_m = \frac{W}{LI_o}\sqrt{1 - B/b_{c2}};$$
$$t_c = k\left(\frac{LI_0 S}{W}\right)^2; \quad h_o \sim 0.005 - 0.01 \mathrm{W/cm^2} K, \ \Delta T_o \sim 5K, \ \mathrm{k} \sim 80 \ \mathrm{kJ/Ohm\,cm^4} \tag{2}$$

AUTOMATIC POWER SUPPLY

The superconducting magnet with PCS provides long time operation in persistent mode with rather short time for charging--discharging. This periodic kind of work permits the use of special approaches to optimize the design of an automatic power supply.

We performed full analysis of losses in the regulator element of the power supply to find out the optimal solution. We concluded that as there are three stages of the APS operation with superconducting magnet with PCS (energizing, stabilization + current's drop and discharging (items -- 2, 3+4 and 6 of the operational cycle of PCS), the APS for each stage should be designed and optimized separately.

Therefore, the APS has to consist of three independent power blocks. Each power block is optimized for a certain stage: charging, stabilizing and current drop or discharging. In this case, we were able to develop an algorithm for the automatic control that provides sharp and reliable operation of APS together with the magnet and switch. The control device switches through the three power blocks in accordance with the digital signals provided by comparators that are checking zero voltage on the PCS and the operation current (see figure 6). All work is performed in accordance with two commands "charge magnet" and "discharge magnet". The opportunity to compensate for some malfunctions is included in the algorithm also.

As only two commands are necessary to start the work of APS and then the rest was done automatically, it was easy to operate via remote control, by radio for instance. Remote control was tested in our sea-tests and worked well.

Some solutions used in design of APS and in the algorithm are as follows:

For magnet charging much higher voltage is necessary than for stabilization of the current, when the voltage should overcome the resistance of current buses only. Hence, we used different voltage sources for charging and stabilizing. To charge the magnet, no powerful control element was used to regulate the current. After the current achieves the necessary magnitude, the comparator provides the signal that switches to the source with lower voltage and the regulator element starts to stabilize the current. As the source voltage at this moment is low, the losses in the regulator element are low also and heating is small. When another comparator recognizes that voltage on PCS is zero, after some time it provides the signal that starts current's drop. If the voltage remains zero, the current decreases to zero. If the voltage reappears due to incomplete cooling of the PCS, the stabilization continues some time more, after which the drop attempt may be repeated.

To discharge the magnet, APS provides heating of the switch to allow energy stored in magnet to discharge to the external load. The voltage on the load was adjusted to the optimal V_0 for the given switch (see above), and remained almost constant during discharge time through use of diodes in the load removing the need for extra control for the discharging process.

Due to periodic cycle of APS operation, its blocks were not designed for permanent cooling. Instead, all power blocks were immersed in an oil bath. The oil having high heat capacitance, accumulates the heat during the short time charging period and then during persistent mode the heat was dissipated in the environment. This solution allows a reduction in the mass of the power blocks.

We tested several types of APS having different regulator elements. Some of the electrical schemes are shown in figures 6 and 7 a,b. In the scheme shown in figure 6 transistors were used as the regulator element. Two voltage sources GB1 and GB2 were used for charging and stabilization. The set of diodes was used as load trough which the magnet was discharged.

In the scheme in figure 7 a, the rheostat was used as the regulator element. It was possible because batteries that do not provide ripples were used as the energy source. The rheostat was made from the stainless steel tubes and was insensitive to sufficient overheating. That provides the high reliability of the APS.

Any ripples may be eliminated by a superconducting choke in series with the magnet and PCS. We used this solution in the APS shown in figure 7 b. The power source was the simplest transformer with the rectifier controlled by the voltage on the primary winding of the transformer. All ripples provided were filtered by the choke inside the cryostat. This APS had record low mass and really looks to be simplest of the all possible choice!

All APS described participated in outdoor--sea tests under severe conditions and worked perfectly. The tests proved the reliability and soundness of the design.

CONCLUSION

A method of optimization and design of PCS and APS was developed. Several PCS and APS were produced and tested along with superconducting magnets for MCM in severe outdoor--sea conditions. The tests show the soundness and reliability of the technical design solutions.

ACKNOWLEDGMENTS

This work was completed by a big team and we are very sorry that there is no room to list all of them. The authors are very grateful to everyone who participated in the work and assure them that we remember their big contribution in our common effort.

REFERENCES

1. M.A.Green, *IEEE Trans. Apll. Superconductivity,* 3:582, (1993)
2. R.C.Whitestone, G,Green, *Advances in Cryog.Engin.*,39:317, (1994)
3. "The issues on the Applied Superconductivity", *Proc. of the P.N. Lebedev Physical Institute,* 121, Moscow, NAUKA, (1980), (in Russian)
4. "The Transient Processes in the Superconducting Magnet Systems", *Proc. of the P.N. Lebedev Physical Institute,* 150, Moscow, NAUKA, (1984), (in Russian)
5. "The Physical Processes in the Superconducting Devices", *Proc. of the P.N. Lebedev Physical Institute,* 205, Moscow, NAUKA, (1991), (in Russian)
6. Stekly Z.J.J, Zar J.L., *IEEE Trans. NS,* NS-12:367, (1965)
7. A.L.Gurevich, R.G.Mints, A.L Rakhmanov, "The Physics of Composite Superconductors", Moscow, NAUKA, (1987), (in Russian)

A LIGHT-WEIGHT RUGGED CONDUCTION-COOLED NbTi SUPERCONDUCTING MAGNET FOR U.S. NAVY MINESWEEPER APPLICATIONS

M. Heiberger[1], A. Langhorn[2], T.W. Hurn[1], D.G. Morris[1], J.D. Walters[3], L. F. Aprigliano[3], D. Waltman[3], and G.F. Green[3]

[1]General Atomics, P.O. Box 85608, San Diego, CA
[2]Startech, Inc., P.O. Box 573, Solana Beach, CA 92075
[3]Naval Surface Warfare Center, Carderock Division, Annapolis Detachment
Annapolis, MD 21402

ABSTRACT

A superconducting magnet using an aluminum-stabilized NbTi coil cooled by two Gifford-McMahon rare earth (GM-RE) cryocoolers has been built and successfully tested in support of the U.S. Navy's efforts to develop the technology for shallow water magnetic mine countermeasures. The 3.4 T, 0.5 meter diameter coil is supported in a vacuum vessel with a series of re-entrant filament-wound cylinders made from E-glass, that are designed to withstand large shock and vibration loads. This high load combined with the need to minimize conduction heat leak to the coil has motivated the need for the lighter weight conductor, thus reducing the stuctural burden. The Navy's additional requirement to minimize the overall system weight has resulted in a compact re-entrant cylinder cold-to-warm support (CTWS) structure design with integrated thermal intercepts and shielding. Structural joints at E-glass and metal transitions in the CTWS use a combination adhesively bonded and riveted scarf joint configuration. Low impedance thermal interfaces between the cryocoolers and the coil, shield, and current leads were developed to ensure that coil temperatures remain well below the NbTi critical transition temperature during operation.

INTRODUCTION

The U.S. Navy is exploring the use of superconducting magnets that can generate magnetic fields high enough to detonate shallow water magnetic mines at an adequate standoff distance. Requirements have been established that prohibit the use of liquid cryogens for practical and logistics reasons. In addition, the system must be designed for transient multi-directional loads as high as 100 g and continuous vibration loads up to 2 g. These are the main design drivers dictating a light weight structurally robust configuration that can be maintained cold enough using only cryocoolers. Coincident with these

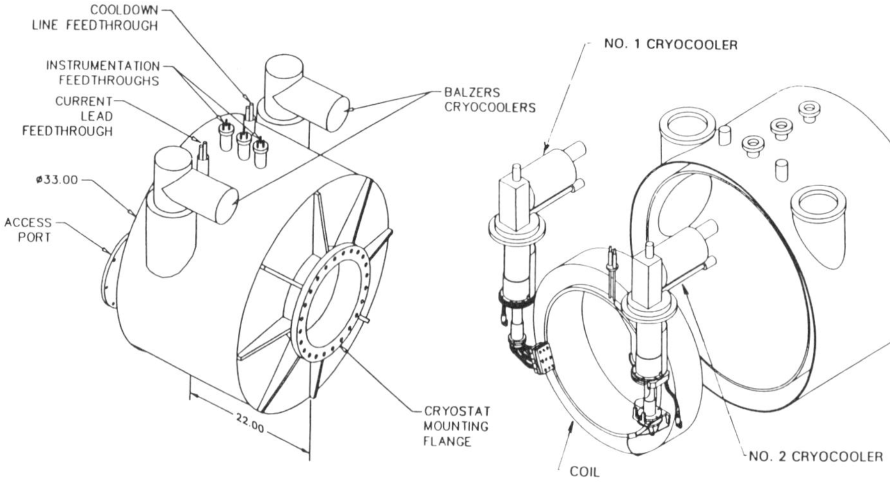

Figure 1. Magnet cryocooler arrangement and interfaces with cold mass components.

requirements is the need to achieve a specified magnetic dipole moment while maintaining strict adherance to a weight limit traceable to the planned method of deployment.

The ruggedness, lightweight, and quench tolerance required for this application motivated the selection of aluminum-stabilized NbTi conductor. The need to maintain the NbTi conductor temperature near 5 K during operation has necessitated the development of low impedance cryocooler thermal interfaces and a CTWS structure design that provided the lowest conduction heat leak given the inertial loads imposed on it by the coil and other cold masses.

A subscale test article mine countermeasures superconducting magnet that embodies these key features has been designed and built at General Atomics, and initial tests that demonstrate feasibility for larger devices have been conducted. Cold temperatures are maintained using two GM-RE cryocoolers manufactured by Balzers, designated as the Kel-Cool 4.2GM. The second stage nominal performance of each cryocooler is rated at 0.5 W at 4.2 K. One cryocooler is thermally interfaced to the coil via a flexible copper link and the other is connected to the coil's 90 A current leads (Figure 1). In this configuration, the coil temperature is maintained at just under 4 K at zero current, and about 5 K at full

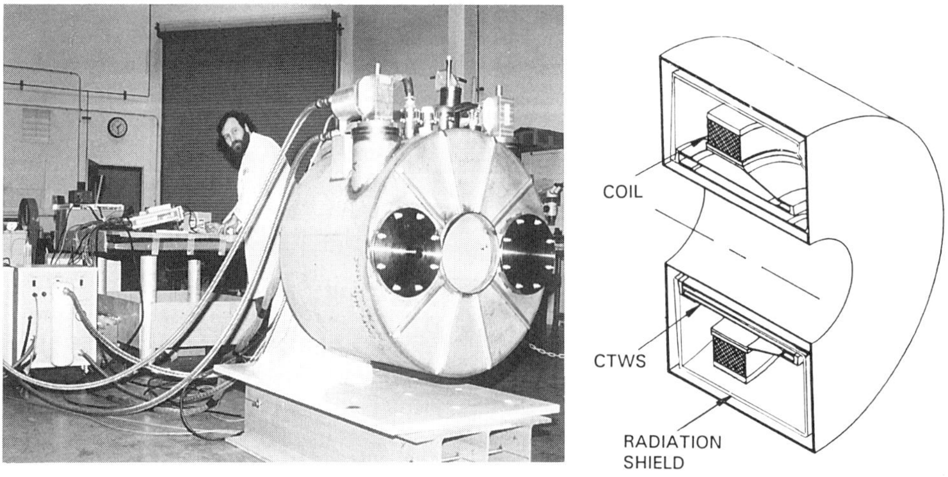

Figure 2. Magnet assembly and cross section showing CTWS, shield and coil configuration.

current. The conductor and coil are designed so the superconductor's critical transition temperature is slightly above 7 K, resulting in a 2 K temperature margin that accommodates coil temperature transients during charging, cryocooler variance, and energy dissipated in the structure from externally imposed vibrations.

Both first stages of the cryocooler are thermally interfaced with an 1100 alloy aluminum radiation shield via flexible copper links, which maintain the shield temperature at 30-32 K. All thermal interfaces between the cryocooler and the cold masses are flexible enough to permit relative motions between them that will occur under shock and vibration loads. The cryocoolers are mounted rigidly to the vacuum vessel, which has different stiffnesses and resonant frequency characteristics compared to the CTWS and its attached cold masses. The shield assembly is also integrated into the CTWS structure in a manner that serves as a thermal intercept to the cryocooler's first stage. Figure 2 shows a photo of the magnet assembly and cross section illustrating its re-entrant cylinder CTWS structure, shield, and coil configuration. The magnet assembly mass is 352 kg.

DESIGN

The magnet was designed with certain key parameters in mind that would be preserved for larger scale designs being considered by the Navy for minesweeper applications:

1. The conductor and coil must be designed so a suitable thermal margin is maintained between the superconductor's steady state operational temperture (T_{op}) and the critical transition temperature (T_c), in order to ensure stable operation under conditions which will cause the coil's temperature to fluctuate. The design margin chosen (T_{op} - T_c) is 2 K.
2. The Lorentz stresses and strains in the coil and its interfaces for the test article magnet should be comparable to larger scale designs in order to demonstrate the basic mechanical and magnetic stability so larger devices can be built confidently. Designs have been selected to limit conductor hoop strains to about 0.1%.
3. The heat loads and thermal gradients at interfaces must be accurately predictable and low enough to be compatible with cryocooler technology for both subscale and full scale designs.
4. The CTWS structure and its interfaces must be designed for 100 g shock loads from any direction and up to 2 g vibration inputs out to 20 Hz. The CTWS conduction heat leak for both subscale and full scale designs must minimize the cryocooler heat burden. In addition, the design approach must have a minimum impact on overall weight.

Experimental efforts at the component level were designed to focus on the above criteria in order to characterize important aspects of the design in greater detail, in addition to predicting the performance of the as-built device. Some of these are described below.

Conductor and Coil

Aluminum stabilized NbTi conductor was supplied by Supercon Inc, Shrewsbury, MA. The conductor was manufactured by coextruding high purity aluminum over a round NbTi-Cu core to produce a final rectangular shape. The conductor was then covered with 2 mil thick Formvar^R. The conductor's nominal specifications are listed in Table 1.

Using the length of conductor available, a coil geometry and operating current density (and consequent peak conductor field, B_p), was selected. The conductor's critical transition temperature, T_c , was then predicted using the following equations [1]:

$$T_c = T_{c(B)} - [j_{op}/j_c (T_{c(B)} - 4.2)], \qquad (1)$$

Table 1. Conductor specifications

Composition by volume	Al (61.5%) - Cu (18.5%) - NbTi (20.0%)		
Bare conductor size (mm)	1.085 x 0.584		
Number of NbTi filaments	294		
Short sample critical current densities at 4.2 K (kA/cm^2)	$j_{c\text{-conductor}}$	$j_{c\text{-NbTi}}$	B (T)
	27.3	136.5	7
	36.2	181.0	6
	47.0	235.0	5
	73.9	369.5	3

where,

$$T_{c(B)} = 9.2 (1 - B_p / 14.5)^{0.59} . \qquad (2)$$

Values of j_c are curve fitted from the Table 1 short sample data. The coil design chosen produces a peak field of 3.41 T at an operating current of 90 A (j_{op} = 14.2 kA/cm^2). From this, the value of T_c is 7.1 K. The coil design parameters are summarized in Table 2. The coil bulk quench temperature value listed assumes all of the energy is dissipated into the conductor's stabilizer. Computer code analysis indicates that the peak temperature during the quench is only a few degrees higher.

The coil was fabricated by "wet winding" it directly on to the outside of an 1100 alloy aluminum bobbin which served as the structural interface with the CTWS and the thermal interface with the cryocooler. The epoxy used was a room temperature cure two-part mix with silica filler, designated as CTD-521[R], available from Composite Technologies Development Inc (CTD), Boulder CO. Aluminum-to-aluminum conductor splices were made using soldering techniques. After the coil was wound and cured, it was overwrapped with a 0.9 cm thick layer E-glass roving impregnated with epoxy to reduce the peak hoop stresses in the coil. The combination of this overwrap and the conductor winding tension establishes the necessary compressive radial interface pressure between the coil and the bobbin after energizing to full current.

Coil Stability

A key attribute of a coil constructed of high purity aluminum stabilized conductor is potentially higher resistance to quenches from external energy disturbances: a chief concern in minesweeper applications where the operational environment can involve shock and vibration loads. The expectation of this higher quench tolerance is based on the higher thermal diffusivity of the aluminum compared with typical superconductor copper. This is true where high residual resistance ratios (RRR) values apply for the aluminum stabilizer, typically ranging from 500 to 1000. In the case of the Table 1 conductor, measurements made by the Naval Surface Warfare Center (NSWC), Annapolis, Maryland and Supercon

Table 2. Coil design parameters

Coil inside diameter	0.5 m	Nominal current	90 A
Coil thickness	5.6 cm	Peak conductor field	3.41 T
Coil length	10.95 cm	Conductor current density	14.2 kA/cm^2
Conductor length	12,100 m	NbTi current density	71.0 kA/cm^2
Number of turns	6933	NbTi transition temperature	7.1 K
Coil mass with epoxy	40 kg	Coil operating temperature	5.0 K
Coil packing factor (A_{cond}/A_{coil})	0.72	Coil stored energy	147 kJ
Coil inductance	36.4 H	Coil bulk quench temperature	72 K

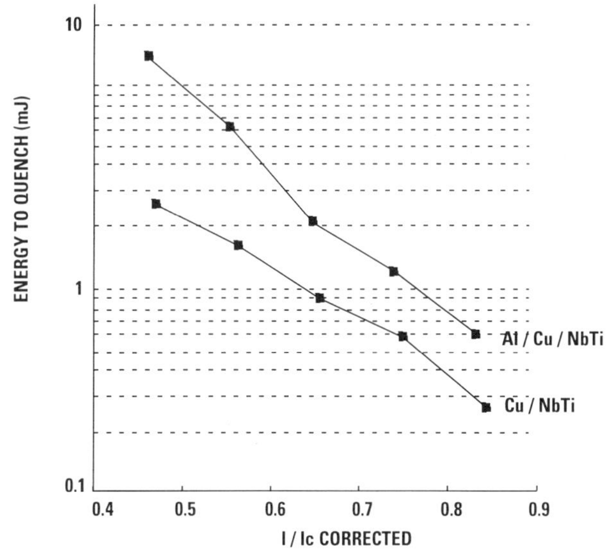

Figure 3. Copper and aluminum stabilized coil energy to quench comparison.

Inc. indicated values that tended more towards the upper limit. NSWC also measured the effect of cyclic strains up to 0.2% on RRR value, with results showing that no significant reduction occurs once the RRR value drops down to about 500.

Tests performed by the NSWC on small test coils compared the quench tolerance of high RRR aluminum stabilized NbTi conductor to that of all-copper stabilized conductor. Separate test coils were wound from aluminum-stabilized conductor (Table 1) and all-copper stabilized conductor with the same dimensions and NbTi percentage. The energy-to-quench value was measured by controlling a heater on the coils while they were energized. Test results are presented in Figure 3, suggest that the aluminum stabilized conductor is significantly more tolerant to quenches from external energy disturbances.

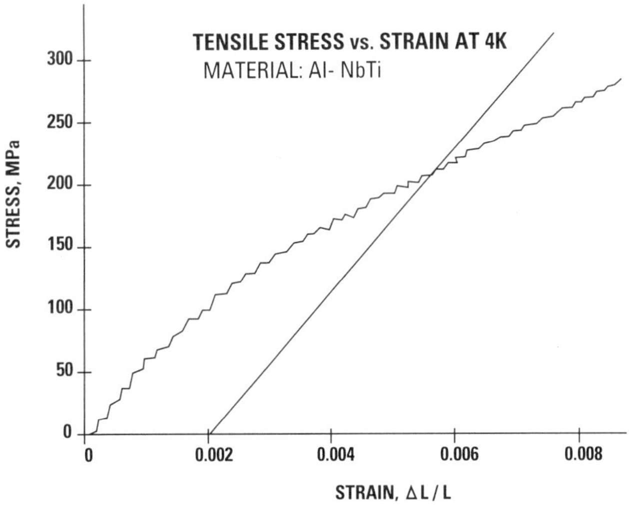

Figure 4. Stress-strain curve for aluminum stabilized conductor.

Table 3. Predicted steady state cryocooler heat loads (W)

SOURCE	FIRST STAGE (30-35K)	SECOND STAGE (4-6K)
CTWS conduction	1.7	0.03
Radiation	1.6	0.02
Current leads	7.0 (0 A), 12.0 (90 A)	0.54 (0 A), 0.97 (90 A)
Instrumentation leads	1.9	0.16
Total	**12.2 (0 A), 17.2 (90 A)**	**0.75 (0 A), 1.18 (90A)**

Mechanical stability is also important to ensure that conductor performance is not degraded under operational stresses and strains. Aluminum-stabilized conductor mechanical tests produced the stress-strain characteristic shown in Figure 4. Based on this result, the coil was designed so stresses are limited to 70 MPa in order to remain in the linear region.

Thermal Management

Predicted heat loads for the superconducting magnet assembly are listed in Table 3. The largest is due to the 90 A conduction-cooled copper current leads. It was therefore important to design an electrically isolated current lead thermal intercept with a high thermal conductance that would interface with the cryocooler's first and second stages. The second stage interface is more critical. For this, two designs were investigated. One used a 1 mm thick beryllium-oxide wafer with copper bonded on both sides that serves as a solder interface for the No. 2 cryocooler copper "cold foot" and the copper current leads. The other used a 1 mm thick sapphire disc sandwiched between the cryocooler and the current leads under a clamping pressure of 16 MPa. Both sides of the sapphire disc were coated with a thin layer of Apiezon-N^R grease. These interface methods were tested by NSWC for thermal conductance with results shown in Figure 5. The sapphire disc was chosen since it did not require soldering to the cryocooler, a process that could entail damage from heating. The current leads were thermally interfaced at the cryocooler's first stage by sandwiching two layers of one mil mylar between them with fastener pressure.

In order to ensure that the coil is maintained at a temperature as low as possible, a low impedance flexible thermal link fabricated from OFHC copper braid was bolted to the coil's 1100 aluminum alloy bobbin at one end, and the No. 1 cryocooler's second stage at the other end. Both the aluminum bobbin and the cryocooler's surfaces were plated with indium, and a one mil thick indium foil was sandwiched between the surfaces as they were bolted tight. A heat gun was applied after assembly to soften the indium followed by fastener re-torquing. The thermal conductance of the complete interface was measured by NSWC, with test results in Figure 5.

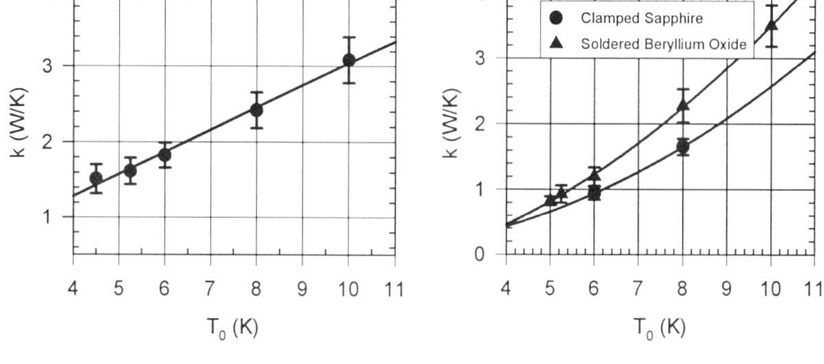

Figure 5. Measured thermal conductance of interfaces: coil-to-cryocooler (left), lead-to-cryocooler (right).

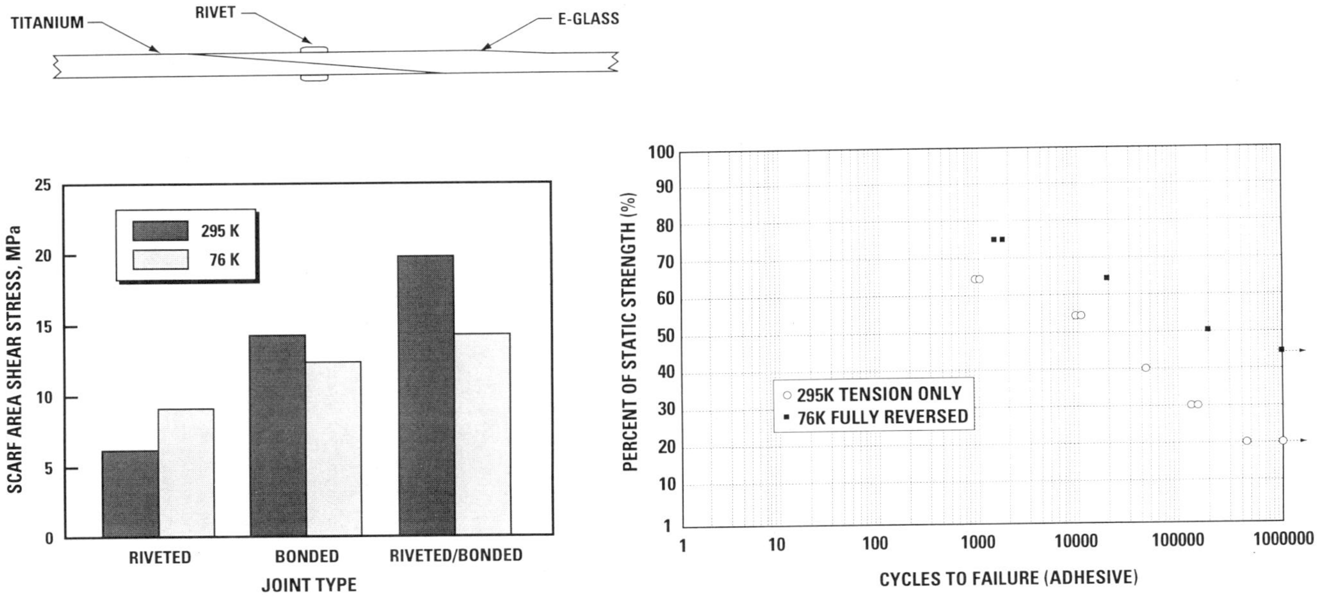

Figure 6. CTWS scarf joint static load performance (left), and fatigue performance (right).

Table 4. Magnet calorimetry data

CURRENT	NO. 1 CRYOCOOLER (coil)			NO. 2 CRYOCOOLER (current leads)		
	2nd Stage (K)	Coil (K)	Est. Heat (W)	2nd Stage (K)	Current Lead (K)	Est. Heat (W)
0 A	3.61	4.14	0.60	3.88	5.36	0.56
91A	4.52	5.13	0.87	4.91	6.45	0.92

Flexible OFHC copper braid links interface both cryocooler first stages with the 1100 alloy aluminum shield. The connections are made via lugs which are bolted with sandwiched indium foil gaskets. The aluminum shield also serves as a conduction thermal intercept to the CTWS. This interface was achieved by brazing the aluminum shield to a titanium ring which joins two re-entrant E-glass cylinders (Figure 2).

The 30 K radiation shield is covered with 60 layers of double-sided aluminized mylar interleaved with spun-bond polyester. 10 layers are used between the coil and the shield.

Cold-to-Warm Support Structure

The E-glass re-entrant cylinders that support the cold mass are sized to yield a minimum factor of safety of four for bending, shear, and buckling stresses that could arise from 100 g shock loads applied from any direction. The cylinders were filament wound with epoxy in a fiber layup that is optimized for strength when these multi-directional loads are considered. The fiber lay-up at the ends of the cylinders are modified to closely match the coefficient of thermal expansion (CTE) of the titanium structures they are bonded to. The cylinders were manufactured by the Brunswick Corp., Lincoln NE. The outer cylinder supports the coil bobbin, which is adhesively bonded to the cylinder's outside diameter. The cylinder is thicker in this region with filaments wound primarily in the hoop direction. The cylinders and joints are instrumented with strain gages, and three accelerometers are mounted on the coil bobbin that will used to obtain data during shock and vibration tests.

Special consideration applies to joints where E-glass cylinders are connected to the metal structural elements at support and re-entrant locations (Figure 2). A riveted and adhesively bonded 5° scarf joint configuration is used at these interfaces (Figure 6). The adhesive used was a two-part epoxy designated as Hysol EA-9330^{R}. The fastener is an A-286 steel rivet. The combined CTE of the E-glass and the titanium adherants closely matches that of steel in the "squeeze" direction. Test results on scarf joint coupons have shown that this combination is stronger than either one alone. Fatigue tests were also performed to characterize the joint strength. The static and fatigue test results at room temperature and 76 K are shown in Figure 6. The test results indicate that the E-glass-to-metal joints used in the test article superconducting magnet may have factors of safety as high as 10 under 100 g shock loads, and when fatigue is considered for vibration loads.

TESTING

The magnet was cooled down on cryocooler power and reached operating temperature in about 95 hrs. At zero current, the coil temperature reached 3.95K with a No. 1 cryocooler second stage temperature of 3.7 K. The magnet was charged to full current at 10 volts in 6 minutes, during which time the coil temperature peaked at 6.5 K due to eddy current and hysteresis losses. No training quenches ever occurred. After about 4 hrs, the coil temperature stabilized at 5.1 K. The shield temperature was maintained at 30-32 K. Temperature gradients measured across the two cryocooler second stage interfaces indicate heat flows when correlating them to the conductance curves shown in Figure 5. These are listed in Table 4. The total indicated heat flow into both cryocoolers is within a factor of about 1.5 of the original Table 3 estimates. The measurements indicate that some of the heat from the leads flows into the coil instead of being intercepted at No. 2 cryocooler, thus shifting some load to No. 1 cryocooler. We also noted that although both cryocoolers were identical, relative performance varied, as indicated by their second stage temperatures and heat flows. Temperatures were measured using ruthenium oxide resistors. The coil sensor was mounted on the outside diameter, 180° from the cryocooler interface.

The conductor's critical transition temperature was measured by shutting down No. 1 cryocooler and allowing the coil to heat up while running at 90 A. Eight minutes later the magnet quenched when a sensor on the aluminum bobbin indicated 7.1 K, while another sensor mounted on the face of the coil indicated 6.9 K. This appears to be a very good verification of the critical transition temperature prediction of 7.1 K described earlier. At the onset of the quench, the peak field region of the coil was likely closer to the 7.1 K bobbin temperature. The coil temperature rose to 65 K after the quench and cooled back down to 5 K on cryocooler power in 4 hrs.

CONCLUSION

An all-conduction cooled superconducting magnet using aluminum-stabilized NbTi conductor has been successfully demonstrated, and larger scale magnets appear feasible. The predicted critical temperature margin of 2 K was achieved, and the magnet was stable under transient charge conditions and cryocooler variances. Additional margin is possible by reducing the current lead heat input through improved thermal intercept design or by using shock-tolerant high-Tc leads.

Scarf joint assembly methods used in the CTWS have proven to be reliable in tests at cryogenic temperatures. The magnet will be vibration and shock-load tested at NSWC to validate the overall CTWS structural integrity in addition to testing conductor stability.

REFERENCES

1. R.K. Maix, D. Salathe, "Practical Scaling Formulas for the Determination of the Critical currents in NbTi Superconductors," Proceedings at the 9th International Conference on Magnet Technology, Zurich, Switzerland, 1985.

PERFORMANCE TESTING OF VARIOUS CRYOCOOLERS FOR CONDUCTION COOLING OF A U.S.NAVY SUPERCONDUCTING MAGNETIC MINESWEEPING SYSTEM

E.Roth,[1] G. Green,[2] and J.N. Chafe[2]

[1] Vector Research Company
Rockville, MD, 20852
[2] CDNSWC-Annapolis Detachment
Annapolis, MD, 21402

ABSTRACT

The U.S. Navy is investigating superconducting magnet systems for use in minesweeping operations. Because of the logistical problems associated with carrying and handling the required liquid cryogens aboard a Navy vessel a cooling system utilizing a mechanical refrigerator is being pursued. Refrigerators with capacities between 1 and 6 watts at temperatures from 4.5 to 10 kelvin are being considered to conductively cool the superconducting magnet system. This paper reports the results of refrigeration capacity tests of three different refrigerators from commercial vendors. The refrigerators tested include a 2-stage G.M and two 2-stage G.M.'s with a J.T. third stage.

INTRODUCTION

The United States Navy is developing a superconducting magnetic minesweeping system under the Advanced Lightweight Influence Sweep System (ALISS) program.[1] This system is designed to detonate shallow water mines by the presence of a magnetic field. In order that strong magnetic fields from a compact system be attainable, a superconducting magnet will be utilized. However, the superconducting materials being considered for this magnet will need to be cooled to temperatures near that of liquid helium (4 K). Although magnets of comparable size and strength anticipated for this project have been constructed for other purposes, they have usually been cooled with liquid helium. Also, these magnetic minesweeping systems will be used at sea so they will have to be portable and require a minimum of service. For these reasons, the Navy has elected to pursue cooling with a cryogenic refrigerator. However, conduction cooling of superconducting magnets is a recent technology. There are various refrigerators available, but no one type at their present level of development is best in every criteria for this project. Refrigeration loads for the minesweeping system are estimated to be 1 to 6 watts at temperatures from 4.5 to 10 kelvin. The refrigeration capacity of three different coolers was tested: 1) a 2-stage G.M built by CryoMech; 2) a 2 stage G.M with a J.T. third stage also built by CryoMech; and 3) a 2 stage G.M's with a J.T. third stage built by APD Cryogenics. Besides thermal performance, also required is simplicity, reliability, and ruggedness. Since the cooler would be used at sea aboard some type of Navy vessel, performance and reliability in a vibration environment must be demonstrated. Some of these coolers have been designed or modified by the manufacturers to withstand the anticipated vibration environment, however extensive testing of performance under vibration conditions has not been performed yet on all the coolers.

Cryocooler Designs

APD GM/JT The cryocooler tested was essentially part of a commercial unit (except as noted below) built by APD to reliquify helium and cool a thermal radiation shield for hospital MRI systems. The modular system consisted of two water cooled compressors and two coldheads; one each for the two-stage Gifford-McMahon (GM) portion and the Joule-Thomson portion (the 3rd stage). A photograph of the APD coldhead is shown in Fig.1. The GM rotary compressor and coldhead were models HC-8C (4.5 kW) and DE-208L respectively and the JT rotary compressor was a HC-4MK1 (3.0 kW); the system as a whole is called a MS-12. These two coolers had their own separate helium gas flow loops but were thermally coupled via heat exchangers. Along with providing cooling to the thermal shield, the two stages of the GM provide precooling of the helium gas in the JT loop via heat exchangers.

The expansion valve in the JT stage was of fixed size which was optimized for flow and performance at about 4.5 kelvin. Though this is a simplifies the operating procedures it causes longer cooldown time and does not allow one to fine tune the performance when at off design conditions.

Fig. 1 Photograph of the APD 3 stage GM/JT cryocooler.

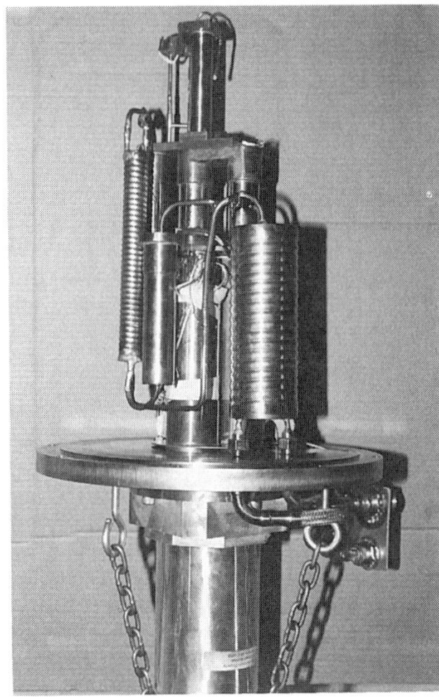

Fig. 2. Photograph of the CryoMech 3 stage GM/JT cryocooler.

The JT compressor had a manually adjustable bypass valve between its supply and return ports. This valve let one set the pressure on the return side and thus the temperature of the 3rd stage (JT). This valve was also important in that it allowed one to always maintain the compressor's return pressure just above atmospheric pressure, thus helping prevent ambient air from entering the compressor.

Fig. 3. Photograph of the Cryomech 2 stage GM cryocooler.

The modifications made by APD were to ruggedize the cooler coldhead to withstand a shock and vibration spectrum of 3 g's from 1 to 1000 Hz. The modifications were to strengthen the stainless steel cold sleeve in the GM coldhead and brace and support the copper tube counterflow heat exchangers of the JT loop. The 1st stage regenerator is phosphor bronze screens and the 2nd stage is lead spheres. Neither compressor was modified.

CryoMech GM/JT. Though CryoMech has no standard line for GM/JT's they had some experience with several prototype units; this one was constructed specially for the Navy and is the largest capacity one they have built. A photograph of the CryoMech system is shown in Fig. 2. The GM portion of this system is a GB 37 similar to but not identical to the modified one described below. It has stationary regenerators and pneumaticlly driven displacers. The 1st stage regenerator was phosphor bronze screens and the 2nd stage was lead spheres. The JT portion had counterflow finned tube heat exchangers and an adjustable JT valve. There was a hydrogen thermo-syphon thermally connecting the 2nd and 3rd stage to assist in cooldown. The compressor was a 19 kW, reciprocating 6 cylinder unit that by staging the individual cylinders had different return pressures for the GM and JT loops. Thus the compressor was able to drive both the GM and the JT loops at different pressure ratios but with the same pressure at the compressor output.

CryoMech 2 Stage GM. A photograph of the CryoMech system is shown in Fig. 3. It is essentially their model GB 37 coldhead. The modifications made to the stock unit were 1) the originally all lead 2nd stage regenerator was replaced with an all neodymium or a lead/neodymium regenerator 2) the originally 100% phoshor bronze 1st stage regenerator now contained 10% lead and 3) the stroke of the 1st and 2nd stage displacers were modified. The details of the these modifications are stated on the performance plots in the results section. The compressor used is the CryoMech's model CP 25 (11 kW) which is the standard compressor for the GB 37.

Instrumentation. All three refrigerator systems had silicon diode temperature sensors and heaters on the coldheads of each stage. The refrigerators had a MLI wrapped copper radiation shield thermally anchored to the 1st stage to intercept radiation heat leak to the lower stages. A mass flow meter was installed in the helium gas lines of the JT loop for both GM-JT units. Pressure sensors were installed on the CryoMech GM-JT at the inlet and exhaust of the JT valve.

Test Procedure

The GM/JT tests consisted of recording the temperature of all three stages while heat loads to the various stages are applied. A 3rd stage (JT) refrigeration maximum capacity data point was determined by the power that could be applied to the 3rd stage while its temperature remained stable and below about 5 K. Generally, once the 3rd stage got above 5 K the temperature began to diverge. This is contrary to the behavior of a GM refrigerator which tends to be stable and have increased capacity at the higher temperatures. For the APD unit the JT compressor bypass valve was adjusted so as to maintain a return pressure of about 1 atmosphere (15 PSIA) and it generally only needed to be readjusted when power levels to the 1st or 2nd stage changed substantially. 15 PSIA at the JT compressor return gave a temperature of about 4.5 K at the JT. The CryoMech unit had the return pressure set at about 12 PSIA.

Helium mass flow in the JT loop was monitored to determine whether any obstructions occurred. This was done by noting any abrupt changes in mass flow data. The mass flow on the APD was not adjustable and was about 48 std l/min. but the CryoMech was set to 170 std l/min.

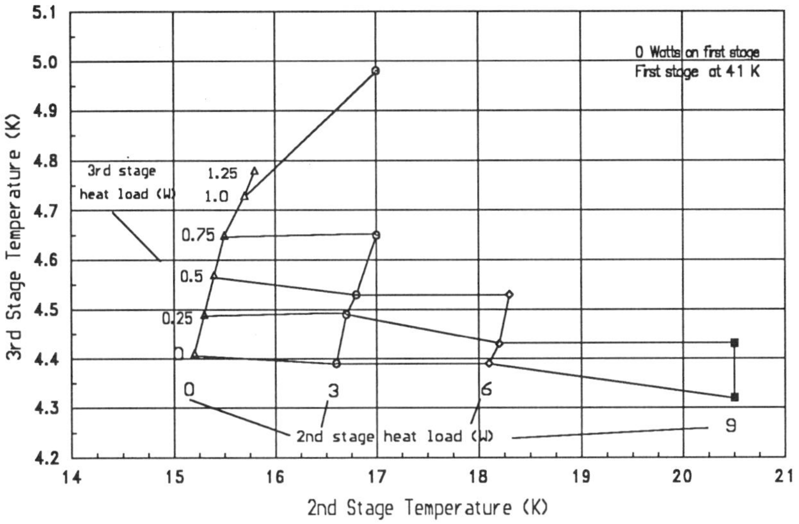

Fig. 4. Performance map of the 3 stage APD GM/JT cryocooler. 0 W on the 1st stage.

Table 1

Heat applied simultaneously to all stages.	1st stage (GM)		2nd stage (GM)		3rd stage (JT)	
	Temp (K)	Watts	Temp (K)	Watts	Temp (K)	Watts
APD Tests	70	30	17	3	4.8	1.2 (max.)
Navy Tests	65	30	17.5	3	4.5	0.5 (max.)
CryoMech GM/JT	59	50	15	6	4.5	3.4
CryoMech GM	67	30	9.1	5	-	-

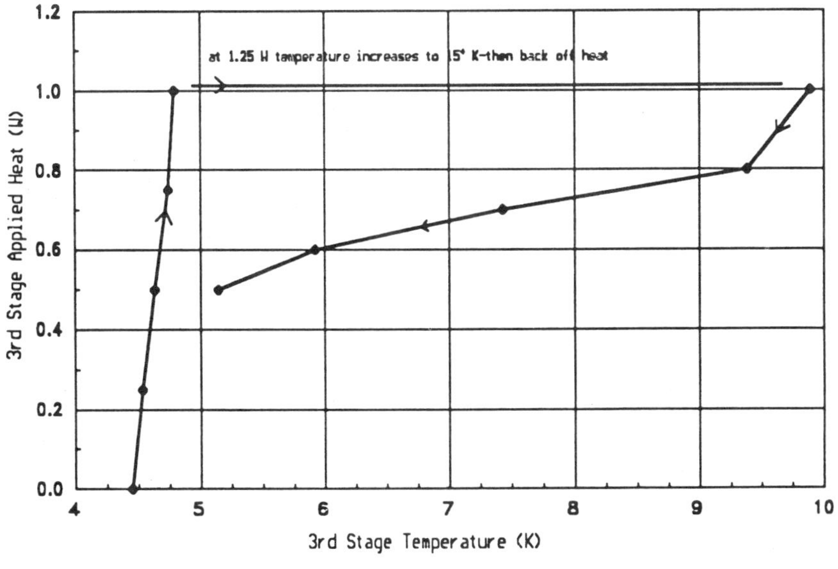

Fig. 5. Observed hysteresis in the performance curve of the 3rd (JT) stage of the APD.

Results

APD GM/JT. Figure 4, shows the performance map for a fixed 1st stage heat load of 0 W. The 3rd stage performance is noticeably affected by the heat loads on the other two stages, particularly loads on the 2nd stage. For example, in Fig. 4 the maximum third stage load is 1.25 W with 0 W on the 1st and 2nd stage, while the 3rd stage has 0.25 W capacity with 0 W on the 1st when 9 W is applied to the 2nd stage. The performance design goals sought by the Navy and some measurements by APD are shown in Table 1. It appears that our performance data does not measure up to APD's data. The most likely reason is due to the significantly smaller helium mass flows we were able to achieve in the JT loop. The data set plotted and discussed above are for helium mass flow rates of about 48 standard liters per minute. This flow is about 75% of the 65 standard liters per minute flow rate that APD had measured (with no loads on any of the stages and a 3rd stage temperature of 4.5 K). At times, flow rates of only 50% of "normal" were occurring. Furthermore, the flow could change unpredictably. For example, if heat was applied the 1st stage, thus warming one of the heat exchangers coupled to the JT loop, the flow would often change abruptly. It appeared as though contaminants would freeze onto the cold surfaces of the heat exchangers or the JT's expansion valve and restricting the flow. The addition of heat would dislodge or melt the contaminants and be flushed out to less critical locations in the system. Though this scenario seems reasonable, it is not clear what the contaminants are and how they got there. There did not appear to be any obvious leaks in the system. Also, purging of the JT compressor, gas lines, adsorber, and heat exchangers was attempted a few times with mixed results. This mode of performance degradation is apparently quit common in JT type refrigerators and similar problems occurred for the CryoMech GM/JT during preliminary testing but we were able to get the system cleaned up satisfactorily for these tests.

Another characteristic discovered is that a JT refrigerator has a qualitatively different refrigeration capacity map than the GM type. This JT had a fairly flat temperature vs. capacity map over a narrow temperature range, then when an upper temperature of about 5K is reached the capacity drops off abruptly. Further, all helium JT's have no cooling above about 25 K as opposed to GM's whose capacity generally get better at the higher temperatures. Fig. 5 shows the diminished capacity of this JT at temperatures above 5 K.

An interesting phenomena observed was that the performance of the JT stage not only depended on its temperature but from which direction the temperature was approached - that is it exhibited hysteresis. Figure 5 shows an example of the APD JT stage's capacity as a function of temperature both for temperatures increasing from 4.5 K and temperatures decreasing from about 15 K. The reduced performance for the path in which the temperature is decreasing is significant because this is the path which the system will take during cooldown. It is *speculated* that one of the reasons for this is that in the volume where the liquid is being produced and evaporated there is either nucleate or film boiling occurring. These two modes of boiling result in significantly different heat transfer coefficients between a surface (eg. the heater) and the liquid helium. It is well known that the transition from nucleate boiling to film boiling or the reverse exhibits a marked hysteresis. Similar results occurred for the CryoMech GM/JT.

Another important aspect related to the refrigerators performance is how quickly the system will cool down from room temperature. For the ALISS system a fast cooldown is desirable. The APD took over 9 hours to cool itself from room temperature to 4.5 K with no added heat load.

CryoMech GM/JT. Figure 6, shows the performance map for a 1st stage heat load of 0 W. The maximum capacity at 4.5 K is 4 W. The capacity drops to 3 W when 15 W is added to the second stage.

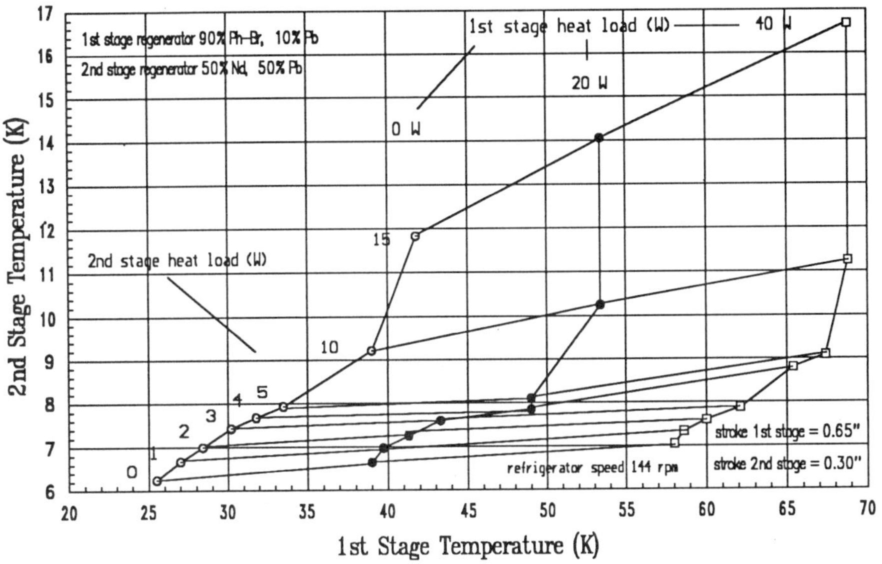

Fig. 6. Capacity map of modified CryoMech GB 37.

The cooldown time was about 2 hours with the aid of the hydrogen heat pipe. A hysteresis effect similar to the APD was noted in its performance.

CryoMech Two Stage GM. This modified GB 37 was tested under a variety of configurations but only the results from the best performing configuration will be reported here. The refrigerator had a no load minimum temperature of 6.2 K on the second stage and 25 K on the first stage. The full performance map is shown in Fig. 7. Also, this unit was operated while being vibrated along its vertical axis. Though a vibration spectrum has been specified for the ALISS project the spectrum tested here had a slightly narrower

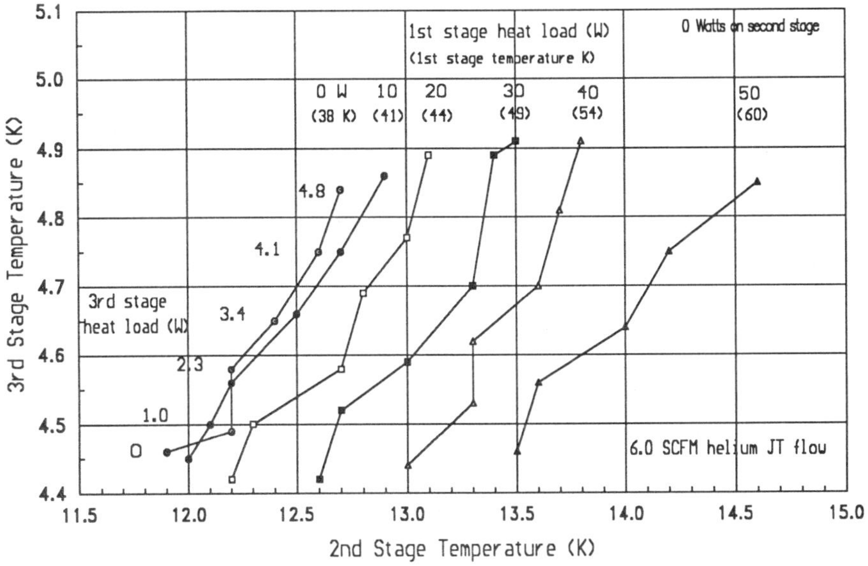

Fig.7. Performance map of CryoMech GM/JT.

frequency span. The cooler was vibrated from 10 to 500 Hz at up to 2 g's with no discernable change in performance; it appears that this refrigerator will not have any problems in performance or reliability when subjected to the full test.

Summary

This paper reports the test results of 3 cryocoolers for possible use in a Navy minesweeping system. With the APD GM/JT stage we were able to obtain only about 75% (1.25 W) of what the manufacturer had tested; this appeared due to partial blockage of the JT helium flow path. The CryoMech GM/JT was able to achieve about 5 W on the 3rd stage and its performance was consistent over the one day of testing. However, during preliminary testing done by CryoMech blockage did occur. It is not clear how this freezeout problem in either cooler can be remedied reliably enough for Navy use at sea. In general, helium JT's provide no cooling from 300 to 25 K, but then their capacity slowly increases to its rated value as the temperature decreases from 25 to about 4.5 K; this is significant in that it could complicate and prolong cooldown. Also, for both units clogging of the JT loop was a problem. The two stage GM ran consistently and trouble free. It had a minimum second stage temperature of 6.2 K but had 5 W at 8 K.

REFERENCES

1. M.E. Golda, J.D. Walters, and G. Green, Applications of Superconductivity to Very Shallow Water Mine Sweeping, *Naval Engineers Journal*, vol. 104, pp 53-64, (1992).

SPACEFLIGHT CRYOCOOLER INTEGRATION ISSUES AND TECHNIQUES

L. M. Sparr,[1] S. H. Castles,[1] I. S. Banks,[2] E. F. James,[2] W. J. Gully,[3] E. J. Krolichek,[4] N. W. Madden,[5] C. P. Cork,[5] R. H. Pehl,[5] P. N. Luke,[5] D. N. Mallone,[5] and R. A. Hopkins,[3]

[1]Cryogenics, Propulsion and Fluid Systems Branch
NASA/Goddard Space Flight Center
Greenbelt, MD, 20771, USA
[2]McDonnell Douglas Corp.
Seabrook, MD, 20706, USA
[3]Ball Aerospace Corp.
Boulder, CO, 80306, USA
[4]Swales Inc.
Seabrook, MD, 20706, USA
[5]Lawrence Berkeley Laboratory
Berkeley CA, 94720, USA

ABSTRACT

NASA/GSFC personnel in cooperation with other researchers are currently investigating a number of issues regarding the integration of cryocoolers into space flight missions. These issues include (I) integration of space flight cryocoolers with detector assemblies, (II) cryocooler-to-spacecraft interactions and vibration suppression with space flight and tactical cryocoolers, and (III) developments in cryogenic capillary pump loop technology to interface multiple cryocoolers to a common load.

INTRODUCTION

This paper discusses three separate investigations related to interfacing cryocoolers with flight sensors and the attendant compatibility issues. A cryocooler developed by Ball Aerospace with funding from NASA/Goddard Space Flight Center (GSFC) was integrated with a germanium gamma-ray detector module developed by Lawrence Berkeley Laboratory (LBL). The integration was performed to determine whether noise induced by the cryocooler would significantly degrade the detector resolution.

NASA/GSFC personnel have investigated the effects of vibration from cryocoolers on a full size, one tenth inertia model of the Earth Observing System (EOS) AM spacecraft. EOS-AM will be launched later this decade. Measurements were made of the effect of the cryocooler's residual vibration on the performance of the laser pointing system incorporated into this spacecraft mock-up. Attempts were made to control and minimize cryocooler axial vibrations. Results of this testing effort are presented.

Under the direction of NASA/GSFC personnel, a cryogenic capillary pump loop (CCPL) was designed and fabricated by SWALES Inc. The CCPL was integrated with a mechanical cooler and underwent thermal vacuum testing at NASA/GSFC's cryocooler test bed. The CCPL technology may be valuable to instrument integrators of long duration space flight missions where multiple cryocoolers are needed to meet mission reliability requirements. This paper reports on initial testing and characterization of this CCPL, including thermal performance, start up and normal operation, 1-g orientation restrictions, and parasitic loading. Conclusions and plans for future developments are also reported.

CRYOCOOLER INTEGRATION & TEST WITH A GAMMA RAY DETECTOR

During March of 1995, an engineering demonstration model of a 30 Kelvin, two-stage, low vibration, long life cryocooler[1] was successfully integrated with a germanium gamma-ray detector module. The tests were conducted over a period of a week at LBL and the results were very favorable, demonstrating the feasibility of using a mechanically cooled germanium gamma-ray detector on a long life space mission[2].

A two-stage Stirling cycle cryocooler was used for this test. Its design reflects special attention to long life and space system compatibility. In the past, liquid and solid cryogens have been used to cool these detectors. This test was run with a cryocooler because flight experiments of a duration exceeding the practical lifetime of a stored cryogen are now being considered. In principle, a cryocooler can operate as long as electrical power is available. Actual usefulness depends upon whether current machines have an acceptable lifetime, and whether they are sufficiently free of mechanical vibration and electrical noise to allow the detector to perform properly.

Figure 1. Test fixture.

The experimental test setup is shown in Figure 1. The germanium detector under study is located in the center of the open test chamber in the figure. This chamber is unique in that the detector is simultaneously connected to two coolers, which enabled testing with and without the cryocooler operating. The first cooler was a gravity-fed LN_2 reservoir located above the chamber which provided cooling through a coldfinger that extended into the chamber through the vacuum sleeve shown atop the chamber. The second cooler was the Ball cryocooler connected to the detector from underneath via a compliant, high thermal conductance strap, visible in the figure at the bottom of the chamber.

The evaluation device was a large volume (68mm dia. X 69mm long) reverse electrode, closed end coaxial germanium detector, designed and built by LBL. The detector was captured in a flight-like module which was previously developed for high altitude balloon-borne detector arrays.

The cooler consists of a compressor module, a displacer module, and its drive electronics. The compressor is in the dome in the foreground in Figure 1. The displacer mechanism is located directly under the vacuum chamber. Its coldfinger protrudes into the chamber and is connected to the detector by an S-shaped thermal link[3]. The "S-link" is designed for very high mechanical compliance and thermal conductance. Because of its shape, the link has similar compliance in all directions. It is ideal for accommodating assembly tolerances and differential contractions between various parts as well as isolating the detector from cryocooler vibrations. It enabled the system to be quickly integrated. The link itself is a bundle of thin aluminum foils and has a thermal resistance of approximately 2 Kelvin per watt. It was indium soldered to the cryocooler coldtip through a sapphire disk (for electrical isolation) on one end, and bolted to the detector module using an indium gasket.

The cooler drive electronics, in breadboard form for this experiment, operates on 28 Vdc provided by a commercial linear power supply. It has a microcomputer for autonomous control and two high power H-bridge drivers to run the compressor linear motors. In this experiment a "ground plane" was created by stringing a heavy braid from the vacuum chamber to the cooler housing, and to the chassis of both the drive and detector electronics modules. The 28Vdc return was tied to the ground plane at a single point.

The primary goal of the experiment was to evaluate the degradation of the width of a reference gamma-ray line caused by electromagnetic interference (EMI) and microphonics induced by vibration from the cryocooler. Given the disparity between the cryocooler's power and the energy detected by the crystal, there are many potential sources of noise coupling from the cryocooler to the detector. Among them are magnetic radiation, chassis currents, electric radiation, and microphonic noise. The magnetic radiation is caused by mutual couplings between the detector and the compressor motors and cables. Chassis currents are a common feature of the pulse width modulated H-bridges that are used to drive the compressor motors. These currents usually appear as unbalanced currents on the input power lines from the 28 Vdc power supply. If they are not properly contained, they often run in among the detector electronics and create unwanted signals. The electric radiation is a common feature of high speed digital controllers like the one that autonomously controls the cooler. Microphonics is the blanket term that describes electronic signals induced by mechanical vibrations, typically in wires, stemming from the operation of the cryocooler.

The test consisted of an evaluation of the noise performance of the detector as power to the cryocooler was increased. The figure of merit was the FWHM (in KeV) of an electronically injected pulser and the spectral line of a gamma ray from a reference Cobalt 60 source. At each level of cryocooler operation, the line width as a function of the pulse shaping amplifier's time constant was explored. The cryocooler operating configurations are shown in Table 1 and the experiment results are summarized in Table 2.

Table 1. Operating conditions

configuration	mechanical cooler status	LN_2 status
1	all equipment off	N2 cooling
2	electronics on only	N2 cooling
3	on and operating at low amplitude	N2 cooling
4	on and operating at full stroke	N2 cooling

Table 2. Experiment results

configuration	pulser (KeV)	1.33 MeV (KeV)	1.17 MeV (KeV)
1	1.03	2.11	2.01
2	1.04	2.20	2.03
3	1.06	2.12	2.04
4	1.08	2.18	2.07

In summary, we were able to operate noise free over a range of conditions that are representative of a final detector/cooler system. No significant electrical pickup from the cooler was detected.

CRYOCOOLER INTEGRATION & TEST WITH EOS STRUCTURAL MODEL

NASA/GSFC personnel visited NASA/Langley Research Center (LaRC) in November, 1994 to conduct cryocooler vibration testing in LaRC's structural dynamics test laboratory. For this test an EOS AM-1 representative testbed was used. This testbed has full scale geometry and 1/10 scale inertia.

GSFC personnel transported two cryocoolers to LaRC; one was a long-life cryocooler manufactured by TRW for space applications and the second was an inexpensive tactical cryocooler manufactured by Texas Instruments (TI).

The TRW cryocooler is an example of a low vibration mechanical cryocooler for space applications with a typical lifetime goal of five years or more[4]. This cryocooler is a small integral Stirling cycle cooler with one compressor, one expander and an active balance motor mounted in a common housing on a single axis. All moving components are supported using flexure springs such that under normal circumstances minimal touch contact occurs. The cryocooler has a nominal 500 milliwatts of cooling capability at 80 Kelvin with a compressor input power of 13 to 18 watts (dependent on the heat sink temperature). The cryocooler mass excluding electronics is about 2.3 kg. Cool down time to no load is about 30 minutes.

For short duration space missions NASA/GSFC has investigated the possible use of tactical cryocoolers manufactured for the Department of Defense[5] and commercial applications. These cryocoolers are readily available at comparatively low cost with typical mean time to failure (MTTF) of about 4000 hours. The Texas Instruments cryocooler provides 1.0 watt of cooling at 80 Kelvin with a compressor input power of 50 watts and a MTTF greater than 4000 hours. Cool down time is about 10 minutes. Twin back-to-back compressors are mounted in a common housing on one axis. A single expander and a GSFC designed balance motor are located in a second housing on a different axis. The compressor and expander assemblies are connected by a 13 cm transfer line. The cryocooler mass excluding electronics is 1.8 kg.

There were several reasons to undertake this testing effort. Previously reported testing of cryocooler vibration control systems has either involved force measurements on massive, stiff force dynamometers or acceleration measurements with the cryocooler suspended from bundgy cords. Neither of these arrangements is a good simulation of expected cryocooler/spacecraft system dynamics. Therefore, this test was conducted to demonstrate that the vibration control algorithms, which must meet certain stability requirements would work equally well on a representative model of a real spacecraft. Also, this testing was necessary to determine if the residual vibration from the cryocooler interfered with the laser pointing system of the spacecraft. In addition, this test investigated the vibration control performance differences between systems using force and acceleration feedback signals. Finally, the test determined whether nulling force and/or acceleration at one instrument location might cause vibration problems at another instrument location.

LaRC personnel were able to get good data regarding cryocooler interaction with the structural platform. This interaction did not cause any detectable degradation of the performance of the laser pointing system, which was able to maintain the actual platform orientation within a 2 arc second radius of the desired direction with the cryocooler running at full stroke, even without vibration suppression.

A working vibration suppression system for the TRW cryocooler was not available for this test so the TRW cooler was used as a baseline vibration signature. Only the first five harmonics were controlled on the TI cryocooler. It was demonstrated that the control system could reduce accelerations at any location on the spacecraft structure by closing the control loop with an accelerometer positioned at that location; however accelerations at other locations on the structure were also affected. The energy from the cryocooler caused increased accelerations at other locations on the structure, which, though they did not interfere with the pointing system, might interfere with other instruments located on the structure.

CRYOGENIC CAPILLARY PUMP LOOP: INTEGRATION & TEST WITH A CRYOCOOLER

Integrating mechanical cryocoolers into spacecraft has proven to be a difficult task. Even with reasonably good vibration control systems, cryocoolers typically vibrate with up to 0.05 Newtons axial and/or radial force over a frequency range from 40 to 400 Hz. For some cryocoolers, the cryocooler cold finger can only withstand about 2 Newtons of static radial loading. Use of conventional copper braided straps transmits excessive vibration from the vibrating cryocooler to the vibration sensitive load. For many applications redundant cryocoolers are required to obtain adequate system reliability; the use of redundant cryocoolers requires the development and use of heat switches. Otherwise, each cryocooler must provide considerable excess cooling capacity to overcome the parasitic heat load from the idled cryocooler. Cryogenic capillary pump loops (CCPL) are able to thermally connect redundant cryocoolers to a single remote load without moving parts or complex controls. CCPLs eliminate the need for thermal switches and provide mechanical isolation at the same time

Under a new technology initiative in CCPL technology which began in 1994, NASA/GSFC and SWALES Inc. have successfully completed the design, development and test of an engineering development model (EDM) miniature cryogenic capillary pump loop. Demonstrations of the closed cycle EDM CCPL working at cryogenic temperatures under steady state thermal loads up to 5 watts have been completed. The CCPL is relatively lightweight (approximately 1 kg) and acts like a heat diode when the cryocooler is idled. Therefore, heat switches are not required to isolate idled cryocoolers when using a CCPL. Unlike heat pipes which must remain horizontal to operate in a 1-g ground test environment, CCPLs have pumping heads of up to 150 mm. Thus, if the length of the CCPL is short, the CCPL will not interfere with instrument-level or payload-level ground testing.

The capillary pump loop (CPL) is a two-phase heat acquisition/transfer device capable of transferring heat efficiently utilizing small temperature differentials. The initial ambient CPL concept was first proposed in the mid-1960s. Over the past fifteen years, extensive ground testing and several flight experiments have been performed, and the potential of the CPL as a reliable and versatile thermal transport system for space applications has been demonstrated. The ambient CPL has been developed to a high state of technology readiness and is currently part of the baseline design of the thermal control system of the Earth Observing System (EOS-AM) and the Commercial Experiment Transporter (COMET)[6].

The basic design of a CPL includes a condenser and an evaporator connected by liquid and vapor transport lines. When heat is applied to the evaporators, liquid evaporates from the saturated wick and flows through the vapor transport line to the condenser zone, where heat is removed. Initially, the condenser zone contains high-velocity vapor and a liquid wall film that eventually condenses into slug flow. The liquid subsequently returns to the evaporators via a subcooled liquid return line, which collapses any remaining vapor bubbles. In the CPL, as in a conventional heat pipe, the meniscus adjusts naturally to establish the pumping needed to match the flow losses associated with the applied heat load and the heat sink condition. The CPL is different from conventional heat pipes in that the wick structure is only required in the evaporator section.

The construction of the CPL offers a number of design and performance features that make it suitable for a variety of applications[7]. It is a device capable of transporting from a few watts to 25 kW or more with no moving parts and relatively high heat transfer coefficients. The CPL can automatically distribute heat to various heat sinks, switch between heat sinks and provide diode action whenever a heat sink exceeds the operating temperature of the CPL.

Any number of combinations of single or multiple evaporator and condenser elements can be used as required to suit a given application. Because of its unique passive heat distribution and management capabilities, there is no need for a one-to-one correspondence of heat acquisition and heat rejection layouts. CPLs are systems of distinct components that can be plumbed together with simple smooth walled tubing. The interconnecting tube can be sized and configured to provide thermal/mechanical isolation and/or accommodation of flexible sections.

The CPL design options discussed above offer a number of potential advantages for cryogenic detector cooling applications. The ability of a CPL to automatically distribute heat to various heat sinks, switch between heat sinks and provide diode action whenever a heat sink exceeds the operating temperature of the CPL, can be used to advantage in cryogenic applications. For example, redundant cryocoolers can be linked to a single source (detector) via a CPL. With no moving parts and no more sophisticated controls than a simple thermostatic heater, the CPL will automatically "switch in" the operating cooler and "switch out" the inoperative ones. Because of the strong diode action intrinsic to CPLs, the parasitic heat leak from inoperative coolers is limited to conduction along thin (about 2 mm diameter) lines.

Startup of a CPL from the supercritical state is achieved by the temperature cycling of the cold reservoir with a thermostatically controlled heater. During initial cool down, working fluid in the gaseous state from the hot reservoir is condensed in the condenser and cold reservoir until they are completely filled with liquid. The cold reservoir is then temperature cycled with a thermostatically controlled heater which results in the injection of liquid into the evaporator. The hot reservoir acts as an expansion volume during this part of the cycle. The cold reservoir is then cooled and the cycle is repeated until the entire

CPL is cooled to below the critical point. Normal CPL operation is initiated below the critical point of the cryogenic working fluid.

During normal operation, the cold reservoir acts as an accumulator that can be temperature regulated or unregulated for operation in variable conductance mode or fixed conductance mode. During non-operating conditions, the hot reservoir provides an expansion volume that minimizes pressure containment problems at high ambient temperatures.

Parasitic heat leaks from the environment can be substantial in any cryogenic system. The CCPL employs small diameter fluid lines which help minimize the magnitude of the heat leaks. The liquid return line of a CPL, however, must be maintained at or below operating temperatures to avoid nucleation in the liquid phase returning to the evaporator. The liquid cooled shield (LCS) in combination with a two-stage condenser eliminates this problem. Condensed fluid from the first condenser stage is passed through a tube attached to the LCS where parasitic heat leaks are absorbed. Vapor bubbles formed in the process are then swept into the second condenser where the energy is removed and the condensed fluid is subcooled. The subcooled fluid is then returned in the center of the shield to avoid any additional parasitic heat leaks.

The CCPL was tested in the cryocooler test facility thermal vacuum chamber at GSFC to demonstrate startup and steady state operation. The CCPL was attached to a modified Sunpower cryocooler[8] which provided cooling to the condenser. This system was then instrumented with 30 thermocouples, three cryogenic diodes, one pressure transducer, and two heaters which were attached to the evaporator and cold reservoir. It was then mounted on G10 standoffs and wrapped with MLI to reduce parasitics. Temperature and pressure data were recorded on a computer-based data acquisition system.

The CCPL was evacuated with an oil-free molecular drag pump to remove potential contaminants such as non-condensable gases. It was then charged to 265 psig with high purity nitrogen. The cryocooler was turned on and within 5 hours the condenser reached 80 Kelvin and the pressure had dropped to 65 psig, indicating that much of the nitrogen in the warm reservoir had transferred into the condenser and cold reservoir. At this time a 10 W heat load was applied to the cold reservoir. This was followed by a rapid rise in system pressure and cooling of the evaporator, indicating that LN_2 had been forced from the cold reservoir and toward the evaporator. The heater power was then turned off. This cycling was repeated once the system pressure indicated that the cold reservoir was sufficiently filled with LN_2. This procedure was repeated 5 times at which point the CCPL was fully primed.

Once the CCPL was primed and operating at 85 Kelvin, a load of 0.5W was applied to the evaporator. After 30 minutes the load was increased by 0.25W, this cycle was repeated until the CCPL deprimed at a 3.0W load. Subsequent testing demonstrated stability both with loads as low as 0.2W and with a 50 mm pumping height.

The CCPL performed well, demonstrating that it is a viable option for future detector-to-cryocooler interfacing. Future plans include the design and manufacture of a CCPL with lower force transmissibility, through the use of thinner lines and geometric configuration, as well as with greater pumping height, through the use of ceramic or metal wicks.

Concluding Remarks

During the past year: with support from GSFC personnel, Ball Aerospace and Lawrence Berkeley Laboratory have successfully demonstrated the feasibility of using a mechanically cooled germanium gamma ray detector on a long life space mission; the capability of existing vibration control methods to provide acceptable performance on a spacecraft structure has been demonstrated by GSFC and LaRC; and GSFC and Swales have demonstrated the viability of CCPL technology. Currently GSFC is ground testing and integrating a space flight TRW miniature pulse tube cryocooler with a long wavelength infrared (LWIR) detector for characterization purposes. This cryocooler and detector assembly will be integrated into the LEISA Instrument and will be placed into orbit on the LEWIS spacecraft during 1996.

References

1. L. M. Sparr et al., NASA/GSFC cryocooler test results for Ball and Lockheed spaceflight cryocoolers, presented at the Cryogenic Engineering Conference, Columbus, Ohio, 1995
2. W. J. Gully et al., Germanium gamma-ray detector operation with a flight-like mechanical cooler, presented at the 1995 Cryogenics Space Workshop, to be published in: "Cryogenics"
3. R. F. Arentz et al., A Verified Broadly Applicable Design for Interfacing Cryocoolers to Sensors or Devices, presented at the Cryogenic Engineering Conference, Columbus, Ohio, 1995
4. L. M. Sparr et al., NASA/GSFC cryocooler test program results for FY93/94, in: "Cryocoolers 8," R.G. Ross, ed., Plenum, New York (1994), p.221
5. I. S. Banks et al., Adaptation of tactical cryocoolers for short duration space-flight missions, in: "Cryocoolers 8," R.G. Ross, ed., Plenum, New York (1994), p.163
6. J. Ku, Overview of capillary pumped loop technology, National Heat Transfer Conference, 1993
7. B. Cullimore et al., Cryogenic capillary pumped loops, in: "Cryocoolers 8," R.G. Ross, ed., Plenum, New York (1994), p.719
8. E. F. James et al., Test results for a commercial, low vibration, linear, stirling cycle cryocooler, presented at the Cryogenic Engineering Conference, Columbus, Ohio, 1995

A VERIFIED, BROADLY APPLICABLE DESIGN FOR INTERFACING CRYOCOOLERS TO SENSORS OR DEVICES

R. F. Arentz, R. A. Hopkins, W. C. Kiehl, D. A. Payne, and J. F. Siebert

Ball Aerospace Systems Division
Boulder, Colorado 80306-1062, U.S.A.

ABSTRACT

Critical components have been designed and verified for efficiently connecting sensors or devices to the cold tips of cryocoolers, especially linear Stirling-cycle coolers. The goals of an optimized interface are high thermal conductance, low parasitic heating, easy integration, isolation of cold tip micromotions, and thermal disconnect of back-up coolers. These goals are met by using (1) a flexible thermal link having low mass, high thermal conductances, high compliance, large range, and a precisely settable null position, (2) high-conductance, dematable joints, and (3) a compact, heat-shrink thermal switch having a very high switching ratio. Designs of these items were developed for use with split, linear, Stirling coolers, but they are also useful for other types of cryocoolers. This paper presents results of testing to verify thermal, mechanical and operating characteristics of a second generation thermal switch and two sizes of the flexible thermal link. With these components an optimized interface can be provided for a variety of cryocooler and sensor applications.

INTRODUCTION

There has been increased interest over recent years in using cryocoolers, rather than stored cryogens or passive radiators, to cool spaceborne sensors and devices. Thermal environments, long mission lifetimes and low temperature requirements may result in dewars or radiators that are unacceptably large and costly. Consequently, many organizations have been developing long-lifetime, sensor-compatible cryocoolers, most notably linear Stirling-cycle and pulse tube coolers. The performance and compatibility issues involved in a sensor cooling application are much the same with both these types of cryocoolers. The component technologies needed for optimal interfacing between the cryocooler cold tip and sensor have received limited attention. This paper discusses the designs and test verifications of component technologies that were developed specifically for cooling applications using long-lifetime, low-vibration Stirling coolers, and that are also useful for pulse tube cooler applications.

Advances in Cryogenic Engineering, Vol. 41
Edited by P. Kittel, Plenum Press, New York, 1996

Design of an optimal interface must address many requirements, as shown in Figure 1 for the case where a back-up cryocooler is used. Key performance issues are sensor vibration isolation, thermal efficiency, ease of integration, and, especially in the case of a clearance-seal Stirling cooler, cold tip side loading.

Although the cooler induced-vibration is transmitted to the sensor primarily through the cooler mounting interface and sensor structure, in some cases it may be necessary to have vibration isolation in the cold tip interface. Thermal efficiency is achieved by simultaneously minimizing thermal impedance of and parasitic heating to the interface assembly. Integration is facilitated by a design that avoids the need for tight tolerances between the sensor and cooler cold head and allows easy attachments. Further, when using a clearance-seal Stirling cooler, the cold tip side loading must be limited to avoid deflection that would cause rubbing of the regenerator. External reinforcement of the cold tip might be used to allow greater side load, but additional parasitic heating would result. In the early machines a side load of only ~0.5N would cause rubbing. Ball Aerospace has developed a fixed-regenerator design with NASA funding that tolerates side loads greater than 2 N without rubbing.[1] While the pulse tube cooler does not have the cold tip mechanical sensitivity of the clearance seal designs, compliance is still needed in the interface to facilitate integration by accommodating generous assembly tolerances and to accommodate relative motions between the sensor and cold head during launch.

DESIGN DISCUSSION

To satisfy typical integration needs, the components comprising the interface must include a flexible thermal link and at least one routinely dematable joint. The flexible link may also have to withstand relatively large motions between the cold tip and the sensor during launch. If a back-up cryocooler is used, the interface may also include a thermal switch, and perhaps a rigid, thermally isolating support for the switch. By eliminating the parasitic heat load of the back-up cooler, a thermal switch can typically save 10-30 W of cryocooler power consumption, depending on cryocooler efficiency. Finally, thermal control features for limiting parasitic heating to the interface assembly can significantly lower the heat load to the cold tip. Ball Aerospace designed and verified through prototype testing a Stirling cooler cold tip interface, including all these components and features, for the XRS Cryogenic Subsystem[2].

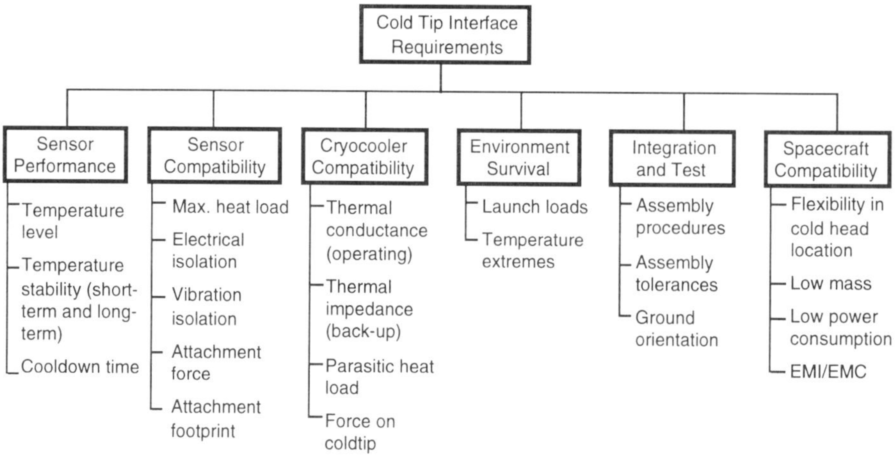

Figure 1. Cold tip interface design requirements come from many sources.

The two key elements of the interface were the flexible thermal link and the thermal switch. More recently, we have made similar thermal links in different sizes for other space hardware applications, and have measured their thermal conductance over the temperature range 5-265 K. We have also developed and completely qualified a second-generation thermal switch.

The thermal link, shown in Figure 2, is made of several hundred aluminum foils with clamped end fittings, which are metalized for soldering. It is called an S-link because of its shape, which provides similar compliance in all directions. Use of aluminum provides minimal mass and a low surface emissivity. Typical characteristics of an S-link are a thermal conductance of 0.5 W/K, a length of 5 cm, a mass of 30 g, a compliance of 1 cm/N, an amplitude capability of 0.3 cm, and a null position tolerance capability of 0.03 cm.

The heat-shrink thermal switch, shown in Figure 3, consists of a 2.5 cm diameter molybdenum disc supported within an aluminum cup by Vespel pins, and a heater assembly, using standard platinum resistance elements, capable of forcing the switch open while it is still connected to the cold sensor. The figure also shows an electrically isolating sapphire disc with both faces metalized for soldering to the aluminum cup and the cryocooler cold tip. A thermal impedance of about 2 K/W, which can be provided by an appropriately designed S-link or insulating shim, is needed between the switch and the sensor to limit heat loss to the sensor during switch opening and avoid an excessive heater power requirement (e.g., >50 W). The resistance of the open switch is maximized by gold coating the molybdenum disk to reduce its emittance and by the very low thermal conductance of the Vespel pins. The switch weighs about 120 g, low enough to be mounted directly on the cold tip of the Ball fixed-regenerator displacer without causing rubbing during horizontal ground operation. The gap between the disk and cup is sized to produce complete contact at a temperature of about 180 K. Hoop stress makes the contact resistance negligible at temperatures below about 100 K. When the necessary impedance (for heater operation) between the switch and sensor is included, the switching ratio is about 1000.

FLEXIBLE THERMAL LINK PERFORMANCE

Environmental survivability of the S-link and the apparatus used to measure its spring rate were previously reported.[2] This apparatus uses a tilt table to apply a given force to the S-link, and then measuring the deflection allows the spring rate to be calculated. Table 1 gives spring rates for two similar S-links, both having characteristic dimensions of about 5 cm. Many spring rate measurements were made, and a hysteresis of about 20 percent was observed, apparently due to stiction between the foils. The values shown in the table represent the highest spring rates measured in each direction.

To understand how compliant the S-link is, consider that a weight of only 5g would deflect it about 1 mm. Further, the spring rate of a compact thermal link made of copper braid and having about the same mass and thermal conductance as an S-link was measured.[3] The result was a spring rate about 100 times greater than the values shown in Table 1 for the S-link.

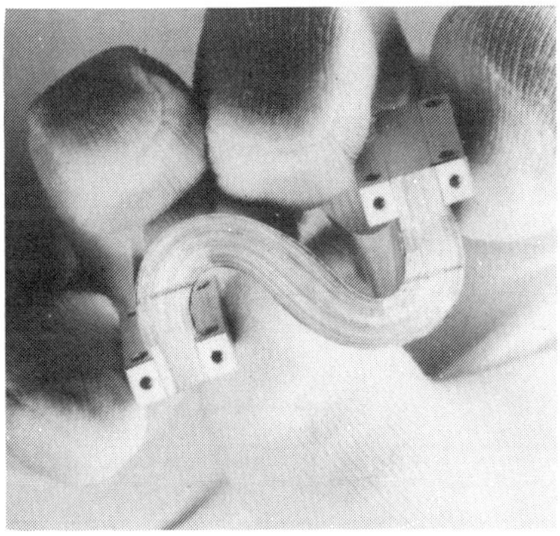

Figure 2. The S-link configuration provides low, omni-directional compliance.

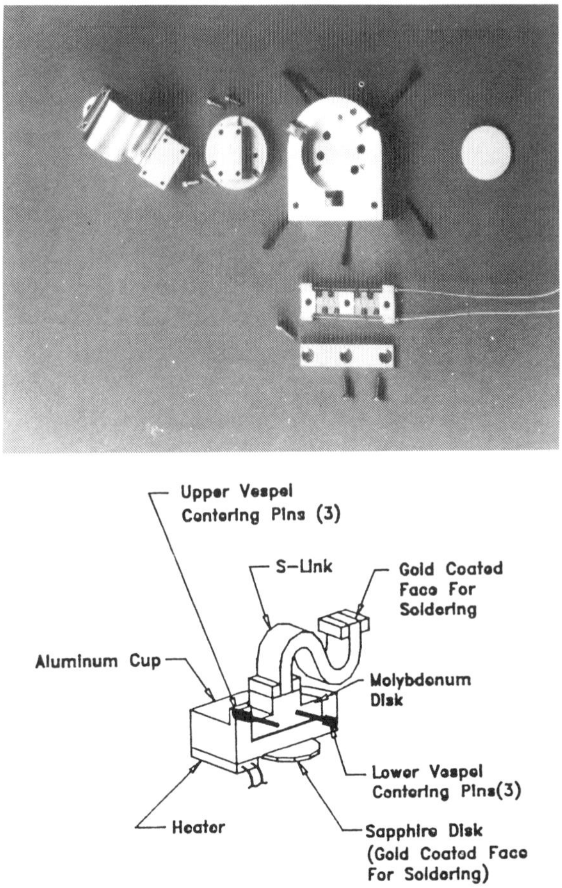

Figure 3. The thermal switch is compact and highly efficient.

Thermal conductance was measured by mounting a link in an evacuated can which was then submerged in either liquid nitrogen or liquid helium (Figure 4). To simplify the setup, a link that had not yet been bent into the S shape was used. A heater on the free end of the link was used to create a several kelvin temperature drop across the link, which was measured by silicon diodes. Parasitic heating by radiation and conduction through sensor wires was negligible compared to the large heat flow through the link. Since the thermal resistance is almost solely in the foils, data are shown as thermal conductivity calculated

Table 1. Spring rate measurements show the S-link has high compliance in all directions

	S-link # 1 (A/L = 0.212 cm)		**S-link # 2 (A/L = 0.167 cm)**	
AXIS	**Force = 0.09 N**	**Force = 0.18 N**	**Force = 0.09 N**	**Force = 0.18 N**
+X	34 N/m	27 N/m	27 N/m	22 N/m
-X	49	55	48	55
+Y	55	68	44	51
-Y	68	78	58	61
+Z	55	73	44	55
-Z	55	73	55	73

Figure 4. The apparatus used to measure S-link thermal conductivity over the temperature range of 5–265 K.

from the measured heat flows and temperature drops and the area-to-length ratio of the link (Figure 5). Results showed the apparent thermal conductivity of the foils to be nearly equivalent to 99.5 percent pure aluminum, as expected.

THERMAL SWITCH PERFORMANCE

After a brief performance verification in the apparatus discussed below, the thermal switch, including the S-link, was soldered via the sapphire disk to a plate, which was then bolted to a shaker. The loose end of the S-link was supported by a bracket attached to the plate. The assembly was then subjected to the following environments:

1. 0.25 g sine sweep for modal survey (all modes are above 80 Hz).
2. Ten cycles of 30 g/30 Hz sine burst for strength verification.
3. Random vibration for 60 s in each axis:

20-80 Hz:	+3 dB/octave
80-800 Hz:	0.02 g^2/Hz
800-2000 Hz:	-3 dB/octave

4. Repeat of the sine sweep to verify mode shapes unchanged.

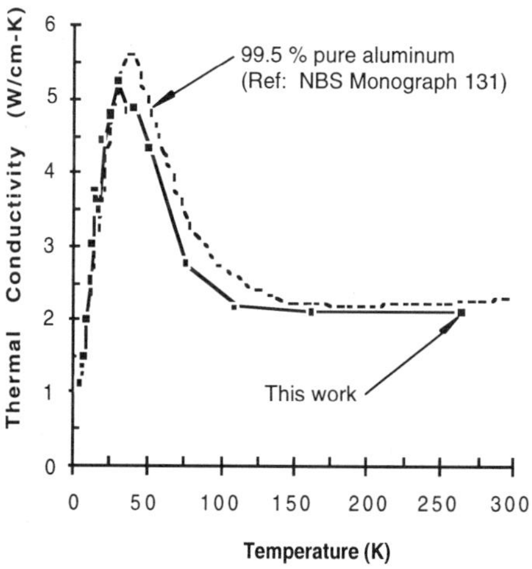

Figure 5. Thermal conductivity of the S-link foils approximately matches 99.5 percent pure aluminum.

Visual inspection following environmental testing showed no apparent damage to the switch assembly, and an electrical resistance check showed the switch was not contacting.

The thermal switch performance was then demonstrated in a cubical cryostat having removable plates for complete access on four sides (Figure 6). The displacer of a split, Stirling-cycle cooler was mounted to the top plate with its cold tip protruding downward into the cryostat. In the figure, the cold tip is wrapped in multilayer insulation. The switch assembly was soldered via the sapphire disk to the cold tip. On the other side the switch was connected by an S-link to a nitrogen-cooled block (i.e., sensor simulator), which was supported by small stainless-steel tubes through which the nitrogen flowed. A stainless steel shim, having approximately the same conductance as the S-link, was placed between the S-link and cold block to reduce heat loss to the cold block during switch opening so that a 28V power supply could produce sufficient heater power. Silicon diodes were used to measure temperatures.

The test consisted of four operational phases:

1. Cooling the block to 77 K with liquid nitrogen while the cryocooler was off, to show that the switch would stay open under this condition without heater power.
2. Turning the cryocooler on, while the block was cooled with liquid nitrogen, to demonstrate switch cooldown and closing.
3. Allowing temperatures to stabilize with the cryocooler running and no liquid nitrogen cooling, to demonstrate the switch closed conductance.
4. Cooling the block with liquid nitrogen, turning the cryocooler off, and powering the heater to demonstrate switch warm-up and opening.

Temperature data through the four test phases are shown in Figure 7. With the cryocooler off and the liquid nitrogen flow on, the cold tip temperature stabilized at about 15 K lower than the displacer body temperature. Based on this, we estimate the heat flow from

Figure 6. Switch operation was characterized in a vacuum chamber using a split Stirling cooler.

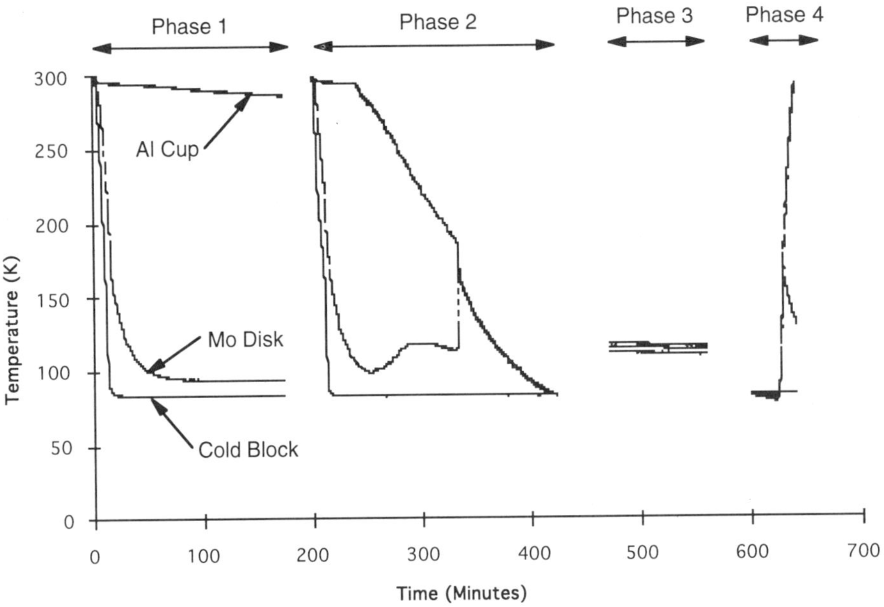

Figure 7. Temperature profiles during thermal switch operation.

the cold tip and through the open switch to be only a few percent of the approximately 400 mW that would flow to the cold tip if it was held at 77 K. This measurement is consistent with the predicted switch open resistance of 5000 K/W and proves the open switch isolates the cold tip from the cold block very well. Measurement of electrical continuity between the molybdenum disk and aluminum cup was used to show the start of contact during the cooling phase. Partial contact was typically made when the aluminum temperature reached about 230 K, but full contact was not achieved until this temperature reached about 180 K. Then, the strong thermal conductance of the interface quickly cooled the switch to the cold block temperature. With the cooler running and the liquid nitrogen flow off, the temperature drop through the switch assembly (from the aluminum cup to the cold block) stabilized at 5.0 K, with a mean temperature of 115 K. About half this temperature drop was in the switch, and half was in the combined S-link and stainless steel shim. The cooler heat lift for this condition is about 1.0 W, giving a switch assembly closed conductance of 0.2W/K. Finally, with the cryocooler turned off and the heater powered by 28 Vdc, the switch was forced open while the cold block temperature was maintained at 77 K. The figure shows the switch starting to open at 180 K after the heater was powered for 5 minutes. In 10 more minutes, the aluminum cup reached 300 K, and the heater was turned off. Heater power peaked out at 35 W. Cycling of the switch is still in progress with a goal of performing 100 open/close cycles.

CONCLUSIONS

The interface between a cryocooler cold tip and the sensor of a spaceborne cryogenic instrument must satisfy many design constraints and performance requirements. Two key components of the interface are a flexible thermal link, needed for nearly all applications, and a thermal switch, which reduces power consumption if a redundant cooler is used. We have verified the excellent compliance and thermal conductance of a flexible thermal link that is being used in flight hardware; its design would be adaptable to many applications. Testing is planned to measure the vibration isolation characteristics of the link. A prototype thermal switch, small enough to mount directly on the cold tip of some cryocoolers and having excellent switching ratio, has been thoroughly tested and is ready for flight implementation.

REFERENCES

1. D. Berry, H. Carrington, and W.J. Gully, "Two-Stage Cryogenic Refrigerator for High Reliability Applications," presented at 1995 CEC Conference, Columbus, Ohio, USA (July 1995).
2. R.A. Arentz, F.A. Eriksen, R.A. Hopkins, S.J. Nieczkoski, D.A. Payne, and J.F. Siebert, "Design and Verification of Stirling Cooler Interfaces Suitable for Long-Lifetime, Space-Borne Sensor Systems," CYOCOOLERS 8 (1995).
3. G.L. Mills and T.K. Kelly, "Integration of Thermoelectric Coolers into a Solid Nitrogen Dewar," presented at 1995 CEC Conference, Columbus, Ohio, USA (July 1995).

INTEGRATION OF THERMOELECTRIC COOLERS INTO A SOLID NITROGEN DEWAR

G.L. Mills[1], T. K. Kelly[1], and B. Brown[2]

[1]Ball Aerospace Systems Division
Boulder, Colorado 80306-1062, U.S.A.
[2]Marlow Industries
Dallas, Texas 85238, U.S.A.

ABSTRACT

The Near Infrared Camera and Multi-Object Spectrometer is a second generation Hubble Space Telescope Instrument. It uses a dewar with solid nitrogen to maintain optics and focal planes at 58 K. The dewar life is extended by a factor of two by using four thermoelectric coolers (TECs) which refrigerate thermal shields. The TECs remove the heat from the shields through cold fingers which use a new high-conductance flexible thermal link technology. The heat from the TECs is rejected by radiators on the outside of the instrument. The design and design tradeoffs of the TECs, cold finger and radiator assembly are discussed. Test data is presented on the thermal performance of the prototype and flight TEC and cold finger assemblies.

INTRODUCTION

The Near Infrared Camera and Multi-Object Spectrometer (NICMOS) is a Hubble Space Telescope (HST) second generation axial science instrument. NICMOS will be launched aboard a space shuttle and installed into the HST by astronauts during an extravehicular activity. NICMOS adds unique near infrared capabilities not present among the current first generation HST instruments or the other second generation HST instruments.[1] NICMOS cameras will detect light in the 1 μ to 2.5 μ wavelengths.

Cryogenic System Design Options

A temperature of 58 K was required for the NICMOS detectors and filters and drove the cryogenic system design options. Using thermoelectric coolers (TECs) to directly cool the detectors and filters is an attractive option because of low mass, low vibrational disturbance and high reliability. However, current TEC technology cannot achieve 58 K. Another low mass option was closed cycle mechanical driven refrigerators. To meet the low induced vibration requirement for HST, the cooler has to run in a carefully balanced opposed pair configuration, and consequently require 80 W power, well in excess of the 35 W power limit. Pure open cycle cryogen systems are too massive. The approach reached was a hybrid TEC-radiator/open cycle cryogen system.

The supercritical or liquid cryogens are less mass efficient than solid sublimers. Also the possible sloshing liquids would present problems in the pointing and control of the HST. Of the solid sublimers, a two stage CO_2/N_2 system performs about the same as a two stage CH_4/NH_3 system and avoids the hazards associated with handling methane and ammonia. To reduce cost, a single stage N_2 design was chosen.

TEC SYSTEM DESIGN

Figure 1 shows the overall design of the NICMOS dewar. At the core of the dewar are the three infrared detector arrays mounted to a cold optics bench. This bench is housed inside a cold well of the cryogen tank. The 120-liter cryogen tank contains 108 kg of solid nitrogen which sublimes at 58 K at a pressure of 30 torr. Surrounding the tank are three thermal shields which reduce the parasitic heat loads reaching the cryogen. The inner most shield is cooled by vented nitrogen vapor. The other two shields are each cooled by a set of two thermoelectric coolers. These thermoelectric coolers reject their heat to radiators mounted on the instrument enclosure. The dewar vacuum shell mounts to the NICMOS main optical bench.

The two TECs attached to each shield provide redundancy and reduce the thermal gradient in the shields. The nominal lifetime of the cryogen with all TECs operating is 4.6 years, based on a computer model of the entire dewar. The failure of one of the outer shield TECs at the start of the mission reduces the cryogen lifetime to 4.4 years. The failure of one of the inner shield TECs at the start of the mission reduces the cryogen lifetime to 3.5

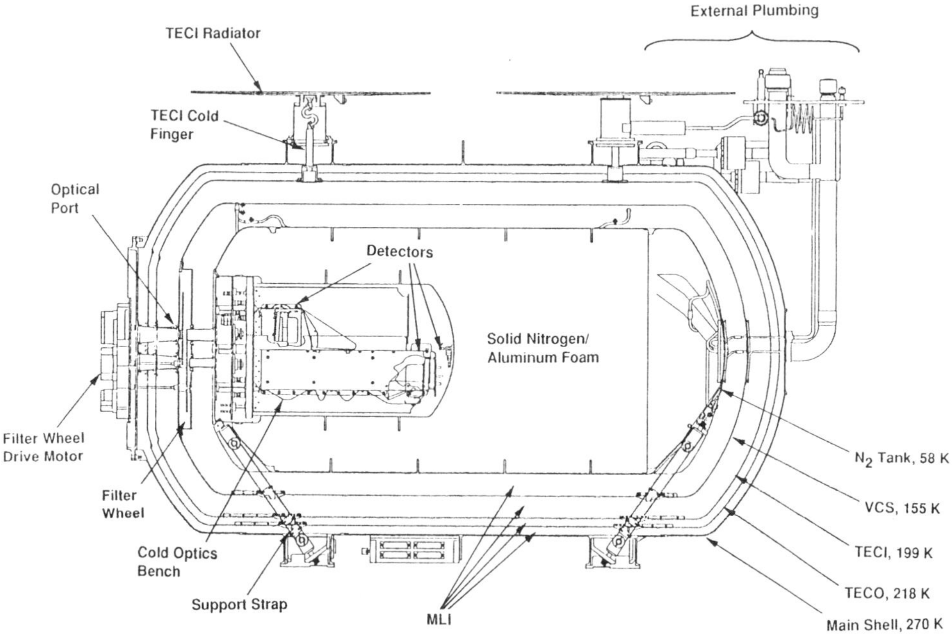

Figure 1. Cross section view of NICMOS solid nitrogen dewar.

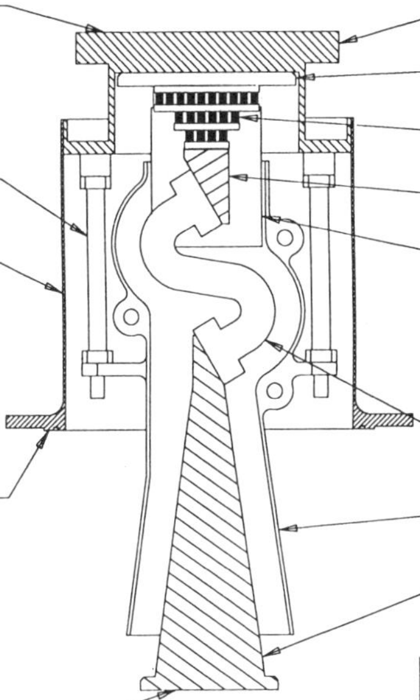

Figure 2. TEC and interface assembly cross section view.

years. The failure of all four TECs at the start of the mission reduces the cryogen lifetime to 2.5 years.

Each of the outer shield TECs is supplied with 2.64 volts of DC electrical power and consumes 2.8 W of power while removing 0.62 W of heat at 232 K. Each of the inner shield TECs is supplied with 5.28 volts of DC electrical power and consumes 13.3 W of power while removing .25 W of heat at 192 K. The total power supplied to the TECs is 32.2 W. This power is supplied from the 28 volt HST electrical bus through a DC to DC power supply which results in a total consumption by the TEC system on the 28 volt bus of 42 W.

Each of the TECs weigh 15 grams. The total weight of the all of the TECs and associated interface assemblies and radiators is 4.45 kg.

Figure 2 shows the assembly required to interface one of the TECs into the dewar. The "warm" end (heat rejection end) of the TEC is soldered to a beryllia (beryllium oxide) interface disk which is screwed to a cup shaped radiator mount. A radiator is screwed to the outside of each radiator mount and radiates the heat rejected by the TEC to the inside of the HST aft shroud. Indium foil is used in these screwed joints to enhance the thermal conductance.

The TEC is attached to the shield it cools by a cold finger assembly. The cold end of the TEC is soldered to a wedge shaped part made from beryllia. This wedge provides the correct angle for attaching the S-link to the cold end of the TEC. The S-link is a flexible, high conductance link between the TEC and wedge and the cold finger. The S-link accommodates the relative motion between the TEC and the shield which it cools, while providing a good thermal link. The S-link is soldered to the cold finger, which is made of high conductivity aluminum and provides the thermal link to the shield being cooled.

The wedge, S-link and wedge are surrounded by an aluminum radiation shield which reduces the radiative parasitic heat input into the cold finger assembly. The radiation shield is supported by four fiberglass struts that screw into the radiator mount. An aluminized polyimide shield is bonded to the first stage of the cooler and reduces radiative heat leak into the TEC and wedge.

The radiator mount is supported from the dewar outer shell by a thermal standoff which is a thin-walled stainless steel tube that surrounds the cold finger assembly and radiation shield. The standoff tube and isolates the radiator from the dewar outer shell, which has to be precisely temperature controlled. Along with the radiator mount, the thermal standoff forms part of the dewar vacuum shell. The thermal standoff tube is welded to the radiator mount and is mounted to the dewar with an indium wire seal.

TEC Design

A common TEC design was developed by Marlow Industries for both the inner and outer shield coolers. It consists of a three stage cooler optimized for this application. Originally, a four stage design was tried, but was abandoned when the top stage fractured during a vibration test with simulated launch loads applied.

The bismuth telluride thermoelectric materials for each stage were optimized for the intended operating temperature by semiconductor doping techniques. The element dimensions, (length and width) were optimized for each stage's operating temperature and material grade. The design of the top stage was off optimum because the cross section of the thermoelectric elements is oversized 20% to insure structural adequacy.

Beryllia was used as the substrate material used between stages and at the top and bottom of the cooler because it has high thermal conductivity, no electrical conductivity and high strength. Copper pads were bonded to the substrates to provide the necessary electrical connections between the elements. The elements were soldered to the copper pads.

Special precautions were taken to maximize the reliability of the TECs. The top and bottom surfaces of the coolers were soldered to beryllia parts to provide an exact match of the coefficient of thermal expansion. Extensive qualification tests were performed by Marlow and Ball Aerospace on coolers from the same manufacturing lot as the flight coolers. The qualification TECs were subjected to sine and random vibration, thermal shock and thermal performance tests after a burn -in period. Each of the flight coolers were subjected to a burn in period and to a vibration test. The electrical resistance of the coolers was checked after each test.

Flexible Thermal Link Design

Structural analysis showed that there could be up to 0.30 cm. of relative motion between the dewar outer shell to which the TECs are mounted and the shields which the TECs cool. A flexible thermal link was a critical part of the cold finger assembly. This flexible link must be partly supported by the cold end of the TEC which has low strength. Therefore requirements for the flexible link are high thermal conductance, high flexibility and low mass. The low mass is needed to reduce stress on the TEC due to the acceleration of the suspended mass.

The first design for this flexible link consisted of 20 braided copper wires soldered between two copper interface plates (**Figure 3**). The wires were 23 AWG 0(.057 cm. in diameter) by 2.5 cm long and were made up of 128 wires of 44 AWG (0.005 cm. in diameter). The interface plates had studs that projected radially to facilitate the soldering of the wires. Prototypes of this first design were built

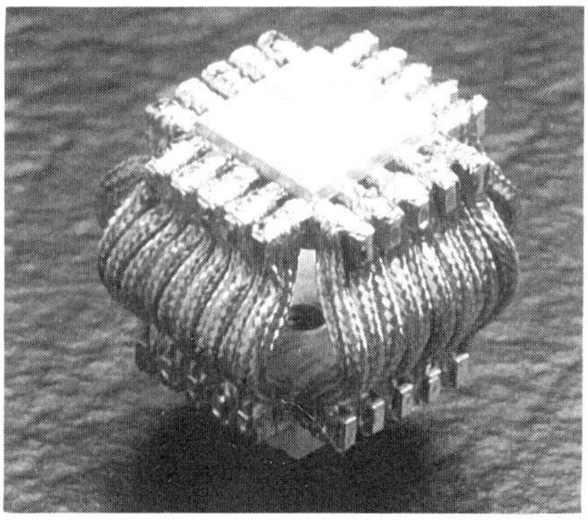

Figure 3. Copper wire flexible thermal link.

and they were found to have a mass of 35 grams and an axial stiffness of 5800 N/m. The conductivity of this design was estimated to be 0.40 W/K.

In support of on work another program, Ball developed a flexible thermal link of another design.[2] This thermal link consists of many layers of aluminum foil, bent in the shape of an "S". The foils are held together at each end with screws between aluminum blocks, but are otherwise free to move independently. This results in a link that is very flexible, has low mass and high conductivity. This design proved to be superior to the copper wire design and was selected for use in the NICMOS TEC cold finger. It can be seen in Figures 2 and 4.

Since the NICMOS design acceleration is 9.7 Gs, it was found that the S-link mass was the design driver, rather than its stiffness. The NICMOS S-link was made as short as possible, since for a given conductance, the mass of a thermal link varies as the square of the length. The thickness and width were adjusted to achieve the most flexible S-link that would meet the conductance requirement of 0.50 W/K. The final S-link has a length of 5.77 cm, a width of 2.29 cm, and a thickness of 0.64 cm.

The completed S-link exceeds all of the requirements with a mass of 26 grams, a maximum measured stiffness of 27 N/m and a measured thermal conductance of 0.55 W/K, near room temperature . It also is clean enough to use in the NICMOS dewar vacuum with low particulate and volatile residues.

Soldering of Cold Finger Assembly

There are five parts that make up the TEC cold finger assembly (see Figure 2): the interface disk, the TEC, the wedge, the S-link and the cold finger. All of these parts were joined to make one assembly by soldering. Soldering was selected over glued or screwed joints, because it was reversible and produced the highest thermal conductance for the least mass.

However, soldering entailed resolving numerous design details. All of the parts were either aluminum or beryllia, which are materials which normally cannot be soldered. In the case of the TEC beryllia interface surfaces, Marlow Industries has developed a proprietary method for nickel plating and solder tinning these surfaces. All of the other surfaces were made solderable by vacuum depositing a thin film of nickel on the surface.

There was only a narrow temperature range to achieve the soldering of these five parts. The TEC was soldered together, in an already developed process, using a solder with a melting point of 135 °C. The dewar is be baked out for vacuum acquisition at a temperature of 80 ° C, so any solder used needs a melting temperature significantly above this.

Several solder sequences and tooling arrangements were tried and a reliable sequence and set of tooling was finally developed. It consisted of first soldering the beryllia interface plate to the TEC using indium/tin eutectic solder (melting point 118°C). The beryllia wedge was soldered to the S-link using pure indium solder (melting point 156°C). The beryllia wedge was then soldered to the TEC using the indium/tin solder while the temperature was carefully monitored to make sure that the parts never got close to the 135°C melting temperature of the TEC solder joints. The final step was to solder the S-link to the cold finger with indium/tin solder, while cooling the TEC-wedge interface joint to prevent desoldering.

Radiator Design

The TEC radiators are critical part of the overall system dewar performance, since the TEC performance depends on the hot side temperature which depends on the performance of the radiators in radiating heat produced by the TECs. A one degree decrease in hot side temperature of the inner TECs resulted in a 13-day increase in dewar life at the final design point. The radiators comprise 70% of the mass of the TEC system.

Table 1. Comparison of predicted and measured performance of prototype TEC

	Milliwatts cooling/watt of power at 0.25 W cooling load at hot side temperatures (K)				Milliwatts cooling/watt of power at 0.62 W cooling load at hot side temperatures (K)			
Cold side temperature	240	260	270	290	240	260	270	290
190 K, predicted	71.9	33.5	22.6	0.0	103.3	0.0	0.0	0.0
190 K, measured	79.1	38.0	27.6		111.3			
200 K, predicted	133.9	60.4	43.0	22.3	193.1	92.5	65.0	0.0
200 K, measured	144.2	67.3	48.3	27.1	201.4	99.4	74.6	
210 K, predicted	258.4	103.4	72.1	38.7	348.4	159.5	114.7	61.7
210 K, measured	272.0	114.4	79.5	44.7		163.0	127.8	71.9
220 K, predicted	558.7	179.5	119.3	61.2	653.6	266.0	186.6	101.4
220 K, measured	612.5	200.6	131.8		650.7	276.7	201.2	111.5

The radiator design had numerous design constraints. The total mass allowed for the four radiators was only 3.0 kg. The radiators could have a maximum thickness of .64 cm. because of system envelope constraints. The inner TEC radiators could have maximum dimensions of 41.9 cm by 41.9 cm and the outer TECs could have maximum dimensions of 22.6 cm by 44.8 cm because of instrument enclosure constraints. The radiators had to be stiff enough that they would not deflect so far that they hit the enclosure panels next to them during launch vibration. It was desirable to keep the natural frequency of the radiators above 100 Hz, to reduce vibration testing.

The shape of the radiators was optimized using the thermal analysis computer program, TAK II[3] . It was first found that the performance of the radiator was strongly driven by its area, so the maximum possible dimensions were used. It was found that making the radiators thicker in the middle and thinner at the edges provided 15 to 17 day improvement in cryogen life of radiators of the same shape, material, and mass but constant thickness. The taper was made linear to simplify machining, but the shape of the area of maximum thickness in the middle of the radiator was optimized.

Several different materials were considered for the radiators. These included pure aluminum, beryllium S-200F, AlBeMet 162 (an alloy of aluminum and beryllium), high conductivity graphite/epoxy composite and high conductivity graphite/pure aluminum composites. Radiator systems of each material were analyzed for thermal and structural performance. Pure aluminum was rejected because its low modulus would have resulted in violating the envelope constraints when it was made thick enough to meet the vibration requirements. This was true even if a pure aluminum radiator was tapered and ribbed to increase its relative stiffness. Radiators made from the all the other materials had adequate

Table 2. Final thermal model prediction of thermal performance of TEC assembly performance in flight dewar

	Inner TEC assembly (average values)	Outer TEC assembly (average values)
TEC parasitic heat leak, watts	0.035	0.024
Cold finger assembly parasitic heat leak, watts	.076	.066
Dewar shield heat load, watts	0.217	0.516
Total TEC cooling load, watts	0.328	0.606
Dewar shell temperature, K	270	270
Radiator average temperature, K	276	277
TEC hot end temperature, K	281	280
TEC cold end temperature, K	194	220
Dewar shield temperature, K	195	224

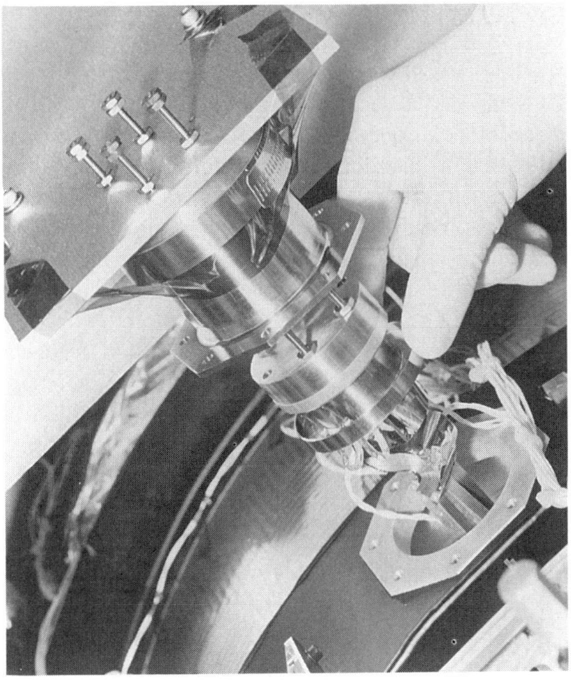

Figure 4. An outer TEC assembly being put into the flight dewar. The radiator mount is suspended from a tooling fixture that attaches to the dewar.

structural performance and similar thermal performance. A radiator system made from the graphite/aluminum composite had the best thermal performance, but this resulted in only 4.7 days of additional cryogen life over the AlBeMet radiators. Since the cost of the graphite/aluminum radiators would have been approximately twice that of the AlBeMet radiators, they were judged not to be cost effective. Similarly, the even smaller additional performance resulting from the use of beryllium S-200F and graphite/epoxy composite did not justify their much higher cost. AlBeMet 162 was selected as the radiator material.

TEC AND COLD FINGER TEST DATA

As a design aid, Marlow industries provided Ball with a prediction of the performance of the final TEC design. Marlow then provided Ball with prototypes of the final design. Ball performed thermal tests on one of these prototypes in a vacuum chamber in which the NICMOS operating conditions were simulated. The results are shown in **Table 1**. The actual performance was better than the prediction by an average of 9.2 % and with a standard deviation of 4.4%. Limited thermal performance testing was done on the flight TECs and the results appeared to be very close to the prototype test results.

Once the TEC installation had been designed, a detailed computer thermal model was written using TAK II. Since the radiative conductances are significant and complex, a TRASYS[4] model was written to calculate these interactions. The TRASYS model was based on the known geometries and estimates of emissivity of the various surfaces. A prototype TEC was assembled into a prototype cold finger assembly the assembly was thermally tested under the NICMOS conditions. From this data the thermal model was updated with a "best fit" model. Finally, thermal testing was performed on two of the flight assemblies and a final update of the thermal model was done. The results are shown in **Table 2**. The parasitic radiative heat loads on the TEC and cold finger were a small portion of the heat being removed from the simulated shield.

INSTALLATION OF FLIGHT HARDWARE ON DEWAR

The installation of the flight TECs into the cold finger assembly was very successful in that none of the TECs failed due to this installation. The subsequent vibration testing caused the electrical resistance of one of the TECs to increase more than the 5% allowed. Two of the TECs failed due to handling mishaps during thermal testing. These failures were not complete operational failures but were an increase of more than 5% in electrical resistance, which is Marlow Industries standard for rejection. These failures emphasized the need for careful handling of TEC assemblies and the wisdom of using a reversible method of bonding the rest of the cold finger to the TECs. There were no problems encountered in replacing the failed TECs.

The installation of the TEC and cold finger assemblies into the dewar went smoothly. Special tooling was built to hold the TEC, cold finger assembly, radiation shield and radiator mount during installation. **Figure 4** shows one of the flight assemblies being installed in the dewar. The electrical resistance of the TECs was checked after installation and was found to be within the allowed 5% change from previous readings. Since their installation in the flight dewar, the TECs have been powered up briefly to verify their operation. The rate at which they cooled the dewar shields was as expected. A complete thermal test of the dewar will not take place until May of 1996, after the dewar has been integrated into the NICMOS instrument.

CONCLUSIONS

1. TECs can be successfully used to significantly increase cryogen life in a dewar.
2. While the installation of the TEC into a dewar appears simple, their are numerous critical details in the interfaces, soldering, and handling.
3. The Ball S-link was critical to making installation the TEC installation in the NICMOS dewar possible. A thermal link made from copper wires is stiffer, more massive, and has lower thermal performance.
4. The thermal performance of TECs can be accurately predicted.

ACKNOWLEDGEMENTS

The NICSMOS instrument is being built by Ball Aerospace and Communications Group, Electro-Optics/Cryogenics Division for the University of Arizona under contract P172984. This work is being supported by the NASA Goddard Space Flight Center through funding to the University of Arizona.

REFERENCES

1. R. Thompson, "NICMOS: Near Infrared Camera and Multi-Object Spectrometer", Space Science Reviews, 61, pages 19-93, 1992.
2. R. F Arentz, R. A. Hopkins, D. A. Payne, J.F. Siebert; "A Verified, Broadly Applicable Design for Interfacing Cryocoolers to Sensors and Devices"; CEC/ICMC 1995.
3. Thermal Analysis Kit II, K & K Associates, 10141 Nelson St. Westminster, CO 80021
4. Thermal Radiation Analyzer, version P22, NASA JSC-22964

DESIGN OF THE XRS HELIUM INSERT

Susan R. Breon[1], Howard D. Branch[1], Garcia J. Blount[2],
Michael L. Jackson[1], Robert F. Boyle[1], and James G. Tuttle[3]

[1]Cryogenics, Propulsion, and Fluid Systems Branch
Code 713
NASA/Goddard Space Flight Center
Greenbelt, MD 20771
[2]Mechanical Engineering Branch
Code 722
NASA/Goddard Space Flight Center
Greenbelt, MD 20771
[3]Hughes STX
4400 Forbes Blvd
Lanham, MD 20706

ABSTRACT

The X-Ray Spectrometer (XRS) instrument has gone through numerous iterations, first as an instrument on NASA's Advanced X-Ray Astrophysics Facility (AXAF), then on AXAF-S, and now scheduled to fly on the Japanese Astro-E satellite. The Astro-E XRS is a high precision x-ray spectrometer with better than 20 eV resolution for x-ray energies from 0.3 to 10 keV. The requirement to obtain a lifetime greater than two years within the weight constraints of Astro-E has presented quite a challenge in the design of the cryogenic system. The design of the superfluid helium insert is described, with emphasis on innovative approaches taken to meet the requirements.

INTRODUCTION

Astro-E is the third host satellite designated for the XRS instrument[1,2]. Astro-E is being built by the Japanese Institute of Space and Astronautical Sciences (ISAS). In addition to the XRS, it will carry a Hard X-Ray Telescope and four Charge-Coupled Device (CCD) cameras. It will be launched from Kagoshima Space Center in January or February, 2000 aboard an M-V rocket. The M-V rocket is currently under development and will be launched for the first time in 1997.

The Japanese/American collaboration extends beyond the Astro-E spacecraft into the design of XRS itself. XRS (see Figure 1) consists of a solid neon dewar–built by Sumitomo Heavy Industries (SHI) under contract to ISAS; a helium insert–built by NASA's Goddard Space Flight Center (GSFC) and comprised of a helium cryostat, an Adiabatic Demagnetization Refrigerator (ADR), detectors, and JFET amplifiers; and associated electronics. The detectors and JFET amplifiers are housed in the Front End Assembly (FEA). To achieve the desired energy resolution, the detectors are cooled by the ADR to a temperature of 0.065 K. When the ADR is cycled, its heat is deposited in the liquid helium bath at a temperature of 1.3 K. The 17 K solid neon provides thermal shielding for the helium and a heat sink for a small thermal shield enclosing the JFETs, which operate between 80 and 120 K. The lifetime requirement for XRS is 2 years, with a goal of 2.5 years. A detailed description of the design of each of the XRS cryogenic systems is given in References 3-6.

DESIGN STATUS

The neon dewar design is a modification of the Infrared Telescope in Space (IRTS) helium dewar[7] which was launched in March, 1995 aboard the Space Flyer Unit (SFU) spacecraft. The IRTS dewar contained 100 L of superfluid helium and lasted approximately one month. The XRS dewar will contain approximately 120 L of solid neon and must last two years or more. The neon tank is suspended from the main shell by S-glass/epoxy straps. Both the suspended mass and the launch loads are higher for XRS than they were for IRTS. These changes present a challenge to SHI to create a long-life neon dewar within the 300 kg allocation[3].

Both the ADR and the FEA have been under development for a number of years. In the FEA, the emphasis has been on improvement and characterization of the detectors and increasing the reliability and repeatability of the detector processing steps. The work on the ADR has focused on providing a long "hold time," or time spent at the low temperature part of the cycle and on making the ADR flightworthy. A low-current superconducting magnet has been developed which provides a field of 2 Tesla with a current of only 2 A. Moving to the Astro-E spacecraft has not resulted in many changes to either the ADR or the FEA, although it has taken advantage of improvements in the thermal performance particularly of the ADR over its original specifications.

Of all the subsystems on XRS, the one that has existed for the shortest amount of time is the helium cryostat. Although both AXAF and AXAF-S XRS configurations had hybrid superfluid helium/mechanical cooler systems, the requirements on the thermal performance were not as demanding: the heat loads on the helium were in the neighborhood of 10 - 20 mW. The thermal budget for the Astro-E/XRS helium tank is ≤800 μW, or more than an order of magnitude lower than the AXAF and AXAF-S configurations. This required more effort than simply downscaling the design. It required rethinking the whole approach and

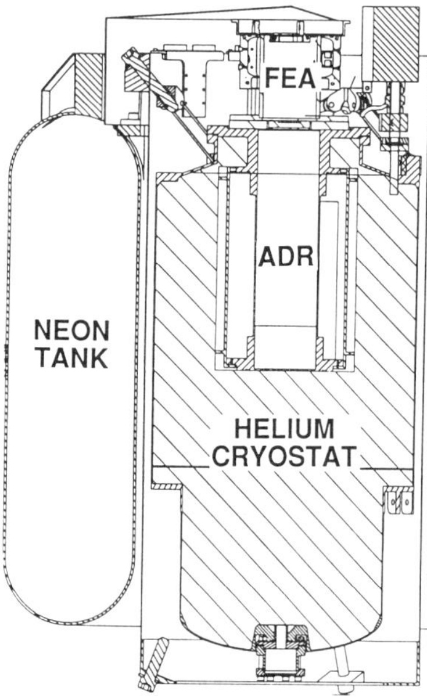

Figure 1. XRS configured for Astro-E. Hybrid cryogenic system provides cooling stages at 17 K (solid neon), 1.3 K (superfluid helium), and 0.065 K (ADR). JFETs in FEA are at 80-120 K.

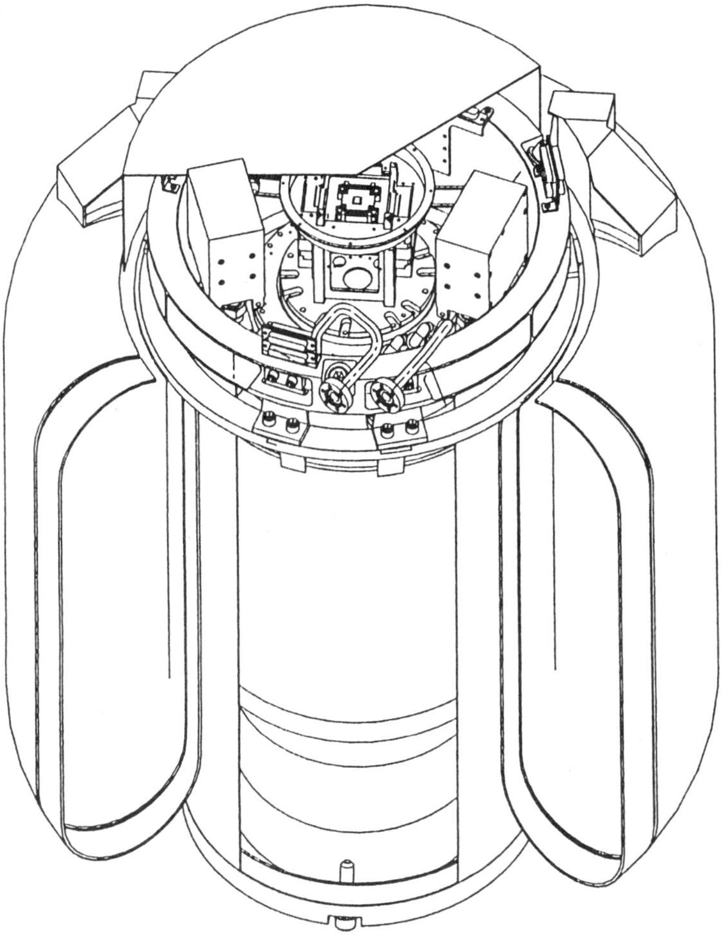

Figure 2. XRS helium insert.

learning not to consider every 10μW as negligible. Numerous trade-offs and configuration studies were conducted in the confusion following the cancellation of AXAF-S in the fall of 1993. Once the dust settled, the present concept was accepted in the summer of 1994. Preliminary design of the cryostat was started in the fall of 1994. A prototype is being fabricated and will be tested by the end of 1995. A mechanical/thermal model of the insert will be delivered to Japan in June 1996 for test. The flight cryostat will be tested and available for integration with the ADR and FEA in November 1996. The flight helium insert and electronics will be delivered to Japan at the end of 1997. Launch is early 2000.

DESIGN APPROACH

As can be seen in Figures 1 and 2, the helium insert is a complicated piece of hardware. The weight budget for the insert, which comprises the cryostat, ADR, and FEA, is 45 kg. The insert is constrained to fit within the center of the annular neon tank, which has a diameter of 356 mm. It is attached to a ring near the neon tank's upper dome. It is nearly impossible to visualize the required geometry in only two dimensions. Since the schedule does not allow a mock-up to be built to determine all the mechanical interfaces, we have instituted the use of the 3-D CAD program Pro/Engineer to design the insert.

The first part of the design to be tackled was the support of the helium tank from the neon tank. Recognizing that it would be difficult for the GSFC and SHI engineers to meet frequently, it was desired to keep the mechanical interfaces between the insert and neon tank as simple as possible. For this reason, the insert is cantilevered from a ring near the front dome of the tank. There are no attachment points at the aft end.

Initial concepts focused on two possibilities. Both incorporated a support ring or truss which bolted to the neon tank at three points. One concept connected the tank to the support truss using six thin-walled graphite epoxy struts, approximately 15 cm long, 1.3 cm in diameter with a wall thickness of about 0.3 mm. This approach was weight-efficient and provided a maximum helium tank volume. The second concept used 12 graphite-epoxy straps, six fore and six aft, attached to a cylinder bolted to the support ring. The ratio of the total cross-sectional area to the length was the same for the straps as for the struts. The advantage was that strap-supported cryogenic systems had been built and flown before. Ultimately, the amount of time needed to develop the techniques for using thin-walled struts and the uncertainty in how to design appropriate end fittings removed the strut-support concept from further consideration. The support cylinder for the strap-supported design is made of titanium, both because of its high strength/weight ratio and because its smaller coefficient of thermal expansion reduces the loss of pre-tension in the straps on cooldown. The titanium cylinder is connected to the aluminum neon tank by six inconel flexures which accommodate the differential thermal contraction between the neon tank and the cylinder. Each pair of flexures is separated by a steel tension bolt used to tighten the straps.

The sizing of the straps is heavily dependent on the launch loads. ISAS had provided loads specifications for Astro-E subsystems, including a random vibration specification. Since the M-V rocket has not yet been built, the loads were derived from the smaller M-III ocket, using conservative extrapolations as is appropriate for an untried launch vehicle. Using a simple, conservative approach to estimating the design loads from the random vibration specification resulted in a design load of 50 g. This was felt to be too conservative: once NASTRAN models were available and a coupled loads analysis was done these numbers should drop dramatically. GSFC chose to design to 25 g loads and developed the concept to meet a 3-year lifetime. A more detailed analysis was done, although still not a coupled loads analysis, which showed that the expected loads would be 75 g. Sizing the straps to handle 75 g loads resulted in a lifetime less than 1 year. These results were presented at the GSFC preliminary design review in March, 1995.

By this time, NASTRAN models had been developed for both the neon dewar and helium insert. Combining these models resulted in 40 g axial loads and 25 g lateral loads. Examination of the details of the analysis showed that the loads carried by the fore straps were much larger than those carried by the aft straps: the center of gravity of the suspended mass is towards the top of the tank, where the ADR and FEA are located. Consequently, the aft straps could be thinner and angled more sharply in the axial direction. The fore straps are vapor-cooled to reduce their heat load; the aft straps are not readily accessible so reducing the aft strap size allowed the lifetime to be raised back to approximately 2.5 years. The present design loads are 35 g axial and 20 g lateral. The total cross-sectional area of each of the fore straps is 0.053 cm^2. They are 13.7 cm long. The aft straps are 0.041 cm^2 by 16.5 cm.

The complex strap geometry has been laid out in the 3-D model. This has allowed interferences between the straps, the strap end fittings, and the helium tank or titanium structure to be resolved relatively quickly. Some of the more difficult pieces have already been sent to fabrication to meet the schedule for development and test of the prototype insert.

While determining the strap layout is important, the power of the 3-D model truly comes into play in packaging the helium insert plumbing and wiring. If the plumbing is poorly designed, it can be a major contributor to the heat load on the helium tank. Even though there is little space, the effective length of the plumbing must be on the order of 20 to 30 cm using small diameter (0.95 cm) thin-wall (0.25 mm) stainless steel tubing. To achieve this effective length, bellows are inserted in both the fill and vent lines between the valves and the conflat flanges. The insert uses two stepper-motor valves similar to the ones used on SHOOT[8], one on the fill and one on the vent line. The routing of the plumbing is further complicated by the need to operate on the ground both vertically and horizontally. This places restrictions on the location of the fill and vent lines as well as the porous plug. Furthermore, any segments of the vent line that do not point generally upwards in either of these orientations will be thermally shorted by convective currents during ground testing and operations.

A plumbing layout has been designed that meets the requirements for ground operations as well as for flight. Short sections of the vent line will be thermally shorted during ground tests, but the additional heat load should not overwhelm the other parasitic heat loads. The majority of the effective length of the plumbing is in the bellows: the bellows in the vent line generally point upwards in both horizontal and vertical orientations. The thermal model will be modified for the ground test configuration to allow a true comparison between the predicted and measured thermal performance of the helium insert.

To complete the design of the insert, the wires and the thermal straps still must be packaged. There are slightly more than 300 wires in the helium insert. Most of them carry very low current, <1 mA, and are 0.071 mm (42 AWG) diameter stainless steel wires. A few wires carry somewhat higher currents, 1 mA $\leq$ I $\leq$ 0.1 A, and are made of larger stainless steel wires. The gauge of these wires is optimized for the combined conductive, radiative, and Joule heating thermal loads. Wires which carry these higher currents only during ground operations are sized so they can carry full current without burning out. In the thermal model all the stainless steel wires are assumed to be 25.4 cm long and heat sunk only at the connectors on each end. We are in the process of determining if these lengths can be achieved without additional support for the wires.

The highest currents, on the order of 1 - 2 A for the ADR magnet and the valves, are carried by YBaCuO high-temperature superconducting (HTC) lead assemblies. Thin wires of YBaCuO are embedded in epoxy. Copper wires are attached at the warm end and NbTi wires with the copper cladding stripped off are attached at the cold end of the assembly. A copper heat strap is also attached at the cold end of the assembly and connected to the cold heat exchanger on the helium vent line. There are two lead assemblies, each with 12 conductors. The concept for mounting the HTC lead assemblies is to connect them to Kevlar ropes and suspend the ropes from attachment points within the insert. This concept will be tested prior to assembly of the insert.

Heat straps are used within the insert to thermally anchor the fore straps, the fill line, and the cold end of the HTC assemblies to copper heat exchangers on the helium vent line. The warm heat exchanger is approximately 8.5 K and the cold approximately 3.5 K. The support straps will have small copper clamps epoxied and bolted at two places on each leg of the strap with the heat straps attached to these clamps. The heat straps are copper wires on the order of 1 to 2 mm in diameter. The heat straps are sized to keep the temperature differences along their length to $\leq$0.01 K. The routing of the heat straps has not yet been determined.

SUMMARY

The design of the XRS helium insert presents numerous challenges in achieving a 2.5 year lifetime within the weight and volume constraints. An additional challenge is posed by the aggressive schedule for delivery of two test units and one flight unit within two years of the start of design work. These challenges are being met by combining new technology, e.g., 3-D CAD modelling and high temperature superconductors, with a wealth of heritage and experience gained previously on COBE, SHOOT, and the precursors to the Astro-E/XRS configuration. Ultimately, the effort will pay off in an unprecedented look at black holes and the formation of new stars at the centers of galaxies, as seen with x-ray vision.

Despite the detailed analytical work, there are many questions that remain about how the insert will perform. The only way to answer these questions is through a thorough test program. Tests of a few components will be conducted over the next few months. The most critical testing of a complete helium insert will be performed on the prototype, which will be a fully functional insert built to the same specifications as the flight design. It will undergo thermal performance testing and structural flight qualification testing to demonstrate that the design meets its requirements. It will also be used in conjunction with engineering unit electronics to demonstrate that the part of the system GSFC is responsible for functions correctly. A mechanical/thermal model of the insert is being fabricated. It will be a fully functional helium cryostat, but will have mass models for the ADR, ADR magnet, and FEA. The mechanical/thermal model will be mounted in an engineering unit neon dewar to verify mechanical, structural, and thermal interfaces. Finally, a flight unit will be built, tested, and delivered to Japan for integration into Astro-E/XRS.

ACKNOWLEDGMENTS

This work is supported by the National Aeronautics and Space Administration Office of Space Science (USA) and the Institute of Space and Astronautical Sciences (Japan).

The authors would like to thank Dr. Michael DiPirro, Ms. Judith Gibbon, Dr. Aristides Serlemitsos, Dr. Peter Shirron, and Dr. Stephen Castles for their contributions to the design of the XRS helium cryostat.

REFERENCES

1. S.R. Breon, R.A. Hopkins, and S.J. Nieczkoski, The X-Ray Spectrometer--A cryogenic instrument on the Advanced X-Ray Astrophysics Facility, *Adv in Cryo Engr* 37:1193 (1992).
2. S.R. Breon and R.A. Hopkins, The new X-Ray Spectrometer--A revamped cryogenic instrument for the restructured AXAF-S mission, *Adv in Cryo Engr* 39:187 (1994).
3. S.M. Volz, K. Mitsuda, H. Inoue, and Y. Ogawara, M. Kyoya, M. Hirabayashi, The X-Ray Spectrometer (XRS)–A multi-stage cryogenic instrument for the Astro-E x-ray astrophysics mission, To be presented at the 1995 Space Cryogenics Workshop.
4. S. R. Breon, J.A. Gibbon, M.J. DiPirro, B.A. Warner, and J.G. Tuttle, Thermal design of the XRS helium cryostat, To be presented at the 1995 Space Cryogenics Workshop.
5. A.T. Serlemitsos, M. SanSebastian, E.S. Kunes, J. Geagan, and J.E. Marketon, Performance of the XRS/Astro-E engineering model ADR, To be presented at the 1995 Space Cryogenics Workshop.
6. R.L. Kelley, R.J. Brekosky, J.D. Gygax, and S.D. Murphy, The microcalorimeter detector assembly for the Astro-E x-ray astrophysics mission, To be presented at the 1995 Space Cryogenics Workshop.
7. G. Fujii, S. Tomoya, M. Kyoya, M. Hirabayashi, M. Murakami, H. Okuda, H. Murakami, and T. Matsumoto, Thermal and mechanical performance of superfluid helium dewar for IRTS, *Adv in Cryo Engr*, 37:1359 (1992).
8. Ralph H. Haycock, A cryogenic vacuum valve for use in space, Report CSE 87-007, National Technical Information Service (1987).

PERFORMANCE IMPROVEMENT OF THE CHeX FLIGHT CRYOSTAT

T.S. Luchik, U.E. Israelsson, D. Petrac, and S. Elliott

Jet Propulsion Laboratory, California Institute of Technology
Pasadena, CA 91109, USA

ABSTRACT

The JPL flight cryostat last flew on the Space Shuttle in October, 1992 in support of the Lambda Point Experiment. A new experiment, the Confined Helium Experiment (CHeX), now in development will reuse this cryostat. An improvement to the cryostat performance was necessitated by the CHeX experiment having a longer mission requirement and stricter requirements imposed by NASA with respect to a launch-scrub turnaround scenario. The parasitic heat load reduction necessary to relieve both constraints was about 15% or 1 liter/day. The techniques implemented to achieve this goal, and subsequent results are presented along with a thermal model used during the analysis of the cryostat.

INTRODUCTION

The superfluid helium cryostat which provides the thermal environment to the JPL Low Temperature Platform (LTP) has flown on two previous missions, SpaceLab 2 in 1985 and the Lambda Point Experiment (LPE) in 1992 (part of the first United States Microgravity Payload, USMP-1). For these missions, the 10 day cryogen lifetime was sufficient to meet the requirements of the experiments it was supporting, given the launch-hold constraints of the time. This cryostat is next planned to be used to provide the cryogenic environment for the Confined Helium Experiment (CHeX) which is part of USMP-4 currently scheduled for launch in October, 1997. This experiment will measure the specific heat of helium in the vicinity of the lambda point in a geometry confined to two dimensions (50 μm spaces).

This cryostat is described in detail elsewhere and will not be described here.[1,2] The essential features of this cryostat is that it has allowance for an 20 cm diameter x 90cm long instrument and holds up to 90 liters of liquid (defined by porous plug location) with an instrument of the size described above installed.

For LPE, the cryostat was "grandfathered in" and allowed to have a 48 hour launch-hold. This was significant since the requirement at the time of the mission was 96 hours.

Launch hold is defined as the period of time between the nominal launch and the final launch attempt for a launch sequence. This launch hold value directly relates to the maximum time between last servicing of the cryostat and the subsequent servicing following a scrubbed launch. For LPE, this was 160 hours which the cryostat could meet with 8 hours

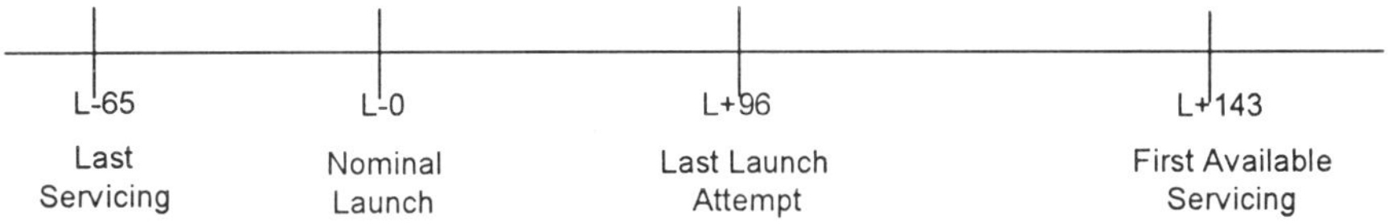

Figure 1. Launch-scrub-turnaround scenario for launch abort after 3 launch attempts.

of contingency. CHeX, however, will have to meet the full 96 hour launch hold criteria. Thus the maximum time between servicing in the event of a launch scrub will be 208 hours. As inherited from LPE the LTP cryostat would be unable to meet this requirement while keeping the instrument cooled to 2K. The Launch-Scrub-Turnaround scenario is shown schematically in Figure 1. The CHeX science requirement for a "Minimum Mission" is for 7 days on orbit compared to the LPE requirement of 5 days. So that even in the event that the cryostat were grandfathered in and allowed the 48 hour launch hold, it would still require an improvement in performance to meet the science requirement of the new experiment, CHeX. The Minimum Mission scenario is shown schematically in Figure 2.

APPROACH

Any improvement to the cryostat was constrained to be a low risk, non-intrusive improvement. This meant that any hardware modification had to be accessible without significant disassembly of the cryostat and limited the possible modifications to those accessible though the neck of the cryostat (the end of the cryostat where the science instrument is inserted).

In its simplest form, the problem of cryogen lifetime is constrained by two values, the maximum fill volume and the parasitic heat rate. However, this problem had an additional parameter and that is the minimum volume of helium required to maintain the instrument at 2K while on the ground. This value is a balance of the amount of heat input to the instrument during a servicing operation and the cooling power to the instrument of the helium in the cryostat bath. The issues of maximum fill and minimum level are related to servicing techniques, while that of the parasitic rate is flight hardware related and subject to the aforementioned modification.

SERVICING ISSUES

A plumbing schematic of the LTP cryostat is shown in Figure 3. On LPE, the path for helium to flow for pre-cooling the external plumbing during a servicing operation was through V2, V4, and V6. During the initial flow of helium there was a significant heat input

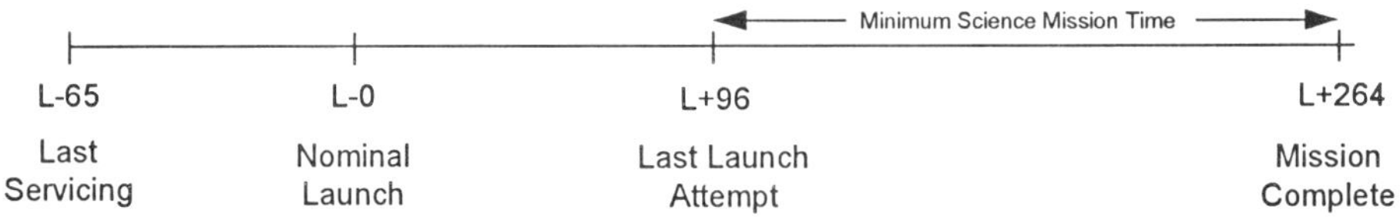

Figure 2. Minimum science mission scenario for CHeX.

to the instrument, since the external plumbing, initially at 300K, had to be cooled prior to liquid flow resulting in some backflow of warm gas through the porous plug heating the portion of the instrument not covered by liquid. Parts of the instrument was seen to heat to a maximum temperature of about 5K in most cases. However, it was required that 25 liters of superfluid be in the cryostat prior to the start of an operation. Otherwise the instrument could get warmer than 5K and put the superconducting shields on the high resolution thermometers of the instrument at risk of going normal.

To minimize the amount of heat input to the instrument during the initial phase of a servicing operation, we have simply chosen a separate pre-cool path. The path is through valves V2, V8, V7 and V6 shown on Figure 3. Since the CHeX instrument has not yet been installed in the cryostat it is impossible to say the exact effect of doing this. However, worst case estimates with an instrument thermal-mechanical model mounted in cryostat have shown that the minimum level that should be tolerable is 20 liters. Several tests showed as little as 17 liters as minimum level. Using the new value of 20 liters, a 5 liter improvement over LPE is achieved. In terms of launch hold at the LPE parasitic rate of 7.5 liters per day this is a 16 hour improvement in hold time simply due to a procedural change.

The other parameter that can be affected by servicing is maximum fill. On LPE, the maximum fill of superfluid helium at 2K was 78 liters (the maximum volume of the dewar given the porous plug configuration is 90 liters). This was achieved by pumping the supply helium down to about 100 torr and allowing it to flow at a maximum pressure differential of about 50 torr. The result was 90 liters of helium at about 50-60 torr which when pumped down to 2K yielded about 78 liters. We have only extrapolated this technique. By pumping the supply dewar down to just above the lambda transition and transferring helium at lower pressures we get a slightly higher fill fraction. When repeated serving operations are performed just after pumping the cryostat down to 10 torr and the supply dewar to 40 torr one can take advantage of thermal stratification in the dewar and achieve a significantly higher fill fraction of superfluid at 2K. We have found fill volumes as high as 85 liters and as low as 80 liters at 2K by performing as many as 3 repeated operations and few as 1 extra operation. The more conservative value of 80 liters gives a 2 liter improvement in fill volume or about a 7 hour improvement in lifetime both on the ground and for flight, using the LPE parasitic rate. Each of these "improvements" were completely procedural and therefore posed no new risk to the hardware.

HARDWARE MODIFICATIONS

Visual inspection of the neck end of the cryostat was the first step in determining the appropriate course of action. The inspection revealed that there was very poor thermal anchoring of all of the thermometry wiring from the warm interface of the cryostat to the thermometer. In some cases, the wires were attached to the vent line at a single point by Kapton tape. The inspection also revealed the use of copper wires for the current leads on a cryostat bath heater and on two superconducting liquid level sensors. It was also noted that the attachments of the vent line to the vapor cooled shields were suspect. Finally, there was a clear 300K to 4K radiation leak for each of cold valves (V3, V4, and V5 on Figure 3).

Although the visual inspection was promising, the magnitude of the improvement that might be expected was not known. A simple thermal model of the cryostat as it existed for LPE was built. A node diagram, with predicted heat flows, of the model for the LPE dewar is shown in Figure 4. The model parameters were fit to the various measured temperatures and parasitic rates from LPE. Data for two outer shell temperatures with the bath at 2K and a 300K outer shell temperature with the bath at 4K were available and matched to the model. Areas which were not within the scope of the improvement were the MLI and the thermal sinking of the cryostat support straps.

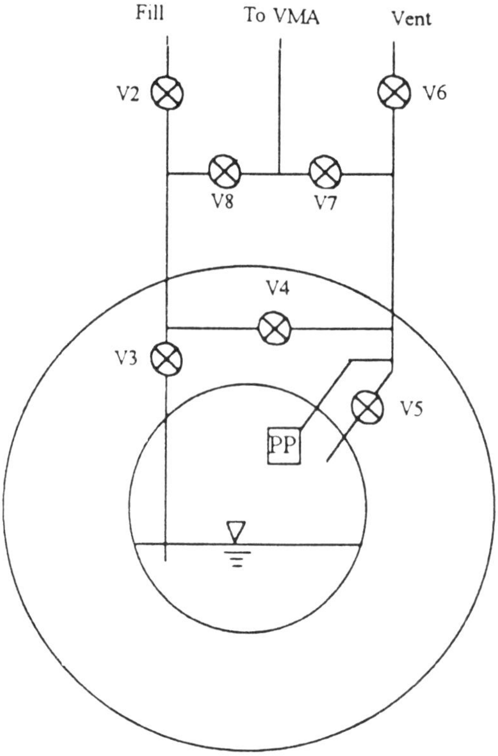

Figure 3. CHeX dewar plumbing schematic.

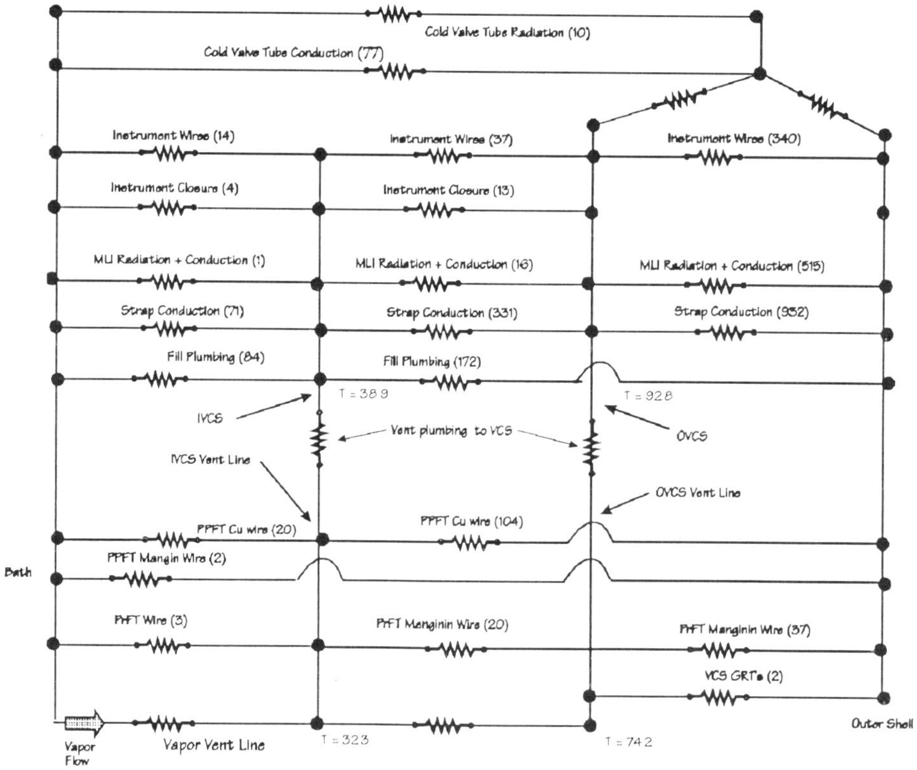

Figure 4. Heat flow diagram for the LTP cryostat in the LPE configuration.

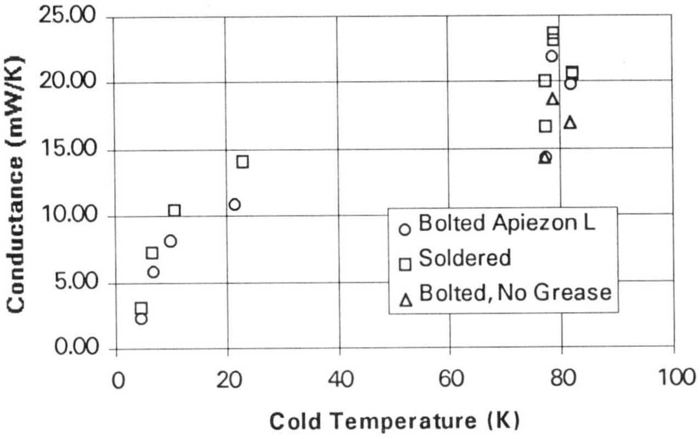

Figure 5. Conductance for various joints and cold temperatures.

The model showed that about 20mW of heat was being transported by the copper wires used for the cryostat heater and liquid level sensors. It was also noted that part of this heat leak was due to the poor thermal anchoring of the copper wires. All wiring was changed to 5 mil manganin wire with the following exceptions. The current leads on the cryostat heat were changed to 2-10 mil manganin wires in parallel and the current leads for the liquid level sensors were changed to 10 mil manganin. The wires were thermally sunk to the vapor vent line at six locations using copper spools soldered to the vent line and wrapping the wires to the spools. The wires were attached to the spools using epoxy.

The model also showed a significant temperature difference between the vapor vent line and the vapor-cooled shields. These two were connected thermally by a series of 0.76 cm wide copper straps that are soldered on the vent line side and a bolted joint on the shield side. It is well known that bolted joints are thermally inefficient. We ran several tests on dry-bolted, greased-bolted and soldered joints. The soldered joints were made using indalloy solder with no flux and an ultrasonic soldering iron. Thermal conductance was measured at powers up to 100mW across the joint and the results of the test are shown in Figure 5. Clearly, the soldered joint out perform the bolted joints over a wide range of temperature. However, in lieu of removing all of the bolted joints, we chose to simply add as many straps between the VCS's and the vent line as possible. The ultrasonic solder technique was used where the strap attached to the aluminum VCS and tin-lead solder was used where the copper strap was attached to the stainless steel vent line. Great care was taken in managing the flux and cleaning all surfaces after the operation was complete so as not to impact the performance of the MLI.

It was noted from the model that the guide tubes for the warm actuators of the cold valves V3, V4 and V5 conduct a significant amount of heat. The tubes are made of G-10 and serve no other purpose other than to insure that these warm actuators do no snag the MLI in the cryostat when they are moved from their stored position to where they engage the cold valve (during servicing operations). Thus it was determined that a reduction in the conduction heat load would be achieve by simply slotting the guide tube with no increase in risk to the cryostat. This modification was made and resulted in a small reduction in parasitic heat load.

The thermal model used for the LPE system was modified for the changes described above with no instrument installed. The model was run and compared to experimental data. The details of the prediction are shown in Figure 6. A comparison of Figures 4 and 6 show a significant increase in heat transport from the vent line to the VCS's due to the additional thermal strapping added. A comparison of selected predicted results with experimental data

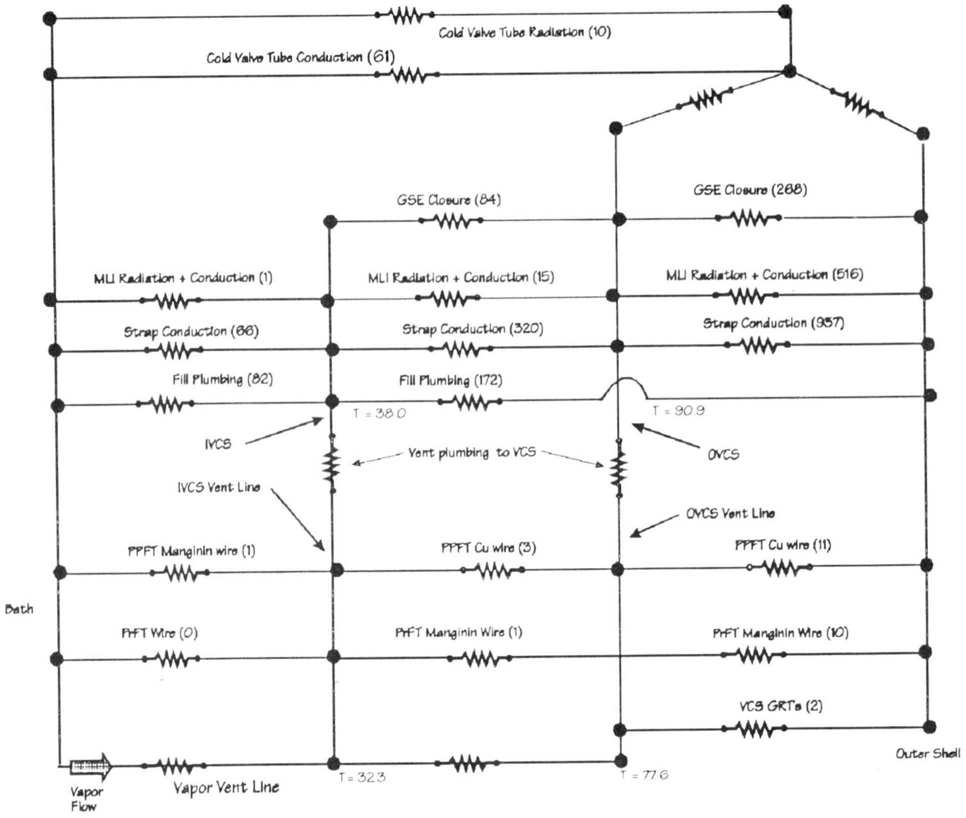

Figure 6. Thermal model with non-intrusive enhancements and no instrument.

are shown in Table 1. About a 19% improvement was predicted while about 14% was realized. However, these data are with no instrument installed. Again using the thermal model of the cryostat, the thermal model of the instrument from the LPE simulation was added to the cryostat model and run. These results are also presented in Table 1. Clearly there is a significant parasitic load from the instrument which is not surprising given the wiring complexity and the more complex installation.

SYSTEM ANALYSIS

These results have not yet answered the question of whether this cryostat enhancement will meet the requirements for CHeX. For launch scrub turnaround the number of hours is given by:

$$LST = (\text{Maximum Fill - Minimum safe level})/\text{parasitic rate}$$

Substituting 80 liters for maximum fill, 20 liters for minimum safe level and 6.7 liters per day for the parasitic rate, 215 hours is obtained which gives 7 hours of contingency on the 208

Table 1. Predicted and measured result with outer shell of cryostat at room temperature

	Measured, No Instrument	Predicted, No Instrument	Predicted with Instrument
Boil off (liters/day)	6.2	5.8	6.2
T_{IVCS} (K)	32	38	39
T_{OVCS} (K)	105	91	93

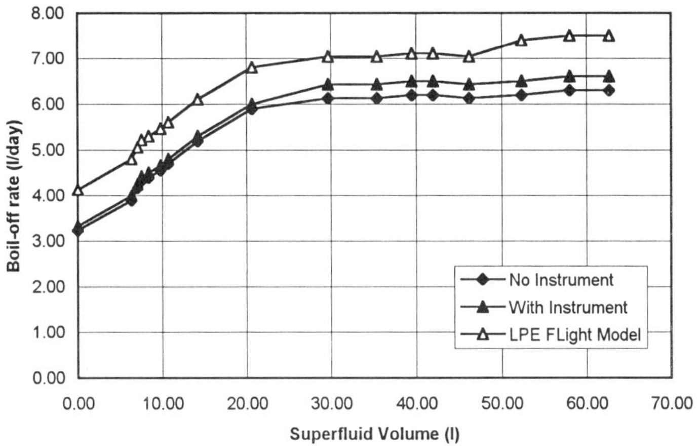

Figure 7. Parasitic rate as a function of liquid volume.

hour requirement. Although this is very little contingency, it represents a worst case analysis since the cryostat may be filled to a higher fill fraction and the minimum safe level for the instrument may be reduced.

The science requirement is for 7 days of experiment after sitting on the ground for 156 hours. However it is inappropriate to use the nominal ground boil-off rate for the entire volume of helium. While on the ground, with the tank moderately full there is an apparent steady state parasitic rate. As the volume of helium in the cryostat lessens, the parasitic rate reduces. This is shown in Figure 7. Also it was noted during LPE flight that the outer shell of the cryostat was at a reduced temperature (273K) compared to the nominal ground condition (300K). The present thermal model predicts a reduction in parasitic load of about 0.3 liter/day due to this temperature reduction. Combining the two effects (Figure 7 and the reduced outer shell temperature) yields a predicted lifetime for the LPE flight of 12.1 days and is shown in Figure 8. The measured value was 12 days, 2 hours.

Figure 7 also shows boil-off data for the cryostat with the CHeX instrument installed. This was obtained by using the delta-boil-off due to the instrument from the thermal model

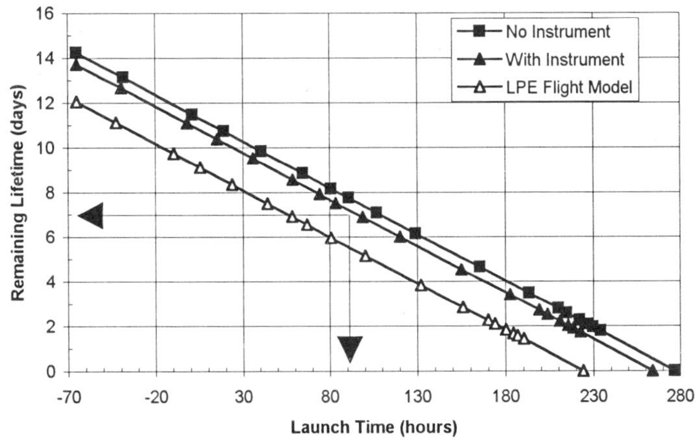

Figure 8. Cryogen lifetime versus launch time.

and adding it to the measured values obtained from the cryostat with no instrument present. For CHeX, the outer shell of the cryostat is expected to cool to 273 K as was seen on LPE. For the flight portion (time greater than the 96 hour launch-hold), the appropriate parasitic rate is the value shown in Figure 7, reduced by 0.3 liters/day due to the colder outer shell. The results of combining these data in this fashion are shown in Figure 8. The arrow indicate the remaining lifetime in the cryostat after 96 hours of launch-hold. The value is 7 days with no contingency.

CONCLUSION

The enhancement to the LTP cryostat presented in this paper were twofold: procedural and hardware modifications. Combined it has been shown that this cryostat can meet the two modified requirements for the Confined Helium Experiment. The launch-scrub scenario, which required an additional 48 hours between servicing, is easily met with 8 hours of cryogen contingency. The science requirement for minimum mission, which required an additional 72 hours of on-orbit cryogen lifetime has been met, although with little to no contingency. Since the likelihood of actually launching the shuttle after a 96 hour launch hold is very small, the project is not concerned with the small contingency. These data are based on experimental results with no instrument, and a predictive model for the cryostat and instrument. Final testing to verify these predictions will occur during the CHeX integration and test period scheduled to begin in January, 1996.

ACKNOWLEDGMENT

This work was carried out by the Science and Technology Section of the Jet Propulsion Laboratory, California Institute of Technology, under contract to the National Aeronautics and Space Administration. The work was funded by NASA Microgravity Science and Applications Division.

REFERENCES

1. D. Petrac, U.E. Israelsson, D. Otth, L.L. Simmons, J. Staats, and A. Thompson, "Technical Challenges Involved in Supporting the Lambda Point Experiment," *Cryogenics*, 30, September supplement (1990).
2. D. Petrac, U.E. Israelsson, and T.S. Luchik, "The Lambda Point Experiment: Helium Cryostat, Cryo-Servicing, Functions and Performance," *Advances in Cryogenic Engineering*, 38, (1993).

CRYOGENIC AND THERMAL ASPECTS OF THE SIRTF WARM LAUNCH CONCEPT

M. J. Lysek, U. E. Israelsson, R. D. Garcia, and T. S. Luchik

Jet Propulsion Laboratory/California Institute of Technology
4800 Oak Grove Drive
Pasadena CA, 91109

ABSTRACT

A fundamentally new concept is being explored for cryogenic operation of the Space Infrared Telescope Facility (SIRTF). In this concept, the helium dewar and instruments would be launched at liquid helium temperature while the telescope and external radiation shields are at room temperature. Once on-orbit, the telescope and the radiation shields would be cooled radiatively to about 70 K. Helium vapor is then used to bring the telescope down to its final operating temperature of 5.5 K. We present a design which meets SIRTF requirements and discuss its merit.

INTRODUCTION

The Space Infrared Telescope Facility[1] (SIRTF) has recently defined an innovative new mission implementation. This "warm launch" architecture for cryogenic space telescopes was developed by Dr. F. J. Low (University of Arizona), the SIRTF Facility Scientist. The principle of this approach will be described by Dr. Low separately (in preparation). This paper presents a detailed application of these principles to the baseline SIRTF design.[2]

In traditional "cold launch" infrared telescope designs (IRAS and ISO), the mirror and instrument package are enclosed within an open-ended cryogenic dewar and are cooled to 2 K when the dewar is filled with liquid helium before launch. In this design, the dewar vacuum vessel must enclose the entire facility, requiring a large, heavy structure for a telescope the size of SIRTF (0.85 m diameter primary mirror).

The current "warm launch" design for SIRTF offers a great savings in mass by reducing the size of the vacuum vessel. In this design, only the instrument package and helium tank are enclosed within the dewar. The primary and secondary mirrors are outside of the dewar and at room temperature when the facility is launched. A thin outer shell and vapor cooled radiation shield surround the telescope and provide thermal isolation when in space. The outer shell is shielded from solar radiation by a solar panel, which allows the outer shell to radiate its heat to space and cool to about 50 K soon after launch. The primary and secondary mirrors (telescope assembly) are initially cooled to about 50 K through a heat switch connected to the outer shell. Then the heat switch is opened, and cold vapor from the helium dewar cools the telescope to below 5 K within a few weeks.

A detailed mechanical and thermal design study demonstrates the feasibility of the warm launch design and highlights some technical challenges to be overcome.

Advances in Cryogenic Engineering, Vol. 41
Edited by P. Kittel, Plenum Press, New York, 1996

OBSERVATORY CONFIGURATION

The configuration of the observatory is shown in figure 1. Low-thermal conductivity struts mount the Cryo-Telescope Assembly (CTA) to the top of the spacecraft which is temperature controlled near 300 K. The solar panel and solar panel shield are supported by low-conductivity struts from the top and bottom of the outer shell. The solar panel shield has a radiator at the top of the observatory. Radiation to space keeps the shell shield at about 180 K, allowing the outer shell to reach 60 K or less. The spacecraft shield, between the spacecraft and the CTA, includes a radiator to cool it below 130 K.

The internal configuration of the CTA has two main parts: the telescope and the dewar. The telescope is mounted to the top of the dewar vacuum shell on struts. The Dewar contains the instrument chamber and the liquid helium tank. A bolted flange near the middle of the dewar outer shell permits disassembly. The telescope and dewar are thermally connected.

The instrument chamber is an aluminum enclosure containing the science instruments. The instrument chamber is supported from the helium tank by struts. Copper straps thermally couple the instrument chamber and the 1.4 K helium bath. The helium tank is supported from the dewar vacuum vessel by struts made of gamma alumina/epoxy composite. Another set of gamma-alumina struts support the dewar from the bottom of the CTA.

The vacuum cover or "cork" is located on the top of the dewar vacuum vessel between the instruments and the telescope. After launch it is opened to admit light to the instrument chamber. The cover may incorporate a window to permit visible light testing of the focal plane while the cover is closed. The vacuum cover must be vacuum tight at room temperature in order to maintain the integrity of the dewar vacuum vessel before launch. Once in space,

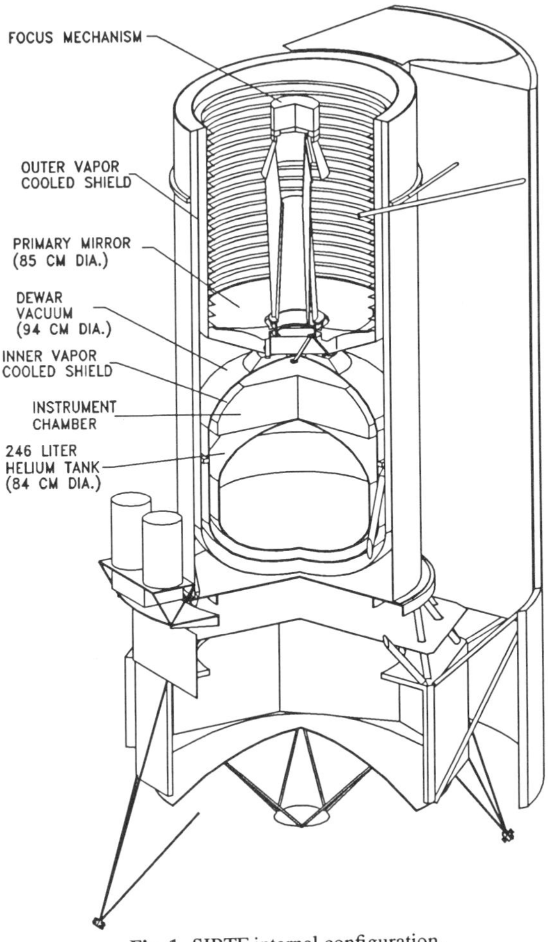

Fig. 1. SIRTF internal configuration.

the vacuum cover needs only to prevent the instrument from becoming contaminated before the telescope cools.

The telescope and dewar are surrounded by a lightweight outer shell. The outer shell is closed at the bottom. Before launch, the outer shell is closed at the top by a dust cover, and the observatory is kept under an intert gas such as dry nitrogen to prevent atmospheric contamination of the mirrors and low emissivity surfaces.

The observatory has two vapor-cooled shields, plus the dewar vacuum shell, which also acts as a vapor-cooled shield. The inner vapor-cooled shield (IVCS) is mounted between the helium tank and the dewar shell. The outside of the IVCS is wrapped with multilayer insulation (MLI). The outer vapor-cooled shield (OVCS) is mounted between the dewar and the outer shell. The OVCS is supported from the outer shell by gamma-alumina struts. All surfaces from the dewar shell to the outer shell are gold coated for low emissivity.

A vapor vent line is connected between the helium tank and vent valves on the outer shell. In orbit, helium vapor flows through the vent line and cools the IVCS, the dewar vacuum shell and telescope, the OVCS, the dewar support struts, the instrument wires, and the outer shell. On the ground, the vapor flow cools the IVCS and instrument wires.

Observatory Cooldown

Before launch, the outer shell, the OVCS, the dewar vacuum shell, and the telescope are at room temperature. During the initial cool-down in space, a heat switch conducts heat from the dewar vacuum vessel and the telescope assembly to the outer shell. This heat switch may be either a removable metal link or helium gas purged into the assembly before the “cork” is opened. The heat switch accelerates the cooldown of the telescope and dewar vacuum shell to about 70 K. After this initial cooldown, the thermal connection between the telescope and the outer shell is broken. Vapor cooling is then used to bring the observatory to its operating temperature.

THERMAL DESIGN

Requirements

There are three driving thermal requirements: The helium bath must be at a temperature of 1.4 K or less, the telescope must be at 5.5 K or less, and the liquid helium must last at least 2.5 years.

The current SIRTF mission plan places the facility in a heliocentric orbit trailing the earth that eliminates the Earth as a heat source and allows a thermal design that puts the observatory behind a solar shield.

Internal Thermal Design

The liquid helium tank has a low-emissivity gold coating and the IVCS is covered with MLI to reduce the heat load on the dewar prior to launch. Gamma-alumina struts are used throughout the design to provide high stiffness supports with low thermal conductivity. The outside surface of the telescope baffle, both sides of the OVCS and the inner surface of the outer shell have low emissivity gold coatings. The instrument wires and various supports are thermally coupled to the vapor vent line at several places to reduce the heat load into the telescope assembly.

A porous plug liquid-vapor separator in the vent line contains the liquid helium while maintaining the instrument temperature at 1.4 K. A particular challenge is to design a porous plug that will operate at both the high flow rate encountered prior to launch, and at the low flow rate after the initial cooldown.

External Thermal Design

To maximize the helium lifetime, the observatory outer shell temperature must be as low as possible. The sources of heat to the outer shell are the spacecraft and the solar panel. A shield is mounted between the spacecraft and observatory and another shield is mounted between the solar panel and the outer shell. The shields are covered with MLI to decrease the emissivity of their surfaces. The wedge shaped solar panel shield curves away from the outer shell in order to improve its view factor to space and to improve the thermal isolation between it and the outer shell.

The cold side of the outer shell (the side away from the sun) has a high-emissivity coating. The outer shell temperature is set by the thermal balance between the heat adsorbed from the heat sources and the heat radiated to space from the cold side. The temperature difference between the hot side and the cold side is about 5 K.

The spacecraft is maintained near 300 K by louvers, MLI, heaters and the electronics heat dissipation. The solar panel has 86 percent of its area covered with solar cells and 14 percent covered with optical solar reflectors. The OSRs reduce the solar panel temperature, typically 350 K.

THERMAL MODEL

The SIRTF baseline design has been developed with the aid of an integrated Microsoft Excel spreadsheet model that couples a mechanical configuration spreadsheet to a thermal balance calculation, enabling many parametric trade studies to be performed quickly. Further static and transient thermal calculations were performed with the same equations and parameters using *Mathematica.*

The thermal design solves for the temperatures of the twelve nodes, assuming that the helium bath temperature is 1.4 K. All radiative and conductive heat flow calculations, except heat transfer to space, were increased by a factor of 1.2 to allow for unexpected design changes. Heat flow through MLI at low temperature was modeled using the data from Spradley et al.[3]

Static Thermal Performance

The temperatures and heat flow data inside the observatory are shown in figure 2, assuming gold coatings with 0.015 emissivity. Note that the heat flow into the helium tank is dominated by the instrument power dissipation, 8 mW. The IVCS would not be necessary in this design except to provide sufficient performance when the dewar outer wall and telescope assembly are at 300 K before launch. The helium flow rate during this time is approximately 13 mg/s, compared with 0.4 mg/s when in space. This design presents a difficult requirement on the porous plug liquid-vapor separator, which must either operate over a very wide dynamic range, or include a valve to switch between two porous plugs of different areas. Future designs may improve the ground performance of the system in order to reduce the required porous plug dynamic range.

The current baseline design achieves a lifetime of 2.5 years using 250 liters of LHe. This is primarily due to the low outer shell temperature, about 50 K. Figure 3 shows the temperatures and heat flow data outside the outer shell. Note that most of the heat is intercepted by the solar panel shield and the spacecraft shield.

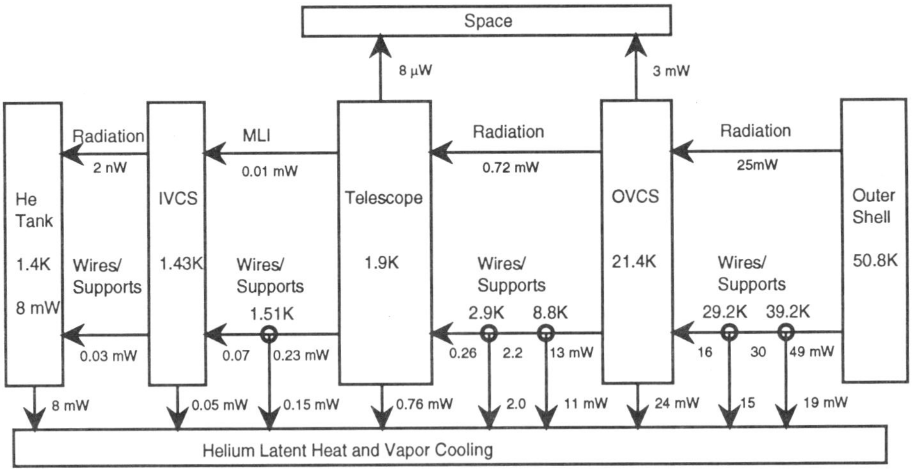

Fig. 2. SIRTF internal temperatures and heat flows configuration.

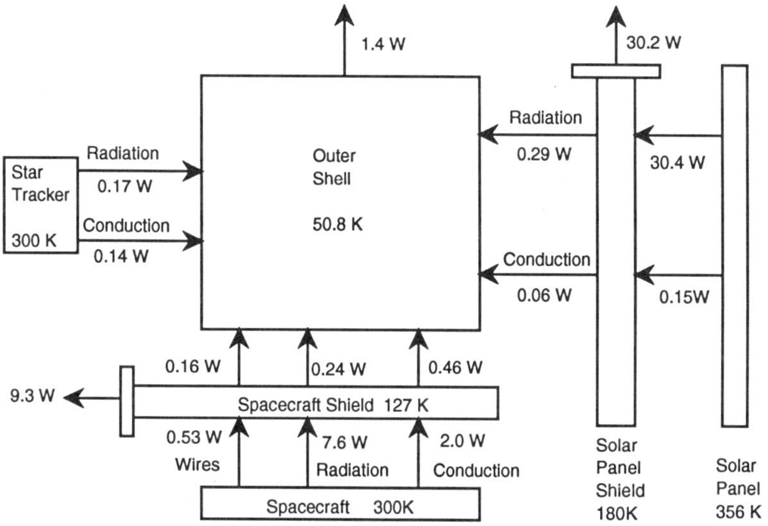

Fig. 3. SIRTF external temperatures and heat flows. View factors for the radiative heat flow calculations were derived using the NEVADA Monte-Carlo ray tracing program.

The telescope and dewar outer wall is primarily cooled by the helium vapor. The boil-off rate is mainly determined by the instrument power dissipation. Figure 4 shows a plot of the telescope temperature vs. instrument power for instrument power dissipations from 0 to 16 mW. The telescope is below the required temperature (5.5 K) when the instrument power is above 5 mW.

The performance of the baseline design depends strongly on the ability to fabricate surfaces with low emissivity coatings. Figure 5 shows the telescope temperature for several values of emissivity. Note that adequate performance is achieved for emissivities less than 0.025, an achievable goal.

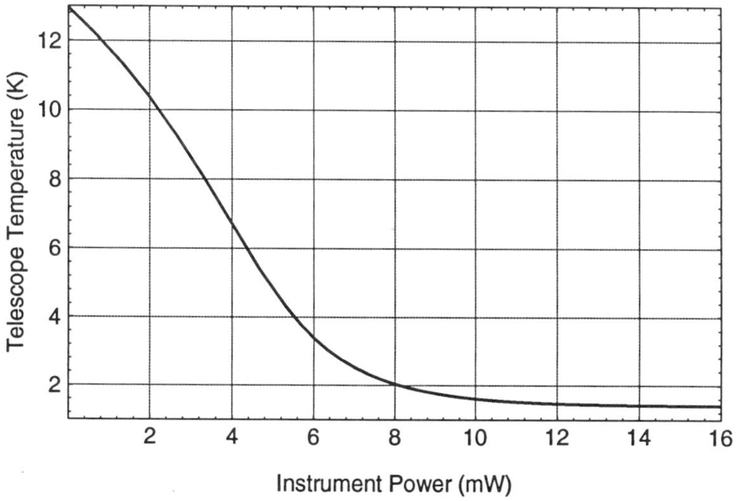

Fig. 4. Telescope temperature vs. Instrument power with the low e surfaces having emissivity values of 0.015.

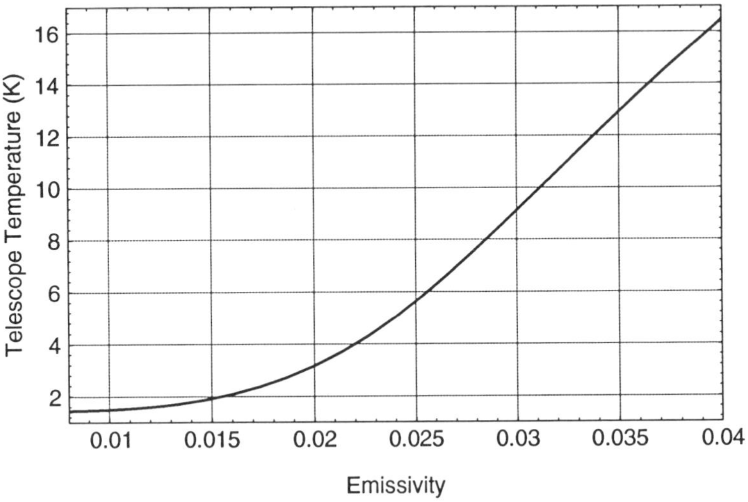

Fig. 5. Telescope temperature vs. emissivity with the instrument power fixed at 8 mW.

Transient Thermal Performance

The warm launch design depends on radiation to space to cool the telescope to below 50 K. The first 30 days of operation were modeled assuming the nominal values for emissivity and instrument power. Figure 6 shows the various temperatures as a function of time. The outer shell temperature drops rapidly to about 50 K within several days of launch. After seven days, the telescope has reached 50 K, and the heat link to the outer shell is opened. At that point, it takes about two weeks for the helium vapor to cool the telescope to its operating temperature.

Figure 7 shows the Helium boil-off rate and figure 8 shows the helium consumption as a function of time for the initial 25 days in space. Approximately 16 L of liquid helium are expended before the telescope cools. Another 8 L of liquid helium will be consumed during the 24 hour pre-launch hold, requiring about 250 L of liquid helium for the 2.5 year mission.

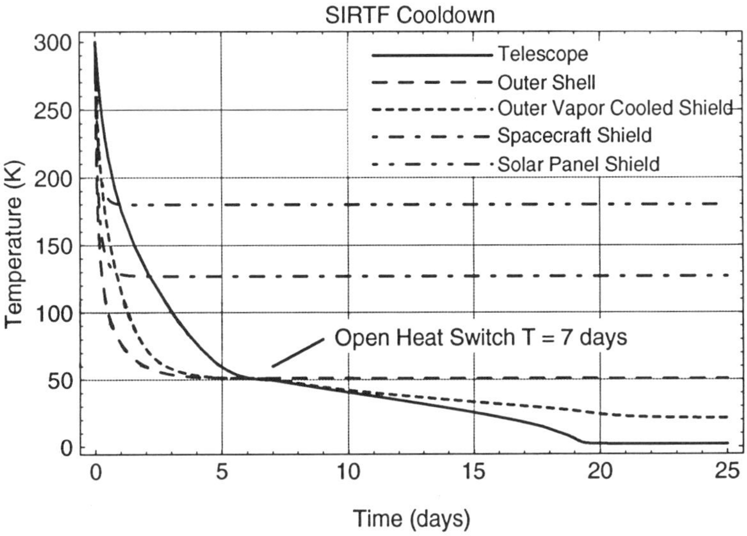

Fig. 6. Telescope temperature vs. time for the nominal instrument power (8 mW instrument power and 0.015 emissivity surfaces).

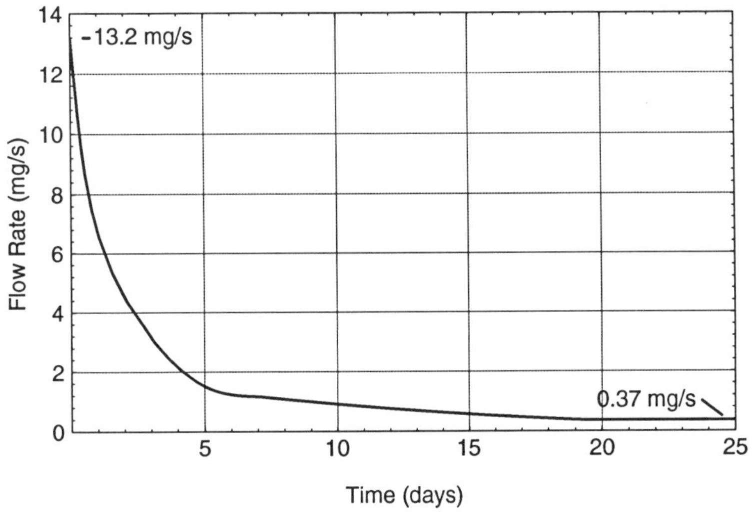

Fig. 7. Helium boil-off rate for the first 25 hours in space assuming 8 mW instrument power and 0.015 emissivity surfaces

BENEFITS AND CHALLENGES OF THE WARM LAUNCH DESIGN

There are several advantages of the warm launch design. Placing the telescope outside of the cryogen tank permits the aperture of the telescope to be increased without increasing the size (and mass) of the dewar, which should allow a larger telescope to be launched on a given launch vehicle. Because the telescope is launched warm, a cryogenic shake test of the telescope is no longer required; it appears that there are other potential savings and simplifications in test and integration as well. The warm-launch architecture is a step towards the future, because it can readily be adapted to telescope apertures larger than the 0.58 m size of SIRTF and its cold-launch predecessors.

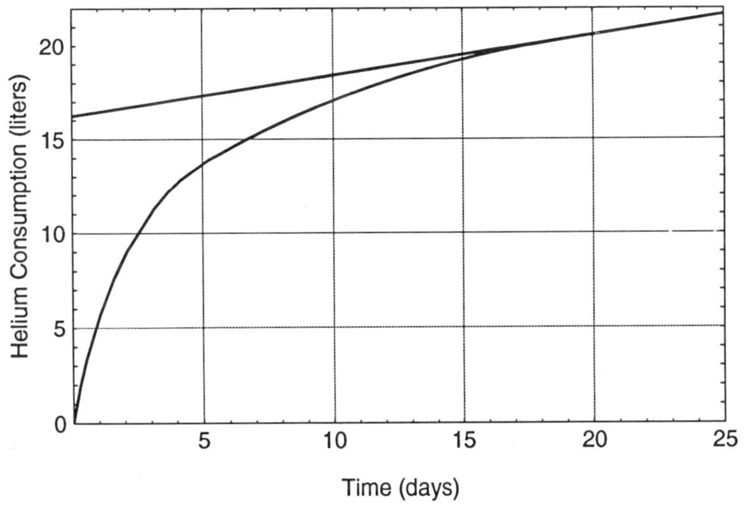

Fig. 8. Lower Curve shows the helium consumption vs. time for the first 25 days in space with the nominal instrument power (8 mW) and emissivity (0.015). The upper line has a slope representing the steady state helium boil off rate after the initial transient has settled.

The warm launch design also presents several challenges. The porous plug liquid-vapor separator will need to operate over a wide dynamic range. The vacuum cover, or cork, must seal a large diameter vacuum opening in the dewar at low temperatures. Low emissivity surfaces must be fabricated and maintained in order for the telescope temperature to be low enough. A particular challenge to the optical design is to maintain alignment between the instrument chamber and the telescope assembly over a wide range of telescope temperatures.

Other design concepts under consideration include locating the instrument chamber outside of the dewar vessel under the telescope assembly where it would be at room temperature during the launch. The instruments would then been cooled by superfluid helium provided by a fountain effect pump. This design further reduces the mass of the overall system but makes pre-launch instrument testing far more difficult.

ACKNOWLEDGMENTS

Dr. Frank Low contributed greatly to the development of the warm launch concept. Many members of the SIRTF design team at JPL contributed to the thermal design including D. Elliott and J. Fanson. The SIRTF project is grateful for the work of several industry partners. This work was performed at the Jet Propulsion Laboratory under a contract with the National Aeronautics and Space Administration.

REFERENCES

1. D.P. Cruikshank, M.W. Werner and D.E. Blackman, SIRTF - Capabilities for the study of planetary systems, *Astrophys. Sp. Sci.*. 212(1-2):407 (1994).
2. "SIRTF Delta 7920 Baseline Design," NASA/Jet Propulsion Laboratory, JPL-D12357, Pasadena (1994).
3. I.E. Spradley, T.C. Nast and D.J. Frank, Experimental studies of MLI systems at very low boundary temperatures, in: "Advances in Cyrogenic Engineering," Plenum Press, New York (1990), v. 35 p. 477.

CRYOGENIC TELESCOPE TEST FACILITY

Thomas S. Luchik, Robert G. Chave,
Ulf E. Israelsson, Melora Larson, Mark Lysek,
Alfred E. Nash, and James Hardy

Jet Propulsion Laboratory, California
Institute of Technology, Pasadena, California
91109, USA

ABSTRACT

An optical test Dewar has been constructed with the unique capability to test mirrors of diameter ≤ 1 m, f ≤ 6, at temperatures from 300 to 5 K with a ZYGO Mark IV/V interferometer. The facility possesses extensive thermometry throughout for characterization of the test chamber thermal environment and Dewar performance. Optical access is controlled with cryogenically cooled shutters. The entire Dewar is vibration isolated by 40 dB where the fundamental resonances of the Dewar structure are highest. The facility has been brought on line for its first user, the Infrared Telescope Technology Testbed (ITTT) for the Space Infrared Telescope Facility (SIRTF) at JPL. The design requirements for this facility and the resultant design and implementation experiences and challenges will be presented.

INTRODUCTION

In the astronomical community there is clearly the desire for large spaceborne optical systems. Due to NASA's need to produce smaller, lighter, lower cost spacecraft, there exists an effort to develop large lightweighted optics. In order to mitigate the risk of using these new lightweight technologies, it was necessary to build a facility to permit the testing of these systems. In keeping with NASA's desire to simplify and reduce the cost of flight operations, the Jet Propulsion Laboratory (JPL) undertook the task of a quick development, low cost facility for the optical interferometric testing of mirrors of diameter ≤ 1 m, f ≤ 6, at temperatures from 300 to 5 K. The project was constrained by an allocation of one year in which to complete the design, fabrication, and optical qualification test, with funding commensurate with the allotted schedule. The system design performance was traded off against cost to minimize the resources necessary to develop the system yet still meet the requirements for cryogenic optical testing. The Dewar, instrumentation, vibration isolation and support equipment were all treated in the trade space. The facility requirements were that: the system environment be stable enough for λ/40 rms interferometric measurements with a helium neon laser interferometer, the Dewar permit disassembly for mounting the test apparatus, and that all surfaces in the test chamber reach 8K. Two goals of the facility were to reach equilibrium in no more than 72 hours and that cryogen hold time be at least 72 hours.

The approach was the use of a small, dedicated team with an industry partner (Janis Research Co., Inc.) for the Dewar fabrication. Specific expertise was procured on an as needed basis. Another part of the approach was to use, to the extent possible, hardware inherited from previous JPL projects.

GENERAL FEATURES

Dewar structure

The Dewar has a vertically oriented inner cylindrical test chamber with a diameter of 1.4 m and an internal height of 2.3 m. The experiment mounting surfaces, the upper and lower surfaces of the chamber, are the exterior surfaces of two liquid helium tanks. The upper liquid helium tank may be positioned at any height within the cylindrical wall of the test chamber to facilitate the testing of optics with various focal lengths. The cylindrical wall of the test chamber is conductively cooled by two 300 liter capacity tanks. All exterior helium cooled surfaces (tank walls and shroud) are covered with a single layer of aluminized mylar to raise the surface emissivity[1]. Surrounding the test chamber is an aluminum dome covered with a 10 layer multi-layer insulation (MLI) blanket made of layers of aluminized mylar with spacer material[2]. This dome is conductively cooled through a bolted contact joint with a 300 liter capacity liquid nitrogen tank (which also has a 10 layer MLI blanket) directly below the helium cooled test chamber. The outer vacuum shell of the Dewar is constructed of an aluminum dome that makes an O - Ring seal with a stainless lower shell surrounding the lower cryogen tanks. Structural and thermal separation between the outer vacuum shell, the nitrogen tank, and the lower helium tank is provided by titanium struts

Figure 1. SIRTF Telescope Test Facility (STTF)

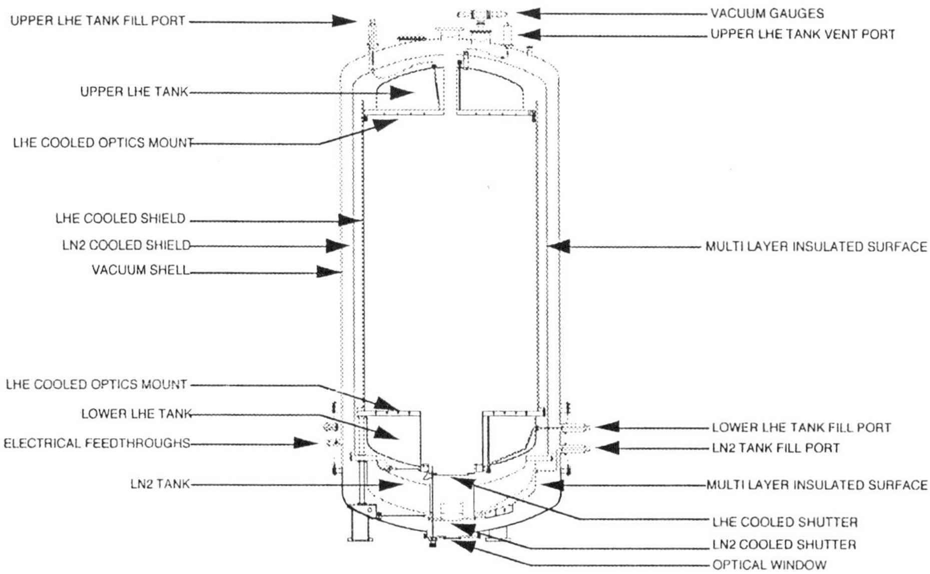

Figure 2. STTF Dewar structure.

and a stainless tension/compression system necessary for earthquake safe operation of the facility in southern California. The entire Dewar is designed to withstand 1.0 g in the vertical and 0.5 g in the horizontal axis without serious hazard to human safety. Optical access is provided through a window at the bottom of the Dewar, with cryogenically cooled baffles and shutters in the apertures through the lower cryogen tanks. The Dewar was fabricated through a partnering arrangement with Janis Research Company, Inc.

Instrumentation

The facility possesses 61 resistance thermometers with multiplexed "quasi" 4 - wire dc readout for characterization of the test chamber thermal environment and Dewar performance. All three cryogen tanks have resistive elements inside with multiplexed "quasi" 4 - wire dc readout for detection of liquid nitrogen level. Both helium tanks have superconducting helium level sensors. The Dewar guard vacuum is monitored with two pairs of cold cathode and Pirani sensors. All instrumentation control and data acquisition is accomplished with a GPIB interfaced Pentium based personal computer. Graphical user interface software written with LabView permits real-time monitoring and analysis of facility parameters.

Cryogenic Optics

A two axis optical gimbal mechanism[3] for aligning 1 meter diameter telescope primaries and test flat mirrors at temperatures from 300 to 4.2 K was constructed for use in the STTF. This mechanism consists of an aluminum frame, pivoting on a monoball bearing, and driven in tip and tilt by tungsten di-sulfide lubricated lead screws with external room temperature drive motors. Flexures decouple the optical support frame from stresses generated by differential rates of cooling. A second set of flexures decouples the mirror mechanically and thermally from distortion in the gimbal mechanism. The mechanism provides arc-second resolution in either axis, while designed to limit the heat leak to less than 100 mW at 4.2 K. Linear variable differential transformers (LVDT's) were used at temperatures from 300 to 4.2 K as readout for the gimbal mechanism. The full system was qualified by obtaining $\lambda/40$ interferograms of a well-characterized beryllium mirror at room temperature, 77K and 6K. These tests proved to be successful and have been presented elsewhere.[4]

Facility Infrastructure

The Dewar was mounted on an existing triangular vibration isolation frame that was modified to provide additional strength to meet earthquake safety requirements and to provide additional stiffening to improve stability for optical measurements. The vibration isolation frame was supported by three pairs of Newport I-2000 pneumatic isolators that have been measured to give the entire facility up to 40 dB of isolation from vibrations in the floor from 5 to 20 Hz (the dominant vibration coupling to the Dewar is through acoustic coupling to the walls of the room).[4]

The attachment and optical alignment of the interferometer components is achieved on standard optical breadboards bolted to the vibration isolation frame. Accelerometer spectra indicate that the vibration isolation frame, the Dewar, and the optical component mounting surfaces respond as a single unit to external vibration.

CRYO-SERVICING AND SYSTEM PERFORMANCE

Two system tests have been performed. The first test was to qualify the facility for optical testing using a ZYGO Mark IV interferometer and a 0.5 m diameter beryllium

Figure 3. Upper helium tank with gimbal and 0.5 m beryllium mirror center of the figure.

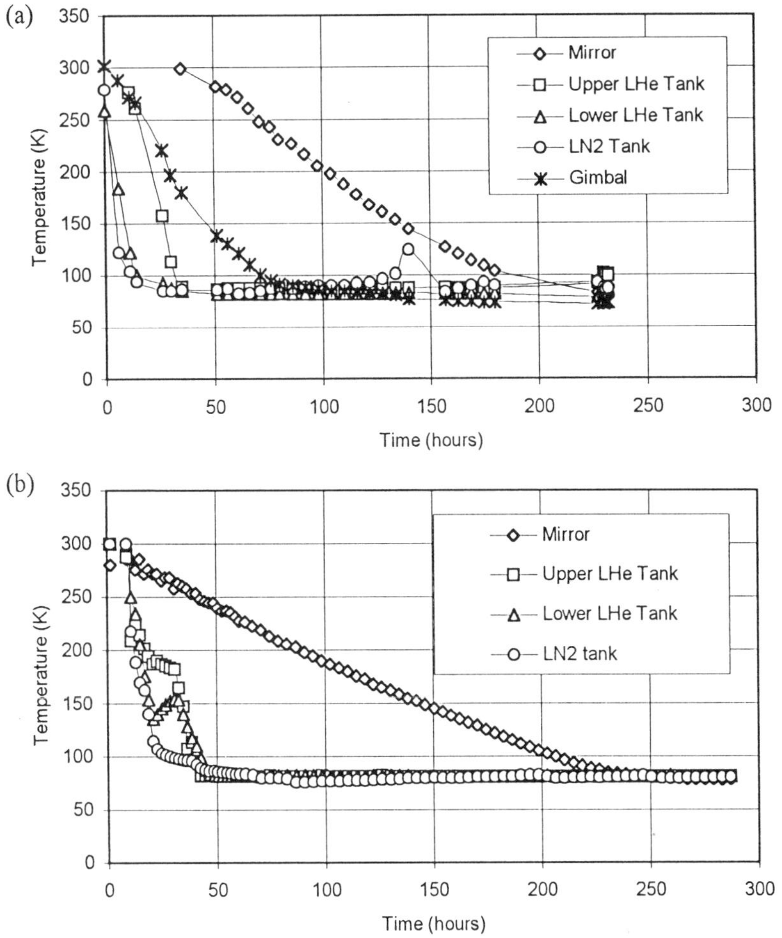

Figure 4. LN_2 cooldown data for (a) first test and (b) second test of the STTF

mirror. The second test was performed after procedural and hardware improvements identified from the first test were implemented. The second test also utilized a new interferometer (ZYGO Mark V).

Liquid nitrogen cool down

After achieving a guard vacuum of better than 10^{-4} Torr in the Dewar, the cool down begins with the introduction of liquid nitrogen into all three cryogen tanks simultaneously. Cool down is achieved by first cooling the tank walls adequately to accumulate liquid inside and fill the tanks, with the conductively cooled surfaces then relaxing to equilibrium with a longer time constant. Results of the LN_2 cooldown are shown in Figure 4. Cooldown of the dewar itself took about 72 hours in the first test and 50 hours in the second test. This met the cooldown requirement of less than 72 hours for cooling to 77K for the dewar. However, the cooldown of the mirror took much longer, 225 hours in each test. The mirror cooling rate was set by the thermal impedence between the cryogen tanks and the mirror. The mirror was thermally attached to the upper tank by 6 copper braid straps bolted at 6 locations to the tank and 3 locations on the mirror. One improvement, that is evident in

Figure 4, is that the upper tank thermal attachment to the LN_2 tank was made much more sound for test #2 based on the result of the first test. This is shown by the more uniform temperature distribution at times greater than 50 hours.

Liquid helium cool down

To cool the test chamber below liquid nitrogen temperatures, any remaining liquid nitrogen is purged from both liquid helium tanks by over pressurizing the tanks with helium gas. Liquid helium is then transferred into both tanks simultaneously. Cool down was achieved by first cooling the tank walls adequately to accumulate liquid inside and fill the tanks, with the conductively cooled surfaces then quickly relaxing to equilibrium. Cool down to steady state took about 24 hours and required approximately 1200 liters of liquid helium when transferring at an average rate of 1 liter/minute into each tank. At steady state the guard vacuum was approximately 10^{-6} Torr in the Dewar. Results similar to Figure 4 for the helium cooldowns of tests 1 and 2 are shown in Figure 5. Figure 6, shows most clearly the effect of the additional thermal strapping added between the upper tank and the helium shroud (twenty two 3.8 cm wide, 0.6 cm thick copper straps). This figure shows the axial

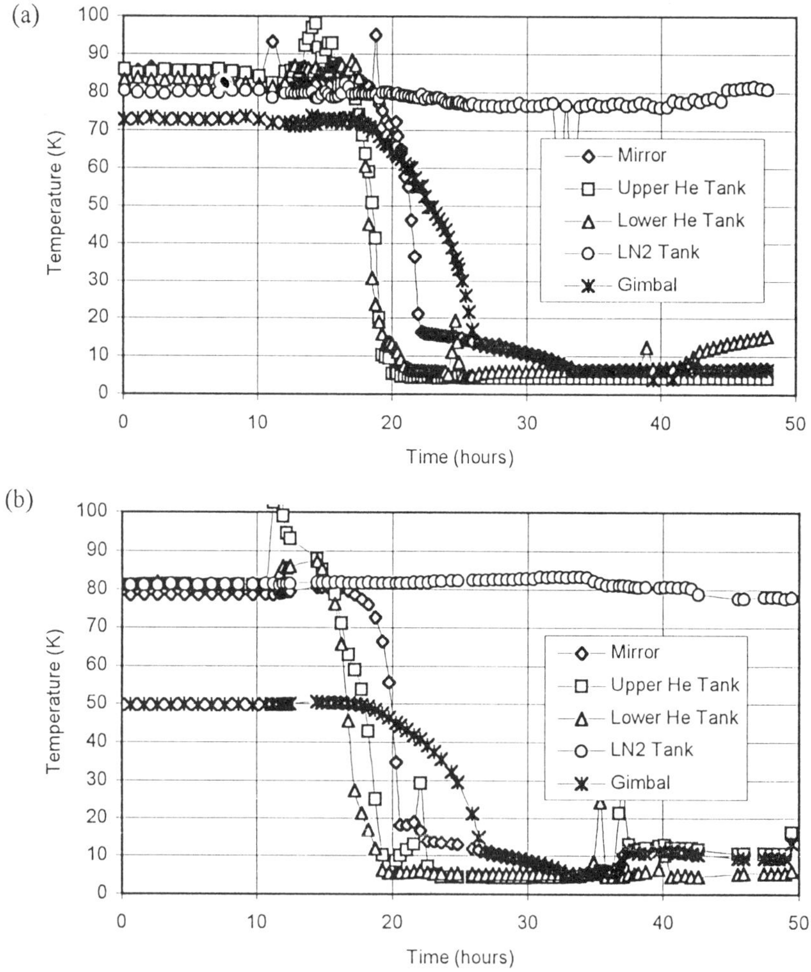

Figure 5. LHe cooldown for (a) first helium test and (b) second helium test.

Table 1. Heat loads on helium and nitrogen tanks.

	First Test		Second Test	
	LN_2	LHe	LN_2	LHe
Upper He Tank	2.4 W	2.2 W	3.7 W	NA
Lower He Tank	2.8 W	9.7 W	2.5 W	NA
LN_2 Tank	99 W	57 W	78 W	NA

temperature variations of that surface when at helium temperatures. In the first series of tests, there was up to 5K variation along the surface. This improvement was critical to the facility as later tests are planned for the Space Infrared telescope Facility which have a requirement that no helium cooled surface have a temperature greater than 6.5K. The excursions in temperature in Figure 6(b) at 35 hours are due to one of the helium tank (the upper tank) running out of helium.

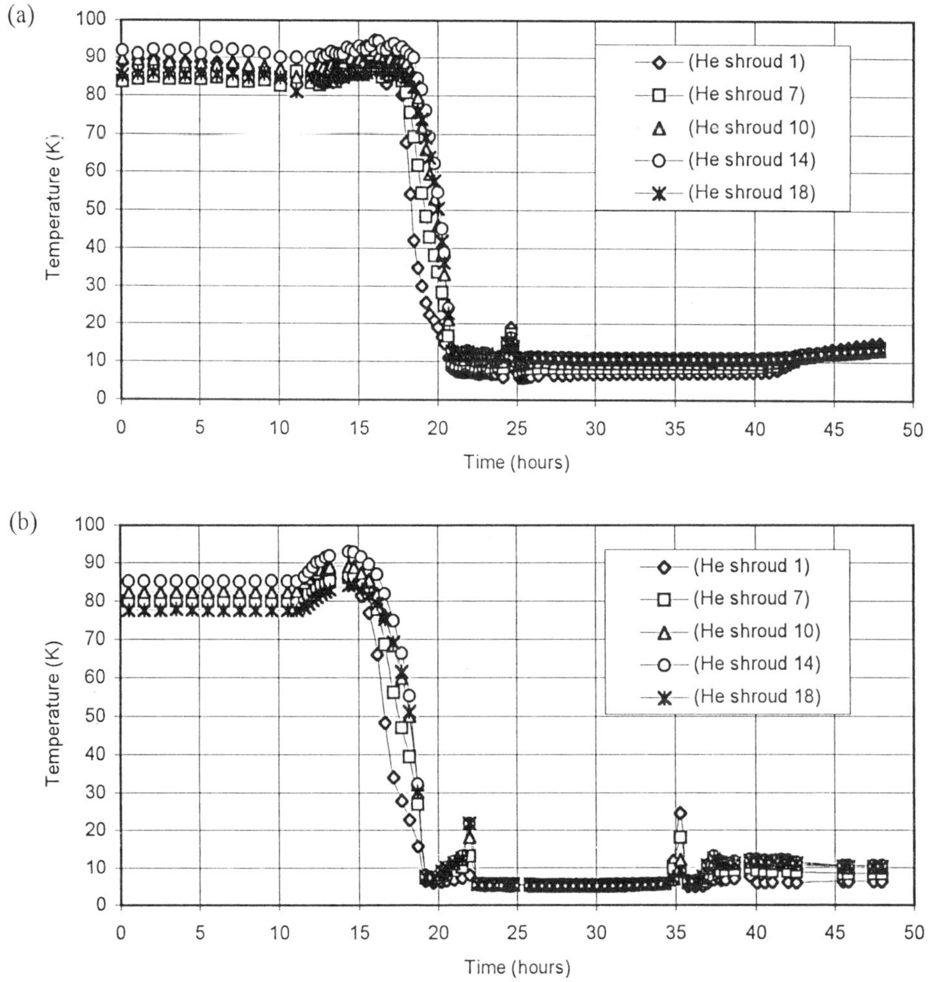

Figure 6. Helium shroud axial temperature variation for (a) first test and (b) second test.

Steady State Performance

Steady state parasitic heat rate data for both tests of the test facility are presented in Table 1. The poor thermal connection between the upper helium tank and shroud in the first test is again evident in this data, particularly at helium temperature. Unfortunately, helium temperature data for the second test were not gotten due to a problem encountered late in the test. However, an increased heat load on the upper helium tank is evident even at nitrogen temperatures. We expect that the parasitic rate at helium temperatures would have roughly a 60-40 split of the total parasitic load, lower tank to upper tank. The overall thermal performance of this system is not particularly good from a cryogenics perspective. Due to resource limitations of the project, thermal performance was forsaken for savings in cost and delivery schedule. The system does, however, meet all of the requirements for cryogenic optical testing

A thermal model of the system[5] indicates that the majority of the heat load is through radiative transport. If a thermal improvement were necessary, it could most likely be achieved by adding 5 to 10 more layers of well laid MLI between the outer shell of the system and the nitrogen shroud.

SUMMARY

Developed by the Jet Propulsion Laboratory's Low Temperature Science and Engineering Group, the SIRTF Telescope Test Facility (STTF) was completed, from conceptual design through construction and evaluation testing, in less than one year and within the allocated budget. Two system tests have been performed and the system has been fully qualified as meeting the performance criteria both cryogenically and optically. The facility has the unique capability to conduct interferometric tests of mirrors of diameter $\leq$ 1 m, f $\leq$ 6, at temperatures from 300 to 5 K. The facility includes instrumentation for characterization of the test chamber thermal environment and Dewar performance.

ACKNOWLEDGMENT

This work was performed by the Jet Propulsion Laboratory, California Institute of Technology under contract to NASA. Special thanks to Dr. Dan Coulter of the Jet Propulsion Laboratory for his continued support during this development.

REFERENCES

1) I. E. Spradley, T. C. Nast, and D.J. Frank, "Experimental studies of MLI systems at very low boundary temperatures," *Recent Advances in Cryogenic Engineering*, Vol. 35, pp. 477 - 485, 1990.
2) T. C. Nast, "A review of multilayer insulation, theory, calorimeter measurements, and applications", *Recent Advances in Cryogenic Engineering*, Vol. 267, pp. 29 - 43, 1993.
3) R. G. Chave, A. E. Nash, J. Hardy, "Gimbal mechanism for cryogenic alignment of 1 meter diameter optics", *Preseented at SPIE's 1995 International Symposium on Optical Science, Engineering and Instrumentation, Optomechanical and Precision Instrument Design Conference*, San Diego, 1995.
4) T.S.Luchik, U.E. Isralesson, R.G. Chave, A.E. Nash and J. Hardy, "SIRTF Telescope Test Facility", *Preseented at SPIE 1995 Internaltional Symposium on Optical Science, Engineering and Instrumentation, Optomechanical and Precision Instrument Design Conference*, San Diego, 1995.
5) A.E. Nash, "Simple Spreadsheet Thermal Models For Cryogenic Applications, "*to be presnted at the Cryogenic Engineering Conference*," Columbus, OH, 1995.

EFFECT OF SPACE FACTORS ON PHYSICO-MECHANICAL PROPERTIES OF MATERIALS OF SPACE VEHICLE OF MARS'96

V.V.Abraimov[1], N.N.Agashkova[1], L.Bohne[2], I.V.Budniak[1], N.I.Velichko[1], A.V.Krevsun[1], V.I.Kostenko[3], F.Lura[2], and A.M.Markus[1]

[1]B.Verkin Institute for Low Temperature Physics and Engineering, Kharkov, Ukraine
[2]German Aerospace Research Establishment, Berlin, Germany
[3]Institute for Space Researches, Moscow, Russia

ABSTRACT

A comprehensive investigation into physico-mechanical properties of four types of the light-absorbing coating materials has been performed under the conditions closely approximating those of outer space. It is shown that a wide variety of physico-chemical processes occurring in irradiated materials result from a variety of the radiation types . These processes are defined by its energy and dose as well as the structural and chemical composition of polymer. Based on the experimental results, the Kapton has been proposed as a base for the light-absorbing coatings as well as for other composites intended for use in outer space. In addition to the maximum resistance to radiation, it features plasticity at 4.2 K in contrast with another polymer materials. The Kapton plasticity is supposed to be of quantum nature at T=4.2 K.

INTRODUCTION

Intensive development of space equipment caused a variety of complicated and complex problems of space material research [1,2] to be solved, such as development of new promising structural materials with pre-set properties, study of their physico-mechanical characteristics, issue of scientifically-grounded recommendations on use of these materials in elements of space vehicle designs, study of their serviceability under "severe" and extreme space conditions and prediction of behaviour of these materials in outer space during long operation (5-10 years). The main space factors are as follows: deep vacuum (10^{-6}..10^{-7} Torr), solar electromagnetic radiation (SEM) including VUV one in the range of 5..2500 nm, proton and electron fluxes of natural and artificial radiation Earth zones, multiple thermocycling in a wide range (4.2..400 K) etc. The problem of reliability and long service life of new structural materials becomes especially pressing when they are to be used in space vehicles of new generation which are now being developed under the international programs "MARS'96" [3], "REGATTA", etc. In these vehicles, new composite materials will be widely used, in particular, black light-absorbing coatings of high absorption coefficients (ε^*=0.95..0.97) in the visible and IR spectra (A_s= 0.4..25 mm). These materials are supposed to be made of the polymer (mylar, fluroplastic, Kapton and their compositions) films as a base and organic (adhesives, lacquer) and non-organic (e.g., low-vacuum Al condensate, LVC Al) films as a coating. The materials used should met severe requirements, such as high stability of mechanical, optical and adhesive properties after irradiation, mass loss of these materials during their service life should not be more than 1.5 %.

The materials under consideration (i.e., composites) are of special interest since an influence of the space factors on physico-mechanical properties has been studied previously (see in Refs. [4-7] only in one-phase systems (pure polymers). In the present work, an effort has been made to study a behaviour of more complicated objects fabricated from two-phase (metal-polymer and polymer-polymer layered structures under the same conditions. It is naturally to expect that phase boundaries of the objects can affect the radiation-physical and other processes occurring in them on their exposure to space factors that results in a change of their physico-mechanical properties. The paper reports the results of the comprehensive investigation into mechanical, optical and structural characteristics of a number of promising composites exposed to some space factors in a wide temperature range 4.2..300 K. The investigations performed made it possible to suggest the degradation mechanisms of these materials and to issue the scientifically-grounded recommendations on application of the materials in a new generation of space vehicles

EXPERIMENTAL TECHNIQUE AND OBJECTS STUDIED

In reaching the above goal, a number of techniques has been developed aimed at accelerated simulation of the following space factors (SF) under terrestrial conditions:

(i)electromagnetic solar radiation including VUV one the range from 5 to 2500 nm by equivalent doses for 1 year. This type of action has been simulated earlier neither in the countries of the former Soviet Union nor in the ESA [1]. The simulator developed enables simultaneous exposure of samples to SEM (λ=200..2500 nm) and VUV radiation (λ=5..200 nm). A feature of this simulator is that it enables one to make investigations of physico-mechanical properties of the materials directly under the beam in vacuum (in situ), e.g. mechanical and electrical-physical properties, mass loss, gas-evolution (outgassing) processes. Fig.1 shows the general view of the simulator consisting of the following main systems: the solar radiation simulator (1), VUV-radiation source (7), vacuum chamber (3), forevacuum system (9), cryogenic system (8,11), deformation unit (2,6) intended for obtaining the hardening curves $\sigma(\varepsilon)$ directly under the beam, sample feeding system (10) and mass-spectrometer of the type ROMS-4 (5). The apparatus of the artificial Sun, IS-160, has been used as a radiation source in the range from the IR to the neighbouring UV (200..2500 nm). The materials to be irradiated are exposed to light flux with intensity of 14 W/cm^2 that corresponds to the solar intensity unit. To simulate the solar radiation in the VUV range (5-200nm), a modification of the VUV simulator has been used based on a gas-jet light source and described in details in Ref. [4]. The intensity of VUV-radiation (I) in the range of 5..200 nm has been measured to be (1..4)$\cdot 10^{-2}$ W/cm^2; the radiation spectrum of the VUV-radiation simulator has been made as close to that of the solar radiation as possible. The simulator has been provided with a variety of equipment intended for carrying out the physical measurements and studies, in particular, a set of the deformation machines

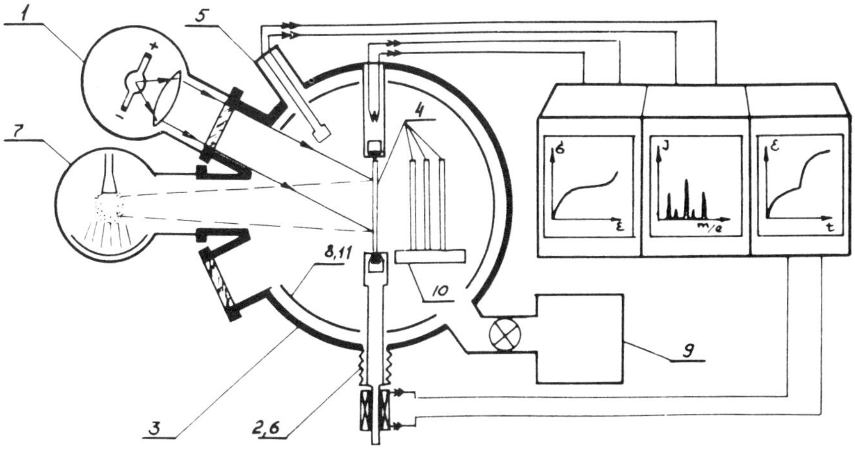

Figure 1. Schematic diagram of simulator designed for simultaneous SEM and VUV irradiation: 1,Sun simulator IS-160; 2,deformation unit; 3,vacuum chamber; 4,samples; 5,mass-spectrometer ROMS-4; 6, attachment designed for studying creep process of materials; 7,VUV-radiation source GIS-1 in the range 2..200 nm; 8,cryogenic pumping system; 9,pumping system (initial vacuum, high vacuum); 10,sample feeding machine; 11, cold shield (77-4.2 K).

of MRK-1 and E-71 type for studying the mechanical properties of materials exposed to active deformation and creep in the temperature range 4.2..300 K [7].

(ii)Irradiating (up to 1 year) of the materials with equivalent dose of protons (p^+) and electrons (e^-) of natural and artificial earth radiation zones with the energy 160 keV, i.e. with a complete fluence of the particles $D=10^{16}$ part./cm^2 provided by a proton-electron injector PEI-1.

(iii)The thermocycling of materials in vacuum 10^{-6} Torr has been carried out by the method described in Ref. [9] and at the temperatures ranging from 4.2 to 300 or from 77 to 300 K depending on the cryogen used (LHe or LN_2).

In investigations of the surface structure of materials, optical and scanning microscopy and microscopes MIM-8, MBS-9, REM-200U have been used. Thermal radiometers TRM-I and spectrophotometers FM-59 and FO-1 [5] have been used for studying the optical and thermal-radiation characteristics.

Table 1 .Influence of different types of radiation upon optical characteristics and mass loss of composites

Type of material	Type of radiation	ε^*	A_s	$\Delta m/m_0$,%
Composite 1 (Kapton + LVC Al)	initial state	0.82	0.97	0
	SEM +VUV (100 h.)	0.82	0.97	0
	VUV (1 year)	0.82	0.97	0
	$p^+ + e^-$ 10^{16} part./cm^2	0.82	0.97	0
Composite 2 (Mylar + Carbon)	initial state	0.82	0.96	0
	SEM +VUV (100 h.)	0.82	0.96	0
	VUV (1 year)	0.84	0.96	0
	$p^+ + e^-$ 10^{16} part./cm^2	0.82	0.97	0
Composite 3 (Kapton + Arimide fabric+Carbon)	initial state	0.81	0.96	0
	SEM +VUV (100 h.)	0.85	0.96	2.60
	VUV (1 year)	0.88	0.96	0.26
	$p^+ + e^-$ 10^{16} part./cm^2	0.77	0.97	2.01
Composite 4 (Kapton + Carbon)	initial state	0.88	0.96	0
	SEM +VUV (100 h.)	0.91	0.96	0
	VUV (1 year)	0.93	0.96	0
	$p^+ + e^-$ 10^{16} part./cm^2	0.88	0.96	0

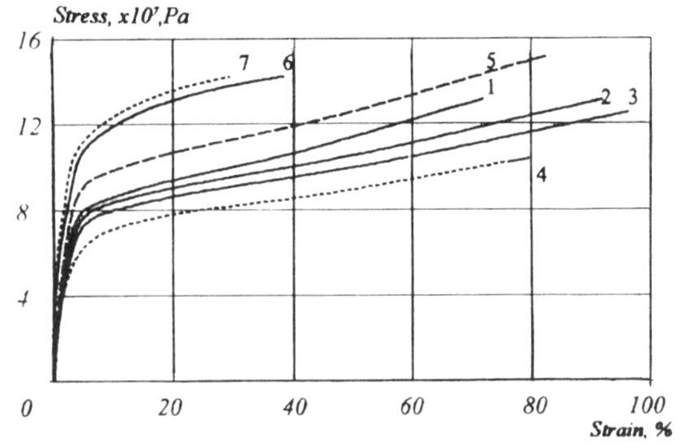

Figure 2. Stess-strain curves for Composite 1.

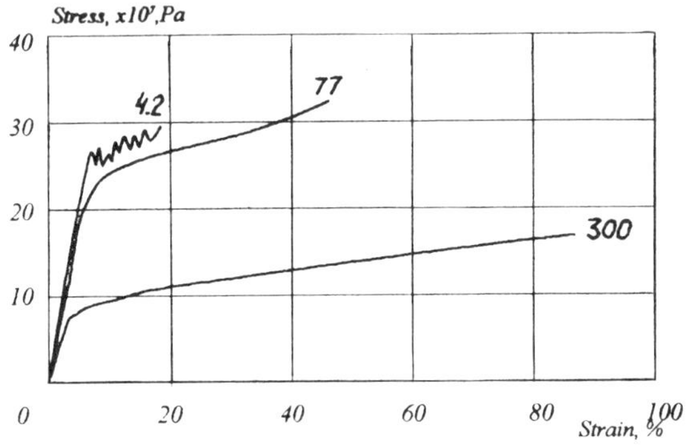

Figure 3. Poor Kapton stress-strain curves for 4.2 K; 77 K and 300 K temperatures.

As the objects of investigations, four types of composite materials with light-absorbing coatings as well as pure Kapton films PM-A 40 μm thick have been chosen:

(i)Composite 1, Kapton film PM-20 20..40 μm thick with a bilateral non-organic coating made of the LVC Al with a thickness of 2 μm which represents a mixture of pure Al and oxide Al_2O_3 with a developed porous surface providing a high absorption coefficient in visible region (A_s = 0.97..0.98);

(ii)Composite 2, Mylar film (PETP) 20 μm thick with a bilateral black coating made of finely-dispersed carbon + adhesive, 5μm thick;

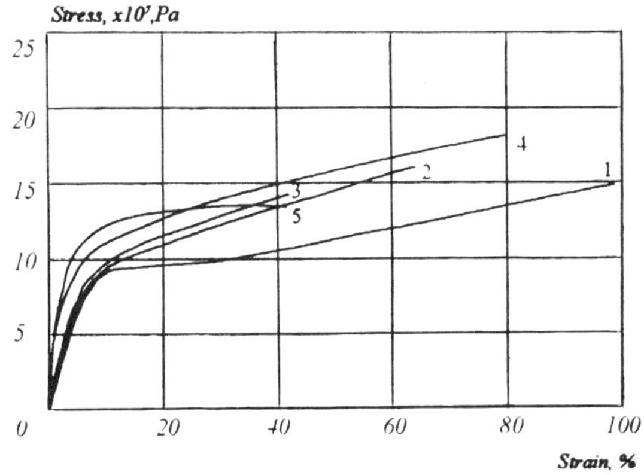

Figure 4. Stess-strain curves for Composite 2.

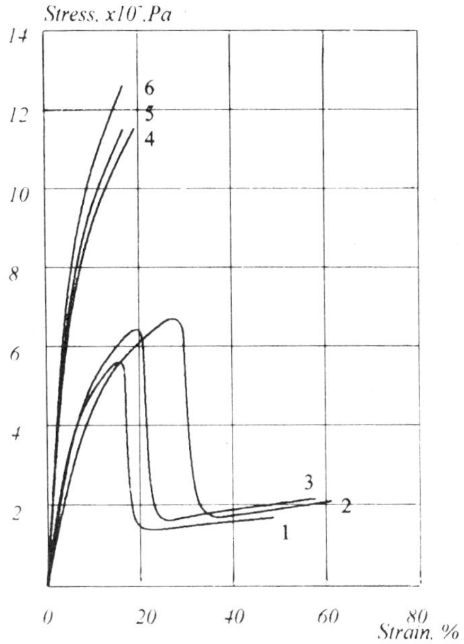

Figure 5 . Stess-strain curves for Composite 3 .

(iii)Composite 3, Kapton films PM-1-3A, reinforced with arimide fabric with bilateral coating of the (finely dispersed carbon + adhesive) type;

(iv)Composite 4, Kapton films PM-a 40 μm in thickness with a coating made of 30% finely-dispersed carbon + polyimide lacquer PAC-1 10 μm in thickness.

These materials have been exposed to above space factors. The following physical properties have been studied before and after the exposure.

1.Mechanical properties. The stress-strain curves $\sigma(\varepsilon)$ have been obtained evidencing effect of the space factors on the main parameters of plastic deformation, where σ_y is the yield stress; σ_u, ultimate strength; ε_Σ, deformation threshold (maximum extent of the deformation of materials followed by their failure); E, Young modulus of the materials.

2.Optical properties. The blackness coefficient ε^* and absorptivity have been measured in the visible range A_s.

3.Structural properties. Using the optical and scanning electron microscopy, a changes of the surface structure of coatings have been studied after exposure to space factors.

EXPERIMENTAL RESULTS AND DISCUSSION

Figs. 2-6 show the typical stress-strain curves of materials as a function of temperature (4.2, 77, 300 and 340 K). Each curve has been plotted on the measurements made on 4-5 samples. In the Figures, the measurement temperatures, type and dose (time) of radiation are shown. In the Table 1 optical (A_s, the absorptivity in the visible range) and thermoradiation (ε^*, the blackness coefficient) characteristics are given as well as the mass loss of the light-absorbing coatings. Let us analyse the data obtained for each material studied.

Composite 1

When comparing the curves 1,2 and 3 shown in Fig.2, it can be seen that electromagnetic irradiation of the material used results in its hardening which manifest itself in the decreased yield stress and strain stresses along the whole curve $\sigma(\varepsilon)$. Note, that after the exposure to SEM and VUV radiation the deformation threshold increases, i.e. this type of irradiation and, perhaps, annealing at high temperatures may cause a plastic deformation of the Kapton film. The stress-strain curves $\sigma(\varepsilon)$ change their shape. After irradiation one-stage stress-strain curve $\sigma(\varepsilon)$ for Kapton typical for the partially crystalline polymers changes to the S-like curve characteristic of the amorphous polymers thus permitting a supposition that the electromagnetic radiation (SEM+VUV) increases the concentration of the amorphous phase of material. From the comparison of curve 6 with curve 7, it becomes evident that increasing dose of the proton-electron irradiation from 10^{15} part./cm^2 to 10^{16} part./cm^2 causes the hardening of the material under consideration that points to a cross-linking of the Kapton. In this case, deformation threshold sharply decreases which is evidence for considerable embrittlement of Kapton arising perhaps from the destruction of the polymer. It is also evident that after irradiation the deformation threshold changes more than the ultimate strength. The greater sensitivity of ε to irradiation may be due to the fact that the both processes caused by irradiation (break of chains and their cross-linking) reduce elasticity of the polymer whereas their effect on the strength is opposite, i.e. break leads to a reduction and a cross-linking to an increase in the strength, respectively. Comparison of the $\sigma(\varepsilon)$ curves obtained at room and nitrogen temperatures reveals considerable hardening (and simultaneous embrittlement) of Kapton with increasing irradiation dose at T=77 K. It should be emphasised that a major property of Kapton, differing it from the other polymer materials, is its plasticity at low temperatures. For example, at T=77 K the deformation threshold amounts to 8..40 % depending on the dose of preliminary irradiation, whereas most of the polymers exhibit a brittle fracture at this temperature.

Similar results have been obtained for pure Kapton films. Fig.3 shows the hardening curves for the Kapton PMA 20 μm thick and at temperatures 300, 77 and 4.2 K, respectively. As can be seen from the curve 3 there exists a region of plastic deformation in the Kapton even at helium temperatures. The values of yield strength obtained for this material at T=77 K and T=4.2 K differ little from each other and are slightly greater at T=300 K (20 kg/mm^2 and 22 kg/mm^2 as compared with 6 kg/mm^2), i.e. at $T \rightarrow 0$ the tendency toward the saturation of the curve is observed . This suggests that at low temperatures the plasticity of the polymer material studied may be caused by the quantum mechanisms. In literature (see in Ref.[11]), there is some information about the role of quantum mechanisms in the low temperature plasticity of metals and their alloys. The present work is the first attempt made to reveal the similar mechanisms of plasticity in the polymer materials.

At helium temperatures, the curve $\sigma(\varepsilon)$ also changes its shape: at T=4.2 K the serrated yielding observed (curve 3, Fig.3) testifies that the mechanisms of the deformation processes are changed. It is known from the theory of metals plasticity [12] that these mechanisms include twinning, dislocations peeling of the Cottrell clouds as well as structural changes (i.e. a turn of the intercrystalline boundaries). For the partially crystalline polymer, as is the case of Kapton containing equal quantities of amorphous and crystalline phases, namely, 50 %, the probability that the turns of separate crystal regions about the deformation axis, relative to each other or to the regions of amorphous phases are such mechanisms is very high.

Composite 2

It is seen from Fig.4 (curves 2 and 3) that electromagnetic irradiation (VUV or SEM) causes a sharp decrease in the extent of deformation of the samples exposed (about 3 times decrease in ε_Σ after 100-hour SEM irradiation) and some increase of their hardening. With decreased temperature and increased irradiation dose, the ductile fracture of the PETF films is changed to the brittle one. Corpuscular irradiation (p^+

and e^-) with the dose $D=10^{16}$ part./cm^2 causes a complete loss of the mechanical stability of the material, i.e. its brittle fracture at null load (point 6 in Fig.4). Comparison of two curves, 2 and 3, shows that combined (VUV and SEM) irradiation of the material leads to a significant non-additive (in respect to energy contribution) decrease of its deformation threshold which correlates with the data on the non-linear variations of physico-mechanical properties of pure polymeric materials, including PETP films, exposed to irradiation of several types 6. This effect may be ascribed to additional absorption channels of optical radiation (SEM) formed after the exposure of the samples to more hard VUV-radiation. Absorption of VUV-radiation by molecular systems whose quantum energy (5-150 eV) exceeds considerably the bonding energy (5 eV) of main chains in polymeric materials is accompanied by a variety of new formations, i.e. intermediate active particles, excited molecular states (singlet, triplet, exciton, etc.); charged particles (electron-cation pair, free and trapped electrons, etc.), free radicals as well as stable products. It has been shown in Ref.[14] that the absorption peaks of new formations are lying in the UV-, visible and in the near-IR regions that makes possible for the polymeric material an optical absorption in a wide spectra.

This suggests that on the exposure of the polymer to VUV radiation there form new channels permeable to short-wave radiation (UV, light) which cause new photochemical processes in polymers .[14,15]

Degradation of optical characteristics and mass loss of coatings after irradiation has not been found (see, in Table). However, VUV radiation changes the surface topology. The surface becomes more developed and wavy that is probably caused by the etching of the coating surface in the VUV radiation field. The data given in Table show that the change of surface structure correlates with the emissivity of the coatings which increases due to radiation from 0.82 to 0.84. The ε^* value is known to depend on two factors: (i) fundamental IR absorption and (ii) the coating surface development. The radiation modifies the surface relief towards its higher non-uniformity, which, in its turn, increases the emissivity factor of the coating (see, in Ref. [5]). This finding may be interpreted as an irradiation-induced modification of the thermal radiation properties of the coating.

Composite 3

A study of mechanical properties (Fig.5) and analysis of the samples made after their deformation have shown that at of 300 K a deformation of two components of the composite (fabric and Capton) is taking place, which continues up to the maximum stress (ε_Σ=35-45%). After passing through the maximum, a failure and further deformation of the fabric have been observed up to ε=60-100%. Decreasing temperature to the nitrogen one leads to a significant increase of yield strength and strain stresses. A simultaneous study of mechanical and structural properties during the deformation process has revealed cracking of the black coatings irradiated by the proton and electron fluxes even at 5 % deformation. As the deformation progresses, the number of cracks is intensively increased. An analysis of the optical characteristics of the material considered (Table 1) measured before and after irradiation has shown that SEM and, especially, VUV-irradiation increases considerably the coefficient ε^* (from 0.81 to 0.85-0.88). Corpuscular irradiation gives an opposite effect: it reduces the coefficient down to 0.77. The mass loss of the coatings has been measured to be 2-2.6 %.

Composite 4

It is seen from the curves shown in Fig.6 that the action of radiation on the mechanical properties of the material is temperature-dependent. At T=300 K,

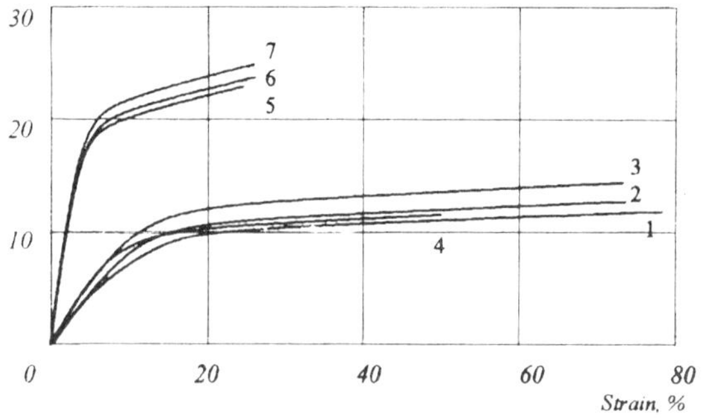

Figure 6. Stess-strain curves for Composite 4.

hardening of the samples has been observed after SEM and VUV irradiation: sy, su are increased and the deformation threshold is somewhat decreased (on p^+ and e^- fluxes irradiation it is decreased much more, by 40 %. At nitrogen temperatures, slight softening of the samples has been observed after their exposure to electromagnetic radiation.

However, in both cases only slight changes have been observed, so there are no reasons to expect a substantial degradation of mechanical properties of the material after SEM and VUV-irradiation. It should be emphasised that the Capton base of this composite as well as that of the Composite 1 retains its plasticity. No degradation of optical characteristics of the coating studied has been observed after SEM and VUV-irradiation. Conversely, electromagnetic radiation increases ε^* from 0.88 to 0.93 after VUV-irradiation. Data obtained with electron microscope of high magnification testify that there appear on the surface of SEM-irradiated coatings crater-like formations and pores of size 40..80 μm (Fig.7), i.e. the relief becomes more developed that can truly cause an increase in ε^*.

a

b

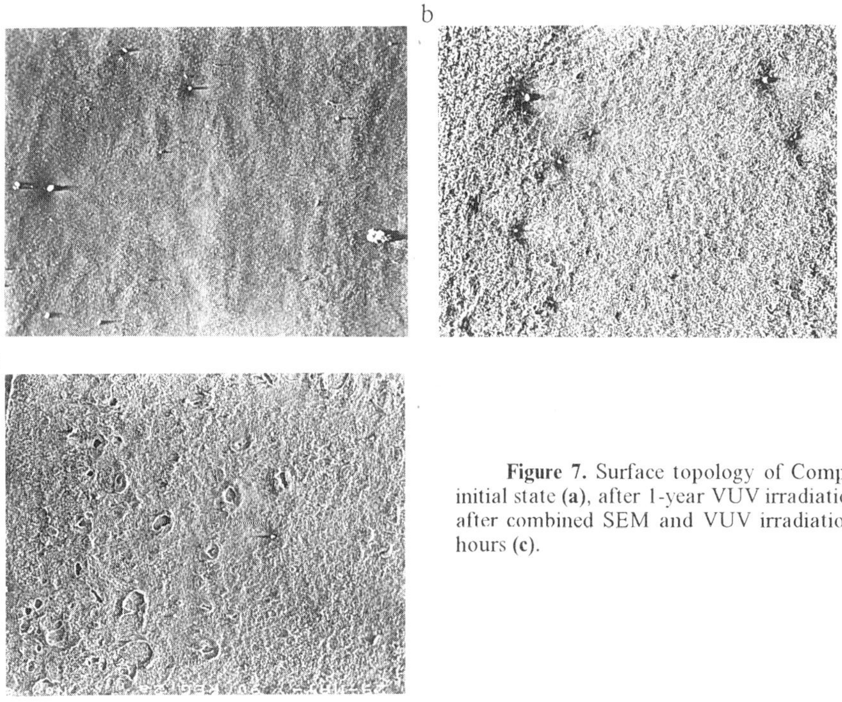

c

Figure 7. Surface topology of Composite 4 in initial state (**a**), after 1-year VUV irradiation (**b**) and after combined SEM and VUV irradiation for 100 hours (**c**).

Now, consider the results obtained from a standpoint of existing concepts of radiation-substance interaction.

Comparison of changes of the mechanical properties of the composites exposed to electromagnetic (VUV-, SEM-irradiation) and hard charged particles (p^+) irradiation shows that the parameters of the plastic deformation of the materials are more sensitive to corpuscular radiation, which is in a satisfactory agreement with the previously obtained data [14,15,16] on high radiation resistance of pure materials (Mylar, Capton). This is probably due to specific processes occurring in or near hard particle tracks, wherein a local absorption of energy will raise the concentration of the initial chemically active particles. While comparing the findings of an investigation into radiation effect upon composites obtained by the authors with the experimental data [4,5,15,16] on radiation-induced changes of pure materials, one arrives at a conclusion of a similar character of the radiation degradation mechanisms in the materials of the two types. This conclusion is of great importance as protons and VUV-radiation are absorbed mainly by a thin near-surface layer of the composite at the absorption length $\xi = 1..2\ \mu m$, i.e. by a coating. Nevertheless, formation of radiation defects in the near surface layer of the composite and their propagation through the substance affect the macroproperties of the whole sample. In other words, the coating-polymer phase boundary has only a minor effect on the propagation of radiation energy, i.e. on accumulation of irreversible radiation defects with big relaxation times (stable radiolysis products) and their diffusion into the substance.

Analysis of experimental data has shown that of all the materials studied, the composites with a Kapton base (Composites 1,4) exhibit the greater resistance to an action of all types of radiation, whereas in the composite with a Mylar base (Composite 2) a complete degradation of mechanical properties has been observed after 1-year dose of corpuscular irradiation. This can be explained by the difference in the chemical structure of these materials. As is known [17], the polymers containing aromatic cycles (Kapton, polystyrene, etc.) display high radiation resistance which is perhaps, associated with a fast dissipation of the energy absorbed by the resonant structures of aromatic cycles ($t = 10^{-15}$ s). These structural elements have a great number of energy levels, so the energy absorbed is not localized but is distributed along the polymer molecule not causing the breaking of chemical bonds, i.e. the irreversible changes of physico-mechanical properties of Kapton.

CONCLUSIONS

1. A comprehensive investigation into physico-mechanical properties of four types of promising composites has been carried out under the conditions closely approximating those of outer space. It is shown that the physico-chemical processes occurring in the irradiated materials are distinguished mainly by the type of radiation and depend on its energy, dose and duration as well as the structure of polymer.

2. A correlation made between the results obtained by the authors for the composites based on Kapton and Mylar and previous data on high radiation resistance of pure materials indicates that in both cases irradiation affects the physico-mechanical properties in a similar way although in the first case the radiation is absorbed by a coating instead of polymer base. Thus, the polymer-coating phase boundary has only a minor effect on the propagation of energy through the substance.

3. Study of the deformation of Kapton and composites based on it at $T = 4.2$ K has shown that these polymer materials retain their plastic properties down to helium temperatures. In connection with this, a tendency has been observed to a saturation of the yield stress vs temperature curve at $T \to 0$ K that confirms the supposition of the quantum character of capton plasticity. The stepwise character of deformation can be explained by reorientation of crystalline regions about the deformation axis and relative to the regions of amorphous phase.

4. Based on the analysis of a great deal of experimental data, a conclusion has been made that Kapton material offers the greatest promise as a base for the light-absorbing coatings as well as other composites intended for using in outer space under the action of the solar, VUV-radiation, proton and electron fluxes of radiation Earth belts as, in addition to high radiation resistance, at temperature 4.2 K it displays the plastic properties in contrast to most of other polymer materials.

5.Comparison studies of four types of composites with light-absorbing coatings has shown that Composite 4 (Kapton film with coating made of Kapton lacquer (as a base) doped with finely-dispersed carbon) exhibits the best optical and mechanical properties and it can be recommended as a candidate for the materials to be used in a new generation of space vehicles.

REFERENCES

1. "Space Model", S.N. Vernov ed., V.2, Simulation of Space Environment Effect on Materials and Facilities of Space Vehicles, Scientific Research Institute of Nuclear Physics Moscow State University, Moscow (1983) (in Russian)
2. "Space Materials Science, Introduction into Scientific Principles of Space Technology", B. Foyerbacher, G. Hamacher, and R.J. Naunan ed., Mir, Moscow (1989) (in Russian)
3. "Mars-94, Mission of Automatic Space Vehicle to Mars", Space Research Institute Russian Academy of Sciences, Moscow (1992) (in Russian)
4. V.V. Abraimov, K.Sh. Bocharov, I.V. Budniak and V.V. Danovski, VUV Induced Creep of Some Polymer Materials, Physics and Chemistry of Materials' Treatment, 1:39 (1993) (in Russian)
5. V.V. Abraimov, N.N. Agashkova, I.V. Budniak, A.A. Zakharova, N.I. Velichko, and A.M. Markus, Effect of Low Temperature, Irradiation and Deep Vacuum on Physico-Mechanical Properties of Polymer Light-Absorbing Coatings, Physics and Chemistry of Materials' Treatment, 2:26 (1992) (in Russian)
6. V.V. Abraimov, F. Lura, L. Bohne, I. Budniak, and V. Danovski, Investigation of Non-Linear and Non-Reversible Degradation of Physico-Mechanical Properties of Materials Under Effect of Space Factors, 1.0 Deutcher Luft und Raumfart Kongress DGLR, Band, Berlin (1993),p.80 (in German)
7. K.Sh. Bocharov, U.S. Stroilov, V.F. Udovenko, M.V. Zinoviev, and V.M. Momot, Deformation Facility MRV-1M for Tests of Polymer Materials in Vacuum and Gas Environment, Strength of Materials, 11:108 (1976) (in Russian)
8. A-.M. Markus, V.F. Udovenko et. al., Facility for Complex Research Work for Radiation Physics of Solids, Devices and Equipment for Experiment, 3:211 (1986) (in Russian)
9. V.V. Abraimov, N.N. Agashkova, V.N. Soloviev, I.P. Zaritskiy, and V.F. Udovenko, Thermal Firmness of lIght-Absorbing Coatings in 4.2-300 K Temperature Interval, Physics and Chemistry of Materials' Treatment, 6:83 (1989) (in Russian)
10. V.S. Ivanov, "Radiation Chemistry of Polymers", "Chemistry", Leningrad (1988) (in Russian)
11. V.I. Startsev, V.P. Soldatov, V.D. Natsik, and V.V. Abraimov, Role of Quantum Mechanisms and Thermal Heating in Low-Temperature Creep of Metals, Phys. Stat. Solidi. 59(a):377 (1980)
12. V.I. Startsev, V.D. Natsik, and V.V. Abraimov, Dislocations and Strength in: "Metals at Very Low Temperatures, Dislocation in Solids, V.6", North-Holland Publishing Company, Amsterdam (1983)
13. "Radiation Firmness of Organic Materials. Reference-book", V.K. Milinchuk and A.I. Tupikov ed., "Energoatomizdat", Moscow (1986) (in Russian)
14. N.P. Khlapova, "Alteration of Structure and Properties of Amorphous-Crystal Polymers Under High Energy Irradiation," Ph.D. Thesis, Kharkov State University, Kharkov (1992) (in Russian)
15. B.I. Verkin, V.F. Udovenko, V.V. Abraimov, and K.Sh. Bocharov, Effect of Electromagnetic Radiation in the Range of 0.01-25 mm on Mechanical Properties of Some Polymer Materials, Space Science and Technology, 3:54 (1988) (in Russian)
16. V.V. Abraimov, K.Sh. Bocharov, V.V. Danovskiy, and V.F. Udovenko, Research of Electromagnetic Radiation in the Range of 5-2500 nm Effect on pHysico-Mechanical Properties of Polymer Materials, in: "Applied Cryogenic and Vacuum Material Science", Naukova Dumka, Kiev (1991), p.103 (in Russian)
17. J. Guilett, "Polymer Photophysics and Photochemistry", Mir, Moscow (1988) (in Russian).

PHENOMENA OF BLISTERING AND FLAKING IN MATERIALS OF SPACE OPTICS CAUSED BY SPACE FACTORS

V.V.Abraimov[1], F.Lura[2], L.Bohne[2], N.I.Velichko[1], A.M.Markus[1], N.N.Agashkova[1], A.V.Krevsun[1], and L.A.Mirzoeva[3]

[1]Institute for Low Temperature Physics and Engineering of National Academy of Sciences of Ukraine, Kharkov, Ukraine
[2]German Aerospace Research Establishment, Berlin, Germany
[3]All- Union Research Centre, S.I.Vavilov State Optical Institute S.Peterburg, Russia

ABSTRACT

Effects of space factors such as proton and electron radiation (E=160 keV), extra-atmospheric solar (and VUV) radiation at λ = 5 - 2500 nm upon the reflectivity of mirrors in scale model IR instruments are investigated experimentally. The phenomena of blistering and flaking - complex defective structures on the mirror surface after exposure to protons (E=160keV) are described. These are either round craters of 2- 5 μm in diameter (flaking) or blisters (detachment of the Al and Ag films from the mirror surface) of 10 - 50 μm in size (blistering). Blistering and flaking develop in a jump - like way during short times ($\delta\tau$=1-2 sec) when the radiation dose becomes critical, $D_{cr}\approx 6\cdot 10^{15}$ particles/cm^2 at T=340K. The defect density is much dependent on the radiation dose and the temperature of the mirrors. The phenomena are related to the formation of hydrogen in the process of proton-electron recombination ($p^{+}+e^{-} \rightarrow H$) in the material. Hydrogen accumulates in blisters in a thin surface layer of glass, its thickness being equal to the proton mean free path ε_{p+}=2μm. When the hydrogen pressure in the blisters exceeds the ultimate strength σ_F of the space optics material, brittle failure occurs in the glass covered with mirror and protective coatings (flaking) and the plastic films (Al or Ag) peel off the mirror surface (blistering). This specific "space erosion" of the mirror surface appreciably impairs the reflectivity of mirrors in a wide spectral range. The experimental findings are interpreted within current models of blistering in metallic materials of nuclear reactors due to He^{+} and D^{+} ions and are in qualitative agreement with the theory of blistering developed by Yu.V.Martynenko [1,2].

INTRODUCTION

The rapid progress in space engineering has brought in a number of problems connected with development of new space system intended for operation in the Earth's radiation belts (250km<H<80000km) where they are exposed to space factors such as simultaneous proton and electron radiation in a wide energy range 30keV<E<10meV, extra-atmospheric solar (and VUV) radiation (λ=50-2500nm), temperature cycling within 77-400K (sometimes 4.2-400K), deep vacuum (10^{-7}Torr), atomic oxygen in lower (200km<H<800km) orbits (E=5eV) and others. There are some problems in space materials science too. Development of new space systems proceeds mainly in the following trends: (I) the lifetime of 5-7 (sometimes 10) years; (II) non-hermetic constructions; (III) a wide use of new composite materials. Materials for space optics set a special group: they are used in astrophysical instruments operating in open space. After a certain period of service in the Earth's radiation belts some IR instruments show degraded reflectivity of the mirrors, which considerably impairs the sensitivity of the whole system. The problems of materials and prediction of their behaviour in space are presently solved in two traditional ways: (I) exposure of samples and materials on orbital stations and subsequent ground-based studies of their physico-mechanical properties; (II) constructions of facilities for simulation of space factors and further laboratory investigations. There is however the third way - to study the impact of space factors on a scale model of space instrument. The latter way was exercised in this work on a simplified scale model IR telescope comprising an objective, a blind (of different materials) and a framework. The work was aimed at (I) finding physical mechanisms of drastic degradation of the reflectivity of the mirrors; (II) revealing factors affecting the reflecting surface of the objective mirrors; (III) separating the contributions of different factors to the material and defect formation under proton, electron, electromagnetic solar (and VUV) radiation; (IV) estimating quantitatively mass loss of the blind material - the AK-243 enamel and arimidic fabric; (V) detecting mechanisms of cryogenic precipitation on the reflecting surface of the mirrors; (VI) estimating the impact of space factors on the material of the model space optics (pure glass and Al, Ag films).

EXPERIMENTAL TECHNIQUE AND OBJECTS

The experimental objects were scale model IR telescopes for studies of the Earth's ecology from space (schematically shown in Fig.1) including a mirror (1) fixed in a copper sleeve (2) within a blind (3) made of arimidic fabric and a rigid component covered with the AK-243 enamel on the inside. The initial blackness coefficient of the blind was ε=0.9. The model was 200mm long, the diameter of the front section of the blind was 80mm, its side surface was $S=2\cdot10^3 mm^2$, the mirror diameter was 60mm. In each scale model the objective mirror was made of a composite material and included a beryllium washer (60mm in diameter, 10mm thick) and a 0.5mm thick glass layer sintered on it. A reflective Al layer (100nm thick) and protective SiO_2 (100nm) layer were evaporated on the glass. The initial reflection coefficient of the mirror in the interval 1-3μm was 83%. The other objects were glass washers 35mm in diameter and 4mm thick. Ag and Al films (100nm) were deposited on half the glass area by evaporating 99.99% pure metals in vacuum 10^{-6} Torr. The shortcut techniques of space simulation developed at the Department of Space materials of the Institute for Low Temperature physics and Engineering (Ukraine) were used. Electromagnetic radiation of the Sun (EMRS) (full spectrum) was produced with an IS-160 Artificial Sun simulator (200-2500nm), a GIS-1 gas jet

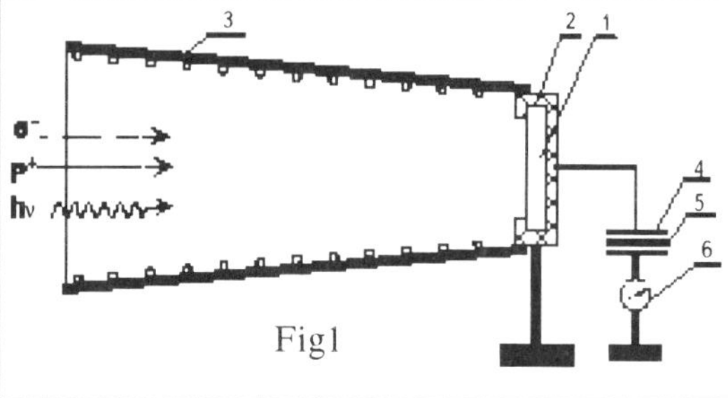

Fig1

light source provided VUV radiation (see Ref.1). The technique enabled us to study the mechanical, electro-physical properties and the mass loss of the material (sensitivity 10^{-4}g) directly in the field of cosmic radiation, and to carry out mass-spectrimetric investigations [3-6] of gas emission from the materials. To irradiate materials and scale models, a PEI-1 proton-electron accelerator was used, whose combined beams (50-200 keV) were directed to a vacuum chamber ($P=5\cdot10^{-6}$ Torr) containing a scale model or a mirror with an Al (Ag) layer on it. Combined and separate (only protons) corpuscular radiation was applied in all experiments. The proton and electron energies were E=160 keV, the current density was j=0.11 μA/cm^2, the radiation dose varied within D= $1\cdot10^{15}$-$2\cdot10^{17}$ particles/cm^2. The beams were registered by the zero-current technique. A microampermeter (6) connected the samples to "ground" and recorded the I=0 current in the process of irradiation. The secondary electron emission from the sample was suppressed using a metallic calibrating aperture with the potential U=250 V. The beams scanned the mirror with the help of a scanning system. Combined proton-electron irradiation was necessary to simulate the electron-neutral plasma which exists in space. The experiments studying the space influence on the structural perfection and topology of the mirror surface can be divided into three groups. In Series 1 the scale model was grounded, no condenser (4) and dielectric material (5) were used (Fig.1), i.e. charges did not accumulate on the scale model. This traditional irradiation technique was used earlier to simulate the Earth's radiation belts [7]. After a fixed $p^+ + e^-$ dose the topology of the mirror surface was studied by the methods of optic and scanning election microscopy. The reflection coefficients R(λ) of the mirrors in the interval 0.75-3 μm and the mass loss of the blind material were also measured. The kinetics of mass loss of the arimidic fabric and the AK-243 enamel was studied at 60-140°C. In other series proton-electron (or only proton) irradiation was performed with charge accumulation on the scale model (Fig.1). This increased the potential relative to the ground as $\delta U = \delta Q/C$, where δQ is the excess charge, δC is the capacity of the potential, whose maximum value is U=20 keV, a discharge device was used between the plates of the condenser (4). The moment of breakdown was registered by a change in the current flowing through the block (6) (Fig.1). This technique of p^+ - e^- irradiation with charge accumulation was used to simulate the kinetics of electrophysical processes similar to those in dielectric materials-glasses, polymers, enamels etc. These materials are noted for high electric resistance ($\rho_v=10^{14}$-10^{15} Ohm·cm) when exposed to quasi-neutral plasma of the Earth's radiation belts. After combined p^+ - e^- (or only p^+) irradiation and investigation of the surface properties, the scale model IR telescope and the mirrors were exposed to combined EMRS and VUV radiation for 50-100 hours with normal incidence of light quanta. Again the structural perfection of the mirrors was examined after exposure to fixed EMRS (and VUV) doses and also reflection coefficients R(λ) at 0,75-3μm and the kinetics of mass loss of the material of the object and the entire scale model were measured. In all experiments temperature was measured at four points of the scalemodel with semiconducting GaAs thermometers (sensitivity 0,1 K). In Series 3 the mirrors (with Al and Ag films) and the initial glasses were exposed only to protons (E=160 keV, D =$1\cdot10^{15}$-$2\cdot10^{17}$ particles/cm^2). Then the topology of the irradiated surface and the reflection coefficients were studied in the same region of the IR spectrum. The mass loss of the arimidic fabric and AK - 243 enamel was measured in the vacuum $P=10^{-7}$ Torr at the constant temperature T=140°C for eight hours (sensitivity 10^{-4} g).

EXPERIMENTAL RESULTS

Series 1

The irradiated objective lens holder(2) was grounded, i.e. charges ran off and the potential was uniform over the entire surface of the scale model. The initial structure of

Fig. 2a

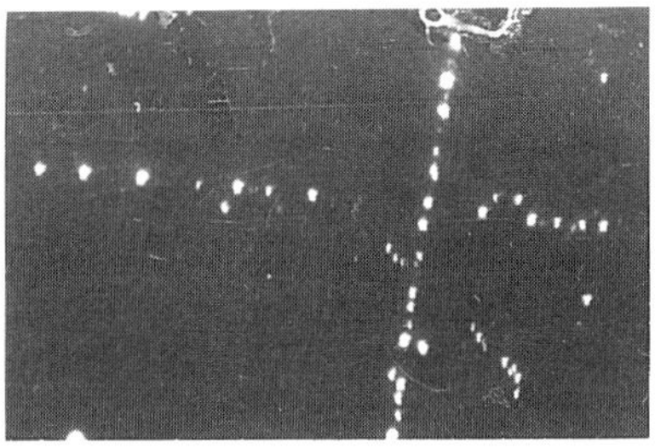

Fig. 2b

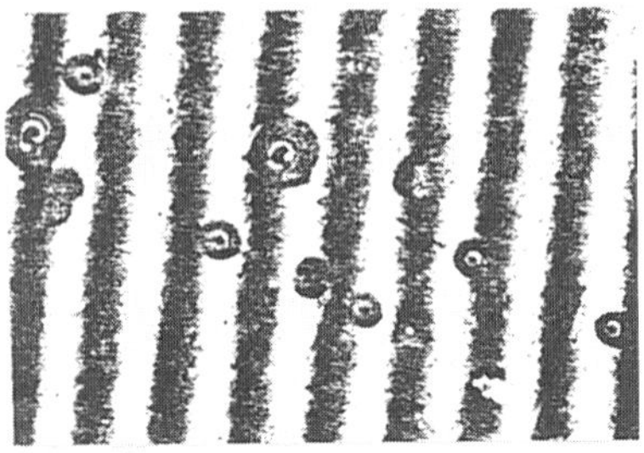

Fig. 2c

Figure 2. Changes in mirror surface made of a composite material exposed simultaneously to protons and electrons D = 10^{16} particles/cm^2 at T = 340 K. a) Initial structure (x 130); b) Characteristic regions with flaking defects (x 130); c) Interference patterns of defects (craters) of semispheric shape.

the mirror is shown in Fig.2a. The photographs of the typical irradiated regions (D = 10^{16} particles/cm^2) are presented in Figs.2b. The mirror surface has defects - round craters 2-5 μm in diameter, which appeared after exposure to corpuscular radiation. The defects are semispherical recesses 2-10 μm in diameter (Fig.2c). In all the cases Al and SiO_2 films were observed to peel off the mirror surface. The surface defects appearing after p^+ and e^- irradiation (D = 10^{16} particles/cm^2) may be interpreted as effects of "blistering" and "flaking" in dielectric materials for space optics, in our case glass with evaporated Al and protective films.

Series 2

Proton-electron irradiation of the model was performed with accumulation of charges (up to 20 kV), i.e. most approached the real condition on a spacecraft in the

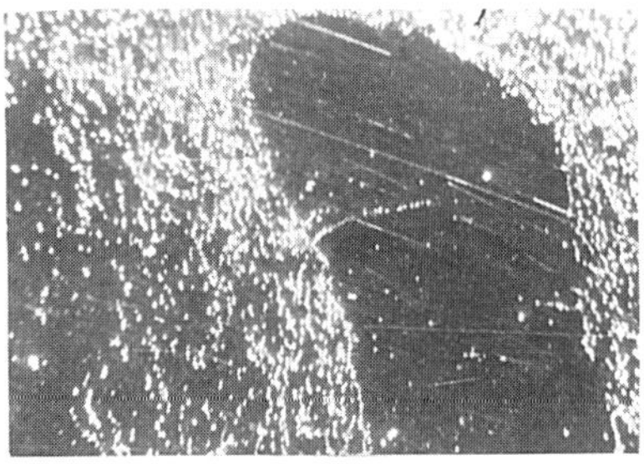

Fig. 3a

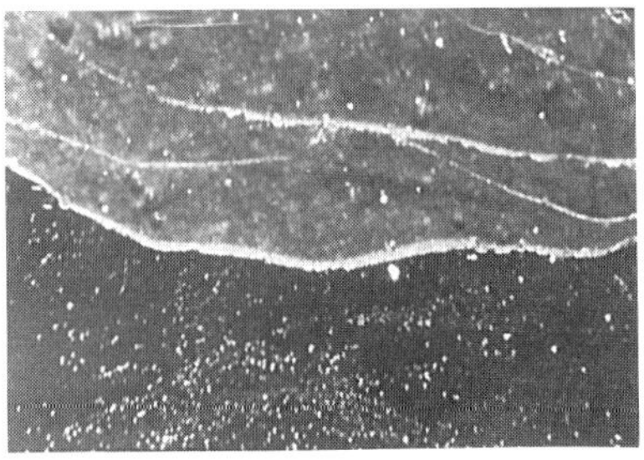

Fig. 3b

Figure 3. a) Typical defective structures due to flaking and internal breakdowns in the mirror after combined proton-electron irradiation, D = 10^{16} particles/cm^2 (x 130); b) Detachment of Al layer along mirror perimeter (top: structure with peeled off Al, the relief of glass and Be is seen; x 130).

Earth's radiation belts. At this voltage a breakdown occurred in the PTFE (5) of the condenser (4) (Fig.1). The scale model was scaled against the Earth. The typical defective structures on the mirror surface after combined p^+ - e^- irradiation ($D=10^{16}$ particles/cm^2) are shown in Figs. 3a,b. As in Series 1, blistering and flaking defects show up as round craters in the glass (Fig. 3a) or the peeled off Al and protective films (Fig. 3b). The defect density was $\rho = 10^4$-10^5 cm^2. There were also new types of defects - tracks, or wide lines of breakdown cascades in the glass. Figure 3a shows the formation of "avalanches" and "cascade breakdowns", and charges running off the centre of the mirror towards its periphery, i.e. the highest electric field strength are seen in Fig.3a. The Al layer peeled off completely on more than one third of the mirror perimeter forming a ring 4-5 mm wide (Fig.3b). A layer of glass and beryllium texture can be seen with a microscope (Fig.3b, top). The characteristic picture preliminary to peeling off can be observed: the defect density inside the "avalanches" is much higher at the edge of the mirror than in the central part (Fig. 3a). After p^+- e^- irradiation ($D=10^{16}$ particles/cm^2) the scale model was exposed to extra-atmospheric solar (and VUV) radiation for 100 hours. Cryogenic precipitation appears on the mirror surface and decreases the reflection coefficient of the objective lens. The structure analysis performed after proton, electron and EMRS radiation shows intensive cracking of the glass surface where the Al film peeled off (along the mirror perimeter). Cracking is caused by the large temperature gradient at the reflection boundary of aluminium and the lower Be layer of the mirror. As a result, thermoelastic stresses develop in the boundary region, which lead to cracking. In Series 2 blistering and flaking caused by proton radiation (E = 160 keV, $D=10^{16}$ particles/cm^2) produce a sort of "space erosion" of the dielectric material surface.

Series 3

Glass samples covered with Al and Ag films were exposed to protons. The experiments were performed to study the results of Series 1 and 2. A proton beam (S = $4cm^2$) was directed so that the mirror surface covered with an Al (Ag) film and the pure glass could be irradiated simultaneously. The beam parameters were E = 150 keV, the current density on the sample j = 2-3 $\mu A/cm^2$, $D = 1 \cdot 10^{15}$ - $2 \cdot 10^{17}$ p^+/cm^2. As in Series 1, the secondary electron emission from the sample surface was suppressed by the negative potential U = 250 V applied to the calibrating aperture in front of the sample. The irradiation conditions were the vacuum $P = 10^{-6}$ Torr, T = 340-490 K. The typical defects on the mirror surface with an Al (Ag) film are shown in Figs. 4,5. The topology of the pure glass surface after fixed doses of protons is shown in Fig. 6. Two typical classes of defects are observed in Series 3. True "blistering" of the Al and Ag films over the glass occurred the mirror surface (Figs.4). The defect sizes and densities were much dependent on the radiation dose and the target temperature. The critical dose D_{cr} was found for T = 340K: on reaching D_{cr} first blisters started to appear in a jump - like way during short times (several seconds) (Fig. 4). The critical p^+ dose which triggered blistering varied within $D_{cr} = 5 \cdot 10^{15}$ - $7 \cdot 10^{15}$ particles/cm^2 for different glass samples of one batch. This dose coincides with a year's integral proton fluence onto a spacecraft in a geostationary (H = 36000 km) orbit ($D = 6 \cdot 10^{15}$ particles/cm^2). The blisters ($D=10^{16}$ particles/cm^2, T_{sample} = 340 K) spread uniformly throughout the irradiation field. They were nearly identical in diameter 30-60 μm .

Their density was high: $\rho = 10^4 cm^2$ at $D=2 \cdot 10^{16}$ p^+/cm^2 and $\rho = 3 \cdot 10^4$ cm^2 at $D= 2 \cdot 10^{17} p^+/cm^2$, T = 340 K. The blisters were observed to coalesce when the radiation dose and temperature were increased. The typical blisters produced by $D= 2 \cdot 10^{17} p^+/cm^2$ at T_{target} = 490 K are shown in Fig.5. The blisters density on the mirror surface was $\rho = 1.4 \cdot 10^5$ particles/cm^2. The change from T = 340 K to T = 490 K with the same fluence thus leads to nearly an order-of-magnitude increase in the blister density. This experimental finding agrees with current theoretical concepts of blistering whose rate is dependent on the diffusion mobility of hydrogen atoms. It is shown experimentally that the critical proton dose $D_{cr} = 6 \cdot 10^{15}$ particles/cm^2 applied to pure

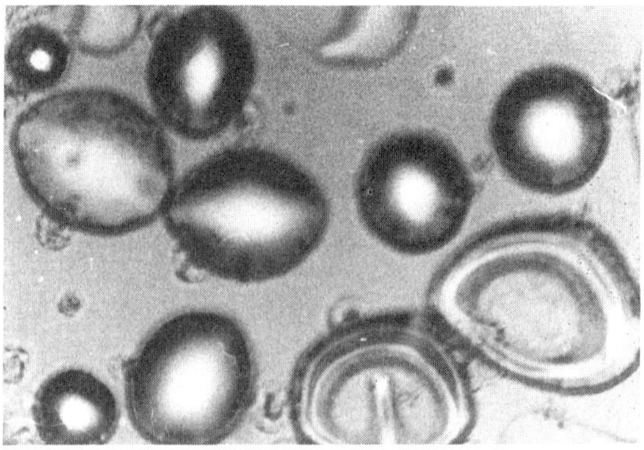

Figure 4. Blistering on the mirror surface: typical bulging of Al (Ag) films after exposure to proton dose D = 10^{16} particles/cm^2 , T = 340 K (x 260).

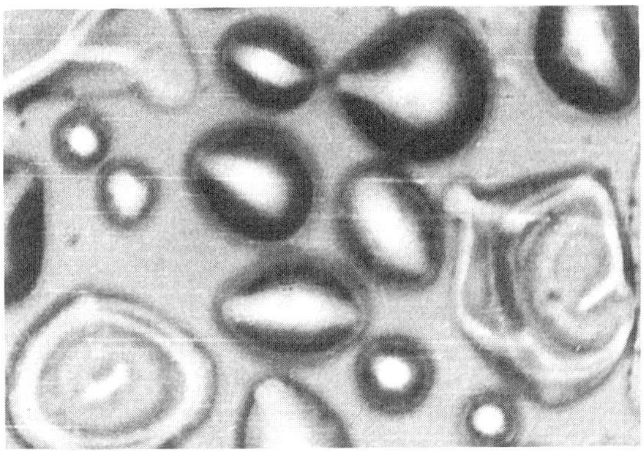

Figure 5. Blistering on mirror surface (Al, Ag film bulging) after exposure to proton dose D = $2 \cdot 10^{17}$ particles/cm^2, T = 490 K (x 260).

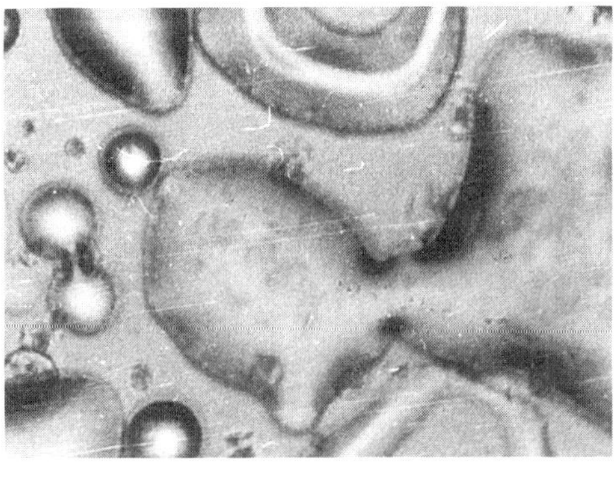

Figure 6. Flaking defects in pure glass after exposure to proton dose D = 10^{16} particles/cm^2 (x 250).

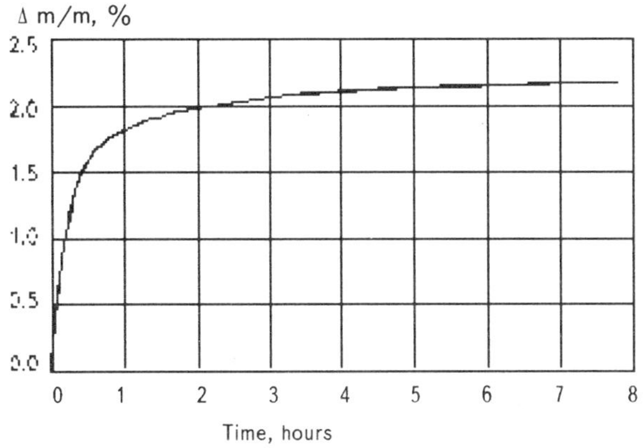

Figure 7. Kinetics of mass loss of arimidic fabric in vacuum 10^{-7} Torr at T = 140°C

glass produces defects similar to those observed on the surface of the composite material of the mirrors in Series 1 semi-spheric craters 2-5 μm in diameter (Fig.6, arrows A). The defect density builds up sharply when the radiation dose and the target temperature are increased.

The kinetics of the mass loss of the arimidic fabric in the vacuum 10^{-7} Torr at T = 140°C after eight hours exposure is shown in Fig.7. The maximum value $\delta m/m_0$ is 2%. The main mass loss occurs during 1 - 1.5 hour.

DISCUSSION

The main mechanisms responsible for development of defective structures on the mirror surfaces are as follows: Group1 - electrophysical mechanisms possible under proton and electron radiation (Fig.8); Group 2 - cryogenic precipitation on the mirror surfaces due to gas liberation from the blind materials under extra-atmospheric solar plus VUV radiation and formation of polymerized films caused by VUV quanta.

Blistering and flaking effects on glass surfaces exposed to protons of the Earth's radiation belts

Let us analyse the mechanisms of Group 1. Protons of the Earth's belts develop defective structures on the glass surface (blistering and flaking). The formation of blisters was reported in Ref.10: blisters were detected on metal surfaces (Cu,Ni,Al) exposed to helium (He^+) and deuterium (D^+) ions. This phenomenon was called "blistering". It was extensively studied in connection with nuclear reactors. The materials used in the reactor were exposed simultaneously to helium, deuterium and tritium. When the radiation dose was increased, the blisters broke up and their caps fell off. This phenomenon was called "flaking". From this standpoint we interpret our experimental results [8,9]. Under proton radiation (penetration depth ε_{p+} = 2μm) an electron adds to a proton $p^+ + e^- \rightarrow H$ (Fig.8a). This process takes place in a thin layer of the material. As a result, a hydrogen atom is formed, which can combine with the vacancies produced by ionic radiation and form blisters. Within a blister the hydrogen atoms form a molecule - H_2. Such blisters of 1-10 nm were observed on He+ bombardment of metals with a dose to Evans's model [8], the pressure in a blister can grow very high P = 0.1 E (E is the Young's modulus of the material) because of poor gas solubility. Such critical dose can cause the material fracture and cracking between the blisters at the depth of the mean free path of the particle, i.e. ε_{p+} = 2 μm, when the highest stress σ_{max} produced by hydrogen pressure in the blisters becomes equal to the ultimate strength of the material σ_F (glass in our case). Being a brittle material, glass, unlike metals, suffers no blistering. The pressure of hydrogen can cause brittle failure (parts of spheres are seen experimentally) in some areas of the surface layer $\varepsilon_{p+} = 2$ μm thick. This process is assisted by rather high internal stresses in the surface layer of ε_{p+} thickness. The stresses develop when hydrogen gets into glass. The material failure is accompanied by detachment of pieces of surface layer of glass (Fig. 8a), together with its layered Al + SiO_2 coating (flaking) (Series 1, 2), Fig. 8c. The picture of defect

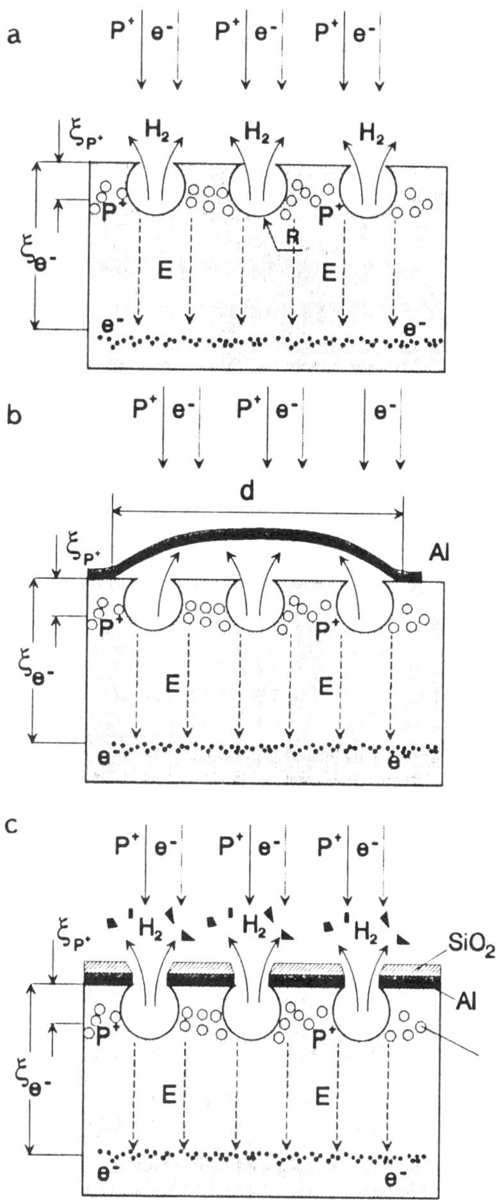

Figure 8. Shematical presentation of defect formation in brittle dielectric materials by mechanisms of -flaking (a,c), - blistering (b), -spatial separation opposite charges (a,b,c).

formation on mirror surface due to proton radiation is somewhat different in Series 3 (Fig. 8b). As fracture occurs in the surface layer, the hydrogen formed in the cavities resulting from flaking does not escape into free expanses but comes to the boundary space between the continuous Al (Ag) film and the glass surface. Since Al and Ag films are plastic materials, the internal pressure of hydrogen first strips the metal film from the glass in the vicinity of craters (primary defects of ϕ= 2-5 μm) and then produces plastic deformation of the films. Microscopically, such defects show up as bulging of the Al (Ag) film over the glass surface. "Pure" blisters, similar to those in metals, are also observed in the glass - Al composite and their caps are plastic too. The equilibrium state of blisters is dependent on the balance of forces - the force of the internal hydrogen pressure, which is in turn dependent on the radiation dose (Hatom concentration) and the adhesive force between the Al film and the glass. This better illuminates the results of Series 1 and 2, i.e. the presence of craters, rather than blisters, in the glass and the composite material of the mirrors (Fig. 8c). It is also clear why experimental IR instruments suffered a drastic degradation of the reflectivity of the mirrors after a 79 month's (sometimes a year's) space service in the Earth's radiation belts. In the latter case the reflecting Al surface (100 nm) was covered with a protective layer of brittle SiO_2 which impeded blistering, i.e. bulging of aluminium. Instead, the mechanisms of flaking was in operation (Fig.8a) producing craters, what just occurs in pure glass. This suppressed the reflectivity of the mirrors. Studies on metals show [9] that subsequent generations of blistering and flaking defects are possible as a new critical radiation dose D_{cr} is applied. In the experiments reported here the critical dose producing flaking in the glass and the mirror was $D_{cr} = 6 \cdot 10^{15}$ particles/cm^2. Blistering in plastic materials and flaking in brittle substances occur as the critical dose

$$D_{cr} = \frac{\sigma_F}{H\sqrt{\delta R^2(E)}} \qquad (1)$$

is reached (σ_F is the ultimate strength of the glass, $\sigma_F \sim 4$ kg/mm^2; H is the energy of the hydrogen atom implantation into the glass, H ~ 1.5 eV; σR^2 shows the rms spreads of paths of protons with E = 160 keV in the glass). Numerical substitution to Eq. (1) gives $D_{cr} = 3 \cdot 10^{15}$ particles/cm^2, which agrees with the experimental dose for blistering $D_{cr} = 6 \cdot 10^{15}$ particles/cm^2. The experimental value is equivalent to an annual proton dose (D = 1 year) incident on a spacecraft in a geostationary orbit. In hard-radiation orbits of spacecraft's (300 km < H < 1000 km) the annual radiation dose is $D = 10^{16}$ particles/cm^2, i.e. the proton (or gas ion) fluence becomes critical for blistering-flaking even after six months of the flight in the Earth's radiation belts. This should be taken into account while predicting the behaviour of materials for long-term service in space. It is found that blistering and flaking are much inhibited in porous materials which may be used in space systems.

Electrostatic effects

There is one more mechanism inducing defects in dielectric materials under simultaneous proton-electron radiation: it initiates spatially separated regions of opposite charges in the glass layer having high resistivity $\rho_v = 10^{14} - 10^{15}$ Ohm·cm^2. Estimation gives the penetration depth ξ_{p+} - 2 μm for protons with E = 160 keV and ξ_{e-} = 100-150 μm for electrons. Hence, there appear two spatially separated regions of opposite charges in the glass. They excite a high electric field in between $E = \sigma/\varepsilon = 1.6 \cdot 10^9$ V/cm (Figs. 8a, b, c) (σ is the surface charge density, ε the dielectric constant of glass). This value much exceeds the breakdown voltage in glass ($E_{br} = 3 \cdot 10^5$ V/cm). The additional defects observed in the material of the objective and the mirrors (Series 1, 2) may be caused by internal breakdowns in the glass, which is accompanied by brittle failure in some regions of the objective, subsequent detachment of the Al = SiO_2 layers and ejection of the substance after an electric breakdown. This mechanism of defect formation permits us to explain why the defect density in Series 2 (with accumulation of electric charges on the objective) grows with distance from the centre of the glass towards its edge, where the electric field has the highest strength in the bulk and on the surface of the mirror. Electrostatic mechanisms operate both in the bulk of the mirror material and in the thin surface layers. They develop due to the difference of potentials between the components of the unit - the objective and the elements of the blind. The electric field strength is the highest at the mirror edge, and the avalanche breakdown is more intensive along the mirror perimeters. The avalanche breakdowns may produce high local heating in the surface layers of the mirrors (a high rise of the local temperature). Glass and aluminium have widely differing thermal expansion coefficients: $\alpha_{gl} = 10^{-7}$ deg^{-1}, $\alpha_{al} = 5\cdot 10^{-6}$ deg^{-1}. Assuming that in the regions of large avalanches the local temperature rises by several tens of degrees, the stresses at the glass-aluminium boundary due to the different α- coefficients is

$$\delta\sigma \sim \frac{\alpha_2 - \alpha_1}{\delta T \cdot E} \sim 10 \div 15 \frac{kg}{mm^2}$$

where E is Young's modulus of glass. This value much exceeds the adhesive strength of the Al coating on the glass. In Series 3 (real conditions of instruments operation in space) three basic electrophysical processes are likely, which significantly affect the thin Al layer and the glass producing defects by various mechanisms-blistering, flaking, internal breakdowns in the glass owed to spatial charge separation and external electric breakdowns between the objective and the blind.

Cryogenic precipitation on the mirror surface

Another group of mechanisms producing defects on the mirrors is related to cryogenic precipitations. The loss of mass of structural materials can be caused by different factors such as direct incidence of the solar emission on the surface of the materials, protons and electrons of the Earth's radiation belts, combined impact of solar emission and ionising radiation of solar flares, gas release due to rising temperature. The mass loss of the arimidic fabric and subsequent condensation if the gas release products lead to cryogenic precipitation uniform over the mirror surface (60 mm in diameter), the thickness of the precipitated film being $\phi_{cr} = 10\mu m$. Considerable degradation of the mirror reflectivity can be expected when the cryoprecipitated film has the thickness comparable with the incident IR wave length (or more exactly $\lambda/2$, i.e. at $\phi_{cr} = 1.5$ μm). In the field of VUV radiation the layer precipitated on the surface of the objective is intensively polymerised. This affects the optical properties and thus impairs the reflectivity of the mirrors in space.

CONCLUSIONS

The degradation of reflectivity of mirrors in IR instruments operating in space is caused by defective structures formed on mirror surfaces. The main mechanisms of defect formation include the following:

1.Complex defective structures develop on the composite material of the mirror surface. They look like round craters , ϕ = 2-5 μm (flaking) or Al (Ag) film bulging (10-50 μm) over the mirror surface (blistering).

2. Blistering and flaking develop in a jump-like way in short times $\delta\tau$ = 1-2 s as a certain critical dose of proton radiation D_{cr} is achieved; at T = 340 K $D_{cr} = 6\cdot 10^{15}$ particles/cm^2. The density of blistering and the target temperature (increases sharply with the temperature).

3. Blistering and flaking can be triggered by hydrogen resulting from proton-electron recombination in the glass [8, 9]. Hydrogen forms blisters in a thin

surface layer of glass at mean free proton paths ξ_{p+} - 2 μm. As the hydrogen pressure in cavities exceeds the ultimate strength of the material σ_r, the glass suffers brittle failure. Brittle failure may also occur in the Al film and the protecting layer and is accompanied by cratering (flaking). One more phenomenon is observed - bulging of the plastic Al (Ag) films over the mirror surface (blistering). In both the cases we observe a kind of "space erosion" on the mirror surface.

4. The critical dose $D_{cr} = 3 \cdot 10^{15}$ particles/cm^2 for blistering estimated in terms of Yu.V. Martynencko's theory [9] shows that this model agrees with our experiment which gave $D_{cr} = 6 \cdot 10^{15}$ particles/cm^2.

5. One more mechanism of defects in dielectric materials of space optics on simultaneous proton-electron irradiation is the formation of spatially separated regions of opposite charges. These appear due to different penetration depths of the particles into the material, which results in internal breakdowns in the glass at different levels.

6. The defects of group 2 are related to cryogenic precipitation films which are formed with condensation of products of gas release from the materials of space instruments.

7. Corpuscular proton-electron radiation produces a heavy impact on development of defects on the mirrors of space optics (by blistering and flaking mechanism). The extra-atmospheric solar emission (including VUV radiation with quanta energies 5-100 eV) does not give any appreciable contribution to blistering and flaking. But it can produce cryogenic precipitation films (in the process of gas release on thermodestruction of the materials) and subsequent condensation of gas products on cold surface of mirrors. Blistering and flaking can be suppressed considerably or completely in porous (e.g., sintered) materials: at growing pressure gas freely escapes through pores. This group of materials may include sintered Al or Be powders, composites (as a base for mirrors) of fine-grain carbon with fixing polymeric varnish or resins. All-aluminium mirrors are also promising for space optics taking into account their high electric conductivity (no charge accumulation and uniform potential over the unit). The critical dose of blistering for Al is $D_{cr} \approx 10^{18}$ particles/cm^2, i.e. about 50-70 years of service in space [2] though this conclusion calls for further investigation.

ACKNOWLEDGMENT

The authors are indebted to Academician V.G.Baryakhtar and Prof. I.M.Nekludov for helpful discussions.

REFERENCES

1. M.I. Guseva and U.V. Martynenko, Radiation Blistering, The Successes of Physical Sciences 135: 671 (1981)
2. U.V. Martynenko, The Theory of Blistering, Moscow, The Kurchatov Institute of Nuclear Energy (1979).
3. V.V. Abraimov, N.N. Agashkova, I.V. Budniak, A.A. Zakharova, N.I. Velichko, A.M. Markus. Effect of Low Temperature, Irradiation and Deep Vacuum on Physico-Mechanical Properties of Polymer Light-Absorbing Coatings, Physics and Chemistry of Materials' Treatment, 2: 26 (1992) (in Russian)
4. V.V. Abraimov, F. Lura , L. Bohne, I. Budniak, and V. Danovski, Investigation of Non-Linear and Non-Reversible Degradation of Physico-Mechanical Properties of Materials Under Effect of Space Factors, Deutcher Luft und Raumfart Kongress DGLR, Band, Berlin (1993), p.80 (in German)
5. V.V. Abraimov, N.N. Agashkova, V.N. Soloviev, I.P. Zaritskiy, and V.F. Udovenko. Thermal Firmness of Light-Absorbing Coatings in 4.2-300 K Temperature Interval, Physics and Chemistry of Materials' Treatment, 6:83 (1989) (in Russian)
6. V.V. Abraimov, K.Sh. Bocharov, I.V. Budniak and V.V. Danovski, VUV Induced Creep of Some Polymer Materials Physics and Chemistry of Materials' Treatment, 1:39 (1993) (in Russian)
7. "Space Model", S.N. Vernov ed., V.2, Simulation of Space Environment Effect on Materials and Facilities of Space Vehicles, Scientific Research Institute of Nuclear Physics Moscow State University, Moscow (1983) (in Russian)

8. J.H. Evans, Formation of Blisters in Mo Bombarded with Helium, Nature, 256: 299, (1975)
9. S.K. Das, M. Kaminsky, et al. Correlation Between Blister Skin Thickness the Maximum in the Damage - Energy Distribution, and Projected Ranges of He+ Ions in Metals, Appl. Phys. Lett., 27: 521 (1975)

THEORY ANALYSIS AND TEST STUDY OF COLD BLACK TARGET FOR SPACE RADIATIVE COOLING ENVIRONMENTAL SIMULATOR

Lu Yan , Han Jun , and Gong Tiequan

Lanzhou Institute of Physics
P. O. Box 94 , Lanzhou , 730000
P. R. C.

ABSTRACT

Based on the principle of the cold black target used in the radiative cooling environmental simulator , an integrated theory has been given for the effective emissivity calculation of the blackbody composed of honeycomb , and every heat load accepted by the cold black target was analyzed in the thermal environmental simulation test of radiant coolers . These provided a theoretic basis for the thermodynamic design and the structural design of the cold black target . By using these theories , we designed a target used in the thermal environmental simulation test of the radiant cooler we are studying . The target is made of five plates to which are glued Aluminum honeycomb and cooled by two G-M cryocoolers . The temperature of the target is below 30K and the effective emissivity is greater than 0.98. We used the simulator with this target to test radiant coolers for many times . All the test results indicated that the target can simulate the thermal and vacuum environment in the space very well . The error of the simulation is below 2% . These also show that the methods that we provided in this paper are reasonable and feasible.

INTRODUCTION

Cosmic space is a cold black and vacuum environment that has a temperature of 3~4K . Space radiative cooling is a refrigeration method that makes use of a special structure radiating heat to the space environment to get low temperature . This method can be widely used in an infrared remote system to cool infrared detectors to get high sensitivity . The cooler must be reliable . Its properties directly affect the performance of the infrared remote system . It must be strictly tested on the ground before the spacecraft is launched . In all of these tests , the most important is the thermal test that examines the thermal properties of the cooler and verifies the calculation of the thermal design and the reasonabilness of the structure and technology .

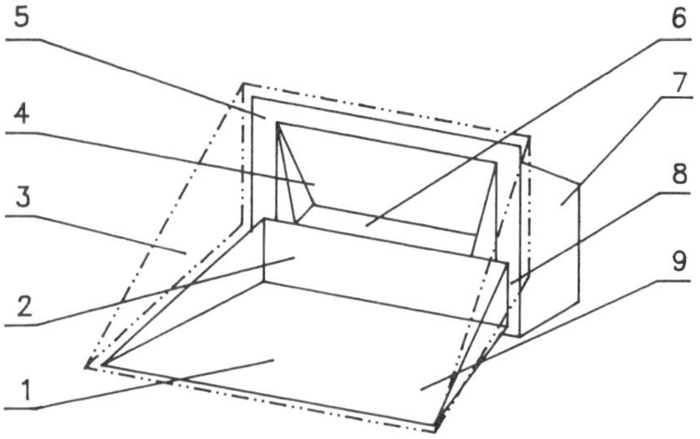

Figure 1. The sketch of coupling relationship between the radiant cooler and the target. 1 is the shield of the cooler. 2, 5 and 6 separately are the first stage radiator, the housing radiator, the second stage radiator. 3 and 8 are the side plate of the target. 4 is the cone of the cooler. 7 is the housing of the cooler. 9 is the base plate.

The test is done in a space environmental simulator that can simulate the cold black and vacuum environment. In this simulator, a cold black target is used to simulate the emissivity and the temperature of cosmic space. The target must have enough refrigeration capacity to meet the needs of radiant cooler thermal testing. The performance requirements of the target are shown as follows:

1. The emissivity of the target must be as high as possible.
2. The temperature distribution of the target must be even.
3. The refrigeration capacity must meet the needs of the thermal test.
4. The blackbody must be continuous and large enough to cover over the radiant cooler and achieve a light tight cavity with the cooler.

Experience has told us only when the temperature of the target is below 30K and the effective emissivity of the blackbody is higher than 0.98 that the test error caused by the temperature and the emissivity is below 2%. To achieve as high emissivity as possible with a large area continuous blackbody, a honeycomb covered with black paint is used. This paper describes the method used to calculate the effective emissivity of honeycomb.

As usual, the target is cooled by a helium or neon circuit using the reverse Brayton cycle. We successfully used a SV cryocooler to cool down the target for the first time in 1985. The chamber has a diameter of 500mm. The temperature of the target is below 20K and the refrigeration capacity is 8W. To meet the needs of

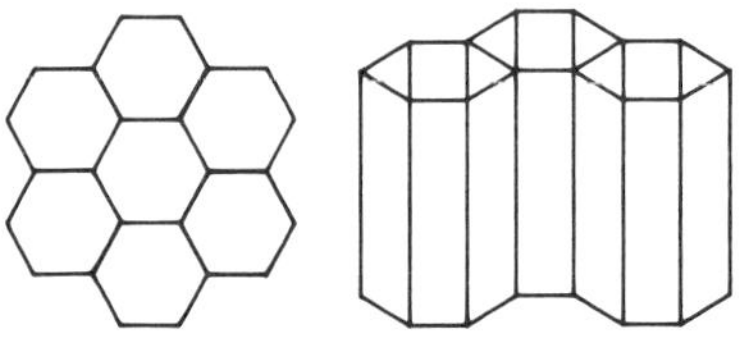

Figure 2. Thin wall honeycomb.

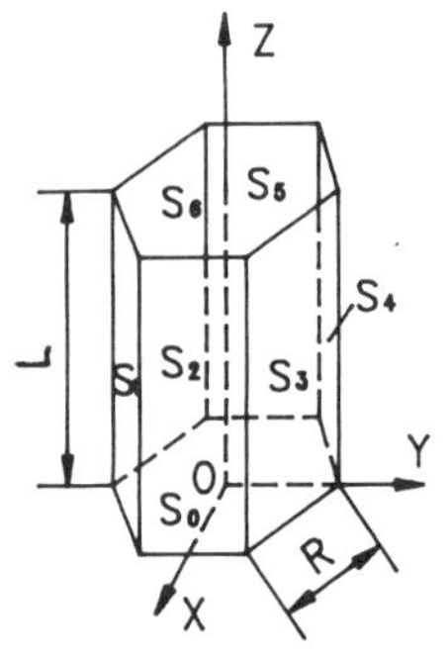

Figure 3. hexagonal prism.

a larger radiant cooler, another simulator that has the chamber diameter of 1000mm was developed in 1990. Because the target is cooled down by a cryocooler, the refrigeration capacity must be exactly calculated. In this paper we also give an analysis method for the heat absorbed by the target and present the design results and test results.

In the thermal simulation test of the radiant cooler, the position relationship between the cooler and the target is shown in Fig. 1. The black target consists of one base plate 9 and two side-plates 3, 8 that all have honeycomb on them. The honeycomb surfaces are covered with black paint that has high emissivity. The effective emissivity of the honeycomb may approach 1 and the blackbody is large and continuous. In the thermal test of radiant cooler, the black target is cooled down by mechanical refrigerator. The temperature of the blackbody must be below 30K and the cooling capacity of the refrigerator must meet the needs of absorbing the heat from the radiant cooler and making the radiant cooler cool down. The radiant cooler just is entirely covered by the target. The distance between these two is very close such that they constitute an enclosed cavity and the scatter by edges can be ignored. The heat from the second stage radiator, the first stage radiator, the housing radiator and other surfaces of the radiant cooler that see the target can all be absorbed by the target.

EFFECTIVE EMISSIVITY CALCULATION OF THE TARGET

Because the blackbody of the target is composed of honeycomb, the effective emissivity is always less than 1. The target doesn't entirely absorb the radiation heat from the radiant cooler and must have a little reflection. This produces errors in the thermal test. To estimate the test error caused by the emissivity, we must first calculate the effective emissivity of the blackbody.

Because the honeycomb is composed of thin walled regular hexagonal prism array (see Fig. 2), the effective emissivity is just the weighted average value of the effective emissivity of a single cavity and the edge emissivity. This can be given by Eq. (1).

$$\varepsilon_{eh} = (1 - \frac{\Delta S}{S}) \cdot \varepsilon_{ec} + \frac{\Delta S}{S} \cdot \varepsilon_0 \tag{1}$$

where

ε_{eh} is the effective emissivity of the honeycomb;
ε_{ec} is the effective emissivity of a single cavity;
ε_0 is the emissivity of the cavity wall;
ΔS is the area of the edge which relate to the thickness of the cavity wall;
S is the total area of the honeycomb plate.

The Effective Emissivity Calculation for a Hexagonal Prism

For a hexagonal prism (see Fig.3), the effective emissivity is given by:

$$\varepsilon_{ec} = \frac{1}{A_0} \iint\limits_S \varepsilon_a(n) \cdot F_{n-0} dS \tag{2}$$

where

A_0 is the area of the cavity opening;
S is the total area of cavity wall;

Table 1. Effective emissivities of hexagonal prism for different ε_0 and L/2R

ε_0 \ L/2R	1	1.5	2	3	4	5
0.80	0.94216	0.94872	0.95300	0.96165	0.97249	0.98591
0.85	0.95818	0.96353	0.96753	0.97637	0.98777	1.00000

ε_0 \ L/2R	1	1.5	2	2.5	3
0.884	0.96847	0.97309	0.97692	0.98110	0.98591
0.891	0.97054	0.97501	0.97881	0.98299	0.98783
0.92	0.97890	0.98279	0.98647	0.99069	1.00000

F_{n-0} is the view factor from the infinitesimal area dS_n to the opening;
$\varepsilon_a(n)$ is the apparent emissivity of an infinitesimal area dS_n obtained by:

$$\varepsilon_a(n) = \varepsilon_0 + (1 - \varepsilon_0) \iint_S \varepsilon_a(p) \cdot \frac{dF_{p-n}}{dS_n} dS \qquad (3)$$

where

ε_0 is the emissivity of the cavity wall;
$\varepsilon_a(p)$ is the apparent emissivity of an infinitesimal area dS_p on the cavity wall;
dF_{p-n} is the view factor from dS_p to dS_n .

We can use numerical methods to solve Eq. (3) , and use numerical integration to calculate ε_{ec} . When the calculation point is on the intersection line/point of two/ three walls , the distribution of the apparent emissivity is disconnected . These points are often named strange points . At these points , the numerical integration on the neighboring wall is given by:

$$\sum_{(S_p)}\sum \cdots\cdots = \begin{cases} \frac{1}{4}\varepsilon_a(\xi,\zeta) & \text{(on the intersection line between two walls)} \\ \frac{1}{2}\varepsilon_a(\zeta,\xi) & \text{(on the intersection line between wall and bottom)} \\ \frac{1}{3}\varepsilon_a(\zeta,\xi) & \text{(on the intersection point of three surfaces for wall)} \\ \frac{7}{24}\varepsilon_a(\zeta,\xi) & \text{(on the intersection point of three surfaces for bottom)} \end{cases} \qquad (4)$$

where

$\varepsilon_a(\zeta,\xi)$ is the apparent emissivity on the strange point;
(ζ,ξ) is the coordinate value of these strange points .

Table 2. Calculation and test results of effective eimissivity for honeycomb

ε_0	L/2R	Test results	Above Method	Sparrow [3]	Campanaro [4]	Treuenfels[5]
0.884	1	0.973	0.96636	0.96317	0.97610	0.96757
	1.5	0.975	0.97086	0.96395	0.98573	0.96757
	2	0.975	0.97456	0.96395	0.99079	0.96757
0.891	1	0.977	0.96855	0.96539	0.97764	0.96969
	1.5	0.979	0.97291	0.96583	0.98665	0.96969
	2	0.979	0.97661	0.96583	0.99134	0.96969

The Calculation Results for a Hexagonal Prism Cavity

For the length R=2mm of the bottom side, different heights of hexagonal prism and different emissivities of cavity wall, we calculated the effective emissivities. The calculation results are shown in Table. 1.

Calculation Results and Test Results for Honeycomb

We used Eq.(1) to calculate the effective emissivities for different honeycombs. The calculation results are given in Table 2. We also used other approximate methods to calculate the effective emissivities and made practical tests with ε_0=0.884, 0.891 and L/2R=1, 1.5, 2. The results are all shown in Table 2. Table 2 also gives out the comparison between all the calculation values and the test results.

HEAT LOAD ANALYSIS FOR COLD BLACK TARGET

Based on the principle of black target in the thermal test of " W-style" radiant cooler and the instruction of the space environment simulator, we know the heat loads of the target not only come from the radiant cooler but also come from the radiation of LN_2 shield. To improve the vacuum of the chamber, we stuck active carbon on the back of the target that can adsorb survival gases when the target is cooled down. The another heat load on the target is the adsorption heat from the active carbon.

Heat from Radiant Cooler

In thermal test of "W-style" two-stage radiant cooler, the cooler and the target constitute an enclosed cavity (see Fig.1). In this cavity, the intermediate stage reflector and the shield are specular reflective surfaces, and the others are diffuse surfaces. It is very difficult to calculate the heat transformation between two surfaces in such a cavity. It is necessary to simplify the radiative interchange. Based on the design of the radiant cooler, we can assume that:

1. The energy from the target to the cone all reaches the cold stage after the first reflection by the cone.
2. The energy from the target reflected by the shield all reaches the first stage radiator.
3. The radiation from the cold patch to the cone all reaches the bottom of the target after only one reflection.
4. The radiation from the first stage radiator to the shield all reaches the bottom of the target after only one reflection.
5. The temperature of the target (include No.3, 8, 9 surfaces in Fig. 1) is well distributed and the temperature of the cone and the first stage radiator are equal.

We ignore any other heat from the radiant cooler. The thermal radiation heat transformation in the cavity can be calculated. In calculation, the temperature of every surface is known. So we can easily obtain the heat quantity from the radiant cooler Q_R.

Radiation Heat from LN_2 Shield

In the space environment simulator, a LN_2 shield is used to reduce radiation heat from the chamber. The temperature of the shield is still higher than the target, thus radiation heat can not be ignored. This heat is given by Eq.(5).

$$Q_S = \frac{\sigma \cdot A_t \cdot (T_S^4 - T_t^4)}{\frac{1}{\varepsilon_t} + \frac{1}{\varepsilon_S} - 1} \tag{5}$$

where

A_t is the area of the back of the target to front of the LN_2 shield.

T_S is the temperature of the LN_2 shield.

T_t is the temperature of the target .

ε_t is the emissivity on the back of the target , i.e. the emissivity of the active carbon there .

ε_S is the emissivity of the inner surface of the LN_2 shield .

σ is the Stefan-Boltzman constant .

Adsorption Heat

The adsorption heat of the active carbon is relative with the quantity of the adsorbent . This is given by

$$Q_a = q_a \cdot M_a \tag{6}$$

where

q_a is the unit adsorption heat .This is decided by the type of the active carbon.

M_a is the total mass of the adsorptive gases .

Then the total heat load absorbed by the cold black target is

$$Q = Q_R + Q_S + Q_a \tag{7}$$

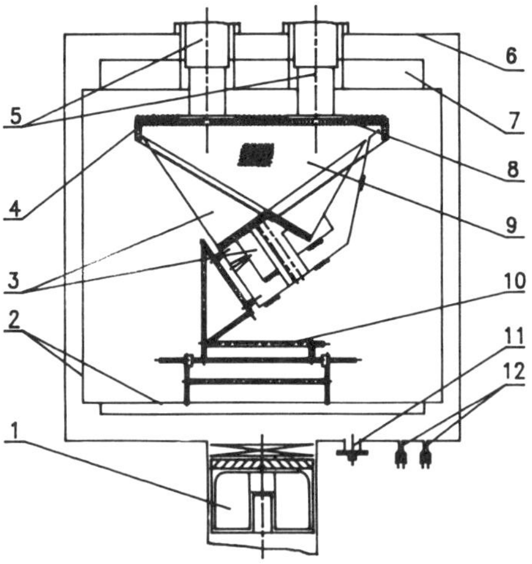

Figure 4. Sketch of the environment simulator used for radiant cooler test. 1 is the cryogenic bump. 2, 7 is the LN_2 heat sink . 3 is the radiant cooler . 4 is active carbon . 8, 9 are plates of the target . 5 is G-M refrigerator . 6 is the vacuum chamber . 10 is the carriage for mounting radiant cooler . 11 is the connector for the electric wire . 12 is the vacuum gage.

The cryocooler that makes the target cool down must provide a refrigeration capacity larger than the heat absorbed by it. The cold target can balance at a specified temperature only under this condition to meet the needs of the thermal test of radiant coolers.

RESULTS OF DESIGN AND TEST

Results of Design

Based on the theory above, a small space environment simulator used for radiant cooler tests was designed. The sketch of the structure is shown in Fig.4. The black target can be cooled down by two G-M coolers. The refrigeration capacity is more than 10W for every cooler. The effective emissivity of the target is greater than 0.98.

Results of Test

For such a simulator, when it runs without radiant cooler, the vacuum of the chamber is better than $7X10^{-4}$Pa. After the LN_2 heat sink was full and the two G-M coolers running, the coolers and the target can be cooled down within two hours. The vacuum of the chamber is less than $1X10^{-4}$Pa, and the temperature of the target is below 18K. When run with a radiant cooler, the radiant cooler must be first heated under the vacuum of $1X10^{-3}$ Pa for 96 hours to prevent contamination. After the LN_2 heat sink was full and the two G-M refrigerators running, the target and the radiant cooler were cooled down simultaneously. The temperature of the target was stabilized at 23.6K after 8 hours. At the same time, the vacuum of the chamber was up to $2X10^{-4}$ Pa. The radiant cooler was continuously cooled down, after 60-80 hours the temperature of the second stage was balanced at the lowest. After examining the cooling capability of the radiant cooler and testing the stability of the temperature control system, the experiment was ended. In the test, some temperature measurement points were set up on the target. All the results of the temperature testing show that the evenness of the temperature is better than ± 1K.

CONCLUSION

1. The method of the effective emissivity calculation for the honeycomb and the method for the heat analysis of the cold black target used in the space environmental simulator of radiative cooling is useful to the simulator design.

2. Using the methods, we designed a simulator for the "W-style" radiant cooler. We did many thermal tests of radiant coolers. All the results of the tests are reasonable and show that the structural design for the simulator using the methods in this paper are feasible and reliable.

3. When the heat absorbed by the target is not very large, using crycooler to cool down the target is a good idea. This can save the running fee and all the operation is simple.

ACKNOWLEDGMENT

The authors wish to thanks for the cooperation of Mr. Pan Yanpin , Mr. Jiang Jishan , Mr. Hong Guotong , Mr. Pang Wei , and all the members in the radiative cooling research group in Lanzhou Institute of Physics.

REFERENCES

1. E. M. Sparrow, R. D. Cess , "Radiation Heat Transfer", 1970.
2. R. E. Bedford, J. Opt. Soc. Am., Vol. 64, No.3, p.339-349(1974).
3. E. M. Sparrow, Trans. ASME J. Heat Tansfer, Vol. 84, No. 1(Feb. 1962).
4. P. Campanaro, J. Opt. Soc. Am., Vol. 57, No. 1, p. 48-50(1967).
5. Ernst W. Treuenfels, J. Opt. Soc. Am., Vol. 53, No. 7, p.1162-1163(1963).

THE DEPENDENCE OF CHOKED FLOW AND BREAKTHROUGH ON PORE SIZE DISTRIBUTION IN VAPOR-LIQUID PHASE SEPARATION OF He II USING POROUS MEDIA

S.W.K. Yuan[1], D.J. Frank[1], and Chris Lages[2]

[1] Research and Development Laboratory
Lockheed Martin Missiles and Space Co.
Palo Alto, CA 94304, U.S.A.

[2] Stanford University
Stanford, CA 94305, U.S.A.

ABSTRACT

Liquid breakthrough and choked flow are important phenomena in the operation of phase separators in space. Breakthrough results in loss of liquid which can reduce the life time of, and potentially terminate a mission. Choked flow limits the amount of liquid boiloff which passes through the phase separator and can result in the warming of the bath temperature. Since the capillary effect plays an important role in both the above phenomena, the distribution of pore sizes in the phase separator governs the critical limits of breakthrough and choked flow. In this paper, the pore size dependence of breakthrough and choked flow in vapor-liquid phase separators will be discussed in detail. The experimental data of various phase separators tested for the Relativity Mission will also be compared to present theory.

INTRODUCTION

Vapor-Liquid phase separation (VLPS) using porous media was studied extensively during the late 70's and early 80's. After the Infrared Astronomical Satellite (IRAS) mission[1] there has been limited effort in the research of the VLPS. The COBE mission[2] raised some uncertainty in the operation of the phase separator as temperature at the downstream of the porous plug was seen to fluctuate with time. The Superfluid Helium On-Orbit Transfer (SHOOT) experiment[3] demonstrated the operation of a high vent rate VLPS system. Recently, the Jet Propulsion Lab conducted experiments[4] on the phase separator to be used for the SIRTF mission. Despite these efforts, there still exists the discrepancy among various theories. Choked flow and hysteresis were not detected in a number of early studies on the VLPS. Whether these investigators overlooked the choked regime is not clear. Moreover, with the same permeability (which has been accepted as a more accurate parameter to characterize VLPS) and dimensions, some plugs tend to choke at higher flowrates than others. We hope that the present paper might shed some light on the theory behind the operation of the phase separator, in particular the pore size dependence of choked flow and breakthrough will be discussed in detail.

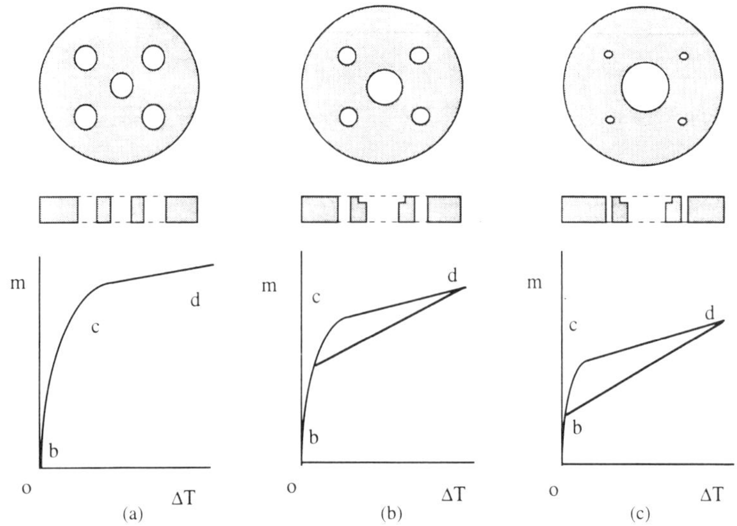

Figure 1. Schematic diagram of the operation of three porous plugs with the same permeability but different distribution of pore size.

THEORY

Vapor-Liquid Phase Separation of He II was studied extensively. The normal operating region or generally know as the linear regime is depicted in Figure 1 as section b-c. The governing equation in this linear regime can be written as[5]

$$m = \frac{\rho^2 S^2 T K_p A \nabla T}{(\lambda + ST)\eta_n} \tag{1}$$

where ρ is the density, S is the entropy, λ is the latent heat of vaporization and η_n is the viscosity of the normal fluid. K_p is the permeability and A is the area. The choked flow regime was first explored experimentally by Murakami et al[6,7] and Hendricks[8]. DiPirro[9] offered a simplified model for the choked flow. Lages[10] postulated a more complete theory

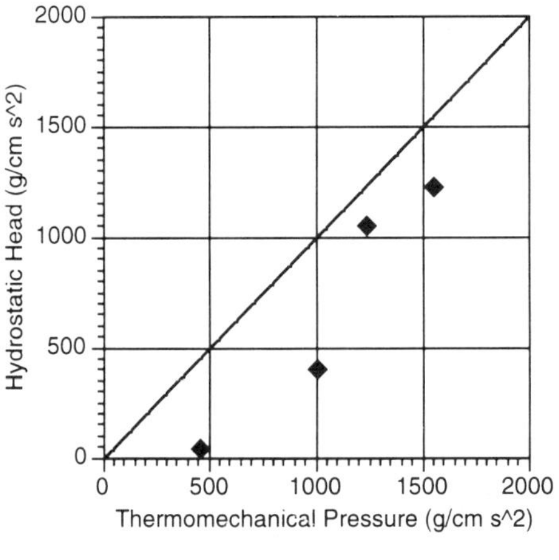

Figure 2. Thermomechanical pressure versus hydrostatic head.

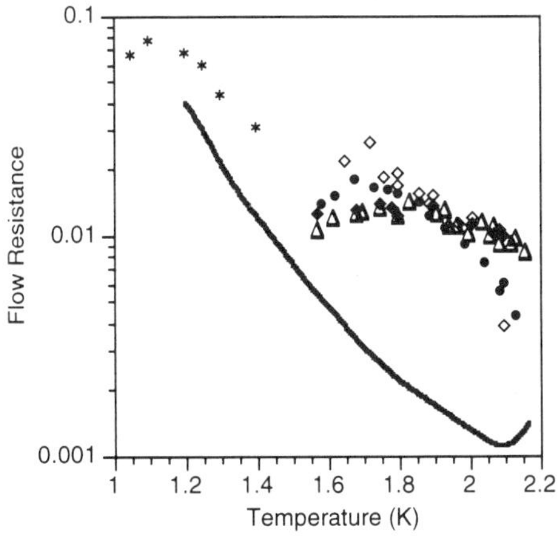

Figure 3. Flow resistance: solid line - Equation 1, * JPL data (Ref. 4), the rest of the data from Ref. 6.

for choked flow based on the Gibbs free energy analysis, which boils down to DiPirro's equation[9], for pore sizes on the order seen in VLPS. According to DiPirro, the temperature across the porous plug in this regime can be expressed as

$$\Delta T = \frac{\dfrac{4\sigma}{D_c} + \rho g h}{\rho S - \left.\dfrac{dP}{dT}\right|_{svp}} \tag{2}$$

where σ is the surface tension, h is the height of the liquid and D_c is the equivalent capillary diameter.

RESULTS AND DISCUSSION

From the experimental data obtained from the Relativity Mission phase separator[11], one learns the following.

The Breakthrough Regime

In this regime, the thermomechanical pressure is not strong enough to overcome the static pressure (e.g., the hydrostatic head and vapor pressure) across the porous plug. This is represented by section o-b in Figure 1. Three plugs were tested. The first plug has a permeability (K_p) of 3.55×10^{-9} cm^2. The second plug has a K_p of 1.7×10^{-9} cm^2. And the third plug has a K_p of 3.8×10^{-10} cm^2. As expected, the first two plugs are more susceptible to breakthrough. For the the last plug, breakthrough was not detected for flowrates higher than 0.05 mg/s-cm^2. Figure 2 is a plot of the thermomechanical pressure against the hydrostatic head for the last plug. The data represent the onset of breakthrough. The fact that the thermomechanical pressure is always larger than the hydrostatic head tends to indicate that liquid breakthrough should not have taken place. The liquid detection at the downstream is due to the presence of a thick film.

The Normal Phase Separation Regime

This regime is generally known as the 'linear regime' (represented by b-c in Figure 1). Murakami first reported a discrepancy between the experimental data and the linear equation (Eq. 1). Elliott[4] also pointed out that Equation 1 does not fit the experimental data of JPL for the SIRTF plug, and that an effective viscosity has to be used in place of the actual viscosity. On the other hand, the NASA GSFC group[12] reported that the COBE test data at small flowrates agree with the linear equation. Figure 3 is a comparison between the experimental data in the literature and Equation 1. The flow resistance is defined as

$$R = \frac{m}{\nabla P K_p A} \tag{3}$$

Note that the resistance of the above equation is unitless. The fact that most data fall above the curve tends to suggest larger flow resistance than the laminar transport (or linear equation). In Figures 4a and 4b, the mass flow is plotted against the temperature difference on a log-log plot. In both of these plots, the slope of unity is denoted by the dashed line, and the slope of 1/2 is shown by the solid line. They are drawn to guide the eye. Since the Murakami data[6] (Fig. 4a) deviate from the slope of unity, it indicates that the flow is in the transition regime. These data show that the transport in the phase separation is not always in the laminar (or linear regime). Due to the distribution of pore size (as discussed in the next section), some phase separators go from the laminar regime directly into choked flow. Others go from laminar to transition, to turbulent before choking. The 1/2 slope of the JPL data[4] (Fig. 4b) tends to indicate an Ergun type transport as suggested in[5]

$$\nabla P_T = \frac{v_n \eta_n}{K_p} + 1.75 \frac{\rho v_n^2}{(150 \varepsilon^3 K_p)^{1/2}} \tag{4}$$

where ∇P_T is the thermomechanical pressure, v_n is the approach normal fluid velocity, η_n is the normal fluid viscosity, and ε is the porosity.

Choked Flow Regime:

Choked flow is characterized by the recession of the vapor-liquid interface into the porous plug. This regime is represented by section c-d in Figure 1. Since the Relativity Mission requires a large range of flowrates, the authors went through painstaking efforts in searching for the appropriate material to be used for the phase separator. Due to the stringent requirements on the bath temperature control, operation in the choked regime is not recommended. Therefore the maximum massflow of the phase separator is limited by the

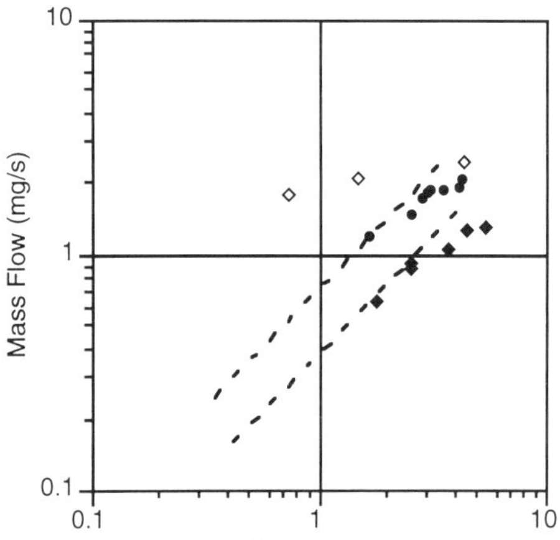

Figure 4a. Data of Murakami[6].

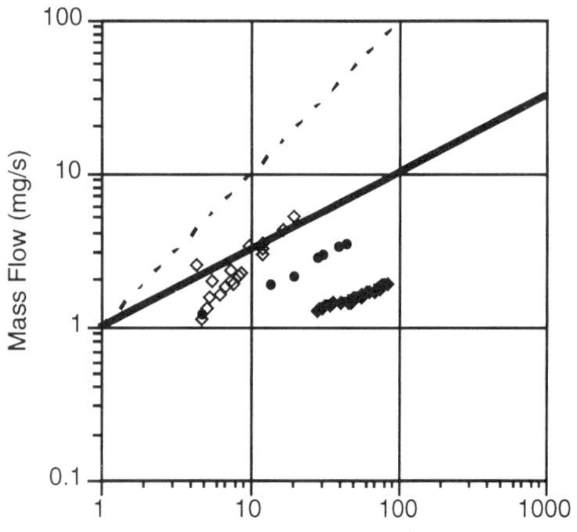

Figure 4b. Data of JPL (Ref. 4)

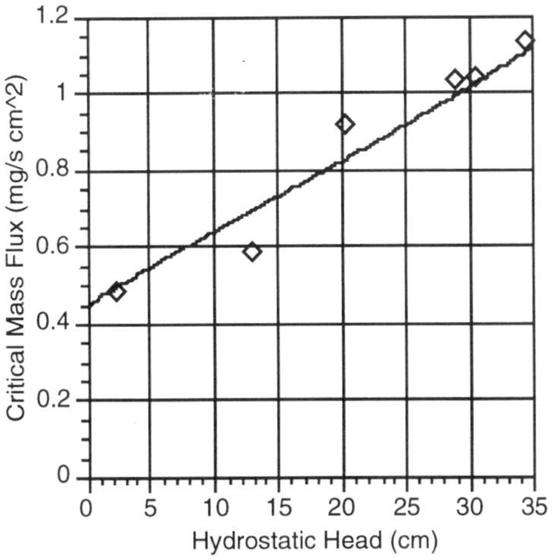

Figure 5a. Lockheed's data (present work)

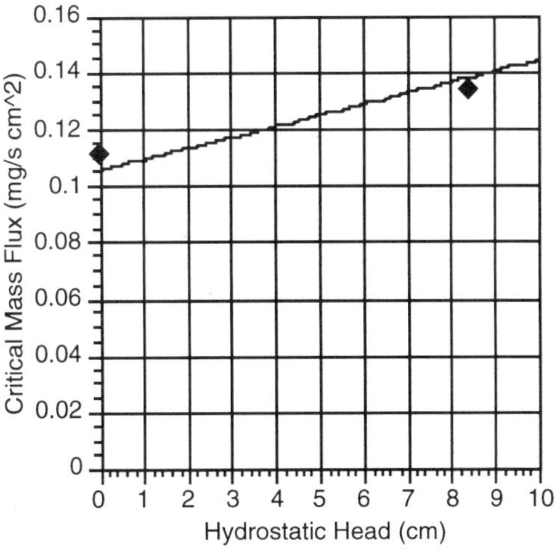

Figure 5b. JPL data (Ref. 4).

critical massflow or better known as the knee[9]. The following lessons are learned from the selection process. Permeability alone is not sufficient in characterizing the critical mass flow (unless the plug has a uniform distribution of pore size). Let's take a look at Figure 1. All three plugs depicted in the figures (a, b, and c) have the same permeability but vastly different distribution of pore size. The plug in Figure 1a has uniform pore size. The transport could go from laminar to turbulent before choking. Also there is little or no hysteresis. This is supported by the data of DiPirro[9] with a glass capillary plug which is supposed to have a uniform distribution of pore size. The plugs in Figures 1b and 1c have progressively larger distribution of pore size. The larger the distribution, the larger the hysteresis loop and the smaller the critical massflow. (The mechanism for hysteresis is very complicated. It is believed to be caused by the varying pore size along the length of the plug which results in the different surface tension during choke and unchoke. Although interconnected pores and superfluid film might also play an important role.) Since the choked flow occurs at much lower flowrates for large distribution of pore size, the transport might choke before the flow turns turbulent or even reaching the transition regime.

The critical mass flux (at the knee) of the phase separator was found to be a strong function of the hydrostatic head. Figure 5a is a plot of the critical mass flux of one of the materials tested for the relativity mission. . This material has a permeability of 3.8 x 10^{-10} cm^2. This is obtained by equating Equation 1 and 2 and solving for the critical mass flow as a function of the hydrostatic head. An equivalent capillary diameter D_c= 3.5 microns is used to correlate the data. Since the plug tends to choke at the larger pores, this value of D_c obtained experimentally gives us a measure of the large pores. On the other hand, the average pore size can be deduced from the analog between the Hagen-Poiseuille equation and the Darcy Law

$$D_k = (32K_p)^{1/2} \tag{5}$$

The present material has an average diameter (D_k) of 1.1 microns. From the difference between D_c and D_k, one can get a feel of the distribution of the pore size. The larger the difference, the larger the distribution and vice versa. Figure 5b shows that a D_c of 3.1 microns is needed to correlate the JPL data, indicatng that the JPL plug (D_k=1.3 microns) has a smaller pore size distribution than the plug tested for the Relativity Mission. Since the bubble point test gives the diameter of the largest pores in a plug, this might be a good way in measuring D_c in Equation 2. While the permeability or D_k is important for the prediction of the transport in the normal operating regime (e.g., Equation 1 or 3), D_c is valuable for calculating the critical mass flow (at the knee) and the size of the hysteresis loop. Note that D_c and D_k could be obtained from the size of the hysteresis loop; D_c on the increasing flow part and D_k on the decreasing flow part.

CONCLUSIONS

The following conclusions can be drawn.

1) Breakthrough Regime:

a) Plugs with large permeability are more susceptible to breakthrough.
b) A thick film is present at the downstream of the phase separator which might set off the liquid level detector to indicate the presence of liquid helium.

2) Phase Separation Regime:

a) The operation in this regime is not always linear
b) The transport in this regime sometimes changes from laminar to turbulent before choking.
c) The Ergun type turbulent transport was detected in some phase separators.

3) Choked Flow Regime:

a) Hysteresis results from a distribution of pore size in the phase separator. Large distribution gives small critical mass flow (at the knee) and large hysteresis loops and vice versa.
b) Like breakthrough, this regime is a strong function of the permeability. The equivalent capillary diameter D_c in Eq. 2 gives a quantitative measure of the large pores in the plug. Because these are the pores that choke first. On the other hand, the $D_k=(32K_p)^{1/2}$ gives the average pore size. If the difference between D_c and D_k is small, the distribution of pore size is small also (Fig. 1a); if the difference between D_c and D_k is large, the distribution of pore size is also large (Fig. 1c).
c) In this regime, permeability alone is not sufficient in predicting the critical mass flow (or the knee). Since the bubbles usually form at the large pores of the plug during a bubble point test. This might be a good way of estimating D_c. A correlation between D_c and the bubble point pressure should be established to characterize plugs used for phase separation.

ACKNOWLEDGMENT

This work was performed under contract to Stanford University, Contract PR4660.

REFERENCES

1) D. Petrac and P.V. Mason, Adv. Cryo. Eng., 29: (1983).
2) S.M. Volz, M.J. DiPirro, S.H. Castles, M.G. Ryschkewitsch, and R. Hopkins, Adv. Cryo. Eng., 37b:1183 (1992).
3) J.G. Tuttle, M.J. DiPirro and P.J. Shirron, Adv. Cryo. Eng., 39:121 (1994).
4) D. Elliott, JPL report, JPL D-11412 (1994).
5) S.W.K. Yuan, Ph.D. Thesis, University of California, in Los Angeles (1985)
6) M. Murakami et al., Proc. Nineth ICEC, Butterworth, UK, 190 (1982).
7) M. Murakami et al., The (Japanese) Institute of Space and Astronautical Science Report No. 612 (1984)
8) J.B. Hendricks and G.R. Karr, Proc. of the Nineth ICEC, Butterworth, UK, 197 (1982).
9) M.J. DiPirro and J. Zahniser, Adv. Cryo. Eng., 35:173 (1990).
10) C.R. Lages, R.H. Torii, and D.B. Debra, Cryogenics 35:33 (1995).
11) D.J. Frank and S.W.K. Yuan, to be presented at the 1995 Cryogenic Engineering Conference, Columbus, Ohio.
12) M.J. DiPirro, private communication.

EXPERIMENTAL RESULTS OF TESTS PERFORMED ON SUPERFLUID HELIUM PHASE SEPARATORS FOR THE RELATIVITY MISSION

D. J. Frank, S. W. K. Yuan

Research and Development Laboratory
Lockheed Martin Missiles and Space Co.
Palo Alto, Ca. 94304, U.S.A.

ABSTRACT

Tests were performed to select a Phase Separator (porous plug) material to be used with the NASA/Stanford University Relativity Mission. The porous plug is used to retain superfluid helium (SFHe) in the 2400 liter tank of the Dewar. Operation of this plug is unique in that it is used to control SFHe temperature, operate over a wide range of flowrate, and provide gas for spacecraft attitude control.

This paper summarizes data taken during evaluation of various materials made of sintered stainless. Data was taken during normal and choked flow conditions. Tests were performed on increasing the flowrate capability by using a heater on the downstream side of the plug. Critical breakthrough parameters were determined. These tests led to the selection of a material for use in the Dewar.

INTRODUCTION

The primary requirements that affected the design of the porous plug were: 1) the SFHe tank temperature has to be controlled to +/- 0.010K, 2) operate over a range of 4-16 mg/s without choking, and 3) operate with a sufficient back pressure so that the vented gas can be used for the spacecraft attitude control system. The predicted nominal flowrate during science phase of the mission is 6.5 mg/s. The higher flowrates are required during the early phase for orbit trim.

These requirements make it unique in that the downstream flow conductance of the plumbing is not constant such as other space borne Dewars. The Relativity Mission utilizes a heater in the tank and a variable flow conductance valve in the vent line to provide the temperature control and flowrates required for attitude control. Increases in venting are achieved by decreasing the downstream temperature of the porous plug and maintaining the bath temperature by turning on a heater. The vent line provides flow to 16 thrusters. The SFHe is to operate at 1.8 K.

A number of different materials were tested. Most of the data presented in this paper reflect the material that was selected. The intention of this paper is not to deal with the theory of phase separators but to present some interesting data that is available for the theorists to use in evaluating modeling techniques. A good review of the theory can be found in references 1-3.

Table 1. Characteristics of porous plugs

Plug #	Thickness (cm)	Dia. (cm)	Bubble Point (cm Hg)	Permeability (cm2)
1	1.27	3.3	7.8	3.55×10^{-9}
2	1.27	3.3	7.9	1.7×10^{-9}
3	0.635	4.06	10.9	3.8×10^{-10}
4	0.635	6.86	19.3	3.8×10^{-10}

MATERIAL CHARACTERISTICS

The materials tested were all porous 316L sintered stainless steel porous media purchased from Mott Metallurgical Corp. The first two plugs had a thickness of 1.27 cm. This thickness was picked due to early concerns of breakthrough while operating at the lower end of a wide range of flowrates. Plug #1 was fabricated in disc form. After preliminary results showed that this plug was very susceptible to breakthrough, a second plug, Plug #2, was assembled with some available 0.318 cm sheet material. Discs were fabricated from the sheet and four discs were stacked to get the 1.27 cm thickness. Results of this plug showed that breakthrough was no longer a concern and subsequent plug thicknesses were reduced to 0.635 cm and fabricated in disc form. The first two materials were purchased by specifying a filtration grade of 0.5 micron. Mott's manufacturing criteria to determine if a material has been fabricated to the correct micron grade is to perform a bubble point test.

Since the two first materials had large difference in permeability for the same filtration grade, subsequent materials were purchased by specifying a permeability. Microscpopic examination showed much different surface characterisitcs between the two materials which are probably due to fabrication differences between discs and sheets. Material for Plug #3 was ordered with the specification of having a permeability comparable to plug #2. The material received had a permeability much lower than that specified and therefore, the diameter was too small. Tests proceeded with this material and once it was well characterized, new material for the final Plug #4 was ordered with the same permeability as #3. Table 1 is a summary of the characteristics of these materials.

Figure 1 is a summary of a number of Mott stainless porous materials that were tested at Lockheed for permeability and bubble point with 2-propanol. It can be seen that materials with comparable permeability can have much different bubble point pressures. Based on Mott's criteria, the material of plug #3 is 0.3 micron and #4 is 0.1 micron. The higher bubble point of #4 would indicate a lower distribution of pore sizes then #3 which had the same permeability.

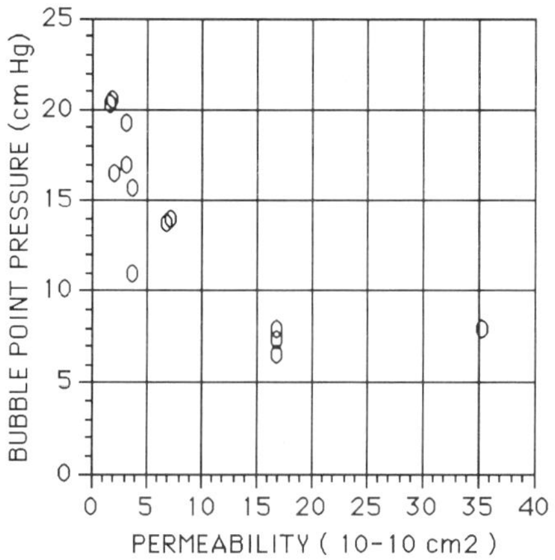

Figure 1. Permeability vs bubble point of sintered stainless material .

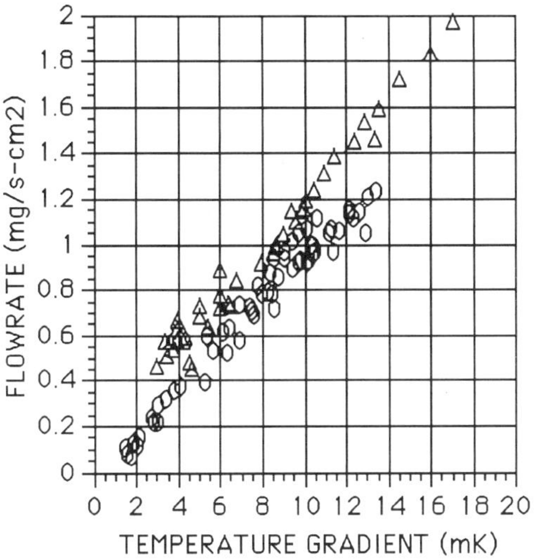

Figure 2. Effect of permeability on the normal operating characteristics at 1.8 K (Δ - K_p=1.7 x 10^{-9} cm^2, 0 - K_p=3.8 x 10^{-10} cm^2).

TEST APPARATUS AND POROUS PLUG DESIGN

In the test apparatus, the porous plug assembly is installed at the bottom of a helium tank such that the inlet of the plug is submerged until the liquid level drops to the inlet of the plug. The downstream side of the plug is facing down. A 100 torr baratron type warm pressure transducer is installed in the warm section of the plumbing. No attempt was made to measure the pressure just downstream of the plug. The transducer was put in the warm section to simulate the location in the actual Dewar vent plumbing. The pressure reading taken at this point is referred to as the "back pressure". The tank contains a superconducting liquid level sensor, to measure the amount of liquid remaining in the tank, and a heater to vary the heat load. During most of the tests, the vent line of the porous plug was hooked up to a spacecraft thruster whose flow capacity was enlarged in order to handle the full flow of all the thrusters. A pump was installed on the outlet of the thruster.

The plug assembly consisted of the porous media with two thermometers on both the upstream and downstream side, a liquid point sensor (LPS) and a 0.28 watt heater on the downstream side. The LPS generates a substantial amount of heat (0.018 W) when activated and therefore was only pulsed on when a set of data points was taken.

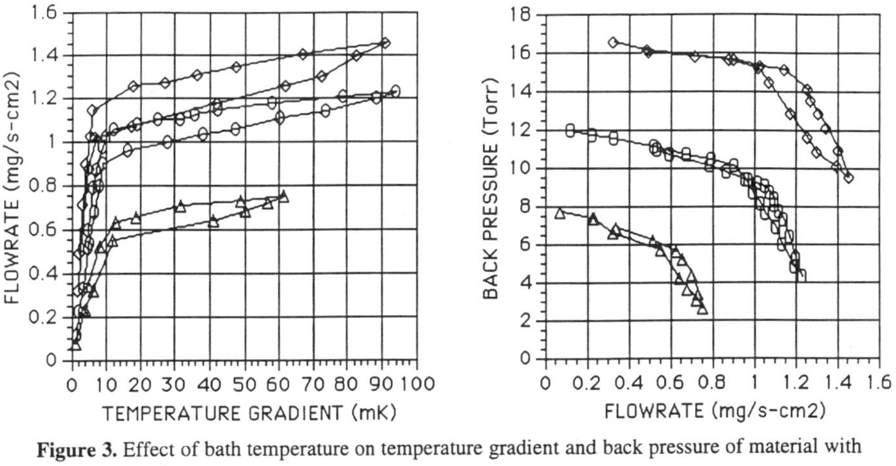

Figure 3. Effect of bath temperature on temperature gradient and back pressure of material with K_p=3.8 x 10^{-10} cm^2 (Δ 1.7K, O 1.8K, ∇ 1.9K).

Thermometers are GR-200B-1500 germanium resistance (GRT) sensors by Lake Shore Cryotronics calibrated by the vendor. The lead wires of the GRTs are epoxied to the surface of the media in order to get a good reading of the surface temperature. In addition to the LPS on the downstream side of the plug, one is installed in the plumbing line a few inches downstream. These point sensors were purchased from American Magnetics. The heater heats a coarse mesh copper screen that covers the downstream side of the media.

NORMAL OPERATION

During normal operation of the plug, the liquid/vapor interface resides on the downstream surface. Figure 2 shows the data taken for plugs #2 and 3. There were thermometry problems with plug #1 and therefore the temperature gradients are not reported. It is reminded that plug #2 was twice as thick as #3. The interesting result is that the slope of the data does not appear to be directly proportional to permeability as some of the theories suggest[4].

CHOKING CHARACTERISTICS

The performance of plugs #3 and #4 are very typical. As the flowrate is increased by decreasing the back pressure, the temperature gradient increases until the plug achieves a choked condition. The flowrate at this point is referred as the critical flowrate. Above this point, as the flowrate is increased and subsequently decreased, the temperature gradient would go through a hysterisis loop. Figure 3 shows typical temperature gradient and back pressure profiles for various bath temperatures. As expected, the slope of flowrate verses temperature gradient in the normal mode and the critical flowrate increase with bath temperature

The most interesting observations are the difference in gradients across the plug before and after going through a hysterisis loop and readings of the LPS on the downstream surface. Upon returning to normal mode, while decreasing the flowrate, the gradients would be larger than those observed while increasing the flowrate at beginning of the loop. This is indicated by larger temperature gradients and lower back pressures.

While operating in the normal mode, the LPS would indicate liquid. This confirms that the liquid/vapor interface resides on the surface of the plug. The LPS is in contact with the copper mesh screen that covers the surface. As the plug was subjected to a hysterisis loop, this sensor would continue to indicate liquid until the plug went into the choked mode. The temperature gradients at the point when the LPS would begin to indicate vapor were 13.0, 31.5, and 27.4 mK for temperatures of 1.7, 1.8, and 1.9K respectively in the three cases shown. The hydrostatic heads were 21, 30, and 24 cm respectively. The sensor would

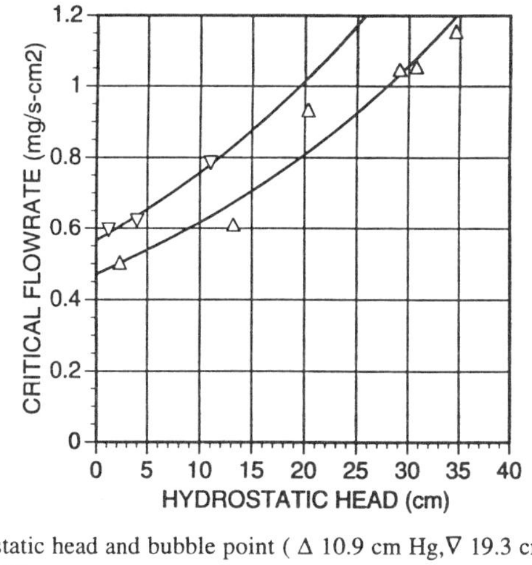

Figure 4. Effect of hydrostatic head and bubble point (Δ 10.9 cm Hg,∇ 19.3 cm Hg) on critical flowrate of material with K_p=3.8 x 10^{-10} cm^2 operating at 1.8K.

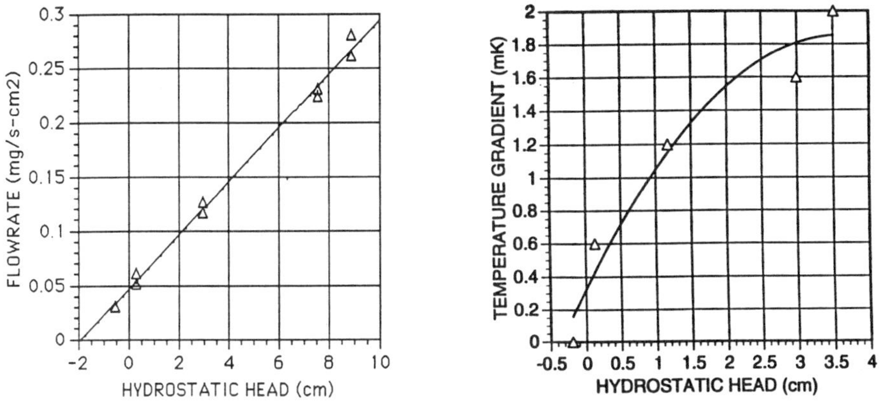

Figure 5. Effect of hydrostatic head on breakthrough of material with K_p=3.8 x 10^{-10} cm^2 operating at 1.8K.

indicate vapor until the plug returned to the normal mode and then sense liquid. The LPS in the plumbing would always indicate vapor. In Plug #4, the LPS was set slightly off the copper mesh and it did not indicate liquid in the normal mode.

Figure 4 shows the effect of hydrostatic head on the critical flowrate of plugs #3 and #4. The interesting observation is the difference which is most likely due to the different bubble point pressures of the two plugs. Based on the results of this test, the 1.8 K critical flowrate at zero hydrostatic head for the material of plug #3 is 0.47 mg/s-cm^2 and 0.57 mg/s-cm^2 for #4.

The behavior of plug #2 in the choked mode did not follow a typical hysterisis loop. While the plug was choked, the temperature gradient would increase, then suddenly decrease and then further increase as the flowrate was increasing. It is speculated that this behavior was due to the fact that the plug was made of four discs stacked on each other and the location of the liquid/vapor interface was effected by this. This behavior was very repeatable. Even though it is intended to only operate in a normal mode, it was decided not to build a unit with a stack of discs because of this behavior while choked.

BREAKTHROUGH CHARACTERISTICS

Plug #1 was very susceptible to liquid breakthrough. This was evident in runs where temperature control of the bath was progressing smoothly for a few hours when suddenly breakthrough would occur. During these operations, the back pressure was staying constant within a few hundredths of a torr. The next plugs were not susceptible to breakthrough.

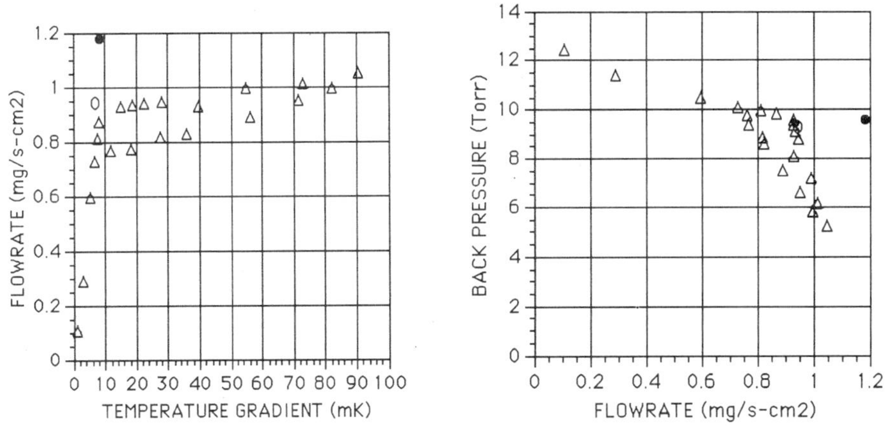

Figure 6. Operation of downstream heater to increase flowrate above critical point (0 start point-heater off, ■ end point-heater on).

A number of tests were performed with Plugs #3 and #4 to characterize breakthrough. The test sequence was such that the flowrate was decreased in small increments and the LPS in the vent line was monitored. As soon as it indicated liquid, the flowrate would be increased, setting the plug in a choked condition and heater in tank turned up to maintain bath temperature. Once the LPS showed vapor, the plug would be unchoked and the test repeated at a lower hydrostatic head. The results are shown in figure 5 which indicate that operation below a flowrate of 0.05 mg/s-cm^2 would cause breakthrough. It is possible that indications of liquid by the LPS was not actual breakthrough but rather a build-up of a large film of SFHe. In either case this would not be desirable due to the possibility of the film conducting heat back to the tank. The hydrostatic head is measured from the inlet surface of the porous media. Both plugs #3 and #4 showed the same characteristics which would indicate that the difference in bubble point does not effect breakthrough.

DOWNSTREAM HEATER TESTS

Tests were performed with the heater on the downstream side of the porous media to determine if it can be used to achieve a flowrate above the critical flowrate without choking. Results of this test are shown in figure 6. The plug was first put through a hysterisis loop to show the critical point. The flowrate was then brought up the critical point at a flowrate of 0.94 mg/s-cm^2 with a temperature gradient of 7.2 mK (0 point in figure). The heater was then turned on and the flowrate increased to 1.18 mg/s-cm^2 while plug temperature gradient only increased to 8.6 mK (■ point-in figure). The plug operated at the higher flowrate (25 % above the critical point) while maintaining a normal mode of operation. The hydrostatic head on the plug during this test was 19 cm. All of the heater power went into vaporizing liquid.

SUMMARY

A porous material for phase separation of SFHe has been succesfully tested. Based on results of these tests, the flight plug (#4) was selected to be 0.635 cm thick with a diameter of 6.86 cm. The flight plug for the Relativity Mission has been assembled, gone through vibration testing, and performance testing. Based on the measured breakthrough point of 0.05 mg/s-cm^2 and a critical point of 0.57 mg/s-cm^2 at zero hydrostatic head, the flight plug operates in a normal mode over the range of 1.8 to 21 mg/s, meeting the requirements of 4-16 mg/s.

A downstream heater is available for increasing the flow beyond it's predicted critical point without choking, but is not planned to be used at this time. Data is presented showing the effect of hydrostatic head on the breakthrough and critical point of the material. In addition results show that the bubble point of the material appears to have an effect on the critical point.

ACKNOWLEDGMENT

This work was performed under contract to Stanford University, Contract PR4660. Authors would like to acknowledge support received by J. Vanden Beukel (Lockheed Martin spacecraft) and C. Lages (Stanford University) in testing and discussions of the results. In addition, we thank B. Imai who assembled the porous plug assemblies and testing apparatus.

REFERENCES

1. M.J. DiPirro and J. Zahniser, The Liquid/Vapor Phase Boundary in a Porous Plug, Adv. Cryo. Eng., Vol. 35 (1990).
2. S. W. K. Yuan, The Non-Newtonian Heat and Mass Transport of He-II in Porous Medium Used for Vapor-Liquid Phase Seperation, Phd dissertation (1985).

3. D. G. Elliot, Porous Plug Tests and Flow Modeling for the Space Infrared Telescope Facility (SIRTF), JPL D-11412 (1994).
4. S. W. K. Yuan, D. J. Frank, and Chris Lages, Size Dependence of Choked Flow and Breakthrough in Vapor-Liquid Phase Separation of HeII Using Porous Media, presented at the 1995 Cryogenic Engineering Conference, Columbus, Ohio.

1995 CEC/ICMC Conference: Jim Peeples (right) presenting the 1995 Russell B. Scott Memorial Award for the best Cryogenic Engineering Research paper to Karl Gschneidner (left)

PASSIVE CRYOPUMP FOR RELATIVITY MISSION SATELLITE

F. C. Chilese

Lockheed Martin Missiles and Space
Orgn. 92-10, Bldg. 253
3251 Hanover Street
Palo Alto, California, 94583, USA

ABSTRACT

The Relativity Mission (GP-B) requires the earth-orbiting science mission instrument to operate at < 2.5 K at a pressure of < 1.3×10^{-9} Pa (10^{-11} torr). The pressure at the orbital altitude is approximately 10^{-7} Pa (10^{-9} torr); thus the pressure within the experiment must be significantly lower than the space environment. In addition, the maximum disturbances allowed preclude the use of any active components. The ultra-high vacuum is achieved using a passive adsorption device, called a cryopump, composed of discrete porous titanium plates welded to form a structural unit. The porous plates provide a large surface area while meeting the requirements for low magnetic susceptibility and level-100 cleanliness. The cryopump functions by being heated to 8 K to drive off all adsorbed helium while the instrument is open to space. When the instrument is closed, the cryopump cools to 2 K. The remaining gas within the instrument will be adsorbed by the cryopump, producing the desired vacuum. This paper describes a sub-scale test performed to validate performance of the device. A comparison is made between the theoretical surface area of the plates (as determined by BET analysis) and the actual performance-derived effective surface area. The full-scale cryopump device is shown and details of fabrication are discussed.

INTRODUCTION

The Relativity Mission is an earth-orbiting experiment that tests the extent of the warping of space-time caused by the earth, and the dragging of space-time caused by the earth's spin. The experiment uses four (4) gyroscopes whose spin is aligned with a guide star whose proper motion is known. During the experiment (approximately 16.5 months on orbit), the axis of spin of those gyroscopes tilts 6.6 arcseconds/year due to local warping of space-time, and 0.042 arcseconds/year in another direction due to the dragging of time space. To observe changes in tilt of this magnitude, non-relativistic effects such as drag and thermal vibration must be reduced to very small magnitudes.

One of the requirements of the environment within which these gyroscopes lie is that the pressure must be less than 1.3×10^{-9} Pa (10^{-11} torr). The pressure at the orbital altitude is approximately 10^{-7} Pa (10^{-9} torr); thus the pressure within the experiment must be significantly lower than the space environment. In addition, the very low maximum allowed disturbances preclude the use of any active components.

The ultra-high vacuum is achieved using a passive adsorption device, called a cryopump, composed of discrete porous titanium plates welded to form a structural unit. The porous plates provide a large surface area while meeting the requirements for low magnetic remanence and level-100 cleanliness.

FLIGHT CONFIGURATION / OPERATION

Flight Experiment Details

The flight instrument is contained in a probe approximately 2 meters (6 ft) long, and 0.25 meters (10 inches) in diameter. A sketch of the probe is shown in Figure 1. The entire probe is evacuated, and is housed within a dewar filled with superfluid helium at 1.8 K. Once on orbit, two 15 cm (6 inch) valves are opened to space. Then the gyroscopes are spun up to their operating rate using gaseous helium. Most of the helium that was used to spin-up the gyros vents to space, and the interior of the probe equilibrates to 10^{-8} Pa (10^{-9} torr), the pressure of space at the probe's orbital altitude of 650 km.

Heaters on the cryopump (and several other metallic components that are normally colder than 2 K) are turned on to raise the temperature of these items to approximately 8 K. This heating drives off nearly all the adsorbed helium from these surfaces. The 15 cm valves are then closed, and all heaters are turned off. The heated items, including the cryopump, return to their normal operating temperature of less than 2 K.

The remaining free helium within the probe adsorbs on the surface of the cryopump, lowering the pressure within the probe. Given some time and the large surface area of the cryopump, a very low pressure may be obtained. The pressure within the probe must be less than 1.3×10^{-9} Pa (10^{-11} torr) for the 16.5 month duration of the flight.

Flight Cryopump Details

The surface area of the probe that is normally colder than 2 K during flight is 6.6 m^2 not including the surface area of the cryopump. Only a small amount of this area is heatable to provide a cryopumping effect. To provide a "large" amount of surface area, the cryopump was sized so that the amount of heatable surface area that is normally colder than 2 K is > 90% of the total surface area that is normally colder than 2 K. This 90% design requirement translated to providing a cryopump with a surface area greater than 44 m^2.

A porous material was selected to provide a large amount of surface area for a given volume and weight of the cryopump. Charcoal, often used for its high specific surface area,

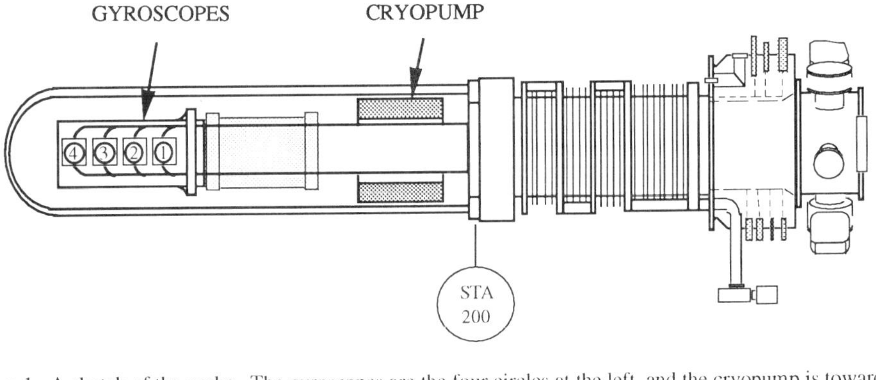

Figure 1. A sketch of the probe. The gyroscopes are the four circles at the left, and the cryopump is toward the center. The valve that opens to space is not shown, but would appear on the right side of the sketch. Everything to the left of "station 200," which is labeled "STA 200" is maintained below 2 K.

was not a viable candidate due to particulation. Any free-floating particles within the probe could cause a gyro failure, and thus failure of the entire mission.

The gyroscopes require a very low magnetic field environment; so low that a significant amount of the Relativity Mission has been devoted to finding and testing ultra-low magnetic materials. One of the few metallic materials that is allowed near the gyroscopes is pure titanium. A vendor who could create and shape porous pure titanium was found. Thin plates of this porous material (pore size of 5 μm) were chosen to minimize the distance the free helium would need to penetrate to find clear surface area. A plate thickness of 0.8 mm (0.032 inches) was chosen so that each plate would retain structural integrity.

The plates were placed side-by-side with a separation distance of 0.5 mm (0.020 inches), and welded to titanium rings for structural rigidity. The allowable volume for the cryopump within the probe determined the maximum size, and the cryopump was designed to fill that volume. The cryopump is shaped as an annulus formed by 360 porous plates and an outer diameter of 210 mm (8.4 inches), an inner diameter of 175 mm (7.0 inches) and a length of 150 mm (6.0 inches). A photo of the cryopump is shown in Figure 2.

The cryopump is constructed using electron-beam welds, with 12 individual welds per porous plate, over 4300 welds in total.

Cryopump Helium Adsorption Capability

A BET analysis of samples of the porous plates found a specific surface area of 0.033 m^2 per gram of plate (average of three tests of different porous plates). There are 1940 grams of porous plates in the cryopump, which equates to a surface area of 64 m^2.

The cryopump operates at a nominal temperature of 2.1 K. At that temperature, the surface should be capable of adsorbing one monolayer of helium before becoming saturated. One monolayer translates to a surface density of helium of 0.37 standard cm^3 per m^2. Multiplying this by the total surface area of the cryopump yields a theoretical adsorption capability of the cryopump of 24 standard cm^3 of helium.

After gyro spinup and the closing of the two 15 cm valves, the interior of the probe has a pressure of approximately 10^{-7} Pa (10^{-9} torr). The temperature of the probe varies linearly from 220 K near the valves (on the right in Figure 1) down to 2 K at station 200. Everything to the left (as shown in Figure 1) of station 200 is at 2 K.

Virtually all of this pressure should be caused by helium, because the probe is very well evacuated on earth before launch and only helium is used in the gyro spinup process. The total volume of the helium in the probe after the two 15 cm valves close should be less than 0.0001 standard cm^3.

The pressure within the probe should be proportional to the volume (in standard cm^3) of the free-floating helium within the probe. Given that the cryopump theoretically can adsorb far more helium than is in the probe, the pressure in the probe should drop to an immeasurably small value while the cryopump remains far from saturated.

This drop in pressure supposes that all the surface area of the porous titanium will be used equally, each part of the plate adsorbing helium to the saturation point. It was unknown how well the helium would penetrate through the thickness of the porous plates and if the area in the center of each plate would be useful. A test was required to determine the effective surface area of the cryopump. In addition, the cryopump will need to adsorb helium as it leaks into the probe both before and after launch.

TEST CONFIGURATION

A small test probe and dewar were used to simulate the flight probe and dewar. The surface area of the test probe which would be colder than 2 K was calculated to be 0.18 m^2, compared to surface area of the actual probe colder than 2 K of 6.6 m^2. The test probe's cold surface area was 1/36th that of the actual probe, so a cryopump 1/36th the size of the actual cryopump was needed. The test cryopump was constructed by electron-beam welding 10 porous titanium plates to two titanium plates. The porous plates were spares remaining after building the actual cryopump, and the spacing between the plates was the same as the average spacing of the actual cryopump. A photo of the test cryopump is shown in Figure 3.

Figure 2. A photograph of the flight cryopump. There are 360 porous titanium plates welded to titanium rings. A segmented heater is bonded to the center ring to elevate the temperature from 2.1 K to 8 K during operation. Three flexures made of beryllium copper mount to the threaded standoffs on the center ring to provide a thermal standoff so that the cryopump may be heated independently from the rest of the probe.

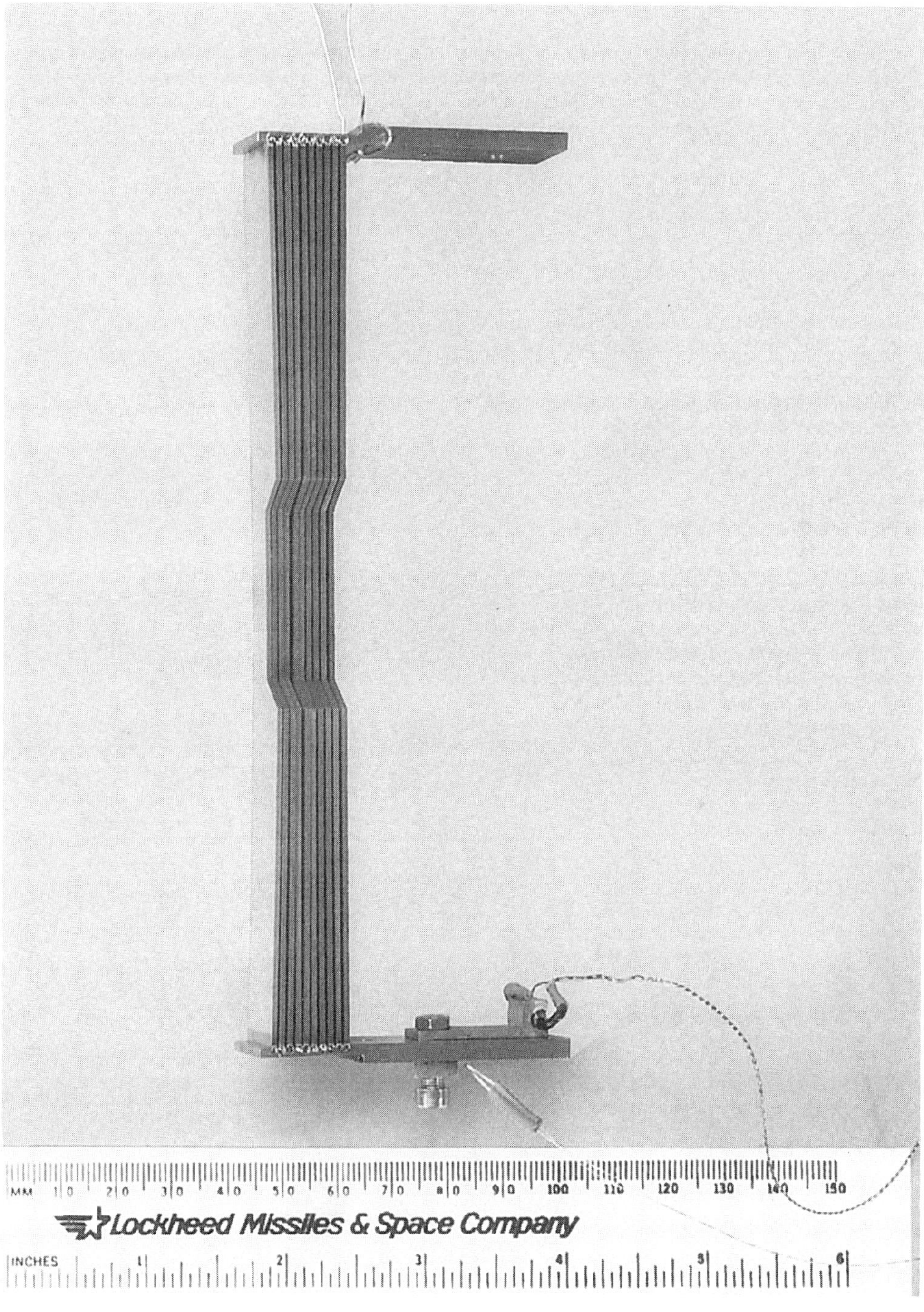

Figure 3. A photograph of the test cryopump. There are 10 porous titanium plates welded to titanium endplates. A resistive heater is screwed to the bottom plate to elevate the temperature from 2 K to 8 K during operation. The cryopump was suspended by the copper wires attached to the top plate.

TEST PROCEDURE

Pressure reduction

The test cryopump was outfitted with a heater and a thermal sensor. It was suspended by a wire whose thermal conduction represented that of the beryllium copper flexures. The probe was inserted into the dewar, and the dewar was filled with liquid helium that was pumped down until a temperature of 1.8 K was achieved. The probe was evacuated, baked out, filled with helium, and then evacuated again.

A vacuum pump was allowed to run for several days to evacuate the probe, and a pressure of 10^{-7} Pa (10^{-9} torr) was achieved. This represented the condition within the probe after the gyros are spun up and the 15 cm valves are open to space.

Next the heater on the cryopump raised its temperature from 2 K to 8 K, driving off any adsorbed helium. Then the valve to the pump was closed and the heater was turned off, representing the closing of the two 15 cm valves and the cooling of the cryopump. The pressure was monitored throughout this process with a residual gas analyzer with an electron multiplier to measure lower pressures. Because the pressure was monitored at the warm end of the probe but it is the pressure near the cryopump that is of interest, the pressure readings were corrected for temperature (i.e., divided by the square root of 300 K / 2 K). The compensated data are reported herein.

The graph of the pressure as a function of time is shown in Figure 4. That graph shows the spike in pressure as the cryopump drives off the adsorbed helium (which is drawn off by the vacuum pump), and the rapid decrease in pressure once the cryopump begins adsorbing helium in the sealed probe.

Total helium adsorption

The test was repeated with a change to determine the adsorption capability of the cryopump. Once the cryopump had been heated to drive off the adsorbed helium, measured amounts of helium were introduced at the top of the test probe. The pressure inside the probe was measured after each introduction of helium. It was expected that the equilibrium pressure

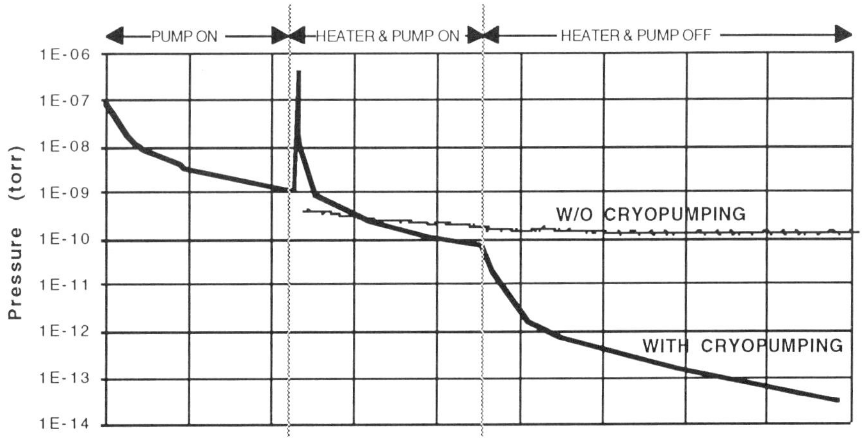

Figure 4. A graph of the results of the first test. Pressure is given in torr and time is given in hours. From 0 to 24 hours, the system is being evacuated by a vacuum pump. At 24 hours, the heater on the cryopump is turned on. At 48 hours, the heater is turned off, the vacuum pump is turned off, and the probe is sealed with a gate valve. From 48 hours to the end of the test, the cryopump is adsorbing helium, lowering the internal pressure of the probe. Data from a test where the cryopump heater was never turned on (labeled "w/o cryopumping") are included to show baseline system performance.

within the probe would stabilize at a higher and higher pressures after each helium introduction. When the equilibrium pressure stabilized above 10^{-11} torr, the cryopump was saturated to the point where it was no longer able to meet the pressure requirement. The total helium adsorbed at that point could then be compared to the theoretical saturation helium volume to determine an "effective" surface area of the cryopump.

Given the specific surface area of 0.033 m^2 per gram of plate and the 54 grams of porous plate used in the test cryopump, there was a total of 1.8 m^2 of theoretical surface area to be used. One monolayer translates to a surface density of helium of 0.37 standard cm^3 per m^2. Multiplying this by the total theoretical surface area of the test cryopump yields a theoretical adsorption capability of the cryopump of 0.67 standard cm^3 of helium.

Helium was added in increments of less than 0.04 standard cm^3, about 5% of the theoretical maximum helium adsorption. After each helium introduction, the helium pressure within the probe was monitored. If it became apparent that the temperature-corrected pressure within the probe would fall below 10^{-11} torr (indicating that the cryopump was still capable of meeting its requirements), another increment of helium was introduced. After a total of 0.32 standard cm^3 had been added, the pressure within the probe did not fall below 10^{-11} torr, indicating that the cryopump was saturated to the point where it could no longer meet its requirements.

TEST RESULTS

The first test, where the pressure was allowed to drop for four (4) days, showed that the cryopump performed as it was intended to perform. The pressure at the cryopump dropped from 10^{-10} torr (the pressure the system could achieve when it was used with no cryopump) to below 10^{-13} torr. When the data taking was concluded, the pressure was still falling at a significant rate. This test shows that the flight cryopump is more than capable of providing the pressure environment required in the flight probe.

The second test, where metered amounts of helium were introduced, showed that the cryopump provides an effective surface area of about 50% of the theoretical maximum, assuming one monolayer of helium yields saturation. Scaled to the full cryopump, this test indicates that the flight cryopump is capable of adsorbing 12 standard cm^3 of helium.

The specified maximum allowable leak rate into the probe is 10^{-6} atm cc/sec of helium, and although there is no equivalent specification for outgassing, the probe is designed to incorporate very little material which can outgas. A test of the prototype probe yielded a combined outgas plus leakage rate of about 10^{-8} atm cc/sec of helium. On orbit, there will be virtually no pressure difference to drive a leak, so all of the helium intrusion should be from outgassing. A conservative assumption is to say all of the measured leakage into the prototype probe was from outgassing. Dividing 12 standard cm^3 by 10^{-8} atm cc/sec yields a pressure lifetime of 40 years on-orbit. Given that the experiment is only sixteen months long, the cryopump provides a significant performance margin.

ACKNOWLEDGEMENT

This work was supported through Stanford University, through subcontract PR4660 and NASA contract NAS8-36125 with the NASA, Marshall Space Flight Center.

The author would like to acknowledge the support of the technicians at Stanford University, Chris Gray and Brian DiDonna, who performed the tests on the cryopump, and of Dr. Sasha Buchman, also of Stanford University, who provided extensive technical advice.

PRODUCTION OF CHARGED AND NEUTRAL LIQUID HELIUM DROPS

C. G. Paine, D. Petrac, and W. K. Rhim

California Institute of Technology
Jet Propulsion Laboratory
Pasadena, California, USA

ABSTRACT

We report on two means of producing drops of liquid helium for injection into an apparatus designed to hold individual drops isolated from material walls. Using a small thermomechanical pump operated in a pulsed mode, we have produced uncharged drops of the superfluid; charged drops of both the normal fluid and superfluid have been produced through electrohydrodynamic effects of a sharply pointed conductor within the bulk liquid, which caused charge-carrying drops to be ejected from the liquid surface.

INTRODUCTION

The capacity to confine and observe isolated drops of liquid for long periods of time would be of value to studies of the properties and behavior of fluids: in particular, behavior which is affected by the fluid surface condition may be best studied in free drops. We are interested in the nucleation of quantized vorticity in superfluid helium, which in contained samples is believed to be catalyzed by interactions at the container surface. Such interactions would be absent in an isolated drop; thus we are developing the capacity to generate individual drop of liquid helium, capture, levitate, and rotate a drop. Operation of the trap requires that the drop carry an electric charge; the drop may be charged immediately upon injection into the trapping region, or it may already carry charge prior to injection.

NETURAL DROP PRODUCTION: MINIATURE FOUNTAIN PUMP

Fabrication of miniature fountain pumps

We have exploited the close-packed grain structure of sintered alumina ceramic to function as a superleak in the fabrication of small fountain pumps for production of superfluid drops. Figure 1 is a schematic representation of such a miniature fountain pump. The body is sintered aluminum oxide [1], 0.4 cm O.D., 0.08 cm wall thickness, 1.5 cm in length, with average pore size approximately 0.2 μm. The lower end is closed with a short piece of alumina rod of 0.18 cm diameter. A 5 cm length of 0.008 cm diameter manganin wire, to function as a resistance heater of approximately 5 Ω resistance, enters and exits the interior volume through two 0.05 cm holes through the length of the lower end plug. Liquid exits the fountain pump through a quartz nozzle of 0.2 cm O.D., approximately 1.5 cm length, and 0.05 cm inner diameter, modified to have a smaller opening at the exit end. This nozzle orifice is fabricated according to a technique described by Switzer [2]. Fountain pumps with nozzle orifice diameters from 80 μm to 230 μm were fabricated and operated. Assembly of the fountain pump consisted of winding the manganin wire around the shank of the nozzle and securing it with a small amount of GLPT varnish, inserting the wire ends through the body and bottom plug, positioning all parts in the desired position and applying a rapid-curing epoxy to seal all access to the interior save that through the orifice; the epoxy also completely covered the ceramic end plug.

Operation of miniature fountain pumps

To evaluate the operating characteristics, the fountain pumps were located in a glass dewar, which provided optical access for a video camera with telescopic lens. Liquid helium at 1.5 K covered the fountain pumps to approximately the level of the vertically-directed exit nozzle. An electric current heated the manganin wire, depositing heat in the liquid.

The behavior of a miniature pump with 170 μm nozzle orifice will be described; behavior of all devices was qualitatively similar, with the major difference being the diameter of the exit stream, which in steady operation appears the same diameter as the nozzle orifice. All tests were performed with the surrounding bath held at 1.5 K.

Constant power in the heater of $2x10^{-4}$ W yielded an exit stream height of 5 cm.; stream height varied roughly linearly with heater power below this range. The stream did not appear to break up into drops until reaching apex.

Operating in pulsed mode, power was dissipated in the heater in a time of order milliseconds, generating flows lasting tens of milliseconds. All of the energy dissipated in the heater is transferred to the liquid in milliseconds or less. For the above device, an energy of $8x10^{-3}$ J dissipated in 0.01 s produced a liquid flow reaching a height of 0.8 cm , continuing for approximately 0.09 s; ejected volume was approximately $1.4x10^{-3}$ cm^3. Viewed with lighting from a stroboscope, the liquid jet is seen to break into drops soon after exiting the nozzle; the drops coalesce near the maximum height, then fall. Considerable deformation from spherical is obvious at all times. Figure 2 shows such a sequence.

The pump repeatedly produced jets of approximately equal volume when operated at pulse rates up to 3 Hz; at higher rates the output decreased. Increasing the energy deposited eventually resulted in boiling of the liquid in the pump. Deposition of less energy resulted in shorter flows reaching lesser height, as expected. There was observed to be a minimum liquid volume flow which will separate from the nozzle, corresponding to an energy deposition of about $3x10^{-3}$ J, which did not depend upon the the power. This limit is a function of the internal volume of the pump and the thermal efficiency of the device.

Simple analysis of fountain pump

Efficient operation of a fountain pump requires that the superleak material have pores small enough to effectively prevent the motion of the normal component while allowing nearly unrestricted flow of the superfluid component, and that the superleak material be a good thermal insulator, so that a temperature gradient can be maintained across it with minimum power.

A simple analysis of the steady-state behavior of the fountain pump proceeds as follows: We ignore all thermal conduction not occuring through the liquid, and we ignore viscous drag on liquid moving in the helium vapor. We then equate the energy density due to the pressure difference across the pump body with the potential energy density of the liquid at the height h to which the liquid rises upon exiting the nozzle, which is located at the level of the surrounding liquid bath. Employing energy conservation we then find

$$\Delta P = \rho \, \sigma \, \Delta T = \rho \, g \, h = 1/2 \, \rho \, v^2 , \qquad (1)$$

where the first two terms are the London equation for a fountain pump, the third term is the potential energy density of the liquid at the stream apex, the last term is the liquid kinetic energy density upon exiting the nozzle, ρ is liquid density, σ is specific entropy, and v is the initial velocity of liquid exiting the nozzle. Observing the height to which a liquid stream rises allows calculation of the temperature difference between the bath and pump, and of the

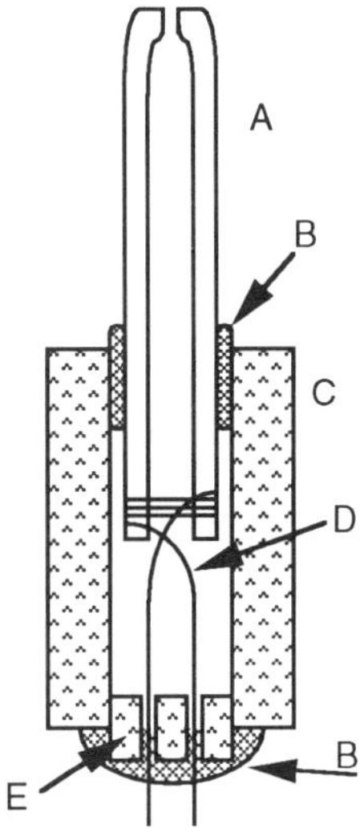

Figure 1
Cross-section of miniature fountain pump
A: quartz nozzle
B: epoxy seal
C: alumina body
D: heater wire
E: bottom plug.

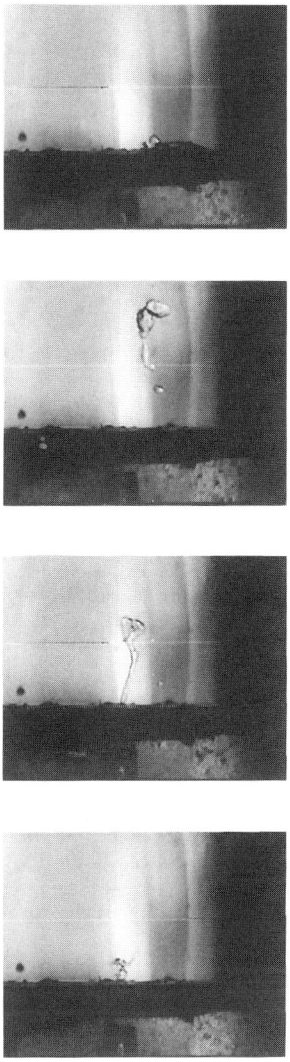

Figure 2
Pulsed operation of fountain pump
Image interval = 33 msec
Earliest image is at bottom
Pulse width = 20 msec, energy = 8 mJ.

mass flow rate through the pump. Knowledge of the liquid specific heat at the operating temperature then yields the power required to maintain that temperature difference as

$$W = C_{sat}\,\Delta T\;v\;A\,, \qquad (2)$$

where C_{sat} is the specific heat of the liquid and A is the area of the nozzle orifice.

For operation as a drop generator, the fountain pump is operated in a pulsed mode; a fixed amount of energy is deposited in the heater in a short time, and the resulting pressure rise expels liquid from the nozzle until the temperature returns to the initial value. This operation is described by

$$Q = C_{sat}\,\Delta T\;V\,, \qquad (3)$$

where Q is the energy deposited in the heater and V is the volume of liquid heated, which is the internal volume of the pump available to the liquid. In a perfect fountain pump V is also the amount of liquid expelled from the nozzle.

Comparison of the observed device behavior with the predictions of this simple model indicate that the fountain pumps are rather inefficient transducers. For the steady-state operation described above, Equation. (3) predicts that this device would require $8.2x10^{-6}$ W to produce the same behavior, rather than the observed $2x10^{-4}$ W. Similarly, the volume of fluid ejected in pulsed operation is 0.02 of the volume predicted by Equation. (4). The disparity between model and actual device is easily understood but harder to remedy; clearly the model is simplistic in its treatment of thermal transport. The amount of normal flow through the pump body is difficult to quantify; normal flow would significantly degrade the thermal characteristics of the devices.

CHARGED DROP PRODUCTION: ELECTROHYDRODYNAMIC EFFECTS

It is well known that the interaction of an electric field with free charges in an insulating fluid medium can produce motion of the fluid, as the charges are accelerated by the field and impart momentum to the fluid via mutual friction: this phenomenon is known as electrohydrodynamic (EHD) flow [3]. For the effect to occur in an insulating medium requires the presence of free charge in the medium. We have produced free charge in bulk liquid helium via field ionization; the resulting fluid motion resulted in charge-carrying drops being ejected from the surface of the liquid.

Field ionization of liquid helium

Field ionization results from the tunneling of electrons through the potential barrier at the interface of a metal and the helium fluid, under the influence of an electric field. A greatly simplified model for the potentials [4], which nonetheless shows the important functional dependences, yields

$$P \propto f(F,I) \exp [\, C_2\, I^{3/2}\, F^{-1}\,], \qquad (4)$$

where P is the tunneling probability, F is the electric field, I is the liquid ionization potential, C_2 is a characteristic constant, and f(F,I) is a slowly varying function of its arguments. The tunneling rate, which determines the charge available to drive fluid transport, is seen to be strongly dependant upon the field F, which is determined by the electrode geometry and applied potentials.

For field ionization to occur in helium requires an electric field in excess of 5 V/A . As a practical matter, this limits means of producing the field to sharp tips or edges held at potentials of a few thousand to several thousand volts, with field ionization occuring in close proximity to the tip or edge. The field cannot be increased arbitrarily; desorption of atoms of the metal occurs at sufficiently high fields [4].

Charged drop production

Tungsten appears to be the material of choice for field ionization electrodes; methods for fabrication of sharp tips are well-known, and tungsten has a field desorption limit [4] comfortably above the ionization field required for helium. We have fabricated electrodes of 0.025 cm tungsten wire by an electrochemical etching process as described elsewhere [5]. Radius of curvature at the electrode tips appears to be less than 0.5 μm, as measured optically; the onset of field ionization below 2.5 KV supports this measurement.

Individual electrodes were positioned in a glass dewar containing liquid helium such that the electrode height relative to the liquid level was adjustable. Insulating materials were used for all structural parts, and the required conductors kept as far away from the electrodes as possible, except when below the level of the liquid. Heavy-wall glass tubing insulated the conductors which carried potential from outside the dewar to the electrodes; this was particularly important when the liquid helium vapor pressure was small (4 torr at 1.5 K) to avoid corona discharge in the vapor. The potential on the electrode could be varied ±20 KV by a Trek 620A Power Supply/Amplifier, controlled either manually or through a HP 8013B Pulse Generator.

A flat electrode of 1.5 cm diameter was located 2 cm above the electrode tip; its potential could be varied ±20 KV independent of the ionization electrode, by another Trek 620A and HP 8130B. This potential could be applied at a predetermined interval after the application of a potential pulse on the ionization electrode. Pulse widths and relative timing were displayed on an oscilloscope.

Flow of the liquid helium was observed, at temperatures from 4.2 K to 1.5 K, resulting from ionization of the liquid and the subsequent coupling of ion motion to the surrounding liquid. No qualitative or quantitative difference was apparent in the onset, magnitude, or behavior of the fluid flow over this temperature range. Typical observations are described below.

With the electrode tip, oriented vertically, located 0.2-0.5 cm below the surface of the liquid helium, raising the potential on the electrode to +2.5 KV induced a flow of liquid above the electrode: the flow was reminiscent of that from a garden hose held below the surface of a pool. The flow was maintained as long as the potential was applied. Increased potential resulted in increased height and volume of flow. In operation the potential never exceeded 10 KV, for safety reasons; potentials above 7 KV appeared to have a detrimental effect upon further operation of the ionization electrode, consistent with the field desorption limit of tungsten.

If the potential were increased sufficiently, the flow would reach the flat electrode, remaining attached at roughly the same point. This flow was surprisingly collimated, usually having a smaller diameter at the upper electrode than at the liquid surface from which it exited. Figure 3 is an image of such a flow.

Qualitatively similar results were obtained when a negative potential was applied to the electrode. The liquid flow was initiated at a lower potential, typically -700 V, consistent with the smaller barrier to field emission [4]. The flow also appeared to be less collimated.

Short pulses of positive potential applied to the ionization electrode resulted in the ejection from the liquid surface of jets of liquid. Generally, the exit stream breaks very quickly into multiple drops, some of which coalesce near the apex of trajectory.

By switching the (upper) flat electrode from ground to a negative potential shortly after the ionization pulse, we have determined that at least some of the ejected drops carry electric charge. In one sequence of tests, 0.12 s after the end of the ionization pulse of +5.5 KV of duration 0.02 s, the upper electrode was switched from ground to -2 KV. Drops of diameter 5×10^{-3} cm were observed to begin to fall after reaching apex, then turn and accelerate upward when the field was applied. This conclusively demonstrates that these drops carried electric charge. Calculations indicate that the surface charge density was approximately 0.3 the Rayleigh charge limit [6] for a spherical drop. Shorter, higher potential pulses were generally more effective at producing well-separated drops than longer, lower potential pulses which produced the same net fluid flow.

All charged drops observed carried positive charge. We did not attempt to produce drops carrying negative charge in this series of tests.

Figure 3
Flow from liquid helium surface to conducting plate.
Separation of liquid surface from conducting surface = 1.1 cm
V = +3.4 KV duration = 20 msec.

We were unable to measure the current flowing through the ionization tip in this apparatus; current was dissipated through various ground paths. We are proceeding to determine if the quantity of charge produced on a drop can be varied with the potential or pulse characteristics, and to reduce the number of individual drops produced by a single pulse. Much work remains to make this a useful technique for controllable charged drop generation.

CONCLUSIONS

We have demonstrated methods for the generation of charged drops of normal or superfluid helium, and neutral drops of liquid the superfluid. This is the first report of which we are aware of EHD effects producing flow in liquid helium, although similar techniques have been used to produce charge in the liquid [7].

Acknowledgement

This work was supported by a grant from NASA under NRA-91-OSSA-24.

REFERENCES

1 Coors Ceramics Company, Golden, CO
2 G. L. Switzer, Rev. Sci. Inst. **62**, 2765 (1991)
3 J. H. Crowley, G. S. Wright, J. C. Chato, IEEE Trans. Ind. App. **26**, 1 (1990)
4 R. Gomer, Field Emission and Field Ionization, Harvard University Press, pp 9 and 70
5 nano tips **1**, 2 (1989) published by Digital Instruments Inc, Santa Barbara, CA
6 Lord Rayleigh, Phil. Mag. **14**, 184 (1882)
7 C. F. Barenghi, C. J. Mellor, J. Meredith, C. M. Muirhead, P. K. H. Sommerfeld, W. F. Vinen, Phil. Trans. R. Soc. London A **334**, 1633 (1991)

BEHAVIOR OF HELIUM II BUBBLE DURING SPIN-UP AND SPIN-DOWN MOTIONS WRAPPING AROUND ROTATING DEWAR WELL

R. J. Hung, H. L. Pan and G.J. Zu
The University of Alabama in Huntsville
Huntsville, Alabama 35899, USA

ABSTRACT

Time dependent animation of spherical bubble spin-up from rest, elongate and wrap around the dewar well; and also, the fully wrapping bubble around rotating dewar well spin-down at an equilibrium rotating speed are numerically studied and simulated. Time dependent deformation of a full wrapping bubble around the rotating dewar well with and without unwrapping during the process of spin-down motion are investigated. Critical rotating speed for bubble to unwrap concerns the possibility of creating the problems of asymmetry in the imbalance liquid-vapor distribution due to sloshing dynamics perturbations in the liquid-vapor interface. Some similarity parameters are considered for how to promote the rotating well during the spin-down motion. Examples are given to illustrate the problems for dewar with various liquid-filled levels. Results show that the degree to assure the rotating bubble with and without unwrapping around the dewar well increases with increasing of the dewar rotating speed, Weber number, and also the size of the bubble.

INTRODUCTION

For the purpose to carry out scientific experiments, some experimental spacecraft use superconducting sensors for gyro read-out and maintain a very low temperature for mechanical stability. The approaches to both cooling and control involve the use of superfluid liquid helium. For example, the Gravity Probe-B (GP-B) Spacecraft adopts the boil-off from the cryogenic liquid helium dewar as a coolant and propellant to maintain the attitude control and drag-free operation of the spacecraft[1,2]. The potential problems for cryogenic liquid in the dewar container could be due to asymmetry in the imbalance of liquid-vapor distribution due to sloshing dynamics perturbations in the liquid-vapor interface.

For the case of the GP-B spacecraft, cryogenic liquid helium II, at a temperature of 1.8K, is used as the propellant. With its superfluid behavior, there is no temperature gradients in the liquid helium. In the absence of temperature gradient along the surface which drive Marangoni convection, the equilibrium shape of the liquid-vapor interface is governed by a balance of capillary, centrifugal and gravitational forces. In particular, the capillary force governed sloshing dynamics plays a major role in the stability analysis of fluid systems in a microgravity environment. Determination of liquid-vapor interface based on computational experiments can uncover details of the flow which can not be easily visualized or measured experimentally in a microgravity environment[3-9].

Time-dependent dynamical behavior of partially-filled rotating fluids in reduced gravity environments was simulated by numerically solving the Navier Stokes equations subject to the initial and the boundary conditions[10-13]. At the liquid-vapor interface, both the kinematic surface boundary condition, and the interface stress conditions for components tangential and normal to the interface, were applied[5-7]. The initial conditions were adopted from the steady-state formulations developed by Hung et al[5-7] with an example applicable to the geometry of the GP-B Spacecraft[1]. Some of the steady-state formulations of interface shapes were compared with the available experiments carried out by Leslie[14] in a free-falling aircraft (KC-135). The experiments carried out by Mason et al[15] showed that the classical fluid mechanics theory is applicable for cryogenic liquid helium in large containers[16-18].

As to the effect of centrifugal force on the liquid helium and helium vapor with an example of the GP-B spacecraft, a spin rate of up to about 1 rpm will be imposed for instrument calibration in the early stages of the experiment. After calibration, the rotation rate will be reduced to its operational value of approximately 0.1 rpm.

The spacecraft is not spinning around its rotating axis when it is deployed to its designated orbit. To give an example, the GP-B dewar container is partially filled with liquid helium II and the ullage is filled with a helium vapor. Initially, the spherically shaped bubble is sitting at the location between the outer and inner walls of the dewar at rest. As the dewar starts to spin-up from rest to certain angular velocities of interest, the bubble shape deforms, elongates and encircles around the inner well of the dewar. It takes a certain period of time for a bubble with a high liquid-filled level to accomplish a complete wrapping around the inner well of the dewar with a rotating velocity of 1.0 rpm and below if liquid helium II with an extremely low viscosity is chosen as the working fluid. In this study, the dynamical behavior of the evolution of how the bubble starts at rest with a spherical-shape then elongates and encircles around the dewar well during the spin-up rotation is investigated. Bubble spin-down from a complete wrapping with steady state is also investigated. Some imaginary and fictitious working fluids with false material constants of certain physical parameters are adopted to elaborate how to accomplish a complete encirclement of the bubble around the rotating dewar well.

BUBBLE SPIN-UP BEHAVIOR IN DEWAR

Thermodynamically, the equilibrium configuration of the bubble is governed by the status of the minimum Gibb's free energy of the bubble. In other words, at the interface between the fluid outside the bubble and the fluid inside the bubble, there is a pressure discontinuity which is governed by Laplace's formulas

$$P_i - P_o = \sigma \nabla \cdot \hat{n} \tag{1}$$

where P_i denotes pressure inside the bubble; P_o, the pressure outside the bubble; σ, the coefficient of surface tension; ∇, the divergent vector of the interface; and $\hat{n}$, unit normal vector pointing outward from the interface.

For the case of bubble, the expression of the divergent of unit normal vector pointing outward from the bubble interface becomes[5]

$$\nabla \cdot \hat{n} = \frac{1}{R_1} + \frac{1}{R_2} \tag{2}$$

where R_1 and R_2 denote the radii of the curvature along the major and minor coordinates on the bubble interface at the point of interest.

In general, P_i and P_o are determined from the full Navier-Stokes equations. For the special case of partially filled liquid inside the rotating dewar, the pressure difference ($\Delta p = P_i - P_o$) is a function of dynamic force of fluid flows induced by the rotating effect of the dewar. That is[5]

$$\Delta p \sim \rho u^2 \tag{3}$$

where u and ρ denote the velocity, and density of fluid inside the dewar, respectively. As the fluid flows inside the dewar are actuated by the rotating container throughout the viscous effect of momentum transfer, u^2 is proportional to $r^2\omega^2$, i.e.,

$$u^2 \sim \alpha(\mu) r^2 \omega^2 \tag{4}$$

where ω and r show the angular velocity, and radial distance of the fluid element from the axis of rotating dewar, respectively. Here, α is the slipping factor of fluid which is a function of viscous coefficient.

The purpose of the present study is to investigate the dynamical behavior of a bubble as the dewar spin-up from rest to ceratin angular velocities of interest and as the dewar spin-down from a higher rotation speed to lower.

This behavior can be characterized by dimensionless Weber number, which is[11]

$$We = \frac{\rho\omega^2 r^3}{\sigma} = \frac{\textit{centrifugal force}}{\textit{surface tension force}} \tag{5}$$

In the study of sloshing dynamics related to similarity rules, Hung et al.[11], indicate that among the physical parameters, such as gravity (g), surface tension (σ), density (ρ), angular velocity (ω), viscous coefficient (μ) and radius of container (r), can be combined and formed into the following similarity factors: Bond number (ratio of gravity to surface tension forces), dynamic capillary number (ratio of viscous to surface tension forces), rotational Reynolds number (ratio of centrifugal to viscous forces) and Weber number (ratio of centrifugal to surface tension forces). In that study[11], it was concluded that Weber number characterizes the amplitudes of slosh wave excitation during the physical process of sloshing dynamics driven by the capillary effect along the liquid-vapor interface in the low gravity environment. In the present study, this conclusion is further extended to cover the effectiveness of the completion of bubble wrapping around the rotating dewar well during the spin-up motion.

Detailed descriptions of the governing equations, boundary and initial conditions, and computational algorithm applicable to cryogenic fluid management under microgravity are given earlier[19-22]. In this study, to show a realistic example, a full scale GP-B spacecraft dewar with an inner radius of 12 cm and an outer radius of 68 cm and a height of 145 cm has been used in the numerical simulation. The dewar is 95% filled with cryogenic liquid helium and the ullage is filled with helium vapor. The temperature of cryogenic helium is 1.8 K. In this study the following data were used: liquid helium density = 0.146 g/cm^3, helium vapor density = 0.00147 g/cm^3, fluid pressure = 1.66×10^4 dyne/cm^2, surface tension coefficient at the interface between liquid helium and helium vapor = 0.353 dyne/cm, liquid helium viscosity coefficient = 1.2×10^{-4} cm^2/s; and contact angle = 0°. The initial profiles of the liquid-vapor interface for fluid with at rest is a sphere which is sitting between the inner and outer walls of the dewar container.

A staggered grid for the velocity components is used in this computer program. The marker-and-cell method of studying fluid flows along a free surface is adopted[23-26]. The formulation for this method is valid for any arbitrary interface location between the grid points and is not limited to middle point interfaces[25]. An algorithm for a semi-implicit method[26] was used as the procedure for modeling the flow field. The time step is determined automatically based on the size of the grid points and the velocity of flow fields. A detailed description of the computational algorithm applicable to microgravity fluid management governed by sloshing dynamics is illustrated in our earlier studies[27-29]. Figure 1 shows the computational algorithm for sloshing dynamics.

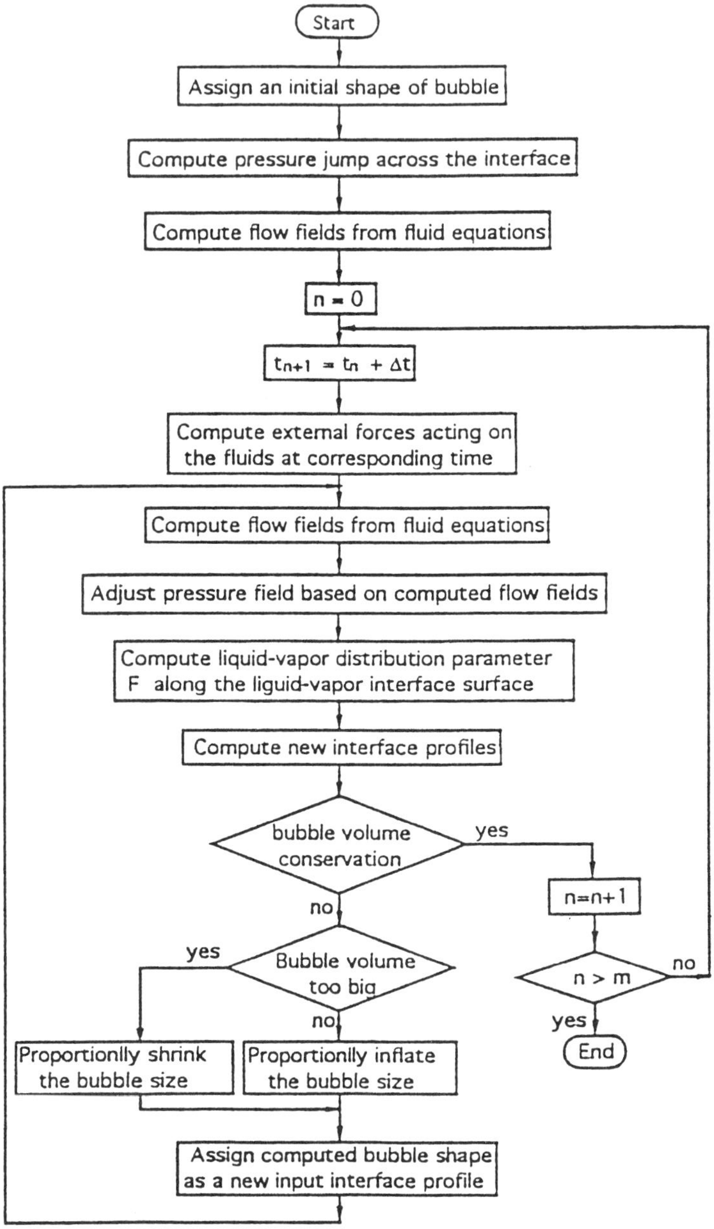

Fig.1 Computational Algorithm for Sloshing Dynamics.

In this study, non-inertial frame mathematical model is employed. In other words, Coriolis, centrifugal and angular accelerations have been applied to the fluid system throughout the entire process of spin-up motion. For the purpose to check the results of numerical simulation, dimensionless time scale of $E_k^{-1/2}$ is used to estimate time required for the development of viscous boundary layer[30]. Parameter E_k is Ekman number which is defined by[30]

$$E_k = \frac{\nu}{\omega (\Delta R)^2} \tag{6}$$

Here $\Delta R = (r_o - r_i)/2$ where r_o and r_i denote outer and inner radii, respectively, of the dewar. In reality, Ekman number is nothing but the reverse of Reynolds number for rotating system (R_e^{-1}). Estimated time required for the fluid to attain the angular velocity of new steady state is on the order of $E_k^{-1/2}/\omega$ during spin-up motion[30]. To show examples, estimated time required for cryogenic helium to reach steady state rotating speed of 0.4 rpm spin-up from rest is on the order of 10^4 s while that of the rotating speed of 0.2 rpm spin-up from rest is on the order of 1.4 X 10^4 s. This order of magnitude estimation of time required to attain steady state is agreeable with the numerical simulation shown in this review.

Weber number also characterizes the time required for bubble wrapping to reach equilibrium shapes. In other words, the higher the Weber number, the shorter the time required for the bubble to reach equilibrium shapes. It shows that the equilibrium bubble profiles with liquid fill level of 95% for bubble wrapping can be resulted with (a) We=56.65, ω=0.2 rpm, and time required to reach equilibrium at t=13,100 s; (b) We=14.18, ω=0.1 rpm, and time at t=34,200 s; and (c) We=3.54, ω=0.05 rpm, and time at t=87,000 s. Degrees of bubble wrapping with different Weber numbers for liquid fill level of 95% are illustrated in Figure 2 (B). It shows that higher Weber number (higher rotating speed) suggests a higher degree of bubble wrapping. This figure shows that the minimum rotating speed for a completion of a full wrapping is 0.2 rpm with liquid fill level of 95%. In other words, bubble wrapping with a rotating speed less than 0.2 rpm will be ending at an incomplete bubble wrapping. Figure 2(B) also shows various angles of wrapping with liquid fill level of 95% for an incomplete bubble wrapping corresponding to different rotating speeds with ω<0.2 rpm. Angles of incomplete wrapping are 53°, 124°, and 178° for bubble wrapping with rotating speeds corresponding to 0.15, 0.1, and 0.05 rpm, respectively, with liquid fill level of 95%. Figure 2 (A) shows the time sequence evolution of the three dimensional dynamical behavior of bubble evolution through the processes of elongation, encircling around the inner well of dewar, and finally a completion of the encirclement around the inner well during the spin-up motion of dewar.

With reference to Figure 2(A) for bubble spin-up from rest to rotating speed of 0.4 rpm with liquid fill level of 95%, Figure 3(A) shows the time evolution of bubble mass center fluctuations during the process of bubble wrapping around the dewar well. The figure shows the bubble mass center time evolution x_c, y_c and z_c along the x, y and z axes, respectively, in non-inertial bubble frame. This figure provides the following conclusions for bubble mass center fluctuations: (a) At time t = 0, the bubble with the spherical-shape was sitting at x_c = -38.2 cm, while y_c and z_c are basically located at a central position. (b) As the dewar container starts to spin-up, the bubble mass center along the x-axis gradually moves toward a central location while the bubble mass center along the y-axis shifts slightly from a central location with small amplitude oscillations and bubble mass center along the z-axis with negligible amplitude of oscillation. (c) The major driving forces for the oscillations of bubble mass fluctuations along the x- and y-axes are due to the centrifugal and Coriolis forces while none of these forces apply to the z-axis. (d) At the very beginning, x_c oscillates from - 38.2 cm toward the central position while y_c fluctuates from the central position toward negative y position. (e) Oscillation periods and amplitudes for bubble mass center fluctuations along the y-axis are such that the initial mode with a shorter periods (2000 s) and a greater amplitude (5 cm) and subsequent second mode with longer period (5000 s) and a smaller amplitude (3 cm). In other words, the bubble mass center oscillations along the y-axis change from a shorter period to a longer period, and from a greater amplitude to a smaller amplitude, and then gradually dies out. (f) Similar characteristics of the variation of a period and an amplitude for bubble mass center fluctuations and oscillations along the x-axis are shown which is similar to that of the figures along the y-axis.

Furthermore, one can study the time evolution of bubble mass center fluctuations with and without a completion of wrapping around the dewar well driven by the spin-up motion. Figures 3(B), 3(C), and 3(D) show bubble mass center evolution with rotating speeds of 0.2, 0.15, and 0.05 rpm,

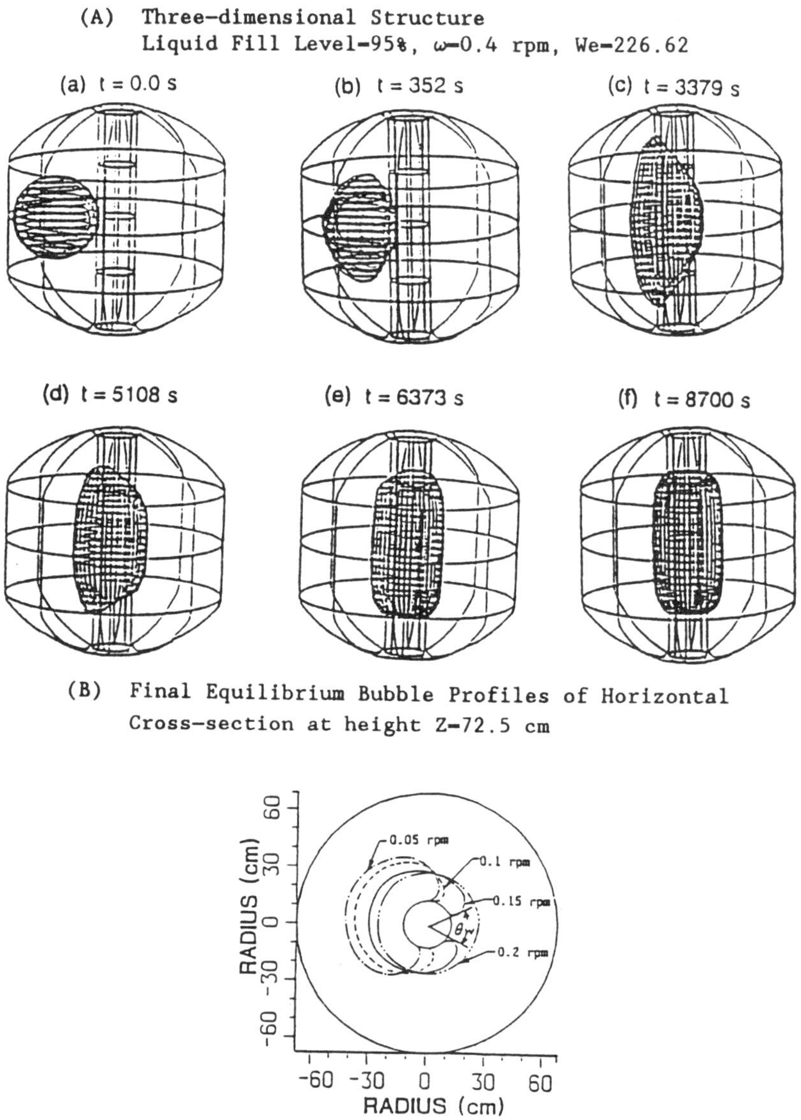

Fig.2 Time Evolution of Cryogenic Helium Bubble Spin-up.

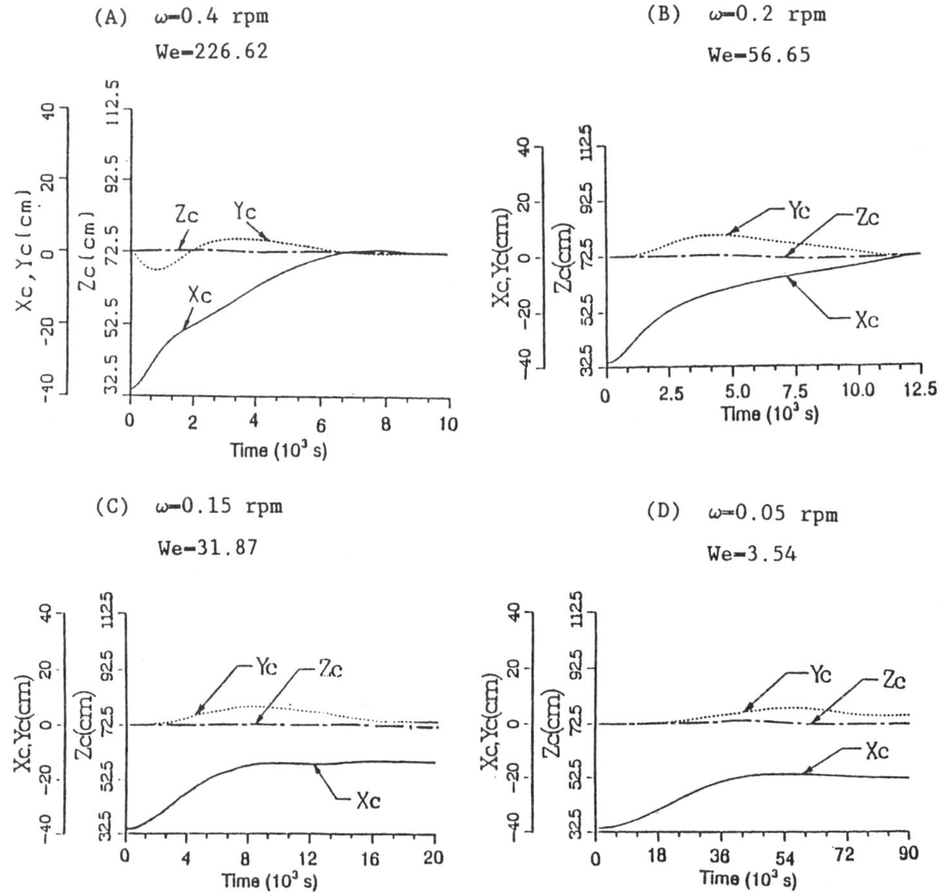

Fig.3 Bubble Mass Center Evolution During Spin-up With Various Container Rotating Speeds.

respectively. These figures provide the following conclusions for bubble mass center fluctuations in addition to the conclusions given for Figure 2(A): (a) The bubble mass center oscillations along the x-axis finally move to a central location for the container rotating speeds $\omega \geq 0.2$ rpm while bubble mass center oscillations are ending at some equilibrium positions near but not at the central position for the container rotating speeds $\omega < 0.2$ rpm. (b) A greater gap between bubble equilibrium locations and central location along the x-axis is shown for the lower container rotating speeds or lower We number in comparison with $\omega = 0.2$ rpm and We=56.65 for minimum values required for the bubble to accomplish a complete wrapping process for the container liquid fill level of 95%.

BUBBLE SPIN-DOWN BEHAVIOR IN DEWAR

Bubble spin-down from a complete wrapping with steady state is also investigated. At time t=0, the dewar container is spinning with 0.2 rpm with a full bubble wrapping around the dewar well for liquid-fill level of 95%. Figure 4(A) shows the time sequence evolution of the three dimensional bubble configuration and select times for spin-down motion of dewar with rotating speeds from 0.2 to 0.15 rpm. Bubble behaves dynamical evolution through the processes of deformations in terms of shrinking in the direction longitudinal to, and stretching in the direction transverse to the rotating

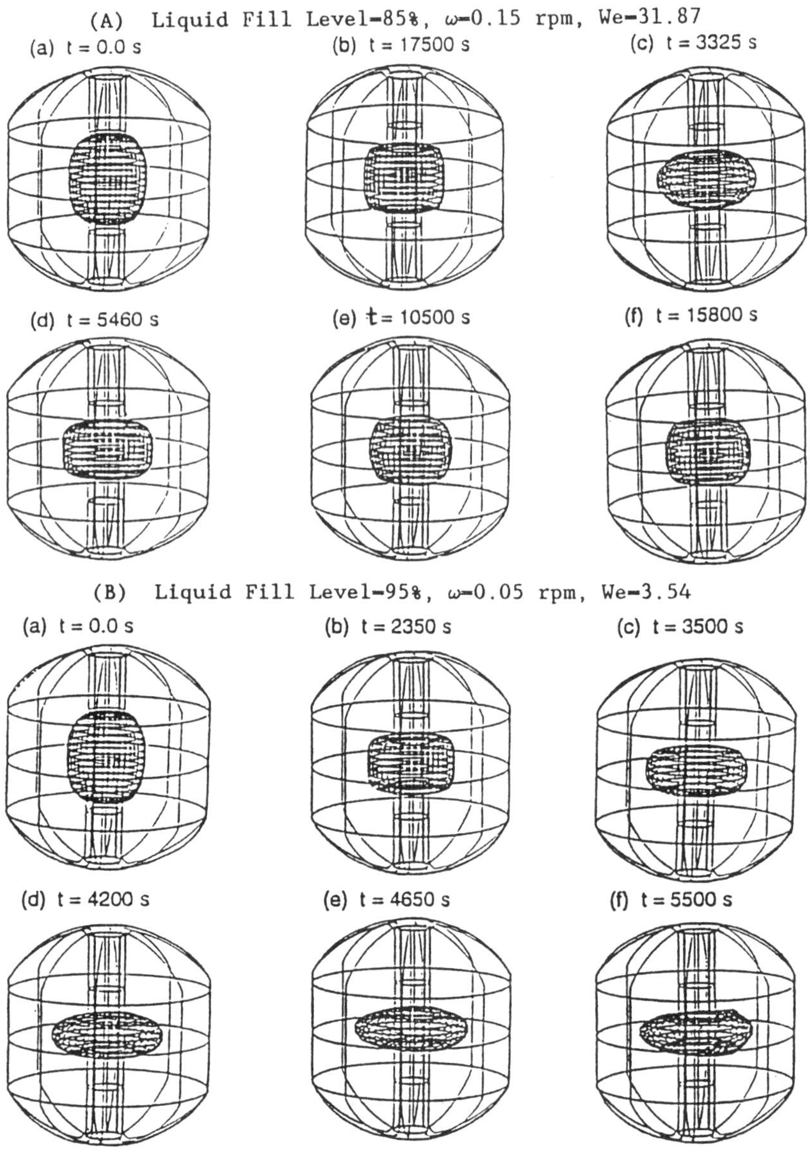

Fig.4 Time Evolution Cryogenic Helium Bubble Spin-down .

axis, jointly, without unwrapping around the dewar well during spin-down motion.

Further decreasing of rotating speed during the spin-down motion than that illustrated in Figure 4(A) eventually causes the motion of bubble unwrapping around the dewar well for dewar with liquid fill level of 95%. Figure 4(B) shows the time sequence evolution of the three dimensional configuration at some select times for spin-down motion of dewar with rotating speeds from 0.2 to 0.05 rpm when bubble starts to unwrap around dewar well with liquid fill level of 95%. Figure 4(B) also clearly shows that the symmetric rotating bubble wrapping around the dewar well starts to create an asymmetric ripple along the surface of the bubble at t=4,650 s. These ripple perturbations also quickly enhance and propagate. Finally the weakening centripetal forces resulted during the spin-down motion are unable to maintain the bubble wrapping around the dewar well. Thus, asymmetry in the imbalance liquid-vapor distribution is then starting to create.

There are some basic differences between the realistic practice and the theoretical computation for the problems of bubble spin-down motion in rotating dewar. In the theoretical computation, a symmetrical profile of distribution with respect to the rotating axis for centrifugal and Coriolis forces together with angular acceleration, which are all nothing but the function of angular velocity of rotating speed, shall be assumed during the time period of spin-down motion. In the realistic practice, a pronounced asymmetric force will be induced in these weakening forces of spin-down motion and initiates a cracking at this particular location of a symmetric force activating on the wrapped bubble which is the major cause of crack propapation leading to the final unwrapping of the rotating bubble around the dewar well. Thus, the symmetrical wrapping bubble will end up in unwrapping status during the spin-down motion. In the theoretical computation, one can never know where and when an asymmetric force shall be initiated unless this force is added into the computation artificially and arbitrarily. This is because there is no mechanism in the theoretical modelling to initiate such an asymmetric force to unwrap the rotating bubble to balance the weakening centrifugal forces during the spin-down motion. In other words, the basic differences between the realistic practice and computation are (a) the bubble will be unwrapped and still maintain in one piece or even two pieces encircling around the rotating dewar well depending upon the strength of surface tension which has a strong intention to maintain a single piece of rotating bubble for the case of realistic practice, and (b) the bubble will be broken into pieces because of the axial symmetric forces acting on the bubble with weakening magnitude of centrifugal force which is unable to ensure bubble wrapping around the rotating dewar well for the case of theoretical computation[51].

The Weber numbers considered in the cases of bubble evolution, shown in Figures 4(A) and 4(B) are 31.87 and 3.54, respectively. It shows that the Weber number characterizes the possibility of the bubble with and without unwrapping around the dewar well. Results indicate that the rotating bubble can maintain its symmetric configurations without unwrapping around the dewar well if the Weber number is greater than 3.54 with dewar liquid fill level of 95%. Otherwise, the centripetal forces are too weak to maintain the rotating bubble without unwrapping around the dewar well.

CONCLUSIONS

A comparison of rotating dewar with liquid fill levels of 85 and 95% shows the results with the following conclusions: (a) The higher the liquid fill level of the rotating dewar, the higher the rotating speed required for a bubble to complete a full wrapping around the dewar well during spin-up motion (0.2 and 0.08 rpm rotating speeds are required for bubble to complete a full wrapping around dewar well with liquid fill levels of 95 and 85%, respectively.) (b) The higher the liquid fill level of the rotating dewar, the higher the rotating speed required for the bubble to maintain unwrapping around the dewar well during the spin-down motion (0.05 and 0.02 rpm rotating speeds are required to maintain a full wrapping bubble without unwrapping around the dewar well with liquid fill level of 95 and 85%, respectively). (c) The higher the liquid fill level of rotating dewar, the shorter the time required for the rotating bubble to accomplish a full wrapping around the dewar well (t=13,100 and 30,400 s are required for the rotating bubble to complete a full wrapping around the dewar well with liquid fill levels of 95 and 85%, respectively). (d) The higher the liquid fill level of the rotating dewar, the shorter the time required for the rotating bubble during spin-down motion with full wrapping to become unwrapped from the minimum rotating speed required to accomplish a full wrapping during spin-up motion (t=4,650 and 10,600 s are required for the rotating bubble with full wrapping to become unwrapped from the minimum rotating speed required to accomplish a full wrapping around the dewar well with liquid fill levels of 95 and 85%, respectively, during the spin-down motion).

Comparison between minimum rotating speeds required for bubble to complete a full wrapping during spin-up motion, and minimum rotating speeds required for bubble to maintain a full wrapping without unwrapping around the dewar well during spin-down motion is quite different. It shows that a minimum speed to accomplish a full wrapping during spin-up motion, and a minimum speed to maintain a full wrapping without unwrapping during spin-down motion are 0.2 and 0.05 rpm, respectively, with liquid fill level of 95%, while the minimum rotating speeds for similar spin-up and spin-down motions are 0.08 and 0.02 rpm, respectively, with liquid fill level of 85%. These discrepancy in two figures between spin-up and spin-down motions are due to the fact that it needs additional energy to change any status of the most stable shape with minimum surface area required by the condition of minimum Gibbs free energy. The situation of the existence of energy barrier between these two motions which are very much similar to the motion of electron such as the thermionic-diode generator of the space-charge-barrier energy between cathode and anode.

REFERENCES

1. Wilkinson, D. T., Bender, P. L., etc., Physics Today, 39, 43, 1986.
2. Proceedings of Society of Photo-Optical Instrumentation Engineers, 619, 1, 1986.
3. Homicz, F. H., and Welch, J. E., Physics Fluids, 8, 2182, 1965.
4. Hung, R. J., and Leslie, F. W., Journal of Spacecraft and Rockets, 25, 70, 1988.
5. Hung, R. J., Tsao, Y. D., Hong, B. B., and Leslie, F. W., International Journal for Microgravity Research and Applications, 11, 81, 1989.
6. Hung, R. J., Tsao, Y. D., Hong, B. B., and Leslie, F. W., Acta Astronautica, 19, 411, 1989.
7. Hung, R. J., Tsao, Y. D., Hong, B. B., and Leslie, F. W., Journal of Spacecraft and Rockets, 26, 167, 1989.
8. Veldman, A. E. P., and Vogels, M. E. S., Acta Astronautica, 11, 641, 1984.
9. Kitchens, C. W. Jr., AIAA Journal, 18, 929, 1980.
10. Hung, R. J., Lee, C. C., and Leslie, F. W., Journal of Guidance, Control and Dynamics, 15, 817, 1992.
11. Hung, R. J., Lee, C. C., and Leslie, F. W., Journal of Fluids and Structures, 6, 493, 1992.
12. Hung, R. J., Shyu, K. L., and Lee, C. C. Journal of Propulsion and Power, 8, 778, 1992.
13. Hung, R. J., Shyu, K. L., and Lee, C. C., Journal of Spacecraft and Rockets, 29, 523, 1992.
14. Leslie, F. W., Journal of Fluid Mechanics, 161, 269, 1985.
15. Mason, P., Collins, D., etc., Proceedings 7th International Cryogenic Engineering Conferences, Surrey, England, Science and Technology Press, 1978.
16. Hung, R. J., Pan, H. L., and Long, Y. T., Cryogenics, 34, 641, 1994.
17. Van Sciver, S. W., Helium Cryogenics, pp. 429, Plenum Press, New York, 1986.
18. Donnelly, R. J., Quantized Vortices in Helium II, pp. 328, Cambridge University Press, Cambridge, 1991.
19. Hung, R. J., and Pan, H. L., Transactions of the Japan Society for Aeronautical and Space Sciences, 36, 153, 1993.
20. Hung, R. J., and Pan, H. L., Fluid Dynamics Research, 14, 29, 1994.
21. Landau, L. D., and Lifshitz, Fluid Mechanics, Pergamon Press, London, 1959.
22. Hung, R. J., and Pan, H. L., Acta Mechanica Sinica, 9, 298, 1993.
23. Harlow, F. H., and Welch, F. E., Physics of Fluids, 8, 2182, 1965.
24. Spalding, D. B. International Journal of Numerical Methods in Engineering, 4, 551, 1972.
25. Patanker, S. V., and Spalding, S. D., International Journal of Heat Mass Transfer, 15, 1787, 1972.
26. Patanker, S. V., Numerical Heat Transfer and Fluid Flow, Hemisphere- McGraw-Hill, New York, NY, pp. 197, 1980.
27. Hung, R. J., Lee, C. C., and Leslie, F. W., Acta Astronautica, 21, 309, 1990.
28. Hung, R. J., Lee, C. C., and Leslie, F. W., Acta Astronautica, 25, 523, 1991.
29. Hung, R. J., and Lee, C. C., Journal of Spacecraft and Rockets, 31, 1107, 1994.
30. Greenspan, H.P., The Theory of Rotating Fluids, Cambridge University press, 327, 1968.
31. Hung, R. J., and Pan, H. L., Journal of Guidance, Control and Dynamics, 17, in press, 1994.

CRYOGENIC ROTATING HELIUM IN RESPONSE TO LATERAL AND AXIAL IMPULSES IN MICROGRAVITY

R. J. Hung and Y. T. Long

The University of Alabama in Huntsville
Huntsville, AL 35899, USA

ABSTRACT

Sloshing dynamics for partially filled liquid of cryogenic superfluid helium II in a rotating dewar container in response to lateral and axial impulses are investigated. The study including how rotating bubble of superfluid helium II reacts to the impulse accelerations in microgravity, how amplitudes of bubble mass center fluctuate with growth and decay of disturbances, and how these disturbances in fluid systems feed back to the system in terms of force actuating on the rotating dewar through the dynamics of sloshing waves are investigated. The numerical computation of sloshing dynamics will be based on the non-inertial frame container bound coordinate, and solve time-dependent, three-dimensional formulations of partial differential equations subject to initial and boundary conditions. Results of simulations of mathematical modelling are illustrated.

I. INTRODUCTION

In order to carry out scientific experiments, some experimental spacecraft use cryogenic cooling for observation instrumentation and telescope, superconducting sensors for gyro read-out and maintain very low temperatures near absolute zero for mechanical stability. The approaches to both cooling and control involve the use of helium II. In this study, the response of cryogenic systems to sloshing dynamics associated with impulsive oscillations is investigated. Some spacecraft adopts the cooling and boil-off from the cryogenic liquid helium dewar as a cryogen and propellant to maintain the cooling of instruments, attitude control and drag-free operation of the spacecraft. Potential fluid management problems may arise due to asymmetric distribution of liquid helium and vapor or to perturbations in the free surface. Baffle-boards may need to install inside the dewar container to reduce the degree of asymmetry and damps disturbances in the liquid-vapor distribution.

Cryogenic liquid helium II, at a temperature of 1.8 K, is assumed to be used as the coolant and propellant. With its superfluid behavior, there is very small temperature gradient in the liquid helium. In the negligibly small temperature gradient along the surface which drive Marangoni convection, the equilibrium shape of the free surface is governed by a balance of capillary, centrifugal and gravitational forces. The nstability of the liquid-vapor interface can be induced by the presence of longitudinal and lateral accelerations. Thus, slosh waves are excited, producing high and low frequency oscillations in the liquid propellant. The sources of the residual accelerations include effects of the Earth's gravity gradient and g-jitter. A recent study[1] suggests that the high frequency accelerations may be unimportant in comparison with the residual motions caused by low frequency accelerations.

The time-dependent behavior of partially-filled rotating fluids in reduced gravity environments was simulated by numerically solving the Navier Stokes equations subject to the initial and boundary conditions[2-5]. At the interface between the liquid and the gaseous fluids, both the kinematic surface boundary condition and the interface stress conditions for components tangential and normal to the interface were applied[2-5]. The initial conditions were adopted from the steady-state formulations developed by Hung et al[4]. Some of the steady-state

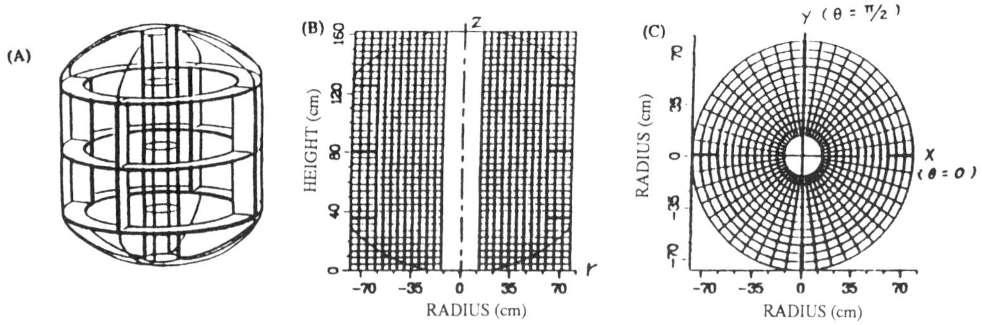

Figure 1 (A) Three-dimensional configuration of container with baffle, (B) distribution of grid points in the radial-axial plane, and (C) in the radial-circumferential plane.

formulations of interface shapes were compared with the available experiments carried out by Leslie[6] in a free-falling aircraft (KC-135). The experiments carried out by Mason et al[7] showed that the classical fluid mechanics theory is applicable to cryogenic liquid helium in large containers.

At temperatures close to absolute zero, quantum effects begin to be of importance in the properties of fluids. At a temperature of 2.17 K, liquid helium has a λ-point (a second-order phase transition); at temperatures below this point, liquid helium (helium II) has a number of remarkable properties, the most important of which is superfluidity. This is the property of being able to flow without viscosity in narrow capillaries or gaps. At temperatures other than zero, helium II behaves as if it were a mixture of two different liquids. One of these is a superfluid and moves with zero viscosity along a solid surface. The other is a normal viscous fluid. The two motions occur without any transfer of momentum from one to another for velocities below a critical velocity[8-13]. For the components of normal and superfluid velocities above a critical velocity, the two fluids are coupled[8-13].

The key parameter of critical velocity to distinguish one-fluid from two-fluid models is a function of fluid temperature and container size. In other words, in considering the dynamical behavior of helium II in a large rotating cylinder, a mixture of the superfluid and the normal fluid without separation of the two fluids is accounted for in the model computation. Density concentration of superfluid is a function of temperature, which is also true for the surface tension and viscous coefficient for helium II[11,12]. In this study, the theory of viscous Newtonian fluids is employed with modification of transport coefficients adjusted by normal and superfluid density concentration which is a function of temperature.

In this paper, the response of superfluid helium II to impulsive disturbances is studied. In order to reduce the degree of asymmetry in the liquid-vapor distribution and damp its associated disturbances, a number of baffle boards are installed inside the dewar[14]. The difference in response for dewar with and without the baffle is also investigated.

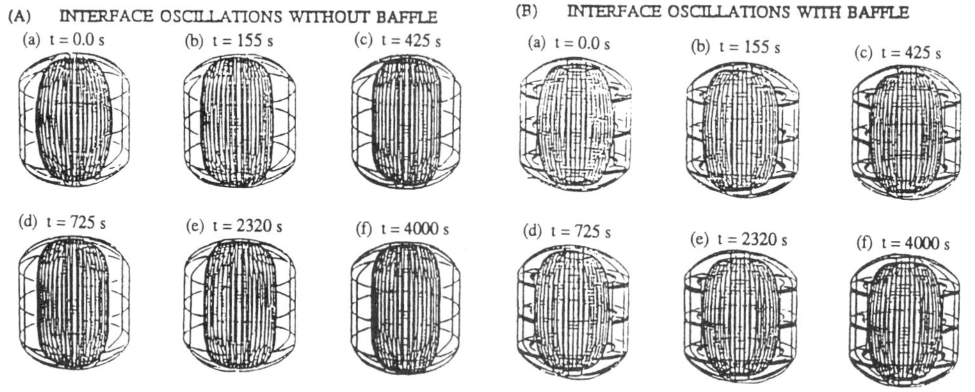

Figure 2 Time sequence evolution of three-dimensional liquid-vapor interface oscillations for rotating dewar in response to lateral impulse for container without and with baffle.

II. NON-INERTIAL FRAME MATHEMATICAL FORMULATION OF FUNDAMENTAL EQUATIONS

Consider a closed circular dewar partially filled with helium II while the ullage is filled with helium vapor. The whole fluid system is spinning in the axial direction z of cylindrical coordinates (r, θ, z), with corresponding velocity components (u, v, w). The governing equations for non-inertial frame spacecraft bound coordinates spinning along its z-axis has been given in our recent studies[15-17].

For the purpose of solving sloshing dynamic problems of liquid propellent systems in orbital spacecraft under a microgravity environment, one must solve the governing equations[15-17] accompanied by a set of initial and boundary conditions. A detailed illustration of these initial and boundary conditions concerning the sloshing dynamics of fluid systems in microgravity was precisely given in Hung and Pan[15,16]. The computational algorithm applicable to cryogenic fluid management under microgravity is also given in Hung et al[18]. In this study, a dewar with an outer radius of 78 cm, and an inner radius of 13.8 cm, top and bottom radius of 110 cm and a height of 162 cm has been used in the numerical simulation. The dewar tank is 60% filled with cryogenic liquid helium and the ullage is filled with helium vapor. The temperature of cryogenic helium is 1.8 K. In this study the following data were used: liquid helium density = 0.146 g/cm^3, helium vapor density = 0.00147 g/cm^3, fluid pressure = 1.66 x 10^4 dyne/cm^2, surface tension coefficient at the interface between liquid helium and helium vapor = 0.353 dyne/cm, liquid helium viscosity coefficient = 9.61 x 10^{-5} cm^2/s; and contact angle = 0°. The initial profiles of the liquid-vapor interface for the rotating dewar are determined from computations based on algorithms developed for the steady state formulation of microgravity fluid management[19].

A staggered grid for the velocity components is used in this computer program. MAC (marker-and-cell) method[20,21] of studying fluid flows along a free surface is adopted. VOF (volume of fluid) method[22] is used to solve finite difference equations numerically. Approximate flow velocity is calculated from the explicit approximation of momentum equations based on the results from the previous time step. Computation of pressure and velocity at the new time step are, thus, obtained from iteratively solving pressure equation through conjugate residual technique[23-26]. A detailed description of the computational algorithm applicable to microgravity fluid management is illustrated in our recent works[15-17]. As for the volume conservation of liquid, a deviation of less than 1 % error of volume is always guaranteed before a move to the next time step. Figs. 1(A), 1(B), and 1(C) show three-dimensional configuration of baffle installation, the distribution of grid points for the dewar tank with baffle boards in the radial-axial plane and radial-circumferential plane, respectively, in cylindrical coordinates. The baffle-boards are installed with three parallel plates perpendicular to the rotating axis and four radial plates aligned with the rotating axis. The inner radius, outer radius and thickness of the three parallel baffle-boards chosen in this computation are 60, 77.48 and 0.1016 cm, respectively. These three baffles are installed at locations z_1, z_2, z_3 of 38.74, 80.94, and 84.4 cm, respectively. Four radial baffle-boards are located at 90° each apart with the same inner and outer diameters and thickness as that of the three parallel baffle-boards.

III. MICROGRAVITY SLOSHING DYNAMICS IN RESPONSE TO LATERAL AND AXIAL IMPULSES

Two examples are given to illustrate sloshing dynamics in response to lateral and axial impulses during the time period of guidance and/or attitude controls. Dewar is spinning with a rotating rate of 0.1 rpm during its normal operation. In this study, with the following forms of impulses are assumed:

(A) Lateral Impulse:

$$\vec{a} = [a_x, a_y, a_z] = [1, 0, 0]\ 10^{-2} g_o \text{ in cartesian coordinates}$$

$$\vec{a} = [a_r, a_\theta, a_z] = [\cos\theta, -\sin\theta, 0]\ 10^{-2} g_o \text{ in cylindrical coordinates} \qquad \text{for } 0 < t \leq 10^{-2}\,\text{s}$$

$$\text{and } \vec{a} = [0, 0, 0] \quad \text{for } t > 10^{-2}\,\text{s}$$

(B) Axial Impulses:

$$\vec{a} = [a_x, a_y, a_z] = [0,0,1]\ 10^{-2} g_o \text{ in cartesian coordinates}$$

$$\vec{a} = [a_r, a_\theta, a_z] = [0,0,1]\ 10^{-2} g_o \text{ in cylindrical coordinates} \qquad \text{for } 0 < t \leq 10^{-2}\,\text{s}$$

$$\text{and } \vec{a} = [0,0,0] \quad \text{for } t > 10^{-2}\,\text{s}$$

where g_o is the Earth gravitational acceleration which is 9.81 m/s^2.

(III-A) Microgravity Sloshing Dynamics in Response to Lateral Impulse

(III-A-a) Bubble Oscillations in Rotating Dewar Without Baffle

The time evolution of the liquid-vapor interface without a baffle is first investigated. Figure 2(A) shows the time sequence evolution of three dimensional liquid-vapor interface oscillations. The figure shows bubble oscillations at times t=0.00, 155, 425, 725, 2320, and 4000 s. It shows that the bubble was deformed in response to the impulse and gradually recovered back to the equilibrium shape at the end of the 4000 s simulation.

Figure 3(A) shows the time sequence of the liquid-vapor interface profiles in the vertical r-z plane at $\theta =$ 0° and 180°. It indicates that the vertical cross-section of the doughnut-shape bubble was first pushed to the left (liquid to the right), then to the right (liquid to the left).

As impulse is pointing toward the eastern (positive-x) direction, the liquid is pushed toward the eastern direction while the bubble is pushed toward the western (negative-x) direction in the r-z plane. Back and forth oscillations are then induced. The oscillations are finally decaying out due to the viscosity effect. This scenario reflects the evolution of the decaying bubble oscillations in response to lateral impulse.

Figure 4 shows the time evolution of growth and decay of the fluid mass center fluctuations with broken lines for rotating dewar without baffle in response to the lateral impulse. Figures 4(A), 4(B), and 4(C) show the time evolution of the locations of the fluid mass center fluctuations in x_c, y_c, and fluctuating locations in the radial axis of the cylindrical coordinates R_c [$= (x_c^{2} + y_c^{2})^{1/2}$], respectively, for the container without baffle. It is due to the effect of Coriolis forces which carry over the lateral fluctuations from x- to y-axes even though the impulse was activating on the x-dire... n only in the rotating dewar along the z-axis. This means that the fluid mass center fluctuations are primarily in the x- and y-axes and not in the z-axis. The locations of maximum absolute values of fluid mass center fluctuations for the container without baffle in the radial direction are Max($|x_c|$, $|y_c|$)=(2.50, 1.96) cm; while the fluctuating values of the maximum fluid mass center disturbances in radial coordinates are Max(Δx_c , Δy_c)=(3.71, 3.41) cm. It shows $\Delta x_c > \Delta y_c$ for fluctuating fluid mass center disturbances in rotating dewar

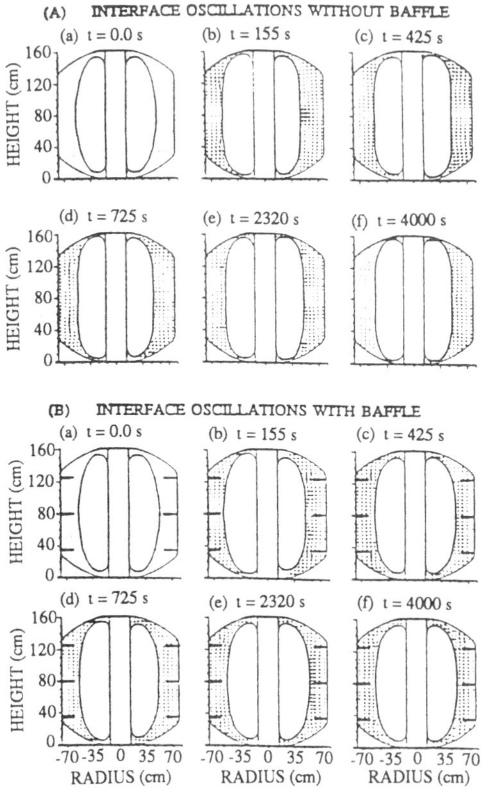

Figure 3 Time sequence evolution of liquid-vapor interface oscillations for rotating dewar in r-z plane at $\theta = 0^{o}$ and 180°, in response to lateral impulse for container without and with baffle.

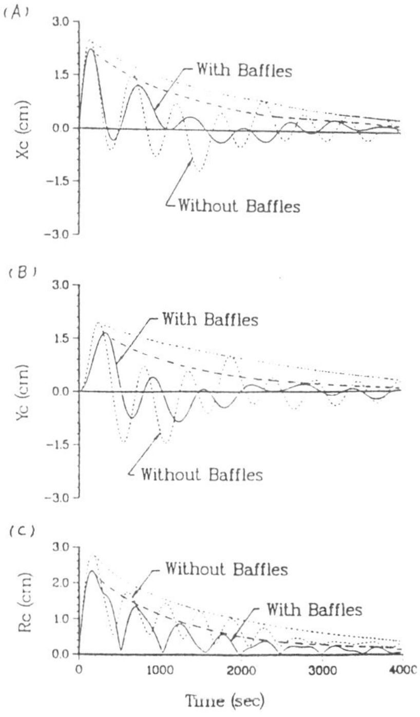

Figure 4 Time sequence evolution of fluid mass center fluctuations in response to lateral impulse for container with and without baffle. (A) Fluid mass center fluctuations in x-axis, (B) Fluid mass center fluctuations in y-axis, and (C) Fluid mass center fluctuations in absolute values of radial direction.

in response to the impulse along the x-axis. In Figure 4 , it shows that fluid mass center fluctuations reach to its peak value at time t=82 s which is almost 82 s after the ending of impulse activating on the fluid systems. It also indicates an exponentially decaying tendency of the fluid mass center fluctuations for the container without baffle.

A comparison of Figures 2 and 4 illustrates some peculiar behavior of cryogenic helium fluids with temperature below λ-point (2.17 K) in which helium demonstrates a number of remarkable properties of superfluidity such as extremely low viscous and surface tension coefficients in response to lateral impulse in a microgravity environment. It can be concluded as follows: (a) Impulse started to actuate on the fluid system at time t = 0 with peak value and end at t = 10^{-2} s. However, fluid mass center fluctuations reacted gradually in response to the impulse with maximum amplitude of fluid mass center fluctuations in x-axis at time t = 80 s and in y-axis at t = 90 s for action transferring from the x- to the y-axes due to the effect of Coriolis forces. (b) Impulsive force was applied to the fluid element in the positive x-direction which pushed the liquid in the positive x-direction and the bubble in the negative x-direction. The displacement of the fluid mass center was sinusoidally shifting continuously even long after the ending of the impulse. This displacement of the fluid mass center fluctuations reached to the maximum value at 82 s after the ending of the impulse. Then the fluid mass center started to oscillate back and forth due to the effect of surface tension forces, in addition to the reflection and bouncing of flows from the wall. (c) The oscillations of the fluid mass center fluctuations continued for a long period of time even after the ending of impulse due to the extremely low viscosity of helium II fluids. (d) It shows exponentially decaying tendency of fluid mass center fluctuations in dotted lines for the container without baffle.

(III-A-b) Bubble Oscillations in Rotating Dewar With Baffles

In this study, the time evolution of the liquid-vapor interface oscillations for the container with baffle due to the sloshing dynamics driven by the same impulse as that actuating on the dewar without baffle has been investigated. Figure 2(B) shows the time sequence evolution of the three dimensional liquid-vapor interface oscillations activated by the impulse on the fluid systems with baffle. It shows that the bubble was deformed in response to the impulse and was gradually recovered back to the equilibrium shape at the end of the 4000 s simulation. In this case, the bubble was oscillating in the middle of the dewar without contacting the edge of the baffles. A comparison of Figures 2(A) and 2(B) shows that less asymmetric behavior of oscillations are characterized for bubble oscillations for dewar with baffle than that for dewar without baffle.

Figure 3(B) shows the time sequence of the sloshing dynamics governed liquid-vapor interface profiles in the rotating dewar in response to impulse along the positive x-direction in the vertical r-z plane at $\theta = 0^{\circ}$ and 180° for dewar with baffle. It shows the following conclusions: (a) the vertical cross-section of the doughnut-shape bubble behaves in the similar manner of leftward and rightward movement of oscillations for the containers with and without baffle; (b) the container with baffle demonstrates a smaller amplitude of fluctuations than that for container without baffle; and (c) it shows a different flow profile for container with baffle in which the flow circumvents around the baffle that is not shown for flow without baffle.

Figure 4 shows the time evolution of the growth and decay of the bubble mass center fluctuations with solid lines for the rotating dewar with baffle in response to the impulse. Figures 4(A), 4(B), and 4(C) show the time evolution of the locations of the fluid mass center fluctuations in x_c, y_c, and fluctuating locations in radial axis of cylindrical coordinates R_c [= $(x_c^2 + y_c^2)^{1/2}$], respectively, for the container with baffle. Similar to the container without baffle, the effect of Coriolis forces which carry over the fluctuations from x- to y-axes also applies to this case. This means that the fluid mass center fluctuations are primarily in the x- and y-axes and not in the z-axis. The locations of maximum absolute values of the fluid mass center fluctuations in the radial direction are $\mathrm{Max}(|x_c|, |y_c|)=(2.24, 1.67)$ cm; while the fluctuating values of the maximum fluid mass center disturbances in the radial coordinates are $\mathrm{Max}(\Delta x_c, \Delta y_c)=(2.64, 2.52)$ cm. Similarly, it also shows $\Delta x_c > \Delta y_c$ for fluctuating fluid mass center disturbances in rotating dewar with baffle in response to the impulse along the x-axis.

A comparison of Figures 2 and 4 between the container with and without baffle also illustrates some peculiar behavior of cryogenic helium fluids with temperature below l-point (2.17 K) in which helium demonstrates a number of remarkable properties of superfluidity such as extremely low viscous and surface tension coefficients in response to impulse in microgravity environment. It can be concluded as follows: (a) Impulse started to actuate on the fluid system at time t = 0 with peak value and end at t = 10^{-2}s for both the container with and without baffle. However, the fluid mass center fluctuations reacted gradually in response to the impulse with a maximum amplitude of fluid mass center fluctuations in the x-axis about the same time as that between the container with and without baffle at time t = 80 s, there is a 15 s delay of maximum amplitude fluid mass center fluctuations in the y-axis for the container with baffle than that for the container without baffle. (b) Displacement of fluid mass center was sinusoidally shifting continuously for both the container with and without baffle long after the ending of the impulse. This displacement of fluid mass center in the radial direction for both cases reached the maximum value at 82 s after the ending of impulse. Then the fluid mass center started to oscillate back and forth with a longer period of oscillations for the container with baffle than that for container without baffle due to the effects of surface tension forces and viscous forces for the container with baffle, in addition to the reflection and bouncing of flows from the wall. (c) The oscillations of fluid mass center

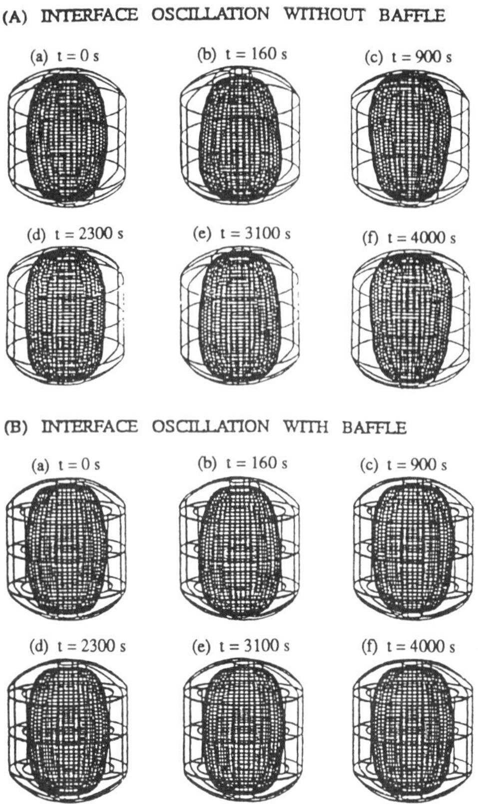

Figure 5 Time sequence evolution of three-dimensional bubble oscillations for rotating dewar in response to axial impulse for container without and with baffle.

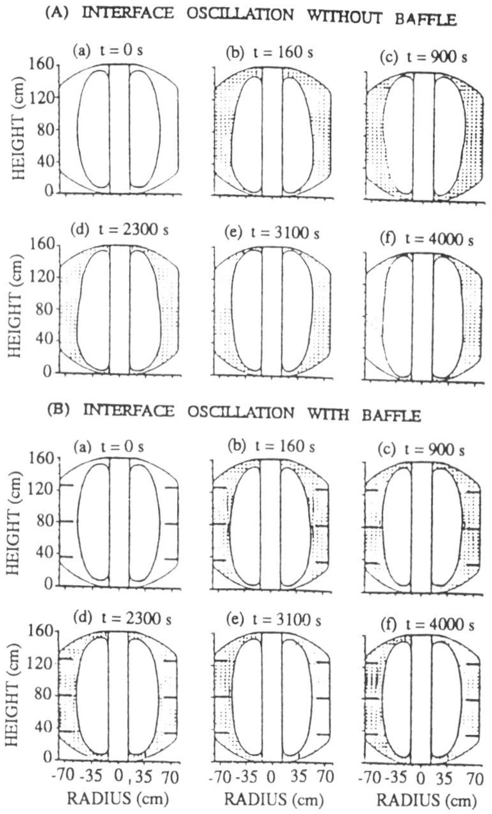

Figure 6 Time sequence evolution of bubble oscillations for rotating dewar in r-z plane at $\theta = 0^\circ$ and 180°, in response to axial impulse for container without and with baffle.

fluctuations continued for a long period of time after the ending of impulse due to extremely low viscosity of helium II fluid which eventually decays the mass center fluctuations. (d) As for the effect of viscous force between the liquid and solid interface in the container with baffle, it contributes a greater damping effect of mass center fluctuations. A comparison of the container with and without baffle shows that the container with baffle can reduce 13% of the maximum value of fluid mass center dislocation and decrease 25% of the maximum amplitude of fluid mass center fluctuations than that for the container without baffle.

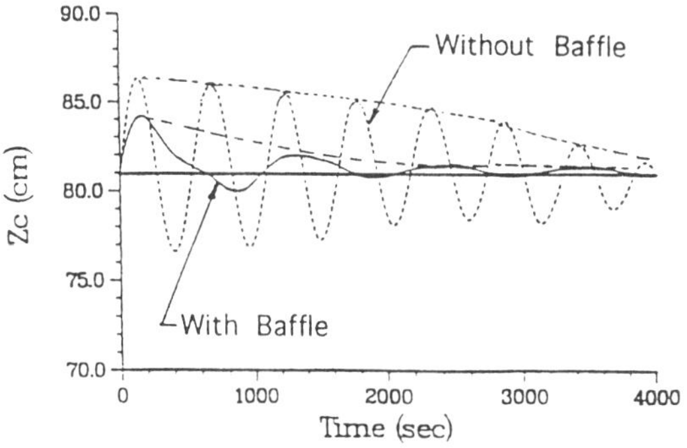

Figure 7 Time sequence evolution of fluid mass center fluctuations in response to axial impulse for container with and without baffle.

(III-B) Microgravity Sloshing Dynamics in Response to Axial Impulse

(III-B-a) Bubble Oscillations Without Baffle

Figure 5(A) shows the time sequence evolution of three dimensional bubble oscillations activated by the axial impulse without baffle in the container. The figure shows time sequence bubble oscillations in response to the axial impulsive force at time t=0.00, 160, 900, 2300, 3100, and 4000 s. It shows that the bubble is inflated at the lower half of the bubble as liquid is pushed upward along the axial direction at the very beginning; and then the bubble is inflated at the upper half of the bubble as the liquid is pushed downward along the axial direction during the up and down oscillations activated by axial impulsive thrust. Finally the bubble is gradually recovered back to the equilibrium shape at the end of 4000 s simulation.

Figure 6(A) shows the time sequence of the sloshing dynamics governed bubble profiles in response to axial impulse along the positive z-direction in the vertical r-z plane at $\theta = 0^{\circ}$ and 180°. It clearly indicates an up and down motion of bubble along axial direction during the animation of decaying bubble oscillations.

Figure 7 shows the time evolution of the growth and decay of the fluid mass center fluctuations with broken lines in the rotating dewar without baffle in response to the axial impulse. The fluid mass center at time t = 0 is located at $(x_c, y_c, z_c) = (0, 0, 80.9)$ cm. The highest amplitude fluctuations of fluid mass center is located at Max $(x_c, y_c, z_c) = (0, 0, 86.3)$ cm at time t = 138 s, while the lowest amplitude fluctuations of fluid mass center is located at Min $(x_c, y_c, z_c) = (0, 0, 76.6)$ cm at time t = 406 s for dewar without baffle. In this simulation, it shows that fluid mass center fluctuations reach to its maximum values at 138 s after the ending of the impulse actuating on the rotating dewar. The wave period of fluid mass center fluctuation is about 500 s for dewar without baffle.

A comparison of Figs. 5 and 7 illustrates some peculiar behavior of cryogenic helium fluids with temperature below l-point (2.17 K) in which helium demonstrates a number of remarkable properties of superfluidity such as extremely low viscous and surface tension coefficients in response to axial impulse in a microgravity environment. It can be concluded as follows: (a) The axial impulsive acceleration started to actuate on the fluid system at time t = 0 with peak value and end at t = 10^{-2} s. However, fluid mass center fluctuations reacts gradually in response to the impulse with a maximum amplitude of fluid mass center fluctuations in z-axis at time t = 138 s and with minimum amplitude at t = 406 s also in z-axis. (b) Impulse was applied to the fluid element in the positive z-direction which pushed the liquid in the positive z-direction and the bubble in the negative z-direction. The displacement of the fluid mass center was sinusoidally shifting continuously even long after the ending of the impulse. Then the fluid mass center started to oscillate up and down due to the effect of surface tension effects, in addition to the reflection and bouncing back of flows from the top wall in response to the impulse. (c) The oscillations of the fluid mass center fluctuations continued for a long period of time even after the ending of impulse due to the extremely low viscosity of helium II fluids and also the damping of pressure fluctuations across the liquid-vapor interface. (d) It shows that the exponentially decay tendency of fluid mass center fluctuations in dotted lines for the container without baffle.

(III-B-b) Bubble Oscillations With Baffle

Figure 5(B) shows the time sequence evolution of the three dimensional bubble oscillations activated by the impulse on the container with baffle. It shows that the bubble was deformed in response to the impulse and was gradually recovered back to the equilibrium shape at the end of 4000 s simulation. In this case, the bubble was oscillating in the middle of the dewar without contacting the edge of the baffle. A comparison of Figs. 5(A) and (B) shows that less asymmetric behavior of oscillations are characterized for bubble oscillations with baffle than that without baffle.

Figure 6(B) shows the time sequence of the sloshing dynamics governed liquid-vapor interface profiles in the rotating dewar in response to impulse along the positive z-direction in the vertical r-z plane at $\theta = 0^{\circ}$ and 180° for the container with baffle. It indicates an up and down motion of bubble which behaves in the similar manner of upward and downward movement of oscillations for the container with baffle as that for the container without baffle at a smaller amplitude of fluctuations. Figure 7 shows the time evolution of the growth and decay of the fluid mass center fluctuations with solid lines in the rotating dewar with baffle in response to the impulse. The highest amplitude fluctuations of fluid mass center is located at Max $(x_c, y_c, z_c) = (0, 0, 84.2)$ cm at time t = 161 s, while the lowest amplitude fluctuations of fluid mass center is located at Min $(x_c, y_c, z_c) = (0, 0, 80)$ cm at time t = 863 s for dewar with baffle. In this simulation, it shows that fluid mass center fluctuations reach to its maximum at t = 161 s after the ending of the impulsive thrust actuating on the rotating dewar. The wave period of fluid mass center fluctuation is about 1000 s.

A comparison of Figures 5 to 11 between the container with and without baffle also illustrates some peculiar behavior of cryogenic helium fluids with temperature below l-point (2.17 K) in which helium demonstrates a number of remarkable properties of superfluidity such as extremely low viscous and surface tension coefficients in response to impulse in microgravity environment. It can be concluded as follows: (a) The axial impulsive acceleration started to actuate the fluid system at time t = 0 with peak value and end at t = 10^{-2}s for both the container with and without baffle. However, the fluid mass center fluctuations reacts gradually in response to the impulse with a maximum amplitude of fluid mass center fluctuations in the z-axis at t = 138 s for dewar without baffle, and at t = 161 s for dewar with baffle, a 23 s delay for dewar with baffle to that for dewar without baffle. (b) The oscillations of fluid mass center fluctuations continue for a long period of time even after the ending of impulse due to extremely low viscosity of helium II fluid which eventually decays the mass center fluctuations. (c) As for the effect of drag force between the liquid and solid interface in the container with baffle, the installation contributes greatly in the increasing of wave period oscillation for the dewar with baffle, which shifts the wave period from about 500 s for dewar without baffle to about 1000 s for the dewar with baffle. (d) The baffle also contributes to the reduction of fluid mass center fluctuations. Results show that the container with baffle can reduce 40% of the maximum amplitude of fluid mass center fluctuations and decrease 70% of the minimum amplitude of fluid mass center fluctuations than that for the container without baffle.

IV. CHARACTERISTICS OF SLOSH WAVES ACTIVATED SLOSH REACTION FORCE

In this study, characteristics of the slosh wave activated fluid reaction forces fluctuations exerted on the dewar container in response to lateral and axial impulses are investigated. The mathematical formulation of the fluctuations of slosh reaction forces exerted on the dewar container are illustrated precisely in the paper given by Hung et al[21]. With reference to Section III for sloshing dynamics driven by impulse together with mathematical formulations[3,16], one can calculate the slosh reaction force and its associated moments acting on the dewar container.

(IV-A) Slosh Reaction Force Driven by Lateral Impulse

Figure 8 shows the computed time variation of the fluctuations of slosh reaction force, with broken lines for the container without baffle and solid lines for the container with baffle, exerted on the dewar container in response to the lateral impulse. Figures 8(A), 8(B), and 8(C) illustrate the time evolution of the fluctuations of slosh reaction forces exerted on the container in x and y components and the total resultant force activating on the container, respectively. As the impulse was applied laterally along the x-axis in container bound coordinates, the slosh reaction forces actuating on the container were in the radial direction along the x- and y-axes, and there was no force in the z-axis.

There are some peculiar data differences shown between the results of the computer plots illustrated in Figure 8 and the computer data file. In the computer plotted figure, shown in Figure 8, the maximum absolute values of slosh reaction forces exerted on the container without baffle are $\mathrm{Max}(|F_x|, |F_y|, |F_z|)_n$= (190, 165, 195) dynes, and those of the slosh reaction forces exerted on the container with baffle are $\mathrm{Max}(|F_x|, |F_y|, |F_z|)_b$=(130, 120, 142) dynes. On the other hand, in the computer data file, the maximum absolute values of slosh reaction forces exerted on the container without baffle are $\mathrm{Max}(|F_x|, |F_y|, |F_z|)_n$=(1328, 0.41, 1329) k-dyn, and those of the

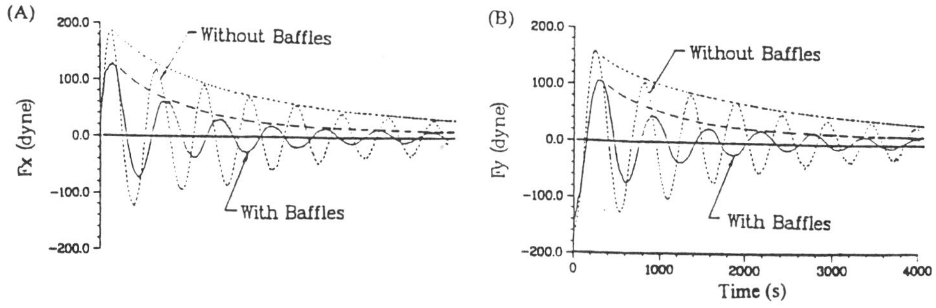

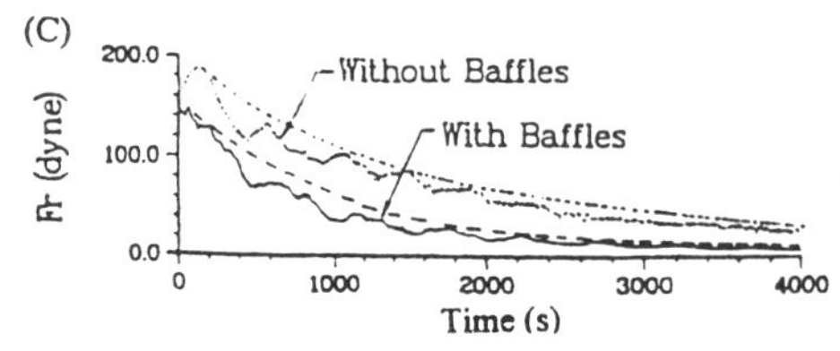

Figure 8 Time sequence of the fluctuations of slosh reaction forces exerted on the dewar in response to lateral impulse for container without and with baffle. (A) Slosh reaction force fluctuations along x-axis, (B) Slosh reaction force fluctuations along y-axis, and (C) Absolute values of total slosh reaction force exerted on the container.

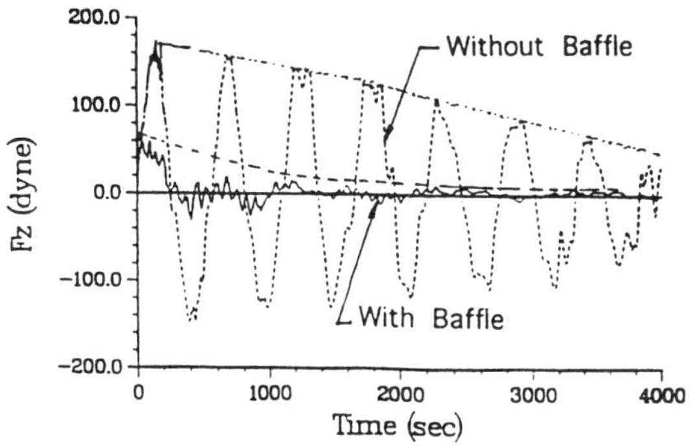

Figure 9 Time sequence of the fluctuations of slosh reaction forces exerted on the dewar in response to axial impulse for container without and with baffle.

slosh reaction forces exerted on the container with baffle are $Max(|F_x|, |F_y|, |F_z|)_b$=(1411, 0.23, 1412) k-dyn. Clearly, some information is missing in the computer plotted figures, in particular during the initial time period of one second when the impulse was applied in a time period of 10^{-2} s. To recover this missing information, the accumulative impulse and the time average of the impulse in response from the cryogenic fluid to the container are computed. The time-average slosh reaction impulsive force from time t=0 to t_K is

$$I_{d,i}^{k} = \frac{\int_0^{t_K} F_i \, dt}{\int_0^{t_K} dt} \qquad (1)$$

Some selected data of the time-average slosh reaction impulsive forces for the first 100 ms exerted on the container without baffle are 1226, 655, 429, 334, 263, 220, 191, and 122 k-dyn for the time period from t=0 to 10, 20, 30, 40, 50, 60, 70, and 100 ms respectively, while those of the time-average slosh reaction impulsive forces exerted on the container with baffle are 1284, 687, 477, 372, 299, 249, 214, and 136 k-dyn for the similar time period from t=0 to 10, 20, 30, 40, 50, 60, 70, and 100 ms respectively. After t>1 s, a drastically great damping effect causes the time-average slosh reaction impulsive force finally reducing to 39 dynes for the container with baffle in comparison with that of 76 dynes for the container without baffle for the time period from t=0 to 4000 s. Obviously this information is not properly illustrated in the first one second of the plot because this data coincides with the vertical axis of force in Figure 8.

A comparison of Figures 4 and 8 can draw the following conclusions: (a) There is a great difference between fluid mass center fluctuations and slosh reaction force for the time period in response to the impulse. Fluid mass center fluctuations of cryogenic helium are gradually built and reach to its peak value at t=82 s after the ending of impulse while the instantaneous and immediate response is reacted from the cryogenic helium to the container through the production of slosh reaction forces. (b) A large slosh reaction force with a value of 1412 k-dyn is immediately produced in response to the impulse and acting on the container with baffle, while a similar value of slosh reaction force is also immediately induced in response to the impulse and acting on the container without baffle. After time t>1 s, the slosh reaction force for the container with baffle becomes smaller than that for the container without baffle because of the viscous damping effect activating between the fluid-solid interface for dewar with baffle. (c) A similar result was concluded for the time-average slosh reaction impulsive force acting on the container. A higher decay rate of slosh reaction impulsive force is concluded for the container with baffle than that for the container without baffle due to viscous damping on baffle at time t>1 s. (d) The major slosh reaction forces acting on the container are in the x-y plane and have nothing to do with the z-axis because the lateral impulse is activating on the x-axis and transferring to the y-axis through the effect of Coriolis forces in the rotating container. (e) The time evolution for the amplitude fluctuations of slosh reaction forces activating on the container with baffle is smaller than that for the container without baffle and slowly decays exponentially due to viscous damping for time t>1 s to the end of t=4000 s. It shows 28% reduction of maximum slosh reaction force fluctuations acting on the container with baffle than that to the container without baffle for time t>1 s. (f) Wave period characteristics for slosh reaction force fluctuations activating on the container between dewar with and without baffle are quite different. The container with baffle prolongs the wave period and makes the amplitude of the fluctuations smaller than that of the container without baffle for time t>1 s.

(IV-B) Slosh Reaction Force Driven by Axial Impulse

Figure 9 shows the computed time variation of the fluctuations of fluid reaction force, with broken lines for the container without baffle and solid lines for the container with baffle, exerted on the dewar in response to the axial impulse. In addition to the fluid reaction forces exerted on the container in response to the axial impulse, the exponentially decay tendency of these fluid feedback forces are also plotted.

There are some peculiar data differences shown between the results of the computer plots illustrated in Figure 9 and the computer data file. In the computer plotted figure, shown in Figure 9, the maximum absolute values of fluid reaction forces exerted on the container without baffle are $Max(|F_x|, |F_y|, |F_z|)_n$= (0, 0, 165) dynes, and those of the fluid reaction forces exerted on the container with baffle are $Max(|F_x|, |F_y|, |F_z|)_b$=(0, 0, 63) dynes. On the other hand, in the computer data file, the maximum absolute values of fluid forces exerted on the container without baffle are $Max(|F_x|, |F_y|, |F_z|)_n$=(0, 0, 1003) k-dyn, and those of the fluid forces exerted on the container with baffle are $Max(|F_x|, |F_y|, |F_z|)_b$=(0, 0, 1085) k-dyn. Clearly, some information is missing in the computer plotted figures, in particular during the initial time period of one second when the impulse was applied in a time period of 10^{-2} s. To recover this important information, the accumulative impulse and the time average of the impulse in response from the cryogenic fluid to the container are computed from Equation (1). Some selected data of the time-average impulsive forces for the first 90 ms exerted on the container without baffle are 704, 428, 293, 223, 180, 151, 129, and 101 k-dyn for the time period from t=0 to 10, 20, 30, 40, 50, 60, 70, and 90 ms respectively, while those of the time-average impulsive forces exerted on the container with baffle are 996, 530, 356, 270, 221, 185, 160, and 124 k-dyn for the similar time period from t=0 to 10, 20, 30, 40, 50, 60, 70, and 90 ms respectively. After t>1 s, a drastically great damping effect causes the time-average impulsive force finally reducing to 9.95 dynes for the container with baffle in comparison with that of 74.7 dynes for the container without baffle for the time period from t=0 to 4000 s. Obviously this information is missing from the data in the first one second of the plot because this data coincides with the vertical axis of force and is not clearly shown for the plot illustrated in Fig. 9.

A comparison of Figs. 7 and 9 can draw the following conclusions: (a) There is a great difference between the fluid mass center fluctuations and fluid reaction force for the time period in response to the impulse. Fluid mass center fluctuations of cryogenic helium are gradually built and reach to its peak value at t=138 s for dewar without baffle and at t=161 s for dewar with baffle after the ending of impulsive acceleration while the instantaneous and immediate reaction is made with a great value of shock in response from the cryogenic helium to the container. (b) A tremendously large reaction force with a value of 1085 k-dyn of fluid reaction force is immediately created in response to the impulse and acting on the container with baffle, while a value of 1003 k-dyn of fluid reaction force is also immediately induced in response to the impulse and acting on the container without baffle. The fluid reaction force for the container with baffle is greater than that for the container without baffle in the first 1 s. After t>1 s, the fluid reaction force for the container with baffle becomes smaller than that for the container without baffle because of the drag damping effect activating between the fluid-solid interface for dewar with baffle. (c) A similar result was concluded for the time-average impulsive force acting on the container. The maximum time-average value of impulsive force at time t=10 ms for the container with and without baffle are 996 and 704 k-dyn, respectively. It shows a greater amplification of both fluid reaction force and fluid time-average impulsive force acting on the container with baffle than that for the container without baffle at the time period within the first 100 ms. It decays with a higher rate for the container with baffle than that for the container without baffle due to drag damping on baffle at time t>1 s. (d) The major reaction forces from the sloshing dynamics modulated fluid system acting on the container are in the z-axis and have no component in the x-y plane because the axial impulse is acting on the z-axis. (e) The time evolution of fluid reaction force fluctuations activating on the container in response to impulse for dewar with baffle is smaller than that for dewar without baffle. These fluctuations decay slowly due to the drag damping effect at time t>1 s and continue to the end at t=4000 s. The result shows a 60% reduction of maximum reaction force fluctuations from the fluid system to the container walls for the container with baffle than that for the container without baffle at t>1 s. (f) Wave period characteristics for fluid reaction force fluctuations activating on the container between the cases for the dewar with and without baffle are quite different. The baffle prolongs the wave period from 500 s for dewar without baffle to 1000 s for dewar with baffle. (g) The baffle also significantly reduce the time-average impulsive force from 74.7 dynes for dewar without baffle to 9.95 dynes for dewar with baffle. It results in 85% reduction from dewar without baffle to dewar with baffle for time-
average impulsive force.

V. DISCUSSION AND CONCLUSION

Sloshing dynamics of fluid system disturbances in response to lateral and axial impulses due to the activation of guidance and/or attitude controls of spacecraft operation[3,26] in microgravity have been investigated. In this study, the effect of surface tension on partially-filled rotating dewar (liquid helium and helium vapor) activated by the time-dependent impulse in the lateral x-direction and axial z-direction with and without baffle have been carried out. Study is based on the computation of three-dimensional non-inertial frame Navier-Stokes equations subject to the initial and boundary conditions[16,17]. The initial condition for the liquid-vapor interface profiles were adopted from the steady-state formulation of rotating dewar developed by Hung and Leslie[19].

In this study, the time evolution of the sloshing dynamics governed liquid-vapor interface oscillations in response to lateral and axial impulses have been investigated. The simulation of fluid mass center fluctuations shows that the amplitude of bubble mass center reached the maximum values at a time around or more than 100 s after the ending of the impulse. In the meanwhile, a great value of slosh reaction force was reacted immediately right at the time of the peak value of the impulse imposed on the fluid systems. At the very beginning, the magnitude of slosh reaction force activating on the container with and without baffle is about the same value. After time t>1 s, slosh reaction force activating on the container with baffle becomes smaller than that for the container without baffle because of the viscous damping between the fluid-solid interface for baffle. The maximum slosh reaction force fluctuations activating on the container without baffle are greater than that for the container with baffle.

Because of the peculiar physical properties of cryogenic liquid helium[8-13], extremely low values of surface tension and viscosity coefficient make oscillations of bubble and sloshing dynamics modulated fluctuating reaction forces actuating on the container in response to the impulse continue for a long period of time and gradually decay exponentially. This paper also discusses how baffle installed in the container can help to damp out these fluctuations. In this research, it has demonstrated how the amplitude of cryogenic helium sloshing dynamics disturbances grow and decay in response to the lateral and axial impulses in microgravity. A full understanding of the time history of growth and decay of disturbances is important for the design of guidance and attitude controls of spacecraft utilizing cryogenic liquid systems[3,8-13,26].

ACKNOWLEDGEMENT

The authors appreciate the support received from the National Aeronautics and Space Administration through the NASA Grant NAG8-938 and NASA contract NAS8-39609/Delivery Order No. 103. They would like to express their gratitude to Richard A. Potter of NASA/Marshall Space Flight Center for the stimulating discussions during the course for the present study.

REFERENCES

1. Kamotani, Y., Prasad, A., and Oastrach, S., AIAA Journal, 19, (1981) 511.
2. Hung, R. J. and Shyu, K. L., Transactions of the Japan Society for Aeronautical and Space Sciences, 36, (1994) 217.
3. Hung, R. J.; Lee, C. C.; and Leslie, F. W., Journal of Guidance, Control and Dynamics, 15, (1992) 817.
4. Hung, R. J.; Lee, C. C.; and Leslie, F. W., Journal of Fluids and Structures, 6, (1992) 493.
5. Hung, R. J.; Long, Y.T.; Pan, H.L., Transactions of the Japan Society for Aeronautical and Space Sciences, 37, (1994) 217.
6. Leslie, F. W., Journal of Fluid Mechanics, 161, (1985) 269.
7. Mason, P.; Collins, D.; Petrac, D.; Edeskuty, F.; Schuch, A.; and Williamson, K., Proceedings 7th International Cryogenic Engineering Conferences, Surrey, England, Science and Technology Press, (1978).
8. Van Sciver, S. W.; Helium Cryogenics, Plenium Press, New York, (1986).
9. Donnelly, R. J.: Quantized Vertices in Helium II, Cambridge University Press, Cambridge, (1991).
10. Hung, R. J.; Pan, H. L.; and Long, Y. T., Cryogenics, 34, (1994), 641.
11. Wilks, H.: The Properties of Liquid and Solid Helium, Clarendon Press, Oxford, (1967).
12. Hoare, F. E.; Jackson, L. C.; and Kurt, N.: Experimental Cryogenics: Liquid Helium II, Butterworths, London, 1961.
13. Donnely, R.J., Scientific American, November (1988) 100.
14. Hung, R.J.; Lee, C.C., Canadian Aeronautics and Space Journal, 40, (1994) 185.
15. Hung, R. J.; Pan, H. L., Transactions of the Japan Society for Aeronautical and Space Sciences, 36, (1993) 153.
16. Hung, R. J. and Pan, H. L., Fluid Dynamics Research, 14, (1994) 29.
17. Hung, R.J.; Pan, H.L.; Leslie, F.W., Z. Fluwiss. Weltraumforsch., 18 (1994) 195.
18. Hung, R. J.; Lee, C. C.; and Leslie, F. W., Transactions of the Japan Society for Aeronautical and Space Sciences, 35, (1993) 187.
19. Hung, R. J. and Leslie, F. W., Journal of Spacecraft and Rockets, 25, (1988) 70.
20. Harlow, F. H.; Welch, F. E., Physics of Fluids, 8, (1965) 2182.
21. Hirt, C. W.; Nichols, B. D., Journal of Computational Physics, 39, (1981) 201.
22. Salvadori, M.G.; Baron, M.L.: Numerical Methods in Engineering, Prentice-Hall, Inc., Englewood Cliffs, NJ, (1961).

23. Hageman, L. A.; Young, D. M.: Applied Iterative Methods. Academic Press, New York (1981).
24. Young, D., Transactions of American Mathematical Society, 76 (1954) 92.
25. Pantanker, S.V.: Numerical Heat Transfer and Fluid Flow, Hemisphere-McGraw-Hill, New York, NY, (1980).
26. Hung, R.J.; Pan, H.L.: Journal of Guidance, Control, and Dynamics, 18 (1995) in press.

THE REFRIGERATION OF HIGH TEMPERATURE SUPERCONDUCTORS BETWEEN 25K AND 65K

R.N.Richardson, R.G.Scurlock and A.C.R.Tavner

Institute for Cryogenics and Energy Systems
University of Southampton
Southampton
SO17 1BJ

ABSTRACT

The present state of the art indicates that acceptable j - H characteristics for power applications of the new high T_c superconductors will only be achieved using materials at temperatures below liquid nitrogen temperature. A boiling point of 27.1K and high specific cooling capacity make neon an eminently suitable choice of refrigerant at these temperatures. A cryostat has been constructed which employs a two stage Gifford - McMahon cooler to liquefy neon gas. The cryostat contains up to 5 litres of liquid neon which can be used for 'in-situ' experiments or transfer to another cryostat. Another set of cryostats are being used with liquid nitrogen/oxygen mixtures at reduced pressure for temperatures down to 50K. All these cryostats provide a core facility for characterising and operating high T_c superconductors at Southampton.

INTRODUCTION: The need for temperatures below 77K

For power engineering applications, the new high temperature ceramic superconductors need to have current densities (j) in the superconducting core of a wire or tape of up to 10^5 Acm^{-2}, in magnetic fields (H) of up to 10 Tesla. At present, these j-H characteristics only appear to be attainable with BSCCO materials at temperatures significantly below 77K. Above about 40K, the j-H characteristics decrease monotonically with increasing temperature, due to thermal deactivation of the flux pinning centres and flux flow effects. This behaviour is illustrated in Figure 1 which shows how the performance of a particular PBSCCO 2223 sample (1) improves markedly as the temperature falls from 77 to 65K. At temperatures above about 70K, the current densities in BSCCO 2223 tapes decrease rapidly with applied field, and therefore the use of liquid nitrogen as a refrigerant at 77K and 1 bar is not a viable option at present. There are indications however that YBCO materials offer satisfactory performance with high current densities at liquid nitrogen temperatures; but the production of long lengths of high current 10 - 100A tapes or wires is proving difficult and it seems unlikely that they will be available in an engineering form for some time.

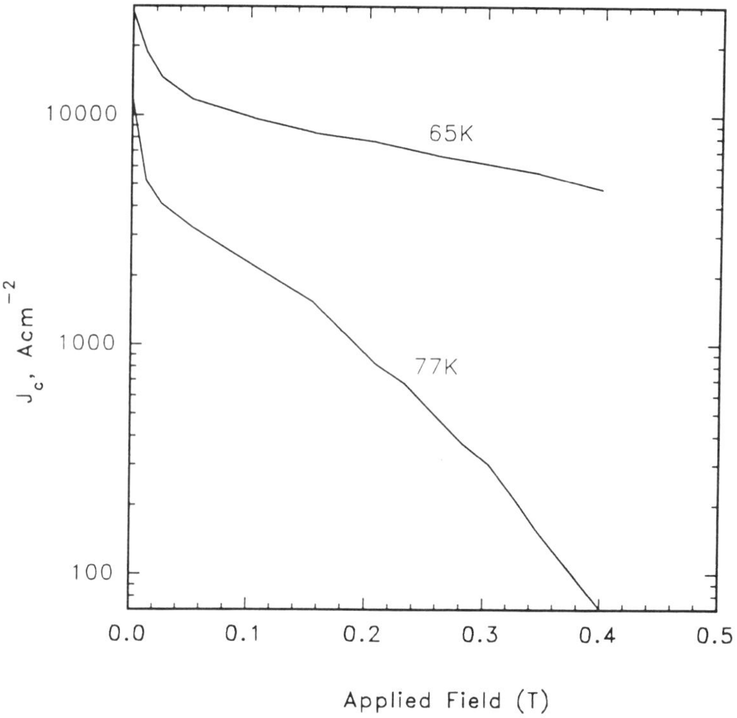

Figure 1. Typical j-H characteristic of PBSCCO 2223 superconductor at 65 and 77K.

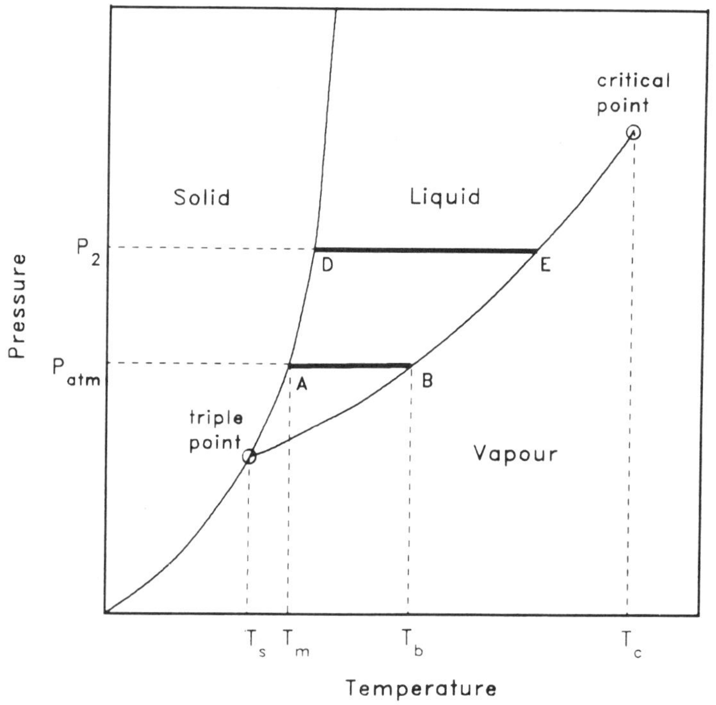

Figure 2. Typical P-T diagram of cryogen (excluding ^3He and ^4He).

CRYOGENIC ENGINEERING BELOW 65K : THE USE OF SUBCOOLED LIQUIDS

Power engineering applications are likely to achieve a large scale very quickly, and an engineering approach is needed to attain temperatures below 77K on a reliable and economic basis. While temperatures below 77K can be easily attained on a small scale in laboratory cryostats with the aid of rotary vacuum pumps,this is not very suitable on a larage scale.

One major way forward for consideration is the use of subcooled refrigerants at atmospheric pressure, or elevated pressures up to critical pressure. Subcooled liquid nitrogen is suitable down to 63.2K, while subcooled nitrogen/oxygen mixtures may be used down to 50.1K. Liquid neon is particularly suitable in the range 25 to 35K.

The technique of subcooling is understood by reference to the P-T phase diagram of a cryogenic fluid (Figure 2). The saturated vapour-liquid equilibrium temperature range over which a liquid refrigerant can be used is generally determined (apart from helium 3 and 4) at the lower end by the triple point T_S, which is the lowest temperature which can be reached by pumping away the vapour and thereby self- cooling the liquid to the solid state,

At temperatures below the triple point, only solid refrigerant is available and this has poor heat transfer characteristics compared with that of the liquid and is generally unsatisfactory for attaining controlled refrigeration. There is little experience with the use of solid cryogens at temperatures below 77K, apart from slush hydrogen at 12 - 14K. At higher temperatures, solid carbon dioxide at 195K and ice at 273K are widely used at 1 bar, but usually in contact with a liquid phase when good heat transfer is required.

The P-T phase diagrams can also be regarded as thermodynamic state diagrams in which the fluid temperature and pressure are separable parameters. For example, refering to Figure 2, the line AB represents the thermodynamic state of the liquid at 1 bar at temperatures from the normal boiling point (T_b) extending down to the melting point (T_m). The liquid is in a sub - cooled state with respect to T_b and can be used very conveniently as a non - isothermal refrigerant at 1 bar pressure along AB. At an elevated pressure P_2, liquid can be used as a sub - cooled refrigerant along the thermodynamic state line DE, but in this case the minimum temperature is that set by the melting point D which will be slightly higher than that at A.

Now, in general the density of subcooled liquid is greater than that of the saturated liquid at the same pressure. This property enables a layer of saturated liquid to stratify on top of the sub cooled liquid with very little heat transfer between the 2 layers owing to (i) the stable density gradient preventing convection and (ii) the observed natural thickening of the liquid inter - layer leading to a small temperature gradient and poor thermal conductance.

Both the concepts of sub - cooling and stratification can be used advantageously in the cryogenic engineering of superconductors below 77K.

THREE MODES OF REFRIGERATION

Three modes of refrigeration are available for cooling the superconductor and its associated composite support structure (a) under steady state power dissipation from AC losses, current leads and joints (b) transient heat pulses thereby providing thermal stabilisation and (c) during a quench when there is rapid dissipation of magnetic energy within the superconductor.

1. **Wet cooling** in which the superconductor is cooled by direct wetted contact with a refrigerant via immersion in (a) a liquid bath, or (b) a conduit.

Very high rates of boiling heat transfer can be achieved in a liquid bath of saturated refrigerant. If required, these rates can be substantially increased by the use of enhanced

Table I. Particular refrigerants

	T_m (melting or triple pt)	T_b (normal boiling pt)	T_c (critical pt)
Oxygen	54.8 K	90.2 K	155 K
Argon	84.0	87.3	151
Nitrogen	63.2	77.4	126
Air 21% O_2 79% N_2	57.0	78.8 bubble pt 81.8 dew pt	132
Eutectic 78% O_2 22% N_2	50.1	85.0 bubble pt 88.5 dew pt	149
Neon	24.6	27.1	44.4
Hydrogen (p)	13.8	20.28	32.9
Helium 4	-	4.214	5.201

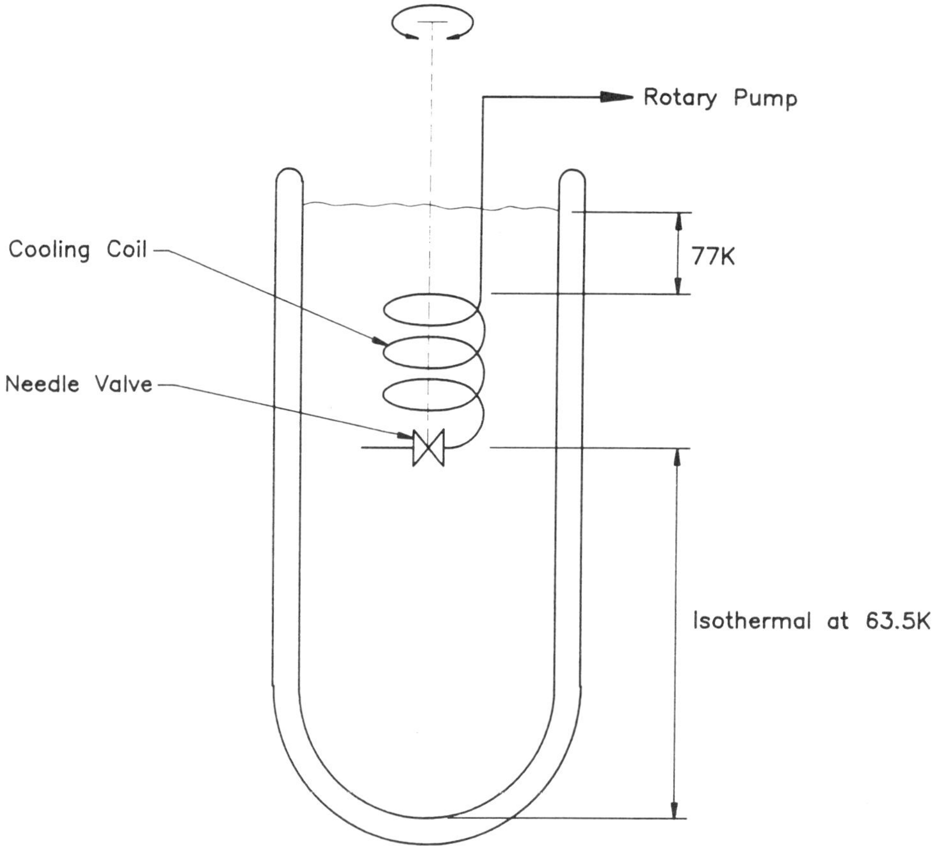

Figure 3. Subcooled liquid nitrogen bath at 1 bar.
Liquid nitrogen admitted through the needle valve is cooled thereby providing the necessary refrigeration for producing an isothermal bath below the coil.

boiling surfaces such as porous metal surfaces. In a conduit, boiling is not advised and heat transfer to the absorbed liquid is by liquid convection only. At this stage of HTS technology, wet cooling by saturated or sub - cooled liquid is the preferred method for quench protection in power engineering applications. One particular reason in favour of wet cooling is the high volumetric heat absorption capacity presented by the latent heat of evaporation as well as the high volumetric heat capacity of a liquid.

2 **Gas cooling** in which the superconductor is cooled by (a) direct contact with a cold gas as the refrigerant or (b)gas acting as a heat exchange medium for indirect heat absorption by (i) a liquid or solid refrigerant through its latent and /or sensible heats, or (ii) by the cold head of a refrigerator.

In general, heat transfer coefficients with natural convective gas cooling are several orders of magnitude less than with convective liquid cooling, although this difference can be reduced by use of high pressure gas and /or forced flow convection. The volumetric heat capacity of a gas is very low compared with that of a liquid refrigerant. While gas cooling is simple and attractive for the steady state refrigeration duties it does not appear to be adequate for quench protection, even when a separate high volumetric capacity cold store of cryogen or high thermal conductivity metal is used.

3. **Dry cooling** in which the system is cooled by thermal contact and / or thermal conduction to high thermal conductivity metals which are in turn cooled by contact with a refrigerant or refrigerator cold head. This cooling and may be satisfactory for the steady state refrigeration duties in the case of small superconducting magnets and electronic devices where no quench protection is needed i.e. where the stored energy, to be dissipated in the event of a quench, is relatively small in comparison with the heat capacity of the superconductor.

PARTICULAR REFRIGERATION SYSTEMS

Let us now examine particular refrigerants (2) and refrigeration systems, with reference to Table 1.

Sub-cooled nitrogen 65-63.2K

Liquid bath cooling with liquid nitrogen offers a very simple refrigeration system. A subcooled liquid nitrogen bath at 1 atmosphere can be easily achieved by using a cooling coil
near the surface of a deep pool as in Figure 3. Liquid nitrogen is admitted through a needle valve into the cooling coil which is connected to a rotary vacuum pump. An isothermal bath down to 63.2K the triple point is easily achieved. Lower temperatures cannot be achieved owing to solid formation downstream of the needle valve. This technique parallels the λ - plate cooler for producing subcooled superfluid helium at 1 bar pressure described originally by Roubeau in 1971 (3).

Solid nitrogen 63.2 - 45K

Solid nitrogen down to about 45K can be achieved by pumping a bath of liquid nitrogen. However, the solid has extremely poor heat transfer characteristics and it has proved very difficult to control temperatures in the presence of a refrigeration load. Solid nitrogen is therefore not recommended as a refrigerant.

Liquid nitrogen / oxygen mixtures 65 - 50K

Liquid air 65 - 57K

Temperatures below 63K can be obtained using subcooled liquid nitrogen/oxygen mixtures at 1 atmosphere, but with a large capacity rotary pump lubricated with oxygen compatible vacuum pump oil. For example, temperatures down to 57K can be attained by subcooling liquid air.

Nitrogen/oxygen eutectic 65 - 50.1K

Lower temperatures can be achieved with greater oxygen concentrations. Liquid baths down to 50.1K are possible where the 22% nitrogen 78% oxygen mixture cools to an eutectic solid at 160 Pa [4]. Further pumping of the eutectic solid will yield temperatures down to 45K, but this is not recommended since the solid is a poor thermal conductor. In practice, a compromise is needed between pumping speed and refrigeration load, and the lowest reasonable working temperature with liquid nitrogen/oxygen mixtures may be around 55K.

Quench protection by sub-cooled LIN/LOX mixtures

These subcooled baths at 1 atmosphere pressure provide extremely safe quench protection, since (a) they vent freely to the atmosphere during a quench (b) they have very high latent heat cooling capacities per unit volume, greater than 200 kJ/ℓ, compared with 2.6 kJ/ℓ for liquid helium (c) the volume of gas produced during a quench is 75 times less than that met in a quench of a liquid helium cooled superconductor.

Liquid neon 40 - 24.6K

Below 40K the use of subcooled liquid neon is very attractive. A particular advantage of neon is its very high evaporative cooling capacity per unit volume of 103 kJ/ℓ and that of helium (latent + sensible heat to 27.1K) of 18Kj/ℓ.

Liquid neon at its normal boiling point, or subcooled to 24.6K is therefore a most attractive refrigerant for use with high temperature superconductors (5). While the cost of neon gas is high at present the cost of unit refrigeration 27.1K is slightly cheaper with neon than with liquid helium. An increase in market interest would bring the price down significantly since the technology for extracting neon from the atmosphere is available at little capital cost to the gas separation industry.

Table II. Thermal impedances (Degrees K per unit heat flux in W/cm^2)

Gas film by convection	100-500
Bulk liquid contact, LIN, 77K	0.2-1.0
10mm thickness of HTS at 77K	100
1mm thickness of HTS at 77K	10
30mm of copper at 77K	0.5
Thermal contact copper-copper at 77K (45 kg load; impedance independent of area of contact)	1.0-3.0 K/W

The use of solid neon below 24.7K is not recommended. Thermal contact with the solid is extremely low , leading to poor control of the refrigeration system. Experience at Southampton suggests that inadvertent production of solid neon by overcooling or over pumping can lead to explosive evaporation of the solid when the temperature is next allowed to rise above the triple point.

Cryocoolers

A number of well designed and reliable cryocoolers are available on the market, and these can be used to attain any temperature level from 77K down to 4K. The cooling power is limited, particularly at the lower temperatures and they are therefore only suitable for cooling small devices, relatively small magnet systems and thermal screens. A major problem with cryocoolers is that associated with their moving parts is a magnetic field. This motion can distort the fields, induce eddy current heating in the cryocooler and provides a source of mechanical vibration. The use of non - metallic moving parts, of split cycle coolers, and good mechanical damping can alleviate these problems. Cryocoolers are, of course, extremely convenient and easy to use.

THERMAL IMPEDANCE MATCHING

The large disparity should be noted between the various thermal impedances associated with the refrigeration of high temperature superconductors. These are illustrated in Table II.

The major thermal impedance is associated with the high temperature superconducting materials themselves, although gas cooling presents a still higher impedance. These impedancies can be reworked in terms of heat extraction during a quench, and the internal temperature gradients required under steady state heating. Various conclusions can be drawn, in particular :

(a) For effective steady - state heat transfer, the superconductor should be within 1 mm of bulk liquid contact, or included in a composite structure of thin (< 1mm thick) superconducting laminae interleaved with high thermal conductivity material such as copper or aluminium.

(b) No quench protection can be achieved in bulk superconductors during a rapid quench - it may not be necessary if the magnetic energy dissipation is less than the the line enthalpy gain.

(c) In large systems where the magnetic energy dissipation is larger thn the limiting enthalpy gain, heat dumping into a liquid coolant will have to be employed via the use of thin superconducting laminae.

ACKNOWLEDGEMENTS

We wish to acknowledge sub- cooling experiments carried out by our colleagues in the Institute of Cryogenics, including Dr C Beduz and Mr D Neate (EPSRC research student) in which temperatures down to 50.1K have been obtained by sub - cooling liquid nitrogen/oxygen mixtures, while temperatures down to 45K have been attained by pumping nitrogen solids.

REFERENCES

1. R.F.Barron. Cryogenic Systems, OUP (1985).
2. R.B.Stewart, R.T.Jacobsen, S.G.Penoncello, Thermodynamic Properties of Refrigerants, ASHRAE (1986)
3. Roubeau, Comptes Rendus (1971) 273 B, 581 - 583
4. M. and B.Ruhemann, Low Temperature Physics, Cambridge University Press (1937) 98 - 101
5. R.N.Richardson, R.G.Scurlock and A.C.R.Tavner, Cryogenics 34 Supplement 293 - 96 (1994).

CLOSED CYCLE CRYOCOOLERS WITH A PHASE-CHANGE SYSTEM FOR THE COOLING OF HIGH-T_C SQUIDS

G. Kaiser, M. Thürk, and P. Seidel

Institut für Festkörperphysik, Friedrich-Schiller-Universität
Jena, D-07743, Germany

ABSTRACT

The use of high-T_C sensors like SQUIDs for practical applications requires user friendly cryogenic subsystems those are easy to handle and have a high reliability[1]. Small scale closed cycle cryocoolers meet these requirements, however, they produce additional noise during the operation. One way to solve this problem is the use of a miniature cryocooler coupled with a phase-change system (PCS) as a cold accumulator for intermittent low noise operation. In this paper we describe our research about two different cryocooler-PCS concepts operating with liquefied gases like nitrogen or neon and solidified hydrocarbons like propane, respectively. Theoretical investigations and experimental results will be presented. Successful high-T_C SQUID operation on an experimental system will be shown.

INTRODUCTION

Small scale closed cycle cryocoolers (CCC) meet the requirements to be a user friendly cooling technique. There are some more advantages, too. Using liquid nitrogen as a coolant it is not possible to reach the optimum temperature range[2] between 45 and 55 K of the mostly used YBCO-SQUID. Furthermore the coefficient of performance of those machines within this temperature interval is quite high and the cooling power requirements of these sensors are a few tenths of a Watt. So, it is possible to get miniaturized cooler solutions consuming only a few Watts of input power[3].

The use of CCC for high-T_C SQUID applications always lead to problems with additional noise coming from the cooler subsystem. One can distinguish mainly three noise influences, which are temperature fluctuations, magnetic noise and mechanical vibrations. For a successful SQUID operation it is necessary to decouple these noise influences from the highly sensitive superconducting sensor. One method to do this is time separation by the use of an intermittent cooling and measuring technique. The CCC acts as a cold generator charging its phase change system (PCS)[4, 5] in the way of a cold accumulator. After that it is possible to switch off the CCC and one can perform SQUID measurements at noise levels equal to those of dewar based systems[6]. The cold is provided by the PCS with a high temperature stability without electronic control.

PCS THEORY

PCS materials

In order to construct a cold accumulator for high-T_C SQUID operation one can use phase transitions exhibiting a sufficiently high amount of volume related latent heat and a good temperature stability at a useful temperature. Liquid-gas or solid-liquid phase transitions are most interesting. Table 1 shows the most interesting candidates with their transition temperatures and volume related specific latent heats.

For the liquid-gas transition the use of neon is preferred because of its low transition temperature and high latent heat. The transition temperatures of solid propane or ethane are in both cases too high. Using an eutectic mixture made of propane (22 %), ethane (25 %) and methane (53 %) it is possible to get a useful transition temperature of 66 K[7].

PCS size

For useful SQUID operation the PCS has to be designed to enable about 90 to 120 minutes of low noise operation. The cold accumulated in the PCS has to compensate the device heat load and the thermal losses. To suppress the conduction losses from the PCS to the coldfinger it is useful to separate the PCS during the low noise operation from the coldfinger. So, we only have calculated radiation losses of about 300 mW and a device heat load of 200 mW mainly caused by the heat conduction of the supporting wires (four-terminal SQUID measuring configuration and the wires for the temperature sensor and the magnetic field coil). The power dissipation of the device in the range of a few tenths of a mW can be neglected. The total heat load is 500 mW which requires 3.6 kJ of accumulated cold for 120 minutes of low noise operation. Table 2 shows the necessary volumes to accumulate this amount of cold for the different cryogens.

It is an advantage of a solid-liquid system to operate with a small closed vessel at the triple point of the PCS material. The operation of a liquid-gas system requires a pressure stabilized reservoir for large amounts of gas. One milliliter of liquid neon generates about 1.4 liters of neon gas at room temperature.

Table 1. PCS materials

liquid-gas transition			solid-liquid transition		
PCS material	temperature [K]	spec. latent heat [J/ml]	PCS material	temperature [K]	spec. latent heat [J/ml]
nitrogen	77.4	110	propane	85.5	57
neon	27.1	106	ethane	90	56

Table 2. PCS volumes for 3.6 kJ of latent heat for 120 minutes of low noise operation

liquid-gas transition		solid-liquid transition	
PCS material	volume [ml]	PCS material	volume [ml]
nitrogen	33	propane	63
neon	34	ethane	64

Table 3. Maximum impurity concentration for 1 mK temperature drift

PCS material	impurity	max. concentration [mol%]
nitrogen	oxygen	0.004
neon	helium	0.07

Table 4. Maximum pressure variations for 1 mK temperature fluctuation

PCS material	max. pressure variation [Pa]
nitrogen	14
neon	40

Temperature stability

During the low noise operation there is no additional magnetic noise or vibration caused by the CCC. The dominating noise influences are now temperature fluctuations. In order to operate a high-T_C SQUID successfully for biomagnetic measurements a maximum temperature fluctuation of about 1 mK is essential[8, 9]. The temperature stability of the phase transition can be influenced by impurities and pressure fluctuations. If the cryogen contains impurities it behaves like a mixture. The phase transition happens in a temperature interval either between the dew point and the boiling point for the liquid-gas transition or between the liquidus point and the solidus point for the solid-liquid transition. This leads to a long term temperature drift during melting or evaporation. For the investigation of the maximum impurity concentration for 1 mK temperature fluctuation, we have considered the PCS materials as binary mixtures of the cryogen and the dominating impurity. Table 3 shows the results of our calculations for liquid-gas systems.

The transition temperature for a PCS material in the thermal phase equilibrium state strongly depends on the pressure due to its vapor- or melt- pressure curve. The steepness of the melt-pressure curve is usually much higher than that of the vapor-pressure curve. The melting temperature variation due to pressure fluctuations is orders of magnitude less compared to the evaporation case. Therefore we have only considered the liquid-gas transition. The results of our calculations for a 1 mK temperature fluctuation are shown in Table 4.

Temperature fluctuations due to boiling

We have performed detailed theoretical investigations in order to understand the boiling process and to get information about the magnitude of the temperature fluctuations due to boiling. To generate a bubble with a radius r one must superheat the liquid to overcome its hydrostatic pressure and the internal pressure shift of the bubble due to the surface tension σ of the surrounding liquid.

The hydrostatic pressure shift Δp_h is

$$\Delta p_h = \rho g h \quad , \tag{1}$$

defined by the liquid density ρ, the earth acceleration g and the height h of the liquid in the vessel. The internal pressure shift Δp_s due to surface tension can be described with the following equation:

$$\Delta p_s = \frac{2\sigma}{r} \quad . \tag{2}$$

The pressure shift corresponds with a temperature shift due to the vapor pressure curve of the PCS material. We have considered the temperature shift as a hint of the temperature fluctuations due to boiling.

Table 5. Properties of some PCS materials[10]

PCS material	liquid density [kg/m³]	surface tension [mJ/m²]	dT/dp [μK/Pa]
nitrogen	804.2	8.8	70
neon	1204	4.8	25
helium	124.8	0.09	14

Table 6. Hydrostatic pressure shift and temperature shift for some PCS materials

PCS material	pressure shift [Pa]	temperature shift [mK]
nitrogen	785	55
neon	1175	29
helium	122	1.7

For the calculation of the hydrostatic effect we have considered a liquid height of 10 cm. Table 6 shows the results using the material properties listed in Table 5. Figure 1 shows the results of our calculations for the radius dependence of the temperature shift due to surface tension for nitrogen and neon in comparison with helium. The temperature shift in boiling nitrogen or neon is about 2 orders of magnitude higher than in boiling helium considering the same bubble size. We have performed measurements in boiling nitrogen using a thermocouple with a low heat capacity (copper-constantan, wire diameter 0.05 mm). The temperature fluctuation was about 10 mK peek to peek confirming the results measured by A. Reichmann[11] years ago. He also measured the temperature fluctuation in boiling helium[12] obtaining about 0.1 mK peek to peek. These fluctuations are in good agreement with the normally used surface tension model under consideration of a typical bubble radius of about 0.1 mm.

Why doesn't the hydrostatic effect contribute to the temperature fluctuation? The hydrostatic pressure shift Δp_h acts in the bubble as in the liquid. So, there is no temperature difference between the gas in the bubble and the liquid if one measures at the same height where the bubbles were generated. The pressure shift due to bubble surface tension Δp_s, however, only acts in the bubble what leads to a temperature shift between the gas in the bubble and its surrounding liquid as shown in Figure 1.

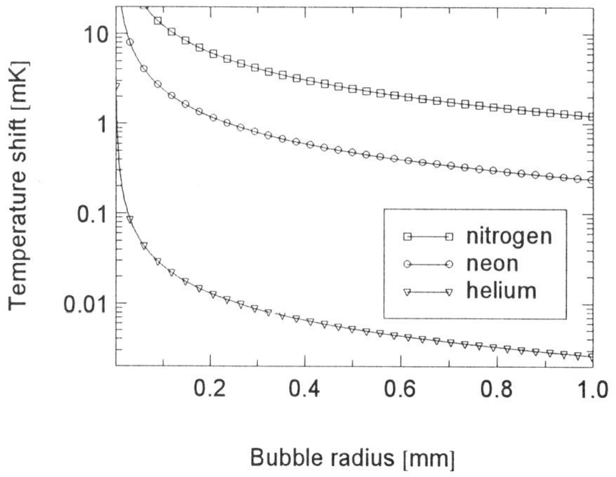

Figure 1. Temperature shift due to the surface tension for some cryogens in dependence on the bubble radius.

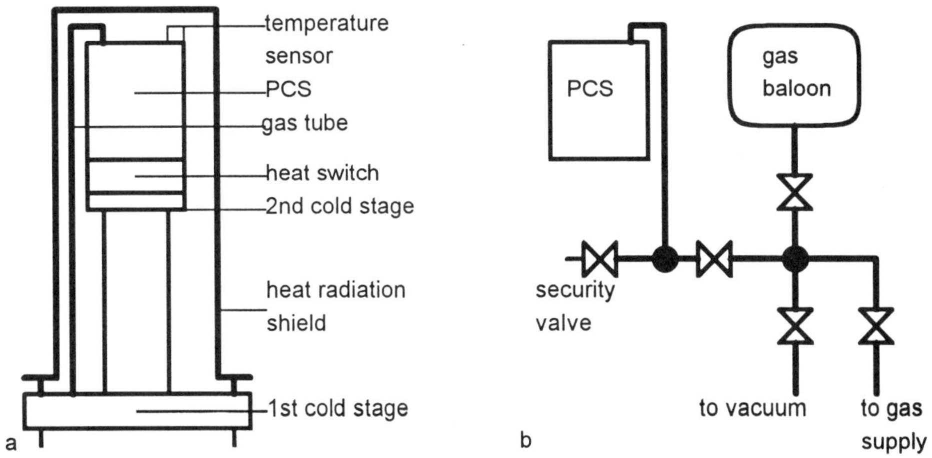

Figure 2. Construction sketch of the PCS (a) and its periphery (b).

EXPERIMENTS

Experimental setup

We have investigated a liquid-gas PCS/CCC concept experimentally. The CCC was a 2-stage Gifford-McMahon cooler (RGD 210, Leybold) with a cooling power of 2 Watt at 20 K. The PCS consits of a cylindrical copper vessel (40 mm diameter, 50 mm long) with a gas inlet at the top side. The vessel contains an additional star shaped copper cold plate of 620 cm^2 to improve the thermal contact to the cryogen. The construction is shown in Figure 2a. A mechanically driven heat switch was located between the cold finger and the PCS to disconnect it during low noise operation from the cold finger. These parts were covered by an aluminium heat radiation shield connected to the first cold stage.

The gas inlet of the PCS was connected to a stainless steel tube leading to a gas supply consisting of a 0.5 m^3 baloon and a system for evacuation or gas filling. The system sketch is shown in Figure 2b. We have operated the system with either nitrogen or neon. A temperature sensor diode (DT 470, Lake Shore) located at the PCS was used for temperature monitoring during cooling and warming of the machine.

Experimental results

Before operation the baloon was filled with the gas and hermetically sealed. The PCS was set in contact to the coldfinger. During the cooling of the CCC the temperature of the PCS was measured. Figure 3a shows the cooling curve of the PCS for operation with nitrogen. In part 1 of the curve the temperature of the PCS falls continuously until condensation starts. After subcooling the temperature at the coldfinger stabilizes some degrees below the condensation temperature (part 2) due to the thermal resistance between the coldfinger and the PCS. When the PCS is completely filled with cryogen the temperature will fall again (part 3). This is the moment to disconnect the PCS from the coldfinger and to switch off the CCC.

The warming curve is shown in Figure 3b. After the disconnection of the PCS from the CCC the temperature of the PCS will rise (part 1) and stabilize a few tenth of a degree above the condensation temperature (part 2) mainly due to heat irradiation. Low noise measurements can be performed for about 110 minutes. Figure 4 shows the modulation curve of a high-T_C dc-SQUID recorded during this time interval. The curve shows 0.5 μV low frequency noise because the SQUID was dc-coupled to the preamplifier.

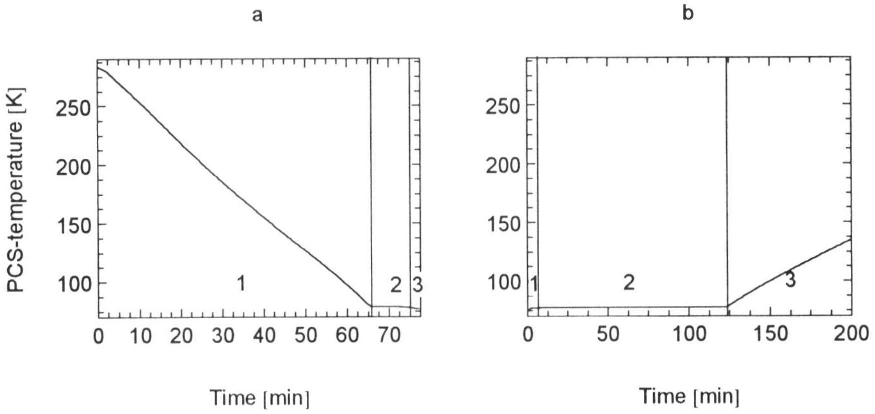

Figure 3. Cooling (a) and warming (b) of the PCS.

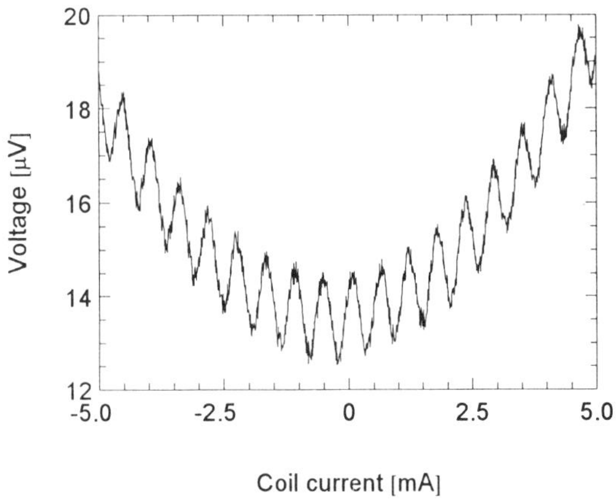

Figure 4. Modulation curve of a YBCO dc-SQUID operated on a CCC/PCS at 27.4 K.

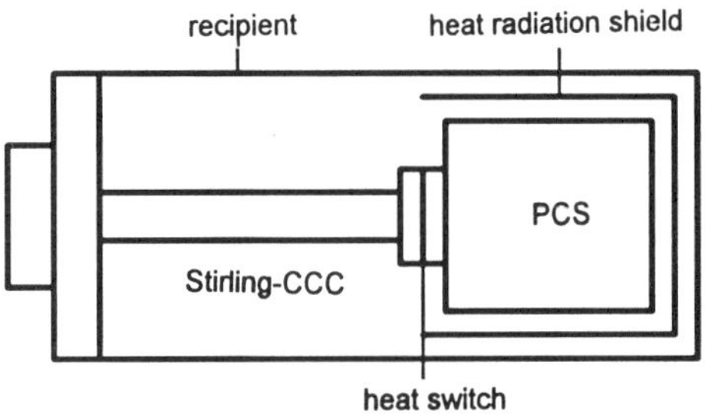

Figure 5. Construction sketch of a miniature solid-liquid CCC/PCS concept.

The PCS was operated with neon at a temperature of 27.4 K. When the cryogen is completely evaporated the temperature will rise again (part 3). During the low noise interval a temperature drift (0.14 K in nitrogen, 0.04 K in neon) has been observed caused by heat irradiation of the warming heat radiation shield. The short term fluctuation was 10 mK peek to peek independent on the PCS material.

FURTHER INVESTIGATIONS

Investigations on a miniature solid-liquid CCC/PCS concept driven by a 1 W Split-Stirling machine are in preparation (Figure 5). The first prototype will operate with solid propane in a pressure resistant vessel designed for 120 minutes of low noise operation. A heat radiation shield floating in the radiation equilibrium will be used to prevent a temperature drift due to variations of heat irradiation to the PCS.

ACKNOWLEDGMENT

Acknowledgment to the German BMBF for financially supporting our work (contract numbers 13N5924A and 13N6618).

REFERENCES

1. M. Nisenoff, Cryocoolers and high temperature superconductors: advancing toward commercial applications, *Cryocoolers* 8: 913 (1995).
2. J.W.M. Hilgenkamp, F.J.G. Roesthuis, and S. Hoogeveen, Monolithic flux transformer coupled high-T_c dc SQUID magnetometer, *Proc. of HTS-workshop on applications and new materials, 8-10 May 1995, Enschede, The Netherlands*: in press (1995).
3. T.H. Clynne, J.R. McCoy, P.E Kimak, M. Kimak, Commercial low cost cryogenic packaging for linear micro-coolers, *Cryocoolers* 8: 949 (1995).
4. G. Kaiser, P. Seidel, M. Thürk, Noise reduction of cryo-refrigerators, *Adv. in Cryog. Eng.* 39: 1281 (1993).
5. B.G. Williams, J.C. Batty, Incorporating a mechanical refrigerator with a refreezable cryogen in space applications, *Cryocoolers* 7: 1043 (1993).
6. N. Kasai, K. Sasaki, S. Kiryu, Y. Suzuki, Thermal magnetic noise of Dewars for biomagnetic measurements, *Cryogenics* 33, No2: 175 (1993).
7. L.B. Robinson, Phase equilibria of cryogenic mixtures: part II, *Cryocoolers* 8: 559 (1995).
8. G. Kaiser, J. Blaurock, R. Hackenberger, M. Thürk, P. Seidel, Closed cycle cryocoolers for HTSC-SQUIDs, *Proc. of HTS-workshop on applications and new materials, 8-10 May 1995, Enschede, The Netherlands*: in press (1995).
9. W.A.M. Aarnink, M. ter Brake, J. Flokstra, H. Rogalla, O. Dössel, High-T_c dc-SQUIDs for multichannel heart scanner operating in an unshielded environment, *Proc. of HTS-workshop on applications and new materials, 8-10 May 1995, Enschede, The Netherlands*: in press (1995).
10. H. Frey, R. A. Haefer, "Tieftemperaturtechnologie", VDI-Verlag, Düsseldorf (1981).
11. A. Reichmann, Physikalisch-technische Aspekte der Kühlung von Infrarotdetectoren, *Proc. of 17. Internationales Symposium Tieftemperaturphysik und Kryoelektronik, 4.-8. Nov. 1985, Georgenthal, Germany*: 126 (1985).
12. A. Reichmann, Temperaturschwankungen im Heliumbad, *Proc. of 17. Internationales Symposium Tieftemperaturphysik und Kryoelektronik, 4.-8. Nov. 1985, Georgenthal, Germany*: 145 (1985).

STUDIES ON A MODIFIED ERICSSON CYCLE WITH NEON AS REFRIGERANT

O. Fredrich, C. Haberstroh, and H. Quack

Technische Universität Dresden
Lehrstuhl für Kälte- und Kryotechnik
Dresden, D-01062, Germany

ABSTRACT

The classical Ericsson cycle consists of two isotherms and two isobars. With an ideal gas as refrigerant a reversible process can be realized. To maintain reversibility with a real fluid as refrigerant, the high pressure isobar is being replaced by a combination of isobaric and isentropic steps.

If the cycle includes the two-phase region, the low temperature isotherm can be obtained by isobaric evaporation without the need for simultaneous work extraction.

This so-called modified Ericsson cycle is still a reversible process. With neon as refrigerant, the heat exchange and work extraction requirements are being studied. In a second step the influence of non-ideal heat exchange and non-ideal work extraction on the total cycle efficiency are being investigated.

INTRODUCTION

The Carnot cycle (Fig. 1) is the "mother" of all reversible refrigeration cycles. It does not contain any internal heat transfer. Therefore it can theoretically be realized with any kind of fluid, even with a real gas with pressure and temperature dependent thermal properties. But the Carnot cycle is impractical in cryogenic applications, because it leads to very high pressure differences.

More practical are the Stirling and the Ericsson cycle, which work with internal heat transfer to bridge the temperature distance between the ambient temperature T_{amb} and the refrigeration temperature T_0. To be reversible, the Stirling and the Ericsson cycle need refrigerants with equal specific heats of the high and the low pressure stream. Therefore one indication of the reversibility is, that the two curves with inner heat exchange are equidistant in the T-s diagram (Fig. 1).

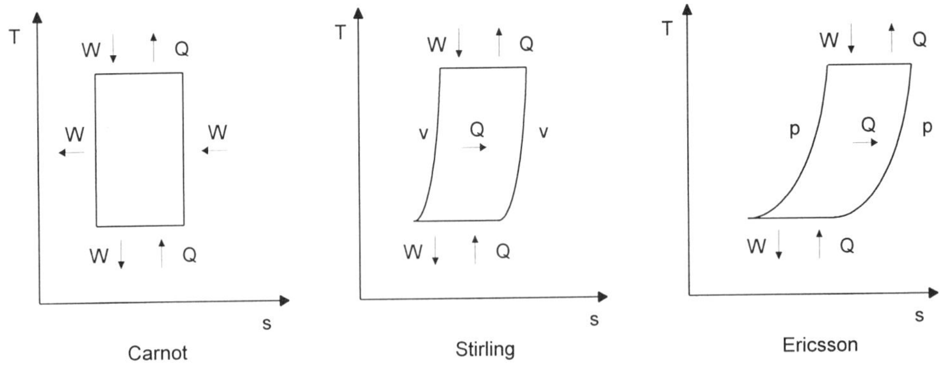

Figure 1. Reversible refrigeration cycles.

Real fluids depart from the requirement of pressure independent specific heats, the more the cycle approaches the vapor dome.

Two years ago we reported on the search for the most efficient cycle for refrigerators at the liquid helium temperature level [1]. We found the cycle shown in Fig. 2 to be very promising. It has two isotherms at T_{amb} and T_0 and an isobar at the low pressure side. Because the specific heats of the high and low pressure stream are not equal, heat exchange alone can not secure the equidistance. This can only be reached by a combination of heat transfer with many small isentropic expansions of the high pressure stream. So the high pressure stream is not isobaric. The pressure is decreasing with decreasing temperature.

Because the cycle shown in Fig. 2 has many similarities with the Ericsson cycle, we call it a modified Ericsson cycle. One significant difference is that work extraction occurs only in the high pressure stream and not along the isotherm at T_0, if one chooses to produce refrigeration below the critical temperature of the refrigerant.

In the following we study the change of state on the high pressure side of the cycle.

REQUIRED TEMPERATURE REDUCTION BY EXPANSION

We investigate a small element of the cycle, where the low pressure stream is warmed up by the temperature step ΔT_L.

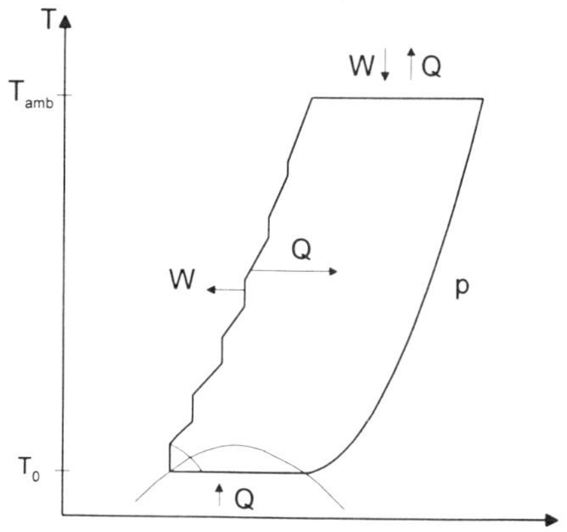

Figure 2. Modified Ericsson cycle with a real fluid.

Assuming the mass flow rates in the high and low pressure streams to be equal, the transferred heat is able to cool the high pressure stream by

$$\Delta T_{H,hx} = \frac{c_{p,L}}{c_{p,H}} \cdot \Delta T_L \tag{1}$$

To obtain an overall temperature reduction of the high pressure stream equal to the temperature rise of the low pressure stream, an additional temperature reduction of the high pressure stream must be obtained by work extracting expansion:

$$\Delta T_{H,e} = \Delta T_L - \Delta T_{H,hx} = \left(1 - \frac{c_{p,L}}{c_{p,H}}\right) \cdot \Delta T_L \tag{2}$$

Which pressure drop by expansion is necessary to produce this temperature reduction?

PRESSURE REDUCTION BY WORK EXTRACTING EXPANSION TO OBTAIN THE DESIRED TEMPERATURE REDUCTION

The modified Ericsson cycle is a "real fluid" process, i.e. it has to take into account the deviations from the ideal gas approximation. We make use of the following thermal properties of the fluid, whose values are pressure and temperature dependent:

$\mu = \left(\frac{\partial T}{\partial p}\right)_h$	Joule Thomson coefficient
$c_p = \left(\frac{\partial h}{\partial T}\right)_p$	Specific heat capacity
v	Specific volume

We start with the total differential of $T = T(h,p)$:

$$dT = \left(\frac{\partial T}{\partial h}\right)_p dh + \left(\frac{\partial T}{\partial p}\right)_h dp = \frac{1}{c_p} dh + \mu\, dp \tag{3}$$

From the 1st and 2nd law one may write:

$$Tds = dh - vdp \tag{4}$$

For an isentropic expansion one obtains:

$$\left(\frac{\partial h}{\partial p}\right)_s = v \tag{5}$$

That means for a small isentropic expansion step (Fig. 4):

$$\Delta h_s = v \cdot \Delta p_e \tag{6}$$

With the help of the definition of the isentropic efficiency

$$\eta = \frac{\Delta h_e}{\Delta h_s} \tag{7}$$

the enthalpy reduction by a non-isentropic expansion step is described by:

$$\Delta h_e = \eta \cdot \Delta h_s = \eta \cdot v \cdot \Delta p_e \tag{8}$$

With change from the differential to a finite, but small difference, Eq. (3) is written:

$$\Delta T_{H,e} = 1/c_{p,H} \cdot \Delta h_e + \mu \cdot \Delta p_e \tag{9}$$

Introducing Eq. (8) into Eq. (9), we get:

$$\Delta T_{H,e} = \frac{\eta \cdot v}{c_{p,H}} \cdot \Delta p_e + \mu \cdot \Delta p_e \tag{10}$$

or

$$\Delta p_e = \frac{1}{\eta \frac{v}{c_{p,H}} + \mu} \Delta T_{H,e} \tag{11}$$

Δp_e is the pressure drop of the work extracting expander and $\Delta T_{H,e}$ is the associated temperature drop. This equation is valid for a small expansion step, where the thermal properties v, c_p and μ can be considered constant.

Using the prescribed temperature step of the low pressure stream ΔT_L instead of the temperature drop $\Delta T_{H,e}$ (see Eq. 2 and Fig. 3):

$$\Delta p_e = \frac{1 - \frac{c_{p,L}}{c_{p,H}}}{\eta \frac{v}{c_{p,H}} + \mu} \Delta T_L \tag{12}$$

While this equation takes into account the inefficiency of the expander, it does not account for a possible inefficiency of the heat exchanger.

NON-IDEAL HEAT TRANSFER

The heat transfer from the high to low pressure gas requires a "driving" temperature difference (δT) between both streams. It can be described by the number of transfer units

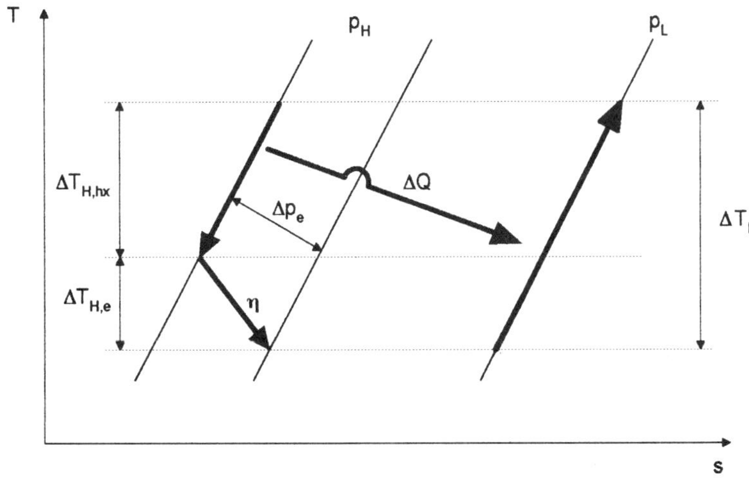

Figure 3. Small element of cycle with heat transfer and expansion .

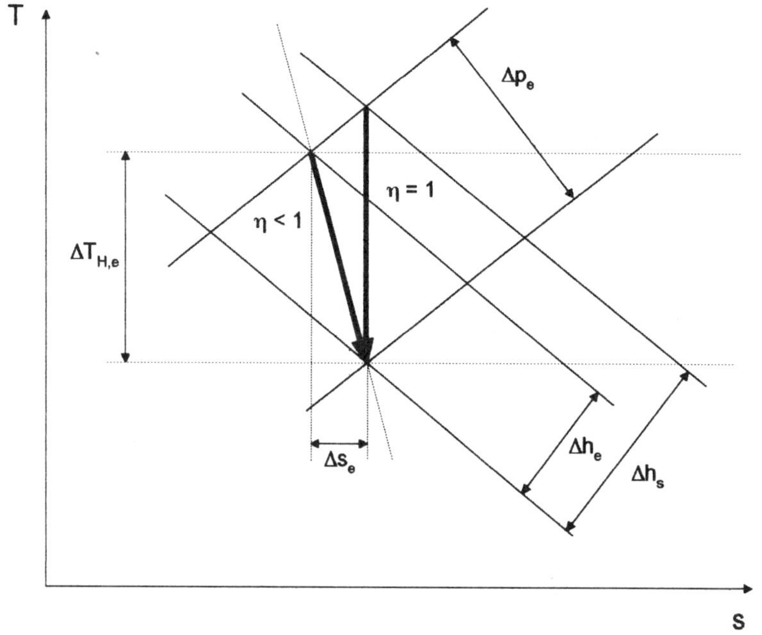

Figure 4. Isentropic (η= 1) and non-isentropic (η< 1) expansion steps shown in a T-s Diagram.

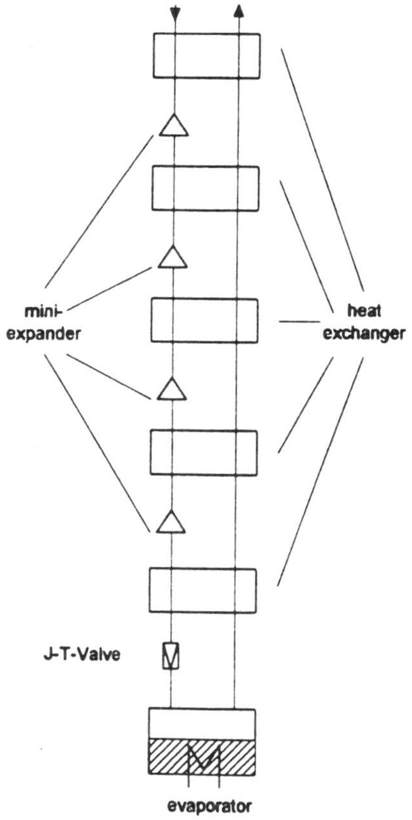

Figure 5. Cycle with 4 mini-expanders in series and a final Joule-Thomson valve.

(NTU) of the counter-current heat exchanger. Thermodynamically it is advantageous, if with a given NTU one chooses a small δT at the cold end, and a larger δT at the warm end of the heat exchanger. The optimum temperature difference distribution for a heat exchanger with a constant specific heat fluids is given by [2]:

$$\delta T = T_H - T_L = \frac{\ln\left(T_{L,w}/T_{L,c}\right)}{NTU} \cdot T_L \tag{13}$$

$T_{L,c}$ and $T_{L,w}$ are the cold inlet and warm outlet temperatures of the low pressure stream. Along a small process element similar to the one shown in Fig. 3, the temperature difference changes by

$$\Delta(\delta T) = \frac{\ln\left(T_{L,w}/T_{L,c}\right)}{NTU} \cdot \Delta T_L \tag{14}$$

This requires an additional temperature reduction, which has to produced by the expansion of the high pressure gas (see Eq. 2 and 14):

$$\Delta T_{H,e} = \Delta T_L - \Delta T_{H,hx} + \Delta(\delta T) = \left(1 - \frac{c_{p,L}}{c_{p,H}} + \frac{\ln\frac{T_{L,w}}{T_{L,c}}}{NTU}\right) \cdot \Delta T_L \tag{15}$$

Therefore the associated pressure drop by expansion (see Eq. 11 and 12) has to increase, too.

$$\boxed{\Delta p_e = \frac{1 - \frac{c_{p,L}}{c_{p,H}} + \frac{\ln\left(T_{L,w}/T_{L,c}\right)}{NTU}}{\eta \frac{v}{c_{p,H}} + \mu} \Delta T_L} \tag{16}$$

This equation is describing the pressure drop over consecutive heat exchange and expansion steps of the high pressure side of the modified Ericsson cycle with both irreversibilities (non-isentropic expansion and non-ideal heat transfer) included.

CALCULATION OF THERMODYNAMIC LOSSES

We study this question again in a process element as shown in Fig. 3, but with an additional temperature difference δT between high and low pressure stream (see Eq. 13).

The entropy production due to the heat flow from the high to the low pressure side is given by [1]:

$$\Delta s_{hx} = \frac{c_{p,L} \cdot \delta T}{T_L(T_L + \delta T)} \cdot \Delta T_L \tag{17}$$

and with Eq. (13) for δT, one obtains for the total losses connected with the heat transfer, assuming the specific heat of the low pressure gas to be constant over a wide range of temperatures:

$$\sum \Delta s_{hx} \approx c_{p,L} \cdot \frac{\left[\ln\left(T_{L,w}/T_{L,c}\right)\right]^2}{NTU + \ln\left(T_{L,w}/T_{L,c}\right)} \tag{18}$$

The entropy generation Δs_e shown in Fig. 4 due to the non-ideal work extracting expansion with the enthalpy drop Δh_e is determined by [1]:

$$\Delta s_e = \frac{\Delta h_e}{T_H} \cdot \frac{1-\eta}{\eta} \tag{19}$$

The total losses by these expansion steps are:

$$\sum \Delta s_e = \frac{1-\eta}{\eta} \cdot \int_{T_c}^{T_w} \frac{\Delta h_e / \Delta T_L}{T_H} dT_L \tag{20}$$

The enthalpy drop Δh_e can be divided into two terms, one due to the pressure dependent specific heat of the refrigerant and a second term due to the non-ideal heat transfer, which produces the optimum temperature difference δT:

$$\Delta h_e = \frac{1 - \frac{c_{p,L}}{c_{p,H}}}{\frac{1}{c_{p,H}} + \frac{1}{\eta} \cdot \frac{\mu}{v}} \cdot \Delta T_L + \frac{\frac{\ln\left(T_{L,w} / T_{L,c}\right)}{NTU}}{\frac{1}{c_{p,H}} + \frac{1}{\eta} \cdot \frac{\mu}{v}} \cdot \Delta T_L \tag{21}$$

So it is possible to distinguish between the different thermodynamic losses.

NEON AS A REFRIGERANT FOR THE 40 K LEVEL

The 40 K temperature level is of interest for applications of high temperature superconductivity. Neon could be a suitable refrigerant for this purpose. Therefore we try this new cycle with neon. An example for the cycle with four mini-expanders in series and a final throttle valve (Joule-Thomson valve) is given in Fig. 5.

To obtain a compact refrigerator, we are using a low pressure level of 20 bar (2 MPa), which corresponds to a refrigeration temperature of 42.2 K. We choose a rate of refrigeration of 33.6 J/ g. In entropy terms this is:

$$\delta s_0 = \frac{33.6 \text{ J/g}}{42.2 \text{ K}} \approx 0.8 \text{ J/g K} \tag{22}$$

In the "lower left corner" we deviate somewhat from the previously described modified Ericsson cycle and include a small Joule-Thomson stage with isobaric heat transfer followed by a throttle valve. the reason is that below 44 K the specific heat of the high pressure (4 MPa) stream is lower than that of the low pressure stream. The final expansion in the J-T valve from 40 bar (4 Mpa) down to 20 bar and the heat transfer in the J-T heat exchanger are connected with a small entropy generation (0.07...0.08 J/ g K).

We are inserting the real properties of neon into above equations for the thermodynamic losses. The results of the numerical calculation with many small process steps, determined by the temperature step ΔT_L, are summarized in Figure 6.

There is shown the entropy span at ambient temperature δs_{amb} in function of 1/ NTU with the efficiency η as parameter.

The efficiency of the total cycle is obtained by dividing the useful entropy (δs_0) by the value of δs_{amb} shown on Fig. 6. This does not include losses of the room temperature compression.

For e.g. NTU=28 and $\eta = 0.6$ one obtains a cycle efficiency of

$$\frac{\delta s_0}{\delta s_{amb}} \approx \frac{800 \quad \text{J/kg K}}{1450 \quad \text{J/kg K}} = 0.55.$$

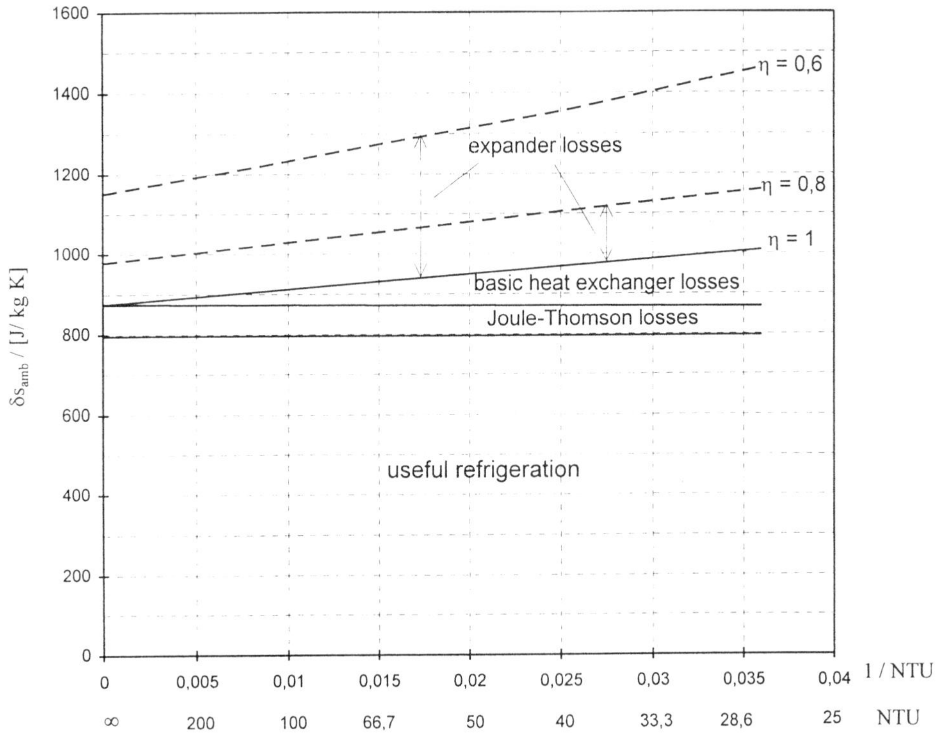

Figure 6. Entropy span δs_{amb} in function of 1/ NTU with the expander efficiency η as parameter.

So this cycle promises to give good performance even with moderately efficient components.

SUMMARY

A new refrigeration cycle, as was mentioned already in [1], with consecutive expansion (work extracting) and heat transfer steps of the high pressure stream, has been identified as a modified Ericsson cycle.

The thermodynamic equations have been derived for an estimation of such a cycle.

With neon as a refrigerant for the 40 K temperature level, the influence of losses (non-ideal heat transfer and non-ideal expansion) on the cycle efficiency is shown. Even with moderately efficient components it is possible to get a good cycle performance.

On the base of the described modified Ericsson cycle a new cryocooler with mini-expanders should be developed.

NOMENCLATURE TABLE

NTU		Number of transfer units
Q	kJ	Heat
T	K	Temperature
W	kJ	Work
c_p	kJ/ kg K	Specific heat capacity @ constant pressure
h	kJ/ kg	Specific enthalpy

p	Pa	Pressure
s	kJ/ kg	Specific entropy
v	m^3/ kg	Specific volume
Δ		Small difference for a variable of state of one stream
δ		Difference of a variable of state between two streams
η		Isentropic efficiency of the expansion
μ	K/ Pa	Joule-Thomson coefficient

Subscripts

H	high pressure stream
L	low pressure stream
amb	ambient temperature level
c	cold end
e	expansion step
h	isenthalpic change of state
hx	heat exchange step
p	isobaric change of state
s	isentropic step
w	warm end
0	refrigeration temperature level

ACKNOWLEDGMENT

This research project is financially supported by the BMBF, contract number 13 N 6412.

REFERENCES

1. H. Quack. Maximum efficiency of helium refrigeration cycles using non-ideal components, in "*Advances in Cryogenic Engineering*", Vol. 39, Part B, (1993), p. 1209-1216
2. P. Grassmann and J. Kopp, Zur günstigsten Wahl der Temperaturdifferenz und der Wärmeübergangszahl in Wärmeaustauschern, *Kältetechnik* 9 (1957) 10, p. 306-308.

CONTINUOUS FLOW SYSTEMS WITH SAMPLE COOLING BY PRESET TEMPERATURE VAPOR BLOW - OVER

A.I.Belyaeva and V.I.Silaev

Institute for Low Temperature Physics & Engineering
National Academy of Sciences of Ukraine
310164 Kharkov,Ukraine

ABSTRACT

The purpose of this paper is to draw attention to merits of continuous flow cryostats for a wide range of laboratory studies involving appreciable heat influx to the sample. Especially significant in this case is heat removal from the sample. The authors faced this problem when developing optical methods for low temperature investigation of solids. Therefore,the continu ous flow systems for the case of appreciable heat influx to the sample will be illustrated by cryostats for spectral and magneto - optical studies in the temperature range from 3,5 to 300K. The method of heat removal from the sample in this way is defined as a special method : pre-set temperature vapor blow-over . It has become the basis of our own optical continuos flow cryostats with smooth sample temperature variation. The advantage of this method is that it provides better (and therefore deeper) cooling of the sample and a comparatively weak dependence of the gas-sample temperature difference on the gaseous coolant temperature.

INTRODUCTION

Continuous flow cryostats (CFC) have become a common and important instrument well suited for variable temperature in low temperature (LT) short time experiments (up to 10 h). At present there are many types of continuous flow systems, where the sample is placed in a vacuum or in an exchange gas [1,2]. The use of cryogenic equipment in optical experiments creates a number of difficulties connected with cooling the objects under investigation and the reliable measurement of their temperature [3].The effect of radiation heating is pronounced not only in cryogenic infrared [2] but in visible optical experiments also [1].Two cryostats design for spectral and magneto - optical studies representing continuous flow system with sample cooling by preset temperature vapor blow-over are described. The characteristic features and advantages are discussed. It is shown that among present-day heat transfer methods this one is the most efficient for studies involving appreciable heat influx to the cooled sample. The resour ces of the small-size CFC for magneto-optical studies are demonstrated on the case of specific phase transition for $RbMnCl_3$.

Advances in Cryogenic Engineering, Vol. 41
Edited by P. Kittel, Plenum Press, New York, 1996

SPECIFIC FEATURES OF THE OPTICAL EXPERIMENT

The most characteristic feature of the cryogenic optical experiment is the continuous intence heat influx to the sample from the light source. Light sources in spectral and magneto-optical studies are various types of lamp of powers ranging mostly from 30 to 1000 W. Such high-power lamps are required by the nature of optical studies of magnetic materials. In spect ral investigations it is desirable to have high - power lamps because of the small light intensity of high resolving power instruments. Also,the experimental specimens are normally small and are made of materials with high light absorption.In the case of visual microscopic observation high-power light sources are used to ensure sufficient intensity of illumination and best possible contrast of the sample picture,so that details can be easily seen (e.g., in visual observation of do main structures by magneto-optical methods using crossed polarizers).

The specimens,as a rule, are plates approximately 30 μm to 5 mm thick and the developed plane diameter between 1 and 10 mm,made of materials with high light absorption (ie,ferrogarnets, orthoferrites). The developed plane area for real specimens is normally 0.2-1.0 cm^2 , and the surface density of the incident energy is as high as several $W\ cm^{-2}$. For microscopic observation, especially in crossed polarizers, it may be large still. Thus, in a LT optical setup an appreciable part of the light source power enters the cryostat and it accumulated by the specimen. As is estimated,almost half of the light energy that reaches the ferrite sample remains there,causing a temperature increase with appreciable gradient over its surface.So even when the samples are under cryogen, their temperatures are not always equal. Its temperature under liquid helium (He^4) may rise above the helium boiling point (4.2 K) by more than 20 K [1],while if helium is pumped to below the λ-point, the temperature difference is within 0.5 K. Thus, in the case of appreciable heat influx, the real sample temperature can differ from the liquid coolant temperature, and when it cannot be measured directly (as in spectral and magneto-optical investigations) its estimation may involve a large error.

An analysis of spectral and magneto-optical LT studies attended by intense heat influx to the sample permits formation of the basic requirements on its cooling and the optical cryostat design with variable temperature. The cooling must provide good heat transfer from the sample over the whole required temperature range. If the thermometer may not be attached to the sample, its temperature needs to be determined remotely,by having the sample and thermometer temperatures very close or by knowing the temperature difference. The particular cooling requirements due to the specific features of magneto-optical experiments are as follows : since cold window cryostats cannot be used for polarized light experiments (because of inherent uncontrollable depolarization) the cryostat for precise polarization measurements must have no cooled air- tight windows. Also, sample mounting for precise spectral and magneto-optical investigations must be free, so that there are no stresses in the material.

PRESET TEMPERATURE VAPOR BLOW-OVER COOLING OF SAMPLES

When considering the whole variety of existing cryostats, one finds the following methods of heat transfer from the sample: through radiation, through cooling gas, through the mechanical contact and by blow-over with gas.

In most continuous flow cryostats the two latter methods are normally used, that is the sample is either fixed to a heat exchanger or blown with cold vapor, when the sample is cooled with gas flow, and its temperature is monitored through the coolant mass flow rate. Fig. 1a . shows a typical diagram of the gas flow rate versus sample temperature. The efficiency of heat transfer from the sample, and hence the temperature difference between the sample and the gas depends on the flow rate of the gas flowing past the sample. This dependence makes it diffi-

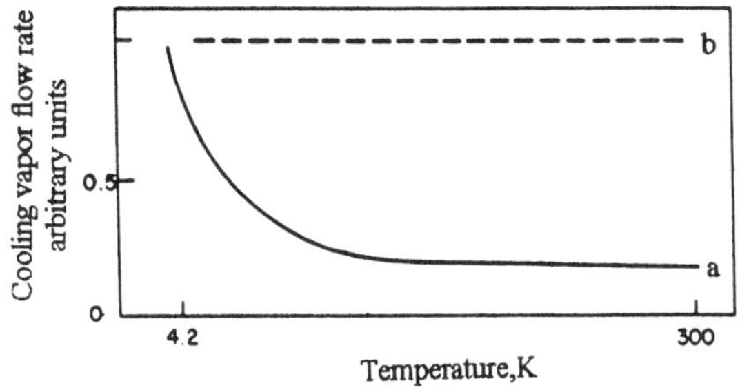

Figure 1. Dependence of the cooling vapor flow rate on the sample temperature.

cult to measure the sample temperature remotely, i.e. by measuring the gas temperature and estimating the temperature difference between the gas and the sample. Another weak point of the method is the difficulty of changing the heat transfer efficiency at a fixed temperature.

A special method, preset temperature vapor blow-over cooling of the sample, has become the basis of our optical continuous flow cryostats with smooth sample temperature variation in the range of 3,5 to 300 K, for intense heat influx to the sample (using a single cryo gen, viz liquid helium). The differences and the advantages of this method is followings:

1. The sample temperature was specified not by the vapor flow rate, but by the vapor temperature.
2. A powerful heater was attached to the cryostat, which were designed to monitor the tempe rature

The gas flow rate as a function of temperature for the new design is shown in Fig. 1b. The coolant mass flow rate depends on the required efficiency of sample cooling, specified by the tolerable temperature difference between the gas (thermometer) and the sample.

CONTINUOUS FLOW CRYOSTAT DESIGNS

Fig. 2 shows a design of continuous flow cryostat based on preset temperature vapor blow-over cooling of the sample.

The cryostat is designed for the precise polarization magneto-optical, optical and spectral studies of transparent samples in conjunction with a solenoid or a small electromagnet. In order to reduce the body outer diameter, the part of the cryostat in the region of the external magnetic system is made in the form of coaxial tubes. The liquid helium from a standard storage dewar is fed into a tube, warmed to the required temperature by the heater and reaches the sample.

The sample is mounted in a holder which is inserted into the cryostat through annular packing, enabling specimen rotation in the plane perpendicular to the optical axis. As shown in Fig. 2 , the cryostat has no internal cold windows, while the external windows in the walls of the gas - filled section containing the sample are kept at room temperature. So there is practically no stress in the optical elements, an important feature for precise polarization measurements.

Samples mounting is similar to that shown in Fig. 3. The sample is fixed to a thin washer by means of thin soft wire legs. The thermometer is placed in the vapor stream in close proximity to the sample.

Temperature below 4.2 K may be achieved by careful adjustment of the needle valve on the transfer tube.

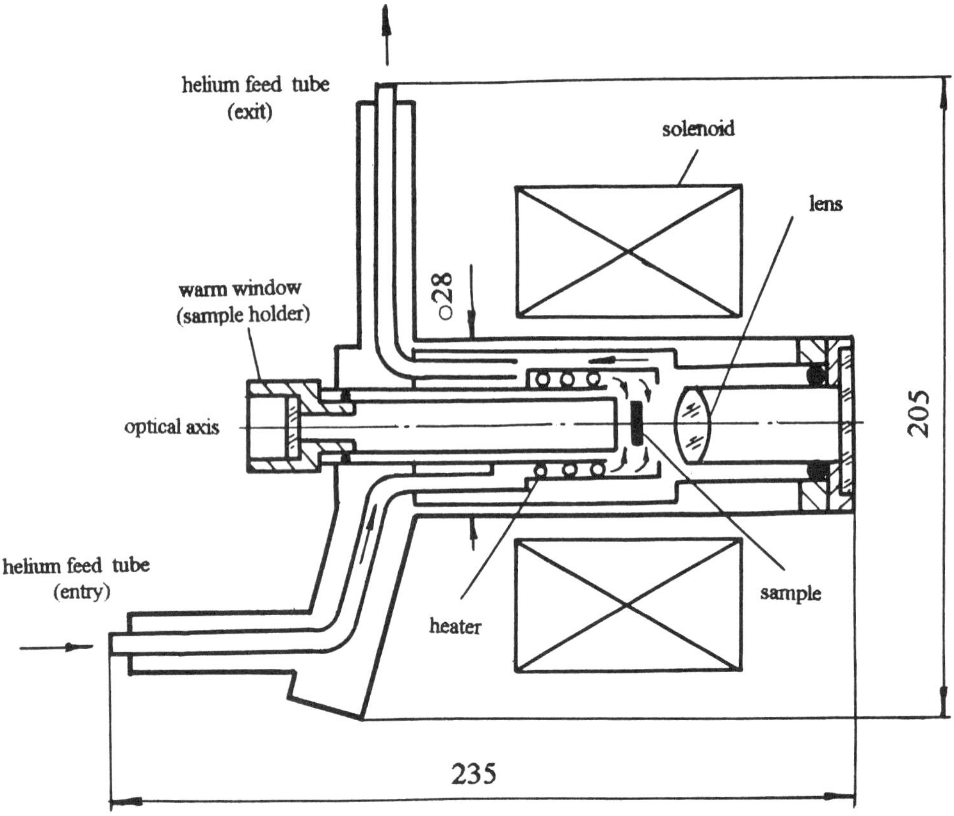

Figure 2. Design of small - size continuous flow optical cryostat. The arrows show the gas flow.

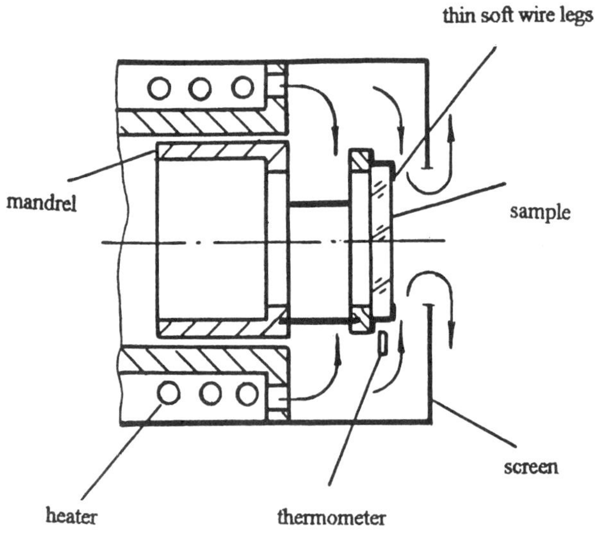

Figure 3. Sample mounting scheme. The arrows show the gas flow.

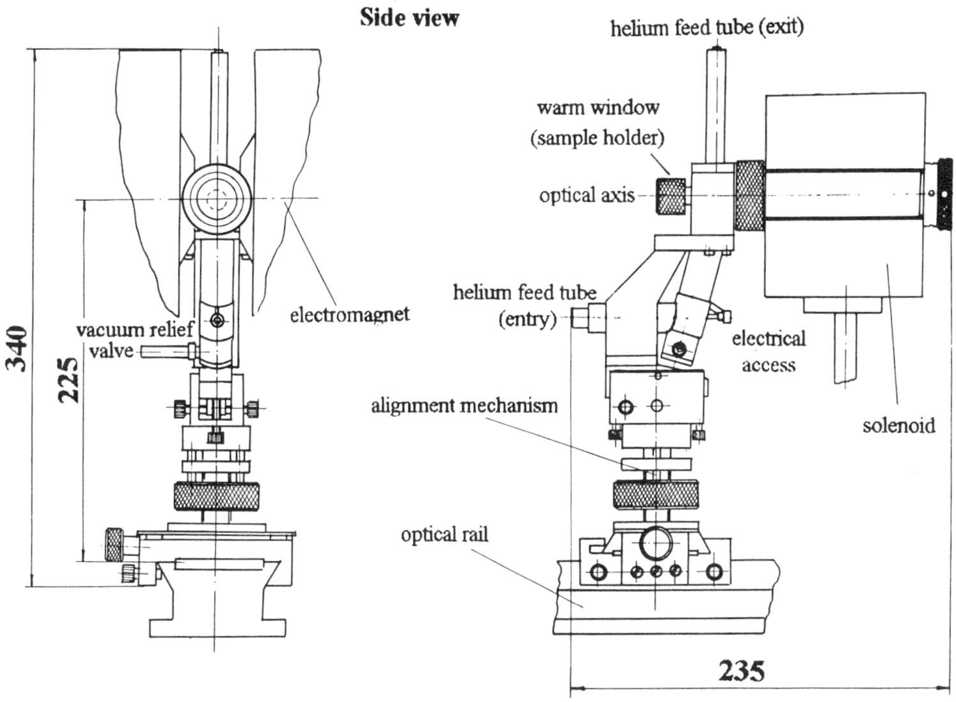

Figure 4. Schematic drawing of magneto-optical experimental arrangement .

Fig. 4 shows a side view of the small-size continuous flow cryostat with the special alignment mechanism and solenoid. This is a schematic drawing of magneto - optical (spectral) experimental arrangement.

The continuous flow cryostat provides the following benefits:

- efficient sample cooling with illumination of up to several W cm^{-2} the whole required temperature range
- a comparatively weak dependence of the gas-sample temperature difference on the gase-

Table 1 contains the main specifications for a small-size continuous flow cryostat

Table 1. Summary of specifications for CFC

Sample space	(mm)	8×8×5
Temperature range	(K)	3,5÷500
Temperature stability	(K)	± 0,1
Sample holder vibration	(μm)	0.05
Cool down time (with transfer tube cold)	(min)	10
Helium consumption (under the heat influx up to 0,5 W cm^{-2}	(l/hr)	
at 4,2K		3,8
at 10K		3,5
at 150K		<0,5
Sample change time	(min)	15
Weight	(kg)	0,5

ous coolant temperature. So the thermometer may be placed in the vapor stream in the close to the sample

- reduction of the temperature gradient along the sample surface despite the high heat influx
- unstressed warm optical windows in the optical channel
- compact, light weight and portable
- easily integrated into spectrometers,microscopes,solenoid or electromagnet
- use with either liquid helium or liquid nitrogen
- economical experimental runs
- operation in any orientation

Fig. 5.1 – 5.7 shows the ferroelastic domain structures (DS) formation for $RbMnCl_3$ at high rate temperature variation (~10 K/s). At low rate (≤1 K/s) DS is absent, but the crystal is optically anisotropic (Fig.5.8). The stability of the image (Fig.5) taken down to 3,5 K is indicative of the low level of vibration of the cryostat system during operation.

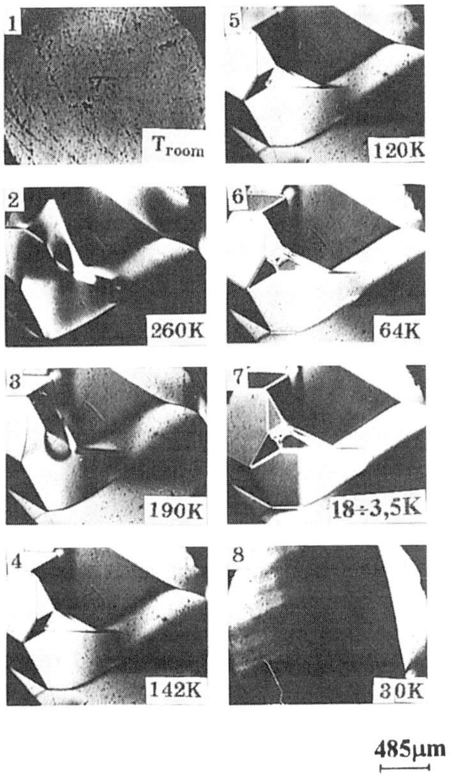

Figure 5. Temperature dependence of the ferroelastic domain structures for $RbMnCl_3$ at high rate of temperature variation (~10 K/s) - (1-7); 8 - optical anisotropy crystal after low rate temperature variation ≤1 K/s down T=30 K.

The cryostat shown schematically in the Fig. 6 is intended for precise magneto-optical studies of transparent objects placed in the field of a superconducting solenoid, with parallel or mutually perpendicular directions of light and magnetic field.

In the lower part of the liquid helium cryostat which cools the superconducting coils of the magnetic system,there are two horizontal mutually perpendicular through channels,the axis of one of them coinciding with the magnetic system axis. Inserts in the optical and continuous flow cooling systems are tightly fitted into these channels.The sample is attached to the optical insert and may be rotated in the plane perpendicular to the insert axis by control 1. By turning control 2, the angle between the sample axis and the optical insert 1 axis is varied within 75° to 105°.

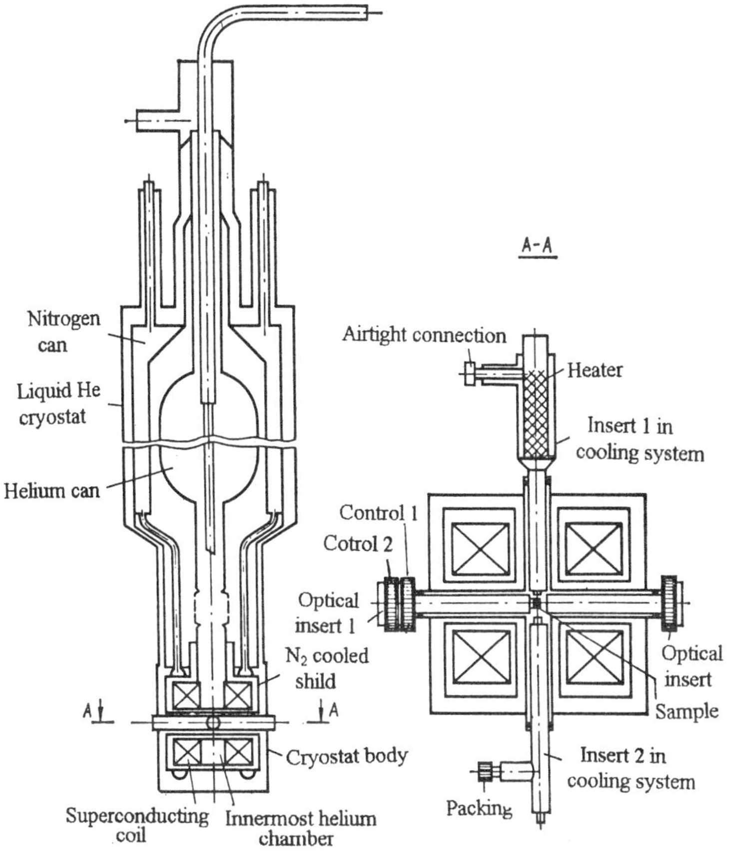

Figure 6. Composite cryostat for magneto-optical studies.

The coolant from the transport dewar is delivered into insert 1, mixed with warm vapor (its delivery line is not shown); then the mix is fed through the heater and passed to the sample. The heater leads are brought out through an air-tight connection.The used gas is discharged through insert 2 which also carries the thermometer of the temperature measurement and stabilization system. The thermometer leads are brought out through the packing.

The lower section of the cryostat represents three horizontal coaxial cylindrical vessels of which the innermost is the helium chamber accommodating super conducting coils, the middle one is a liquid nitrogen - cooled shield, and the outer one a cryostat body. The vessels are fixed with wire braces. Each of the two mutually perpendicular channels is a system of three coaxial tubes soldered into the end walls of the appropriate cylindrical vessel. As is shown in fig.6, the channel interior is not connected with the cryostat jacket, which ensures rapid heating and replacement of the sample with full helium and nitrogen cans. One cryostat charge (3,5 l of liquid helium) is sufficient for continuous operation during 10 h.The helium flow rate in the cooling system depends on the temperature to be maintained, the heat evolved by the sample and the required maximum tolerable sample-thermometer temperature drop.

TEMPERATURE ADJUSTMENT AND STABILIZATION IN CFC

CFC are used in conjunction with certain subsidiary devices which form, together with the cryostat, the continuous flow cryostatic system. Temperature variation and stabilization in the cryostat to be provided by such a system amounts to adjustment and stabilization of the coolant temperature [1]. Liquid coolant is fed from the storage dewar,through the overflow siphon with a needle valve,to the heater and then blown over the sample.The gas issued from the cryostat, through the valve, air valve and manostat and is drawn away by a mechanical pump. The valve ensures smooth variation if the gas flow entering the air valve, essentially a conventional diaphragm manostat for pressure stabilization in vapor pumping systems. As the gas flow rate changes,which is attended by a change in the pressure drop across the valve,the diaphragm in the valve enlarges or reduces the valve flow area, thus providing the required stabilization.This system is reliable when used in conjunction with an electronic system of temperature stabilization using heater.

CONCLUSION

CFC well suited for comparatively shot time laboratory experiments requiring wide temperature variation range (3,5 to 300 K) and involving intense heat influx to the sample are described.Its have warm optical windows and a gas-filled test chamber and may be used for precise investigations in polarized light. These CFC have allowed novel visual polarization magneto-optical techniques to be developed for the study of the physical properties of solids over a large span of low temperatures. In particular, the cryostat shown in fig.2 has become the major element of an arrangement for magneto-optical investigation of the submicrometer size domain structure and basic magnetic parameters of materials in the temperature range 3,5 to 400 K and in various magnetic fields. CFC have enabled visual optical and magneto-optical techniques to be used for investigations of domain structures in magnetic materials in the case of weak magneto-optical effects and high-accuracy microspectral techniques to be applied over a wide low temperature range to the study of the nature of low-temperature magnetic phase transitions.

The continuous flow systems based on forced sample blow-over with preset temperature vapor flow are the most promising not only in optical experiment,but also in other fields of physics,especially where high heat influx to the sample is involved.

ACKNOWLEDGMENT

The research described in this publication was made possible in part by Grant No U9K000 from the International Science Foundation.

REFERENCES

1. A.I.Belyaeva, V.I.Silaev and Yu.E.Stetsenko"Protochnye Criostaty Dlya Laboratornykh Issledovanii (Flow Cryostats for Laboratory Studies),Naukova Dumka,Kiev,USSR (1987).
2. A.I.Belyaeva and V.A.Sirenko "Kriogennye Mnogosloynye Pokrytya" (Cryogenical Multilayer Coatings), Naukova Dumka,Kiev,USSR (1991).
3. A.I.Belyaeva,T.G.Litvishkova,S.N.Marushko et.al."Problems of cooling and temperature measurement on objects in cryogenic optical studies",Cryogenics 30:56 (1990).

DESIGN, DEVELOPMENT AND TESTING OF THE CRYOGENIC SYSTEM FOR THE 45-T HYBRID

S. W. Van Sciver,[1,2] S. J. Welton,[1] K. Bartholomew,[1]
M. Gorbunov, [1]J. R. Miller,[1] G. E. McIntosh,[3] G. Bon Mardion,[4]
F. Viargues,[4] D. Balcer[5] and K. Kreinbrink[5]

[1] National High Magnetic Field Laboratory
Tallahassee, FL 32306 USA
[2] ME Department, FAMU/FSU College of Engineering
Tallahassee, FL 32310-2175
[3] Cryogenic Technical Services Inc.
Boulder, CO 80301 USA
[4] Commissariat a L'Energie Atomique
France
[5] CVI Inc.,
Columbus, OH 43216 USA

ABSTRACT

The cryogenic system for the 45-T Hybrid superconducting magnet employs an innovative design primarily driven by the requirement that it be a user-friendly, reliable facility. To achieve this requirement, the superconducting magnet is being built with highly stable cable-in-conduit conductors, operating at 1.8 K and internally cooled with static He II. Further, the magnet is housed in a separate cryostat connected to the refrigeration system through a services duct to the supply cryostat. This design allows the magnet cryostat to be free of penetrations that may interfere with the magnet user. The 45-T Hybrid cryostat was delivered to the NHMFL in March of 1995. The present paper reports on the design, development, fabrication, installation and preliminary testing of this unique system.

INTRODUCTION

The National High Magnetic Field Laboratory in collaboration with the Francis Bitter Magnet Laboratory is developing a new hybrid magnet system which will, when complete, provide the world's highest steady magnetic fields to the physics and materials research community.[1] This system consists of two separate magnets: a 14 T, 616 mm warm bore superconducting outsert magnet and a 31 T, 33 mm bore resistive insert. The outsert magnet is supported by a cryogenic system operating at 1.8 K with static, pressurized He II.

Various aspects of the design and development of the 45-T Hybrid cryogenic system have been previously published.[2-5] The present paper all further describe progress on this development with, in this case, the emphasis on results of recent thermal performance tests. The first portion of the paper summarizes the design and analysis of the superconducting magnet and cryogenic system. Subsequent sections report on measurements of heat leak and other performance parameters based on testing performed in June of 1995.

Table 1. General specifications of the superconducting outsert magnet

Field contribution by outsert magnet	14 T
Magnet cold bore	710 mm
Winding height (outer coil)	1028 mm
Total energy stored	120 MJ
Total cold mass	12 t

OVERVIEW OF MAGNET DESIGN

Table 1 lists the general specifications of the superconducting outsert magnet. This magnet consists of three nested coils, the inner two (coils A&B) use Nb_3Sn conductor and the outer (coil C) uses NbTi. All three coils employ a Cable-in-Conduit Conductor (CICC) technology. Coils A&B are layer wound, while coil C is a double pancake wound coil. All three coils run in series at a nominal current of 10 kA.

The CIC Conductor has several advantages that make it the ideal choice for the 45-T Hybrid outsert magnet. With a CICC, the mechanical structure is distributed throughout the windings allowing a more effective utilization under electromagnetic loading. This approach also allows the fragile, strain-sensitive Nb_3Sn conductor to be in a rather strain-free state. Also, the containment of the helium within the conduit ensures better turn-to-turn insulation than can be achieved with bath cooled magnets. However, of possibly greatest interest to the present audience, the approach allows improved cooling of the conductor and consequently enhanced stability at the high current densities required to achieve the goals of the 45-T Hybrid design.

One particularly unique feature of the present CICC superconducting magnet design is that the conductor is cooled by static He II (superfluid helium) at 1.8 K. This novel approach to cooling was primarily selected to improve the performance of the conductor in the high field region. Not only is the current density in the superconductor higher at the lower temperature operation, but the conductor stability or resistance to external disturbance is also much better. A high degree of stability is required for this magnet system, for certain fault modes, particularly the rapid discharge of the resistive insert magnet, produce large AC losses which must not quench the superconducting magnet. He II cooling of the CICC should ensure this level of stability. Static He II is preferred over forced flow for its passive nature and ease of operation.

CRYOGENIC SYSTEM DESIGN

The cryogenic system for the 45-T Hybrid must provide the environment for safe and reliable operation of the superconducting magnet. This goal establishes a set of general requirements, listed in Table 2, for the cryogenic system.

In the final analysis, the 45-T hybrid magnet system will have its success dependent on its ability to supply a magnetic field to the users. Thus, the system must be reliable and accessible to the researchers. Reliable operation means that the outsert magnet should be at field during Laboratory operations (8:00-20:00, 5 days/week) and in standby mode (at 1.8 K or 4.2 K) during other times with shut down and warm up only for maintenance. It is highly desirable for the users to be able to access the bore of the magnet at any time, even when the magnet is discharged. Thus, a separation of the operations from the user should be a goal of the cryogenic system design.

The cryogenic system for the 45-T Hybrid was fabricated by CVI and delivered to the NHMFL in March of 1995. A photo of the system is shown in Fig. 1.

Table 2. General requirements of the 45-T Hybrid cryostat

Cooldown time should not exceed 7 days
Ramp time to full field in 1 hour
Steady operation with available refrigeration
Fast recovery from major disturbance such as insert trip or quench
Thermally insulating, structural support for EM loading of outsert

Figure 1. Photo of the completed 45-T Hybrid cryogenic system .

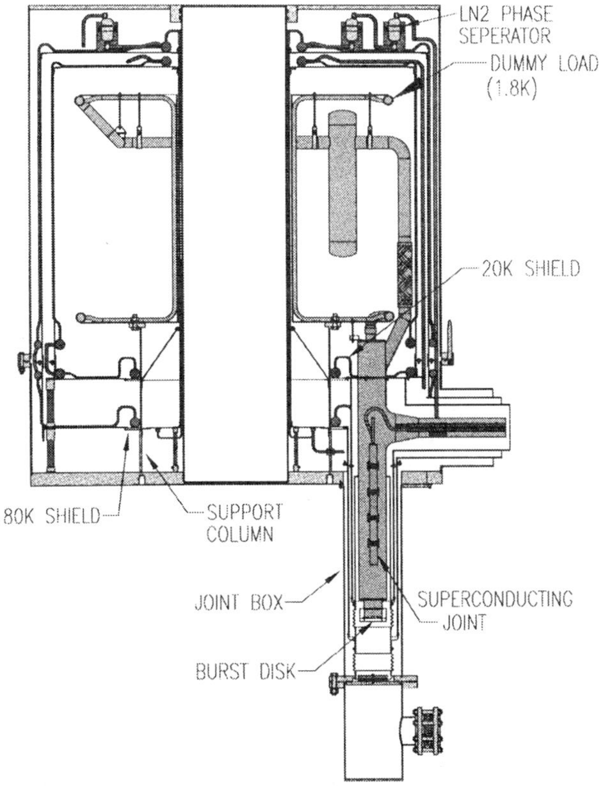

Figure 2. Cross section of magnet cryostat with dummy load installed .

The cryogenic system for the 45-T Hybrid consists of two major components: magnet cryostat and supply cryostat. The magnet cryostat has the sole purpose of supporting and insulating the superconducting magnet. However, since the cryogenic system is installed at the NHMFL before completion of the superconducting magnet, three features in the magnet cryostat design are required. First, the magnet cryostat must be easily opened for installation of the superconducting outsert magnet. Second, the bottom of the cryostat must be located four meters above the floor to allow installation of the resistive insert magnet. Finally, in the interim, the cryostat must be equipped with a dummy load to allow testing of the system before the magnet arrives. To meet this last requirement, we have equipped the magnet cryostat with a temporary vessel instrumented with a heater and temperature sensors, as shown in Fig. 2.

Certainly, one of the most important components in the magnet cryostat is the support column, which transfers the electromagnetic and gravitational load to the base plate. The required load bearing capacity of this column is +/-6.2 MN. Further, the support column must be thermally insulated. To meet these requirements, we have built the column as a stepped 304 SS cylindrical shell with discrete cooling stations at 20 K and 77 K. These cooling stations are approximately equally spaced and selected to minimize the cooling power requirements, particularly at the lower temperatures. Heat loading on the column at the three temperatures was calculated to be: 480 W at 80 K; 24 W at 20 K and 1.3 W at 1.8 K.

Another important subcomponent in the magnet cryostat is the joint box. This assembly, which is located below the cryostat, contains the demountable low loss resistive joint between the conductor bus and magnet. Note also that the bottom of the joint box contains the burst disk which is set to rupture at 0.5 MPa. This component provides the last resort protection in the event of a major disturbance that can not be discharged through the primary relief path.

The supply cryostat contains all the interfaces for operation of the magnet. A cross sectional view of this part of the system is shown in Fig. 3. The supply cryostat consists of an upper and lower helium reservoir, which are maintained at 4.5 K and 1.8 K, respectively, during normal operation. The supply cryostat supports a number of demountable components necessary for the operation of the system. These include: two JT refrigerators (only one shown in the figure), two current leads, a cooldown JT valve and

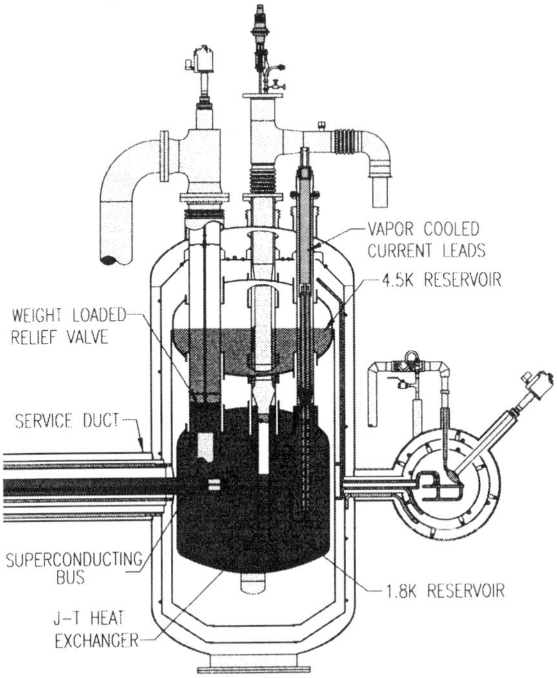

Figure 3. Cross section of the supply cryostat .

Table 3. Steady state total heat load estimate for the 45-T hybrid cryostat during normal, full current operation

T(K)	Heat load	Main contributions
1.8 K	6.7 W	support column, leads
4.5 K	52 l/hr + 15 W	VCR, He II supply
20 K	36 W	support column
80 K	632 W (4.2 l/hr)	support column

Table 4. Performance specifications of a single refrigerator/liquefier unit with LN_2 precooling

	Mode 1	Mode 2	Mode 3
Liquefaction Rate	60L/h	56 L/h	70 L/h
5-K Refrigeration		20 W (1.6 g/s)	
20-K Refrigeration	60 W (1.8 g/s)		

three-way valve, a fill JT valve, instrumentation probe and quench relief. The cooldown JT valve and three-way valve are designed to supply high pressure helium to the magnet during cooldown or in the closed position provide thermal insulation between the two reservoirs. During normal operation, the cooldown JT and three-way valve are closed so that liquid helium can be supplied to the upper reservoir through the fill JT valve. Cooldown and quench relief are facilitated using a weight loaded relief valve (WLRV) between the upper and lower reservoirs in the supply cryostat. This device has a conical stainless steel plug and seat that can be actuated from outside the cryostat. During cooldown from 300 K this valve is open to allow return of the helium to compressor suction. It also must be periodically opened during the cooldown from 4.5 K to 1.8 K to refill the lower reservoir due to condensation and densification of the lower reservoir inventory. During reduced temperature operation, the WLRV must be closed to minimize the heat leak to the lower reservoir. A more detailed discussion of the design, development and operating characteristics of the JT refrigerators is given in a companion paper by S.J. Welton, et.al.[6]

The magnet vessel is connected to the supply cryostat by the services duct which is a five meter long, horizontal double conduit section. The larger conduit connects to the bottom of the magnet vessel and contains the conductor bus and half of the magnet voltage tap cables. As currently configured, the conductor bus is configured with a short circuit in the joint box to allow powered testing of the bus in advance of magnet installation. The

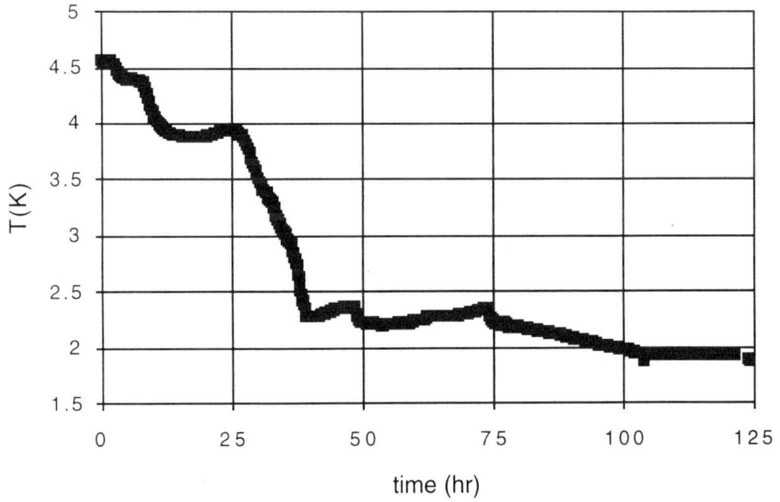

Figure 4. First cooldown of the 45-T Hybrid cryostat .

smaller conduit contains the other half of the voltage tap cables cables and is connected to the top of the magnet vessel. During cooldown, the cold helium is forced to flow to the top of the load and returned through the larger duct to the supply cryostat.

The 45-T Hybrid cryostat heat load has been calculated and minimized as part of the design process. The results of these calculations are listed in Table 3. These values are compared, in the next section, to measured heat leaks obtained during the preliminary testing.

The refrigeration system that supplies cooling to the 45-T Hybrid cryostat was purchased from Process Systems International. The system consists of two Model 1630 cold boxes, each with two screw compressors. Each of these units has a capacity as specified in Table 4.

Two Kinney vacuum pumps, each consisting of a 1000 CFM at 10 Torr roots blower backed by a liquid ring pump, supply the He II refrigeration by reducing the vapor pressure of the He II reservoir in the JT refrigerators. Combined with the JT heat exchangers, the maximum capacity of each He II refrigerator is 18 W at 1.8 K.

The philosophy behind the design of the hybrid refrigeration system is to provide redundancy to increase reliability during operation. One refrigeration unit and vacuum pump should be able to supply the capacity sufficient to maintain the cryostat during steady state operation, constant current to the magnet. Under extraordinary conditions such as cooldown, magnet charge or recovery from quench, it will be necessary to use both refrigerators operating in parallel.

PERFORMANCE ANALYSIS & TESTING

The 45-T Hybrid cryostat was first cooled to helium temperature in June of 1995. The cooldown procedure involved first establishing vacuum in the isolation space, followed by filling of the LN_2 system and starting the liquefiers. Analysis has indicated that system including the magnet should cooldown in approximately 160 hours.[5] However, due to a number of technical difficulties, the first cooldown to 4.5 K actually took about two weeks.

Once the supply cryostat was full of liquid at 4.5 K, the cooldown JT was closed and the fill JT opened to supply liquid to the upper reservoir. The weight loaded relief valve was then closed and pumping initiated on the JT heat exchangers. Cooldown proceeded to operating temperature, around 1.8 K. The recorded cooldown curve from 4.5 to 1.8 K is shown in Fig. 4. This procedure required several days since it consisted of both a reduction of temperature, liquid filling and condensation of vapor in the magnet cryostat. In the present case, the process is primarily limited by the liquefaction rate. Since the magnet cryostat heat load can not vent except through the supply cryostat, heat generated in the load increases the pressure and expels the contained liquid. Thus, above T_λ, the stable condition is with the load void of liquid even with the lower reservoir of the supply cryostat full. On the other hand, below T_λ the high conductivity of the He II allows easy transport of heat between the two cryostats without the presence of vapor.

Once the cryostat reached operating temperature (about 1.8 K), testing was initiated. The test plan included the following tasks: 1) determination of the steady state heat leak; 2) measurement of the steady heat removal capability of the He II refrigeration system; 3) investigation of recovery from a major thermal disturbance deposited in the load; and 4) testing of the current bus. Although all four tasks were, to some degree, carried out during this phase of the testing program, the data obtained are still preliminary and dependent on future measurements for confirmation.

The LN_2 temperature heat leak was determined by measuring the nitrogen gas evolution. The measured value was 482 W which is about 76% of that predicted by analysis (632 W). Although the reasons for the lower LN_2 consumption are unclear at this time, it may be due to inadequate heat stationing on the support column or that the lower end of the column is at some temperature below 300 K.

The steady state heat loads at 20 K and 4.5 K could only be estimated from these preliminary results. The 20 K heat load was determined by measuring the mass flow and enthalpy change from the inlet to the outlet of the refrigerator that supplies the 20 K loop. The estimated value was in the range of 50 W, which is slightly higher than the calculated 36 W; however, improved measurements will be required to confirm these differences. An estimate of the 4.5 K heat load was also difficult to obtain from these tests. The only

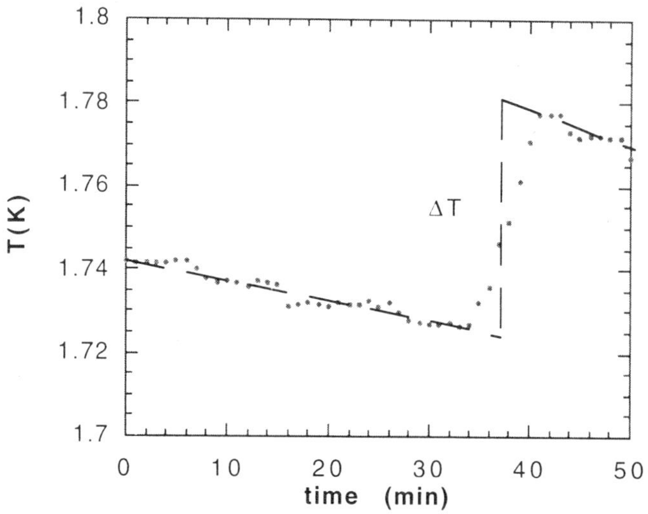

Figure 5. Temperature increase in the He II reservoir resulting from a 24 kJ heat pulse.

quantitative statement that can be made at this time is that one liquefier (70 liters/hr) is sufficient to maintain the supply cryostat full with a 16 W heat load into the 1.8 K reservoir and zero bus current. The 1.8 K heat load was measured by a calorimetric method as described below.

The first task of the 1.8 K heat load measurement was to make an accurate determination of the He II volume within the cryostat. This was accomplished by the heat pulse method. Typical data are shown in Fig. 5. With the temperature of the cryostat stable or cooling slowly, a heat pulse of 24 kJ was applied to the load. The corresponding temperature increase is recorded by the temperature sensor in the lower reservoir. By a simple extrapolation technique, the net temperature change, ΔT, and enthalpy change of the liquid are measured. For the data in Fig. 5, this calculation corresponds to Δh = 142 J/kg and m = 169 kg (1150 liters) of He II, which is consistent with estimates based on cryostat geometry.

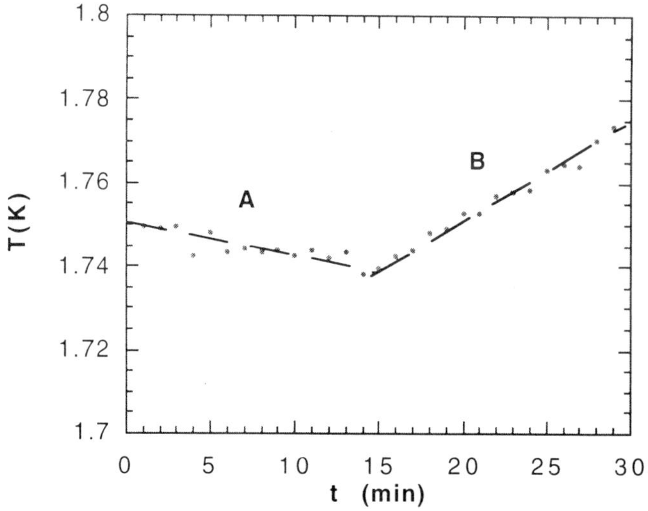

Figure 6. Heat leak measurement for He II space.

Once the volume of He II is known, the overall 1.8 K heat leak was measured by a similar calorimetric technique. With the system cooling slowly, the heat exchanger JT valve and the vacuum pump valve were closed, stopping refrigeration to the He II reservoir. The corresponding rate of temperature increase gives the heat leak from

$$Q_c = m\,C_p\,dT/dt \tag{1}$$

where m is the helium mass. A typical trace for this process is shown in Fig. 6: **A** corresponds to steady cooling; **B** is the temperature increase without cooling. Based on these data, we estimate the 1.8 K heat load to be 16 W. This value is approximately a factor of three higher than that obtained from analysis. The reason for the relatively high 1.8 K heat leak is unclear at this time; however, this result may at least be partially due to a loss of cooling on half of the 20 K shields and on one side of the support column that occurred before the 1.8 K heat load measurement.

The steady cooling capacity of the JT heat exchangers was investigated as part of the testing program. For most measurements, only one HX valve was open, as each heat exchanger was designed to extract 20 W at 1.8 K. The total measured cooling capacity at 1.8 K with two vacuum pumps operating was 25 W. At this temperature, the measured overall heat transfer coefficient was about 1.8 kW/K. For higher temperatures, the capacity increased due to the higher density of the helium vapor. In general, the heat exchanger performance was within the requirements set for the operation of the 45-T Hybrid. A more detailed report on the design and performance of this component is given by Welton, et al[6].

An important design constraint on the operation of the cryostat is to ensure recovery from a major disturbance that might occur during magnet operation. Examples of such a disturbance might include: AC losses in the coil due to magnet charge; AC losses in the coil due to trip of the resistive insert; and in the extreme case, quench of the outsert magnet. In an effort to understand the performance of the cryostat under such conditions, a series of tests were performed consisting of applying a heat pulse to the dummy load heater and monitoring the equilibration of the temperatures in the two cryostats. An example of this measurement is shown in Fig. 7, which is a time base plot of the temperature on the dummy load and in the lower reservoir of the supply cryostat associated with a 24 kJ heat pulse applied to the load. The variation of the lower reservoir temperature during this same interval can be seen in Fig. 5.

Note that there is an initial temperature difference (T_l-$T_{He\ II}$) of around 0.25 K, which may be caused by the load temperature measurement made on the surface of the vessel or

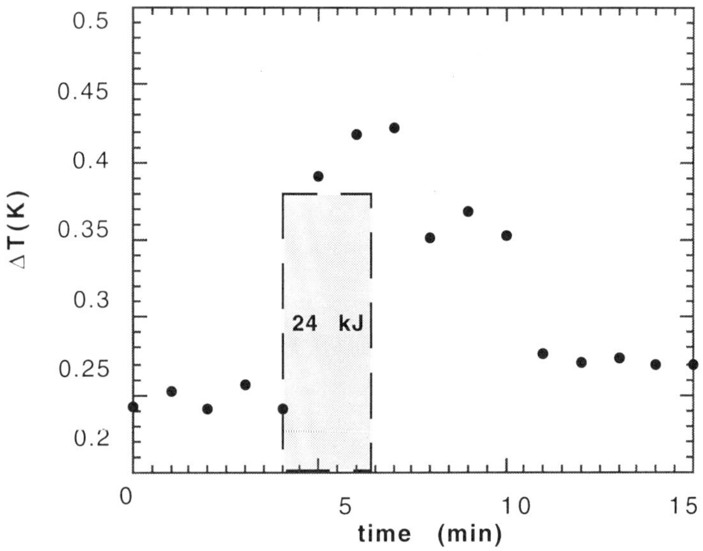

Figure 7. Temperature difference between load and He II reservoir during heat pulse

possibly a calibration effect. However, the important observation is the roughly 0.2 K increase in the temperature difference caused by the pulse. This difference diminishes with time such that after 6 to 7 minutes, the temperature difference returns to a steady value. Thus, short duration transient disturbances that occur in the load must depend on the stored He II for heat absorption.

Preliminary tests of the superconducting bus simply involved the application of current in 1-kA increments, while monitoring voltage drops across vapor cooled leads (VCLs), buses and joints, until 10 kA was reached. At the 10-kA level, helium vapor flows were adjusted to achieve reasonably stable VCL voltages in the 50 - 100 mV range. An indication of VCL stability at 10 kA is the observation that the voltage drop across each easily remained within the 50 - 100 mV range for more than 20 minutes with no adjustments to the flows. Another indication comes from observations of the rate of rise of VCL voltage following a complete and abrupt stoppage of flow. Starting from an initial flow that resulted in 50 mV across a lead, the time to exceed 100 mV after flow stoppage was approximately 1 minute for each lead, and for each lead, intervention to reestablish flow at the end of 1 minute brought the voltage back below 100 mV within about 10 s.

The vertical and horizontal busbars within the He II space remained superconducting throughout the bus-system tests, and the resistances of joints in the bus system were far too small to measure accurately, the upper limit being in the range of 10^{-11} ohms.

FUTURE PLANS

At the conclusion of these tests, the system was warmed to room temperature. Further testing of the 45-T Hybrid cryostat as configured is scheduled throughout 1995 and until the superconducting magnet is delivered to the NHMFL. At that time, the magnet cryostat will be opened and the superconducting magnet installed. First testing of the entire system is scheduled for the end of 1996.

CONCLUSION

The cryostat for the NHMFL 45-T Hybrid superconducting magnet has been developed and subjected to preliminary testing. Overall, the cryostat achieved most of its design requirements, with the only significant exception being the 1.8 K heat leak. Work will continue to better measure and improve the cryostat performance characteristics in anticipation of eventual operation with the outsert magnet.

ACKNOWLEDGMENT

Research supported by the State of Florida and the National Science Foundation through NSF Cooperative Grant No. DMR 9016241.

REFERENCES

1. J.R. Miller et al., An Overview of the 45-T Hybrid Magnet System for the New National High Magnetic Field Laboratory, , IEEE Trans. on Magnetics 30:1563 (1994).

2. S. W. Van Sciver, J. R. Miller, W. J. Kenney, H. J. Schneider-Muntau and G. E. McIntosh, Design of the Cryogenic System for the 45 Tesla Hybrid Magnet , 9th Intersociety Cryogenic Symposium, Houston TX , January 31-Feb 2, 1993.

3. S. W. Van Sciver, J. R. Miller, S. Welton, H.J. Schneider-Muntau and G. E. McIntosh, Cryogenic System for the 45 Tesla Hybrid Magnet , Advances in Cryogenic Engineering. 39A: 375 (1994).

4. A. Bonito-Oliva, M.B. Gorbunov, K. Iyengar, J. Miller, S.W. Van Sciver, S.J. Welton, Thermal Analysis of the Superconducting Outsert of the NHMFL 45-T Hybrid Magnet System, Cryogenics 34 (Supplement):713 (1994).

5. S. W. Van Sciver, A. Bonito-Oliva, M. Gorbunov, and S. Welton, Thermal Hydraulics of the 45-T Hybrid Superconducting Magnet System, in: "Proc. International Symp. on Safety of Superconductors", Kyushu Univ., (1994).

6. S.J. Welton, S.W. Van Sciver, K. Bartholomew and M. Gorbunov, Design, Development and Testing of the JT Refrigerators for the 45-T Hybrid, Paper WE-PE2b-3 -1995 CEC/ICMC.

DESIGN, DEVELOPMENT AND TESTING OF THE JT REFRIGERATORS FOR THE 45-T HYBRID MAGNET

S. J. Welton,[1] S. W. Van Sciver,[1] K. Bartholomew,[1] M. Gorbunov,[1]
G. E. McIntosh,[2] G. Bon Mardion,[3] F. Viargues [3]

[1]National High Magnetic Field Laboratory / Florida State University
Tallahassee, FL 32306-4005, U. S. A.

[2]Cryogenic Technical Services Inc., Boulder CO 80301, U. S. A.

[3]Commissariat a L'Energie Atomique, France

ABSTRACT

The 45-T Hybrid Facility requires 1.8 K refrigeration for extended periods of time. The JT refrigeration system design is fully redundant to guarantee continuous cooling of the super-conducting magnet. Each refrigerator is equipped with a removable instrumentation probe and JT valve plug. These removable components make service and repair possible at 4.2 K. In June of 1995, preliminary testing of the JT refrigerators was performed during the initial cool down of the Hybrid Cryostat. Details of the design, development, and fabrication will be presented along with the results of these early tests.

INTRODUCTION

When the 45-T Hybrid Facility comes on line in 1996, it is expected to operate continuously for eleven months out of the year (with one month reserved for maintenance and service of equipment). This facility will provide a 45 T field for the research community. The normal operating point (and idle mode) for the super-conducting magnet will be 1.8 K. Maintaining the magnet at this temperature requires a dependable 1.8 K refrigeration system. The scope of this paper is to discuss the design and development of the JT refrigerators, describe some of the preliminary testing, and present the test data. For more details on the Hybrid Cryogenic System itself, the reader is referred to Ref. 2.

DESIGN AND DEVELOPMENT

System Overview

Flow Description and Operation. The JT Refrigeration circuit for the 45-T Hybrid Cryostat consists of two JT Heat Exchangers, an electric gas heater, two vacuum pumps, four compressors and two liquifiers. A simplified flow schematic is shown in Figure 1. Liquid He, at approximately 4.6 K and 1.4 bar, flows from the upper reservoir of the

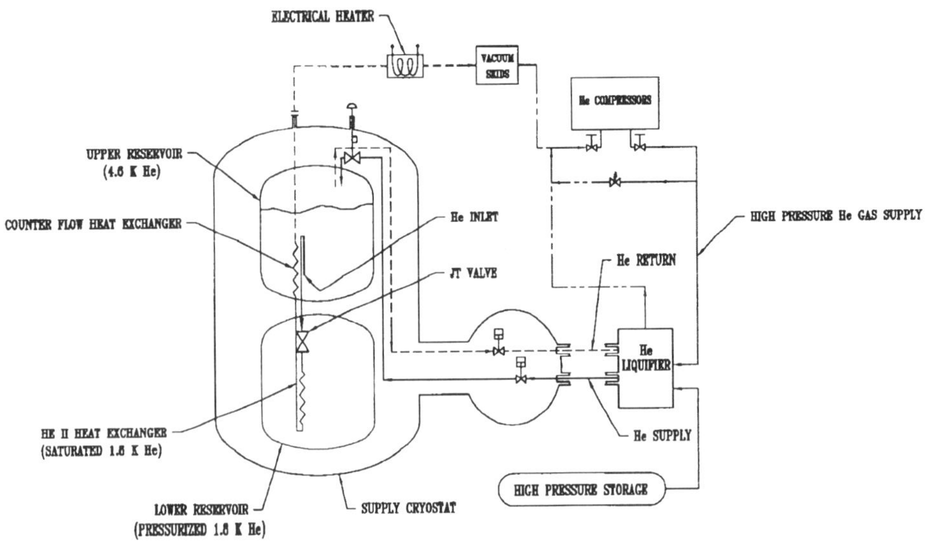

Figure 1. Simple flow schematic of the JT Refrigeration circuit for the 45-T Hybrid Cryogenic System.

Supply Cryostat, through a 15 μm inlet filter, into the counter-flow heat exchanger where it is cooled (prior to entering the JT valve) to near T_{λ}. After the JT expansion, the liquid fills the saturated bath of He II in the bottom of the refrigerator (See Figure 2). The saturated He bath pressure is maintained by two Kinney Vacuum Pumps each capable of 472 l/s volumetric flow at 13.3 mbar (10 torr). The Kinney Vacuum Pumps are two stage systems consisting of a positive displacement blower and a liquid ring pump. Prior to entering the pumps, the cold He gas is warmed to 300 K by an electric heater. The outlet from the vacuum pumps goes directly to compressor suction. The rotary screw (RS) compressors

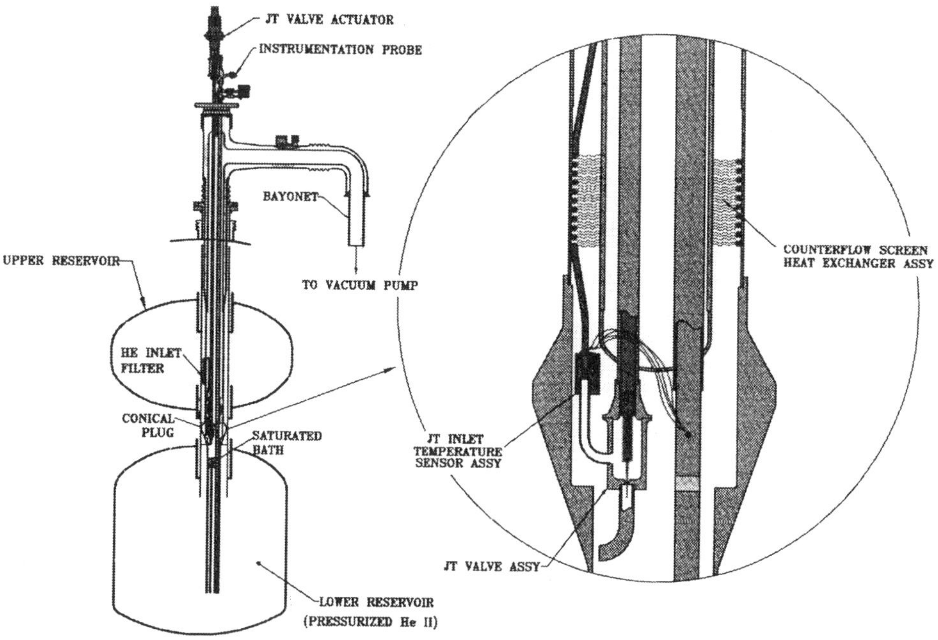

Figure 2. Cross-section of JT Refrigerators with blow-up of the Counter Flow Heat Exchanger and JT valve assembly.

provide He gas at 17.6 bar, 300 K, to the liquifiers. Each Model 1630 PSI liquifier is capable of making 56 to 60 l/hr into the upper reservoir of the supply cryostat during normal operating conditions.

Heat is removed from the pressurized He II bath (lower reservoir) across the copper heat exchange surface of the refrigerators. The liquid level in the saturated bath of the refrigerators is controlled by adjusting the position of the JT valves. These valves are pneumatically actuated and will be automated once testing has been completed and the facility comes on line. As liquid is used in the JT Refrigeration System it is replaced in the upper reservoir by the liquifiers.

JT Refrigerator Design

The design of the JT refrigeration system is fully redundant in order to ensure reliable performance. In the event of a total failure, fully removable components, make repair possible. Figure 2 shows the details of a JT refrigerator. The main components are: the counter flow heat exchanger, the JT valve assembly, the He II heat exchanger, and the instrumentation probe. Two other important pieces of the design are the stainless steel conical plug and the bayonet assembly on the vacuum line.

The seal between the He I (normal fluid) upper reservoir and the He II (superfluid) lower reservoir in the cryostat is accomplished by the stainless steel conical plug, which seats in a mating receptacle in the cryostat. The purpose of this conical plug (and the bayonet fitting on the vacuum line) is to make removal of the refrigerators from the cryostat possible. If necessary, this can be accomplished at 4.2 K, saving many days in warm up and cool down time. Once removed, the refrigerator is replaced with a plug assembly which was fabricated for this purpose. Only one refrigerator is required to maintain the superconducting magnet at 1.8 K.

The instrumentation probe and JT valve rod assembly are also removable at 4.2 K. Back up temperature sensors and liquid level indicators (LLI) are mounted on the probe to ensure long term reliability.

The JT refrigerators were designed at NHMFL and fabricated by CDM Machining and Fabricating in Commerce City, CO.

JT Refrigerator Components

Counter flow heat exchanger. The counter flow heat exchanger cools the liquid entering the JT valve with the gas evaporated from the saturated bath. The heat exchanger was designed by the Commissariat a L'Energie Atomique (CEA) and manufactured by Societe Nouvelle de Labo-Soudure (SNLS) both of France.

The heat exchanger consists of two concentric stainless steel tubes. The annular space between the two tubes is fitted with the Cu heat exchanger screen and tube assembly. The cold He gas flows through the screen assembly which acts as cooling fins for the coiled copper tube. The efficiency of this heat exchanger is given later in Table 2.

JT valve assembly. The Joule-Thompson (JT) valve assembly consists of the valve seat and exit tube, the valve plug and rod assembly, and the rod actuator assembly. The JT valve seat and exit tube are both stainless steel. The exit tube was rotated during installation to cause incoming liquid to flow down the inner wall of the heat exchanger. The diameter of the seat was machined to 1.56 mm. The JT valve plug was machined from threaded brass rod to prevent galling with the seat. The plug is tapered slightly (approximately 0.57°) for maximum flow control. The plug is threaded into a long G-10 rod. The JT rod assembly is sealed from the outside by an O-ring flange at room temperature. This flange makes it possible to remove the rod assembly. A bellows allows the rod assembly to be actuated without breaking the vacuum.

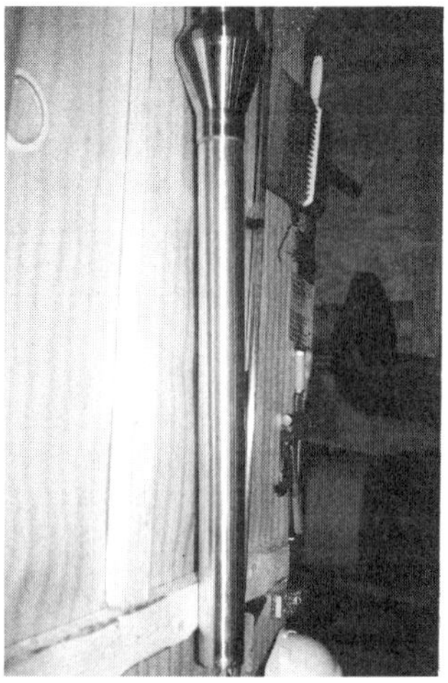

Figure 3a. The outside surface of the copper heat exchange tube is shown. Transition piece is welded to bottom of conical plug.

Figure 3b. View of bottom surface of He II heat exchanger. Note re-entrant tube arrangement. Gap in pattern is location of instrumentation probe.

He II Heat Exchanger. The heat exchange surface, between the saturated and pressurized He II baths, consists of 7 re-entrant copper tubes inside of one outer copper tube (see Figure 3a). The diameter of the outer tube is 79.4 mm. The re-entrant tubes consist of one central tube (ø25.4 mm) surrounded by six smaller (ø19.1 mm) tubes. As shown in Figure 3b, there is a gap in the re-entrant tube arrangement where the instrumentation probe is inserted. The entire copper heat exchanger assembly (everything below the conical plug) is brazed together. At the top of the heat exchanger there is a stainless transition piece to which it is brazed and which is welded to the stainless steel conical plug.

Design Calculations and Specifications

The He II heat exchanger was sized to limit the ΔT across the copper surface to 50 mK or less. Conservative estimates of the heat transfer coefficient, h (see Eq. (1a) below), were

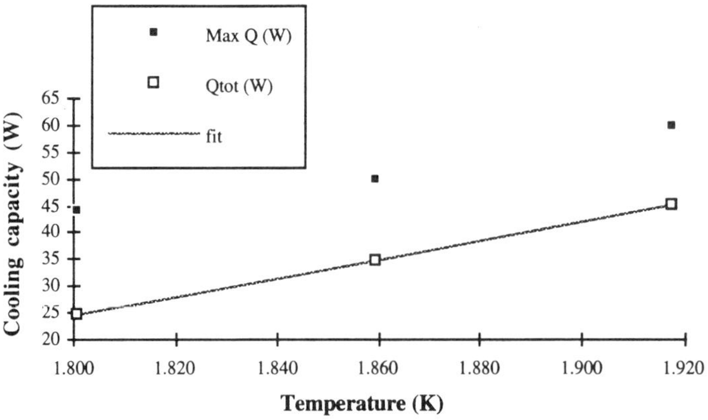

Figure 4. Total refrigeration power, Qtot, measured during initial cool down test of 45-T Hybrid Cryostat. Maximum refrigeration power, Max Q, calculated from Eq. (4).

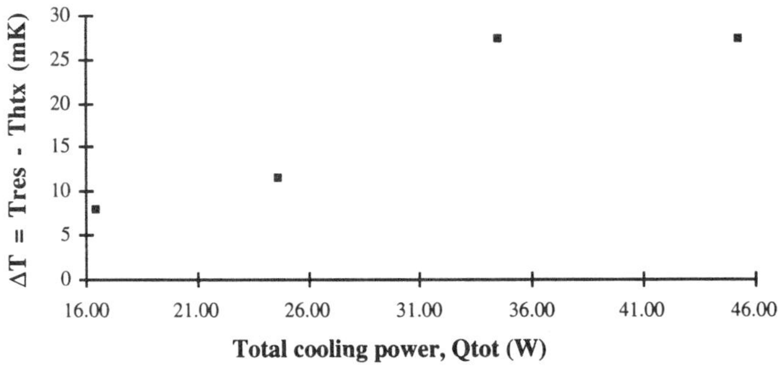

Figure 5. Temperature difference across the walls of the He II heat exchanger due to Kapitza Resistance and gradients in saturated bath. Tres = temperature of the pressurized He II reservoir, Thtx = Temperature of the saturated bath in the heat exchanger.

used to give a worst case estimate for the ΔT due to Kapitza resistance of 34 mK. Because of the good condition of the copper surface initially, the heat transfer coefficient is expected to be closer to that predicted by Eq. (1b)[1].

$$h = 0.4\ T^3 \text{ (in kW/m}^2\text{) - dirty surfaces} \tag{1a}$$

$$h = 0.9\ T^3 \text{ - clean surfaces} \tag{1b}$$

The value of h (heat transfer coefficient) at various temperatures was calculated from experimental data and is compared to values generated from Eq. (1b). This data will be given later in Table 1.

The temperature difference in the saturated liquid, from the bottom of the heat exchanger to the liquid surface, was determined by integration of the heat transport equation as shown below[1].

$$\int f^{-1}(T)\, dT = \int q(x)^3\, dx \tag{2}$$

$f^{-1}(T)$ is the heat conductivity function for He II which is a function of temperature and pressure; $q(x)$ is the heat flux within the He itself. An iterative solution to Eq. (2) predicted a ΔT in the bath of 19 mK at 1.8 K. Thus, the maximum ΔT from the pressurized bath to the saturated bath was conservatively estimated at 53 mK.

EXPERIMENTAL METHOD AND RESULTS

Refrigeration power (Q). The refrigeration (or total cooling) power, Q_{tot}, of the JT refrigerators is given by the following equation:

$$Q_{tot} = Q_{htr} + Q_{hl} \tag{3}$$

Q_{htr} is heat added by a resistive heater in the He II bath, and Q_{hl} is the heat leak into the He II system from the surroundings. The heat leak into the He II bath was measured to be about 16 W. The method for making this measurement is described in another paper[2]. Adding power to the He II bath from the heater caused the temperature of the bath to rise. After a finite time interval, the temperature in the bath stabilized (i.e. dT/dt = 0). The refrigeration power at this stabilized temperature was then determined by Eq. (3). The value obtained from Eq. (3) was then compared to the theoretical maximum cooling power which is obtained from Eq. (4) below. These results are shown in Figure 4 on the following page.

$$Q_{max} = (1-x)\,\rho\, V'\,\lambda_l \quad (4)$$

Q_{max} is the theoretical cooling capacity of the refrigerators, (1-x) the mass fraction of liquid exiting the JT valve, ρ is the density of the saturated vapor, V' is the volumetric flow rate through the pumps, and λ_l is the latent heat of the saturated liquid.

Temperature difference across heat exchange surface. The temperature difference between the pressurized He II bath and the saturated He II bath in the refrigerator was measured in the following manner. A calibrated temperature sensor in the pressurized bath was used to measure the temperature outside the heat exchanger while the pressure over the saturated bath was used to calculate the temperature inside the heat exchanger. The difference between these two temperatures (ΔT) is plotted as a function of refrigeration power (Q_{tot}) in Figure 5.

Heat transfer coefficient (h). From the data in Figure 5, the heat transfer coefficient was calculated from Newton's law of cooling.

$$h = Q_{tot} / A_s\,\Delta T_s \quad (5)$$

It was assumed that ΔT_s in Eq. (5) is 1/2 of the total ΔT measurement from Figure 5. Table 1 below gives the heat transfer coefficient calculated for three points. The total area of the heat exchange surface is 0.582 m^2.

Table 1. Heat transfer coefficient (h) at surface of He II heat exchanger

T (K)	Qtot (W)	h (Eq. 1b)	h (W/m^2-K)
1.800	24.7	5249	**7391**
1.859	34.6	5782	**4338**
1.917	45.3	6340	**5659**

Efficiency of Counter flow Heat Exchanger. The efficiency, η, of the counter flow heat exchanger is defined in the following equation:

$$\eta = \Delta h_{act} / \Delta h_{max} = (h_{T_{in}} - h_{T_{out}}) / (h_{T_{in}} - h_{T_\lambda}) \quad (6)$$

The temperature of the gas exiting the heat exchanger, T_{out}, was measured by the sensor on the JT inlet tube. The pressure drop from the upper reservoir of the cryostat to the JT inlet was assumed to be negligible and isenthalpic expansion was assumed for the JT valve. The efficiency of the heat exchanger was calculated for the same three data points as in Table 1, and the results are given below in Table 2.

Table 2. Counter flow Heat Exchanger Efficiency (η)

T (K)	Tin (K)	Tout (K)	η
1.800	4.631	2.752	**0.742**
1.859	4.706	2.979	**0.700**
1.917	4.719	3.065	**0.682**

DISCUSSION OF RESULTS AND FUTURE TESTING

The results presented here are preliminary and future testing of the cryostat and JT refrigerators is planned for some time in the fall of 1995. The cooling capacity of the JT refrigeration system, though adequate for the Hybrid Magnet, is less than expected. Figure 3 shows that the actual cooling capacity of the refrigerators is considerably less than the expected maximum. The ratio of Q_{tot} / Q_{max} ranges from 0.56 at 1.8 K to 0.75 at 1.92 K.

By extrapolating a linear fit of Q_{tot} to the temperature at which Q_{tot} equals the heat leak (16 W), that temperature is 1.754 K. During testing of the cryostat, however, the minimum temperature achieved was approximately 1.710 K. The reasons for this discrepancy and for the lower than expected refrigeration are not clear at this time. It is hoped that future testing will verify and/or clarify these initial findings.

The maximum ΔT across the heat exchanger was measured to be 27.5 mK which is well within the design specifications. This measurement is believed to be conservative as it neglects the pressure drop through the counter flow heat exchanger screen. Improvements to the thermometry in the saturated bath are required before more testing is performed. Difficulties with the sensors in this bath made the temperature measurements unreliable. As a result, the pressure over the bath was used to determine the saturation temperature of the liquid.

The heat transfer coefficient values from Table 1 show fairly good agreement with the predicted theoretical values of Eq. (1b). It is difficult to know the accuracy of these values because of the uncertainty of the surface area across which the heat is transferred. The outside of the heat exchanger is constantly immersed in the pressurized bath, but the liquid level inside the heat exchanger fluctuated.

It was observed during testing, that some cooling was achieved with no liquid registering on the LLI. This cooling was a result of the liquid exiting the JT valve and spraying down the inner surface of the copper tube. As this liquid evaporated off of the tube walls, cooling was achieved without actually collecting liquid in the bottom of the heat exchanger. The effect of this "empty heat exchanger cooling" is to increase the cooling surface area.

Only one JT refrigerator was filled with liquid during these tests. The other was empty with the JT valve closed. However, it is doubtful that the closed valve provided a leak tight seal. Increasing the spring force on the valve to close it tighter, would reduce the leakage by the valve, but it would also increase the possibility of the valve needle seizing in the seat. Thus, the "leaking" JT valve provided additional cooling surface, as described above, and made accurate calculation of the heat transfer coefficient difficult. For future tests, both heat exchangers will be used simultaneously (full of liquid) for calculation of the heat transfer coefficient.

CONCLUSIONS

The cooling capacity of the JT refrigerators are sufficient for operation of the Hybrid Superconducting Magnet, though their efficiency was less than expected. The ΔT across the heat exchanger due to Kapitza resistance is better than the design specification. Additional testing is required to verify the preliminary data presented in this paper.

ACKNOWLEDGMENT

* Research supported by the State of Florida and the National Science Foundation through NSF Cooperative Grant No. DMR 9016241.

REFERENCES

1. S. W. Van Sciver, "Helium Cryogenics", Plenum Press, New York (1986).
2. S. W. Van Sciver et. al., Design, Development and Testing of the Cryogenic System for the 45-T Hybrid, this conference.

A THROTTLE CYCLE REFRIGERATOR OPERATING BELOW 77 K

Ajay Khatri

APD Cryogenics
Allentown, PA 18103 USA

ABSTRACT

A cryocooler has been developed that uses a unique refrigerant in a throttle cycle to achieve temperatures below the normal boiling point of liquid nitrogen. 70 K was achieved at no load condition and continuous operation with a 1.5 W heat load below 77 K has been demonstrated. The refrigerator consists of a compressor package that is similar to a standard air conditioning system, a cold end which has no moving parts that consists of a heat exchanger with a cold plate, and two gas lines connecting the compressor and the cold end. The system requires no maintenance. Several prototypes have been built, tested and delivered to a customer for application in the field of high temperature superconducting filters for cellular phone communications. The cold end has the design flexibility to be customized for different applications. System performance can also be customized by designing different refrigerants that can be used without the need to modify the hardware.

INTRODUCTION

A closed cycle throttle refrigerator system has been tested that provides continuous refrigeration below 77 K. The basic system[1] shown in fig. 1 consists of a compressor assembly, a cold end assembly and two gas lines that connect the compressor to the cold end. The compressor is a single stage oil lubricated compressor typically used in house hold air-conditioning systems. The other major components are an after-cooler, oil separator and adsorber. The cold end consists of a coiled counter flow heat exchanger, a throttling device and a cold plate. The system does not require maintenance. The ability to distribute refrigeration to the area to be cooled gives this system its unique flexibility. This relatively simple system works with high efficiency and has low levels of vibration and noise. The fact that the refrigerant can be customized for the temperature range of 70 K to 120 K without changing the hardware makes the system very suitable for a variety of commercial applications as well as for research applications. Reliability is high due to the fact that the compressor is the only active mechanical component and has commercially proven reliability.

Advances in Cryogenic Engineering, Vol. 41
Edited by P. Kittel, Plenum Press, New York, 1996

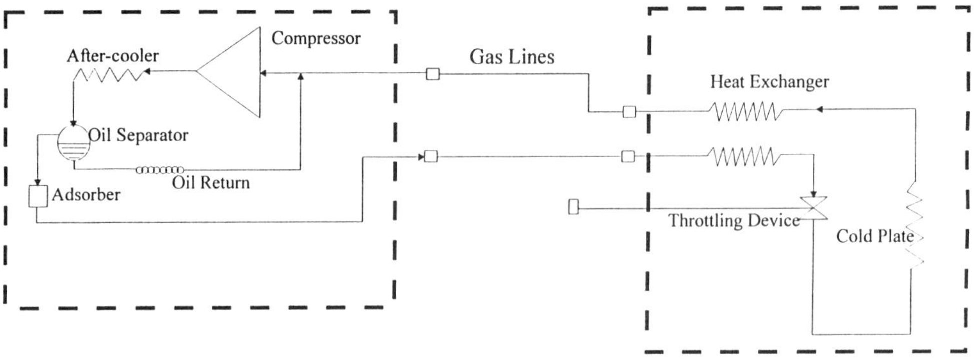

Figure 1 Closed-Cycle Throttle Refrigerator Schematic.

SYSTEM DESIGN AND OPERATION

The design of the system emphasizes minimum complexity in fabrication and operation. The gas is compressed by the oil lubricated compressor. The gas is typically discharged at 18 to 20 atm. with suction pressure of 1 to 3 atm[(2)]. The gas then flows through an air cooled after-cooler before entering the oil separator. The oil from the oil separator is returned to the compressor via an oil return line. The high pressure gas then flows through ambient temperature gas lines to the cold end. The cold end is tolerant to a low level of oil carryover. In the cold end the gas passes through a counter flow heat exchanger where the incoming high pressure gas is cooled by the returning low pressure gas. The gas then passes through a throttling device producing cooling effect on the cold plate. Constituents of the gas mixture start condensing in the heat exchanger as the cooldown progresses. As the system temperature approaches the steady state operating condition i.e. when the cold plate is cooled down to cryogenic temperature (less than 120

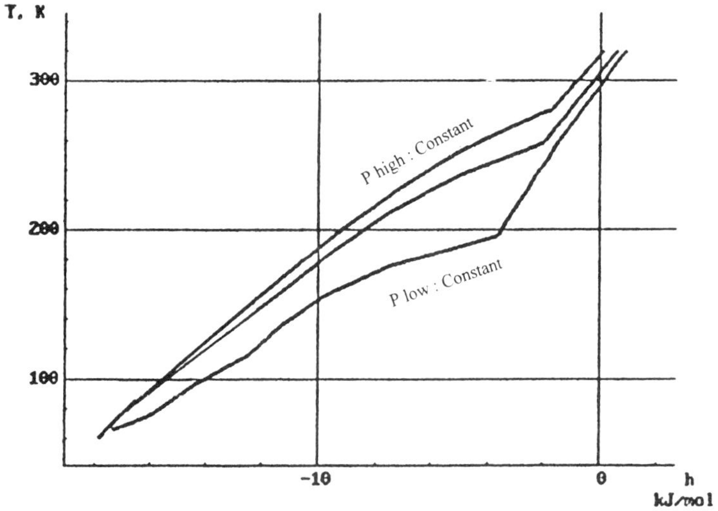

Figure 2 Temperature-Enthalpy Plot for A Mixed Refrigerant.

K), the high pressure supply gas is mostly liquefied before entering the throttling device. The throttling effect does not produce a significant temperature drop from the high to the low pressure sides of the throttle device. Refrigerant is mostly liquid before and after throttle valve.

The refrigerant gas mixtures[3] in the system consist of some combination of inert gases like helium, neon, nitrogen, argon, and hydrocarbons such as methane, ethane, propane, butane and R14, R23, R22 etc. Flammability of the gas mixture will depend on the gases used and their concentration in the mixture. The freezing point of the mixture is typically lower than that of the pure constituent components except neon and helium.

A typical refrigerant mixture has a temperature- enthalpy plot[4] as shown in fig.2. The minimum cold end temperature achieved depends on the gas mixture as well as the return pressure in the evaporator. These are important design parameters that define system performance. The heat exchanger design and system analysis is based on the properties of the gas mixture in steady state operating conditions. The gas mixture circulating in the system during the steady state conditions is significantly changed from the gas mixture that is used to charge the warm system. This is due to several factors including solubility of several components of the mixture in the oil as well as liquefaction and accumulation of liquid components in the heat exchanger and the evaporator.

PERFORMANCE

Primary focus of the tests was to investigate the performance of the system below 77 K. Good system performance at liquid nitrogen temperature would significantly increase the opportunities for applications in the field of high temperature superconductivity, detector technologies, medical applications and many more.

Several parameters affect the performance. Compressor pressure ratio, cold end heat exchanger design, throttle valve setting and refrigerant mixture are the primary variables. The effect of variation of these parameters and the results are discussed below.

Fig.3 is a plot showing the temperature vs. applied heat load at the cold plate. Tests were run with the same compressor and cold end setup but different composition of the gas mixture. The manual throttle valve setting was kept the same. Refrigerant "Mix1" was tested first and performance was mapped. Then the system was evacuated and recharged with "Mix2" and the performance was mapped again. No load temperature was lowered from 76.5 K to 70.9 K for the same setting of the throttle valve. Applied heat load was limited to 3 W maximum. At 3 W applied heat load the cold plate temperature was lowered from 82.8 K to 77.3 K. The return pressure was 1.0 - 1.5 atm.

Fig.4 shows a plot of temperature vs. applied heat load when the refrigerant mixture as well as the throttle valve setting is changed. Refrigerant "Mix3" was used for the test and the throttle valve was opened by one full turn from the original setting. This helps not only to lower the temperature with a special refrigerant mixture but also increase the mass flow rate and maintain high total heat load capacity at lower temperature. Specific cooling capacity is low for this low temperature refrigerant mixture so it is important to increase the mass flow rate to keep the total heat load capacity high. Alternately the temperature can be lowered by reducing the evaporator pressure by closing the throttle valve further but it also reduces total heat load capacity which is not desirable. The no load temperature was lowered to 68 K and with 3W applied heat load 76.2 K was achieved.

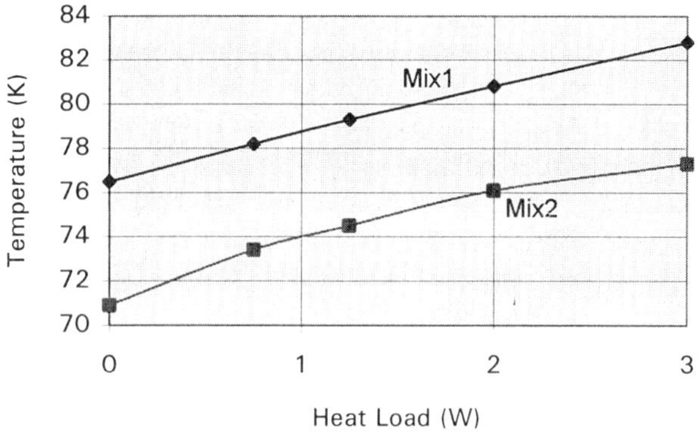

Figure 3. Temperature vs. Heat Load for refrigerants "Mix1" and "Mix2". Mixture "Mix2" is designed to produce minimum temperature below 77 K.

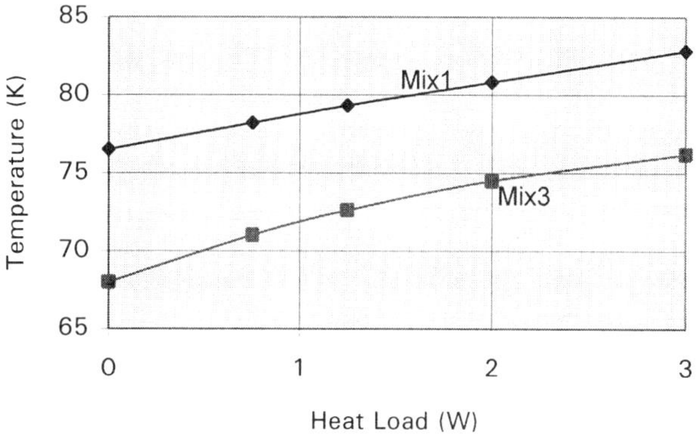

Figure 4. Temperature vs. Heat Load for refrigerants "Mix1" and "Mix3". Refrigerant "Mix3" allows higher flow through throttle valve and still achieve temperature below 77 K.

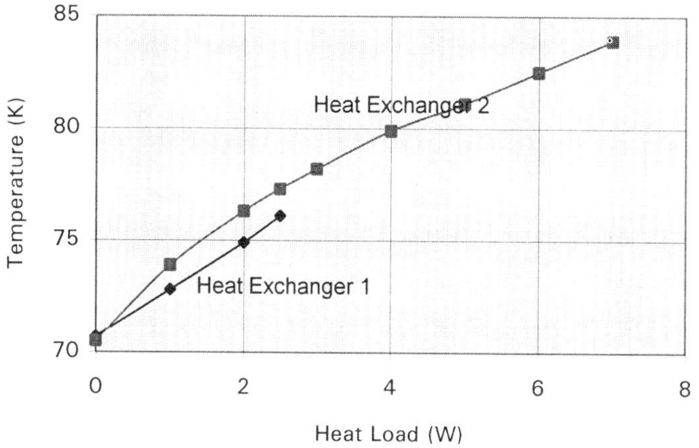

Figure 5. Temperature vs. Heat Load for heat exchangers "HE1" and "HE2". HE2 allows higher heat load carrying capacity for the same refrigerant gas mixture "Mix4".

Fig.5 shows a plot of temperature vs. applied heat load when the same refrigerant mixture is used in two different designs of the cold end. Refrigerant "Mix4" was used for the testing. Heat Exchanger #1 (HE1) was the same one used in the tests described above. Heat Exchanger #2 (HE2) was designed for higher efficiency. Both heat exchangers had similar flow rates through the throttling devices. Improved efficiency of the heat exchanger allowed HE2 to carry 7 W with "Mix4" while HE1 could carry only 2.5 W maximum. The HE2 delivered higher total heat load carrying capacity due to higher efficiency design. The HE2 carried 10 W at about 91 K and 20 W at 120 K with properly designed refrigerant gas mixtures in each case. This shows that the design of the heat exchanger is an important parameter along with the refrigerant design.

Compressor has about 400 W power consumption while delivering 10 W at 91 K. The system efficiency relative to Carnot at this point is 5.8%. With higher compressor efficiency and heat exchanger efficiency this should improve.

TEMPERATURE CONTROL

The operating temperature is set by the refrigerant pressure in the evaporator and the heat load. As the heat load is applied, the liquid refrigerant evaporates and pressure in the evaporator increases, which increases the cold plate temperature. The rise in temperature takes place over a limited range. If the heat load is applied beyond the maximum that is sustainable then the unit will continue warming up as long as that heat load is maintained. If the heat load is reduced or taken off then the unit will cooldown again. The operating temperature range can be shifted by increasing or reducing the flow through the throttling device. A temperature controller with a heater element can be used to maintain a constant temperature if so desired.

SUMMARY

This type of closed cycle refrigeration system is the first commercialization of mixed gas refrigerant technology at or below 77K. It provides high reliability, no maintenance, long life, no moving parts in the cold end, possibility of customized cold end configuration, separation of cold end and compressor and the possibility of low cost in high volume production. It is also possible to design multiple stage configuration to significantly increase the refrigeration capacity at less than 80 K. Various applications are possible in different fields where liquid nitrogen is presently required to achieve cryogenic temperature.

ACKNOWLEDGMENT

Author would like to acknowledge analytical help received from Dr. M. J. Boiarski and Dr. B. Yudin from MEI, Moscow, as well as from Dr. R.C.Longsworth at APD Cryogenics, Inc. Stephen Harold provided significant help during the setup and testing of the systems

REFERENCES

1. Longsworth, R.C. Boiarski, L.A. Klusmier, "80 K Closed Cycle Throttle Cycle Refrigerator", presented at the International Cryocooler Conference, Vail Co., June 1994.

2. Longsworth R.C., "Cryogenic refrigerator with Single Stage Compressor", US Patent 5,337,572 Aug. 1994.
3. Alfeev V.N., Brodyanski V.M., Yagodin V.M., Nikolsky V.A., Ivantsov A.V., "Refrigerant for a Cryogenic Throttling Unit", UK Patent 1,336,892 (Nov. 1973).
4. R.C.Longsworth and M.J.Boiarski, "New Smart Gaseous Materials for Cryogenic Applications", presented at Institute of Materials, "Materials Issues in Smart Structures" Conference London,UK, April 1995.

CONTINUOUS FLOW CRYOGEN SUBLIMATION COOLER

R. C. Longsworth and A. Khatri

APD Cryogenics Inc
Allentown, PA 18103-4783 U.S.A.

ABSTRACT

A new technique has been developed that permits temperatures below the triple point to be maintained continuously when liquid cryogen is fed to a special cold stage that vents to vacuum. The technology has been demonstrated using hydrogen which has been liquefied by a Displex GM refrigerator and fed to the cold stage at pressures ranging from 0.1 to 0.5 MPa. With no heat load applied the temperature remained constant at 9.7 K independent of pressure. The temperature and sublimation rate increased when heat was applied. This paper describes the device that was tested and presents a theoretical model which explains the results.

INTRODUCTION

A closed cycle refrigerator that utilizes hydrogen adsorbing hydrides to pump hydrogen has been demonstrated to produce temperatures below 10 K on a periodic basis, [1]. This system uses a Stirling cycle refrigerator to maintain a thermal block at about 65 K which serves as a heat sink to cool H2 that is circulated from a gas storage bottle at an initial pressure of about 10 MPa through a JT heat exchanger to cool a reservoir and collect liquid at about 25 K while returning gas to a Vanadium hydride adsorber. After filling the reservoir with liquid the gas flow from the bottle is stopped, the line to the Vanadium hydride adsorber is valved off, and a direct vent line from the reservoir to a ZrNi hydride adsorber is opened. The ZrNi adsorber can pump H2 at $<$ 2 Torr so it pumps the about one third of the liquid H2 that has collected to a temperature below 10 K and maintains that temperature until all of the solid H2 that formed has sublimed. The ZrNi adsorber is then valved off and the adsorbers heated in sequence to compress the gas back into the storage bottle.

The successful operation of the hydride sorption compressor system motivated an investigation of a means to produce solid H2 at 10 K on a continuous basis.

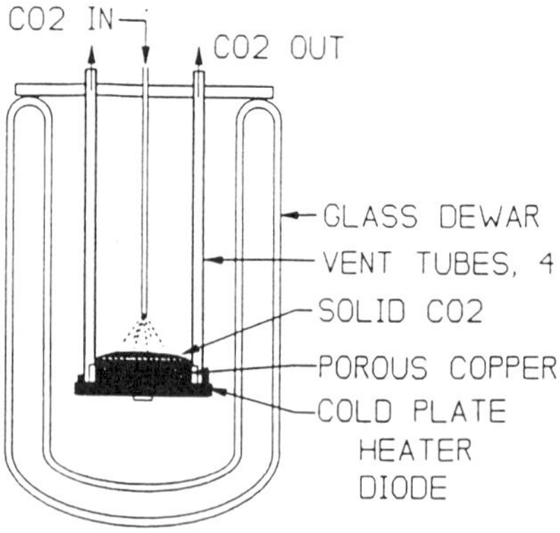

Figure 1 Test Cell in Clear Glass Dewar to Observe Collection and Sublimation of CO2 Tests.

EXPLORATORY TESTS WITH CO2

Initial tests were done with CO2 in a clear glass dewar, as shown in Figure 1, so that the formation of the solid could be observed. Liquid CO2 was fed from a supply bottle at 3.5 MPa through a 3 mm diameter tube which terminated in a 0.15 mm ID capillary tube that was directed at a porous copper disc attached to a copper cold plate with a heater and temperature sensing diode. Solid CO2 snow from the cold jet impinged on the surface of the porous copper while the gas could either flow through the porous copper or vent directly to the inside of the dewar which was covered but open to atmosphere. Solid CO2 was observed to collect on the porous copper and periodically collapse onto the surface as the solid near the surface sublimed. With the heat and diode connected to a temperature

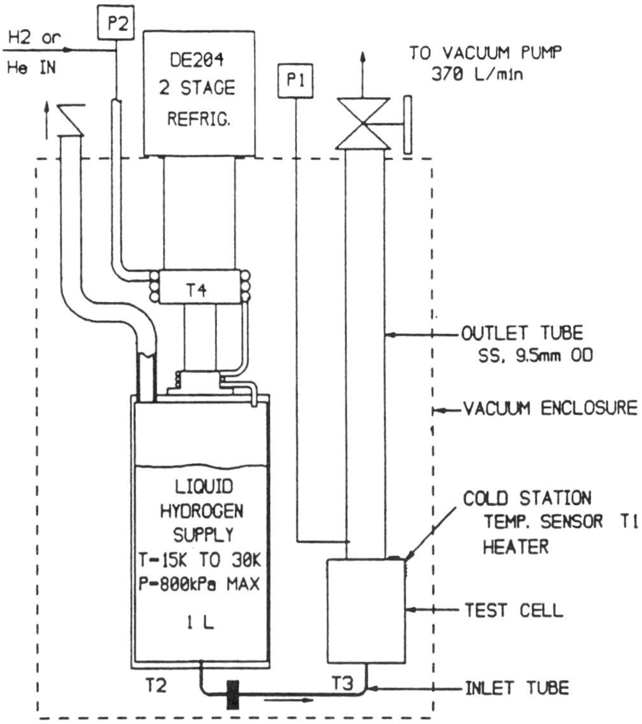

Figure 2 Schematic of Continuous Flow H2 Sublimation Cooler Test Arrangement

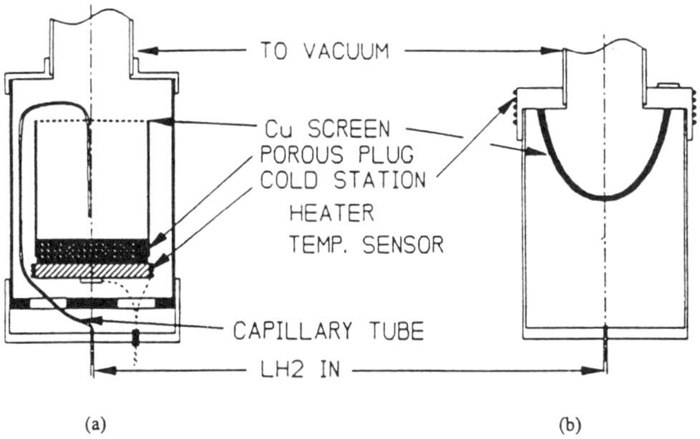

Figure 3 SH2 Test Cells. a) Modified CO2 Cell. b) Cell Designed for SH2.

controller it was easy to maintain a constant temperature between 196 K and 210 K (the NBP of CO2 is 194.7 K). Excess snow was observed to blow off the heat station and vaporize in the dewar. Flow from the capillary tube was steady indicating that there was no freezing of the liquid CO2 in the capillary tube.

TESTS WITH HYDROGEN

Tests with H2 were run using the small liquefier shown schematically in Figure 2. A model DE204 Displex® expander is used to condense H2 at a rate of about 0.5 L/hr which collects in a 1.5 L accumulator attached to the cold stage. A line from the bottom of the accumulator connects to the test cell which is supported by a vent tube that connects to a vacuum pump with a displacement of 370 L/m. Hydrogen can be supplied at pressures up to 650 kPa (100 psi). It is precooled on the first stage at a temperature of about 75 K (T4) then condensed on the second stage at the saturation temperature corresponding to the pressure (P2). Temperature sensor T2 measures the temperature at the bottom of the accumulator which may be less than the temperature at the surface because cold LH_2 is denser than warm LH_2, but is never greater. This happens when the supply pressure is increased, for example. The LH_2 can also be pressurized with He and cooled by the refrigerator to 15 K. With this system LH_2 can be fed to the test cell at pressures up to 650 kPa and liquid temperatures down to 15 K.

The experimental phase separator cell is mounted along side the LH_2 accumulator. The pump-out tube has a copper wire soldered to it that is also connected to the second stage of the refrigerator to intercept heat leak and to help cool the test cell from 300 K to 25 K. A separate tube is connected to the cold end of the pump-out tube to measure the pressure of the SH_2 (P1). A separate pressure gauge at the inlet to the vacuum pump is used to calculate the flow rate.

The test cell used for the tests with CO2 was modified as shown in Figure 3a. The capillary tube was sized on the assumption that 8 mg/s of LH2 would be supplied at a temperature of 24 K and pressure of 265 kPa while the vacuum pump would maintain an exhaust pressure of 1.6 kPa corresponding to a temperature of less than 12 K.

The first test was run by cooling down the accumulator and test cell, with some gas flowing through it, then partially filling the accumulator with LH2 at 24 K. The valve to the vacuum pump was then opened. The initial results were disappointing because the LH2 in the capillary tube went through repeated freeze thaw cycles. Subsequent tests showed that freezing occurs when the exhaust pressure drops below the triple point pressure. After a number of freeze thaw cycles we were surprised to find that the temperature dropped to about 9.5 K and stayed there until a heat load was applied. Turning a heat load of 50 mW on and off caused the temperature to cycle in the range of 18 K to 10 K. Figure 4 picks up the test 2 hours after first cooling to 14 K and shows some of temperature excursions. At about 650 s into this plot the H2 vent pressure was increased to atmospheric pressure and the cell warmed to the triple point and stayed there for 16 minutes. This indicates that a significant amount of SH2 had formed in the cell. When the vacuum was reestablished the temperature dropped below 10 K and stayed there for 20 minutes at which time the test was terminated.

This test indicated that it should be possible to maintain 10 K continuously, but the mechanism for doing it is entirely different from the method that worked for CO2. It was conceptualized that in the case of H2, solid was forming in the cell where it was trapped by the screen on the top and the porous copper on the bottom and subliming at the surface. We were not sure however whether or not the SH2 was being replaced.

The test was repeated with a larger capillary tube. This time the cell cooled below 10 K in about 1 minute and stayed at 9.5 K until the test was stopped 17 minutes later. During the test the pressure in the accumulator was increased from 160 kPa to 545 kPa with no change in temperature or vent pressure.

A new test cell was then made as shown in Figure 3b. It has several layers of copper screens to trap the SH2 and transfer heat from the heater to the subliming SH2. It was

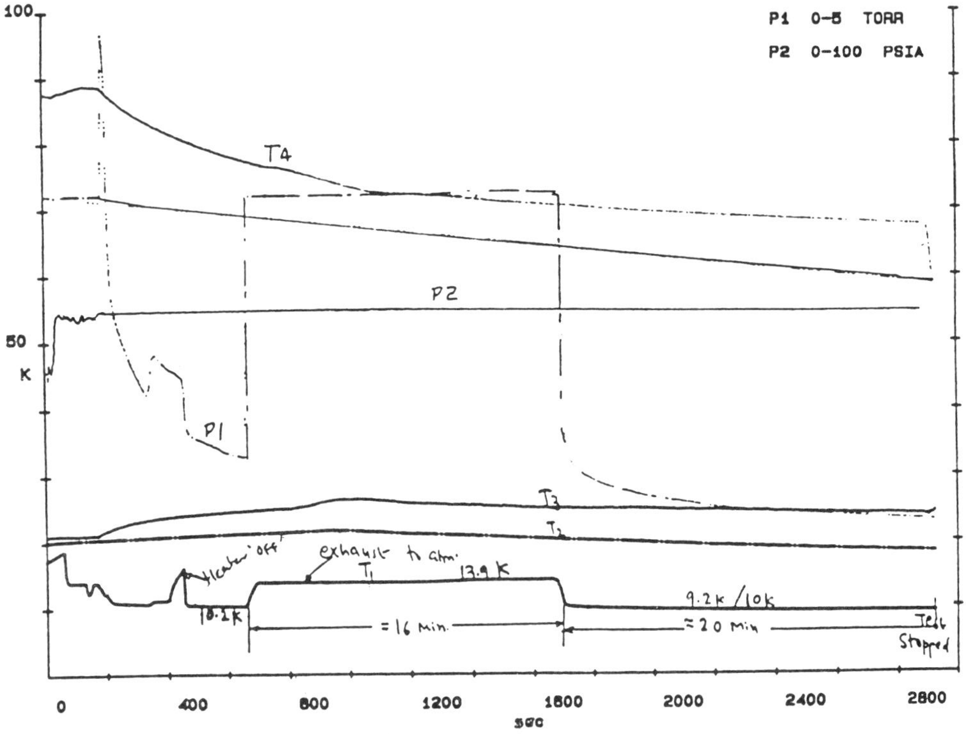

Figure 4 Test Plot with LH2 Using Modified CO2 Cell. Plot Starts 2 Hours after First Reaching 10 K.

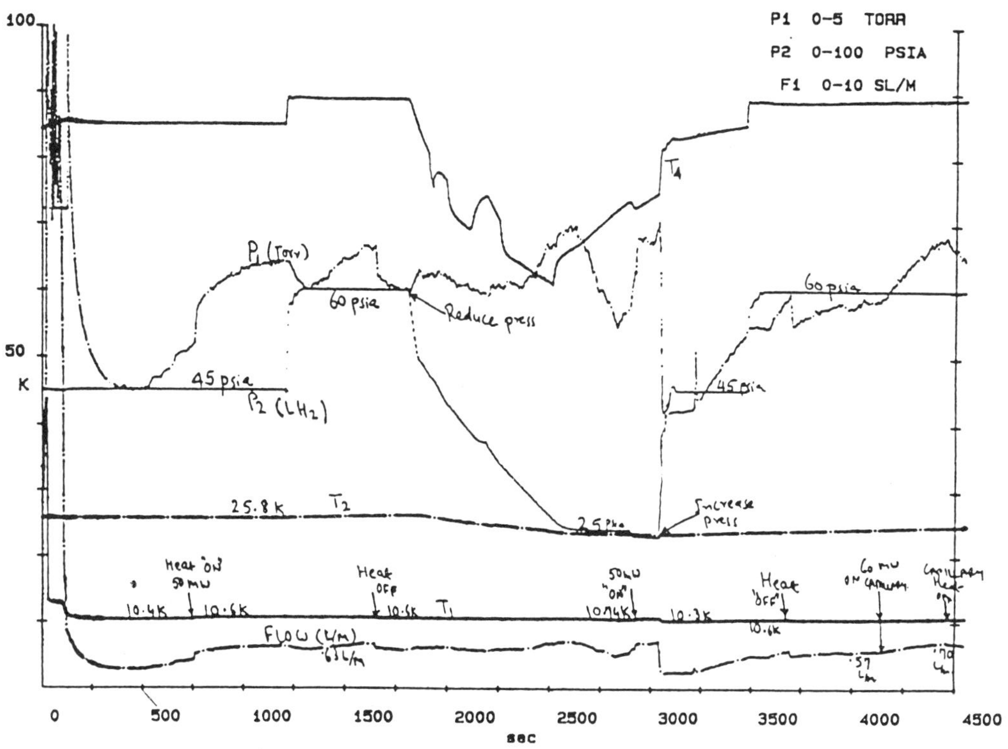

Figure 5 Test Plot with the Cell Designed for SH2.

designed to test our theory that the SH2 subliming at the surface of the porous copper could be continuously replenished in a through flow design in which the supply capillary tube is separated from the subliming SH2 so that the liquid will not freeze in the tube. The pressurized LH2 also provides a force that keeps the solid in contact with the copper screen. The capillary tube is sized so that the vacuum pump can maintain a pressure below the triple point while the cell cools down and solid forms on the copper screen. After solid has formed the sublimation rate drops and the vacuum pump reduces the temperature.

Figure 5 is one of the test plots with the cell shown in Figure 3b. Heater wire was wrapped around the capillary tube to see what the effect would be of heating the tube. The cell cooled below 11 K in about 2 minutes and stayed near 10.5 K for 75 minutes during which time a heat load of 50 mW was turned on and off twice, the supply pressure was changed from 160 kPa to 545 kPa, and 60 mW was applied to the capillary tube during different periods of time as noted on the plot. The rate at which the SH2 sublimed changed in response to changes in heat load with some delay, and the changes in supply pressure caused some transient changes in the vent rate. A flow rate of .63 L/min (at room temperature and pressure) was measured when 50 mW was applied. This corresponds to a total heat load at 10.5 K of 270 mW.

THEORETICAL MODEL

Figure 6 shows a model of a simple cylindrical cold cell in which LH2 feeds in under pressure at one end and vacuum is maintained beyond a porous heat conducting plug at the other. If we assume an orientation with the liquid above the solid then there is no convection

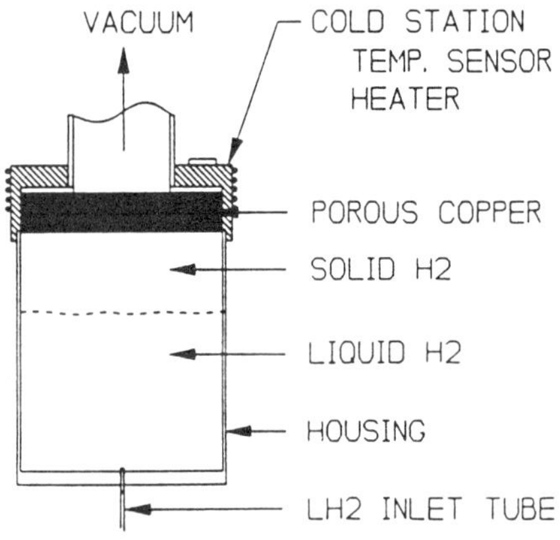

Figure 6 Model of a Simple Cylindrical Cell.

in the cell. Heat flows into the porous plug from an external source and from the LH2 / SH2 by conduction. Assuming the applied heat load to be 250 mW and the LH2 supply temperature to be 25 K then 135 mW must flow from the LH2 / SH2 to cool it from 25 K to 10 K. A total of 385 mW of cooling is provided by the sublimation of SH2 at the rate of 1.27 mg/s. At this flow rate a vacuum pump with 370 L/min displacement will have an inlet pressure of 0.26 kPa which corresponds to a temperature of just under 10 K. Figure 7 is a plot of the heat flow rate through the H2 versus temperature as a function of temperature for the same example. The step increase at 14 K is due to the heat of fusion that has to be removed to solidify the LH2. Figure 8 is a plot of the temperature versus distance from the subliming surface for a diameter of 18 mm as derived
from the heat flow rate of Figure 7 and the thermal conductivity of LH2 / SH2. The density of SH2 continuously increases as it cools below the triple point so it does not tend to stick to the walls of the cylinder.

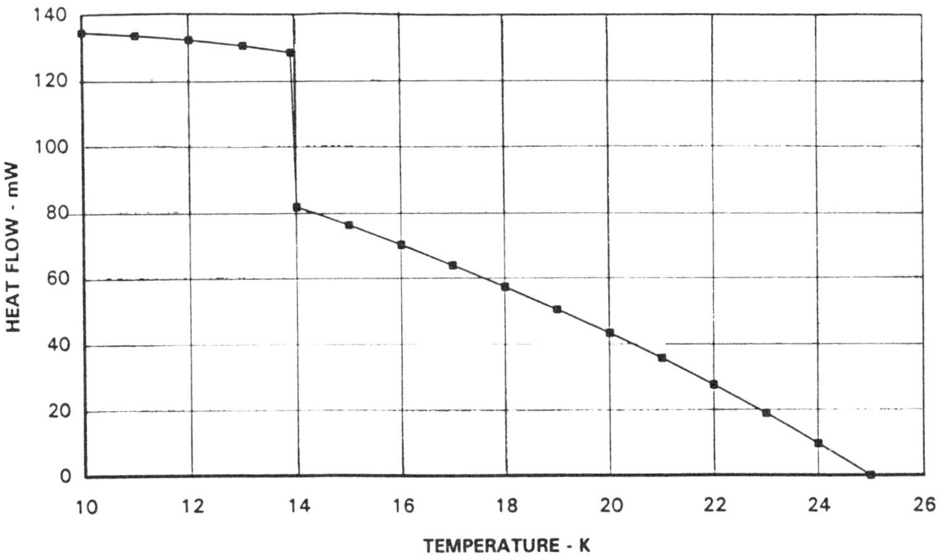

Figure 7 Heat Flow Rate Vs Temperature Through SH2 / LH2 Needed to Cool it from 25 K to 10 K to Produce 250 mW at 10 K.

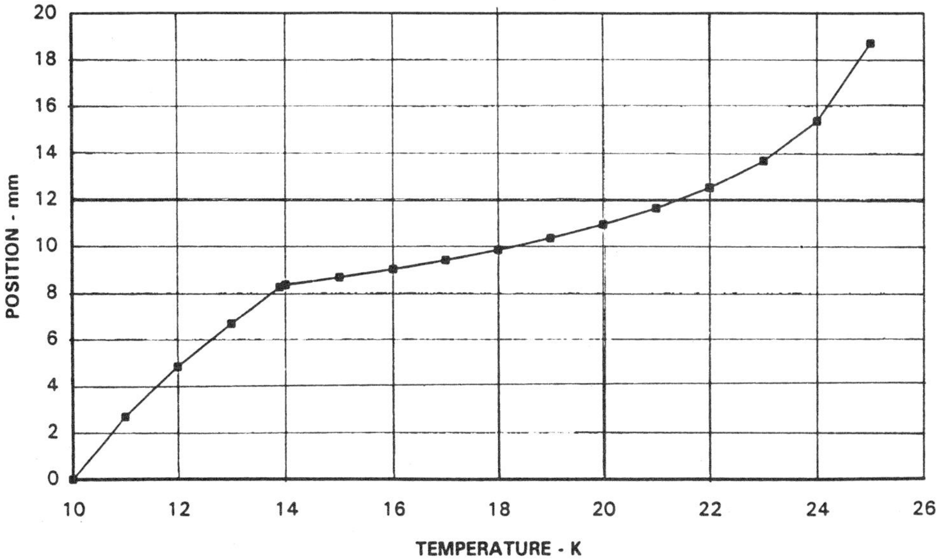

Figure 8 Position Vs Temperature in SH2 / LH2 when Cooling it from 25 K to 10 K to Produce 250 mW at 10 K in an 18 mm Diameter Cell.

Had a larger diameter or a lower heat load been assumed then a greater length would be needed to avoid the possibility of the inlet tube freezing. It is clear that there are limitations on the relationships between heat load, cell size, cell shape, supply temperature, and orientation for cooling to be produced continuously at a near constant temperature.

OTHER CRYOGENS

This cooling method is applicable to other cryogens. Table 1 lists the most common cryogens that are the apt to be used and the temperature they would operate at if a vacuum of 0.13 kPa (1 Torr) were maintained. Liquid nitrogen for example is produced at an efficiency of more than 35 % in a large plant so it may be more economical to produce refrigeration at 50 K to 60 K by pumping on LN2 than by using a small refrigerator to provide the cooling. This cooling method can also be vibration free and provide cooling in a small remote space.

Table 1 Properties of Cryogens that are Suitable for Continuous Sublimation Cooling

Gas	NBP -K	Triple Point Press. - kPa	Triple Point Temp. - K	Temp. At .13 kPa (1 Torr)
Hydrogen	20.3	7.0	13.9	9.5
Neon	27.1	44.3	24.5	15.8
Nitrogen	77.3	12.5	63.2	47.0
Oxygen	90.2	0.15	54.4	54.1
Argon	87.3	68.8	83.8	54.4
Methane	111.7	11.7	90.7	67.3
Krypton	120.0	73.0	116.2	74.8

ACKNOWLEDGMENT

This work was funded by Aerojet General Corporation.

REFERENCES

1 J. J. Wu, S. Bard, W. Boulter, J. Rodriguez and R. C. Longsworth, "Experimental Demonstration of a 10 K Sorption Cryocooler Stage", Advances In Cryogenic Engineering, Vol. 39, Plenum Press, New York, 1994

2 R. C. Longsworth and A. N. Khatri, "Continuous Flow Cryogen Sublimation Cooler", U. S. Patent 5,385,027, Jan. 31, 1995

DEVELOPMENT OF A PERIODIC 10 K HYDROGEN SORPTION CRYOCOOLER

P. Karlmann, S. Bard, and J. J. Wu

Jet Propulsion Laboratory
California Institute of Technology
Pasadena, CA 91109

ABSTRACT

The design and fabrication of a small, closed-cycle, periodic 10 K sorption cryocooler laboratory system is described. The cooler is designed to provide cooldown from 60 to < 10 K in under 2 minutes, maintain a simulated detector heat load of 100 mW below 10 K for over 10 minutes, and have the ability to be recycled in under 6 hours. A $LaNi_{4.8}Sn_{0.2}$ hydride sorbent bed is used to compress and circulate hydrogen refrigerant in the 24-30 K Joule-Thomson intermediate refrigeration stage. A ZrNi hydride sorbent bed is used to provide the low vacuum pressure (< 2 torr) needed to solidify and cool the hydrogen to 10 K. Thermally cycling the sorbent beds between 290 and 550 K produces a compression ratio of nearly 10^6. A vacuum-pumped liquid nitrogen loop is used to simulate a mechanical cooler or radiator for the upper 60-80 K stage. Performance predictions are presented, results are compared with ground test data from the flight-qualified Brilliant Eyes Ten-Kelvin Sorption Cryocooler Experiment (BETSCE), and plans for using the cooler as a test bed for future reliability physics experiments are described. Performance and reliability test results, combined with BETSCE flight validation data, will enable early insertion of this long-life (> 10 years), low-vibration refrigeration technology into precision-pointing military surveillance, earth-observation, and astrophysics space satellite applications.

INTRODUCTION

Recent advances in sorption cooler technology have shown a significant potential for the application of sorption coolers in a variety of areas[1,2]. Thermally powered sorption compressors with no moving mechanical parts have a fundamental advantage over conventional compressors due to their low vibration and potential for long-life reliability. Cryogenic applications for this technology include astrophysics, earth-observation, and military surveillance satellite systems. For those applications which require only intermittent or periodic cooling, sorption technology is particularly attractive due to its quick cooldown capability and its low average power consumption.

This paper describes a small, closed-cycle periodic 10 K sorption cryocooler that is being developed to conduct performance and reliability studies. Test results are expected to lead to improved understanding and enhanced performance of both periodic and continuous 10 - 25 K sorption coolers.

Periodic 10 K Sorption Cooler Concept

The concept of a periodic 10 K sorption cooler was originally developed by Johnson and Jones in 1990.[3, 4] A simplified schematic of the 10 K system is shown in Figure 1. The basic principal is to utilize metal hydride powders to compress and recirculate hydrogen gas within a closed-cycle Joule-Thomson system for periodic cooling of sensors at 10 K.

The cycle begins with the tank initially charged to approximately 10.3 MPA with hydrogen gas and the cryostat upper stage maintained at a temperature of 60 - 80 K. Cooling for the upper stage could be provided by a number of currently available Stirling, pulse-tube, or sorption coolers operating in the 60 K temperature range. For this particular cooler, liquid nitrogen near its triple point is used to obtain temperatures from 65 to 80 K.

When the command is given, a solenoid valve is opened, releasing hydrogen from the tank, through the J-T loop of the cryostat, through a check valve, and into a $LaNi_{4.8}Sn_{0.2}$ hydride sorbent bed. Within the J-T loop of the cryostat, the hydrogen gas is precooled and expanded, producing liquid hydrogen at the cold head. The $LaNi_{4.8}Sn_{0.2}$ hydride sorbent bed absorbs the in-coming hydrogen gas and maintains the vapor pressure of liquid produced in the coldhead between 24 to 30 K.

After enough liquid has been accumulated in the coldhead (typically 1 to 2 grams), the solenoid valve is closed, halting gas flow. Immediately, another solenoid valve is opened, exposing the coldhead to the low pressure ZrNi hydride sorbent bed. The ZrNi hydride absorbs the gaseous hydrogen in the coldhead, reducing the vapor pressure to below 266 Pa. This reduction in vapor pressure solidifies the liquid hydrogen and cools the coldhead to approximately 10 K. As the solid hydrogen sublimates due to the applied heat load, the low pressure sorbent bed maintains the temperature of the coldhead at 10 K by absorbing the sublimated gas. This temperature is maintained until the solid hydrogen is completely sublimated, or until the ZrNi hydride is entirely saturated.

To recharge the tank with hydrogen and prepare the system for another cooldown, the low pressure bed is first heated to a temperature of approximately 550 K. This drives the hydrogen gas present in the ZrNi hydride from the low pressure bed at about 1.01 kPa, through a check valve, and into the high pressure bed where it is absorbed by the $LaNi_{4.8}Sn_{0.2}$ hydride. Next, the high pressure bed is heated to a temperature of approximately 545 K. This releases hydrogen gas at about 10.3 MPA. The high pressure gas then flows through a check valve from the high

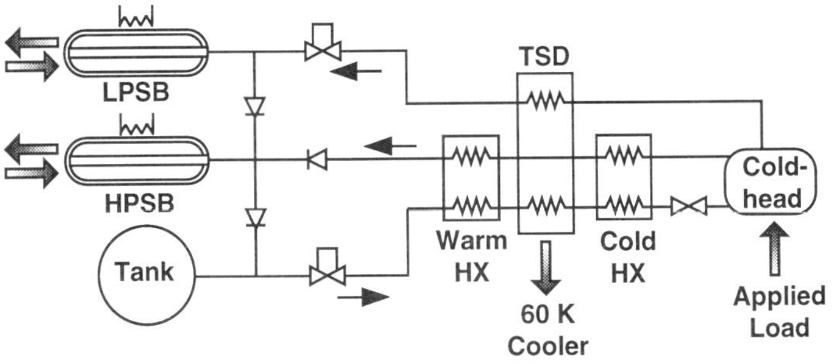

Figure 1. Periodic 10 K sorption cryocooler simplified fluid schematic.

Table 1. Cooler specifications & design characteristics

Design Parameter	**Description**
Cooldown time from 60 K to 10 K	< 2 minutes
Cooling duration below 10 K with 100 mW load	> 10 minutes
Recycle time	< 6 hours
Quantity of LaNi4.8Sn0.2 in high pressure sorbent bed	626 g
Hydrogen capacity, high pressure sorbent bed	8.01 g
Quantity of ZrNi in low pressure sorbent bed	251 g
Hydrogen capacity low pressure sorbent bed	1.87 g
Avg. hydrogen flowrate during cooldown from 60 to 10 K	0.1 g/s
Total hydrogen consumed during cooldown from 60 to 10 K	7.3 g
Coldhead mass	78 g
TSD mass	2.9 kg

pressure sorbent bed back into the tank. Once the sorbent beds have cooled back down to ambient temperature, the cycle is complete and the system is ready for the next cooldown operation.

JPL Sorption Cooler Development Program

A series of proof-of-principle experiments were conducted in 1991 which successfully demonstrated the key technologies of a 10 K sorption stage.[5] Based on the success of these experiments, the Brilliant Eyes Ten-Kelvin Sorption Cryocooler Experiment (BETSCE) was developed as a spaceflight demonstration of the critical 10 K sorption technologies.[2] BETSCE is a space shuttle payload side-wall-mounted experiment which represents the first closed cycle 10 K sorption cryocooler ever built. Extensive ground testing of BETSCE has already been conducted, and the scheduled shuttle launch is set for April, 1996.

The periodic 10 K sorption cooler described in this paper represents the next step for the technology. In developing this new cooler, it was desired to produce a greatly simplified and compact laboratory version of BETSCE which would incorporate many of the lessons learned during ground testing of the spaceflight experiment. Important changes incorporated in this new design include the following: a greatly reduced size and mass, use of two rather than three sorbent beds, use of passive check valves as a replacement for active solenoid valves, and a completely redesigned Joule-Thomson expansion device, coldhead, and thermal storage device. In addition, the gas-gap heat switches for the sorbent beds were eliminated to simplify the laboratory cooler.

HARDWARE DESIGN

Off-the-shelf technology was incorporated into the cooler design whenever practical. Emphasis was placed on minimizing the size of the cooler while balancing the additional goals of facilitating fabrication and minimizing costs. System robustness, reliability, and ease of operation were also important design considerations.

Displayed in Table 1 are the principle design specifications used for the development of this cooler and some of the critical overall design characteristics resulting from these specifications. In general, the cooldown time and the cooling duration are the critical specifications that determine flowrates, capacities, and sizes of the various cooler components. In most cases, BETSCE design characteristics and performance data were used as initial reference points for the design of this cooler.

Although fundamentally identical to Figure 1, the actual cooler includes a number of pressure transducers, service valves, and relief valves. An additional pneumatic valve was placed between the inlet and outlet lines of the J-T cooldown loop to quickly vent gas from the high pressure side to the low pressure side prior to opening the low pressure sorbent bed. Also, a saran carbon coldtrap was incorporated into the recharge cycle between the high pressure sorbent bed and the tanks.

The physical layout of the cooler is shown in Figure 2. At the center of the upper level is the 15.2 x 30.5 cm cylindrical vacuum dewar that houses the cryostat. On left and right sides of the vacuum housing are the low pressure and high pressure sorbent beds, respectively. Underneath the sorbent beds are located two 1-liter tanks which hold the 10.3 MPA of hydrogen prior to a cooldown. Scattered over the remainder of the upper and lower levels are a number of pneumatic valves, check valves, service valves, relief valves, and pressure transducers.

Sorbent Bed Design

The low pressure and high pressure sorbent beds circulate the hydrogen gas in a closed-cycle system and thermally compress it from a pressure of 266 Pa to 10.3 MPA. In general, powdered metal hydride materials such as ZrNi and $LaNi_{4.8}Sn_{0.2}$ absorb hydrogen gas at low temperatures, and desorb it at higher temperatures and pressures. During chemical absorption, the materials reject heat and absorb mass. During desorption, the materials absorb heat and reject mass. From a design perspective, sorbent beds can be viewed as pressure vessels designed to transfer heat and mass to and from the powdered materials contained within them.

The overall design of the low pressure and high pressure sorbent beds are shown in Figure 3. As was previously described, the basic function of the low pressure sorbent bed is to maintain a vacuum of < 266 Pa on the collected hydrogen in the coldhead, thereby solidifying the hydrogen and reducing the temperature to below 10 K. To achieve this low pressure for the ten minute duration, 162 grams of powdered ZrNi hydride alloy was used for an estimated reversible hydrogen capacity of 1.87 grams.

The primary functions of the high pressure sorbent bed are to absorb a large quantity of hydrogen in a short period of time during the initial cooldown at low pressure, and then to re-pressurize the tank by desorbing the hydrogen at high pressure. This sorbent bed uses 626 grams of powdered $LaNi_{4.8}Sn_{0.2}$ hydride for an estimated reversible hydrogen capacity of 8.01 grams.

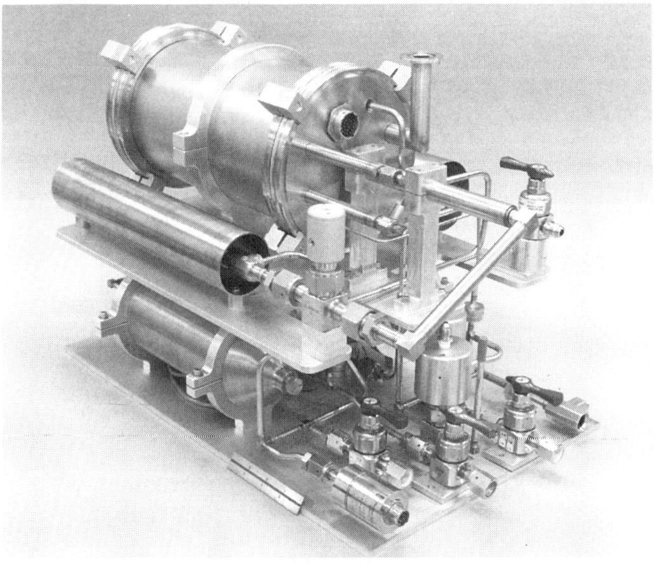

Figure 2. Periodic 10 K sorption cryocooler.

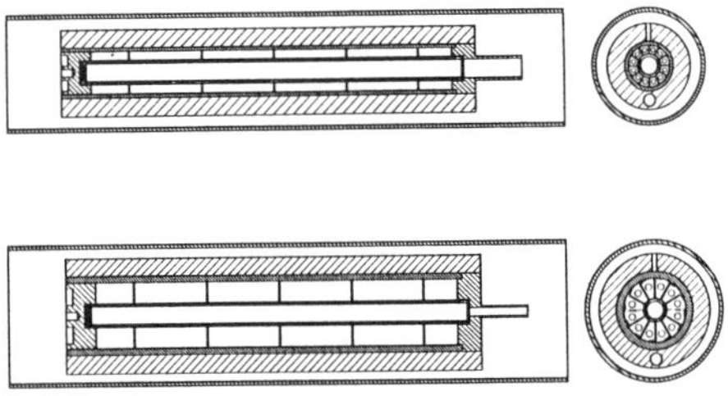

Figure 3. Low pressure and high pressure sorbent beds.

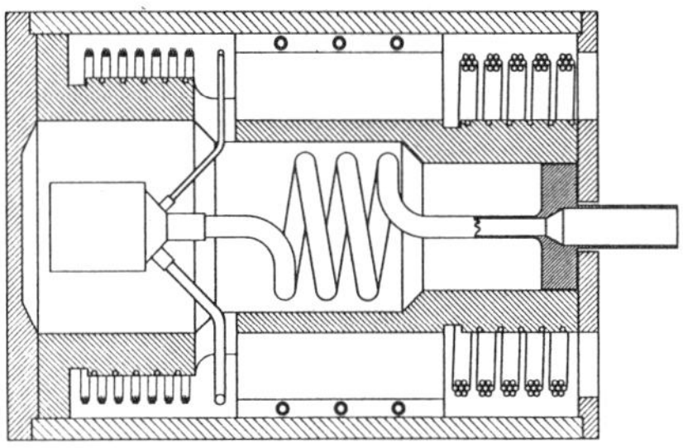

Figure 4. Cryostat assembly.

In general, the designs of both sorbent beds are identical except for the outer diameter, inlet tube size, and the type of hydride material contained within. The primary housing for the powdered hydride material is a stainless steel cylinder with welded endcaps. To facilitate free molecular flow into the low pressure sorbent bed with minimal pressure drop, a larger diameter inlet tube was selected. From the inlet tube, through the center of the sorbent bed, runs a sintered stainless steel 0.5 micron filter. This filter allows for gas flow into and out of the sorbent bed while preventing migration of the powdered metal hydride throughout the system. Pressed within the annulus between the filter and outer wall, occupying the same area as the powdered metal alloy, is a series of inter-locking copper fins. These fins are in direct contact with the metal powder and aid in the transfer of heat to and from the hydride material.

Wrapped around the stainless steel housings of the sorbent beds are thick aluminum sleeves. Along the base of these sleeves are drilled holes for insertion of 500 W rod heaters. During gas absorption, the aluminum sleeves help to draw the generated heat from the powdered metal alloy thereby minimizing temperature rise of the hydride material. During desorption, the sleeves enable more uniform heating of the sorbent beds.

Cryostat Design

The overall cryostat assembly is depicted in Figure 4. The basic structure of the cryostat is provided by the almost 2.9 kg of solid aluminum comprising the thermal storage device (TSD) and radiation shield. This aluminum structure was designed to be maintained at 60 to 80 K by means of a liquid nitrogen cooling loop operated near the triple point, and its large mass is meant to buffer the system from the high heat flux present during a rapid cooldown from 60 to 25 K. This uniform temperature TSD would be replaced by a segmented TSD for a flight system. The segmented TSD serves as a single stream regenerator, and results in a > 50% reduction in TSD mass.

The liquid nitrogen cooling loop consists of five cylindrical canisters connected in series which are directly inserted into the TSD. To facilitate the inclusion of two saran carbon coldtraps, a system bakeout heater, and the liquid nitrogen cooling assembly, a series of eight cylindrical clamps were machined directly around the circumference of the TSD.

At the top and bottom ends of the TSD, around the circumference, is machined a spiral groove in which a 3.2 mm stainless steel tube is epoxied in place. This tube, in close thermal contact with the aluminum TSD, makes up the 2.49 m of TSD heat exchanger and connects the warm and cold heat exchangers. Floating freely between the TSD heat exchanger and outer radiation shield of the cryostat, the warm and cold heat exchangers are thermally isolated from the 60 to 80 K cold sink. Although not explicitly shown in Figure 4, the J-T expander is located at the end of the cold heat exchanger and is also free floating.

At the bottom of the cryostat, within the center cavity of the TSD, is located the coldhead and low pressure vent tube. The low pressure vent tube is thermally and structurally attached at

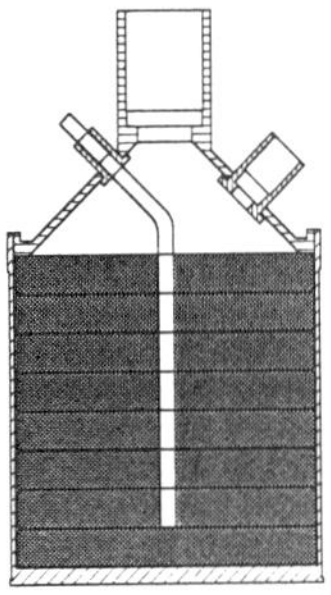

Figure 5. Coldhead assembly.

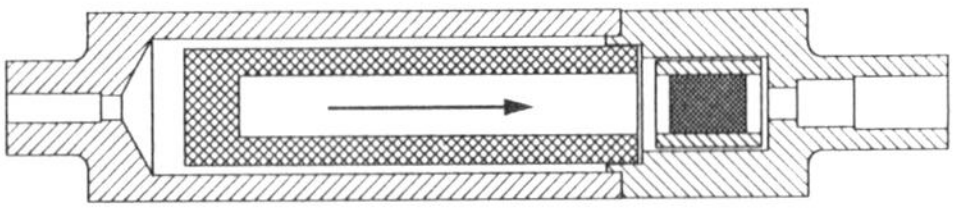

Figure 6. Porous metal Joule-Thomson expansion device.

the top of the TSD and coiled to a length of 38.1 cm to minimize heat leakage. Upon exiting the TSD, the vent tube transitions to 12.7 mm diameter tubing to minimize pressure drop and to provide a rigid support for the cryostat within the vacuum housing. The total length of 12.7mm diameter tubing between the TSD and the outer vacuum housing is approximately 26.7 cm. Estimated conduction parasitics along the vent tube between the vacuum housing and TSD and between the TSD and coldhead are 0.5 W and 12 mW respectively. Radiation parasitics between the coldhead and TSD are estimated at 4 mW.

Coldhead. The primary function of the coldhead is to collect and retain the liquid/solid hydrogen used for cooling the applied heat load. In designing the coldhead, emphasis was placed in the following areas: (a) minimizing thermal mass to reduce gas consumption and thus sorbent bed size, (b) improving conductivity between the collected hydrogen and the applied heat load, (c) developing a phase-separation and wicking mechanism to improve liquid and solid hydrogen retention, and (d) maintaining structural integrity at operating pressures up to 11 MPA.

Figure 5 illustrates the overall design of the coldhead. The main body of the coldhead is a cylindrical cup with a conical endcap machined from beryllium copper. Pressed inside the main body of the coldhead are eight, 3.18 mm thick, sintered OFHC copper disks. These disks provide the wicking necessary to retain the liquid hydrogen and also provide an excellent heat path from the center of the coldhead to the outer shell.

In addition to the passive wicking provided by the sintered medium, it is also desirable to prevent liquid in the coldhead from being blown out by the incoming gas-liquid mixture. To partially remedy this problem, gas-liquid phase separation was incorporated in the coldhead design. By placing holes at specific positions in the sintered metal disks, it is possible to induce a helical flow. While the gas phase follows the helical path provided by the holes, the liquid phase is thrown out radially into the wicking medium where it is retained by capillary forces.

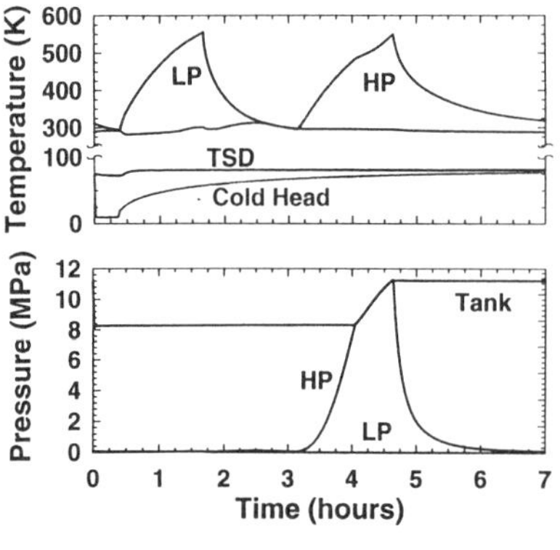

Figure 7. Sorbent bed performance predictions.

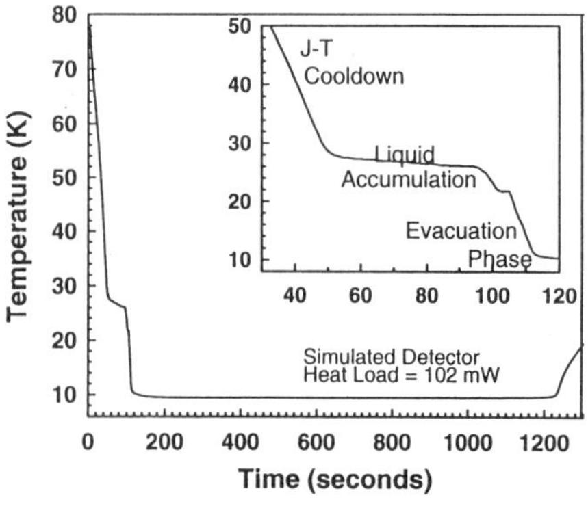

Figure 8. Example cooldown curve from BETSCE.

Heat Exchangers. Within the J-T loop of the cryostat are two high-effectiveness counterflow soldered-tube heat exchangers. During a quick cooldown, these heat exchangers function as recuperators which extract the heat from the high pressure incoming gas by using the low temperature, low pressure, exhaust gas from the coldhead. In addition to minimizing pressure drop and maximizing effectiveness, size and ease of fabrication were also important design considerations.

The warm heat exchanger consists of six 1.59 mm OD stainless steel tubes soldered around a central tube. Through the center tube flows the high pressure incoming hydrogen gas and through the six outer tubes flows the low pressure exiting gas. The overall length of the heat exchanger is 2.41 m. Once soldered together, the warm heat exchanger was coiled into a 10 cm diameter helix.

The seven tubes of the cold heat exchanger are .813 mm OD stainless steel tubes soldered together in the same configuration as the warm heat exchanger. The total length of the cold heat exchanger is 1.78 m.

Joule-Thomson Porous Plug. The design of the expansion device used to produce the Joule-Thomson (J-T) effect in this cooler was significantly influenced by the desire to reduce the cooler's sensitivity to contamination while also maintaining the necessary system performance. Instead of conventional orifice or capillary tube methods, a sintered metal porous plug was chosen to induce the desired pressure drop and set the required flowrate. A schematic of the J-T assembly is shown in Figure 6.

The actual porous plug is a commercially available, encapsulated flow restrictor, available over a wide range of standardized flow rates. During assembly, the flow restrictor is pressed into a machined stainless steel housing. On top of the flow restrictor is pressed a 0.5 micron sintered stainless steel cup filter with an available surface area of approximately 2.07 cm^2. Surrounding both the filter and flow restrictor is an all-welded stainless steel housing. The total length of the assembly is 2.86 cm.

PERFORMANCE PREDICTIONS

Sorbent Bed Performance

Based on experimental sorbent bed data from BETSCE and from previous material cycling and isotherm studies, it is possible to make accurate performance predictions for both the low pressure and high pressure sorbent beds. Based on the design margins for both heat removal and hydrogen absorption, it is also expected that performance during absorption will not significantly deviate from previous experimental data. Shown in Figure 7 are the predicted temperature and pressure profiles for both sorbent beds for the entire six hour cycle from initial cooldown to final tank recharge based on BETSCE test results.[2]

Cooldown Performance

Using experimental test results from previous proof-of-principal and BETSCE cooldown data, it is possible to predict the cooldown performance of this system. While the heat exchangers, TSD, J-T and sorbent beds are sized for a cooldown flow rate comparable to BETSCE flowrates (approx. 0.1 g/s), the coldhead size and mass are significantly less than that of BETSCE. With this in mind, cooldown time is expected to be slightly faster for this cooler. However, cooling duration below 10 K is predicted to be less than that for BETSCE because of the reduced sorbent bed capacity margins. As an example, a typical cooldown curve for BETSCE is shown below in Figure 8.

SUMMARY AND CONCLUSIONS

This cryocooler is the result of recent efforts at JPL toward the future development of space qualified, low-cost, long-life, high performance 10 K sorption cryocoolers. Novel ideas implemented in this design include a sintered copper cryogen wick with an induced phase separation mechanism, and a highly robust contamination resistant porous plug J-T. Utilizing the same basic technology as the space qualified BETSCE, this laboratory model is designed to provide comparable performance at less than one tenth of the size and weight. By replacing many of the off-the-shelf technologies with components optimized for size and weight, future miniaturization of this type of cooler appears readily attainable. Performance characterization testing of this cryocooler is scheduled to begin in August, 1995. Future studies proposed for this test bed include cooler contamination and long-term cycling tests. Test results will benefit future designs of both periodic and continuous 10 K sorption coolers.

ACKNOWLEDGMENTS

Research carried out by the Jet Propulsion laboratory (JPL), California Institute of Technology under contract with the National Aeronautics and Space Administration, and sponsored by the USAF Space and Missiles Systems Center. The authors would like to thank Capt. Charles Shultz and Lt. Eugene Croft for their support of this effort. The fabrication and test support provided by Tim Connors and Mike Schmelzel is greatly appreciated.

REFERENCES

1. S. Bard, J. Wu, P. Karlmann, C. Mirate, and L. Wade, "Component Reliability Testing of Long-Life Sorption Cryocoolers", *Advances in Cryogenic Engineering,* Vol. 39, Plennum Press, New York, NY, (1994).
2. S. Bard, J. Wu, P. Karlmann, P. Cowgill, C. Mirate, and J. Rodriquez, "Ground Testing of a 10 K Sorption Cryocooler Flight Experiment (BETSCE)," *Proc. Eighth Intl. Cryocooler Conference,* Vail, CO, (1994).
3. A. L. Johnson, and J. A. Jones, "Periodic 10 K Sorpti on Cooling," *USA Patent Pending,* (1992).
4. A. L. Johnson, and J. A. Jones, "Evolution of the 10 K Periodic Sorption Refrigerator Concept," *Proc. Seventh Intl. Cryocooler Conference,* Santa Fe, NM, (1992).
5. J. Wu, S. Bard, W. Boulter, J. Rodriquez, and R. Longsworth, "Experimental Demonstration of a 10 K Sorption Cryocooler Stage," *Advances in Cryogenic Engineering,* Vol. 39, Plennum Press, New York, NY, (1994).

PERFORMANCE OF A MAGNETIC REFRIGERATOR OPERATING BETWEEN 2 K AND 10 K

A. Kashani,[1] B. P. M. Helvensteijn,[2] F. J. McCormack,[3] and A. L. Spivak[3]

[1]Atlas Scientific
Sunnyvale, CA 94086
[2]Sterling Federal Systems
Palo Alto, CA 94303
[3]Trans-Bay Electronics
Richmond, CA 94804

ABSTRACT

A prototype of an adiabatic demagnetization refrigerator (ADR) designed for space applications, operating between 2 K and 10 K, has been built and tested. The refrigerant for the ADR consists of thick slices of single-crystal Gadolinium Gallium Garnet (GGG) that are sandwiched together with strips of high-purity copper. The copper strips are to transfer heat between the GGG and the 2 K He II-gap and the 10 K helium gas-gap heat switches. In a typical test when the ADR is cycled from 9.6 K to 1.8 K it provides 30 mW of cooling for 150 s. The cycle period is 690 s and the peak field is 6.5 T. The figure of merit for this cycle is 25 %. Methods for improving the performance of the ADR are discussed.

INTRODUCTION

Some future astrophysics missions planned by NASA require 2 K cooling. An adiabatic demagnetization refrigerator (ADR) operating between 2 K and 10 K is being developed at Ames Research Center as the last stage of a multi-stage cooler to achieve 2 K. The magnetic refrigerant selected for operating in the 2 - 10 K temperature range is Gadolinium Gallium Garnet (GGG). During the ADR cycle, thermal contact has to be made between the GGG and a 2 K heat source or between the GGG and a 10 K heat sink. Liquid- and gas-gap heat switches are employed to eliminate the need for moving parts, thus increasing reliability.

The heat switches are configured as two matching surfaces which are separated by a narrow gap. A switch is on when its gap is filled with helium and is off when the gap is emptied. This is accomplished by an activated carbon pump (ACP). The carbon adsorbs helium when cooled and desorbs it when heated. The 2 K heat switch is a liquid-gap heat switch to reduce the required heat transfer surface area, taking advantage of the high thermal conductivity of superfluid helium. The 10 K heat switch is a gas-gap heat switch since the thermal conductivity of gaseous helium is sufficiently high at 10 K.

The heat switches are thermally coupled to the GGG by strips of high purity copper. The copper strips are sandwiched between thick slices of GGG. To enhance thermal conductance high purity indium foil is placed at the interfaces between the copper and the GGG.

A prototype of the ADR described above has been built and tested.[1] The ADR consists of three main components; namely, the Cu/GGG sandwich, the 10 K heat switch, and the 2 K heat switch. The two heat switches are described elsewhere[2,3] and are briefly described here. This is followed by a description of the ADR. Finally, results of tests performed on the ADR are presented.

10 K GAS-GAP HEAT SWITCH

The 10 K heat switch consists of two finned copper cylinders, as shown schematically in Fig. 1. The fin patterns on the two cylinders are a mirror image of one another. When put together the fins are separated by a gap of width 0.010 cm. A 0.012 cm-wall SS support tube holds the two pieces together and seals the gap volume. The fins are 2.54 cm long and 0.142 cm thick. The diameter of the heat switch is 2.45 cm. The effective heat transfer area of the heat switch is approximately 82 cm^2.

The ACP for the 10 K heat switch has an internal volume of 4.5 cm^3 and contains 0.02 g of activated carbon. It is charged to 5.3 kPa at room temperature with helium gas. For the switch to fully turn on the ACP temperature is raised to 20 K. To turn the switch off a copper wire links the ACP to the heat sink.

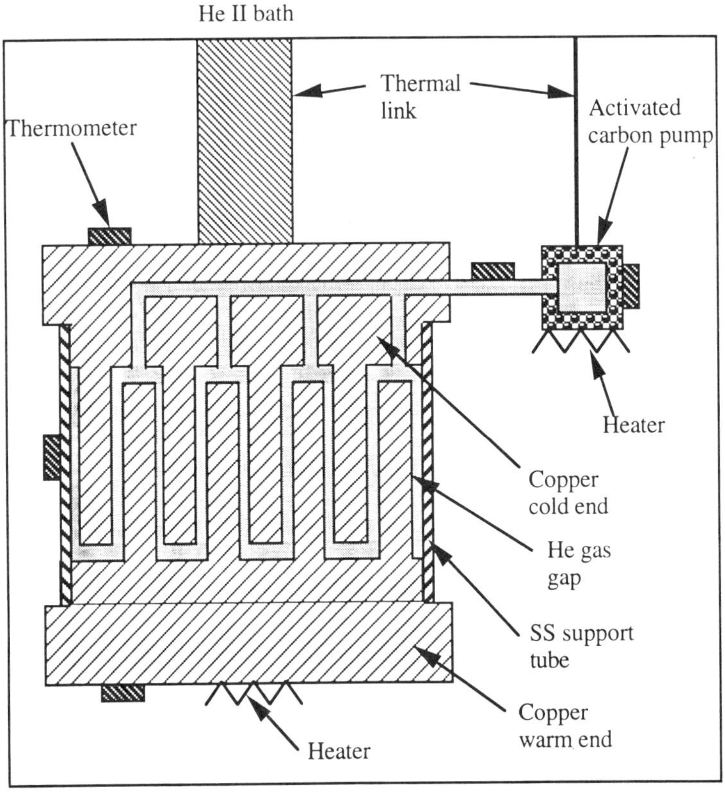

Figure 1. Schematic of the 10 K heat switch.

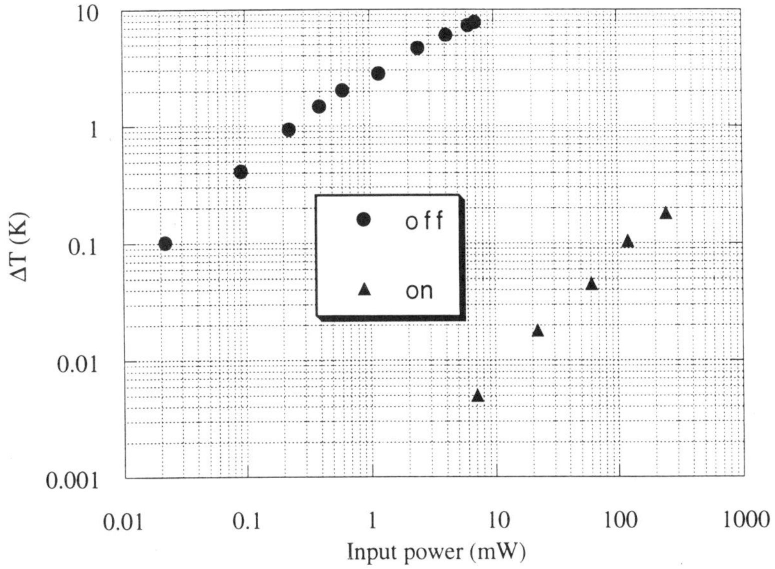

Figure 2. Temperature gradient across 10 K heat switch vs. input power.

The temperature gradient across the heat switch as a function of applied power is shown in Fig. 2 for both the on- and the off-mode. The on- to off-conductance ratio of the heat switch at an input power of 7 mW is 8000. The turn-on time of the switch is on the order of 30 s; whereas, its turn-off time is nearly one minute.

2 K LIQUID-GAP HEAT SWITCH

The 2 K liquid-gap heat switch consists of two high purity copper cylinders that are positioned end to end separated by a 0.010 cm-wide gap. The copper cylinders are annealed after machining and their RRR is estimated to be 800. The two cylinders are held together by a SS support tube which has a wall-thickness of 0.012 cm and is 2.5 cm long. The seal between copper and SS is made by Stycast 2850 epoxy. An annular gap, 0.16 cm in width, is maintained between the SS tube and the longer cylinder as shown in Fig. 3. The diameter of the gap between the two copper cylinders is 1.27 cm providing a heat transfer area of 1.27 cm^2.

The ACP for the 2 K heat switch has a volume of 11.3 cm^3 and contains 0.63 g of activated carbon. It is charged to 4 atm with ^4He gas at room temperature. This amount of gas is sufficient to fill the gap between the two copper cylinders with He II. For the switch to fully turn on at 2 K the ACP temperature is raised to 40 K. The switch is turned off when its ACP is cooled via a thermal link connecting it to the heat sink.

In Fig. 4, the temperature gradient across the heat switch is plotted as a function of the input power for both the on- and the off-mode of the heat switch. The on- to off-conductance ratio of the heat switch for an applied power of 10 mW is 6900. The turn-on time of the heat switch is less than 40 s. However, for the switch to fully turn off it takes over three minutes.

THE PROTOTYPE ADR

The prototype ADR, shown schematically in Fig. 5, consists of the Cu/GGG sandwich and the two heat switches. Five slices of single crystal GGG are sandwiched together with four strips of

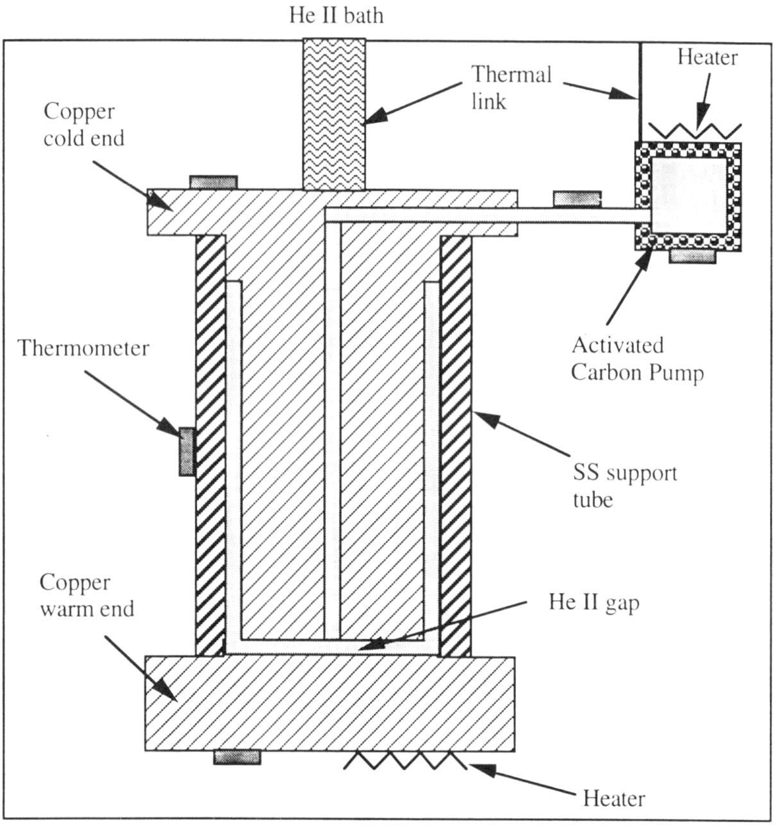

Figure 3. Schematic of the 2 K liquid-gap heat switch.

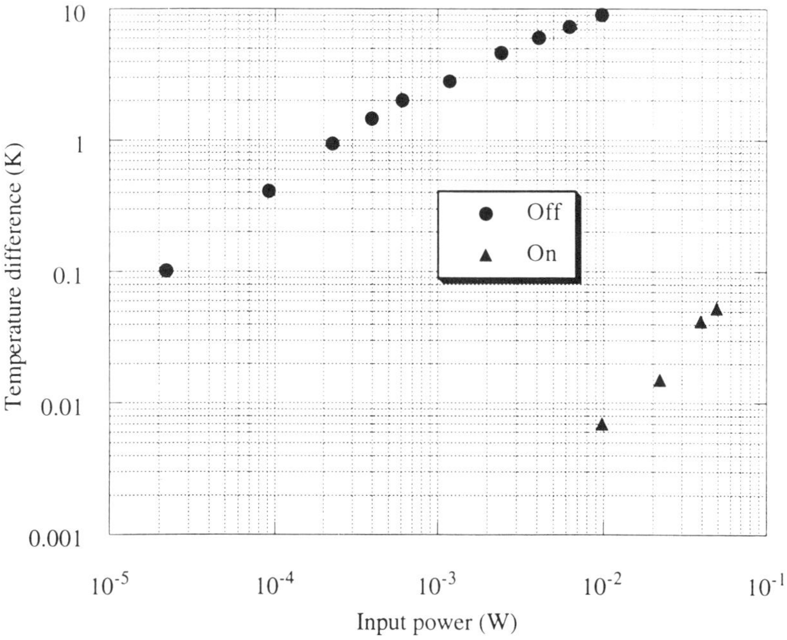

Figure 4. Temperature gradient across 2 K heat switch vs. power input.

high purity copper. Indium foil is used at interfaces between GGG and Cu to improve thermal conductance .[4] The GGG slices are 2.5 cm wide and have an overall thickness of 1.9 cm. The total volume of the GGG is approximately 46 cm^3. The copper strips have a RRR of 800. They are 2.5 cm wide and 2 mm thick. The sandwich is held under compression by Kevlar strands which are tensioned by a SS draw-bar mechanism.

The copper strips allow heat transfer between the GGG and the heat switches. Two of the strips are connected to the 2 K heat switch; while, the other two are connected to the 10 K heat switch. The overall length of copper strips running to the 2 K heat switch is 14 cm; whereas, the ones attached to the 10 K heat switch are 18 cm long. At both heat switches the strips are clamped down to one end of the heat switch with high purity indium foil placed at the interface.

The ADR is thermally anchored to the cold plate of a helium cryostat. As shown in the schematic, the end of the 10 K heat switch not in thermal contact with the GGG is used for this purpose. A control heater is mounted at this location to maintain the temperature at 10 K with the aid of a temperature controller. This end of the heat switch, referred to as the 10 K heat sink, is where the thermal links from both ACPs are terminated.

A heater is mounted on the end of the 2 K heat switch, not in thermal contact with the GGG, to simulate the 2 K heat source. Thermometers are positioned at several locations on the ADR to determine the temperature drop between the 2 K heat source/10 K heat sink and the GGG.

The magnet used in the tests is a 10 H superconducting magnet that is rated to 7 T at 4 K. The field homogeneity is within 5 % over the entire length of the Cu/GGG sandwich.

TEST PROCEDURE AND RESULTS

Tests were conducted to characterize the performance of the prototype ADR at two different heat sink temperatures; 8.5 K and 10 K. Figure 6 shows a typical test cycle for the ADR. In the test the heat sink temperature is stabilized at 8.5 K while the power required to maintain this temperature is monitored. The magnet is ramped up to a field of 6.5 T. When the temperature of GGG rises to 6.5 K the 10 K heat switch is turned on. Complete magnetization and thermal equilibrium at 10 K take 160 s after which the switch is turned off. This takes an additional 60 s. Next, the ADR is demagnetized by discharging the magnet energy through five power diodes. Below a GGG temperature of 3 K the 2 K heat switch is turned on. At this time the demagnetization rate is reduced by discharging the magnet current through only one diode. When the temperature of the GGG is below 1.75 K (120 s after demagnetization is initiated), 30 mW is applied at the 2 K heat source. The heat source temperature is maintained at 1.85 K, by regulating the magnetic field discharge, for 210 s until the superfluid helium in the heat switch begins to evaporate. This is marked by an increase in the temperature of the 2 K heat source. Finally, both the 2 K source heater and the 2 K heat switch are turned off. It takes nearly 190 s for the 2 K heat switch to fully turn off.

The temperature difference between the 2 K heat source and the GGG, when the source heater is turned on, is mainly due to: a) the temperature drop across the 2 K heat switch, b) the temperature drop across the clamped joint between the heat switch and the copper strip, and c) the temperature drop across the Cu/GGG interface. With 30 mW applied at the 2 K heat source, these are measured to be 24 mK, 36 mK, and 20 mK, respectively. Thus, a temperature difference of 80 mK is measured between the 2 K source and the GGG. The temperature drop across the GGG slices adds an additional 72 mK to the overall temperature drop. This contribution to the overall temperature drop can be decreased using a single slice of GGG in the middle of the sandwich.

The figure of merit (FOM) for the ADR is calculated based on the following relation:

$$FOM = (Qc/Qr)_{actual} \;/\; (Qc/Qr)_{theoretical}$$

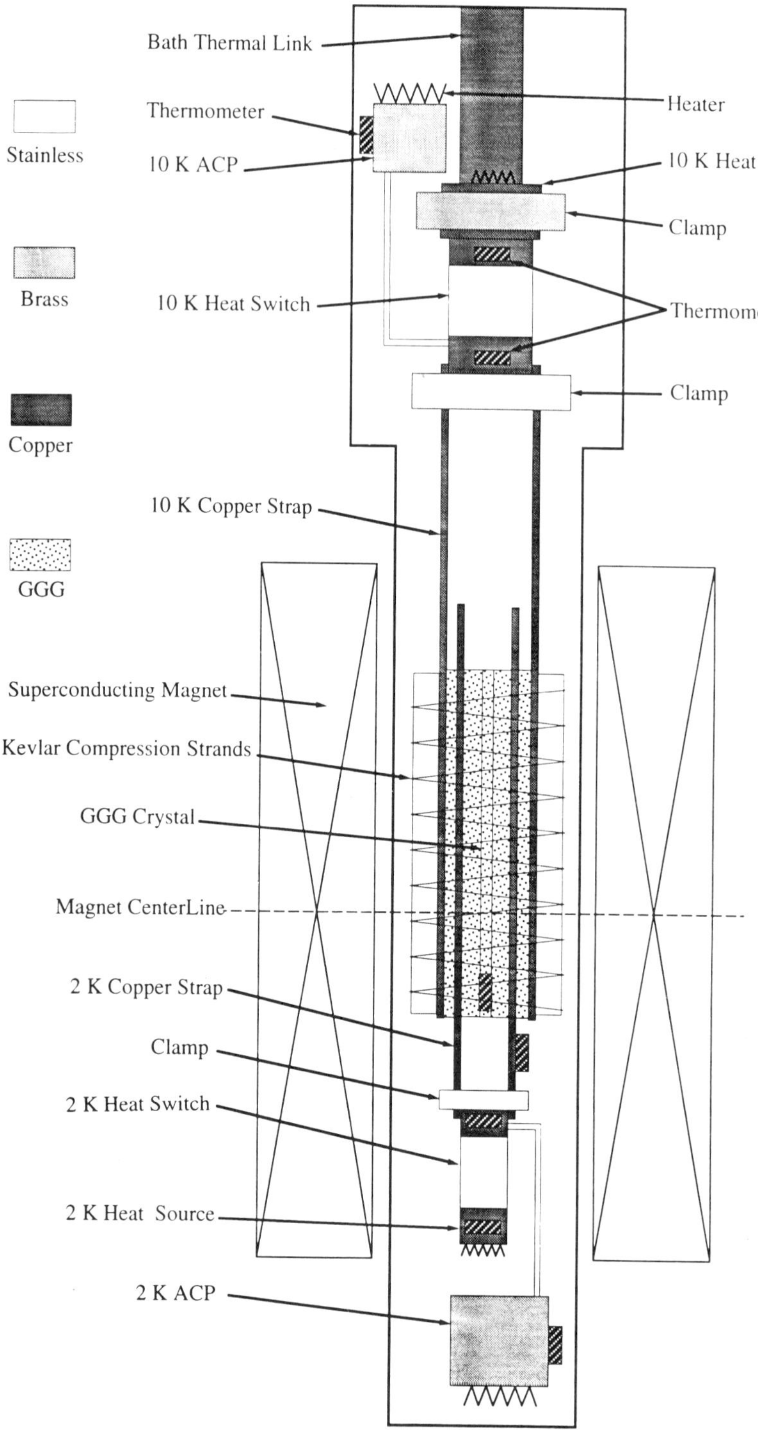

Figure 5. Schematic of the prototype ADR.

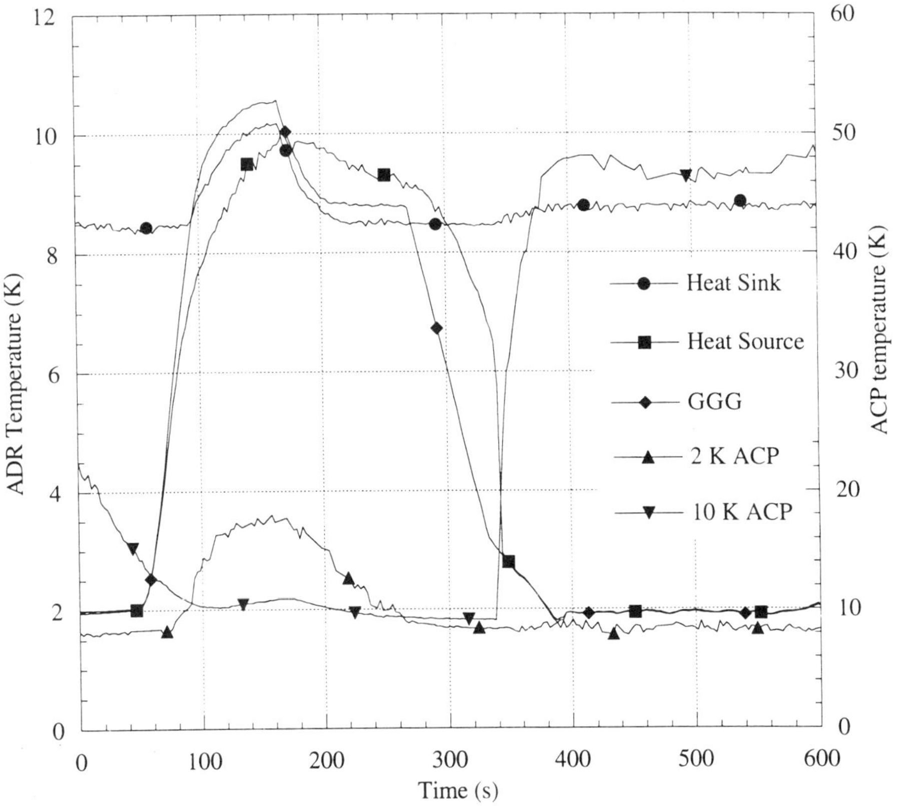

Figure 6. Temperature profiles during the prototype ADR cycle.

Table 1. Experimental results

Cycle temperature	1.85 K - 8.5 K	1.81 K - 9.6 K
GGG temperature	1.72 K - 9.0 K	1.70 K - 10.2 K
Cooling energy at 2 K	6.3 J	4.5 J
Heat rejected at 10 K	97.9 J	96.9 J
No load temperature	0.93 K	1.05 K
Cycle period	740 s	690 s
Peak field	6.5 T	6.5 T
FOM	29 %	25 %

where Qc is the cooling energy at 2 K and Qr is the heat rejected at 10 K. From the power applied at the heat source and its duration the cooling energy of the prototype ADR is calculated. The heat rejected at 10 K is obtained from the decrease in the power applied to 10 K control heater. These values along with the calculated FOM are listed in Table 1 for two different heat sink temperatures.

When the 2 K heat switch is on heat is conducted from its ACP to the 10 K heat sink through the thermal link, reducing the cycle efficiency. A gas-gap heat switch can be incorporated in the thermal link to isolate the ACP when the switch is on. This also allows for a more conductive thermal link to achieve a faster turn-off time.

An analytical investigation has also been undertaken to model the ADR design presented in this study. Results of the model are compared to the experiment in a separate paper.[5]

CONCLUSIONS

A prototype of an ADR operating between 2 K and 10 K has been built and tested. Test results indicate that the ADR is capable of providing 30 mW of cooling at 1.8 K for 150 s. The total cycle lasts nearly twelve minutes and the peak field is 6.5 T. The figure of merit for this cycle is 25 %. The measured temperature drop between the 2 K heat source and the center of the GGG is 0.15 K.

The efficiency of the system can be improved by the addition of a small gas-gap heat switch in the thermal link of the 2 K heat switch ACP. Further improvement can be made by eliminating the interfaces within the refrigerant.

REFERENCES

1. A. Kashani, B.P.M. Helvensteijn, F.J. McCormack, and A.L. Spivak, Development of a magnetic refrigerator operating between 2 K to 10 K, Proc. of ICC, 8:637 (1994).
2. A. Kashani, B.P.M. Helvensteijn, F.J. McCormack, and A.L. Spivak, Helium liquid- and gas-gap heat switches for applications at 2 K, Proc. of CEC , 39:1657 (1994).
3. A. Kashani, B.P.M. Helvensteijn, F.J. McCormack, and A.L. Spivak, Helium liquid- and gas-gap heat switches, Proc. of ICC, 7:355 (1993).
4. B.P.M. Helvensteijn, A. Kashani, and F.J. McCormack, Methods for improving the transfer of heat across the copper and gadolinium gallium garnet interface, Proc. of CEC, 39:1303 (1994).
5. B.P.M. Helvensteijn, A. Kashani, and P. Kittel, Efficiency calculations for a magnetic refrigerator operating between 2 K and 10 K, Presented at CEC 1995.

EFFICIENCY CALCULATIONS FOR A MAGNETIC REFRIGERATOR OPERATING BETWEEN 2 K AND 10 K

B. P. M. Helvensteijn [1], A. Kashani [2], and P. Kittel [3]

[1] Sterling Federal Systems
Palo Alto, CA 94303
[2] Atlas Scientific
Sunnyvale, CA 94086
[3] NASA, Ames Research Center
Moffett Field, CA 94035

ABSTRACT

An adiabatic demagnetization refrigerator (ADR) operating between 2 K and 10 K has been built and tested. The refrigerant in the ADR is Gadolinium Gallium Garnet (GGG). Absorption of heat at 2 K and heat rejection at 10 K in this fully static refrigerator is made possible by the incorporation of 2 K and 10 K heat switches. The physical layout and experimental results are presented in a parallel paper, whereas the current paper presents the computational point of view. The performance of the as-built ADR is assessed taking into account the dimensional features of the ADR components, the thermal characteristics of the components (heat capacity, latent heat, thermal resistance), the GGG entropy diagram and thermal loss mechanisms as they occur during the refrigeration cycle (eddy current heating, temperature gradients). The results are summarized in terms of the ADR efficiency.

INTRODUCTION

The vision at NASA for future astrophysics missions includes the deployment of satellites requiring 2 K cooling. For long duration flights adiabatic demagnetization refrigerators (ADRs) may be chosen in the final refrigeration stage reaching down to 2 K. The benefits of the ADR are mainly a high efficiency and the lack of wear if built without moving parts. The ADR being developed at Ames Research Center (ARC) fulfills this requirement.

The 2 - 10 K ADR prototype, built and tested at ARC, provides refrigeration at 2 K and rejects heat to an 8 - 10 K heat sink (cryocooler). The magnetic pill consists of Gadolinium Gallium Garnet (GGG). During the isothermal stages of the ADR cycle, the pill is in thermal contact with either the 2 K heat source or the 10 K heat sink by means of a liquid- and a gas-gap heat switch, respectively. Operation of the heat switches takes place by activation of electrical heaters. A detailed description of the actual ADR prototype is part of a previous publication[1] and also of a parallel paper[2] which discusses the ADR's thermal

Advances in Cryogenic Engineering, Vol. 41
Edited by P. Kittel, Plenum Press, New York, 1996

performance as measured. In order to assess how well the experimental results compare to the performance potential of the ADR prototype the present paper takes a computational approach. Reported here are the results of our calculations of the heat flowing into and out of the GGG pill. The computer program used in the calculations traces the thermodynamic cycle of the ADR's GGG pill on a diagram of the entropy of GGG vs. temperature and magnetic field. Taken into account are all the significant features of the experimental layout and the associated heat load on the GGG pill. We will start off with an exploration of the entropy of GGG. This is followed by the discussion of the computational approach to traversing the ADR cycle. The final section contains the results of the calculations.

GGG ENTROPY

The evolution of the entropy S during the refrigeration cycle plays a pivotal role in the assessment of the magnetic refrigerator's efficiency. The heat absorbed by a magnetic substance in each time interval dt is given by the product of the temperature T and the change dS in the GGG entropy. The entropy is a function of the magnetic field H and the temperature and may be approximated by the following equation[3]:

$$\frac{S}{R} = \frac{x}{2J}\coth\frac{x}{2J} - \frac{(2J+1)}{2J}\coth\frac{(2J+1)x}{2J} + \ln\left(\frac{\sinh\frac{(2J+1)x}{2J}}{\sinh\frac{x}{2J}}\right) + S_{nuc} + S_{latt} \quad (1)$$

the symbols representing: R gas constant; $x=\beta gJB/kT$; β Bohr magneton; g Lande factor (2.0); J total angular momentum quantum number (7/2 for Gd^{3+}); k Boltzmann's constant; $B = \mu_o H$ magnetic induction (μ_o is the permeability of free space); the nuclear magnetic entropy and the lattice heat capacity[4] are to be accounted for as: $\Delta S = \int C/T\, dT$; $C_{nuc} = 5.11*10^{-4}\, B^2\, T^{-2}$ J moleGGG^{-1} K^{-1}; $C_{latt} = 2.36*10^{-4}\, T^3$ J moleGGG^{-1} K^{-1} taking e.g. T = 0.5 K as a reference point.

Comparison with published data[4] indicates that eqn. (1) merely represents the trends in the entropy of GGG. A two-dimensional correction factor (c = c(B,T)) is needed to bring eqn. (1) into agreement with the data. In previous work[1] we applied a two-dimensional quadratic interpolation to the correction factor in order to attain the entropy at any point within the range of interest (0 - 7 T; 1 - 15 K). Close inspection of this algorithm shows it is troubled by small discontinuities occurring when either the temperature or the magnetic field passes through a tabulated value. Subsequently, an algorithm has been formulated creating a smoother approximation to S(B,T) using two-dimensional cubic spline curve fitting. However, it has been found that the results obtained by this method depend on the path taken on the B-T plane towards desired points. Although problems (such as walk-out and infinite loops) associated with two-dimensional quadratic and cubic spline interpolation have been surpassed by relaxing convergence criteria locally, these algorithms are no longer part of our calculations. The entropy calculations in our work now employ an altogether different method merely summarized below since it is described in detail by Gallagher[5].

The foundation to Gallagher's interpolation scheme is the thermodynamic expression of the entropy for a magnetically active system, expressed in terms of the exact differential:

$$dS = \left(\frac{\partial S}{\partial T}\right)_H dT + \left(\frac{\partial S}{\partial H}\right)_T dH \quad (2)$$

Rewriting the first partial differential in terms of the specific heat at constant magnetic field and using the Maxwell relations on the second partial, eqn. (2) is rewritten as:

$$dS = \left(\frac{C_H(T)}{T}\right)_H dT + \mu_o V \left(\frac{\partial M}{\partial T}\right)_H dH \quad (3)$$

in which V is the molar volume and M the magnetization (in our calculations the entropy reflects the magnetic pill size and is expressed in $J\ pill^{-1}\ K^{-1}$; for graphs the entropy is converted to $J\ kgGGG^{-1}\ K^{-1}$).

The first term in eqn. (3) is integrated using the specific heat at zero field as quoted by Fisher et al. However, the second term is more problematic since the magnetization data[4,6] do not facilitate straightforward integration to any given field and temperature. The following scheme was chosen as a solution. Manipulating various combinations of variables Gallagher found the following fourth order polynomial to give a most satisfactory overall approximation of the data ($T \leq 4.2$ K: Fisher et al.; $4.2\ K \leq T \leq 20$ K: Hashimoto):

$$\frac{\mu_o H}{VM} = c_0(T) + c_1(T)\,\mu_o H + c_2(T)\,(\mu_o H)^2 + c_3(T)\,(\mu_o H)^3 + c_4(T)\,(\mu_o H)^4 \qquad (4)$$

Each of the coefficients ($c_{0\text{-}4}$) is a polynomial in the temperature (see below). Equation (4) clearly demonstrates the strong dependence of the magnetization on both the magnetic field and the temperature, thus allowing simple integration to the applied field only under isothermal conditions. Therefore, integration of the second term in eqn. (3) is approximated by Gauss-Legendre quadrature:

$$\int_0^{\mu_o H} V \left(\left(\frac{\partial M}{\partial T}\right)_H \right)_T d\mu_o H \approx \frac{\mu_o H}{2} \sum_{k=1}^{12} \omega_k V \left(\frac{\partial M(T \bullet x_k)}{\partial T} \right)_H \qquad (5)$$

with: $x_k = (1-\xi_k)\mu_o H/2$. The zeros ξ_k and weights ω_k of the 12th degree Legendre polynomials are listed by Kopal[7].

The entropy of GGG calculated with the method outlined above is graphically displayed in figure 1. An arbitrary offset of $50\ J\ kgGGG^{-1}\ K^{-1}$ is added avoiding $S < 0$.

FEATURES OF THE COMPUTATIONS

The Gallagher algorithm has been applied in a computer program which traces the location of the GGG pill on the entropy diagram for one complete ADR cycle. The program walks through a cycle by introducing small steps in time which, depending on the part of the cycle being traversed, may or may not involve a change in the magnetic field. Synchronized with the progress made along the ADR cycle, heat switches are gradually turned off and on, allowing for the exchange of heat between the pill and its environment (2 K load and 10 K heat sink). Applying triangulation, the state of the GGG pill on the entropy diagram is determined by the following simple relation:

$$\Delta Q = T_{low}\,\Delta S + \frac{1}{2}\,\Delta T\,\Delta S \qquad (6)$$

In this equation ΔQ is the amount of heat flowing into the pill in a given step during which the GGG changes its temperature by ΔT and its entropy by ΔS. T_{low} is the lower GGG temperature for each step.

The simplicity of equation (6) is rather misleading. The heat load to the pill depends for the main part on the in-series thermal resistance of the heat switches[1] and the stacked layers of copper, indium foil and GGG[8]. Both these resistances are a function of the imposed temperature gradients while each is subjected to the same heatflow. The thermal resistance of the heat switches depends also on the amount of time that has past since the on-off position was changed. Allowing the temperature gradients to change during each step adds complexity to the problem turning nearly all of the variables in Eqn. (6) into moving targets. This problem is slightly aggravated by the addition of other heat loads to the pill such as the heat capacity of the copper bus bars, the latent heat of helium condensing when closing the 2 K heat switch (evaporating during magnetization), and the eddy currents occurring in the copper bus bars whenever the magnetic field changes.

The manner in which the program solves the problem of locating a new position on the entropy diagram is as follows. Whenever the field is changing the program resets the field by a fixed amount. At the new field value, adjustments are made iteratively to the temperature. Given a change in field, the first task in the program is to find the temperature for an isentropic step. This temperature is found using the relative change in the entropy:

$$T_m = T_i \bullet (1.0 - (\pi \bullet \frac{(S_i - S_o)}{(S_i + S_o)})) \quad (7)$$

Original values before a step (in time and field) have assigned to them the index "o"; endpoint values after a step change have the index "n"; initial values before an iteration have the index "i"; intermediate values ending each iteration have the index "m". A non-unity prefactor, effectively $\pi / 2$, is employed accelerating the convergence towards the shift in temperature $T_m - T_o$.

Once the isentropic temperature shift is found, this shift is altered in order to find the change in temperature $\Delta T = T_n - T_o$ corresponding to the shift in entropy $\Delta S = S_n - S_o$ such that Eqn. (6) holds to within the convergence criterion. Convergence is declared when the left and right hand sides of eqn. (6) match to a relative error in ΔQ of 0.01%. This process is time-consuming since most of the relevant variables depend on T_n and T_o. In each iteration an interval in temperature surrounding T_m is subdivided into three parts. Comparisons are made between the nodes for deviations to Eqn. (6). Successively smaller deviations are found with the temperature intervals shrinking in size until the convergence criterion for Eqn. (6) is satisfied. This scanning of the temperature scale was found to be necessary in order to avoid striking down on local minima and to remedy walk-out (divergence). The program includes resetting the interval being scanned, enabling recovery whenever the convergence ends at a local minimum not satisfying Eqn. (6).

Several stages exist in the ADR cycle for which the scheme outlined above needs slight modifications. At the end of the isentropic demagnetization when nearing the target load refrigeration temperature T_r (2 K) the steps in time and field are refined such that the

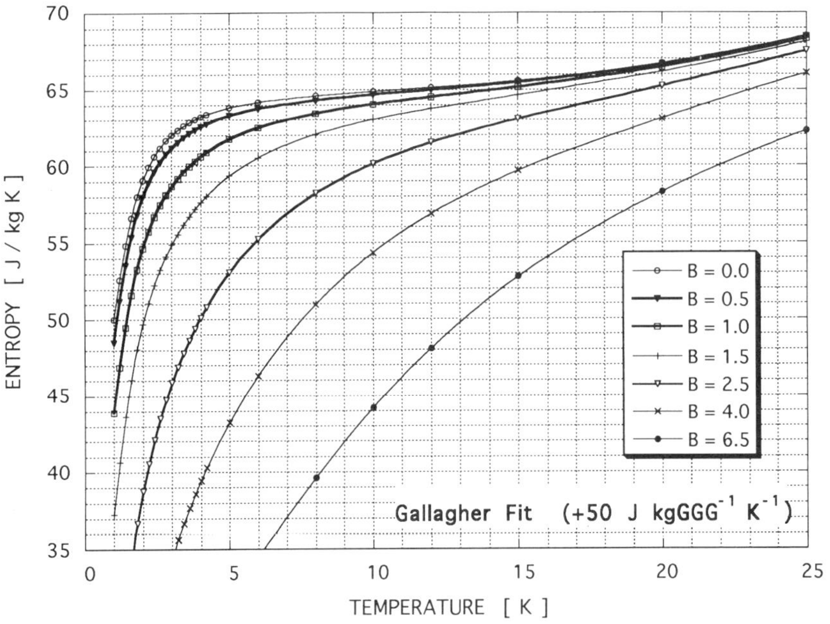

Figure 1. Entropy of GGG as a function of temperature and magnetic field with $B=\mu_0 H$ (in Tesla).

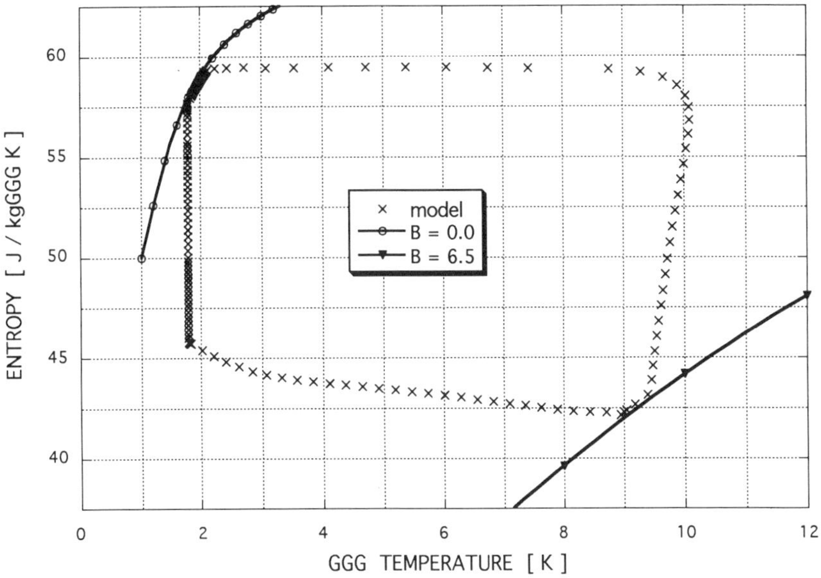

Figure 2. The calculated thermodynamic cycle for the ADR prototype.

thermal load (to the GGG) enters the cooling phase at $T_{ld} = T_r \pm 0.001$ K. In this short pre-cooling phase the load is kept at T_r continuing the demagnetization thus requiring the application of a small amount of heating power q to the load. Closing in on the target temperature is accomplished by variation of both the time step and the field step. Then, when q reaches the required level the cooling phase is entered. At this time the temperature drop over the 2K heat switch and the Cu-In-GGG stack is fixed as well as the heater power and the temperature of the GGG pill.

After demagnetization, the 2 K heat switch requires some recovery time before sufficient isolation exists to warrant magnetization without causing significant heat flow to the load. During this episode all the parts in the ADR are attempting to equilibrate which is prohibited mainly by the thermal isolation provided by the 10 K heat switch. A similar stage in the cycle happens after heat rejection at 10 K when the 10 K heat switch is opened.

COMPUTATIONAL RESULTS

The ADR cycle as calculated is depicted in figure 2, where the entropy is converted into J $kgGGG^{-1}$ K^{-1} (as opposed to J $pill^{-1}$ K^{-1}). The program cycle starts off with the GGG fully magnetized at 8.9 K in a field of 6.4 T (S = 42.2 J $kgGGG^{-1}$ K^{-1}). At this point the pill is nearly in equilibrium with the heat sink ($\Delta T = 0.4$ K; $T_{hs} = 8.5$ K). The demagnetization is started decreasing the field until the GGG cools down to below 2 K. At 4 K an algorithm is started to gradually activate the 2 K heat switch. The condensation of helium into the switch is apparent by the rise in entropy below 2 K up to roughly 46 J $kgGGG^{-1}$ K^{-1}. The remainder of the demagnetization occurs as a temperature controlled ramp-down of the field until the field becomes 0.25 T near 57.5 J $kgGGG^{-1}$ K^{-1} (the program becomes unstable below 0.125 T; this may relate to crossing constant field lines at low magnetic fields in the entropy tables by Fisher et al.). From this point on the ADR idles until the conductance of the 2 K heat switch becomes sufficiently low. This occurs near 2.1 K and 59 J $kgGGG^{-1}$ K^{-1}. The magnetization begins here, then near 7 K the algorithm starts the gradual activation of the 10 K heat switch. The magnetization rate is the same rate as in the near-isentropic demagnetization stage. This rate is not slowed down until stopped at full field (6.4 T, 43 J $kgGGG^{-1}$ K^{-1}). From hence on the ADR idles until the isolation by the 10 K heat switch becomes sufficient and the demagnetization may be started again.

Figure 2 shows an overshoot in temperature of the GGG during the isothermal phase at 10 K simulating a timing error in turning on the 10 K heat switch as might occur in an actual experiment. Clearly, this would increase the load to the heat sink while also increasing the duration of the subsequent equilibration at maximum field (exhibited by the short trace along the 6.4 T constant field line.

In addition to the graphics of figure 2 the following statistics result from the program:

– temperature GGG when load at 2.00 K:	1.79 K
– temperature heat sink:	8.50 K
– refrigeration at 2 K:	5.98 J
– heat rejection at 10 K:	48.8 J
– eddy current heating; isentropic demagnetization:	0.88 J
– eddy current heating; isothermal demagnetization:	0.06 J
– eddy current heating; magnetization:	1.14 J
– latent heat load:	0.15 J
– vapor cooling load:	1.08 J
– heat capacity loading:	0.67 J
– heat leak 10 K heat switch (when cooling):	0.78 J
– heat leak 10 K heat switch ($T_{GGG} < T_{heatsink}$):	4.27 J

The listing above results in a figure of merit for the refrigeration attained with the GGG of:

$$FOM_{GGG} = \frac{\left(\frac{5.98}{48.8}\right)}{\left(\frac{1.79}{8.5}\right)} = 58\ \%$$

However, the additional heat loads to the heat sink due to the operation of the heat switches amount to roughly 50 J, bringing the figure of merit for the cooler down to:

$$FOM_{2K\ COOLER} = 29\ \%$$

Modifications to the design of the carbon pumps controlling the heat switches, which are discussed in the parallel paper, could bring this figure into closer agreement with the FOM_{GGG}.

The computer program has also been applied to establish a no-load bottom temperature for the GGG, i.e. no heat is applied to the pill while being demagnetized down to zero field. The temperature for this case bottomed out at 1.01 K.

CONCLUSIONS

The operation of the prototype ADR has been formulated successfully in a computational model. The model assesses the FOM of the cold stage at 58 % not including the manipulation of the heat switches. The FOM changes to 29 % when the heat switch losses are included. It should be noted that these losses could be reduced drastically with a redesign of the controls for the heat switches.

COEFFICIENTS FOR EQUATION (4)

Below are listed the coefficiènts to equation (4) as established by Gallagher[5]. Gallagher's model and the coefficients below are set up to attain the entropy in units of J $gGGG^{-1}$ K^{-1}. In our work we convert to J $pill^{-1}$ K^{-1} or J $kgGGG^{-1}$ K^{-1} for graphs.

$c_0(T) = ((-0.0025072373 \cdot T + 0.1278728) \cdot T + 2.718523) \cdot T + 13.67388;$

$c_1(T) = (((9.8684995 \cdot 10^{-6} \cdot T + 0.0011088194) \cdot T - 0.073461875) \cdot T + 0.9144735) \cdot T - 2.980799;$

$c_2(T) = (((4.3787923 \cdot 10^{-6} \cdot T - 0.0010840200) \cdot T + 0.048334930) \cdot T - 0.6364962) \cdot T + 3.166695;$

$c_3(T) = ((1.3868879 \cdot 10^{-4} \cdot T - 0.0069979527) \cdot T + 0.090944469) \cdot T - 0.3461423;$

$c_4(T) = ((-6.8436516 \cdot 10^{-6} \cdot T + 3.3840450 \cdot 10^{-4}) \cdot T - 0.0041968897) \cdot T + 0.013560383;$

REFERENCES

1. A. Kashani, B.P.M. Helvensteijn, F.J. McCormack, and A.L. Spivak, "Development of a magnetic refrigerator operating between 2 K to 10 K", Proc. of ICCC , (1994).
2. A. Kashani, B.P.M. Helvensteijn, F.J. McCormack, and A.L. Spivak, "Performance of a magnetic refrigerator operating between 2 K and 10 K", Adv. in Cryog. Eng., (CEC, Columbus, Ohio, 1995).
3. O.V. Lounasmaa, "Experimental principles and methods below 1K", Academic Press (1974) p.87.
4. R.A. Fisher, Brodale, G.E., Hornung, E.W., Giauque, W.F., "Magnetothermodynamics of Gadolinium Gallium Garnet. I. Heat capacity, entropy, magnetic moment from 0.5 to 4.2K, with fields to 90 kG, along the [100] Axis", Journal of Chemical Physics Vol. 59 (1973) pp. 4652-4663.
5. G.R. Gallagher, "Analysis of a magnetically active regenerator", M.S. Thesis, Massachusetts Institute of Technology, 1986.
6. T. Hashimoto, personal communication to Y. Iwasa (Gallagher's thesis advisor) from S. Kuriyama of Toshiba Corporation. Data collected by Hashimoto at Tokyo Institute of Technology.
7. Z. Kopal, "Numerical Analysis", John Wiley & Sons Inc., New York, 1955.
8. B.P.M. Helvensteijn, A. Kashani, F.J. McCormack, "Methods for improving the transfer of heat across the copper and Gadolinium Gallium garnet interface", Adv. in Cryog. Eng.Vol.39B, (1993) pp. 1303-1308.

A NEW METHOD FOR PRODUCING PERFORATED PLATE RECUPERATORS

John B. Hendricks

Alabama Cryogenic Engineering, Inc.
P.O. Box 2470
Huntsville, Alabama 35804 USA

ABSTRACT

Perforated plate heat exchangers, consisting of a series of parallel, high thermal conductivity perforated plates, that are separated by a series of spacers that have low thermal conductivity, can provide large values of heat transfer per unit volume. For plates with circular cross-section flow passages, the compactness factor can approach that of gap geometries so that heat exchangers using plates with circular holes can have high performance. The design of low pressure drop heat exchangers is straightforward. Design principles for perforated plate heat exchangers will be given. Experimental results on a prototype system will be presented.

INTRODUCTION

The perforated plate heat exchanger was conceived in the 1960's as an approach to high performance heat exchanger construction. The early models used thin sheets of a high thermal conductivity material, such as copper, with the perforations either punched or photoetched[1,9]. Alternately, sheets of metal screen could be employed as perforated plates [4,8]. In these early models, the spacers between the perforated plate were fabricated from some type of plastic or composite, and the plates and spacers were bonded together using an epoxy or other adhesive. This use of plastics made the reliability of the devices suspect, but the units did demonstrate that perforated plate heat exchangers could provide high effectiveness with low pressure drop. More recent work has focused on all-metal, higher reliability structures utilizing perforated plate material that has well defined flow passages[2,5,7]. In this paper, design principles for high-performance perforated plate heat exchangers will be presented. Experimental results that support the design are also included.

An example of a perforated plate heat exchanger is shown in Figure 1. In this case, rings of stainless steel are diffusion bonded or brazed between the perforated plates. These rings separate the central, high pressure, flow channel from the annular, low pressure flow channel. The outer shell, which is necessary to contain the low pressure flow, is not shown.

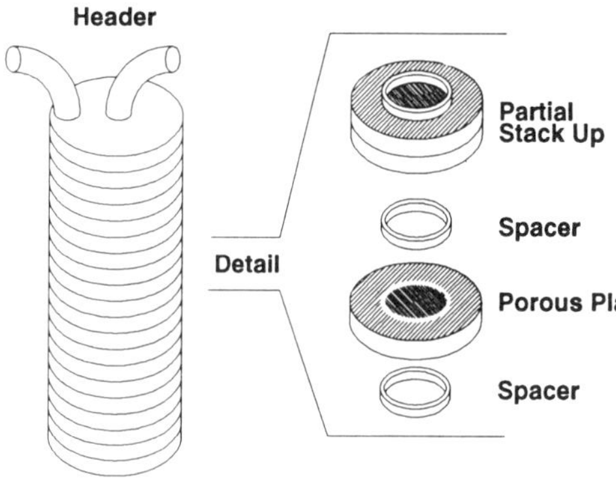

Figure 1. Schematic of a concentric, two-passage perforated plate recuperator. The outer shell has been removed to show the details of the inner construction.

Design Principles for Perforated Plate Heat Exchangers

The questions to be answered in the design process include:

1. What passage diameters and porosities are needed?
2. What number of plates and what plate thickness is required?
3. What are the spacer dimensions?

The input parameters include:

1. The temperature span of the heat exchanger
2. The desired pressure drop.
3. The desired effectiveness.

The design considerations will be discussed separately.

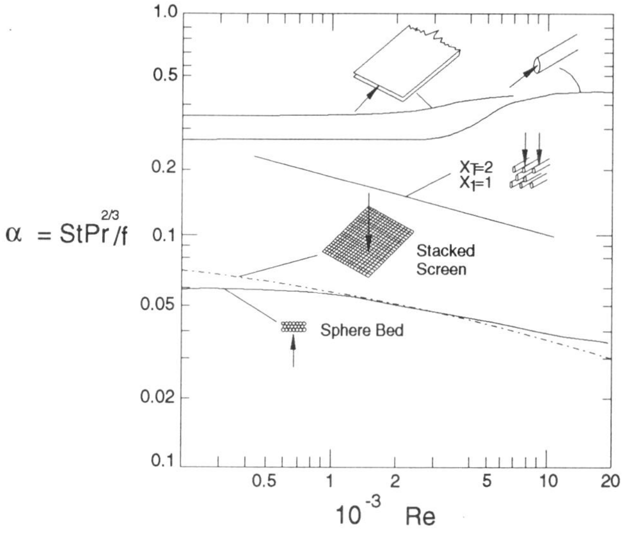

Figure 2. The compactness factor, α, for several heat transfer matrices. This chart is a modified version of Figure 2-41 of Kays and London.[3]

Choice of Flow Passage Geometry

The compactness factor is defined as:

$$\alpha = St\,\mathrm{Pr}^{2/3}/f \tag{1}$$

Where St is the Stanton number, Pr is the Prandtl number and f is the friction factor. The compactness factor is a measure of the "cost" in pressure drop that is necessary to achieve a certain heat transfer. The compactness factor, for several heat transfer geometries, is shown as a function of the Reynolds number in Figure 2. The perforated plate system is ideally a system of parallel channels with identical , uniform flow passages. Therefore, internal tube flow should describe the perforated plate system when the length to diameter (L/D) ratio is large. On a compactness factor basis, matrices such as gaps and perforated plates are preferred over stacked screens and sphere beds. We believe that thin perforated plates with L/D<1 will act more like screens, and not like internal tube flow.

Pressure Drop Calculations

Relatively small cryogenic heat exchangers with fine passages will operate in the laminar flow regime. Therefore, the pressure drop can be calculated from:

$$\Delta P = \frac{128\mu\dot{m}L}{\pi\rho N_h D_h^4} \tag{2}$$

Where μ and ρ are the fluid viscosity and density, $\dot{m}$ is the total massflow, and N_h and D_h are the number and diameter of the flow passages. This assumes that all the passages are identical and that the total massflow divides equally between the passages. The presence of the gaps between the plates allows the flow to re-distribute, so flow maldistribution is minimized. The equation can be modified using the definition of porosity:

$$pD_p^2 = N_h D_h^2 \tag{3}$$

Where p is the porosity and D_p is the equivalent diameter of the area of flow passages:

$$A_f = \frac{\pi}{4} D_p^2 \tag{4}$$

For simplicity, the calculations are usually carried out with the thermophysical qualities μ and ρ evaluated at the mean temperature of the heat exchanger.

Heat Transfer Calculations

If we make an assumption of laminar flow, then the Nusselt number, Nu, is constant, and the heat transfer coefficient, h, can be calculated from:

$$h = Nu\,k/D_h \tag{5}$$

where k is the fluid thermal conductivity and D_h is the passage diameter. The hA product can then be calculated for one side of the heat exchanger:

$$hA = \pi\,Nu\,k\,N_h\,L \tag{6}$$

with the quantities defined as before. The overall UA value can be calculated from:

$$UA = \left[\frac{1}{(hA)_1} + \frac{1}{(hA)_2}\right]^{-1} \tag{7}$$

However, $(hA)_1 = (hA)_2$, so the NTU is

$$NTU = \frac{(UA)}{2\dot{m}C_{p,\min}} \tag{8}$$

Plate Conduction Effects

The heat transfer path in a perforated plate heat exchanger consists of the two wall-fluid interfaces in series with the plate conduction. The thermal resistance of the plate will create a temperature difference, ΔT_p. This temperature difference will contribute directly to the ineffectiveness of the heat exchanger. In order to calculate this thermal resistance, the effect of the holes must be included. An approximate relation that holds for p<0.68 is:

$$\lambda_e = (1 - 1.16p)\lambda_p \tag{8}$$

where λ_p is the thermal conductivity of the plate material and λ_e is the effective value corrected for the presence of the holes.

In order to calculate the heat transfer over the plate, we use Fourier's Law of heat transfer from the middle to the edge of the plate:

$$\int_o^{r_2} \frac{\dot{Q}(r)}{A(r)} dr = -\int \lambda_p dT_p$$
$$= -\lambda_p \Delta T_p \tag{9}$$

To solve this equation, we must break the area into three regions (See Figure 1 for the geometry).

1. The central region -- the heat transferred is zero at the center and maximum at the outer circumference.
2. The spacer region -- the heat transferred is approximately constant. The spacer is assumed to be very thin compared to the plate diameter.
3. The outer region -- the heat transferred is maximum at the inner circumference and zero at the outer rim.

The free design parameter that lowers the temperature difference is the total length of the heat exchanger, $L = N_p t_p$. increasing the length increases the heat transfer area and thus reduces ΔT_p. The solution of Equation (9) can be reduced to:

$$L = \frac{0.159\dot{m}C_p\varepsilon}{\lambda_p(1-1.6p)(1-\varepsilon)} \frac{r_2^2}{r_2^2 - r_1^2} \ln\frac{r_2}{r_1} \tag{10}$$

where ε is the desired effectiveness and r_1 and r_2 are the radii of the inner and outer regions,

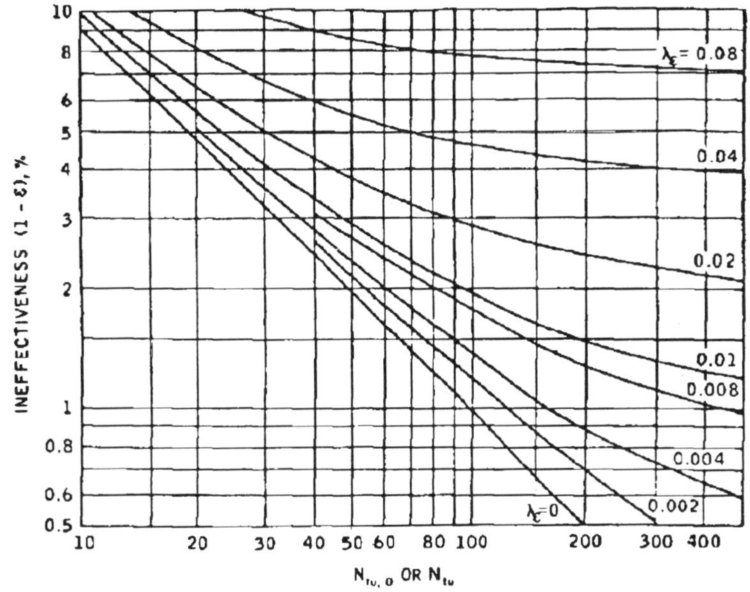

Figure 3. The effect of the longitudinal thermal conductivity parameter, λ_ε, on the performance. This chart is Figure 2-39 in Kays and London.

respectively. The spacer width is assumed to be small enough to be neglected in this calculation.

The number of plates, N_p, in,

$$L = N_p t_p \tag{11}$$

is subject to an additional constraint[6]. For a series of isothermal surfaces, the maximum NTU_p per plate is:

$$NTU_p \leq 1 \tag{12}$$

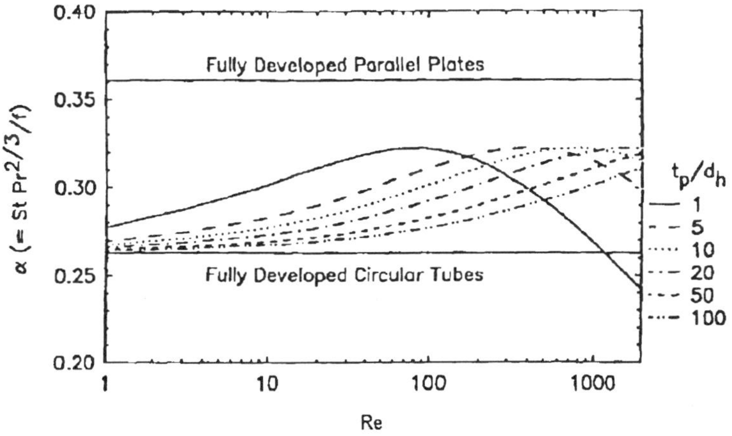

Figure 4. The compactness factor of internal tube flow as a function of the Reynolds number. The parameter t_p/d_h, is the length to diameter ratio of the flow passages.

TABLE 1 HIGH EFFICIENCY HEAT EXCHANGER DESIGN SUMMARY

Neon, Operating Range = 265 K TO 65 K
m_t = 1.27 g/s ΔP = 0.23 psi Copper Plates, Stainless Steel Spacers

	ε	n_p	d_h (µm)	t_p (mm)	OD (mm)	OL (mm)
$w_s/t_s = 2$	.99	427	148	.13	87	88.1
	.985	258	136	.14	77	56.1
	.98	180	128	.15	70	41.0
$w_s/t_s = 1$	.99	241	170	.23	76	73.3
	.985	150	156	.24	67	47.9
	.98	107	146	.25	62	35.4
$w_s/t_s = .5$	.99	164	187	.345	69	75.8
	.985	104	170	.350	61	49.7
	.98	76	160	.367	57	36.9

w_s/t_s = spacer width-to-thickness ratio
ε = heat exchanger effectiveness
d_h = diameter of the passages
n_p = number of plates
t_p = plate thickness
OD = overall diameter
OL = overall length

so that the total NTU_t is,

$$NTU_t \leq N_p \tag{13}$$

Longitudinal Thermal Conduction

The heat transferred through the spacers and outer wall of the heat exchanger can significantly degrade the performance of the device. Kays and London[3] define the longitudinal conductivity parameter as:

$$\lambda\varepsilon = \frac{(\lambda_s A_s / L_s)}{\dot{m}C_p} = \frac{\delta\varepsilon}{\varepsilon} \tag{14}$$

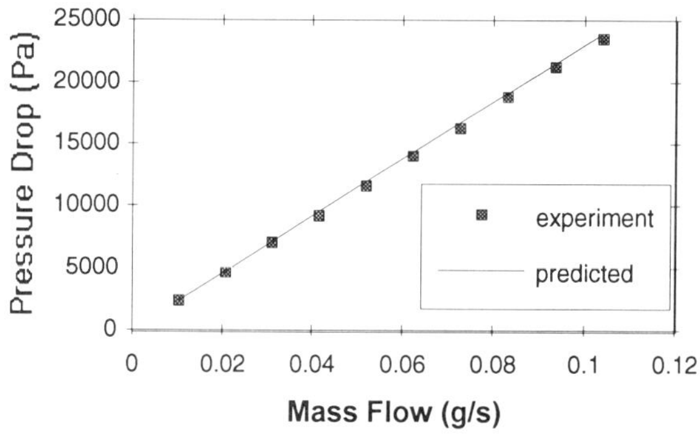

Figure 5. Measured and predicted pressure drop of a stack of 19 perforated plates.

where λ_s is the thermal conductivity of the spacer material and A_s and L_s are the cross-sectional area and total length of the spacers. The necessary NTU to achieve a desired ineffectiveness (1-ε) can be taken from Figure 3. If the area, A_s, and the spacer material is selected, then this approach can be used to find the minimum value of L_s that will provide the desired performance.

Effect of Finite Length to Diameter Flow Passages

The earlier calculations of pressure drop and heat transfer assumed fully developed laminar flow, with an infinite L/D ratio. However, with a series of finite L/D ratio plates that do not have the holes from plate to plate exactly aligned, the flow must re-develop at the surface of each plate, and this can significantly alter the parameters. Using the methods of Kays and London[3], the skin-function coefficient data and the Nusselt number data for the entrance region can be fit with equations, and the compactness factor, α, can then be calculated. Figure 4 shows the compactness factor of a perforated plate system as a function of the Reynolds number, with the L/D ratio of the individual plates as a parameter. Since α is higher for values of Re<1000, the effect on heat transfer must be greater than the effect on pressure drop. Therefore, a large number of plates can improve the overall performance.

Design Method for Perforated Plate Heat Exchangers

The principles given in the previous section can be applied by solving the series of equations that represent the principal effects on performance. Due to the non-linear nature of many of the relations, this cannot be done in closed form; however, a computer can be used to solve the system. Alabama Cryogenic Engineering, Inc., has developed a computer program to carry out this design process. The required input variables are the operating temperatures, plate and spacer material, porosity, desired effectiveness, pressure drop, massflow rate, spacer area, and the working fluid parameters (ρ, μ, k, C_p). The program will produce a design that meets the requirements. The output parameters include the diameters of the high and low pressure flow regions, the hole size, the number of plates and the thickness of the plates. A series of design results are given in Table 1. We note the very strong effect of the spacer width to thickness ratio, w_s/t_s. Clearly the longitudinal thermal conductivity is a critical parameter in the design of perforated plate heat exchangers.

Model Comparison

The easiest measurement to perform on perforated plate heat exchangers is the pressure drop. Figure 5 shows the predicted and measured pressure drop for a series of nineteen (19) plates. The agreement is excellent. We expect that the heat transfer effects, which are directly related to the pressure drop through the compactness factor relations, should show similar agreement. However, high accuracy heat transfer measurements have not been performed on this system.

HEAT EXCHANGER FABRICATION

The perforated plate manufacturing process that Alabama Cryogenic Engineering, Inc., has developed is based on extrusion technology. A sacrificial material fills the flow passages through processing, and this maintains their shape. Slices are cut from the finished composite rods, and the sacrificial material is then removed by use of a suitable chemical etch. The perforated plate is fully characterized by its matrix material (e.g. OFE copper), its porosity, the number and diameter of passages and the diameter and thickness of the plate. The ratio of the passage diameter and the plate thickness gives the L/D ratio of the passage which is

Figure 6. A sub-scale prototype of a perforated plate heat exchanger. This model has two flow passages with equal areas and the same passage dimensions.

constrained to less than roughly 5,000 with current fabrication techniques. As shown in Figure 1, it is possible to have regions on the perforated plate with different passage diameters. Each region is then separately characterized as above.

We have fabricated a number of perforated plate heat exchangers using vacuum brazing. The systems are leak free and can withstand repeated thermal shocks. The close thermal match of copper and stainless steel should minimize any differential expansion problems. The completed heat exchangers have been pressure tested, and performed as expected. This confirms that the vacuum brazing was continuous in the joint regions.

A short prototype perforated plate heat exchanger is shown in Figure 6. This exchanger is made up of 19 OFE copper perforated plates. The outer diameter is 61 mm and the overall length is 183 mm. This exchanger has equal areas in both flow channels, and was constructed as a demonstration of the techniques required.

SUMMARY

Perforated plate heat exchangers can offer a number of advantages. These include:

1. They can be very compact for a given set of requirements.

2. It is relatively easy to produce a low pressure drop, high effectiveness design.

3. They can be fabricated as an all-metal system which can minimize any impurity contribution to the flow.

ACKNOWLEDGMENT

Substantial contributions by Dr. Myron E. Calkins, Dr. Michael L. Dingus and Dr. Michael J. Nilles are acknowledged.

The work was partially funded by the Department of Energy under Grant DE-FG05-90ER80865.

REFERENCES

1. R.B. Fleming, "Advances in Cryogenic Engineering-Volume 14," Plenum Press, New York (1969), p. 197.
2. R.H. Hubbell, High effectiveness perforated plate heat exchangers for cryocoolers, *Proceedings 4th Interagency Meeting on Cryocoolers-DTRC-91/003* (1991), p. 31.
3. W.M. Kays and A.L. London, "Compact Heat Exchangers -3rd Edition," McGraw-Hill, New York (1984).
4. R.C. Lins and M.A. Elkan, "Advances in Cryogenic Engineering-Volume 20," Plenum Press, New York (1975), p. 283.
5. M.J. Nilles, M.E. Calkins, M.L. Dingus and J .B. Hendricks, Heat transfer and flow friction in perforated plate heat exchangers, *Experimental Thermal and Fluid Science*, 10: (1995), p. 238.
6. S. Sarangi and J.A. Barclay, "Cryogenic Processes and Equipment-1984" American Society Mechanical Engineering, New York (1984), p. 37.
7. W.L. Swift, "Cryocoolers 8," Plenum Press, New York (1995), p. 499.
8. G. Vonk, "Advances in Cryogenic Engineering-Volume 13," Plenum Press, New York (1968), p. 582.
9. C.E. Witter, "Advances in Cryogenic Engineering-Volume 11," Plenum Press, New York (1966), p. 107

ETCHED FOIL REGENERATOR

R. Yaron,[1] S. Shokralla,[1] J. Yuan,[2] P. E. Bradley,[3] and R. Radebaugh[3]

[1] Yaron Consulting
Los Altos, California 94024, USA
[2] University of Wisconsin
Madison, Wisconsin 53706, USA
[3] National Institutes of Standards and Technology
Boulder, Colorado 80303, USA

ABSTRACT

A new regenerator for cryocoolers has been designed and tested. This regenerator improves performance of existing cryocoolers. The regenerator promises to allow regenerative cryocoolers to run at higher speeds. The regenerator can be fabricated in numerous shapes and from a variety of materials, including materials with high heat capacity at low temperatures. Thermodynamic analysis of the regenerator with a computational fluid dynamics model and test results in a pulse tube cryocooler are described.

INTRODUCTION

Regenerator analysis[1,2] suggests that the ratio of local convective heat transfer to pressure drop is much higher for parallel plates than for stacked screens or packed spheres.

However, measurements[3] have shown that solid parallel plates do not perform as well as expected in cryocooler regenerators.

The poor performance of the solid parallel plates may be a result of a slight flow maldistribution that becomes enhanced during the cold blow. In the regions where the flow is higher during the cold blow, the gas temperature will be lower than in other regions because of imperfect heat exchange. Because the viscosity is lower at the lower temperature, the flow rate will increase further in those regions. Thus, there is an intrinsic flow instability during the cold blow that can cause nonuniform flow and lower regenerator effectiveness. During the hot blow the temperature dependent viscosity tends to make the flow uniform. Because the flows at any location in the regenerator are not the same in the two directions, there is the need to transfer heat between different locations, which degrades the regenerator performance even further.

By introducing slits in the foil that are perpendicular to the fluid flow, the fluid flow is interrupted periodically, which can stabilize the flow. In addition, the slits interrupt the conduction path and reduce conduction heat flow in the axial direction.

The regenerator[4] is fabricated from a single piece of foil material. Holes are etched into the foil in precise patterns, opening highly defined flow passages. Figure 1 illustrates a regenerator matrix formed from a roll of etched foil. As shown, the horizontal strips in the background are about half the thickness of the original material. The vertical connection straps in the foreground are also approximately half the original thickness. Only at the places where the vertical and horizontal elements overlap does the material still retain its original thickness.

Foils from 30 μ to 127 μ thick have been etched successfully. A sample of stainless steel etched foil is shown in Figure 2 and Figure 3. Materials used to date include stainless steel, beryllium copper and antimony lead. No difficulty is anticipated in etching phosphor bronze. Preliminary analysis suggests that neodymium foil can also be etched successfully.

ANALYSIS

Flow in cryocooler regenerators is not turbulent; typical Reynolds numbers are less than 200. Neither is the flow compressible; typical velocities are well below Mach 1, thus avoiding the problems inherent in modeling shock waves. These circumstances suggest

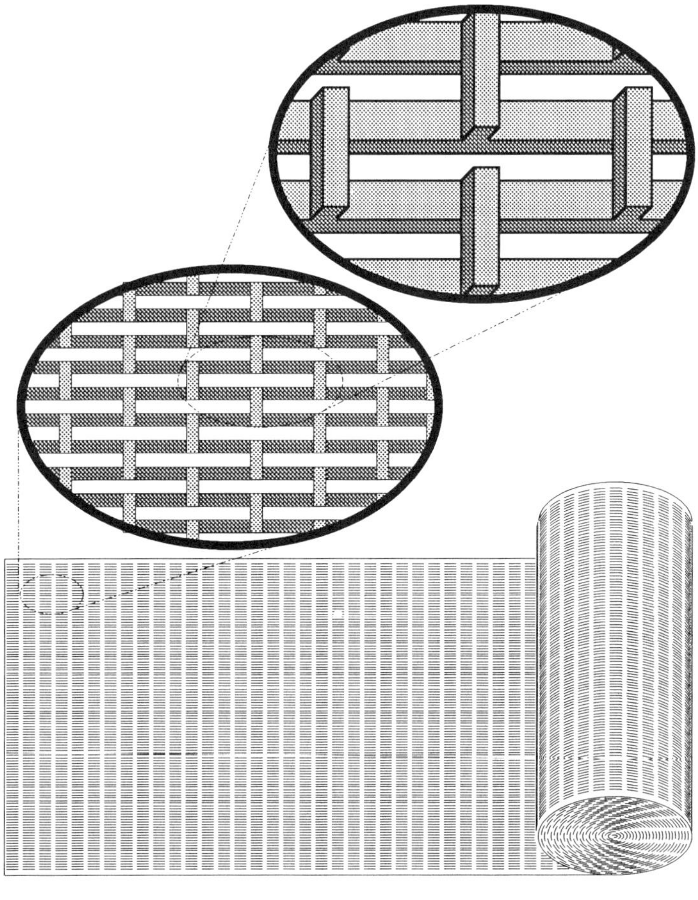

Figure 1. Etched foil regenerator.

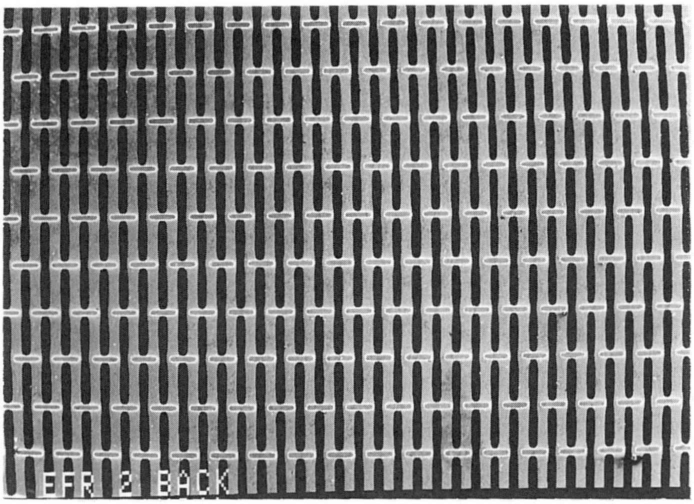

Figure 2. SEM photo of stainless steel etched foil. (Courtesy of Lockheed Martin Corp., Palo Alto, California.)

Figure 3. SEM photo of the stainless steel etched foil single element. (Courtesy of Lockheed Martin Corp., Palo Alto, California.)

that solving the complete Navier-Stokes equations (including energy) would model the regenerator adequately. However, solution of those equations requires an exact specification of flow path geometry. Happily, the geometry of the etched foil regenerator can be precisely defined. The Navier-Stokes equations were solved using an existing, proven, easy-to-use CFD code. Using a proven commercial code eliminated many interesting questions regarding the validity of the results that might have been obtained with a new, custom-made computer program.

Figure 4 illustrates the computed ratio of heat transfer to pressure drop for etched foil regenerator as compared with measured values for screens and spheres[1]. In the range of relevant Reynolds numbers, the computed ratio for the foil regenerator is more than three times the ratio measured for screens.

TESTS

Pressure Drop Measurements

First tests with the new etched foil regenerator were conducted at the National Institutes of Standards and Technology at Boulder, Colorado in 1994. Pressure drop tests were run on two regenerators designed to fit the NIST small pulse tube test rig. One was the original screen regenerator. The other was a foil regenerator sized to fit the rig.

To measure pressure drop, a regenerator was installed in a pressure drop test rig shown schematically in Figure 5. The pressure regulator "PR" and the adjustable valve "⊗" were simultaneously adjusted until inlet pressure to the regenerator stabilized at 4.0 MPa. Then mass flow rate "dm", pressure "P" and pressure drop "dP" readings were taken. Measurements were repeated for different mass flow rates.

Pressure Drop Results

Figure 6 shows measured pressure drop as a function of flow rate for the foil and screen regenerators and theoretical pressure drop from the CFD code, as well as calculated pressure drop based on friction factors obtained from Kays and London[1]. The measured porosity of the tested screens was 0.640 which is slightly higher than the porosity of 0.602 of Kays and London. The theoretical values of pressure drop for the foil regenerator from the CFD code are slightly lower than those measured at NIST. The measured results show a sevenfold decrease in pressure drop through the foil regenerator as compared to the screen regenerator at steady-state flow conditions with no temperature gradient.

TESTING IN THE SMALL PULSE TUBE CRYOCOOLER TESTING FACILITY

Performance tests with the new etched foil regenerator were conducted at NIST in 1994 using a small, single-inlet, orifice pulse tube test rig. The testing facility consisted of an instrumented pulse tube cryocooler equipped with temperature and pressure sensors at various locations. While tests were running, the measured data were digitally presented in real time. When needed, the data were downloaded to a computer for further analysis.

The screen regenerator, the foil regenerator and the small pulse tube were all about ∅ 5 mm by 50 mm long. After etching, the thickness of the remaining elements of the foil was about the same as the wire diameter of the screens.

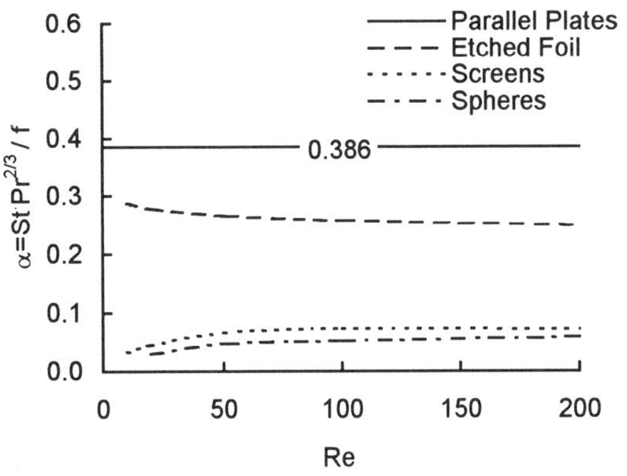

Figure 4. Ratio of heat transfer to friction factor as function of Reynolds number; computational for foil, experimental for screen and spheres.

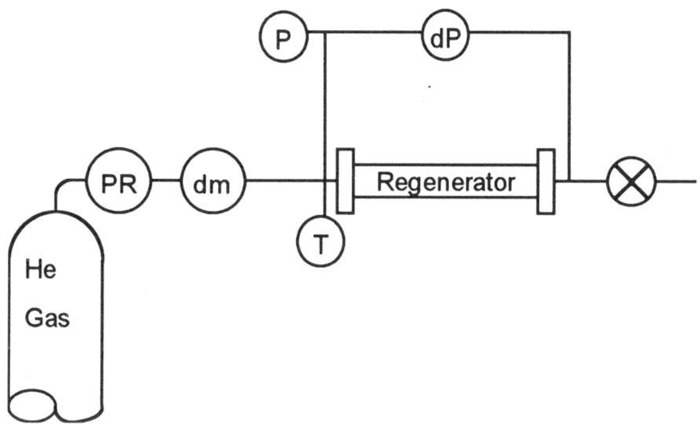

Figure 5. Pressure drop measurement rig.

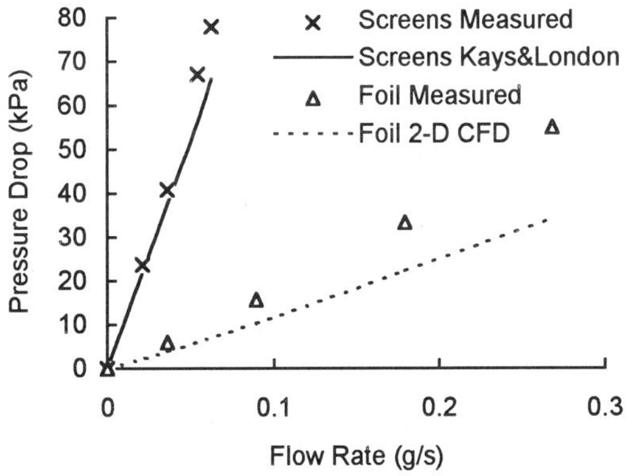

Figure 6. Pressure drop vs. flow, with He at 295 K and 4.0 MPa. pressure, under steady-state flow conditions with no temperature gradient.

1994 Small Pulse Tube Setup

In the 1994 tests, a rotary 1 cc compressor was used. The 1994 tests were run at frequencies from 20 to 25 Hz. At that time, the pulse tube had a single inlet.

1995 Small Pulse Tube Setup

Subsequent testing was conducted in 1995 using a faster and larger compressor. The new compressor was a linear type which allowed variation of volumetric displacement from 0 to 3.5 cc and speeds from 40 to 70 Hz. The test rig was also modified; a secondary inlet was added to the original pulse tube.

Figure 7 shows schematics of the test rig. Dash-dot lines represent the secondary inlet valve used in the 1995 test.

1994 Pressure Drop Results in the Small Pulse Tube Rig

In addition to heat loads and temperature measurements, other readings were also taken during the 1994 pulse tube performance tests described below. Pressure was measured in real time and logged for four different locations; 1) at the compressor; 2) at the regenerator ambient side; 3) at the pulse tube ambient side; and 4) at the reservoir volume. These measurements were made for both screen and foil regenerators. With compressor running at 23 Hz and delivering peak-to-peak pressure swing of 1.0 MPa, peak-to-peak pressure drop through the etched foil was about .055 MPa while in the screen regenerator it was about 0.4 MPa. These results, under actual cryocooler operating conditions, validate the 1 to 7 ratio in pressure drop between the foil and the screen regenerator as measured separately under steady-flow conditions in the regenerator flow measurement facility.

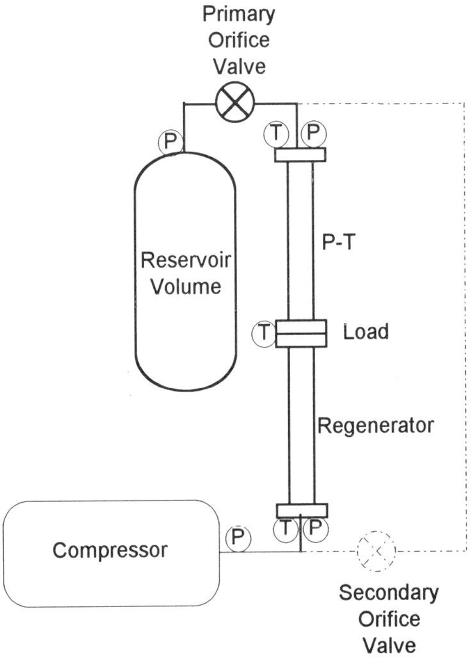

Figure 7. NIST small pulse tube test rig.

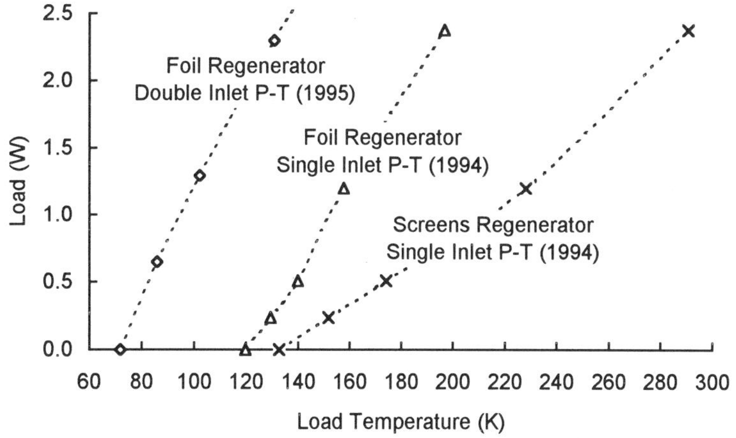

Figure 8. Load curves of the NIST pulse tube cryocooler.

Small Pulse Tube Results

The pulse tube tested in 1994, when equipped with the screen regenerator at optimized operating conditions (frequency of 23.8 Hz and average pressure of 3 MPa), reached a no-load temperature of 132.7 K; with the new etched foil regenerator at similar compressor input power (at frequency of 22.55 Hz and average pressure of 4 MPa), it reached 119.8 K.

The 1995 tests were run at speeds of 40-70 Hz as compared with 20-25 Hz speed range used in the 1994 tests. Comparison between the foil and screen regenerators was not possible because the pulse tube rig equipped with the original screen regenerator did not approach cryogenic temperatures at those frequencies.

Figure 8 contains load curves for the tested configurations. While operating at frequency of 50 Hz, average pressure of 1.2 MPa, pressure ratio of 1.57 and at a rejection temperature of 350 K, the double inlet pulse tube equipped with the foil regenerator stabilized at a no-load temperature of 71.81 K.

CONCLUSION

Under the operating conditions tested, particularly at higher frequencies, the small pulse tube cryocooler equipped with the etched foil regenerator reached lower temperatures, and delivered more cooling power, than when equipped with the screen regenerator.

ACKNOWLEDGMENTS

To Fumio Kuriyama of Ebara Corporation and Dick Rawlings of Texas Instruments for their part in the 1995 tests. To Matt Mitchell, who did most of the writing.

Research supported in part by Ballistic Missile Defense Organization under contract F29601-93-C-0153, managed by United States Air Force, Air Force Materiel Command, Phillips Laboratory, Kirtland AFB, New Mexico.

REFERENCES

1. W. M. Kays and London, in: "Compact Heat Exchangers", McGraw-Hill, New York (1984).
2. R. Radebaugh and B. Louie, A simple, first step to the optimization of regenerator geometry, in: "Proceedings of the Third Cryocooler Conference", NBS Special Publication 698, U.S. Government Printing Office, Washington D.C. (1985), p.177.
3. W. Rawlings, K.D. Timmerhaus, R. Radebaugh and D.E. Daney, Measurement of performance of a spiral wound polyamide regenerator in a pulse tube refrigerator, in: "Advances in Cryogenic Engineering," R.W. Fast, ed., Plenum Press, New York, (1992), p. 947.
4. U.S. Patent 5,429,177.

USEFUL SCALING PARAMETERS FOR THE PULSE TUBE

J. M. Lee[1], P. Kittel[1], K. D. Timmerhaus[2], and R. Radebaugh[3]

[1]NASA Ames Research Center
Moffett Field, CA 94035-1000, USA

[2]University of Colorado, Boulder
Boulder, CO 80303, USA

[3]National Institute for Standards and Technology
Boulder, CO 80303, USA

ABSTRACT

A set of dimensionless scaling parameters for use in correlating performance data for Pulse Tube Refrigerators is presented. The dimensionless groups result after scaling the mass and energy conservation equations, and the equation of motion for an axisymmetric, two-dimensional ideal gas system. Allowed are viscous effects and conduction heat transfer between the gas and the tube wall. The scaling procedure results in reducing the original 23 dimensional variables to a set of 11 dimensionless scaling groups. Dimensional analysis is used to verify that the 11 dimensionless groups obtained is the minimum number needed to describe the system. We also examine 6 limiting cases which progressively reduce the number of dimensionless groups from 11 to 3. The physical interpretation of the parameters are described, and their usefulness is outlined for understanding how heat transfer and mass streaming affect ideal enthalpy flow.

INTRODUCTION

Experimentalists must often decide upon the most efficient way to correlate a large amount of laboratory data that will give a clear understanding of the results. This eventually leads to scaling the differential fluid equations, and then determining the number of mutually independent dimensionless groups using dimensional analysis..

As an example of the usefulness of dimensional analysis, consider heat transfer from a hot wire submerged in a moving fluid. The heat transferred per unit length of wire, Q/l, is a function of the temperature difference, θ, wire diameter, d, fluid velocity, u; and the fluid density, ρ, heat capacity, C_p, viscosity μ, and thermal conductivity, κ. The number of data points required to measure these 8 variables a minimum of 3 times per variable is 3^8.

However, by using dimensional analysis, the above 8 variables can be reduced to 3 independent dimensionless groups

$$\frac{Q}{\kappa l \theta} = f\left(\frac{\rho u d}{\mu}, \frac{C_p \mu}{\kappa}\right),$$

which are the Nusselt, Reynolds and Prandtl Numbers where $Q = hA\theta$, with h as the heat transfer coefficient and $A \propto d \cdot l$ as the heat transfer area. Thus only 3^3 independent measurements are required.

Limited progress has been made on modeling the Pulse Tube (PT) through solution of the fluid equations. Most models are based upon ideal one-dimensional (1D) flows[1]. To account for diffusion heat transfer and viscous effects, lumped-parameter corrections to the 1D models have been employed[2]. Direct solution of the two-dimensional (2D) fluid equations to account for heat transfer and viscosity have been made earlier in acoustics for boundary layer flows[3]. A recent 2D analysis has examined thermal and viscous diffusion and second order steady mass streaming for the case of negligible axial temperature gradient[4].

These models are not completely satisfactory. One-dimensional models cannot account for steady mass streaming, and solving for the temperature profile in 2D models is difficult. The limited progress is due to the complexity of the coupled differential equations for mass, momentum and energy conservation. Fortunately, some relief is available from dimensional analysis. Dimensional analysis gives the experimentalist the ability to correlate data without resorting to a complete solution of the differential equations. A recent dimensional analysis for thermoacoustics illustrates the usefulness of this approach[5].

The purpose of this paper is to present a formal dimensional analysis that suggests a set of useful dimensionless groups for correlating PT data. The dimensionless groups are shown to be consistent with the dimensionless groups obtained by scaling the 2D axisymmetric fluid equations. The analysis, being 2D, contains transverse heat transfer and viscous effects, which are the transport processes associated with departures from ideal 1D models. The set of dimensionless groups are then reduced for 6 special limiting cases of pulse tube operation.

SCALING

In this first section we scale the governing fluid equations that describe the flow dynamics of the tube. A sketch of the system is shown in Figure 1. Two problem domains are considered. The gas domain extends from $r^* = 0$ to $r^* = r_w^*$ and $z^* = 0$ to $z^* = L^*$ (starred variables are *dimensional* quantities). The tube wall domain extends from $y^* = 0$ to $y^* = l^*$ and $z^* = 0$ to $z^* = L^*$, where l^* is the tube wall thickness. Adiabatic conditions exist for the outer wall surface, and continuity of temperature and heat flux must exist between the gas and the tube wall interface. The velocity boundary conditions are of small amplitude and periodic so that time is described using complex notation: at $z^* = 0$, $u = U_0^* e^{i\omega^* t^*}$, and at $z^* = L^*$, $u = U_L^* e^{i(\omega^* t^* + \phi_U)}$ where $\omega^* = 2\pi f^*$ is the angular frequency, f^* is the frequency; t^* is time; ϕ_U is the velocity phase angle between the tube ends; and U_0^* and U_L^* are the velocity amplitudes at each end. The energy boundary condition at the outer tube wall is adiabatic, with the temperature at $z^* = 0$ and $z^* = L^*$ taken as $T^* = T_c^*$ and $T^* = T_h^*$, respectively. In this paper, we will refer to the z-direction as the axial direction, and the r-direction as the transverse direction. We will also use the following notation for partial derivatives: $\chi_{,\eta} = \partial\chi/\partial\eta$.

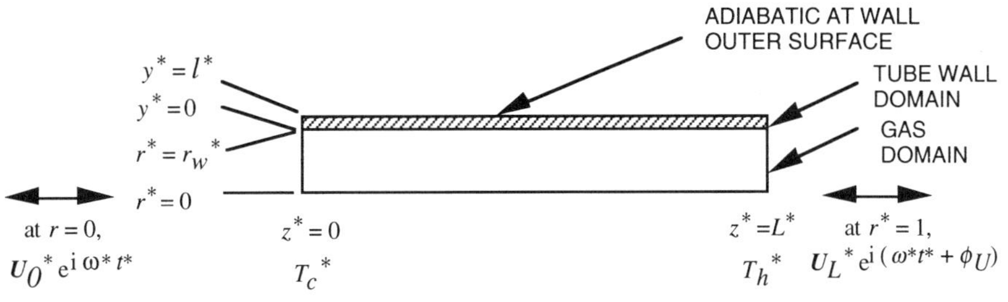

Figure 1. Axisymmetric system examined where $r_w^*/L^* \ll 1$.

The fluid equations[6] are reduced for our system using the following simplifying assumptions: axisymmetric cylindrical geometry; ideal gas; constant transport properties; Stokes' assumption for the second viscosity; and $r_w^{*2}/L^{*2} \ll 1$ (implying that $\partial p^*/\partial r^* \approx 0$ so that the r-momentum equation can be decoupled from the rest of the problem, and so that axial viscous transport is negligible in the z-momentum equation). The reduced fluid equations for mass conservation, equation of motion, energy conservation, and equation of state become, respectively,

$$\rho^*_{,t^*} + \frac{(\rho^* \upsilon^* r^*)_{,r^*}}{r^*} + (\rho^* u^*)_{,z^*} = 0; \tag{1}$$

$$\rho^*\left[u^*_{,t^*} + \upsilon^* u^*_{,r^*} + u^* u^*_{,z^*}\right] = -p^*_{,z^*} + \frac{\mu^*}{r^*}\left(r^* u^*_{,r^*}\right)_{,r^*} \tag{2}$$

$$\rho^* C_p^*\left[T^*_{,t^*} + \upsilon^* T^*_{,r^*} + u^* T^*_{,z^*}\right] = -\left(p^*_{,t^*} + u^* p^*_{,z^*}\right) + \frac{\kappa^*}{r^*}\left(r^* T^*_{,r^*}\right)_{,r^*} + \kappa^* T^*_{,z^*z^*} + \mu^* u^{*2}_{,r^*} \tag{3}$$

$$p^* = \rho^* R^* T^*. \tag{4}$$

The energy conservation equation for the tube wall domain for $l^* \ll r_w^*$ is

$$\rho_t^* C_{pt}^* T^*_{,t^*} = \kappa_t^*\left(T^*_{,y^*y^*} + T^*_{,z^*z^*}\right). \tag{5}$$

The time-averaged enthalpy flow is of primary interest since it represents refrigeration,

$$\overline{H}^* = 2\pi\omega^* \oint_{1/\omega^*} \left(\int_0^{r_w^*} \rho^* u^* C_p^* T^* r^* dr^* \right) dt^*, \tag{6}$$

where the overbar represents time-averaged values over a cycle. The kinematic velocity components in the z^* and r^* direction are u^* and υ^*; the thermodynamic gas variables p^*, ρ^*, T^* are pressure, density and temperature, respectively; the density of the tube wall material is ρ_t^*; gas properties, μ^*, κ^*, C_p^*, are the dynamic viscosity, thermal conductivity and heat capacity, and the tube wall properties, κ_t^* and C_{pt}^*, are the thermal conductivity and heat capacity of the tube wall.

The above dimensional equations are next scaled (normalized) to dimensionless form so that the resulting dimensionless variables range from 0 to $\boldsymbol{O}(1)$ (order 1). The variables are scaled as follows: r^* is scaled with r_w^*, z^* is scaled with L^*, y^* is scaled with the tube wall thickness l^*, and t^* is scaled with the angular frequency ω^*; u^* is scaled with the axial boundary condition velocity U_0^*; υ_0^* is scaled with $(U_0^* r_w^*/L^*)$; p^*, ρ^*, T^*, are scaled with mean pressure p_0^*, reference density ρ_0^*, and reference temperature T_0^*. The transport properties μ^*, κ^*, C_p^*, κ_t^* and C_{pt}^*, and the tube wall density ρ_t^* are taken as constant. These scaling parameters are substituted into the dimensional Eqs. (1) to (6) and rearranged to give the corresponding dimensionless form (unstarred variables are *dimensionless*) for mass conservation, equation of motion, energy conservation, equation of state, tube wall energy conservation and time-average enthalpy flow, respectively:

$$\rho_{,t} + \varepsilon\left[\frac{(\rho \upsilon r)_{,r}}{r} + (\rho u)_{,z}\right] = 0 \tag{7}$$

$$\rho\left[u_{,t} + \varepsilon\left(\upsilon u_{,r} + u u_{,z}\right)\right] = -\frac{\varepsilon}{\gamma \mathrm{M}^2} p_{,z} + \frac{1}{\mathrm{V}^2}\frac{(r u_{,r})_{,r}}{r} \tag{8}$$

$$\rho\left[T_{,t} + \varepsilon\left(\upsilon T_{,r} + u T_{,z}\right)\right] = \frac{\gamma - 1}{\gamma}\left(p_{,t} + \varepsilon u p_{,z}\right) + \frac{1}{\mathrm{P}^2 \mathrm{V}^2}\left(\frac{(r T_{,r})_{,r}}{r} + \frac{r_w^{*2}}{L^{*2}} T_{,zz}\right) + (\gamma - 1)\frac{\mathrm{M}^2}{\mathrm{V}^2} u_{,r}^2 \tag{9}$$

$$p = \rho T \tag{10}$$

$$T_{,t} = \frac{1}{\mathrm{F}^2}\left(T_{,yy} + \frac{l^{*2}}{L^{*2}} T_{,zz}\right). \tag{11}$$

$$\overline{H} = \pi r_w^2 \omega \oint_{1/\omega} \rho u T dt \tag{12}$$

The enthalpy flow of the gas, Eq. (12), is normalized by the *sum of the axial heat conduction of the gas, the axial heat conduction of the tube wall, and the enthalpy flow of the gas at some reference point, all averaged over the transverse area* (which includes the gas and the tube wall). Since this quantity is constant in the axial direction when averaged over the transverse area (recall the adiabatic boundary conditions at the outer tube wall, Fig. 1), we take the reference enthalpy flow, $\overline{H}_{ref}^*$, to be at the cold end (this will be convenient for illustrating a later point in the section on dimensional analysis). The above set of equations identifies 9 dimensionless scaling groups. Two additional dimensionless groups, ϕ_U and U_L, enter through the boundary conditions. These groups are listed in Table 1 along with their physical meanings. The relative magnitudes of the groups provide an understanding of the importance of the various dynamic effects (friction, heat transfer, compressibility, etc.).

DIMENSIONAL ANALYSIS

The Buckingham-Pi Theorem[7] states that the minimum number of dimensionless groups, Π, is determined from the expression

Table 1. Dimensionless Scaling Groups

	Name	Definition	Physical Meaning
ε	expansion parameter	$U_0^*/(\omega^* L^*) = d^*/L^*$	ratio of displacement length, d^*, to tube length
γ	heat capacity ratio	C_p^*/C_V^*	ratio of constant pressure to constant volume heat capacities.
V^2	Valensi Number	$r_w^{*2}\omega^*/\nu^*$	ratio of tube inner radius to viscous diffusion length
P^2	Prandtl Number	ν^*/α^*	ratio of viscous to thermal diffusion length scales
M	Mach Number	$U_0^*/\sqrt{\gamma R T_0^*}$	ratio of velocity at z = 0 to speed of sound
F^2	inverse Fourier Number	$\alpha_t^*/\left(\omega^* l_t^{*2}\right)$	ratio of thermal diffusion length to tube wall thickness
r_w^*/L^*	gas domain length ratio		ratio of tube radius to tube length
l^*/L^*	tube wall length ratio		ratio of tube wall thickness to tube length
$\overline{H}$	normalized enthalpy flow	$\overline{H}^*/\overline{H}_{ref}^*$	ratio of enthalpy flux to reference enthalpy flux $\overline{H}_{ref}^* = \rho_c^* T_c^* U_0^* C_p^* \pi r_w^{*2}$
U_L	velocity ratio	U_L^*/U_0^*	ratio of velocity amplitude at z = 1 to amplitude at and z = 0
ϕ_U	velocity phase angle		velocity phase angle at z = 1 relative to z = 0

$$\Pi = \text{m} - \text{n}$$

where m is the minimum number of independent variables (such as velocity, pressure, thermal diffusivity, etc.) and n is the number of primary dimensions. For the pulse tube, we take time, length, mass and temperature to be our primary dimensions.

A helpful procedure used for obtaining the minimum number of independent variables (and therefore the minimum number of dimensionless groups) has been outlined by Krantz[8] and is summarized as follows:

1. List all equations necessary to solve for the quantity of interest.
2. List the boundary conditions.
3. Simplify the equations listed in steps 1 and 2 through use of any additional information. For example, neglecting higher order terms to reduce the number of variables, or using explicit solutions that can relate one variable in terms of others.
4. List all variables and group-of-variables that are mutually independent by examining the set of relations listed in 3. This typically requires finding natural groups within the equations and from the boundary conditions. The number of independent variables is "m"
5. List all fundamental dimensions (such as mass, length, time, temperature). The number of fundamental dimensions is "n".
6. The minimum number of independent dimensionless groups, Π, is m – n.
7. Combine the independent variables to form the minimum number of independent dimensionless groups.

In keeping with the methodology outlined above, the first step in dimensional analysis is to identify the quantity of interest. This is the time and (transverse) area averaged enthalpy flow, $\overline{H}^*$, with C_p^* taken as constant. In general, it is a nonlinear quantity, the time-averaged product between mass flux and temperature (Eq. 6) or velocity and pressure,

$$\overline{H}^* = 2\pi\omega^* \frac{\gamma}{\gamma - 1} \int_0^{r_w^*} \left(\oint_{1/\omega^*} p^* u^* dt^* \right) r^* dr^*. \tag{13}$$

Table 2. Full Complement of Dimensional Variables

Dimensional Variable	Symbol	Dimensional Variable	Symbol
enthalpy flow	$\overline{H}^*$	leading order pressure, gas	p_0^*
tube inner radius	r_w^*	leading order density, gas	ρ_0^*
tube length	L^*	leading order temperature, gas	T_0^*
tube thickness	l^*	oscillating pressure, gas	p_1^*
angular frequency	ω^*	oscillating density, gas	ρ_1^*
oscillating radial velocity	v_0^*	oscillating temperature, gas	T_1^*
oscillating axial velocity	u_0^*	dynamic viscosity	μ^*
axial velocity at z=0	U_0^*	thermal conductivity, gas	k^*
axial velocity ratio	U_L^*	heat capacity, gas	C_p^*
axial phase angle	ϕ_U	density, tube	ρ_t^*
mass ideal gas constant	R	thermal conductivity, tube	k_t^*
		heat capacity, tube	C_{pt}^*

Determining $\overline{H}^*$ requires solving for p^* and u^* over the domain. This generally requires solving the full 3D set of fluid equations. However, we can immediately reduce the equations to 2D axisymmetric as given by Eqs. (1) to (5).

The next step is to identify the boundary conditions. The velocity can be of different amplitude (U_0^* and U_L^*) and phase (ϕ_U) as indicated in Figure 1. The temperature (energy) boundary condition at the outside of the tube is adiabatic. Temperature and heat flux must be continuous across the gas/tube interface.

The equation set can be further simplified by linearizing if the parameter ε is small (see Eqs (7), (8) and (9)). An asymptotic series solution in the small parameter ε, with $M \ll \varepsilon$, can then be used to obtain an anelastic linear set of leading order equations with second order corrections. Details of this expansion are given elsewhere.[9]

The full complement of 23 dimensional variables relevant to the problem is listed in Table 2. These are not all mutually independent. The Pi Theorem requires the minimum number of variables, that is, the number of variables that are mutually independent. The independent variables are obtained by considering additional information. In particular, the number of variables listed in Table 2 can be reduced by finding "natural groups" within the equations, by using information from the boundary conditions, and by using explicit equations relating variables. The following discussion makes use of this additional information.

Thermal Diffusivity for the Tube Wall. The energy equation for the tube wall is simply the unsteady diffusion equation, with the thermal diffusivity, $\kappa_t^*/\left(\rho_t^* C_{pt}^*\right)$ being a natural group that describes the relative importance between unsteady effects and diffusion. The result of defining this natural group is to combine 3 variables and reduce them to a single independent group. Thus the number of variables is reduced by 2.

Ideal Gas Relation. From linearization, The zeroth and first order dependence of pressure, density and temperature is described by the corresponding zeroth and first order equation of state for an ideal gas. This constitutes 6 unknowns (zeroth order: p_0^*, ρ_0^*, T_0^*, and first order: p_1^*, ρ_1^*, T_1^*) in two equations ($p_0^* = \rho_0^* T_0^*$, $p_1^* = \rho_0^* T_1^* + \rho_1^* T_0^*$), hence there are 4 degrees of freedom and thus the number of variables is reduced by 2 (= 6 – 4).

Adiabatic Boundary Conditions. The adiabatic boundary conditions at the outer tube wall requires that the gas enthalpy flow *plus* conduction for the total (gas and tube wall) system to be constant. This relation reduces the number of variables by 1.

Analytic Solution for u_0^*. An analytic solution for u_0^* in z^* is obtained directly from the linear momentum equation[9]

$$u_0^* \propto \Re\left[\left(U_L^* e^{i\phi_U} - U_0^*\right)z^* + U_0^*\right] F(r^*)$$

where $F(r^*)$ is the functional radial dependence and $\Re$ represents the real part. Thus u_0^* (which is the quantity of interest to us) contains the effects of three variables, U_0^*, U_L^* and ϕ_U. The number of variables is reduced by 3.

The total number of variables has been reduced from the original 23 to 15, hence m = 15. We now take time, length, mass and temperature to be our fundamental dimensions, so that n = 4. Finally we arrive at the minimum number of independent dimensionless groups, which is 11.

The set of variables are now formed into 11 dimensionless groups. With the help of the previous scaling discussion, we choose the dimensionless groups listed in Table 1. These groups are mutually independent, hence, these are the parameters to vary when conducting experiments (all other things being equal, i.e., regenerator, heat exchanger and compressor performance being constant). The 11 groups, in conjunction with experimental data, quantify how heat transfer and secondary mass streaming influence the enthalpy flow and the temperature difference between the tube ends.

LIMITING CASES

This dimensional analysis has resulted in reducing the original 23 variables to 11 independent dimensionless groups. The 11 dimensionless groups listed in Table 1 are reiterated in Table 3 as Case 0. It is the most complex case for correlating data based on our reduced problem (Eqs. (1) to (6) and the boundary conditions of Fig. 1). Mapping the entire parameter space for these 11 groups is still a formidable task. Fortunately, there are several limiting cases for the PT that further reduce the number of dimensionless groups. This is accomplished by expanding for small values of a particular dimensionless group, which implies negligible effects for that group (at leading order). We present 6 cases.

Case 1: Small Mach number limit and constant γ. The pulse tube operates at the small Mach number limit (typically, $M = O(10^{-3})$). This reduces the number of dimensionless groups from 11 to 10 and effectively reduces the problem to an anelastic one with negligible (energy) viscous dissipation. In addition, for practical purposes, we can simply take γ as large as possible, since a larger γ results in larger temperature oscillations (for an isentropic process), hence larger enthalpy flows. Helium, a monatomic molecule whose γ is 5/3, is usually used as the working fluid. Constant γ further reduces the number of groups from 10 to 9.

Case 2: Isothermal wall. The isothermal tube wall limit implies that the thermal inertia of the wall is infinite, i.e., the temperature of the gas is pinned at the wall. This eliminates the effect of the wall domain, groups (l^{*2}/L^{*2}) and (F^2), and so reduces the number of groups from 9 to 7.

Case 3: Adiabatic wall. The adiabatic limit for the tube wall implies that the tube wall has no effect on the temperature oscillations of the gas (negligible heat capacity of the

Table 3. Suggested Dimensionless Groups for Various Limiting Cases

Case 0	Case 1	Case 2	Case 3	Case 4	Case 5	Case 6
full 2D	M<<1 and γ = constant	Case 1 plus isothermal wall	Case 1 plus adiabatic wall	Case 3 plus no viscous flow	Case 4 plus $T_{,z}$=constant	Case 5 plus $\phi = 90°$
M						
γ						
l^{*2}/L^{*2}	l^{*2}/L^{*2}					
F^2	F^2					
P^2	P^2	P^2				
$Ұ^2$	$Ұ^2$	$Ұ^2$	$Ұ^2$			
r_w^{*2}/L^{*2}	r_w^{*2}/L^{*2}	r_w^{*2}/L^{*2}	$r_w^{*2}/L^{*2}\,P^2Ұ^2$	$r_w^{*2}/L^{*2}\,P^2Ұ^2$		
ϕ_U	ϕ_U	ϕ_U	ϕ_U	ϕ_U	ϕ_U	$\phi_U = 90°$
ε	ε	ε	ε	ε	ε	ε
U_L	U_L	U_L	U_L	U_L	U_L	U_L
$\overline{H}$	$\overline{H}$	$\overline{H}$	$\overline{H}$	$\overline{H}$	$\overline{H}$	$\overline{H}$

wall). The adiabatic boundary condition at the outer tube wall is directly seen by the gas, and is characterized by the limit in which the dimensionless groups for the tube wall domain go to zero, $(l^{*2}/L^{*2}) \to 0$ and $(l^{*2}/L^{*2}F^2) \to 0$. At this limit, there is no heat flux at the gas/wall interface, which effectively eliminates the tube wall domain. There is also no heat flux at the centerline due to symmetry. These temperature boundary conditions eliminate transverse thermal diffusion thus requiring the temperature to be constant in the radial direction (see Eq. (9)). All this leads to a reduction of the number of dimensionless groups from 9 to 6. Axial heat conduction remains and is now characterized by the group $r_w^{*2}/\left(L^{*2}P^2Ұ^2\right)$.

Case 4: Adiabatic wall with no viscosity. This is the same as Case 3 but with the additional constraint of $1/Ұ^2 \to 0$. This further reduces the number of groups from 6 to 5 and effectively eliminates all transverse diffusion effects, hence, the problem becomes one-dimensional.

Case 5: Adiabatic wall with no viscosity and constant axial temperature gradient of the gas. This is the same as Case 4 with the additional assumption of a constant axial temperature gradient (axial linear temperature profile). A constant temperature gradient eliminates the effect of axial heat conduction scaling group $r_w^{*2}/\left(L^{*2}P^2Ұ^2\right)$(see Eq. 9). In real systems, though, this may not be the case.

Case 6: Adiabatic wall with no viscosity, constant axial temperature gradient of the gas, and ideal integrated flow through the orifice. This is the same as Case 5 with the additional requirement that the phase angle between the velocities at the two ends is 90° (or equivalently, the phase angle between pressure in the tube and mass flow through the orifice is 0°). This is usually the case in ideal models, however, it is not necessarily the case in which there is active control of the hot end mass flow, such as for moving plug systems or systems in which the orifice is replaced with solenoid valves.

As an example of how these dimensionless groups may be used, consider the simplest case of ideal enthalpy flow, Case 6 in Table 3. Case 6 contains 3 independent dimensionless parameters, ε, U_L, and $\overline{H}$. These parameters represent the following measurable quantities: ε represents the dynamic pressure oscillation relative to the mean pressure, or equivalently, the pressure ratio (ε can be written as d^*/L^* where d^* is the (imaginary) piston displacement at one end of the tube and L^* is the tube length); $U_L = U_L^*/U_0^*$ represents the mass flow ratio between the two ends; and $\overline{H}$ is the normalized enthalpy flow which is determined by the amount of heat rejected at T_h for a given T_c. Measurements of the dimensional data (U_0^* may be a bit difficult), for the prescribed conditions of $\{l^{*2}/L^{*2}\}$, $\{l^{*2}/(L^{*2}\mathrm{F}^2)\}$ and $\{1/Ꝟ^2\}$ being small, and the axial temperature profile being reasonably linear (which may not necessarily be the case) will allow a correlation between the three groups. These groups are also indirectly suggested by Storch and Radebaugh, et. al.[10] (in their Eqs. (2-39) and (2-40)) who give a 3 parameter functional relation that can be rewritten in terms of ε, U_L, and $\overline{H}$.

CONCLUSION

A set of 11 dimensionless scaling groups have been identified for use in correlating pulse tube data. The 11 groups allow for transverse heat transfer to the tube wall and viscous effects. For ideal one-dimensional enthalpy flow in which the phase angle between the pressure in the tube and mass flow through the orifice is in phase, and the axial temperature is reasonably linear, the number of dimensionless groups reduce to 3: pressure ratio, velocity ratio between the two ends, and normalized heat rejected at the hot end.

REFERENCES

1 R. Radebaugh, A review of pulse tube refrigeration, in: "Adv. Cryo. Eng.", Plenum Press, New York (1990), 35, pg. 1191-1205.

2 C. Wang, P. Wu, and Z. Chen, Numerical modeling of an orifice pulse tube refrigerator, *Cryogenics*, 32 (1992) pg. 785.

3 N. Rott, Damped and thermally driven acoustic oscillations in wide and narrow tubes, *Z. Angew Math. Phys.*, 20 (1969), pg. 230-243.

4 J. M. Lee, P. Kittel, K. D. Timmerhaus and R. Radebaugh, Steady secondary momentum and enthalpy streaming in the pulse tube refrigerator, in: "Cryocoolers 8", Plenum Press, New York (1995), pg. 359-369.

5 J. R. Olson and G. W. Swift, Similitude in thermoacoustics, *J. Acoust. Soc. Am.*, 95 (3) 1994, pg. 1405.

6 R. B. Bird, W. E. Stewart, E. N. Lightfoot, "Transport Phenomena", Wiley (1960).

7 P. W. Bridgman, "Dimensional Analysis", Yale University Press, New Haven (1931).

8 W. A. Krantz, Course lecture notes on transport phenomena, University of Colorado, Boulder, CO (1992).

9 J. M. Lee, Ph. D. Thesis to be published.

10 P. J. Storch, R. Radebaugh and J. Zimmerman, Analytical model for the refrigeration power of the orifice pulse tube refrigerator, Nat. Inst. of Standards and Technology, Tech Note 1343 (1990), pg. 28-29. For an ideal gas, the grouping $\left(\dot{m}_h^* T_h^*\right)/\left(\dot{m}_c^* T_c^*\right)$ of Eq. 2-39 can be rewritten as

$$\frac{\dot{m}_h^* T_h^*}{\dot{m}_c^* T_c^*} = \frac{\left(\rho_h^* U_L^* A\right) T_h^*}{\left(\rho_c^* U_0^* A\right) T_c^*} = \frac{U_L^*}{U_0^*} \frac{\left(\rho_h^* T_h^*\right)}{\left(\rho_c^* T_c^*\right)} = \frac{U_L^*}{U_0^*} \frac{p_0^*/R}{p_0^*/R} = U_L .$$

THEORETICAL ANALYSIS OF A PULSE TUBE REGENERATOR

Pat R. Roach[1], Ali Kashani[2] and J. M. Lee[1]

[1]NASA Ames Research Center
Moffett Field, CA 94037-1000
[2]Atlas Scientific
Sunnyvale, CA 94086

ABSTRACT

A theoretical analysis of the behavior of a typical pulse tube regenerator has been carried out. Assuming simple sinusoidal oscillations, the static and oscillatory pressures, velocities and temperatures have been determined for a model that includes a compressible gas and imperfect thermal contact between the gas and the regenerator matrix. For realistic material parameters, the analysis reveals that the pressure and velocity oscillations are largely independent of details of the thermal contact between the gas and the solid matrix. Only the temperature oscillations depend on this contact. Suggestions for optimizing the design of a regenerator are given.

CONSERVATION EQUATIONS

The conservation equations for an element of gas undergoing three-dimensional flow and exchanging heat with its surroundings, ignoring gravitational forces, can be written as follows[1] (we adopt the notation that variables with an asterisk are the normal unscaled variables; later we will introduce scaled, dimensionless variables, which will not have an asterisk):

Mass conservation:
$$\frac{\partial \rho^*}{\partial t^*} = -\nabla \cdot \rho^* \mathbf{v}^* \tag{1}$$

Momentum conservation:
$$\frac{\partial \rho^* \mathbf{v}^*}{\partial t^*} = -\nabla \cdot \rho^* \mathbf{v}^* \mathbf{v}^* - \nabla P^* - \nabla \cdot \boldsymbol{\tau}^* \tag{2}$$

Energy conservation:
$$\frac{\partial}{\partial t^*}(\rho^* U^*) = -\nabla \cdot \rho^* U^* \mathbf{v}^* - P^* \nabla \cdot \mathbf{v}^* - \nabla \cdot \mathbf{q}^* - \boldsymbol{\tau}^* : \nabla \mathbf{v}^* \tag{3}$$

where $\mathbf{v}^*$ is the vector velocity, $\boldsymbol{\tau}^*$ is the stress tensor, $\mathbf{q}^*$ is heat flow, and U^* is the internal energy density.

In the case of one-dimensional flow in the z-direction, eqs. (1), (2) and (3) become (using eq. (1') to arrive at the form of eqs. (2') and (3')), in cylindrical coordinates:

$$\frac{\partial \rho^*}{\partial t^*} = -\frac{\partial(\rho^* v_z^*)}{\partial z^*} \tag{1'}$$

$$\rho^* \frac{\partial v_z^*}{\partial t^*} = -\rho^* v_z^* \frac{\partial v_z^*}{\partial z^*} - \frac{\partial P^*}{\partial z^*} + \mu\left(\frac{\partial^2 v_z^*}{\partial r^{*2}} + \frac{1}{r^*}\frac{\partial v_z^*}{\partial r^*} + \frac{4}{3}\frac{\partial^2 v_z^*}{\partial z^{*2}}\right) \tag{2'}$$

$$\rho^* C_p^* \frac{\partial T^*}{\partial t^*} - \frac{\partial P^*}{\partial t^*} = -\rho^* v_z^* C_p \frac{\partial T^*}{\partial z^*} - k\frac{\partial^2 T^*}{\partial z^{*2}}$$

$$+ v_z^* \frac{\partial P^*}{\partial z^*} + Q + \mu\left[\left(\frac{\partial v_z^*}{\partial r^*}\right)^2 + \frac{4}{3}\left(\frac{\partial v_z^*}{\partial z^*}\right)^2\right] \tag{3'}$$

where U^* has been replaced by $C_p T^* - P^*/\rho^*$ and the $\nabla \cdot \mathbf{q}^*$ term has been split up into an axial conduction term, $\frac{\partial}{\partial z^*}(k \frac{\partial T^*}{\partial z^*})$, and a term, Q, for heat transfer from the matrix. In eqs. (2') and (3') the stress tensor, $\boldsymbol{\tau}^*$, has been expressed in terms of its velocity gradient components [1].

The terms involving $\partial v_z^*/\partial z^*$ in eqs. (2') and (3') can be neglected and in most regenerator calculations the term $\rho^* \partial v_z^*/\partial t^*$ in eq. (2') is also very small compared to the pressure gradient and the viscous terms. In that case eq. (2') can be used to evaluate the term, $v_z^* \partial v_z^*/\partial t^*$, in eq. (3') and one obtains[2]:

$$\rho^* C_p \frac{\partial T^*}{\partial t^*} - \frac{\partial P^*}{\partial t^*} = -\rho^* \frac{q^*}{f} C_p \frac{\partial T^*}{\partial z^*} + k\frac{\partial^2 T^*}{\partial z^{*2}} + Q \tag{4}$$

where v_z^* has been replaced by $q^*/f = u^*_{av}$ by averaging v_z^* over the cross section of the pore; q^* is the effective velocity that the fluid would have with no matrix present and f is the void fraction.

The $\partial v_z^*/\partial r^*$ terms in eq. (2') are generally replaced by a Darcy permeability expression for a porous matrix. Then eq. (2') becomes:

$$\frac{\rho^*}{f}\frac{\partial q^*}{\partial t^*} = -\frac{\partial P^*}{\partial z^*} - \frac{\mu q^*}{K_p} \tag{5}$$

where K_p is the Darcy permeability.

To convert the variables to dimensionless form we use the following scaling parameters: P_0^*, average pressure in the system, T_0^*, temperature at hot end of regenerator, ρ_0^*, density at hot end of regenerator, q_0^*, gas velocity at hot end of regenerator, L_0^*, length of regenerator and τ^*, period of oscillation. These scaling parameters can be combined into dimensionless parameters as follows:

$\varepsilon = \tau^* q_0^*/L_0^*$; $\gamma = C_p/C_v$, M = Mach number = $q_0^*/\sqrt{\gamma R T_0^*}$, where R is the gas constant, Va = Valensi number = $K_p \rho_0^*/\mu \tau^*$; $\lambda = f/2\pi \mathrm{Va}$, where f is the void fraction. These parameters are convenient when expressing the previous equations in dimensionless form:

Scaled mass conservation equation, eq. (1'):

$$\frac{\partial \rho}{\partial t} + \frac{\varepsilon}{f}\frac{\partial}{\partial z}(\rho q) = 0 \tag{6}$$

Scaled ideal gas law:

$$P = \rho T \tag{7}$$

Scaled momentum conservation equation, eq. (5):

$$\frac{\rho}{f}\frac{\partial q}{\partial t} = -\frac{\varepsilon}{\gamma M^2}\frac{\partial P}{\partial z} - \frac{q}{\mathrm{Va}} \tag{8}$$

Scaled energy conservation equation, eq. (4):

$$\rho_0^* C_P \left(\frac{\partial T}{\partial t} + \frac{\varepsilon q}{f} \frac{\partial T}{\partial z} \right) \rho - R \rho_0^* \frac{\partial P}{\partial t} = \frac{\tau^* k}{L_0^{*2}} \frac{\partial^2 T}{\partial z^2} + \frac{\tau^*}{T_0^*} Q \tag{9}$$

The heat transferred between the gas and the matrix (per unit volume of gas) is: $Q = h A_s (T_m^* - T^*) / V_{gas}$ where h is the heat transfer coefficient, A_s is the area of the gas–matrix interface and V_{gas} is the gas fraction of the total volume. The hydraulic radius is $r_h = V_{gas}/A_s$ so $Q = h (T_m^* - T^*) / r_h = h T_0^* (T_m - T) / r_h$ in dimensionless variables. Therefore, eq. (9) becomes:

$$\rho_0^* C_P \left(\frac{\partial T}{\partial t} + \frac{\varepsilon q}{f} \frac{\partial T}{\partial z} \right) \rho - R \rho_0^* \frac{\partial P}{\partial t} = \frac{\tau^* k}{L_0^{*2}} \frac{\partial^2 T}{\partial z^2} + \tau^* \frac{h}{r_h} \left(T_m - T \right) \tag{10}$$

A similar energy conservation equation can be written for the matrix:

$$\rho_m^* C_m (1 - f) \left(\frac{\partial T_m}{\partial t} \right) = \frac{\tau^* k_m (1 - f)}{L_0^{*2}} \frac{\partial^2 T_m}{\partial z^2} + \tau^* f \frac{h}{r_h} \left(T - T_m \right) \tag{11}$$

where ρ_m and C_m are the density and heat capacity of the matrix, T_m is the scaled temperature of the matrix and k_m is the thermal conductivity of the matrix. The term describing the heat flow from the gas (per gas volume) has been multiplied by f, the gas fraction of the total volume, and the terms involving just the matrix are multiplied by (1 – f), the matrix fraction of the total volume.

SOLUTION OF EQUATIONS

These equations will be solved in a perturbation treatment that is valid when the level of oscillation of the gas is small. The perturbation parameter is ε which is proportional to q_0^*, the amplitude of the velocity oscillation at the hot end of the regenerator. Further, it is assumed that all the important features of the system operation can be described with only fundamental frequency terms and that higher harmonic terms can be neglected. These assumptions lead to the following expansions for the dimensionless variables in the above equations:

$$\begin{aligned}
P &= P_0 + \varepsilon \left(P_a + P_d e^{i 2 \pi t} \right) + \varepsilon^2 \left(P_{2a} + P_{2d} e^{i 2 \pi t} \right) \\
T &= T_0 + \varepsilon \left(T_a + T_d e^{i 2 \pi t} \right) + \varepsilon^2 \left(T_{2a} + T_{2d} e^{i 2 \pi t} \right) \\
T_m &= T_0 + \varepsilon \left(T_{ma} + T_{md} e^{i 2 \pi t} \right) + \varepsilon^2 \left(T_{m2a} + T_{m2d} e^{i 2 \pi t} \right) \\
\rho &= \rho_0 + \varepsilon \left(\rho_a + \rho_d e^{i 2 \pi t} \right) + \varepsilon^2 \left(\rho_{2a} + \rho_{2d} e^{i 2 \pi t} \right) \\
q &= q_a + q_d e^{i 2 \pi t} + \varepsilon \left(q_{2a} + q_{2d} e^{i 2 \pi t} \right)
\end{aligned} \tag{12}$$

Subscripts a refer to average values and subscripts d refer to the amplitude of oscillating or dynamic variables.

When eqs. (12) are substituted into eqs. (6), (7), (8), (10) and (11) the lowest order equations describing the oscillating-term coefficients are:

$$\frac{1}{f} \frac{\partial (\rho_0 q_d)}{\partial z} + 2 \pi i \rho_d = 0 \tag{13}$$

$$P_d = \rho_d T_0 + \rho_0 T_d \tag{14}$$

$$\frac{2 \pi i}{f} \rho_0 q_d = - \frac{q_d}{Va} - \frac{\varepsilon^2}{\gamma M^2} \frac{\partial P_d}{\partial z} \tag{15}$$

$$\rho_0^* C_p \left[\frac{\rho_0 q_d}{f}\frac{\partial T_0}{\partial z} - 2\pi i \left(\frac{R}{C_p} P_d - \rho_0 T_d\right)\right] = \frac{\tau^* h}{r_h}(T_{md} - T_d) + \frac{\tau^* k}{L_0^{*2}}\frac{\partial^2 T_d}{\partial z^2} \tag{16}$$

$$(1-f)\,\rho_m^* C_m [\,2\pi i\, T_{md}] = \frac{f\,\tau^* h}{r_h}(T_d - T_{md}) + \frac{(1-f)\,\tau^* k_m}{L_0^{*2}}\frac{\partial^2 T_{md}}{\partial z^2} \tag{17}$$

Equation (17) is easily solved if the last term, $k_m \partial^2 T_{md}/\partial z^2$, is neglected and if the oscillating matrix temperature is assumed to be proportional to the gas temperature oscillations, but with a phase shift. Then eq. 16 becomes (neglecting $k\, \partial^2 T_d/\partial z^2$):

$$\frac{\rho_0 q_d}{f}\frac{\partial T_0}{\partial z} - 2\pi i \left(\frac{R}{C_p} P_d - \rho_0 T_d\right) = \frac{-2\pi i \Gamma}{1 + i\alpha} T_d, \tag{18}$$

where $\alpha = \dfrac{2\pi(1-f)\rho_m^* C_m r_h}{f\,\tau^* h}$ and $\Gamma = \dfrac{(1-f)\,\rho_m^* C_m}{f\,\rho_0^* C_p}$.

The variables ρ_0, P_0 and T_0 are determined by lower order equations and by static boundary conditions. The results are $P_0 = 1$, $T_0 = 1 + \Delta T\, z$ and $\rho_0 = 1/T_0$, where ΔT is the temperature difference from the hot to the cold end of the regenerator. Equations (14), (15) and (18) can be used to express the three oscillating-term coefficients, q_d, T_d and ρ_d in terms of P_d and eq. (13) can use these expressions to arrive at the differential equation[2] for P_d:

$$0 = \frac{\partial^2 P_d}{\partial z^2} + \left[\frac{(1+\alpha i)\Gamma T_0 + \alpha^2 + 1}{\left([\Gamma T_0 + 1]^2 + \alpha^2\right) T_0} - \frac{\lambda(\lambda T_0 - i)}{\lambda^2 T_0^{\,2} + 1}\right]\frac{\partial T_0}{\partial z}\frac{\partial P_d}{\partial z}$$
$$+ \frac{4\pi^2 M^2 (\lambda i T_0 - 1)\left\{(\gamma - 1)[(1+\alpha i)\Gamma T_0 + \alpha^2 + 1] - \gamma\left[(\Gamma T_0 + 1)^2 + \alpha^2\right]\right\} P_d}{\varepsilon^2\left[(\Gamma T_0 + 1)^2 + \alpha^2\right] T_0} \tag{19}$$

With specified boundary conditions, eq. (19) can be solved for P_d by numerical methods and the result used to find q_d, T_d and ρ_d. The boundary conditions will be provided by the requirements of matching pressures and mass flows at the two ends of the regenerator--at the compressor (hot end) and at the pulse tube (cold end).

There is one problem, however. When average enthalpy flow (time-averaged over one cycle) in the regenerator is calculated from these oscillating parameters, one finds:

$$\langle \dot{h} \rangle = C_p \langle \dot{m}\, T\rangle = \frac{C_p}{f}\langle \rho\, q\, T\rangle = \frac{\varepsilon C_p}{2f}\rho_0\, T_d\, q_d^{cc} \tag{20}$$

using the relationships $\rho_0 q_a = 0$ and $\rho_a q_a + \rho_0 q_{2a} + \frac{\rho_d}{2} q_d^{cc} = 0$ that come from the requirement that there is no average mass flow over a cycle. $\langle\,\rangle$ denotes a time average over one cycle and q^{cc} refers to the complex conjugate of q; $T_d\, q_d^{cc}/2$ is the zero frequency term arising from multiplying two fundamental–frequency expressions.

In general this is not constant with z in the regenerator. This means that there must be some conversion of enthalpy flux into heat that is conducted axially through the gas and/or conducted into the matrix and conducted axially through the matrix. For this to happen there must be a non-linear static temperature profile in the gas and/or the matrix. But T_0 is linear in z and it can be shown[2] that neither T_a nor T_{ma} can provide the necessary temperature profile correction. Therefore, the correction to the linear T_0 profile must come from T_{2a}.

Using higher order versions of the conservation equations, and taking the gas axial conduction term, $k\, \partial^2 T_{2a}/\partial z^2$, to be negligible, one can show that

$$\frac{\rho_0^* C_p}{2}\frac{\partial(\rho_0\, q_d^{cc}\, T_d)}{\partial z} = \frac{(1-f)\,\tau^* k_m}{L_0^{*2}}\frac{\partial^2 T_{m2a}}{\partial z^2} \tag{21}$$

which can be integrated to find T_{m2a} and

$$\frac{\rho_0^* C_p}{2f}\frac{\partial\left(\rho_0 q_d^{cc} T_d\right)}{\partial z} = \frac{\tau^* h}{r_h}\left(T_{m2a} - T_{2a}\right) \tag{22}$$

which can be used to find T_{2a}.

We now have the corrected static temperature profile and the lowest order oscillating terms, P_d, T_d, ρ_d and q_d. In addition, we can calculate the average enthalpy flux and it obeys the second order energy conservation equations.

BOUNDARY CONDITIONS AT COMPRESSOR

In the compressor, using scaled parameters:

Pressure: $P_c = 1 + \varepsilon P_{cd} e^{i2\pi t}$ (scaled by P_0^*) (23a)

Mass in the compressor: $m_c = 1 + \varepsilon m_{cd} e^{i2\pi t}$ (23b)

(scaled by m_{ca}^*, the average mass in the compressor)

Distance above the piston: $\zeta_c = \zeta_{ca} + \varepsilon \zeta_{cd} e^{i2\pi t}$ (scaled by L_0^*) (23c)

Temperature: $T_c = T_{ca} + \varepsilon T_{cd} e^{i2\pi t}$ (scaled by T_0^*) (23d)

For an isothermal compressor, (essentially the same whether scaled or unscaled)

$$\frac{1}{P_c}\frac{\partial P_c}{\partial t} = -\frac{1}{\zeta_c}\frac{\partial \zeta_c}{\partial t} - \frac{1}{m_c}\frac{\partial m_c}{\partial t} \tag{24}$$

But $P_c = P$ at entrance to regenerator ($z = 0$) and $\partial m_c/\partial t$ is mass flowing into regenerator at $z = 0$. Therefore $\partial m_c^*/\partial t = A_{regen}\,\rho^* q^*$ in unscaled parameters. Applying scaling yields:

$$\partial m_c/\partial t = (F\varepsilon)\left(\rho_0 + \varepsilon \rho_d e^{2\pi i t}\right)\left(q_d e^{2\pi i t}\right) \tag{25}$$

where $F = A_{regen} L_0^* \rho_0^* / m_{ca}^*$.

Using eq. (25) and eq. (23c), keeping only the lowest order oscillating terms, eq. (24) becomes

$$2\pi i \varepsilon P_d(z=0) = -\frac{2\pi i \varepsilon \zeta_{cd}}{\zeta_{ca}} - F\varepsilon \rho_0 q_d \quad (\text{here, } \rho_0 = 1) \tag{26}$$

This equation provides one of the boundary conditions for the solution of eq. (19).

BOUNDARY CONDITIONS AT PULSE TUBE

In the pulse tube, using scaled parameters:

Pressure: $P_p = 1 + \varepsilon P_{pd} e^{i2\pi t}$ (scaled by P_0^*) (27a)

Mass in the pulse tube from z to the warm end:

$m_p = m_{pa} + m_{pd} e^{i2\pi t}$ (scaled by m_{ca}^*) (27b)

Temperature: $T_p = T_{pa} + \varepsilon T_{pd} e^{i2\pi t}$ (scaled by T_0^*) (27c)

Based on the treatment of Radebaugh[3]:

$$\frac{\partial m_r^*}{\partial t^*} = \frac{A_{pt}\left(\Lambda - L_{regen}\right)}{\gamma R T_{cold}^*}\frac{\partial P_p^*}{\partial t^*} + \frac{T_0^*}{T_{cold}^*}\frac{\partial m_o^*}{\partial t^*} \tag{28}$$

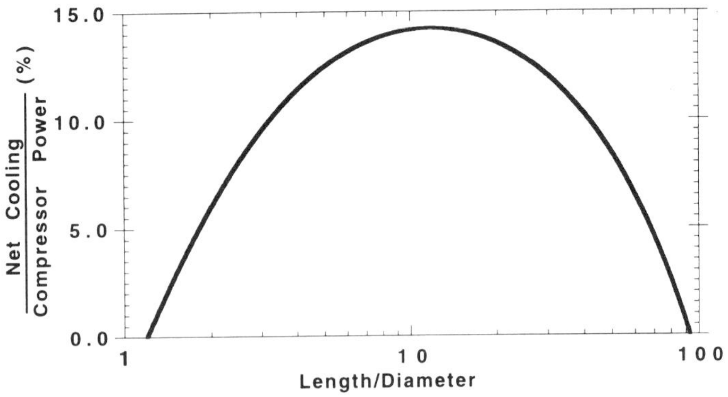

Fig. 1. Cooling power as a function of length-to-diameter ratio for 150-mesh screens.

where $\partial m_r^*/\partial t^*$ is the mass flow from the regenerator into the pulse tube at its cold end, P_p^* is the pressure in the pulse tube, $\partial m_o^*/\partial t^*$ is the mass flow through the orifice at the warm end, A_{pt} is the area of the pulse tube and Λ is the length of regenerator plus pulse tube. Eq. (28) scales to

$$\frac{\partial m_r}{\partial t} = G\frac{\partial P_p}{\partial t} + \frac{1}{T_{cold}}\frac{\partial m_o}{\partial t} \quad \text{where } G = A_{pt} L_{pt} \rho_0^* T_0^*/\gamma m_{ca}^* T_{cold}^* \tag{29}$$

But $\frac{\partial m_r}{\partial t}$ is the mass flow from the regenerator at z = 1:

so
$$\frac{\partial m_r}{\partial t} = F \varepsilon \rho_0(1)\, q_d(1) e^{2\pi i t} \tag{30a}$$

and $\partial m_o/\partial t$ is the mass flow at the orifice (proportional to pressure):

$$\frac{\partial m_o}{\partial t} = k \varepsilon P_{pd}\, e^{2\pi i t} = k \varepsilon P_d(1) e^{2\pi i t} \tag{30b}$$

Therefore,
$$F \varepsilon \rho_0(1)\, q_d(1) = \varepsilon \left(2\pi i G + \frac{1}{T_{cold}} k\right) P_d(1) \tag{31}$$

This equation provides the second boundary condition for the solution of eq. (19).

At the present time, the solution of eq. (19) is carried out with temperature-independent values for thermal conductivity, viscosity and heat transfer coefficient. The effect of a temperature-dependent viscosity will be incorporated in the next improvement of the calculation. We are looking into ways to include the temperature dependence of the

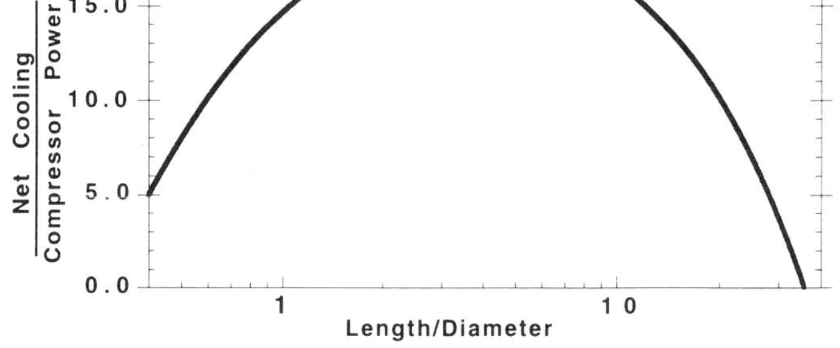

Fig. 2. Cooling power as a function of length-to-diameter ratio for 250-mesh screens.

thermal conductivity and the temperature and velocity dependence of the heat transfer coefficient.

If all the dimensions and parameters of the system are inserted into the equations, the results for cooling power can be calculated. The enthalpy flow in the pulse tube is the fundamental cooling of the system from which various heat leaks must be subtracted. These are: axial conduction in the wall of the pulse tube, axial conduction in the wall and matrix of the regenerator, and enthalpy flux in the gas at the hot end of the regenerator. (This last term redistributes itself to become an enthalpy flux in the gas at the cold end of the regenerator plus an additional axial conduction through the wall and matrix as discussed above; the sum of these two contributions doesn't change, however, as it passes through the regenerator.) The orifice opening parameter is then varied to find a value that maximizes the net cooling power.

RESULTS FROM MODEL

When typical numbers are used for the various parameters in the above equations, it is noticed that the two terms in eq. (19) with denominators $(\Gamma T_0 + 1)^2 + \alpha^2$ are very small, approximately 2 orders of magnitude less than the other terms in the equation. These terms come from the expression for the oscillating temperature. It would not distort the solution appreciably to ignore these terms in the calculation of the pressure. This is plausible since it would not be expected that the thermal behavior of the system would have a large effect on the mechanical behavior (pressure and velocity) of the system. Once the pressure and velocity were found from the simpler approach then the temperature oscillation could be found from the eq. (18). This is not the approach we take but it would certainly be reasonable if circumstances made it advantageous.

It is interesting to ask how the performance of a pulse tube depends on the length-to-diameter ratio of the regenerator. For a constant volume of the regenerator, there should be an optimum length-to-diameter ratio since the axial-conduction heat leak will become very large at short lengths and the pressure drop will become very large at long lengths. We used our model to calculate some performance data for a typical orifice pulse tube operating between 300 K and 100 K using helium gas with γ=1.68. The cooling power we calculate is composed of the enthalpy flow in the pulse tube section reduced by the enthalpy flow through the regenerator toward the cold end and reduced by axial heat conduction through the walls of the pulse tube and regenerator. Figure 1 shows the results for a regenerator filled with 150-mesh screens. The optimum length/diameter ratio is 12 for this case.

When the regenerator screens are made finer by going to 250 mesh, the cooling power is higher at the optimum length/diameter ratio as shown in fig. 2. Because of the more-restrictive screens, the optimum length/diameter occurs at 3.6.

REFERENCES

1. R. Byron Bird, Warren E. Stewart and Edwin N. Lightfoot, *Transport Phenomena,* John Wiley & Sons, New York, (1960).
2. Pat R. Roach, Ali Kashani and J. M. Lee, *Cryogenics* (to be published).
3. Peter J. Storch, Ray Radebaugh and James E. Zimmerman, NIST Technical Note 1343.

EVALUATION OF THE THERMODYNAMIC PERFORMANCE OF PULSE TUBES

Catherine Pinsky, S. Yoshida,
K. V. Ravikumar, T. H. K. Frederking

In collaboration with No Sang Myung, Steven Sapida,
Nicolae Lupu, Nell Papavasiliou, Cindy Chan & SRP Group

Chemical Engineering Department, University of California
Los Angeles, CA 90095-1592, USA

ABSTRACT

The thermodynamic performance of pulse tubes (PT) has been studied with emphasis on second law efficiencies and volume effects. The various pulse tube data of operation are compared to small Stirling coolers, characterized by a parameter set similar to the PT specifications. A modified "thermal shunt" model (modified "Bejan model") is utilized for the comparison. The model predicts a small refrigeration load asymptote of the efficiency proportional to the load itself in partial agreement with pulse tube data.

INTRODUCTION

The importance and advantages of pulse tube cryocoolers[1] are becoming widely evident, e. g. in the proceedings of the 1994 conferences "Cryocoolers-8" and ICEC-15. Among characteristic features recognized are reliable cooling due to lack of moving components in the low temperature region and a trend toward increased efficiencies comparable to other small coolers. Simple construction of the low temperature pulse tube section translates into a relatively small investment cost. Additionally, there exists the potential for miniaturization and upscaling toward elevated temperatures, e.g. non-CFC refrigeration of the future.

This paper focusses on comparing pulse tube performance to other small cryocoolers, in particular the Stirling cooler. Radebaugh has assessed the intrinsic performance of the pulse tube without compressor and without regenerator [2]. Cooling efficiencies (fraction of Carnot) of up to 90 % have been deduced from his model[2]. Thus, improvements in compressor and regenerator contribute toward bringing the pulse tube to an attractive place in the cooler field. Therefore, the present emphasis is on the overall performance, as expressed by the second law efficiency for a particular cooler at its best point (design point). A modified shunt model (Bejan cooler model[3]) is used based on literature results and small cooler experiments[4].

First, the model is outlined noting two results, an efficiency function involving the refrigeration load, and a function of dimensionless variables which does not contain the volume/size influence explicitly. Second, the types of pulse tubes evaluated are discussed, and pulse tube data are presented in the same frame of reference as other cooler data, e.g. Stirling results. Third, thermal conductances and volume effects are considered. Finally, our findings are summarized.

SHUNT MODEL

For a convenient characterization of cooler/refrigerator performance the "second law efficiency", also designated as "thermodynamic efficiency", is evaluated.

$$\eta_{II} = COP/COPc \tag{1}$$

This efficiency is the ratio of the coefficients of performance (COP = ratio of the refrigeration load to the power input required). COPc is the Carnot value of COP (= ratio of the low temperature T_L to the temperature lift). The temperature lift is the difference between the environmental "high" temperature and T_L. For a specified refrigeration load, the Carnot work (W_c = minimum work) is to be augmented by $\Delta W = W - W_c$. Then the second law efficiency is written as

$$\eta_{II} = 1/[1 + \Delta W/W_c] \tag{2}$$

The shunt model is characterized by a thermal link attached as shunt to an ideal Carnot refrigerator operating between T_L and $T_H = T_E$ (Figure 1). The heat Q_{is} transported through the shunt causes an entropy generation rate imposed on the low temperature subsystem. Consequently, the refrigerator has to pump up the gross refrigeration load $Q_L' = Q_L + Q_{is}$. The work required is enlarged compared to the ideal Carnot refrigerator. Thus, the second law efficiency is rewritten as

$$\eta_{II} = 1/[1 + Q_{is}/Q_L] \tag{3}$$

According to reference 3, it is convenient to normalize the thermal conductance in the shunt heat leak rate Q_{is} = product of the thermal conductance [C_T] and the temperature lift. In the search for a suitable normalized shunt conductance, guidance has been based on results for small coolers[4]. A reference thermal conductance of Q_L/T_L (= cold domain entropy difference per time ΔS_L) is very convenient for the simplified version of the shunt model.

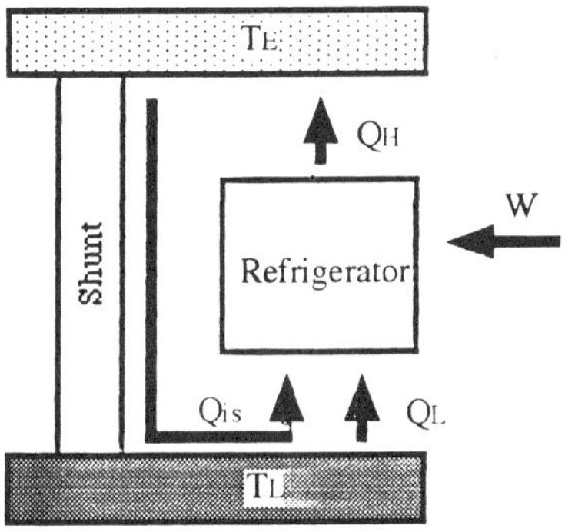

Figure 1. Shunt model, schematically.

Table 1. Pulse Tubes Listed by Category

Type of Pulse Tube	Number of PT Considered
Basic	1
Single Orifice	5
Double Orifice	4
Other	2
Miniature	3

Table 2. Performance Data of Miniature Pulse Tubes & Thermo-Acoustic Systems

Pulse Tube	COPc	η(II)
Haruyama and Inoue	1.857	0.0144
Haruyama and Inoue	2.896	0.023
Kanao et al.	0.657	0.0054
Thermo-Acoustic	COPc	η(II)
T. J. Hofler	4.66	0.126

Table 3. Stirling Cooler Reference List
a. Cryocoolers 8 ["CC8"] (R.G. Ross, Jr., Ed.)[17]

AUTHOR(S)	CC8 PAGE NO.	TYPE/MODEL
B.G. Jones	1	Brit Aerosp. 1-stage split Stirling (St)
Spradley et al.	13	Lockheed/Lucas 2-compressor split St
Horsley et al.	23	Ball/RAL/Oxford 1-stage split St
Kawada et al.	35	ETL-Tsukuba Oxford type split St
Duband et al.	47	Grenoble 2-stage/twin piston St
Nast et al.	55	Lockheed split Stirling/1710-C
Wakugawa et al.	69	Hughes split Stirling/Oxford type
Carrington et al.	85	Ball Aerospace/2-stage Stirling
Stacy et al.	103	30 D diaphragm 2-stage Stirling
Pruitt	113	Hughes twin-piston linear drive St
Rawlings	121	Texas Instrum. linear drive St
Harayan	120	Magnavox dual opposed piston St
Chardin et al.	137	SIGNAAL USFA cooler (St)
Smedley et al.	149	Sunpower/JPL 140 K cooler (St)
Sparr et al.	163	NASA GSFC: 3 cooler tests
Smedley et al.	185	JPL: 5 cooler tests--maximum efficiency
b. Intl. Cryog. Engr. Conference (ICEC-15, 1994)[21]		
Nast et al.	187	Lockheed split Stirling (Lucas/Oxford)
Deac	191	Split Stirling-rotary drive motor
Meromi et al.	195	RICOR split St.-springless expander

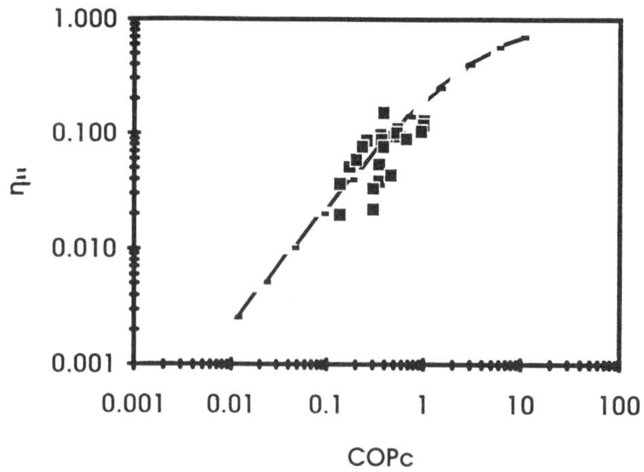

Figure 2. Second law efficiency as a function of COPc for pulse tube coolers.

The resulting second law efficiency (= percent Carnot divided by 100) is

$$\eta_{II} = 1/[\ 1 + C_m^*/COPc\ ] \tag{4}$$

C_m^* is the dimensionless conductance $C_T/\Delta S_L$. Aside from reference 4, small coolers have been considered as reported in the literature on early cooler developments. The value C_m^* = 4.3 has been deduced as a guide characterizing the status of the small cooler category. In "high-T" refrigeration applications, high η_{II} values near 80 % are predicted by equation (4) in agreement with performance results. In the low-T cryocooler area, a linear function is approached: η_{II} = const COPc ; COPc << 1. In the subsequent section, we consider recent small cooler results in relation to equation (4).

COOLER CATEGORIES AND THERMODYNAMIC PERFORMANCE

First, we consider the energetics of the coolers. Second, we address questions of the volumes required. The second law efficiency is evaluated using the preceding equations in the form

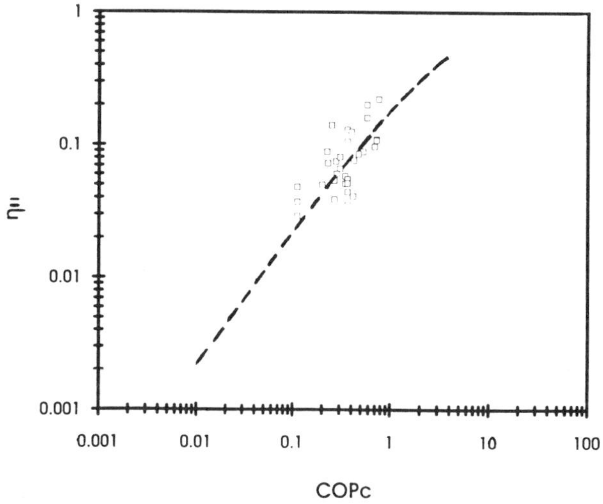

Figure 3. Second law efficiency as a function of COPc for Stirling coolers.

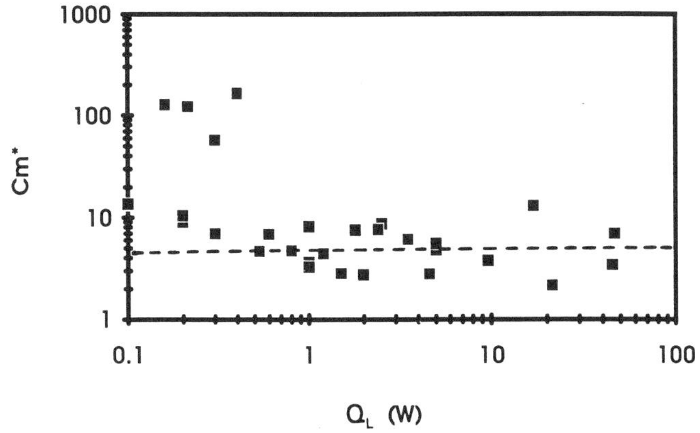

Figure 4. Dimensionless conductance (ratio C_m*) as a function of refrigeration load.

$$\eta_{II} = (Q_L/W)\ (\Delta T)/T_L \tag{5}$$

In some cases, direct information has been given by authors. In other cases only restricted data are available. The cooler categories considered are listed in Table 1. Miniaturized coolers have special characteristics with mostly higher C_m* values compared to other small cooler categories (Table 2). Another category is the high frequency thermoacoustic system, e. g. reference 7.

Figure 2 plots the second law efficiency versus COPc for pulse tubes (references 8-16). Equation (4) is included for C_m* = 4.3. This function serves as a guide for the assessment of the individual cooler.

Figure 3 is a similar plot which includes Stirling coolers taken primarily from contributions contained in the recent "Cryocoolers 8" proceedings, as listed in an overview of these coolers[17] (Table 3). This comparison suggests that Stirling coolers and pulse tubes have reached the same efficiencies region. This confirms the conclusions reached in an early comparison[19]. Equation (4) again is useful. For a specified COPc, i.e. a low temperature, good operation is exhibited by a point above the line, and moderate operation is below the line.

A comparison of Figures 2 and 3 suggests that both pulse tube and Stirling coolers are close to each other in their performance figures.

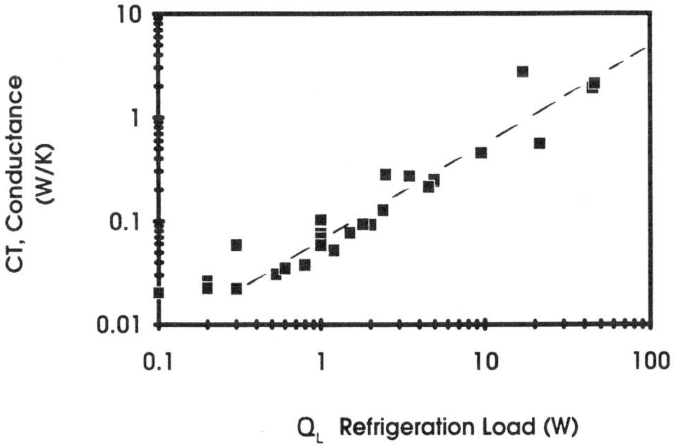

Figure 5. Absolute thermal conductance of the shunt model.

THERMAL CONDUCTANCES AND VOLUME EFFECTS

The shunt model permits determination of the dimensionless conductance and of the absolute conductance. It is emphasized that these quantities constitute the only "free" parameter used for the overall characterization of the cooler operation.

The dimensionless conductance of the shunt model is obtained from data for a pulse tube cooler from our shunt model for steady cooler operation.

$$C_m^* = COPc\,[\eta_{II}^{-1} - 1] \tag{6}$$

The shunt thermal conductance is determined using the entropy difference (per time) at the cold end $\Delta S_L = Q_L/T_L$.

$$C_T = C_m^*\,\Delta S_L = C_m^*\,Q_L/T_L \tag{7}$$

The C_m^* - values of pulse tube coolers are shown in Figure 4 as a function of the refrigeration load Q_L. The value $C_m^* = 4.3$, deduced from early cooler data and from reference 4, is included in Figure 4 as dashed (horizontal) line. There is considerable scatter of the dimensionless conductance C_m^* in general. Despite this wide range, there is a trend toward high C_m^* values at low loads. This effect is pronounced for "mini"-pulse tubes.

The absolute thermal conductances are displayed in Figure 5 versus Q_L. According to equation (7), we expect for a specified T_L that the conductance is a linear function of the refrigeration load provided pulse tube cooler similarity is retained for a class of pulse tubes. Figure 5 indicates indeed a monotonic variation of the conductance with refrigeration load.

The cooler volume varies monotonically with the refrigeration load aside from the relative increase at low Q_L as the cooler undergoes miniaturization. Thus, both Figures 4 and 5 indicate that the dimensionless conductance C_m^* is increased as the volume is decreased. This implies that coolers are similar, at least in major parameters of geometry and operation.

In Figure 5 the low temperature $T_L = 80$ K is included as a dashed line based on equation (7) for the value of $C_m^* = 4.3$. It is noted that C_m^* represents an overall loss value for the entire cycle, not only for a single component carrying heat. This is reflected in Figures 4 to 6.

Figure 6 shows the volume (V) of pulse tubes versus Q_L . V is the volume of the pulse tube proper. For similar geometries, V is comparable to the regenerator stack volume. Only

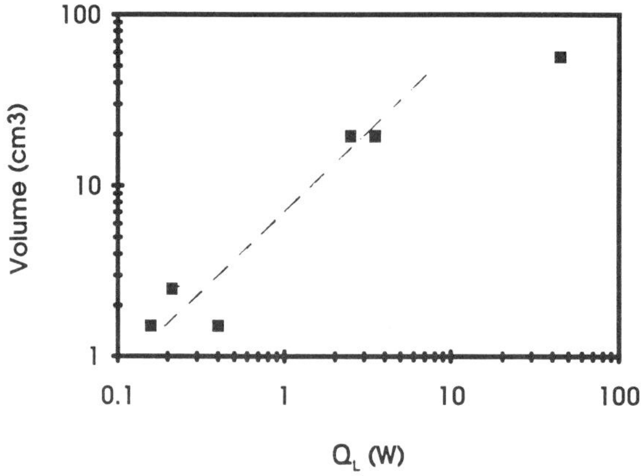

Figure 6. Volume of pulse tube versus refrigeration load.

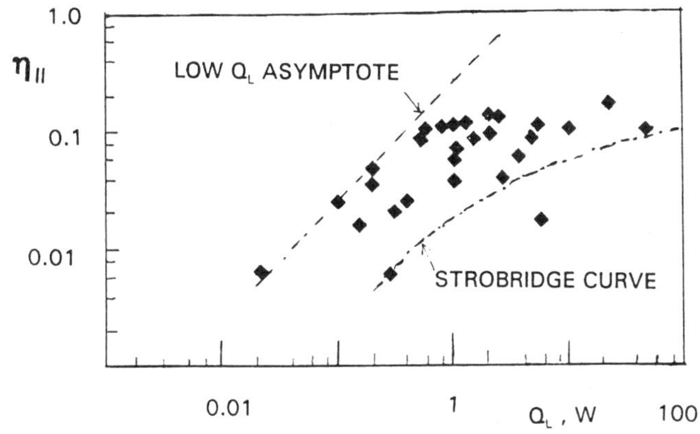

Figure 7. "Strobridge function" with linear, low load asymptotes of the modified shunt model for constant leak rates Q_{is}.

a small pulse tube data set is available with complete geometry information and parameters of operation. Ths set in Figure 6 indicates indeed a nearly linear variation of volume with refrigeration load consistent with the preceding discussion.

Our shunt model incorporates an efficiency asymptote for small refrigeration loads which is linear in Q_L. For similar coolers we reconsider equation (4) in the following form

$$\eta_{II} = 1/\left[1 + C_T\, \Delta T/Q_L\right] \tag{8}$$

The efficiency versus Q_L function has been known as "Strobridge diagram" [20]. Our shunt model predicts a special asymptote for the small cooler range. From equation (8) we obtain for relatively large shunt heat leaks, $Q_{is} = C_T\, \Delta T >> Q_L$, a second law efficiency η_{II} which varies linearly with refrigeration load Q_L.

$$\eta_{II} = Q_L/\, Q_{is} \tag{9}$$

This linear function for constant Q_{is} - values has been entered in the "Strobridge diagram" [20] (Figure 7). Part of the data support this equation, i.e. similarity may exist for pulse tube coolers located close to equation (9) for a constant "shunt leak" representing the overall cycle loss parameter. In this context it is noted that not all pulse tube data considered represent optimum (maximum) efficiency points. Often the minimum temperature has been emphasized more than the maximum η_{II} location of a particular pulse tube cooler.

In addition, the (mean) "Strobridge function" of efficiency versus Q_L is included in Figure 7. The pulse tube data show significant improvements made over the years with respect to the original (mean) Strobridge function.

SUMMARY AND CONCLUDING REMARKS

The present modified "shunt model" for small coolers addresses usual "balance" results, i.e. an Euler control volume system for steady operation and overall characterization purposes. This is in contrast to Lagrangian approaches, often considered in pulse tube discussions[2,22,23]. Our modified Bejan model has been found to agree with some data trends for small coolers, i.e. low refrigeration loads. A cooler has a low dimensionless conductance $C_m{}^*$ when thermodynamic losses are relatively small, and high conductances $C_m{}^*$ when losses are relatively large.

The early phase of this work has been presented at the 10th Intersociety Cryogenics Symposium, Houston Texas, March AIChE Meeting.

ACKNOWLEDGMENTS

Seed funding of the pulse tube work by Hughes Aircraft Co. is acknowledged with great appreciation. Dr. Frithjof Mastrup has been most instrumental in facilitating a rapid transition from transport phenomena toward the applied cooler thermodynamics. Numerous experts gave helpful comments and advice.

Three students, the first author, Steven Sapida and No Sang Myung, acknowledge receiving a Summer-Industrial Intern Hughes Scholarship. We like to thank all supporters of the present work.

REFERENCES

1. G. Walker and E.R. Bingham, "Low-capacity Cryogenic Refrigeration", Clarendon Press, Oxford (1994).
2. R. Radebaugh, LT-18, *Jap. J. Appl. Phys.* 26S: 2076 (1987).
3. A. Bejan, "Advanced Engineering Thermodynamics", Wiley, New York (1988).
4. S. Yoshida et al., *Adv. Cryog. Eng.* 39B: 1263 (1994).
5. T. Haruyama and H. Inoue, "Cryocoolers 8", R.G. Ross, Jr., ed., Plenum Press, New York (1995), p. 353.
6. K. Kanao, N. Watanabe and Y. Kanazawa, *Cryogenics* Suppl. 34:18 (1995).
7. T.J. Hofler, Ph.D. Thesis, University of California, San Diego, 1986.
8. C. Wang et al., *Cryogenics* Suppl.33:648 (1993).
9. J. Wang et al., *Cryogenics* Suppl.30:267 (1990).
10. Y. Ishizaki and E. Ishizaki, "Cryocoolers 7", Proc. Vol. 1, Santa Fe (1993), p. 140.
11. Y. Matsubara and J. Gao, *Cryogenics* Suppl.34:115 (1995).
12. S.C. Soloski and F.N. Mastrup, "Cryocoolers 8", R.G. Ross, Jr., ed., Plenum Press, New York (1995), p. 321.
13. W.W. Burt and C.K. Chan, "Cryocoolers 8", R.G. Ross, Jr., ed., Plenum Press, New York (1995), p. 313.
14. Y. Zhou and Y.J. Han, "Cryocoolers 7", Proc. Vol. 1, Santa Fe (1993), p. 147.
15. T. Kuriyama et al., "Cryocoolers 7", Proc. Vol. 1, Santa Fe (1993), p. 157.
16. C.K. Chan et al., "Cryocoolers 7", Proc. Vol. 1, Santa Fe (1993), p. 113.
17. R.G. Ross, Jr., "Cryocoolers 8", Plenum Press, New York (1993).
18. B.G. Jones and S.R. Scull, *Cryogenics* 32:850 (1992).
19. P. Kittel, *Cryogenics* 32:843 (1992).
20. K.D. Timmerhaus and T.M. Flynn, "Cryogenic Process Engineering", Plenum Press, New York (1989).
21. C. Rizzuto and C. Ferdeghini, eds, Proc. ICEC-15, Cryogenics Supplement: 34 (1994).
22. G. W. Swift, *Journ. Acoustical Soc. America* 84:1145 (1988).
23. A. Tominaga, *Cryogenics* 35:427 (1995).

PRESSURE HEAT PUMPING IN THE ORIFICE PULSE-TUBE REFRIGERATOR

P.C.T. de Boer

Sibley School of Mechanical and Aerospace Engineering
Cornell University, Ithaca, NY 14853, USA
e-mail ptd1@cornell.edu facsimile 607 255-1222

ABSTRACT

The mechanism by which heat is pumped as a result of pressure changes in an orifice pulse-tube refrigerator (OPTR) is analyzed thermodynamically. The thermodynamic cycle considered consists of four steps: 1) the pressure is increased by a factor π_1 due to motion of a piston in the heat exchanger at the warm end of the regenerator; 2) the pressure is decreased by a factor π_2 due to leakage out of the orifice; 3) the pressure is further decreased due to motion of the piston back to its original position; 4) the pressure is increased to its value at the start of the cycle due to leakage through the orifice back into the pulse tube. The regenerator and the heat exchangers are taken to be perfect. The pressure is assumed to be uniform during the entire cycle. The temperature profiles of the gas in the pulse tube after each step are derived analytically. Knowledge of the temperature at which gas enters the cold heat exchanger during steps 3 and 4 provides the heat removed per cycle from this exchanger. Knowledge of the pressure as a function of piston position provides the work done per cycle by the piston. The pressure heat pumping mechanism considered is effective only in the presence of a regenerator. Detailed results are presented for the heat removed per cycle, for the coefficient of performance, and for the refrigeration efficiency as a function of the compression ratio π_1 and the expansion ratio π_2. Results are also given for the influence on performance of the ratio of specific heats. The results obtained are compared with corresponding results for the basic pulse-tube refrigerator (BPTR) operating by surface heat pumping.

NOMENCLATURE

A	cross-sectional area of tube
c_p	specific heat at constant pressure
COP	coefficient of performance
h	specific enthalpy
k	ratio of specific heats
L	physical length of pulse tube, including high-temperature heat exchanger
L_c	non-dimensionalized stroke of virtual piston
L_h	non-dimensionalized stroke of orifice piston
$\dot{m}$	mass flow rate
p	pressure
Q	amount of heat flowing per cycle, non-dimensionalized by $A\,L\,p_1$
R	gas constant
T	temperature
W	net work per cycle, non-dimensionalized by $A\,L\,p_1$
x	axial coordinate, non-dimensionalized by L
x_p	position of piston
z	$x + kL_h$; non-dimensional axial coordinate
z_h	$1 + kL_h$

Greek symbols

π_1 p_2/p_1, compression ratio

π_2 p_2/p_3, expansion ratio

π_4 p_4/p_1

ρ density

Subscripts

1-4 conditions before steps 1-4, resp.

a after compression or expansion
b before compression or expansion
c refers to (cold) heat exchanger
h refers to (hot) heat exchangers at both ends
he heat exchanger
p refers to virtual or orifice piston
pr refers to driving piston
pt pulse tube

INTRODUCTION

Shortly after Gifford and Longsworth introduced the concept of pulse-tube refrigeration[1], they described the underlying mechanism in terms of *surface heat pumping*.[2]. Their device, now known as the Basic Pulse-Tube Refrigerator (BPTR), had one end of the pulse tube closed. When the pressure is increased at the open end, each gas element in the tube is compressed. The compression results in heating, as well as in motion towards the closed end. The heated gas element then can give off its heat to the wall. Upon a subsequent decrease of the pressure at the open end, the gas element gets expands. The expansion results in cooling, as well as in motion away from the closed end. The cooled gas element then can take up heat from the wall, at a location farther removed from the closed end than the location where it deposited its heat. Each gas element thus "pumps" heat toward the closed end, by means of thermal interaction with the wall surface. A detailed thermodynamic analysis of the BPTR without regenerator has recently been presented by de Boer[3]. The influence of the regenerator was investigated in a subsequent paper[4]. Using control volume analysis to determine enthalpy flows, and control mass analysis to determine heat flows associated with individual gas elements, it was found that the addition of the regenerator yields significant improvements in the coefficient of performance, the heat removed per cycle, and the refrigeration efficiency.

A significant improvement to pulse-tube refrigeration was made by Mikulin et al.[5], who provided the closed end of the tube with an orifice-reservoir assembly. The resulting device has become known as the orifice pulse-tube refrigerator (OPTR), and has received a great deal of attention.[6-21] The orifice-reservoir assembly acts like a piston, the motion of which is out of phase with the piston driving the motion. The OPTR thus resembles a Stirling cryocooler consisting of a regenerator with a piston at each end. It is generally understood that Stirling cryocoolers do not rely on surface heat pumping. This raises questions about the mechanism underlying the operation of the OPTR. It might be argued that this mechanism still must involve surface heat pumping, because otherwise there would be an apparent violation of the second law of thermodynamics. The argument leading to this apparent violation is based on consideration of the gas elements that spend part of the cycle in the cold heat exchanger, and the remaining part in the pulse tube. Without surface heat pumping, these gas elements are in thermal contact with only one heat reservoir, i.e. the cold heat exchanger. One of the corollaries of the second law states that the elements then cannot remove net heat per cycle from the exchanger. This would preclude any cooling effect. If surface heat pumping is present, the gas elements are in contact with two heat reservoirs, and cooling of the heat exchanger can occur. One of the purposes of the present paper is to resolve this apparent paradox. The resolution lies in the circumstance that while the gas elements just considered indeed provide heat to the cold heat exchanger in the absence of surface heat pumping, there is a separate mechanism that removes heat from the exchanger. This separate mechanism requires the presence of the regenerator. As shown in Ref. 4 and further discussed below, heat is pumped from the heat exchanger at the cold side of the regenerator to the heat exchanger at the warm side of the regenerator. The amount of heat removed this way is larger than the amount of heat provided by the gas elements considered above. There remains the question of how to find the net heat removed from the cold heat exchanger. As was also shown in Ref. 4, this amount follows from determining the temperature of the gas elements entering the cold heat exchanger. The temperature of an entering gas element is lower than the value T_c at which the element has left the heat exchanger, and the decrease is due to adiabatic cooling caused by pressure

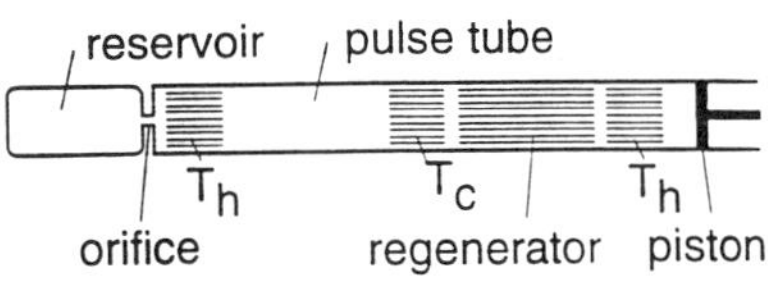

Figure 1. Layout of the OPTR

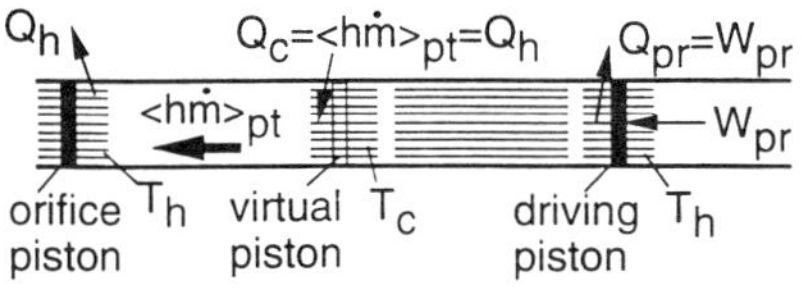

Figure 2. Simplified model of OPTR.

changes. Therefore, it appears appropriate to call the corresponding mechanism *pressure heat pumping*. While pressure heat pumping clearly is more effective than surface heat pumping, both mechanisms will occur simultaneously in actual OPTR's. The next section considers the pressure heat pumping mechanism in detail, and neglects surface heat pumping.

THERMODYNAMIC ANALYSIS

Formulation of the Problem

The layout of an OPTR is shown in Fig. 1. The heat exchangers at the left and right are maintained at temperature T_h by appropriate cooling methods. Heat is removed from the cold environment by the heat exchanger at T_c. In order to make the problem tractable, the OPTR is treated in terms of the simplified model shown in Fig. 2. The driving piston is taken to move within the heat exchanger on the right. The action of the orifice-reservoir assembly is represented by a piston moving within the heat exchanger on the left; this piston will be referred to as the *orifice piston*. The concept is introduced of a *virtual piston* moving within the cold heat exchanger. The following simplifying assumptions are made. All frictional effects are neglected, as are all conduction effects within the gas and within the walls, the heat exchangers and the regenerator. The pressure is taken as uniform during the entire cycle, which requires that the piston velocities are much less than the speed of sound. The temperature of a gas element in a heat exchanger equals the temperature of the exchanger. The temperature of a gas element in the regenerator equals the local temperature of the regenerator. The assembly is filled with an ideal gas having constant specific heats.

The thermodynamic cycle is taken to consist of the following four steps:

1. The virtual piston moves towards the left, while the orifice piston is stationary at the entrance of the heat exchanger on the left. The pressure increases from p_1 to p_2.
2. The orifice piston moves towards the left, while the virtual piston is stationary at the entrance of the cold heat exchanger. The pressure decreases from p_2 to p_3.
3. The virtual position returns to its original position by moving to the right, while the orifice piston remain stationary. The pressure decreases from p_3 till p_4.
4. The orifice position returns to its original position while the virtual piston remains stationary. The pressure returns from p_4 till its original value p_1.

In order to keep the virtual piston in place during steps 2 and 4, the driving piston must move in the direction opposite to that of the orifice piston during these steps. As shown in Ref. 4, it follows from conservation of mass between the driving piston and the virtual piston that the ratio of the cyclic work W_{pr} done by the former to the cyclic work W_p done by the latter is equal to the temperature ratio:

$$W_{pr} / W_p = T_h / T_c. \tag{1}$$

As also shown in Ref. 4, the amount of cyclic heat Q_{pr} removed by the heat exchanger on the right equals the work W_{pr} done by the driving piston (see Fig. 2). This heat is removed by the part of the heat exchanger within the range of the piston. At locations of the heat exchanger outside the range of the piston, the gas provides a cyclic heat flow Q_{pr} from left to right. Simultaneously, cyclic flow work W_{pr} is done on the gas from right to left, resulting in zero cyclic enthalpy flow. Similarly, in the cold heat exchanger there is cyclic heat flow Q_p from left to right , and cyclic flow work W_p from right to left. The differential cyclic flow work $W_{pr} - W_p$ is converted to heat within the regenerator, and this heat Q_c removed from the cold heat exchanger equals the cyclic enthalpy flow $< h \dot{m} >_{pt}$ in the pulse tube, which in turn equals the cyclic heat Q_h removed by the heat exchanger on the left. All these quantities are indicated in Fig. 2. The net work done by the gas on the orifice piston is taken to be dissipated into heat, which is included in the cyclic heat Q_h.

The cyclic heat Q_c equals the heat taken up by gas entering the cold heat exchanger from the pulse tube.[4] Calculating Q_c therefore requires knowledge of the temperature of the gas just before it enters the heat exchanger. This temperature can be found by following the cyclic temperature history of the column of gas that is expelled from the cold heat exchanger during step 1. After the temperature of the entering gas is found as a function of pressure, the determination of Q_c is straightforward. Similarly, once the piston position x_p is found as function of pressure, the determination of the cyclic work W_p is straightforward. These determinations are carried out in the following subsections. The basic independent parameters are taken to be the compression ratio $\pi_1 = p_2/p_1$ and the expansion ratio $\pi_2 = p_2/p_3$; both of these ratios are larger than one.

Basic thermodynamic relations

There are only a few basic thermodynamic relations that are needed for the determination of the temperature profiles. These are summarized below; the derivation can be found in Refs. 2 and 3. The first one pertains to the isentropic compression or expansion of a gas column of length x in the pulse tube. As the pressure is changed from p to p + dp, the length of the column is changed by an amount dx = - x dp/(k p). Integration yields

$$x_a / x_b = (p_b / p_a)^{1/k} \text{ (column in pulse tube)} \tag{2}$$

where subscripts a and b indicate after and before the change, respectively. This result holds for any temperature distribution within the column. An element originally at position x_b with temperature T_b ends up at position x_a with temperature T_a given by

$$T_a / T_b = (p_a / p_b)^{(k-1)/k} \tag{3}$$

Analogously, a column of length x in a heat exchanger has its length changed by dx = - dp/p as the pressure changes by dp. Integration yields

$$x_a / x_b = p_b / p_a \text{ (column in heat exchanger)} \tag{4}$$

Another basic relation that is needed is the change in length that a gas element suffers when it enters a heat exchanger. The ratio of the length dx_{pt} of an element just before it enters to its length dx_{he} just after is has entered equals the ratio of temperature T_{pt} of the element just before entering to the temperature T_{he} of the heat exchanger:

$$dx_{pt} / dx_{he} = T_{pt} / T_{he} \tag{5}$$

Gas elements leaving the heat exchanger retain their length while exiting.

Temperature profiles

The motion of the virtual piston during step 1 is indicated schematically in Fig. 3a). Use of the results given in the previous subsection directly yields the temperature profile and the length of the gas column that was expelled during this step from the cold heat exchanger. The results are shown in Fig. 3b). The non dimensional axial coordinate x varies from 0 at the entrance of the heat exchanger on the left till 1 at the entrance of the cold heat exchanger. The length L_c over which the piston moves follows from the equation expressing conservation of mass of the column expelled from the heat exchanger:

$$\frac{p_1}{RT_c} L_c = \int_{x_i}^{1} \frac{p_2}{RT} dx, \text{ yielding } L_c = (\pi_1 - 1) / k \tag{6}$$

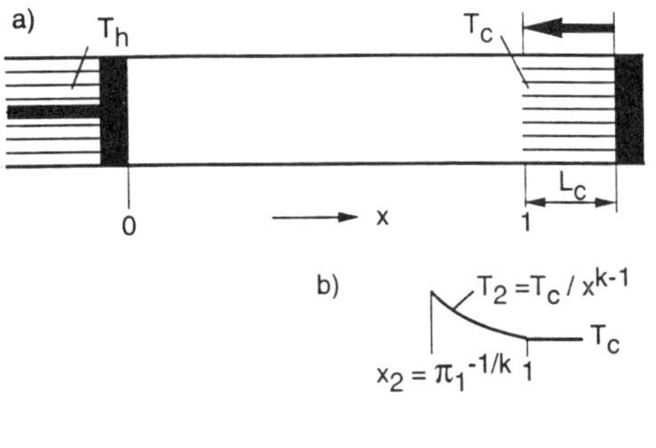

Figure 3. a) Motion of virtual piston during step 1; b) Temperature profile of column expelled in step 1.

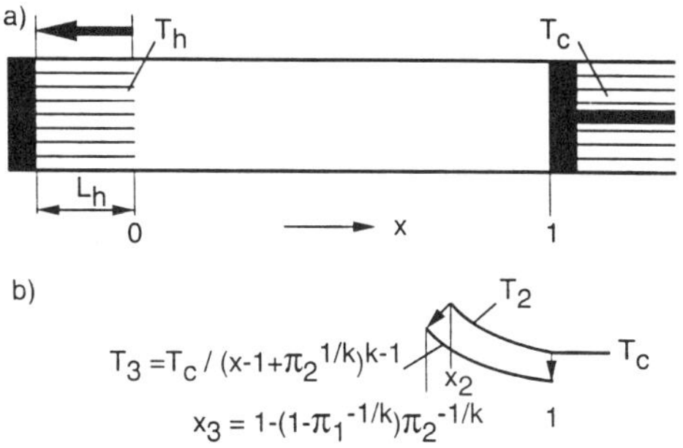

Figure 4. a) Motion of orifice piston during step 2; b) Temperature profile of column considered.

The motion of the orifice piston in step 2 is indicated schematically in Fig. 4a). In deriving results for this step, it is convenient to introduce a temporary axial coordinate $y = 1 - x$. It is found that the gas column considered in step 1 now attains the temperature profile and length shown in Fig. 4b). The distance L_h over which the orifice piston travels during this step depends on the temperature profile of the gas entering the heat exchanger on the left. This profile is unknown at this stage (it can be determined later, if desired). Therefore, the distance L_h is an unknown of the problem.

Fig. 5 indicates the motion of the virtual piston during step 3. The piston position is denoted by x_p, and use is made of the axial coordinate $z = x + kL_h$. After some algebra, the temperature profile during step 3 is found to be

$$T / T_c = \left[z + (p_3 / p)^{1/k}\left(\pi_2^{1/k} - z_h\right)\right]^{-k+1} \tag{7}$$

where $z_h = 1 + kL_h$. In order to find the pressure p_4 after this step in terms of the variables already defined, consider a change in pressure by dp (< 0). Let the distance over which the piston moves during this change be dx_p. The length of the gas element dx_{pt} entering the heat exchanger equals $- z_h\, dp/(k\, p)$. Its length after entering is $dx_p + x_p dp/p$. Using Eqs. (5) and (7), it follows that

$$d\left(p x_p\right) = -\left[z_h + (p_3 / p)^{1/k}\left(\pi_2^{1/k} - z_h\right)\right]^{k-1}(z_h / k)\,dp \tag{8}$$

Integrating this equation from $p = p_3$ to $p = p_4$ yields, after algebra,

$$z_h = \frac{1 - \left[1 - \pi_4\left(1 - 1/\pi_1\right)\right]^{1/k}}{1 - \left(\pi_2 \pi_4 / \pi_1\right)^{1/k}}\,\pi_2^{1/k} \tag{9}$$

where $\pi_4 = p_4/p_1$, and where use was made of Eq. (6) for the distance L_c over which the piston moves during this step. Eq. (9) relates the two unknowns π_4 (replacing p_4) and z_h

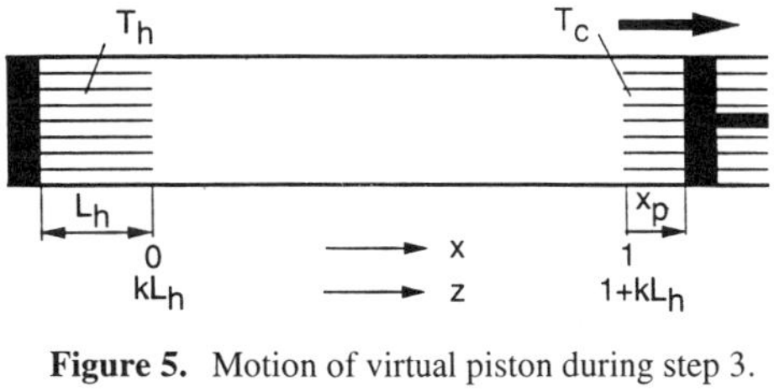

Figure 5. Motion of virtual piston during step 3.

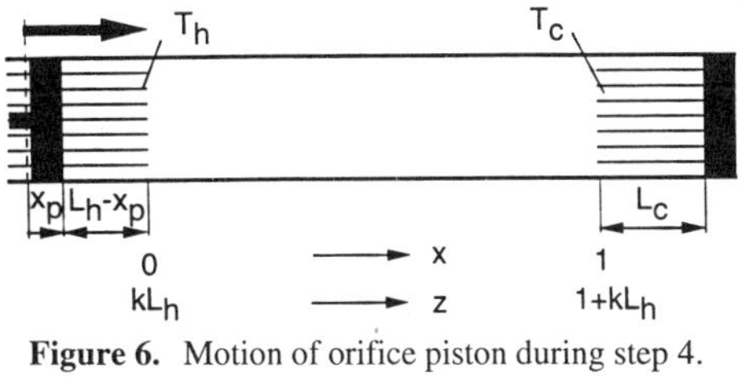

Figure 6. Motion of orifice piston during step 4.

(replacing L_h). The temperature of the gas entering the cold heat exchanger during step 3 is given by Eq. (7) with $z = z_h$.

The piston motion during step 4 is indicated in Fig. 6. The piston position is again denoted by x_p. The temperature profile $T(z, p_4)$ at the start of step 4 follows from Eq. (7) with $p = p_4$. The temperature of a gas element entering the cold heat exchanger at pressure p is given by

$$T(z_h, p) = T(z_4, p_4)(p / p_4)^{(k-1)/k} \tag{10}$$

where $z_4 \equiv x_4 + k\,L_h$. is the position and $T(z, p_4)$ the temperature of the element at the start of the step. The position z_4 follows from conservation of mass of the gas that has entered the cold heat exchanger when the pressure reaches the value p. The equation expressing this conservation is

$$\int_{x_4}^{1} \rho_4 dx = \int_{x_4}^{1} \frac{p_4}{RT_4} dx = \frac{p - p_4}{RT_c} L_c \tag{11}$$

where subscript 4 indicates quantities at the start of step 4. Integrating this equation yields

$$\left[z_4 + (p_3 / p_4)^{1/k}\left(\pi_2^{1/k} - z_h\right)\right]^k = \left[z_h + (p_3 / p_4)^{1/k}\left(\pi_2^{1/k} - z_h\right)\right]^k + (1 - p / p_4)kL_c \tag{12}$$

As the pressure increases from p to $p + dp$, the length of the gas element entering the heat exchanger is $dx_p - (L_h - x_p)\,dp/p - dp/(k\,p)$. Its length after entering is $L_c dp/p$. Application of Eq. (5) yields

$$d\left(p x_p\right) = \left[z_h + kL_c T(z_h, p) / T_c\right] / k \tag{13}$$

Integrating this equation from $p = p_4$ till $p = p_1$ and using Eqs. (12), (7) and (11) finally results in

$$F(\pi_4) \equiv 1 - z_h \pi_4 + (\pi_1 - 1)^{1/k} B\pi_4 \int_{1/B}^{1/(B\pi_4)} (-1 + 1/s)^{(k-1)/k} ds = 0 \tag{14}$$

where

$$B = 1 + \left[z_h + (\pi_1 / \pi_2 \pi_4)^{1/k}\left(\pi_2^{1/k} - z_h\right)\right]^k / (\pi_1 - 1) \tag{15}$$

Since z_h is given in terms of π_4 by Eq. (9), Eq. (14) is a single equation for the single unknown π_4. Given k, π_1 and π_2, this equation can be easily solved using a computer program such as MATLAB. The value of z_h then follows from Eq. (9).

Computation of heat removed and work done

The non-dimensional amount of heat Q_c removed per cycle follows from integrating over all elements dx entering the cold heat exchanger during steps 3 and 4:

$$Q_c p_1 = \int_{p=p_3}^{p_4} \rho c_p\left[T_c - T(z_h, p)\right] dx + \int_{p=p_4}^{p_1} \rho c_p\left[T_c - T(z_h, p)\right] dx \tag{16}$$

Carrying out the integrations results in

$$Q_c = \pi_1 (1 - z_h / \pi_2) / (k - 1) \tag{17}$$

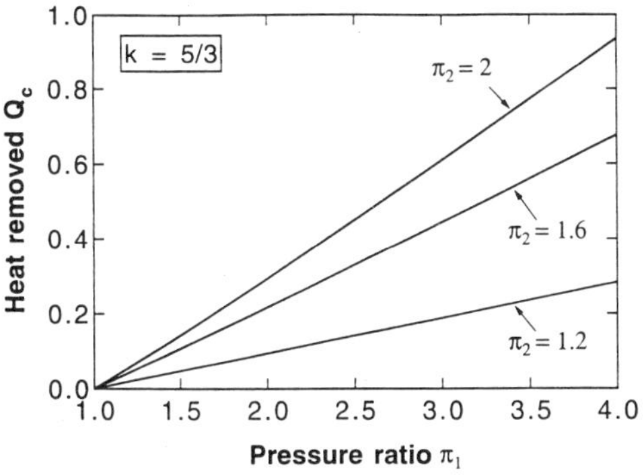

Figure 7. Heat removed as function of compression. ratio π_1, for various values of expansion ratio π_2.

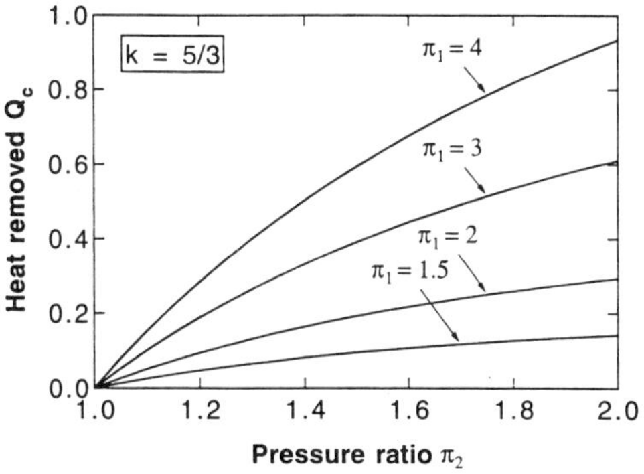

Figure 8 Heat removed as function of expansion ratio π_2, for various values of compression ratio π_1.

The non-dimensional work W_{p1} done by the virtual piston during step 1 is

$$W_{p1} = \int_0^{L_c} \frac{p}{p_1} dx_p = \int_0^{L_c} \frac{\pi_1}{\pi_1 - k x_p} dx_p = \frac{\pi_1 \ln \pi_1}{k} \tag{18}$$

In order to find the work W_{p3} (< 0) done by the virtual piston during step 3, the piston position x_p must be known as a function of pressure. The relation between x_p and p can be found by applying Eq. (5):

$$-L_h\, dp / p - dp / (k\, p) = \left(dx_p + x_p dp / p\right) T(z_h, p) / T_c \tag{19}$$

Use of Eq. (7) with $z = z_h$ and integrating yields

$$x_p = -\left[1 - (p_3 / p)^{1/k}\left(1 - \pi_2^{1/k} / z_h\right)\right]^k z_h^k / k + \pi_2 p_3 / (k\, p) \tag{20}$$

Using the differential of this equation leads to

$$W_{p3} = -\int_0^{L_c} \frac{p}{p_1} dx_p = \frac{\pi_1}{\pi_2}\left(z_h - \pi_2^{1/k}\right)^k \int_{\left(1-\pi_2^{1/k}/z_h\right)^{-1}}^{(\pi_2\pi_4/\pi_1)^{1/k}\left(1-\pi_2^{1/k}/z_h\right)^{-1}} \frac{(s-1)^{k-1}}{s} ds + \frac{\pi_1}{k} \ln \frac{\pi_2 \pi_4}{\pi_1} \tag{21}$$

This expression can again be easily evaluated by using a computer program such as MATLAB. The cyclic work done by the driving piston follows from Eq. (1) with $W_p = W_{p1} + W_{p3}$.

RESULTS AND DISCUSSION

As can be seen from Eq. (17), the cyclic heat Q_c depends on the compression ratio π_1 and the expansion ratio π_2 only, for given value of k. The independence of Q_c on the temperature ratio T_c/T_h is a result of the circumstance that the pressure p_1 has been used to non-dimensionalize Q_c. Keeping p_1 constant as T_c decreases causes the density of the gas in the cold heat exchanger to increase. The product of this density and the temperature difference $T_c - T(z_h, p)$ of the gas entering the cold heat exchanger thus remains constant, resulting in constant Q_c (cf. Eq. 16). Similarly, Eqs. (18) and (21) show that the cyclic work done by the virtual piston also depends on π_1 and π_2, only, for given value of k.

This simplifies presentation of the results obtained in graphical form. Figure 7 shows the results for the cyclic heat Q_c as a function of the compression ratio π_1, for $k = 5/3$ and various values of the expansion ratio π_2. Figure 8 is similar to Figure 7, with the roles of π_1 and π_2 interchanged. It is seen that Q_c increases nearly linearly with π_1, but less than linearly with π_2. Figure 9 shows Q_c for two values of k, as a function of π_2 and for $\pi_1 = 2$. The result for a diatomic gas ($k = 7/5$) in this case is more favorable than that for a monatomic gas ($k = 5/3$).

The cyclic work done by the driving piston equals the cyclic work done by the virtual piston times the temperature ratio T_c/T_h (see Eq. 1). As a result, the COP is proportional to T_c/T_h, and the product $COP*T_h/T_c = (Q_c/W_{pr})*T_h/T_c$ is independent of T_c/T_h. Figure 10 shows this product as a function of the expansion ratio π_2, for two values of k. It was found that the dependence of $COP*T_c/T_h$ on π_1 is very small (it decreases only a fraction of a percent as π_1 increases from 1.5 till 12, for given values of π_2 and k). This dependence is too small to be seen in Fig. 10. The result for a diatomic gas in this case again is more favorable than that for a monatomic gas. The maximum value of $COP*T_c/T_h$ is seen to be equal to one, and to occur at $\pi_2 = 1$. In this limiting case the orifice piston is stationary and no heat is being removed from the cold heat exchanger. The occurrence of maximum value of one can be understood by noting that in absence of the regenerator, heat would be added to the cold heat exchanger (see the discussion of this point in the Introduction). Considering the virtual piston/pulse tube/orifice piston as a separate assembly (Figures 3-6), it follows that part of the cyclic work W_p done by the virtual piston is transferred into heat added to the cold heat exchanger. The work W_p therefore is larger than Q_h, which equals Q_c (see Figure 2). As a result, the value of Q_c/W_p is less than one in the general case. In the limiting case with stationary orifice piston the assembly operates reversibly. No heat then is transferred to the cold heat exchanger, resulting in $Q_c/W_p = 1$. Since $COP = (Q_c/W_p)*(T_c/T_h)$, the COP is always less than one in the general case $\pi_2 > 1$, $T_c/T_h < 1$. reaches a maximum value of less than one for $T_c/T_h = 1$. This is a limitation to the performance of OPTR's. The limitation derived here is slightly more severe than the result $COP = T_c/T_h$ derived by Kittel.[15]

It is of interest to compare the OPTR performance characteristics derived here with those of the BPTR considered in Ref. 4. The two devices have a number of independent parameters in common. These are the compression ratio, (π_1 in the case of the OPTR and π in the case of the BPTR), the temperature ratio T_c/T_h, and the ratio of specific heats k, However, the remaining independent parameter is not the same for the two devices. For the BPTR this is the non-dimensional length L_h of the heat exchanger at the closed end,[4] while for the OPTR it is the expansion ratio π_2. In the latter case, there also occurs a parameter L_h, representing the stroke of the orifice piston (see Fig. 4). It seems reasonable

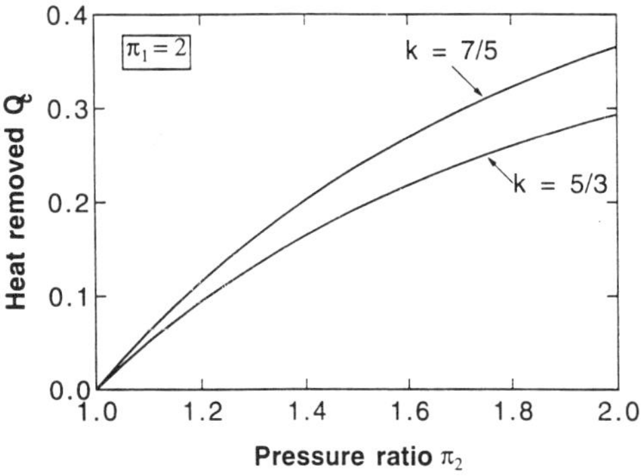

Figure 9 Heat removed as function of expansion ratio π_2, for two values of k.

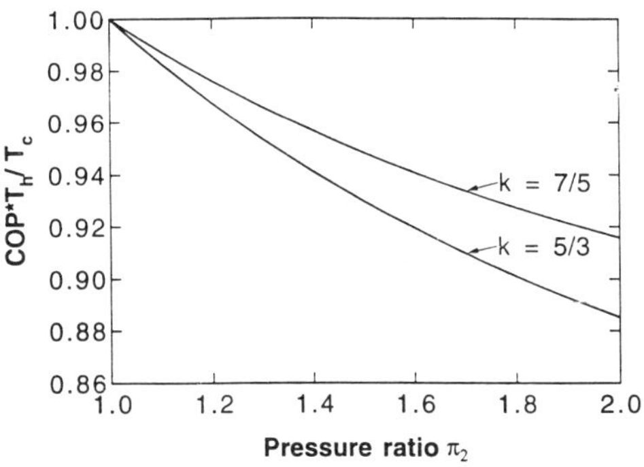

Figure 10. COP times temperature ratio as function of expansion ratio π_2, for two values of k.

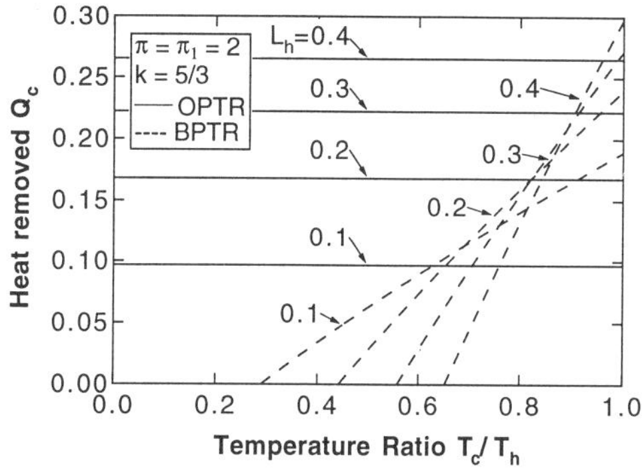

Figure 11. Comparison between OPTR and BPTR of heat removed as function of temperature ratio.

Table 1. Values of π_2 and some other parameters as function of L_h, for $\pi_1 = 2$

L_h	π_2	π_4	Q_c	COP^*T_h/T_c
0.1	1.2055	0.9220	0.0966	0.9664
0.2	1.4124	0.8536	0.1679	0.9396
0.3	1.6201	0.7935	0.2223	0.9176
0.4	1.8282	0.7405	0.2651	0.8993

to make the comparison in terms of identical values of L_h. The values chosen for L_h are shown in Table 1, together with corresponding values of π_2 and some other OPTR parameters. Results for Q_c with $\pi = \pi_1 = 2$ and $k = 5/3$ and various values of L_h are shown in Fig. 11. The BPTR has a minimum temperature below which it no longer functions as a refrigerator. As mentioned previously, the OPTR will maintain its refrigeration capability even at very low temperature ratios T_c/T_h. It is clear from Fig. 11 that the OPTR provides greater cooling power over almost the entire range of T_c/T_h. The only exceptions occur at values of T_c/T_h near unity. Fig. 12 shows the COP as function of the temperature ratio T_c/T_h, for the cases considered. The BPTR results in this case are independent of π,[4] while the OPTR results are essentially independent of π_1 and only slightly dependent on L_h and hence on π_2 (cf. Table I). Also shown in this figure is the Carnot result $1/(-1+T_h/T_c)$, applicable to the ideal case of a reversible refrigerator. At low values of T_c/T_h, the COP of the OPTR is quite close to the Carnot curve. However, since its value of COP cannot exceed unity, the two curves differ widely for larger values of T_c/T_h. The COP of the BPTR is seen to be significantly higher than that of the OPTR over the full range of BPTR operation.

The refrigeration efficiencies of the OPTR and BPTR are shown in Fig. 13. The BPTR results again are independent of the compression ratio π, while the OPTR results are essentially independent of π_1 and only slightly dependent on π_2. Both sets of results decrease linearly with T_c/T_h. In accordance with the result shown in Fig. 12, the efficiency of the BPTR is considerably higher than that of the OPTR over the entire range of BPTR operation.

Finally, it should be noted that the results presented are obtained within the context of the usual thermodynamic idealizations. While the results may be expected to predict

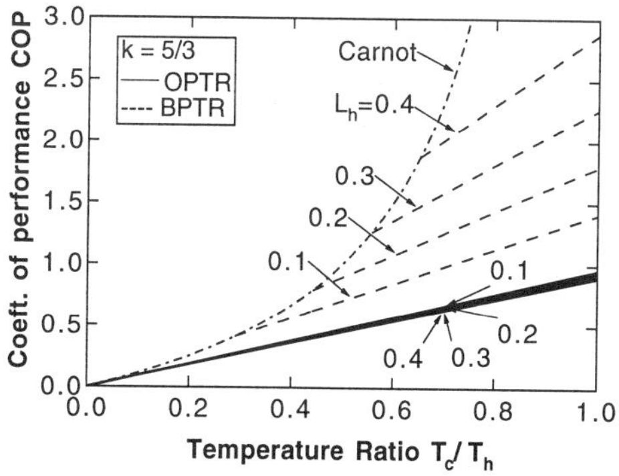

Figure 12. Comparison between OPTR and BPTR of COP as function of temperature ratio.

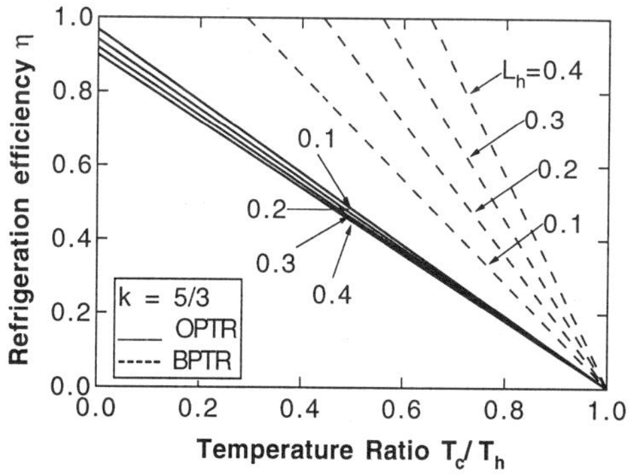

Figure 13. Refrigeration efficiency of OPTR and BPTR as function of temperature ratio.

various important trends correctly, quantitatively accurate predictions will require that the various loss mechanisms be taken into account. Inclusion of the loss mechanisms in the model described is considered outside the scope of this paper.

REFERENCES

1. W.E. Gifford and R.C. Longsworth, Pulse-tube refrigeration, *Trans. ASME, J. Eng. Ind.* 86 Series B:264 (1964).
2. W.E. Gifford and R.C. Longsworth, Surface heat pumping, in: "Advances in Cryogenic Engineering," Vol. 11, Plenum Press, NY (1966), p.171.
3. P.C.T. de Boer, Thermodynamic analysis of the basic pulse-tube cryocooler, *Cryogenics* 34:699 (1994).
4. P.C.T. de Boer, Analysis of the basic pulse-tube refrigerator with regenerator, in AES-Vol. 33, "Thermodynamics and the Design, Analysis and Improvement of Energy Systems," R.J. Krane, ed., Am. Soc. Mech. Eng. (1994), p.277. Accepted for publication in Cryogenics.
5. E.I. Mikulin, A.A. Tarasov, and M.P. Shkrebyonock, Low temperature expansion pulse tubes, in: "Advances in cryogenic engineering", Vol. 29, Plenum Press, New York (1984), p. 629.
6. R. Radebaugh, J. Zimmerman, D.R. Smith, and B.Louie, A comparison of three types of pulse tube refrigerators: New methods for reaching 60 K, in: "Advances in Cryogenic Engineering," Vol. 31, Plenum Press, NY(1986), p. 779.
7. P.J. Storch, R.Radebaugh, and J.E. Zimmerman, Analytical model for the refrigeration power of the orifice pulse tube refrigerator, NIST Technical Note 1343 (1990).
8. Radebaugh, R., Zimmerman, J., Smith, D.R., and Louie, B., A comparison of three types of pulse tube refrigerators: New methods for reaching 60 K, in "Advances in Cryogenic Engineering," Vol. 31, Plenum Press, NY, (1986), p. 779.
9. Radebaugh, R. and Hermann, S., "Refrigeration efficiency of pulse tube refrigerators," Proceedings of the 4th International Cryocooler Conference, Easton, MD. Edited by G. Green, G. Patton and M. Knox, David Taylor Naval Ship and Development Center, Annapolis, MD 21402-5067. NTIS accession number AD-A211 760/4/xAB, pp. 119-133 (1986).
10. Hermann, S. and Radebaugh, R., "Measurements of the efficiency and refrigeration power of pulse-tube refrigerators," NBS Technical Note 1301. See also Radebaugh, R. and Herrmann, S., "Refrigeration efficiency of pulse-tube refrigerators," Proceedings of the 4th International Cryocooler Conference, Boulder, CO (1986).
11. R.N. Richardson, Pulse tube refrigerator - an alternative cryocooler?, Cryogenics 26:331 (1986).
12. Radebaugh, R., Chowdhury, K., and Zimmerman, J., "Optimization of a pulse tube refrigerator for a fixed compressor swept volume," Proceedings of the 5th International Cryocooler Conference, Naval Postgraduate School, Monterey, CA (1988).
13. M. David and J.-C. Maréchal, An experimental investigation of pulse tube refrigeration heat transfer rates, in: "Advances in Cryogenic Engineering," Vol. 35, Plenum Press, NY (1990), p. 1231.
14. M.J.A. Baks, A. Hirschberg, B.J. van der Ceelen and H.M. Gijsman, Experimental verification of an analytical model for orifice pulse tube refrigeration, *Cryogenics* 30:948 (1990).
15. P. Kittel, Ideal orifice pulse tube refrigerator performance, *Cryogenics* 32:843 (1992).
16. S.F. Kral, D. Hill, J. Restivo and J. Johnson, Test results of an orifice pulse tube refrigerator, in: "Advances in Cryogenic Engineering," Vol. 37, part B, Plenum Press, NY (1992), p. 931.
17. M. David, J.-C. Maréchal and P. Encrenaz, Measurements of instantaneous gas velocity and temperature in a pulse tube refrigerator, in "Advances in Cryogenic Engineering," Vol. 37, part B, Plenum Press, NY (1992), p. 939.
18. M. David, J.-C. Maréchal, Y. Simon and C. Gulpin, Theory of ideal orifice pulse tube refrigerator, Cryogenics 33:154 (1993).
19. W. Rawlins, R. Radebaugh, P.E. Bradley and K.D. Timmerhaus, Energy flows in an orifice pulse tube refrigerator, in "Advances in Cryogenic Engineering," Vol. 39, Plenum Press, NY (1994), p. 1449.
20. H. Mirels, Linearized theory for pulse tube cryocooler performance, *AIAA J.* 32:1662 (1994).
21. Y. Ohtani, H. Hatakeyama, T. Kuriyama, H. Nakagome, Y. Matsubara, H. Okuda and H. Marakami, High efficient pulse tube refrigerator with linear-motor drive compressor, in: "Advances in Cryogenic Engineering," Vol. 39, Plenum Press, NY (1994), p. 1441.

A BLIND TEST ON THE PULSE TUBE REFRIGERATOR MODEL (PTRM)

S.W.K. Yuan[1] and Ray Radebaugh[2]

1 Research and Development Division
Lockheed Martin Missiles and Space Co.
Palo Alto, CA 94304, U.S.A.

2 Thermophysics Division
National Institute of Standards and Technology
Boulder, CO 80303, U.S.A.

ABSTRACT

The Stirling Refrigerator Performance Model (SRPM) has been validated extensively against the Lockheed built Stirling Coolers and various units in the literature. This model has been modified to predict the performance of the Pulse Tube Coolers (PTCs). It was successfully validated against a Lockheed in-house-built PTC. The results are to be published elsewhere[1]. In this paper, the validation of PTRM against a NIST (National Institute of Standards and Technology) orifice pulse tube cooler is reported. Dimensions and operating condition of the PTC were obtained from NIST without prior knowledge of the performance. In other words, this is a 'blind test' on the PTRM with the help of the National Institute of Standards and Technology. Good correlation was found between the test data and the prediction. PTRM is a generic model that gives accurate performance prediction of the pulse tube coolers.

INTRODUCTION

The Pulse Tube Performance Model (PTRM) was modified from the Stirling Refrigerator Performance Model (SRPM) which is a third order model that has been validated extensively against various Stirling coolers in the literature. They include the Lucas-Lockheed 60K unit[2], the NASA/Philips Magnetic Bearing unit[3], the Oxford refrigerators[4], and the Astronomic Infrared Sounders (AIRS) units A, B, and C. A detailed description of the model can be found in Reference 5. In this report, the PTRM model is validated against a large laboratory PTC built by the National Institute of Standards and Technology.

THEORY

The equations and assumptions used in the PTRM model were discussed elsewhere[5]. The model breaks up the pulse tube cooler into a number of nodes. The number of nodes in each section depends on the value of the state variables.

For examples, more nodes are required in the regenerator because of the large temperature difference and large pressure drop in the axial direction. Conservation of energy, momentum and mass are solved until the solutions converge. Equation of states and empirical equations for pressure drop and heat transfer are also used. No fudge factors are used in the program.

Figure 1 is a schematic diagram of the nodal network used in the Pulse Tube Refrigerator Model (PTRM).

This model was modified from the Stirling Refrigerator Performance Model. The expansion space of the SRPM was replaced by the pulse tube with an orifice and the surge volume. The volumetric variation and the flow passage (to the displacer motor) at the hot end of the regenerator (of the Stirling model) were also eliminated. The gas transport in the pulse tube is modeled as unidirectional laminar or turbulent flow, depending on the Reynolds number. Heat transfer in the axial direction is modeled as enthalpy flow whereas the radial heat transport is predicted by the forced flow heat transfer coefficient (for both laminar and turbulent regimes). The transport across the orifice is modeled by the discharge flow coefficient.

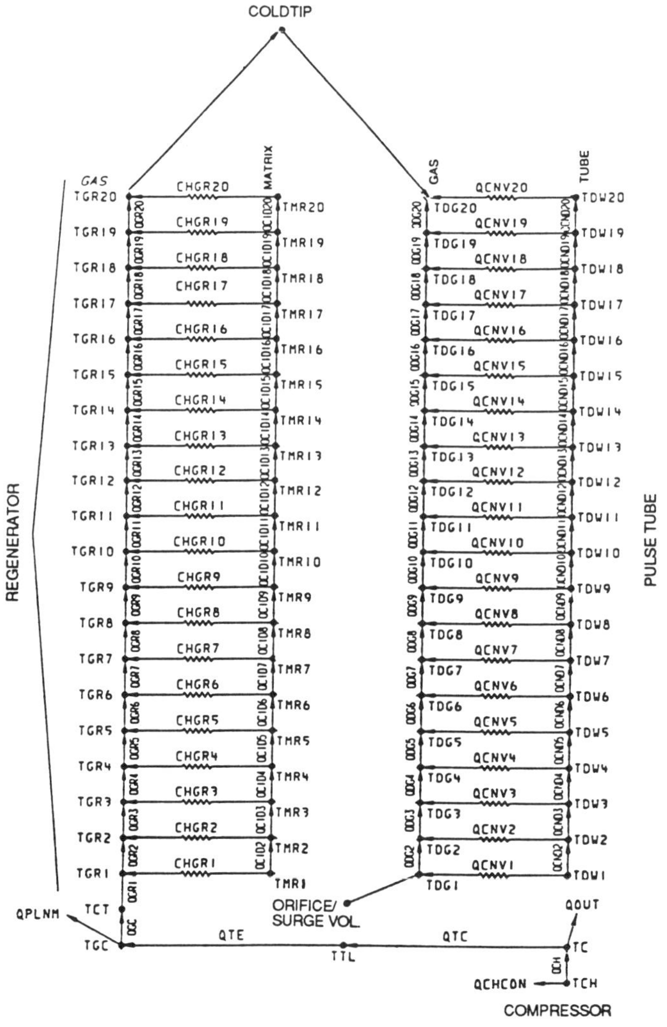

Figure 1. Schematic of the nodal network diagram.

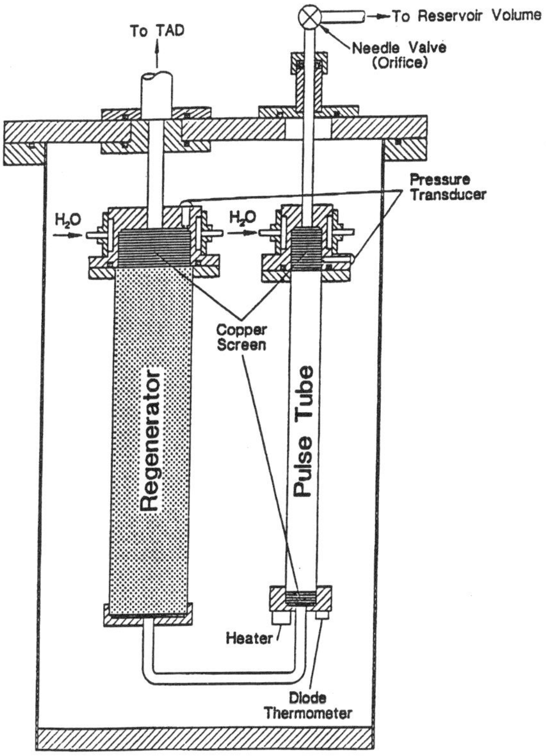

Figure 2. Cross-section NIST large pulse tube cooler.

EXPERIMENTAL SETUP

The pulse tube refrigerator was originally designed to be used with a thermoacoustic driver (TAD) that operated at a frequency of 30 Hz, a pressure ratio of 1.1, and an average pressure of 3.0 MPa. After some measurements in 1990, NIST changed the size of only the pulse tube and operated the pulse tube refrigerator with a mechanical compressor. The system achieved an efficiency of about 15% of Carnot when referred to input PV power and is one of the most efficient pulse tube refrigerator ever measured, in spite of the fact that it was not optimized for the conditions that it experienced with the mechanical driver. The measurements on this pulse tube refrigerator were made on January 1991.

Geometry

Figure 2 shows a cross-section of the pulse tube refrigerator. Not all of the dimensions are to scale on this drawing. The compressor and the connecting line are not shown in this figure. They would be connected in place of the TAD. A commercial dry-lubricated compressor was used. The top head with inlet and outlet valves was removed and replaced with a steel plate with a threaded hole for the output connection. Only one of the two cylinders was used to generate the oscillating pressure. This single piston/cylinder has a swept volume of 347.5 cm^3. At the outlet of the compressor was an aftercooler that consisted of seven copper tubes in parallel which were surrounded by water. Each of these copper tubes had an inside diameter of about 0.125 inch and a length of about 22 inches. At the exit of the aftercooler the working helium gas passed through 18 inches of 0.25 in. ID copper tubing before entering the 2 inch length of 0.285 inch ID stainless steel tube just inside the vacuum can. This stainless steel tube then connects on to the second aftercooler as shown in Fig. 2.

Table 1. Dimensions of three heat exchangers

	Aftercooler	Cold HX	Warm HX
Diameter (mm)	39.5	25.4	19.1
Length (mm)	19.0	8.4	19.1
Mesh (copper)	#80	#80	#80
Wire Diameter (mm)	0.140	0.140	0.140
Porosity	0.65	0.65	0.65
Gas volume (cc)	15.1	2.77	3.56

The dimensions of this aftercooler, the cold heat exchanger, and the pulse tube warm heat exchanger are given in Table 1. The dimensions of the regenerator and the pulse tube are given in Table 2.

The connecting tube between the regenerator and the pulse tube is stainless steel with a length of about 6 inches and an inside diameter of 0.230 in. The connecting tube between the pulse tube warm end heat exchanger and the needle valve is 4 in long and 0.23 in ID. The 11-turn needle valve has an orifice diameter of 3.18 mm. It is a Nupro needle valve, model #SS-4LA, with a full open C_v of 0.15. A few measurements were made with the secondary valve partially open, but the improvement in performance was barely noticeable. The reservoir volume is 2700 cc.

Operating Parameters

Table 3 lists the operating parameters and the performance of the pulse tube refrigerator when the second orifice was closed. The orifice settings used for these data are approximately equal to the optimum values.

RESULTS AND DISCUSSION

The prediction on the NIST PTC was performed without prior knowledge of the performance of the pulse tube. Inputs to the model were supplied by NIST on May 10, 1994. The prediction was completed on July 14, 1994 and the experimental performance was obtained from NIST and compared with the analysis on July 19, 1994.

Table 2. Dimensions of regenerator and pulse tube

	Regenerator	Pulse Tube
Outside diameter (mm)	41.28	25.4
Wall thickness (mm)	0.89	0.51
Length (mm)	208	208
Mesh	#200 stainless	NA
Wire Diameter (mm)	0.0533	NA
Porosity	0.69	NA
Gas volume	175.8 cc	97.1 cc

Table 3. Operating parameters and performance of pulse tube refrigerator

T_c (K)	37.5	43.0	58.9	78.4
T_h (K)	298.9	304.9	310.2	315.8
P Average (MPa)	1.739	2.551	2.504	2.488
P Ratio (pulse tube)	1.255	1.301	1.334	1.384
P Ratio (regenerator)	1.337	1.370	1.394	1.453
P Ratio (compressor)	1.404	1.424	1.445	1.491
Frequency (Hz)	4.6	4.5	4.5	4.5
Orifice turns	7.0	7.0	7.0	7.0
PV Work (W)	454	653	642	619
Net Cooling (W)	0	0	14.5	29.9

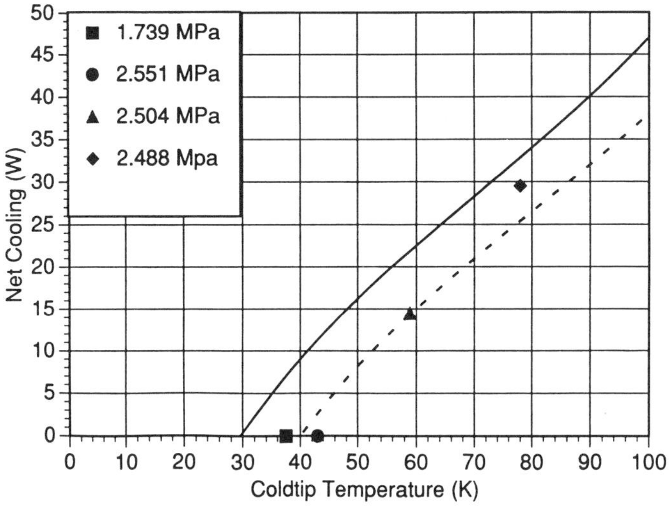

Figure 3. Net cooling versus coldtip temperature.

Figure 3 is a plot of the net cooling versus the coldtip temperature at two different fill pressures. (In both Figures 3 and 4, the solid line represents prediction at 1.739 Mpa average pressure; the dashed line represents prediction at 2.551 MPa). Excellent agreement is found between the prediction and the experimental data. Figure 4 shows the specific power (net refrigeration / PV power) of the unit versus the coldtip temperature. The agreement at low temperature is remarkable. The experimental data at 80K tends to give better performance than the prediction. The discrepancy is mainly due to the difference in the compressor PV work between the experimental data and the prediction. The predicted PV work is about 50% higher. This might have been caused by the uncertainty in knowing where the PV work was measured experimentally, and the unaccounted dead volumes in modeling the pulse tube system.

It should also be mentioned that the experimental data being compared here were taken with a fixed orifice setting (of 7 turns). As for the PTRM prediction, better net cooling was found with a large valve opening at high

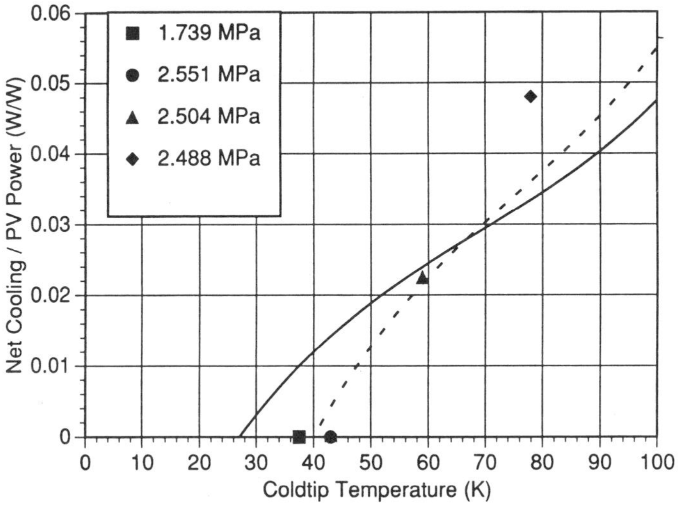

Figure 4. Specific power versus coldtip temperature.

temperatures, and a small valve opening at low temperatures. Unfortunately, at large orifice openings, the PV work is also much larger. This might also have contributed to the discrepancy between the experimental data and the prediction. Since the predicted net cooling at 80K is higher than the experimental data, one might be able to run the model at a smaller orifice opening to reduce the PV work and get even better correlation. In other words, running PTRM at a small orifice opening (within the limit) results in better specific power performance, but the net cooling is worse accordingly.

CONCLUSIONS

The Pulse Tube Refrigerator Model (PTRM) was validated against a NIST pulse tube cooler without prior knowledge of the performance. Good agreement was found between the prediction and the experimental data. The PTRM was proven to be a generic model that is invaluable for the analysis and design of pulse tube coolers.

REFERENCES

1) S.W.K. Yuan, Validation of the Pulse Tube Refrigerator Model Against a Lockheed Pulse Tube Cooler, to be presented at the Space Cryogenic Workshop, Maryland, July 1995.

2) S.W.K. Yuan, I.E. Spradley, P.M. Yang and T.C. Nast, Computer Simulation Model for Lucas Stirling Refrigerators, Cryogenics 32:143 (1992)

3) S.W.K. Yuan and I.E. Spradley, Validation of the Stirling Refrigerator Performance Model Against the Philips / NASA Magnetic Bearing Refrigerator, Proc. 7th Int. Cryocooler Conf., Vol. 1, Philips Lab, USA, p 280 (1993).

4) S.W.K. Yuan and I.E. Spradley, Validation of the Stirling Refrigerator Performance Model Against the Oxford Refrigerators, Adv. Cryo. Eng, 39B:1359 (1994).

5) S.W.K. Yuan and I.E. Spradley, A Third Order Computer Model for Stirling Refrigerators, Adv. Cryo. Eng, 37B:1055 (1992).

NUMERICAL ANALYSIS AND EXPERIMENTAL VERIFICATION OF MULTI-BYPASS PULSE TUBE REFRIGERATORS

Chao Wang, Jinghuei Cai, Yuan Zhou, and Suqing Wang

Cryogenic Laboratory, Chinese Academy of Sciences
P.O.Box 2711, Beijing 100080, China

ABSTRACT

Numerical analysis and experiment were performed to study the effects of the middle-bypass valve on the performance of multi-bypass pulse tube refrigerators (MPTR). Numerical predictions reveal the energy transportation and two refrigeration processes occurring in MPTR. The compared results verified that the multi-bypass version improved the performance of the refrigerator in a low temperature range. A lowest temperature of 23.8K was obtained with the MPTR of one middle-bypass tube. A MPTR of two middle-bypass tubes was first constructed and reached a temperature of 24K.

INTRODUCTION

The pulse tube refrigerator contains no moving parts at its cold head, so it has considerable system advantages over most other types of cryogenic refrigerators in reliability, lifetime, vibration, and cost. However, the initial interest in it weakened due to its poor refrigeration capacity. Pulse tube refrigerator has been developed rapidly since an important innovation called orifice pulse tube was introduced by Mikulin et al. in 1984.

The development and innovations achieved in recent years enable pulse tube refrigerator to have a comparable refrigeration capability with Stirling or G-M machines [1,2,3]. More and more laboratories around the world are interested in pulse tube refrigerator because of its promising application potentials.

The multi-bypass pulse tube refrigerator was first suggested by a research group headed by professor Zhou in Cryogenic Laboratory, Chinese Academy of Sciences. Its configuration is shown in Fig.1. A bypass tube and an orifice connect the middle of the pulse tube and the regenerator. There is mass flow in or out the pulse tube owing to the pressure drop in the regenerator. The mass flow rate is controlled by the valve 11. The improvement of the middle-bypass to the pulse tube refrigeration has been initially verified in reference [1]. However, there has little theoretical and experimental analysis published concerning this version up to now.

Our objects in the present study are to investigate the refrigerative mechanism of multi-bypass pulse tube refrigerator by numerical simulation and experimental test, and to improve the performance of the refrigerator.

Advances in Cryogenic Engineering, Vol. 41
Edited by P. Kittel, Plenum Press, New York, 1996

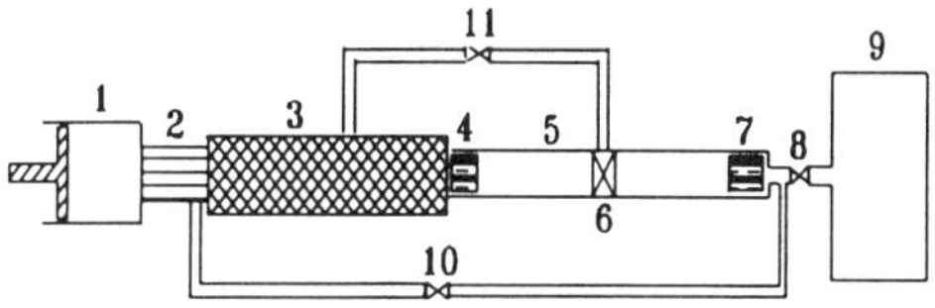

Fig.1. Schematic of a multi-bypass pulse tube refrigerator.
1.compressor; 2.aftercooler; 3.regenerator; 4.cold head; 5.pulse tube;
6.middle heat exchanger; 7.hot heat exchanger; 8.orifice; 9.reservoir;
10.double-inlet valve; 11. middle-bypass valve.

NUMERICAL ANALYSIS

Numerical model of a multi-bypass pulse tube refrigerator has been developed and the equations of continuity,momentum and energy (for both gas and solid) are solved numerically. The mass flow rate through the bypass valve 10 or 11 is determined by the pressure difference between either side of it. The part 6 in the Fig.1 is assumed as a heat exchanger of constant temperature. The basic assumptions and the numerical method are very similar to that suggested in references [4,5]. In the present study, if the part 6 is eliminated and the bypass valve 11 closed, the model is a conventional orifice version. The structure parameters of the refrigerator for typical calculations are given in the descriptions of the experimental apparatus. The working fluid is helium.

The calculated energy transportation in the refrigerator is shown in the Fig.2. Where Hr is the enthalpy flow in the regenerator, Hp the enthalpy flow through the middle-bypass valve, Qc the heat flow in the cold head, Hc the enthalpy flow in the pulse tube, Hd the enthalpy flow through the double-inlet valve, etc. There is the enthalpy flow Hp from the regenerator to the pulse tube at the middle-bypass part. That is mean that refrigerative capacity is generated at this point. Therefore, the part 6 filled with screens as a heat exchanger in Fig.1 is necessary for this configuration (experimental results verified it in Fig.7). There is a Hd from the hot end to the aftercooler. The double-inlet tube acted as a expander in the refrigeration process. The gas temperature variations (see Fig.3) show that two expansion processes occur in the pulse tube. The heat exchanger 6 is cooled by the expansion gas in the right part of the pulse tube and heated by the gas in the left part.

Obviously, it can be concluded that two refrigeration processes occur in the multi-bypass pulse tube refrigerator. However, the refrigerative processes of the MPTR have distinctions with the two-stage pulse tube refrigerator described in reference [6] since the expansion process at the middle-bypass part is restricted by the valve 11. So, the performance comparisons of the MPTR with a two-stage pulse tube refrigerator are carrying on in our laboratory.

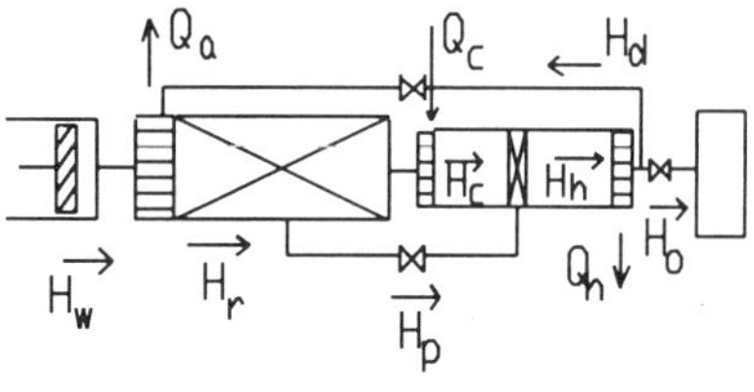

Fig.2. Energy flow in a Multi-bypass pulse tube refrigerator.

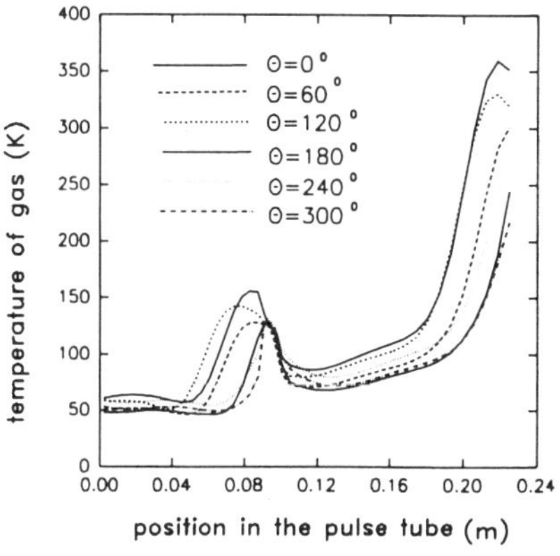

Fig.3. Gas temperature variations in the pulse tube.

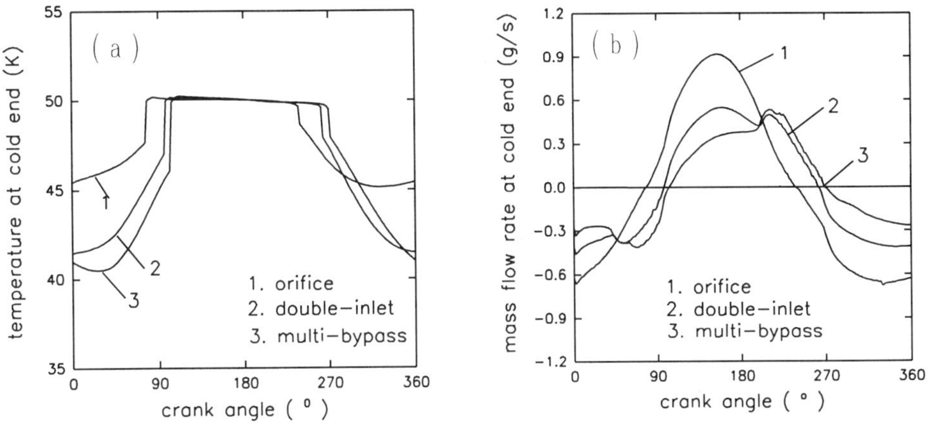

Fig.4. Variations of the temperature and the mass flow rate at cold end in a cycle.

The transient gas temperatures and the mass flow rates of three kinds of the refrigerator at the cold end of the pulse tube were compared in Fig.4 (a) and (b). The gas temperature of the MPTR was lower and the mass flow rate smaller than the others. It is the main reason why the multi-bypass configuration could reach a lower refrigerative temperature (see Fig.7). The effects of the valve 11 on the amplitude and the phase of the gas parameters are very similar to that of the valve 10.

The predicted pressure oscillations in the compressor and the pulse tube were compared with the experimental datum as shown in Fig.5. The results from the analysis and the experiment are in good agreement.

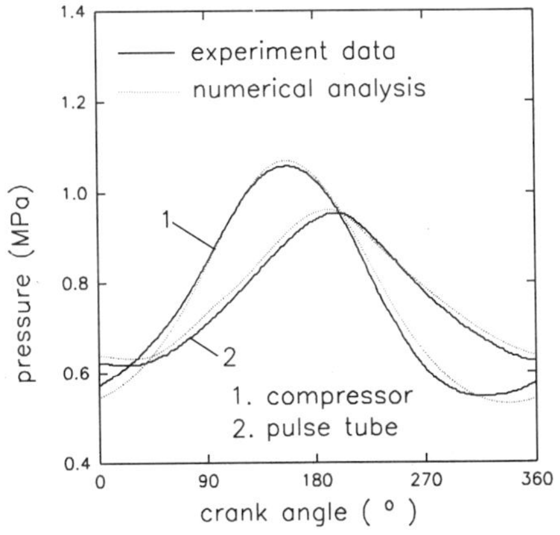

Fig.5. Comparison of numerical predictions with experiment data.

EXPERIMENTAL APPARATUS

The configuration of the experimental apparatus is shown in Fig.6. The refrigerator consists of a valveless compressor, regenerator, middle-bypass valve, vacuum chamber, reservoir, orifice, double-inlet valve, pulse tube and cold head. The main structural parameters of the pulse tube refrigerator are as follows: the swept volume of the compressor is 55.6 cm^3; the reservoir volume is 300 cm^3. The pulse tube and the regenerator are separated into two parts respectively. The pulse tubes were 11.5mm i.d. x 120mm long and 9.5mm i.d. x 90mm long stainless steel tubes. The regenerators were 17mm i.d. and 14mm i.d. stainless steel tubes and were packed with phosphor bronze screens and stainless steel screens.

Two pressure transducers measured the transient gas pressures at two points: before the regenerator (after the compressor) and after the regenerator (at the hot end of the pulse tube). The cold head temperature was measured by a gold(+0.07At%Fe)-Chromel thermal couple. The wall temperatures of the pulse tube and the regenerator were measured by ten pairs of copper-constantan thermal couple. All thermal couples were calibrated in the center for cryogenic metrology in our laboratory.

In the experiment, the test rig was an orifice version(OPTR) or a double-inlet version(DPTR) if there are no middle-bypass and screens in the pulse tube. The test of a multi-bypass version is in the situation of the three valves 2,6 and 7 opened.

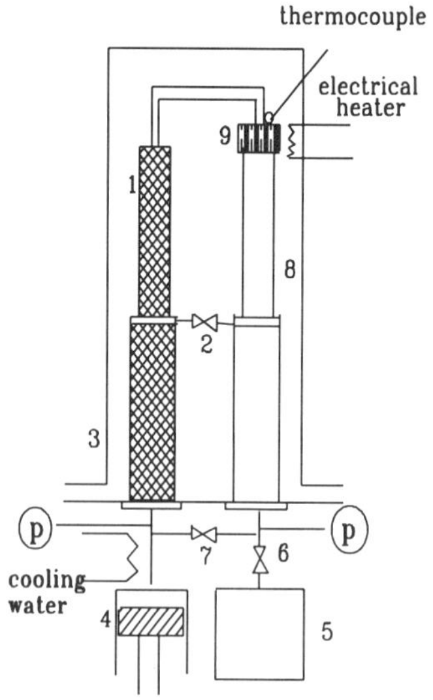

Fig.6. Experimental apparatus, 1.regenerator; 2.middle-bypass valve;3.vacuum chamber;3.vacuum chamber;4.compressor; 5.reservoir;6.orifice;7.double-inlet valve; 8.pulse tube;9.cold head .

EXPERIMENTAL RESULTS AND DISCUSSION

1) Comparing the performances of the multi-bypass version with the double-inlet version

At first, it is important for using the multi-bypass version that whether it is necessary to add the metal screens as a gas flow straightener or a heat exchanger in the connecting part in the pulse tube. The comparing results of adding screens and no screen were shown in Fig.7. The open of the middle-bypass valve led to lightly higher temperature of the cold head for the no screen case. The present authors reasoned that this negative effect is due to gas disturbance and elimination of a heat exchanger in the pulse tube. There is a large temperature difference between the regenerator and the pulse tube (about 50-80°C, see Fig.8). For the case of adding screens in the pulse tube, the temperature was lowered with the open of the middle-bypass valve. The refrigerative temperature with the MPTR is 10K lower than that with the DPTR. It was noteworthy that adding screens in the DPTR had negative effects on the performance and led to a higher refrigerative temperature.

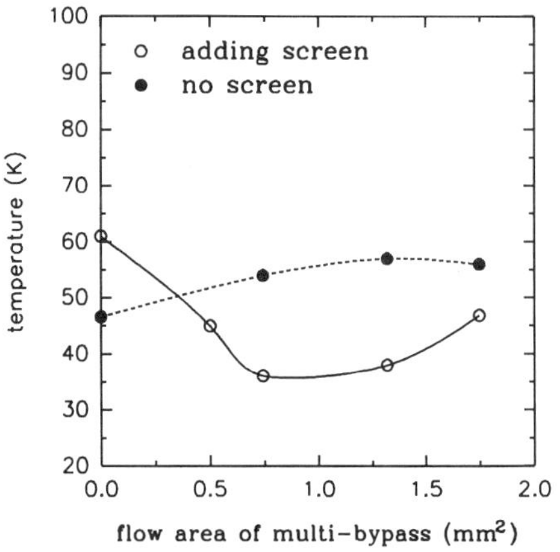

Fig.7. Comparisons of the cold head temperature.

The temperature distributions on the walls of the pulse tube and the regenerator in the DPTR are shown in Fig.8 and that in MPTR shown in Fig.9. All wall temperatures were recorded at the optimum values of the valves. The temperature at the connecting point on the pulse tube increased and that on the regenerator decreased with the open of the middle-bypass valve. It is caused by a amount of enthalpy flowing into the pulse tube and the middle-bypass part becoming a second refrigerative part in the MPTR. It agrees with the numerical results.

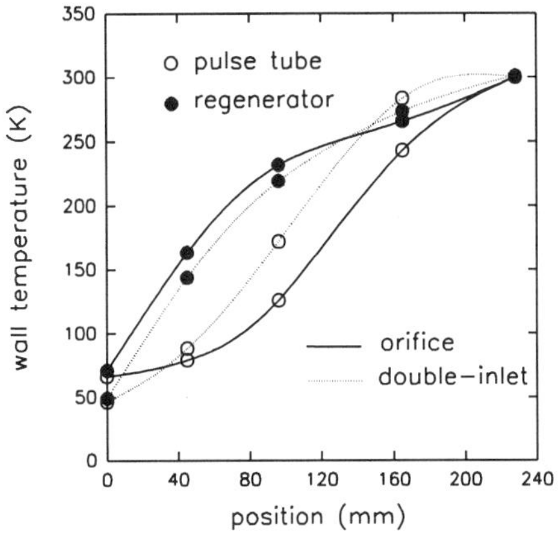

Fig.8. Wall temperature distributions in DPTR .

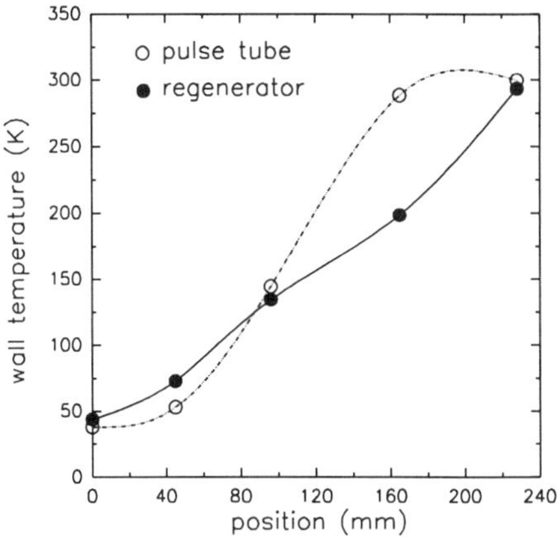

Fig.9. Wall temperature distributions in MPTR.

2) Performance of the refrigerator

By further optimizing the operating and the structure parameters of the refrigerator, a lowest temperature of 23.8K was obtained in the present experiment with system pressure of 1.05MPa and at a frequency of 13Hz. Fig.10 shows the variation of the cold head temperature with operating frequency.

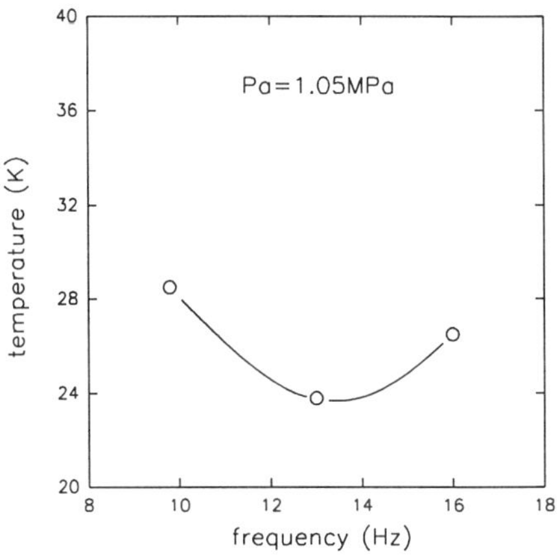

Fig.10. Temperature vs frequency.

Considering whether two or more middle-bypass tubes can further lower the refrigerative temperature, a MPTR of two middle-bypass tubes was constructed and tested. Fig.11 is a photograph of it. The optimum values of the two middle-bypass valves were both 0.74mm^2. The refrigeration capacities of the MPTR of two middle-bypass tubes (see Fig.11, No.2 MPTR), the MPTR of one middle-bypass tube (No.1 MPTR) and the DPTR (see Fig.6) were compared in Fig.12. The multi-bypass version improves the performance of the pulse tube refrigerator in a very low temperature range, but it failed in a high refrigerative temperature range.

A lowest temperature of 24K was reached with the No. 2 MPTR. However, the refrigeration power of it is lower than that of the No.1 MPTR due to the less mass flow rate through the cold head. A cooling capacity of the No.1 MPTR at 30K is 0.49W and at 77K is 3.6W. That of the No.2 MPTR at 30K is 0.41W and at 77K is 2.8W.

The pulse tube refrigerator of two middle-bypass tubes operated successfully in the present study. It is expected by the authors that a lower temperature could be reached with the No.2 MPTR or the MPTR of more middle-bypass tubes by optimizing its structure parameters.

CONCLUSIONS

Numerical analyses and experiments revealed that the two refrigeration processes occur in the multi-bypass pulse tube refrigerator, and verified the multi-bypass version improved the performance of the refrigerator in a low temperature range.

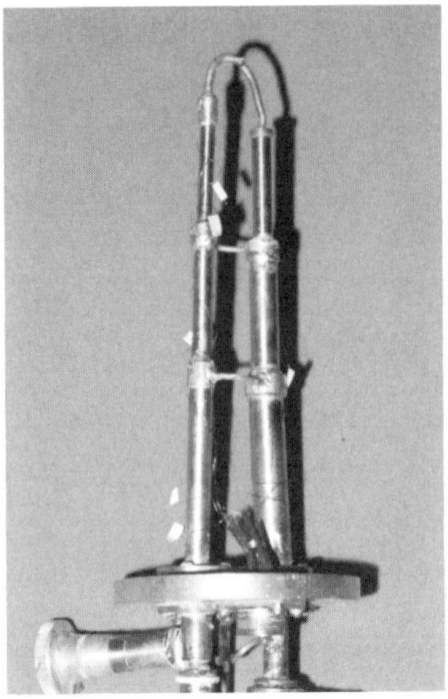

Fig.11. Photograph of the No.2 MPTR.

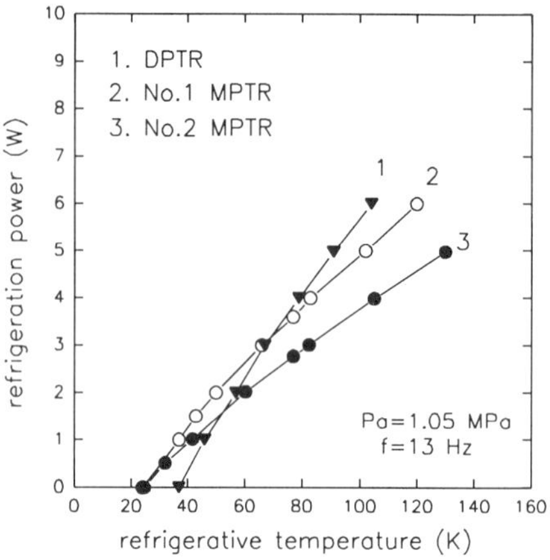

Fig.12. Refrigeration vs temperature.

A lowest temperature of 23.8K and a cooling capacity of 0.49W at 30K were obtained with the MPTR of one middle-bypass tube.

A MPTR of two middle-bypass tubes was constructed and reached a temperature of 24K. It is expected to reach a lower temperature by optimizing its structure parameters.

ACKNOWLEDGEMENTS

Research supported by the National Natural Science Foundation and the Postdoctoral Science Foundation of China. We wish to thank Ms. W.X. Zhu and Mr. P. Zhang for their help in building the test rig.

REFERENCES

1. Cai,J.H., Wang,J.J. and Zhu,W.X., Zhou,Y., Experimental analysis of the multi-bypass principle in pulse tube refrigerator, Cryogenics(1994), Vol.34, P.713
2. Chan,C.K., Jaco,C.B., Raab,J., et al, Miniature pulse tube cooler, Proc. of the 7th International Cryocooler Conference, Santa Fe, New Mexico, Nov., 1992, P.113
3. Matsubara,Y., Tanica,K., et al, Four-valve pulse tube refrigerator, Proc. of JSJS-IV on Cryocoolers and Concerned Topics, Beijing, China, Oct., 1993, P.54
4. Wang,C., Wu, P., and Chen, Z., Numerical modelling of an orifice pulse tube refrigerator, Cryogenics(1992), Vol.32, P.785
5. Wang,C., Wu,P., and Chen, Z., Numerical analysis of a double-inlet pulse tube refrigerator, Cryogenics(1993), Vol.33, P.526
6. Matsubara,Y., Gao, J., Novel configuration of three-stage pulse tube refrigerator for temperature below 4K, Cryogenics(1994), Vol.34, P.259

TEMPERATURE OSCILLATIONS OF WORKING GAS INSIDE A BASIC, AN ORIFICE AND A DOUBLE INLET PULSE TUBE REFRIGERATORS

Kazuya Seo[1], Masao Shiraishi[2], and Masahide Murakami[1]

[1]Institute of Engineering Mechanics, University of Tsukuba
Tennodai 1-1-1, Tsukuba 305 Japan
[2]Mechanical Engineering Laboratory, MITI
Namiki 1-2, Tsukuba 305 Japan

ABSTRACT

The effect of the driving frequency on the oscillating behavior of working gas in pulse tube refrigerators is experimentally studied by directly measuring the pressure and temperature of the gas for three types of pulse tube refrigerators, which are a basic, an orifice, and a double inlet types. The condition for the optimum frequency, at which the minimum wall temperature is reached, is discussed from the point of view of the dependence of the temperature oscillation on the driving frequency. It is found that the characteristic feature of the temperature oscillation at lower frequencies than the optimum frequency can be clearly distinguished from that at higher frequencies for every type. In the case of the basic type, the time for the compression process specified by the compressor-valve system is nearly equal to the relaxation time for gas temperature, defined as a period for the temperature rise by gas compression process at the optimum frequency. In the case of the orifice and the double inlet types, the phase difference between the pressure and temperature oscillations is nearly zero around the optimum frequency, and it shifts from negative to positive with the increase of the driving frequency.

INTRODUCTION

Since the invention of the original type of pulse tube refrigerator by Gifford and Longsworth[1], which is now called the basic type, several variations of pulse tube refrigerators, such as an orifice type[2, 3] and a double inlet type[4], have been developed. Recently, it has been demonstrated that a pulse tube refrigerator had a comparable refrigeration capability with a Stirling and a G-M refrigerators. The lowest temperature ever reached is as low as 4K.[5] It has got to be recognized that the pulse tube refrigerator will find various potential applications in future not only as a practical cryocooler but also as a new heat pump device.

The performance of the pulse tube refrigerator depends on the driving frequency, the mean gas pressure, and the dimension of the refrigerator. Effects of these parameters have been explained qualitatively with the aid of several analytical models.[6, 7, 8] On the other hand, a few experimental studies[9] have been reported, in which the effects of the parameters were investigated systematically. As to the dependence on the driving frequency, it was reported that the performance of a pulse tube refrigerator depended on the driving frequency for every type, and there existed an optimum frequency at which the minimum wall temperature was reached.[10, 4] Nevertheless there remains open questions why the performance depends on the driving frequency, and how the oscillating behavior of working gas varies with the driving frequency.

In this study, the variation of the oscillating behavior of working gas with the driving frequency is examined by measuring the pressure and temperature oscillations to investigate the dependence of the driving frequency on the performance in a basic, an orifice, and a double inlet types, and the condition for the optimum frequency is discussed.

EXPERIMENTAL APPARATUS AND PROCEDURE

A schematic of the experimental apparatus is shown in Fig. 1. It is designed primarily for the measurement of the pressure and temperature of working gas at several points along the pulse tube refrigerator. For the sake of easy wiring of thermocouples and pressure sensors, the pulse tube and the regenerator are only wrapped with polyurethane thermal insulator instead of installing them in a vacuum chamber for thermal isolation from the room temperature environment. A commercially available helium compressor unit for a Gifford-Macmahon refrigerator (SRD-208, Sumitomo Heavy Industries) and a rotary valve are employed to generate prescribed pressure oscillation. The frequency of the pressure oscillation can be varied from 0 to 20 Hz. The pressure ratio between the high and the low pressures can be changed from 1 to 3.4 by adjusting the opening of the bypass valve connecting the high pressure exit to the low pressure inlet of the compressor. The regenerator is composed of a hollow cylindrical frame made of bakelite with dimensions of 35 mm in diameter and 160 mm long, and a pile of 865 pieces of circular stainless steel screen

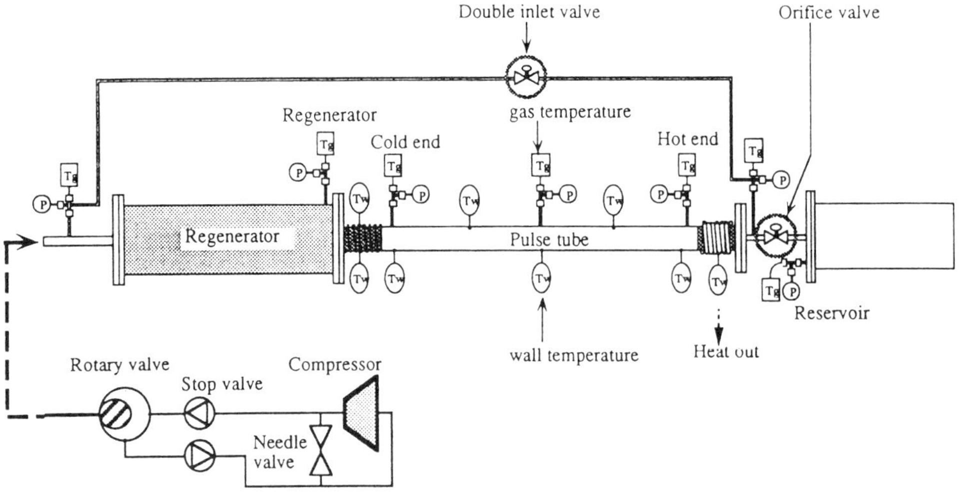

Fig.1 Experimental apparatus.

of 100 mesh size filled up into the frame. The pulse tube is made of stainless steel, 13 mm in inner diameter, 1.5 mm in wall thickness and 242 mm long. Two short pieces of copper tube with a length of 33 mm are soldered to the stainless steel tube at both ends as heat exchangers. One at the hot end is kept at the room temperature with a forced flow of cooling water. The rate of heat out from it is evaluated from the temperature rise of cooling water and the mass flow rate. The other is at the cold end. Two capillary tubes are connected to the hot end of the pulse tube, one leading to the reservoir having a volume of 308 cm^3 through the orifice valve, and the other returning to the entrance of the regenerator through the double inlet valve. A particular type of the pulse tube refrigerator can be selected by opening a corresponding valve. The branch tubes, 2 mm in inner diameter by 40 mm long, are also welded to the refrigerator as feedthroughs for a thermocouple and a pressure tap as shown in Fig. 1.

The pressure is measured with a strain gauge type pressure transducer and the temperature by an E-type (Chromel/Constantan) sheathed thermocouple of 0.25 mm in diameter, which is inserted to the center of the pulse tube. The response times of the pressure transducer and the thermocouple are about 0.25 msec (4 KHz) and 5 msec (200 Hz), respectively. However, a practical response time of the thermocouple is estimated to be 15 msec (70 Hz) in gas. T-type (Copper/Constantan) thermocouples are also soldered on the surface of the pulse tube wall to measure the wall temperature at eight points.

The experimental procedure is as follows. After evacuation of the whole system, helium gas is charged at the pressure level of 1.3 MPa. Cooling water is circulated through the heat exchanger at the hot end. Then, the compressor unit is started, and the driving frequency and the pressure ratio are adjusted to prescribed values. After the wall temperature at the cold end and the heat out from the hot end reach steady values, data are acquired. Experiments are carried out in the range of the driving frequency from 1 to 18 Hz and the pressure ratio from 1.8 to 3.4 for the basic, the orifice, and the double inlet pulse tube refrigerators. A typical result for the case of a compression ratio of 2.2 is presented here.

EXPERIMENTAL RESULTS

Measurement of the wall temperature and the heat out

Shown in Fig. 2 is the results of the steady values of the wall temperature at the cold end and the heat out in the case with no net refrigeration power as a function of the driving frequency in the case of a basic, an orifice, and a double inlet types. The openings of the orifice and the double inlet valves are fixed at the optimum opening at which the minimum wall temperature is reached. It is seen that the wall temperature and the heat out depend on the driving frequency, and there exists an optimum range of frequency at which the minimum wall temperature is attained. Typical frequencies in the optimum range, 2 Hz, 5 Hz, and 5 Hz, are selected for the basic, the orifice, and the double inlet types, respectively. On the other hand, the heat out increases with the driving frequency, and is the largest for the orifice type among three types at all driving frequencies except 1 Hz. The reason why the minimum wall temperature appears is considered by investigating the result shown in Fig. 2. Up to the optimum frequency, the wall temperature at the cold end drops and the heat out increases with the driving frequency, because P-V work per unit time increases proportionally with the driving frequency and so does the heat pumping rate. Above the optimum frequency, the wall temperature rises but the heat out still increases with the frequency because the frictional loss due to viscosity increases with the frequency and thus the net heat pumping rate decreases.

Pressure and temperature oscillation of working gas

The pressure and temperature oscillations of working gas are shown in Figs. 3-a, b, c for a basic, an orifice, and a double inlet types operated at the optimum frequency, where the pressure oscillation is indicated

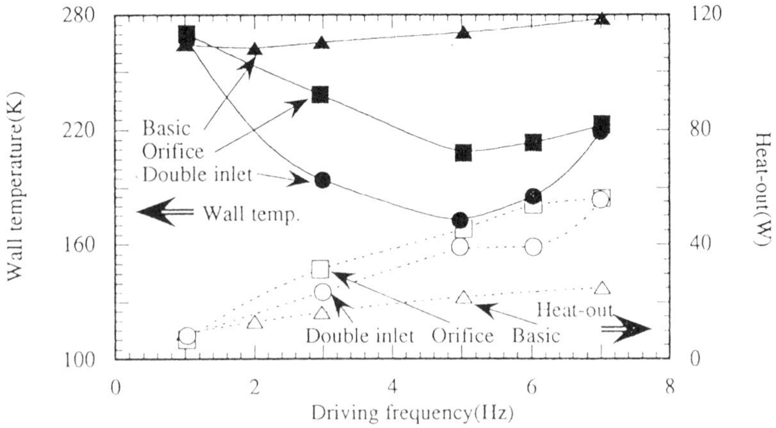

Fig. 2. The wall temperature at the cold end and the heat out as a function of the driving frequency. Closed triangles, squares, and circles denote the wall temperature at the cold end for the basic, the orifice, and the double inlet types. Open triangles, sruares, and circles denote the heat out for the basic, the orifice, and the double inlet types.

by the dotted line and the temperature oscillation is by the solid line. Since every pressure oscillation data is quite similar throughout the refrigerator except in the reservoir for three types, the data measured at the hot end is shown as a typical example together with that in the reservoir. On the other hand, the temperature oscillation differs from location to location along the pulse tube. In the case of the basic type (Fig. 3-a), the profile of the temperature oscillation at the cold end is similar to that of the pressure oscillation, but the data measured at the hot end is far from a square wave. There exists some phase difference between the pressure and temperature oscillations. The temperature oscillation is slightly delayed with respect to the pressure. The amplitude of the temperature oscillation at the hot end is three times as large as that at the cold end. In the case of the orifice type (Fig. 3-b), the profile of the temperature oscillation in the pulse tube is nearly triangular. The temperature oscillation proceeds the pressure oscillation. The amplitude of the temperature oscillation at the hot end is as large as that at the cold end unlike the basic type. The profile of the temperature oscillation in the double inlet type is close to that in the orifice type as seen from Fig. 3-c.

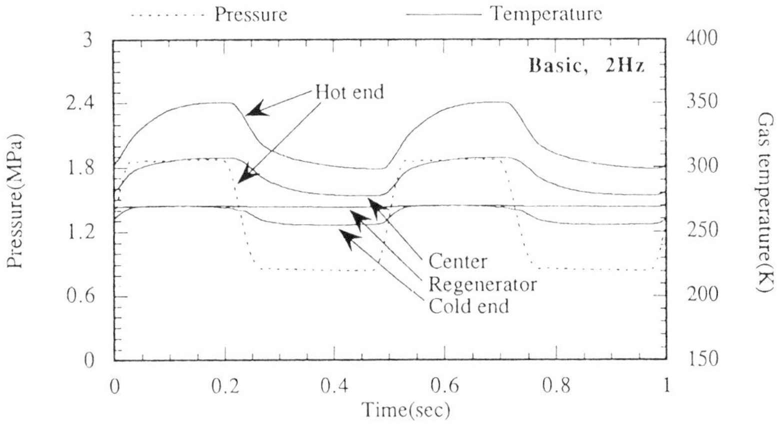

Fig. 3-a. Basic type, 2 Hz.

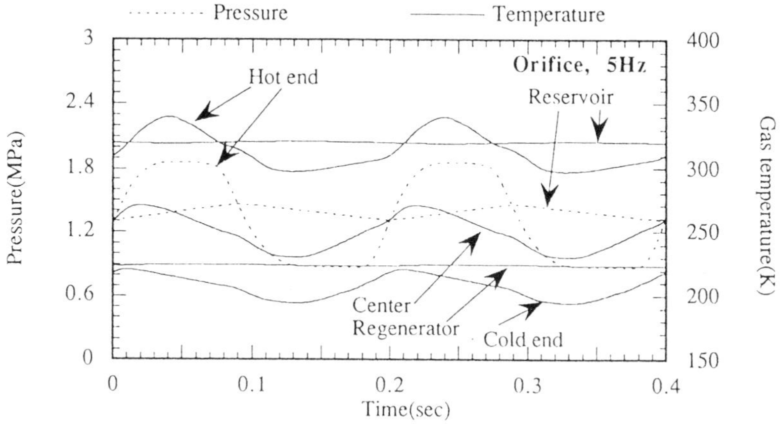

Fig. 3-b. Orifice type, 5 Hz.

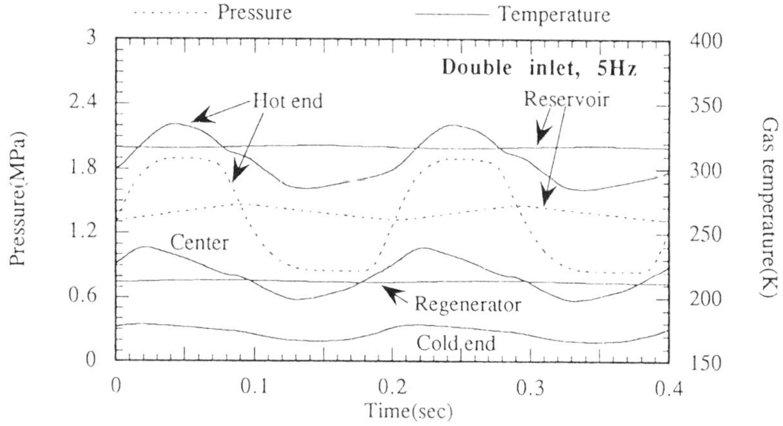

Fig. 3-c. Double inlet type, 5 Hz.

The amplitude of the temperature oscillation at the cold end in the double inlet type is slightly smaller than that in the orifice type, though the amplitude at the hot end in the double inlet type is larger than that in the orifice type. The difference in the profiles of the temperature oscillations between the orifice and the double inlet types is hardly found though the wall temperature at the cold end is evidently different.

Shown in Figs. 4-a, b, c are the variation of the temperature oscillation at the hot end with the driving frequency in the basic, the orifice, and the double inlet types at typical three driving frequencies, that is the optimum frequency and the lower (1 Hz) and the higher (7 Hz) frequencies than the optimum frequency. The origin of the abscissa is taken as the onset of the compression process. The pressure oscillation at the hot end is shown in every figure and that at the reservoir is also shown in Figs. 4-b, 4-c. It is seen that the temperature oscillation at lower frequencies can be distinguished from those at higher frequency than the optimum frequency for every type. In the case of the basic type (Fig. 4-a), the temperature at 1 Hz slowly drops with the laps of time during the high pressure plateau while it still rises until the onset of the expansion process at higher frequencies than the optimum frequency. The peak of the temperature at 1 Hz appears before the onset of the expansion process while that at the frequencies higher than the optimum one appears in phase with or after the onset of the expansion process. In the case of the orifice type (Fig. 4-b), the profiles of the temperature oscillation at 1 Hz and 5 Hz are quite different from that of the pressure. The temperature starts to drop during the high pressure plateau phase and does to rise during the low pressure plateau phase. This variation of the temperature can be well explained by considering the gas flow between the reservoir and the pulse tube through the orifice valve. The flow is driven by the pressure difference between them caused by the flow impedance of the orifice valve, producing some phase difference (about 90 degree) of the pressure variations between the reservoir and the pulse tube. In this situation, after the temperature rise by the compression the temperature drops by the cold inflow of gas from the regenerator during the high pressure plateau phase because of the pressure difference between the high pressure in the pulse tube and the low pressure in the reservoir, and then after the temperature drop by the expansion the temperature rises by the hot inflow of gas from the reservoir during the low pressure plateau phase. At 7 Hz the amplitude of the pressure oscillation in the reservoir is so small to induce the large scale inflows. As a result of the negligible inflow the profile of the temperature oscillation in the orifice type at 7 Hz becomes similar to that in the basic type. Moreover, the phase angle from the origin of the abscissa to an angle where the pressure difference between the reservoir and the pulse tube vanishes are about 8 degree for 1 Hz and 40 degree above the optimum frequency. The inflow from the reservoir continues for a while after the onset of the compression process at higher frequencies than the optimum frequency. It is seen from Fig. 4-b and Fig. 4-c that there seems almost no difference in the pressure and temperature oscillation profiles between the double inlet and the orifice types at any frequencies.

A number of characteristic times defined below are introduced for detail discussion on the temperature variation around the optimum frequency in the basic type as shown in Fig. 5. Here, τ_c is defined as the time difference between the onsets of the compression and the expansion processes, which are externally specified by the compressor-rotary valve system, in fact being nearly equal to 1/(2 fd), and τ is defined as the time from the instance at which the temperature rises by gas compression to the temperature peak measured at the hot

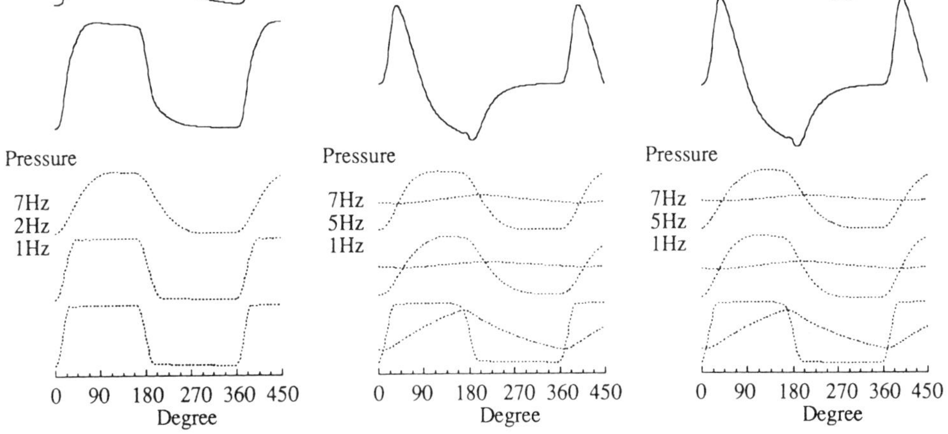

Figs. 4. The pressure and temperature oscillations in the basic (4-a), the orifice (4-b), and the double inlet (4-c) types.

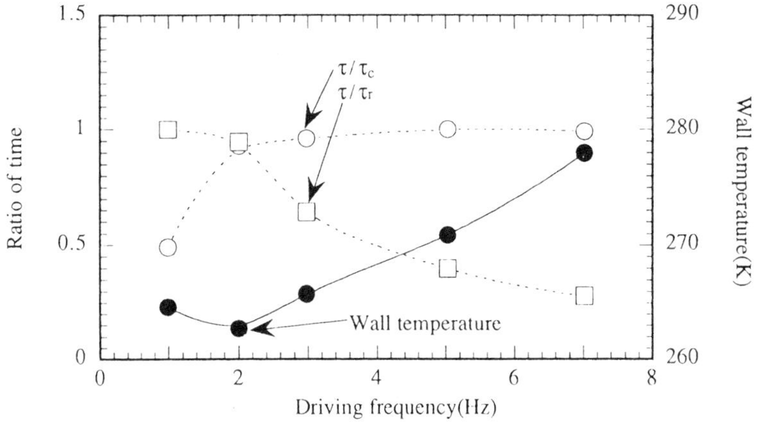

Fig. 5. Ratio of characteristic times and the wall temperature at the cold end as a function of the driving frequency in the basic type.

end. τ_r is defined as the relaxation time, that is the period required for the full temperature rise by gas compression, which can typically be evaluated at the lowest frequency, 1 Hz. It is found that the time for the compression process, τ_c, is nearly equal to the relaxation time, τ_r, at the optimum frequency (2 Hz), and is too long compared with the relaxation time at lower frequencies, and is too short compared with the relaxation time at higher frequencies.

Fig. 6 shows the results of the phase difference between the pressure and temperature oscillations at the hot end plotted as a function of driving frequency in the case of the basic, the orifice, and the double inlet types. The phase of each oscillation profile is obtained with respect to the fundamental mode. The phase difference is defined as the phase of the temperature oscillation minus that of the pressure oscillation. In the case of the basic type, the phase difference is always positive and increases with the driving frequency. In the case of the orifice and the double inlet types, it shifts from negative to positive values with the increase of the driving frequency and is nearly zero around the optimum frequency. The reason why it becomes negative in the range of low frequencies is that the temperature oscillation varies owing to the gas dynamic process when it takes and the flow between the reservoir and the pulse tube while the pressure variation is very quick in both compression and expansion processes as specified by the quick action of the rotary valve.

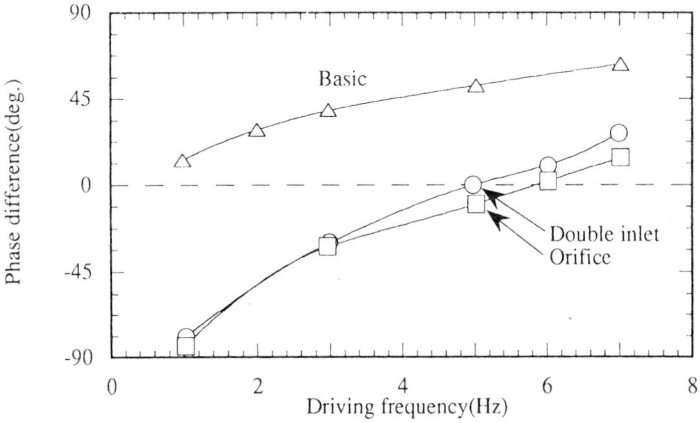

Fig. 6. The phase difference between the pressure and temperature oscillations at the hot end as a function of the driving frequency.

CONCLUSION

The following conclusions are drawn from the investigation of the effect of the driving frequency on the pressure and temperature oscillations inside a basic, an orifice, and a double inlet type pulse tube refrigerators.

1. The profile of temperature oscillations can be clearly distinguished between lower and higher frequencies than the optimum frequency for every type.

2. In the case of the basic type, the time for the compression process is nearly equal to the relaxation time defined as the period of the temperature rise by the compression at the optimum frequency.

3. In the case of the orifice and the double inlet types, the phase difference between the pressure and temperature oscillations is nearly zero around the optimum frequency, and it shifts from negative to positive with the increase of the driving frequency.

ACKNOWLEDGMENT

This research was partly supported by the Japan Society for the Promotion of Science.

REFERENCES

1. W.E.Gifford and R.C.Longsworth, "Pulse -Tube Refrigerator", ASME paper, (1963) No.63-WA-290.
2. E.I.Mikulin, A.A.Tarasov, and M.P.Shkrebyonock, "Low Temperature Expansion Pulse Tubes", Advances in Cryogenic Engineering, Vol.29, Plenum Press, New York, (1984) p.629.
3. R.Radebaugh, J.Zimmerman, D.R.Smith and B.Louie, "A COMPARISON OF THREE TYPES OF PULSE TUBE REFRIGERATORS : NEW METHODS FOR REACHING 60 K", Advances in Cryogenic Engineering, Vol.31, Plenum Press, New York, (1986) p.779.
4. S.Zhu, P.Wu, Z.Chen, W.Zhu, and Y.Zhou, "A single stage double inlet pulse tube refrigerator capable of reaching 42 K", Cryogenics, Vol.30 September Supple., (1990) p.257.
5. Y.Matsubara and J.L.Gao, "Novel configuration of three-stage pulse tube refrigerator for temperatures below 4 K", Cryogenics, Vol.34 (1994) p.259.
6. N.Rott, "Damped and Thermally Driven Acoustic Oscillations in Wide and Narrow Tubes", Z. Angrew. Math. Phys. Vol.20 (1969) p.230.
7. A.Tominaga, "Basic notion of Thermoacoustic theory and some results of simulations", Proc. of 4th Joint Sino-Japanese Seminar on Cryocoolers and Concerned Topics, Beijing (1993) p.79.
8. J.H.Xiao and X.T.Wang, "Regenarator irreversibility analysis based on thermoacoustic theory", Proc. of 4th Joint Sino-Japanese Seminar on Cryocoolers and Concerned Topics, Beijing (1993) p.242.
9. R.C.Longsworth, "AN EXPERIMENTAL INVESTIGATION OF PULSE TUBE REFRIGERATION HEAT PUMPING RATES", Advances in Cryogenic Engineering, Vol.19, Plenum Press, New York, (1967) p.608.
10. R.N.Richardson, "Pulse tube refrigerator - an alternative cryocooler ?", Cryogenics, Vol.26 (1986) p.331.

VIBRATION CHARACTERISTICS OF A 100 Hz TADOPTR

D. L. Patelzick, E. L. Hershberg, Y. K. Kwong, and K.M. Godshalk

Systems Integration Laboratory
Technology Group - Video and Networking Division
Tektronix Inc.
POB 500, M/S 50-470
Beaverton, OR 97077

ABSTRACT

Tektronix Inc., Los Alamos National Laboratory and the National Institute of Standards and Technology have successfully collaborated in building and testing a 100 Hz thermoacoustic driver (TAD) and orifice pulse tube refrigerator (OPTR) that demonstrated a 10 W cooling capacity at 110 K. The mating of the TAD and the OPTR resulted in some surprising vibration behavior that had to be understood and addressed before further work could be done. In this paper we present the results of our analysis and a description of the solution. We also present evidence that an extension mode in the TAD with an unusually low frequency was excited. The tools and techniques used to understand what was taking place in the TADOPTR structure and to determine how the acoustic energy was coupled to the structure are also described. This case study underscores the importance and value of even first order structural analysis of experimental machines so their response to stimuli can be included in the design process.

BACKGROUND

This report results from a project that is part of an Advanced Research Projects Agency (ARPA) program entitled "Development of Novel Cryosystems Technology and Demonstration of a System Utilizing a Functioning LSI Superconductive Microcircuit." One objective of this program was to extend the previous work of Radebaugh, et al,[1] from 27 Hz to 100 Hz as part of a larger program to miniaturize the ThermoAcoustically Driven Orifice Pulse Tube Refrigerator (TADOPTR).

The TAD has an overall length of 3.6 m. At each end there is a thermoacoustic assembly consisting of a hot heat exchanger, thermoacoustic stack and cold heat exchanger. Each assembly is about 71 cm long, with a mass of 68 kg. Separating these assemblies is a 2.18 m long, 88.9 mm diameter tube that functions as the resonator. The entire assembly is supported on simple jack stands with a rubber damper between each stand and the TAD. The OPTR is mounted on a rigid platform and is enclosed in a heavy vacuum can. During

Advances in Cryogenic Engineering, Vol. 41
Edited by P. Kittel, Plenum Press, New York, 1996

initial tests of the 100 Hz TAD, before the OPTR was attached, there were no detrimental levels of vibration. When the TAD and the OPTR were mated with a thick wall stainless steel tube, the system vibrated so violently that we were prevented from applying full power to the TAD due to safety concerns. Initially the worst case amplitude was 0.25 mm (8.3 g) on the OPTR support structure with approximately 50 % power applied to the TAD. The vibrations coupled through the support stands to the floor and propagated throughout the five story building that houses our laboratories.

The obvious solution was to decouple the motion of the TAD from the OPTR. All the modeling done on this project had concentrated on the gas dynamics of the TAD and the OPTR, but little work had been done on the structural dynamics. We wanted to ensure no new problems were created with the "fix." In complex dynamic systems there are sometimes surprising effects that may go undetected for long periods of time. We were concerned that simply inserting a flexible coupling would solve the immediate vibration problem, but set the stage for other problems in future designs. We decided to compare the observed mechanical vibration behavior of the TADOPTR with a numerical model of its structure that would provide a clearer understanding of what happens during the operation of the TADOPTR. An accurate model could also provide a way to test the effects of dampers and other vibration reducing measures prior to fabricating parts or modifying the TADOPTR.

PROBLEM APPROACH

Analytical modeling using finite elements was performed using COSMOS/M[2]. The first model considered a system with a thick wall stainless steel tube between the TAD and OPTR. The second model considered a flexible tube. We used simple modal analysis of the structure. The spectrum of the pressure waveform of the TAD was recorded and compared to the mode shape and frequency to determine if one of the modes could be excited.

In order to quantify the effect of the redesigned structure, a transducer was used to measure vibration amplitudes at various points along the TADOPTR and the support structure before and after the rigid coupling was replaced with a flexible coupling. The vibration phenomenon was measured and analyzed under steady state conditions. The mass of the transducer was extremely small relative to the mass of the system, so moving it from place to place had a minimal effect on the vibration characteristics of the system.

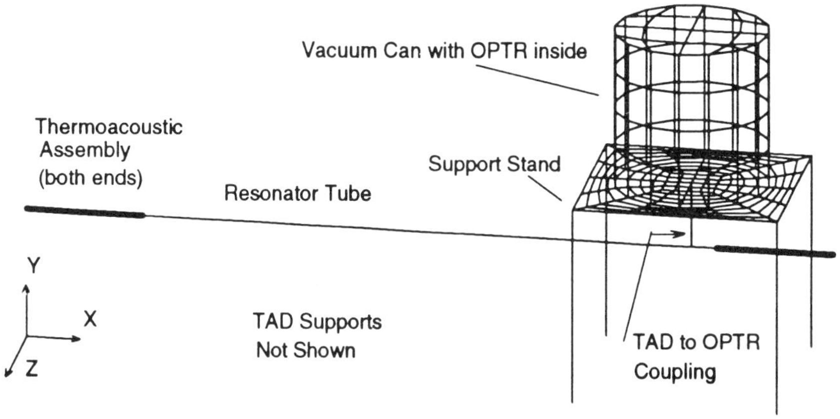

Figure 1. Finite Element Model of TADOPTR with vacuum can and stand.

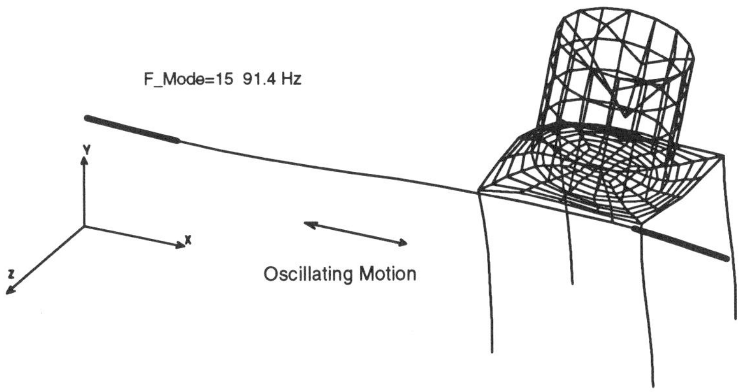

Figure 2. TADOPTR mode shape at 91.4 Hz. with stiff coupling between the TAD and the OPTR.

Finite Element Model

The finite element model shown in Figure 1 is a combination of plate elements, beam elements and pipe elements. Plate and beam elements are used to emulate the vacuum can surrounding the OPTR and the support stand. Pipe elements are used to model the TAD and the tube connecting the TAD to the OPTR. For the TAD, the center portion of the model accurately represents the resonator duct as far as dimensions and materials are concerned. The two ends represent the drivers. These have the same length as the actual system, but different diameters and material properties, so that inertial properties are accurately emulated. The material for the connecting tube can be readily changed to observe mode shapes and frequencies for the TADOPTR system. The stands that supported the TAD were not modeled as there was no evidence that the TAD was behaving any differently than a freely supported system. This was later shown to be a reasonable assumption.

Modal Analysis Results

The analysis of the TAD rigidly coupled to the OPTR gives results that seem reasonable. The 100 Hz TAD actually operates at 96 Hz, and it is the motion of the compressed gas in the TAD at this frequency that provides the forcing function for the system vibrations. The model we developed identified a structural resonance mode at 91.4 Hz that involves a complicated flexure mode of the support stand under the vacuum can coupled through the legs of the stand, and driven by the motion of the TAD parallel to the TAD axis. The motion of the assembly is horizontal and parallel to the axis of the TAD as shown in Figure 2, and it includes some flexure of the tad with a node at the attachment point of the rigid coupling. The model identified many modes at many different frequencies for this complex structure, but there were no others close enough in frequency to the TAD 96 Hz operating frequency to be of consequence. The one remarkable thing about this situation is that it is driven entirely by the oscillatory motion of a compressed gas in a closed system.

The model results for the system when the TAD was flexibly coupled to the OPTR are quite interesting. The closest mode to the forcing frequency was 93.3 Hz, but it was entirely different from the rigidly coupled case. With a flexible coupling, the vacuum can and the stand were predicted to oscillate normal to the axis of the TAD instead of parallel to it. This limits the effect of the TAD on the overall system and effectively isolates the TAD from the OPTR; greatly reducing the transfer of energy from the TAD to the OPTR, the OPTR stand and vacuum can.

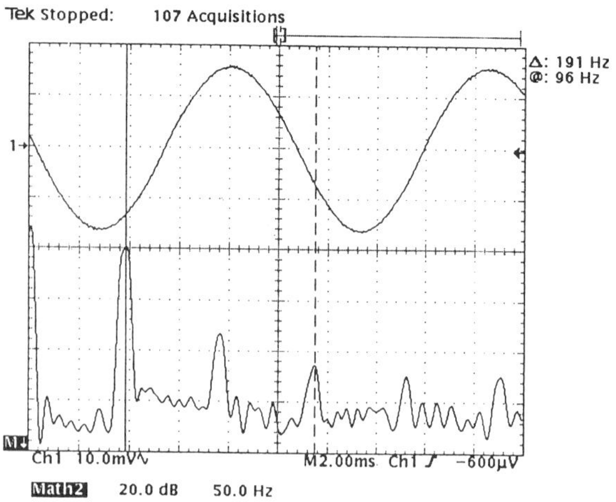

Figure 3. TAD gas pressure data at 40% input power - top trace: output from the pressure transducer, bottom trace: FFT of pressure waveform.

The model predicts a mode at 189 Hz for both the rigidly and flexibly coupled configurations. In this mode shape the drivers oscillate about the center of the TAD exactly 180 degrees out of phase with each other and parallel to the long axis of the tube. This results in limited motion of the OPTR, the OPTR stand and vacuum can. Essentially the resonator tube acts as a long spring holding the mass of the drivers apart. This frequency is almost exactly twice the operating frequency of the TAD and it would be possible that higher order harmonics in the pressure waveform could excite this mode, although the coupling would have to be very efficient.

Pressure Measurements

A thermoacoustic driver is designed to convert thermal energy into acoustic energy consisting of a standing resonant oscillating pressure wave that is confined to a closed tube. The 100 Hz TAD was built with a pressure transducer already located near the coupling to the OPTR. The upper waveform in Figure 3 shows the output of the transducer and the lower waveform is the fast fourier transform (FFT) of the pressure transducer output. The

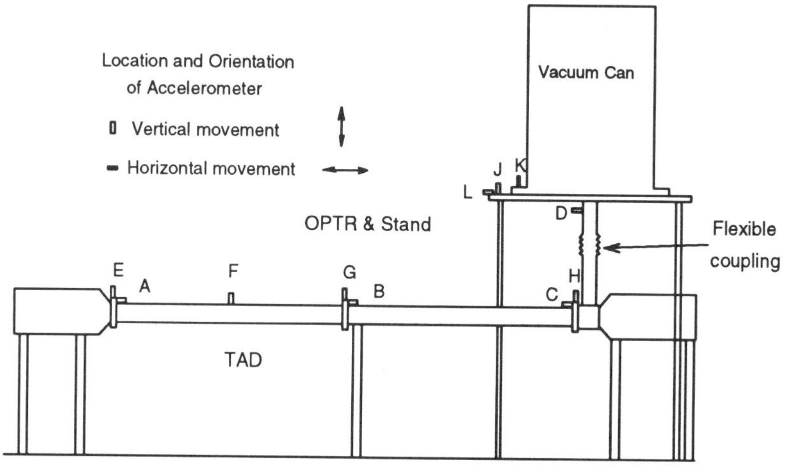

Figure 4. Transducer locations on the TADOPTR.

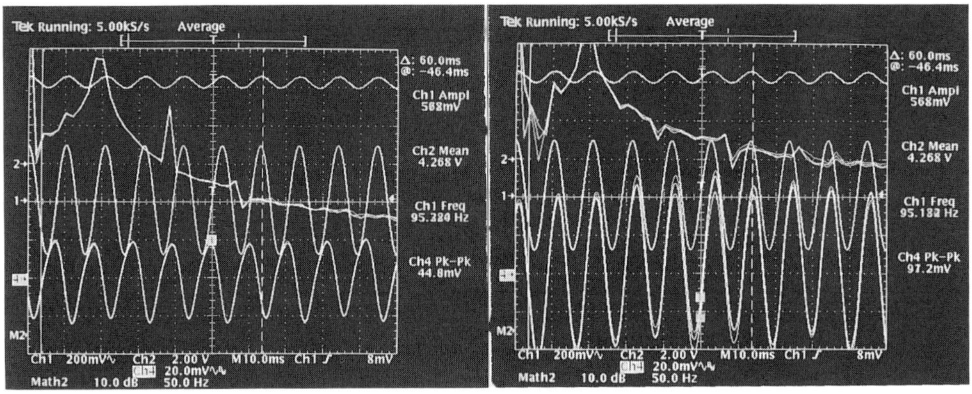

Figure 5. Displacement transducer output on for a 100 Hz TADOPTR with rigid coupling. LEFT: Horizontal position "A" on the TAD. RIGHT: Horizontal position "L" on OPTR stand. Traces from top to bottom: output of the pressure transducer (DC coupled, 2 V/div.), FFT of the displacement transducer signal (10 dB/div.), Output of the pressure transducer (AC coupled, 200 mV/div.) and the displacement transducer signal (20 mV/div.). Horizontal axis 10 ms/div. and 50 Hz/div. for FFT.

FFT is computed by the oscilloscope we used for collecting the transducer data and is produced from the actual time series acquired by the acquisition system. The FFT of this pressure waveform reveals some second and third harmonic components on what appears to be a perfect sinusoidal waveform. The amplitude of the second harmonic is 40 dB lower than the fundamental, and is thus quite small. These second and third harmonics were not originally expected to have a large effect on the vibration of the TADOPTR.

Vibration Measurements

We characterized the vibration of the TADOPTR system with a displacement transducer[3] and a piezo-electric accelerometer[4]. The displacement transducer was sensitive to motion normal to it's primary axis so we used the accelerometer for much of the characterization. In Figure 4 we have shown the locations on the TADOPTR at which the

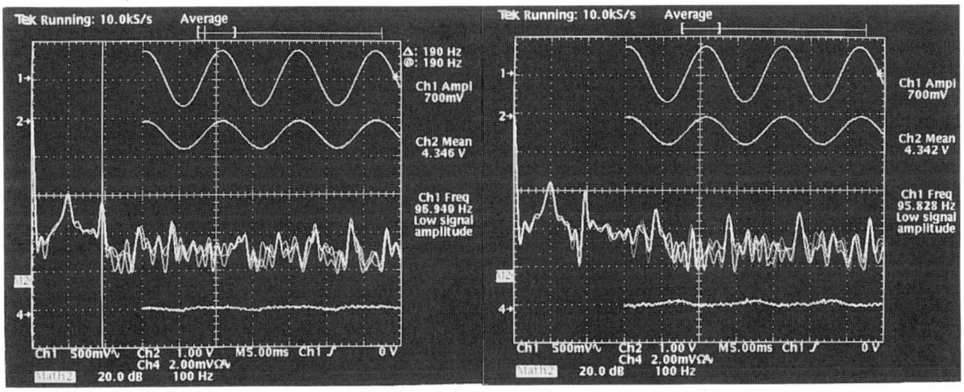

Figure 6. Accelerometer transducer output for a 100 Hz TADOPTR with flexible coupling LEFT: Horizontal position "L" on the OPTR stand. RIGHT: Vertical position "J" on OPTR stand. Traces from top to bottom: output of the pressure transducer (AC coupled, 500 mV/div.), output of the pressure transducer (DC coupled, 1 V/div.), FFT of the accelerometer transducer signal (20 dB/div.),and the accelerometer transducer signal (2 mV/div.). Horizontal axis 5 ms/div. and 100 Hz/div. for FFT.

two transducers were placed. The accelerometer captured data in the horizontal plane at locations A through D, and in the vertical plane at locations E through K. The displacement transducer was used to capture data in the horizontal plane parallel to the axis of the TAD. The transducers were bonded to the apparatus at each location with a thin layer of microcrystalline wax. At each location a set of data was gathered using a digital oscilloscope[5] with built-in FFT and data analysis capabilities. The calibration of the displacement transducer is 10 mV / 0.025 mm and that of the accelerometer is 5.0 mV/g. The fast fourier transform (FFT) using rectangular filtering[6] was performed on the accelerometer data while it was taken. Each peak of the FFT represents a component frequency of the input waveform.

Vibration characteristics of a 100 Hz TAD and OPTR connected by a rigid coupling

The rigidly coupled system was characterized with both pressure and displacement measurements. Peak displacements on the TAD of 0.11 mm were observed at point A, leading to accelerations of 4.2 g. The vacuum can support stand showed the greatest displacement at point L. This was 0.25 mm at a frequency of 95.2 Hz, yielding 9.0 g accelerations. This was at only 50 % applied power to the TAD. The wave forms and of the pressure and displacement transducers are shown in Figure 5, along with the FFT of the displacement data taken at these two locations. At these vibration amplitudes we did not apply any greater input power to the TAD due to concerns of structural damage to the higher stress components. Our data are consistent with the predictions made by the FEA model. The resonant frequency is very close to the drive frequency and the greatest accelerations are on the deck of the support stand under the vacuum can.

Vibration characteristics of a 100 Hz TAD and OPTR connected by a flexible coupling

Figure 6 shows the data obtained after the rigid coupling was replaced with a flexible coupling. A very low mass piezoelectric accelerometer was used to characterize the TADOPTR vibrations after the flexible coupling was installed. This permitted us to place the transducer at difficult to reach locations, and allowed us to obtain measurements without regard to the orientation of the transducer. The measurements shown in Figure 6 were made while the TAD was operating at 100 % of its design input power, which emphasizes how significant a reduction was observed in the vibration amplitudes. Clearly the vacuum can support stand is very well isolated from the TAD by the flexible coupling. Even at maximum input power, measurements on either side of the coupling showed accelerations of less than 0.04 g. This result is also consistent with the model predictions. The flexible coupling isolated the TAD from the OPTR and provided for a shift in mode frequencies that led to complementary motions of the OPTR and vacuum can support stand that further reduced the vibration coupling from the TAD to the OPTR.

Unexpected extension mode vibration identified in a 100 Hz TADOPTR

Our measurements before and after the installation of the flexible coupling uncovered the consistent appearance of a vibration signature at roughly twice the TAD drive frequency. This initiated a search for the potential sources and mechanism. As mentioned previously, modal analysis predicted a resonance at about 190 Hz due to the extension mode along the TAD axis. This is an unusual mode, but one that is easy to identify by direct measurement due to it's unique characteristics. Figure 7 shows the accelerometer measurements at two points on the TAD parallel to the TAD axis. They show peak accelerations of 0.8 g at 100% input power. Measurements made parallel to the TAD axis

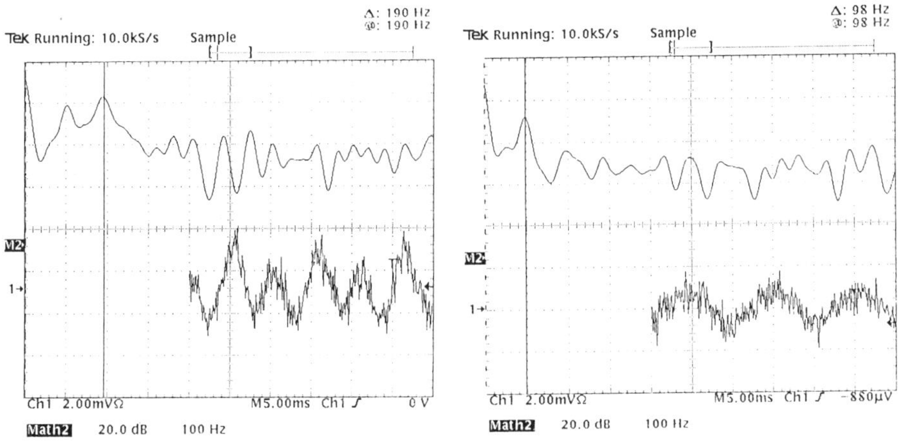

Figure 7. Accelerometer transducer output for a 100 Hz TAD - LEFT: horizontal position "A" on the TAD. RIGHT: horizontal position "B" on the TAD. Traces from top to bottom: FFT of the accelerometer transducer signal (20 dB/div.), and the accelerometer transducer signal (2 mV/div.). Horizontal axis 5 ms/div. and 100 Hz/div. for FFT. The accelerometer output at position "B" shows the operating frequency of the TAD at 98 Hz. and the essentially no motion in the axial direction.

but at the midpoint between the two ends showed the operating frequency of TAD and the frequency of the extension mode down 16 dB from the operating frequency. This is completely consistent with an excited extension mode. There are no off axis modes that will show this characteristic. Thus, the high vibration amplitudes at the lower frequency of the TAD drive were coupled through the structures to the higher frequency extension mode. This characteristic effect was seen in the FFT of the TAD pressure waveform when the TAD and OPTR were separated and when they were both rigidly and flexibly coupled. The frequency of the extension mode resonance at almost exactly double the excitation frequency of the TAD was completely fortuitous. Had the resonant tube connecting the two driver assemblies been made of steel instead of aluminum, the extension mode would have shifted in frequency and the effect would not have been seen.

A good check on the interpretation of the FFT of the accelerometer waveform as seen in Figure 7 is to compare it to a waveform constructed by adding two sinusoids. Figure 8 shows a composite waveform derived from the addition of two sinusoids in which the second sinusoid is twice the frequency and 2.5 times the amplitude of the first. The FFT of

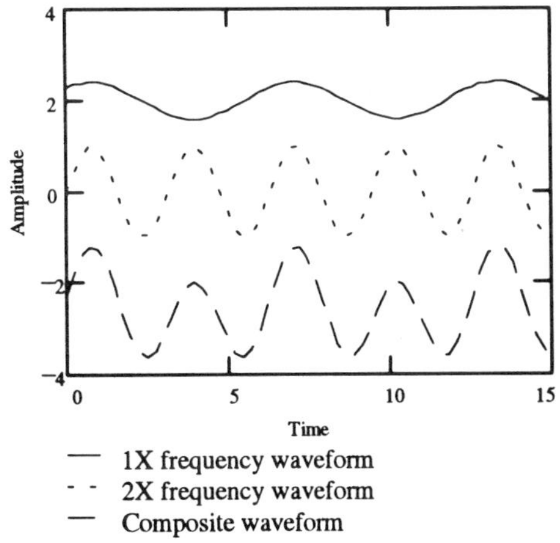

Figure 8. Composite waveform.

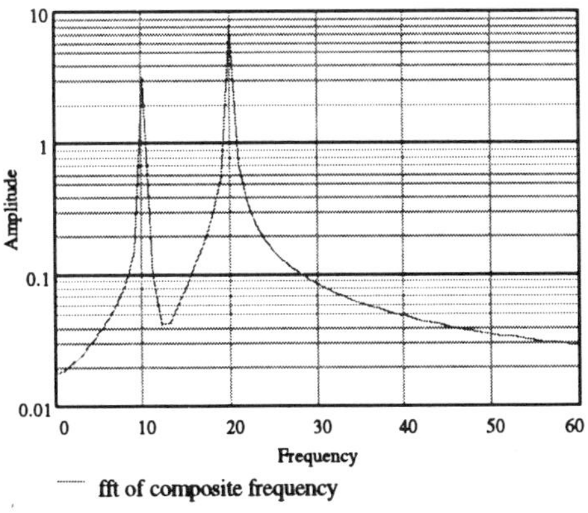

Figure 9. FFT of composite waveform.

the composite waveform is shown in Figure 9 and is strikingly similar to the FFT shown in Figure 7. This example leads to a significant insight about this system. The component of the accelerometer waveform from the resonant extension mode centered at twice the driving frequency of the TAD is significantly larger than the component from fundamental. The second peak of the two component frequencies in Figure 7 is 4 dB higher than the first which means the 190 Hertz amplitude is 2.5 times the amplitude of the TAD motion induced by the fundamental frequency of the oscillating gas. In this case the apparent coupling coefficient to a higher order harmonic is greater than one.

DISCUSSION AND SUMMARY

No significant noise and vibration accompanied the initial tests of the TAD by itself. After rigidly coupling the TAD and OPTR together there was significant vibration that prevented application of full power. Replacing the rigid coupling with a flexible coupling reduced the noise and vibration to a tolerable level that was only a fraction of what was initially encountered. In addition, it is now apparent that the second harmonic of the pressure oscillation in the TAD, along with efficient coupling of the TAD oscillation to the fundamental structural resonances is capable of exciting a resonant extension mode of the mechanical structure of the TAD. Sufficient damping exists in the present system to limit the oscillatory amplitude at the resonant point thereby preventing the TADOPTR from being damaged.

We have developed a structural model of a system that confirms the presence of resonant modes at the first and second harmonic of the pressure oscillation in the TAD. The understanding of the boundary conditions and the gas dynamics derived from this experience provided a path to analyze another system with an operating frequency of 350 Hz. The 350 Hz system was analyzed prior to fabrication and the subsequent operation confirmed the validity of the analysis.

ACKNOWLEDGMENT

This work is supported by the Advanced Research Projects Agency, Contract Number MDA972-929C-0011.

REFERENCES

1 R. Radebaugh, K.M. McDermott, G.W. Swift, R.A. Martin, Development of a Thermoacoustically Driven Orifice Pulse Tube Refrigerator, Proc. of Fourth Interagency Meeting on Cryocoolers, 205 (1991)

2 Structural Research & Analysis Corporation

3 Tektronix Inc., Horizontal Vibration Transducer, P/N 015-0167-00

4 PCB Piezotronics Inc., Quartz accelerometer P/N 309A & Transducer power unit model 480D06.

5 Tektronix TDS 540 with FFT option.

6 "TDS Family Option 2F Advanced DSP Math," Published by Tektronix Inc., 1992.

CHARACTERIZATION OF 350 HZ THERMOACOUSTIC DRIVEN ORIFICE PULSE TUBE REFRIGERATOR WITH MEASUREMENTS OF THE PHASE OF THE MASS FLOW AND PRESSURE

K.M. Godshalk[1], C. Jin[1], Y.K. Kwong[1], E. L. Hershberg[1], G. W. Swift[2], R. Radebaugh[3]

[1]Tektronix, Inc.
Beaverton, OR 97077
[2] Los Alamos National Laboratory
Los Alamos, NM 87545
[3] National Institute of Standards and Technology
Boulder, CO 80303

ABSTRACT

The world's first 350 Hz thermoacoustic driven orifice pulse tube refrigerator (TADOPTR) has been designed and built by Tektronix, Inc., in cooperation with Los Alamos National Laboratories (LANL) and the National Institute of Standards and Technology (NIST). This highly instrumented system includes hot wire anemometers and pressure sensors for measuring the phase of the mass flow and pressure at all key locations in the TADOPTR, permitting for the first time detailed comparison to analytical models developed by LANL and NIST. Characterization results for velocity and pressure phase, pressure amplitude, and enthalpy flow show good agreement with the simulations. We have also demonstrated a new design method that uses the inertance of the pulse tube at 350 Hz to achieve the desired phase between the mass flow and pressure, rather than the usual double inlet design. We have designed and characterized single stage and two stage 350 Hz TADOPTRs.

INTRODUCTION

TADOPTR's are of interest because they have no moving parts and thus should have a long lifetime. The first TADOPTR was built by NIST and LANL and operated at 27 Hz[1]. Tektronix, LANL, and NIST have jointly designed and constructed 100 Hz and 350 Hz TADOPTRs.

We describe here the design and characterization of single stage and two stage 350 Hz OPTRs. The design of the both OPTRs uses the inertance of the pulse tube to control the phase angle of the velocity at the cold end of the pulse tube. We are able to design the OPTR to obtain phasing of the mass flow similar to that found in a Stirling cycle[2,3]. The two stage system is designed to be a testbed for validating our design methodology. In order to verify the design prediction for velocity phasing, we included hot wire anemometers at key locations in the two stage TADOPTR. Temperature, pressure and heat rejection are also measured, allowing detailed comparison to analytical models developed by LANL and NIST.

DELTAE[4,5] and REGEN3.1[6] are modeling tools developed by LANL and NIST, respectively. DELTAE (Design Environment for Linear ThermoAcoustic Engines) can simulate thermoacoustic systems that are combinations of segments such as ducts, heat exchangers, regenerators or thermoacoustic stacks, and impedances. Among its predictions at each segment are: pressure, volume velocity, enthalpy flow and work flow. REGEN3.1 solves for the performance of many types of regenerators. Since the stacked screen regenerator segment for DELTAE was only recently completed, REGEN3.1 was used for the initial design of the stacked screen regenerators, and was used to verify the results of the new DELTAE segment.

A 350 Hz thermoacoustic driver (TAD) designed and constructed by LANL is used to drive the OPTRs. All characterization has been done with helium as the working fluid; most tests have been done at a mean pressure of 3.1 MPa, but some experiments have been performed at lower mean pressures.

SINGLE STAGE 350 HZ OPTR

The design of the 350 Hz single stage pulse tube is based on simulation results from DELTAE and REGEN3.1. The DELTAE simulation shows that in the design optimized for the best coefficient of performance, the phase angle of the mass flow at the two ends of the regenerator is shifted by about 90 degrees, and this shift is approximately evenly split around the pressure oscillation, as seen in Figure 1. This arrangement is very similar to that of Stirling machines, where one has the freedom to adjust the mass flow phase angles for optimal performance. It is possible to obtain this phasing relationship without pistons by making use of the inertance of the pulse tube. This is feasible at 350 Hz due to the shorter wavelength of sound in helium at this pressure.

A schematic of the single stage OPTR is shown in Figure 2. The regenerator is 3.6 cm in diameter and 5.5 cm long; it is packed with #325 stainless steel mesh and has a

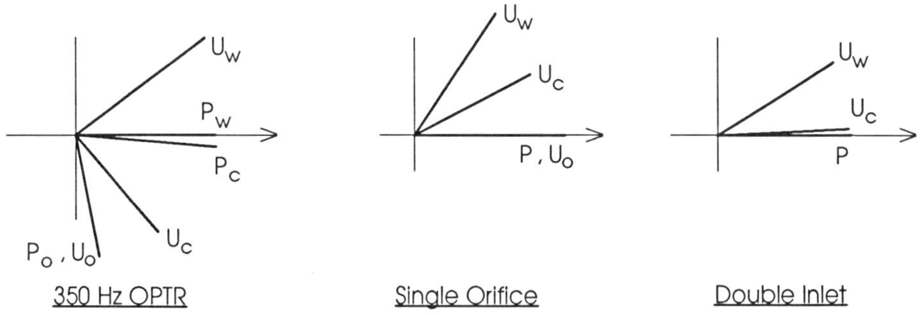

Figure 1. Phasor diagrams for the 350 Hz single stage OPTR design, and for conventional single inlet and double inlet orifice pulse tubes. U denotes the velocity, P denotes the pressure, and the subscripts w, c, and o denote the warm end of the regenerator, the cold end of the regenerator, and the orifice (the warm end of the pulse tube). Phases are relative to the pressure at the warm end. In the 350 Hz OPTR, the velocity lags the pressure at the cold end; the phase of the pressure changes also, whereas in conventional low frequency OPTRs, the change in the phase of the pressure is usually negligible.

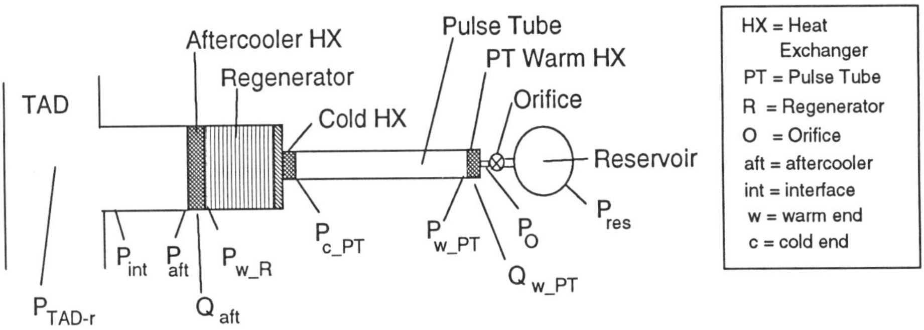

Figure 2. Schematic of the single stage OPTR, with locations of components and pressure sensors indicated.

measured porosity of 73%. The pulse tube is 0.84 cm in diameter and 20.0 cm long. The reservoir is 150 cc, and the orifice valve is 7.1 mm in diameter. The orifice valve is completely open for these experiments. The connecting duct between the TAD and OPTR is 20 cm long and 3.6 cm in diameter.

The single stage OPTR reached a cold temperature of 147 K with a mean pressure of 3.1 MPa and a pressure amplitude of 0.190 MPa (6.1%) at the entrance to the OPTR. The heat rejected to the aftercooler heat exchanger was 494 W, which is a measure of the power input to the OPTR. Figure 3 shows the cold temperature of the OPTR as a function of the normalized pressure amplitude squared at the TAD. Although we were limited in the pressure amplitude obtainable by limitations in the hot end temperature of the TAD heaters, we came close to achieving the desired performance. This result is a very strong confirmation of the utility and power of the design tools, and the overall design approach to this new concept for high frequency orifice pulse tube refrigerators.

The pressure amplitude at several locations in the OPTR is shown in Figure 4 as a function of the pressure amplitude at the warm end of the regenerator. Also shown is the DELTAE model prediction. The simulation uses the geometry of the OPTR and uses the observed pressure amplitude at the warm end of the regenerator as an initial condition. The impedance between the pulse tube warm end and the reservoir is modeled as an 8 cm long 7.6 mm diameter duct, which is approximately the total length of the duct between the reservoir and pulse tube. There are no adjustable parameters in the model aside from the initial condition. The observed pressure amplitude agrees very well with the model simulation.

The simulation also gives the enthalpy flow through the regenerator, and the heat rejected at the aftercooler heat exchanger. For the operating point mentioned above (6.1% pressure amplitude at the TAD), the measured regenerator loss is 32 W, and the simulation result is 36 W. The measured heat rejected at the aftercooler was 494 W and the prediction was 288 W (42% discrepancy). The system was designed for the velocity phase at the cold end to lag the pressure by 41°, and the simulation indicates that the velocity lags the pressure by 29°. Although the basic performance agrees well with the predictions, several factors need more investigation. The temperature at the exit of the aftercooler heat exchanger was lower than its cooling water temperature. This behavior was unexpected, and the disagreement between the predicted and observed aftercooler heat is puzzling given the agreement for other parameters. The system was also designed to reach 110 K with 5 W cooling power, at 10% pressure amplitude at the TAD; extrapolations of the data shown in Figure 3 to higher pressure amplitude indicate that this performance would probably not be realized. Although the observed performance is consistent with a simulation that uses the observed initial conditions for pressure amplitude, in order to efficiently design the final system it is necessary to be able to construct a system that reaches its targeted operating condition.

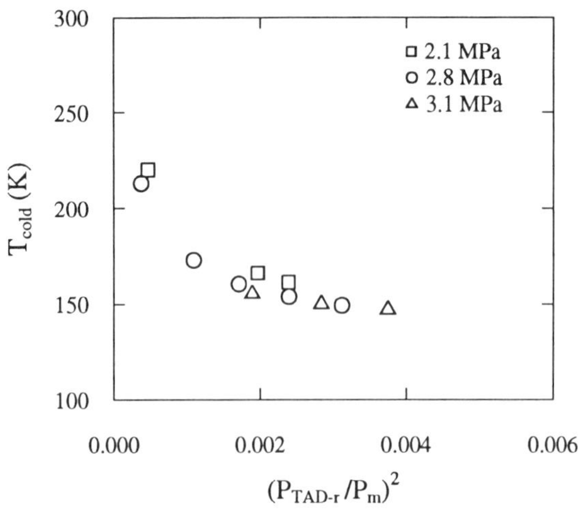

Figure 3. The cold temperature for the single stage OPTR as a function of the square of the normalized pressure amplitude at the TAD. Data are shown for three different mean pressures. We are prevented from going to higher pressure amplitude due to limitations on the hot end temperature of the thermoacoustic driver.

350 HZ TWO STAGE OPTR

We applied the design method developed for the single stage OPTR to the design of the two stage OPTR. The two stage system was designed primarily to give a detailed evaluation of our design methods and tools, in preparation for design and construction of the final system. As a result, instrumentation rather than absolute performance was of primary importance. A more in depth characterization and comparison to theory was desired for the two stage device than was obtained for the single stage device. This was due to the added complication of designing the branch between the two OPTRs to preserve the desired phasing scheme, and because of the unexpected behavior of the aftercooler heat exchanger in the single stage OPTR. As mentioned in the introduction, hot wire anemometers were included to measure the phase of the velocity, allowing us to make direct comparisons between the prediction and results for the velocity phasing.

The design of the two stage OPTR was again based on DELTAE and REGEN3.1. Two separate models for the first and second stages were used. The branch connecting the two stages was modeled as a load on the first stage, with the branch impedance being

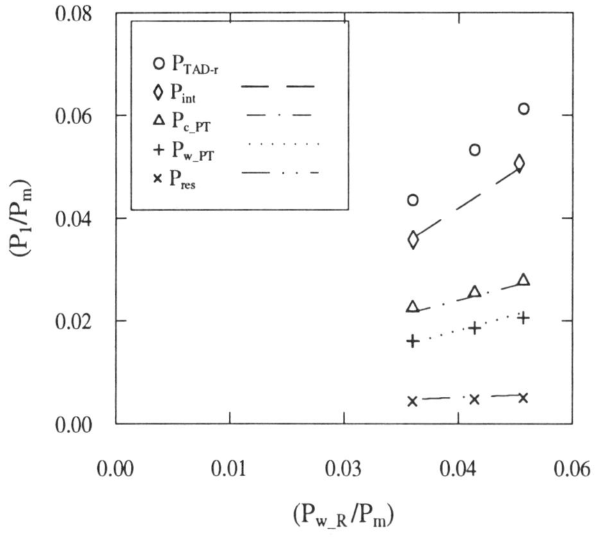

Figure 4. The pressure amplitude P_1 at various locations in the single stage OPTR as a function of the pressure amplitude at the warm end of the regenerator. The mean pressure P_m is 3.1 MPa. Experimental data are shown with symbols, and DELTAE model predictions are shown with lines. The model uses the observed pressure amplitude at the warm end of the regenerator as an initial condition, but requires no other adjustable parameters.

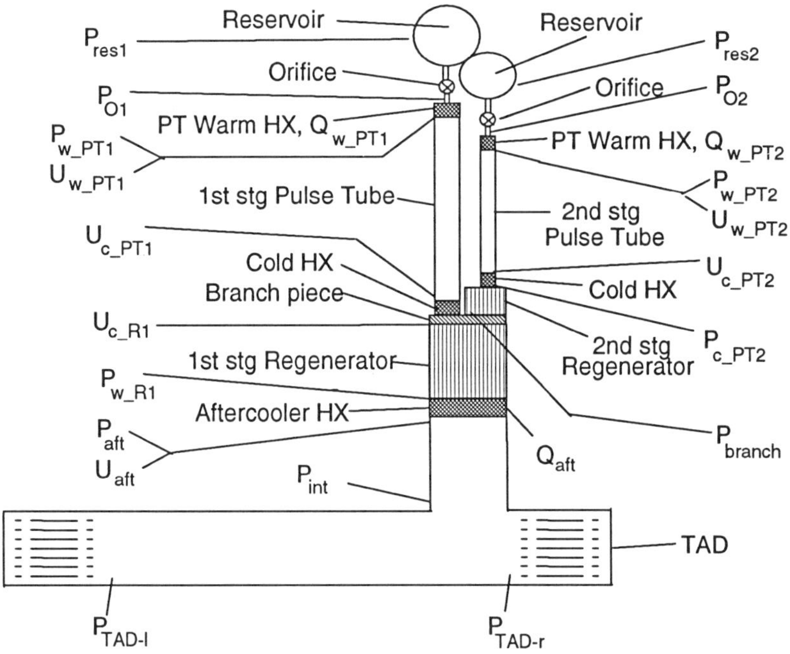

Figure 5. A schematic of the 350 Hz two stage TADOPTR system, showing location of pressure sensors (P), and anemometers (U). The anemometers are located at the entrance to the aftercooler heat exchanger, at the cold end of the first stage regenerator (before the branch to the second stage), at the cold end of the first stage pulse tube (after the branch to the second stage), at the warm end of the first stage pulse tube, and at the cold and warm ends of the second stage pulse tube. The heat Q rejected at the aftercooler and first and second stage warm end heat exchangers is determined by measuring the cooling water flow rate and temperature rise.

determined from the second stage model. A schematic of the two stage OPTR is shown in Figure 5. The first stage regenerator is 4.58 cm in diameter and 3.0 cm long; it is filled with #325 stainless steel mesh with 0.028 mm wire diameter, and has a measured porosity of 71%. The second stage regenerator is 3.19 cm in diameter and 3.0 cm long; it is filled with #325 bronze mesh calendered to 0.061 mm thickness, and has a measured porosity of 61%. The first stage pulse tube is 0.93 cm in diameter and 22.0 cm long; the second stage pulse tube is 0.57 cm in diameter and 11.0 cm long. The first and second stage reservoirs are 300 cc and 150 cc, respectively. The orifice valves are 7.1 mm and 4.4 mm in diameter for the first and second stage, respectively. The duct between the TAD and OPTR is 13 cm long.

The design of the first stage regenerator emphasized a low pressure drop in order to insure that adequate pressure amplitude was available at the second stage. As a result of this, the performance of the first stage is poor, and we only reached a cold temperature of 255 K at the first stage for a 4.30% pressure amplitude at the TAD. The second stage reached a cold temperature of 180 K at this operating point.

The anemometers used are TSI 1260-T1.5[7]. Since we only desire phase information, it was not necessary to calibrate the anemometers. Figure 6 shows typical oscilloscope traces for one pressure sensor and the six hot wire anemometers in the system. The anemometer signal is at 2ω, where ω is the frequency of the pressure oscillation. A lockin amplifier referenced to the pressure oscillation in the TAD is used to measure the phase of the anemometer signal.

Model comparisons are made by first fixing the value of the pressure amplitude in the model at the warm end of the regenerator to be the same as the experimental data, and then adjusting the model value for the orifice impedance until the simulated pressure amplitude in the reservoir matches the data. This sets the mass flow at the warm end of the pulse tube to be the same as observed in the experiment. The second stage data are analyzed first, and the impedance at the warm end regenerator is calculated from the pressure and mass flow. The second stage impedance is then entered into the branch segment for the first stage model, and the first stage data are analyzed with the above method.

Pressure amplitude, pressure phase and velocity phase data for the first stage and second stage are shown in Figure 7 as a function of position in the OPTR, for a pressure amplitude of 4.30% at the TAD. The agreement between observed and predicted pressure amplitude is good, as seen in (a) and (b) of Figure 7. For the phase relationships in the first stage, the mass flow precedes the pressure amplitude at the warm end of the regenerator, is in phase with the mass flow just before the branch to the second stage, and lags the pressure amplitude at the cold end of the first stage pulse tube (just after the branch to the second stage). This is what we expected to see based on our original design. The agreement

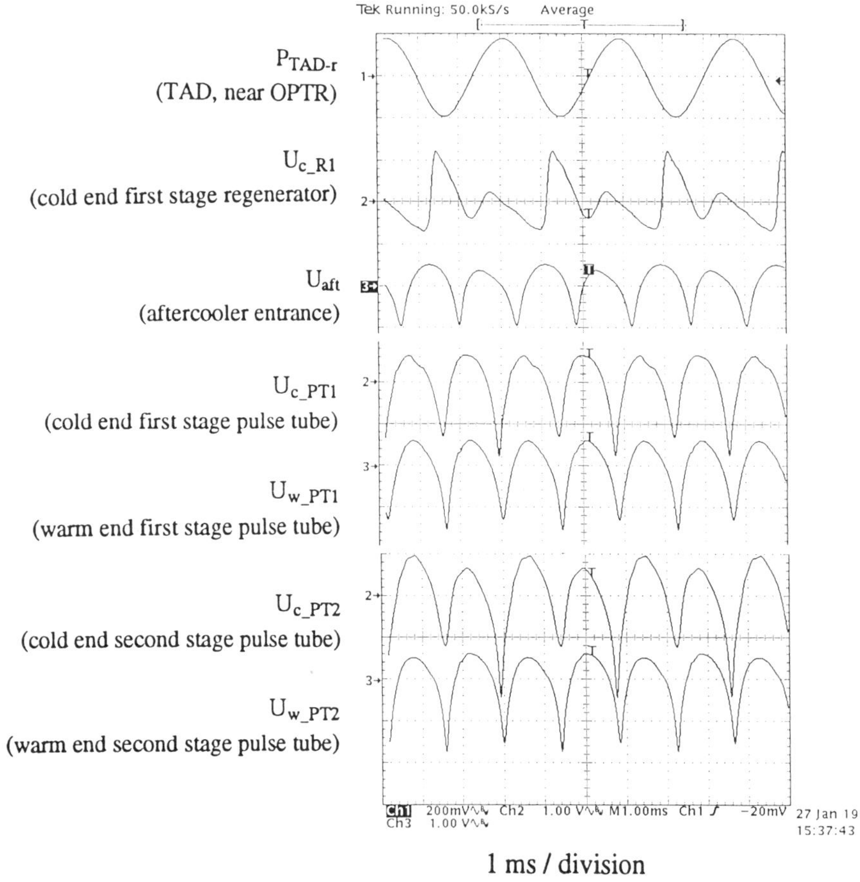

Figure 6. Oscilloscope traces for the anemometers in the 350 Hz two stage system. The locations of the anemometers are shown in Figure 5. The asymmetric traces for the cold end first stage regenerator and cold end second stage pulse tube indicate differing temperatures for the incoming and outgoing gas. The vertical scale is for reference only.

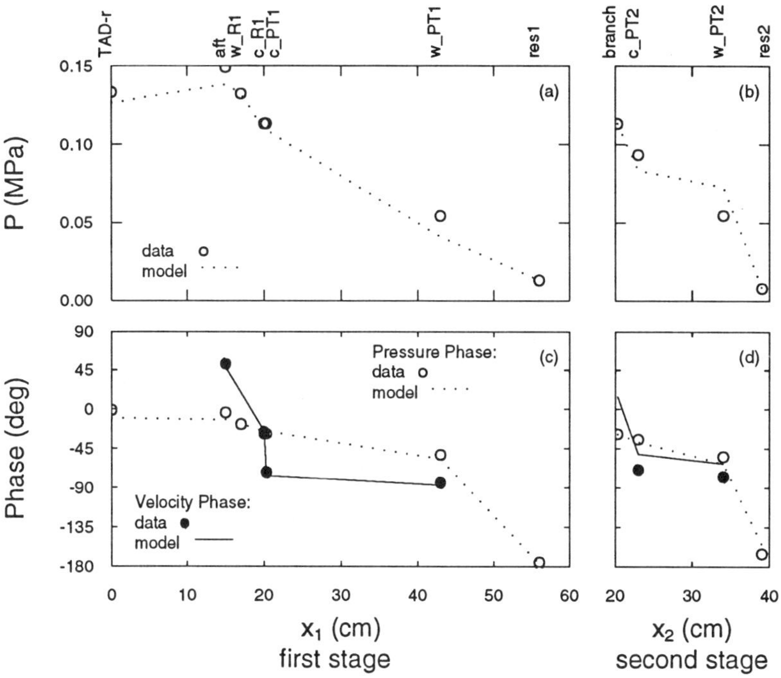

Figure 7. The pressure amplitude as a function of position in the OPTR for the first stage (a), and second stage (b). The model prediction is shown with a dotted line. The phase of the pressure and velocity are shown in (c) and (d) for the first and second stage, respectively. The dotted and solid lines give the model predictions. The top axis gives the location of the OPTR components, and is keyed to Figure 5. As seen in Figure 5, the locations c_R1 and c_PT1 are on either side of the branch, and therefore have the same pressure amplitude but different velocity phases. Mean pressure $P_m = 3.1$ MPa.

between observed and predicted phases is very good for the first stage, as seen in Figure 7c. The agreement for the velocity phase is not quite as good for the second stage. This may be due to sensitivity of the system to the geometry between the warm end of the pulse tube and the reservoir. A significant result is our confirmation that the mass flow lags the pressure at the cold end of the second stage pulse tube. These measurements provide an important quantitative verification of the design method and the concept of using the inertance of the pulse tube to shift the phase of the mass flow.

The regenerator loss for the first stage regenerator was 45 W when the pressure amplitude at the TAD was 4.30%; the model prediction was 28 W. The agreement for the regenerator loss is much better for a low pressure amplitude point: 19.4 W measured regenerator loss and 19.5 W predicted at a 2.84% pressure amplitude at the TAD. The regenerator loss for the second stage regenerator was 11.3 W at the 4.30% operating point, and the model prediction was 9.5 W. The heat rejected to the aftercooler at this point was 269 W, and the prediction was 192 W (29% discrepancy). The mass flow into the first stage reservoir was 8.8 g/s, and 2.7 g/s into the second stage reservoir.

In the results for the single stage 350 Hz OPTR, we mentioned that we saw unexpected behavior at the aftercooler heat exchanger, i.e. the gas temperature at the exit of the aftercooler was lower than the cooling water temperature. This behavior was not seen in the two stage OPTR. We observed instead that the exit gas temperature from the aftercooler was slightly above the cooling water temperature, as expected. The two systems used the same aftercooler heat exchanger, so other system parameters are affecting the aftercooler performance.

The poorer agreement between simulation and experiment for the phase in the second stage OPTR may be due to sensitivity of the system to the geometry at the orifice valve. The best design would eliminate any ducts going to and from the orifice valve, or use a carefully characterized capillary tube or porous plug as the impedance. Since it is difficult in practice to eliminate ducts going to and from a valve, the capillary or porous plug might be the best choice. The system is also sensitive to the design of the heat exchangers at either end of the pulse tube. Typical OPTR designs have used copper screen heat exchangers which are the same diameter as the pulse tube. We found that the small diameter of the pulse tubes in these designs caused an unacceptably high pressure drop in the heat exchangers. In order to reduce the pressure drop, we used copper screen heat exchangers of a diameter larger than the pulse tube, in combination with flow straighteners.

CONCLUSION

We have presented experimental results from single stage and two stage 350 Hz TADOPTRs that use the inertance of the pulse tube to shift the phase of the mass flow. In addition to pressure sensors, the two stage system included anemometers at key locations in the TADOPTR for measurement of the phase of the mass flow. The comparison of experimental results to simulations indicated good agreement in general, especially at low pressure amplitude. The agreement for the pressure amplitude, pressure phase and velocity phase are very good, especially for the first stage of the two stage system, and demonstrated that the design method using the inertance of the pulse tube to shift the phase of the mass flow is successful. This quantitative comparison to simulation results gives us confidence that other parameters determined by the simulation, such as mass flow, are also correct. As far as we know this is the first direct confirmation of the internal operation and phase relationships in an OPTR at key locations along its length. At this point we have confidence in our design tools and methods for designing orifice pulse tubes at frequencies up to 350 Hz.

ACKNOWLEDGMENT

This work is supported by the Advanced Research Projects Agency, MDA972-92-C-0011.

The Los Alamos contributions to this work were partially supported by the US Department of Energy's Technology Transfer Initiative.

REFERENCES

1. R. Radebaugh, K.M. McDermott, G.W. Swift, and R.A. Martin, Development of a Thermoacoustically Driven Orifice Pulse Tube Refrigerator, Proc. of Fourth Interagency Meeting on Cryocoolers, 205 (1991).
2. A. Kirk, On the Mechanical Production of Cold, Proc. Ins. Civil Eng. **37**, 244, London (1874).
3. J.W.L. Kohler, The Stirling Refrigeration Cycle, Sci. Am. **212**(4), 119 (1965).
4. W.C. Ward and G.W. Swift, Design environment for low amplitude thermoacoustic engines, J. Acoust. Soc. Am. 95, 3671 (1994). Fully tested software and users guide available from Energy Science and Technology Software Center, US Department of Energy, Oak Ridge Tennessee. For a beta-test version, contact ww@lanl.gov (Bill Ward) via Internet.
5. G.W. Swift and W. C. Ward, Simple harmonic analysis of stacked-screen regenerators, submitted to J. Thermophys. Heat Trans.
6. J. Gary and R. Radebaugh, An improved numerical model for calculation of regenerator performance (REGEN3.1), Proc. of Fourth Interagency Meeting on Cryocoolers, 165 (1991).
7. TSI Inc., 500 Cardigan Road, P.O. Box 64394, St. Paul, MN 55164.

MINIATURIZATION OF A CO-AXIAL PULSE TUBE COOLER WITH LINEAR MOTOR DRIVE COMPRESSOR

Chao Wang, Jinghui Cai, Wenxiou Zhu and Yuan Zhou

Cryogenic Laboratory, Chinese Academy of Sciences
P.O.Box 2711, Beijing 100080, China

ABSTRACT

In order to minimize a pulse tube cooler for practical use, a co-axial version with linear motor drive compressor is designed and investigated. The numerical predictions and experiments show that the heat transfer between the regenerator and pulse tube retards the performance. Experiments verify that a modified co-axial version suggested by the present authors greatly improves the performance. A temperature of 68.5K is achieved by the modified cooler with system pressure of 1.6MPa and frequency of 50Hz. The swept volume of the compressor is about 2 cm^3. This compact design is very similar to a split Stirling cooler and commercially available for pulse tube coolers.

INTRODUCTION

Pulse tube cooler is very attractive for cryogenic cooling because no moving part at its cold head leads to long life, high reliability, less vibration and lower cost. The innovations and developments achieved in recent years enable the pulse tube cooler to have comparable refrigeration capacity and efficiency to Stirling or G-M cooler. More and more laboratories in the world intend to develop miniature pulse tube cooler for space application etc.

There are three configurations for the design of pulse tube cooler, which are as follows:

(1) the pulse tube and regenerator are arranged in parallel (Figure 1 a), called U type pulse tube;

(2) the pulse tube and regenerator are connected at a line (Figure 1 b), called line type pulse tube;

(3) co-axial design in which the pulse tube and regenerator are arranged concentrically (Figure 1 c), called co-axial type pulse tube.

Obviously, the co-axial pulse tube cooler is very compact and convenient for practical applications. The penalty of it with the conventional configuration is lower refrigeration efficiency.

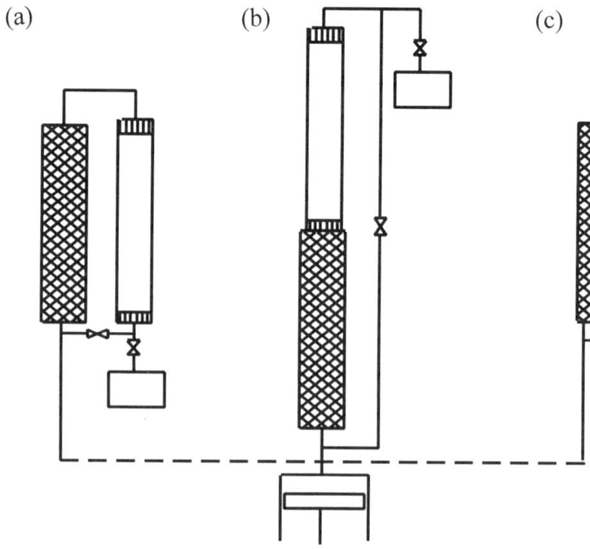

Fig.1. Schematic diagram of various pulse tube arrangement.

It is encouraging that a line type pulse tube cooler with linear motor drive compressor developed by C.K.Chan et al [1] reached a cooling power of 0.53W at 80K for an input power of 17.8W to the compressor. The lowest temperature of 98K was obtained by K.Kanao et al [2] with a U type pulse tube cooler with linear compressor.

To meet the needs of application for small size, low weight and high efficiency, a practical co-axial pulse tube cooler with linear compressor is designed and studied in this paper. The effects of co-axial design on the performance are analysed and compared by numerical simulation and experiments. Many efforts are undertaken to improve the efficiency of the cooler.

ANALYSES OF CO-AXIAL PULSE TUBE CONFIGURATION

The numerical modelling of pulse tube cooler was developed by the present authors and described in references [3,4]. The line type pulse tube for the typical design is simulated. The swept volume of the compressors is $2cm^3$. The temperature of the cold head is 80K and that

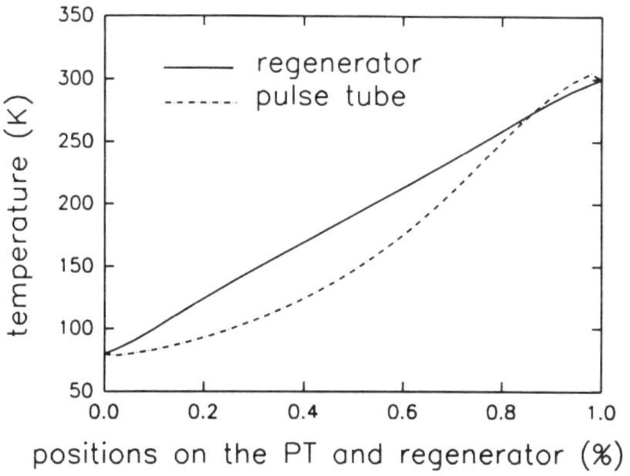

Fig.2. Temperature distributions on PT and regenerator.

Table 1. Predicted results of a line type and a co-axial type pulse tube cooler

	P-V work (W)	Hr (W)	Hc (W)	Qc (W)	η(%)
line type	17.92	0.66	2.47	1.58	8.82
co-axial type	17.96	0.67	2.12	1.22	6.79

where Hr is the enthalpy flow in the regenerator; Hc the enthalpy flow at the cold end of the pulse tube; Qc the refrigeration power; η the efficiency. The performance of conventional co-axial configuration is lower than that of the line one due to the effects of the heat transfer between the pulse tube and regenerator.

of the hot end is 300K. The operating frequency is 50Hz. The predicted temperature distributions along the regenerator and pulse tube are shown in Figure 2. The temperature distributions of them are sufficiently different. So, there is heat transfer between the pulse tube and regenerator for the co-axial configuration.

In order to investigate the effects of co-axial design, the numerical modelling for this configuration is developed in this paper. The heat transfer between the pulse tube and regenerator is considered in this model. The predicted results of the line type and co-axial type pulse tube coolers are shown in table 1.

CO-AXIAL PULSE TUBE DESIGN

A photograph of the practical design of co-axial pulse tube cooler is shown in Figure 3 and the cooler cross section in Figure 4. The compressor design utilizes a dual opposed pistons configuration in order to minimize output vibration. Each piston in the compressor is directly coupled to the moving coil of the linear motor in a permanent magnet system. The total effective swept volume of the compressor is about 2 cm^3.

Fig.3. Photograph of the co-axial pulse tube cooler.

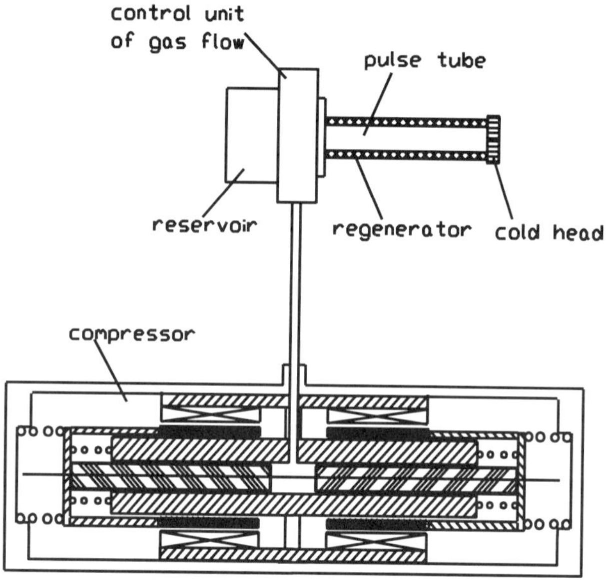

Fig.4. Cross section of the cooler.

The outside diameter of the cold head is 13mm. The pulse tube is placed within the annular regenerator. The flow channels of the pulse tube and regenerator are connected by the cold heat exchanger. The parameters of the regenerator and pulse tube are optimized by the numerical predictions.

The double-inlet orifice version is used in the present design. The hot ends of the pulse tube and regenerator are amounted on a gas flow control unit. A reservoir is amounted on the backside of it. The gas flow from the pulse tube to the reservoir and the gas flow through the bypass tube are controlled by this unit. It is apparent that the outside configuration of the designed miniature pulse tube cooler is very similar to that of split Stirling cooler. This compact design is commercially available for pulse tube coolers.

In order to compare the effects of the heat transfer between the pulse tube and regenerator, two pulse tubes with the same inner parameters are designed and fabricated. One pulse tube is made of stainless steel. Another is made of nylon. The experimental results are compared in Table 2.

The refrigerative temperature with nylon pulse tube is 10K lower than that with stainless steel one. The axial conduction loss of the stainless steel tube is about 0.14W larger than that of nylon one. So, the main part of loss in the cooler with stainless steel tube is caused by the heat transfer between the pulse tube and regenerator. It is concordant with the numerical predictions.

Table 2. Performance comparisons between SS pulse tube and nylon pulse tube

	f (Hz)	Pa (MPa)	Rc	Tc(K)
SS pulse tube	50	0.9	1.42	107
nylon pulse tube	50	0.9	1.42	97

where f is the operating frequency, Pa the average pressure, Rc the pressure ratio in the compressor, Tc the cold head temperature.

Table 3. Performance comparisons between the conventional version and modified version

	f (Hz)	Pa (MPa)	Tc (K)
nylon pulse tube	50	0.9	97
modified pulse tube	50	0.9	78

Fig.5. Photograph of the modified co-axial pulse tube cooler.

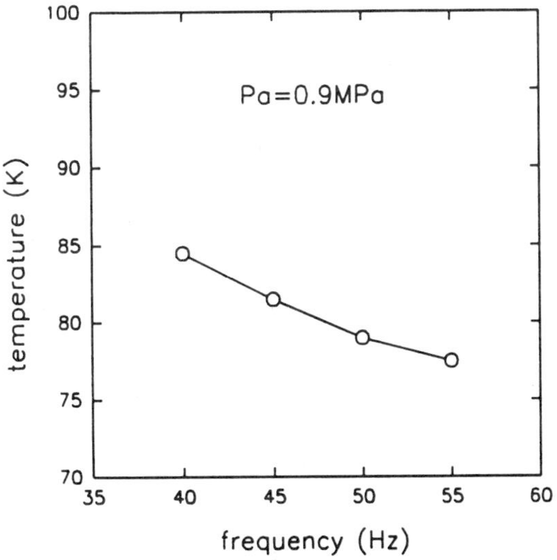

Fig.6. Temperature variations with frequency.

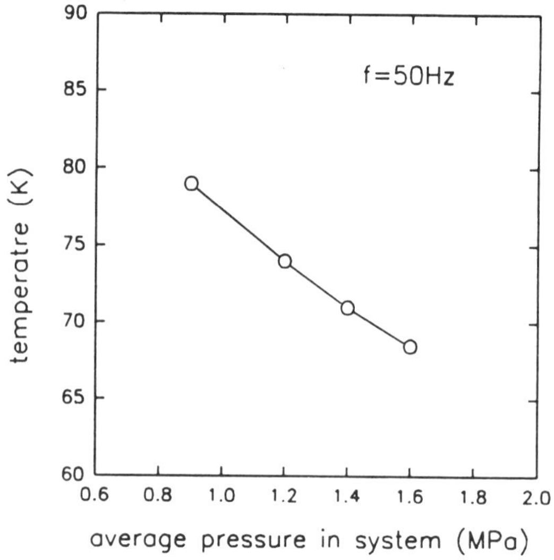

Fig.7. Temperature variations with pressure.

PERFORMANCE OF A MODIFIED CO-AXIAL PULSE TUBE COOLER

According to the above analyses and experiments, a modified configuration is used in present study to improve the performance of the co-axial pulse tube cooler. A photograph of the modified configuration is shown in Figure 5. The outside diameter of the cold head is only 8mm. The designed mini cold head is very convenient for packing with instruments.

The comparisons between the modified co-axial version and the conventional co-axial version are shown in Table 3. The structure parameters of them are almost the same.

The modified version greatly improves the performance of the cooler. The temperature of it is 19K lower than that of the conventional co-axial version.

The variations of the temperature with operating frequency are shown in Figure 6 and that with system pressure in Figure 7. The lowest temperature of 68.5K is obtained with the system pressure of 1.6MPa and frequency of 50Hz.

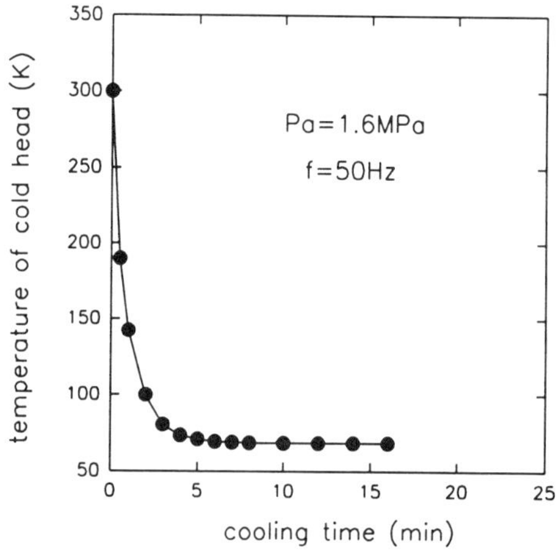

Fig.8. Cool-down curve of the co-axial pulse tube cooler.

It would be noticed from Figure 6 and Figure 7 that the operating parameters have not been optimized. The optimization is limited by the performance of the linear compressor used in the present study. The regenerator is made of stainless steel. We estimate that the conduction loss through the regenerator may be as large as 0.46W. This loss can be reduced by 50% by using titanium in place of stainless steel. Therefore, it is expected that much lower temperature could be reached by improving the performance of the compressor and using proper materials for the regenerator.

The comparisons between the double-inlet version and the conventional orifice version show that the double-inlet version still have great positive effects on the performance of our miniature pulse tube cooler. The temperature of the double-inlet version is about 10-20K lower than that of the conventional one in the present study.

The typical cool-down curve of the co-axial pulse tube cooler is shown in Figure 8. It takes about 8 minutes to reach the lowest temperature.

CONCLUSIONS

A miniature co-axial pulse tube cooler with linear motor drive compressor is designed and its compact configuration is commercially available. The numerical predictions and experiments show that the heat transfer between the pulse tube and regenerator retards the performance of the cooler.

Experiments verify that a modified co-axial pulse tube cooler suggested by the present authors greatly improves the performance. The lowest temperature of 68.5K is achieved with this modified version. The swept volume of the compressor is about 2 cm^3.

Better performance is expected to be obtained by improving the performance of the linear compressor and using proper materials for the regenerator.

ACKNOWLEDGMENT

Research supported by the National Natural Science Foundation and Postdoctoral Science Foundation of China.

REFERENCES

1. Chan,C.K., Jaco,C.B., Raab,J. et al. Miniature pulse tube cooler, Proc. of the 7th International Cryocooler Conference, Santa Fe, New Mexico, Nov., 1992, P.113
2. Kanao,K., Watanabe, N., and Kanazawa,Y., A miniature pulse tube refrigerator for the temperature below 100K, Cryogenics (1994),Vol.34, ICEC Supplement,P.164
3. Wang,C., Wu, P., and Chen, Z., Numerical modelling of an orifice pulse tube refrigerator, Cryogenics (1992), Vol.32, P.785
4. Wang,C., Wu,P., and Chen, Z., Numerical analysis of a double-inlet pulse tube refrigerator, Cryogenics (1993), Vol.33, P.526

COOLING CHARACTERISTICS OF A MODIFIED MINIATURE PULSE TUBE REFRIGERATOR

T. Haruyama and H. Inoue

KEK, National Laboratory for High Energy Physics
Oho 1-1, Tsukuba, Ibaraki 305, Japan

ABSTRACT

The cooling performance of a split type miniature pulse tube refrigerator with a flexure spring compressor was investigated. The diameter and length of the miniature pulsc tube were 5 mm and 60 mm, respectively. A stack of stainless steel mesh of #300 was used as a regenerator. An Oxford type compressor was driven at frequencies from 40 to 55 Hz with a swept volume of 1-2 cm^3. The lowest temperature of 99.6 K with no load was achieved for an input power of 30 W at a filling pressure of 1.32 MPa. The compressor back-side volume was used as a reservoir and capillary tubes were used instead of double inlet and orifice needle valves. The influence of orifice capillary tube length on the cooling performance of the miniature pulse tube refrigerator was investigated.

INTRODUCTION

In a previous report[1] on a miniature split type pulse tube refrigerator, a cooling power of 0.16 W at 195 K for an input power of 6 W was reported. For approaching the final target of 1 W at 80K for 30 W input power, a refrigerator was redesigned mainly to reduce the void volume in a system. Two needle valves for double inlet and orifice valves were removed and capillary tubes were used[2] for less void volume and more effective phase shifting between pressure and displacement of the working gas. Compressor back-side volume was used for a buffer reservoir as in the previous system.

The cooling performance of an orifice and double inlet pulse tube refrigerator was studied by varying frequencies from 40-55 Hz in a filling pressure range of 1.3-1.8 MPa. The compressor piston position was well regulated by adding DC bias current so as to apply an input power of over 30 W. The lowest temperature dependencies on the length of capillary tube, frequency and filling pressure were investigated.

MINIATURE PULSE TUBE REFRIGERATOR

A so-called Oxford type compressor which was designed and fabricated in our laboratory[3] was used for this experimental research. Main parameters of the refrigerator are listed in Table 1. More detailed parameters of the refrigerator were mentioned in a previous paper, and information on hardware and operating conditions are also listed in Table 2.

Table 1. Main parameters of the miniature pulse tube refrigerator

Parameters	Details
Compressor	Linear drive with flexure spring
Resonance frequency	49 Hz at P = 1.5 MPa
Comp. swept volume	1-2 cc
Pulse tube	5 mmΦ i..d. X 60 mm, V = 1.2 cc
Regenerator	8 mmΦ i..d. X 55 mm, Vvoid = 1.8 cc #300 SUS mesh X 700

Table 2. Hardware and operating conditions

Hardware	Details
Double inlet	Capillary tube (0.7 mmΦ i..d. X 40 mm)
Orifice	Capillary tube (0.4 mmΦ i..d. X 25, 50 cm)
Buffer	400 cc (compressor back volume)
Operating condition	
Frequency	40- 60 Hz
Pressure	1.3-1.8 MPa
Pressure ratio	1.2-1.3
Input power	~30 W
Piston position control	DC bias current

Figure 1 shows a schematic diagram of the refrigerator. A compressor, regenerator and pulse tube are connected by fine copper tubes of 1/8 inch in diameter and 0.5 mm in thickness. Compressor discharge pressures were monitored by pressure sensor (PMS30H bulk-type semiconductor) with an accuracy of a few percent. The displacement of the compressor piston was monitored by a Linear Variable Differential Transformer (LVDT) position sensor. Cold head temperature was monitored by a silicon diode thermometer (SMDT[4]) with an accuracy of ±0.5 K. There is no radiation shield between the cold head and a vacuum housing. During the experiment, a vacuum housing was connected to the high vacuum pump. Figure 2 shows a detailed draw of pulse tube and regenerator. Figure 3 shows a photograph of the compressor and cold head with a vacuum jacket.

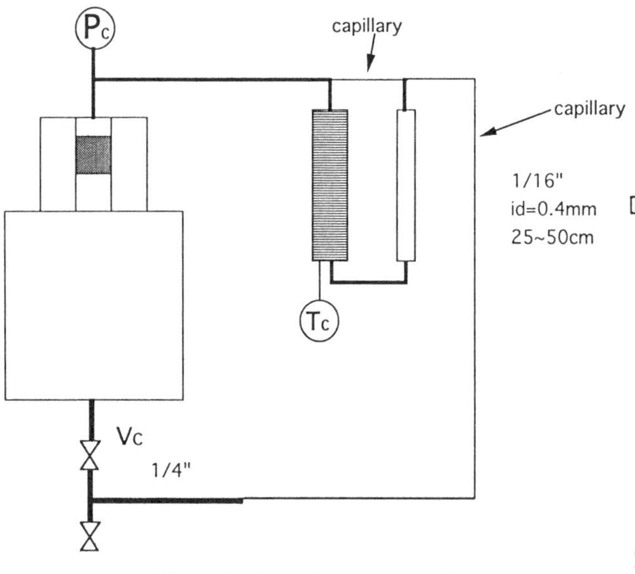

Figure 1. General arrangement of the miniature pulse tube refrigerator.

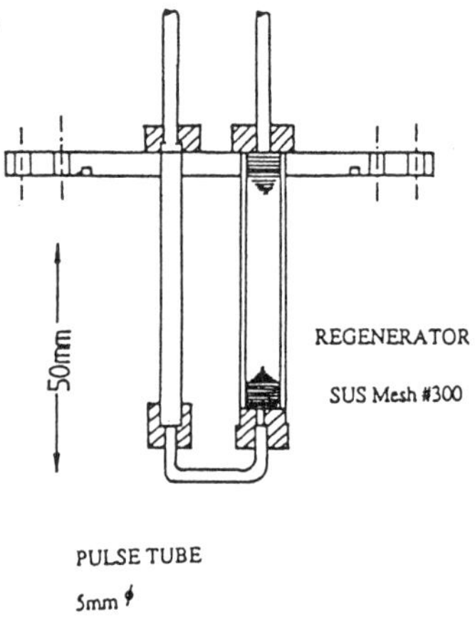

Figure 2.
Miniature pulse tube and regenerator.

Figure 3. Photograph of the miniature pulse tube refrigerator with a flexure spring compressor.

COOLING PERFORMANCE

The system was evacuated and filled with grade-A pure helium gas up to 1.3-1.8 MPa. The compressor was then operated at frequencies around 50 Hz. The input power to the compressor, Q_{IN}, was kept around 30 W. The compressor p-v work on working gas, Q_{PV}, was obtained by calculating the net area of the compressor p-v plane diagram. As the inner diameter and length of double inlet capillary was fixed, three parameters were varied. Those are frequency, filling pressure and the length of orifice capillary tube.

Cool down

Figure 4 shows the typical cool down characteristics when the refrigerator was operated at 50 Hz with a filling pressure of 1.32 MPa. Cooling speed is also shown in the figure. These data were obtained with a double inlet and orifice configuration using capillary tubes. Detailed information about the capillary tubes is given in Table 1.

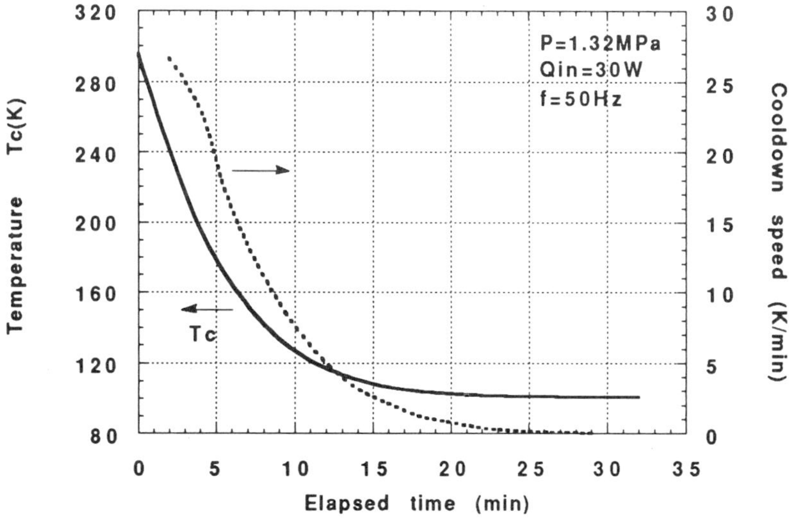

Figure 4. Cool down characteristics.

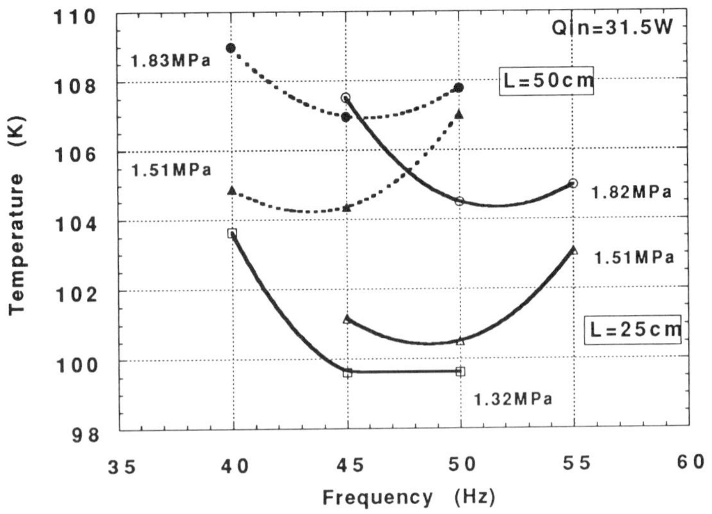

Figure 5. Frequency dependence of lowest temperature at Qin = 31.5 W. Filling pressure and length of orifice capillary are varied(25 and 50 cm) to investigate those influences.

Frequency dependence

The relationships between frequencies and the lowest temperature achieved are shown in figure 5. The influence of orifice capillary tube length and filling pressure on achieved temperature are also shown in the figure. Compressor input power was kept at 31.5 W, but the piston stroke varied from 4.6 to 5.8 mm. Its variation depended on filling pressure and frequency when the length of orifice capillary tube was same. Increasing the filling pressure results in a decreasing piston stroke. Also, increasing operating frequency results in a decreasing the piston stroke.

The lowest temperature was obtained at a filling pressure of 1.32 MPa with an operating frequency of 50 Hz and when the length of orifice capillary was 25 cm. However, when the compressor was driven at an input power of 15 W, the lowest temperature was obtained at a filling pressure of 1.51 MPa, not at 1.32 MPa even with the same capillary length. The typical wave form of the compressor delivery pressure, piston displacement and swept volume of compressor are shown in Figure 6. The compressor p-v diagram at 45 Hz, 50 Hz and 55 Hz is also shown in the figure.

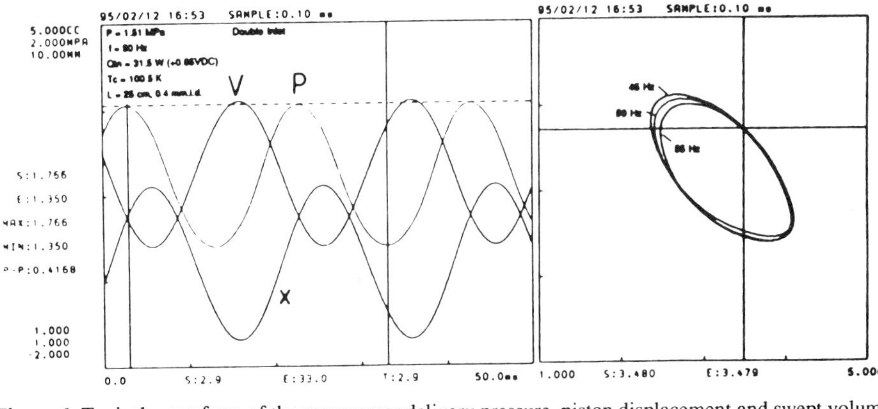

Figure 6. Typical wave form of the compressor delivery pressure, piston displacement and swept volume of compressor in a p-v diagram.

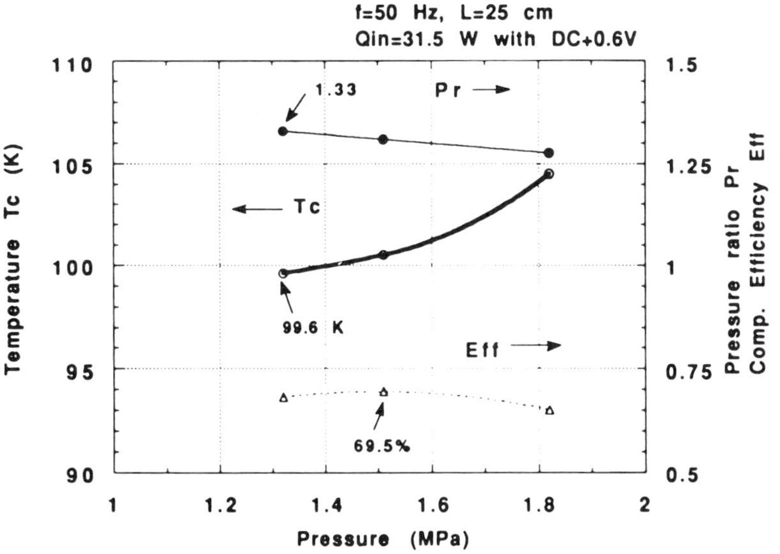

Figure 7. Filling pressure influence on the lowest temperature. The compressor pressure ratio and compressor efficiency are also shown.

Filling pressure dependence

Figure 7 shows filling pressure influence on the lowest temperature. The compressor pressure ratio, defined as P_{MAX}/P_{MIN}, and compressor efficiency, defined as Q_{PV}/Q_{IN}, are also shown in the figure. The operating frequency was 50 Hz, the length of orifice capillary was 25 cm and compressor input power was kept at 31.5 W. A much lower temperature was achieved when the filling pressure was decreased. The maximum pressure ratio was 1.33 at a filling pressure of 1.31 MPa, and the maximum compressor efficiency of 69.5% was obtained at a filling pressure of 1.51 MPa.

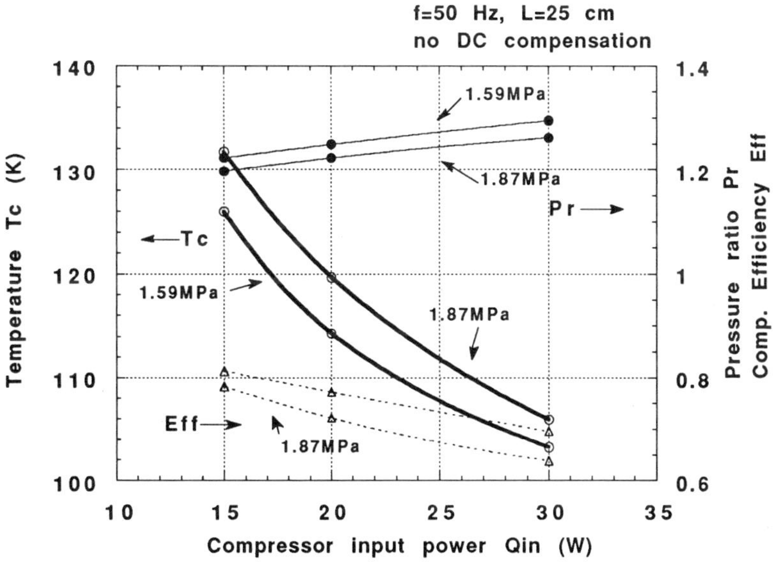

Figure 8. Achieved lowest temperature versus compressor input power at filling pressures of P=1.59 and 1.87 MPa.

The filling pressure was decreased to 1.18 MPa. However, noise came from the compressor and an input power of more than 28 W could not be applied. The lowest temperature for this condition was 103 K.

Input power dependence

Figure 8 shows the compressor input power dependence on the lowest temperature at filling pressures of 1.59 and 1.87 MPa. The operating frequency is 50 Hz, and the orifice capillary length is 25 cm with a double inlet configuration.

Pressure ratio dependence

Figure 9 shows the pressure ratio dependence on the lowest temperature. For all cases, the refrigerator configuration is the same (capillary double inlet tube and orifice capillary tube of 25 cm in length). Various pressure ratios were obtained by changing the filling pressure, input power and piston position control by using DC bias current. Data for the Stirling refrigerator using the same compressor are also shown in the figure for comparison.

In the case of the miniature pulse tube refrigerator, the temperature could not be lowered even if the pressure ratio was increased from 1.26 to 1.33. Under the same compressor operating conditions using the same compressor system, the Stirling refrigerator achieved a lowest temperature of 50 K, which is 50 degrees lower than that of miniature pulse tube refrigerator.

SUMMARY

Cooling characteristics of a miniature pulse tube refrigerator, a split-type with a flexure spring compressor, and with capillary tubes for double inlet and a orifice valves were investigated. The lowest temperature of 99.6 K was achieved for a compressor input power of 30 W. The optimum length of orifice capillary tube was found to be 25 cm for 0.4 mm$^{\phi}$ i.d. at a frequency of 50 Hz and a filling pressure of 1.32 MPa. A simple DC bias current method worked well for piston position control for large piston displacement. Even when the pressure ratio is close to that of Stirling refrigerator, the achieved lowest temperature by the pulse tube refrigerator is 50 degrees higher than the Stirling refrigerator. Thus, cooling mechanism and losses peculiar to the miniature pulse tube refrigerator should be investigated.

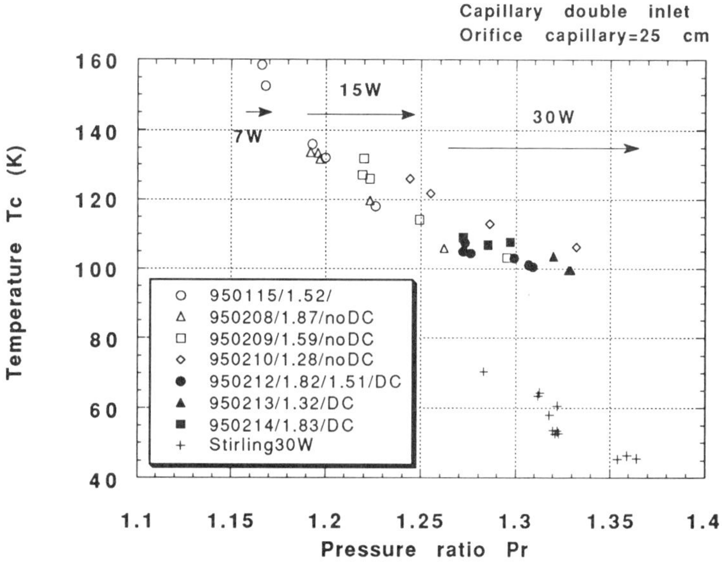

Figure 9. Lowest temperature versus the pressure ratio of pulse tube refrigerator. Results of Stirling refrigerator by same compressor is also shown.

ACKNOWLEDGMENT

The authors wish to thank Prof. S. Iwata, the Director of Physics Department, for his continuous encouragement of this work.

REFERENCES

1. T. Haruyama and H. inoue, Cooling performance of a prototype miniature pulse tube refrigerator with a flexure spring compressor, Presented at the 8th International Cryocooler Conference, Colorado (1994)
2. K. Kanao, N. Watanabe and Y. Kanazawa, A miniature pulse tube refrigerator for temperature below 100K, *Cryogenics* Vol. 34 ICEC Supplement, 167-170 (1994)
3. T. Haruyama. and H. Inoue, Design and fabrication of a prototype Stirling Cryocooler for space borne applications, *Adv Cryog Eng* Vol. 39, 1367-1373 (1994).
4. M. G. Rao, R. G. Scurlock and Y. Y. Wu, Miniature silicon diode thermometer for cryogenics, *Cryogenics* Vol. 23, 635-638 (1983)

MINIATURE DOUBLE-INLET PULSE TUBE CRYOCOOLER: DESIGN BY THERMOACOUSTIC THEORY COMPARED WITH PRELIMINARY EXPERIMENTAL RESULTS

J. H. Xiao*,[1] J. H. Yang,[1] and Z. D. Tao[2]

[1]Cryogenic Laboratory, Chinese Academy of Sciences
PO Box 2711, Beijing 100080, P. R. China
[2]Institute of Acoustics, Chinese Academy of Sciences
PO Box 2712, Beijing 100080, P. R. China

ABSTRACT

A miniature double-inlet pulse tube cryocooler was designed using the thermoacoustic theory for regenerative cryocoolers developed by the author. The available pressure oscillator determined the operating parameters used in the design. These parameters were: a filling pressure of 1.2 MPa, operating frequency of 48 Hz, and a swept volume of 4 cm^3. This pressure oscillator, recently designed and built for use with studies of pulse tube cryocoolers, utilized a linear motor, clearance seal, and linear-arm flexure bearings. The pulse tube and regenerator were fabricated using the U-tube configuration, which was split from the pressure oscillator by about 280 mm. A lowest temperature of 78 K was achieved with this cooler at present. Our theoretical and experimental results show that the double-inlet concept yields good performance at this high frequencies. The lowest temperature achieved by introducing the double-inlet valve is about 23 K lower compared with that of the single-orifice type. This paper will report on the design approach and the preliminary experimental results.

INTRODUCTION

An important aspect of the development of a practical compact cryocooler for cryo-electronics devices is that of the increased operating frequency. The compactness, efficiency, reliability and long life time are critical in placing a cryocooler into practical use. Pulse tube cryocoolers have several advantages over other regenerative type cryocoolers relating to some of these characteristics, especially when they are used to cool infra-red

*Present address: Cryogenic Technology, NIST, 325 Broadway, MS 838.09, Boulder, CO 80303, USA

detector, superconducting electronics devices and other cryo-electronics devices. Pulse tube cryocoolers have made great progress in recent years. Despite its intrinsic simplicity and reliability, the novel pulse tube cryocoolers have achieved an efficiency comparable to Stirling and G-M cryocoolers in many cases[1-4].

Basically, pulse tube cryocooler and other regenerative cryocoolers are physically similar in principle. So the pulse tube cryocooler and regenerative cryocoolers of other type can share the same theoretical fundamentals and experimental data, even the manufacturing experiences. Furthermore, in many cases they can share most of the mechanical parts or components, such as the compressor, the regenerator, etc.. All of these factors made the pulse tube cryocooler a favorite candidate for a cryocooler of next generation.

However, the understanding of the working mechanism of regenerative cryocoolers is not entirely adequate, especially a detailed understanding of the flow dynamics, heat transfer processes, energy converting mechanism, sources of irreversibility, and the mutual interacting of the oscillatory flow and heat transfer processes of the cryocooler. We must face this problem in the design of a new cryocooler using an existing theoretical model or a computer code. Otherwise, we will not obtain a satisfactory optimized design.

The first author has being involved in developing a new theoretical background for regenerative cryocoolers (based upon thermoacoustic effects) for several years. Having used the thermoacoustic theory for regenerative cryocoolers for several case studies and being satisfied with the results, we are attempting to design a pulse tube cryocooler using this theory. The so called 'high frequency' miniature pulse tube would be a good candidate for this work, because we have no experience with a pulse tube cryocooler working around 50 to 60 Hz. The purpose of this paper is to report on our design approach of a 50 Hz pulse tube cryocooler with the preliminary experimental results.

DESIGN OF THE PULSE TUBE COOLER BY THERMOACOUSTIC THEORY

Design of a new pulse tube cryocooler by using a theoretical model is different from calculating the performance of an existing pulse tube cryocooler using the same model. When one is involved in a work of calculating the performance of a known pulse tube cryocooler, he knows every detail of the configuration, geometry, thermophysical properties and operating parameters of the cooler. In the design of a cryocooler, the geometry and even some of the operating parameters of the cryocooler remain unknown, which is just what we need to decide.

At present, there are basically three classes of theoretical method of modeling regenerative cryocoolers: thermodynamic cycle analysis method, nodal analysis method and thermoacoustic method.

The thermodynamics cycle analysis method uses simple uniform lumped thermodynamic parameters to characterize the cooler. Making a lot of assumption to make the analysis possible, this model establishes the relation among the parameters of the cryocooler by using the total mass balance and energy balance, finally a very rough estimation of the performance of the cryocooler could be obtained. Good examples are the isothermal model and adiabatic model for Stirling cryocooler analysis. As this kind of model is too idealized, we can not expect it to yield a good geometry for design work.

The nodal analysis method[5-6] is a kind of numerical method, it uses one-dimensional differential equations of fluid dynamics and thermodynamics, it makes quasi-steady assumption so as to allow the use of available steady flow friction factor and heat transfer coefficient correlation data. To solve these one-dimensional, quasi-steady partial differential equations, the cryocooler is spatially discretized into a large number of segments and the time is divided into a series of small enough time steps. These procedures convert the differential equations in to a large number of finite-difference equations and can be solved with proper boundary and initial conditions. Despite the doubtful application of the steady flow experimental data of friction factor and heat transfer coefficient, especially with high operating frequencies, this method requires a long computation time to achieve a reasonable convergence. We can expect this kind of modeling to turn out a good design if the operating frequency is not too high to break the quasi-steady assumption, but the computation time could be months, for design work needed to compare hundreds, or even thousands of different geometries to achieve an optimized design suggested by the model.

Thermoacoustic analysis method is a kind of perturbation method. Generally speaking, it is a two-dimensional and transient analysis method, but it is a locally linearized harmonic analysis method at present. Nonlinear terms could possibly be included if further attempts to treat the nonlinear effects involved can be completed successfully. For the theoretical treatment of the regenerator, the most important component in a regenerative cryocooler, this method, and the nodal analysis method, use one-dimensional equations of fluid dynamics and thermodynamics to describe the flow dynamics and thermal characteristics of the regenerator, with variables representing spatial averages over volumes large enough to smooth over the details of the screen mesh or porous materials at each position. However, we can obtain the correlation equations for characterization of the flow and heat transfer processes from the capillary models of porous material. Although these correlation equations are only from a theoretical base now, and not from the experimental data (no transient oscillating flow friction and heat transfer experimental data are available at this time), we believe that the thermoacoustic analysis method for regenerative cryocooler design could be a good approach, especially at higher frequencies, because this method is physically a self-consistency transient model.

Transient modeling of regenerative cryocoolers provides the theoretical background to develop higher frequency cryocoolers. The author believes that further development of cryocoolers working around the liquid nitrogen temperature range will also be characterized by increased operating frequencies, because the cryocooler should be very compact to cool cryo-electronics devices. Perhaps we can consider the thermoacoustic theory for regenerative cryocoolers to be the initial stage of such transient modeling, since it seems to meet some of the needs for theoretical modeling.

In addition, using thermoacoustic theory to calculate the performance of an existing cryocooler is time saving. The time duration of computer calculations of performance of a regenerative cryocooler based on thermoacoustic analysis method is orders of magnitude faster (seconds or minutes) than that of the nodal analysis method (hours or days). This makes it possible to find out the optimal geometry suggested by the theoretical model in a reasonable time on a 486 or 586 computer.

Having started the theoretical work since the end of 1980's, a special effort has been made in recent years to develop thermoacoustic modeling and a computer code for the analysis and design of conventional and novel orifice pulse tube cryocoolers, including the double-inlet type and multi-bypass type pulse tube cryocoolers. As the fundamentals and the main elements of the thermoacoustic theory for regenerative cryocoolers have been published elsewhere[7-13], we only give the outline of the design approach here.

The computer program for design work based on thermoacoustic theory is mainly composed of four parts: the first part is a subroutine for definition of the structure, geometric configuration, and operation parameters of the pulse tube cryocooler, with adjustable geometrical dimensions; the second part is a subroutine for the thermophysical properties of the solid media and fluid medium; the third part is a subroutine to solve the thermoacoustic equations which determine the flow dynamics and thermal characteristics of the cooler, and the fourth part is a subroutine to optimize the geometry of the cryocooler, with the refrigeration power at a given cold end temperature as the target function for optimization. At the very beginning, we give an initial design of the pulse tube cryocooler to put into the computer code, then the computer program will automatically adjust the geometry of each component to achieve a higher refrigeration power until the highest is achieved, at which time the corresponding geometry represents the optimal design.

For our present practical design work, the operating parameters and the geometry of the pressure oscillator (or compressor) is given. This determines the main parameters of the design. These parameters were: a filling pressure of 1.2 MPa, operating frequency of 48 Hz, and a swept volume of 4 cm^3. The pressure oscillator, designed and built for use with studies of pulse tube refrigerators, utilize a linear motor, clearance seal and linear-arm flexure bearings.

Even with a given pressure oscillator, we should first decide the configuration of the pulse tube cryocooler, and furthermore there are still many dimensions to be determined. A double-inlet, split type, U-shape configuration is appropriate in this work, as shown in Figure 1. We further decided to specify all the other geometry of the cooler beside that of the regenerator and pulse tube, with the regenerator matrix material made of screen with a variable mesh size. Other important geometric parameters are as follows: the volume of the gas reservoir is 225 c.c.; both the hot end heat exchanger of regenerator and pulse tube are 15 mm in diameter, 15 mm in length, both filled with 80 mesh brass screen with wire diameter of 112 μm; the cold end heat exchanger are 9 mm in diameter and 10 mm in length which is filled with 200 mesh brass screen with wire diameter of 51 μm. A copper connecting tube is used to split the cooler from the pressure oscillator, the inner diameter of which is 2 mm with wall thickness of 1 mm and length of 280 mm. There also is a copper connecting tube at the cold end to connect the cold end heat exchanger and the pulse tube, the inner diameter of which is 2 mm with wall thickness of 0.5 mm and length of 70 mm.

The remaining geometry needed be designed are the diameter and length of the pulse tube, the diameter and length of regenerator and the lengths of the three types of stainless steel screen mesh, e.g. 150 mesh number with wire diameter of 61 μm, 200 mesh number with wire diameter of 51 μm and 300 mesh number with wire diameter of 31 μm, which will be used to fill into the regenerator as matrix material.

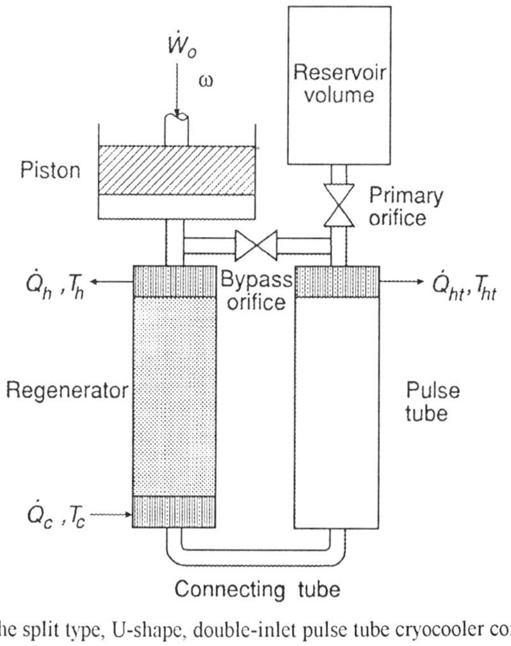

Figure 1 The split type, U-shape, double-inlet pulse tube cryocooler configuration.

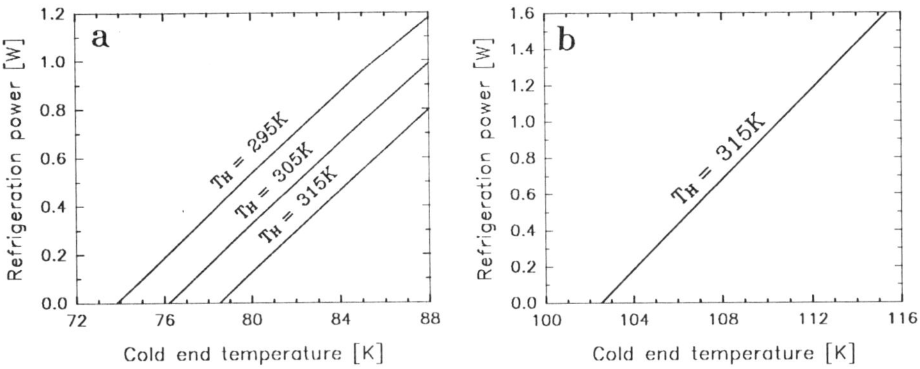

Figure 2 Performance of the designed pulse tube cryocooler.
a. Double-inlet type b. Single-orifice type

The results of optimization turns out to be reasonable. The diameter and length of the regenerator is 9 mm and 69.87 mm, respectively. The lengths of the screen mesh are 30.64 mm of 150 mesh screen, 15.36 mm of 200 mesh screen and 23.87 mm of 300 mesh screen. The diameter and length of the pulse tube is 6 mm and 79.68 mm, respectively. Here the diameters of regenerator and pulse tube are values rounded up to the nearest available tube size.

The calculated performances of the designed pulse tube cryocooler are shown in Figure 2. Here the performance of the orifice type was obtained by closing the double-inlet valve. The results show that the lowest temperature of the orifice type is 103 K for a hot end temperature of 315K, and the lowest temperature of the double-inlet type is 74 K for a hot end temperature of 295 K and 78.6 K for 315 K.

PRELIMINARY EXPERIMENTAL RESULTS

The designed pulse tube cryocooler was built and tested at a preliminary stage.

Further details of the geometry relating to the fabrication and setup of the pulse tube cryocooler are given here. The regenerator is comprised of a SS tube of 9 mm inner diameter, 70 mm in length, and wall thickness of 0.2 mm, filled with 360 disk of 150 mesh screen, 250 disk of 200 mesh screen and 320 disk of 300 mesh screen. The cold end heat exchanger is of 9 mm inner diameter, 8 mm in length, filled with 60 disk of 200 mesh brass screen. The pulse tube is a SS tube, with 6 mm inner diameter, 75 mm in length, and a wall thickness of 0.24 mm. The warm end heat exchanger of both regenerator and pulse tube is fabricated from a copper plate, that had a 15 mm diameter hole, 10 mm in length, and filled with 180 disks of 80 mesh copper screen.

Table 1 shows the results of three experiments. Experiment A was conducted only with the first stage mechanical vacuum pump, the lowest vacuum achieved was 2.2 Pa (2×10^{-2} torr), a lowest temperature of 87.8 K was obtained . Experiment B was conducted by turning on the second stage of the vacuum pump, a vacuum of 1×10^{-4} Pa was achieved, a lowest temperature of 82.6 K was obtained. In experiment C, a flow straighter, which was made by a 0.8 mm thick copper plate with 1 mm diameter fine holes, was installed at the cold end of the pulse tube, a better lowest temperature of 77.8 K was achieved.

Figure 3 shows the cooling down process of experiment B. At the first 20 minutes we left the double-inlet valve being closed, and the orifice needle valve was adjusted to its best opening, a lowest temperature of 106.5 K was achieved. From that time on, we further

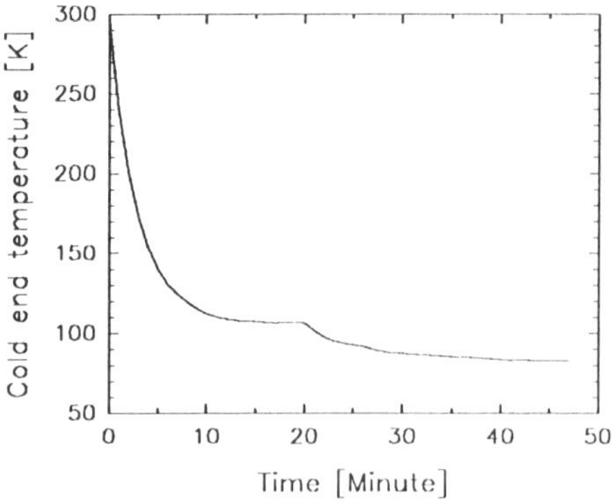

Figure 3 Cooling down characteristics of experiment B.

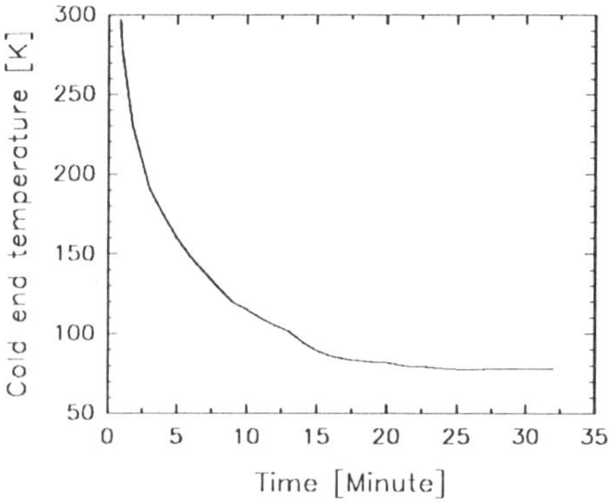

Figure 4 Cooling down characteristics of experiment C.

Table 1 Preliminary experimental results

Experiment	Vacuum	Lowest temperature	Remarks
A	2 Pa	88 K	poor vacuum
B	10^{-4} Pa	83 K	good vacuum, orifice 106.5 K
C	10^{-4} Pa	78 K	with flow straighter

adjusted the double-inlet valve and the orifice valve to their best opening, then the lowest temperature of 82.6 K was achieved.

Figure 4 shows the cooling down characteristics of experiment C with the flow straighter installed at the cold end of the pulse tube. This time we let both the orifice valve and double-inlet valve adjusted to their best opening, a lowest temperature of 77.8 K was achieved at the cold end of the regenerator.

Figure 5 and Figure 6 show the piston displacement, the pressure fluctuation at the middle of the connecting pipe at the warm end and the P-V diagram of the pressure oscillator, respectively, while cold end was at the lowest temperature in experiment B.

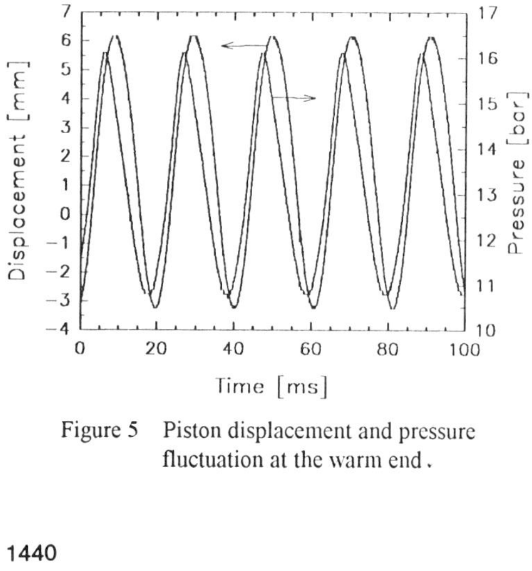

Figure 5 Piston displacement and pressure fluctuation at the warm end.

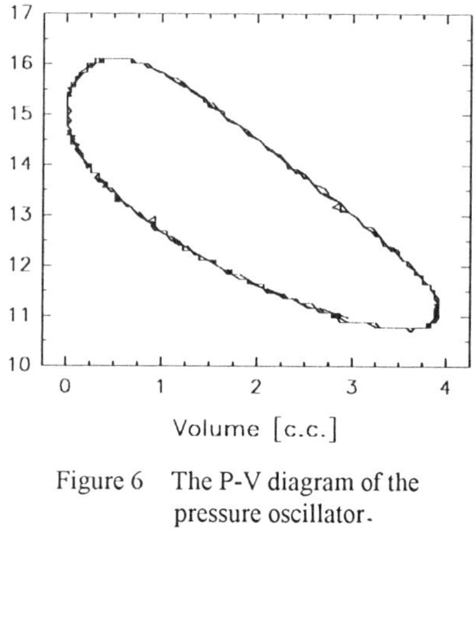

Figure 6 The P-V diagram of the pressure oscillator.

CONCLUSIONS

The attempt to design a pulse tube cryocooler by using thermoacoustic theory for regenerative cryocoolers is basically successful. The following conclusions can be drawn from the present work.

Thermoacoustic theory for regenerative cryocoolers is a useful tool for the design of new pulse tube cryocoolers geometries. It takes reasonable computation time and the results are within acceptable accuracy compared with that of the preliminary experimental results.

Both theory and experiment shows that the double-inlet concept performs excellent at a frequency around 50 Hz.

Vacuum level has a noticeable influence on the lowest temperature of the pulse tube cooler.

The flow straighter at the cold end of pulse tube also has a noticeable positive influence on the performance of the pulse tube cryocooler operating around 50 Hz.

ACKNOWLEDGMENTS

The authors would like to thank Wen Xiu Zhu and Prof. Yuan Zhou for their active and very helpful assistance in conducting the experiments. The authors also wish to express their acknowledgment to Dr. Ray Radebaugh of NIST, for his kind help in preparing and critical reading of this paper. This research was supported by the National Science Foundation of China and the Chinese Academy of Sciences.

REFERENCES

1. Radebaugh, R., A review of pulse tube refrigeration, "Advances in Cryogenic Engineering", Plenum Press, New York (1990), Vol. 35, p1191
2. Zhu, S. W., Wu, P. Y. and Cheng, Z. Q., Double-inlet pulse tube refrigerators: an important improvement *Cryogenics* 30:514 (1990)
3. Chan, C. K., Jaco, C. B., Raab, J., Tward, E. and Waterman, M., Miniature pulse tube cooler "Proceedings of the 7th International Cryocooler Conferences", Air Force Report No. PL-CP-93-1001 (1993), p113
4. Matsubara, Y. and Gao, J. L. Novel configuration of three-stage pulse tube refrigerator for temperature below 4 K *Cryogenics* 34:259 (1994)
5. Wang, C., Wu, P. Y. and Cheng, Z. Q., Numerical analysis of double-inlet pulse tube refrigerator *Cryogenics* 33:526 (1993)
6. Gary, J., O'Gallagher, A., and Radebaugh, R., A numerical model for regenerator performance, "Technical Report", NIST, Boulder (1994)
7. Xiao, J. H., Thermoacoustic heat transportation and energy transformation. Part I: Formulation of the problem *Cryogenics* 35:15 (1995)
8. Xiao, J. H., Thermoacoustic heat transportation and energy transformation. Part II: Isothermal wall thermoacoustic effects *Cryogenics* 35:21 (1995)
9. Xiao, J. H., Thermoacoustic heat transportation and energy transformation. Part III: Adiabatic wall thermoacoustic effects *Cryogenics* 35:27 (1995)
10. Xiao, J. H., Thermoacoustic theory for cyclic flow regenerators. Part I: Fundamentals *Cryogenics* 32:895 (1992)
11. Xiao, J. H., Thermoacoustic theory for regenerative cryocoolers: A case study for a pulse tube refrigerator "Proceedings of the 7th International Cryocooler Conferences", Air Force Report No. PL-CP-93-1001 (1993), p305
12. Xiao, J. H., and Wang, X. T., Regenerator irreversibility analysis based on thermoacoustic theory "Proceedings of the 4th Joint Sino-Japanese Seminar on Cryocoolers and Concerned Topics", Beijing, China, (1993), p242
13. Xiao, J. H. and Yang, J. H., Computation of the performance of novel orifice type pulse tube cryocooler using thermoacoustic theory, "Proceedings of the 6th National Cryocooler Conference of China", Tianjing, China (1994) (in Chinese)

MULTIPLE COLD FINGER CRYOCOOLER WITH VOLTAGE ISOLATION

J.M. Pfotenhauer and J. Yuan

Applied Superconductivity Center
University of Wisconsin - Madison
Madison, WI 53706

ABSTRACT

A design is presented for a single Orifice Pulse Tube (OPT) cryocooler to conductively cool a pair of high temperature superconducting current leads which are at voltage potentials differing by $\pm$ 3.5 kV. A single compressor, regenerator, and ballast is combined with two identical pulse tubes, the cold ends of which are metallically connected to the two current leads. The necessary voltage insulation is incorporated in the cryocooler structure at the regenerator - pulse tube connection, and in the cold end to warm end construction of the pulse tube itself. Voltage insulation calculations account for voltage breakdown in the helium gas (working fluid) and voltage creep along the insulating members. This approach provides a general method of optimizing the cooling power to devices at various voltage potentials since the usual problem presented by the Wiedeman Franz law is circumvented by the heat transfer through the pulsing helium gas flow.

INTRODUCTION

Of the various methods used for cooling high temperature superconducting current leads, the option of using an intermediate thermal intercept to absorb heat from the upper conventional stage offers one of the best methods for minimizing the room temperature refrigeration power associated with the current lead heat load. In many superconducting systems the thermal intercept is typically accomplished by a cryocooler. The obvious requirement of a single cryocooler used to cool both current leads is that it be electrically isolated from the leads so that the magnet circuit will not be shorted at that point. Various methods have been developed to address this problem including a liquid nitrogen heat pipe[1] and, more commonly, the use of an electrically insulating member between the cryocooler cold finger and the current lead. Designs incorporating Al_2O_3 or G-10 members have been reported in the literature[2,3]. For the μSMES system, the necessary voltage insulation between the cryocooler and current lead is substantial, with design insulation of $\pm$ 3.5 kV required between each current lead and ground.

The disadvantage of using electrically insulating members within the thermal conduction path between the cold finger of the cryocooler and the current lead is the unavoidable accompanying thermal insulation across that member. The thermal resistance becomes significant for current leads carrying current on the order of 1 kA or larger, because these will have a relatively large associated heat load at the intercept point. In general, the heat load at the thermal intercept point will depend on the intercept temperature and the method of cooling used for the upper conventional stage. However for example, a pair of 1.5 kA current leads whose upper stages are conduction cooled and operating between room temperature and 70 K will have an associated heat load of 120 watts. In such a situation the temperature drop across the insulating pieces can easily be as large as 10 K. A decreased cryocooler efficiency will be associated with this thermal resistance since the cryocooler must operate at a temperature 10 K colder to provide the 70 K intercept temperature. The added thermal resistance represents a decrease in efficiency for a Carnot refrigeration cycle from 0.30 to 0.25, or a minimum increase of 22% in the equivalently required room temperature power. It is therefore of significant advantage to eliminate this thermal resistance.

Although the necessary conductive mode of heat transfer between the cryocooler cold finger and the current leads is best accomplished by a metallic path which does not simultaneously allow electrical isolation, heat transfer within cryocoolers themselves is most typically accomplished by the flow of inert helium gas. Thus, the method of heat transfer within cryocoolers offers the opportunity both to transfer heat and to provide voltage insulation at the same time. The details of exploiting this possibility with an orifice pulse tube (OPT) cryocooler are the subject of the present paper. Means of exploiting the same possibility with Gifford-McMahon cryocoolers are being developed and will be presented in a future publication.

DESIGN APPROACH

The arrangement of a twin cold finger OPT cryocooler used to intercept heat from a pair of HTS current leads is shown in figure 1. In this application, the use of a single regenerator combined with two pulse tubes provides a space savings on the dewar top flange as compared to an alternate possibility of using two separate regenerator - pulse tube systems. Admittedly, the occupied space of the option in figure 1 is more than that required by a single regenerator - single pulse tube arrangement. However the refrigeration efficiency achieved by eliminating insulation pieces between the cold finger and the current leads is the gained advantage.

Voltage insulation within the body of the cryocooler is provided between the regenerator and the cold end of each pulse tube, and between the cold and warm ends of each pulse tube. This is accomplished by inserting a section of standard ceramic tubing between the regenerator and pulse tubes, and by replacing a section of the pulse tube wall (typically stainless steel) with an insulating material. A solid metallic (copper) conduction path can then be installed between the cold end of the pulse tube and the thermal intercept point on the current lead. When the superconducting magnet is charged or discharged, the cold ends of the pulse tubes become electrically "hot", but the remainder of the cryocooler stays at ground potential. For reasons of safety and reliability it is of advantage to minimize the portions of the cryocooler which become electrically charged.

The requirements for the electrically insulating pieces are detailed in the following paragraphs.

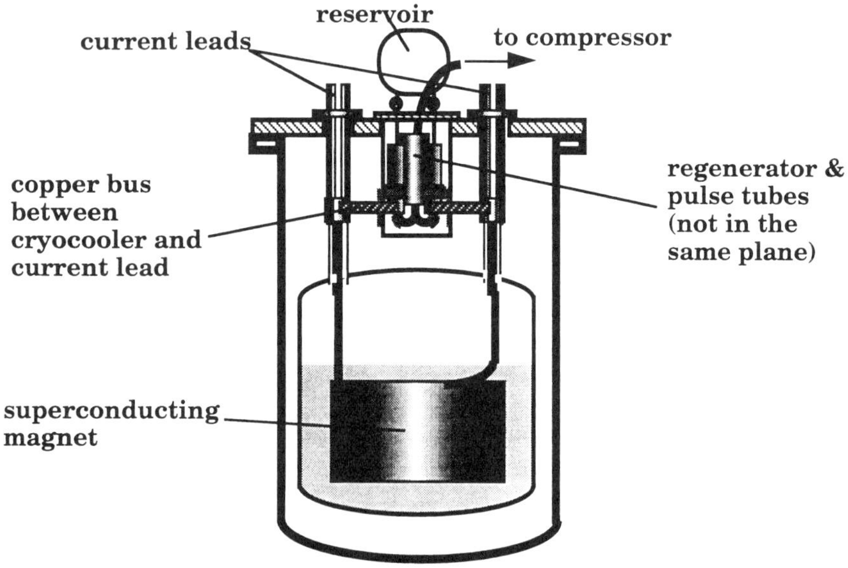

Figure 1. Arrangement of voltage insulated OPT for cooling HTS current leads.

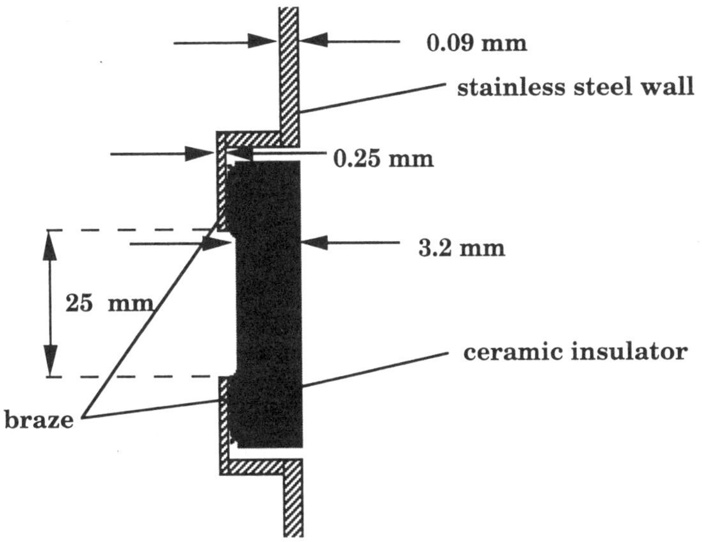

Figure 2. Stainless steel to ceramic joint in pulse tube wall.

Voltage insulation

The first requirement of the insulating pieces is that they provide a voltage stand-off of at least 3.5 kV. Provision for electrical insulation is a well studied issue in the field of superconducting systems. For the pulse tube application, voltage breakdown must be avoided in the helium gas inside the tube as well as in the vacuum space outside the tube. Schwenterly[4] graphically presents the breakdown voltage limits for various gases, including helium, as defined by Paschen's curve. In general, the breakdown voltage in gases is a monotonically increasing function of the product of the gas density and the separation distance between the surfaces of differing electrical potential. The lowest possible gas density therefore defines the minimum allowable separation without a voltage breakdown. In the pulse tube, the lowest density occurs at the warm end of the tube and at the minimum of the pressure oscillation. For the conditions of room temperature and 1.5 MPa pressure the required separation is only 1.3 mm. Considerations of voltage creep recommend a multiplier of at least 2.5 to this value.

Structural requirements

Because cryocoolers typically operate at moderately high pressures, the insulating member in the pulse tube wall must be able to support the pressure differential between the peak operating pressure and the vacuum on the outside. This requirement is also easily met by most candidate insulation materials. For example, with a peak pressure of 3 MPa and for wall dimensions of 40 mm diameter by 3 mm thickness, the minimum yield strength allowed is 20 MPa. In comparison, ceramics can be easily obtained which have tensile yield strengths on the order of 100 MPa, while that of G-10 is well in excess of 100 MPa.

During the lifetime of the pulse tube, it will be subjected to repeated cycling between room temperature and cryogenic temperatures. The joint between the insulating material and the stainless wall must be able to survive this cycling. Placement of the insulating joint near the room temperature end of the pulse tube reduces this concern, but for the earlier mentioned goal of minimizing the surface area of the cryocooler which becomes electrically charged, the joint should be located near the cold end of the pulse tube. Minimizing thermal stresses is accomplished as shown in figure 2. Here the ceramic material is brazed to a thin-wall section of stainless steel which provides a degree of mechanical flexibility to compensate for the differential thermal contraction of the joint. In fact this same feature is crucial for the success of the brazing process and an initial leak tight seal demonstrates the success of the joint.

Additional requirements

All cryocoolers are structurally fabricated to provide maximum thermal insulation between the hot and cold ends of the cooler. The electrical insulation section must therefore also provide

Table 1. Dimensions of 10 watt OPT components

	Regenerator	10 watt pulse tube	twin 5 watt pulse tubes
inner diameter (cm)	3.95	3.03	2.08
height (cm)	19.7	14.9	14.9
wall thickness (cm)	0.09	0.09	0.07

equivalent thermal insulation as the wall section which it replaces. This requirement is nominally achieved since the thermal conductivity of non-crystalline ceramics are typically lower than that for stainless steel by a factor of two or more. A thicker wall required for the ceramic insulator offsets this advantage somewhat, however the net thermal resistance per unit length remains essentially the same.

In addition to the vacuum requirements implied above for the joint, it is necessary to ensure that the insulation material is also not porous to helium gas. Construction of dewars which use G-10 material typically include a metal foil within their structure to eliminate the diffusion of helium gas through the G-10[5]. Thus, in spite of the advantages afforded by G-10 for insulation, strength and low thermal conductivity, it is not acceptable for this application because of its porosity to helium gas.

DESIGN SPECIFICS

An experimental program has been initiated to test the above design ideas using an orifice pulse tube cryocooler with total cooling capacity on the order of 10 watts at 70 K. A Corkin model D91 single stage, single cylinder compressor is used to generate the pressure wave. The swept volume of the compressor is 290 cm^3 and it is configured to operate at a frequency of either 400 rpm or (preferred) 825 rpm. Design operation is for an average pressure of 2 MPa and a pressure ratio of 1.5. Considerations of mass conservation were combined with the design pressure ratio of 1.5 to determine an approximate size of 150 cm^3 for the dead volume in the regenerator, and 110 cm^3 for the internal volume of a single 10 watt pulse tube. Subsequently the programs REGEN, REGEN2 and REGEN3[6,7] were used to define the appropriate regenerator aspect ratio.

The final dimensions of the regenerator, the 10 watt pulse tube, and the twin 5 watt pulse tubes are given in table 1. The configuration for the twin pulse tube system is shown in figure 3. An alumina ceramic with approximately 6 % silicone impurities is being used for the insulation pieces in the pulse tubes, and these are installed to allow a 25 mm gap between the stainless steel wall sections above and below.

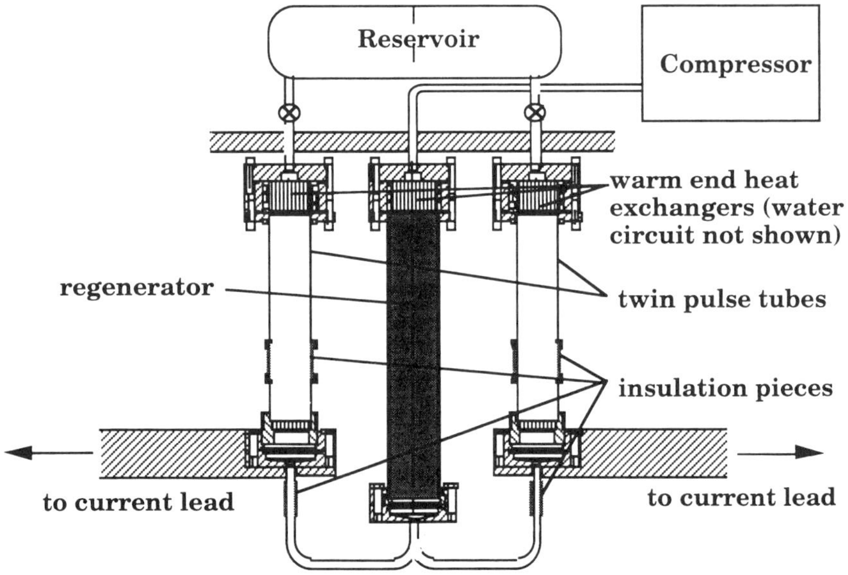

Figure 3. Insulated Orifice Pulse Tube details.

Present activities are occupied with fabrication and operation of a 10 watt uninsulated version to provide a baseline efficiency for the system. To date the pulse tube has been operated down to a temperature of 100 K, but various practical operation details are still being resolved and lower temperatures are anticipated.

Beyond the ability to develop and demonstrate a voltage insulated cryocooler, a number of other interesting questions are being pursued with this project. These include the question of whether the twin 5-watt pulse tube will be more or less efficient than the single 10 watt pulse tube, whether the parallel flow arrangement of the twin pulse tube will provide a stable or unstable fluid flow, and whether it is possible to operate multiple pulse tubes of differing capacities off of a single regenerator. Activities are underway to address each of these questions in the very near future.

ACKNOWLEDGMENT

This study was funded by the U.S. Defense Nuclear Agency under DNA MIPR 92-719, work unit CD:00014, RCC: 7010.

REFERENCES

1. M.A. Daugherty, F.C. Prenger, D.D. Hill, D.E. Daney, and K.A. Woloshun, "HTS Current Lead using a Composite Heat Pipe," *IEEE Trans. on Applied Superconductivity* 5(2):773-776 (1995).

2. P.F. Herrmann, E. Beghin, G. Bottini, C. Cottevieille, A. Leriche, T. Verhaege, and J. Bock, "Test Results of a 1 kA (2 kA) 20 kV HTSC current lead model," *Cryogenics* 34(6):534 (1994).

3. R.C. Niemann, J.D. Gonczy, T.H. Nicol, Low-thermal-resistance, high-electrical-isolation heat intercept connection, in "Advances in Cryogenic Engineering 39B," P. Kittel , ed., Plenum, New York (1994) p. 1665.

4. S.W. Schwenterly, Design and testing of electrical insulation for superconducting coils, in "Advances in Cryogenic Engineering 33," R.W. Fast, ed., Plenum, New York (1988) p. 271.

5. private communication with D. Gummer of Precision Cryogenics Inc.

6. V. Arp and R. Radebaugh, "Interactive Program for Microcomputers to Calculate the Optimum Regenerator Geometry for Cryocoolers (REGEN and REGEN2) US Air Force report # AFWAL-TR-87-3040, (1987)

7. J.Gary, A. O'Gallagher, and R. Radebaugh, "A Numerical Model for Regenerator Performance (REGEN3), publication available upon request to authors.

THE INFLUENCE OF CHECK VALVE WITH A FILTER ON THE PERFORMANCE OF PULSE TUBE REFRIGERATORS

Y.Zhou, Y.L.Ju, C.Wang, R.C.Luo and J.J.Wang

Cryogenic Laboratory, Academia Sinica
P.O.Box 2711, Beijing, 100080, China

ABSTRACT

A simple, high performance scheme with a check valve and a filter for pulse tube refrigerator is suggested in this paper. Results show that the scheme is succcssful to eliminate the contamination of the working gas. The experimental results and refrigeration performance of the pulse tube refrigerator are also presented.

INTRODUCTION

Interest in the pulse tube refrigerator has grown rapidly in the last few years because of the merits of mechanical simplicity, high reliability and low vibration. This type of cryogenic refrigerator has potential application in the cooling of infrared devices and sensors, superconducting electronic components and semiconducting devices. The developments achieved in recent year enable pulse tube refrigerator to have a comparable refrigeration capability with stirling machine [1]. Stirling refrigerator has two moving parts, the piston and the displacer, while the pulse tube refrigerator contains only one moving part at ambient temperature, no moving part at its cold head, so it has considerable system advantages over most other types of cryogenic refrigerator in reliability, lifetime, vibration and cost.

In order to achieve high reliability and long lifetime, we need to investigate systematically this moving part, improvement of the lifetime of the spring supporting piston and adoption of the frictionless configuration between the piston and cylinder with ensuring perfect of the seals are the most importance things. According to practice experience, we know that the gas in refrigeration system keep purity and no contamination is one of the key problems of improving the performance of the refrigerator. Obviously, with the operating time increasing, the high purity helium originally would be contaminated continuously and be liquidized and solidified to pellet resulting in blocking up the transport of regenerator partially, therefore, increasing the flow resistance and degrading heat transfer efficiency, so the performance of the refrigerator decreased not to satisfy the requirements of cooling devices and sensors. The object of this present paper is to employ a new technological process to eliminate the gas contamination.

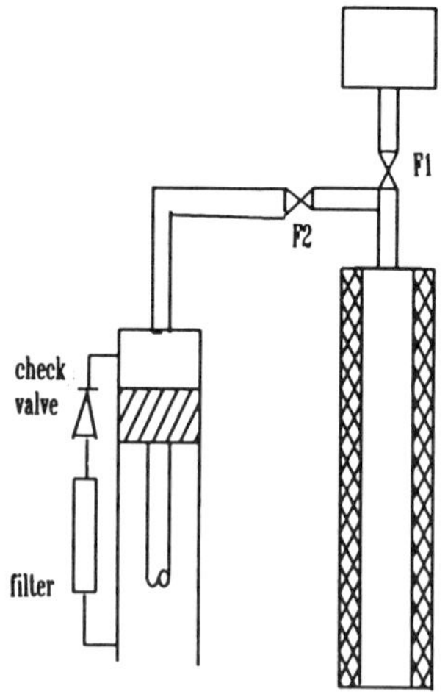

Figure 1(a). The scheme of orifice pulse tube refrigerator with a check valve and a filter.

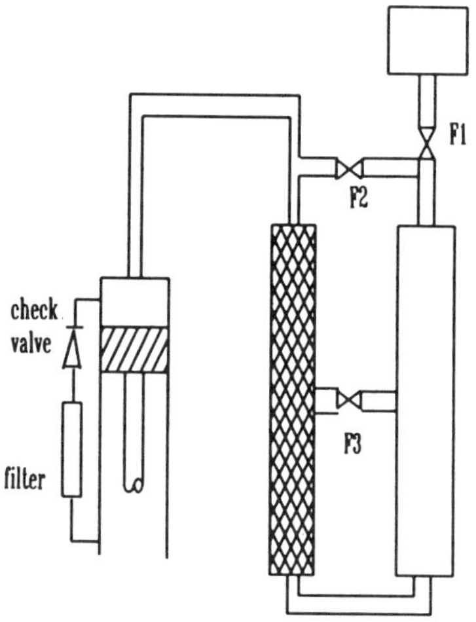

Figure 1(b). The scheme of double-inlet and multi-bypass pulse tube refrigerator with a check valve and a filter.

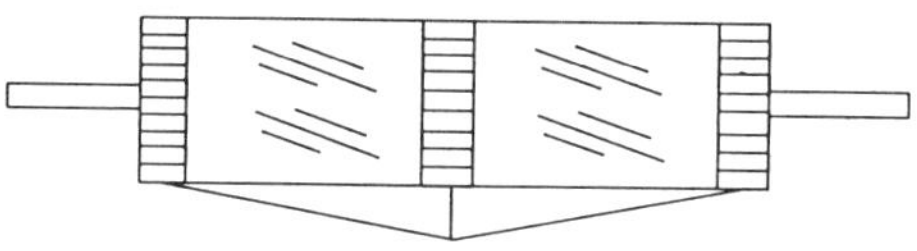

Figure 2. The schematic of the filter.

THE SOURCE OF IMPURITY GAS

It should be noted that the electrical motor of the oilless valveless compressor is often directly connected with the refrigeration system, the impurity gas adsorbing in electrical coil or exhaling from the decomposition of enamel-insulated wire will send off slowly and seep into helium continuously due to the electrical motor coil operating, the gases are almost chemical compounds and will be solidified into pellet in regenerator under -60^{0}C and decrease the performance of the refrigerator.

When the system is vacuumed and fulled in the high pure helium, the non working gases will be adsorbed on the wall of components in system, the gases are almost air.

The motion between the components of refrigeration system has friction resulting in producing the harmful impurity gases.

In order to eliminate the contamination of the above impurity gas, we should not only select the material and technological process to make all best to achieve frictionless and avoid the overheat of system components, but also master the technological process of manufacture, installing, debugging to decrease the residual impurity gas.

However, the impurity gas may still seep into the working gas, high purity helium, after long time operating. In this paper, we design a technological process to pure the working gas in long time operating resulting in keeping the performance of the refrigerator and achieving long lifetime and high reliability.

THE SCHEME OF THE FLOW CHART

The scheme of technological process is shown in figure 1, figure 1(a) is the technological process of orifice pulse tube refrigerator, and figure 1(b) is one of double-inlet and multi-bypass pulse tube refrigerator. In above two schemes, we mounted a check valve between the space of the valveless compressor and the crank box, a filter is mounted on the bottom of the check valve, the gas can't go through the filter from the crank box into the compressor space until the piston moving back (the compressor space being larger), thus the gas can only get into the regenerator and the pulse tube. The configuration of the filter is very simple, it can be seen in figure 2, there are packed with 400 meshes stainless steel screen at top, bottom and middle in the cylinder to prevent the solid pellet from crank box getting into the refrigeration system, the cylinder is composed of two parts, whose bottom part is packed silica gel and top is active carbon. It is obviously that the adsorbent is used to adsorb the impurity gases of system after long time operating. From previous experiment, the impurity gases are almost chemical compounds sent off from heating of the coil, the adsorbent should be heated to achieve activation before vacuuming and filling in the high pure helium.

EXPERIMENTAL RESULTS AND DISCUSSIONS

The test apparatus in figure 1(a) is a co-axial pulse tube refrigerator which is used as cryotherapy apparatus. The sweep volume of the valveless compressor is 30cm^3, the pulse tube and the regenerator are made of stainless steel, both of the length are 110mm, the pulse tube is 7mm inside diameter, the regenerator installed on outside of pulse tube is 14mm in outside diameter, the connection pipeline between the valveless compressor and the pulse tube is 4mm in diameter by 1800mm length. Compressor is adopted the type of oilless crank, the motor is connected directly with crank box, the motor coil was not heated well before installing, so the outgassing is serious. Figure 3 shows the experimental result, we can see that the temperature

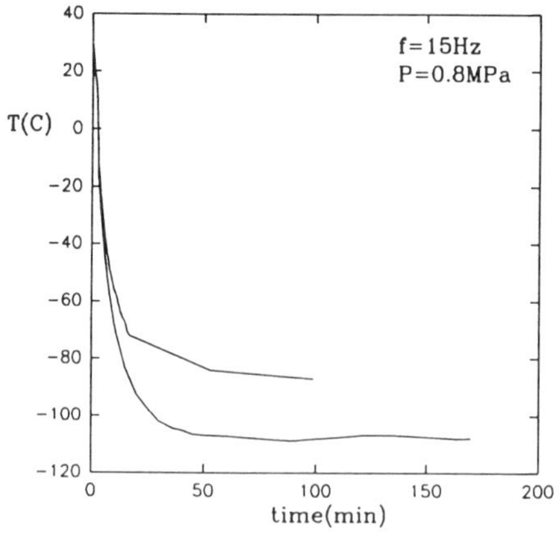

Figure 3. Cold-down processes of the pulse tube refrigerator.

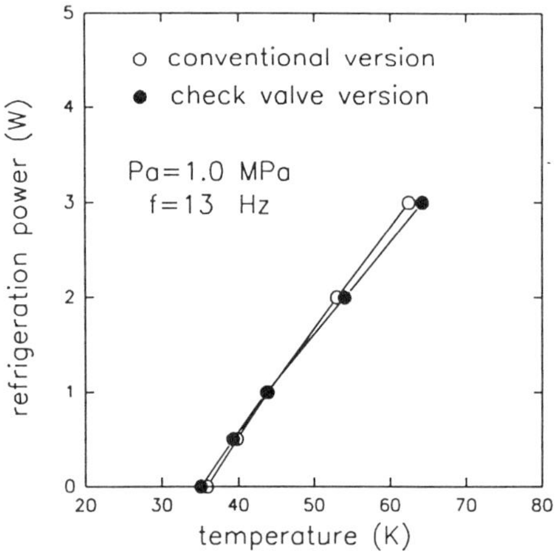

Figure 4. Cooling performance.

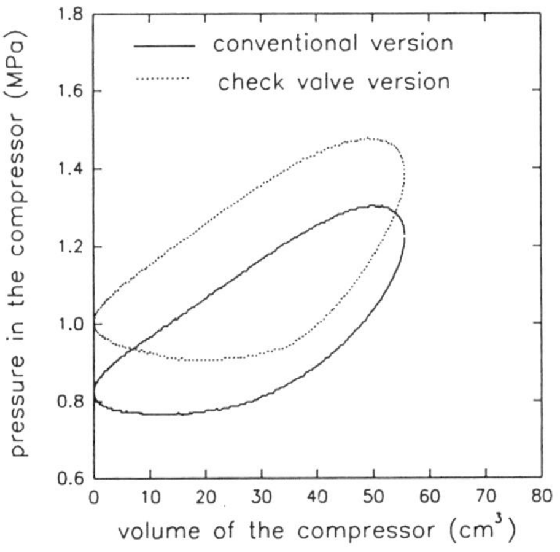

Figure 5. Variation of the compressor P-V work.

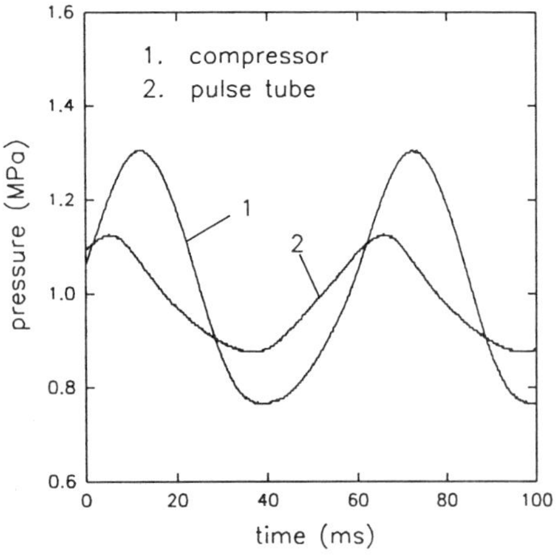

Figure 6. Typical pressure wave.

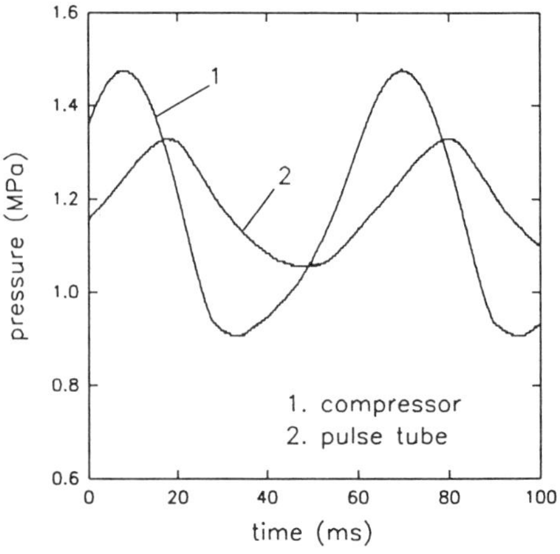

Figure 7. Typical pressure wave.

of cold head insulated by cotton needs long time to drop down to -80^{0}C without check valve and filter. Being restoration, make another experiment, we find that the performance of the refrigerator is worse than upper time. However, if we exhaust off the gas of system and vacuum out the air and full in helium again, the curvature of the temperature is the same as that in figure 3(1). The curvature 2 in figure 3 is the performance of the refrigerator whose cold head is insulated by cotton with cheek valve and filter, the lowest temperature drops up to -100^{0}C and maintains a long period, this machine has being worked half year. This shows that using the filter to pure the impurity gas is available.

Figure 1(b) shows the scheme of double-inlet and multi-bypass pulse tube refrigerator, the valveless compressor is adopted the type of oilless crank, the sweep volume of the compressor is 50cm^3, the average pressure is 1.0MPa, the frequency is 13Hz, the working gas is helium. The pulse tube and the regenerator are arranged in U shape, the pulse tube is 210mm length and its outside diameter is 10mm, the regenerator is 170mm length and its outside is 18mm. The experimental results are shown in figure 4,5,6,7. It can be seen in figure 4 that the ultimate temperature of the cold head and the refrigerate power are almost the same with the check valve and filter or not for the double-inlet and multi-bypass pulse tube refrigerator, that is to say, the influence on the performance of the refrigerator is small whether adding check and filter or not, unfortunately, the comparison of reliability of long lifetime doesn't make. Figure 6 shows the shape of pressure wave without check valve and filter at entrance of the compressor and the pulse tube, figure 7 shows that with one, comparison of the two figures, we can obtain the following results:

Figure 5 shows the P-V diagram of the compressor space, we can see that the input power of pulse tube refrigerator with a check valve and a filter is 10 percent higher than that without one. For the amplitude of pressure in the compressor space increasing, parts of gases go continuously from crank back box into compressor space by the check valve and filter, and come back into the crank box by the gap between piston and cylinder from the compressor space, resulting in composing a gas cycle, but no influence on the performance of the refrigerator.

The amplitude of the pressure oscillation at the entrance of the compressor in figure 7 is larger than that in figure 6, however, the pressure difference in pulse tube are almost the same. This demonstrates that although the amplitude of the pressure in compressor space increases, the relative resistance between the piston and the cylinder is lower than that in regenerator for large gap, therefore, the gas flowing through the valve into compressor space doesn't get into the refrigerator but transport into the crank box by the gap between the piston and the cylinder, so constructing a gas cycle, the effect of the cycle is to eliminate the impurity gas.

The average difference of pressure in figure 7 is larger than that in figure 6, this shows that the average pressure in pulse tube with check valve and filter is higher than that without one, according to the phase shift theory [2] [3] and thermacoutics theory [4], the refrigerate power should increase with the pressure increasing, however, the experimental results show that the lowest temperature and the refrigeration power are keep unchanged, this indicates that the configuration and the dimension of regenerator don't achieve to optimize, the multi- bypass is not adjusted to optimum opening.

The technological process of the pulse tube refrigerator with check valve and filter still have potential application in improving the performance of the refrigerator.

CONCLUSION

1. In order to eliminate the contamination of the impurity gas produced by various facts in pulse tube refrigerator, the scheme with a check valve and a filter is suggested and testified successful, the power input is 10 percent larger.
2. It needs to optimize the configuration and dimension of components to improve the performance of the pulse tube refrigerator for adapting to new technological process.

ACKNOWLEDGEMENT

Financial support provided by the National Natural Science Foundation of China is greatly appreciated. The authors are also grateful to W.X.Zhu, J.H.Cai and L.W.Yan for their help in this paper.

REFERENCES

1. Y.Matsubara and J.L.Gao, "Novel Configuration of Three-Stage Pulse Tube Refrigerator for Temperatures Below 4K," Cryogenics, 34(1994), p.259.
2. R.Radebaugh and S.Herrmann, "Refrigeration Efficiency of Pulse Tube Refrigerator," Proc. 4th Int. Conf. Cryocoolers, (1986), p.119.
3. R.Radebaugh, "A Review of Pulse Tube Refrigeration," Adv. Cryog. Eng., 35B(1990), p.1191.
4. A.Tominaga, "Phase Control for Pulse Tube Refrigerator of the Third Generation," Cryog. Eng., 27(1992), p.146

START-UP BEHAVIOR OF PULSE TUBE REFRIGERATORS

M. Shiraishi[1], Y. Kim[1], K. Seo[2] and M. Murakami[2]

[1]Mechanical Engineering Laboratory
Namiki 1-2, Tsukuba, Ibaraki, 305 Japan
[2]University of Tsukuba
Tennodai 1-1, Tsukuba, Ibaraki, 305 Japan

ABSTRACT

A start-up behavior of a pulse tube refrigerator is experimentally investigated to gain a better understanding of the basic refrigeration mechanism. Two typical types of pulse tube refrigerators, which are a basic and an orifice pulse tube, are used. The pressure oscillation is introduced suddenly into the pulse tube under the uniform temperature, and the transient changes of gas temperature, wall temperature and heat out at the hot end are obtained as a time history from an onset of the pressure oscillation to an ultimate state that can be obtained after a long time. It is found that after only two cycles from the onset of pressure oscillation, the difference of gas temperature between the cold and the hot ends appears and it reaches to 65 % level of the difference at the ultimate state. The heat out becomes stable after several hundreds of seconds. It is also found that the change of temperature at the very beginning of the oscillation is not adiabatic. Significant differences in the start-up behavior between a basic and an orifice pulse tube did not appear.

INTRODUCTION

A cryocooler generates refrigeration at temperature below 120K [1] and its application fields are expanding with the years.[2] The practical cryocooler with high reliability promises to play an important role in future technology for the cooling of high Tc superconductors. A pulse tube refrigerator, more simply called a pulse tube hereafter, has only been recently introduced. The pulse tube is classified in the same group of Stirling, Gifford-McMahon as the regenerative refrigerator.[3] The great advantage of the pulse tube is its mechanical simplicity because of having no moving part. This promises to provide high reliability compared to other refrigerators. Moreover, it is of interest to apply the refrigeration mechanism of the pulse tube to a refrigeration system in a low temperature range and a new heat pump system. Because of a ban on CFC refrigerants in near future, it is becoming increasingly important to develop new environmentally acceptable refrigeration and heat pump systems.

Of late years, refrigeration power and the lowest achievable temperature have been improved with the modification of original basic pulse tubes [4] to orifice pulse tubes [5] and double-inlet pulse tubes [6,7] and there exist several analytical models.[8-13] Although these models are convenient for making approximate predictions of basic trends or dependencies of performance, a clear explanation of a detail refrigeration mechanism based on the thermofluiddynamic phenomena inside the pulse tube has not been given yet. For obtaining more accurate analytical model, it should be required to gain a better understanding of internal oscillating flow phenomena, especially thermofluid phenomena in one cycle from compression to expansion processes.

The objective of this study is to investigate the start-up behavior of a pulse tube refrigerator by measuring pressure and temperature oscillations, and to discuss a basic refrigeration mechanism based on thermofluiddynamic phenomena in the pulse tube in more detail for a better understanding of the refrigeration mechanism.

EXPERIMENTAL SETUP AND PROCEDURE

Experimental Pulse Tube Apparatus

Figure 1 shows a schematic of the apparatus. The test apparatus is designed to provide detail and convenient measurements of pressure and temperature oscillations at many locations through the pulse tube refrigerator so less attention is paid to obtaining good refrigeration performance. A pulse tube and regenerator section are insulated from the room temperature environment by using a polyurethane foam insulation, while the orifice and reservoir sections are not done. The pressure oscillation is generated by using a combination of a helium compressor and a rotary valve. To introduce a pressure oscillation suddenly, two valves are installed at the entrance and the exit of the rotary valve as shown in Fig.1. The regenerator is composed of the stainless steel housing and a bakelite tube, 35 mm in diameter by 160 mm long packed with 865 discs of 100-mesh stainless steel screen. The pulse tube is composed of a stainless steel tube, 13 mm inside diameter by 242 mm length, brazed with a copper tube at both ends as heat exchangers having same diameter as the stainless steel part and a length of 33 mm. The whole length of the pulse tube is 308 mm long. The heat exchanger at the hot end is kept at the room temperature by circulating cooling water and the rate of heat extraction from the hot end is derived from a mass flow rate and a temperature increase of circulating water. The reservoir having a volume of about 7.5 times of the pulse tube volume is connected to the main body of pulse tube at the hot end through the orifice valve.

A probe is devised to measure the inside pressure and temperature variations along the apparatus. The probe consisted of a strain gauge pressure transducer and a type E(Chromel/Constantan) sheathed thermocouple of 0.25 mm in diameter with exposed cold junction. The nominal 63 % response time of the thermocouple is 5 msec(200 Hz) so that a practical response time is estimated to be 20 msec(50 Hz) at most. The response time of the pressure

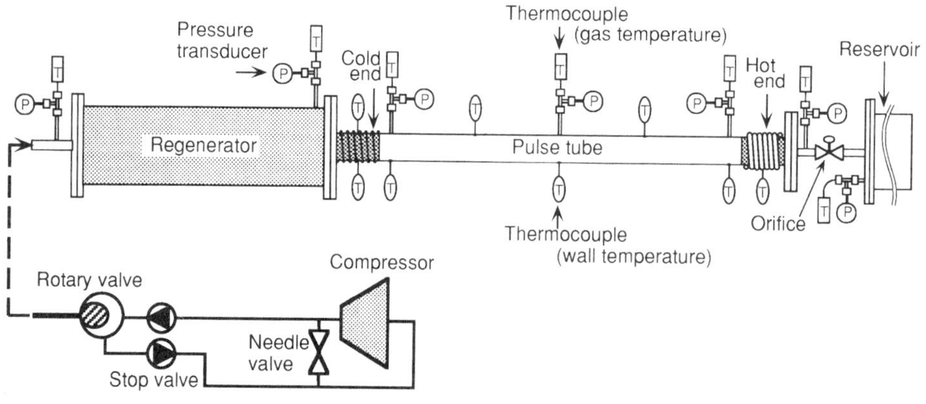

Figure 1. Schematic of the experimental apparatus.

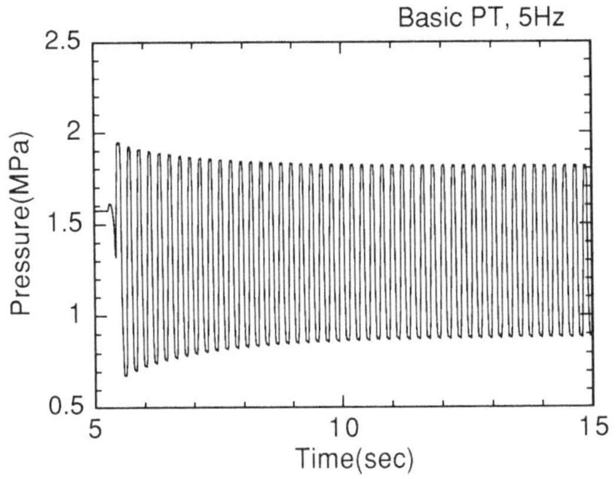

Figure 2. The behavior of the pressure oscillation in the beginning of the oscillation.

transducer is about 0.25 msec(4 kHz). An introducing pipe having an inside diameter of 2 mm and a length of 40 mm is welded to apparatus at seven locations as shown in Fig. 1, and the probe is attached by T-shaped fitting to the pipe. The thermocouple is installed at the center of the pipe at each location. Type T(Copper/Constantan) thermocouples are also soldered to the pulse tube wall at eight positions (see Fig. 1).

Experimental Procedures

After evacuation of the apparatus, helium gas is charged at the pressure of 1.3 MPa. The type of refrigerator configuration is set by the orifice valve. Upon circulating cooling water the temperature distributions of inside and wall surface of the pulse tube are measured to confirm the uniform temperature distribution. After both temperature distributions have settled down to a uniform level within 1 K, the compressor and rotary valve begin to work at previously adjusted conditions. Then the two valves are opened immediately and the pressure oscillation is introduced suddenly to the pulse tube. The changes of the temperature distributions, temperature oscillations of gas and heat out are obtained as a time history. Measurements are taken at the frequencies of 1, 5 and 10 Hz and the compression ratio at 2.0 for both the basic and the orifice pulse tube configurations. A typical result of a basic pulse tube at a frequency of 5 Hz is presented here.

EXPERIMENTAL RESULTS AND DISCUSSION

Because an oscillatory behavior of gas pressure is quite similar at all positions through the pulse tube,[14] the pressure oscillation at the cold end is presented as a typical one. Figure 2 shows the transient pressure oscillation before and after opening the valve suddenly. Although the amplitude just after the onset of the oscillation is about 35 % larger than the one at the ultimate state, it rapidly decreases with time and settles down within 6 % of the ultimate one after about 3 sec(15 cycles) from the onset of oscillation.

Figure 3 shows the temperature oscillation which corresponds to the pressure oscillation in Fig.2. It is noteworthy that the temperature difference between the cold end and the hot end appears simultaneously with the onset of the pressure oscillation, more precisely after only two cycles. Moreover, the gas temperature at the cold end decreases suddenly from the initial level. That indicates the beginning of refrigeration. The amplitude at the very beginning of the oscillation is larger compared to the one at the ultimate state corresponding to the pressure oscillation as shown in Fig.2. The temperature inside the regenerator, which is obtained near the cold end of the regenerator, does not show a change from the initial level in

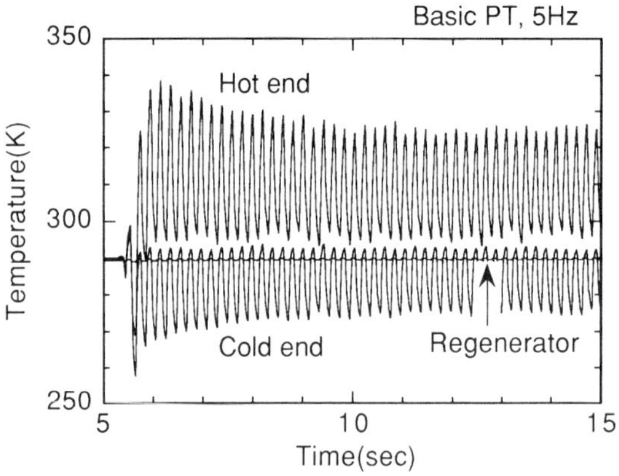

Figure 3. The behavior of the gas temperature oscillation in the beginning of the oscillation corresponds to the pressure oscillation in Fig.2.

about the first 20 sec(100 cycles). After that, the temperature begins to decrease with time and the temperature gradient develops inside the regenerator gradually. The delay of the temperature change is due to the large heat capacity of the regenerative materials.

Figures 4 and 5 show the behaviors of gas and wall temperatures during the period from the beginning to the time when the heat out reaches the ultimate level, that is about 800 sec. The gas temperatures except inside the regenerator change transiently at the beginning, and the temperature difference between at the cold end and at the hot end increases suddenly over about 65 % level of the difference at the ultimate state. Although this difference at the beginning is very large, it takes about 900 sec to reach 90 % level of the ultimate one. It can be considered that the reason for the temperature change with time is that just after the beginning, thermofluiddynamic effects mainly affect the temperature change. After that the wall temperature distribution as well as the thermofluiddynamic effects influences the observed changes.

Although both the wall temperature and the heat out change with gas temperature early in time, their behavior is more gradual compared to the change of the gas temperature. Concerning the change of the wall temperature, the times to reach 90 % levels of the ultimate state are about 420 sec for the hot end and 1540 sec for the cold end. The time for the cold end is

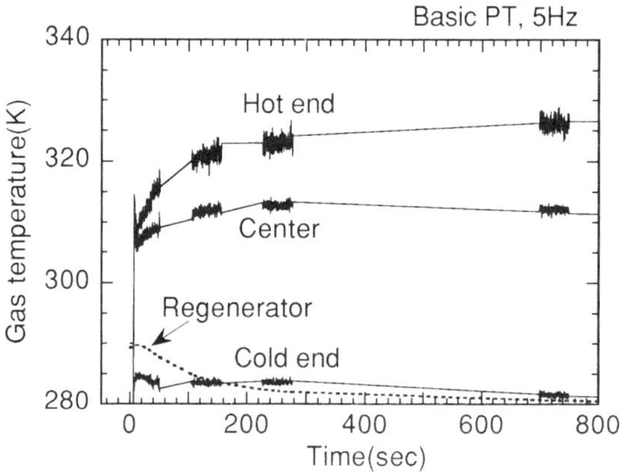

Figure 4. The behavior of the gas temperature during the cool-down process for the basic pulse tube.

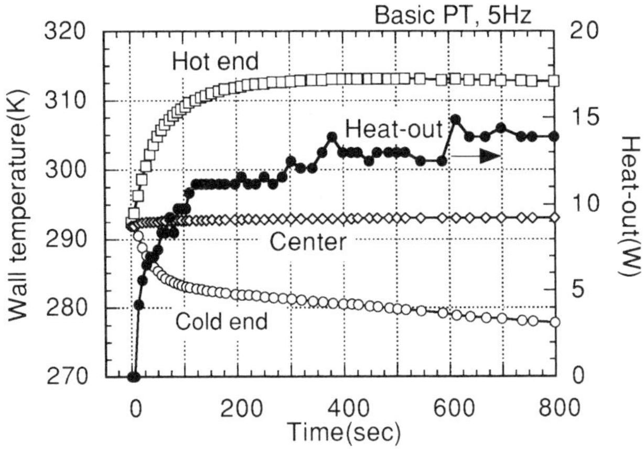

Figure 5. The behaviors of the wall temperatures and heat out during the cool-down process corresponds to Fig.4.

over three times as long as for the hot end. The heat out reaches 90 % of the ultimate state after about 530 sec. That is only 110 sec late after the hot end. The heat out is roughly determined by the level of the wall temperature at the hot end so that the behavior of the heat out is similar to that of the wall temperature at the hot end. Moreover, since the refrigeration power is related to the heat out, its behavior is also similar to the behavior of the heat out. Since the ultimate state is obtained after 10000 sec in this experiment, it can be considered that the main part of the cool-down process completes in the very beginning of the process.

Figure 6 shows the start-up behavior in the case of an orifice pulse tube. The basic start-up behavior is the same as that explained for the case of the basic pulse tube except for the change at the hot end in the beginning.

It is remarkable that a large temperature difference between the cold end and the hot end appears after only two cycles from the beginning of oscillation. Because this temperature change of gas occurs very fast, it is assumed that heat transfer to the wall during the pressure change is negligible and the phenomena occur adiabatically. Figure 7 shows the comparison between the experimental results and the prediction from the well-known thermodynamic relation,

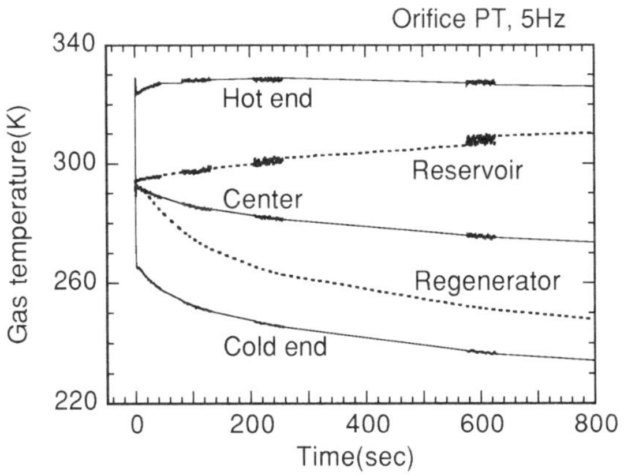

Figure 6. The behavior of the gas temperature during the cool-down process for the orifice pulse tube.

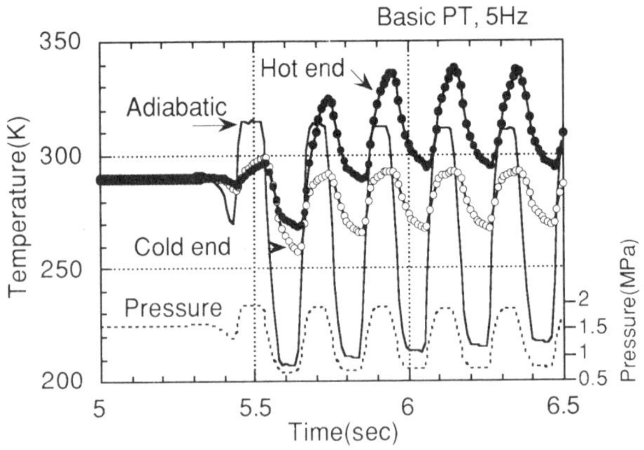

Figure 7. Comparison of the gas temperature oscillation during several cycles at the beginning of the oscillation with the calculation from Eq.(1).

$$T = T_0 (P/P_0)^{(\gamma-1)/\gamma} \quad (1)$$

in which P_0 and T_0 denote the initial conditions before the onset of oscillation and γ is the ratio of specific heats. The prediction doesn't agree well with the experimental results. This discrepancy suggests that the phenomena can not be explained well by an adiabatic process even though is the state just after the onset of oscillation.

CONCLUSION

To gain a better understanding of a basic refrigeration mechanism, the start-up behavior of a pulse tube refrigerator is experimentally investigated by measuring the transient temperature oscillation and its process is discussed. The results can be summarized as follows:

In the first several cycles, the remarkable temperature change appears in gas inside the pulse tube and the temperature difference between the cold end and the hot end reaches over about 65 % of the ultimate temperature difference. After that, the gas temperature changes gradually with time. The wall temperature distribution does not affect significantly the refrigeration process at the beginning.

The gasdynamic processes inside the pulse tube are not assumed to be adiabatic even if the state is adiabatic just after the onset of oscillation.

REFERENCES

1. G.Walker, "Cryocoolers Part 1: Fundamentals, Plenum Press, New York, 1983.
2. R.G.Ross, Jr., "Cryocoolers 8", Plenum Press, New York, (1995).
3. A.Radebaugh,"A review of pulse tube refrigeration", Advances in Cryogenic Engineering, Vol. 35B, Plenum Press, New York, (1990), p.1191.
4. W.E.Gifford and R.C.Longsworth, "Pulse Tube Refrigeration", ASME paper No.63-WA-290, (1963).
5. E.I.Mikulin, A.A.Tarasov and M.P.Shkrebyonock,"Low Temperature Expansion Pulse Tubes", Advances in Cryogenic Engineering, Vol.29, Plenum Press, New York, (1984), p.629.

6. S.Zhu, P.Wu, Z. Chen, W. Zhu and Y. Zhou, "A single stage double inlet pulse tube refrigerator capable of reaching 42 K", Cryogenics, Vol.30 September Supple., (1990), p. 257.
7. E.C.Luo, J.H.Xiao and Y.Zhou, "A simplified thermoacoustic modeling for pulse tube refrigerator", Proceedings of Fourth Joint Sino-Japanese Seminar on cryocoolers and concerned topics, Academia Sinica, Beijing, (1992), p.94.
8. G.W.Swift, "Thermoacoustic engines", J. Acoust. Soc. Am., Vol. 84(4), (1988), p.1145.
9. P.J. Storch, R. Radebaugh and J.E. Zimmerman, "Analytical Model for the Refrigeration Power of the Orifice Pulse Tube Refrigerator", NIST Technical Notes 1343, (1990).
10. A.Tominaga, "Basic notion of Thermoacoustic theory and some results of simulations", Proceedings of Fourth Joint Sino-Japanese Seminar on cryocoolers and concerned topics", Academia Sinica, Beijing, (1992), p.79.
11. J.H. Xiao, "Thermoacoustic heat tranportation and energy transformation, Part 1: Formulation of the problem", Cryogenics, Vol. 35, (1995), p.15.
12. J.H. Xiao, "Thermoacoustic heat transportation and energy transformation, Part 2: Isothermal wall thermoacoustic effect, Cryogenics, Vol. 35, (1995), p.21.
13. J.H. Xiao, "Thermoacoustic heat transportation and energy transformation, Part 3: Adiabatic wall thermoacoustic", Cryogenics, Vol. 35, (1995), p.27.
14. M.Shiraishi, K.Seo and M.Murakami, "Pressure and temperature oscillations of working gas in a pulse tube refrigerator", Cryocoolers 8, R.G.Ross, Jr. ed., Plenum Press, New York, (1995), p.403.

1995 CEC/ICMC Conference: Jim Peeples (right) presenting the 1995 Russell B. Scott Memorial Award for the best Cryogenic Engineering Applications paper to Kurt Schaubel (left)

PULSE TUBE REFRIGERATOR FOR HIGH-T_c SQUID OPERATION

G. Thummes, R. Landgraf, F. Giebeler, M. Mück, and C. Heiden
Institut für Angewandte Physik, Justus–Liebig–Universität
D–35392 Giessen, Germany

ABSTRACT

We report on a pulse tube refrigerator (PTR) designed for cooling an YBCO RF—SQUID electronic gradiometer and on measurements of the magnetic flux noise of the system. To reduce the size of the cooler, a coaxial configuration of regenerator and pulse tube is used. For higher performance the PTR can be operated in orifice and double–inlet configuration. A motor–driven rotary valve serves to control the helium gas flow between compressor and refrigerator inlets. In order to reduce vibrations and interference noise at the cold head, the rotary valve and refrigerator inlets are connected by a flexible tube. Using a tube with a length and inner diameter of 4 m and 3 mm, respectively, an average pressure of 18 bar, a frequency of f = 2 Hz and a high to low pressure ratio of 1.5, a minimum temperature of 53 K has been achieved. The gradiometer with a base line length of 45 mm consists of two YBCO microwave RF–SQUIDs which are attached to a mounting block connected with the cold heat exchanger of the PTR. The two SQUIDs were successfully operated in flux locked loop in laboratory environment without any magnetic shielding. The flux noise spectrum exhibited interference signals at f and f/2, which decreased rapidly upon increasing the distance between cold head and the rotary valve. At a distance of 2 m and more, a white noise of the order of $6 \times 10^{-5}\ \Phi_0/\sqrt{\text{Hz}}$ was measured in gradiometer configuration.

INTRODUCTION

The application potential of highly sensitive cryoelectronic sensors such as high–T_c SQUIDs could be considerably increased if appropriate refrigeration devices were available. For continuous operation of a SQUID by use of a cryocooler, the system has to meet severe requirements that arise from the high magnetic flux sensivity of the sensor. Magnetic fields originating from the cryocooler must be kept as small as possible. Also mechanical vibrations of the cold head should be reduced to a minimum, since vibration of the sensor in an external magnetic field will produce a corresponding magnetic flux signal.

Compared to the traditional regenerative cryocoolers such as the Stirling– and Gifford–McMahon refrigerator, the pulse tube refrigerator (PTR) offers the advantage of having no moving parts at low temperatures. This feature results in mechanical simplicity and reduced vibrations of the cold head, thus making the PTR very attractive for cooling SQUID sensors.

The original version of the PTR has only a rather poor refrigeration efficiency with lowest temperatures above 120 K achieved in a single stage unit[1]. The performance of the PTR has been greatly improved within the last decade. In 1984 an orifice and a

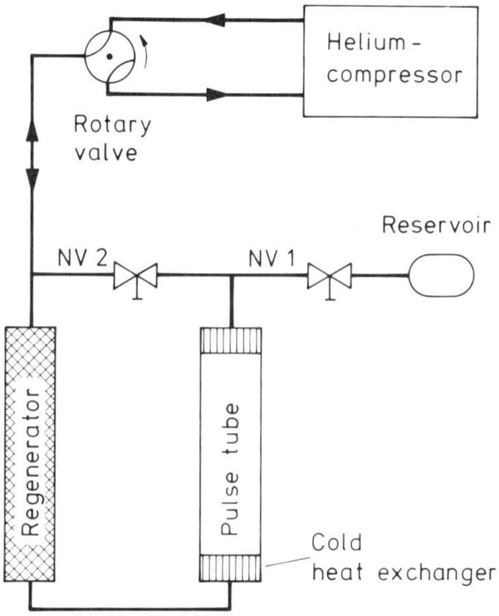

Figure 1. Schematic of a double–inlet pulse tube refrigerator.

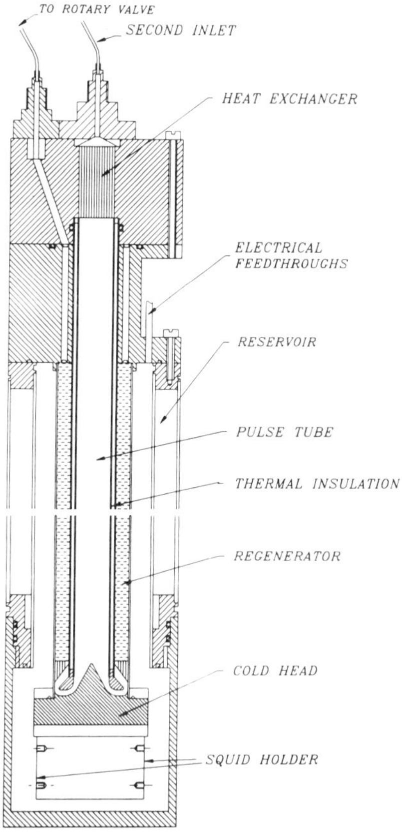

Figure 2. Cross section of the coaxial PTR for SQUID operation.

buffer volume (reservoir) were added to the hot end of the pulse tube[2] leading to the orifice PTR (OPTR). In 1990, a modified version of the OPTR, the double–inlet PTR (DIPTR), was suggested[3], in which an additional flow impedance connects the pressure wave generator and the hot end of the pulse tube. A lowest temperature of 28 K has been reported for a single stage DIPTR.[4] Even lower temperatures have been obtained with other versions, such as the two–piston type PTR of Ishizaki.[5]

Figure 1 shows a schematic of a valved PTR for operation in orifice and double–inlet modes. A helium compressor provides the high pressure helium gas as working fluid. The pressure wave p(t) is generated by a rotary valve which connects periodically the high and low pressure side of the compressor to the refrigerator inlet. The regenerator and pulse tube are usually made of thin–walled stainless steel tubes and the regenerator matrix consists of a stack of wire screens. Two heat exchangers are connected to the hot and cold end of the pulse tube. Orifice–type operation is achieved by opening the needle valve NV1. Additional opening of NV2 corresponds to the DIPTR mode. A review on pulse tube refrigerators and the operation principles can be found in Ref. 6.

In a previous study[7], we have developed a single–stage DIPTR, which is capable of reaching a minimum temperature of 42 K and has a net cooling power of 4 W at 80 K. The low noise level of the cooler was demonstrated by operating a single YBCO RF–SQUID attached to the cold head[8]. On the basis of these results, we constructed a PTR designed for cooling a high–T_c SQUID gradiometer consisting of two YBCO RF–SQUIDs.

COOLER DESIGN

Practical operation of a SQUID gradiometer by use of a PTR requires a compact design of the cooler and of the vacuum vessel and also a reduction of mechanical and electromagnetic noise from the rotary valve and the compressor.

A cross section of the PTR designed for SQUID operation is shown in Figure 2. In order to reduce the transverse dimension, a coaxial configuration of regenerator and pulse tube is used. Pulse tube and regenerator are made from stainless steel tubes with lengths of 250 and 205 mm and inner diameters of 13.4 and 29 mm, respectively. The regenerator matrix consists of a stack of 1600 phosphorus bronze screens of 220 mesh and 400 stainless steel screens of 220 mesh. The screens, which are annular in shape, are placed concentrically around the pulse tube.

One problem in the case of coaxial configuration arises from the different temperature profiles along regenerator and pulse tube, since these cause radial thermal conduction losses. In order to reduce such losses, a double wall structure is employed. This structure consists of an additional tube with inner diameter of 17 mm, which is welded to the ends of the pulse tube. The space between both tubes is filled with polyurethane foam. The part of the refrigerator, which is located outside the vacuum insulation, and to which the pulse tube and regenerator are attached, is fabricated of two cylindrical copper blocks containing the vacuum flange, gas inlets, the warm end heat exchanger, and electrical feed–throughs. Axial grooves are machined into the outer surface of the copper blocks in order to increase the cooling by natural air convection. The warm end heat exchanger consists of a copper cylinder with thin axial flow channels providing heat exchange and flow straightening. The inner surface of the cold head (made of copper) is conically shaped and serves for heat transfer and guidance of the gas flow. The design of the cold head is similar to that described in Ref. 5.

The vacuum vessel with an outer diameter of 68 mm consists of two pieces: the upper part, made from stainless steel, contains the reservoir in form of an annular tank; the lower part, surrounding the cold head with SQUID holder, is fabricated from acrylic glass. The total length of the cooler with vacuum vessel in Fig. 2 is 34 cm.

The whole cooler assembly is as shown in Fig. 1. A standard compressor for Gifford–McMahon coolers is used, and a motor–driven rotary valve serves to control the helium gas flow between the compressor and refrigerator inlets, as illustrated in Fig. 1 and described in more detail in Ref. 7. To reduce magnetic noise and mechanical vibrations, the refrigerator is connected to the rotary valve by means of a flexible polyimide tube with inner diameter of 3 mm and with lengths between L = 1 and 4 m.

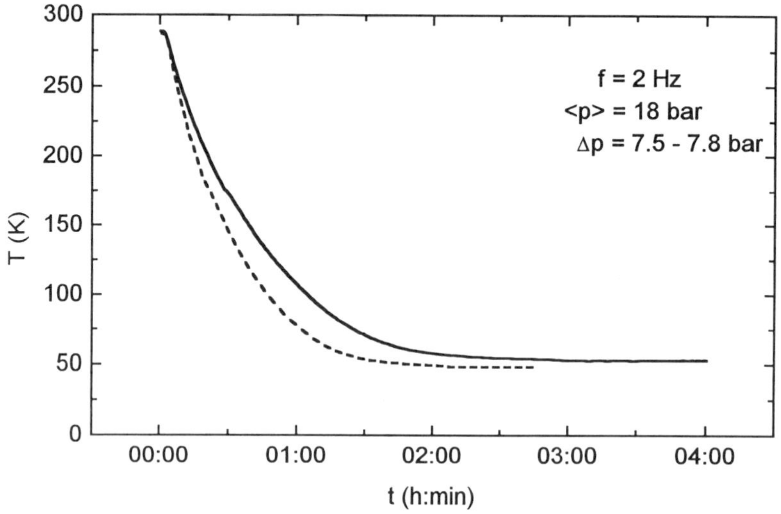

Figure 3. Cool–down curves of the coaxial PTR without radiation shielding of the cold head. Dashed line: without SQUID gradiometer, L = 1 m; solid line: with SQUID gradiometer, L = 2.2 m.

The two YBCO RF–SQUIDs forming the electronic gradiometer are mounted on the opposite sides of an aluminium block, which is bolted to the cold head and is designed to give a base line length of 45 mm, see Fig. 2. To reduce thermal Johnson noise, the aluminium block in a later version will be replaced by one made of sapphire.

Each SQUID is placed in the center of an S–shaped microwave strip resonator, which is prepared from a laser–deposited YBCO film on a $LaAlO_3$ substrate. Inductive coupling to the room temperature electronics is made by means of two semi–rigid coaxial cables with outer diameters of 1.2 mm. The operating frequency is 3 GHz. A more detailed description of the SQUID preparation and the electronic readout system can be found in Refs. 9 and 10.

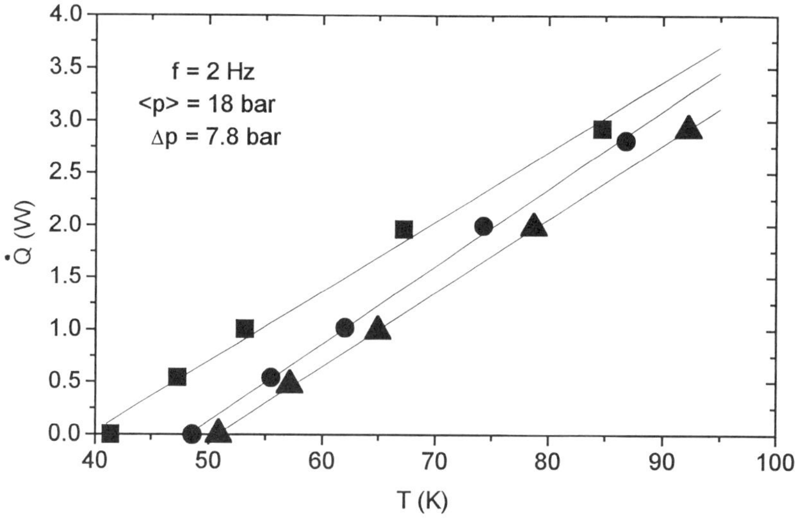

Figure 4. Net cooling powers vs cold end temperature for L = 1m. Squares: without gradiometer and with radiation shielding (r.s.) of the cold head; circles: without gradiometer and without r.s.; triangles: with gradiometer and without r.s..

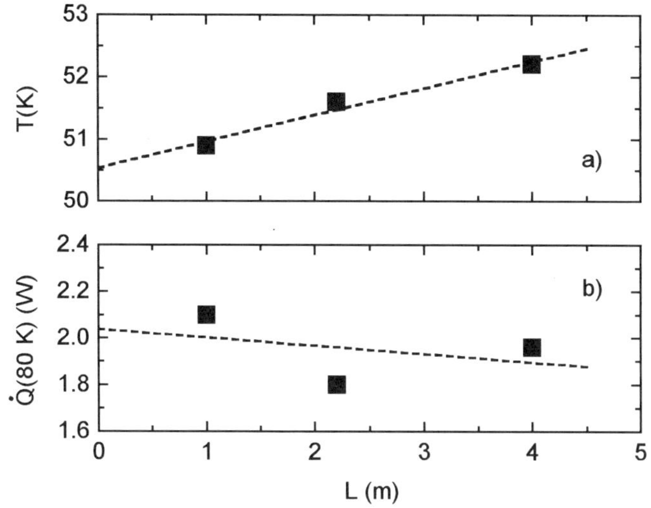

Figure 5. Variation of a) minimum temperature and b) net cooling power $\dot{Q}$(80 K) with length L of the flexible tube; gradiometer installed.

EXPERIMENTAL RESULTS

Cooler Performance

Best performance of our coaxial PTR is achieved in double–inlet configuration at a frequency of 2 Hz for the pressure wave and with the needle valves NV1 and NV2 (Nupro, type M) set to optimum flow coefficients of c(NV1) = 0.0053 and c(NV2) = 0.0116. The average pressure in the cooler is typically 18 bar. The peak to peak pressure difference at the cooler inlet is $\Delta p = 7.8$ bar for a flexible tube lengths of L = 1 m and decreases to $\Delta p = 7.2$ bar for L = 4 m. The pressure variation in the 250 cm^3 – reservoir (see Fig. 2) at the optimum setting of the orifice valve NV1 is $\Delta p_0 = 0.6$ bar.

Fig. 3 shows the cool–down curves of the PTR with and without the SQUID gradiometer installed. In both cases no superinsulating foil was used in the acrylic glass region of the vacuum vessel. With the gradiometer installed, the lowest temperature increases from 48.6 to 52 K and a longer cool–down time of about 2.5 hours is needed. The rather slow cooling down in both cases is caused by the large heat capacity ($\simeq$ 125 J/K) of the copper cold head with the aluminium SQUID–holder attached.

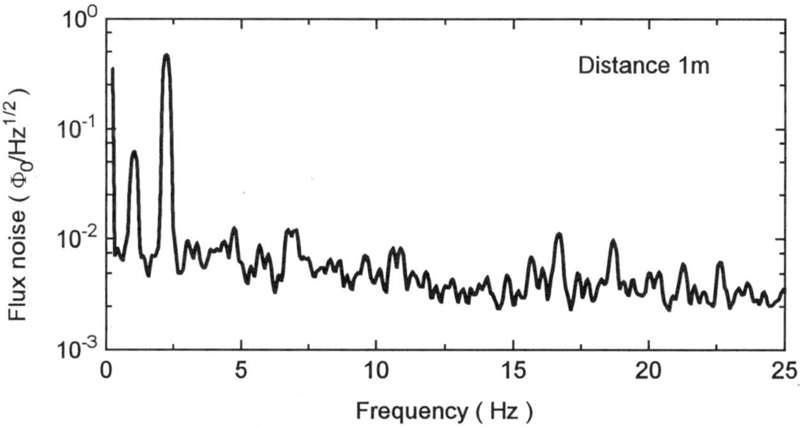

Figure 6. Flux noise spectrum of a microwave YBCO RF–SQUID operated in flux locked loop. Cold head temperature T = 52 K; distance to rotary valve: 1 m.

The measured net cooling powers dQ/dt as a function of cold end temperature for L = 1 m with and without gradiometer are displayed in Fig. 4. The thermal load was introduced by supplying a current to a resistive heater fixed to the SQUID–holder. The linear variation of dQ/dt with T is explained in Ref. 7. With proper radiation shielding of the cold head, a minimum no–load temperature of 41 K and dQ/dt = 2.8 W at 80 K are attained. Upon removal of the superinsulation and installation of the gradiometer the no–load temperatures increase to 49 K and 51 K, respectively. The slope of dQ/dt, however, remains nearly unaltered. The vertical offset of $\simeq$ 0.2 W between the two lower lines in Fig. 4 is given by the heat load from the gradiometer, which is mainly due to thermal conductance along the microwave cables.

Fig. 5 displays the minimum temperature and the net cooling power at 80 K as a function of the length L of the flexible tube, in both cases with the gradiometer installed. Increasing the tube length in the range L = 1 – 4 m is found to increase the minimum temperature by about 0.4 K/m, while the net cooling power at 80 K decreases roughly by $\simeq$ 0.1 W/m.

In stationary operation at 54 K and with L = 4 m, the temperature of the convection–cooled copper radiator (see Fig. 2) is only $\simeq$ 308 K compared to the ambient temperature of $\simeq$ 297 K. This finding is different from that in the work of Ishizaki[11], where a strong pulse heating of the air–cooled radiator to $\geq$ 335 K was observed for L $\geq$ 2 m. We relate this difference to the cooling power of our valved PTR, which is smaller than that of the two–piston type PTR of Ishizaki[11], but makes possible the use of larger tube lengths between cooler and rotary valve.

SQUID Gradiometer Test

In order to detect any magnetic interference from the cooler, measurements of the flux noise spectrum of the RF–SQUIDs have been performed. In these experiments the SQUIDs were operated in flux–locked loop and in an unshielded environment. A flexible tube length of L = 2.2 m was used. The temperature of the cold head with SQUID holder was 52 K.

The typical flux noise spectrum at frequencies below 25 Hz of a single SQUID operating as a magnetometer is depicted in Fig. 6. In this measurement the cold head was placed a distance of 1 m apart from the rotary valve. In addition to a white noise of about $5 \times 10^{-3}\ \Phi_0/\sqrt{\text{Hz}}$ the spectrum in Fig. 6 exhibits two peaks located at 1 Hz and 2 Hz. These peaks can be related to the rotation of the valve shaft (partly made of steel) that occurs with a frequency of 1 Hz.

At the same distance of 1 m but in electronic gradiometer configuration, the white noise and the two interference signals are reduced by about two orders of magnitude, as illustrated in Fig. 7 a). With increasing distance, these interference signals are reduced and at a distance of 2 m they are almost buried in the white background noise, which is of the order of $6 \times 10^{-5}\ \Phi_0/\sqrt{\text{Hz}}$, see Fig. 7 c). With a field to flux transfer coefficient of $\simeq$ 100 nT/Φ_0 of the used SQUIDs, this white flux noise corresponds to the gradient noise of our laboratory being equivalent to a field noise of a few pT/($\sqrt{\text{Hz}}$ cm).

The observed dependence of the two interference signals on the distance between SQUIDs and the rotary valve (at a fixed flexible tube length) clearly shows that the rotary valve is the main source of interference in our system.

There are two other possible sources of magnetic noise a contribution of which cannot be totally ruled out at the present stage of our work. Firstly, residual vibrations that arise from the 2 Hz pressure oscillation in the regenerator, to which the cold head and SQUID holder are attached, could generate a disturbance. For a pressure variation of Δp = 7.8 bar the resulting length variation of the regenerator tube is calculated to be about 12 μm. In a strong field gradient, this length change can cause a flux signal in the gradiometer. Such a gradient could originate from a remanent magnetization of the stainless steel parts and welded joints of the cooler in the vicinity of the SQUIDs. Secondly, a flux signal can arise from an oscillation of the SQUID temperature. As has been found in Ref. 12 for low–T_C as well as high–T_C SQUIDs, the magnetic flux signal due to a temperature variation is of the order of $10^{-4}\Phi_0$/mK. This temperature effect is probably related to the movement of vortices in the SQUID structure. In our experiments no oscillation of the temperature of the SQUID holder could be detected within the experimental resolution of 10 mK. This finding is consistent with the

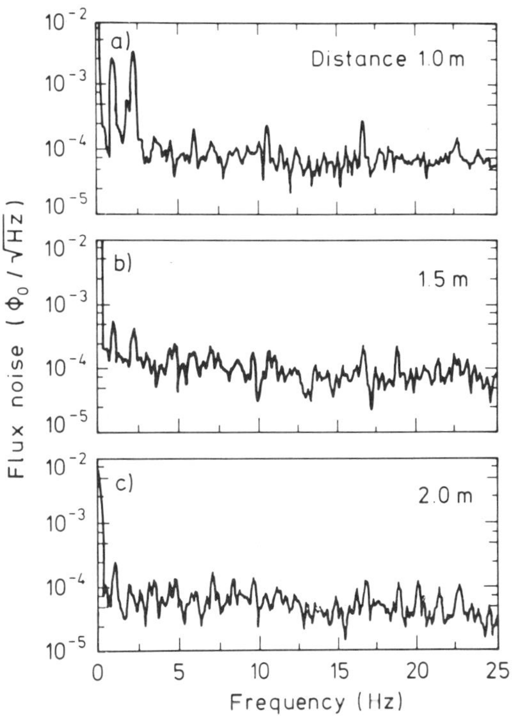

Figure 7. Flux noise spectrum of a microwave YBCO RF—SQUID electronic gradiometer (base line: 45 mm) cooled by the coaxial PTR. The distance between the rotary valve and the cold head is a) 1 m, b) 1.5 m, and c) 2 m. Cold head temperature T = 52 K.

calculated temperature variation ΔT in the SQUID holder arising from the oscillation of gas temperature during one cycle. A calculation on the basis of the Fourier equation yields $\Delta T < 7$ mK at the surface of the holder. The oscillation of the SQUID temperature is estimated to be lower by more than a factor of ten, because of the thermal contact resistance between holder and substrate.

CONCLUSIONS

The coaxial pulse tube refrigerator described in this paper is capable of cooling an YBCO 3 GHz–SQUID gradiometer to an operating temperature of 50 – 55 K. The compact size of the cooler and the use of a flexible tube to connect the cooler inlets to the pressure–wave generating rotary valve allows for easy handling of the cold head, as is required for practical applications. Flexible tube lengths up to 4 m have been tested successfully, and even larger tube lengths seem to be possible with only slight degradation of the performance.

Measurements of the flux noise spectrum of the electronic SQUID gradiometer without any shielding and at small distances between the SQUIDS and the rotary valve reveal magnetic interference signals which are ascribed to the rotating shaft of the valve. At a distance of 2 m these signals disappear and an average white flux noise of $\simeq 6 \times 10^{-5}\ \Phi_0/\sqrt{Hz}$ is visible, which corresponds to the magnetic field gradient noise of our laboratory. At present, additional pressure–wave induced magnetic interference from residual vibrations and from temperature oscillations cannot be totally ruled out, but their magnitude should be less than about $2 \times 10^{-4}\Phi_0/\sqrt{Hz}$, as can be seen from Fig. 7 c).

These results clearly demonstrate, when compared to other closed–cycle refrigerators, the great advantage of the PTR in cooling of highly sensitive high–T_c sensors. Further improvements of the present PTR system consist in making the rotary valve and the cooler completely out of non–magnetic materials, increasing the mechanical rigidity of the cooler, and in replacing the aluminium SQUID holder by a sapphire block in order to reduce electronic noise.

ACKNOWLEDGEMENTS

The authors thank Y. Zhang and A.I. Braginski (KFA Jülich) for supplying the microwave SQUIDs. Financial support by a BMBF grant (No. 13N6176 0) is gratefully acknowledged.

REFERENCES

1. W.E. Gifford and R. Longsworth, Pulse tube refrigeration progress, in: "Advances in Cryogenic Engineering", Vol. 10 B, Plenum Press, New York (1965), p. 69.
2. E.I. Mikulin, A.A. Tasarov, and M.P. Shkrebyonock, Low temperature expansion pulse tubes, in: "Advances in Cryogenic Engineering", Vol. 29, Plenum Press, New York (1984), p. 629.
3. S. Zhu and P. Wu, Double inlet pulse tube refrigerators: an important improvement, Cryogenics **30**: 514 (1990).
4. A. Ravex, P. Rolland, and J. Liang, Experimental study and modelisation of a pulse tube refrigerator, Cryogenics **32** ICEC 14 Suppl.: 9 (1992).
5. Y. Ishizaki and E. Ishizaki, Experimental performance of modified pulse tube refrigerator below 80 K down to 23 K, in: "Proc. 7th Int. Cryocooler Conf.", Santa Fee (1992), p. 140.
6. R. Radebaugh, A review of pulse tube refrigeration, in: "Advances in Cryogenic Engineering" Vol. 35, Plenum Press, New York (1990), p. 1191.
7. G. Thummes, F. Giebeler, and C. Heiden, Effect of pressure wave form on pulse tube refrigerator performance, in: "Cryocoolers 8", R.G. Ross, Jr., ed., Plenum Press, New York (1995), p. 383.
8. F. Giebeler, G. Thummes, and C. Heiden, Aufbau und Betriebsverhalten eines Pulsröhrenkühlers, in: "Supraleitung und Tieftemperaturtechnik: zum Statusseminar 13. — 14. Juni 1994", VDI Verlag, Düsseldorf (1994), p. 248.
9. M. Mück, Progress in RF—SQUIDs, IEEE Trans. Appl. Supercond. **AS—3**: 2003 (1993).
10. M. Mück, A readout system for 3 GHz RF—SQUIDs, Rev. Sci. Instrum. **63**: 2268 (1992).
11. Y. Ishizaki and E. Ishizaki, Prototype of pulse tube refrigerator for practical use, in: "Advances in Cryogenic Engineering", Vol. 39 B, Plenum Press, New York (1994), p. 1433.
12. S. Schöne, M. Mück, and C. Heiden, Response of superconducting quantum inter— ference devices to temperature variation, to be submitted to Appl. Phys. Lett.

PERFORMANCE CHARACTERIZATION OF THE TRW 35K PULSE TUBE COOLER

S. A. Collins, D. L. Johnson, G. T. Smedley, and R. G. Ross, Jr.

Jet Propulsion Laboratory
California Institute of Technology
Pasadena, CA 91109

ABSTRACT

The TRW 35K pulse tube cooler is configured as an integral cooler, with the pulse tube attached perpendicular to a pair of compressors operating into a common compression chamber. The cooler was optimized for 35K operation and has a nominal cooling capacity of 850 mW at 35 K with a cooler input power of 200 W. It also provides 2 W of cooling at 60 K for 90 W of input power. The cooler was extensively characterized by JPL, measuring the thermal performance and the cooler-generated vibration and EMI as a function of piston stroke and offset position. The thermal performance was found to be quite sensitive to the piston offset position. The pulse tube parasitic conduction levels were also measured and shown to have a strong angular dependence relative to gravity. Magnetic shielding studies were performed to examine radiated magnetic emission levels from compressors with and without shielding.

INTRODUCTION

As part of JPL's cryocooler characterization program sponsored jointly by the Air Force Phillips Laboratory and the NASA Earth Observing System AIRS Instrument Project, the TRW 35K pulse tube cooler was recently characterized in terms of its refrigeration performance, generated vibration, and electromagnetic compatibility. The results of these performance measurements are summarized in this paper.

The TRW 35K pulse tube cooler is an integral cryocooler that consists of a pair of flexure-bearing compressors operating on a common compression chamber ported to a pulse-tube cold head. The cold head is mounted perpendicular to the compressor axis in the center of the compressor assembly, as shown in Figure 1. Additional details can be found in Reference 1.

REFRIGERATION PERFORMANCE

Refrigeration performance measurements were conducted in the JPL thermal-vacuum test facility that is used to simulate conditions in space and to provide a highly stable thermal test environment; the high level of environmental stability allows the accurate and repeatable measurements that are necessary to illuminate subtle and important performance sensitivities. The cooler's heatsink temperature was regulated by attaching a pair of copper heat exchangers with a temperature-controlled fluid loop to the mechanical flats on either side of the cooler's centerplate. The heat exchangers were thermally isolated from the vacuum housing and also served as the cooler's structural supports. The cooler's coldblock was outfitted with a cryodiode to measure the coldblock temperature and a wire-wound heater to apply a heat load. The coldblock was then wrapped in several layers of aluminized Kapton MLI to reduce any parasitic radiation heat load to negligible levels, and the vacuum was maintained below 10^{-6} torr to avoid gaseous conduction effects.

The cooler was driven using a low-distortion audio amplifier with a sinusoidal voltage waveform. The power to the cooler was monitored using a true-RMS power meter.

Sensitivity studies were conducted with respect to key operational and environmental parameters such as compressor stroke, piston offset, drive frequency, and heatsink temperature. No attempt was made to optimize the cooler's performance in any one test; the parameter values were selected to allow sufficient variation to observe performance sensitivity while restricting the input power to less than 200 W.

Thermal Performance

The thermal performance of the TRW 35K cooler is presented in a multivariable format that depicts the interrelationship between the input power, the coldblock load, the specific power, the coldblock temperature, and a sensitivity parameter. For example, Figure 2 maps the

Figure 1. TRW 35K pulse tube cooler.

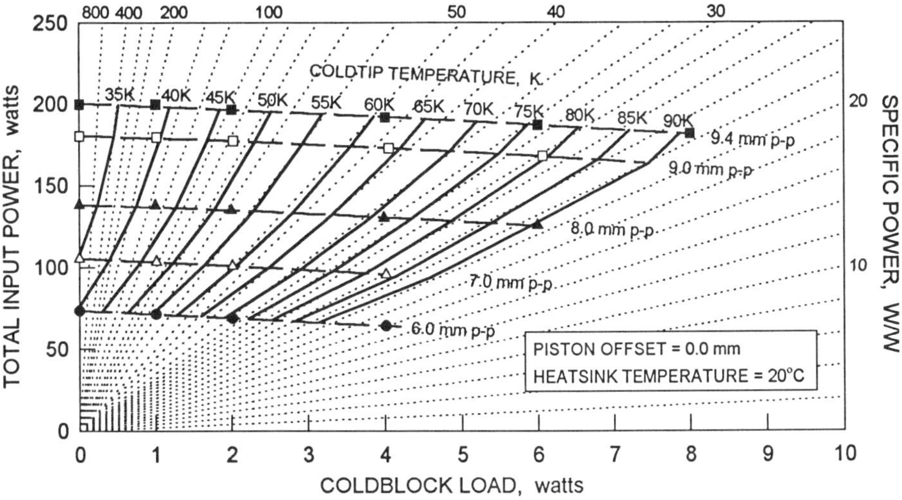

Figure 2. Sensitivity of thermal performance to compressor stroke.

performance as a function of compressor stroke. At each compressor stroke, measurements were made at several coldblock loads. A curve fit through these data yields a constant-stroke load line. Interpolation is then used to find the performance at a particular coldblock temperature, allowing isotherms to be developed across the sensitivity parameter, *e.g.* the compressor stroke. The slope of the isotherms may be compared to the specific-power grid lines to determine whether a particular change in the sensitivity parameter yields better or worse performance with regard to overall refrigeration efficiency.

Figure 2 shows the sensitivity of the cooler's thermal performance to compressor stroke with no piston offset and with a heatsink temperature of 20°C. The input power is strongly dependent on the compressor stroke and weakly dependent on the coldblock load. It is worth noting that very good performance is achieved over much of the parameter space: the isotherms are parallel to the specific-power grid lines over a wide range of coldblock loads, coldblock temperatures, and compressor strokes. Thus, at a particular coldblock temperature, the compressor stroke can be adjusted to accommodate a range of coldblock loads with little change in refrigeration efficiency.

The efficiency of the cooler may be improved by reducing the compression chamber's dead volume. Figure 3 shows the sensitivity of the thermal performance to piston offset with a compressor stroke of 9.0 mm_{pp} and a heatsink temperature of 20°C. As the offset increases from -1.0 mm to +2.0 mm, which moves the pistons toward one another, more input power is required to maintain a given stroke and the cooling capacity at a given temperature is increased. The cooler is, however, operating more efficiently at the higher offsets: notice that the isotherms are less steeply sloped than the specific-power grid lines.

Cooler Efficiency

An important figure-of-merit for cryocoolers is the thermodynamic coefficient of performance (COP) of the refrigerator expressed as a percentage of the ideal Carnot COP, which is uniquely defined in terms of the coldblock temperature and the heatsink temperature. The thermodynamic COP is a measure of the cooler's ability to convert work done on the gas into net cooling power. In the data presented here, the work done on the gas is approximated as the input electrical power minus the drive-motor i^2R losses.

Figure 4 shows the measured % Carnot COP as a function of the coldblock temperature; the sensitivity to compressor stroke is also shown. Notice that larger strokes achieve colder no-

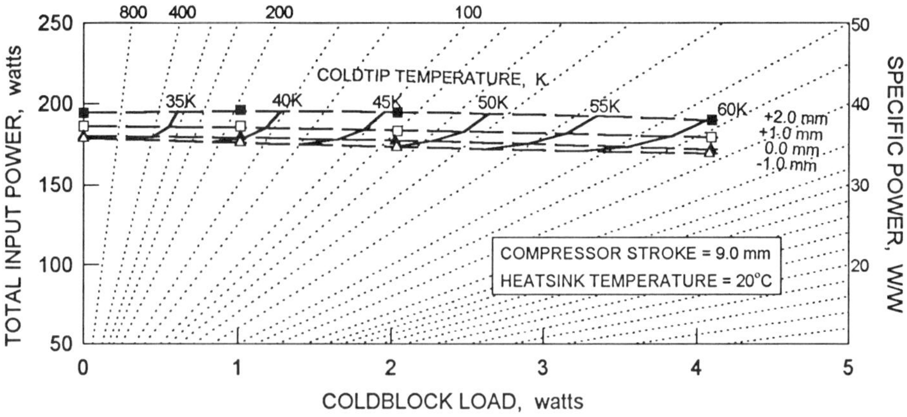

Figure 3. Sensitivity of thermal performance to piston offset.

load temperatures but appear to peak at lower % Carnot COP values than do the smaller strokes, as is also seen with Stirling coolers.[2]

The performance of the cooler drive motors was assessed: the motor efficiency was approximately 85% and the power factor was near 0.96. This performance was typical over a broad range of operational parameters, which contributes to the cooler's high refrigeration efficiency at a variety of operating points.

Off-State Conductance

Cold-head conductive losses can significantly reduce the amount of useful cooling work that a cooler can perform. These losses can also represent a major heat load in cryogenic applications that include redundant or non-operating coolers. The off-state conductance was measured using an equilibrium technique described in Reference 3. The 35K pulse tube cooler is physically quite large, and the coldblock is located in the middle of the coldhead, resulting in a relatively high conductive heat load of 500 mW at 35 K.

The off-state heat load can be significantly enhanced under certain orientations of the cold head with respect to gravity, as the hollow pulse tube becomes a source of convective helium gas flow. This is shown dramatically in Figure 5. Off-state heat loads in excess of 2000 mW were measured when the cold head was horizontal. The convective effect was highly sensitive to

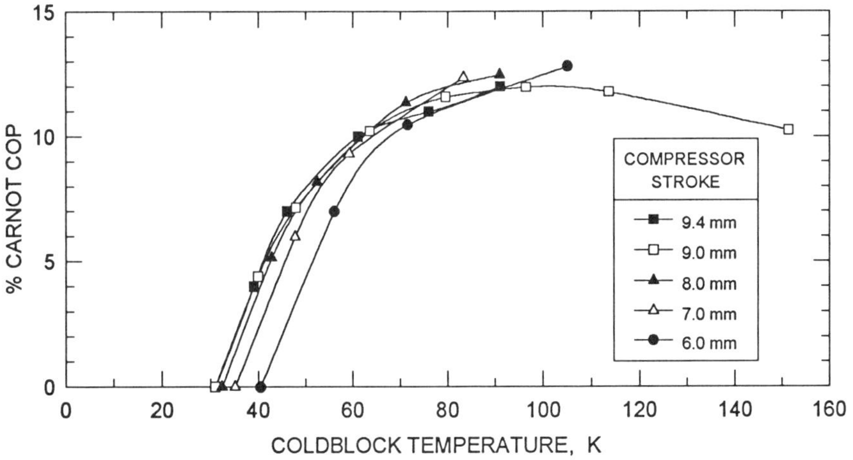

Figure 4. Sensitivity of % Carnot COP to compressor stroke.

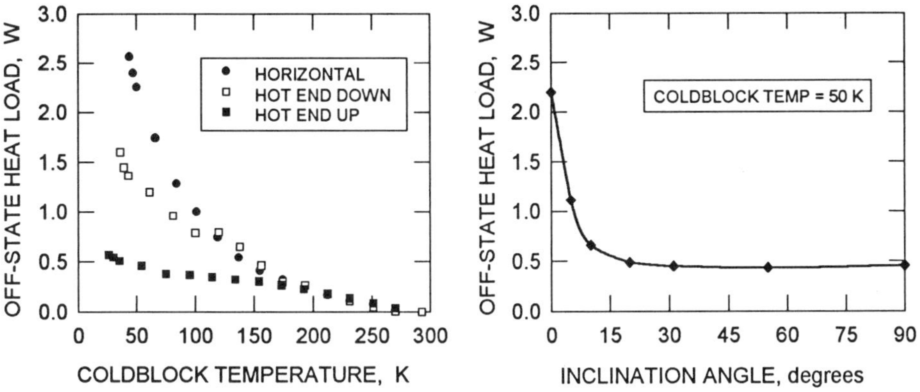

Figure 5. Sensitivity of off-state conductance to orientation.

coldhead inclination, with 20° of inclination being sufficient to eliminate the enhanced heat flow. This is an important consideration when planning the ground-testing of cryogenic systems with non-operating cryocoolers.

GENERATED VIBRATION

Measurements of the vibration generated by the TRW 35K pulse tube cooler were conducted in the JPL cryocooler vibration characterization facility using a special-purpose six-degree-of-freedom dynamometer. This dynamometer has a frequency range from 10 to 500 Hz and a force sensitivity from 0.005 N to 450 N full scale. During operation, the generated forces are simultaneously recorded in real time using a spectrum analyzer.

Integral Cooler Forces

As this cooler features a back-to-back compressor configuration, the vibration along the compressor axis at the drive frequency was minimized by trimming the stroke of one of the two compressors. The resulting forces are shown in Figure 6. The z-axis is the compressor axis, the x-axis is the cold-head axis, and the y-axis completes the orthogonal set. Most of the harmonics have amplitudes of 0.1 to 1 N, and many of them display little sensitivity to compressor stroke. Notice that the axial (F_z) and the lateral (F_x and F_y) vibrations are of similar amplitudes. These vibration levels are comparable to those generated by back-to-back Stirling coolers.[4]

Split Cooler Forces

In order to gain greater insight into the vibration generated by the pulse-tube cold head, a split cooler was also tested on the JPL dynamometer. The compressors and the cold head were provided by TRW and were similar to the 35K cooler, but they were not mated in the integral configuration. Figure 7 depicts the coldhead-generated forces as a function of the pressure-wave amplitude. Notice that the cold head is not free of vibration, although the levels are significantly smaller than the integral cooler's vibration.

Coldblock Motion

In addition to the vibration generated by the compressors and the cold head, the motion of the coldblock itself can act as a disturbance. This motion was measured with the integral cooler mounted on the dynamometer, providing a rigid mount for the cooler. A set of three

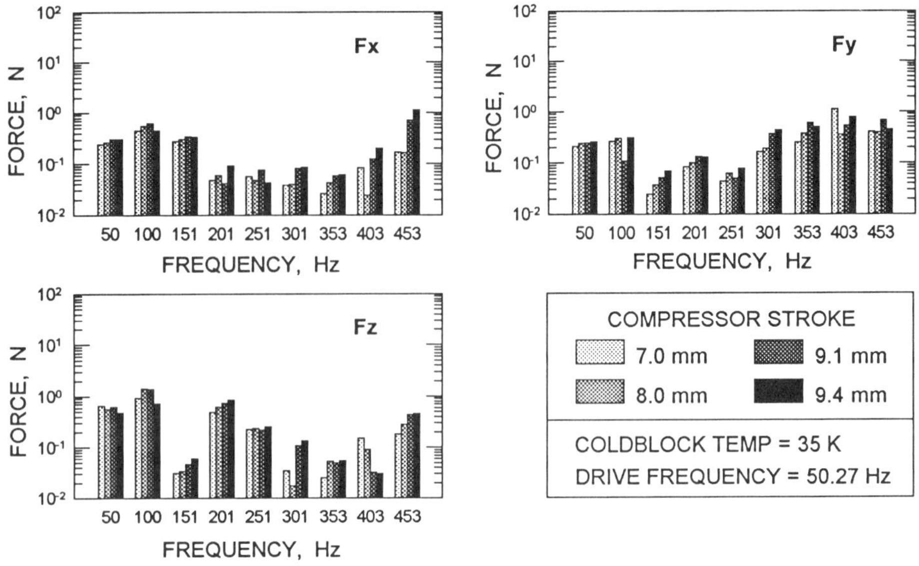

Figure 6. Sensitivity of generated vibration to compressor stroke, where the z-axis is the compressor axis and the x-axis is the cold-head axis.

accelerometers was then attached to the coldblock and aligned to the same axes as the force sensors. The measurements are shown in Figure 8. Notice that the motion is significantly greater along the cold-head axis than it is in the other two directions. These displacements are comparable in amplitude to the coldtip motion of Stirling coolers.[5]

ELECTROMAGNETIC COMPATIBILITY

EMC measurements were made with the cooler placed in a steel RF-shielded room and grounded to a copper-laminated table. During the testing the drive electronics were placed in an

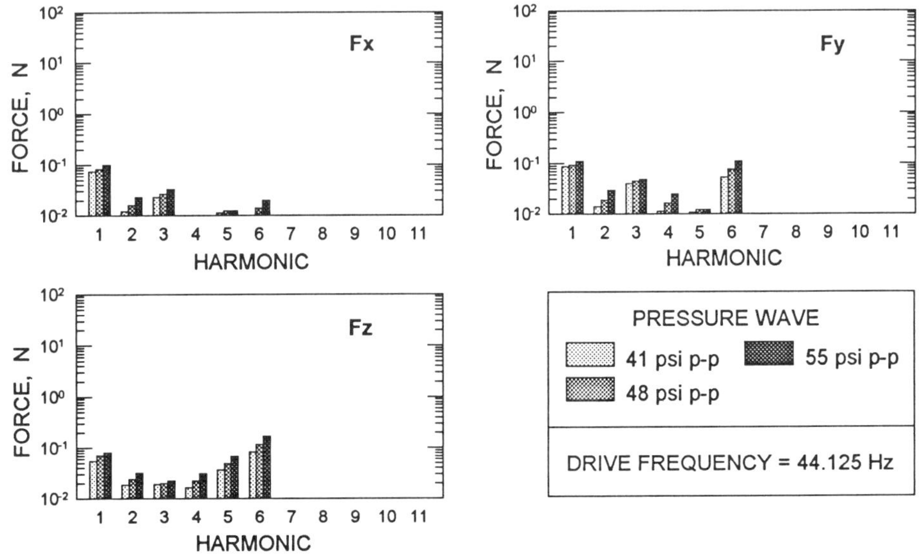

Figure 7. Sensitivity of cold-head vibration to pressure amplitude, where the z-axis is the compressor axis and the x-axis is the cold-head axis.

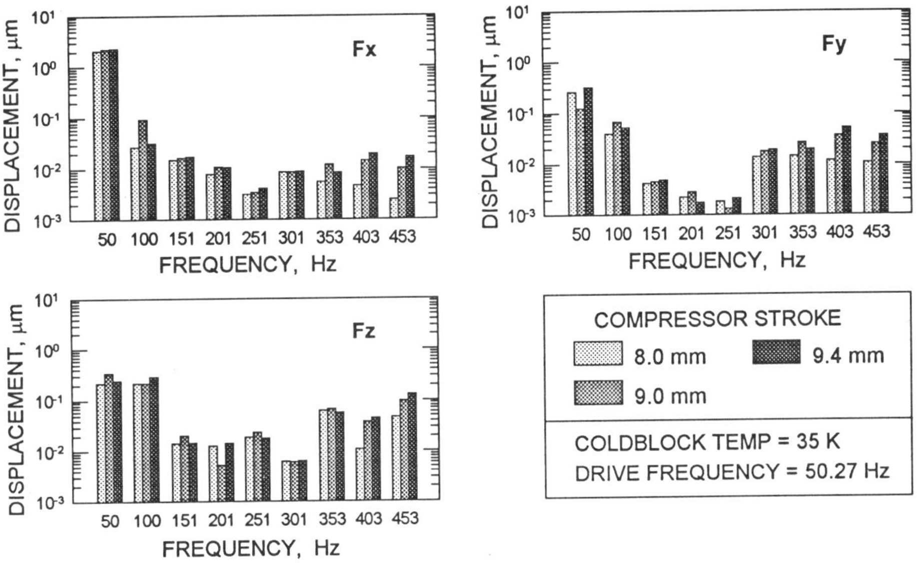

Figure 8. Sensitivity of coldblock motion to compressor stroke, where the z-axis is the compressor axis and the x-axis is the cold-head axis.

adjacent area outside the shielded room with connecting cabling fed through a bulkhead in the wall. The cabling was sheathed in aluminum foil and grounded to the copper table top to minimize any contributing radiation.

Measurements of the cooler-generated AC electric and magnetic field emissions were made. The cooler was operated at full power to create worst-case emissions. The cooler was powered with linear-amplifier power supplies, so the measurements did not contain the high-frequency EMI that is typical of the high-efficiency PWM-type power converters used in spacecraft applications. For the most part, the radiated AC electric field emissions were well below the specification limits of MIL-STD 461C: RE 02. The AC magnetic field emissions were measured at distances of 7 cm and 1 m from the cooler. The specification limits of MIL-STD 461C:RE 01 and RE 04 were exceeded by the first three harmonics of the cooler drive frequency, which is typical for cooler compressors of this size.[6]

The spatial distribution of the AC magnetic field emitted from the compressors was mapped in both the axial and radial directions. Figure 9 depicts the field distribution and strength (in nT) based on the magnitude of the fundamental. The measurements were then repeated with 0.030" thick μ-metal shielding wrapped around each compressor housing. As the figure indicates, the shielding had a significant effect on the intensity of the AC magnetic field emissions.

The DC magnetic field of the cooler was mapped. For these measurements the cooler was disconnected from the drive electronics and was mounted inside a set of DC coils that generate a magnetic field to cancel the earth's magnetic field within the test volume. The magnitude of the cooler's DC magnetic field was found to be 200 milligauss at a distance of 0.5 m.

SUMMARY

The TRW 35K pulse tube cooler was recently characterized at JPL in terms of its refrigeration performance, generated vibration, and electromagnetic compatibility. Sensitivity studies were conducted with respect to key operational and environmental parameters such as compressor stroke, piston offset, drive frequency, and heatsink temperature. The thermal

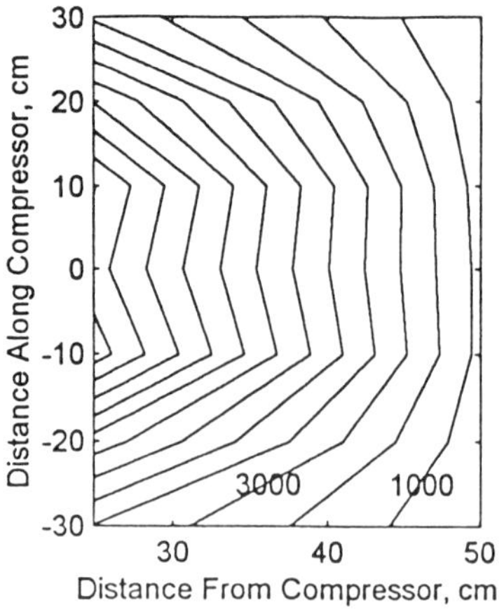

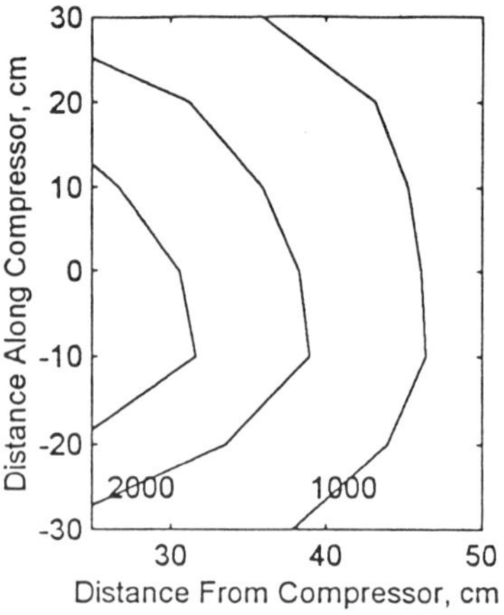

Figure 9. Effect of shielding on AC magnetic field emissions. The contour spacing is 1000 nT.

performance was found to be quite sensitive to piston offset position, and the off-state heat load was shown to have a strong angular dependence relative to gravity.

ACKNOWLEDGMENTS

The work described in this paper was carried out by the Jet Propulsion Laboratory, California Institute of Technology, and was sponsored by the Air Force Phillips Laboratory and the NASA EOS AIRS Project under a contract with NASA. Particular credit is due S. Leland, who provided assistance and support with many of the test setups.

References herein to a particular cooler, or to any specific commercial product, process, or service by tradename, trademark, manufacturer, or otherwise, does not constitute or imply its endorsement by the United States Government or the Jet Propulsion Laboratory, California Institute of Technology.

REFERENCES

1. W.W. Burt and C.K. Chan, Demonstration of a high performance 35 K pulse tube cryocooler, "Cryocoolers 8," R.G. Ross, Jr., ed., Plenum Press, New York (1995), pp. 313–319.
2. G.T. Smedley, G.R. Mon, D.L. Johnson, and R.G. Ross, Jr., Thermal performance of Stirling-cycle cryocoolers: a comparison of JPL-tested coolers, "Cryocoolers 8," R.G. Ross, Jr., ed., Plenum Press, New York (1995), pp. 185–195.
3. V. Kotsubo, D.L. Johnson, and R.G. Ross, Jr., Cold-tip off-state conduction loss of miniature Stirling cycle cryocoolers, "Advances in Cryogenic Engineering," Vol. 37, Part B, Plenum Press, New York (1992), pp. 1037–1043.
4. G.R. Mon, G.T. Smedley, D.L. Johnson, and R.G. Ross, Jr., Vibration characteristics of Stirling cycle cryocoolers for space application, "Cryocoolers 8," R.G. Ross, Jr., ed., Plenum Press, New York (1995), pp. 197–208.
5. R.G. Ross, Jr., D.L. Johnson, and V. Kotsubo, Vibration characterization and control of miniature Stirling-cycle cryocoolers for space application, "Advances in Cryogenic Engineering," Vol. 37, Part B, Plenum Press, New York (1992), pp. 1019–1027.
6. D.L. Johnson, G.T. Smedley, G.R. Mon, R.G. Ross, Jr., and P. Narvaez, Cryocooler electromagnetic compatibility, "Cryocoolers 8," R.G. Ross, Jr., ed., Plenum Press, New York (1995), pp. 209–220.

EXPERIMENTAL INVESTIGATION ON THE ROLE OF ORIFICE AND BYPASS VALVES IN DOUBLE-INLET PULSE TUBE REFRIGERATORS

T.Inada,[1] S.Nishio,[1] Y.Ohtani,[2] and T.Kuriyama [2]

[1]Institute of Industrial Science, University of Tokyo
Tokyo 106, Japan

[2]Toshiba R&D Center
Kawasaki 210, Japan

ABSTRACT

The double-inlet pulse tube refrigerator (DPTR) has a mechanism which consists of an orifice valve, a bypass valve and a reservoir tank at the hot end of the pulse tube. The mechanism brings about a phase angle between pressure and displacement of the working gas. Linearized thermoacoustic theories suggest that the phase angle has a strong influence on the performance of the pulse tube refrigerator (PTR). In this report, to estimate the phase angle, instantaneous mass flow rates of the working gas through the orifice and bypass valves were measured by hot wire anemometers. The experimental results indicate that the phase angle at the hot end of the pulse tube can be over 90° in the DPTR while it is below 90° in the orifice pulse tube refrigerator (OPTR). The experimental results also show that not only the phase angle but the average mass flow rate at the hot end of the pulse tube must be optimized to reach lower temperatures at the cold end.

INTRODUCTION

The most attractive point of pulse tube refrigerators (PTRs) is the absence of low-temperature moving parts. So practical use of the PTRs is anticipated where high reliability is needed. Since Mikulin et al.[1] invented the orifice pulse tube refrigerator (OPTR) which has an orifice and a reservoir tank at the hot end of the pulse tube as a phase shifter, the performance of the PTRs has been improved remarkably by means of phase shifters such as a moving plug[2] and a double-inlet[3].

These PTRs can be classified into three types by the phase angle between the cyclic oscillations of pressure and displacement of the working gas at the hot end of the pulse tube[4]. Denoting the phase delay of the gas displacement ξ_h compared with the pressure P_p at the hot end of the pulse tube by θ_h, the first type is specified as $\xi_h=0$, the second type as $0° < \theta_h \leq 90°$ and the third type as $\theta_h > 90°$. According to the thermoacoustic theory[4,5], the

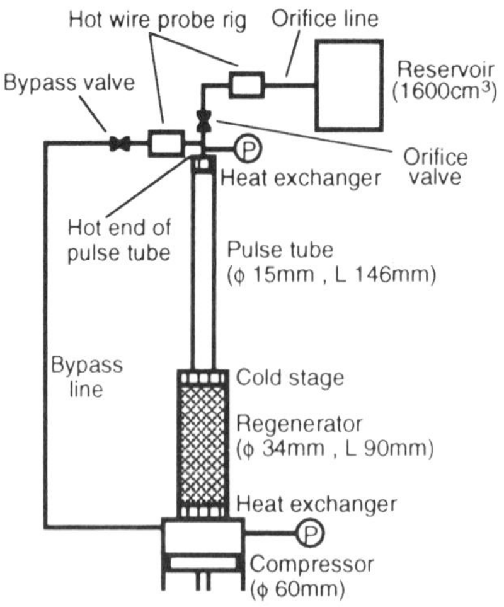

Figure 1. Schematic diagram of double-inlet pulse tube refrigerator used in experiment.

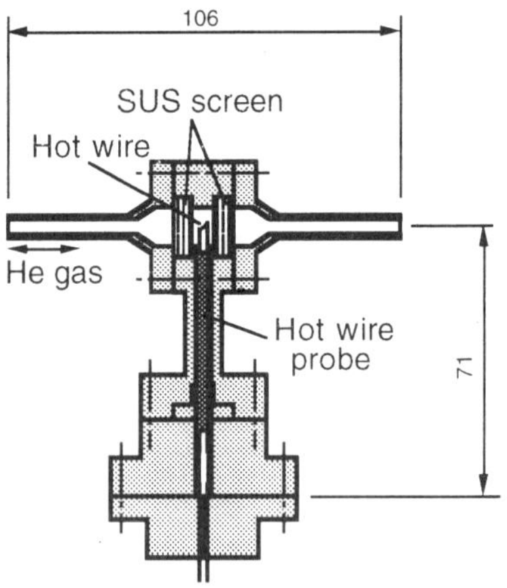

Figure 2. Hot wire probe rig.

third type refrigerator is best among these three types to generate a larger refrigeration power and to reach a lower temperature. The basic pulse tube refrigerator (BPTR)[6] belongs evidently to the first type because the hot end of the pulse tube is closed. It is often said that the OPTR is the second type and the double-inlet pulse tube refrigerator (DPTR) may be the third type. However there are only a few experimental data of the phase angle in PTRs, especially in the DPTR. Cai et al.[7] measured mass flow rates at the hot end of the pulse tube in a DPTR and obtained experimental results supporting the characteristics of the phase angle mentioned above. But in their report, it is not clear whether the phase angle could be over 90° in the DPTR. Yazaki[8] measured mass flow rates at arbitrary points in the pulse tube using LDV and proved that a phase angle over 90° is possible in the arrangement of the DPTR, but their apparatus didn't operate as a refrigerator because their experiments were conducted for very small amplitudes of pressure oscillation.

In this report, instantaneous mass flow rates through the orifice and bypass valves were measured by hot wire anemometers and the phase control function of both the valves were investigated.

EXPERIMENTAL APPARATUS

A schematic diagram of the DPTR used in the experiments is shown in Figure 1. Dimensions of each part are also shown in this figure. The working gas was helium. A linear driving piston was used as the compressor. Stainless steel spheres (ϕ0.30~0.35mm) filled a hotter half volume of the regenerator and copper screens (ϕ34mm,325mesh) another colder half volume. The line which connects the hot end of the pulse tube to the reservoir tank is called as the orifice line in the present paper. Another line which connects the hot end of the pulse tube to the compressor volume is called as the bypass line. The valves inserted to the orifice and bypass lines are called as the orifice and bypass valves respectively. During the experiments, the two heat exchangers shown in Figure 1 were kept at a constant temperature by cooling water. Temperature at the cold stage was measured by a resistance thermometer. Pressure oscillations were monitored at two locations near the two heat exchangers.

Mass flow rates of the working gas in the orifice and bypass lines were measured at the points near the orifice and bypass valves by hot wire anemometers as shown in Figure 1. Constant temperature anemometry was adopted to measure the mass flow rates. Figure 2

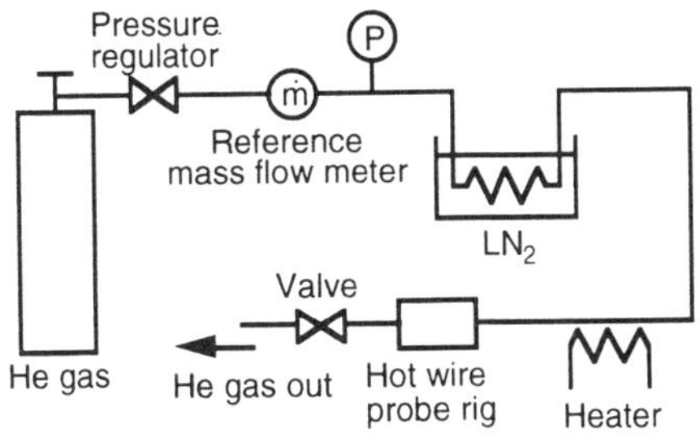

Figure 3. Arrangement to calibrate hot wire anemometer.

shows the cross-sectional detail of the hot wire probe rig. The hot wire was located in a stainless steel tube of 10mm in diameter. Two sets of stainless steel screens (200mesh) were stacked in the pipe at up- and down-streams 3mm away from the hot wire. This is the same method as Rawlins et al.[9] employed. The screens make the flow in the hot wire probe rig laminar and uniform. The diameter of the screens was 15mm. The thickness was 5mm for the orifice line and 3mm for the bypass line.

CALIBRATION OF HOT WIRE ANEMOMETER

The hot wire used in this measurement was a fine tungsten wire of 5μm in diameter and 1mm in length. Such a tungsten wire can be also used as a high response resistance thermometer[10]. Since the gas temperature fluctuates in the orifice and bypass lines and the fluctuation affects the outputs of the hot wire anemometer, it is needed to calibrate the hot wire anemometer to the gas velocity by changing temperature levels of the gas.

Fortunately, in this study, there was no need to get simultaneously the outputs of the anemometer and the gas temperature because the experiments were executed in a periodic steady state and the calibration was executed in a steady state. So one tungsten wire was placed in the hot wire probe rig as both an anemometer and a resistance thermometer by changing in turn an electric circuit.

Figure 3 shows a schematic arrangement for the calibration. The gas pressure was controlled by a pressure regulator. The mass flow rate was adjusted by a valve at the outlet of the gas line. The gas temperature in the hot wire probe rig was controlled by an electric heater attached to the gas line. When the gas was precooled by immersing a part of the gas line into a liquid nitrogen bath, we could make the gas temperature lower than room temperature. The hot wire probe was set in the same rig as Figure 2.

It is known that mass flow rates with pressure and temperature oscillations can be calculated by using fluid temperature and outputs of a hot wire anemometer regardless of pressure level[9]. Therefore the hot wire anemometer was calibrated to the mass flow rate at a constant pressure by changing the gas temperature.

Calibration results of the square of the output E^2 to the gas temperature T at several mass flow rates $\dot{m}$ are shown in Figure 4. They can be approximated by straight lines around room temperature. They are replotted in Figure 5, which represents calibration curves for constant gas temperatures. The abscissa in Figure 5 is chosen as $\dot{m}^{0.45}$ based on the following equation by Collis and Williams[11] for forced convection around fine wires.

$$Nu\,(T_m / T_\infty)^{-0.17} = 0.24 + 0.56\,Re^{0.45} \qquad (0.02 < Re < 44) \qquad (1)$$

In this equation, Re and Nu are the Reynolds number and the Nusselt number respectively. T_∞ is the fluid temperature far from the wire and T_m is the mean temperature of T_∞ and the

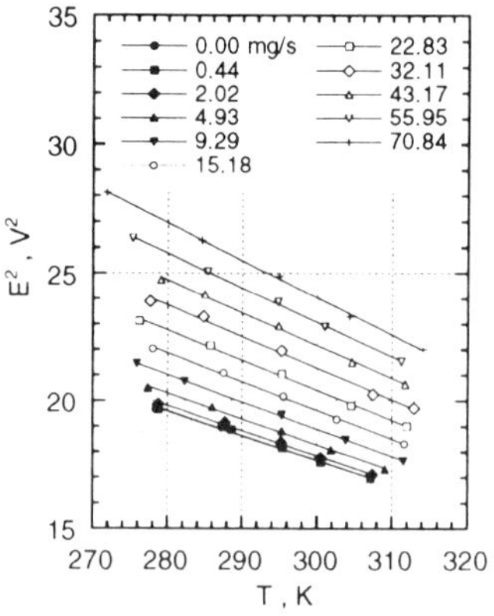

Figure 4. Plot of E² vs. temperature curves at various constant mass flow rates.

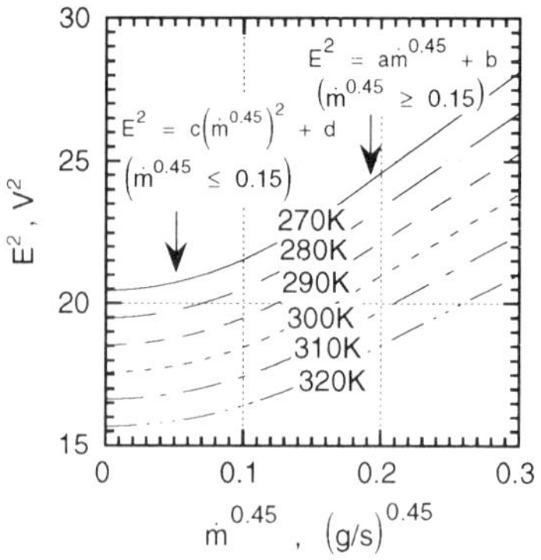

Figure 5. Calibration curves of hot wire anemometer at various temperatures.

wire temperature. The Reynolds numbers around the hot wire were always smaller than 44 in the experiments discussed here. In Figure 5, the correlating equations of E^2 to $\dot{m}^{0.45}$ are shown. The calibration curves in this figure are approximated by straight lines for larger mass flow rates and by quadratic lines for smaller mass flow rates. The coefficients a to d in the correlating equations are linear functions of the gas temperature.

The response time of the anemometer can be examined by supplying a square wave current to the hot wire. The response time of the anemometers used here was less than 20 μsec and it is enough to detect the mass flow oscillation under the operating conditions as mentioned later.

RESULTS AND DISCUSSION

Experiments to measure the instantaneous mass flow rates were executed by changing the opening of the orifice and bypass valves. As an index of opening of a needle valve, the flow coefficient, *Cv*, is often used . The flow coefficient is defined by non-oscillatory steady flows in general, so it is doubtful whether the flow coefficient is an appropriate measure of the opening for oscillating flows discussed here. Hence the absolute value of the mass flow rate through the needle valves averaged during one cycle is introduced as an index of the opening. It is called the average mass flow rate in this report and is represented as $\dot{m}_{av}$.

In these experiments, the refrigerator was operated under the following conditions. The frequency was 2.5Hz, the initial pressure was 1.1MPa and the stroke of the piston was 12.5mm i.e. the swept volume was 35.3cm^3.

Because the compressor efficiency was not optimized in these experiment, we cannot discuss the total efficiency of the refrigerator and the performance of the refrigerator is evaluated by the temperature of the cold stage.

In the following discussion, the term "hot end of the pulse tube" means the top of the heat exchanger shown in Figure 1.

Function of Orifice Valve

At first, in order to investigate the function of the orifice valve, experiments were conducted by changing only the orifice valve opening and keeping the bypass valve closed. Though it was impossible to cut off perfectly the flow through the bypass valve because of

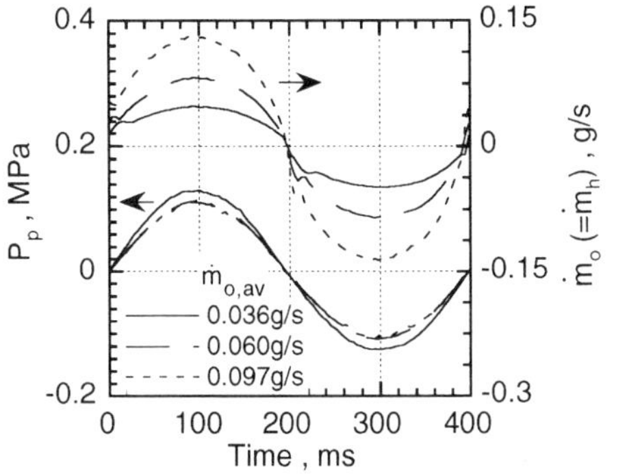

Figure 6. Mass flow rate through orifice valve and pressure at hot end of pulse tube as functions of time for several average mass flow rates through orifice valve.

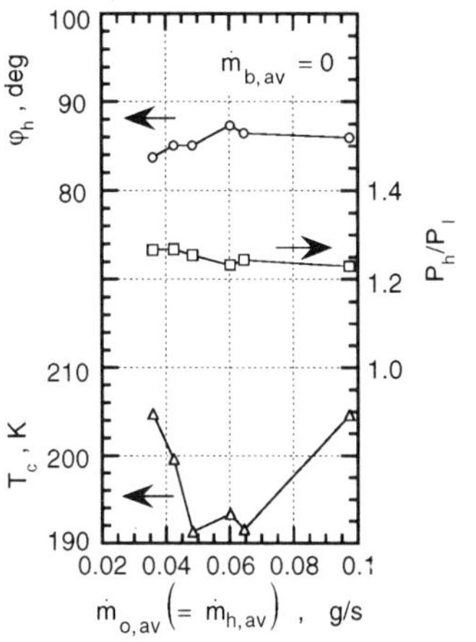

Figure 7. Dependence of φ_h, P_h/P_l, and T_c on average mass flow rate through orifice valve.

the characteristics of the needle valves used here, the mass flow rate through the bypass valve was negligible compared with that through the orifice valve so that these experiments were almost similar to experiments about the OPTR.

Figure 6 shows typical experimental results of the oscillations of the mass flow rate through the orifice valve $\dot{m}_o$ and the pressure at the hot end of the pulse tube P_p for several average mass flow rates through the orifice valve $\dot{m}_{o,av}$. The tube volume between the hot wire and the hot end of the pulse tube was so small that $\dot{m}_o$ was almost equal to the mass flow rate at the hot end of the pulse tube $\dot{m}_h$. In the figure, the flow direction from the pulse tube to the reservoir tank is defined as the positive direction. In these experiments, the flow direction was determined by referring to the pressure oscillation at the hot end of the pulse tube. The abscissa in Figure 6 is the elapsed time after P_p changes to a positive value from negative one.

To discuss the phase angle θ_h mentioned above, first, the mass displacement at the hot end of the pulse tube m_h was calculated by integrating $\dot{m}_h$. Since the measured wave-forms of the pressure and mass flow rate are not pure sine waves, a criterion to determine the phase angle from the experimental results is needed. In this paper, the representative phase angle in each wave-form was chosen as that at which the value changes from a negative value to positive one. Then the phase delay of m_h compared to P_p was defined by that between the representative phase angles and it is denoted by φ_h. Since θ_h is equal to φ_h in the linearized thermoacousitc theory [5], the experimental results are discussed using the phase angle φ_h in this report.

We checked the method to determine the phase angle as follows. The obtained wave-forms were expanded into the Fourier series and the phase of each wave-form was determined as that for the operating frequency of the refrigerator. The comparison showed that there was no meaningful difference between the results obtained by the two methods.

Figure 7 shows the results of the phase angle φ_h, the ratio of the maximum pressure to the minimum one at the hot end of the pulse tube P_h/P_l and the minimum temperature T_c. T_c is strongly influenced by $\dot{m}_{o,av}(=\dot{m}_{h,av})$ and there is an optimum value of $\dot{m}_{o,av}$. The phase shift φ_h is almost independent of $\dot{m}_{o,av}$ and it is constant at about 85° within the operating conditions discussed here. The measured value of the phase angle indicates that the OPTR is

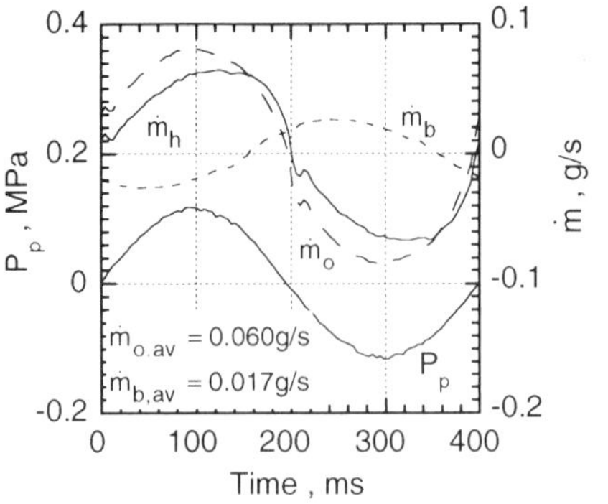

Figure 8. Mass flow rates at various locations and pressure at hot end of pulse tube as functions of time.

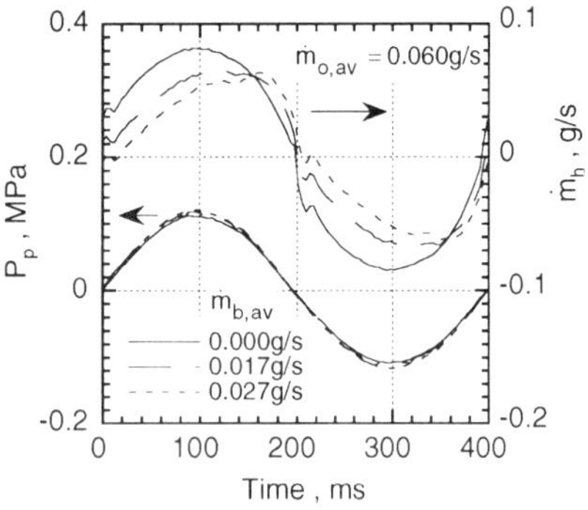

Figure 9. Mass flow rate and pressure as functions of time at hot end of pulse tube for several average mass flow rates through bypass valve.

classified into the second type in which $0°<\varphi_h\leq90°(0°<\theta_h\leq90°)$. Though P_h/P_l decreases as $\dot{m}_{o,av}$ increases, the dependency is relatively weak. Hence the most influential parameter for the minimum temperature in these experiments is probably $\dot{m}_{h,av}$. When $\dot{m}_{h,av}$ becomes smaller, the refrigerator approaches the BPTR and the refrigeration power must be decreased. On the other hand, the increase in $\dot{m}_{h,av}$ deteriorates the regenerator efficiency and, also it may increase some losses related to the gas oscillation in the pulse tube[12,13]. They are the reasons why the optimum $\dot{m}_{o,av}(=\dot{m}_{h,av})$ does exist.

Function of Bypass Valve

In order to look into the function of the bypass valve, experiments were conducted by changing also the opening of the bypass valve.

The tube volume among the two hot wires and the hot end of the pulse tube was so small that the sum of the mass flow rate through the orifice valve $\dot{m}_o$ and that through the bypass valve $\dot{m}_b$ was almost equal to the mass flow rate at the hot end of the pulse tube $\dot{m}_h$. Figure 8 shows the experimental results of $\dot{m}_o$, $\dot{m}_b$, $\dot{m}_h$ and P_p during a cycle for $\dot{m}_{o,av}$=0.060g/s and $\dot{m}_{b,av}$=0.017g/s. In the figure, the positive direction is defined as the direction from the pulse tube to the reservoir tank for $\dot{m}_o$, as the direction from the pulse tube to the compressor through the bypass line for $\dot{m}_b$ and as the direction from the cold end to the hot end of the pulse tube for $\dot{m}_h$. The flow direction was estimated by the pressure oscillations at the hot end of the pulse tube and in the compressor volume. Neither the phases of $\dot{m}_o$ nor $\dot{m}_b$ changed in these experimental conditions even if $\dot{m}_{o,av}$ and $\dot{m}_{b,av}$ were changed. However the phase of $\dot{m}_h$, which was obtained by the superposition of $\dot{m}_o$ and $\dot{m}_b$, changed if the $\dot{m}_{o,av}$ and $\dot{m}_{b,av}$ were changed.

Figure 9 shows the results of $\dot{m}_h$ at some values of $\dot{m}_{b,av}$ for $\dot{m}_{o,av}$=0.060g/s. Reminding the definition of φ_h mentioned already, φ_h increases as $\dot{m}_{b,av}$ increases.

Figure 10 shows the dependence of φ_h, $\dot{m}_{h,av}$, P_h/P_l and T_c on $\dot{m}_{b,av}$. The value of $\dot{m}_{o,av}$ is 0.049g/s in (a), 0.060g/s in (b) and 0.100g/s in (c) respectively. These figures indicate peculiar tendencies for the dependence of T_c on $\dot{m}_{b,av}$ within the range of the valve opening shown in Figure 10. In the case of (a) in which $\dot{m}_{o,av}$ is relatively small, T_c increases monotonously as $\dot{m}_{b,av}$ increases. In the case of (b) in which $\dot{m}_{o,av}$ is somewhat large, $\dot{m}_{b,av}$ has the optimum value that makes T_c minimum. In the case of (c) in which $\dot{m}_{o,av}$ is larger than other cases, T_c decreases monotonously as $\dot{m}_{b,av}$ increases. Though P_h/P_l increase with increasing $\dot{m}_{b,av}$ in all the cases from (a) to (c), the dependence of P_h/P_l on $\dot{m}_{b,av}$ is

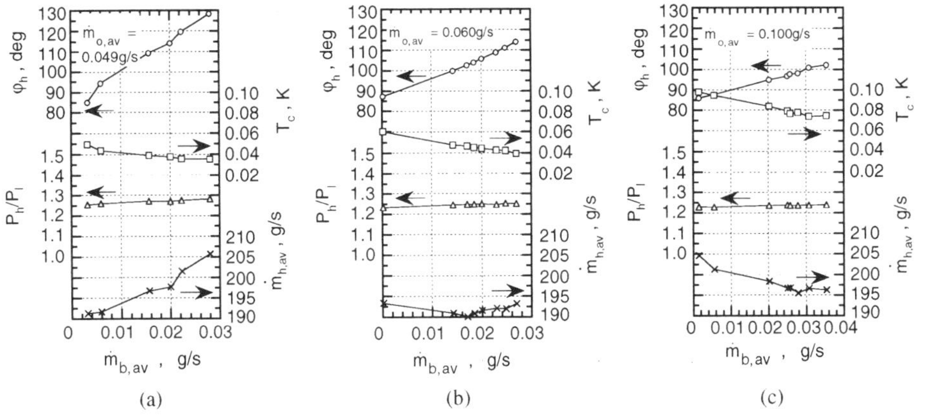

Figure 10. Dependence of φ_h, P_h/P_l, $\dot{m}_{h,\,av}$ and T_c on average mass flow rate through bypass valve.

not so strong and P_h/P_l does not seem to be dominant on T_c. φ_h increases monotonously as $\dot{m}_{b,\,av}$ becomes large and φ_h>90° is realized by opening the bypass valve. Hence the DPTR is classified into the third type in which φ_h>90°(θ_h>90°). $\dot{m}_{h,\,av}$ decreases as $\dot{m}_{b,\,av}$ increases in all the cases of (a) to (c) within the range of the valve opening in these experiments.

In Figure 11, T_c are plotted on φ_h - $\dot{m}_{h,\,av}$ map because P_h/P_l does not have strong influences on the performance of the refrigerator within the experimental conditions discussed here. The contour lines of T_c inferred from the experimental results are also drawn. The optimum value of φ_h providing the minimum T_c exists in the range between 90° and 110°. Because the phase angle φ_h over 90° cannot be realized when the bypass valve is closed, the lowest value of T_c is obtained only by opening the bypass valve. In addition, Figure 11 also indicates that T_c varies with $\dot{m}_{h,\,av}$ even if φ_h is kept constant and that T_c varies with φ_h even if $\dot{m}_{h,\,av}$ is kept constant. Therefore both φ_h and $\dot{m}_{h,\,av}$ must be optimized to get lower temperatures.

CONCLUSION

Mass flow rates of the working gas through the orifice and bypass valves in the OPTR and DPTR were measured by hot wire anemometers to investigate the function of phase

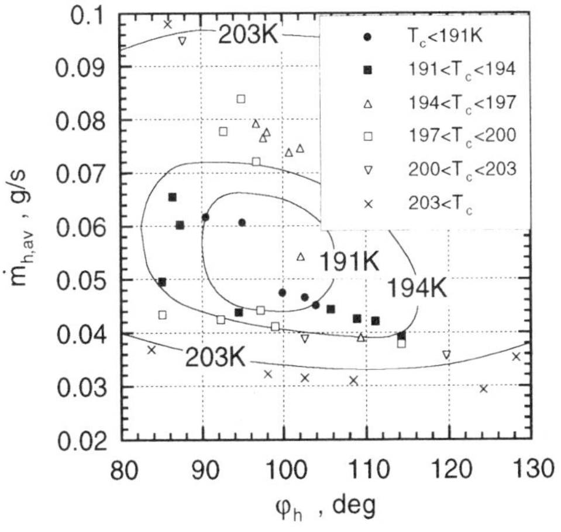

Figure 11. Map of minimum temperature contours drawn by experimental results.

control of these valves, and the following results were obtained. The phase angle φ_h between the mass displacement and the pressure can be over 90° in the DPTR while φ_h is below 90° in the OPTR. The effectiveness of phase control by the bypass valve is ascertained because the optimum value of φ_h exists between 90° to 110° in these experiments. To improve the performance of the refrigerator, both φ_h and $\dot{m}_{h,av}$ which is the average mass flow rate at the hot end of the pulse tube must be optimized.

REFERENCES

1. E.I.Mikulin, A.A.Tarasov, and M.P.Shkrebyonock, Low-temperature expansion pulse tubes, in:"Advances in Cryogenic Engineering," Vol.29, Plenum Press, New York (1984), p.629.
2. Y.Matsubara and A.Miyake, Alternative method of the orifice pulse tube refrigerator, in:"Proc. 5th Intl. Cryocooler Conf.," (1988), p.127.
3. S.W.Zhu, P.Y.Wu, and Z.Q.Chen, Double inlet pulse tube refrigerators: an important improvement, Cryogenics, 30:514 (1990).
4. A.Tominaga, Phase dependences of pulse-tube refrigerators, Cryogenic Engineering, 27:134 (1992) (in Japanese)
5. A.Tominaga, Thermodynamic aspects of thermoacousitc theory, Cryogenics, in press (1995)
6. W.E.Gifford and R.C.Longsworth, Pulse-tube refrigeration, Trans. ASME Ser.B J. Eng. Ind., 86:264 (1964)
7. J.H.Cai, Y.Zhou, J.J.Wang and W.X.Zhu, Experimental analysis of double-inlet principle in pulse tube refrigerators, Cryogenics, 33:3010 (1993).
8. T.Yazaki, in:"Proc. 53rd Meeting on Cryogenics and Superconductivity," (1995), p12. (in Japanese)
9. W.Rawlins, R.Radebaugh, and K.D.Timmerhaus, Thermal anemometry for mass flow measurement in oscillating cryogenic gas flows, Rev. Sci. Instrum., 64:3229 (1993).
10. W.Rawlins, K.D.Timmerhaus, and R.Radebaugh, Resistance thermometers with fast response for use in rapidly oscillating gas flows, Temperature Meas. Control Sci. Ind., 6:471 (1992)
11. D.C.Collis and M.J.Williams, Two-dimensional convection from heated wires at low Reynolds numbers, J. Fluid Mech., 6:357 (1959).
12. J.M.Lee, P.Kittel, K.D.Timmerhaus and R.Radebaugh, Flow patterns intrinsic to the pulse tube refrigerator, in:"Proc. 7th Intl. Cryocooler Conf.," (1992), p.125.
13. J.Yuyama and M.Kasuya, Experimental study on refrigeration losses in pulse tube refrigerator, Cryogenics, 33:947 (1993).

EXPERIMENTAL AND ANALYTICAL STUDY OF TWO STAGE PULSE TUBE REFRIGERATOR

M.Tanaka[1], T.Nishitani[2], T.Kodama[1], T.Araki[2], E.Kawaguchi[3] and M.Yanai[3]

[1] Faculty of science, Osaka City University
Sugimoto 3-3-138, Sumiyoshi-ku, Osaka 558, Japan

[2] Shiga Technology Center,Iwatani International Corp.
Katube-cho 1095, Moriyama-shi, Shiga 524, Japan

[3] Iwatani Plantech Co. Ltd.
Katube-cho 1095 Moriyama-shi, Shiga 524, Japan

ABSTRACT

We have constructed a two stage pulse tube refrigerator equipped with two rotary valve system. The first stage regenerator was packed 1735 stainless steel screens of #200 and the second stage regenerator was packed with 359 g Pb shot of 0.3~0.4 mm diameter. The experiment was carried out at fixed frequency of 2.6 Hz and phase angles of 122 degrees using two compressor units of 40 m^3/h. The performance test was made by changing systematically the volume ratio of first and second pulse tube and their total volumes. The lowest temperature at the second stage cold end was 15.6 K at the optimum operated conditions, when the volume ratio was 1 and the total volume was equal to the total void volume of the regenerators. We tried to simulate the performance using the equivalent P-V work method, and investigated the correlation between the experimental results and the calculated PV work. It is found that the equivalent P-V diagram method is useful for the understanding of the gas behavior in the pulse tube and the performance of the refrigerator.

INTRODUCTION

Since the orifice type pulse tube refrigerator was discovered by Mikulin et al[1], the pulse tube refrigerator has been improved remarkably. The double inlet type and the other modified pulse tube refrigerator[2,3] which has new type phase shifter were developed and have reached to very low temperatures. Recently three stage pulse tube refrigerator was developed and reached below 4 K[3].

We have constructed a two stage pulse tube refrigerator[4] which has two rotary valves, one at the entrance of the regenerator and the other at the hot ends of the first and second pulse tubes and have examined the performance effect of exchanging the materials of the second stage regenerator. The lowest temperature at the second stage cold end of the regenerator was 19 K[4]. After this study, we optimized the total volume and the volume ratio of the pules tubes and we have tried to analyze the gas behavior in the pulse tube to get some idea how to improve the design of the refrigerator using the equivalent P-V diagram method[5].

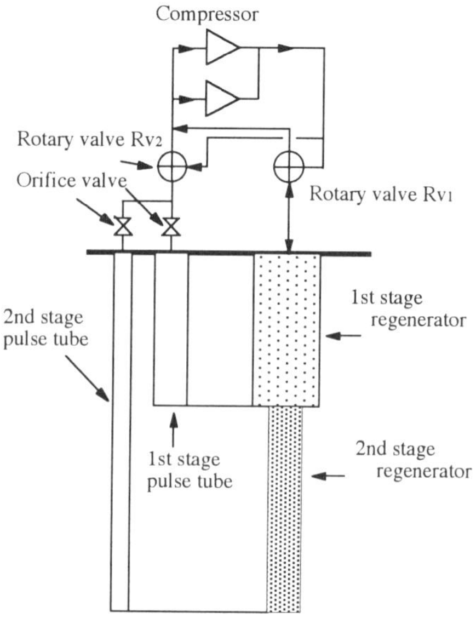

Figure 1. Schematic diagram of two stage pulse tube refrigerator.

In this paper, we report the experimental results about the effect of the pulse tube volume and the results of the simulation for the double inlet type pulse tube refrigerator based on the equivalent P-V diagram method proposed by Matsubara et al[5].

EXPERIMENTAL

Figure 1 shows the schematic drawing of the experimental set-up. The refrigerator consists of two regenerators, two pules tubes, two rotary valve systems(R_{V1} and R_{V2}) and two compressor units. The first stage regenerator was made of 180 mm long, 0.5 mm thickness stainless steel tube of 39 mm OD and 1735 stainless steel screens of 200 mesh. The second stage regenerator was made of stainless steel tube 18 mm OD, 0.3 mm thickness and 250 mm long packed 359 g of lead shots of about 0.3 mm diameter.

Each rotary valve system consists of a three-way rotary valve and a stepping motor mounted in the valve housing. The connections of the three-way valves are shown in the figure. Each compressor has a capacity of 40 cubic meters per hour, their pressure ratio was about 2.3 and they are operated at the mean pressure of 1.6 MPa. The rotation rates of the rotary valves were controlled by adjusting the pulse frequency to the stepping motor. Two pressure cycles were produced by one turn of the rotor. The angular position and the phase angle of the two rotors were monitored by a photo sensor mounted in the valve housing. As the rotary valves can be operated independently, it is possible to set the phase angle difference of the two valves to any value. The amount of the gas which flows to and from each hot end of the pulse tubes was controlled by adjusting the opening of the orifice valves. The temperatures at the first and second stage cold end and the middle point of the second stage regenerator, were measured by the Au-Fe vs chromel thermocouple.

EXPERIMENTAL RESULTS

Experiments were carried out by exchanging systematically the total volume and volume ratio of the first and second pulse tubes for the same regenerator combination.

Measurement were made at the fixed frequency of 2.62 Hz and phase angles of 122 degree. Here the phase angle 0 degree is defined when the entrance of the regenerator is

Table 1. Lowest temperature for various volume of 1st and 2nd pulse tube

Type	Volume and size of the pulse tube V cm^3, (IDmm×Length mm) 1st stage(V1)	2nd stage(V2)	Total volume Vt cm^3(=V1+V2)	V1/V2	Temp. at 2nd stage K
a	127.5,(19×450)	42.5,(11.7×395)	170	3.0	25.7
b	113 ,(19×400)	57 ,(13×425)	170	2.0	19.0
c	85 ,(19×300)	85 ,(17.4×360)	170	1.0	15.6
d	70 ,(19×247)	100 ,(17.4×421)	170	0.7	16.5
e	57 ,(19×201)	113 ,(19×400)	170	0.5	20.5
f	100 ,(19×354)	100 ,(17.4×421)	200	1.0	16.5
g	70 ,(19×247)	70 ,(17.4×295)	140	1.0	19.8
h	100 ,(19×354)	85 ,(17.4×360)	185	1.176	16.5
i	70 ,(19×247)	85 ,(17.4×360)	155	0.82	20.0

connected to high pressure side of the compressor by Rv1 and the hot end of the pulse tube is connected to low pressure side of the compressor by Rv2. Therefore the phase angle θ means the state that Rv1 leads Rv2 by θ/2. The openings of the orifice valves were adjusted to their optimum position for each data point.

Table 1 shows the lowest temperature of the second stage cold end for various volume of the first and second stage pulse tube. Fig 2 shows the temperature of the 1st and 2nd stage cold end as a function of their volume ratio. Type c reaches the lowest temperature of 15.6 K, which had volume ratio of 1 and total volume of 170 cm^3. This value is equal to the total void volume of the 1st and 2nd regenerators.

Figures 3 and 4 show the relation between the heater power of the 2nd stage cold end and the temperature of the 1st and 2nd stage cold end of Type c and g respectively. These figures show the importance of the pulse tube volume for the performance of the refrigerator. When the heater power to the cold end was changed stepwise, the behavior of the temperature change of Type c was most stable and reached the next steady state condition faster than any other types.

It is not clear why the optimum volume ratio of the pulse tubes is equal to 1 and the total volume of the pulse tubes is equal to the regenerator void volume. However, the data shows the existence of the optimum volume ratio.

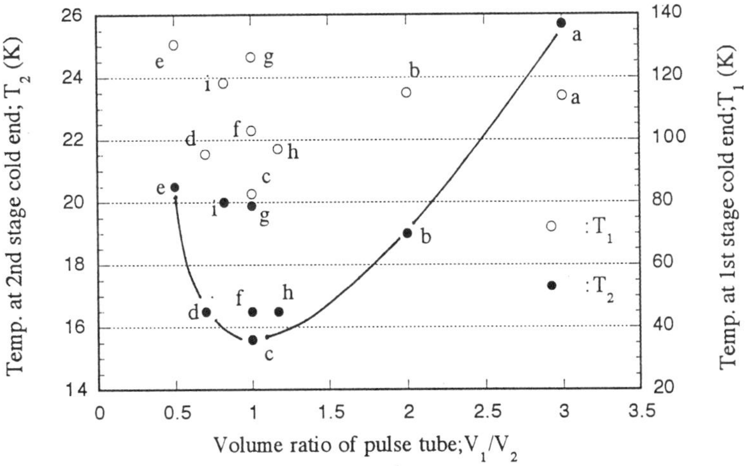

Figure 2. The 1st and 2nd stage cold end temperature for various volume ratio of the pulse tubes. a~i are the type in the table 1.

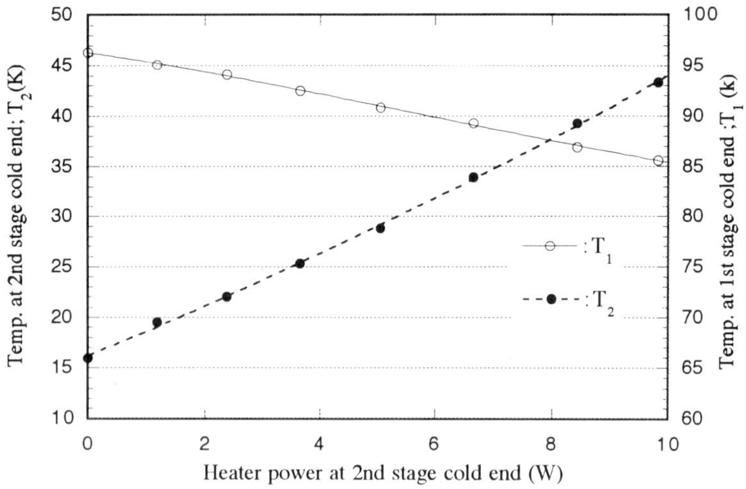

Figure 3. The 1st and 2nd stage cold end temperature as a function of the heater power supplied to the 2nd stage cold end of type c.

EQUIVALENT P-V DIAGRAM

Figure 5 shows a schematic diagram for the equivalent P-V diagram simulation. The pulse tube is divided into three parts; the gas in the cold part flows to and from the regenerator, the gas of the middle part is reciprocated in the pulse tube, the gas in the hot part flows via orifice valve and bypass valve. The regenerator is divided into two parts connected via a flow-resistance. In this model, it is assumed that:

1) the pressure in the pulse tube and the cold part of the regenerator is equal and uniform.
2) the gas temperature in each part of the regenerator and the pulse tube is uniform and constant except in the middle part of the pulse tube.

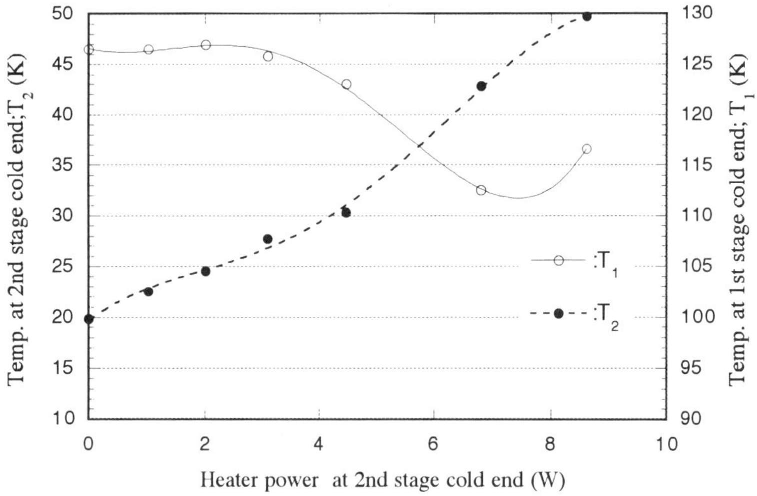

Figure 4. The 1st and 2nd stage cold end temperature as a function of the heater power supplied to the 2nd stage cold end of type g.

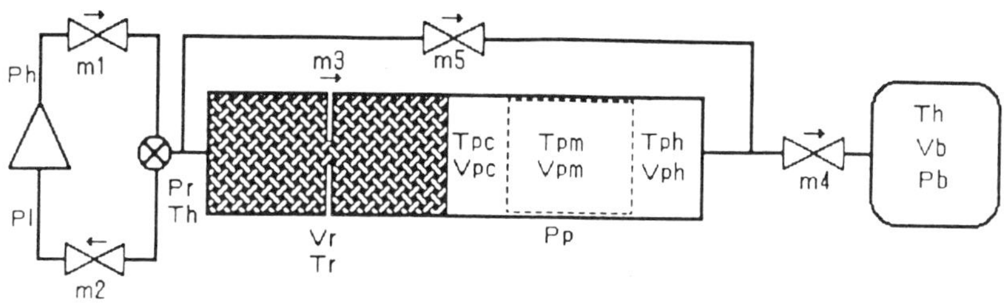

Figure 5. Schematic diagram for the equivalent P-V diagram simulation. m, P, V, T are shown mass flow rate, pressure, volume and temperature respectively.

3) the behavior of the gas column in the middle part of the pulse tube is adiabatic.
4) the mass flow rate is proportional to pressure difference.
5) the working substance is assumed ideal gas.

The governing equations can be formulated as follows using the mass conservation law and state equation of the ideal gas:
the pressure change is

$$\frac{d\,Pr}{dt} = \frac{2RTr}{Vr}(m_1 - m_2 - m_3 - m_5) \tag{1}$$

$$\frac{dPp}{dt} = \frac{RTh(m_5 - m_4) + RTcm_3}{Vp + \frac{Vr}{2Tr} + \frac{1-\gamma}{\gamma}Vpm} \quad (\equiv F) \tag{2}$$

$$\frac{dPb}{dt} = \frac{RTh}{Vb}m_4 \tag{3}$$

where γ is specific heat ratio of the gas, and the mass change in the cold part and hot part of the pulse tube are

$$\frac{dMpc}{dt} = m_3 - \frac{Vr}{2RTr}F \tag{4}$$

$$\frac{dMph}{dt} = -m_4 + m_5 \tag{5}$$

The mass flow rate is assumed as follows;

$$m_1 = \alpha_1(Ph - Pr)$$
$$m_2 = \alpha_2(Pr - Pl)$$
$$m_3 = \alpha_3(Pr - Pp)$$
$$m_4 = \alpha_4(Pp - Pb)$$
$$m_5 = \alpha_5(Pr - Pp)$$

where α is the flow coefficient of valves and the flow resistance in the regenerator.

After all flow coefficients,α, the high and low pressure of the compressor(Ph,Pl) and the temperature of each part were determined, the P-V work at the cold and hot part of the pulse tube was calculated by Runge-Kutta method using the above equations. The amplitude of the gas displacement and the volume change were calculated from Eq.(4)and (5). The phase difference between gas displacement and pressure were determined from the first mode of the Fourier transformed form of the oscillation.

We roughly estimated the heat loss which results from moving the gas piston in the pulse tube. It is assumed that:

1) the temperature gradient along the pulse tube wall is linear and constant for a period.
2) only the gas of cold part in the pulse tube transfers the heat to and from the wall.
3) the gas displacement is sinusoidal.

The heat loss for unit time can be expressed as:

$$Q_{loss} = \frac{3}{4}(Hh - Hc)\frac{Ld\,\rho}{d}\frac{Vpc^2}{Vp} \tag{6}$$

where, *Hh* and *Hc* are enthalpy of the gas in the hot and cold part, ρ is the density at cold part, *d* is the diameter and *Vp* is the volume of the pulse tube, *Vpc* is the volume change of the cold part. *Ld* is the thermal penetration depth of the gas and is given as

$$Ld = \sqrt{\frac{D}{\pi f}}$$

where, *D* is the thermal diffusivity and *f* is the frequency.

We simulated a single stage pulse tube refrigerator of our previous work[4]. Its regenerator was made of stainless steel tube of 18 mm OD, 0.3 mm thickness and the length of 200 mm packed with 1700 brass screen mesh of #200 and pulse tube was 14 mm OD, 0.5 mm thickness and the length of 350 mm made of stainless steel. The lowest temperature of the cold end was 31.7 K at the pressure frequency of 6.5 Hz.

The typical P-V diagram are shown in Fig.6. Figures 7 and 8 show the simulated P-V work, displacement, phase difference at cold part and the net cooling power as a function of the flow coefficient of the orifice valve α_4 for orifice and double inlet types respectively.

For the orifice type refrigerator (Fig.7), the P-V work has a maximum at about α_4=5. As the opening of the orifice valve is increased, the displacement ξ_l increases whereas the phase difference θ_l saturates above $\alpha_4 \approx 10$. At the same time the pressure amplitude decreases. Therefore, the P-V work become maximum at α_4=5.

In the case of double inlet type(Fig.8), the bypass valve is opened and fixed at α_5=5. The value of θ_l is larger and ξ_l is smaller compared to the orifice type for the same opening of the orifice valve. The maximum P-V work increases slightly compared to that of the orifice type at the higher openings of the orifice valve.

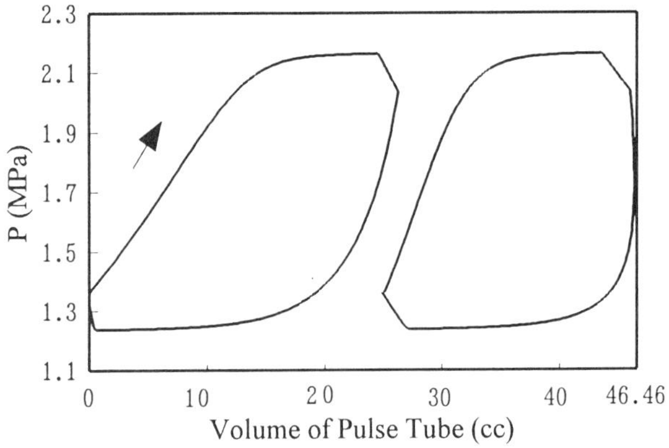

Figure 6. The typical simulated P-V diagram of orifice type pulse tube refrigerator. Where, P_h=2.3 MPa, P_l=1.1 MPa, f=8 Hz, $\alpha_1, \alpha_2, \alpha_3$=100 g/MPa·sec, α_4= 2 g/MPa·sec, α_5=0 g/MPa·sec.

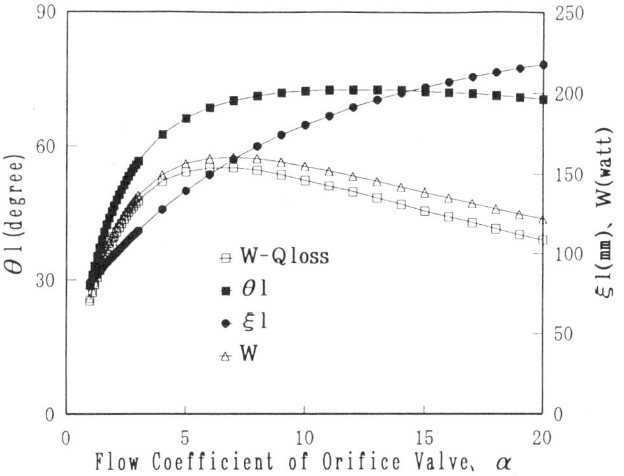

Figure 7. Equivalent P-V work W, displacement ξl, phase difference θl at cold part and net cooling power W-Qloss as a function of the flow coefficient of the orifice valve at a condition of the orifice type. Where, P_h=2.3 MPa, P_l=1.1 MPa, f=8 Hz, $\alpha_1, \alpha_2, \alpha_3$=100 g/MPa sec, α_5=0 g/MPa sec.

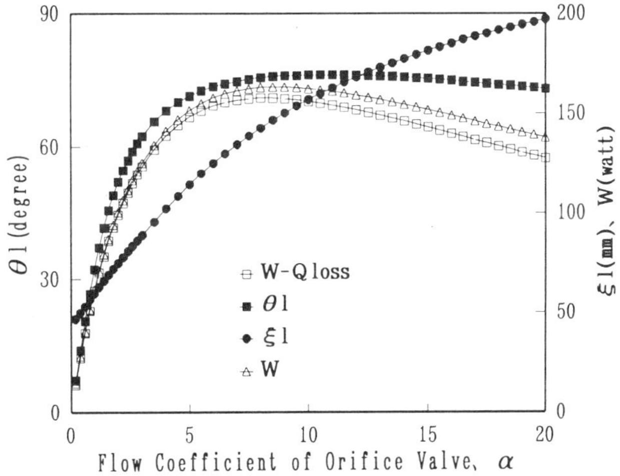

Figure 8. Equivalent P-V work W, displacement ξl, phase difference θl at cold part and net cooling power W-Qloss as a function of the flow coefficient of the orifice valve at a condition of the double inlet type. Where, P_h=2.3 MPa, P_l=1.1 MPa, f=8 Hz, $\alpha_1, \alpha_2, \alpha_3$=100 g/MPa sec, α_5=5 g/MPa sec.

These results show that the gas behavior in the pulse tube and the performance can be understand qualitatively. However, the flow coefficient values are not known for the practical device and the estimation of the heat loss are rough. Improvements of the model are necessary for the practical use of the design of the refrigerator.

CONCLUSION

We have constructed a two stage pulse tube refrigerator and optimized the volume ratio of the pulse tube. When the total volume was equal to the total void volume of the regenerator and the volume ratio of the first and second stage pulse tubes is equal to 1, the temperature at the second stage cold end reached the lowest temperature of 15.6K.

The equivalent P-V work have been calculated based on a simple model and compared with experimental results. These simulation results seem to be helpful for the understanding of the performance and gas behavior in the pulse tube.

REFERENCES

1. E.I.Mikulin, A.A.Tarasov and M.P.Shkrebyonock. Low-temperature Expansion Pulse Tubes, *Adv. Cryog. Eng.*, 29: 629(1984)
2. Z.Zhu, P.Wu and Z.Cheen. A Single Stage Double Inlet Pulse Tube Refrigerator Capable of Reaching 42K , *Cryogenics*. 30 ICEC Supplement: 257(1990)
3. Y.Matssubara and J.L.Gao. Novel Configuration of Three-stage Pulse Tube Refrigerator for Temperature below 4K, *Cryogenics*: 259(1994)
4. M.Tanaka, T.Kodama, T.Nishitani, T.Araki, E.Kawaguchi, and M.Yanai. Two Stage Pulse Tube Refrigerator with Double Rotary Valves, *Cryogenics*. 34 ICEC Supplement; 159(1994)
5. Y.Mattsubara, J.L.Gao, K.Tanida, Y.Hiresaki, M.Kaneko. An Experimental and Analytical Investigation of 4K Pulse Tube Refrigerator, *7th International Cryocooler Conf. Proceeding*. Part1:166(1993)

A HYBRID TWO-STAGE REFRIGERATOR OPERATED AT TEMPERATURES BELOW 4K

J. L. Gao,[1] Y. Hiresaki,[1] and Y. Matsubara[2]

[1] Suzuki Shokan
Kawagoe, Saitama 350, Japan
[2] Nihon University
Funabashi, Chiba 274, Japan

ABSTRACT

Elimination of moving parts at low temperature is the best approach to increase the reliability of refrigerators for cryogenic applications. This paper describes a hybrid two-stage refrigerator with no moving parts at low temperature stage. In this hybrid configuration of a two-stage refrigerator, the first stage is a Gifford-McMahon (G-M) cycle with a solid displacer and the second stage is a double inlet pulse tube refrigerator with no moving parts. 3.2 K is the minimum refrigeration temperature and 0.37 W is the cooling capacity at 4.2 K achieved by this hybrid two-stage refrigerator. With heating load 20 W at 48 K on the first stage, the refrigerator has about 0.2 W of cooling capacity at 4.2 K at the second stage. Measurement results of the heat generation at the pulse tube hot end are presented in order to analyze the intrinsic efficiency of this hybrid 4 K refrigerator.

INTRODUCTION

There are a growing number of superconducting devices and detectors that require cryogenic cooling in the liquid helium temperature region. Recent developments in regenerative refrigerators using magnetic regenerator materials in the last stage have stimulated interest in the production of commercially available 4 K refrigerators as alternatives to liquid helium cooling systems which are often not convenient[1]. In the case of the 4 K Gifford-McMahon (G-M) refrigerator, there still exist some problems to be studied to increase its reliability, such as the low temperature rubbing seal ring, vibration from displacer's movement and magnetic noise due to movement of the regenerator with magnetic materials. Pulse tube refrigeration is one refrigeration method which could meet the reliability requirement for the cryogenic cooling in some applications because of no

Advances in Cryogenic Engineering, Vol. 41
Edited by P. Kittel, Plenum Press, New York, 1996

moving parts at low temperature[2]. The performance of a pulse tube refrigerator is greatly increased by improvement of the phase shifter located at the pulse tube hot end. With the multiple staging configuration for pulse tubes, the refrigeration temperature has been lowered much further for pulse tube refrigerators. In our previous studies, a three-stage pulse tube refrigerator with a valved compressor was developed which gave a 3.6 K minimum temperature and 119 mW of cooling capacity at 4.9 K at the third stage. The input power of the compressor used for this three-stage pulse tube refrigerator was about 5 kW, although the calculated isothermal compression work required for this device was about 2 kW. We have continued to study and develop the 4 K three-stage pulse tube refrigerator to increase its efficiency and to make the machine compact[3]. In an effort to provide a cooling system with a pulse tube refrigerator for actual applications, we developed a hybrid two-stage refrigerator to increase both reliability and performance. This paper first describes the configuration of the hybrid refrigerator and the test apparatus used in this study. Next, the experimental results are presented to show the refrigeration performance of this hybrid two-stage refrigerator. Finally, the refrigeration efficiency of the refrigerator, as well as its great potential to actual applications, is discussed with measurement results of the heat flow at the hot end of the second stage.

CONFIGURATION AND DESCRIPTION

Many efforts have been made to increase the reliability of the two-stage 4 K G-M refrigerator, such as modification of the configuration for the rubbing seal at the second stage, improvement of materials for the cold seal and reduction of rubbing forces on seals. It is best approached by elimination of moving parts at the cold region to increase reliability and to decrease vibration and noise. This work describes a hybrid two-stage refrigerator which has no moving parts in the low temperature stage, as shown in Figure 1. The pressure fluctuation is generated by a valved compressor or by a valveless compressor. The first stage consists of a regenerator and a displacer or a expansion piston. The second stage is a double inlet pulse tube refrigerator, which consists of a regenerator, a pulse tube, a bypass valve, an orifice and a reservoir. The pulse tube hot end is operated at the first stage cold end

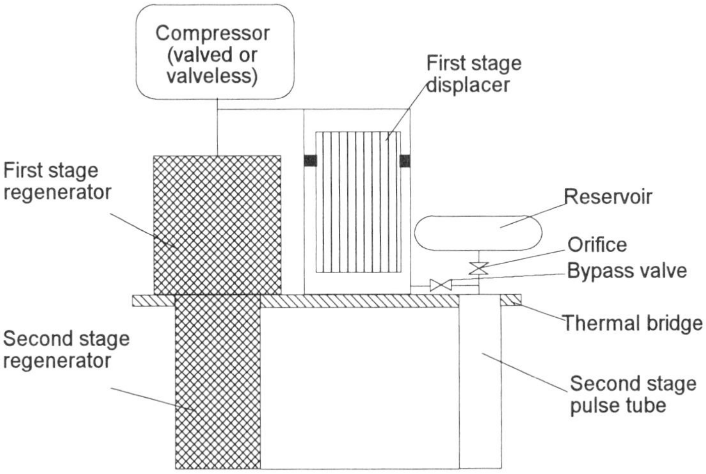

Figure 1. A hybrid configuration of a two-stage refrigerator.

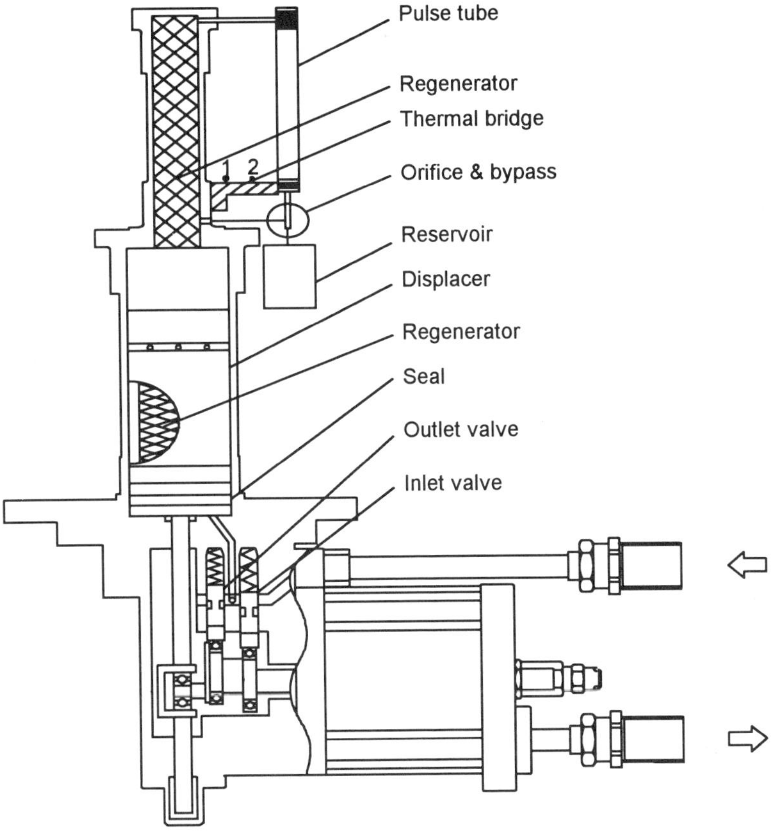

Figure 2. Test apparatus with a valved compressor.

temperature by use of a thermal bridge. As described before, the first stage in this hybrid two-stage refrigerator can use either a G-M cycle with a valved compressor or a Stirling cycle with a valveless compressor. Figure 2 shows the test apparatus used in this study where the pressure fluctuation was generated by a valved compressor which is generally used in commercial G-M refrigerators. The construction of the first stage was then the same as that of the G-M cycle. The displacer which contains the regenerator material is reciprocated by a drive motor with a scotch yoke mechanism. The high pressure and low pressure gases were controlled by slide spool valves having the correct open-close timing with displacer's movement to go through the G-M cycle. There should be almost no problem for reliability and performance from the first stage rubbing seal, since it is operated at room temperature. The double inlet pulse tube configuration was used for the second stage. The pulse tube hot end was thermally connected with the first stage cold end by a copper plate. A reservoir was connected to the pulse tube hot end through a stainless steel capillary tube as an orifice. Another stainless steel tube as a bypass tube was also inserted between the pulse tube hot end and the second stage regenerator hot end. The flow resistance of the orifice and the bypass was controlled by changing the inner diameter of these two stainless steel tubes to obtain the optimum phase shift for maximizing the performance of the second stage. A description of the hybrid two-stage refrigerator is given in Table 1. At both ends of the second stage pulse tube bronze mesh was brazed into the tube for flow straightening and heat exchange. Rhodium-iron resistance thermometers were used to measure the temperatures at the cold ends of the first and second stage. Heating

Table 1. Description of the refrigerator used in this study

Component	Materials	Size (mm)
First stage displacer	Bakelite filled with 250 no. bronze mesh 75% in volume and lead shot 25% in volume	O.D. 70 reciprocating with 31.8 stroke
Second stage regenerator	Stainless steel tube filled with Er3Ni and lead shot (half and half)	$\phi 25 \times 0.5 \times 200$
Second stage pulse tube	Stainless steel tube	$\phi 13 \times 0.5 \times 200$
Orifice	Stainless steel capillary tube	Length 60, I.D. changeable
Bypass tube	Stainless steel capillary tube	Length 40, I.D. changeable
Reservoir	Copper	Volume 0.283 L

resistances were provided at the cold ends of the second and first stage for cooling power measurements. For measurement of the heat flow through the copper plate, the rhodium-iron resistance thermometers were set on the point 1 and 2 on the copper plate in Figure 2 to measure their temperature difference.

EXPERIMENTAL RESULTS

The experiment was done with the hybrid two-stage refrigerator as shown in Figure 2 to measure its refrigeration performance. In order to obtain the best phase shift effect for the second stage pulse tube, the flow resistances of the orifice and the bypass tube were optimized by changing the inner diameters of the stainless steel capillary tubes. At first, the experiment was made with the simple orifice type (without bypass tube) for the second stage pulse tube. Figure 3 shows the effect of the orifice opening on the minimum temperature at the pulse tube cold end. At the optimum orifice opening of 0.6 mm inner diameter, the minimum temperature achieved at the second stage cold end was only about 7.5 K. As we well understand, the simple orifice pulse tube refrigerator has several advantages, but it has one fatal disadvantage that is the low cooling capacity per unit mass flow rate, which results in the larger mass flow rate through the regenerator cold end. The low temperature limit by the simple orifice pulse tube will be determined primarily by the regenerator loss due to larger mass flow rate through the regenerator cold end. Thus with only the orifice as a phase shifter, there is little hope for the pulse tube to reach 4 K. To reach liquid helium temperature with a pulse tube, the first thing that should be considered would be to introduce an efficient second phase shifter. Recent developments in pulse tube refrigerators provide three different second phase shifters; second piston, second inlet (bypass) tube and second valve at the pulse tube hot end, all of which further improve performance. The simplest way to enhance the shift effect of the simple orifice pulse tube is the double inlet pulse tube refrigerator. The bypass tube in the double inlet pulse tube can provide an optimum phase shift effect to increase the cooling capacity per unit mass flow and decrease the regenerator loss by reducing the mass flow rate through the regenerator cold end, if the three friction factors from the regenerator, orifice and bypass tube were optimized. In order to lower refrigeration temperature down to 4 K with the hybrid refrigerator, an experiment was then made using the double inlet pulse tube refrigerator for the second stage. Three different inner diameters of the bypass tubes with different orifices were used in the experiments to optimize their friction factors. Figure 4 shows the refrigeration performance at the cold end of the second stage using 0.3 mm, 0.6 mm and 0.9 mm inner diameters of the bypass tubes with a 0.3 mm inner diameter of the orifice. Figure 5 shows the refrigeration performance at the cold end of the second stage using 0.3 mm, 0.6 mm and 0.9 mm inner diameters of the bypass tubes with a 0.6 mm inner diameter of the orifice. The best refrigeration performance was achieved by the 0.3 mm inner diameter orifice and 0.6 mm

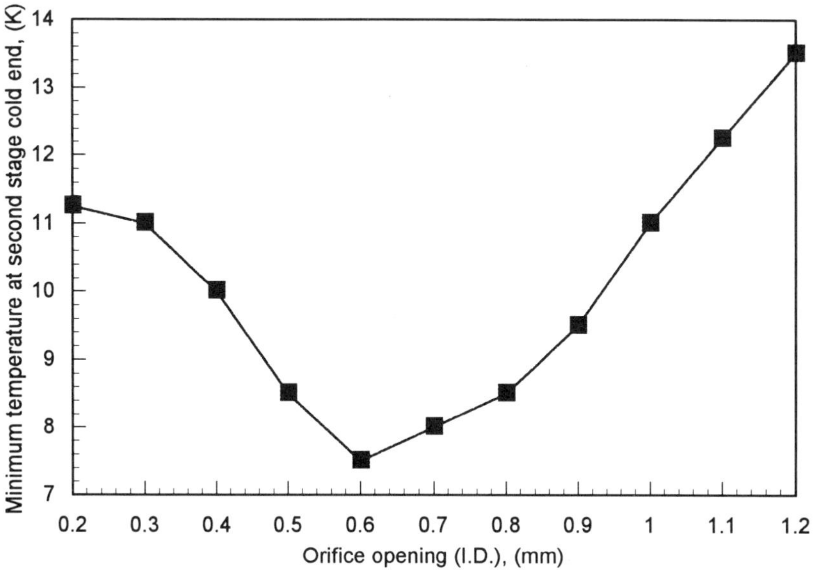

Figure 3. Effect of the orifice opening on the minimum temperature at the second stage cold in the case of the simple orifice type.

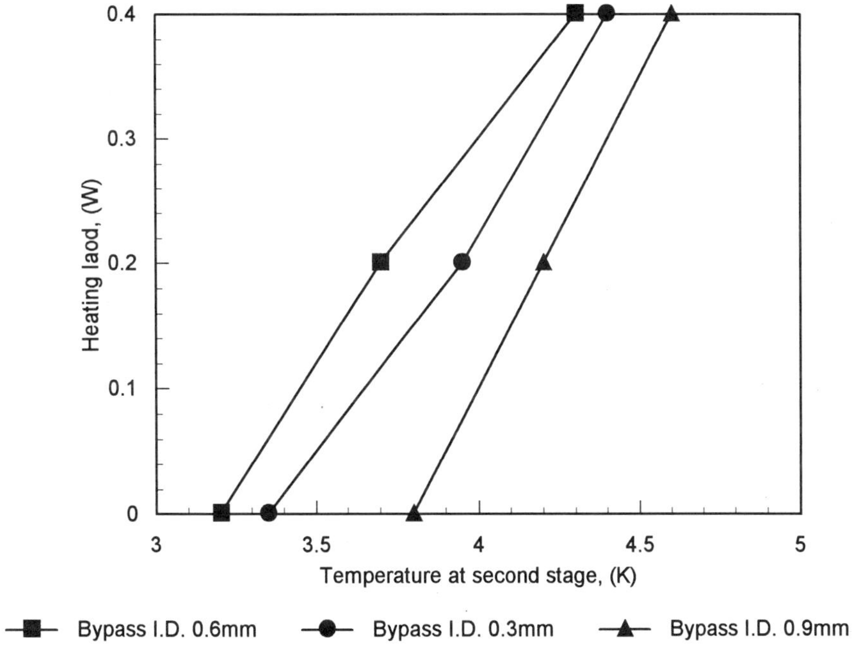

Figure 4. Effect of the bypass opening on the refrigeration performance at the second stage cold end with I.D. 0.3 mm orifice.

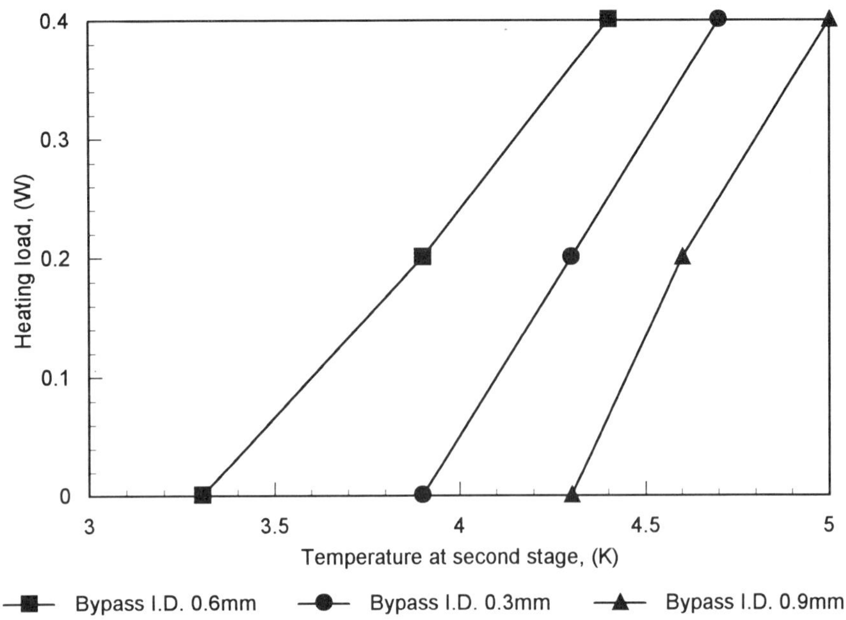

Figure 5. Effect of the bypass opening on the refrigeration performance at the second stage cold end with I.D. 0.6 mm orifice.

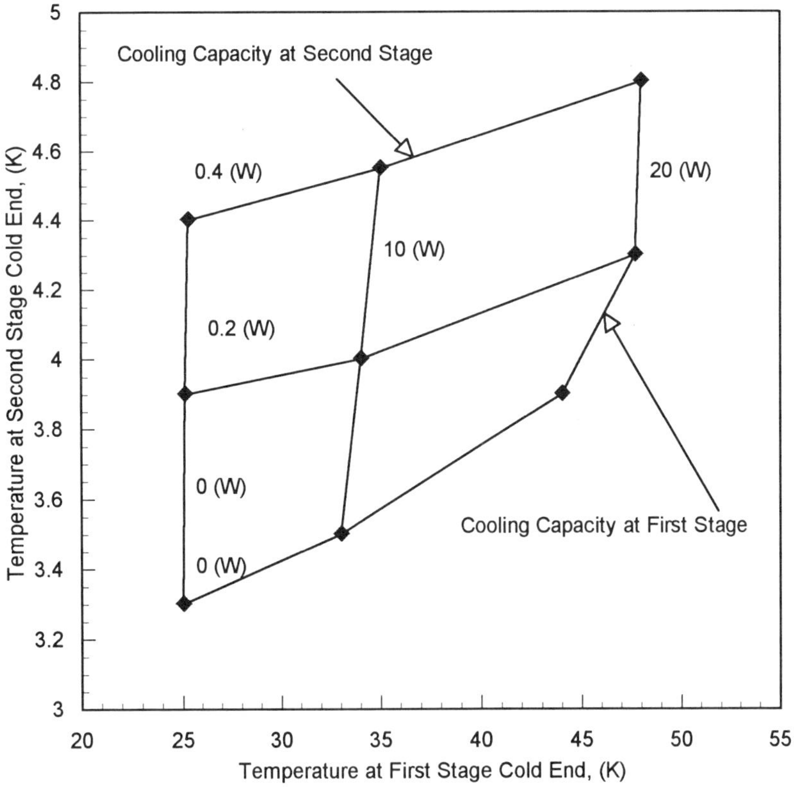

Figure 6. Cooling load map of the hybrid two-stage refrigerator.

inner diameter bypass tube, which gave a 3.2 K minimum temperature and 0.37 W of cooling power at 4.2 K at the second stage. The experiment was also done using a 0.9 mm inner diameter orifice with the three different sizes of the bypass tubes, but the minimum temperature obtained at the second stage cold end was over 9 K. The refrigeration performance at the first and second stage was measured by the heating loads at the cold ends. As an example, Figure 6 shows the cooling load map of this hybrid two-stage refrigerator with 0.6 mm inner diameter for both orifice and bypass tubes. All the experiments were done with the same operating parameters of 1.2 Hz operating frequency, 1.72 MPa mean pressure, 2.06 MPa operating high pressure and 0.68 MPa operating low pressure from the compressor except for changes in the size of the orifice and the bypass tube. Figure 6 also indicated that the hybrid refrigerator could produce about 0.2 W of cooling at 4.2 K at the second stage with a heating load 20 W at 48 K on the first stage.

DISCUSSION

The most attractive advantage of the pulse tube refrigerator, compared with other types of the refrigerators, is the increase in simplicity. The configuration in Figure 2 allows one to make the machine compact for actual applications. For example, the second stage in the hybrid refrigerator could be constructed using a coaxial double inlet pulse tube configuration. The reservoir, decreased in its volume due to low temperature operation, could be designed either inside of the first stage cylinder or around the first stage cylinder. From the view point of thermodynamic efficiency, one problem remained in the configuration of Figure 2, that the heat generated at the second stage pulse tube hot end would be removed by the first stage cooling power, although the first stage could have a large enough cooling power with a G-M cycle. Based on the first law of thermodynamics, the energy balance at the cold end of the first stage in Figure 2 is given by

$$<\dot{Q}_{C1}>=<\dot{W}_{E1}>-<\dot{H}_{R1}>-<\dot{Q}_{L1}>-<\dot{Q}_{HF}>+<\dot{H}_{R2}>+<\dot{E}_{B}> \tag{1}$$

and in the case of the simple orifice (without the bypass tube), it is given by

$$<\dot{Q}_{C1}>=<\dot{W}_{E1}>-<\dot{H}_{R1}>-<\dot{Q}_{L1}>-<\dot{Q}_{HF}>+<\dot{H}_{R2}> \tag{2}$$

where: $<\dot{Q}_{C1}>$ is the actual cooling power at the first stage cold end; $<\dot{W}_{E1}>$ is the ideal expansion work produced by the displacer at the first stage; $<\dot{H}_{R1}>$ is the first stage regenerator loss; $<\dot{Q}_{L1}>$ is the first stage heat loss; $<\dot{Q}_{HF}>$ represents the heat flow through the copper plate connected between the pulse tube hot end and the first stage cold end, which should be included in both the generated heat at the pulse tube hot end by the orifice and the heat flow from the hot end to the cold end of the second stage pulse tube; $<\dot{H}_{R2}>$ is the second stage regenerator loss; and $<\dot{E}_{B}>$ is the energy flow to the pulse tube hot end through the bypass tube. In order to understand how much extra cooling power at the first stage could be required in Figure 2, the experiment was performed to measure the heat flow $<\dot{Q}_{HF}>$ in Eq.(1) and Eq.(2) through the copper plate located between the pulse tube hot end and the first stage cold end. After calibration of the thermal conductivity of the copper plate, the results were used for the heat flow measurements. The measurement was carried out with both the simple orifice type and the double inlet type for the second stage pulse tube. The results are shown in Table 2. It was indicated that in the hybrid 4 K refrigerator the heat dissipated by the orifice was about 6.1 W and the additional heat produced by the orifice due to increase in mass flow rate through the orifice caused by

Table 2 Measurement results of heat generation at the pulse tube hot end

Configuration	Heat flow through the copper plate (W)	Temperature at first stage cold end with no heat load (K)	Temperature at the second stage cold end (K)	Reservoir temperature (K)	Point 1 temperature on the copper plate (K)	Point 2 temperature on the copper plate (K)
Simple orifice	6.10	23.70	7.00 (with no heating load)	33.50	24.70	26.65
Double inlet	7.50	22.65	3.40 (with no heating load)	35.00	23.70	26.05
Double inlet	7.40	24.30	7.50 (with 2 W heating load)	36.50	25.20	27.60

bypass tube was about 1.4 W at the second pulse tube hot end. It required about 7.5 W of extra cooling power for the first stage when the refrigerator in Figure 2 was operated at temperatures below 4 K. To increase the overall performance of this hybrid refrigerator, the pulse tube hot end of the second stage could be extended to room temperature, in which the heat generated at the pulse tube hot end could be dissipated at room temperature without using the first stage cooling power. However, it may bring about a more complicated construction than that of the coaxial configuration as discussed before. Depending on the requirements from applications, either configuration as described above for this hybrid refrigerator could be selected for actual applications.

CONCLUSION

The hybrid two-stage refrigerator configured with the double inlet pulse tube second stage with a G-M cycle first stage was developed to increase both reliability and performance. The refrigeration performance of this hybrid refrigerator with a valved compressor was measured. The refrigerator gave 3.2 K for the minimum temperature and 0.37 W for the cooling power at 4.2 K at the second stage cold end. The input power for the compressor used in this refrigerator was about 5.5 kW. Test results with the simple orifice pulse tube refrigeration for the second stage indicate that it seems to be impossible for the pulse tube to obtain 4 K only with the simple orifice as a phase shifter. Measurement results of the heat flow from the pulse tube hot end to the first stage cold end showed that it required about 7.5 W extra cooling power at 22.65 K for the first stage to remove the generated heat at the pulse tube hot end when the second stage pulse tube cold end was operated at temperatures below 4 K which would decrease the overall performance of the refrigerator. It could be improved by modification of the hybrid configuration. However, it still has a great potential for actual applications when small cooling power required at the second stage and large cooling power at the first stage.

REFERENCES

1. T. Kuriyama et al., "Development of 1 Watt Class 4K G-M Refrigerator with Magnetic Regenerator Materials", Advances in Cryogenic Engineering 39B (1994) P1335
2. Y. Matsubara and J.L. Gao, "Novel Configuration of Three-Stage Pulse Tube refrigerator for Temperatures below 4 K", Cryogenics, 34, 4 (1994) P256
3. R. Radebaugh, "A Review of Pulse Tube Refrigeration", Advances in Cryogenic Engineering 35 (1190) P1191

THREE-STAGED PULSE TUBE REFRIGERATOR CONTROLLED BY FOUR-VALVE METHOD

K. Tanida[1,2], J. L. Gao[2], N. Yoshimura[1] and Y. Matsubara[1]

[1] Nihon University, Funabashi, Chiba 274 Japan
[2] Suzuki Shokan Co., Ltd, Kawagoe, Saitama 350 Japan

ABSTRACT

Our previous study of three-stage pulse tube refrigerators controlled by the double-inlet method shows that cooling temperatures below 4 K can be acheived. However, it required a number of large reservoirs which are not convenient to make the refrigerator compact. Another experimental study of single-stage pulse tube refrigerators, shows that the four-valve method gives better performance than the double inlet method. Based on those results, a three-staged pulse tube refrigerator controlled by the four-valve method was fabricated and tested. The minimum temperature of 2.75 K at the third stage cold end has been obtained at the cycle frequency of 2 Hz. The first and second stage temperatures at this condition were 115 K and 29.5 K respectively. Optimum valve timing for the mass flow at the hot end of each stage pulse tube is discussed.

INTRODUCTION

The three-stage pulse tube refrigerator developed in Nihon University has made 4 K operation possible, which could be applied for NbTi magnet cooling[1~4]. The gas flow in each pulse tube of this device has been controlled by the double inlet method[5]. One of the problems of the double inlet method applied to multi-staged pulse tubes is mutual interaction of phase between the pressure and the mass flow rate at each pulse tube. This leads to the difficulty of getting the optimum phase at each pulse tube. The reservoir volumes required for the double inlet method also make the refrigerator non compact.

The four-valve pulse tube refrigerator does not require reservoir volumes and it could be extensively used for multi-staged pulse tube refrigerator. This paper describes the experimental results of a three-stage pulse tube refrigerator controlled by the four-valve method.

COMPARISON OF DOUBLE INLET AND FOUR VALVE METHOD

As one phase shift mechanism for the pulse tube, the four-valve method was developed in 1992[6]. The early experiments of this method were performed using four different switching valves, two for the regenerator hot end gas flow switching and the other two for the pulse tube hot end gas flow control. With this configuration, it is possible to get the optimum phase between the pressure and the mass flow at the cold end of the pulse tube. A comparative study with the double inlet method, which is known as one of the best phase shift mechanisms, indicated the four-valve method gave similar performance to the double inlet method. After obtaining the optimum switching time of each valve, it was replaced by a rotary valve.

Figure 1 shows a single stage pulse tube refrigerator controlled by a rotary valve connected to the compressor. This valve has two holes in its stator. The large hole is connected to the regenerator directly and the small hole is connected to the hot end of the pulse tube through a needle valve. A similar design was reported by J. Blaurock[7]. This arrangement allows operation in a four-valve mode. The phase angle between the large and small hole was decided by the previous experiment to give the optimum timing of each valve opening time as shown in figure 1 (b). The shadow area indicates the open period. Vph is for high pressure and Vpl is for low pressure. Subscripts 1 and 2 are for regenerator and pulse tube respectively. The minimum temperature of 22 K has been obtained with a pulse tube of 25 mm inner diameter and 220 mm length. The regenerator is made using 200 bronze mesh of 36 mm diameter and 123 mm stack length. The operating frequency is 1.8 Hz and the mean pressure is 1.4 MPa with the pressure ratio of 2. The cooling capacity at 80 K is 17 watts. This result is slightly better than that in our previous study controlled by four different valves.

On the basis of these studies of single stage pulse tubes, the three-stage pulse tube controlled by the four-valve method was constructed. In this case, each pulse tube hot ends must be located at room temperature as shown in figure 2, because the phase shift mechanism at the hot end of the pulse tube includes the mechanical switching valves which are not convenient to use at low temperature. The possibility of this kind of arrangement, however, was already confirmed by our previous study of the three-stage pulse tube refrigerator controlled by the double inlet method[2]. It gave a minimum temperature of 3.6 K at the third stage cold end. Therefore we expect the similar or better results from the configuration of figure 2 (a).

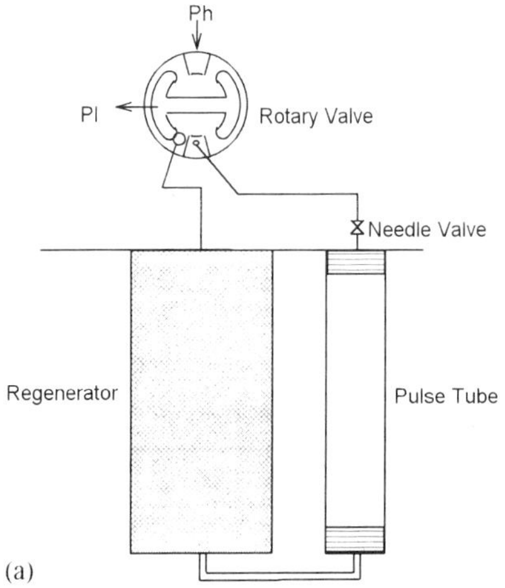

(a)

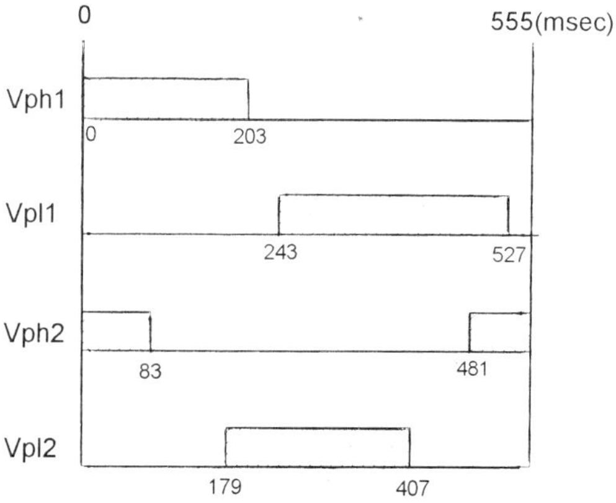

(b) valve timing chart

Figure 1. Single stage pulse tube refrigerator controlled by four-valve method.

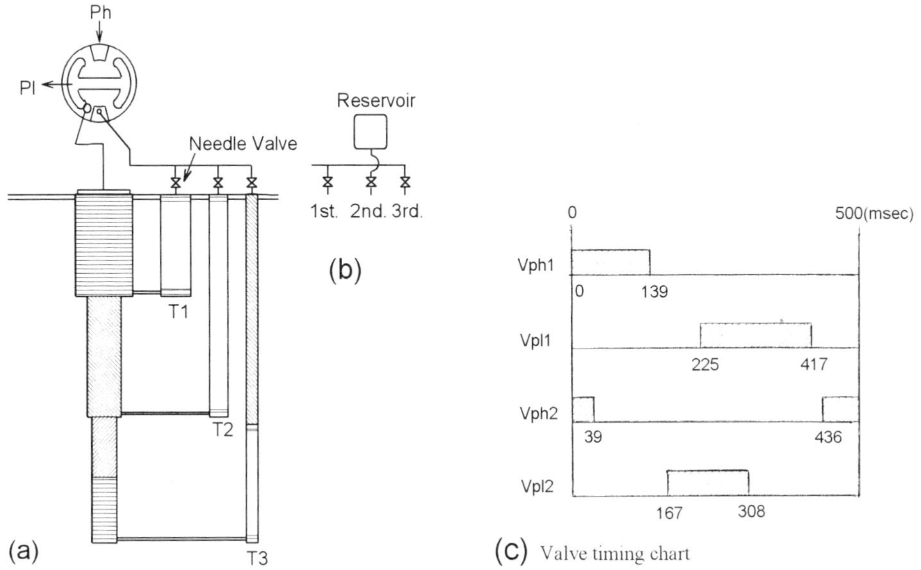

Figure 2. Three-stage pulse tube refrigerator controlled by a rotary valve.

Table 1. Size and materials of the component

	Pulse Tube Size(mm)	Regenerator Size(mm)	Regenerator Material
1st Stage	ϕ 25 × 0.5 × 190	I.D.50.8 × 130	Bronze Mesh (#200)
2nd Stage	ϕ 18 × 0.5 × 210	I.D.29 × 130	Lead shot(0.3mm)
3rd Stage	ϕ 9 × 0.5 × 140	I.D.19 × 150	Lead shot(0.3mm) and Er3Ni shot(0.3mm)
Regenerative Tube		I.D.5 × 300	Lead shot(0.3mm)

PERFORMANCE OF THE TEST DEVICES

The first experiment was done using the arrangement of figure 2 (a) with the same rotary valve used for single stage pulse tube. The sizes of the pulse tubes and the regenerators are given in table 1. The regenerative tube installed between the hot end of the third stage pulse tube and the mass flow controller are also used in this experiment. Radiation shields are connected at the cold end of the first and second stage regenerators. Test runs were done using the valve timing both of figure 1(b) and figure 2(c), however, the liquid Helium temperature was not reached. This unexpected result may be due to the mismatch of the rotary valve to the multi-staging operation.

To overcome this mismatch, the second stage pulse tube hot end was connected to a resevoir of 1 litre volume which operate as a simple orifice pulse tube as shown in figure 2 (b). A third stage temperature of 4.8 K was attained with second stage at 40 K and first stage at 90 K using the valve timing of figure 2 (c). The operating frequency is 2 Hz and the pressure ratio measured at the hot end of the first regenerator is 2.2. A 5 kW compressor from a G-M cooler was used in this experiment.

From these results, it is pointed out that the optimum valve timing for the single stage and the multi-stage may not be same. To find the optimum operating condition of the

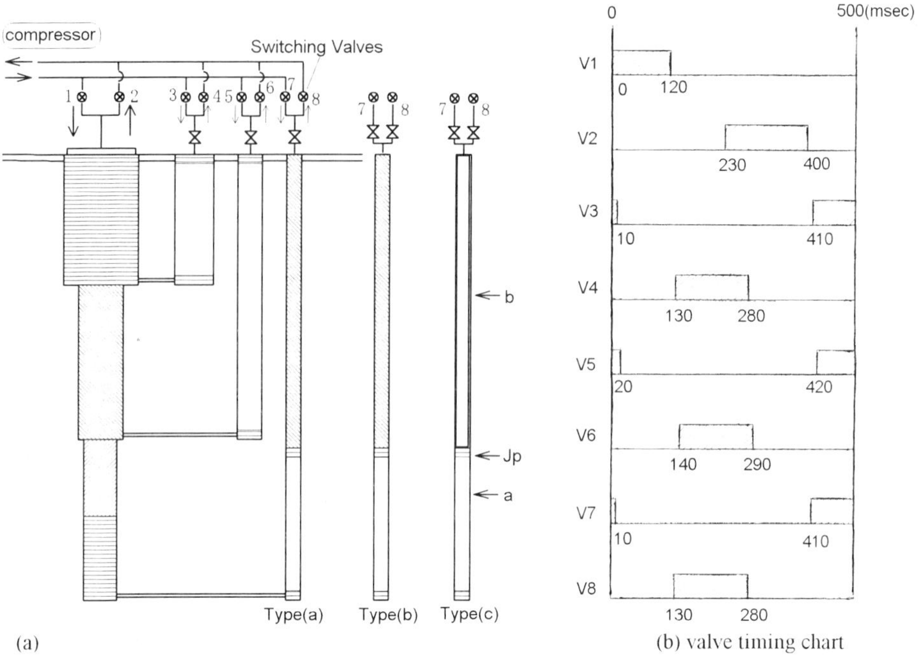

Figure 3 Three-stage pulse tube refrigerator controlled by solenoid valves.

multi-stage pulse tube, the pressure controller consisting of eight solenoid valves replaced the rotary valve as shown in figure 3 (a). Each valve timing can be controlled independently by means of a process controller. A needle valve was installed at the hot end of each pulse tube to adjust the mass flow rate.

With this arrangement, at first, Type (a) was tested and a minimum temperature of 3.9 K and a cooling capacity of 40 mW at 5 K were obtained with the valve timing of figure 3 (b). The measured pressure wave at the first regenerator hot end is given in figure 4. Table 2 shows each stage temperature without heating load (run 1) and with heating load (runs 2,3).

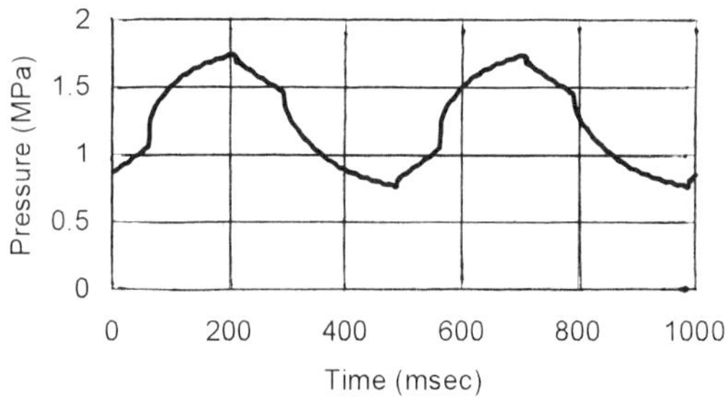

Figure 4. Pressure wave measured at 1st stage regenerator hot end.

Table 2 The cooling capacity of three-staged pulse tube

	Heating load / Temperature		
	(run 1)	(run 2)	(run 3)
1st stage	0 W / 120 K	4 W / 138.8 K	4 W / 139.3 K
2nd stage	0 W /21 K	1 W / 29 K	2 W / 33 K
3rd stage	0 W / 3.9 K	50 mW / 6.3 K	50 mW / 6.7 K

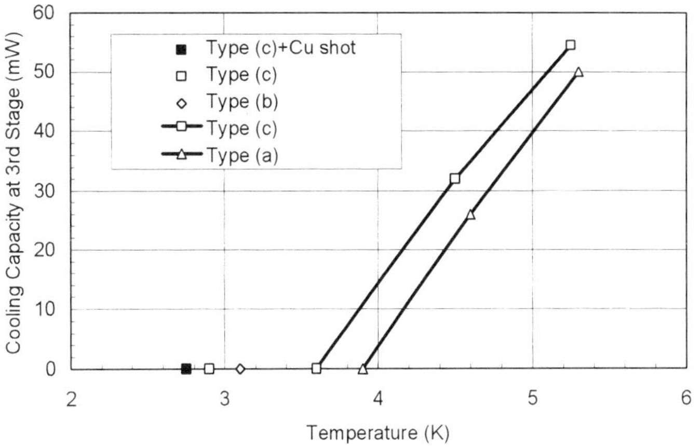

Figure 5. Cooling capasity of 3rd stage pulse tube.

As a general charactaristic of the four-valve method, gas flow streaming will exist. If the low pressure valve of the pulse tube side is opened slightly longer than the high pressure valve, the flow streaming will appear from the regenerator to the pulse tube. Another direction of the flow streaming will also appear by changing the opening times of the valves. This phenomenon indicates that the flow streaming always exists and the exact opening time of each valve must be controlled to reduce the rate of the flow streaming. However, our previous experiment done using the single stage pulse tube shows stable operating condition by controlling the valve timing properly. At this operating condition, the small amount of the flow streaming toward the pulse tube from the regenerator seems to exist because the hot end temperature of the pulse tube never warms up during steady state operating conditions. The regenerator in this condition is similar to the liquefaction mode with unbalanced mass flow rate.

Flow streaming at the third stage pulse tube, however, should be minimized to get much lower temperature. Therefore the connecting line of V_7 and V_8 to the pulse tube was replaced by Type(b) of figure 3 to minimize the flow streaming. By controlling the needle valves at V_7 and V_8 independently, the minimum temperature at the third stage is lowered to 3.1 K with second and first stage temperature of 23 K and 82 K respectively.

The function of the regenerative tube is still not well understood. Therefore the performance of Type (c), which has no regenerative materials in the tube, was tested. Our previous study of the three-stage pulse tube refrigerator controlled by double inlet method did not show any posibility of reaching 4 K with this kind of empty tube instead of the regenerative tube. In the case of four-valve method, however, the minimum temperature at the third stage of 3.6 K has been obtained by Type (c). The cooling capacity at the third stage is given in figure 5 in comparison with that of Type (b). After the measurement of cooling capacity, the minimum temperature of 2.9 K was obtained by further fine controlling the needle valves.

During the experiment using the empty tube, an abnormal quench phenomenon of the temperature distribution along the 3rd stage pulse tube was observed several times. Figure 6 shows an example of the temperature trace during the cool down process. The 1st and 2nd stage temperatures are almost constant and the 3rd stage temperature is decreasing down to 4 K. The wall temperatures along the 3rd stage pulse tube and the empty tube are measured at the position a, jp and b as shown in figure 3. When the 3rd stage cold end temperature comes to 4 K, T_a is decreasing but T_b and T_{jp} are increasing.

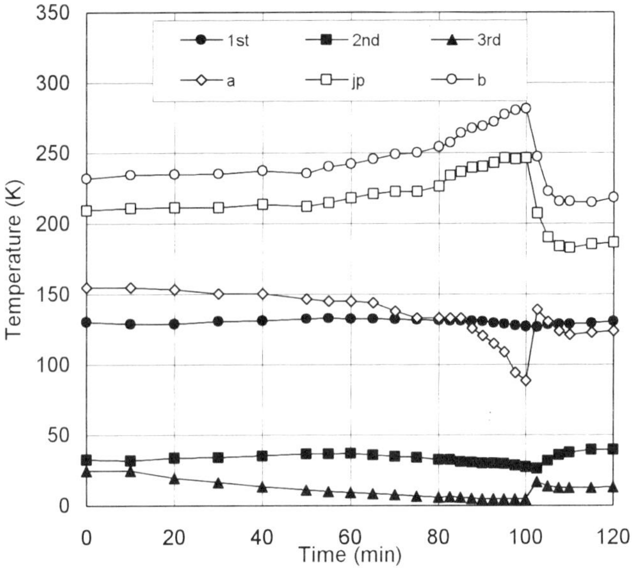

Figure 6. Temperature change during the cool down process.

This tendency, however, is suddenly quenched to a rather normal temperature distribution. Figure 7 shows the comparison of the temperature distributions before and after the quench. The maximum temperature gradient of 5.3 K/mm between T_{jp} and T_a was observed just before the quench. This extremely large temperature gradient may generate an acoustic oscillation which induces additional heat flow from the hot end to the cold end of the pulse tube. This kind of phenomena has never been observed during the experiment of Type (a) and (b).

To avoid this type of instability, the copper shots of 1 mm in diameter were installed in the empty tube and tested again with same operating condition. The result shows a minimum temperature of 2.75 K as a stable operating condition with the 1st stage temperature of 115 K and the 2nd stage temperature of 29.5 K.

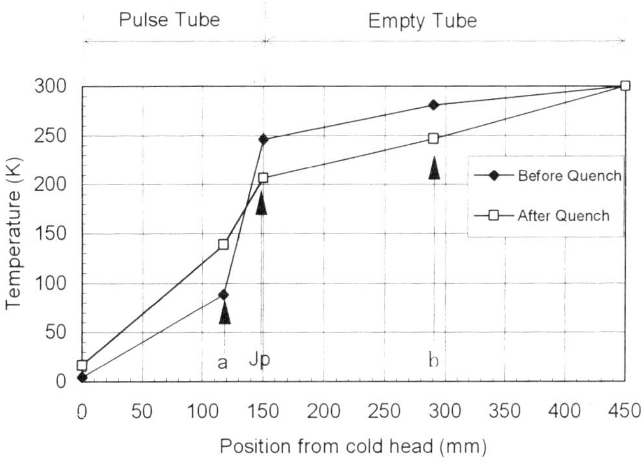

Figure 7. Temperature distribution along the 3rd stage pulse tube.

CONCLUSION

The multi-staging method developed for the double inlet pulse tube refrigerator can be applied to the four-valve pulse tube, because each stage phase controller can be located at room temperature. On the basis of this concept, a three-stage pulse tube refrigerator controlled by the four-valve method was fabricated and tested.

The first experiment done using a rotary valve did not operate well, however, the following experiments using solenoid valves instead of the rotary valve gave a minimum temperature of 2.75 K at the third stage pulse tube cold end. Through these experiments, we found an important configuration for the mass flow controller located at the hot end of the pulse tube, which gives the optimum flow streaming passing through the pulse tube.

To make the refrigerator compact and reliable, the use of a single rotary valve for the multi-stage pulse tubes would be required. To realize Type (b) or (c) of figure 3 by a rotary valve, it is proposed to make two small holes in the stator of the rotary valve for the pulse tube, one is only for switching the high pressure gas and the other is only for switching the low pressure gas. This configuration gives almost the same operating condition of Type (b) or (c) of figure 3.

The empty tube instead of the regenerative tube can be used for the 3rd stage pulse tube in the case of the four-valve method, however, a regenerative tube having relatively small heat transfer area seems to give a much more stable operating condition than the empty tube.

ACKNOWLEDGMENTS

We are grateful to S.W. Zhu for useful discussions and S. Mori for measurements of cooling performance.

REFERENCES

1. J.L. Gao and Y. Matsubara. Experimental Investigation of 4 K Pulse Tube Refrigerator. Cryogenics. 34. 1. 25 (1994).
2. Y. Matsubara and J.L. Gao. Novel Configuration of Three-Stage Pulse Tube Refrigerator for Temperatures below 4 K. Cryogenics. 34. 4. 256 (1994).
3. Y. Matsubara and J.L. Gao. Multi-staged Pulse Tube Refrigerator for Superconducting Magnet Applications. Cryogenics. 34. ICEC Supplement. 155 (1994).
4. Y. Matsubara and J.L. Gao. Multi-Stage Pulse Tube Refrigerator for Temperatures below 4 K. "Cryocoolers 8". R.G. Ross.Jr. Ed.. Plenum. New York. (1995) p. 345.
5. S.W. Zhu. P. Wu and Z. Chen. Double inlet pulse tube refrigerator: an important improvement. Cryogenics. 30. 514 (1990).
6. Y. Matsubara. J.L. Gao. K. Tanida. Y. Hiresaki and M. Kaneko. An experimental and analytical investigation of 4 K pulse tube refrigerator. Proceeding of 7th International Cryocooler Conference: 166 (1993).
7. J. Blaurock. R. Hackenberger. P. Seidel and M. Thurk. Compact four-valve pulse tube refrigerator in coaxial configuration. in: "Cryocoolers 8". R.G. Ross. Jr. Ed.. Plenum. New York (1995). p. 395.

A THREE-STAGE PULSE TUBE REFRIGERATOR FOR TEMPERATURES NEAR 4 K*

P.E. Bradley and R. Radebaugh
Thermophysics Division
Chemical Science and Technology Laboratory
National Institute of Standards and Technology
Boulder, Colorado 80303, USA

ABSTRACT

This paper describes the design and construction of a three-stage pulse tube refrigerator intended for reaching temperatures near 4 K. Until recently little work has been done with pulse tube refrigerators in this temperature range. This refrigerator was designed to be operated with a commercially available helium compressor mated to a rotary valve to provide the necessary pressure oscillation. The refrigerator is designed to operate at a frequency of about 2 Hz with a pressure ratio of about 2 and an average pressure of about 1.5 MPa. At present a low temperature of 10.5 K has been reached with 2 W of refrigeration delivered at 22.5 K. The effects of real gas properties on the design of the third-stage regenerator are discussed. Measurements of the refrigeration power, minimum temperature, and other relevant parameters are presented.

INTRODUCTION

Superconducting magnets for use in the Navy mine-sweeping program may require about 1 W of refrigeration at 4 K. In support of this effort NIST has been working with the Navy to develop a scalable prototype three-stage pulse tube refrigerator capable of reaching 4 K. While the demonstration unit would not be able to withstand the high-g shock loads associated with mine detonation, a sufficiently robust full scale unit could be constructed to withstand this environment.

The pulse tube refrigerator is noted for its high reliability and low cost. Most of the developments in pulse tube refrigeration have been for temperatures above about 50 K.

---

*Research partially sponsored by the Naval Surface Warfare Center, Annapolis Detachment, Carderock Division, Annapolis, MD 21402-5067.

Recently, however, Matsubara and Gao[1] have shown that temperatures of 4 K are possible with a three-stage pulse tube refrigerator. They used an unusual regenerative tube at the warm end of the third-stage pulse tube in order to eliminate the use of cold valves. We discuss here the design and preliminary results for a three-stage pulse tube refrigerator in which a cold valve is used for the third-stage pulse tube. The system was designed for 4 K operation, but 10.5 K is the lowest temperature achieved to date.

SYSTEM DESIGN

The basic design philosophy proceeds as follows: (1) determine the optimum system mean operating pressure, (2) determine the operating conditions at various stages, (3) perform the basic regenerator design at each of the stages, (4) perform a coupled or system approach to balance the boundary conditions for the conservation of work flow and conservation of mass flow at the junction of each of the stages, and (5) determine the size of the pulse tube at each stage.

In order to achieve a relatively low cost cooler to meet the Navy's goal, a commercially available compressor, similar to the type used with Gifford-McMahon (GM) coolers, was chosen and then mated to a rotary valve to provide the pressure oscillation in the system. The rotary valve can be operated from 1 to 10 Hz and is suitable to meet the flow rate required by this system. A low operating frequency of about 2 Hz was chosen to maintain a relatively high effectiveness in the regenerators.

Determine optimum mean pressure. To determine the optimum mean operating pressure for the system, the effects of nonideal gas properties at temperatures below about 15 K must be taken into account. This is accomplished by first looking at the gross refrigeration power in the pulse tube with an ideal regenerator summarized by

$$\langle\dot{Q}\rangle_r = (T\beta Z)RT\langle\dot{m}_c(P_1/P_0)\rangle, \tag{1}$$

where $\dot{m}_c$ is the mass flow at the cold end of both the pulse tube and the regenerator, β is the expansivity, T is the average temperature for the last stage, R is the gas constant per unit mass, Z is the compressibility factor, P_1 is the dynamic pressure, and P_0 is the mean pressure. For an average temperature of 10 K a maximum value of the product $T\beta Z$ occurs at a pressure of 1 MPa. To achieve nearly adiabatic behavior in the pulse tube, the thermal penetration depth should be quite small compared to the diameter of the pulse tube. Therefore, a minimum penetration depth, which leads to a mean pressure of 1.5 MPa is desirable. Using a higher pressure leads to a decrease in the size of the regenerator, pulse tube, and compressor which is better for the upper stage(s).

Determine operating conditions of various stages. Ideally an infinite number of stages would provide the maximum efficiency for any refrigeration cycle, because the interception of losses at the highest temperature minimizes the entropy generation. In fact, an infinite number of stages, or in other words, a continuous staging concept, while preferred, is not practical. Therefore, a finite number of stages should be selected and, in practice, should be minimized. Gifford-McMahon cryocoolers have achieved temperatures of 4 K with the use of two stages. However, because the pulse tube expansion process is not quite as efficient, this cooler utilizes three stages. A high refrigeration power per unit mass flow may be obtained, by using a similar operating frequency of 2 Hz and a pressure ratio of about 2, thus relaxing somewhat the requirements on the regenerators. The system is designed with the following stage temperatures:

Ambient:	300 K		
1st stage:	70 K	Temp. ratio:	4.29

2nd stage:	15 K	Temp. ratio:	4.67
3rd stage:	4 K	Temp. ratio:	3.75

Because the third stage has low heat capacity, the ratio at the third stage is lowered somewhat. The first-stage ratio is also lower because the radiation shield is attached to it requiring some additional refrigeration power.

Perform the basic regenerator design at each of the stages. Once the operating conditions have been determined, the design of the regenerators can begin. This consists of performing the initial optimization of the geometry of the regenerator at each stage by using a simplified analytical model developed by Radebaugh and Louie[2] in the mid-1980's called REGEN in software form. There are two approaches that can be employed. The first is to start with the desired refrigeration power at the coldest stage and calculate the refrigeration power and regenerator geometries for each stage from that stage up to the compressor. The second approach is to start with the available compressor power and proceed toward the coldest stage. In either case the initial optimization of the regenerator geometry for each stage will be completed, and the information can then be used as input to the numerical model called REGEN3.1 developed by Gary and Radebaugh[3,4] to fine-tune the design. REGEN3.1 is a performance code which makes very few assumptions and incorporates nonideal gas properties in the analysis.

In this case, the analysis for this cooler began at the third stage and worked up to the compressor. This compressor can provide approximately 6 g/s continuous flow rate of helium gas for the pressure and pressure ratio required for the pulse tubes. During a half cycle the compressor can provide 12 g/s, but when a sinusoidal flow rate is imposed the amplitude can be as high as 18.8 g/s.

Perform a coupled or system approach to balance the boundary conditions. Boundary conditions must be balanced at the junction of the two stages when we combine two or more stages. This can and should be performed simultaneously while designing the regenerators. First, the total hydrodynamic work flow at the cold end of a regenerator must equal the sum of the hydrodynamic work flow at the cold end of the pulse tube for that stage and the hydrodynamic work flow at the warm end of the regenerator of the next colder stage. This can be summarized by the equation

$$\dot{W}_{c1} = \dot{W}_{h2} + \dot{W}_{pt1} , \tag{2}$$

which assumes that there is no work lost at the interconnections between the various components. Second, the mass flow at the junction must also be conserved and is expressed by

$$\dot{m}_{c1} = \dot{m}_{h2} + \dot{m}_{pt1} , \tag{3}$$

which is a phasor equation that has both real and imaginary components that must be conserved. This coupled approach should be performed in conjunction with the design of the regenerators.

The hydrodynamic work flow is related to the amplitude and the phase of the mass flow by the equation

$$\dot{W}_x = (1/2)\,|\dot{m}_x|\,RT_c(P_1/P_0)\cos\phi_x , \tag{4}$$

where $\dot{W}_x$ is the hydrodynamic work flow at the junction in the element x, $|\dot{m}_x|$ is the amplitude of the mass flow at the junction for element x, T_c is the temperature at the cold end of the pulse tube, P_1 is the pressure amplitude, P_0 is the mean pressure, and ϕ_x is the

Table 1 Design geometries of the pulse tube system

Parameter	1st Stage	2nd Stage Original PT	2nd Stage Large Volume PT	3rd Stage
Regen. Volume Total (cm^3)	437	113	113	77
Reg. Diam. (mm)	73.4	53.9	53.9	24.3
Regen. Length (mm)	103.2	49.5	49.5	165.6
Regen. Material	Stainless Steel	Pb+5%Sb	Pb+5%Sb	Nd
Mesh or Sphere Dia. (μm)	#325	127	127	203
porosity	0.692	0.404	0.404	0.397
Pulse Tube Volume (cm^3)	111.5	13.4	19.8	11.84

phase angle between the mass flow and the pressure. We assume that a phase angle of 0 can be achieved in the pulse tube, such as for a double inlet pulse tube. The gross refrigeration power of the pulse tube is some fraction of the work flow $\dot{W}_{pt}$

$$\dot{Q}_r = \eta_{pt}\dot{W}_{pt} , \tag{5}$$

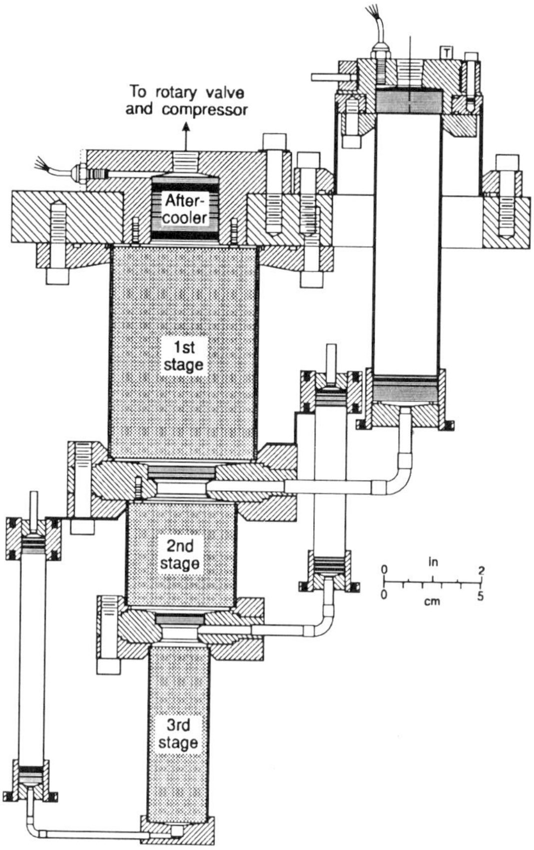

Figure 1 Cross sectional view of three stage pulse tube refrigerator.

where η_{pt} is the pulse tube efficiency. The gross refrigeration power must be sufficiently large to provide for the losses and any net refrigeration required. The necessary mass flow to provide this refrigeration at the cold end of the pulse tube is found from Eqs. (4) and (5).

Determine the size of the pulse tube at each stage. Once this flow rate is determined for each pulse tube, the volume of the pulse tube V_t can be calculated from

$$\frac{|\dot{P}|V_t}{\gamma RT_c} = \frac{2\pi f(P_1/P_0)V_t}{\gamma RT_c} = |\dot{m}_{ct}|\sin\theta \ , \qquad (6)$$

where θ is the phase angle between the mass flows at the hot and cold ends of the pulse tube. For an optimum orifice setting θ is about 30°. The design geometries of the pulse tube system are summarized in Table 1.

CONSTRUCTION OF A THREE-STAGE PULSE TUBE REFRIGERATOR

A three-stage pulse tube refrigerator was constructed using the geometries listed in Table 1. Figure 1 shows a cross sectional view of the refrigerator. As Figure 1 shows, the refrigerator consists of three stages of regenerators that are colinearly arranged, with the corresponding pulse tubes arranged radially about the regenerators. Dark heavy lines are thermal links between components depending upon the selected configuration. A radiation shield (not shown) is attached to the cold end of the first stage as well. Needle valves (not shown) are attached to the warm end of each of the pulse tubes with a reservoir volume (not shown) connected to each of the valves. The orifice valves are adjusted by means of a mechanical linkage (not shown) that is thermally isolated. The pressure oscillation is generated by mating a continuous flow helium compressor to a rotary valve. The valve's operating frequency can be varied from 1 to 10 Hz. Silicon diode thermometers are placed strategically within the system at the cold and warm ends of the pulse tubes. The thermometers placed on the first and second stages have an uncertainty of ±0.5 K at temperatures less than 100 K, while those on the third stage have an uncertainty of ±0.25 K at temperatures less than 100 K. A resistive heater is placed on the cold end of the third-stage pulse tube to measure the refrigeration power. Piezoresistive pressure transducers are placed at the warm end of the first and second stage pulse tubes only when the pulse tubes are anchored to ambient temperature.

REFRIGERATOR PERFORMANCE

Preliminary experiments were conducted with the refrigerator operated in one of two configurations. The first had the warm ends of both the first- and second-stage pulse tubes linked to ambient temperature, while the warm end of the third-stage pulse tube was linked to the cold end of the first-stage pulse tube. The second configuration thermally anchored the warm end of the second-stage pulse tube to the cold end of the first-stage pulse tube while keeping all other anchors the same as in the first configuration.

Table 2 lists the temperatures achieved at each stage for various operating conditions. Configuration I achieves a low temperature of only 25 K at the third stage. For configuration II, the original 13.4 cm^3 second-stage pulse tube achieved a temperature only 10.6 K below that of the first stage, which indicated that the second-stage pulse tube was not sized to provide sufficient refrigeration. Therefore for configuration IIb we installed the larger 19.8 cm^3 pulse tube to increase the refrigeration at the second stage. After we installed the larger second-stage pulse tube, the second stage achieved 47 K and the third

Table 2 Preliminary experimental results

Parameter	Type I	Type II	Type IIb	Type IIc	Type IIc*
1st Stage, Temp.	82.78 K	92.2 K	90.9 K	90.99 K	82.8 K
2nd Stage, Temp.	96.84 K	81.6 K	46.5 K	52.11 K	39.76 K
3rd Stage, Temp.	25.0 K	11.65 K	10.5 K	10.4 K	15.3 K
O_1 , # Turns	0.18	0.18	0.18	0.18	0.18
O_2 , # Turns	1 5/8	1 5/8	2 3/5	2 3/5	2 3/5
O_3 , # Turns	1 1/2	1 1/2	1 1/4	1 1/4	1 1/4
P_{mean}, MPa	1.73	1.86	1.97	1.5	1.5
Press. Ratio	1.62	1.53	1.54	1.7	1.7
Frequency, Hz	2.25	2.61	2.44	2.7	2.66

Note: * The lowest temperature reached was 10.4 K, but a heat leak developed and caused an upward drift to 15.3 K. The load curve performed in any case.

stage a low temperature of 10.5 K. To date this is the lowest sustained temperature achieved by this system.

The data listed in Table 2 are for the nearly optimal operating conditions to achieve the lowest third-stage temperature. The mean pressure varies from about 1.5 MPa to almost 2 MPa with pressure ratios ranging from 1.53 to 1.7. Pressure ratios above about 1.6 were not attainable using the compressor selected due to a larger than expected pressure drop across the rotary valve. Therefore a second compressor was installed in parallel with the original, whereupon a higher ratio of about 1.7 was attained, as denoted for the configuration labeled IIc in Table 2. This ratio is still lower than the design value of 2. The optimum orifice settings were initially determined from early experiments of each configuration and further refined during operation. For an approximately 50% increase in volume of the second-stage pulse tube there is a linear correlation to the optimum orifice setting. The optimum orifice setting for the third stage varies little for the different configurations and is fairly independent of the mean pressure, pressure amplitude, and frequency, as is the first stage.

System cool-down times to achieve the lowest third stage temperature ranged from 4½ hours for configuration I, 6½ hours for configuration II, 9 hours for configuration IIb, and

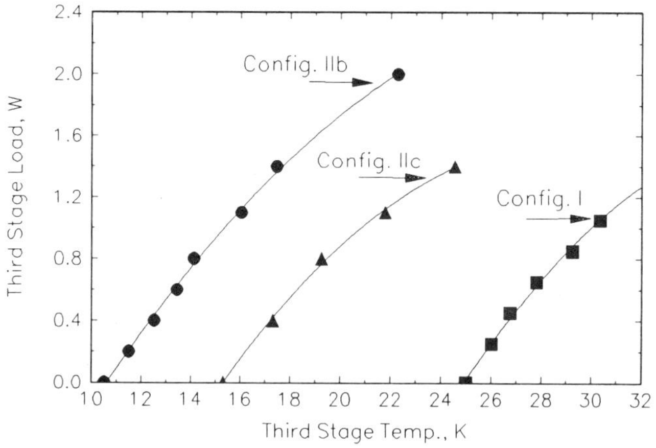

Figure 2 Refrigeration power at the third stage for configurations I, IIb, and IIc.

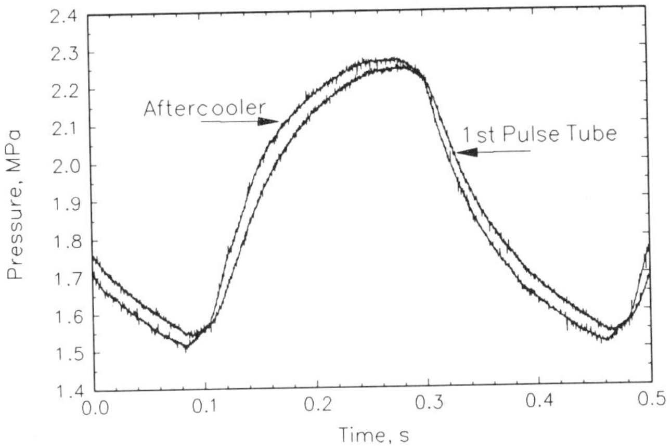

Figure 3 Typical pressure input to system as a function of time.

5½ hours for configuration IIc. The cool-down times for configurations II and IIb were longer than for configuration I because of the additional thermal mass associated with the reservoirs attached to the first stage. The quicker cool down for configuration IIc is due to a higher refrigeration per unit mass flow which is brought about by the higher pressure ratio at a lower mean pressure.

Heat leaks were detected during operation of the system at higher mean pressures, which caused small helium leaks into the vacuum space. These leaks were not significant enough to cause a complete loss in vacuum; however, the vacuum degraded over time as a result. These heat leaks certainly contributed to the poor performance, especially considering that an estimated heat leak of 700 mW was present during the configuration II experiment just before we took data; the low temperature was only 22.5 K. Some of this was due to apparent leaks at the interconnections resulting from a difference in thermal expansion between the stainless steel connecting bolts and the copper interconnection. After the bolts were replaced with those of bronze, the performance improved leading to the low temperature of 10.5 K in configuration IIb.

Figure 2 shows the refrigeration power as a function of temperature for the third stage for configurations I, IIb, and IIc. The 2 W refrigeration at 22.5 K for configuration IIb and 1.4 W at 24.5 K for configuration IIc are below the 25 K lowest temperature achieved by configuration I. Configuration IIb in fact has useful refrigeration of nearly 1 W at 15 K.

Figure 3 shows the pressure as a function of time at the aftercooler and at the warm end of the first-stage pulse tube. This curve shows that the pressure delivered to the system is not very nearly sinusoidal and is characteristic of valved systems. The design of the system assumes a sinusoidal input, which is the standard practice when designing such a system. The curve shows that a rapid change in pressure occurs during the inlet phase of operation, while there is a fairly long period in which the pressure levels before the pressure declines rapidly at the start of the exhaust phase. The increased mass flow during the initial stages of inlet and exhaust generates a higher load within the regenerator, thus reducing the effectiveness of the regenerators. This effect becomes more evident at the lower-temperature stages. Figure 3 also shows that the pressure drop in the first-stage regenerator is small and is about what is expected from the design calculations.

CONCLUSIONS

A three-stage pulse tube refrigerator able to achieve temperatures near 4 K was designed and constructed. This refrigerator reached a low temperature of 10.5 K and can at present provide up to 2 W of cooling at 22.5 K. This low temperature is not as low as desired; however, it may be reduced by improving the third-stage regenerator, modifying the rotary valve to handle higher flows in a more sinusoidal manner, and eliminating vacuum leaks. The design approach presented here is for a sinusoidal pressure oscillation, which certainly was not observed here. Suitable factors need to be employed during the regenerator design to accommodate the more realistic pressure oscillation provided.

ACKNOWLEDGMENTS

The authors wish to thank Christoph Walther for his assistance in conducting experiments and constructing portions of the refrigerator

REFERENCES

1. Y. Matsubara and J.L. Gao, "Multi-Stage Pulse Tube Refrigerator for Temperatures below 4 K", in Cryocoolers 8, R.G. Ross Jr., ed., New York (1995), p. 345.
2. R. Radebaugh and B. Louie, "A Simple, First Step to the Optimization of Regenerator Geometry," NBS Special Publication 698, (1985) p. 177.
3. J. Gary, D. E. Daney, and R. Radebaugh, "A Computational Model for a Regenerator," NBS Special Publication 698, (1985) p. 199.
4. J. Gary and R. Radebaugh, "An Improved Numerical Model for Calculation of Regenerator Performance (REGEN3.1)," Proceedings of the Fourth Interagency Meeting on Cryocoolers, DTRC-91/003 (1991) p. 165.

SUPERFLUID ORIFICE PULSE TUBE REFRIGERATOR BELOW 1 KELVIN

A. Watanabe, G. W. Swift, and J. G. Brisson[1]

Condensed Matter and Thermal Physics Group
Los Alamos National Laboratory, Los Alamos NM 87545

ABSTRACT

Half the moving parts of the superfluid Stirling refrigerator have been eliminated by adopting an orifice-pulse-tube configuration. Our first such device has cooled to 0.64 K with the hot platform anchored at 1.0 K. Performance of the refrigerator is in reasonable agreement with expectations. Two curious features of the superfluid pulse tube are in distinct contrast with features of conventional pulse tubes. First, stability of the ^{3}He-^{4}He mixture against free convection requires that the hot end of the pulse tube must be *below* the cold end. Second, the low heat capacity of metals below 1 K makes heat loss along the pulse tube due to the fluid's oscillatory motion very small.

INTRODUCTION

The superfluid Stirling refrigerator[2] uses the ^{3}He in a ^{3}He/^{4}He mixture as a thermodynamic working fluid. Bellows-sealed superleak pistons work only on the ^{3}He, carrying it through a Stirling refrigeration cycle. The ^{3}He behaves approximately like an ideal gas; the ^{4}He passes freely through the superleaks, and has little effect on the thermodynamics of the cycle.

In the Stirling refrigeration cycle[3,4], the phasing of the cold piston's motion relative to the oscillating pressure is such that the cold piston absorbs work from the working fluid while the pressure is high, and returns work to the working fluid while the pressure is low; the net effect (averaged over each cycle) is that the cold piston removes work from the fluid. As shown in Fig. 1, in orifice pulse tube refrigerators[5] the cold piston of the Stirling refrigerator is replaced by a passive, dissipative structure which maintains this phasing between fluid flow and pressure in the vicinity of the cold heat exchanger, but with no moving parts. The "orifice" is a resistive flow impedance, which passes oscillatory fluid flow in phase with the oscillatory pressure. Work is thereby removed from the fluid, and dissipated into heat; a heat exchanger near the orifice is thermally anchored to the hot temperature of the refrigerator to dispose of this heat. The "pulse tube" is a simple tube full of the working fluid, connecting the cold heat exchanger to the orifice. It thermally isolates the orifice's dissipation from the cold end, while transmitting fluid motion. In other words, the fluid in the pulse tube can be imagined as a sort of long, thermally insulating piston. The cold end of this imaginary piston, in the vicinity of the cold heat exchanger, mimics the former cold piston of the Stirling

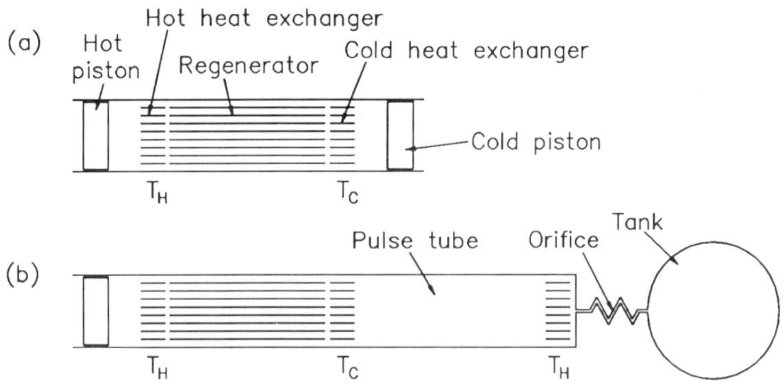

Fig. 1. Schematics of (a) Stirling cycle refrigerator, and (b) orifice pulse tube refrigerator, in which the cold piston of the Stirling cycle refrigerator is replaced by passive components.

refrigerator; the hot end of the imaginary piston delivers net work to the orifice.

In principle, orifice pulse tube refrigerators are less efficient than Stirling refrigerators: Mechanical work is dissipated into heat in the orifice of the pulse tube refrigerator, whereas it is efficiently recovered at the cold piston of the Stirling machine; and the effective thermal conductance of the pulse tube often puts a greater thermal load on the cold heat exchanger than does the heat generated by friction and other losses at the cold piston in the Stirling refrigerator. The lack of cold moving parts is a significant advantage, however, which may outweigh the lower efficiency for some applications. These considerations led us to undertake the work reported here, to explore the application of orifice pulse tube technology to the superfluid Stirling refrigerator. As discussed below, we modified our existing refrigerator, replacing the cold pistons with the simplest possible orifice pulse tube components, and operated it with a 17% mixture. The basic operation of the system is understandable, but a quantitatively accurate understanding will require better knowledge of the thermophysical properties of the mixture. For simplicity, numerical estimates of performance presented here will model the ^{3}He in the mixture using the equation of state and specific heat of a classical ideal gas, and with thermal conductivity $K \sim 0.03$ W/m-K and viscosity $\mu \sim 1.5 \times 10^{-6}$ kg/s-m.

Gravity plays an important role in our pulse tube, both with respect to ordinary convection and by flattening the oscillatory flow profile. Hence ground-based testing will not generally be relevant to the performance of a superfluid pulse-tube refrigerator in zero gravity; a more conventional Stirling configuration of the superfluid Stirling refrigerator may entail less risk for satellite-based applications.

APPARATUS

Our unmodified superfluid Stirling refrigerator, described elsewhere[2] in this volume, essentially comprised two simple Stirling refrigerators as illustrated in Fig. 1a operating thermally in parallel and temporally 180° out of phase from each other. A counterflow heat exchanger served as regenerator, so that the fluid in each of the two refrigerators regenerated that in the other, without need of the usual solid heat capacity.

To convert this apparatus into an orifice pulse tube refrigerator, we removed the two cold pistons, replacing them with two pulse tubes and an orifice as shown in Fig. 2. Because of the 180° temporal phase shift between the two halves of the refrigerator, a single orifice linking the hot ends of the two pulse tubes was used, so that each pulse tube served as the "tank" (cf. Fig. 1b) of the other pulse tube. The heat exchangers at the hot ends of the pulse tubes were thermally anchored to the hot platform.

Choice of dimensions of the pulse tubes was a rough compromise among several considerations. The total fluid volume in the pulse tube should be significantly larger than the volumetric displacement through the pulse tube, so that the "imaginary piston" is not swept entirely out of the pulse tube during operation. The peak-to-peak volumetric displacement of each of our hot pistons was $2V_1 \leq 2$ cm^3, so we expected vol-

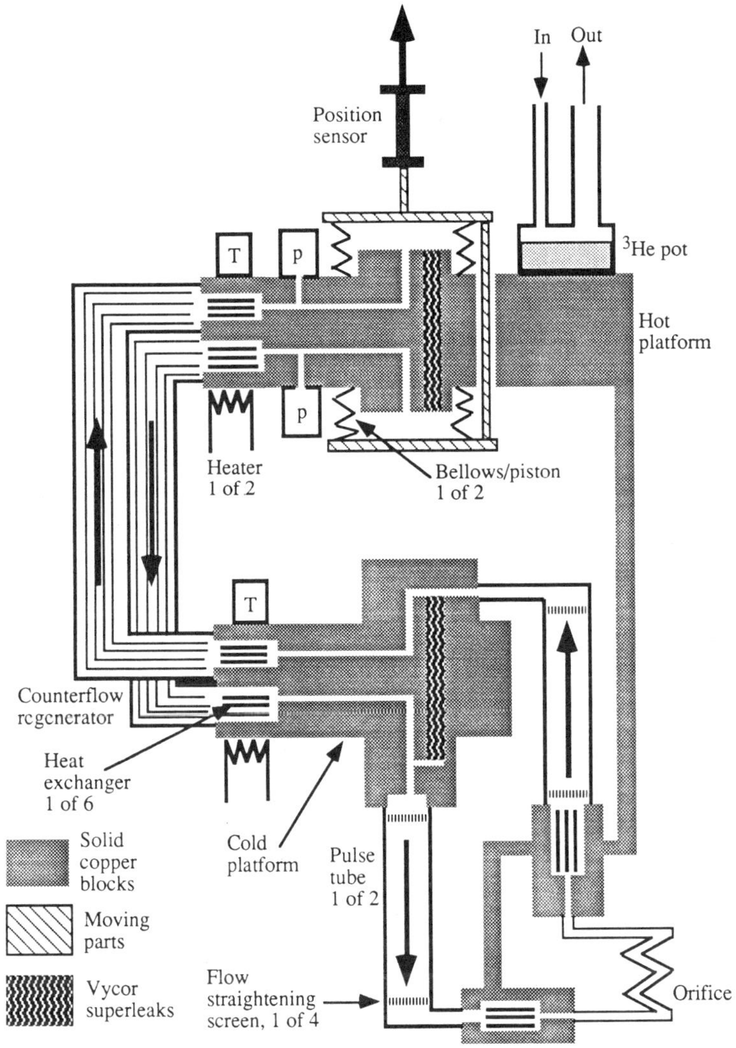

Fig. 2. Schematic of the superfluid orifice pulse tube refrigerator. Pressure sensors are indicated by "p"; thermometers by "T".

umetric displacements in the pulse tubes to be somewhat smaller than that. However, if the pulse tube volume is too large, the pressure amplitude is reduced significantly for a given piston stroke. Erroneously, we had believed that the pulse tube diameter should be significantly larger than the mixture's viscous penetration depth $\delta_\mu = (\mu/\pi f \frac{m*}{m}\rho)^{1/2}$ (where f is operating frequency, ρ is mass of ^{3}He per unit volume, and m^* and m are the effective mass and true mass of a ^{3}He atom), so the "imaginary piston" could move with a velocity that is essentially independent of radial location within the pulse tube. (This consideration will be discussed more fully in the "Gravity" section below.) For the operating frequencies accessible with our drive motor (4 mHz $\lesssim f \lesssim$ 50 mHz), 0.5 mm $\lesssim \delta_\mu \lesssim$ 1.6 mm. The area-to-length ratio of the pulse tube should not be too large, lest ordinary heat conduction along the "imaginary piston" be excessive. Finally, we were constrained by the available space at the bottom of our cryostat. We chose pulse tube inner diameter 6.0 mm and length 7.0 cm, so that the fluid volume in each pulse tube was 2.0 cm^3. The tubes were CuNi with 150 μm thick walls.

Copper screens (60 mesh/inch) at each end of each pulse tube served as flow straighteners, to prevent possible jetting of fluid entering from the heat exchangers or 3 mm diam connecting tubes, in which the Reynolds number was $\sim 10^3$. The connecting tubes contributed 0.7 cm^3 to the volume of fluid in each half of the refrigerator.

Research orifice pulse tube refrigerators usually use needle valves as orifices so that the impedance can be varied widely during the course of an experiment. For our superfluid orifice pulse tube refrigerator, with small displaced volume and low frequency, the required impedance seemed far higher than the range of commonly available valves. Hence we used a fixed impedance, arrived at by trial and error. It consisted of 16 capillary tubes in parallel, with wires inserted to block most of the area. Each tube was 30 mm long, approximately 0.2 mm inner diameter, with 0.15 mm diameter wire. The Mach number (based on second-sound speed) of the flow in this impedance was ~ 0.01; the Reynolds number was ~ 10.

Each of the hot heat exchangers between the orifice and the pulse tubes was simply four holes (0.8 mm diam, 1 cm long) drilled through a copper block. These blocks were bolted to copper bars, which were in turn bolted to the hot platform.

Thermometers and heaters on the hot and cold platforms allowed measurement of temperatures T and cooling power $\dot{Q}$, and regulation at constant temperature when desired. Piston volumetric displacement V was measured with a variable differential transformer, and oscillatory pressure p was measured with flexible diaphragm sensors. These measurements were automated with a personal computer, so that cycle-averaged work $W = \oint p\,dV$ and average power $\dot{W} = fW$ could be computed easily. More details on these sensors may be found elsewhere.[2]

GROSS COOLING POWER

To investigate basic operation of the refrigerator under simple conditions, we regulated the cold platform temperature T_C to be equal to the hot platform temperature T_H, so that temperature was essentially uniform throughout the apparatus, and measured $V(t)$, $p(t)$, and $\dot{Q}$ as functions of f. The results, shown in Fig. 3, can be understood qualitatively by considering a simple model: an oscillating volume $V(t) = V_0 - V_1 \cos(\omega t)$ of gas, weakly linked by a resistive impedance R to a location of constant pressure p_0. (For our dual refrigerator, this location is the midpoint of the orifice.) Viscous pressure drops within V are negligible, so that $p(t)$ is spatially uniform. The mass of gas in the volume is proportional to pV; the rate of change of mass in the volume is $-(p-p_0)/R$. Hence the differential equation describing the system is $d(pV)/dt = -(p-p_0)/R$. Using $p(t) = p_0 + \mathrm{Re}[p_1 e^{i\omega t}]$ and $V(t)$ as given above, we find

$$\frac{p_1}{p_0} = \frac{V_1/V_0}{1 + 1/i\omega R V_0} \quad \text{and} \quad \dot{W} = \frac{p_0}{2R}\left|\frac{p_1}{p_0}\right|^2 . \tag{1}$$

We can easily estimate the parameters in these equations. First, we use $p_0 = 370$ torr, the pressure of a classical ideal gas at 1 K with number density equal to that of a 17% mixture. Second, we use $V_0 = 5.0$ cm^3. The geometrical volume of fluid in each half of the refrigerator is 6.7 cm^3. However, V_0 is somewhat smaller than this, because most of the fluid volume experiences nearly adiabatic pressure oscillations, and hence is stiffer (by a factor $\gamma \sim 5/3$) than we assumed in the isothermal derivation above. Our value of V_0 is based on the estimate that 40% of our geometry should be treated as isothermal and 60% as adiabatic. Finally, we obtain $R = 2.2$ s/cm^3 from the measured 11 s relaxation time of pressure following an abrupt change in piston position.

Using this value of R, the length and number of capillaries, the wire diameters, and the viscosity of the mixture, we estimate that the annular gap between wires and capillaries in the orifice is 7 μm. This is the correct order of magnitude, but a little smaller than would be inferred from the geometry of the capillaries and wires. The flow velocities are of order 10 times the critical velocity for pure ^{4}He in this geometry; perhaps superfluid turbulence plays a significant role in the behavior of the impedance.

With these values for p_0, V_0, and R, Eqs. (1) then yield the lines in Fig. 3, which are in qualitative agreement with the experimental results for the magnitude and phase of p_1 and for $\dot{W}$. (According to Radebaugh's tables[6] for the equation of state, most of the difference between the lines and the measurements is due to our use of the ideal

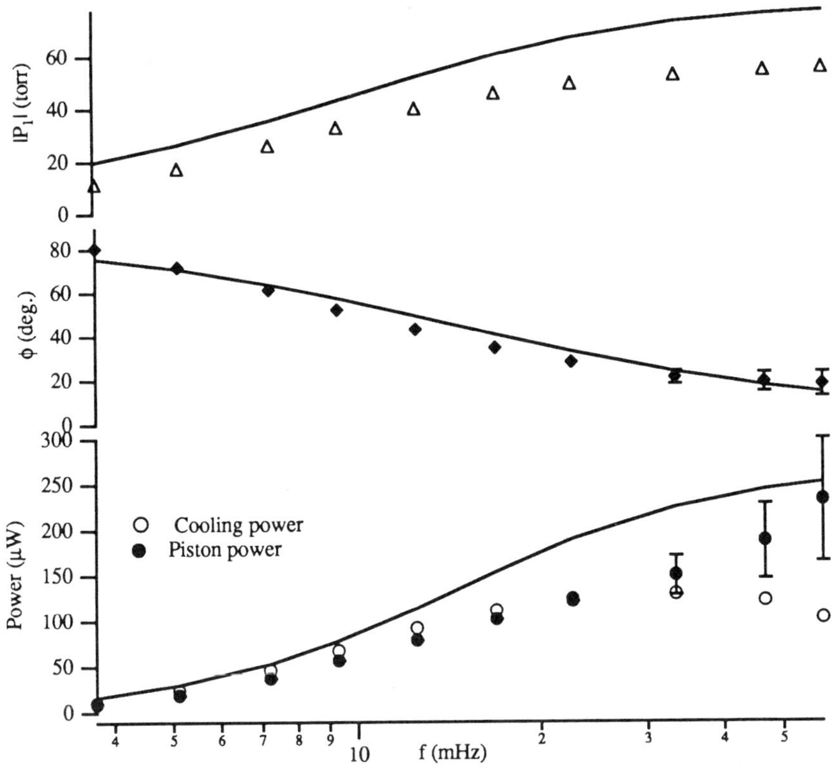

Fig. 3. Amplitude and phase of oscillatory pressure, and cooling power and piston power, vs drive frequency, for $T_C = T_H = 1.0$ K and $V_1 = 0.56$ cm^3. Here, p_1 is the complex amplitude of the pressure difference between the two halves of our refrigerator, and the powers are the sum of those for the two halves, so appropriate factors of 2 have been used relative to Eqs. (1).

gas approximation; but we hesitate to use his values or any others because of the lack of experimental data on osmotic pressure in this concentration range.)

For small regenerator volume and $T_C = T_H$, the volumetric flow rates at the two ends of the regenerator are equal. The work fluxes at the two ends of the regenerator are also equal, because $p(t)$ is spatially uniform. The enthalpy flux through an ideal gas in a perfect regenerator is zero, so the cooling power must equal the time-averaged work flux at the cold end. Hence under these conditions we expect $\dot{Q} = \dot{W}$. At low and mid frequencies, the data agree with this expectation.

We do not fully understand the observed increase of $\dot{W}$ and decrease of $\dot{Q}$ at the highest frequencies. Plausible explanations, in rough agreement with the measurements, include[7] irreversible oscillatory pressure in fluid that is neither isothermal nor adiabatic and nonlinear flow in the impedance.

EFFECTS OF GRAVITY IN THE PULSE TUBE

Initially, we made the mistake of orienting the cold end of the pulse tubes down, following conventional pulse tube refrigerator practice for stability against gravitational convection. With this arrangement, we could not cool below $T_C = 0.95T_H$. Stability against convection in ^{3}He-^{4}He mixtures requires that the cold end be up, because the total (^{3}He + ^{4}He) mass density is lower at the cold end, where the ^{3}He density is higher. To be consistent with our use of the classical ideal gas equation of state for the ^{3}He, we estimate the effective thermal expansion coefficient β_{grav} for purposes of gravitational convection by assuming that the total number density in our mixture of mass-3 and mass-4 atoms is constant, independent of temperature and ^{3}He concentration. Then

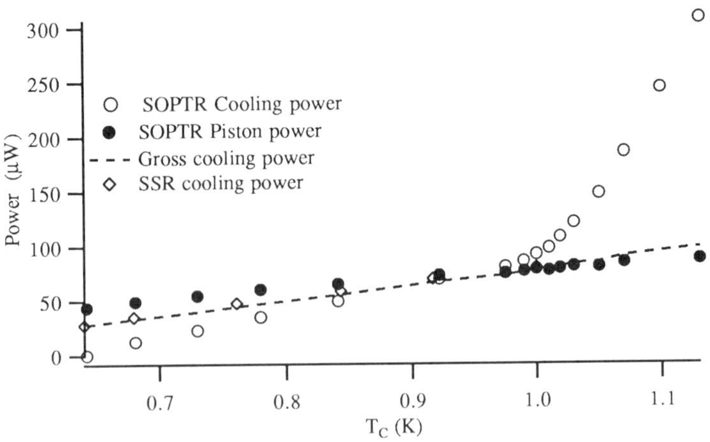

Fig. 4. Powers vs T_C, with $T_H = 1.0$ K and $f = 12.5$ mHz. Circles are measured cooling power and mechanical power with $V_1 = 0.56$ cm^3. Triangles are scaled measured cooling power under similar conditions[2] for conventional Stirling configuration. Dashed line is expected gross cooling power.

it is easy to show that $T\beta_{grav} \cong -1/3$, a third the magnitude and of opposite sign compared to a classical ideal gas. This value is close to tabulations[8] based on a much more complete treatment[9] in the context of Rayleigh-Bènard convection.

All the measurements presented here have the cold end of the pulse tubes up (as shown in Fig. 2). Figure 4 shows typical measurements of cooling power as a function of T_C, with T_H maintained at 1 K. The "cooling power" for $T_C > T_H$ shows a dramatic increase, which we attribute to gravitational convection. For the hottest datum, the Rayleigh number in the pulse tube is of order 10^6.

Gravity also has a profound effect on the oscillatory motion. Consider the x component of the equation of motion of the fluid in the pulse tube:

$$(m^*/m)\rho\left[\partial u/\partial t + (\mathbf{v}\cdot\boldsymbol{\nabla})\,u\right] = -\partial p/\partial x + \mu\nabla^2 u + \rho g_{\text{eff}}, \tag{2}$$

where x is directed along the axis of the pulse tube, and is positive downwards, and $g_{\text{eff}} = -g/3$ with g the ordinary acceleration of gravity. For small, oscillatory motion, $u \rightarrow \text{Re}[u_1(r)e^{i\omega t}]$ and $\rho \rightarrow \rho + \text{Re}[\rho_1 e^{i\omega t}]$, so that Eq. (2) reduces to

$$i\omega(m^*/m)\rho u_1 = -\partial p_1/\partial x + \mu\nabla^2 u_1 + \rho_1 g_{\text{eff}}. \tag{3}$$

Usually the gravitational term can be neglected. Then the scale of the distance near the walls over which u_1 changes from 0 to its free-stream value is determined by the ratio of the viscous and inertial terms, which yields the (square of the) viscous penetration depth δ_μ alluded to in the introduction. In contrast, in the superfluid pulse tube, we cannot neglect g. The oscillatory displacement of the fluid with its temperature gradient, along the temperature gradient, causes temperature oscillations, contributing a spatially dependent $\rho_1(r)$ proportional to $u_1(r)$. (Density oscillations ρ_1 also arise from pressure oscillations, but these are spatially uniform and so have no effect on the shape of u_1). Then Eq. (3) becomes

$$i\omega(m^*/m)\rho u_1 = -\partial p_1/\partial x + \mu\nabla^2 u_1 - (\rho g_{\text{eff}}/T)T_1. \tag{4}$$

with T_1 given by the first-order temperature equation

$$\rho c_p\left[i\omega T_1 + \nabla_x T\, u_1\right] = K\nabla^2 T_1 \tag{5}$$

(again neglecting a spatially uniform term due to pressure oscillations). Combining Eqs. (4) and (5) by eliminating T_1 produces a 4th order differential equation in u_1, whose spatial dependence is governed by an effective viscous penetration depth $\delta_{\mu,\text{eff}}$ given in the low-frequency limit by

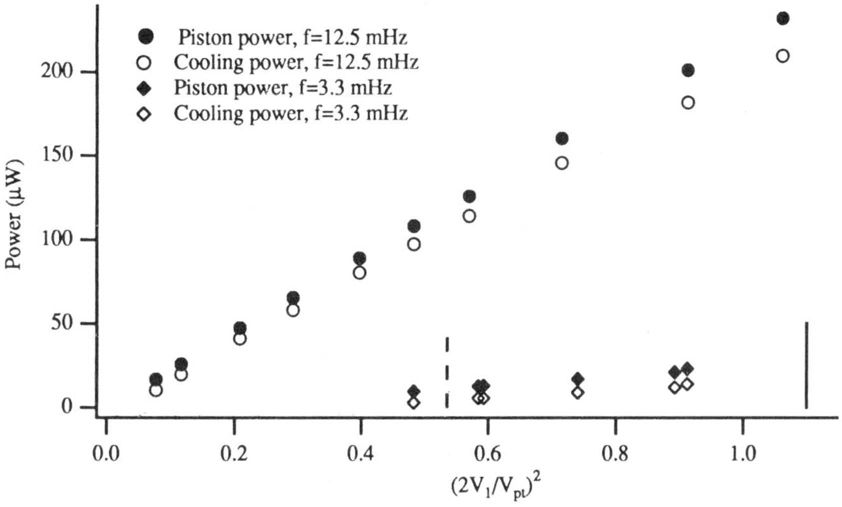

Fig. 5. Cooling power and mechanical power vs piston stroke, with $T_H = 1$ K, $T_C = 0.9$ K, at two different frequencies. Here, we normalize the peak-to-peak piston stroke $2V_1$ by the pulse tube volume $V_{pt} = 2.0$ cm^3, the volume of fluid between the flow straighteners in the pulse tube.

$$\delta^4_{\mu,\text{eff}} = \delta^4_\mu \omega^2 T / 4\sigma \left| g_{\text{eff}} \right| \nabla_x T, \tag{6}$$

where the Prandtl number $\sigma = \mu c_p / K$. This effective viscous penetration depth gives the scale of the distance near the walls over which u_1 changes from 0 to its free-stream value in our pulse tube. For values appropriate to the data shown in Fig. 4, $\delta_\mu = 0.9$ mm and $\delta_{\mu,\text{eff}} = 0.1$ mm: Gravity flattens the velocity profiles dramatically. We believe this effect is negligible in *conventional* pulse tubes ($g_{\text{eff}} \to g$) only when $\omega^2 T / 4\sigma g \nabla_x T \ll 1$. Typical pulse tubes are ~ 10 cm long and have $T_H - T_C \sim T$, implying $f \ll 3$ Hz.

Some indirect evidence of this flattening is shown in Fig. 5, which displays cooling power as a function of piston stroke. We expect that the "imaginary piston" in the pulse tube would disappear for large enough piston strokes, because it would be swept entirely out of the pulse tube as the fluid oscillates, dramatically destroying its thermally insulating quality. The diameter of the "imaginary piston" is approximately equal to the pulse tube diameter minus twice the thickness of the fluid layer held motionless near the pulse tube walls by viscosity[10] ; if gravity could be neglected, this layer would have thickness δ_μ, but the effects of gravity reduce its thickness to $\delta_{\mu,\text{eff}}$. The vertical dashed line in Fig. 5 shows, for $f = 3.3$ mHz, the piston stroke above which the imaginary piston would disappear if δ_μ were the appropriate thickness; the vertical solid line shows the location of this expected transition for $\delta_{\mu,\text{eff}}$ the appropriate thickness. (For $f = 12.5$ mHz, the corresponding lines would be at $(2V_1/V_{pt})^2 = 1.6$ and 2.2, respectively.) The measurements are consistent with the $\delta_{\mu,\text{eff}}$ interpretation (with a heat leak of the order of 10 μW that does not increase dramatically with stroke).

This flattening of the velocity profile increases the viscous dissipation in the pulse tube, because it steepens the velocity gradient near the wall. We estimate this dissipation to be $\lesssim 1$ μW for the conditions of Fig. 4; it varies as $V_1^2 f^{3/2}$.

The flattening of the velocity profile also helps make this an essentially ideal pulse tube. The small value of $\delta_{\mu,\text{eff}}$ compared to the pulse tube diameter makes the "imaginary piston" picture reasonably accurate. The heat capacity in the metal tube wall and the heat capacity in the stationary fluid layer adjacent to the wall are small compared to the heat capacity within the first thermal penetration depth in the moving fluid, so negligible heat is transferred laterally to and from the moving fluid; hence shuttle heat loss is small. The strong stability enforced by gravity and the temperature gradient may suppress streaming-driven convective mixing as well. This leaves ordinary thermal conduction in the x direction as the only significant loss mechanism in this pulse tube.

PULSE TUBE THERMAL LOSS

In the Gross Cooling Power section above, we argued that $\dot{Q} = \dot{W}$ when $T_C = T_H$. For unequal temperatures, the measured cooling power $\dot{Q}$ will be reduced by heat loads to the cold heat exchanger through the regenerator and pulse tube, due both to ordinary conduction and to convection driven by the oscillatory fluid motion. Let us call the cooling power if such heat loads were absent the gross cooling power. For small regenerator volume, the ratio of the volumetric flow rates at the two ends of the regenerator must equal T_C/T_H. Then the ratio of work fluxes at the two ends of the regenerator also equals T_C/T_H, because $p(t)$ is spatially uniform. The enthalpy flux through an ideal gas in a perfect regenerator is zero, so the gross cooling power must equal the time-averaged work flux at the cold end. Hence under these conditions we expect the gross cooling power to be $\dot{W}\, T_C/T_H$, where $\dot{W}$ is the work flux at the hot end, as before. The dashed line in Fig. 4 is this expected gross cooling power.

The difference between the gross and observed cooling powers is the sum of the heat loads due to the regenerator and pulse tube. To estimate what fraction is due to the regenerator, we display in Fig. 4 data taken before[2] we replaced the cold pistons with the pulse tubes, for the same ^{3}He concentration and operating frequency. (To account as well as possible for the slightly different piston stroke used in the earlier data, we multiplied the earlier cooling powers by 0.85 to bring the observed $T_C = T_H$ cooling powers, and hence the gross cooling powers, into agreement.) These data fall much closer to the gross cooling power line than do the data for the pulse tube system, suggesting that the heat load on this system is largely due to the pulse tubes. The data are in rough agreement with our estimate (from geometry and the thermal conductivity of the mixture) that ordinary conduction through the pulse tubes should deliver $(50\ \mu\text{W/K})(T_H - T_C)$. Although many pulse tubes suffer from substantial thermal loss due to streaming- or turbulence-driven mixing, there is no evidence of such loss here.

By using longer, thinner pulse tubes, it will be simple to reduce pulse tube losses by an order of magnitude in future superfluid orifice pulse tube refrigerators, while still keeping $\delta_{\mu,\text{eff}}$ much smaller than the pulse tube diameter and keeping viscous dissipation within $\delta_{\mu,\text{eff}}$ acceptably low. Hence, we expect that a ground-based superfluid pulse tube refrigerator can perform nearly as well as a superfluid Stirling refrigerator, except for the added heat loads imposed on the heat sink at T_H by the orifice dissipation.

ACKNOWLEDGMENT

This research has been supported by the Division of Materials Science in the US Department of Energy's Office of Basic Energy Sciences. We acknowledge helpful conversations with Ray Radebaugh, Mike Hayden, and Bob Ecke.

REFERENCES

1. Permanent address: Mechanical Engineering Department, 41-206, Massachussetts Institute of Technology, 77 Massachusetts Avenue, Cambridge MA 02139.
2. A. Watanabe, G. W. Swift, and J. G. Brisson, Measurements with a recuperative superfluid Stirling refrigerator (1996) elsewhere in this volume; and references therein.
3. G. Walker, "Stirling Engines," Clarendon, Oxford (1960).
4. G. Walker, "Cryocoolers," Plenum, New York (1983).
5. R. Radebaugh, Advances in Cryogenic Engineering 35:1191 (1990).
6. R. Radebaugh, Thermodynamic Properties of ^{3}He-^{4}He Solutions with Applications to the ^{3}He-^{4}He Dilution Refrigerator, National Bureau of Standards Technical Note 362, Boulder CO (1967).
7. A. Watanabe, Studies of superfluid Stirling and pulse tube refrigeration, Master's thesis, Massachusetts Institute of Technology (1995).
8. Y. Maeno, H. Haucke, R. E. Ecke, and J. C. Wheatley, *J. Low Temp. Phys.* 59:305 (1985), Appendix.
9. A. L. Fetter, *Phys. Rev. B* 26:1164 (1982), Eq. (22).
10. Our calculations of the effective diameter of the imaginary piston use the complex Bessel functions of N. Rott, *Z. Angew. Math. Phys.* 20:230 (1969).

MEASUREMENTS WITH A RECUPERATIVE SUPERFLUID STIRLING REFRIGERATOR

A. Watanabe, G. W. Swift, and J. G. Brisson[1]

Condensed Matter and Thermal Physics Group
Los Alamos National Laboratory, Los Alamos NM 87545

ABSTRACT

A superfluid Stirling refrigerator cooled to 168 mK using a 4.9% ^{3}He-^{4}He mixture and exhausting its waste heat at 383 mK. Cooling power versus temperature and speed is presented for 4.9%, 17%, and 36% mixtures. At the highest concentration, a dissipation mechanism of unknown origin is observed.

INTRODUCTION

The superfluid Stirling refrigerator (SSR) uses a ^{3}He-^{4}He liquid mixture[2] as a working fluid in a Stirling cycle. It operates at temperatures below 2 K where the ^{4}He component of the working fluid is superfluid. The ^{3}He component of the working fluid, to first approximation, behaves thermodynamically as an ideal gas in the inert background of superfluid ^{4}He. Using pistons equipped with a superleak bypass, it is possible to compress and expand the ^{3}He solute "gas" alone. The SSR is a Stirling machine equipped with these "superleaked" pistons to take advantage of the properties of the ^{3}He solute to cool below 1 K. The principle was first demonstrated by Kotsubo and Swift[3,4] in 1990; a more practical refrigerator was studied by Brisson and Swift[5-8] beginning in 1992.

The refrigerator used in the present work, shown in Fig. 1, is a modified version of that discussed by Brisson and Swift.[5,6] The major modification is the addition of a ^{3}He evaporation refrigerator that cools the hot platform on which the hot pistons are mounted. With this arrangement, the hot platform temperature can be maintained at between 0.3 K and 1 K. This refrigerator is actually two SSR's operating 180° out of phase with each other, which allows the use of a counterflow heat exchanger to act as a regenerator for each of the SSR's. Our counterflow regenerator,[8] or recuperator, consists of CuNi tubes silver soldered in a hexagonally close packed array with alternating rows corresponding to each "half" of the SSR.

The pistons are made with welded bellows which have wavy segments that nest into one another to minimize the dead volume. The hot platform pistons are rigidly connected together and driven sinusoidally by a rigid push rod from a room temperature drive. The cold platform pistons are similarly driven. Linear position sensors are mounted on a pumped ^{4}He platform maintained at 1.2 K to monitor the positions of the pistons. The hot piston's stroke ranged from 2.41 mm to 4.85 mm, corresponding

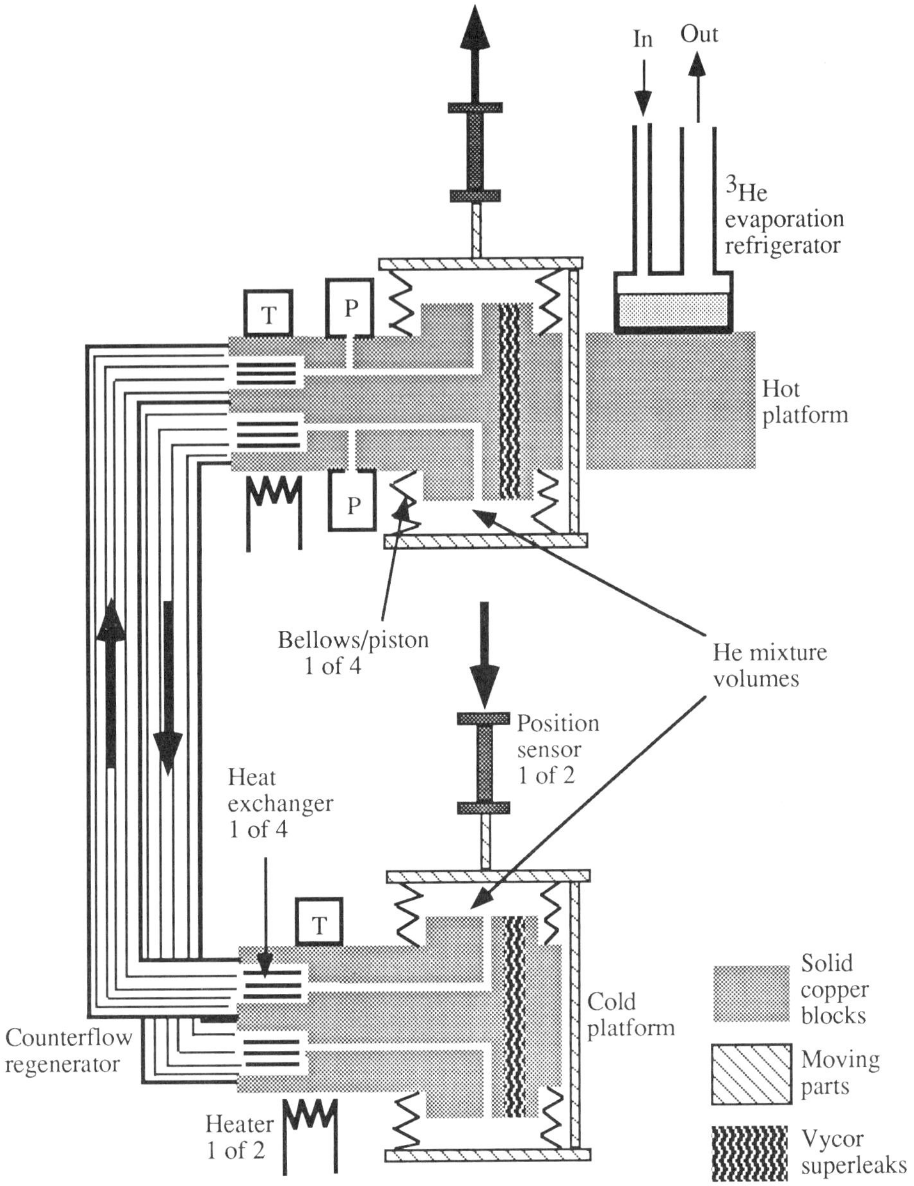

Fig. 1. The SSR. Pressure gauges are delineated with a "P", thermometers with a "T". The pumped ^{4}He platform, which is located above the hot platform, is not shown. The arrows in the regenerator indicate the direction of fluid flow corresponding to the direction of piston motion shown.

to a swept volume of 0.763 cm^3 to 1.53 cm^3. The cold piston's stroke ranged from 2.34 mm to 4.93 mm, corresponding to a swept volume of 0.737 cm^3 to 1.56 cm^3. The superleaks shown in Fig. 1 are cylinders of Vycor glass. These allow the superfluid ^{4}He to freely flow between the halves of the refrigerator during its operation. Flexible diaphragm pressure gauges are connected to each half of the SSR at the hot platform. After loading the SSR with ^{3}He-^{4}He mixture of desired concentration, the two fill lines into each of the SSR halves are sealed with mechanically actuated low temperature valves mounted on the pumped ^{4}He platform. If we did not use these valves, operation of the SSR would cause the mixture to oscillate up and down the fill capillaries and put a significant heat load on the refrigerator. Heaters and calibrated germanium thermometers are mounted on the outside of the hot and cold platforms.

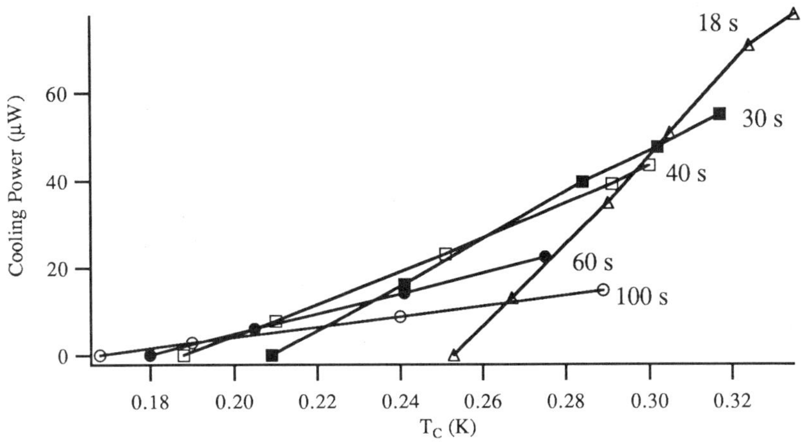

Fig. 2. The SSR's cooling power vs the cold platform temperature for a 4.9% mixture. Here and in the following figures, each set is labeled with the SSR's period, and the lines are only guides to the eye. The hot and cold piston swept volumes were 1.53 cm^3 and 1.56 cm^3, respectively. The hot platform temperature was approximately 0.450 K, 0.417 K, 0.407 K, 0.400 K, and 0.387 K for the period of 18, 30, 40, 60, and 100 seconds, respectively.

RESULTS AND DISCUSSION

The SSR's cooling power shown in Figs. 2 through 5 was measured as a function of the mixture's average concentration, hot platform temperature T_H, cold platform temperature T_C, and the operating speed. T_C was actively controlled by a feedback controller that delivered current to the heater mounted on the cold platform, and the heat delivered averaged over five to ten periods yielded the cooling power. With the exception of the 4.9% mixture, T_H was actively controlled by a combination of the ^{3}He evaporation refrigerator and a feedback controller delivering current to the heater mounted on the hot platform. For the 4.9% mixture, T_H was kept as cold as the ^{3}He evaporation refrigerator allowed. The SSR's minimum period, the time necessary to complete one cycle, was limited by the ^{3}He pot's capability in all of the runs. Running the SSR too fast put a large heat load on the hot platform and resulted in the ^{3}He pot quickly running dry. Concentrations given in this paper are molar per cent of ^{3}He.

Figure 2 shows the SSR's cooling power at various speeds as a function of cold platform temperature for a 4.9% mixture. The SSR reached a minimum temperature of 168 mK with a period of 100 s and hot platform temperature of 383 mK. Figure 2 also shows that there is an optimum operating speed at each temperature to maximize the cooling power. The upper envelope of the lines defines this SSR-^{3}He refrigerator system's maximum cooling power as a function of temperature.

It is of interest to model the mixture's concentration as a function of position in the refrigerator corresponding to the experimental conditions measured above. Because of the temperature difference, the concentration in the cold platform pistons is higher than the "nominal" concentration, and that in the hot platform pistons is lower. For low concentrations of ^{3}He, the mixture should obey[5]

$$P_3(\rho, T) + P_f = \text{constant} \tag{1}$$

throughout the refrigerator, where P_3 is the osmotic pressure of ^{3}He, ρ is the mixture's ^{3}He density, T is the temperature, and P_f is the fountain pressure. Approximate values of $P_3(\rho, T)$ can be obtained using the ideal Fermi gas equation of state.[9] It was assumed that the hot and cold piston volumes are 3.2 cm^3. The regenerator, which has a volume of 1.0 cm^3, was modeled as six discrete volumes with a linear temperature profile. The number of ^{3}He particles in the refrigerator is constant, which means

$$\sum \rho_i(T) V_i = \rho_{avg} V_{tot}, \tag{2}$$

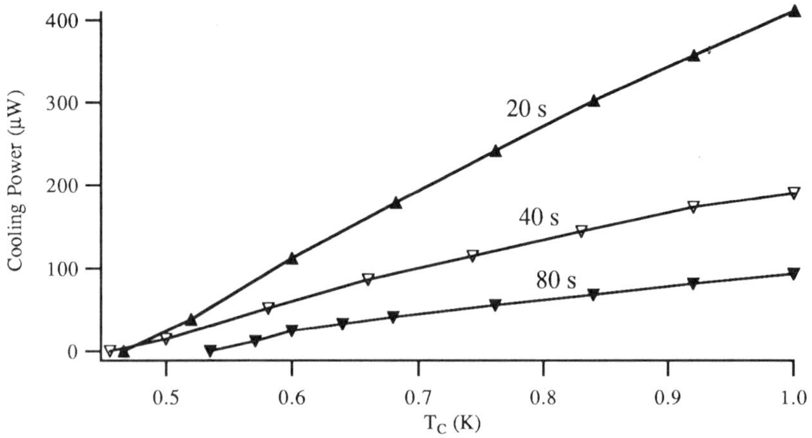

Fig. 3. The SSR's cooling power vs the cold platform temperature for a 17% mixture and T_H = 1 K. The hot and cold piston swept volumes were 0.763 cm^3 and 0.737 cm^3, respectively.

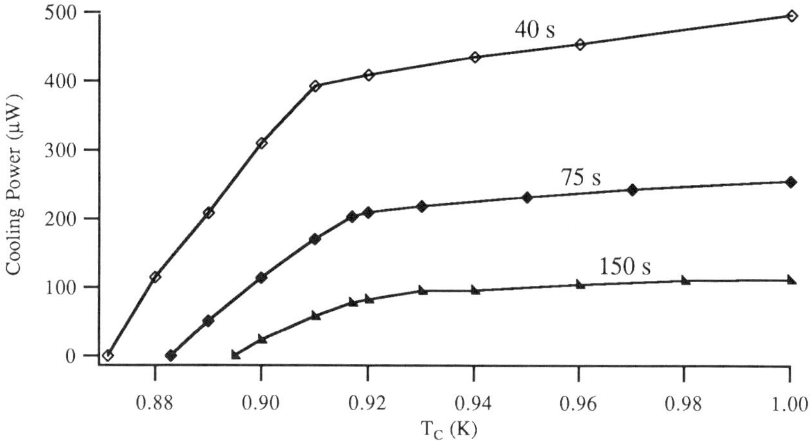

Fig. 4. The SSR's cooling power vs the cold platform temperature for a 36% mixture and T_H = 1 K. The hot and cold piston swept volumes were 0.767 cm^3 and 0.737 cm^3, respectively.

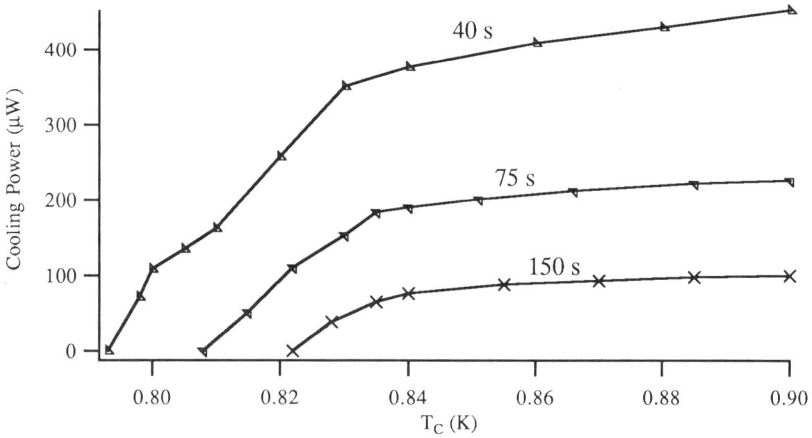

Fig. 5. The SSR's cooling power vs the cold platform temperature for a 36% mixture and T_H = 0.9 K. The hot and cold piston swept volumes were 0.767 cm^3 and 0.737 cm^3, respectively.

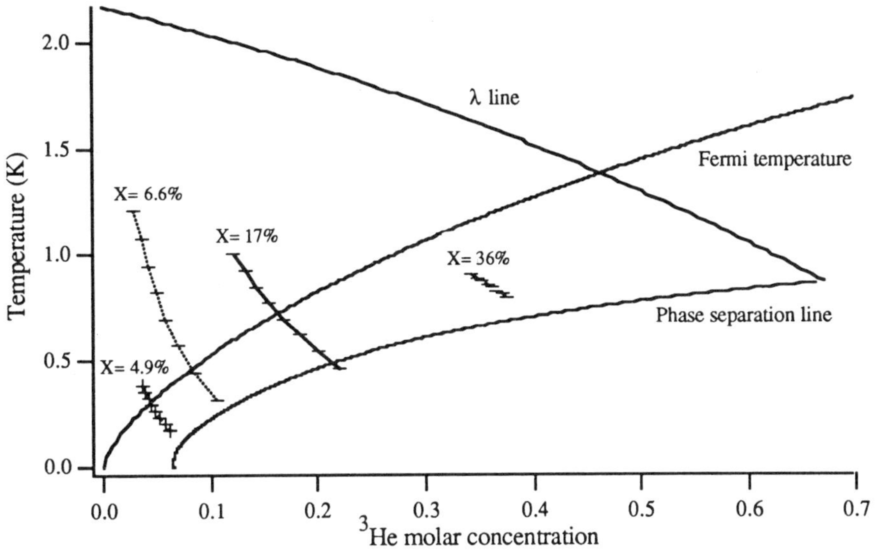

Fig. 6. The mixture's state as a function of position in the refrigerator for four different average concentrations, superimposed on a phase diagram. Each curve is labeled with the mixture's average concentration. Three of the curves are for operating conditions of this paper; the 6.6% curve is for the conditions of Ref. 6. The ^{3}He Fermi temperature as a function of concentration is also shown. In each curve, the top point is the state of the mixture in the hot piston, the bottom point is the state of the mixture in the cold piston, and the six intermediate points represent the mixture states in the regenerator.

where $\rho_i(T)$ and V_i are the ^{3}He density and volume at various points in the refrigerator, ρ_{avg} is the average ^{3}He density in the refrigerator, and V_{tot} is the total volume of the refrigerator. With Eq. (2) as a constraint, we can solve iteratively for the mixture's concentration as a function of position in the refrigerator, given the hot and cold platform temperatures and average concentration. Figure 6 shows the results of the calculation for four sets of operating conditions, superimposed on a ^{3}He-^{4}He phase diagram. The top point in each curve is the state of the mixture in the hot piston, the bottom point is the state of the mixture in the cold piston, and the six intermediate points represent the mixture states in the regenerator.

We can make several observations using Fig. 6. First, for all the conditions modeled the state of the ^{3}He in the cold piston is in the Fermi regime. In addition, when the SSR was loaded with an average concentration of 36%, the entire refrigerator was operated below the Fermi temperature. Our model for the 17% mixture in the SSR suggests that there is phase separation in the cold piston. Unfortunately, we have no direct experimental evidence for phase separation in the 17% data.

The experimental cooling power results for the 36% mixture, Figs. 4 and 5, show a "cutoff" temperature where the cooling power drops off sharply as a function of temperature. Clearly, this is not due to phase separation of the mixture, as a ^{3}He-^{4}He mixture does not phase separate at 0.92 K regardless of the concentration.[10] Yet the sharpness of the transition suggests that it is caused by an abrupt transition in fluid properties. This transition may be caused by a critical velocity. The ^{3}He component may be exceeding a critical velocity in the 0.8 mm diameter capillary connecting the piston to the heat exchanger in the cold platform. As the cold platform gets colder, the concentration in the cold piston increases according to Eq. (1), and the net effect may be to decrease the critical velocity. For a saturated mixture below 100 mK, Zeegers et al.[11] empirically found a critical velocity of

$$v_c = (K/d)\ln(d/d_0) \qquad (3)$$

where K=0.05 cm^2/s, d is the diameter of the capillary, and d_0=15 μm. Although both the temperature and the concentration used to find Eq. (3) are significantly lower than

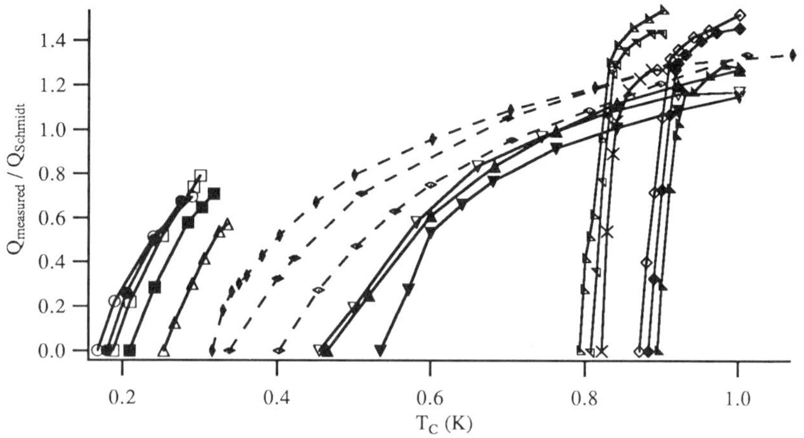

Fig. 7. The measured SSR cooling power normalized by the cooling power calculated from Schmidt analysis vs cold platform temperature. The symbols correspond to those shown in Figs. 2 through 5. The dashed lines represent the normalized cooling power of data of Ref. 6.

those used in our refrigerator, we will use this expression because we are unaware of any other published data for the critical velocity of superfluid ^{3}He-^{4}He mixtures. For a 0.8 mm diameter capillary, this expression yields 2.5 cm/s. In our experiment, the maximum ^{3}He velocity in the capillary during the cycle ranged from around 2.3 cm/s for the period of 150 s to around 8.5 cm/s for the period of 40 s, which is of the same order of magnitude as that given by Eq. (3).

Except for this unusual behavior at 36% concentration, cooling power exhibits remarkable linearity in T_C. This was also true for measurements[6] for a 6.6% average mixture concentration. This linearity can be understood as follows. Assuming equal clearance volumes and swept volumes in the hot and the cold pistons and with 90° phasing between the pistons, the isothermal Schmidt analysis[12] of the Stirling cycle predicts a cooling power in the cold platform of

$$\dot{Q}_{Schmidt} = f\frac{\pi}{2}\left(\frac{V_{sw}}{V_0}\right)^2 MRT_H\left(\frac{\alpha}{1+\alpha}\right)^2 \Big/ \left(1 - \frac{V_r}{V_0}\frac{\alpha\ln\alpha}{1-\alpha^2}\right)^2 \tag{4}$$

to lowest order in swept volume, where f is the frequency, V_{sw} is the piston swept volume, V_0 is the average volume of each piston, M is the mass of ^{3}He in the refrigerator, R is the gas constant per mass of ^{3}He, $\alpha = T_C/T_H$, and V_r is the volume of the regenerator. We have multiplied the standard Schmidt result by twice the operating frequency because our refrigerator comprises two Stirling refrigerators operating in parallel. For a small regenerator volume, the temperature dependent part of Eq. (4) is very nearly linear in T_C for our operating range of $T_C/T_H > 0.3$. We can write

$$\dot{Q}_{measured} = \dot{Q}_{Schmidt} - \dot{Q}_{losses} \tag{5}$$

where $\dot{Q}_{losses}$ accounts for thermal loads on the cold platform due to thermal conduction from the hot platform, imperfect thermal contact of fluid in the regenerator, and other, smaller sources. Thermal conduction and regenerator losses are roughly proportional[8] to $T_H - T_C$. Since both $\dot{Q}_{Schmidt}$ and $\dot{Q}_{losses}$ are approximately linear in T_C, Eq. (5) predicts that $\dot{Q}_{measured}$ will also be approximately linear in T_C.

The measured cooling powers normalized by those given by Schmidt analysis are plotted in Fig. 7. The normalized cooling powers extrapolate to between 1.1 and 1.6 near $T_H = T_C$, where the regenerator losses are predicted to be negligible. Without the regenerator losses, the normalized cooling power should be 1 when $T_H = T_C$. The deviation from this value can be attributed[7], in part, to the deviation of the working fluid's thermodynamic behavior from that of an ideal Boltzmann gas. The deviation is enhanced further by the non-isothermal compression and expansion in the real refrigerator. The latter effect can be estimated by considering the thermal penetration depth δ_κ in the ^{3}He-^{4}He mixture, given by $\delta_\kappa^2 = \kappa/\pi f \rho c_p$, where κ is the thermal conductivity of the mixture and ρ and c_p are the density and the heat capacity of ^{3}He. Assuming $c_p = 5R/2$ and using published figures[13] for κ, δ_κ varies between 1.3 mm and 4.2 mm for our operating conditions at 1.0 K. Since the fluid is as much as 6 mm away from isothermal solid surfaces, part of the fluid behaves more adiabatically than isothermally. This increases the pressure amplitude above that predicted by the Schmidt model, leading to a cooling power enhancement[7] of the order of 30%. Figure 7 nevertheless indicates that the cooling power is roughly proportional to the concentration of the working fluid.

In summary, the SSR has cooled to 168 mK with the hot platform being held at 383 mK, demonstrating the ability to cool to temperatures well below that attainable with a continuous ^{3}He evaporation refrigerator. The SSR has also been operated with mixtures of up to 36% concentration with proportionately higher cooling powers.

ACKNOWLEDGMENT

This work was supported by Los Alamos Laboratory-Directed R&D funds.

REFERENCES

1. Permanent address: MIT, 41-206, 77 Massachusetts Avenue, Cambridge, MA 02139.
2. J. Wilks, "The Properties of Liquid and Solid Helium," Clarendon, Oxford (1967), p. 232.
3. V. Kotsubo and G.W. Swift, in: Proceedings of the Sixth International Cryocoolers Conference, edited by G. Green and M. Knox, David Taylor Research Center, Bethesda, MD (1991), 2:59.
4. V. Kotsubo and G.W. Swift, Superfluid Stirling-cycle refrigeration below 1 Kelvin, *J. Low Temp. Phys.* 83:217 (1991).
5. J.G. Brisson and G.W. Swift, in: Proceedings of the Seventh International Cryocoolers Conference, (Phillips Laboratory, Kirtland AFB, Albuquerque, NM 1992) p. 460.
6. J.G. Brisson and G.W. Swift, A recuperative superfluid Stirling refrigerator, "Advances in Cryogenic Engineering," 39B:1393 (1994).
7. J.G. Brisson and G.W. Swift, High-temperature cooling power of the superfluid Stirling refrigerator, *J. Low Temp. Phys.* 98:141 (1995).
8. J.G. Brisson and G.W. Swift, Measurements and modeling of a recuperator for a superfluid Stirling refrigerator, *Cryogenics* 34:971 (1994).
9. K. Huang, "Statistical Mechanics," Wiley, New York (1963), p. 200.
10. R. Radebaugh, "Thermodynamic Properties of ^{3}He-^{4}He Solutions with Applications to the ^{3}He-^{4}He Dilution Refrigerator", NBS Tech Note 362, 19 (1967).
11. J.C.H. Zeegers, R.G.K.M. Aarts, A.T.A.M. de Waele, and H.M. Gijsman, Critical velocities in ^{3}He-^{4}He mixtures below 100 mK, *Phys. Rev. B* 45:12442 (1992).
12. I. Urieli and D.M. Berchowitz, "Stirling Cycle Engine Analysis," Hilger, Bristol (1984), p. 46.
13. O.V. Lounasmaa, "Experimental Principles and Methods Below 1 K," Academic Press, New York (1974), p. 253.

AN EXACT EXPRESSION FOR SHUTTLE HEAT TRANSFER

H.-M. Chang and J. H. Baik

Hong Ik University, Department of Mechanical Engineering
Seoul, 121-791, Korea

ABSTRACT

A new and general analytical expression is obtained for the shuttle heat transfer rate in displacer-cylinder systems. The shuttle heat transfer occurs when the displacer has a reciprocating motion over the cylinder with an axial temperature gradient. The heat transfer might be important in small cryogenic refrigerators because it represents a loss of refrigeration in addition to the wall conduction. The oscillating temperature distributions of both the displacer wall and the cylinder wall are exactly obtained as functions of time and space from the conduction equations by introducing the complex temperatures. From the cyclic steady state solution of the temperatures, a simple mathematical expression is derived to calculate the shuttle heat transfer. The expression includes the axial temperature gradient, the stroke and the angular speed of the reciprocating motion, the heat transfer coefficient between two walls, and the thermal properties (such as thermal conductivity, density, or specific heat) of two walls. The usefulness of the results is justified by the approximate solutions of previous works and the physical interpretations are presented.

INTRODUCTION

Many cryogenic refrigerators have displacers and cylinders as their components. When there exists an axial temperature gradient along the walls of the displacer and the cylinder, the reciprocating motion of the displacer causes an extra heat transfer from the high to the low temperature region in addition to the wall conduction. This motional heat transfer has been called shuttle heat transfer. The shuttle heat transfer could be important in Stirling, Gifford-McMahon, Vuilleumier or Claude refrigerators because it represents a loss of refrigeration.

The basic principle of the shuttle heat transfer can be explained in Figure 1. While the displacer is moved toward the high temperature side from the center of the stroke, the displacer wall has a lower temperature than the cylinder wall of the same axial position. Therefore heat is transferred from the cylinder to the displacer. Similarly, while the displacer is moved toward the low temperature side, the displacer wall has a higher temperature and heat is transferred from the displacer to the cylinder. Because of the heat capacity of the displacer wall, the oscillations of the displacer temperature and of the displacer movement are not in phase. At a fixed axial position, the enthalpy flow accompanying the displacer motion to the low temperature direction is always greater than that to the high temperature direction. The net enthalpy flow rate from the high to the low temperature region is the shuttle heat transfer.

Several mathematical expressions have been introduced up to date to estimate the amount of the shuttle heat transfer. Zimmerman and Longthworth[1] derived an expression based upon the assumptions that the walls of the displacer and the cylinder have an infinite heat capacity and that the displacer has a square-wave motion. Rios[2] presented an approximate solution by linearization and application of Fourier series. Radebaugh and Zimmerman[3] developed an approximate solution by heat conduction analysis for a semi-infinite plate with a sinusoidal surface temperature distribution. In their analysis, the thermal resistance of gas in the gap between the walls was neglected in the calculation of the wall temperatures, and was included later by an approximate method. Martini[4] summarized these expressions to estimate the shuttle heat transfer. Recently, Nishio et al.[5] presented an approximate solution including the temperature oscillations for both of the displacer and the cylinder. In their work, a thermal resistance model was employed and the results were compared with those of their own numerical simulation.

Recently the authors published a new mathematical solution for the shuttle heat transfer, including the wall properties of both the displacer and the cylinder[6]. In this paper, a simpler and more useful form of the expression is presented and the physical interpretations are discussed.

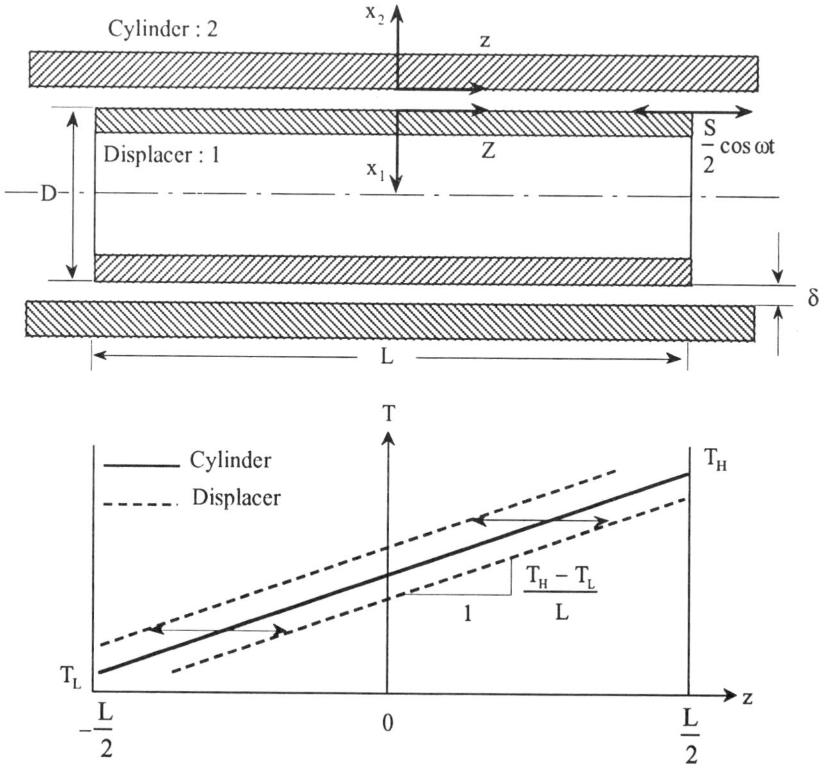

Figure 1. Schematic representation of geometry and temperature distribution for displacer-cylinder system.

FORMULATION AND SOLUTION

Analysis Model

As shown in Figure 1, both the displacer and the cylinder have the same axial linear temperature gradient, $(T_H-T_L)/L$, and the displacer reciprocates with angular speed ω and stroke S. The stationary axial coordinate z is measured from the center of the cylinder and the moving axial coordinate Z is measured from the center of the displacer. The radial coordinates x_1 and x_2 are measured from the facing surfaces of the displacer and the cylinder, respectively. The subscripts 1 and 2 denote the dispalcer and the cylinder, respectively.

For simplicity, the system is assumed be in a cyclic steady state with a sinusoidal motion with angular velocity ω and stoke S. The two axial coordinates are related with time, t, by the following relation.

$$z = Z + \frac{S}{2} \cos \omega t \tag{1}$$

The wall thicknesses of the displacer and the cylinder are assumed to be much smaller than the diameters, but greater than the thermal penetration depths so that the displacer and the cylinder are considered as two semi-infinite flat plates facing each other. These assumptions can be justified in terms of the real situation where the shuttle heat transfer is significant as checked later. It should be noted that the heat capacities of the displacer and the cylinder are not assumed to be infinity and that the temperature of the walls is a function of the axial position (z or Z), the radial position (x), and time (t).

The governing equations for the temperature of the cylinder, T_2, and the temperature of the displacer, T_1, are

$$\frac{1}{\alpha_2}\frac{\partial T_2(t,x_2,z)}{\partial t} = \frac{\partial^2 T_2(t,x_2,z)}{\partial x_2{}^2} \tag{2}$$

$$\frac{1}{\alpha_1}\frac{\partial T_1(t,x_1,Z)}{\partial t} = \frac{\partial^2 T_1(t,x_1,Z)}{\partial x_1{}^2} \tag{3}$$

respectively. It is noted that the axial coordinate, z, of Eq. (2) is stationary while the axial coordinate, Z, of Eq. (3) is moving with the displacer and that α's are the thermal diffusivities.

Two boundary conditions are

$$T_2(t, x_2 = \infty, z) = T_0 + \frac{T_H - T_L}{L} z \tag{4}$$

$$T_1(t, x_1 = \infty, Z) = T_0 + \frac{T_H - T_L}{L} Z \tag{5}$$

where T_0 is the cylinder temperature at $z=0$, and radially far from the surface. It is clear that two temperatures of Eq. (4) and (5) are same when $z=Z$ or the displacer is at the center of the stroke. Two other boundary conditions are obtained from the heat transfer relation between the facing surfaces of the displacer and the cylinder.

$$-k_1 \frac{\partial T_1\left(t, 0, z-\frac{S}{2}\cos\omega t\right)}{\partial x_1} = h\left\{T_2(t,0,z) - T_1\left(t,0,z-\frac{S}{2}\cos\omega t\right)\right\} = k_2 \frac{\partial T_2(t,0,z)}{\partial x_2} \tag{6}$$

where k's are the thermal conductivities of the walls and h, is the heat transfer coefficient between the two surfaces. In Eq. (6), the axial position of displacer, Z, was replaced with a function of z and t via Eq. (1).

Wall Temperatures

The wall temperatures for the displacer and the cylinder are obtained by solving Eq. (2) and (3) simultaneously with the boundary conditions, Eq. (4), (5), and (6), by introducing complex temperatures[6]. For the displacer wall,

$$\begin{aligned} T_1(t,x_1,z) &= T_0 + \frac{T_H - T_L}{L}\left[z - \frac{S}{2}\cos\omega t\right] \\ &+ \frac{\frac{T_H - T_L}{L}\frac{S}{\sqrt{2}}b_1}{\sqrt{(b_1+b_2)^2 + (b_1+b_2+1)^2}} exp\left[-\sqrt{\frac{\omega}{2\alpha_1}}x_1\right] cos\left[\omega t - \sqrt{\frac{\omega}{2\alpha_1}}x_1 + \frac{\pi}{4} - \phi\right] \end{aligned} \tag{7}$$

where

$$tan\phi = \frac{b_1 + b_2 + 1}{b_1 + b_2} \tag{8}$$

$$b_1 \equiv \frac{h}{k_1}\sqrt{\frac{\alpha_1}{2\omega}} \tag{9}$$

$$b_2 \equiv \frac{h}{k_2}\sqrt{\frac{\alpha_2}{2\omega}} \tag{10}$$

The dimensionless numbers, b_1 and b_2, are Biot numbers whose characteristic length is a thermal penetration depth in the walls.

Shuttle Heat Transfer Rate

With the obtained temperature of the displacer wall, the shuttle heat transfer can now be calculated. At an arbitrary axial position, the shuttle heat transfer can be defined as the net enthalpy flow rate from the high to the low temperature side, or negative z direction. A cycle-averaged net enthalpy flow rate of the displacer wall can be obtained by integrating over a period and the displacer wall area.

$$Q = \frac{\omega}{2\pi}\int_0^{2\pi/\omega}\int_0^{\infty}\rho_1 c_1 T_1(t,x,z)\cdot\left(\frac{\omega S}{2}sin\,\omega t\right)\cdot\pi D dx\cdot dt \tag{11}$$

where the parentheses denotes the velocity of displacer in the negative z direction and ρ and c denote density and specific heat of the walls, respectively. The result of integrating Eq. (11) is the new expression for shuttle heat transfer.

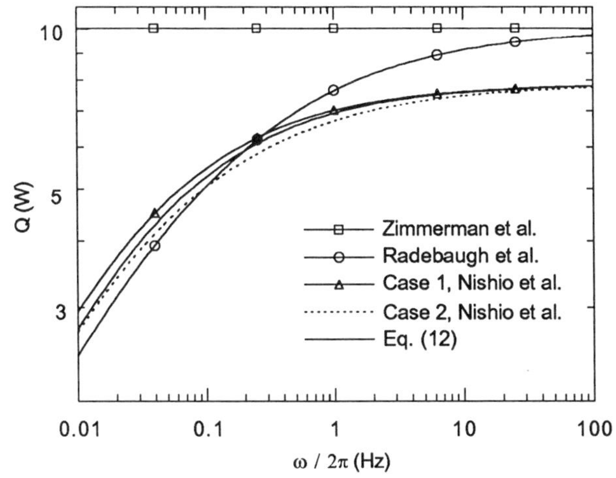

Figure 2. Comparison of current and existing expressions for shuttle heat transfer rate as a function of frequency.

$$Q = \frac{T_H - T_L}{L} \frac{\pi D S^2 h}{8} \frac{b_1 + b_2 + 1}{(b_1 + b_2)^2 + (b_1 + b_2 + 1)^2} \tag{12}$$

It can be noted that the shuttle heat transfer rate is independent of z.

RESULTS AND DISCUSSION

The new expression, Eq. (12), could be compared with the existing data for a specific situation. The materials and the physical dimensions for this calculation are given in Table 1. The shuttle heat transfer rates were calculated for a variable frequency of oscillation by using the existing methods and the new method. In all cases, the heat transfer coefficient is obtained by

$$h = \frac{k_g}{\delta} \tag{13}$$

where k_g and δ are the thermal conductivity of gas and the gap between the walls, respectively.

Figure 2 shows the shuttle heat transfer as a function of the frequency of oscillation for four different cases. The case 1 and the case 2 of Nishio et al. indicate the results without and with temperature oscillation of the cylinder, respectively. The shuttle heat transfer from Zimmerman et al. is constant and much over-estimated due to the infinite heat capacity of the two walls. The result from Radebaugh et al. has greater values of shuttle heat transfer at high frequencies and lower values at low frequencies . The result from the case 1 of Nishio et al. has greater values than those from the case 2 and from Eq. (12). The cause for the difference is very obvious; the phase shift from the temperature oscillation of the displacer to that of the cylinder was over-estimated due to the heat capacity of the cylinder being infinite.

A fairly good agreement is observed between the case 2 of Nishio et al. and Eq. (12). The slight difference is due to the assumption of Nishio et al. that the phase shift of the surface temperature oscillation is neglected. Clearly, Eq. (12) is much more useful than

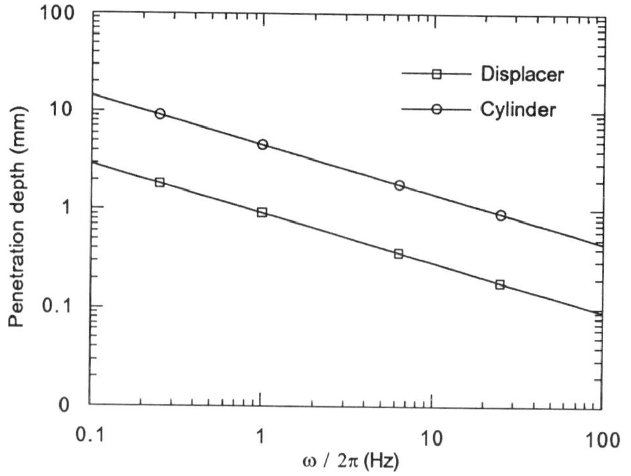

Figure 3. Thermal penetration depths of displacer and cylinder as a function of frequency.

Nishio et al., because the former is an exact solution that satisfies the governing equations and the boundary conditions while the latter is an approximate solution with several simplifying assumptions. In addition to the accuracy, it is also obvious that Eq. (12) is simpler to use.

As mentioned earlier, Eq. (12) is based on the model of two semi-infinite bodies. The validity of the model should be checked by seeing whether the thermal penetration depth is smaller than the thickness of the wall. Figure 3 shows the thermal penetration depth as a function of frequency. In the calculation, the same dimensions given in Table 1 were used, and the thermal penetration depth is defined as the distance from the surface to the point where the amplitude of the temperature oscillation is 1% of the amplitude of the surface temperature oscillation. As can be seen in Eq. (7), the penetration depth is proportional to the square root of α / ω. In this specific problem, the displacer penetration depth is always smaller than the wall thickness for all frequencies. On the other hand, the cylinder penetration depth is smaller than the wall thickness only when the frequency is greater than about 1.2 Hz in the specific situation.

Figure 4 shows the effect of the frequency and the gap between the walls on the shuttle heat transfer. Generally, the shuttle heat transfer increases as the frequency increases or as the gap decreases. It is observed that at low frequencies the heat transfer is strongly dependent on the frequency but is not much dependent on the gap. It is also observed that at high frequencies the heat transfer is strongly dependent on the gap but is not much dependent on the frequency. Quantitatively, the first observation is explained by an asymptotic expression of Eq. (12) for small ω or for small δ.

Table 1. Specifications for sample calculation

Displacer (Bakelite)	Diameter, D	125 mm
	Stroke, S	25 mm
	Thickness	3 mm
Cylinder (Stainless Steel)	Thickness	4 mm
	Temperature gradient	1 K / mm
Gas (Helium)	Temperature	160 K
	Gap, δ	0.4 mm

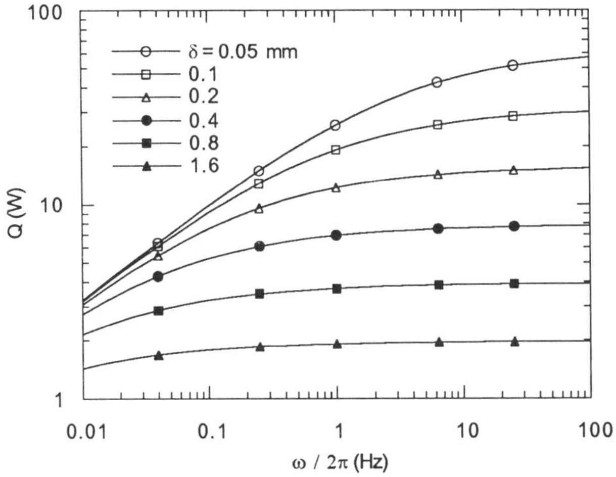

Figure 4. Shuttle heat transfer rate as a function of frequency for various gaps between the walls.

$$Q = \frac{T_H - T_L}{L} \frac{\pi D S^2}{8\sqrt{2}} \frac{\sqrt{\omega}}{\dfrac{1}{\sqrt{k_1 \rho_1 c_1}} + \dfrac{1}{\sqrt{k_2 \rho_2 c_2}}} \tag{14}$$

In words, the shuttle heat transfer is proportional to the square root of the frequency and independent of the gap when the Biot numbers defined by Eq. (9) and (10) are much greater than 1 or $b_1 + b_2 >> 1$. Eq. (14) is useful because it shows explicitly the effects of the wall properties on the heat transfer. Similarly, the second observation is explained by an asymptotic expression of Eq. (12) for large ω or for large δ.

$$Q = \frac{T_H - T_L}{L} \frac{\pi D S^2}{8} \frac{k_g}{\delta} \tag{15}$$

The shuttle heat transfer is independent of the frequency and inversely proportional to the gap when the Biot numbers are much smaller than 1 or $b_1 + b_2 << 1$. It can be noted that Eq. (15) is the same form of expression by Zimmerman and Longthworth[1] except the constant, because the thermal capacities of the two walls were assumed to be infinite in this extreme case.

CONCLUSIONS

A new analytical expression was presented for the shuttle heat transfer rate, including the wall properties of both the displacer and the cylinder. The exactness of the new expression was verified. The effects of the oscillating frequency, the gap between the walls, and the wall properties on the amount of heat transfer were discussed. Most of the phenomena were qualitatively mentioned in the previous works but could be quantitatively and precisely described via the expression. The results are expected to be directly applicable to many cryogenic systems where the shuttle heat transfer is significant.

REFERENCES

1. F.J. Zimmerman and R.C. Longthworth, Shuttle heat transfer, "Advances in Cryogenic Engineering," Vol. 16, Plenum Press, New York (1970) pp.342-351.
2. P. A. Rios, An approximate solution to the shuttle heat transfer losses in a reciprocating machine, *ASME J of Eng for Power*, 4:177 (1971).
3. R. Radebaugh and J.E. Zimmerman, Shuttle heat transfer in plastic displacer at low speeds, NBS Special Publication 508 (1978) pp.67-73.
4. W.R. Martini, "Stirling Engine Design Manual," DOE/NASA/3194-1, NASA CR-168088 (1983) p.112 .
5. S. Nishio, T. Inada and H. Nakagome, Shuttle heat transfer in refrigerators, "Proceedings of 6th International Symposium on Transport Phenomena in Thermal Engineering," Vol.4 (1993) pp.165-170.
6. J.H. Baik and H.-M. Chang, An exact solution for shuttle heat transfer, *Cryogenics*, 35:9 (1995).

OSCILLATING FLOW MODELLING OF A STIRLING CYCLE CRYOCOOLER

Z. F. Sun and C. G. Carrington

Physics Department
University of Otago
Dunedin, New Zealand

ABSTRACT

The operating behavior of an Oxford cryocooler has been modelled using the SIMPLER algorithm of Patankar, and the implicit correction scheme by the present authors, to solve the balance equations.

Kornhauser and Smith's complex Nusselt number correlation for heat transfer is used in the compression and expansion spaces. Armour and Cannon's friction factor and Kays and London's heat transfer correlations are used in the regenerator and connecting pipe. However it has been necessary to modify these correlations using correction factors, which take account of the effects of oscillating flow, constructed on the basis of experimental data. Comparisons with measurements show that the correction correlations can be used successfully to evaluate the friction force and heat transfer under operating conditions and that the model can be used to predict the operating behavior of the cryocooler.

INTRODUCTION

To establish a quantitative theoretical analysis of the performance of a Stirling cryocooler, the mass, momentum and energy balance equations should be solved using appropriate correlations for the heat transfer coefficients and fluid friction factors. Yuan et al[1,2] have recently developed such a model using the correlation of Kornhauser and Smith[3] for heat transfer in the compression and expansion spaces. This correlation, which is expressed in terms of a complex Nusselt number, explicitly includes the influence of oscillating flow. It is well accepted that the limited displacement of fluid elements and the acceleration due to oscillating flow, should have a significant effect on flow friction and heat transfer in Stirling cryocoolers[4–6]. For the regenerator, Yuan et al[1,2] used a correlation obtained from the data of Kays and London[7] for heat transfer, together with the corresponding friction factor obtained by Armour and Cannon[8]. However these correlations, which were derived from steady flow measurements, do not take any account of oscillation effects.

The purpose of this paper is to present a simulation model for a Stirling cryocooler which includes the influence of oscillating flow in the regenerator and connecting pipe, as well as in the compression and expansion volumes. The model represents an Oxford miniature split Stirling cycle cryocooler[9]. In developing the model, the experimental measurements of Orlowska and Davey[9,10] have been utilized to characterize the effect of flow oscillation on heat transfer and fluid friction in the regenerator and connecting pipe. Throughout this paper the notation is the same as that of Slattery[11] , except where indicated otherwise.

MODEL EQUATIONS AND ASSUMPTIONS

The model provides a one-dimensional representation of gas flows in the expansion space, regenerator, warm-end compression space, connecting pipe and compressor of an Oxford cryocooler[9] . The system is divided into control volumes based on component functions; the system dimensions were provided by the Oxford group. The ideal gas equation is used for the working fluid, helium. The temperature dependence of properties, such as viscosity, conductivity, and heat capacity, of both the working gas and the bronze meshes have been included in the model equations explicitly. The wall temperature of the warm end, connecting pipe, and compressor is fixed at a constant value, 298 K, and the wall temperature of the cold end is determined by operating conditions.

When the gravity force is neglected, the governing equations for the gas, which is in turbulent flow, can be written as[11] :

Mass balance:

$$\frac{d}{dt}\int_{V_{(s)}} \overline{\rho} dV = \int_{S_{(ent\ ex)}} \overline{\rho(\mathbf{v}-\mathbf{v}_{(s)})\cdot(-\mathbf{n})} dS, \tag{1}$$

Momentum balance:

$$\frac{d}{dt}\int_{V_{(s)}} \overline{\rho\mathbf{v}} dV = \int_{S_{(ent\ ex)}} \overline{\rho\mathbf{v}(\mathbf{v}-\mathbf{v}_{(s)})\cdot(-\mathbf{n})} dS - \int_{S_{(ent\ ex)}} \overline{\mathsf{P}^{v}\cdot(\mathbf{n})} dS - \overline{\mathbf{F}_f} - \overline{\mathbf{F}_c}, \tag{2}$$

Energy balance:

$$\frac{d}{dt}\int_{V_{(s)}} \overline{\rho\left(\hat{U}+\frac{1}{2}v^2+\frac{P_0}{\rho}\right)} dV = \int_{S_{(ent\ ex)}} \overline{\rho\left(\hat{H}+\frac{1}{2}v^2\right)(\mathbf{v}-\mathbf{v}_{(s)})\cdot(-\mathbf{n})} dS$$
$$+\int_{S_{(ent\ ex)}} \overline{\mathbf{J}_q'\cdot(-\mathbf{n})} dS + \overline{Q_{cv}} - \int_{S_{(ent\ ex)}} \overline{(\mathsf{P}^{v}\cdot\mathbf{v})\cdot(\mathbf{n})} dS - \overline{W_f} - \overline{W_c}. \tag{3}$$

Here, Q_{cv} is the rate of convective heat transfer to the system across the impermeable surfaces of the system; $\mathbf{F}_f$ is the friction force the fluid exerts on the impermeable portion of the system boundary suface; $\mathbf{F}_c$ is the compression force the fluid exerts on the boundary surface; W_f and W_c denote the friction and compression work done by the gas system on the solid surface and boundary surface respectively; and $\mathbf{J}_q'$ and P^v represent the heat conduction flux and the viscous pressure tensor.

The temperature and velocity of solid members (regenerator matrix and walls) are different from those of the helium gas. By including the effect of the gravity force, the momentum and energy balance equations, similar to Eqns.(2) and (3), have been obtained for the solid members.

Eqns.(1), (2), and (3) and the balance equations for the solid members can be used also for the clearance space between the cold finger and the displacer to model the shuttle heat transfer. However, to do this at the same time as the simulation of the momentum and heat transfer processes in the regenerator, connecting pipe, compression and expansion spaces would greatly complicate the modelling of the cryocooler. In order to give a reasonable estimation of these processes, the simplified method of Zimmerman and Longsworth[12] was used to estimate the rate of the shuttle heat transfer.

The implicit correction scheme proposed by the authors[13] has been used to obtain the difference expressions for the governing balance equations. For the systematic solution of the model equations, the SIMPLER algorithm of Patankar was used[14] . To improve the rate of convergence, the results of previous time steps were used to get the initial approximation for the current step using extrapolation. The implicit scheme permits the use of a relatively coarse spatial grid[13] . In most calculations, we divided the regenerator and connecting pipe into 15 and 5 control volumes respectively. The time step was fixed at 3 degrees of crank angle.

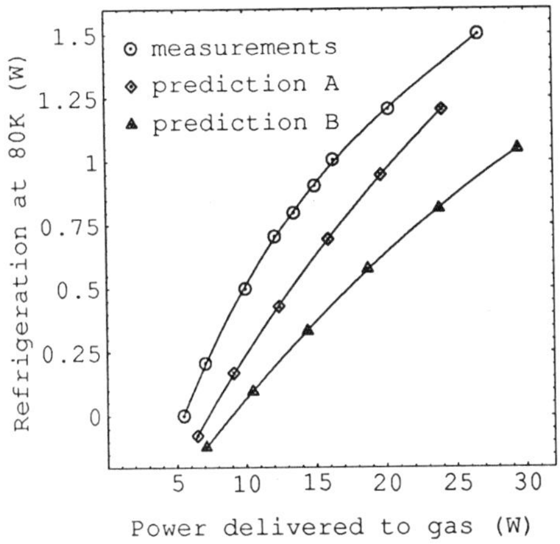

Fig. 1. Comparison of the prediction with the measurements. Prediction A: Armour and Cannon's friction factor and Kays and London's heat transfer coefficient correlations are used. Prediction B: Friction factor correlations are corrected for oscillating flow in the regenerator and connecting pipe. The heat transfer correlations are the same as in A.

HEAT TRANSFER AND FRICTION FACTOR CORRELATIONS

For steady flow in the x-direction[11,15,16] ,

$$(\mathbf{F}_c)_x = \left[\int_{S_{(s)}} [(P-P_0)\mathsf{I})\cdot \mathbf{n}dS\right]_x = [(P-P_0)A]_2 - [(P-P_0)A]_1, \tag{4}$$

$$W_c = \int_{S_{(s)}} [(P-P_0)\mathbf{v}_{(s)}\mathsf{I}]\cdot(\mathbf{n})dS = [(P-P_0)u_bA]_2 - [(P-P_0)u_bA]_1, \tag{5}$$

$$(\mathbf{F}_f)_x = \left[\int_{S_{(s)}-S_{(ent\ ex)}} (\mathsf{P}^v\cdot\mathbf{n})dS\right]_x = f(\frac{1}{2}\rho|u-u_s|^2A_c)\frac{u-u_s}{|u-u_s|}, \tag{6}$$

$$Q_{cv} = \int_{S_{(s)}-S_{(ent\ ex)}} \mathbf{J}_q^{'}\cdot(-\mathbf{n})dS = hA_c(T_s-T), \tag{7}$$

$$W_f = \int_{S_{(s)}-S_{(ent\ ex)}} (\mathsf{P}^v\cdot\mathbf{v})\cdot(\mathbf{n})dS = u_s(\mathbf{F}_f)_x, \tag{8}$$

where the subscripts 'b' and 's' represent the properties of the boundary surface and solid surface respectively. Eqns.(4) to (8) should be modified to take account of oscillating flow on heat and momentum transfer. This is not necessary in the expansion and compression spaces, since the friction force is negligible and the heat transfer correlations suggested by Kornhauser and Smith[3] include the influence of oscillating flow. In the expansion and compression spaces, when $\mathrm{Pe}D_h/X_s > 100$ the power law relationship of Kornhauser and Smith may be used for the real and imaginary parts of the complex Nusselt number, and when $10 < \mathrm{Pe}D_h/X_s < 100$ Lee's model[3] was used instead.

For the regenerator we initially used Armour and Cannon's friction factor correlation[8] , and the following equation for the Stanton number, obtained from Kays and London's data[7] ,

$$\mathrm{St} = 0.63\mathrm{Pr}^{-0.667}\mathrm{Re}^{-0.418}. \tag{9}$$

In the connecting pipe the friction factor and Nusselt number were initially evaluated using well established correlations[7,17,18] for unidirectional flow.

The measured shuttle heat transfer rate and heat conduction rate across the solid members obtained by Orlowska and Davey were used to determine the effective shuttle heat transfer coefficient, the effective conduction coefficient, and the effective conduction heat transfer area of the cold finger and the alloy liner of the displacer.

COMPARISON WITH MEASUREMENTS

To illustrate the effect of using the unidirectional steady flow friction factor and heat transfer correlations in the regenerator and connecting pipe, the model predictions are compared with the measurements by Davey and Orlowska[9,10] .

Figure 1 shows the comparison of measured and calculated refrigeration under the operating conditions: displacer stroke, 3 mm; cold-end temperature, 80 K; filling pressure, 12 bar; mesh size, 250; operating frequency, 38 Hz; phase angle, 90 angle degrees[9] . The calculated power delivered to the working gas is varied by adjusting the compressor stroke. Prediction A in Figure 1 represents the cooling power obtained using Armour and Cannon's friction factor and Kays and London's heat transfer coefficient correlations. It is seen that the calculated cooling power is lower than that measured by approximately 0.3 W, depending on the input power.

Figure 2 shows the frequency dependence of the calculated and measured[10] compressor indicated power when the displacer is kept stationary. The operating conditions are: cold-end temperature, 80 K; filling pressure, 11.5 bar; mesh size, 250. In Figure 2, the parameter E_h represents the enhancement factors for the heat transfer coefficients in the compressor, which will be discussed in detail below. It is seen that there are fundamental differences between the model results ($E_h = 1$) and the measurements. First the measured compressor power varies linearly with frequency, whereas the calculation predicts the quadratically non-linear dependence. Second the measured compressor power is much larger than that calculated. Similar results were obtained using Kays and London's friction factor correlation in the regenerator.

We believe the reason for the discrepancies shown in Figures 1 and 2 is that the friction factor and heat transfer coefficient correlations are only applicable to unidirectional steady flows.

Under the operating conditions in Figure 2 no refrigeration is being produced by the cryocooler. Hence the input power is mainly used as the driving force for the irreversible heat transfer during the expansion and compression in the compressor and for the irreversible flow friction in the regenerator and connecting pipe. The indicated work of the compressor, W_{comp}, can be calculated using the following equation,

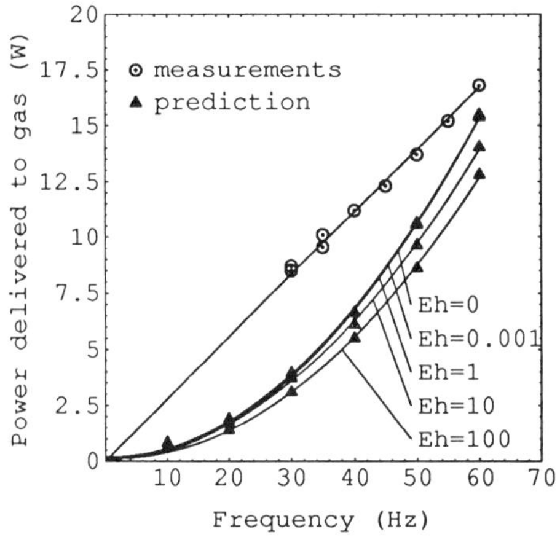

Fig. 2. Indicated compressor power vs. frequency with displacer stationary and 7.1 mm piston stroke. Armour and Cannon's friction factor and Kays and London's heat transfer coefficient correlations are used without correction for oscillating flow.

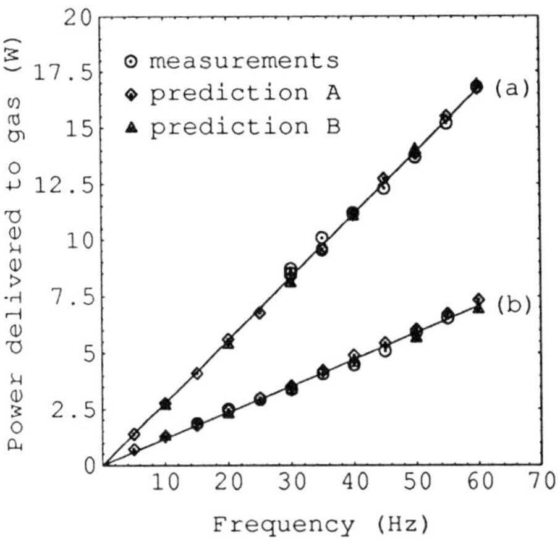

Fig. 3. Indicated compressor power vs. frequency with displacer stationary. Eqns.(11) and (12) are used to correct the friction factors for oscillating flow. Prediction A: Kays and London's heat transfer correlations are used without modification. Prediction B: Heat transfer correlations are corrected using Eqns.(14) and (15). (a) 7.1 mm piston stroke; (b) 4.8 mm piston stroke.

$$W_{comp} = \oint PdV = \int_{V_{min}}^{V_{max}} \Delta PdV = \Delta P_{ave} \Delta V \tag{10}$$

where ΔP is the difference between the pressures during the compression and expansion in the compressor, ΔP_{ave} is the average of ΔP in the sense of the integral mean value theorem, and ΔV is the swept volume. Since ΔV is constant, the discrepancy shown in Figure 2 must be produced by under-estimation of the pressure difference. The main factors which influence ΔP_{ave} are the heat transfer in the compressor and the pressure drop in the regenerator and connecting pipe. Unfortunately no measured heat transfer and pressure drop data of the cryocooler is available to make a comparison of the effect of these two factors. In the following we have tried to resolve this difficulty using the existing measured irreversible compressor power shown in Figure 2.

The effect of oscillating flow on heat transfer in the compressor has been included by using Kornhauser and Smith's complex Nusselt number correlation. The influence of the possible inaccuracy of this correlation has been investigated using enhancement factors for the heat transfer coefficients as illustrated in Figure 2. The difference between the measured and estimated input power for $E_h = 0$ must be the minimum value produced by under-estimation of the friction force. This minimum value must be increased if the heat transfer coefficients are under-estimated. However, it is noted from Figure 2 that the asymptotic value ($E_h = 0$) of the input power will be a good approximation even for a worse case, say $E_h = 100$. Nevertheless, Kornhauser and Smith's complex Nusselt number correlation was validated using the experimental data[3] and was utilized by the previous authors[1,2] . Based on this analysis, we suggest that the discrepancy shown in Figure 2 is produced mainly by under-estimation of the friction force. In order to secure agreement with the measurements of Orlowska and Davey[10] , we propose one empirical correction factor for the friction factors to represent the effects of oscillating flow on the basis of the measured and calculated input power ($E_h = 1$) shown in Figure 2:

$$C_f = \left(\frac{67.6331F}{F^2 + 3.9811} \right)^{1.4863}, \tag{11}$$

where F is the operating frequency. The number, 67.6331, in Eq.(11) can be considered as the characteristic frequency of the cryocooler and the number, 3.9811, is a constant. Hence, C_f is a dimensionless variable. The corrected friction factor, f_c, is given by

$$f_c = C_f f. \tag{12}$$

Eq.(11) has the following feature,

$$C_f = 0 \qquad \text{when } F \to 0 \text{ or } F \to \infty. \tag{13}$$

That is, when the fluid is stationary ($F = 0$) or when it is oscillating at a fixed position ($F = \infty$), the friction factor obtained by Eq.(12) is equal to zero. The correction correlation, Eq.(11), has been established using measured values of F in the range 15 to 60. In the tested range, the magnitude of the correction factor is 9.14 to 1.19.

The validity of Eq.(11) was tested under different operating conditions. Figure 3 shows the frequency dependence of the measured and calculated compressor indicated power obtained using frequency corrected friction factor correlations for the regenerator and connecting pipe. The heat transfer correlations are those of Kays and London without modification. Two cases of different stroke lengths are shown and the calculated indicated input power is in good agreement with the measured values. From Figure 3 it may be inferred that the friction factors are only weakly dependent on the compressor piston stroke, or on the displacement of fluid elements; hence Eq.(11) can be used in the analysis of the cryocooler.

Prediction B in Figure 1 shows the calculated cooling power using the frequency corrected friction factor and Kays and London's heat transfer correlation. We see that the discrepancy between the measured and calculated cooling power is increased by using the corrected friction factors in the regenerator and connecting pipe.

In order to represent the effect of the limited displacement of fluid elements or the compressor stroke on heat transfer, an empirical correlation to correct Kays and London's heat transfer coefficients in the regenerator and connecting pipe is proposed. This correction factor, which is based on the results of prediction B and the measured data shown in Figure 1, is,

$$C_h = 1.0 + \left(\frac{6.6811 X_s}{X_s^2 + 4.3741}\right)^{1.8864}. \tag{14}$$

Here X_s is the compressor stroke in mm. The number, 6.6811, in the equation above can be considered as the characteristic piston stroke of the compressor and the number, 4.3741, is a constant. The corrected Nusselt number, Nu_c, takes the form,

$$\mathrm{Nu}_c = C_h \mathrm{Nu}. \tag{15}$$

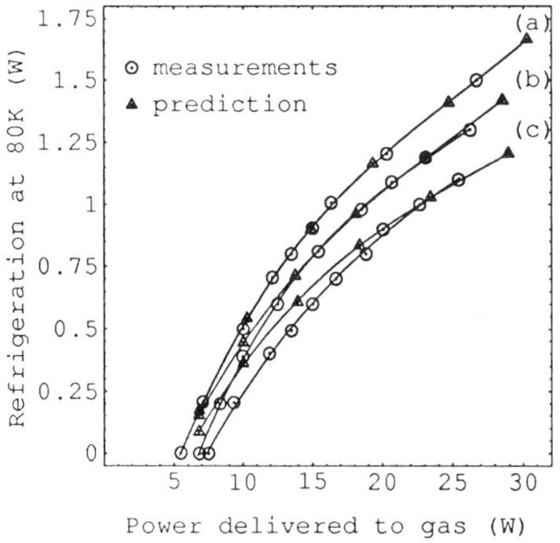

Fig. 4. Variation of net cooling power with compressor indicated power. Displacer stroke 3 mm, Cold end temperature 80 K, $\phi = 90°$. Eqns.(11), (12), (14) and (15) are used. (a) Filling pressure 12 bar, Frequency 38 Hz, Mesh size 250; (b) 12 bar, 38 Hz, 150; (c) 10 bar, 45 Hz, 150.

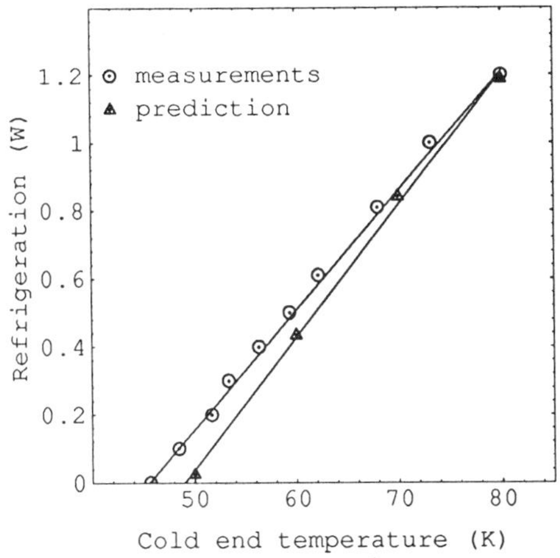

Fig. 5. Influence of cold end temperature on net cooling power at a constant input power of 32.5 W. Displacer stroke 3 mm, Filling pressure 10 bar, Mesh size 250, Frequency 40 Hz, $\phi = 90°$. Eqns.(11), (12), (14) and (15) are used.

The correction factor C_h has the following property,

$$C_h - 1 \qquad \text{when } X_s \to 0 \text{ or } X_s \to \infty. \tag{16}$$

It is seen from Eqns.(14) and (15) that heat transfer in the regenerator and connecting pipe is enhanced by the oscillating flow. Eq.(16) indicates that this enhancement vanishes for static heat conduction ($X_s \to 0$) and for unidirectional steady flows ($X_s \to \infty$). The correction correlation, Eq.(14), has been established using measured values of X_s in the range 3 to 8 mm. In the tested range, the magnitude of the correction factor, C_h is 3.15 to 1.63.

The use of the frequency corrected friction factors was re-examined by using the displacement corrected Nusselt number correlations for heat transfer, as shown by prediction B in Figure 3. Here it is demonstrated that heat transfer in the regenerator and connecting pipe has negligible effects on the friction force or pressure drop in the regenerator and connecting pipe.

Figure 4 shows the comparison of measured and calculated results obtained using the correction factors, Eqns.(11) and (14), for several mesh sizes and cyclic frequencies. Figure 5 shows the measured and calculated cooling power plotted against the cold end temperature. The motor efficiency is assumed to be 70%[9] .

It is seen from Figures 3, 4, and 5 that the model results agree well with the measurements, indicating that the proposed correction correlations, Eq.(11) and Eq.(14), are reasonable to represent the effect of oscillating flow on the friction force and heat transfer in the cryocooler. However curve (c) in Figure 4 appears to indicate that the correction correlation for the heat transfer coefficients should also include the effect of the cyclic frequency. We have not investigated this further due to the lack of sufficient relevant data.

CONCLUSION

A simulation model has been developed for an Oxford cryocooler using the SIMPLER algorithm by Patankar and the implicit correction scheme by the present authors to solve the balance equations.

When friction factor and heat transfer correlations derived from unidirectional flow data are used in the regenerator and connecting pipe, there are fundamental differences between the model results and measurements. Accordingly, based on the measurements of Orlowska and Davey[9,10] , two correction factor correlations, Eqns.(11) and (14), for the friction factors and for heat transfer coefficients have been proposed. The corrected friction factors are strongly dependent on the cyclic frequency and the corrected heat transfer coefficients are mainly dependent on the limited displacement of fluid elements.

The dimensionless correction factors, C_f and C_h, may be understood to be functions of the dimensionless frequency and relative fluid dislacement amplitude, as defined by Seume and Simon[5,6]. Here the correlations for C_f and C_h, Eqns.(11) and (14), have been expressed in terms of the cyclic mean properties, frequency F and compressor stroke X_s. More generally it may be necessary to use instantaneous local parameters to construct correction correlations for the friction factor and heat transfer coefficient. These parameters could include the instantaneous input indicated power, the local temperature difference, the local heat transfer rate, and the local fluid displacement[3-6]. However, comparisons of the model results with the measurement data shown in Figures 3, 4, and 5 indicate that the correction factor correlations for the friction factors and heat transfer coefficients are valid for the cyclic operation prediction of the system, such as the cyclic overall refrigeration and input indicated power.

REFERENCES

1. S.W.K. Yuan, and I.E. Spradley, A third order computer model for Stirling refrigerators, Advances in Cryogenic Engineering, Vol. 37, Part B, Edited by R. W. Fast, Plenum Press, New York, pp.1055-1062, (1992).
2. S.W.K. Yuan, I.E. Spradley, and W.G. Foster, Validation of the Stirling refrigerator performance model against the Oxford refrigerator, Advances in Cryogenic Engineering, Vol. 39, Edited by P. Kittel, Plenum Press, New York, pp.1359-1365, (1994).
3. A.A. Kornhauser, and J.L. Smith, Jr., Application of a complex Nusselt number to heat transfer during compression and expansion, On Flows in Internal Combustion Engines-IV, FED-Vol. 65, Edited by T. Uzkan, The American Society of Mechanical Engineers, New York, pp.1-8, (1988).
4. A.A. Kornhauser, and J.L. Smith, Jr., Heat transfer with oscillating pressure and oscillating flow, Proceedings 24th Intersociety Energy Conversion Conference, August, pp.2347-2353, (1988).
5. J.R. Seume, and T.W. Simon, Flow oscillation effects in tubes and porous material: unresolved issues, Symposium on Fluid Flow and Heat Transfer in Reciprocating Machinery, ASME Winter Annual Meeting, FED-Vol. 62(HTD-Vol. 93), Boston, pp.55-62, (1987).
6. J.R. Seume, L.F. Goldberg, and T.W. Simon, Description of an oscillating flow test program, Proc. 22nd IECEC, paper #879029, (1987).
7. W.M. Kays, and A.L. London, Compact Heat Exchanger, 2nd ed., McGraw-Hill Book Company, New York (1964).
8. J.C. Armour, and J.N. Cannon, Fluid flow through woven screens, AIChE Journal, Vol. 14, No. 3, pp.415-420, (1968).
9. G. Davey, and A.H. Orlowska, Miniature Stirling cycle cooler, Cryogenics, Vol. 27, pp.148-151, (1987).
10. A.H. Orlowska, and G. Davey, Measurement of losses in a Stirling cycle cooler, Cryogenics, Vol. 27, pp.645-651, (1987).
11. J.G. Slattery, Momentum, Energy, and Mass Transfer in Continua, 2nd ed., Robert E. Krieger Publishing Company, Inc., New York (1981).
12. F.J. Zimmerman, and R.C. Longsworth, Shuttle heat transfer, Advances in Cryogenic Engineering, Vol.16, pp.342-351, (1970).
13. Z.F. Sun, and C.G. Carrington, A new numerical scheme for convective dominated heat transfer in a porous medium with strong temperature gradients, Transport Phenomena in Porous Media, in press, (1995).
14. S.V. Patankar, Numerical Heat Transfer and Fluid Flow, McGraw-Hill Book Company, New York (1980).
15. R.B. Bird, W.E. Stewart, and E.N. Lightfoot, Transport Phenomena, Wiley, New York (1960).
16. Z.F. Sun, and C.G. Carrington, Exergy balance for flows in complex configurations, Proceedings of ECOS'92 (On Efficiency, Costs, Optimization And Simulation of Energy Systems) Zaragoza, Spain, June, pp.529-535, (1992).
17. W.M. Kays, and H.C. Perkins, Forced convection, internal flow in ducts, in Handbook of Heat Transfer, ed. by W.M. Rohsenow, and J.P. Hartnett, McGraw-Hill Book Company, New York (1973).
18. L.C. Thomas, Fundamentals of Heat Transfer, Prentice-Hall Inc., Englewood Cliffs, N.J. (1980).

SIMULATION AND SECOND LAW ANALYSIS OF A MINIATURE STIRLING CYCLE CRYOCOOLER

Z. F. Sun and C. G. Carrington

Physics Department
University of Otago
Dunedin, New Zealand

ABSTRACT

The loss mechanisms of an Oxford Stirling cryocooler are examined using a new simulation model together with second law analysis. The simulation employs a new correlation for heat transfer and friction in oscillating flow. The loss mechanisms have been analysed quantitatively using the exergy balance equation based on non-equilibrium thermodynamics. In particular, the variation of exergy destruction with position and crank angle have been evaluated by integrating the exergy balance equations. The significant exergy destructions in the cryocooler, due to viscous dissipation, heat transfer, mixing processes, shuttle heat transfer and heat conduction in the solid members, are evaluated independently.

INTRODUCTION

Orlowska and Davey[1,2] made extensive measurements on the performance of an Oxford split Stirling cryocooler and analyzed the loss mechanisms using this data on an energy balance basis. While they obtained good energy balances, the losses were measured under operating conditions different from those for the normal operation of the cryocooler. For example, the loss due to irreversible compression and expansion in the system was measured with the displacer stationary while other parameters (cold-end temperature, filling pressure and piston stroke) were kept as in normal operation with the displacer running[2] . In fact, the irreversible power obtained by this method represents only the losses due to the irreversible processes in the cryocooler under these special conditions. This measured loss tends to be lower than that for normal operation, since no refrigeration is being produced under these conditions. These difficulties were noted by Orlowska and Davey[2] . They also recognized a significant discrepancy between the total indicated refrigeration power and the total refrigeration power obtained by adding the useful refrigeration and the refrigeration losses. Clearly, appropriate measurements of the real losses in the cryocooler can be obtained only when the cryocooler operates normally. However, it is very difficult to make measurements of the losses under these conditions because of the synergetic linkages between the processes of the Stirling machine.

In order to resolve these difficulties, and to give a reasonable dissection of the contributing losses in the cryocooler, we have analyzed the cryocooler processes using second law analysis methods. The analysis employs the exergy balance equations and exergy destruction formulations based on non-equilibrium thermodynamics[3-5]. By using these equations, we are able not only to determine the location of the exergy losses, but also to identify the causes for the exergy destruction. Since this analysis can not be applied to measured data directly, the simulation model developed by the present authors for the Oxford cryocooler[6] has been used to provide a basis for the second law analysis. In this model, the effect of oscillating flow in the cryocooler has been considered by using correction factors to modify friction and heat transfer correlations derived from unidirectional flow data.

In this paper, the second law analysis results of the Oxford cryocooler under typical operating conditions are presented. This provides a basis for further investigation of the optimum operation and design of the cryocooler.

BALANCE EQUATIONS

The simulation model provides a one-dimensional representation of gas flows in the expansion space, regenerator, warm-end compression space, connecting pipe and compressor of an Oxford cryocooler[1]. The system is divided into control volumes based on component functions; the system dimensions were provided by the Oxford group. The ideal gas equation is used for the working fluid, helium. The temperature dependence of properties, such as viscosity, conductivity, and heat capacity, of both the working gas and the bronze meshes have been included in the model equations explicitly.

When the gravity force is neglected, the governing equations for the gas, which is in turbulent flow, can be written as[6,7]:

Mass balance:

$$\frac{d}{dt}\int_{V_{(s)}} \bar{\rho} dV = \int_{S_{(ent\ ex)}} \overline{\rho(\mathbf{v}-\mathbf{v}_{(s)})\cdot(-\mathbf{n})} dS, \tag{1}$$

Momentum balance:

$$\frac{d}{dt}\int_{V_{(s)}} \overline{\rho\mathbf{v}} dV = \int_{S_{(ent\ ex)}} \overline{\rho\mathbf{v}(\mathbf{v}-\mathbf{v}_{(s)})\cdot(-\mathbf{n})} dS - \int_{S_{(ent\ ex)}} \overline{\mathsf{P}^{v}\cdot(\mathbf{n})} dS - \overline{\mathbf{F}_{v}} - \overline{\mathbf{F}_{c}}, \tag{2}$$

Energy balance:

$$\begin{aligned}\frac{d}{dt}\int_{V_{(s)}} \overline{\rho\left(\hat{U}+\frac{1}{2}v^2+\frac{P_0}{\rho}\right)} dV &= \int_{S_{(ent\ ex)}} \overline{\rho\left(\hat{H}+\frac{1}{2}v^2\right)(\mathbf{v}-\mathbf{v}_{(s)})\cdot(-\mathbf{n})} dS \\ &+ \int_{S_{(ent\ ex)}} \overline{\mathbf{J}_q'\cdot(-\mathbf{n})} dS + \overline{Q_{cv}} - \int_{S_{(ent\ ex)}} \overline{(\mathsf{P}^{v}\cdot\mathbf{v})\cdot(\mathbf{n})} dS - \overline{W_v} - \overline{W_c}. \end{aligned} \tag{3}$$

In addition to the mass, momentum, and energy balance equations above, the exergy balance equation for the working gas in a system in turbulent flow can be written as[4]

$$\begin{aligned}\frac{d}{dt}\int_{V_{(s)}} \overline{\rho b_{Tn}} dV &= \int_{S_{(ent\ ex)}} \overline{\rho b_{Ts}(\mathbf{v}-\mathbf{v}_{(s)})\cdot(-\mathbf{n})} dS + \int_{S_{(ent\ ex)}} \overline{\left(1-\frac{T_0}{T}\right)\mathbf{J}_q'\cdot(-\mathbf{n})} dS \\ &+ \overline{B_{q,cv}} + \int_{S_{(ent\ ex)}} \overline{(-\mathsf{P}^{v}\cdot\mathbf{v})\cdot\mathbf{n}} dS - \overline{W_v} - \overline{W_c} - \overline{\dot{D}_g}. \end{aligned} \tag{4}$$

In Eqns.(1)-(4), Q_{cv} is the rate of convective heat transfer to the system across the impermeable surfaces of the system; $\mathbf{F}_v$ is the viscous force the fluid exerts on the impermeable portion of the system boundary surface due to viscous flow and flow obstacles; $\mathbf{F}_c$ is the compression force the fluid exerts on the boundary surface; W_v and W_c denote the viscous and compression work done by the gas system on the solid surface and boundary surface respectively; $B_{q,cv}$ is the heat exergy current across the impermeable surfaces of the system; $\mathbf{J}_q'$ and P^v represent the heat conduction flux and the viscous pressure tensor.

The temperature and velocity of solid members (regenerator matrix and walls) are different from those of the helium gas. By including the effect of the gravity force, the momentum, energy, and exergy balance equations, similar to Eqns.(2), (3) and (4), have been obtained for the solid members.

The implicit correction scheme proposed by the authors[9] has been used to obtain the difference expressions for the governing balance equations. For the systematic solution of the model equations, the SIMPLER algorithm of Patankar was used[10].

For steady flow in the x-direction[4,8],

$$
\begin{aligned}
(\mathbf{F}_c)_x &= [(P - P_0)A]_2 - [(P - P_0)A]_1, && (5)\\
W_c &= [(P - P_0)u_{(s)}A]_2 - [(P - P_0)u_{(s)}A]_1, && (6)\\
(\mathbf{F}_v)_x &= (\mathbf{F}_{v,f})_x + (\mathbf{F}_{v,o})_x \\
&= f(\frac{1}{2}\rho|u - u_s|^2 A_c)\frac{u - u_s}{|u - u_s|} + e_v(\frac{1}{2}\rho|u - u_s|^2 A)\frac{u - u_s}{|u - u_s|}, && (7)\\
Q_{cv} &= hA_c(T_s - T), && (8)\\
B_{q,cv} &= Q_{cv}\left(1 - \frac{T_0}{T_s}\right), && (9)\\
W_v &= u_s(\mathbf{F}_{v,f})_x + u_s(\mathbf{F}_{v,o})_x, && (10)
\end{aligned}
$$

where f, e_v, and h represent the friction factor due to fluid viscousity, the friction loss factor due to flow obstacles, and the heat transfer coefficient respectively.

The complex Nusselt number correlation for heat transfer suggested by Kornhauser and Smith[11], which has included the influence of oscillating flow, is used in the compression and expansion spaces. The approximate correlations[8] for sudden changes in cross-sectional area are used for friction loss factor. Armour and Cannon's friction factor[12] and Kays and London's heat transfer correlations[13,14] are used in the regenerator and connecting pipe. However, since Armour and Cannon's and Kays and London's correlations were derived from unidirectional flow data, it has been necessary to modify these correlations using correction factors, which take account of the effects of oscillating flow, constructed on the basis of experimental data[6].

In the working gas side of the cryocooler, several irreversible processes contribute to the exergy destruction rate, $\dot{D}_g$. These contributions are: viscous flow of the working gas, $\dot{D}_{v,f}$; flow obstacles and sudden changes in cross-sectional area in the system, $\dot{D}_{v,o}$; convective heat transfer from the solid surfaces to the working gas, $\dot{D}_{q,cv}$; axial heat conduction in the working gas, $\dot{D}_{q,cd}$; and mixing of the working gas elements with different temperatures, $\dot{D}_m$. For a unidimensional flow in the x-direction, these contributions except for $\dot{D}_m$ can be evaluated by using the following equations[4],

$$\dot{D}_{v,f} = T_0(\mathbf{F}_{v,f})_x\left(\frac{u - u_s}{T_f}\right), \tag{11}$$

$$\dot{D}_{v,o} = T_0(\mathbf{F}_{v,o})_x\left(\frac{u - u_s}{T_f}\right), \tag{12}$$

$$\dot{D}_{q,cv} = T_0 Q_{cv}\left(\frac{T_s - T}{T_f^2}\right), \tag{13}$$

and

$$\dot{D}_{q,cd} = T_0 Q_{cd,x}\left(\frac{\Delta T}{T^2}\right)_x. \tag{14}$$

Since the detail of the mixing processes in the cryocooler is unknown, we can not evaluate $\dot{D}_m$ directly. However, the total exergy destruction in the gas side, $\dot{D}_g$, can be calculated by using Eq.(4), the exergy balance equation, and also

$$\dot{D}_g = \dot{D}_{v,f} + \dot{D}_{v,o} + \dot{D}_{q,cv} + \dot{D}_{q,cd} + \dot{D}_m. \tag{15}$$

Hence, $\dot{D}_m$ can be determined by using Eq.(4) together with Eqns.(11)-(15)

For the sake of simplicity, the method of Zimmerman and Longsworth[15] was used to evaluate the the shuttle heat transfer rate, $Q_{shuttle}$. On the basis of this theory, the exergy destruction rate due to the shuttle heat transfer in a control volume is given as

$$\dot{D}_{q,shuttle} = T_0 Q_{shuttle} \left(\frac{\Delta T_{wall}}{T_{wall}^2} \right)_x . \tag{16}$$

In the solid members of the cryocooler, the only significant irreversible process is the axial heat conduction from the warm-end to the cold-end through the wall of the cold finger and the alloy liner for the displacer, since the contact between the mesh disks in the regenerator is negligible and the wall temperatures in the warm-end, connecting pipe, and compressor are fixed at a constant, 298 K[6] . Thus, we have the following equation for the exergy destruction rate, $\dot{D}_s$, in the solid members,

$$\dot{D}_s = \dot{D}_{q,cd,wall} = T_0 Q_{cd,wall} \left(\frac{\Delta T_{wall}}{T_{wall}^2} \right)_x . \tag{17}$$

Since the complex Nusselt number correlations suggested by Kornhauser and Smith were used to account for oscillating flow in the compression and expansion spaces[11] , for the application of Eq.(13) an effective temperature difference between the working gas and solid members in these spaces is defined as,

$$(T_s - T)_{eff} = (T_s - T) - \frac{1}{\omega} \frac{\mathrm{Nu}_i}{\mathrm{Nu}_r} \frac{dT}{dt} . \tag{18}$$

where Nu_r and Nu_i are the real and imaginary parts of the complex Nusselt number[11] .

THE PERFORMANCE OF THE CRYOCOOLER

Using the simulation model established by the present authors[6] , the performance of the cryocooler has been investigated, under the operating conditions: displacer stroke, 3mm; cold-end temperature, 80K; filling pressure, 12 bar; mesh size, 250; operating frequency, 38 Hz; phase angle, 90 angle degrees[1,2] .

Figures 1 (a), (b) and (c) show the pressure and volume variations with crank angle and the *P-V* diagrams of the cold-end expansion space, warm-end compression space, and compressor. It is seen from Figure 1 (a) that the cold-end pressure is higher than that of the compressor when the whole system is expanded and lower than that of the

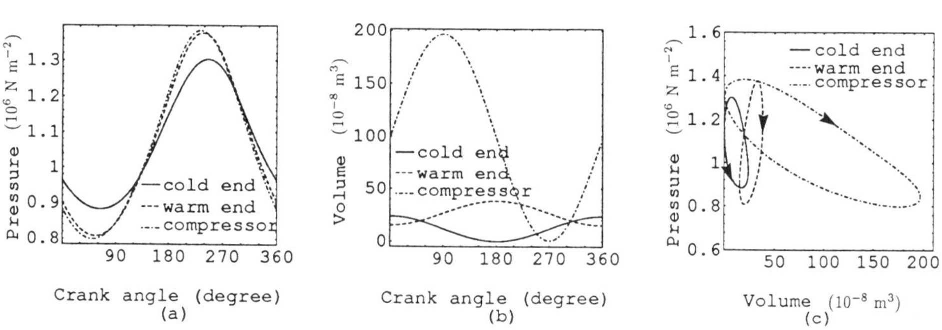

Fig. 1. Pressure and volume variations with crank angle in the cold-end expansion space, warm-end compression space, and compressor. (a) pressure variation; (b) volume variation; (c) pressure-volume diagrams.

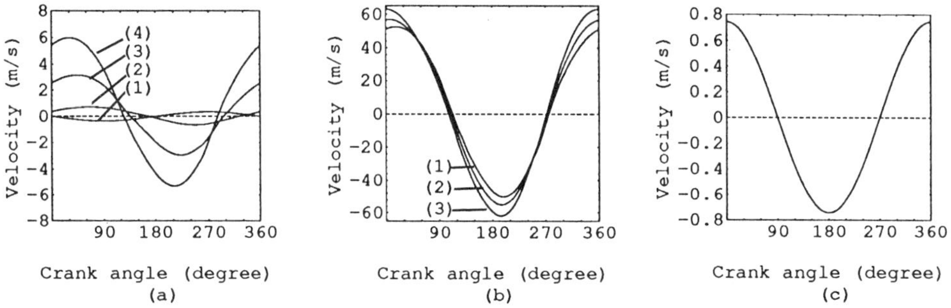

Fig. 2. Velocity variations with crank angle. (a) (1) displacer; (2) working gas in the cold-end of the regenerator; (3) working gas in the middle of the regenerator; (4) working gas in the warm-end of the regenerator; (b) (1) working gas in the warm-end of the connecting pipe; (2) working gas in the middle of the connecting pipe; (3) working gas in the compressor end of the connecting pipe; (c) compressor piston.

compressor when the whole system is compressed. This is different from the assumption that the expansion space pressure is always less than the compressor space pressure by some authors[16] . In Figure 1 (a), the pressure difference between the warm-end and the compressor represents the pressure drop in the connecting pipe, and that between the cold-end and the warm-end denotes the pressure drop in the regenerator. It can be seen that the pressure drop in the regenerator is larger than that in the connecting pipe. Figure 1 (c) indicates that the input indicated power of the compressor and warm-end is much larger than the refrigeration obtained in the cold-end.

Figures 2 (a), (b), and (c) show the velocity variations with crank angle in the regenerator, connecting pipe, and compressor. It can be seen that the working gas velocity decreases very quickly from the compressor to the cold-end, and the velocity propagation from the compressor piston face to the cold-end is delayed by some phase angles from one point to another point. In Figure 2 (a), the velocity difference between the working gas and displacer represents the relative velocity.

Figures 3 (a), (b), and (c) show the temperature variations with crank angle for some typical positions in the regenerator. As shown, the temperature variations in the regenerator are complicated. Here we take Figure 3 (b) as an example to discuss the effect of the working gas velocity, and the compression and expansion of the whole system, on the temperature.

The temperature of the working gas in a control volume is determined by the energy accumulation, energy entering and leaving the control volume by convection and conduction, and heat transfer between the gas and the regenerator, as described by Eq.(3). The working gas temperature is generally lower than that of the meshes in the regenerator during the expansion period and higher than that of the meshes

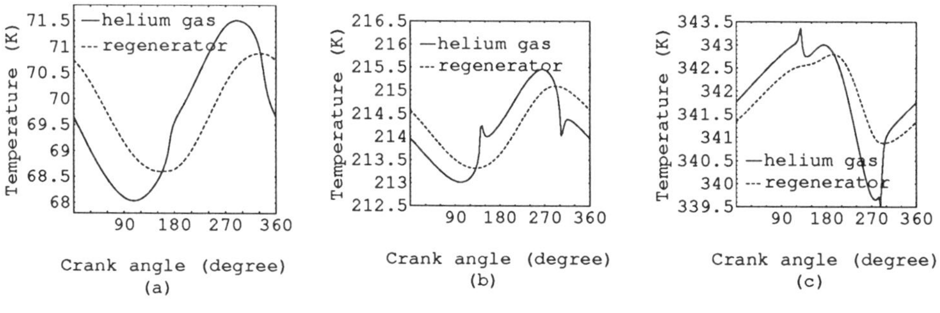

Fig. 3. Temperature variations with crank angle in the regenerator. (a) in the cold-end of the regenerator; (b) in the middle of the regenerator; (c) in the warm-end of the regenerator.

during the compression period. However, because of the velocity delay, the effect of compression and expansion is delayed also from the compression space to the expansion space. During the compression period, the first small peak around 150 degree of crank angle in the temperature curve is due to the very small relative velocity as shown in Figure 2 (a) and hence the small heat transfer between the working gas and the regenerator. After this peak point, the relative velocity increases. Hence, the heat transfer is increased and the gas temperature is decreased. The largest temperature in the temperature-crank angle curve corresponds to the end of the compression period of the compressor as shown in Figure 1 (c). The influence of the compression is much stronger than that of heat transfer around 270 degree of crank angle. During the expansion period, the small valley around 300 degree of crank angle is caused by the small relative velocity and hence the small heat transfer between the working gas and the regenerator. The lowest temperature around 90 degree of crank angle is due to the strong effect of the expansion of the whole system, since the total volume is then maximum.

The working gas temperature in the cold and warm ends of the regenerator is influenced by the cold-end expansion space and warm-end compression space as well as the velocity, and the compression and expansion of the whole system. Due to heat transfer, the temperature of the meshes varies with the temperature of the working gas in a similar way.

The temperature variations in the connecting pipe, which are similar to those in the regenerator shown in Figure 3, are not shown.

Figure 4 shows the temperature of the cold-end space, warm-end space, and compressor. We note here that the gas temperature in the cold-end varies significantly with crank angle as does the gas temperature in the warm-end and compressor, and is not a constant[16] . The wall temperature of the cold finger in the cold-end is fixed at the operating temperature, 80 K. The wall temperatures of the warm-end cylinder and compressor cylinder are also taken to be fixed at 298 K[6] . It is seen that most of the exhausted heat to the environment is transferred through the warm-end and the connecting pipe.

ANALYSIS OF THE EXERGY DESTRUCTIONS

By integrating the exergy balance equations over a time step we can obtain the exergy currents and exergy destructions during this step. The time and space distributions of the exergy destructions caused by different processes are shown in Figure 5. In these figures, control volumes 1, 19, and 25 represent the cold-end expansion space, warm-end compression space, and compressor; control volumes 2 and 18 denote the annular space and connecting ducts which link the regenerator with the cold-end expansion space and warm-end compression space; control volumes 3 to 17, which are uniformly sized, represent the regenerator; and control volumes 20 to 24, which are also uniform, represent the connecting pipe.

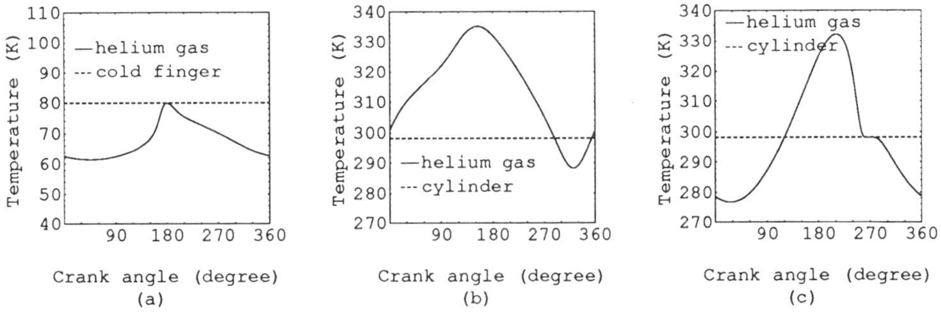

Fig. 4. Temperature variations with crank angle. (a) in the cold-end expansion space; (b) in the warm-end compression space; (c) in the compressor.

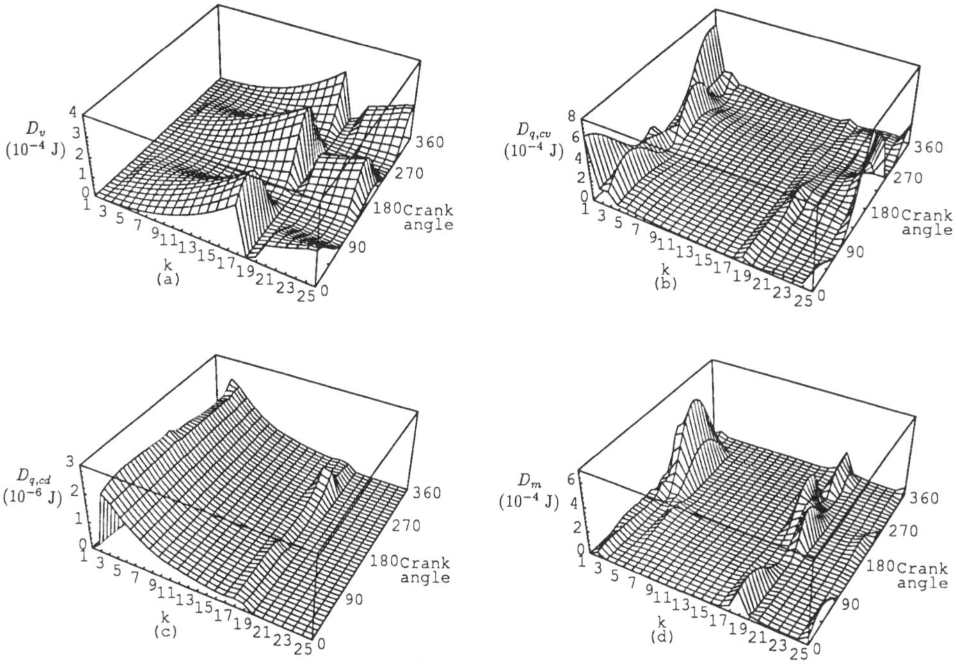

Fig. 5. Time-space distributions of the exergy destructions in the cryocooler. (a) exergy destruction by viscous flow and sudden changes in cross-sectional area; (b) exergy destruction by heat transfer between the working gas and the solid members; (c) exergy destruction by axial heat conduction in the working gas; (d) exergy destruction by mixing processes of the working gas elements with different temperatures.

Compared with losses due to viscous flow, exergy destruction by sudden changes in cross-sectional area in the regenerator and in the connecting pipe is small. Therefore, the exergy destruction by viscous flow and by sudden changes in cross-sectional area are presented accumulatively in Figure 5 (a). In the cold-end space, warm-end space, and compressor, the friction force is neglected. Consequently the exergy losses in these spaces, represented by control volumes 1, 19, and 25 shown in Figure 5 (a), are due to exergy destruction by flow obstacles. It is seen from Figure 5 (a) that the exergy destruction by flow friction is mainly produced in the connecting pipe and in the warm-end region of the regenerator. The reason for this is that the exergy destruction by viscous flow or obstacles is proportional to the cube of the relative velocity, $|u - u_s|$, of the working gas, as described by Eqns.(11) and (12). As shown in Figures 2 (a) and (b), the relative velocity of the working gas in the connecting pipe and in the warm-end region of the regenerator is much larger than that in the cold-end region of the regenerator.

As illustrated in Figure 5 (b), the exergy destruction by heat transfer in the regenerator (control volume 3 to 17) dominates in the cold-end region of the regenerator. The exergy destruction by heat transfer is proportional to the square of the temperature difference and inversely to the square of the film temperature, as described by Eq.(13). From the simulation results shown in Figure 3, we know that the temperature difference in the regenerator increases only slightly from the warm end to the cold end, but the temperatures of the working gas and the meshes in the cold-end region of the regenerator are much lower than those in the warm-end region of the regenerator. The exergy destruction due to heat transfer between the working gas and the walls of the connecting pipe, cold and warm end spaces, and compressor represented by control volume 1, 2, 18 to 25 is also shown in Figure 5 (b). It is seen that the exergy destruction is larger in the cold-end space than in the warm end space and in the connecting pipe.

Table 1. First law and second law analysis results of the cryocooler

	Calculation	Measurement[1,2]
Compressor power, (W)	20.50	20.5
Displacer power, (W)	0.74	<1.0
Useful refrigeration, (W)	1.21	1.2
Exergy of useful refrigeration, (W)	3.31	3.27
Exergy loss due to friction, (W)	6.19	
Exergy loss due to flow obstacles, (W)	0.22	
Exergy loss due to heat transfer (gas and regenerator), (W)	2.18	
Exergy loss due to heat transfer (gas and walls), (W)	4.70	
Exergy loss due to heat conduction in gas, (W)	0.06	
Exergy loss due to mixing, (W)	2.16	
Exergy loss due to shuttle heat transfer, (W)	1.18	
Exergy loss due to heat conduction in solid, (W)	1.29	
First law efficiency, (%)	5.71	5.85
Second law efficiency, (%)	15.57	15.96

The reason is that although the absolute heat flux in the cold-end space is small, the average temperature of the working gas in the cold-end is much lower than in the warm end space and connecting pipe. The exergy destruction due to heat transfer in the compressor is also larger than in the warm-end space and in the connecting pipe. This is due to the lower average temperature and the larger absolute heat transfer flux in the compressor.

The exergy destruction due to axial heat conduction in the working gas is shown in Figure 5 (c). This exergy destruction is mainly produced in the cold-end region of the regenerator, since the gas temperature in the cold-end region is much lower than that in the warm-end region. Compared with the exergy destruction by convective heat transfer, this contribution can be neglected. It is seen that the exergy destruction by axial heat conduction is only weakly dependent on crank angle.

Due to the sharp changes in the flow cross-sectional area and the volume expansions of the cold-end expansion space, warm-end compression space, and compressor, the working gas accumulates in some regions in the cryocooler. This causes the mixing of the working gas elements with different temperatures. Figure 5 (d) shows the exergy destruction due to mixing processes in the cryocooler. It is seen that this mode of exergy destruction occurs mainly in the cold-end expansion space, warm-end compression space, the compressor, and in neighbouring spaces.

The time-space distributions of the exergy destruction caused by the shuttle heat transfer and heat conduction in the cold finger and the alloy liner of the displacer should be similar to that of axial heat conduction in the working gas, and are independent of crank angle. These losses, which have not been drawn here, are dominated by the cold end contributions, similar to the exergy loss due to axial conduction in the working gas.

The results of first law and second law analysis are summarized in Table 1 together with some available measured data[1,2] . This analysis applies to the fixed operating conditions given above; the measured exergy of refrigeration and first and second law efficiency are evaluated on the basis of the useful refrigeration, 1.2 W and the input power 20.5 W, with the displacer power neglected. The significant contributions to exergy destruction in the cryocooler are: flow friction, 35.6%; heat transfer in the regenerator, 12.1%; heat transfer between the gas and the walls, 26.15%, mixing processes, 12.0%; shuttle heat transfer, 6.6%; and axial heat conduction in the solid members, 7.2%.

CONCLUSION

The various contributions to the exergy destruction in the cryocooler have been located and quantified by process under typical operation conditions. This analysis shows the exergy destructions by heat transfer between the working gas and the regenerator, and by heat conduction in the gas and in the solid members, are located primarily in the cold-end region. Exergy destruction by heat transfer between the gas and the solid walls is produced mainly in the cold-end expansion space, warm-end compression space, connecting pipe, and compressor. The exergy loss by viscous dissipation dominates in the warm-end region of the regenerator and in the connecting pipe. Exergy destruction due to mixing processes is important mainly in the cold-end and warm-end spaces, the compressor and their neighbouring spaces. In decreasing order, the significant exergy destructions in the cryocooler are due to viscous dissipation, heat transfer, mixing processes, shuttle heat transfer, and heat conduction in the solid members.

The exergy destruction is determined by both the operating parameters and the geometric configuration of the cryocooler. For example, the relative velocity of the gas, which has a strong influence on the exergy destruction by viscous dissipation and heat transfer, is determined by the phase angle between the displacer and the compressor piston. It is evident that a regenerator consisting of graded mesh disks, fine at the cold-end region and coarse at the warm-end region, would have improved the efficiency of the cryocooler. The operating conditions and geometric configuration for the optimum operation and design of the cryocooler will be investigated further.

NOMENCLATURE

A	cross-sectional area for fluid flow (m^2)
A_c	contact area between fluid and solid (m^2)
B	steady-flow exergy current ($J\ s^{-1}$)
b_{Tn} (b_{Ts})	specific total non-flow (steady-flow) exergy function ($J\ kg^{-1}$)
D	exergy destruction (J)
e_v	friction loss (exergy destruction) factor due to flow obstacles
$\mathbf{F}$	force on solid surfaces (N)
F	cyclic frequency (s^{-1})
f	friction factor
$\hat{H}$	specific enthalpy of fluid, ($J\ kg^{-1}$)
h	heat transfer coefficient ($J\ m^{-2}\ K^{-1}\ s^{-1}$)
$\mathbf{J}_q'$	heat flux, excluding enthalpy of diffusion currents ($J\ m^{-2}\ s^{-1}$)
Nu	Nusselt number for heat transfer
$\mathbf{n}$	unit vector normal to surface, outwardly directed
P^v	viscous part of pressure tensor ($N\ m^{-2}$)
P	thermodynamic pressure ($N\ m^{-2}$)
Q	heat current to the system ($J\ s^{-1}$)
S	surface area (m^2)
T	temperature (K)
t	time (s)
$\hat{U}$	specific internal energy of fluid, ($J\ kg^{-1}$)
u	velocity of unidimensional flow ($m\ s^{-1}$)
V	volume (m^3)
$\mathbf{v}$	velocity vector ($m\ s^{-1}$)
$\mathbf{v}_{(s)}$	velocity vector of closed bounding surface ($m\ s^{-1}$)
v	velocity, $v^2 = \mathbf{v} \cdot \mathbf{v}$, ($m\ s^{-1}$)
W	work done by fluid on solid surfaces ($J\ s^{-1}$)

Greek symbols

ρ	density ($kg\ m^{-3}$)
ω	$2\pi F$ (s^{-1})

Subscripts

0	values for the environmental reservoir
c	compression
cd	conductive heat transfer
cv	convective heat transfer
($ent\ ex$)	the entrance and exit portions of $S_{(s)}$
f	viscous flow friction or film property

g	gas
m	mixing process
o	flow obstacle
q	heat transfer
s	solid surface properties
(s)	closed bounding fluid system
v	viscous
$wall$	solid wall
x	axial direction
$1, 2$	convective entry and exit surface for control volume

Superscripts

$\overline{}$	(over-line) time average quantity in turbulent flow

REFERENCES

1. G. Davey and A.H. Orlowska, Miniature Stirling cycle cooler, Cryogenics, Vol. 27, pp.148-151, (1987).
2. A.H. Orlowska and G. Davey, Measurement of losses in a Stirling cycle cooler, Cryogenics, Vol. 27, pp.645-651, (1987).
3. Z.F. Sun and C.G. Carrington, Application of non-equilibrium thermodynamics in second law analysis, Trans ASME: J. Energy Resources Tech, 113, 33-39, (1991).
4. Z.F. Sun and C.G. Carrington, Exergy balance for flows in complex configurations, Proceedings of ECOS'92 (On Efficiency, Costs, Optimization And Simulation of Energy Systems) Zaragoza, Spain, June, pp.529-535, (1992).
5. C.G. Carrington and Z.F. Sun, Second law analysis of combined heat and mass transfer in internal and external flows, Int. J. Heat and Fluid Flow, Vol. 13, No. 1, 65-70, (1992).
6. Z.F. Sun and C.G. Carrington, Oscillating flow modelling of a Stirling cycle cryocooler, Advances in Cryogenic Engineering, Vol.41, (1995).
7. J.G. Slattery, Momentum, Energy, and Mass Transfer in Continua, 2nd ed., Robert E. Krieger Publishing Company, Inc., New York (1981).
8. R.B. Bird, W.E. Stewart, and E.N. Lightfoot, Transport Phenomena, Wiley, New York (1960).
9. Z.F. Sun, and C.G. Carrington, A new numerical scheme for convective dominated heat transfer in a porous medium with strong temperature gradients, Transport Phenomena in Porous Media, in press, (1995).
10. S.V. Patankar, Numerical Heat Transfer and Fluid Flow, McGraw-Hill Book Company, New York (1980).
11. A.A. Kornhauser and J.L. Smith, Jr., Application of a complex Nusselt number to heat transfer during compression and expansion, On Flows in Internal Combustion Engines-IV, FED-Vol. 65, Edited by T. Uzkan, The American Society of Mechanical Engineers, New York, pp.1-8, (1988).
12. J.C. Armour, and J.N. Cannon, Fluid flow through woven screens, AIChE Journal, Vol. 14, No. 3, pp.415-420, (1968).
13. W.M. Kays and A.L. London, Compact Heat Exchanger, 2nd ed., McGraw-Hill Book Company, New York (1964).
14. W.M. Kays and H.C. Perkins, Forced convection, internal flow in ducts, in Handbook of Heat Transfer, ed. by W.M. Rohsenow, and J.P. Hartnett, McGraw-Hill Book Company, New York (1973).
15. F.J. Zimmerman and R.C. Longsworth, Shuttle heat transfer, Advances in Cryogenic Engineering, Vol.16, pp.342-351, (1970).
16. M.D. Atrey, S.L. Bapat, and K.G. Narayankhedkar, Cyclic simulation of Stirling cryocoolers, Cryogenics, Vol. 30, pp.341-347, (1990).

LOW COST SMALL CRYOCOOLERS FOR COMMERCIAL APPLICATIONS

A. Karandikar and D. Berchowitz

Sunpower, Inc.
6 Byard Street, Athens, Ohio 45701

ABSTRACT

Two models of small free piston cryocoolers with the potential for low cost and high reliability have been developed for emerging commercial applications. Model M223 has been designed for applications ranging from 173 K to 273 K lifting up to 40 W at 223 K. The other, model M77, has been designed for operation from 65 K to 150 K. This unit will lift up to 4 W at 77 K and more than 10 W at 150 K. Both passive and active balance units have been implemented to reduce vibration. Performance data for each model are presented over a wide temperature range for representative machines. A summary of the life and reliability information to date is also included.

INTRODUCTION

The machines described here, the models M223 and M77, are based on a common hardware platform. The entire drive section of the machines is essentially identical. The main differences between the machines are in the design of the cold finger and thermodynamic section. The common hardware approach offers two main advantages, namely, increased sample size for life and reliability assessment and larger production volumes for common components. Another useful result is that modified versions of these machines may be developed for a fraction of the development time and cost of a new machine. Some examples include a coupled version of an M223 which cooled a refrigerator (SOR/F) during Space Shuttle *Discovery's* 10 day mission in February 1994 [1], and another modified M223 installed in a small super insulated domestic refrigerator [2].

The M223 has been optimized to lift 35 W at 223 K with a reject temperature of 313 K and an input of 65 W (Figure 1). The principle application is to cool electronic multichip modules to low temperatures in order to improve clock speeds[3,4]. This application requires a cooler that is highly reliable (> 60,000 hours with no maintenance), low in cost for large volumes (< $100 at 1 million per year), and compact.

The M77 machine has been optimized to lift 4 W at 77 K with a reject temperature of 313 K. Input power is about 90 W. This unit was developed for cooling high temperature superconductivity (HTS) devices (Figure 2). In order for many emerging HTS applications to become commercially viable, a cooler which is highly reliable (> 60,000 hours with no maintenance), compact, and relatively low in cost for medium volumes (< $1500 at 5,000 - 10,000 per year) is required.

To date, 107 machines have been built using the common hardware platform and 10 more M77's are scheduled for completion by the end of August, 1995. While some machines are operating in test environments, a number of others are being used in the development of systems and field trial evaluations.

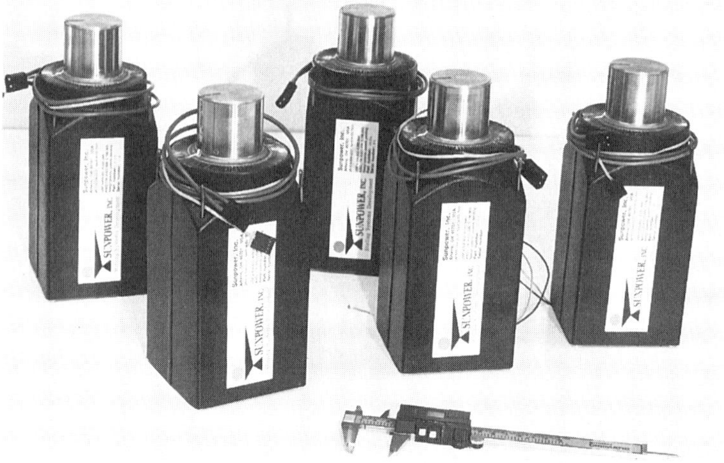

Figure 1. M223 with controller.

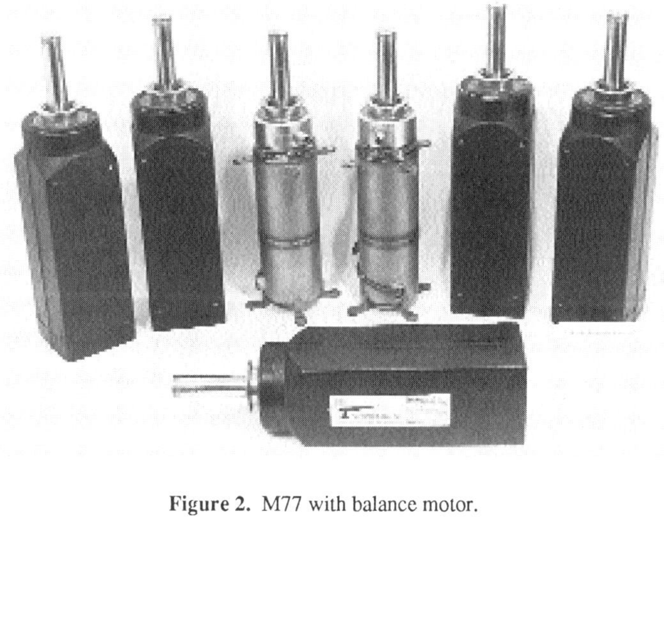

Figure 2. M77 with balance motor.

FEATURES AND SYSTEM DESIGN

Both the M223 and M77 are free piston Stirling cycle machines with the cold finger and compressor assembly integrated into one compact package (integral configuration). The moving components, the piston and displacer, are levitated on gas bearings during operation, thereby virtually eliminating friction and wear. The gas bearings make it possible to achieve long life and high reliability and consume less than 2 W. Additional advantages include modulation by adjusting RMS voltage and balancing of both moving components by the use of only one balancing device.

An electronic controller ramps the machine down from ambient to operating temperature and modulates the machine to maintain cold end temperature. Cold end temperature is adjustable.

A passive balance absorber consisting of a sprung mass is used to reduce casing vibration to within 25 μm amplitude. A balance motor which may be driven by dedicated electronics has been developed for ultra low casing vibration requirements. This results in some additional mass and increased length. The units shown in Figure 2 include this balance motor.

PERFORMANCE DATA

The performance of free piston Stirling coolers may be mapped as shown in Figures 3 and 4. These curves were generated using Sunpower's proprietary Stirling simulation program SAUCE and indicate input and lift as a function of piston amplitude and cold end temperature for a given reject temperature. Figure 4 curves were calibrated at 77 K and tend to underpredict at lower temperatures and overpredict at warmer temperatures.

The maximum input power is limited by either the mechanical limits or the magnetic saturation of the linear motor. When the machine is first turned on, the power consumption for a given piston amplitude is much lower than at low temperatures so the input power is limited by the mechanical stops. As the machine cools down, the input power for a given piston amplitude increases until limited by magnetic saturation of the linear motor.

Sensitivity to reject temperature is shown in Figure 5 for the M77. As may be expected, lift capability is reduced with increasing temperature. The scatter in data is both a function of the variation in performance and instrumentation error. No special care is taken to reduce loading caused by radiation on the cold finger which could be as high as 0.5 W at 77 K.

Figure 6 shows measured COP for the M223, M77, and other custom machines based on the same hardware platform. The rejector temperature in all cases is fixed at 313 K and COP is defined as lift divided by input power to the linear motor. The ideal COP is shown for comparison.

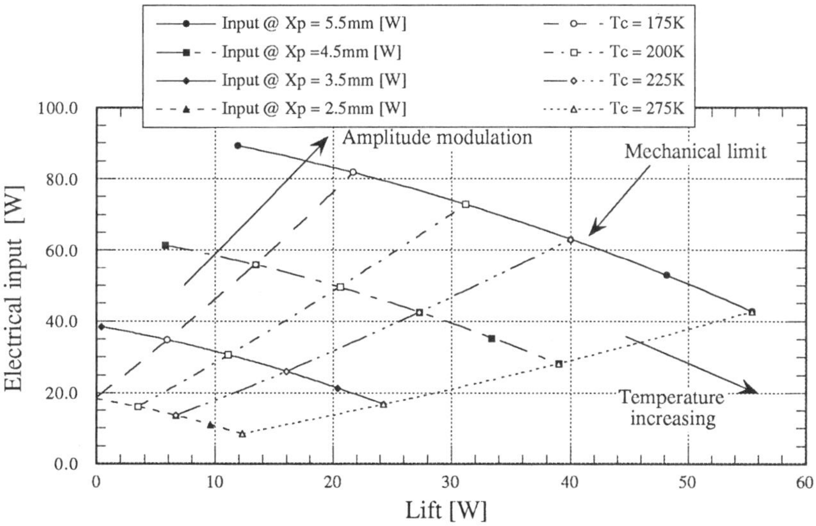

Figure 3. M223 Performance map (xp = piston amplitude, Tc = cold end temperature, T reject = 313 K).

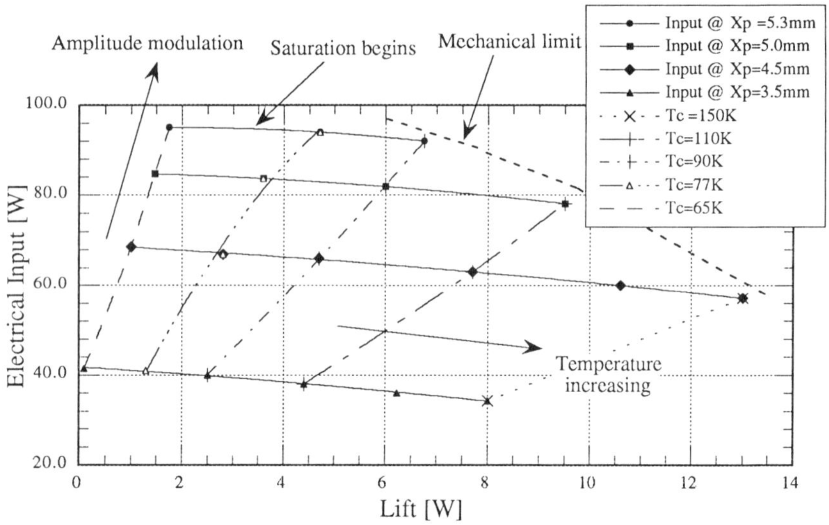

Figure 4. Performance map of M77 (xp = piston amplitude, Tc = cold end temperature, Treject = 313 K).

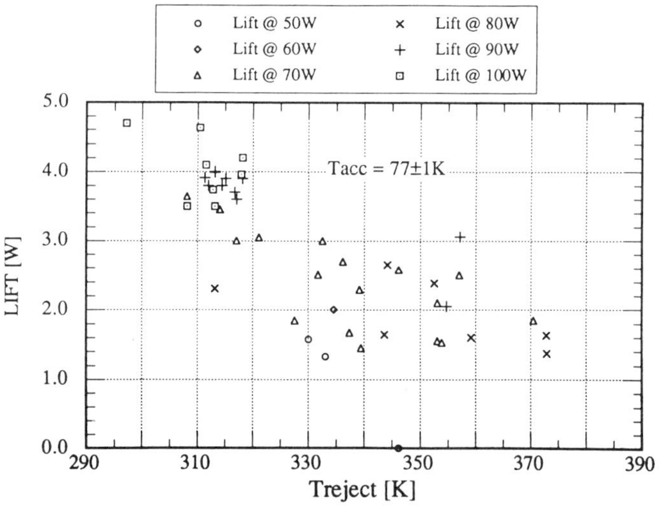

Figure 5. Lift vs reject temperature for M77 at 77 K.

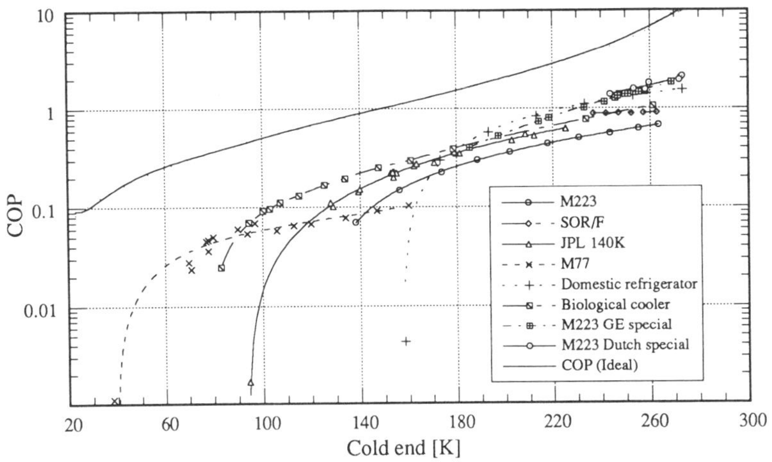

Figure 6. COP vs cold end temperature for different machines at reject temperature of 313 K.

OPERATIONAL CHARACTERISTICS

Tests have been conducted on two machines operating with reject temperatures up to 373 K but testing at low temperature ambients (<288 K) has not been done at Sunpower. No cooler damage is expected over an ambient temperature range of 233 to 373 K. The electronic components used are standard commercial grade and are rated for ambient temperatures from 273-348 K. If required, components are available for operation from 218-398 K.

Custom machines have been designed to tolerate sustained 5g loading (Space Orbiting Refrigerator/Freezer -SOR/F) and momentary 20g without catastrophic destruction. During the Shuttle mission, the SOR/F unit operated during both the ascent and descent stages of the mission. Space vehicle launch vibration testing has been conducted on only one unit (M223); there was no damage or degradation in performance.

EMI tests were conducted by GE Government Services (now Martin Marietta) for space shuttle operation on both the SOR/F custom units and on one M223. These tests showed one exceedence which was not considered to be serious [5]. The main source of EMI radiation was in the power leads to the cooler. This could be reduced by incorporation of line filters.

Voltage surges have not presented problems to date. The electronic drivers deduce the piston amplitude from the back-EMF. Therefore, the drivers may be subjected to voltage variations up to the maximum limit of the electronic components without affecting cooler performance.

LIFE AND RELIABILITY

To date, life and reliability testing has been sparse and incidental. In order to obtain useful life and reliability data, tests need to be conducted on a statistically significant sample size (minimum 100 machines) under appropriate stress conditions. Up to this point, most life data have been anecdotal and gathered through feedback from customers. Some formal reliability testing has been done on the M223.

Since the M223 was initially developed in 1991, a total of 107 machines have been built and delivered, including the M77 and modified versions for various applications and customers. To date, two M77 units have operated continuously for over 12,000 hours and at least two other M77's have operated at customer facilities for over 2000 hours. An M223 operated at Sunpower for 5500 hours at full load with no degradation in performance. A summary of deliveries, failures, and operational hours is shown in Table 1.

Wear is generally considered to be the primary failure mode. In order to investigate the efficacy of the gas bearings to reduce wear, a number of tests have been completed and are included in Table 2. These include rapid stop/start tests, temperature cycling, and machine operation with the gas bearings deactivated. The number of stress cycles induced far exceed

Table 1 Summary of deliveries, failures, and hours on machines

Machine type	Client	Old or new motor?	Number built	Number in operation	# failures	Reasons	Remedy	Hours
M223	Proprietary	old	32	Info. not available	4 known	Magnets delaminated	Redesigned motor/piston	3 units -2500+ 1 units - 100
M223	Various	old	5	Info. not available	-	-	-	Unknown
M223	Proprietary	new	28	Info. not available	-	-	-	Unknown
M223	Proprietary	new	8	8	0	-	-	Unknown
M223	Sunpower	new	1	0	*	*Test apparatus damaged machine	None required	5500
Modified M223 SOR/F	NASA	new	3 opposed pairs	0	1 known	Misalignment of piston	Realigned piston	600+ Survived launch and mission
Modified M223	Mennink	new	2	2	1	Bolt came loose	Eliminated bolt	3600
M77	Proprietary	old	9	Info. not available	Info. not available	-	-	2 units 12,000 + Still running
M77	Proprietary	new	3	Info. not available	Info. not available	-	-	Info. not available
M77B	LLNL	new	2	2	0	-	-	1 unit - 2000+ 1 unit - 1000 +
M77B	NASA Goddard	new	8	8	2	Inadequate water and/or hydrocarbon removal	Better vacuum bakeout procedure	2 units 1880+ to date 2 units 740+ to date 3 units 79-240 to date Reference 7
M77B	UCSB	new	1	1	Info. not available	-	-	Unknown
M77B	LLNL	new	2	2	Info. not available	-	-	Unknown

Table 2 Stress Tests

Machine type	Old or new motor	Stress test type	Number of units	Number of stress cycles Total operation time	Failures
M223	old	Rapid stop/start @1200/hr	4	2.5 million cycles each	1
M223	old	Operational cycling from 293K to 223K @ 12/hour	4	12000 cycles each	1
M223	old	Continuous duty	4	2700+ hours	1
M223	new	Deactivate bearings	1	140 hours	0
M223	new	318K ambient temperature	1	890+ hours ; then shipped to customer	0
M77	new	373K reject temperature	1	50+ hours	0

those seen in normal operation. All 3 failures during these tests were due to magnet delamination and not bearing failure. This failure mode has been corrected with a redesign of the magnet assembly.

Vacuum bakeout procedures have been improved; however, it is unknown whether this procedure is good enough for 60,000 hours of trouble free operation. With the large diameter cold finger, the machines are less susceptible to problems caused by water and other contaminants. There have been two known instances where inadequate bakeout resulted in reduced performance, while in another case the machines have shown no problems in over 12,000 hours of operation.

A large amount of qualitative information has been gathered which has been useful in identifying potential problems. All of the known failures have been dealt with either through design changes or by improved quality control. The number of "infant mortality" issues has dropped to a negligible level. So far there has not been a single gas bearing failure.

MACHINE COST

The M223 and M77 machines were designed to be low cost machines at medium to high volumes. Increasing volume is the primary factor in reducing cost, and the common platform shared by the machines makes it possible to increase volume for a number of parts.

A cost study conducted by General Electric as part of a gas cycle refrigerator project concluded that a 205 W unit of similar proportions could be made for $88 at volumes of 250,000 per year and as low as $30 at volumes of one million per year[6]. Based on this study, using the same labor costs and prorating materials according to mass, an M223 should cost about $20 for volumes of one million per year.

Another cost study for the M223, the details of which are confidential, indicated a manufacturing cost of $363 for volumes of 10,000 per year. These two studies show evidence of meeting the cost targets for the applications under consideration.

CONCLUSION

There are emerging technologies which require low cost, highly reliable cooling machines. The M77 and M223 are two possible candidates for these applications. However, in order to fully realize commercial viability, the following tasks need to be completed:

- obtain statistically significant reliability data
- obtain statistically significant life data
- conduct a more rigorous production and manufacturing evaluation
- establish a medium to large volume market (10,000 to 1,000,000 per year)

The most important of these tasks is to establish a market for medium to high volumes. Work done to date indicates a strong likelihood that cost and performance targets can be achieved.

ACKNOWLEDGEMENT

The following have made significant contributions to the development of these coolers: W. Beale, L. Haas, D. Kiikka, S. McDonald, G. Morris, R. Redlich, and R. Unger.

REFERENCES

1. K. McDonald, D. Berchowitz, J. Rosenfeld, and J. Lindemuth. Stirling refrigerator for space shuttle experiments, *Proc. 29th IECEC*, 4:1807 (1994).
2. D. Mennink and D.Berchowitz. Development of an improved Stirling cooler for vacuum super insulated fridges with thermal store and photovoltaic power source for industrialized and developing countries, *Proc. IIR*, " New Applications of Natural Working Fluids in Refrigeration and Air Conditioning," Hannover (1994).
3. R. Solomon. Opportunities and problems for refrigeration in small computers, *Proc XVIIIth Int. Congr. Refr.* (1991), p. 202.
4. M.J. Deen, C.Y. Chan, and N. Fong. Operational characteristis of a CMOS microprocessor system at cryogenic temperatures, *Cryogenics* 28:336 (1988).
5. K.E. Wahler, Internal Report, GE Government Services, Houston, TX. (June 1992).
6. W.F. Bessler. *Electrical Motor-Driven Non-Phase Change Gas-Cycle Refrigeration; Phase 1: GCR System Study and Design*, Final report, Contract 85X-SJ837C, U.S. Dept of Energy or General Electric Corp. (June 1993).
7. L. Sparr, S. Castles, E. James, and S. Banks. Test results for a commercial, low vibration, linear, stirling cycle cryocooler," *Adv. Cryogenic Engineering* 41: in press (1995).

TEST RESULTS FOR A COMMERCIAL, LOW VIBRATION, LINEAR, STIRLING CYCLE CRYOCOOLER

Leroy Sparr[1], Stephen Castles[1], Edward James[2], and Stuart Banks[2]

[1]Cryogenics, Propulsion and Fluid Systems Branch
NASA/Goddard Space Flight Center
Greenbelt, Md. 20771

[2]McDonnell Douglas Space Systems Co.
Seabrook, Md. 20706

ABSTRACT

NASA/GSFC personnel have evaluated a commercial, low vibration, linear motor driven Stirling cryocooler made by Sunpower Inc. for potential space use on long duration space flight missions. NASA, the U.S. Air Force and industry have invested considerable money during the last decade to develop low vibration mechanical cryocoolers for space applications with typical lifetimes of five years or more. Present space-based cryocoolers typically cost about $1 million per cryocooler. As a potential alternative for one to three year space flight missions, NASA/GSFC has purchased and evaluated eight identical commercial cryocoolers manufactured by Sunpower, Inc. These cryocoolers were designed to have an operational life of greater than 50,000 hours. The cryocooler is an integral Stirling cycle, free piston displacer using linear drive motors for a single compressor and single counterbalancer. Characterization tests include load curves, power measurements, thermal vacuum testing and axial and lateral vibration measurements with and without an active vibration control system. Launch vibration and preliminary life test data will also be presented.

INTRODUCTION

NASA/Goddard Space Flight Center (GSFC) has been actively participating with industry in the development of long life cryocoolers for over twenty five years. This effort has been a combination of contracted cryocooler development, in-house mathematical modeling[1] and in-house testing[2,3]. In addition, Goddard had made a significant contribution to the understanding and implementation of cooler vibration control systems[4,5].

During the last 2 years Sunpower Inc. of Athens, Ohio, has developed inexpensive, single stage, integral Stirling cycle cryocoolers designed for operational lifetimes greater than

50,000 hours. The cryocoolers use a Sunpower proprietary gas bearing/flexure support system to prevent contact between moving parts. NASA/GSFC purchased 8 of these cryocoolers to enter a GSFC testing program designed to investigate the flight worthiness of this cryocooler design for long life space missions.

SUNPOWER CRYOCOOLER & TEST SETUP DESCRIPTION

Sunpower delivered eight cryocoolers to NASA/GSFC during December of 1994, about 8 months after the order was placed. The cryocoolers were an integral Stirling cycle design with one compressor motor and a free piston, moving regenerator expander. At GSFC's request, a second compressor motor was installed as a balancer within the working fluid pressure boundary, in-line with the other moving components. Before shipment to GSFC, each cryocooler was run by Sunpower for about 70 hours to eliminate infant mortality failures.

Sunpower provided a set of laboratory electronics for each cryocooler which were designed to start the machine, cool it down and modulate the stroke as required to maintain a cold tip temperature of approximately 77 Kelvin. Cooldown time to no load temperature is about 10 minutes. The Sunpower supplied electronics did not power the balancer motor and therefore provided no vibration suppression of the cryocooler fundamental or harmonics.

However, the existence of the balancer did permit the cryocoolers to be run with GSFC's vibration suppression electronics. These cryocoolers do not have piston position or fill pressure sensors. Therefore, the GSFC vibration control system closed the loop on the force sensors on the 6-axis force dynamometer on which the Sunpower coolers were mounted.

The Sunpower cryocoolers can provide up to 5 watts of cooling at 80 K with an ambient heat sink temperature and with 100 watts of input power to the compressor. The optimum operating frequency was determined by Sunpower to be 58.6 Hz. All testing at NASA/GSFC was performed at this frequency.

Of the eight cryocoolers delivered, six were air cooled units which included a plastic shroud around the cryocooler and a 4 inch muffin fan attached at one end. A radial copper fin heat exchanger was soldered onto the heat rejection surface of the cryocooler. Ambient

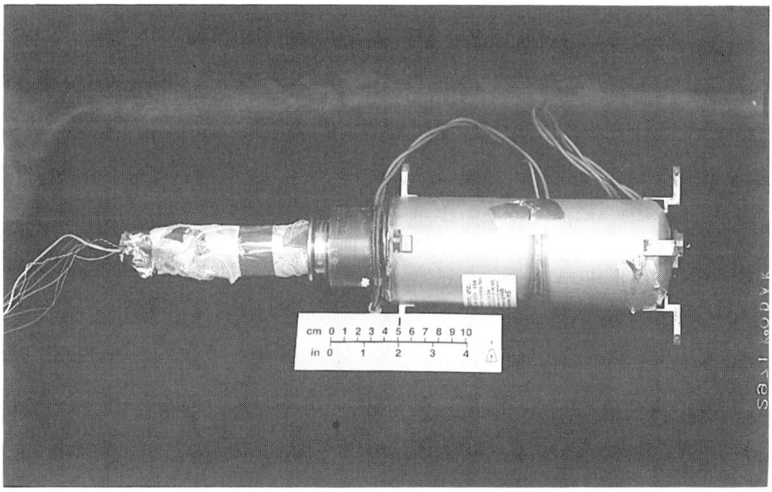

Figure 1. Typical Sunpower 70K Cryocooler.

Table 1. Status of each of the 8 Sunpower Cryocoolers

Cryocooler #	Run time (hrs)	Testing
4	1897	Launch shake, thermal vac
5	1882	Launch shake, thermal vac
6	740	Ambient characterization
7	940	Ambient characterization
8	0	Ambient characterization
11	123	Ambient characterization
12	79	Ambient characterization
13	240	Support capillary pump loop test

air is drawn by the fan between the plastic shroud and the cryocooler pressure boundary and passes through this heat exchanger to cool the machine in a laboratory environment.

GSFC requested that the two remaining cryocoolers to be delivered without the heat exchanger, shroud or muffin fan installed. These two cryocoolers were installed in GSFC designed 6 degree of freedom force dynamometers and installed in the GSFC thermal vacuum cryocooler test bed facility. A GSFC designed liquid heat exchanger was installed on each cryocooler to provide an acceptable heat sink between -40 °C and +30 °C.

The remaining six cryocoolers were also installed in dynamometers of the same design. These cryocoolers also underwent thermodynamic, vibration, and electrical characterization at ambient temperatures and will continue to undergo life testing with periodic re-characterizations. In addition to life testing and characterization efforts, the remaining six cryocoolers will be used for additional analysis and experimentation purposes, such as vibration suppression, cryocooler control, and instrument interfacing studies.

DATA ACQUISITION, ANALYSIS, AND CRYOCOOLER CONTROL

To permit unattended, autonomous operation of the cryocoolers and to insure a high degree of repeatability in thermal and vibration characterization tests, GSFC personnel designed and implemented cryocooler data acquisition and control systems which acquires and stores data simultaneously with the additional ability to autonomously shut down the cryocoolers if any operational restrictions are violated. Two separate systems were implemented. To support the cryocooler test bed thermal vacuum chamber tests, a VXI - B (VME extension) instrumentation system is utilized. It is controlled by software written in LabVIEW and it runs on a Macintosh PC. Communication takes place over a IEEE-488 General Purpose Interface Bus (GPIB).

The VXI -B contains a custom designed plug in control circuit card which is used for autonomous cryocooler control. This system controls all of the coolers operating in the cryocooler test bed thermal vacuum facility.

The second system implemented consists of several stand-alone control systems used to operate each cryocooler separately. Both systems utilize elapse counters and run time of each cryocooler is maintained and summarized above in Table 1.

THERMAL VACUUM / THERMODYNAMIC TEST RESULTS

Two Sunpower cryocoolers underwent thermal vacuum testing with heat sink temperatures of approximately -25° C, 0° C and +30° C. A family of load curves were then generated at each heat sink temperature for cryocooler compressor power levels of 20, 60, and 100 watts. Figure 2. below illustrates the results for +30° C and -25 ° C.

The two Sunpower coolers in the thermal vacuum camber, #4 and #5, were determined to have basically identical load curves. Therefore load line data for Sunpower cryocooler #5 is not presented. Six additional cryocoolers underwent thermodynamic testing on an optical bench under ambient conditions (typically 20° C to 25° C) with results as anticipated.

Two of the 8 Sunpower cryocoolers (cooler #6 and #4) have experienced a temperature degradation problem. After cooler #6 had about 400 hours of operation the cold tip temperature slowly increased approximately 50 Kelvin before an automatic shutdown occurred. Once restarted (about 24 hrs later) the cryocooler again achieved it's specified performance. After approximately 200 hours of run time, degradation started again. Each subsequent occurrence of degradation appeared at shorter and shorter intervals of run time. Cooler has #4 experienced temperature degradation only twice, once after approximately 1000 hrs of run time and again at 1300 hrs. Sunpower Inc. suspects that the helium working fluid within the coolers is contaminated with water. They have indicated that they are aware of this potential condition and that they have already taken steps in their manufacturing process to prevent it in future cryocoolers. GSFC intends to return its coolers to Sunpower to allow them to bake out the coolers and refill them with helium. Additional endurance testing will be performed following this bake-out.

VIBRATION CHARACTERIZATION

The objectives of the vibration characterization tests were to determine axial and radial force generated by the cryocooler under heat sink temperatures of -30° C, 0° C, and +30° C, as well as with different drive electronics operating at various power levels and heat loads. A spectrum analyzer was used to record the axial and radial force spectrum from 0 to 800 Hz.

Each dynamometer was designed to be very rigid with its own resonant frequency above 200 Hz to prevent excitation of the dynamometer by the cryocooler. It is important to understand that each vibration signature in this paper represents the cryocooler dynamometer system response and not the cryocooler by itself.

At full power the vibration levels generated by the Sunpower cryocoolers without their counter balancers operating were so high that the axial force signal exceeded the range of our force transducer amplifier. Therefore, the signals were clipped unless the power level was lowered. The maximum power level that we could operate the cryocoolers

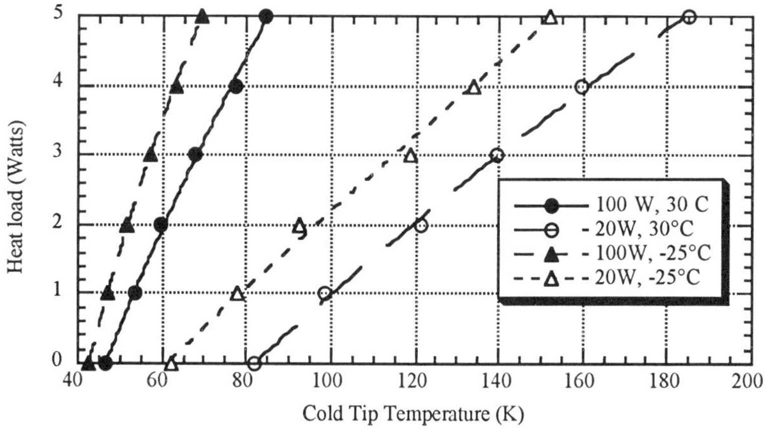

Figure 2. Sunpower #4 load line data at multiple heat sink temperatures.

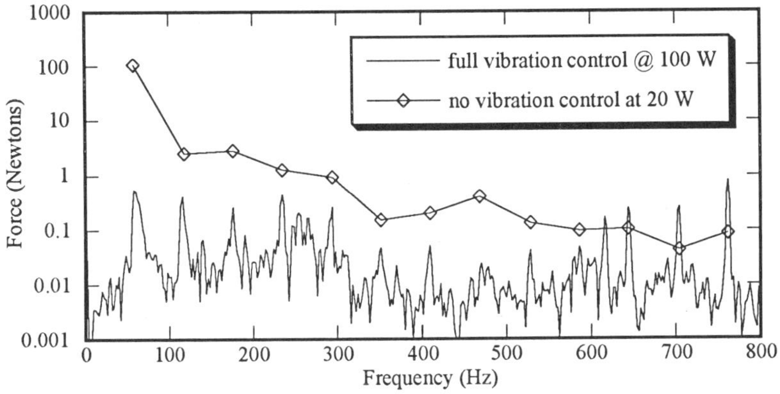

Figure 3. Comparison of vibration levels with and without control.

without driving the counterbalance was about 20 Watts compressor power. Figure 3 illustrates the peak vibration levels at 20 Watts compressor power and is compared to the full spectrum vibration levels with the cryocooler at 100 Watts compressor power with vibration control implemented.

Off-axis residual vibration was also measured. It is reasonably unaffected by axial vibration suppression. The off-axis (radial) force spectrum is shown in Figure 4.

As described below, the operation of the counterbalance with a simple sine wave at the proper phase angle can result in a considerable reduction of the fundamental force component, from 110 Newtons to about 1 Newton at 100 Watts of compressor power. All subsequent harmonics are generally below 5 Newtons even with no counterbalance. Once thermodynamic equilibrium has been reached, the cryocooler vibration is extremely stable in this mode. This mode of vibration control has been implemented on two cryocoolers and it has held constant for several weeks at a time.

CRYOCOOLER ELECTRICAL CHARACTERIZATION

All Sunpower cryocoolers were operated from GSFC electronics. The compressor motor total impedance was unusually low compared to most of the cryocoolers evaluated at

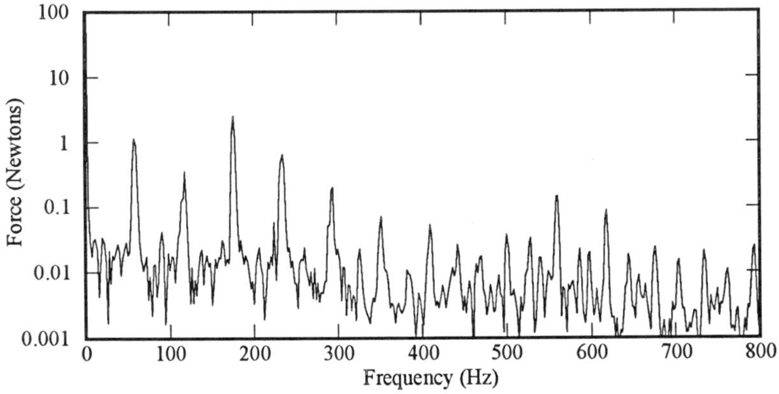

Figure 4. Typical radial vibration at 100 Watts and 0 ° C.

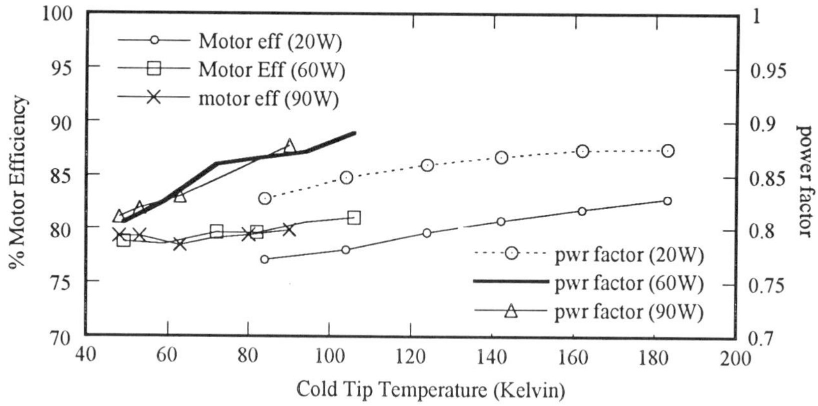

Figure 5. Motor efficiency and power factor at various cold tip temperatures.

GSFC. As a result, the current requirement is very high. Approximately 1.5 ohms total impedance at 60 Hz was measured on each of the Sunpower cryocooler compressors tested. The linear power amplifier utilized for the GSFC drive electronics is the Techron Model 7540 and it is optimized for a 4 ohm load. The maximum power level this amplifier could deliver across the 1.5 ohm impedance of the Sunpower cooler was approximately 100 watts, depending on the power factor. When required, two power amplifiers were configured in parallel to increase current output capability, which allowed cryocooler operation up to 120 watts.

The compressor motor power factor is a function of the working gas temperature. It was observed that the power factor decreased as the cryocooler approached its operating temperature. Figure 5 illustrates the power factor and motor efficiency measured as a function of cold tip temperature. The drop in power factor shown in this figure was observed on all of the Sunpower cryocoolers. A reduction in power factor represents a loss in efficiency.

Sunpower Inc. indicated that the cryocoolers are slightly detuned in order to increase the range of power levels over which the cryocoolers can effectively operate. They indicated that the linear motor deliverable power is limited by the onset of magnetic saturation. Sunpower also indicated that upon request, they would tune the cooler for a specific operating temperature and, as a result, increase the power factor and overall COP at

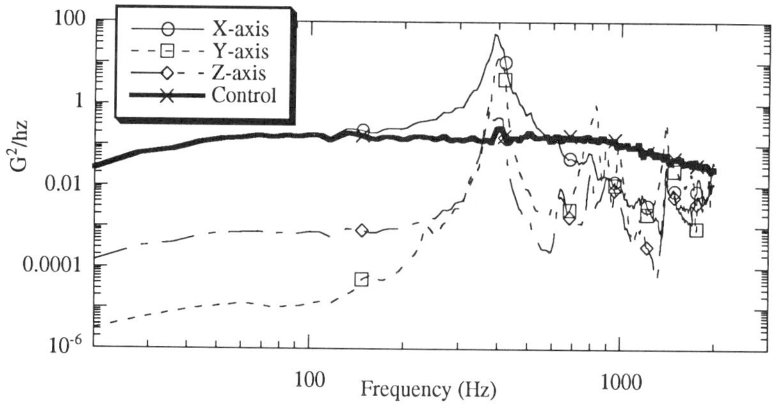

Figure 6. Sunpower Axial Vibration: Full qualification level Control and Response.

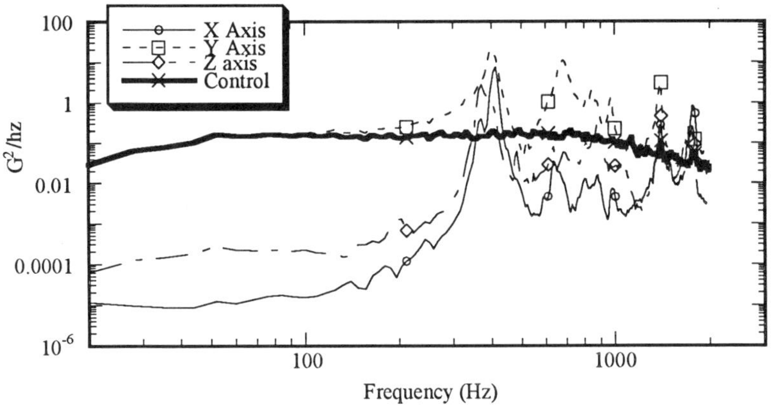

Figure 7. Sunpower Full Qualification level Radial Vibration: Control and Response.

that operating temperature. Figure 5 also illustrates the motor efficiency at three compressor power levels and as function of cold tip temperature. The motor efficiency appears to be relatively flat over a temperature range of 60 K to 150 K and actually increases slightly at higher power levels. The compressor motor dc resistance measured was 0.138 ohms. The motor ac resistance was provided by Sunpower Inc. and is .152 ohms.

The counterbalance motor utilized on the Sunpower cryocoolers require very low power compared to the compressor motor. The balancer motor has an overall impedance of about 150 ohms at 60 Hz, with a current requirement of less than 50 mA. In a continuing attempt to explore methods to simplify cryocooler drive techniques, GSFC explored utilization of only one power amplifier for both motors. This was accomplished by driving the balancer motor with a sine wave from the amplifier for the compressor. A passive network consisting of a small transformer and an RC circuit was used to obtain a sine wave at the proper amplitude and phase angle to null the cooler fundamental reaction force. This scheme nulled the fundamental vibration component to less than 1 Newton.

LAUNCH VIBRATION TEST RESULTS

Two Sunpower cryocoolers underwent random vibration tests. The levels chosen were the generalized component vibration levels as specified in the NASA/GSFC General Environmental Verification Specification (GEVS). The cryocooler motor coils were shorted during these tests. Cryocooler # 5 was subjected to the full qualification level which is 14.1 g's RMS. Cryocooler # 4 was subjected only to the acceptance level of 10.0 g's RMS. Figure 6 and 7 illustrate the acceleration spectral density as a function of frequency in all three axis. Each cryocooler was subjected to axial excitation (Figure 6) as well as radial excitation (Figure 7). In both figures, the X axis is in-line with the excitation.

Once the random vibration tests were complete, the two cryocooler were put back into the cryocooler test bed thermal vacuum chamber, and the earlier thermodynamic test were repeated. Both cryocoolers had test results which indicated no thermodynamic degradation as a result of the random vibration testing.

SUMMARY

In general, the Sunpower cryocoolers have exhibited exceptional performance considering the low cost compared to other long life cryocoolers. The fact that cooler # 5 was subjected to the full random vibration qualification levels without degradation to its thermodynamic performance is very impressive. However, reliability issues do remain a concern. In communication with Sunpower Inc. regarding the results of these characterization tests, it is clear that Sunpower is aware of our concerns and plans to correct them on future coolers.

ACKNOWLEDGMENTS

Special credit is due Vince Arillo (Hughes-STC corp.) for an extremely rapid response to the electronic fabrication requirements necessary to support the effort described in this paper. Also, much credit is due Thomas Gibboney of Orbital Sciences Corporation and John Marketon of Hughes-STC corp. for software development.

REFERENCES

1 S. Castles et al., "NASA/GSFC Cryocooler Development Program", 7th International Cryocooler Conference Proceedings, Pl-CP--93-1001 (1992).
2 L. Sparr et al., "NASA GSFC Cryocooler Test Program Results", 7th International Cryocooler Conference Proceedings, Pl-CP--93-1001 (1992).
3 S. Banks et al., "Adaptation of Tactical Cryocoolers for Short Duration Spaceflight Missions", presented at 8th International Cryocooler Conference, Vail, CO (1994).
4 R. Boyle et al., "Harmonic Vibration Control System for a Cryogenic Refrigerator", 7th International Cryocooler Conference Proceedings, Pl-CP--93-1001 (1992).
5 E. James et al., "Flight Hardware Implementation of a Feed Forward Vibration Control System for Space Flight Cryocoolers", presented at 8th International Cryocooler Conference, Vail, CO (1994).

NASA/GSFC CRYOCOOLER TEST PROGRAM RESULTS FOR BALL AND LOCKHEED SPACE FLIGHT CRYOCOOLERS

Leroy Sparr[1], Stuart Banks[2], Edward James[2], and Stephen Castles[1]

[1]NASA/Goddard Space Flight Center
Greenbelt, MD 20771

[2]McDonnell Douglas Corp.
Seabrook, MD, USA 20706

ABSTRACT

This paper summarizes results from testing efforts performed at NASA/Goddard Space Flight Center (GSFC) on the latest generation of long life cryocoolers produced under contract to GSFC by Ball Aerospace and Lockheed. NASA/GSFC has invested significant resources during the last decade to develop long life, low vibration mechanical cryocoolers for space applications with typical lifetimes of five years or more. These cryocoolers and the technology development they represent are the result of this dedicated effort. Both cryocoolers have undergone extensive characterization testing at NASA/GSFC which includes the measurement of load curves, power input requirements, axial and lateral vibrations measurements (with and without vibration control systems in operation), stiction testing, thermal vacuum performance testing, and alignment verification methods.

CRYOCOOLER TESTING DESCRIPTION

Upon arrival at GFSC, each cryocooler is installed in a custom, dedicated test fixture-dynamometer assembly and undergoes environmental testing in the GSFC cryocooler test bed thermal vacuum chamber. The floor of the thermal vacuum chamber is the upper surface of a 4 meter thick concrete isolation pad and each dynamometer is bolted to the chamber floor to ensure negligible vibration levels from external disturbances.

The GSFC dynamometers are uncalibrated. They are conservatively estimated to be accurate to approximately $\pm 10\%$. This accuracy is considered to be adequate for these tests since the primary goal is to demonstrate the ability to reduce cryocooler induced vibration forces with an active counterbalancer and an existing vibration control algorithm rather than to make absolute comparisons between cryocoolers.

Each dynamometer was designed to be very rigid, with its own resonant frequency above 200 Hz to prevent excitation of the dynamometer by the cryocooler. However, it is difficult to ensure that we have been completely successful. Therefore, it is important to understand that each vibration signature presented in this paper represents the cryocooler/dynamometer system and not the cryocooler by itself. A space flight cryocooler/spacecraft system will not be as stiff as our dynamometer.

The cryocooler test bed vacuum chamber is fitted with a cooling header used to control the temperature of the cryocooler heat sink. The fluid loop temperature is controlled by two remotely located chillers capable of providing operating temperatures between -35° C and +30° C.

Each cryocooler cold finger was enclosed in a vacuum bonnet connected to a vacuum header. An automatic gate valve is used to isolate the vacuum header from the turbo pumping system in the event of a loss of power or vacuum. In order to permit unattended, autonomous operation of the coolers during cryocooler life testing, GSFC personnel have designed and assembled a data acquisition and control system with the ability to safely shut down the coolers if any operational restrictions are violated.

BALL 30K CRYOCOOLER DESCRIPTION & TEST RESULTS

The Ball Engineering Development Model (EDM) cryocooler was developed over a two year period under contract to NASA/GSFC; it was completed during the winter of 1993. This cryocooler produces approximately 400 milliwatts of cooling at 30 Kelvin with a heat sink temperature of 20°C. Total power input to the breadboard electronics is 108 watts. Cooldown time to no-load temperature is about 40 minutes. This cryocooler is a split Stirling cycle cryocooler with twin opposed compressors. A 30 cm transfer tube connects the compressors to a single two-stage, moving regenerator expander with opposed active counterbalancer. It is designed for a minimum operational time of 5 years. All moving components are driven by high efficiency, moving coil linear motors and supported using beryllium copper flexure bearings to eliminate touch contact. The cryocooler is filled to 130 psig with ultra pure helium. Performance from this cryocooler was reported at the 1994 Cryocooler Conference in Vail, CO[1].

During the spring of 1994, Ball proposed to change the expander regenerator from a moving regenerator used in the EDM cryocooler to a fixed regenerator. Although this particular design change posed complicated fabrication challenges, the potential payoffs were significant, namely no touch contact in the displacer, reduced expander axial and radial vibration, and a more robust coldfinger. To determine the feasibility of a flight design, Ball built a breadboard fixed regenerator expander. In the fall of 1994, the compressor from the EDM cryocooler described above was mated to the fixed regenerator expander.

The fixed regenerator, as opposed to the earlier moving regenerator displacer, results in a significant reduction in moving mass in the expander which, in turn, results in less axial and radial vibration and an improved margin to touch contact in the expander under any operating condition, including when the cryocooler is operated horizontally in a 1-g environment. These are significant advantages over earlier moving regenerator designs. Test results for this configuration are reported in this paper.

The cryocooler mechanical assembly weighs approximately 9 kg. This mass includes a compressor custom pressure dome with feedthroughs for the piston radial position sensors. The cryocooler operational frequency is approximately 35 Hz. The breadboard electronics which drives the cryocooler uses the Harris RTX-2000 microprocessor and resides in two 40 cm x 28 cm x 8 cm boxes and one 30 cm x 30 cm x 8 cm box. The user interface to these boxes is via a DOS-based personal computer. The cryocooler electronics also provides a vibration cancellation feature which works well to null axial vibrations through the seventh harmonic (only 7 by design; more memory would be required to

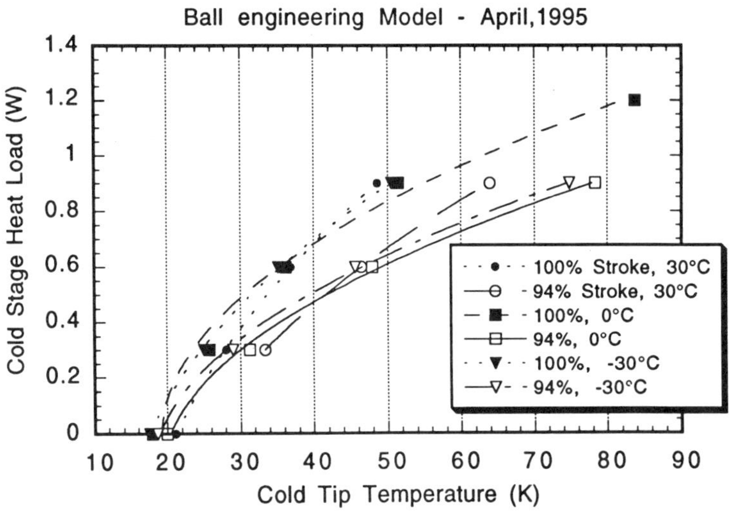

Figure 1 Cooling power as a function of compressor stroke and heat sink temperature.

balance more harmonics). Total power to the cryocooler will be reduced by as much as 12 watts in a flight electronics design. The power measurements assocociated with Figures 1 and 2 reflect total power input into the existing EDM cryocooler and EDM electronics.

Figure 1 shows the heat lifting capability of the cryocooler cold stage as a function of heat sink temperature. Two different compressor strokes were measured.

The maximum possible compressor stroke corresponds with a total input power into the EDM cryocooler electronics of 108 watts at a helium fill pressure of 132 psig and a heat sink temperature of +30°C. This stoke provided 430 mwatts of cooling at 30 Kelvin, which is 33% greater than the original design requirement. The 94% of full stroke data corresponds to a total input power of 75 watts at a fill pressure of 132 psig and a heat sink temperature of +30°C. At this stroke the cryocooler delivered 300 mW of cooling capability at 33 K. The reader must keep in mind the electronics tested here were non-flight (see Reference 1 for a more complete description). The EDM electronics were air cooled and drew a quiescent power of 18 watts.

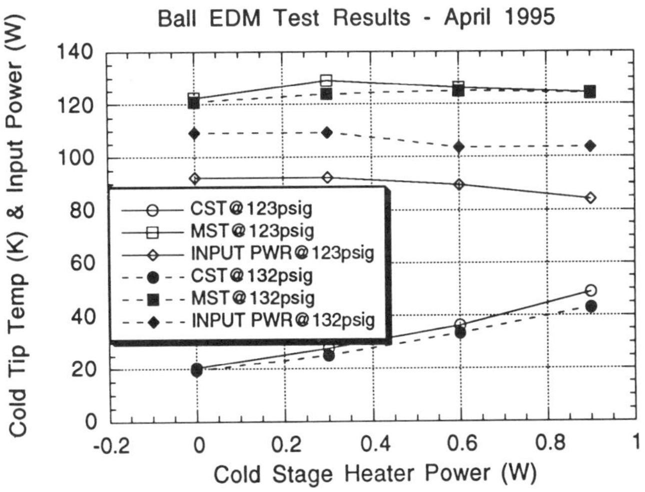

Figure 2. Cooling power and input power as a function of fill pressure.

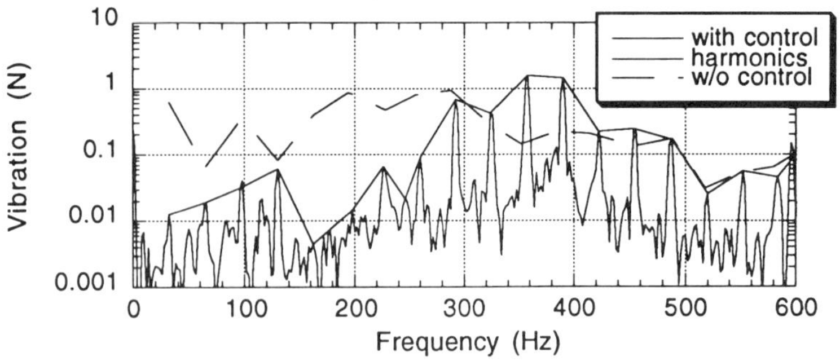

Figure 3. Ball Compressor Axial Vibration Comparison. Solid line is with vibration control on first 7 harmonics. Dashed line is wothout vibration control

Figure 2 shows Ball cryocooler thermodynamic performance for both the cold-stage (CST is cold-stage temperature) and mid-stage (MST is mid-stage temperature) for variations in working fluid pressure. There is a modest increase in the cold-stage temperature with decreases in working fluid pressure. Generally, a drop in input power is also observed.

Initial stiction tests were performed following temperature soaks at -30°C, 0°C and +30°C degrees. No indication of stiction was seen on either the compressor or expander pistons. The cryocooler was then stabilized at normal operating temperature and residual vibration signatures were taken with and without vibration control. Figures 3 and 4 demonstrate the change in axial vibration signatures for the compressor and expander when the vibration control system is engaged. As illustrated in Figure 3 for the case of no vibration control, the lines in these figures connect the vibration levels recorded at each harmonic of the cryocooler operating frequency. At frequencies between these harmonics the residual vibration is generally quite low. Only the first 7 harmonics were controlled. The resulting residual vibration was well below the specification of 0.1 newtons, maximum, at any frequency through the first 7 harmonics.

Figures 5 and 6 depict the residual radial vibration at each harmonic for the cryocooler operating frequency for the compressor and expander. These residual vibration levels are quite low for a cryocooler with the cooling power of the Ball EDM cryocooler. Note that the residual radial vibration is not significantly diminished as the vibration control systems is engaged. This is typical of the linear cryocoolers tested by GSFC.

Based on the successful outcome of the EDM test, Ball is currently fabricating a protoflight two-stage cryocooler with a fixed regenerator. This cryocooler will also incorporate radial position sensors into the compressor and expander assemblies. These

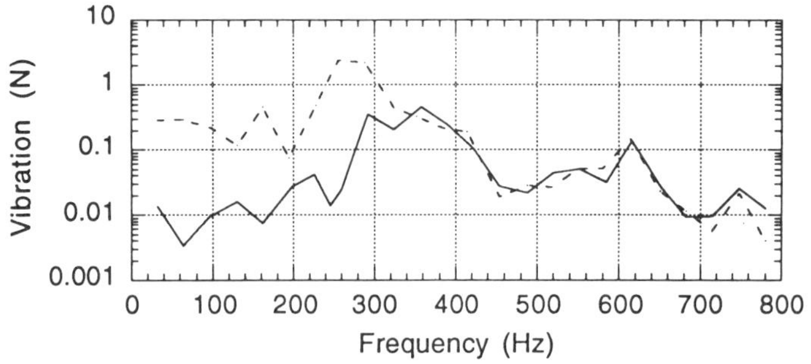

Figure 4 Ball Expander Axial Vibration. Solid line is vibration control on the first 7 harmonics and dashed line is without vibration control.

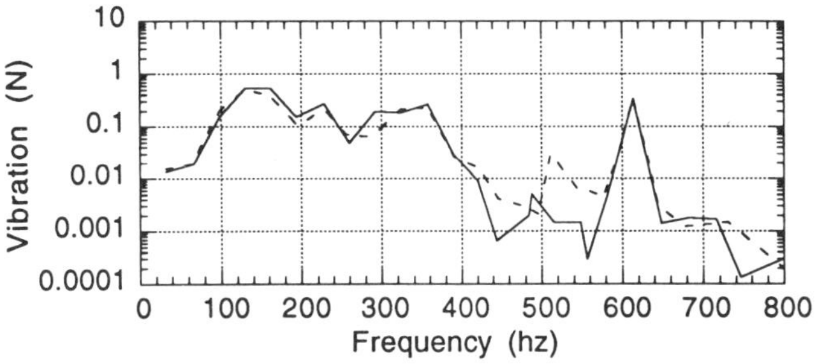

Figure 5 Ball Compressor Radial Vibration Comparison. Solid line is vibration control on the first 7 harmonics and dashed line is without vibration control.

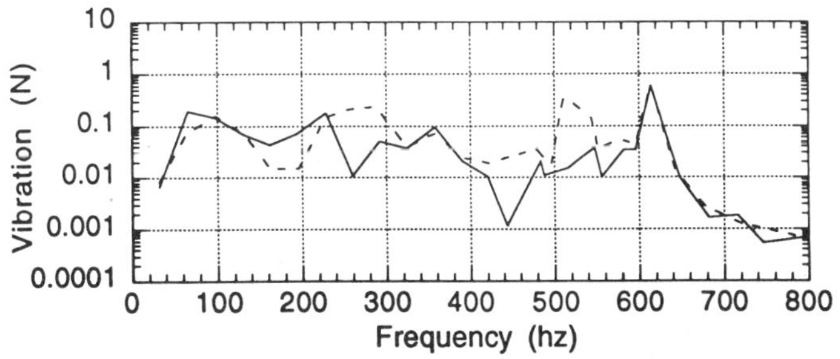

Figure 6 Ball Expander Radial Vibration Comparison (fixed regenerator). Solid line is vibration control on the first 7 harmonics and dashed line is without vibration control.

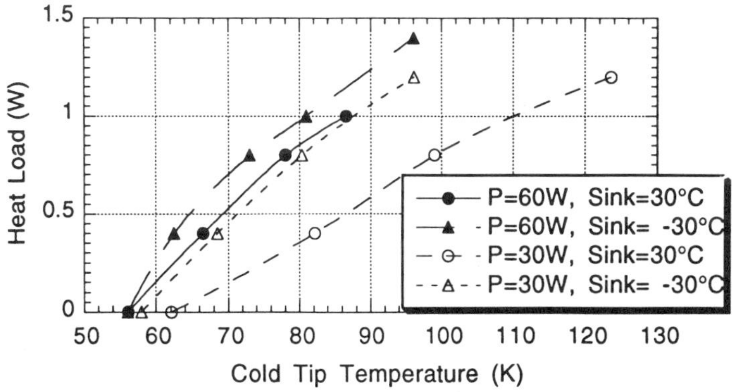

Figure 7 Lockheed 80K Cryocooler Load Curves for ±30°C heat sink temperatures and two input power levels.

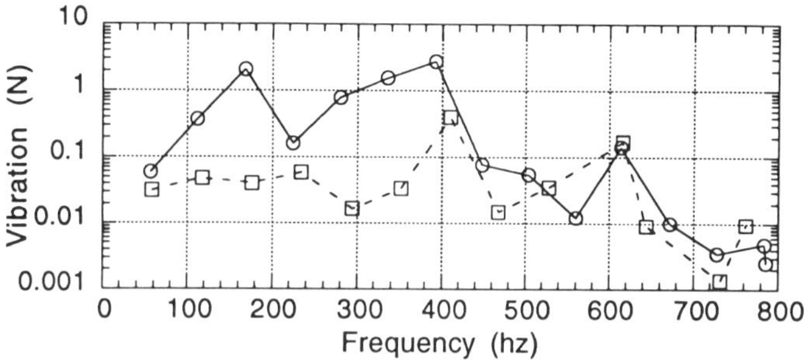

Figure 8 Lockheed Compressor Axial Vibration. Solid line is is without vibration control and dashed line is vibration control on the first 7 harmonics.

sensors are used to assemble and align the compressor and expander pistons within their respective assemblies. The sensors can then be used to measure the radial displacement of the moving components as a function of piston and displacer axial displacement with the cryocooler operating. This feature is of tremendous value in assuring no touch contact within the compressor and expander assemblies in any orientation over the full operating range of heat loads and heat sink temperatures. This alignment system can be used on a flight cryocooler during ground testing and system integration to verify that the no touch contact feature remains intact during payload integration and test. This is particularly important for long duration missions requiring a highly reliable cryocooler, particularly where the mass penalty of flying redundant cryocoolers can not be tolerated.

LOCKHEED 80K CRYOCOOLER DESCRIPTION & TEST RESULTS

The "80K" cryocooler was designed, fabricated and tested under contract to NASA/GSFC during the past three years by Lockheed in Palo Alto, California. Lucas Aerospace of the United Kingdom was the primary subcontractor to Lockheed for the design and fabrication of the mechanical cryocooler piece parts.

The cryocooler is a split Stirling cycle machine with twin opposed compressors connected by a 30 cm transfer tube to a single-stage, moving regenerator expander with opposed active counterbalancer. Each motor is a high efficiency, moving coil, linear motor. The cryocooler is hermetically sealed with a design helium leak rate of less than 1×10^{-9} cc/sec.

The cryocooler mechanical components are made primarily of aluminum and stainless steel. The assembled cryocooler weighs 7 kg. All moving components are supported using flexure bearings to eliminate touch contact. A fiber optic alignment technique is used by

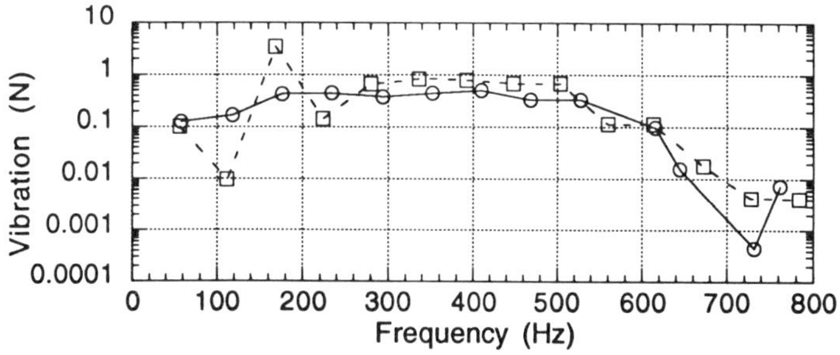

Figure 9 Lockheed Compressor Radial Vibration. Solid line is vibration control on the first 7 harmonics and dashed line is without vibration control.

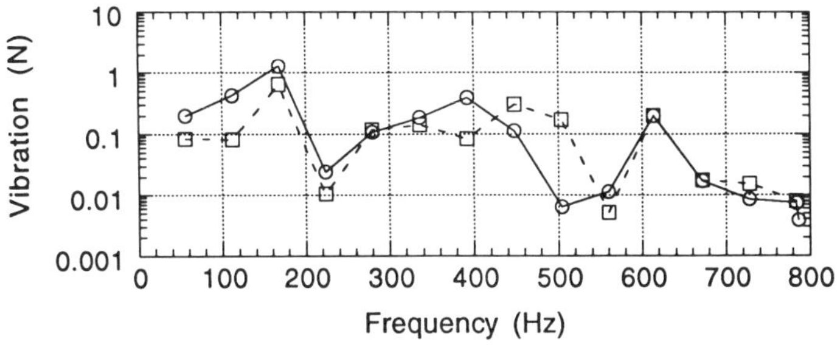

Figure 10 Lockheed Expander Axial/Radial Vibration. Solid line is radial and dashed line is axial.

Lockheed to characterize and assemble the cryocooler moving components to ensure satisfactory alignment of critical moving components with respect to fixed cryocooler parts. The operational frequency of this cryocooler is approximately 56 Hz. The conduction cooled electronics (which weigh 6 kg and occupies .014 m^3) were designed, fabricated and tested by Lockheed. The microprocessor is the United Technologies 69R Risk processor. The cryocooler electronics use a closed loop control system to keep axial vibration levels to within specification. Either force or acceleration signals can be used to close the control loop.

Prior to final delivery to NASA/GSFC this cryocooler and its electronics will complete an extensive flight qualification test program at Lockheed, including launch vibration testing, thermal vacuum testing, and radiated and conducted emission testing. No launch lock power is required for this cryocooler. Shorted coils and inherent natural gas dampening in the cryocooler prevent overstroke conditions from occurring.

Figure 7 presents the thermodynamic performance of the cryocooler with 30 watts and 60 watts of compressor input power and with heat sink temperatures of +30°C and -30°C. Cooldown time to no load temperatures is about 50 minutes because the coldfinger end cap has several ounces of copper attached for mounting instrumentation.

Figures 8 and 9 display the axial and radial compressor vibration signatures with and without vibration control. Figure 10 displays the axial and radial expander vibration signatures without vibration control. The axial vibrations are reduced significantly for both the compressor and expander using the control electronics. Because of space limitations only the compressor data is displayed. There is no significant change in the radial vibration signatures for either the compressor or the expander as vibration control is implemented.

SUMMARY

During the past year, NASA/GSFC has tested Stirling cycle coolers built by Ball Aerospace and Lockheed. These coolers were built to design requirements that meet the needs of future NASA spacecraft.

The Ball cryocooler successfully demonstrates a two stage, 30 Kelvin, fixed regenerator cryocooler with radial position sensors that measure radial clearances under all operating conditions. This feature ensures no touch contact and, therefore, high reliability. The cryocooler has excellent thermodynamic performance and low axial and radial vibration signatures. The cryocooler has maintained these operating characteristics through environmental testing, including launch vibration, and through handling and shipping. Ball is currently fabricating a protoflight unit with this fixed regenerator design for delivery to NASA/GSFC in the fall of 1995.

The Lockheed cryocooler successfully demonstrates a single-stage, moving regenerator, 80 Kelvin cryocooler which exceeds its thermodynamic performance requirements, has no apparent touch contact between moving parts, has low axial and radial vibration signatures and can survive launch vibration and other environmental testing, handling and shipping.

Both cryocoolers will undergo extensive life testing at GSFC during the next 5 years.

ACKNOWLEDGMENTS

The authors would like to acknowledge the contributions of Mr. Vincent Arillo and Mr. John Marketon of Hughes-STX. Their hardware and software contributions assisted the cryocooler testing upon which this paper was written.

Mr. Harold Carrington and Dr. Wilfred Gully of Ball Aerospace and Mr. Larry Naes and Mr. Mark Pryor of Lockheed were instrumental in assisting this testing effort.

REFERENCES

1. L. Sparr et al., "NASA/GSFC Cryocooler Test Results For FY93/94", p221, 8th International Cryocooler Conference Proceedings, Plenum Press, New York, NY (1994)

2. L. G. Naes et al., "NASA/GSFC 80K Longlife, Low Vibration Mechanical Cryocooler Performance Test Program Results", p85, 8th International Cryocooler Conference Proceedings, Plenum Press, New York, NY (1994).

TWO-STAGE CRYOGENIC REFRIGERATOR FOR HIGH RELIABILITY APPLICATIONS

D. Berry, H. Carrington, and W.J. Gully

Ball Aerospace Systems Division
Boulder, Colorado 80306-1062, U.S.A.

ABSTRACT

We have been developing a two-stage Stirling cycle cryocooler specifically for space borne applications. We have made a number of configuration changes to address reliability and vibration issues. In this paper we focus on our change from a moving regenerator to a fixed regenerator displacer design. We describe the change and compare the measured performance of the two designs.

INTRODUCTION

We have been developing Stirling cycle mechanical cryocoolers specifically to meet space borne requirements both with internal funding and under contract to NASA's Goddard Space Flight Center. We intend to develop a reliable multistage cooler that will be relatively easy to integrate into a spacecraft.

We are configuring our cooler to make it as compatible with space hardware as possible. As a first example, we have refined an approach based upon the Oxford variant of linear cooler technology to improve the cooler's reliability by truly achieving non-contacting operation. In this approach, diaphragm springs support the armatures laterally as they oscillate axially. In principle, these springs are stiff enough to maintain the small clearance seal gaps between the pistons and cylinders and these machines can operate without wear. However, when we introduced lateral position sensors into the compressor to monitor these clearances and study the dynamics of this motion, we found that pneumatic destabilizing forces were causing contact and wear in units that were perfectly free statically.[1,2] To achieve non-contacting operation, we had to radically change the construction and alignment of key parts in order to suppress these pneumatic forces. We are currently installing these sensors in our latest displacer mechanism for a corresponding study.

In another example of how we are trying to improve system compatibility, we radically reconfigured the regenerator to make the displacer smaller and more compact while improving its non-contacting characteristics. Our original test displacer, based upon a moving regenerator design created at the Rutherford Appleton Laboratory ([3]), worked well thermally but presented us with a number of fairly intractable system problems as we tried to

Advances in Cryogenic Engineering, Vol. 41
Edited by P. Kittel, Plenum Press, New York, 1996

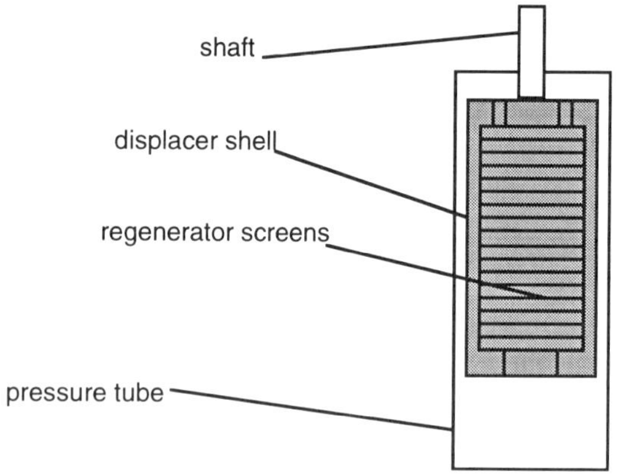

Figure 1. Schematic of a conventional displacer where the regenerator is located within the displacer piston.

modify it to meet other space flight criteria. We will discuss where this conventional design approach led us and then show an alternate approach based upon a more radical fixed regenerator design that helped us circumvent these problems. Finally, we will provide data that illustrates the difference between the two approaches.

ORIGINAL APPROACH

Systems that require small amounts of cooling at low temperatures typically use a moving regenerator displacer like the one illustrated schematically in **Figure 1**. The co-location of the regenerator within the displacer piston results in a compact design that works well for these small, lightweight displacers. As we modified our breadboard displacer to optimize the cooling efficiency at the relatively low temperature of 30K, we found that the cantilevered multiple stages and dense regenerators that gave good thermal performance also made the displacer difficult to support. The structure required to hold the original 30K displacer horizontally in 1G without contact is shown in **Figure 2**. We decided it was too large and heavy, and we set out to improve it using another approach.

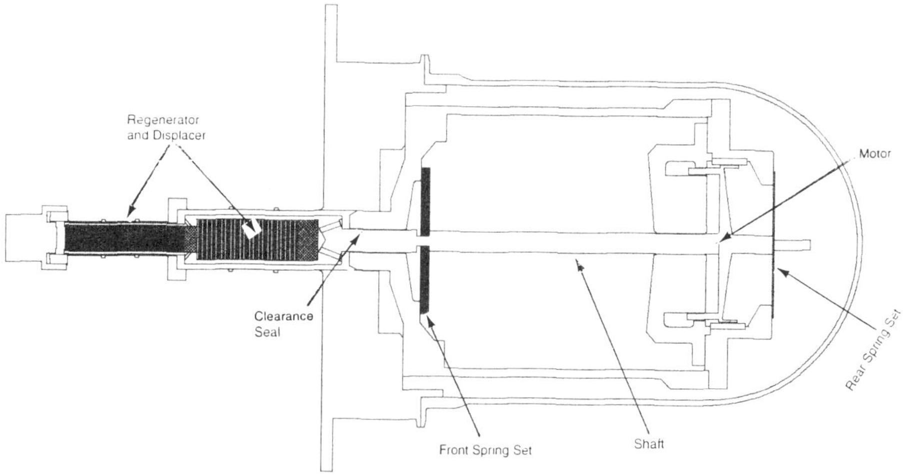

Figure 2. The original moving regenerator displacer. The mechanism must be large and heavy to support the cantilevered mass.

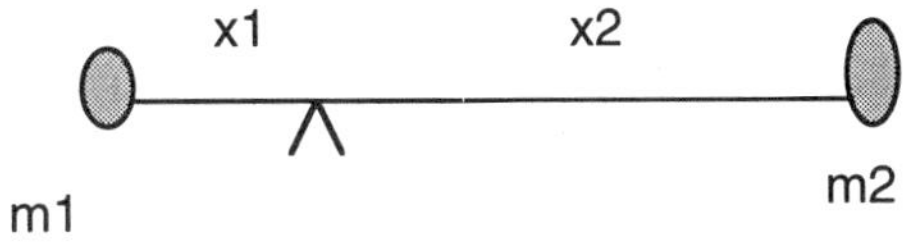

Figure 3. The distribution of mass along the shaft can be understood in terms of a simple lever.

The principal features of the conventional displacer are labeled in Figure 2. The displacer is supported on a shaft, itself supported by two sets of springs located in the large, open chamber. A close tolerance gap seal is required around the shaft to separate the open volume from the working gas around the displacer. The tightest clearances between the moving mechanism and the stationary housing occur at this seal and in the vicinity of the cold tip. The majority of the support springs are located just behind the shaft seal. The few remaining springs are located on the end of the shaft beyond the coil of the loudspeaker type motor. In this particular design we placed an active counterbalance in the open space in the figure between the front springs and the motor stator, but it has been omitted from Figure 2 for clarity.

Non-contacting operation introduces a number of design requirements on the moving mass and its support. Since the cooler has to be thoroughly tested on the ground, there can be no significant static sag due to gravity in any orientation because of the tight clearances at the shaft and along the cold region. And the moving displacer assembly must have good lateral and torsional stability so that it can withstand de-stabilizing dynamic forces that occur during operation. Launch vibration survival adds additional constraints. We have to minimize the moving mass to reduce the threat of damage due to launch vibrations that occur at the beginning of a mission. Lateral accelerations are not a significant problem because the clearances are so tight that not much energy is coupled into the mechanism. But accelerations along the axis of the linear mechanisms work against relatively soft springs in the direction of travel and can cause the displacer to violently strike its stops. To counter this we short out the motor drive coils during launch. An efficient motor, when its coil is shorted, acts as a magnetic brake and protects the armature by "locking" it to the housing and preventing relative motion. For applied harmonic acceleration at frequency w and amplitude of G times the acceleration of gravity, the internal relative displacement of a damped motor is given approximately by:

$$x = \text{weight} * G / (w * (g * g / R))$$

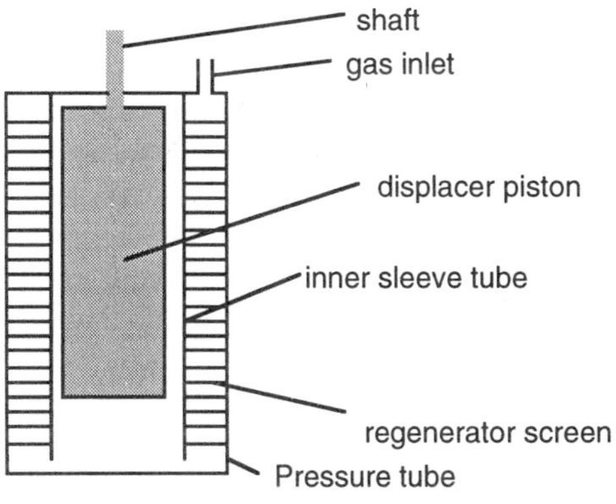

Figure 4. The separation of the displacer and regenerator into separate parts reduces the cantilevered mass.

where g is the force/amp constant of the motor, and R its winding resistance. Given a weight and G level, an efficient (i.e., large and heavy) motor is required to reduce x to acceptable levels. Conversely, it is important to minimize the weight of the moving armature to make the relative amplitude x less than its clearance.

The essence of the design that results from these constraints can be understood with the help of the lever in **Figure 3**. This lever is an abstraction for discussing the weight distribution of the displacer and the impact of the fixed regenerator design. We begin with a given displacer/regenerator of mass m1 located at an effective distance x1 away from the front spring support, which for simplicity we render here as a fulcrum. The rest of the moving mass of the displacer, m2, consisting of the motor, shaft, springs, and sensor parts, is distributed to the right of the fulcrum. To neutrally balance the shaft we would have to locate the CG of the whole displacer at the fulcrum by distributing the mass m2 at acting center x2 such that m2*x2 equals m1*x1. We could, for example, make do with less mass m2 by making x2 longer. To give the displacer some rotational stability about the fulcrum we have to distribute m2 more to the right and add a second outboard spring to support it. A lot of extra weight would be required to pull the CG very far from the fulcrum, so in practice the front springs support the weight and act as the fulcrum, which justifies the model, and the outboard spring provides practically all of the torsional stability of the displacer. A proper design requires a more complex analysis, but this simplified argument suffices to explain the impact of the fixed regenerator approach.

In the design in Figure 2 we tried to minimize the mass m2. We extended the shaft x2 to the point where we could locate the counterweight inside, and we settled for the relatively few springs on the end of the shaft as indicated by the thin packet. But we found that other details increased the mass m2 in unexpected ways. First, we had to make the shaft rigid enough to control the sag at the ends which made it larger and heavier than we had originally planned. And as the mass increased we had to enlarge the braking motor, which in itself further increased the mass on the shaft that had to be braked. We also had to insure that the mode spectrum of the device occurred above the frequency range of operation of the vibration control. The design in Figure 2 was the result of all of these "improvements". It represents the mechanism required to support the thermally efficient 30K moving regenerator displacer subject to our design requirements of non-contacting and launch lock conditions. The mass and size had increased to unacceptable levels, and the full unit shown was never built.

FIXED REGENERATOR APPROACH

Because of the difficulties with the original approach, we returned to the basic design driver, the moment m1*x1 that had to be supported by the shaft. We realized that if we separated the regenerator and displacer, as shown in **Figure 4**, we could make the regenerator a stationary part of the cold end and significantly reduce the amount of mass that had to be supported by the spring system. This has been done before in large coolers, but it was not clear that it could be done successfully in the small size appropriate to our spaceborne applications. By taking the regenerator out of the displacer we have them in parallel and significantly increase the cross sectional area into the cold environment. We also have to come up with a lightweight, insulating, clean replacement for the space vacated.

Our solution was to make a thin wall metal piston, as shown in the photo of **Figure 5**. We chose this over a plastic piston because we found non-metallics to be too dimensionally unstable in these tight quarters, and often to be the source of gaseous contamination in the cooler. And we could evacuate it and weld it shut for the required thermal insulation. The piston in the photo has less than a quarter of the cantilevered mass of the original design. The decrease had a dramatic impact on the design of the support structure and led to the

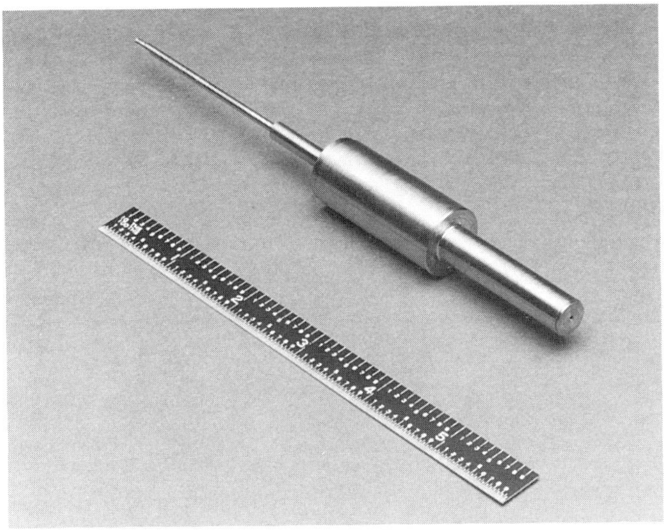

Figure 5. The thin wall metal displacer used in the fixed regenerator breadboard.

much more compact assembly shown in **Figure 6**. We made a very successful breadboard version of this device that we will report on here, and are currently assembling our flight version.

The regenerators in the new design have the same volumes and cross sectional areas as before. But the screens are now punched as rings and are packed around an inner sleeve tube introduced to keep them away from the moving displacer bottle. The critical displacer running clearance is now between the displacer and this tube instead of between the displacer and the outer pressure tube. The heat load increased because of conduction through the enlarged pressure tube, through the new inner sleeve tube, and through any gas that happens to be inside the piston. This degrades the performance. But we expect some offsetting performance improvement because the ends of the stationary regenerators can do double duty as heat exchanger elements because they are now in contact with the load and with the ambient end of the displacer.

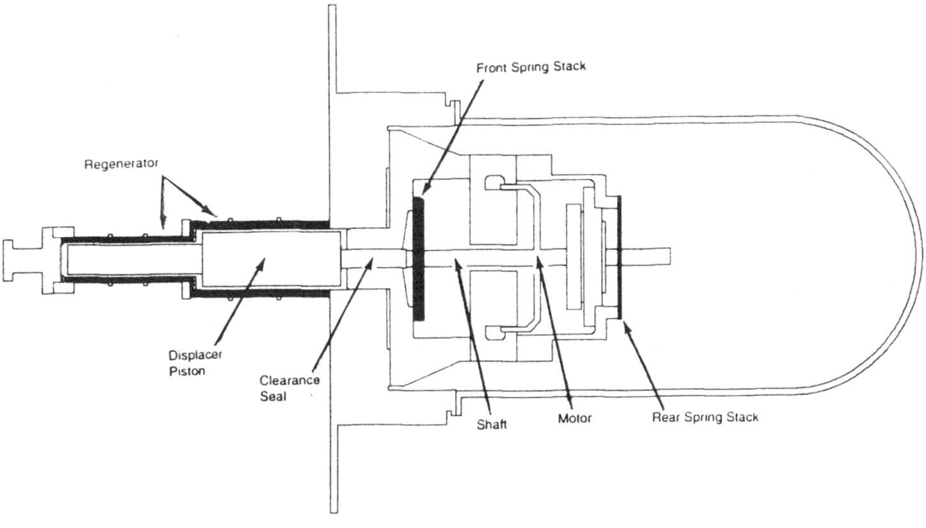

Figure 6. The fixed regenerator approach results in a more compact displacer design.

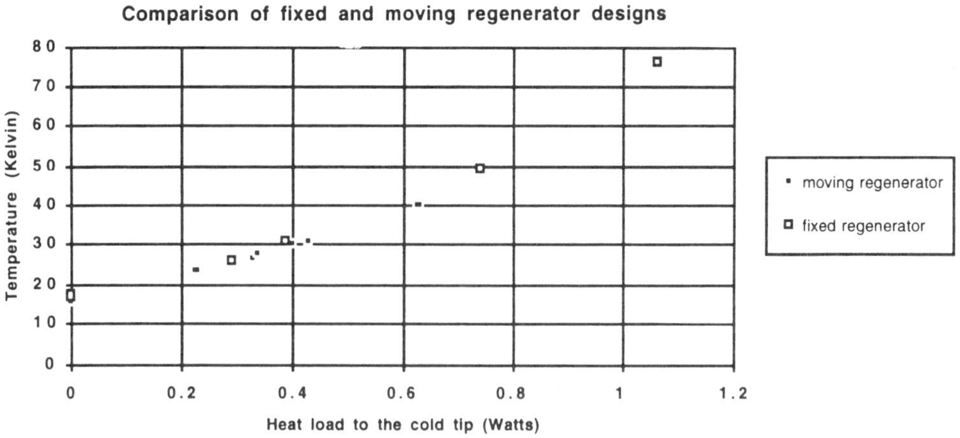

Figure 7. Cold tip performance for the moving regenerator and fixed regenerator designs is virtually identical.

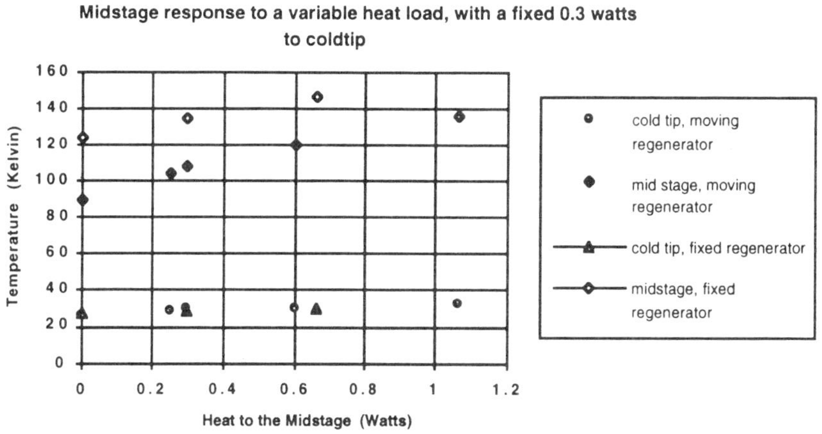

Figure 8. The mid stage temperature is unexpectedly higher for the fixed regenerator design.

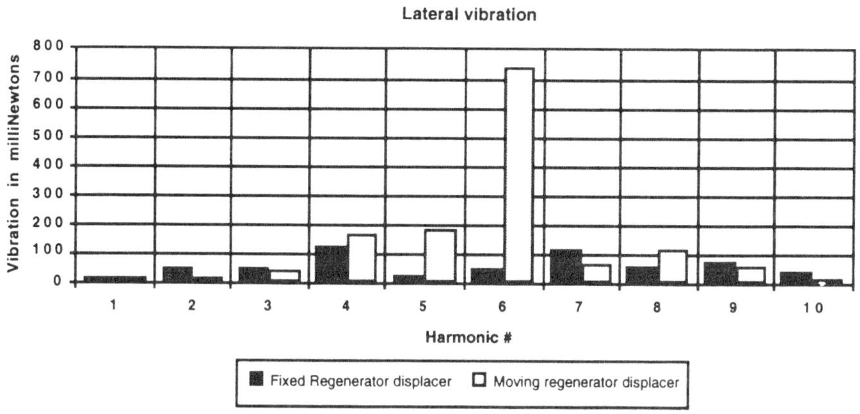

Figure 9. The lateral output forces of the fixed regenerator are much lower than those of the moving regenerator displacer because of the lower mass and inertia of the device.

Figure 10. A photo of the fixed regenerator displacer breadboard cold tip supporting a Ball thermal S-link.

RESULTS

We ran comparison tests using the same compressor and drive electronics with the fixed regenerator and moving regenerator displacer breadboards. Both breadboard displacers were good thermal representations of the more complete versions that were to follow. These flight versions would have the same thermal design but would be sealed and have the non-contacting and launch survivability features. Because of its light weight the fixed regenerator breadboard already was non-contacting, but the moving regenerator breadboard was non-contacting only in the vertical direction. And the fixed regenerator had the final moving mass, whereas the moving regenerator breadboard had only two-thirds the mass of the flight version. Therefore, the vibration output contrast between the flight versions of the displacers would be even more pronounced than what we show for the breadboard versions below. We now discuss various aspects of the performance.

Thermal. The key question is whether the new fixed regenerator version would work as well thermally as the moving regenerator displacer. The results are shown in **Figures 7** and **8**. In Figure 7 we show that the 30K performance is virtually unchanged, and equal to 0.4 watts of heat lift for 75 watts of system input power. The cooler works as well as before for its intended purpose. However, the mid stage performance was another matter. In Figure 8 we show that the mid-stage no-load temperature had apparently absorbed the brunt of the design change and had increased substantially. We cannot account for the increase simply by the extra conduction expected for this design, and are doing auxiliary studies to understand this result.

Vibration output. The lower moving mass had directly observable consequences. We illustrate the change by comparing the lateral vibration spectrum through 10 harmonics for both displacers at a 32.5 Hz operating frequency in **Figure 9**. Both sets of data were taken on the same 6-axis dynamometer at the NASA Goddard test facility. The vibrations at most harmonics are significantly reduced. These harmonics result from an interplay between the drive and the internal modes of the mechanism. Here, the lighter weight and more compact design have reduced the inertia significantly and resulted in a cleaner mode spectrum. We note in closing that no effort had been made in either case to minimize the magnitude of vibration. The data in Figure 9 is only for a relative comparison of the two designs.

Non-contacting operation. Unlike our earlier moving regenerator displacer, we found the fixed regenerator displacer much easier to align because the low mass and stiff assembly made the total sag only a few percent of the gap clearances at all of the critical locations. The usual quasi-static stiction tests showed that it could move through its full stroke at all temperatures in all orientations without touching. We have not yet made dynamic measurements with a final design incorporating the lateral position sensors, but simulations show the design to have good margin under the anticipated conditions, primarily because no margin is lost to static considerations and because of a more favorable torsional spring rate.

Ease of integration. One unexpected benefit of the new design came from having the critical operating clearance deeper within the mechanism, between the inner sleeve tube and the piston, rather than between the pressure tube and the piston. We found that the inner sleeve tube does not move at all when lateral forces of up to 2 Newtons are applied to the tip of the pressure tube, which itself moves by about 10 micron. This relatively small deflection of the outer tube is not coupled to the inner tube by the screens. So the cold finger can easily support the weight of a heat switch or thermal link, like the one shown in **Figure 10,** without eroding the clearance between operating parts in the cooler.

REFERENCES

1. "Multistage cryocooler for Space Applications", H. Carrington, *et al.*, CRYOCOOLERS 8, Plenum., pp. 93
2. "NASA/GSFC Cryocooler Test Program Results for FY93/94", L. Sparr, *et al*, CRYOCOOLERS 8, Plenum, pp. 221.
3. "First results on a prototype two-stage miniature Stirling cycle cooler for space flight applications", T. Bradshaw, Proceedings of the 4th International Cryocooler Conference, Easton Md, 1986.
4. "Design and Verification of Stirling Cooler Interfaces Suitable for Long-Lifetime, Space-Borne Sensor Systems", Arentz, *et al*, CRYOCOOLERS 8, Plenum, p. 855.

DEVELOPMENT OF A COMPUTER MODEL FOR TWO-STAGE GIFFORD-McMAHON CRYOCOOLER

M.D. Atrey [1], H.U Häfner [2], C. Heiden[1]

[1]Institute for Applied Physics
University of Giessen, 35392 Giessen, Germany
[2] Leybold AG
Bonner-Str. 498, D-50968 Cologne, Germany

ABSTRACT

Computer modelling of cryocoolers has emerged as a major research area in the last 4-5 years. In this paper, the authors describe a computer model for two-stage Gifford-McMahon cycle cryocooler. The model is developed based on certain simplifying assumptions. It starts with the computation of ideal refrigeration effect that can be obtained from the cryocooler. The model then computes different losses in the system giving the net refrigeration effect available from the cryocooler. Measurements are performed to get the input data for the model from the machine. The input data are pressure-volume variations, temperature and refrigeration effect at both expansion stages. The model predictions have been found to be very close to the actual results obtained from the machine.

The model is used further to study the effect of different parameters like phase angle between pressure and volume variations, frequency and also of the regenerator materials on the refrigeration available at first and second stage of expansion.

INTRODUCTION

Gifford-McMahon cryocoolers were reported by Gifford and McMahon[1,2] in 1959 and found ready applications in small capacity cryocooleres. These G-M cycle cryocoolers have become quite popular because of their simplicity and high reliability. Methods of computing the performance of the G-M cryocoolers have been reported by several authors [3-6]. Most of the computations are analytical in nature and do not consider the actual pressure and mass flow rate changes in the cycle.

Recently, Tirumaleshwar et al.[7-9] reported heat balance analysis for single and double stage G-M cooler. Although the loss analysis was based on the average mass flow rates and pressure variation, the analysis could give an insight to relative weight of the different losses in the system. Minas and Hualde [10] recently reported a dynamic model of G-M cryocooler which however could represent only the trends of the observed physical phenomena in an accurate fashion. The model does not consider heat transfer processes inside the system.

In this paper, the authors present a computer model for two stage G-M cryocooler. The pressure-displacement charts obtained experimentally from the machines for specific temperatures form the input to the model. The model computes ideal refrigeration effect, mass flow rate of the working fluid in the expansion spaces and losses in the machine. This then gives the net refrigeration effect available at the first and second stage of expansion. Experimental data is collected from the machine for various temperatures in order to compare the model predictions. The predictions of the model are found to be very close to the actual results. The model is further extended to study the effect of different parameters like phase angle between pressure and displacer motion, frequency of the machine and also of the regenerator materials on the performance of the cryocooler.

MODEL DEVELOPMENT

Assumptions

The important assumptions made are listed below :

i) The working fluid, Helium, behaves as an ideal gas.
ii) The process of expansion is isothermal.
iii) The pressure at any instant is constant throughout the machine.

Volume Variations

The volume variations of the expansion spaces with the crank angle are taken directly from the Fig.1. The displacement curve gives the motion of displacer in a cycle. In this analysis, one cycle period (360 degrees) is divided in different intervals (indicated by i) and the corresponding pressure and volumes are computed from this graph. The position of the displacer, Y, at any time is interpolated from the graph. The volume variations, therefore are given as,

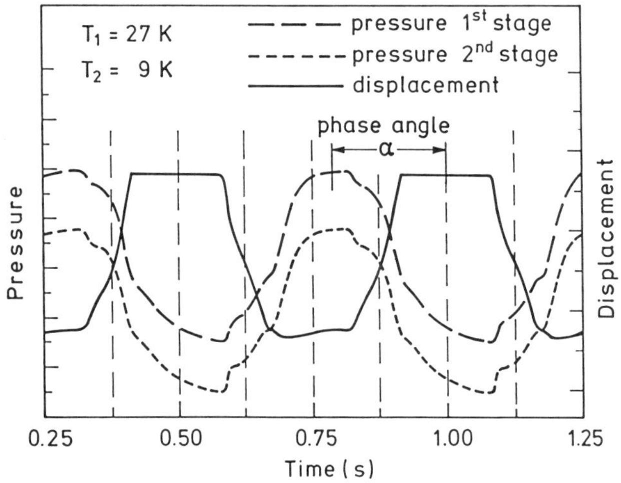

Figure 1. Pressure-Displacement variations.

$$V_E(j,i) = V_{EM}(j) * (S_D - Y(i)) / S_D \quad (1)$$

where

$V_E(j,i)$ is the volume variation of the jth stage of expansion for the ith interval of the cycle. $V_{EM}(j)$ is the maximum expansion space volume for the jth stage. S_D is the stroke of the displacer while $Y(i)$ denotes the instantaneous position of the displacer for the interval i.

Mass Flow Rate

The mass flow rate in the expansion spaces is different because of the regenerators, connecting tube, dead space and temperature changes. To calculate these mass flow rates, the variations of mass of the gas inventory in each space for each interval is calculated. By applying the gas law we get,

$$M(j, i) = P(i) * V_E(j, i) / (R\, T_E(j)) \quad (2)$$

where

$M(j, i)$ is the moles of the gas inventory in the ith interval of the cycle and in the jth expansion space. $P(i)$ is the corresponding pressure at the ith interval of time. R is the universal gas constant and T_E is the gas temperature in the expansion space.

The mass flow rate in an interval therefore can be given by,

$$W_{ES}(j, i) = [M(j, i+1) - M(j, i)]\, M_W\, N / (\phi / 360) \quad (3)$$

where

$W_{ES}(j, i)$ is the mass flow rate in the ith interval and in the jth expansion stage.
M_W is the molecular weight of the working fluid and N is the frequency of the machine. ϕ is the chosen size of the crank angle interval.

Ideal Refrigeration Effect

The ideal refrigeration effect per cycle for the jth stage, $Q_I(j)$ could be given by,

$$Q_I(j) = \int P(i)\, dV_E(j, i) \quad (4)$$

The integration is carried out using the trapezoidal rule.

Loss Analysis

Following losses are considered to determine the net refrigeration effect available from the different expansion stages.

1. Loss due to regenerator inefficiency
2. Loss due to shuttle heat conduction
3. Temperature swing loss
4. Pumping loss
5. P-V loss due to pressure drop
6. Loss due to conduction

The losses are calculated for each interval of the cycle using the available expressions.

EXPERIMENTAL SET-UP

The basic coldhead (expander unit) used for the experimental investigation is a Leybold RGD 5-100-2, a sectional view of which is shown in Fig.2. As the displacer in the present cryocooler is not mechanically driven by the motor, the volume variations in the expansion spaces are not sinusoidal. The displacer is driven pneumatically and this leads to an unique displacement pattern depending on the temperature of the working spaces. A rotating valve, driven by a synchronous motor is operating on the hardened surface of a control disk in such a way that high or low pressure Helium is transferred into the main volume of the coldhead cylinder as well as to the small control volume above the control piston at certain times. The pressure difference between the control volume and the main volume causes displacer movement. The displacer is connected to a metallic rod with magnetic segments, the movement of which is picked up via two induction coils. This whole displacement pickup assembly is locked in a bore of the control disc, and the electrical signal is processed in a computerised system.

The actual pressures in the system are measured using small piezoelectric Kistler pressure transducers at three locations, the warm end of the coldhead and at the cold flanges of the first and second stages. The corresponding signals from the pressure transducers are processed by computer as well.

Both stages of the coldhead are equipped with Silicon diodes (Lake Shore type DT470) for temperature monitoring. Electrical heaters are employed at the cold ends to measure the refrigeration effects for certain temperatures.

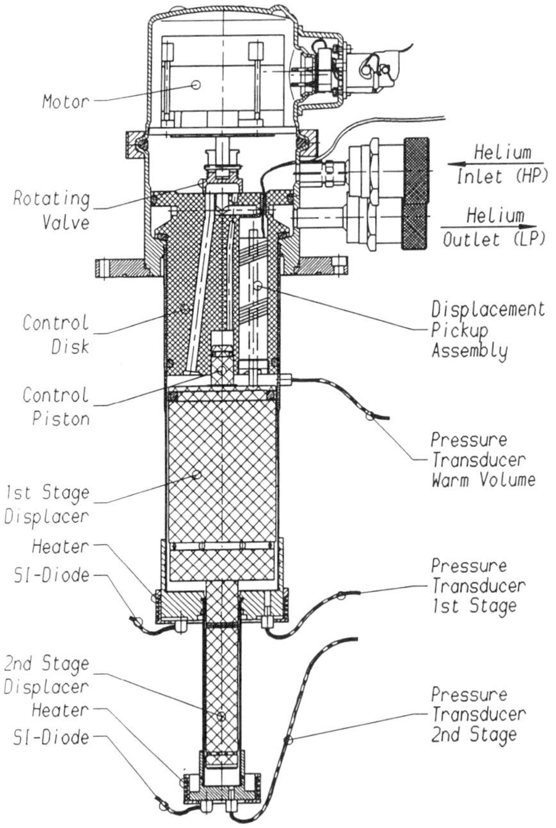

Figure 2. Experimental Set-up for measurements.

Table 1. Comparison of Model Predictions with Experimental Results

Filling Pressure - 1.6 MPa, Frequency - 2.0 Hz

Sr. No.	Temperature (K) Stages		Refrigeration Effect - I Stage (W)		Refrigeration Effect-II Stage (W)	
	I	II	Experi-mental	Model	Experi-mental	Model
1	28	10	0	1.78	0	0.3
2	48	10	50	42.16	0	-0.19
3	80	11	100	104.1	0	0.16
4	48	14	50	38.7	4	4.22
5	81	15	100	98.07	4	4.12
6	28	25	0	-0.84	8	8.05
7	46	23.5	50	53.7	8	8.97
8	80	23.5	100	107.5	8	8.57

MODEL VALIDATION

The model was validated by taking various experimental cases from the actual machine. A correlation was developed between the temperatures and the mass flow of the gas in the expansion spaces to compute the mass flow of the gas in the regenerator. For each case, the pressure displacement diagrams were taken from the actual machine. Table 1 gives the comparison between model predictions and the experimental results. The comparison shows a good agreement between them. The negative values of refrigeration effect obtained in some cases simply means total losses exceeding the ideal refrigeration effect at those temperatures.

EFFECT OF DIFFERENT PARAMETERS ON REFRIGERATION

The model so developed has been then used to study the effect of different parameters on the performance of the cryocooler. Effect of three parameters, namely, phase angle between pressure and displacer motion (indicated in Fig.1), frequency and the regenerator material are studied here.

Effect of Phase Angle

Fig. 3 gives the effect of phase angle on the refrigeration effect for the first and second stage. The curves show that there exists an optimum phase angle for each stage for which

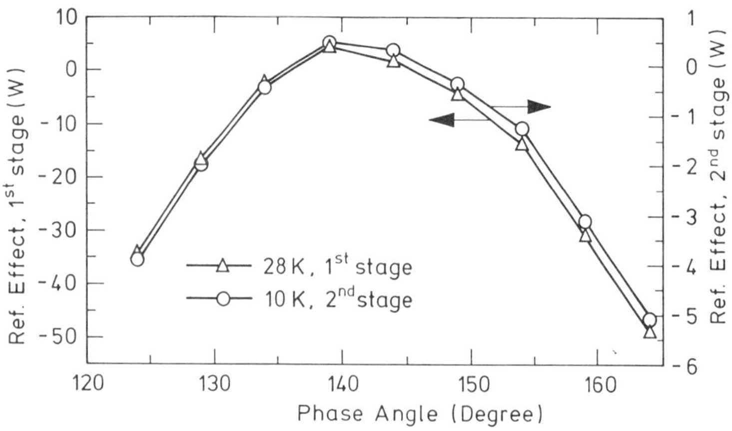

Figure 3. Effect of phase angle on refrigeration effect.

the respective net refrigeration effect is maximum. It is found that the optimum phase angle for the present case lies very close to the existing one in the actual machine. As the displacer is a free one, the phase angle between pressure and displacer movement is also a function of temperature which changes the dynamics of the system.

Effect of Frequency

Figs. 4 and 5 show the effect of frequency on the performance of the cryocooler for different temperatures. Fig. 4 is for higher temperatures 80 K and 20 K as 1st and 2nd stage temperatures resp. while Fig. 5 is for 40 K and 12 K resp. A very interesting fact can be observed from these figures. It is found from Fig. 4 that at higher temperatures, as the frequency increases the refrigeration effect both at 1st and 2nd stages increases. On the other hand, for the lower temperatures as seen in Fig. 5, there exist an optimum frequency at which the refrigeration effect at first and second stage is maximum. In the present case the optimum frequency lies very close to 2 Hz. The difference in the behaviour of these curves is due to the fact that at lower temperatures the increase in the losses at higher frequencies

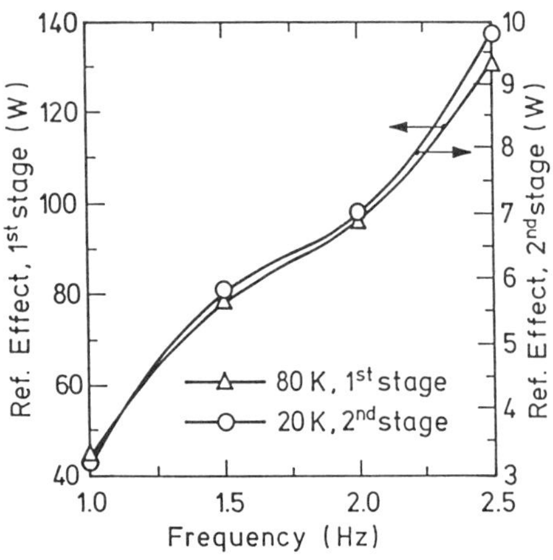

Figure 4. Predicted effect of frequency at higher temperatures.

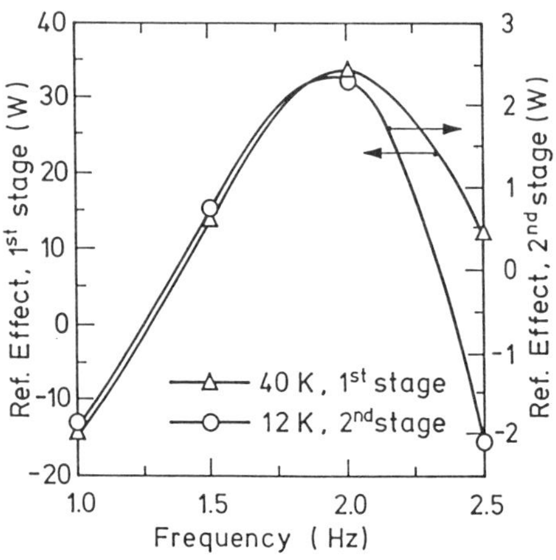

Figure 5. Predicted effect of frequency at lower temperatures.

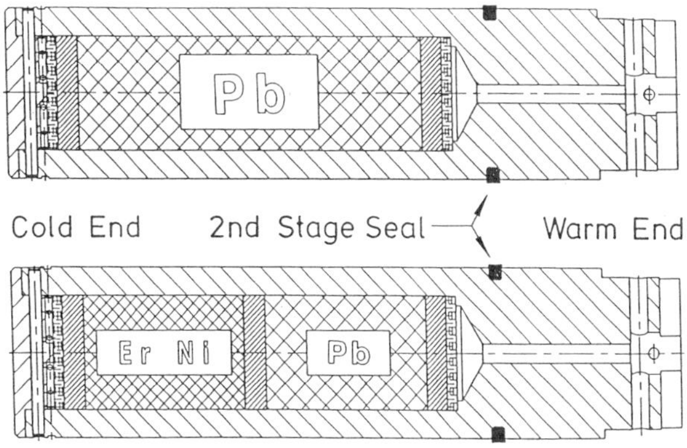

Figure 6. Hybrid Regenerators.

is more pronounced than at higher temperatures. The losses, mainly the regenerator loss and the loss due to pressure drop, show a distinctive increase at higher frequencies and lower temperatures. If the temperatures are still lowered it is obvious that the optimum frequency would decrease.

Hybrid Regenerator

At lower temperatures, the specific heat capacity of regenerator matrix material becomes a vital parameter. Materials like lead can no longer be used, instead, rare earth based materials exhibiting an anomaly in the specific heat may be employed as indicated in the literature[11-12]. The present model can study the effect of these regenerator materials, arranged in a hybrid manner, on the refrigeration effect at the second stage. The important data required for this purpose is the variation of specific heat capacity of different materials like ErNi, Er_3Ni, Nd, etc. at lower temperatures. The present model incorporates the data for six different materials. The temperature distribution across the regenerator is assumed to be logarithmic.

Two types of regenerators are analysed as shown in Fig.6. The first type of regenerator has only lead balls of 0.2 mm diameter as the matrix material and it is found that end temperatures at no load are 28 and 10 K for 1st and 2nd stages resp. A hybrid regenerator is then considered which is filled up with lead balls and ErNi in the equal proportions, separated by a sintered disc or a stabiliser, keeping ErNi at the low temperature side. It is found that in the second case, the actual refrigeration effect at 10 K increased to 2.5 W owing to higher specific heat capacity of ErNi while the model predicted it to be 2.4 W. The study, thus, highlights the importance of hybrid regenerators for obtaining low temperatures and also the role of the model in order to choose the right material.

CONCLUSIONS

The G-M model so developed was tested extensively for different cases and was compared with the actual experimental data obtained from the machine. The model, presently, is being used to predict the performance of the machine using hybrid regenerators. The model can predict the performance of the machine using rare earth materials in the regenerators. The regenerator can have two, three or more materials stacked in its body separated by a disc. The significance of the computer model is demonstrated here. Experimental developments, based on the predictions of the model, are being carried out to realise an optimum hybrid regenerator in practice.

ACKNOWLEDGEMENTS

The authors wish to acknowledge the efforts of D. Sous and Ch. Beckers for collecting the data from the G-M cryocooler for model validation.

REFERENCES

1. W.E. Gifford and H.O. McMahon, A new low temperature gas expansion cycle, part I Adv. Cryo Engg., 5:354 (1960).
2. W.E. Gifford and H.O. McMahon, A new low temperature gas expansion cycle, part II Adv. Cryo Engg., 5:368 (1960).
3. W.E. Gifford, The Gifford-McMahon cycle, Adv. Cryo. Engg., 11:152 (1966)
4. R.A. Ackermann, W.E. Gifford, A heat balance analysis for Gifford-McMahon cryorefrigerator, adv. Cryo. Engg.,16:221 (1971)
5. P.Herycak, Thermodynamic analysis of a new gas refrigeration cycle, Cryogenics 3:23 (1963).
6. W.E. Gifford, and T.E. Hoffman, A new refrigeration system for 4.2 K, Adv. Cryo Engg., 6:82 (1961)
7. M. Tirumaleshwar and S.V. Subramanyam, heat balance analysis of single stage Gifford-McMahon cycle cryorefrigerator, Cryogenics 26:189 (1986)
8. M. Tirumaleshwar and S.V. Subramanyam, Two stage Gifford-McMahon cycle cryorefrigerator operating at 20 K, Cryogenics, 26:547 (1986)
9. M.Tirumaleshwar and S.V. Subramanyam, Gifford-MyMahon cycle - a theoretical analysis, Cryogenics, 26:177 (1986)
10. C. Minas, P.M. Hualde, Dynamic modelling of a Gifford-McMahon cryorefrigerator, Cryogenics, 32:634 (1992)
11. T.Hashimoto, T.Tsukagoshi, H.Nitta, M.Yabuki,T.Kuriyama, H.Nakagome, Excellent character of multi-layer type magnetic regenerator near 4.2 K, Cryocoolers 8, R.G. Ross,Jr.,ed.,Plenum Press, New York (1995),p.677.
12. K.A. Gschneidner,Jr., V.K. Pecharsky, M.Gailloux, New ternary magnetic lanthanide regenerator materials for the low temperature stage of a Gifford-McMahon cryocooler, Cryocoolers 8, R.G. Ross,Jr.,ed.,Plenum Press, New York (1995),p.685

INFLUENCE OF VALVE OPEN TIMING AND INTERVAL ON PERFORMANCE OF 4 K GIFFORD-McMAHON CYCLE CRYOCOOLER

R.Li, A.Onishi, T.Satoh, and Y.Kanazawa

Research and Development Center, Sumitomo Heavy Industries, Ltd., Yuhigaoka, Hiratsuka, Kanagawa, 254, JAPAN

ABSTRACT

The influence of intake/exhaust valve timing on performance of a 4 K Gifford-McMahon(GM) cycle cryocooler has been investigated. The 4 K GM cryocooler is one of the standard two-stage types with magnetic regenerator material, $ErNi_{0.9}Co_{0.1}$, in its second stage regenerator, and delivers more than 1 W cooling at 4.2 K. A reasonably early open timing of intake/exhaust valves not only brings about a cooling capacity above 1 W at 4.2 K on the second stage, but also produces a much larger cooling capacity at the first stage. Under the optimum conditions, 39.4 W cooling at 40 K on the first stage, as well as 1.18 W at 4.2 K on the second stage, was obtained with an input electric power of 7.05 kW. This paper discusses the results of cooling capacity with P-V diagrams, and indicates that the reduction in pressure drop at intake/exhaust valves is important for a 4 K GM cryocooler.

INTRODUCTION

The development of magnetic regenerator material after 1988[1,2,3,4] brought great progress in regenerative cycle cryocoolers. For Gifford-McMahon(GM) cycle cryocoolers especially, the successful use of rare earth magnetic compounds has reduced the no-load temperature down to the superfluid helium region[5]. The hybrid structure regenerators with $ErNi_{0.9}Co_{0.1}$ or $Er_{0.9}Yb_{0.1}Ni$ placed at the colder end make it possible for cryocoolers to deliver about 1 W cooling capacity at 4.2 K[6,7], Furthermore, the 4 K GM cryocooler has been successfully applied to some cryogenic uses, such as SIS mixer cooling[8], helium recondensation in a superconducting MRI magnet system[9], and conduction cooling of small superconducting magnets[10].

In the last several years, a substantial body of literature on applying magnetic materials to regenerative cycle cryocoolers has been published, but the papers describing the improvement in other components of the 4 K cryocoolers are very few. As a matter of fact, 4 K GM cryocoolers are developed from commercially available GM cryocoolers and many of their components are not yet optimized for the cooling at 4 K. It is the same as the advances in magnetic regenerator material that the improvement in other components should be important to a 4 K GM cryocooler.

In many cryogenic systems, sufficient shield cooling by the first stage of a GM cryocooler yields an effective cooling by the second stage. That requires an appropriate

Advances in Cryogenic Engineering, Vol. 41
Edited by P. Kittel, Plenum Press, New York, 1996

balance of cooling capacity between the first stage and the second stage of the cryocooler, and it is also a matter of course for a 4 K GM cryocooler. The previous researches on 4 K GM cryocooler, however, generally focused on the second stage cooling capacity, and almost did not mention that of the first stage.

The intake/exhaust valves in GM cryocoolers are a key component, and contain several important factors, such as valve type, valve structure, valve open timing and valve interval. In order to find an optimum valve for 4 K GM cryocoolers, we investigated the influence of valve open timing and interval on performance of the cryocooler. As a result, it is shown that a much larger cooling capacity is available on the first stage, as well as the cooling capacity on the second stage is more than 1 W at 4.2 K. This paper describes the experimental details, and discusses the results with pressure-volume(P-V) diagrams of both the stages.

EXPERIMENTAL DETAILS

The experimental setup shown in figure 1 is used in the present study. Both the RD-210L4 type cold head of the 4 K GM cryocooler and the compressor unit of CSW-71A type are manufactured by Sumitomo Heavy Industries Limited. In the cold head, the first regenerator has both copper meshes and lead spheres, with the latter arranged at the colder end. The second regenerator is also a hybrid one with lead spheres of 360 g placed at the hotter end and spherical magnetic regenerator material, $ErNi_{0.9}Co_{0.1}$, of 360 g placed at the colder end. The stroke of the displacer is 30 mm, and the inside diameters of the cylinders are respectively 82 mm for the first stage and 35 mm for the second stage.

A rotary valve which consists of a crank, a plate and a body is employed for intake/exhaust in the cryocooler. As is shown in figure 2, the valve open timing and the interval are usually represented by an angular measure, such as -42°/-46° for open timing and 124°/173° for interval. The minus value means that the intake/exhaust valve opens at the angle (=-42°/ -46°) before the expansion volume turns to minimum/maximum (defined as 0°/0°). The valve interval given above, therefore, indicates that the intake interval is 124° and the exhaust one is 173°. Because of the angular expression, the valve open timing will

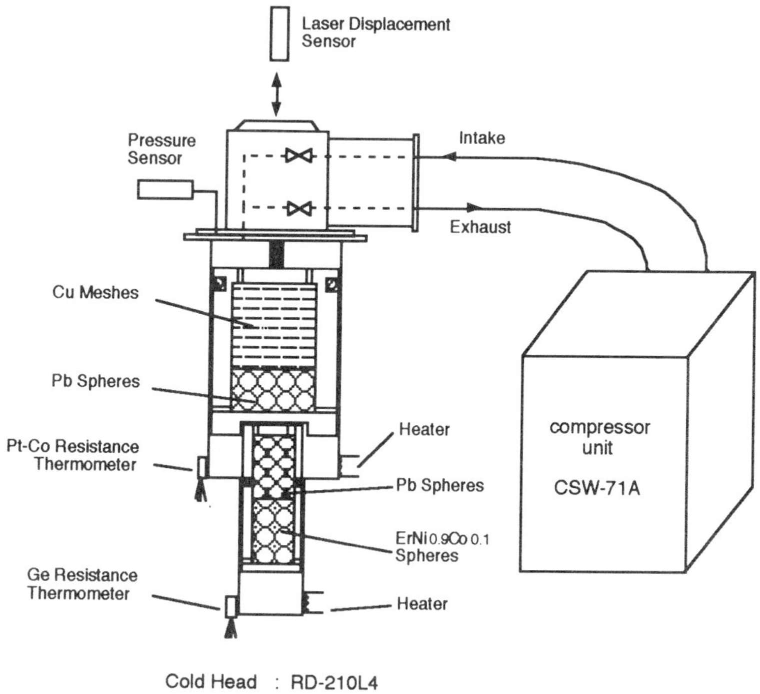

Figure 1. Schematic of experimental setup.

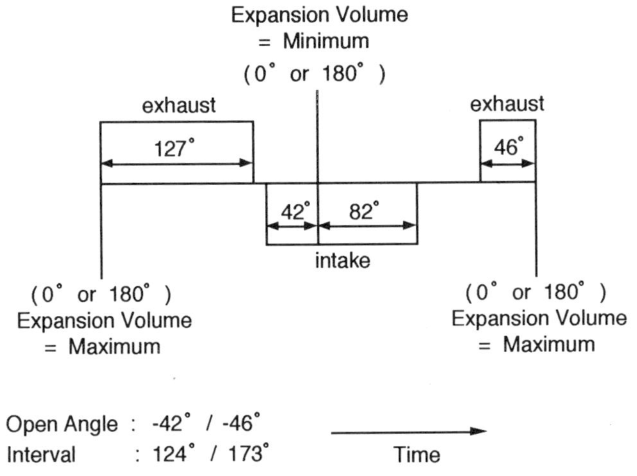

Figure 2. Definition of valve open timing and valve interval by an angular measure.

be also designated as valve open angle as the following. The 18 different valves were prepared for this work. Their open angles are from -12°/-16° to -62°/-66°, and the intervals are from 104°/153° to 134°/183°.

The pressure of working gas in the cold head is measured in the room temperature chamber with a pressure transducer of the strain gauge type. The expansion volumes of both the stages are calculated from the displacer position which is monitored by a laser displacement sensor. A Pt-Co resistance thermometer and a Germanium resistance thermometer are respectively used to obtain the first and the second stage temperatures. The heat loads to both the stages are supplied by electric heaters of manganin wire.

RESULTS AND DISCUSSIONS

Figure 3 shows cooling capacities on both the stages for all of the 18 valves, under the conditions that the temperature of the first stage is 40 K while that of second stage is 4.2 K.

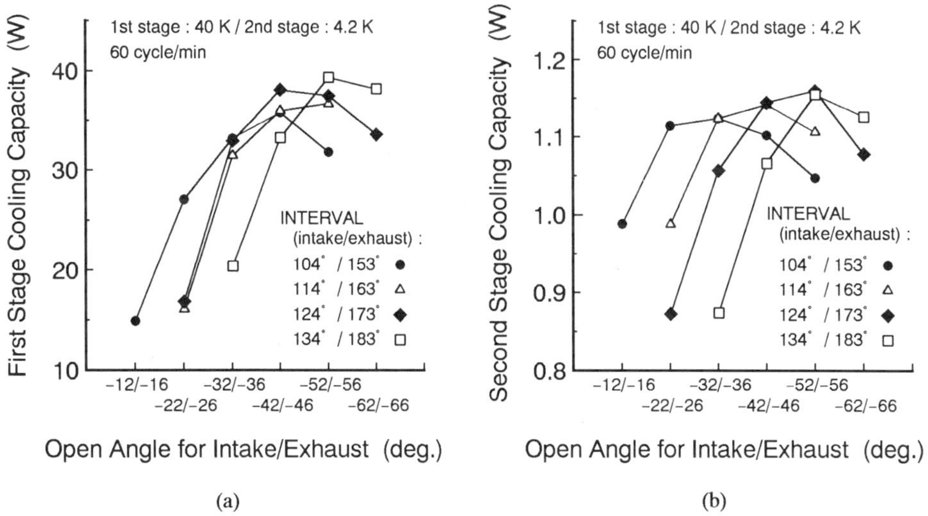

Figure 3. Cooling capacities on (a)first stage and (b)second stage for all of the 18 valves.

It is clear that the performance of the 4 K GM cryocooler is greatly influenced by intake/exhaust valve open timing and interval. For each valve interval, there is an optimum valve open angle to give a maximum cooling capacity on the first stage or on the second stage. Since the larger absolute value in the horizontal axes of Fig. 3 means that the intake/exhaust valve opens earlier, Fig. 3 shows that a reasonably early valve open timing brings about much larger cooling capacities not only on the second stage but also on the first stage. When the valve interval is 124°/173°, for example, the second stage cooling capacity at 4.2 K of the -52°/-56° valve is 1.16 W which is 0.29 W larger than that of the -22°/-26° valve, and the first stage cooling capacity at 40 K of the -42°/-46° valve is 38.03 W which is 21.25 W larger than that of the -22°/-26° valve.

Though the optimum valve open angles for the second stage cooling capacity are much different for each valve interval, there is a strong resemblance in the valve open angle dependence of cooling capacity on both the stages among the four valve intervals. For making discussion easy to understand, therefore, we mainly focus on the results about the interval of 124°/173° in the following discussions.

Figure 4 is the load map for valves with the interval of 124°/173°. Whether the second stage is heated or not, the first stage cooling capacity shows a maximum at an open angle of -42°/ -46°. As can be seen in figure 5, when the first stage has no load, the later the valve opens, the larger the second stage cooling capacity at 4.2 K is. For the condition that the first stage is warmed up to 40 K, the cooling capacity at 4.2 K is maximum for the open angle of -52°/-56°.

The P-V diagrams corresponded to Fig. 4 are given in figure 6. Since the pressure of the working gas is not directly measured at the cold stages, Fig. 6 only describes the pressure changes versus expansion volume approximately, but it is considered that the diagrams are enough to make a comparison among the valves. For an earlier intake/exhaust valve open timing, the P-V diagram is longer and narrower, and the pressure change in expansion volume is larger while the area of the diagram becomes smaller.

In the ideal case of GM cycle, the intake/exhaust valve opens exactly at the moment that the expansion volume is minimum/maximum, and the interval for intake/exhaust is just 180°. As a result, the GM cycle draws a rectangular P-V diagram and delivers a maximum cooling capacity which is equal to the P-V work.

Figure 7 shows the P-V work and the maximum pressure change(P_{max}-P_{min}) in the first stage expansion volume. For the -22°/-26° valve which open timing is closest to the ideal case, the P-V work is certainly largest, but the first stage cooling capacity is smallest as shown as in Fig. 3 and Fig. 4. On the other hand, the maximum pressure change in expansion volume increases when the valve open timing becomes earlier. The increasing maximum pressure change and the decreasing P-V work for an earlier open timing are

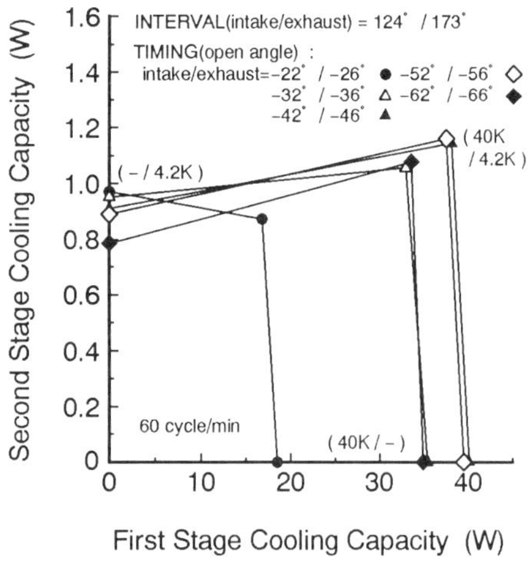

Figure 4. Load map for valves with the interval of 124°/173°.

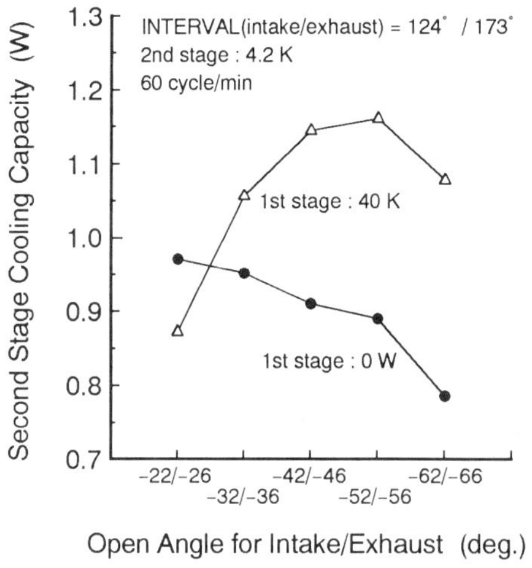

Figure 5. Second stage cooling capacity at 4.2 K for the cases that first stage is 40 K and is on-load.

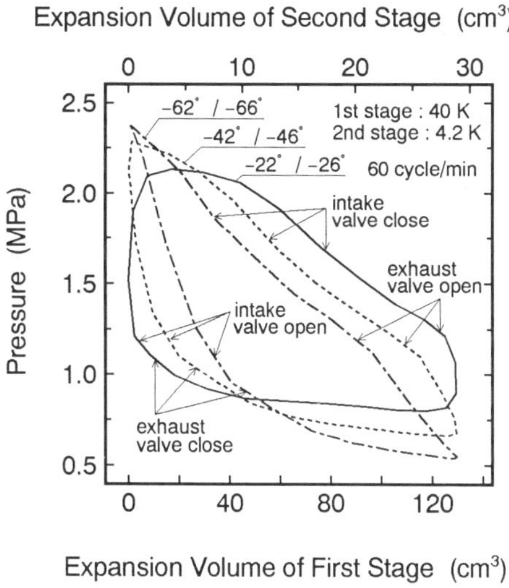

Figure 6. P-V diagrams for valves with the interval of 124°/173°.

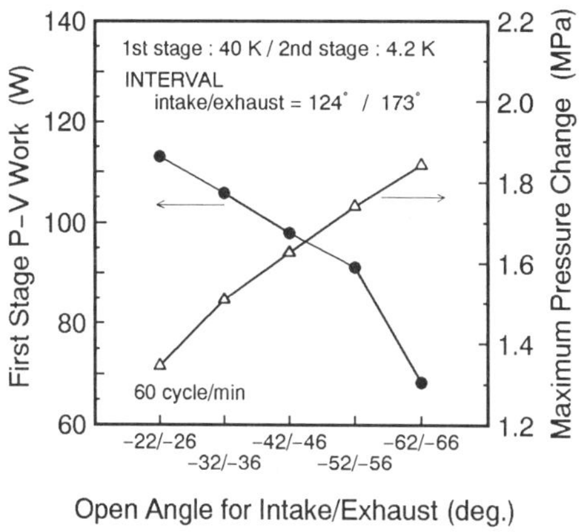

Figure 7. P-V Work and maximum pressure change in first stage expansion volume.

thought to be the reason that the optimum open angle for the first stage cooling capacity is not -22°/-26°, but -42°/ -46°.

In a real GM cryocooler, there is a resistance against the gas flow through valves and displacers. This resistance causes pressure drops at such components. The ideal GM cycle, however, leaves the pressure drop out of consideration, and imagines that the pressure changes from lower/higher one to higher/lower one as soon as the intake/exhaust valve opens. Consequently for a real cold head and a given compressor unit, the ideal valve open timing does not always yield a sufficient intake/exhaust process and does not always produce a cooling capacity in expansion process as large as expected. In other words, it is necessary to take an early intake/exhaust valve open timing for reducing the pressure drop at intake/exhaust valve and getting a powerful cooling capacity on the cold stages.

The reduction in the pressure drop by means of an early valve open timing is shown clearly in figure 8. In the figure, the pressure difference between the supply pressure and the return pressure of compressor unit and the maximum pressure change in the cold head are given as functions of the valve open angle. There are two advantages of taking an early valve open timing. One is the increase in the pressure difference between the supply pressure and the return pressure, another is the decrease in pressure drop at the intake/exhaust valve. Both of these effects contribute to the good performance of the first stage at -42°/-46°.

For the second stage, it is considered to be somewhat different from the first stage. As described above, the open angle dependence of the cooling capacity at 4.2 K is dissimilar for the cases that the first stage is 40 K or no-load (Fig. 5). Figure 9 shows the P-V work and the maximum pressure change of the second stage. When the first stage is warmed from the no-load case up to 40 K, the maximum pressure change in the cold head increases with almost the same degree for every valve open angle. If the maximum pressure change was a dominating factor for the second stage, the open angle dependence of the 4.2 K cooling capacity would be similar whether the first stage is heated or not.

The difference in open angle dependence of the 4.2 K cooling capacity can be explained by the increase in P-V work. Fig. 9 also gives the increment of the second stage P-V work ($\Delta W_{P\text{-}V\ work}$) and that of the maximum pressure change in the cold head (ΔP_{40K}-ΔP_{OW}), for warming the first stage from the no-load case up to 40 K. The increment of maximum pressure change is almost independent of the valve open angle, but that of the P-V work has a maximum at -52°/-56°. When the first stage is no load, the second stage cooling capacity at 4.2 K is a decreasing function of valve open angle and has an analogy with the P-V work. For the condition that the first stage is 40 K, the cooling capacity at 4.2 K is

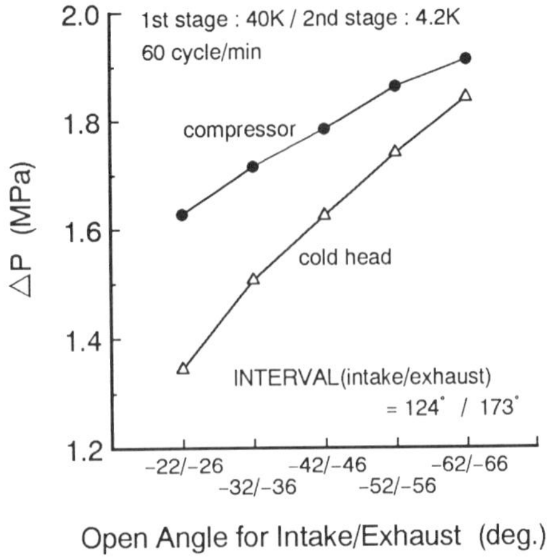

Figure 8. Pressure difference between supply pressure and return pressure of compressor unit and maximum pressure change in cold head.

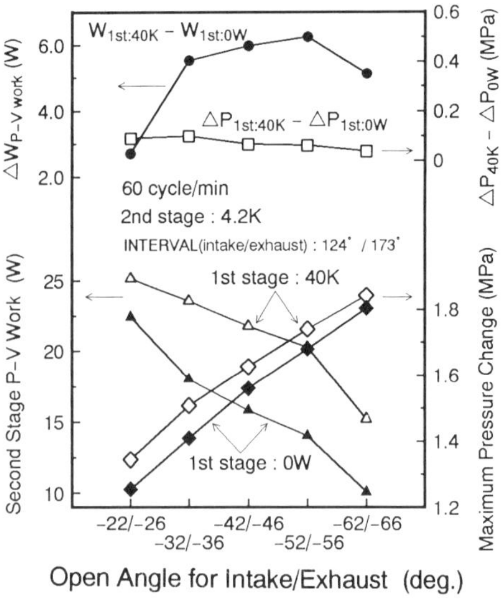

Figure 9. P-V work and maximum pressure change in second stage, and their increments for warming first stage from no-load case up to 40 K.

maximum at the open angle of -52°/-56°, which corresponded to the increment of P-V work shown in Fig. 9.

The effect of reduction in pressure drop by taking an early valve open timing also appears when the cycle speed is changed. Figure 10 gives the cooling capacity variations of both the stages versus the cycle speed. As is shown, by changing the valve open angle from -22°/-26° to -62°/-66°, the optimum cycle speed can be shifted from lower one (<=48 cycle/min) to higher one (>=60 cycle/min).

The first stage P-V works correlated with Fig. 10 are shown in figure 11. The ratios of the maximum pressure change in cold head to the pressure difference between the supply

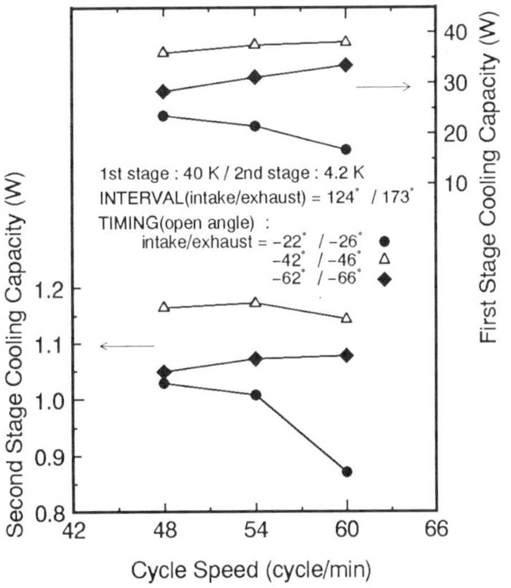

Figure 10. Cooling capacities on both stages as the functions of cycle speed.

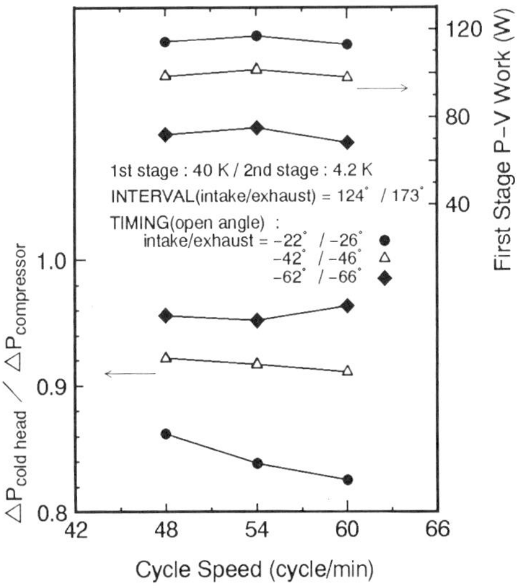

Figure 11. First stage P-V work, and the ratio of maximum pressure change to pressure difference of compressor unit.

pressure and the return pressure of compressor unit are plotted in the figure too. The P-V work of each valve open angle has a different value, but is roughly independent of the cycle speed. This means that the cycle speed dependence of cooling capacity does not result from the P-V work. For the ratio of maximum pressure change to pressure difference, on the other hand, there is a resemblance to the cooling capacities given in Fig. 10. In comparison with the ratio at 48 cycle/min, those at a higher cycle speed increase more rapidly with an earlier valve open timing and become almost the same as the ratio at a lower cycle speed. This fact indicates that an early valve open timing is much more effective in reducing the pressure drop at intake/exhaust valve for a higher cycle speed and that is the reason why the optimum cycle speed shifts to higher one by means of an early valve open timing.

As reported by other researchers the optimum performance of 4 K GM cryocooler is usually achieved at cycle speeds lower than those for conventional GM cryocoolers, however, Fig. 10 suggests a different way to operate a 4 K GM cryocooler with a good performance, though the cycle speeds can be the same as used in conventional GM cryocoolers.

CONCLUSIONS

The influence of intake/exhaust valve open timing and interval on the performance of a 4K GM cryocooler has been investigated. The results indicate that the cooling capacity of both the stages is strongly dependent on valve open timing rather than valve interval. The advantages brought about by taking a reasonably early intake/exhaust valve open timing are summarized as the following.

(1) A much larger first stage cooling capacity, close to 40 W at 40 K, is available.

(2) A cooling capacity above 1.1 W at 4.2 K on the second stage can be obtained even though the first stage is warmed up to 40 K. This feature is favorable to many practical cryogenic uses.

(3) A 4 K GM cryocooler can be operated at usual speeds as the same as those for conventional GM cryocoolers with the optimum performance.

In addition, the optimum conditions found in this work are the valve open angle of -52°/-56°, the valve interval of 134°/183°, and the cycle speed of 54 cycle/min. The cooling capacity under these conditions is 39.4 W at 40 K on the first stage, as well as 1.18 W at 4.2 K on the second stage, with an input power of 7.05 kW to compressor unit.

According to the results and the discussions above, the valve open timing and the valve interval are truly important to a 4 K GM cryocooler, but the authors should like to point out that the open timing and the interval are only two parameters of many in a GM cryocooler and the essence in this work is just how to reduce the pressure drop at intake/exhaust valve.

REFERENCES

1. R.Li, T.Hashimoto, M.Sahashi and Y. Tohkai, in "Proceedings of the Twelfth International Cryogenic Engineering Conference", Butterworth, Guildford, UK (1988), p.423.
2. M.Sahashi, Y.Tokai, T.Kuriyama, H.Nakagome, R.Li, M.Ogawa and T.Hashimoto, Advances in Cryogenic Engineering, Vol.35, Plenum Press, New York (1990), p.1175.
3. R.Li, M.Ogawa and T.Hashimoto, Cryogenics, Vol.30 (1990), p.521.
4. M.Ogawa, R.Li and T.Hashimoto, Cryogenics, Vol.31 (1991), p.405.
5. M.Nagao, T.Inaguchi, H.Yoshimura, S.Nakamura, T.Yamada and M.Iwamoto, in "Proceedings of the Sixth International Cryocooler Conference" Vol.II, David Taylor Research Center Report DTRC-91/002, Ed. by G.Green and M.Knox, Bethesda, MD (1991), p.37.
6. G.Ke, H.Makuuchi, T.Hashimoto, A.Onishi, R.Li, T.Satoh and Y.Kanazawa, Advances in Cryogenic Engineering, Vol.40, Plenum Press, New York (1994), p.639.
7. T.Kuriyama, M.Takahashi, H.Nakagome, T.Hashimoto, H.Nitta and M.Yabuki, Advances in Cryogenic Engineering, Vol.39, Plenum Press, New York (1994), p.1335.
8. R.Plambeck, N.Thatte and P.Sykes, in "Proceedings of the 7th International Cryocooler Conference", Air Force Phillips Laboratory Report PL-CP-93-1001, Kirtland AFB, NM (1993), p.401.
9. M.Nagao, T.Inaguchi, H.Yoshimura, T.Matsumoto, S.Nakagawa, K.Moritsu and T.Watanabe, Advances in Cryogenic Engineering, Vol.39, Plenum Press, New York (1994), p.1327.
10. K.Watanabe, S.Awaji, T.Fukase, Y.Yamada, J.Sakuraba, F.Hata, C.K.Chong, T.Hasebe and M.Ishihara, Cryogenics, Vol.34, ICEC Supplement (1994), p.639.

OPERATION CHARACTERISTICS AND ANALYSIS OF A TWO STAGE G-M CRYOCOOLER

Shimo Li, P. Huaiyu,
C. Yan, L. Qifen,
and Z. Haoping

Cryogenic Lab., Zhejiang University
Hangzhou, 310027 P. R. China

ABSTRACT

This paper consists of two parts: (1) an introduction to the experimental results using rare earth intermetallic compounds in the second-stage regenerator. Our experiment shows that there is a wide range of suitable regenerator matrices when refrigeration temperature is above 15K; (2) a description of the changes in warm-chamber gas pressure, influence of the void space and enthalpy loss during normal operations.

INTRODUCTION

Current theoretical and experimental research in the field of regenerative refrigerator emphasizes the search for new types of regenerator matrices for enhancing regenerator efficiencies. This leads to break-throughs in developing regenerative refrigerators operating at LHe temperatures. For refrigerators working at higher temperatures, the main focus is on the improvement of the structures and usable lifetime. The typical requirement for long-time operation without failure is approximately 5 years and longer.

This paper compares the experimental results using conventional and non-conventional regenerator matrices for operations on a two-stage G-M refrigerator. Changes in warm-chamber pressure in a variable-mass system, void-space effect and temperature effects are also discussed.

Table 1. No-load cooling temperatures for various second stage matrices

Matrix (2-stage)	ratio of volume	T_1(K)	T_2(K)
200g Pb	—	42	7.0
100g Pb + 57g MH①	1 : 1	38.6	7.7
100g Pb + 61g MH②	1 : 1	40.5	7.7
100g Pb + 61g MH③	1 : 1	39.8	7.7
100g Pb + 48g MH④+42g MH⑤	2 : 1 : 1	46.3	10.3
100g Pb + 6.15g active carbon	1 : 1	43.0	9.0
163g Pb + 3.3g active carbon	3 : 1	—	8.7
163g Pb + 30.7g Er_3Ni	3 : 1	40.4	6.4

①MH I ②Mm $Ni_{4.5}Mn_{0.5}$ ③La Ni_5 ④MH I +Pb ⑤Mm $Ni_{4.5}Mn_{0.5}$ Sintered

EXPERIMENTAL PLATFORM AND RESULTS

The main specifications of the two-stage G-M cryocooler are as follows:

Second-stage cooling capacity: 3W/20K
No-load cooling temperature: T_1=45K, T_2=9.4K
Normal expansion rotation speed: 60rpm
Second-stage regenerator height: 60mm
Second-stage regenerator diameter: 22mm

The experimental results for various regenerator matrices are listed in Table 1 and Figure 1. This cryocooler reaches the lowest cooling temperature of 5.6K using Er_3Ni under variable-frequency operating conditions.[1]

CHOICE OF REGENERATOR MATRICES

Regenerator matrices suitable for low-temperature cryocoolers can be roughly divided into the following three types:

1. Debye matrices

These regenerator matrices are suitable for operation above 15K. Their specific heat-temperature relations observe Debye theory, i.e. the lattice specific heat is proportional to

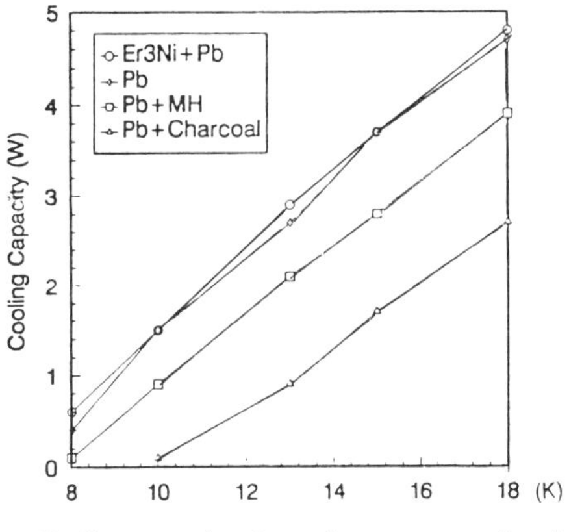

Fig. 1 Cooling capacity (zero first-stage cooling load).

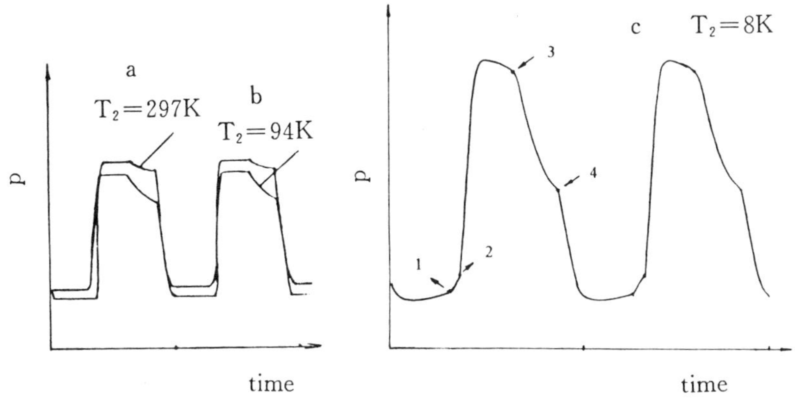

Fig. 2 Warm-chamber pressure variation vs time.

the third power of temperature: $C_{v,l}=464.5\left(\frac{T}{\theta_D}\right)^3$. Since the specific heat of these matrices decrease drastically with decreasing temperature, conventional regenerative cryocoolers usually can operate above 10K only.

2. Adsorption matrices

These are the active carbon, molecular sieve and some metallic substitutes for helium adsorption at low temperatures. The resultant lowest cooling temperature is slightly decreased. In fact, this type of matrices are only suitable for temperature below 8K according to the thermophysical properties of helium. Above 10K the mass-specific heat difference of helium before and after expansion is diminishingly small and the adsorption effect nearly vanishes.

3. Magnetic matrices

Recently developed heavy rare-earth lanthanide matrices[2] and ceramic matrices utilize the magnetic phase transitions in metals as temperature decreases or Schottky effect induced abnormal increase in thermal capacities[3]. The peak magnetic thermal capacity values produced by magnetic phase transitions are usually one order of magnitude larger than lattice thermal capacities. Newly discovered matrices include intermetallic compounds composed of heavy rare-earthes Er, Dy, Ho, Gd, Nb and transition metals Co, Ni etc. Their magnetic thermal capacities are, however, peaked very narrowly so a single material cannot satisfy the requirement for the lowest cooling temperature and cooling with heat load at various operation temperatures. Consequently, layered composite matrix regenerator is more desirable. Matrices with high thermal capacity including both peak magnetic thermal capacity and lattice thermal capacity are also desirable. This type of matrices have lower Debye temperatures.

VARIATION OF WARM CHAMBER PRESSURE

Correct measurement of the pressure variations in the expansion chamber can help analyzing the thermal operating conditions in cryocoolers. Considering that the warm-chamber pressure variation can approximately reflect the pressure changes in the cold chamber, and it is relatively easy to measure pressure in the warm-chamber, we used a BPR-2 type pressure transducer, and an SC-16 optical oscilloscope for measuring warm-chamber pressure.

The warm-chamber pressure variation and valve openings at each sections under varied second-stage temperature T_2 are shown in Fig. 2. With a decreasing T_2, the warm-chamber pressure waves gradually deviate from the square wave form for an ideal G-M cryocooler. This is due to the combined effects of the regularity of the exhaust, phase difference between that of the inlet and outlet and the thermophysical properties of the working helium medium. At $T_2=8K$, the cryocooler operates in a stable state and the pressure variation is close to sinusoidal. If the regenerator impedance is neglected, then the cold-chamber pressure variation should also be sinusoidal. It can be seen from the figure that pressure variation amplitudes between nodes (1) and (2), (3) and (4) increase with increasing temperature difference between the warm and cold chambers. There are virtually no isobaric charge and discharge processes when the cold-chamber temperature is very low. The pressure waves can be explained using the temperature dependence of the helium density. Inlet and outlet valve openings can also be optimized.

The measurement of the second-stage cold-chamber temperature variation corresponding to the first group of experiments in Table 1 is shown in Figure 3. The nodes correspond to those in Figure 2c. The cold end temperature variation can be easily explained from the pressure variation.

VOLUME EFFECT OF THE VOID SPACE IN REGENERATOR

After some neccessary assumptions by convention, the control-volume-gas flow equations can be written according to the variable-mass-system thermodynamics:

$$\frac{\alpha A_f}{L}(T-T_m)+C_p\frac{\partial(\dot{m}T)}{\partial x}+C_p\rho A_c\frac{\partial T}{\partial t}-\rho A_c T\left(\frac{\partial v}{\partial T}\right)_p\frac{\partial p}{\partial t}=0 \qquad (1)$$

where the first term is the heat exchange between the gas and matrix; the second term is the corresponding enthalpy change; the third term represents the energy change due to the time dependent gas temperature in the regenerator; and the fourth term represents the energy change due to the time dependent gas pressure.

The influence of the first and second terms on the regenerator efficiency can be easily understood. The helium specific heat and density are rather large when the temperature in the control volume is very low, hence the third term cannot be neglected. the helium in

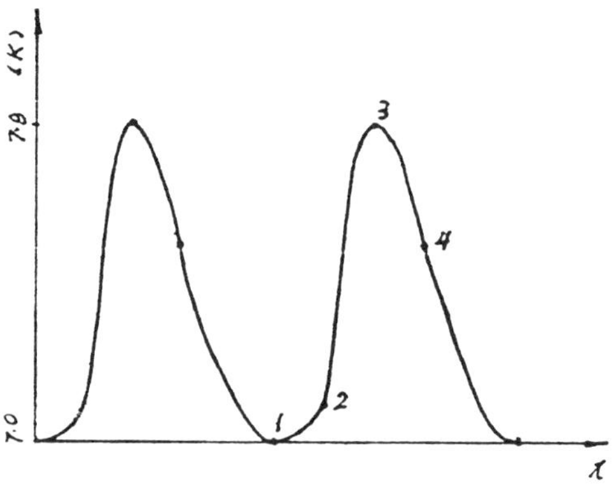

Fig. 3 cold-chamber temperature variation.

Table 2. The adiabatic-expansion temperature difference in the cold-chamber gas

T_i(K)	P_h(MPa)/P_l(MPa)	ΔT_e(K)
9/8/7/6	1.8/0.45	2.5/2.1/1.5/1.1
9/8/7/6	1.8/0.6	1.9/1.7/1.3/0.9
9/8/7/6	1.8/0.8	1.3/1.4/1.0/0.7

the control volume functions roughly as a "matrix", slowing the tendency for decreasing regenerator efficiency due to the decreasing thermal capacity of the solid matrix. The active carbon used in the experiment has low specific heat, but its helium adsorption effect reduces the thermal saturation in the regenerator. The fourth term in the above equation reflects the compression term in the internal energy change. It is a cooling capacity loss component in cryocoolers. From the warm-chamber pressure variations, the compression heat flux corresponding to $\partial p/\partial t$ is

$$\theta_c = \rho A_c T\left(\frac{\partial v}{\partial T}\right)_p \frac{\partial p}{\partial t} = A_c \alpha_p T \frac{\partial p}{\partial t} \quad (2)$$

where $\alpha_p = \frac{1}{v}\left(\frac{\partial v}{\partial T}\right)_p$ is the isobaric expansivity of the gas.

According to the thermophysical properties of helium, it can be derived that there is a specific-heat peak near 10K and its value depends on the temperature and pressure. For temperatures larger than 20K, helium can be treated as an ideal gas and from pV=RT, we obtain:

$$\rho A_c T\left(\frac{\partial v}{\partial T}\right)_p \frac{\partial p}{\partial t} = A_c \frac{\partial p}{\partial t} \quad (3)$$

i.e. the compression heat is no longer temperature dependent. On the other hand, when temperature drops below 6K the compression heat decreases rapidly, thus reducing the cooling capacity loss due to the compression heat. As a result, the regenerator efficiency is even improved. This phenomenon can be used to explain the experimental results in reference.[4]

REGENERATOR EFFICIENCY AND COLD-CHAMBER-EXPANSION TEMPERATURE DIFFERENCE

This section discuss the effect of the regenerator cold-end temperature difference ΔT_r and cold-chamber-expansion temperature difference ΔT_e on cryocooler efficiency. ΔT_r not only depends on regenerator efficiency but also depends on cold-chamber temperature and pressure.[5] This ΔT_r causes the following enthalpy loss:

$$\Delta Q_r = \dot{m} C_p \Delta T_r (= \oint d(\dot{m}h)_{in} - \oint d(\dot{m}h)_{out}) \quad (4)$$

where $\dot{m}$ is the mass flux of the gas, C_p is the gaseous specific heat, and ΔT_r is the temperature difference between the inlet and outlet gas at the regenerator cold end.

Assuming the gas flow in the cold chamber to be Simon isoentropy expansion and knowing the P_h and P_l values, the temperature difference before and after the expansion can be determined. Accordingly, the cooling capacity is

$$Q_c = \dot{m} C_p \Delta T_e \quad (5)$$

If $\dot{m}$ and C_p in eqs. (4) and (5) are approximately equal, then for obtaining net cooling capacity, ΔT_e must be greater than ΔT_r. Because ΔT_e decreases drastically due to decrease in the expansion ratio P_h/P_l and the temperature T_i before expansion, as in Table 2, and that $\triangle T_r$ remains roughly unchanged, one must try to lower $\triangle T_r$ or reduce the enthalpy loss.

CONCLUSIONS

From the above analysis, we can conclude:

1. If we want to lower the refrigeration temperature of a two-stage G-M cryocooler or a Solvay cryocooler to LHe temperatures, the cold-end regenerator matrix in the second-stage regenerator must be a matrix with the corresponding specific heat peak value, and the first stage cooling temperature must be below 30K. Since the lowest cooling temperature of the first stage in our experimental cryoccoler is only 39K, the second-stage refrigeration temperature can reach 5.9K only at 37rpm rotation speed.

2. Our experiment has demonstrated that there are a wide range of regenerator matrices suitable for cryocoolers operated above 10K. rase-earth metallic hydrides can be expected to take the place of lead spheres as the matrix.

3. Under steady operating conditions, the phase difference of the inlet and outlet openings in the expansion chamber can be optimized by analyzing the warm-chamber pressure waves.

REFERENCES

1. S. Li, Y. Mao, H. Zhao, H. Pan, Z. Huang and S. Ge, Theory and Experiment for Rare-Earth Magnetic Regenerative Materials, p625-630 in "Advances in Cryogenic Ergineering-Materials", vol. 40 A, Plenum, New York (1994)
2. T. Hashimoto, M. Ogawa, R. Li, and K. Aoki, Recent progress on rare earth magnetic regenerator materials, p859 in "Advances in Cryogenic Engineering-Materials", vol. 37B, Plenum, New York(1991)
3. L. L. Sparks, Specific heat, p. 47 in "Materials at Low Temperature", R. P. Reed and A. F. Clark, eds., American Society for Metals, Metals Park, Ohio (1983)
4. T. Kuriyama, et al, High efficient two-stage GM refrigerator with magnetic material in the liquid helium temperature region, p1261 in "Advances in Cryogenic Engineering, vol. 35B, Plenum, New York(1990)
5. G. Walker, Chapter 5. p247 in "Cryocoolers, Part 1: Fundamentals", Plenum, New York (1983)

OPTIMIZATION OF OPERATIONAL PARAMETERS FOR A 4 K-GM REFRIGERATOR

T.Kuriyama[1], Y.Ohtani[1], M.Takahashi[1], H.Nakagome[1], H.Nitta[2], T.Tsukagoshi[2], A.Yoshida[2], and T.Hashimoto[2]

[1]Toshiba Research and Development Center
Ukishima, Kawasaki, 210, Japan
[2]Tokyo Institute of Technology
Oh-Okayama, Meguro, 152, Japan

ABSTRACT

This paper describes the experimental results of a 4 K-GM refrigerator, which uses magnetic regenerator materials. The technical point of this study is ascertaining the effects of operational parameters for improving the 4.2 K refrigeration capacity. Compressor capacity and displacer stroke are main parameters in this study. A larger compressor enables a larger refrigeration capacity at a smaller temperature difference in the second regenerator. A smaller compressor, however, is preferable to maintain 4.2 K, when temperature difference is large. Displacer strokes of 12, 20 and 32 mm are investigated. The optimum reciprocating speed for the 4.2 K refrigeration capacity strongly depends on the displacer stroke. The largest refrigeration capacity, however, are almost equal for each of the strokes. In this study, the optimized stroke is found to be 20 mm to obtain refrigeration capacities at the second stage (4.2 K) and the first stage (40 K) simultaneously.

INTRODUCTION

Two-stage GM refrigerator is the most commonly used cryocooler in many cryogenic systems, such as cryopumps and thermal shield coolers for MRI magnets or other superconducting magnets. The lowest temperature for a conventional two-stage GM refrigerator, however, is limited to about 8 K level. Because the specific heat of conventional regenerator material, Pb, decreases rapidly with decreasing temperature and it is not large enough below 10 K level.

The authors, for the first time, succeeded in achieving liquid helium temperature region by conventional two-stage GM refrigerator using a magnetic regenerator material, erbium 3 nickel (Er_3Ni)[1,2]. Thereafter, much effort to obtain lower cooling temperatures and larger refrigeration capacities at 4.2 K has been done[3-5]. Recently, 4 K-GM refrigerators have been applied to some cryogenic systems, such as an SIS receiver for astronomical radio observation[6,7], MRI[8] and a cryocooler-cooled superconducting magnet[9].

Two procedures have been carried out to improve the 4.2 K refrigeration capacity. The first procedure was to select the magnetic regenerator materials and optimize the regenerator structure. Two or three layers of regenerator materials in the second regenerator has improved the 4.2 K refrigeration capacity[10,11]. The second procedure was to change the operating conditions of the refrigerator. The reciprocating speed of the displacer and the valve timing for the supply/exhaust valves had a strong effect on the 4.2 K refrigeration capacity. Since helium gas behavior is non-ideal in liquid helium temperature region, the refrigeration mechanism at 4.2 K is much different from that when helium gas behavior is ideal. Therefore, the optimum operating condition was not equal to that of conventional two-stage GM refrigerator.

In this paper, the experimental results of the second procedure, which changes the operational parameters of the GM refrigerator, are described. Compressor capacity and displacer stroke effects on the improvement of the 4.2 K refrigeration capacity were investigated.

EXPERIMENTAL APPARATUS

A conventional two-stage GM refrigerator was used in this paper. The first and the second cylinder sizes were 70 mm and 32 mm in diameter and 220 mm and 260 mm in length respectively. The displacer stroke was one of the experimental parameters of this study, and the details are described in the next section.

Two kinds of graded regenerators with spherical magnetic regenerator materials were used as the second regenerator. One was two layers of Er_3Ni and $Er_{0.9}Yb_{0.1}Ni$ and the other was three layers of $Er_{0.75}Gd_{0.25}Ni$, Er_3Co and $Er_{0.9}Yb_{0.1}Ni$. Figure 1 shows the volumetric specific heat of these materials. The diameter of these materials were 180 - 450 micro meter.

Compressor capacity is the other experimental parameter of this study. Three types of compressor were used with the cold head mentioned above. Details of these compressors are also described in the next section.

The displacers were driven by an AC induction motor. The reciprocating speed could be

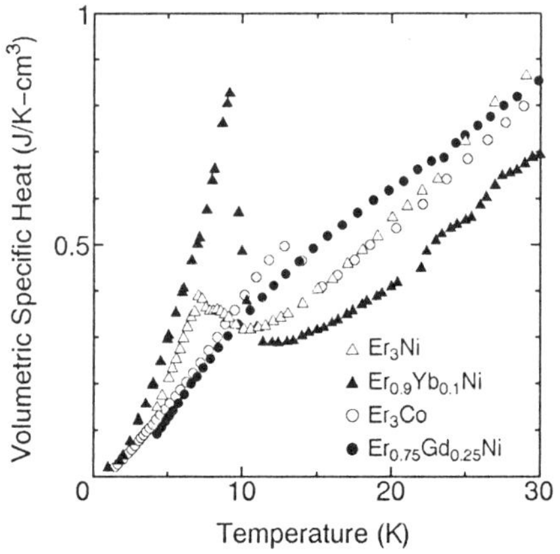

Figure 1. Volumetric specific heat of magnetic regenerator materials.

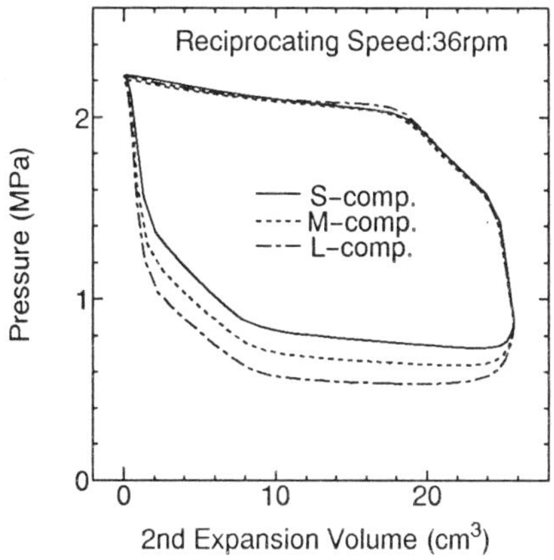

Figure 2. Second stage PV diagrams obtained by three compressors.

varied by changing the supply frequency. A displacement transducer and a pressure transducer were mounted to obtain a Pressure - Volume (PV) diagram for the expansion space. Carbon glass resistance thermometers and electrical heaters were mounted at both refrigeration stages.

RESULTS AND DISCUSSIONS

Effect of Compressor Capacity

The effect of compressor capacity on the 4.2 K refrigeration was investigated using three kind of compressors. Input power to the compressors were 3 kW, 5 kW and 7 kW. In this paper, these compressors are called as S-compressor, M-compressor and L-compressor respectively for the sake of convenience. The second regenerator has a graded structure of Er_3Ni (50 % of higher temperature side) and $Er_{0.9}Yb_{0.1}Ni$ (50 % of lower temperature side).

The displacer stroke of 32 mm was used in this experiment and the expansion volume at the second refrigeration stage was 25.7 cm^3. Figure 2 shows the PV diagrams for the second expansion space obtained by the three compressors. These diagrams are obtained at the reciprocating speed of 36 rpm and second stage temperature of 4.2 K. The PV works for “S”, “M” and “L” compressors were 15.8 W, 17.3 W and 19.3 W, respectively.

Figure 3 shows the relationship between the displacer reciprocating speed and the refrigeration capacity at 4.2 K obtained by "S" and "L" compressors. There was no heat load at the first stage. The first stage temperatures were between 25 K and 35 K. Both results have almost the same tendency and the largest refrigeration capacity is obtained at 30 rpm reciprocating speed. The maximum refrigeration capacities were 0.78 W for L-compressor and 0.59 W for S-Compressor.

Figure 4 shows the relationship between the first stage temperature and the 4.2 K refrigeration capacity obtained by "S" and "L" compressor. The first stage temperature was changed by changing the heat load to the first stage. When the first stage temperature is low, i.e.; a temperature difference in the second regenerator is small, a larger refrigeration capacity is obtained by L-compressor, which provides a larger PV work. The refrigeration capacities for both compressors decrease with increasing the first stage temperature. The result for L-compressor, however, decreases more rapidly compared with that for S-compressor. In spite

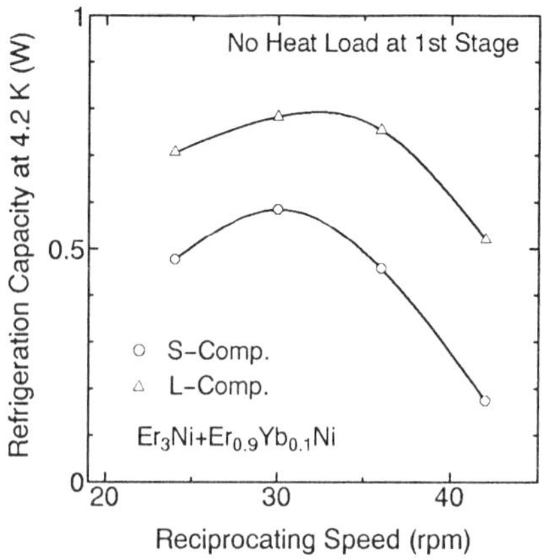

Figure 3. Relationship between reciprocating speed and 4.2 K refrigeration capacity.

of the small PV work of S-compressor, the 4.2 K refrigeration capacity of S-compressor exceeds that of the L-compressor when the first stage temperature is above about 52 K.

The reason why L-compressor cannot maintain 4.2 K for large temperature difference in the second stage is as follows. The 4.2 K refrigeration capacity is determined by an ideal refrigeration and refrigeration losses. The ideal refrigeration depends on a PV work at the expansion space. The dominant refrigeration loss at this temperature is considered to be the regenerator ineffciency. When the temperature difference in the second regenerator is small, the regenerator inefficiency is small. As the result, the refrigeration loss becomes sufficiently small in L-compressor, and a larger refrigeration capacity than S-compressor is obtained.

The regenerator inefficiency increases either with increasing mass flow rate through the regenerator or temperature difference in the regenerator. Assuming that the expansion volume is completely swept by the displacer, the mass flow rate at the cold end of the second regenerator is proportional to a swept volume, a reciprocating speed and helium density for high pressure. The high pressure for both compressors are almost the same as shown in Figure 2. Then, the mass flow rate at the cold end of the second regenerator is almost the

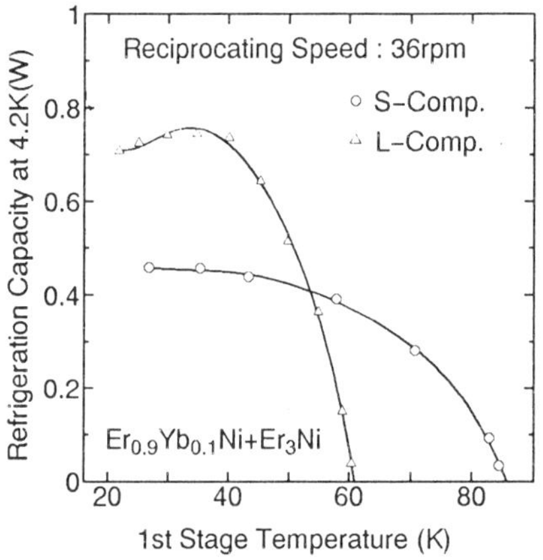

Figure 4. Relationship between first stage temperature and 4.2 K refrigeration capacity.

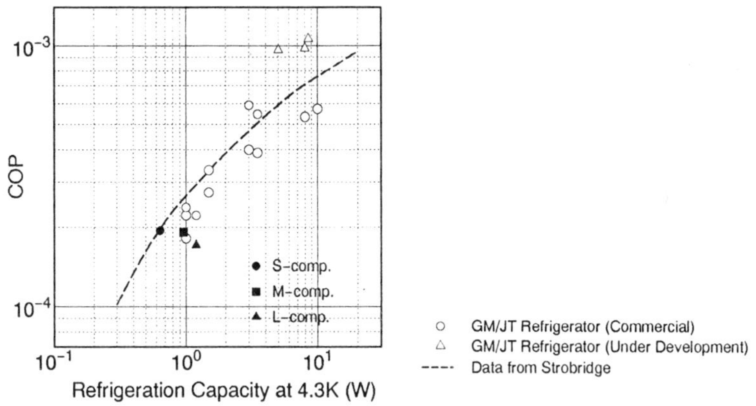

Figure 5. COP for 4K-GM refrigerator.

same for both compressors. On the other hand, the mass flow rate at the warm end of the second regenerator is the sum of mass flow rate to the expansion space and to the dead volume in the regenerator. The former is equal to the mass flow rate at the cold end and those for both compressors are almost equal. The latter is proportional to the dead volume in the regenerator, reciprocating speed and the helium density difference between high and low pressures. Since L-compressor has larger pressure difference than S-compressor as shown in Figure 2, the mass flow rate to the dead volume is much larger than that for S-compressor. As a result, at the warm end of the second regenerator, the mass flow rate for L-compressor is much larger than that for S-compressor. Therefore, when the temperature difference in the second regenerator is larger, the regenerator inefficiency for both compressors increase, especially more rapidly for L-compressor because of its high mass flow rate. That is why the 4.2 K refrigeration capacity for L-compressor decreases more rapidly with an increasing in the first stage temperature, and L-compressor cannot maintain the 4.2 K refrigeration for a larger temperature difference in the second regenerator compared with S-compressor.

The COP values at 4.3 K are calculated for three compressors. The calculated results are shown in Figure 5. The COP values of this study are compared with the data from Strobridge[12] and from GM/JT refrigerators. In the figure, circles indicate the data from commercially available GM/JT refrigerators; triangles are the data from those under development, which use magnetic regenerator materials in the second regenerator. The COP values for the 4 K-GM refrigerator are almost the same as those for commercially used GM/JT refrigerator. The reason why the COP value for L-compressor is the lowest is that the L-compressor is considered too large for the cold head used in this study.

Effect of Displacer Stroke

In this section, the effect of the displacer stroke is discussed. The strokes examined in this study were 12 mm, 20 mm and 32 mm. The expansion volumes at the second stage were 9.7 cm^3, 16.1 cm^3 and 25.7 cm^3. The M-compressor was used in this experiment. The second regenerator has three layers of $Er_{0.9}Yb_{0.1}Ni$ (50 % of lower temperature side) , Er_3Co (25 % of middle part) and $Er_{0.75}Gd_{0.25}Ni$ (25 % of higher temperature side). Other conditions were the same as in the previous section.

Figure 6 shows the PV diagrams at the second expansion space, where the reciprocating speed is 48 rpm and the second stage temperature is 4.2 K. The pressure difference of PV

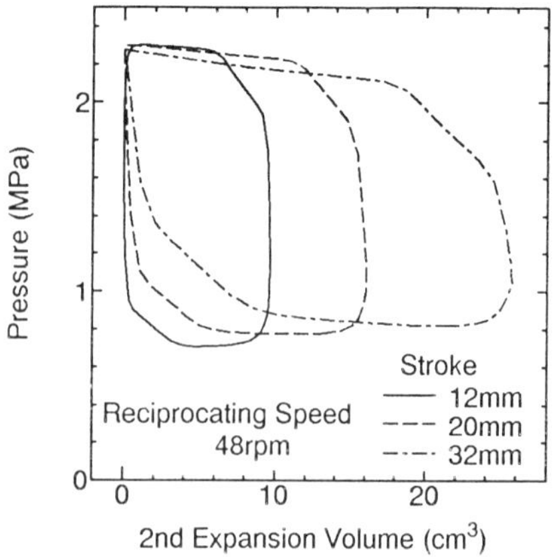

Figure 6. Second stage PV diagrams obtained by various strokes.

diagram decreases with increasing stroke, because the mass flow rate increases with increasing expansion volume, which is determined by the displacer stroke. The PV works for 12, 20 and 32 mm stroke indicated in Figure 6 are 11.0 W, 16.6 W and 22.2 W, respectively.

Figure 7 shows the relationship between the reciprocating speed and no-load temperature at the second stage. The lowest temperatures for each stroke conditions are almost the same at about 2.5 K. However, the dependency against the reciprocating speed is different. Shorter stroke enables to obtain a lower temperature for a wider reciprocating speed range. For example, the no load temperature less than 4 K was obtained by less than 48 rpm for the 32 mm stroke, but the same temperature range was obtained by less than 144 rpm for the 12 mm stroke.

The refrigeration capacity at 4.2 K is shown in Figure 8. No heat load was added to the first stage. As in the case of no-load temperature results, the shorter stroke enables to obtain the 4.2 K refrigeration capacity for a wider range of reciprocating speeds. The reciprocating speed that gives the largest refrigeration capacity for each stroke, increases with decreasing stroke, such as 30 rpm for 32 mm stroke, 42 rpm for 20 mm stroke and 72 rpm for 12 mm

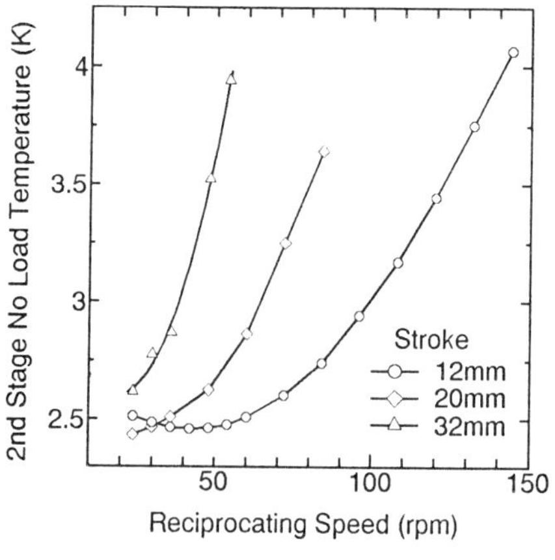

Figure 7. Relationship between reciprocating speed and no-load temperature.

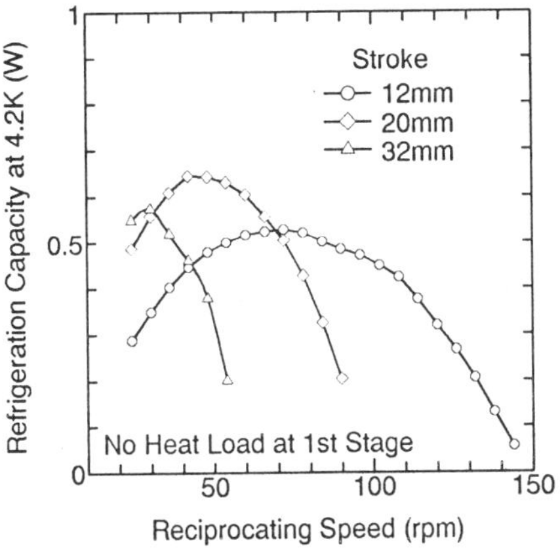

Figure 8. Relationship between reciprocating speed and 4.2 K refrigeration capacity.

stroke. In this experiment, the largest refrigeration capacity for each stroke is between 0.53 W and 0.64 W. The stroke does not have a strong effect on improving the 4.2 K refrigeration capacity.

A two-stage GM refrigerator is often required to have refrigeration capacities both at the first and the second stages simultaneously. For example, the two-stage GM refrigerator, used in a cryocooler-cooled superconducting magnet system, cools a thermal shield and current leads at the first stage and cools a superconducting coil at the second stage. Therefore, the relationship between the first and the second refrigeration capacities are investigated.

Figure 9 shows the 40 K refrigeration capacity at the first stage and the 4.2 K refrigeration capacity at the second stage, which were obtained simultaneously. The reciprocating speed is changed from 24 to 144 rpm for 12 mm stroke, from 24 to 90 rpm for 20 mm stroke and from 24 to 54 rpm for 32 mm stroke. The first stage refrigeration capacity strongly depends on the stroke compared with the 4.2 K refrigeration capacity at the second stage. The largest 40 K refrigeration capacity for the 20 mm stroke is almost twice as much

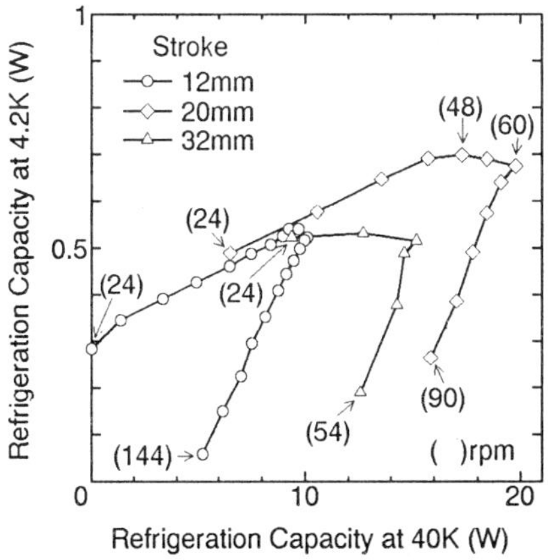

Figure 9. Refrigeration capacities of 4.2 K at second stage and 40 K at first stage.

as that for the 12 mm stroke. The maximum first refrigeration capacity in this experiment was obtained for the 20 mm stroke at 60 rpm. In this condition, the 4.2 K refrigeration capacity at second stage is 0.67 W and is almost equal to the maximum 4.2 K refrigeration capacity of 0.70 W obtained at 48 rpm for the 20 mm stroke. These results indicates that when both the first and the second refrigeration capacities are required, the displacer stroke and the reciprocating speed should be optimized. In this study, the optimum operational conditions of this GM refrigerator are the stroke of 20 mm and the reciprocating speed of 60 rpm. The refrigeration capacities of 0.67 W at 4.2 K for the second stage and 19.8 W at 40 K for the first stage were obtained simultaneously by M-compressor, which input power was 5 kW.

CONCLUSIONS

Effects of operational parameters, such as compressor capacity and displacer stroke are investigated for a two-stage GM refrigerator using magnetic regenerator materials, and the following conclusions are drawn.

The refrigeration capacity at 4.2 K is improved by using a larger compressor, where the temperature difference in the second regenerator is small. The 4.2 K refrigeration, however, can be maintained by a smaller compressor at larger temperature difference in the second regenerator. A smaller compressor is suitable for 4.2 K refrigeration where the temperature difference is large.

The displacer stroke has a strong effect on the optimum reciprocating speed for the 4.2 K refrigeration capacity. A shorter stroke gives a higher optimum reciprocating speed. To improve both the first and the second stage refrigeration capacities simultaneously, the optimum stroke is 20 mm according to this study.

REFERENCES

1. T.Kuriyama et al., Adv. Cryog. Eng.,Vol.35B:1261(1990).
2. M.Sahashi et al., Adv. Cryog. Eng.,Vol.35B:1175(1990).
3. M.Nagao et al., Proc.of the 6th Intl.Cryocooler Conf.Vol.II:37(1991).
4. J.N.Chafe et al.,Proc.of the 7th Intl.Cryocooler Conf.:1157(1993).
5. T.Kuriyama et al., Adv. Cryog. Eng.,Vol.39:1335(1994).
6. R.Plambeck .,Proc.of the 8th Intl.Cryocooler Conf.:795(1995).
7. M.Takahashi et al.,Proc.of the 7th Intl.Cryocooler Conf.:495(1993).
8. M.Nagao et al.,Adv. Cryog. Eng.,Vol.39B:1327(1994).
9. T.Kuriyama et al .,Cryogenics, Vol.34, ICEC Supplement:643(1994).
10. T.Hashimoto et al., Adv. Cryog. Eng.,Vol.40A:655(1994).
11. T.Tsukagoshi et al.,presented at Cryog. Eng. Conf.,(1995).
12. T.R.Strobridge:NBS TN:655(1974).

REFRIGERATION CAPACITY OF A GM REFRIGERATOR WITH MAGNETIC REGENERATOR MATERIALS

T. Tsukagoshi[1], H. Nitta[1], A. Yoshida[1], K. Matsumoto[1], T. Hashimoto[1], T. Kuriyama[2], M. Takahashi[2], Y. Ohtani[2], and H. Nakagome[2]

[1]Tokyo Institute of Technology, Dept. Appl. Phys., Oh-okayama Meguro, Japan

[2]Toshiba R and D Center, Toshiba Co., Ukishima-cho, Kawasaki Kanagawa, Japan

ABSTRACT

An investigation was made of the refrigeration capacity of a two-stage Gifford-McMahon (GM) refrigerator using a layer structural regenerator (multilayer regenerator) with magnetic regenerator materials.

To prove the effectiveness of the multilayer regenerator, we made two kinds of second regenerators with magnetic materials: a triplelayer regenerator with $Er_{0.75}Gd_{0.25}Ni$ (a new magnetic regenerator material), Er_3Co and $Er_{0.9}Yb_{0.1}Ni$, and a doublelayer regenerator with Er_3Co and $Er_{0.9}Yb_{0.1}Ni$. We calculated the regenerator efficiency of the triplelayer regenerator using computer simulation to determine its optimum structure. We then made two kinds of second regenerators and their refrigeration capacities were experimentally compared.

From the experimental results we will discuss the optimum structure of the regenerator.

INTRODUCTION

With the rapid progress now being in applications of metal-superconductors, development of a high efficiency, small size He liquefier for cooling superconductor equipment is urgent. We therefore investigated the refrigeration capacity of a two-stage GM(Gifford-McMahon) refrigerator using magnetic regenerator materials. The magnetic materials are thought to have large heat capacity in the temperature range of liquid He. In this paper, we report on the effectiveness of these materials.

Advances in Cryogenic Engineering, Vol. 41
Edited by P. Kittel, Plenum Press, New York, 1996

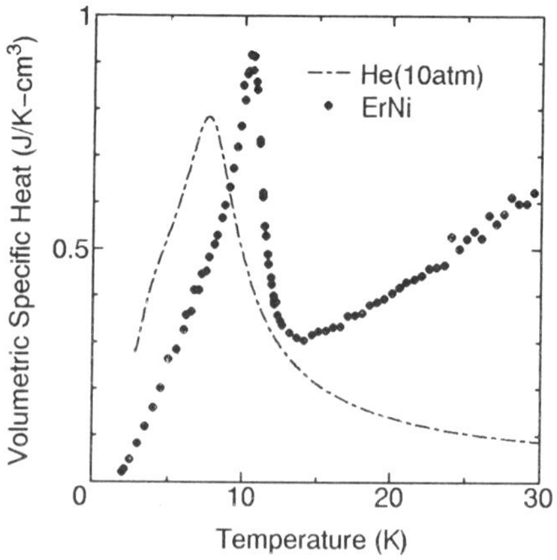

Figure 1. Specific heats of erbium nickel (ErNi) and He gas at 1MPa.

The first point to be discussed is why we chose magnetic materials for the regenerator materials. The specific heat due to thermal agitation of the lattice is very small below ~20K, so non-magnetic materials such as Pb or Cu are useless as regenerator materials in the temperature range of liquid He.

The physical phenomenon producing large heat capacity of a certain material seems to be due to a phase transition. We selected a magnetic phase transition which is easily controlled.

An example of the specific heat, $C(T)$, of a magnetic material is given by erbium nickel (ErNi). Figure 1 shows the $C(T)$ of ErNi near Curie temperature T_C. The $C(T)$ of He gas which is a working gas is also shown for comparison.

Two characteristics are noted in this comparison:

1. Their peak value is the same order of magnitude.

2. The reduced value of half width of the specific heat peak, $\Delta T/T_C$, of ErNi is ~0.35 in contrast to the ~1 of He gas.

This means that the heat capacity of the magnetic material is relatively small in comparison with that of He gas in the temperature range of liquid He. Therefore, it is impossible to cover the peak of the specific heat of He gas by only one kind of magnetic material.

McMahon and Gifford[1] experimentally determined that the specific heat of a regenerator material must be two to four times larger than that of a working gas. Considering the characteristics of the specific heat of magnetic materials, we concluded that a layer

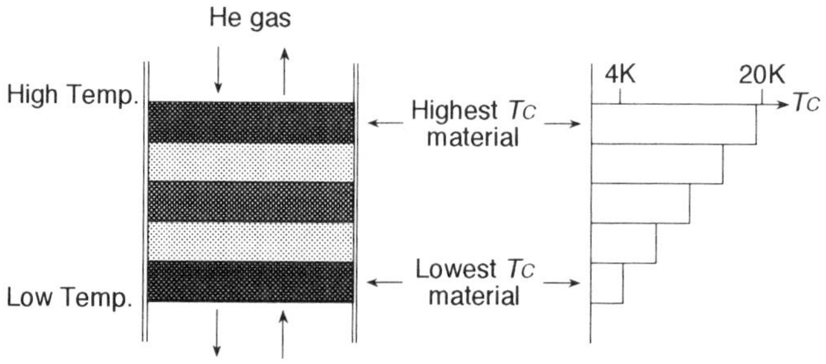

Figure 2. Conceptualized figure of a multilayer regenerator.

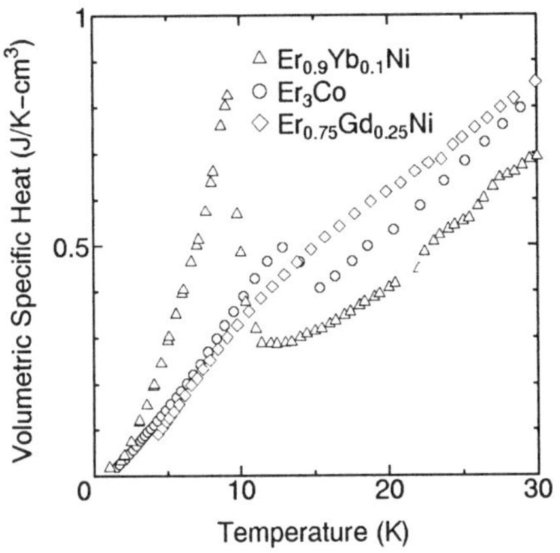

Figure 3. Specific heats of three kinds of magnetic regenerator materials.

structural regenerator (multilayer regenerator) in which the magnetic regenerator materials are placed according to their T_C valueswould be most promising. Figure 2 shows a conceptualized figure of a multilayer regenerator.

In order to determine the effectiveness of the multilayer regenerator with magnetic materials, we made two kinds of second regenerators with these materials. We then compared the refrigeration capacity of the two-stage GM refrigerator with one of these regenerators and that with the other. Finally, we discuss the optimum structure for a regenerator based on the experimental results.

AN INVESTIGATION OF THE MULTILAYER REGENERATOR

The Multilayer Regenerators Used in This Study

We have investigated some kinds of second regenerators[2,3,4] until now. $Er_{0.75}Gd_{0.25}$ Ni was used for the first time as the material to be placed in the high temperature part of the regenerator. Its specific heat and those of others are shown in Figure 3.

To prove the effectiveness of the multilayer regenerator, we made the two kinds of second regenerators with magnetic regenerator materials:

A. A doublelayer regenerator contained $Er_{0.9}Yb_{0.1}Ni$ and Er_3Co[3]. The internal space was divided into two parts. $Er_{0.9}Yb_{0.1}Ni$ was placed in the low temperature part and Er_3Co in the high temperature part.

B. A triplelayer regenerator contained $Er_{0.9}Yb_{0.1}Ni$, Er_3Co and $Er_{0.75}Gd_{0.25}Ni$.

Computer Simulation of the Regenerator[5]

To determine an optimum structure for the triplelayer regenerator (regenerator B) and to verify its superiority over the doublelayer regenerator (regenerator A), we calculated the regenerator efficiency of regenerator B using computer simulation.

Fundamental Equations. Now we consider a one-dimensional model of the regenerator. When effects of the void volume of the regenerator were neglected, the mass flow rate of the working gas, $\dot{m}$, became constant[5]. Finally, two fundamental equations were introduced:

$$\frac{\partial H_g}{\partial x} = -\frac{hA}{L\dot{m}}(T_g - T_m) \tag{1}$$

$$\frac{\partial H_m}{\partial t} = \frac{hA}{M}(T_g - T_m) \tag{2}$$

H_g; Enthalpy of gas
H_m; Enthalpy of regenerator material
T_g; Temperature of gas
T_m; Temperature of regenerator material
M; Total mass of regenerator material
$\dot{m}$; Mass flow rate of the working gas
h; Heat transfer coefficient
A; Net heat transfer area
L; Length of the regenerator

Regenerator Efficiency. The calculation method of the regenerator efficiency, η, is:

$$\eta = \frac{\Delta H_{real}}{\Delta H_{ideal}} \tag{3}$$

where,

$$\Delta H_{ideal} = \dot{m}\tau \min(H(P_H,T_H) - H(P_H,T_C), H(P_L,T_H) - H(P_L,T_C)) \tag{4}$$

$$\Delta H_{real} = \dot{m}\int_0^{\tau}[H(P_H,T_H) - H(P_H,T_{Cout})]dt \tag{5}$$

$$= \dot{m}\int_0^{\tau}[H(P_L,T_{Hout}) - H(P_H,T_C)]dt \tag{6}$$

P_H; Highest pressure of the gas
P_L; Lowest pressure of the gas
T_H; Temperature at hot end
T_C; Temperature at cold end
T_{Hout}; Temperature of the gas as it left from the cold end of the regenerator
T_{Cout}; Temperature of the gas as it left from the hot end of the regenerator
τ; Half period time

Assumptions for the Computer Simulation. The following assumptions were made in calculations using computer simulation:

1. Thermal conductions along the axis of the regenerator material and the working gas were neglected.
2. Generation of heat due to the viscosity of the working gas was neglected.
3. Temperature and pressure of the working gas did not change according with time when the working gas went into the regenerator.
4. Pressure loss in the regenerator was neglected.

The following conditions were assumed: size of the regenerator was 85mm long and 14mm in diameter. Packing factor of the material was 60%. Mass flow rate of the working gas was 2g/s; Highest and lowest pressures of the working gas were 2 and 0.8MPa respectively. The reciprocating speed was 60rpm. The initial conditions were as follows: the temperature at the hot and cold ends of the regenerator was 30K and 4K, respectively, and the temperature distribution in the regenerator was linear.

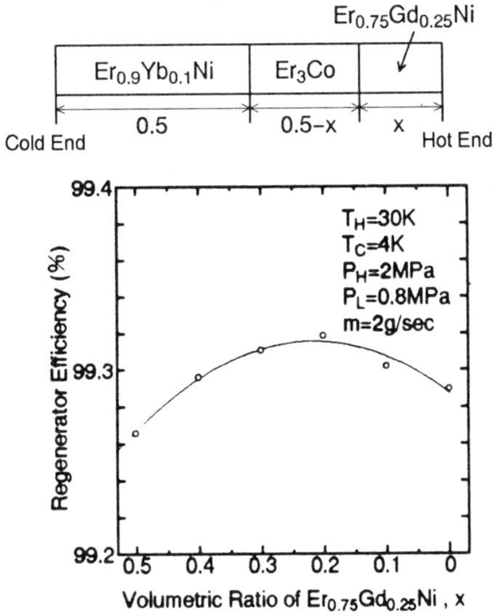

Figure 4. Calculation results of the regenerator efficiency.

Calculation Result of Regenerator Efficiency

To make the triplelayer regenerator (regenerator B), we simply assumed the following. We first divided the internal space into two parts, and the low temperature part was filled with $Er_{0.9}Yb_{0.1}Ni$ particles. A portion of the high temperature part (volumetric ratio to the whole was $(0.5 - x)$) was filled with Er_3Co particles and the rest (volumetric ratio to the whole was x) was filled with $Er_{0.75}Gd_{0.25}Ni$ particles (Figure 4).

In Figure 4 the calculation result of the regenerator efficiency of the triplelayer regenerator as a function of x (volumetric ratio of $Er_{0.75}Gd_{0.25}Ni$ to the whole) is shown. Maximum efficiency was obtained at $x \simeq 0.25$, and the regenerator efficiency of the triplelayer regenerator (regenerator B) was better than that of the doublelayer regenerator (regenerator A).

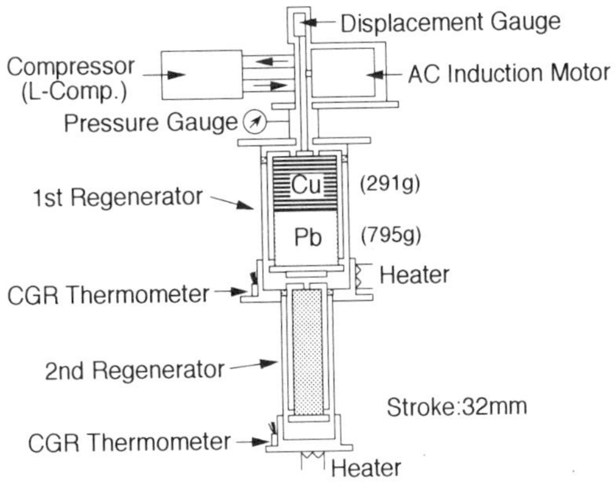

Figure 5. Experimental apparatus of the two-stage GM refrigerator.

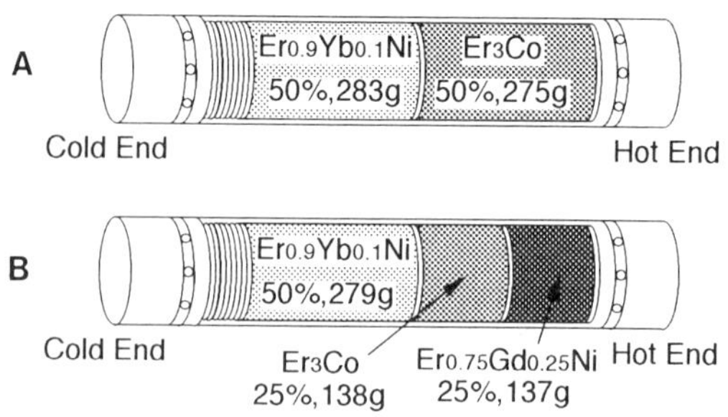

Figure 6. Structures of the two kinds of second regenerators (regenerators A and B).

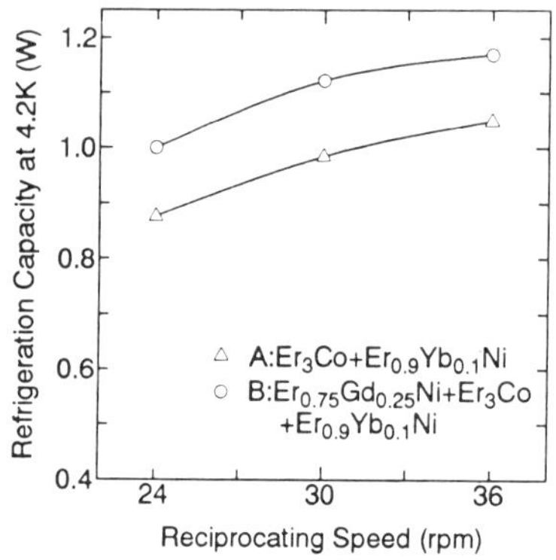

Figure 7. Relationship between the refrigeration capacity at 4.2K and the reciprocating speed of the displacer.

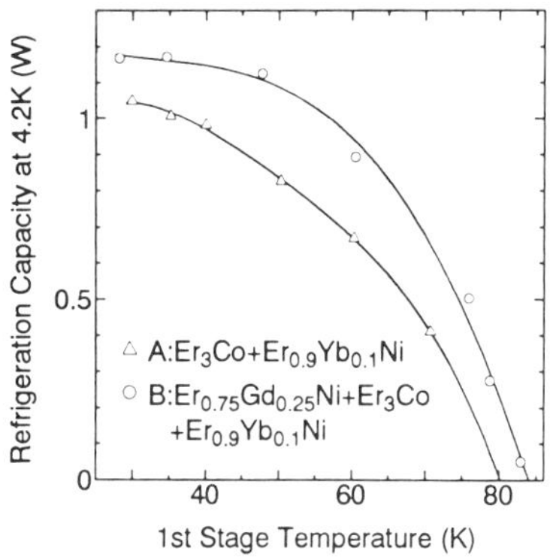

Figure 8. Relationship between the refrigeration capacity at 4.2K and the first stage temperature.

EXPERIMENTAL APPARATUS

Figure 5 shows the experimental apparatus of the two-stage GM refrigerator. L-Comp.[6] was used as the compressor, and the displacer stroke was 32mm.

The size of the first regenerator was 177mm in length and 70mm in diameter, and the internal space was 100mm in length and 50mm in diameter.The internal space was divided into two parts, and Cu mesh (#180) and Pb spheres (200~400μm in diameter) were placed in each part.

The second regenerator was 252mm in length and 32mm in diameter, and had an internal space 157mm in length and 28mm in diameter.

The structures of these two kinds of second regenerators (regenerators A and B) are shown in Figure 6. Regenerator A is the doublelayer regenerator with Er_3Co spheres and $Er_{0.9}Yb_{0.1}Ni$ spheres and regenerator B is the triplelayer regenerator with $Er_{0.75}Gd_{0.25}Ni$ spheres, Er_3Co spheres and $Er_{0.9}Yb_{0.1}Ni$ spheres. Packing factor of these materials was ~60%.

EXPERIMENTAL RESULTS AND DISCUSSION

The experimental results are shown in Figures 7 and 8. Figure 7 shows the relationship between the refrigeration capacity at 4.2K and the reciprocating speed of the displacer. Maximum refrigeration capacities were obtained at 36rpm for regenerator A, 1.05W, and for B, 1.17W.

Figure 8 shows the relationship between the refrigeration capacity at 4.2K and the first stage temperature. The refrigeration capacities decreased as the first stage temperature increased using the two kinds of second regenerators. The capacity using regenerator B was always larger than that using regenerator A, no matter how the first stage temperature changed.

From these results we concluded that the triplelayer regenerator (regenerator B) was superior to the doublelayer regenerator (regenerator A).

CONCLUSIONS

A multilayer regenerator with magnetic regenerator materials was investigated using a two-stage GM refrigerator, and it was experimentally confirmed that the refrigeration capacity at 4.2K was improved by increasing the number of layers in the second regenerator.

Further discussion is required on optimization of the structure of a regenerator with magnetic regenerator materials.

REFERENCES

1. H.O. McMahon and W.E. Gifford, A New Low-Temperature Gas Expansion Cycle part I, *Adv. in Cryog. Eng.*, 5:354 (1960).

2. T. Kuriyama et al., Helium Liquefaction by a Two-Stage Gifford-McMahon-Cycle Refrigerator Using New Regenerator Material of Er_3Ni, *Jpn. J. Appl. Phys.*, 31:L1206 (1992).

3. T. Hashimoto et al., Effect of High Entropy Magnetic Regenerator Materials on Power of the GM Refrigerator, *Adv. in Cryog. Eng.*, 40A:655 (1994).

4. T. Hashimoto et al., Development of Powerful Magnetic Regenerator Materials and Verification of Their Effectiveness, *Cryogenics*, 34:223 (1994).

5. H. Seshake et al., Analysis of Rare Earth Compound Regenerators Operating at 4K, *Adv. in Cryog. Eng.*, 37B:995 (1992).

6. T. Kuriyama et al., Optimization of Operational Parameters for a 4K-GM REfrigerator, *Adv. in Cryog. Eng.*, to be published.

DEVELOPMENT OF 1.5W 4K G-M CRYOCOOLER WITH MAGNETIC REGENERATOR MATERIAL

Toshimi Satoh , Atsushi Onishi , Rui Li ,
Hiroshi Asami, and Yoshiaki Kanazawa

Research and Development Center Sumitomo Heavy Industries , Ltd .
63-30 , Yuhigaoka , Hiratsuka , Kanagawa , 254 , Japan

ABSTRACT

A two-stage 4K Gifford-McMahon (G-M) cycle cryocooler with magnetic regenerator material which has cooling capacity 1.5W at 4.2K has been developed. The hybrid structural second regenerator composed of lead and $ErNi_{0.9}Co_{0.1}$ was used in the cryocooler. $ErNi_{0.9}Co_{0.1}$ has a large specific heat peak at lower temperature than 10K and lead has a larger specific heat in the higher temperature region. The intake/exhaust valve timing was optimized to improve the cooling capacity not only of the second stage but also of the first stage. A larger size second cylinder in diameter than the former one was used to get a larger pressure-volume (PV) work.

INTRODUCTION

The recent development of superconducting technology requires a small 4K cryocooler with high refrigeration performance, high reliability and easy operation. A G-M cryocooler is widely used for many cryogenic systems, such as a cryopump, a shield cooler of an MRI superconducting magnet and a direct cooling system of a superconducting coil because of its high reliability and compactness. If it is possible to achieve 4.2K or lower by a G-M cryocooler, it becomes a desirable 4K cryocooler.

Usually lead is used as a regenerator material in cryogenic temperature region. The lowest obtainable temperature of a typical two-stage G-M cryocooler is about 8 K because the specific heat of lead is lower comparing to that of helium at lower temperature than 10K. The utilization of large magnetic specific heat of rare earth material at lower temperature than 10K was proposed to achieve lower temperature by a G-M cryocooler.

The helium temperature was firstly achieved by a two-stage G-M cryocooler with Er_3Ni[1,2]. But the cooling capacity of the G-M cryocoolers at 4.2K was low for practical uses. The reason of low cooling capacity was thought that the specific heat was still small. To obtain larger cooling capacity, a regenerator material having larger specific heat than that of Er_3Ni below 10K was needed. The magnetic materials such as $ErNi_{0.9}Co_{0.1}$[4] and $Er_{0.9}Yb_{0.1}Ni$[5] are appropriate to this purpose.

The specific heat of these materials below 10K was large, but it was low comparing to, for example, that of lead or Er_3Ni in the higher temperature region than 10K. The second regenerator was expected to work more effective if $ErNi_{0.9}Co_{0.1}$ or $Er_{0.9}Yb_{0.1}Ni$ in the low specific heat region was replaced by other materials having larger specific heat. Then the effect of a hybrid structural second regenerator was investigated. A hybrid regenerator is composed of more than one kind of regenerator materials. The refrigeration performance of a G-M cryocooler was much improved by using the hybrid second regenerator, and the cooling capacity of about 1W at 4.2K was obtained with the two-layer hybrid second regenerator[4,6].

The G-M cryocooler with cooling capacity of about 1W is applicable to many practical uses. But the cooling capacity and also the refrigerator efficiency of a G-M cryocooler are still lower than that of a GM+JT refrigerator. It is needed to develop a high efficient G-M cryocooler with larger cooling capacity than 1W to extend its application area.

1.5W 4K G-M CRYOCOOLER CONSTITUTION

The two-stage 4K G-M cryocooler used in this paper is shown in Fig. 1. The first and second displacers of the refrigerator unit were driven by a synclonous motor. Cycle speed of the motor was varied by an inverter. The technical points of the second regenerator and the cylinder are explained below.

Thermometers and heaters were installed on the refrigeration stages for the cooling capacity measurement. A platinum-cobalt resistance thermometer for the first stage and a germanium resistance thermometer for the second stage were placed, respectively. Manganin wire was wound around each refrigeration stage as electric heater. A displacement transducer and a pressure transducer were also attached to obtain the PV diagram. The pressure transducer was connected to the first expansion space. All these attachments for the measurement are omitted in Fig. 1.

Second Regenerator

The authors developed magnetic intermetallic compound $ErNi_{0.9}Co_{0.1}$ which has specific heat peak $0.65 J/Kcm^3$ at 7K as is shown in Fig. 2, and the cooling capacity 0.95W at 4.2K was achieved by using the two-layer hybrid second regenerator composed of $ErNi_{0.9}Co_{0.1}$ and Er_3Ni. A half of the regenerator volume in the low temperature region was filled with $ErNi_{0.9}Co_{0.1}$ and the remaining volume in the high temperature region was filled with Er_3Ni. The cooling capacity of 0.8W at 4.2K was obtained by the two-stage G-M cryocooler with the regenerator containing only Er_3Ni, and 0.9W with Er_3Ni and $ErNi_{0.9}Co_{0.1}$[4,7]. The refrigeration performance was much improved by replacing Er_3Ni in the low temperature region of a regenerator with $ErNi_{0.9}Co_{0.1}$.

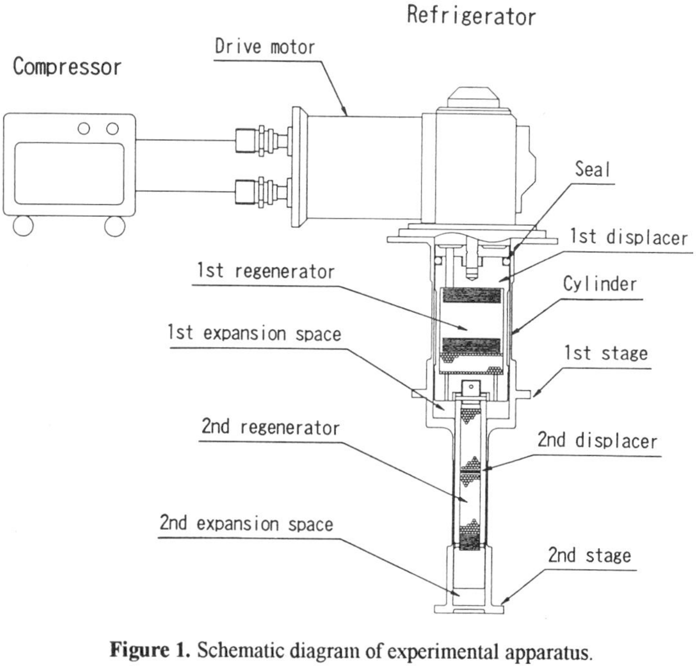

Figure 1. Schematic diagram of experimental apparatus.

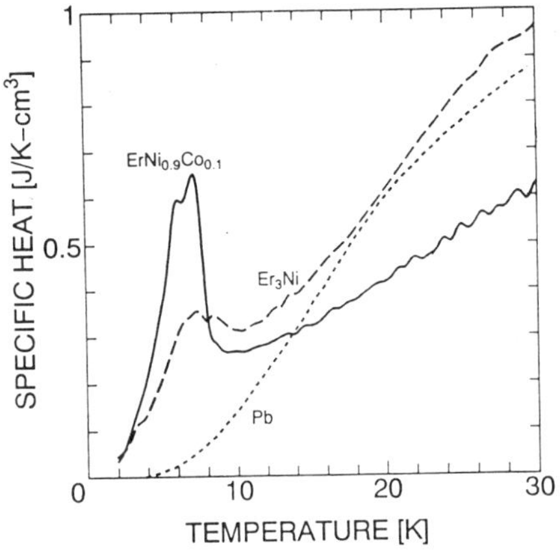

Figure 2. Temperature dependence of volumetric specific heat of $ErNi_{0.9}Co_{0.1}$,Er_3Ni and lead.

In practical uses, the first stage temperature is usually higher than 30K because of the heat load. A material of a two-layer hybrid regenerator in the high temperature region, therefore, must have large specific heat in very wide temperature range from 8K to higher than 30K. According to Fig. 2, lead has nearly same specific heat as that of Er_3Ni above 15K. Then lead is expected to be a potential hybrid regenerator material in the high temperature region, though the specific heat below 15K is small. The authors investigated the refrigerator performance of the two-stage G-M cryocooler with the hybrid regenerator containing lead. If lead is usable, the manufacturing cost of a 4K G-M cryocooler is reduced since the price of lead is much lower than that of rare earth compound.

The cooling capacity of the G-M cryocoolers with three types of regenerators listed in Table 1 was measured to investigate the effect of lead. The experiments were performed at the motor speed 60rpm and the displacer stroke 20mm. The cooling capacity data at 4.2K are shown in Table 2. The cooling capacity of the G-M cryocooler using the hybrid regenerator with $ErNi_{0.9}Co_{0.1}$ and lead was smaller than that with $ErNi_{0.9}Co_{0.1}$ and Er_3Ni, but it was still larger than that with Er_3Ni only. From this experiment the hybrid regenerator with lead and $ErNi_{0.9}Co_{0.1}$ was used for the new G-M cryocooler.

Cylinder

The largest cooling capacity at 4.2K was 1.29W in the experiments so far (Fig. 3). These results were obtained by the refrigerator with the regenerator composed of lead 350g and $ErNi_{0.9}Co_{0.1}$ 350g. The first and the second cylinder sizes were 82mm and 35mm in diameter, respectively.

The cooling capacity is produced by the PV work in the expansion space, therefore it is a effective means to make the PV work larger to obtain larger cooling capacity. As the PV work is proportional to the sectional area, a new cylinder which has larger second cylinder diameter was designed to improve the second cooling capacity. The first and second cylinder sizes of the former G-M cryocooler were 82mm and 35mm in diameter, respectively. The reciprocation stroke was 30mm, then the first and the second maximum expansion volume were 158.4mm^3 and 28.9mm^3, respectively. The new second cylinder diameter was 40mm to obtain the cooling capacity 1.5W at 4.2K. The maximum second expansion volume of the new cylinder was 37.7cm^3. The PV work becomes about 30% larger than the former one.

Table 1. Construction of investigated regenerators

Type	High temperature region		Low temperature region	
	Material	Volume	Material	Volume
A	Er_3Ni	50%	$ErNi_{0.9}Co_{0.1}$	50%
B	Pb	50%	$ErNi_{0.9}Co_{0.1}$	50%
C	Er_3Ni			

Table 2. Cooling capacity with three types of 2nd regenerators.

1st stage temperature		25K	30K	35K	40K	50K
Cooling capacity at 4.2K (W)	Regenerater A	0.87	0.9	0.89	0.91	0.91
	Regenerator B	0.85	085	0.81	0.70	0.43
	Regenerator C	0.81	0.76	0.73	0.43	----

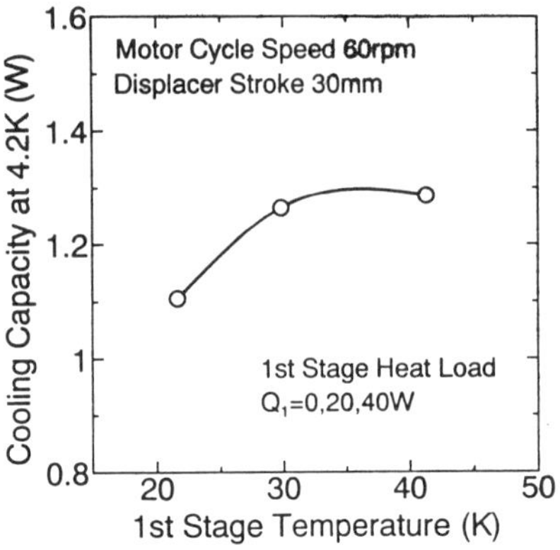

Figure 3. Cooling capacity at 4.2K obtained with the former cylinder as a function of the first stage temperature.

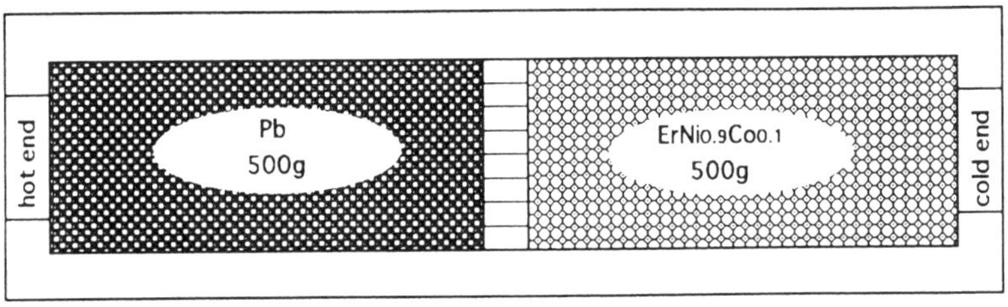

Figure 4. Construction of the two-layer hybrid regenerator with $ErNi_{0.9}Co_{0.1}$ and lead.

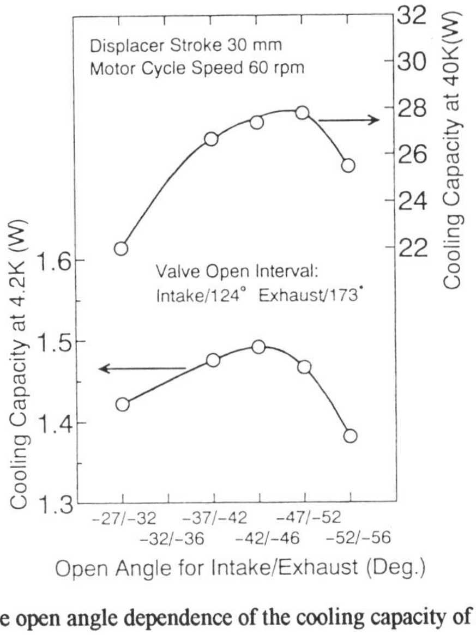

Figure 5. Intake/exhaust valve open angle dependence of the cooling capacity of the second stage at 4.2K and the first stage at 40K.

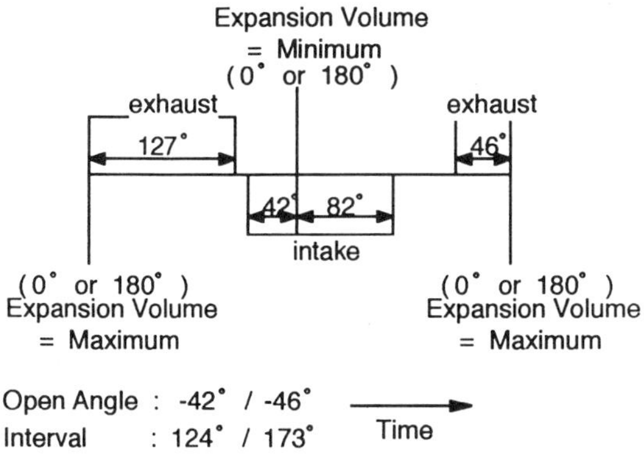

Figure 6. Optimal intake/exhaust valve timing and interval.

EXPERIMENTAL RESULTS

The 1.5W 4K G-M refrigerator was constructed of the above mentioned regenerator and cylinder. The two-layer hybrid second regenerator shown in Fig. 4 was used in the refrigerator. Lead spheres (diameter 0.5 to 0.6mm) were filled in the high temperature region and $ErNi_{0.9}Co_{0.1}$ (diameter 0.2 to 0.3mm) in the low temperature region. The used amount was 500g for each material. The first and the second cylinder sizes were 82mm and 40mm in diameter, respectively. The displacer stroke was 30mm.

Intake/Exhaust valve optimization

Not only the second cooling capacity but also the first cooling capacity of a 4K G-M cryocooler was strongly influenced by the valve timing[8,9]. The intake/exhaust valve timing was determined from the experiment. Fig. 5 shows the experimental results of the first and the second cooling capacity at 60rpm. The valve open intervals expressed in angle were fixed in this experiment to 124 deg. for the intake valve and to 173 deg. for the exhaust valve, respectively. The valve timing was expressed by the valve open position in angle. Minus sign of the open angle of the horizontal axis in Fig. 5 means that the intake/exhaust valve opens before the expansion volume becomes minimum/maximum, respectively.

The cooling capacity of the second stage at 4.2K and of the first stage at 40K are shown against the open angle for the intake/exhaust valve (valve timing) in Fig. 5. The second cooling capacity at 4.2K were largest at the valve timing -42deg./-46deg. from this figure. The usual valve timing was -27deg./-32deg.. The first and the second cooling capacities were much improved by changing the valve timing to -42deg./-46deg., though the first cooling capacity was not largest. The valve timing shown in Fig. 6 was chosen for the experiment.

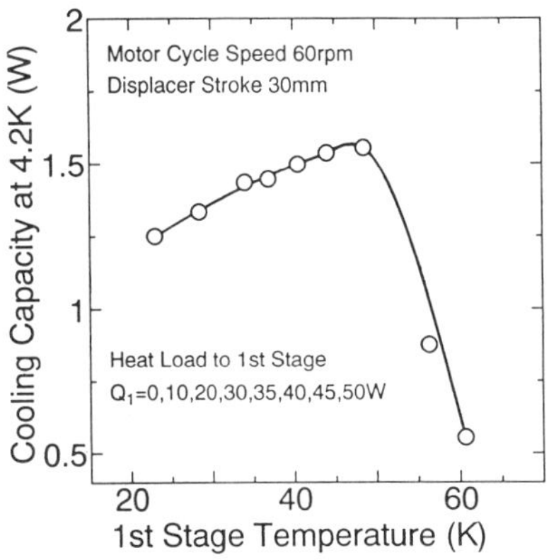

Figure 7. Cooling capacity at 4.2K as a function of first stage temperature.

Table 3. Comparison of cooling capacity with two types of cylinders

1st stage temperature		25K	30K	40K
2nd dia.35mm cylinder	1st heat load(W)	8.7	20.4	38.0
	cooling capacity at 4.2K(W)	1.14	1.24	1.26
2nd dia.40mm cylinder	1st heat load(W)	3.3	12.1	27.4
	cooling capacity at 4.2K(W)	1.30	1.39	1.49

Improved cooling capacity

Fig. 7 shows the relation between the second stage cooling capacity at 4.2K and the first stage temperature at the motor cycle speed 60 rpm. The cooling capacity at 4.2K was going up as the first temperature increased, and the largest value 1.56W was obtained at the first temperature 48.5K (the first heat load 40W). The cooling capacity at 4.2K with no heat load to the first stage was 1.25W. The cooling capacity was improved more than 20% by the increase of the first stage temperature. According to another experiment, the PV work in the second expansion space increased when the first stage temperature went up[9]. This suggests that the increase of the second stage cooling capacity was caused by the increase of the PV work coming from the first stage temperature rise.

The motor cycle speed affects the no-load second stage temperature, and also affects the cooling capacity[1]. The lowest temperature was 2.35K at 36rpm, and 3.3K at 72rpm. The experiment was not performed below 36rpm, since the cooling capacity did not become larger any more. The cooling capacity other than at 60rpm was also measured. The cooling capacity 1.56W at 4.2K was obtained at 54rpm and the first stage temperature 35K (the first heat load 30W). The largest cooling capacity at 54rpm was same as that at 60rpm, but the COP value was better at 54rpm. The cooling capacity at 4.3K was 1.69W and the input power was 7.1kW at 54rpm. Then the COP value at 4.3K was $2.38*10^{-4}$, which is the same order of magnitude as that for GM+JT refrigerators[10].

The cooling capacity with the cylinders having the second diameter 35mm and 40mm is shown in Table 3. At all the first stage temperatures shown, the second cooling capacity with the second cylinder in diameter 40mm (the new cylinder) was larger than that with the second cylinder in diameter 35mm (the former cylinder), but the first cooling capacity with the new cylinder was smaller, though the first cylinder diameter was same. Fig. 8 shows the comparison of the first stage PV diagrams for the two types of cylinders. The second stage temperature was below 4K in this experiment. The measured PV work with the new cylinder was smaller than that with the former cylinder. The decrease of the first cooling capacity of the new cylinder is caused by the decrease of the PV work. The ratio of the difference of the high and low pressures was almost same with the ratio of the PV work. Then the difference of the calculated PV work is related to the difference of the high and the low pressures in the expansion spaces.

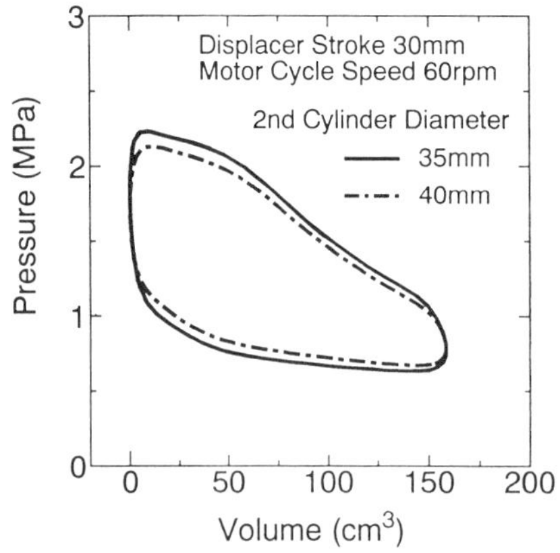

Figure 8. PV diagram at first expansion space with two types of cylinders.

CONCLUSIONS

A two-stage G-M cryocooler with magnetic regenerator material has been developed and its performance was investigated. The results are summarized as follows.

(1)The second stage cooling capacity 1.56W at 4.2K was obtained by using a two-stage 4K G-M cryocooler with a new cylinder in second cylinder diameter 40mm. The COP value for the 4K G-M cryocooler at 4.3K was $2.38*10^{-4}$.
(2)The two-layer hybrid second regenerator composed of $ErNi_{0.9}Co_{0.1}$ and lead was used in the 4K G-M cryocooler. It was shown that lead works effective in a 4K G-M cryocooler as a second regenerator material when it was used hybrid.
(3)The intake/exhaust valve timing was changed to improve not only the second cooling capacity but also the first cooling capacity. The first and the second cooling capacities were much improved comparing to those for the usual valve timing.

REFERENCES

1.T.Kuriyama et al.,High efficient two-stage GM refrigerator with magnetic material in the liquid helium temperature region,in:"Advances in Cryogenic Engineering," vol.35B, Plenum Press, New York(1990),p.1261.
2.T.Inaguchi,M.Nagao and H.Yoshimura,Two-stage Gifford-McMahon cycle cryocooler operating at about 2K,*Proceedings of the Sixth International Cryocoolers Conference*, vol. II :25(1990)
3.M.Sahashi et al.,New magnetic material R_3T system with extremely large heat capacities used as heat regenerators,in:"Advances in Cryogenic Engineering," vol.35B,Plenum Press,New York(1990),p.1175.
4.G.Ke et al.,Improvement of two-stage GM refrigerator performance using a hybrid regenerator,in: "Advances in Cryogenic Engineering," vol.40A,Plenum Press, New York(1994),p.639.
5.T.Hashimoto et al.,Recent progress on rare earth magnetic regenerator material,in:"Advances in Cryogenic Engineering," vol.37B,Plenum Press,New York(1992), p.859.
6.T.Hashimoto,M.Yabuki,T.Eda,T.Kuriyama and H.Nakagome,Effect of high entropy magnetic regenerator materials on power of the GM refrigerator,in:"Advances in Cryogenic Engineering," vol.40, Plenum Press,New York(1994),p.655.
7.A.Onishi et al.,A 4K-GM cryocooler with hybrid regenerator of magnetic materials,*Proceedings of Fourth Joint Sino-Japanese Seminar on Cryocoolers and Concerned Topics,* 44(1993)
8.T.Kuriyama et al.,Development of 1 watt class 4K GM refrigerator with magnetic regenerator materials, in:"Advances in Cryogenic Engineering," vol.39, Plenum Press, New York(1994),p.1335.
9.R.Li,A.Onishi,T.Satoh and Y.Kanazawa,The influence of valve timing and interval on performance of 4K Gifford-McMahon cycle cryocooler,presented at Cryog. Eng. Conf., (1995)
10.T.Kuriyama et al.,Development of a watt-class 4 K GM refrigerator using magnetic regenerator materials, *Proceedings of Fourth Joint Sino-Japanese Seminar on Cryocoolers and Concerned Topics,* 34(1993)

THE THEORETICAL COOLING CAPACITY AND HEAT LOSS ANALYSIS OF A 4.2K G-M REFRIGERATOR

Zheng Zhu* and Liang Zhang

Cryogenic Laboratory
Chinese Academy of Sciences
P.O.Box 2711, Beijing, 100080, China

ABSTRACT

In this paper, a formula is built up for calculating the theoretical cooling power of a 4.2 K G-M refrigerator in a wide temperature range. Numerical analysis of heat losses is presented due to the imbalance of the helium enthalpy between the high pressure and low pressure, and the Simon compression in the void volume of the low temperature regenerator. The comparison between the experimental result and the numerical analysis is also presented.

INTRODUCTION

Since magnetic materials began being used as the regenerator matrix, the experimental study of G-M refrigerators has developed rapidly, the cooling power has reached 0.5~1.0W/4.2K[1,2], and the no load temperature has reached 2.09K[3]. The related theoretical analysis of the heat balance has not been further studied. As helium can not be treated as an ideal gas for the temperatures below about 40K, the generally used formula :

$$Q_{co} = P_H - P_L \tag{1}$$

is no longer suitable for the ideal cooling power of a 4.2K refrigerator. The purpose of this paper is to set up a more satisfactory formula reflecting realistic features of helium in a wide temperature range. Moreover, the heat losses caused by the enthalpy imbalance of helium and Simon compression will be analyzed numerically.

---

*Present address: Kejian institute, Chinese Academy of Sciences, P.O.Box 8712, Beijing , 100080, China

FORMULA DEDUCTION

In this section, we will infer a formula based on an isothermal Simon expansion model.

First, we assume helium is a real gas and all kinds of heat losses are negligible. Thus, the ideal G-M cycle could be expressed by four processes: a) inlet gas with constant high pressure, b) outlet gas with Simon expansion, c) outlet gas with constant low pressure, d) inlet gas with Simon compression. Fig.1 is a T - S diagram for Simon expansion . 2 -3 shows the isothermal Simon expansion, 2 -3' shows the isentropic Simon expansion, point 4 and 2' represent the difference from the ideal expansion caused by enthalpy imbalance and Simon compression. Since Simon expansion is an irreversible process, the only purpose of the processes shown in Fig.1 is to express the relationship of the temperatures and pressures before and after the expansion, and not to represent any mass or energy change during the expansion.

We apply the basic thermodynamic equation of unsteady flow to the four processes of the G-M cycle:

$$dQ - dW = [\, h_{out} dM_{out} - h_{in} dM_{in} \,] + [\, M_2 u_2 - M_1 u_1 \,] \tag{2}$$

where: dQ and dW, heat and work exchange of the system with the environment; h_{out} and h_{in}, unit enthalpy outlet and inlet the system; dM_{out} and dM_{in}, mass outlet and inlet the system; M_2 and M_1 , mass of the system after and before the process; u_2 and u_1, unit internal energy of the system after and before the process.

We choose the cold space of the low temperature stage as the research object and use Q_i (J/cm^3) as the ideal cooling capacity per unit of the cold volume.

After a simple deduction, we obtained:

$$Q_i = P_H - P_L + \rho_L h_L - \rho_H h_H + \int_{\rho_L}^{\rho_H} h \, d\rho \tag{3}$$

where: ρ_H, density of the high pressure helium before expansion ; ρ_L , density of the low pressure helium after expansion; h_H, enthalpy of the high pressure helium before expansion; h_L , enthalpy of the low pressure helium after expansion.

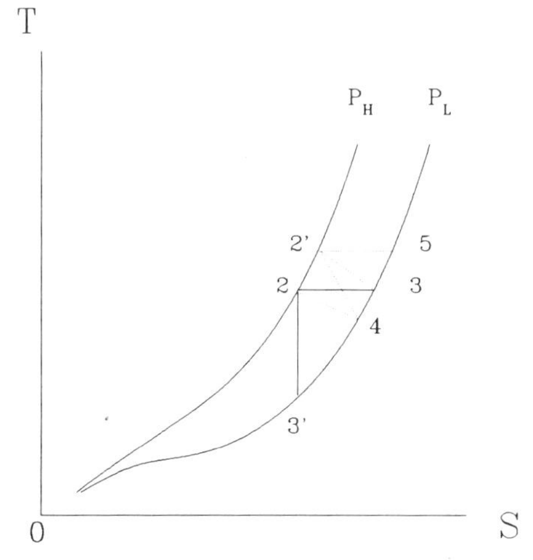

Figure 1. T–S diagram of helium.

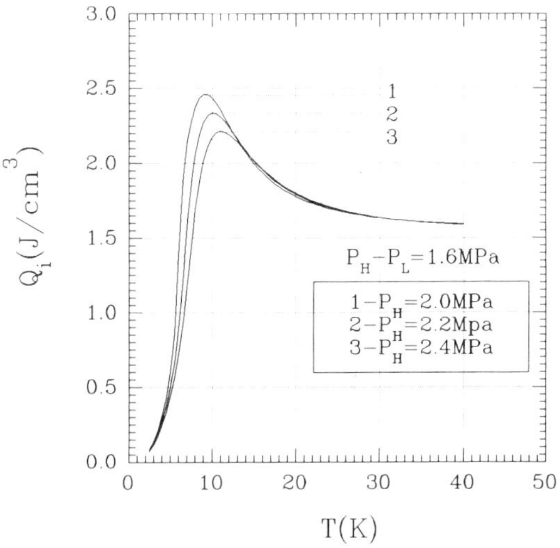

Figure 2. Ideal cooling power of the G–M cycle.

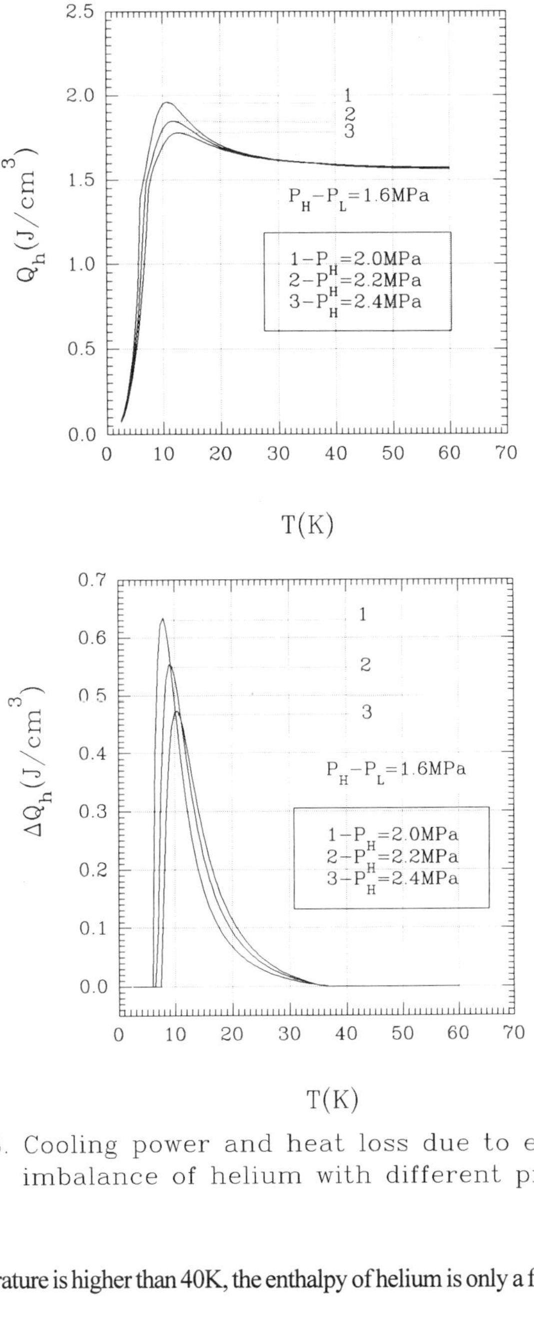

Figure 3. Cooling power and heat loss due to enthalpy imbalance of helium with different pressures.

When the temperature is higher than 40K, the enthalpy of helium is only a function of temperature, then:

$$\rho_L h_L - \rho_H h_H + \int_{\rho_L}^{\rho_H} h \, d\rho = 0 \tag{4}$$

and equation (3) will be transformed to equation (1). This conclusion can also be verified by Fig. 2, which shows a numerical analysis of equation (3), in which the Simon expansion is shown by 2-3 in Fig.1. All helium data used is taken from reference 4. When the temperature is higher than 40K, the cooling powers discussed in this paper are almost the same(Q_{co}), therefore, our discussion below will focus on a temperature range no higher than 40K.

If we consider the expansion process as an isentropic one, we will obtain another formula which shows little difference from isothermal cooling power below 10K, because the entropy change of helium for the expansion decreases very fast with temperature.

THE THEORETICAL COOLING POWER OF THE G-M CYCLE

In this section, the effect of enthalpy imbalance and Simon compression will be considered, also a theoretical cooling power(Q_{th}) quite different from Q_i will be obtained.

Heat Loss of Enthalpy Imbalance ΔQ_h

From the H-T diagram of helium, it is easy to find two curves with different pressures that cross at T_A (about 7K) and overlap after T_B (about 40K), this characteristic yields non-linear temperature distributions in the low temperature regenerator.

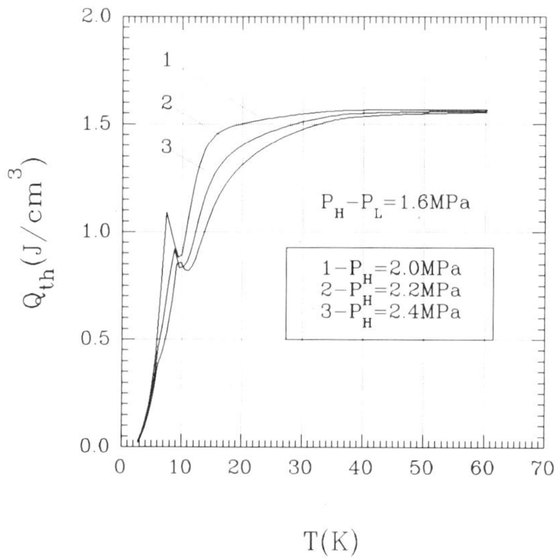

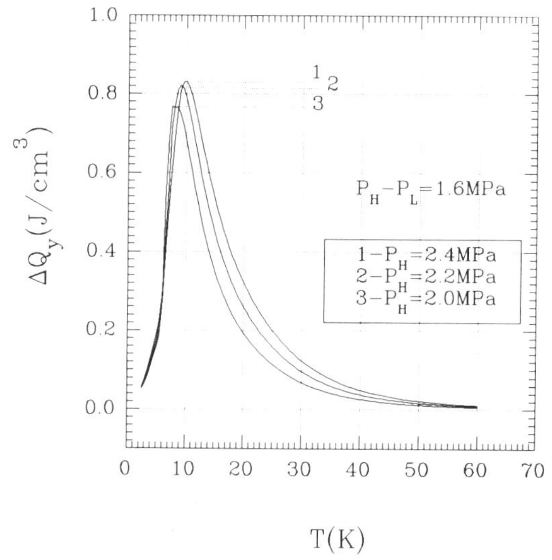

Figure 4. Theoretical cooling power and heat loss of Simon compression.

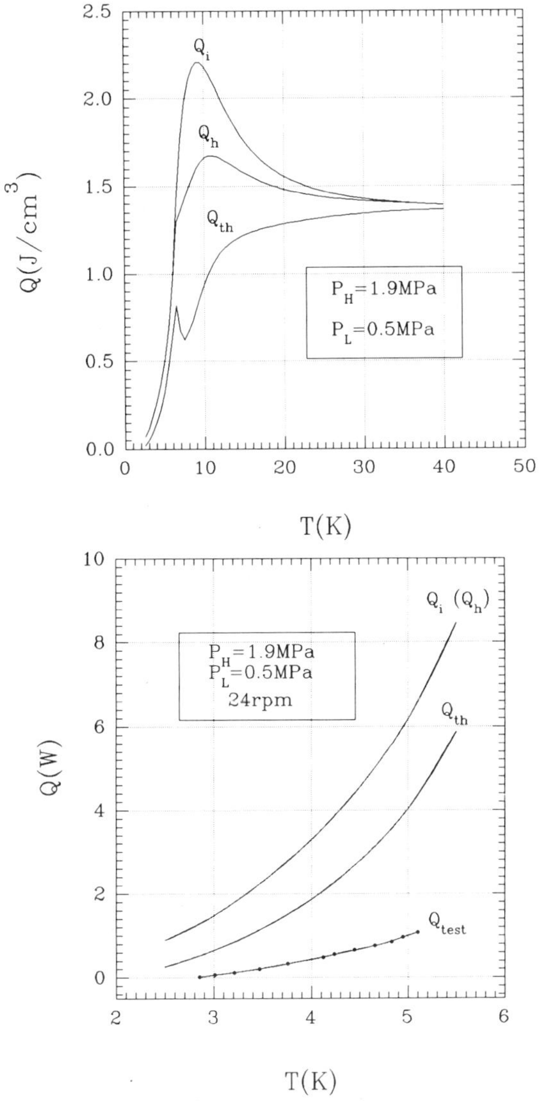

Figure 5. Comparison of the ideal, theoretical and experimental cooling power.

Here we use Q_h as the cooling capacity. T_2 represent the helium temperature before entering the cold space.

When $T_2 > T_A$:

$$Q_h = P_H - P_L + \rho_4 h_4 - \rho_2 h_2 + \int_4^2 h \, d\rho \tag{5}$$

The heat loss is $\Delta Q_h = Q_i - Q_h$. Fig. 3 is the numerical result of equation (5).

Heat Loss of Simon Compression ΔQ_y

Implied in one of the above assumptions is that the regenerator has no porosity. Actually, porosity is not only necessary for the passage of helium in a regenerator, it is also a main part of the void volume in a refrigerator. During the compression process, the low pressure helium remaining in the regenerator is first compressed to the high pressure, then surges into the cold space in a constant high pressure process. The heat produced by compression is absorbed by the helium and the matrix resulting in a temperature jump from T_2 to $T_{2'}$, the jump has been described by Hong[5]:

$$\delta T \cong \frac{p\,Q_S}{p\,C_p + (1 - p)\,C_m} \tag{6}$$

where: p, porosity factor ; Q_S, Simon compression heat ; C_p , specific heat of helium ; C_m , specific heat of matrix.

$T_{2'} = T_2 + \delta T$, we use Q_{th} as the cooling power in this case.
When $T_A < T_2 < T_B$:

$$Q_{th} = P_H - P_L + \rho_4 h_4 - \rho_{2'} h_{2'} + \int_4^{2'} h\, d\rho \tag{7}$$

and when $T_2 < T_A$:

$$Q_{th} = P_H - P_L + \rho_3 h_3 - \rho_{2'} h_{2'} + \int_3^{2'} h\, d\rho \tag{8}$$

The heat loss is $\Delta Q_y = Q_h - Q_{th}$. Fig. 4 shows a numerical result of equation (7) and (8). Fig. 5 shows a comparison of the experimental result and the numerical analysis .

CONCLUSION

The ideal and theoretical cooling capacities of the G-M cycle are different when the temperature is below T_B, the difference comes from two kinds of heat losses, in which ΔQ_h is inevitable, while ΔQ_y could be reduced by improve the specific heat of the matrix or the structure of the regenerator. Other kinds of heat losses could be divided into two parts: one is connected with processes, such as inefficiency of the regenerator, shuttle heat transfer , pressure drop, leak of helium, pumping affect, heat exchange inefficiency. The other is irrelevant with processes, such as axial conduction, residual gas conduction and radiation. Both of these are analyzed in reference 6.

ACKNOWLEDGEMENT

The experimental and theoretical research of a 4.2K G-M refrigerator is supported by the National Natural Science Foundation of China, the National Centre for Research and Development on Superconductivity and the Chinese Academy of Sciences.

Discussion with and help of members of 202 group, Cryogenic Laboratory, Chinese Academy of Science is sincerely appreciated.

REFERENCES

1. T.Kuriyama et al.: "Development of a Watt-class 4K G-M Refrigerator Using Magnetic Regenerator Materials" <JSJS-4> 1993, P34-39
2. L. Zhang et al.: " Development and Analysis of a 4.2K G-M Refrigerator" Cryogenics 1994 Vol.34 ICEC Supplement, P179-182
3. M.Nagao et al.:"Generation of Superfluid Helium by a G-M Cycle Cryocooler" <The 6th - Cryocoolers> Vol.2 1990, P37-47
4. V.D.Arp et al: "Thermophysical Properties of Helium - 4 from 0.8 to 1500K with Pressures to 2000MPa", NIST-Technical Note 1334, Nov. 1989
5. C.S. Hong: "On The Cooling Capacity of Low Temperature G-M Refrigerators" Proc. 3rd National Cryogenic Engineering Conference, P40-43, 1992 (In Chinese)
6. Zheng Zhu: "The experiment Study and Heat Balance Analysis of a 4.2 K G-M Refrigerator" Thesis for master degree, Cryogenic Laboratory, Chinese Academy of Sciences, August, 1994. P32 - 57

STUDY OF EFFECT OF HEAT TRANSFER IN THE COLD HEAD TO THE PERFORMANCE OF A 4.2K G-M REFRIGERATOR

L. Zhang, L.H. Gong and X.D. Xu

Cryogenic Laboratory, Chinese Academy of Sciences
P.O.Box 2711, Beijing 100080 China

ABSTRACT

A ncw prototype of a G-M refrigerator with V_{CO} = 13.3 cm^3 is designed and tested. The experimental result of 2.9 K and 515 mW / 4.2 K has been got at 60 rpm with the cool-down time of 100 minutes. The design principles and more information in detail are given in this paper.

The heat transfer from helium gas to the wall of the cold head, occurring in the gap between a cylinder and a displacer of a Gifford-McMahon refrigerator, is very important to the refrigeration. This process in the colder stage of a 4.2 K G-M refrigerator with magnetic material in the regenerator, is one of the dominant factors to the enhancement of the useful cooling capacity, because the thermal properties of helium in the range of temperature from 20 K to 4.2 K and pressure from 2.0 MPa to 0.7 MPa present dramatic change. A numerical analysis method considering the heat transfer together with the regenerator is suggested. The calculation results and some discussion are presented in this paper.

INTRODUCTION

In recent years one of the most important progress in two-stage Gifford-McMahon refrigeration is the success in reaching the liquid helium temperatures using magnetic materials, exhibiting high specific heat at low temperature, as a regenerator matrix for the colder stage[1.2.3]. This kind of refrigerator with 0.5 W / 4.2 K, running with a compressor of about 7.5 kW input power, is commercially available[4.5]. For getting liquid helium temperatures the rotating speed of the refrigerator is generally reduced from 60 rpm to about 20-30 rpm. The refrigerator is more complex and has an extra operation and also needs about 3 hours to cool down.

A new two-stage G-M refrigerator with Vco = 13.3 cm^3 has been designed and tested. The experimental result of this prototype is satisfactory. The lowest temperature of 2.9 K and a cooling capacity of 515 mW / 4.2 K can be reached at 60 rpm with a cool-down time

Advances in Cryogenic Engineering, Vol. 41
Edited by P. Kittel, Plenum Press, New York, 1996

Table 1. Design Parameters

	First stage		Second stage	
Cylinder				
Diameter	53 mm		26 mm	
Length	159 mm		250 mm	
Stroke	25 mm		25 mm	
Regenerator				
Diameter	40 mm		22 mm	
Length	70 mm		190 mm	
Matrix				
Materials	S.S.	Pb	Pb	$ErNi_2$
Type	screen	ball	ball	grain
Size	250 mesh	0.5 mm	0.18-0.45 mm	
Mass	800 sheets	202 g	150.8 g	284.2 g
Fill factor	31 %	64 %	65.4 %	59.4 %

of 100 minutes. The key points of this design and more detail information will be presented in this paper.

The heat exchange between helium and the wall of the cylinder in the cold head becomes weaker when the refrigeration temperature reaches liquid helium temperature, as the specific heat of helium in volume increases rapidly inversely with temperature. A rough estimation of this effect on the useful cooling capacity has been made[1]. A numerical analysis method considering this heat transfer together with the regenerator is suggested. The calculation results and some discussion are presented in this paper.

DESIGN AND TEST OF THE PROTOTYPE

The ratio of V_R/V_{CO}, where V_R is the volume of the regenerator and V_{CO} is the cold space volume, in the colder stage is a very important parameter[6.7]. The ratio of 5.4, which is a little over the critical datum, is chosen in order to supply a tolerance to test regenerators with different sizes.

Another key point of design is to increase the surface of the heat exchanger in the gap between the cylinder and the displacer in the colder stage for the sake of ameliorating the heat transfer.

The main parameters of this prototype are shown in the Table 1.

The experimental results, which are tested with the refrigerator always running at 60 rpm, are shown in the Fig.1 and Fig.2. The curves in the Fig.1 present the cool-down of this prototype and indicate that the lowest temperature of 2.9 K at the second stage and 37.5 K at the first stage are reached within about 100 minutes with P_H = 2.01 MPa and P_L = 0.73 MPa. The Fig.2 shows the cooling capacity of the second stage while the temperature of the first stage goes from 37.5 K to 50.0 K without power supplied. Curve 1 is for two compressors running in parallel and curve 2 with only one compressor. The electrical power supplied to the compressor(s) is measured by a wattmeter. Two compressors consume 5.47 kW and one compressor needs 2.60 kW. The efficiency of this refrigerator is difficult to be evaluated because the cooling capacity of the first stage has not been measured simultaneously with that of the second stage.

The reason of getting liquid helium temperature without reducing rotating speed is not clear yet. More experiment and analysis are need to be done in the future.

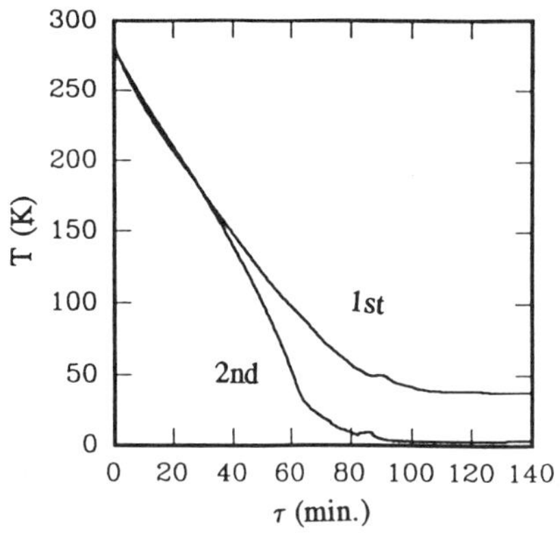

Figure 1. Curves of cooling down.

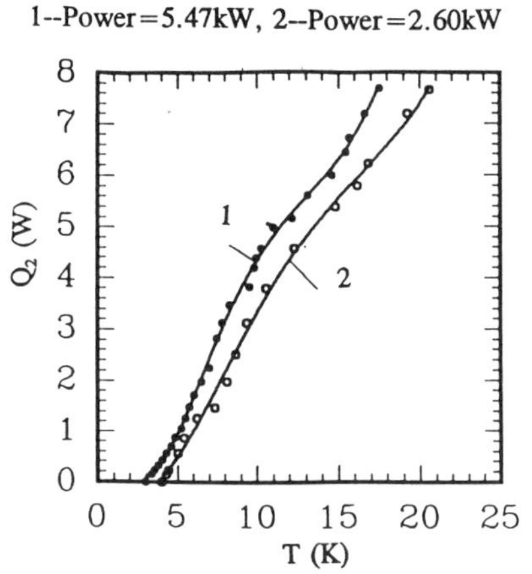

Figure 2. The cooling capacity of the second stage.

THE CALCULATION MODEL OF THE 2ND REGENERATOR AND THE GAP HEAT EXCHANGER

Special attention has been attracted to the heat transfer from the fluid to the cold head in the gap heat exchanger which is the flow passage between the outside of the 2nd displacer and the inside of the cylinder. This heat exchange becomes poor when the refrigerating temperature goes down to liquid helium temperature, because the physical properties of helium change dramatically in this temperature range. The poor heat transfer is caused by low velocity at this gap since the density of helium is very high at the operating condition of the refrigerator. The inefficiency of the heat exchanger leads, but is not equivalent, to a reduction in the useful cooling capacity because the helium at the lower temperature flowing into the regenerator could improve the operating condition of the regenerator if the design of the regenerator is rational.

A numerical analysis method considering the process in the gap and the regenerator together is suggested. The basic equations are as follows:

$$M_{s1}C_p\frac{\partial T_f}{\partial x}+\frac{\alpha_1 A_{\alpha 1}}{\psi}(T_f-T_m)+\rho C_p\frac{\partial T_f}{\partial t}-\beta_T\frac{\partial P}{\partial t}=0 \qquad (1)$$

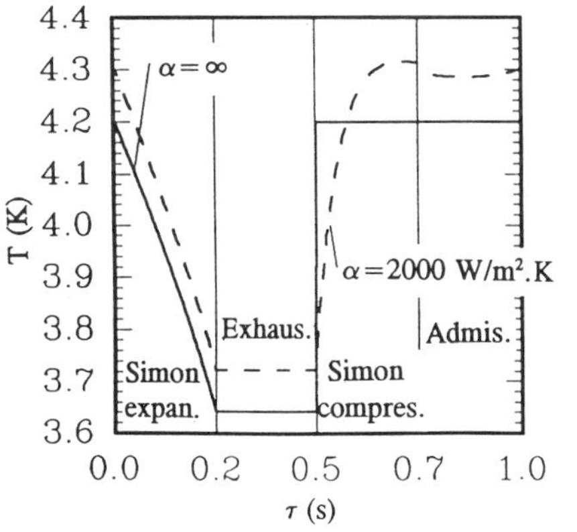

Figure 4. The heat exchange power between helium and the wall of the cold head.

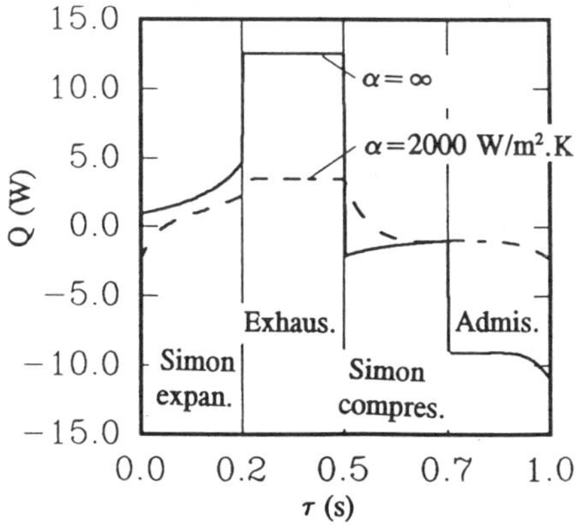

Figure 3. The helium temperature vs. time in the cold chamber.

$$\alpha_1 A_{\alpha 1}(T_f - T_m) - (1-\psi) C_m \frac{\partial T_m}{\partial t} = 0 \tag{2}$$

$$\frac{\partial M_{s1}}{\partial x} + \rho \beta_P \frac{\partial P}{\partial t} - \frac{\rho}{T_f} \beta_T \frac{\partial T_f}{\partial t} = 0 \tag{3}$$

$$M_{s2} C_p \frac{\partial T_f}{\partial x} + \alpha_2 A_{\alpha 2}(T_f - T_c) + \rho C_p \frac{\partial T_f}{\partial t} - \beta_T \frac{\partial P}{\partial t} = 0 \tag{4}$$

$$\frac{\partial M_{s2}}{\partial x} + \rho \beta_P \frac{\partial P}{\partial t} - \frac{\rho}{T_f} \beta_T \frac{\partial T_f}{\partial t} = 0 \tag{5}$$

The processes in the regenerator and the gap are governed by equations (1),(2),(3) and (4),(5) respectively. The relationship of these two processes is connected by equation (6):

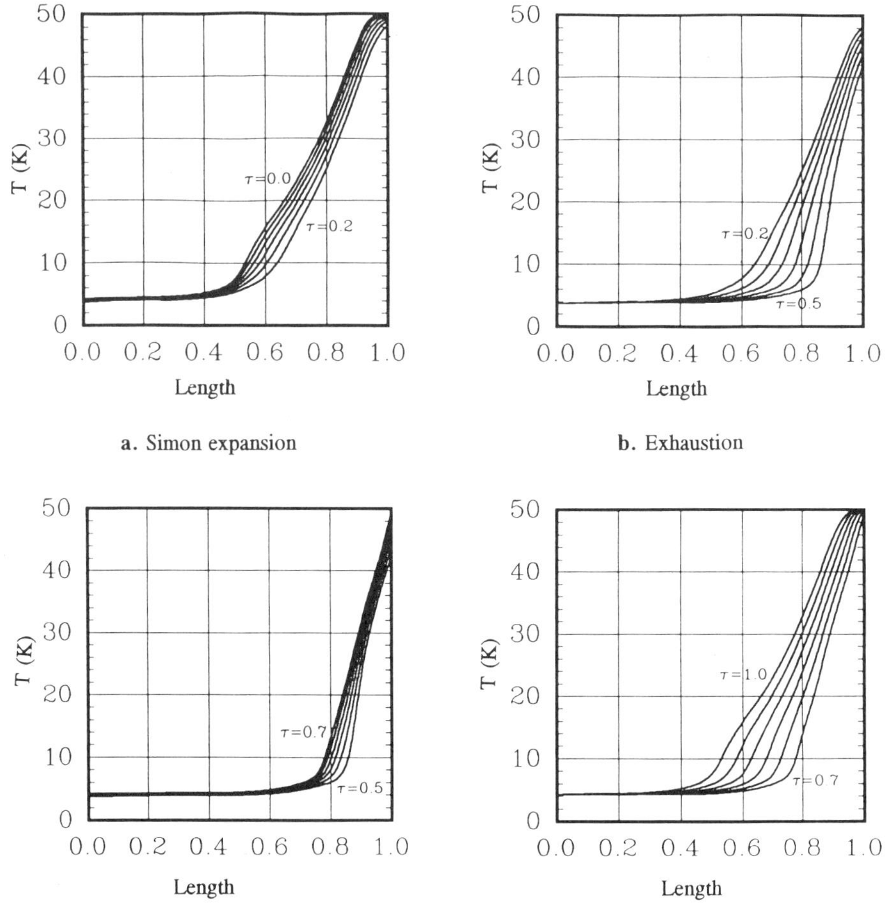

a. Simon expansion

b. Exhaustion

c. Simon compression

d. Admission

Figure 5. The temperature profiles vs. time and the location in the heat exchanger (0.0-0.25 of the length) and the regenerator (0.25-1.0 of the length) with α=2000 W/m^2.K.

$$M_{s1}A_{c1}=M_{s2}A_{c2} \tag{6}$$

The boundary conditions are as follows:

1. Simon expansion

$$x=0,\quad M_{s2}=-\frac{V_{CO}}{A_{c2}}(\rho\beta_P-\frac{\beta_T^2}{C_P T_f})\frac{\partial P}{\partial t}$$

2. Exhaustion

$$x=o,\quad M_{s2}=\frac{\rho V_{CO}}{A_{c2}\tau_2}$$

3. Simon Compression

$$x=0,\quad M_{S2}=0.0 \qquad\qquad x=L,\quad T_f=T_w$$

4. Admission

$$x=0,\quad M_{s2}=\frac{V_{CO}}{\tau_4 h_{CO} A_{c2}}(\rho u+\tau\frac{\partial(\rho u)}{\partial\tau}+P_H) \qquad\qquad x=L,\quad T_f=T_w$$

CALCULATION RESULTS AND DISCUSSION

The calculation parameters are $D_R=0.022$ m, $L_R=0.180$ m, $D_{gap}=0.026$ m, $L_{gap}=0.060$ m, $B_{gap}=0.0002$ m, Er_3Ni matrix, cold head $T_c=4.2$ K; where D_R and L_R are diameter and length of the regenerator and D_{gap}. L_{gap} and B_{gap} are diameter. length and width of the gap heat exchanger, respectively.

The helium temperature vs. time in cold chamber is shown in Figure 3 for $\alpha=2000$ W/m²K and $\alpha=\infty$. During Simon compression process the helium temperature in the cold chamber is that of the cold end of the gap heat exchanger. The helium average temperature in the cold chamber in a refrigeration cycle for $\alpha=\infty$ is about 0.1 K lower than that for $\alpha=2000$ W/m²K.

The heat exchange power vs. time in the gap heat exchanger between helium and the wall of the cold head is shown in Figure 4 for $\alpha=2000$ W/m²K and $\alpha=\infty$. It can be seen in Figure 4 that the heat exchange quantity for $\alpha=\infty$ is much larger than that for $\alpha=2000$W/m²K. The cooling capacity per unit volume of the cold chamber can be obtained by integrating the heat exchange power in the gap heat exchanger vs. time during one refrigeration cycle, which are 0.0561 J/cm³ for $\alpha=2000$ W/m²K and 0.0850 J/cm³ for $\alpha=\infty$. Obviously,the imperfection of heat transfer of cold head will greatly reduce the cooling capacity of a G-M refrigerator. This is mainly due to the cooling quantity not being completely transferred to the cold head by the gap heat exchanger. Hence, enhancement of the heat transfer ability of the gap heat exchanger, for example: increasing the heat exchange area, is one of key factors increasing the cooling capacity of a G-M refrigerator.

Figure 5 presents the temperature profiles vs. time and the location in the gap heat exchanger (0.0-0.25 of the length) and the regenerator (0.25-1.0 of the length) for $\alpha=2000$ W/m²K,including four processes: Simon expansion. exhaustion. Simon compression and admission. Figure 6 presents the temperature profiles vs. time and the location in the gap heat exchanger (0.0-0.25 of the length) and the regenerator (0.25-1.0 of the length) for

$\alpha=\infty$. By comparing Figure 5.b and Figure 6.b we find that the helium outlet average temperature of the gap heat exchanger for α=2000 W/m²K is lower than that (4.2K) for $\alpha=\infty$. It proves that the cooling quantity isn't completely transferred to the cold head.

CONCLUSION

1. A new prototype of a G-M refrigerator with Vco = 13.3 cm³ has been designed and tested. The experimental result of 2.9 K and 515 mW / 4.2 K has been achieved at 60 rpm with a cool-down time of 100 minutes.
2. The imperfection of the heat transfer in the gap heat exchanger in the cold head is one of the key factors to the degradation of the useful cooling capacity of a G-M refrigerator.
3. The improvement of the performance of the gap heat exchanger, for example increasing heat transfer area and coefficient, will increase the cooling capacity of a G-M refrigerator.

NOMENCLATURE

A_α heat transfer area per unit volume
A_c flow area

GREEK LETTERS

α heat transfer coefficient
β_P isothermal compression coefficient

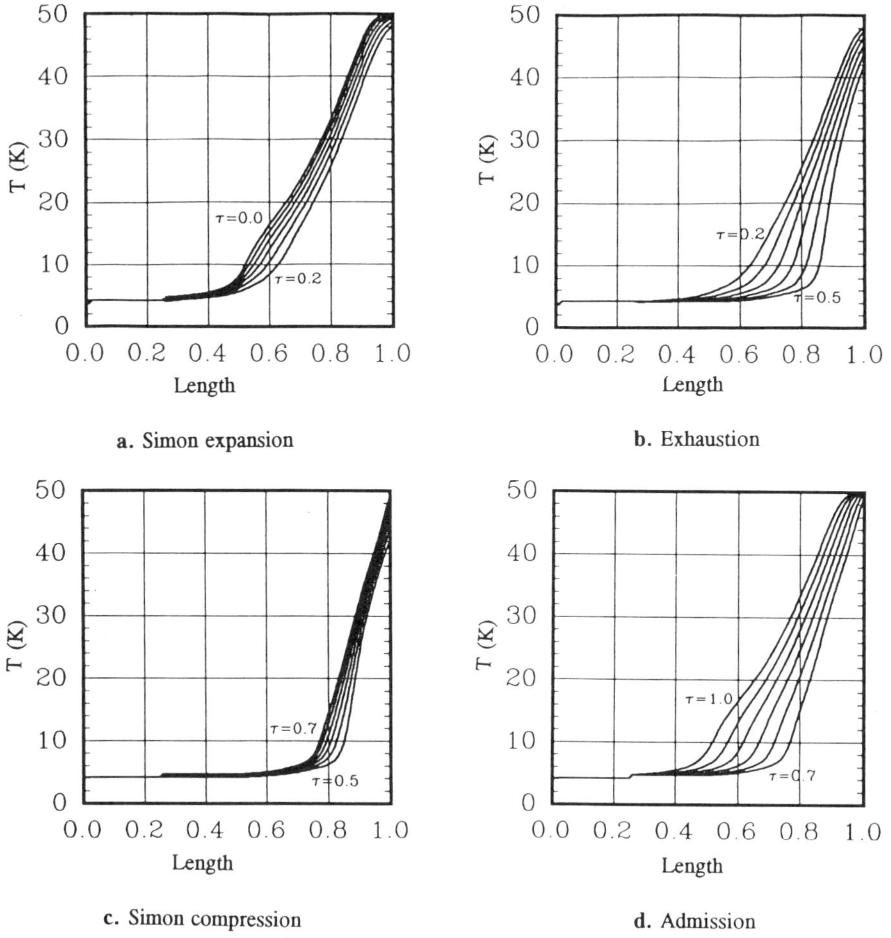

Figure 6. The temperature profiles vs. time and the location in the heat exchanger (0.0-0.25 of the length) and the regenerator (0.25-1.0 of the length) with $\alpha=\infty$.

C_m specific heat of matrix
C_p isobaric specific heat of helium
h_{CO} specific enthalpy in cold chamber
L length
M_s mass flow rate per unit area
P pressure of helium
P_H high pressure
T_f temperature of helium
T_m temperature of matrix
t time
u specific internal energy of helium
V_{CO} cold space volume
x length coordinate along regenerator and gap heat exchanger

β_T isobaric expansion coefficient
ρ density
τ_2 exhaustion period
τ_4 admission period
ψ porosity

SUBSCRIPT

1 in regenerator
2 in gap heat exchanger

ACKNOWLEDGEMENT

This research is supported by the National Natural Science Foundation of China, the National Center for Research and Development on Superconductivity and the Chinese Academy of Sciences.

The authors would like to think Academician S.C. Hong for his useful suggestion, Mr.J.C. Hu for his helpful discussion.

REFERENCES

1. L.Zhang et al. "Development and Analysis of a 4.2 K G-M Refrigerator" Cryogenics ICEC Supplement, Vol 34, P 179 (1994)
2. T.Inaguchi et al. "Two-stage Gifford-McMahon Cycle Cryocooler Operating at about 2 K" Proc. 6th Int. Cryocoolers Conf. Vol 2, P 25 (1990)
3. T.Kyriyama et al. "Regenerator Performance and Refrigeration Mechanism for 4 K GM Refrigerator Using Rare Earth Compound Regenerator Materials" Proc. 7th Int. Cryocoolers Conf. P 429 (1992)
4. Sumitomo Heavy Industries, Ltd. "Technical Specification for 4 K GM Refrigeration System"
5. J.L.Gao, Suzuki Shokan Co., Ltd. Private Communication
6. C.S.Hong et al. "On the Unique Features of the Low Temperature Regenerator in a 4.2 K G-M Refrigerator" Proc. 4th Joint Sino-Japanese Seminar on Cryocoolers and Concerned Topics, P 6, (1993)
7. C.S.Hong et al. "On the Thermodynamic Cycle of the Low temperature G-M refrigerator" Cryogenics ICEC Supplement, Vol 34, P 183 (1994)

THERMODYNAMIC CONSIDERATIONS FOR THE DESIGN OF ACTIVE MAGNETIC REGENERATIVE REFRIGERATORS

J. L. Hall, C. E. Reid, I. G. Spearing, and J. A. Barclay
Cryofuel Systems Group, University of Victoria
P. O. Box 3055, Victoria, B.C., Canada V8W 3P6

ABSTRACT

Recent work has yielded new insights into the thermodynamic operation of active magnetic regenerative refrigerators (AMRR's). Existing theory specifies a linear profile of adiabatic magnetization temperature change (ΔT_{ad}) with temperature (T) of the material in the flow direction of the regenerator. This theory is shown here to be incorrect because an AMRR cannot be represented as a collection of an infinite number of independent refrigerators, as is commonly thought. Furthermore, it is argued that there is no unique $\Delta T_{ad}(T)$ profile for idealized (reversible) AMRR's; however, acceptable profiles must satisfy an integral constraint on the total magnetic work input, as well as satisfying boundary conditions on ΔT_{ad} on the hot and cold edges of the regenerator. It is shown that convex temperature-distance (T(r)) profiles result in minimized entropy generation in an AMRR; however, the magnitude of the magnetic work term appears is too small to produce significant convexity. Also, it is shown that efficient AMRR's require monotonically increasing T(r) profiles in the magnetized region. This condition requires $d(\Delta T_{ad})/dT \geq -1$ for any magnetic material that is used in the regenerator. Finally, some examples are given to illustrate how these new thermodynamic considerations are being used in the design of a rotary AMRR natural gas liquefier.

INTRODUCTION

Magnetic refrigeration is based on the phenomenon that some materials undergo a temperature change when they are adiabatically magnetized and demagnetized. This "magnetocaloric effect" can be combined with heat flows to and from the magnetic material to create a thermodynamic cycle. One common example of this is the magnetic Brayton cycle shown in Fig. 1. The material is adiabatically magnetized in process 1-2, rising to a temperature ΔT_{ad} above the initial temperature T_1. The material is then cooled back down to T_1 in process 2-3, with the transferred heat being rejected to the environment. Adiabatic demagnetization occurs in process 3-4, lowering the material temperature below the initial temperature by ΔT_{ad}. The cooling load is then absorbed in process 4-1, which completes the cycle. Note that the use of the term "magnetic Brayton cycle" is deliberate and underscores the direct analogy between magnetic and gas thermodynamic cycles. In this case, the magnetization processes take the place of gas compression and expansion in the standard Brayton cycle, and the heat transfer operations occur at constant magnetic field rather than constant gas pressure.

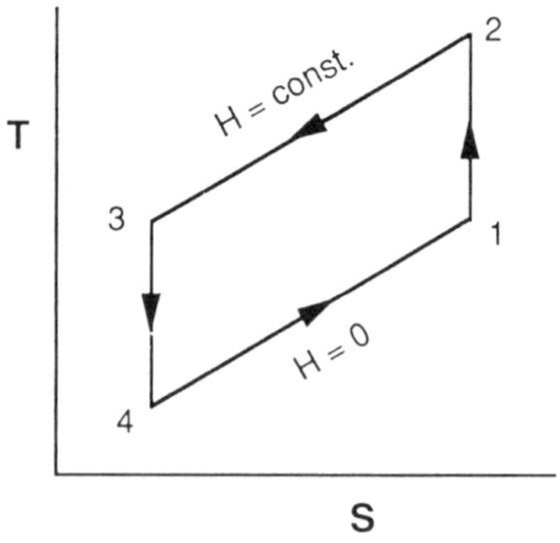

Figure 1. Magnetic Bryton Cycle.

The magnetocaloric effect in most ferromagnetic materials is strongly temperature and field dependent, with a maximum value at the magnetic phase transition temperature. For example, a typical material such as Gadolinium exhibits a caret-shaped profile as illustrated in Fig. 2. These are results from our molecular field model that are in agreement with the experimental data of Benford and Brown.[1] The peak in the profile occurs at the Curie temperature, the point at which Gadolinium undergoes a phase transition from paramagnetism to ferromagnetism. The magnitude of the magnetocaloric effect is small even at the peak, a fact that is generally true for other magnetic materials as well. Therefore, given attainable fields of the order of 10 Tesla with existing superconducting magnet technology, maximum temperature changes of only 15 to 20 Kelvin occur. This contrasts sharply with adiabatic gas compression processes that can produce hundreds of Kelvins of temperature change at room temperature. The relatively small ΔT_{ad}, combined with the narrow temperature range over which it is non-nero, has a major impact on cycle selection for magnetic refrigeration applications. In particular, recuperative cycles are restricted to source-sink temperature spans of the order of ΔT_{ad}

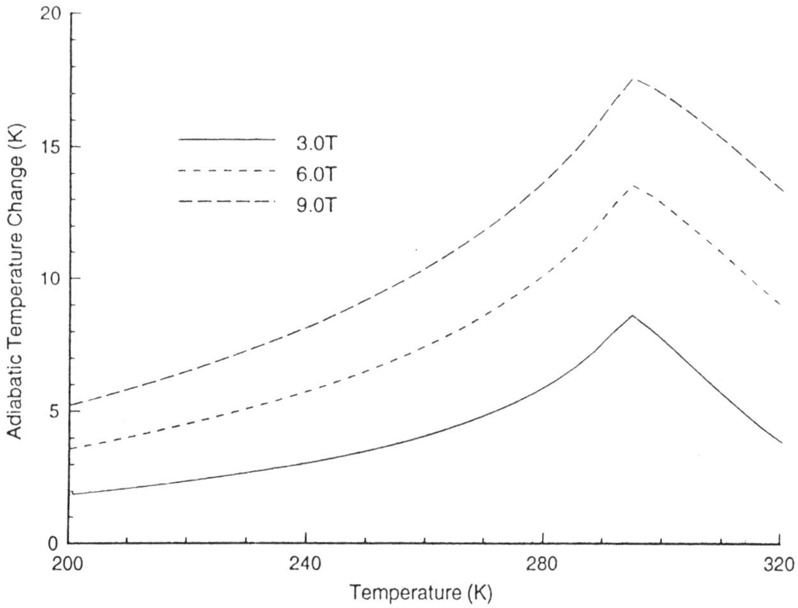

Figure 2. Magnetocaloric Effect in Gadolinium.

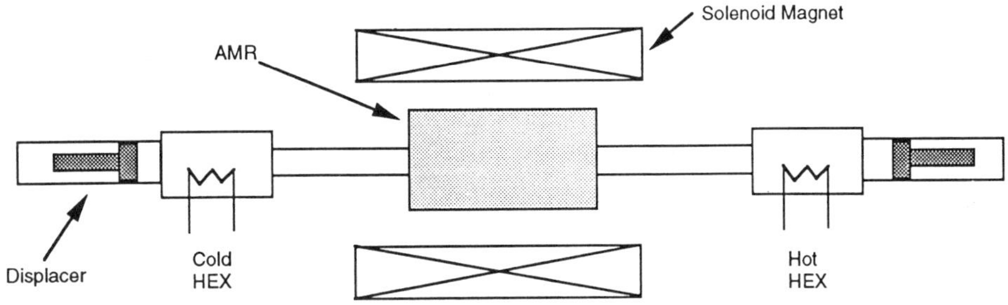

Figure 3. Conceptual Reciprocating AMR Device.

because the magnetic material must have a significant magnetocaloric effect over the entire temperature range it experiences.

Applications involving large temperature spans require regenerative magnetic refrigeration cycles. This approach separates the functions of heat pumping and temperature spanning, allowing the small magnetocaloric effect to be leveraged across a large temperature range. Each element of the regenerator will experience only a small temperature change during steady-state operation; therefore, once can select materials that are optimized for the mean temperature at every location throughout the regenerator. Specifically, the material will be optimized if the peak in the $\Delta T_{ad}(T)$ curve (the Curie point) coincides with the mean temperature at that location.

This idea, while simple in principle, has proven to be very difficult to implement in practice. Considerable effort has been expended by many researchers over the past two decades in attempts to develop and test various regenerative magnetic refrigerators, with modest success at best.[2-12] Much of the difficulty stems from the fact that the working material is a magnetic solid, not a fluid. In particular, heat transfer within or between solid bodies is not particularly good, leading most researchers to use a fluid to thermally connect the magnetic solid to both the load and the environment. This requires that the magnetic material be manufactured in the form of a porous solid through which the fluid can be pumped. The common geometries include particle beds, perforated plates, parallel plates, and wire screens. The need for high heat transfer coefficients typically results in very fine-scale flow passages (< 1 mm), which leads to a host of problems associated with flow pressure drops, manufacturability, and structural integrity. Finally, the combined effects of fluid flow, heat transfer and magnetization result in a coupled physical problem that is very difficult to analyze mathematically.

In an attempt to gain insights into the problem, most researchers have sought analogies between regenerative gas compression cycles and regenerative magnetic cycles. However, unlike the recuperative magnetic Brayton cycle example described in Fig. 1, regenerative analogies are not straightforward, and in fact have often proven to be misleading. One reason for this is the subtlety in defining the thermodynamic system where a solid material does the work, but a circulating fluid transports the heat. Another key reason is that many researchers have employed an active magnetic regenerative (AMR) device that uses magnetic material for both the work input and passive regeneration functions, a coupling that even further blurs the boundary of the thermodynamic system in question. Nevertheless, AMR's appear to have significant advantages over other magnetic cycles[13,14] thereby motivating their continued use despite the attendant analytical complexities.

The present work was motivated by the need to better understand the thermodynamic theory behind active magnetic regenerative refrigerators (AMRR's) in order to properly design a small-scale prototype that will liquefy natural gas. It will be argued that the AMRR is a unique thermodynamic system that cannot be adequately described by analogies to the various recuperative and regenerative gas cycles. We will attempt to show where these analogies break down and how this invalidates some of the theoretical conclusions that have been drawn concerning these devices. Finally, we will present some new insights to the thermodynamics of AMRR's and show how these results are being incorporated into the prototype design. Further details can be found in the recent thesis by Reid.[15]

PROBLEMS WITH EXISTING THEORY

In its simplest configuration, an active magnetic regenerative refrigerator consists of a porous magnetic solid surrounded by a solenoidal magnet that is periodically turned on and off (Fig. 3). The fluid that fills the void space is moved back and forth through the system by a pump or displacer. This fluid accepts heat at the cold heat exchanger and rejects it at the hot heat exchanger. Note that a temperature gradient exists across the regenerator at steady-state operating conditions. The fluid flow and magnetization processes are synchronized in a similar fashion to the magnetic Brayton cycle example described above. First, the fluid is held stationary while magnet is turned on. This raises the regenerator temperature by an amount equal to the magnetocaloric effect. Second, the fluid is pushed from the cold to the hot end, picking up heat from the solid material and rejecting it at the hot heat exchanger. Third, the fluid flow is stopped and the magnet is turned off. The regenerator temperature drops back down. Fourth, the fluid is pushed from the hot to the cold end, depositing heat back into the regenerator and accepting heat from the load at the cold heat exchanger.

A serious problem arises when we try to define the thermodynamic cycle corresponding to this AMRR process. Every part of the system experiences *different* operating conditions. The solid regenerator material exists across a range of temperatures, resulting in a different magnetocaloric effect for each piece. Likewise, the fluid moves back and forth through the device with each infinitesimal element oscillating across a different temperature range depending on its starting location. Many previous researchers have attempted to finesse this non-uniqueness problem by assuming that some kind of composite cycle exists that can adequately describe this thermodynamic process.[4,5,16] This composite cycle approach is typically justified on the basis that the complete regenerator is equivalent to a "cascade" of thermodynamic cycles for each solid element. More specifically, it is postulated that each solid element functions as its own Carnot or Brayton or Ericcson refrigerator, receiving heat from the adjacent lower element and pumping it up to the adjacent higher element. The sum of all of these sub-cycles therefore becomes equivalent to a single composite cycle that spans the complete temperature range of the device. Note that this approach is directly analogous to multi-stage gas compression devices.

However, we argue here that this underlying premise is false. The AMRR does not consist of a cascade of sub-cycles because each solid element of the regenerator does not directly pump heat to its neighbours. All of the solid elements are either accepting or rejecting heat at the same time depending on which way the fluid is flowing. Heat is not pumped from one solid element to the next; instead, it is pumped to the fluid which then carries it across the regenerator. If the fluid exits the regenerator during this half cycle, the heat is rejected to the environment; otherwise, the heat is deposited back into the regenerator where the fluid stops moving. In this very basic sense, there is no direct communication between adjacent solid elements. This is reflected in their thermodynamic cycles plotted on the T-s plane (Fig. 4). Adjacent solid elements in the AMRR will generally have overlapping plots, in contrast to those seen in true cascaded (multi-stage) systems. This is indicative of the fact that those solid elements do not pump heat from one neighbour to the next in a cascaded fashion.

Another argument can be advanced against the cascade theory. Consider two separate AMRR devices connected by an intermediate heat exchanger (Fig. 5). In this case, each AMRR is clearly an independent refrigerator with its own separate fluid loop. There is no need to synchronize magnetization or fluid flows between the stages, and

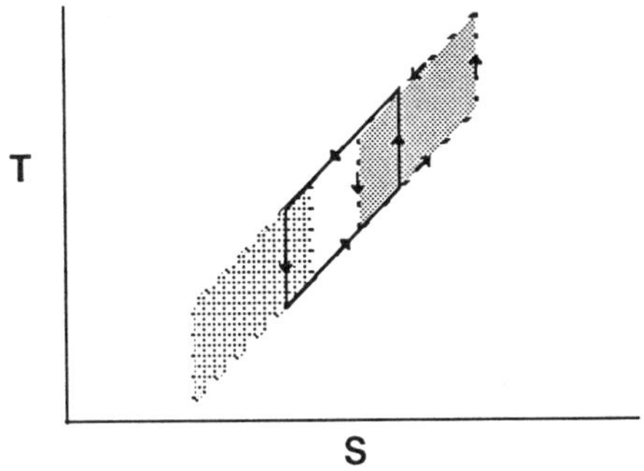

Figure 4. Overlapping Cycles of Adjacent AMRR Solid Elements.

in fact, they may operate at different frequencies altogether. It is also clear that heat from the cooling load is pumped from one stage to the next in the usual cascaded fashion. This operation contrasts sharply with the postulated sub-cycle cascading within a given AMRR device itself. They are not the same.

The preceding arguments suggest that analogies between AMR refrigerators and regenerative gas compression devices either do not exist or simply are not useful. In fact, we think there is a strong case for considering the AMRR to be a unique thermodynamic system altogether. We deliberately avoid the use of the term "cycle" here because there is no unique cycle described by the elements of the AMRR, nor is there a composite cycle composed of cascaded sub-cycles. We will proceed on this basis and describe how this hypothesis leads to some significant changes in the theory of designing AMRR's.

ΔT_{ad} PROFILES

Perhaps the key result of earlier theories was the specification of a linear ΔT_{ad} versus T profile for a given AMRR design.[16] This profile was derived in four steps. First, it was argued on the basis of the sub-cycle cascade model that entropy flows from one element of solid to the next will be equal in an ideal (reversible) device. More specifically,

$$\dot{S} = \frac{\dot{Q}}{T} = constant \tag{1}$$

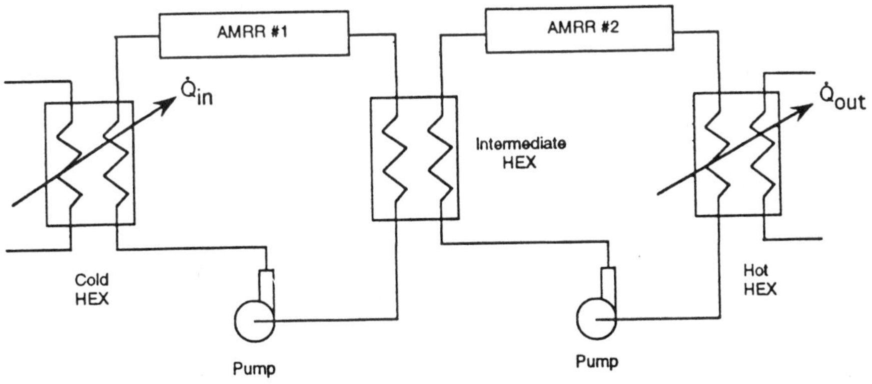

Figure 5. Schematic of a Two-stage AMRR.

where S is the entropy flow, Q is the heat flow and T is the absolute temperature where the heat flow occurs. Second, the net heat flows into the AMRR at the cold and hot ends (Q_c and Q_h) are directly proportional to the magnetocaloric effect. In other words, the heat transfer fluid can only be heated up by an amount $\Delta T_{ad,h}$ above the environment temperature, and cooled by an amount $\Delta T_{ad,c}$ below the load temperature. Therefore,

$$\dot{Q}_c = \dot{m}_f C_{P_f} \Delta T_{ad_c} \tag{2}$$

and

$$\dot{Q}_h = \dot{m}_f C_{P_f} \Delta T_{ad_h} \tag{3}$$

where m_f is the fluid mass flow rate and Cp_f is the fluid heat capacity. The third step is to substitute equations Eqs. 2 and 3 into the entropy balance Eq. (1) to find

$$\frac{\Delta T_{ad_c}}{T_c} = \frac{\Delta T_{ad_h}}{T_h} \tag{4}$$

where a constant specific heat capacity has been assumed. Finally, the fourth step is to draw upon the cascade analogy and argue that Eq. 4 must apply throughout the regenerator and not just at the endpoints. Therefore, this sets the criterion for material selection in the AMRR: the magnetocaloric effect must be a linearly increasing function across the temperature span of the device, with a slope given by T_h/T_c.

Now, since we have argued that the cascade analogy is not valid, then the requirement that Eq. 4 be satisfied across the interior of the active regenerator becomes invalid as well. The ΔT_{ad} versus T profile does not have to be linear in an ideal device. However, Eq. 4 still applies at the boundaries of an AMRR because this simply reflects the need to have the same entropy flow from the load to the environment in a reversible process (Eq. 1).

Although there are boundary conditions for the $\Delta T_{ad}(T)$ profile across an idealized AMRR, there does not appear to be a unique curve for satisfying a zero entropy-producing constraint. Consider the possible entropy producing mechanisms in the AMRR: fluid flow pressure drop, heat transfer across a finite temperature difference between the solid and the fluid, and heat conduction through the solid and fluid due to spatial temperature gradients. Note that none of these mechanisms directly depend on the $\Delta T_{ad}(T)$ profile. In essence, if you can provide a zero viscosity fluid, infinite surface area and zero spatial thermal conductivities, then the AMRR will produce zero entropy for any adiabatic temperature profile that is specified.

However, at least one other constraint on the $\Delta T_{ad}(T)$ profile in addition to the boundary conditions at the hot and cold end. Each infinitesimal element of solid will undergo a process similar to the magnetic Brayton cycle shown in Fig. 1. Note that the area inside the closed curve on the T-s plane is equal to the magnetic work input per cycle. Note also that this area is roughly proportional to the magnetocaloric effect, ΔT_{ad}, although the exact relationship depends on the material properties. The integral of this work over all solid elements must equal the total work input into the AMRR. By definition, this must equal the difference between the heat input and heat rejection,

$$\dot{W} = \dot{Q}_h - \dot{Q}_c \tag{5}$$

where W is the total work per cycle divided by the frequency. This establishes a relationship between the integral of the $\Delta T_{ad}(T)$ profile and the magnetic work input to the AMRR,

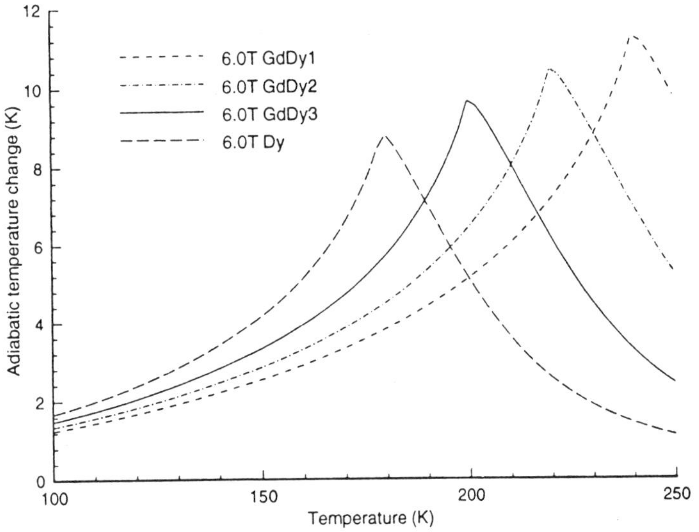

Figure 6. Layered Materials in an AMRR.

$$\dot{W} = \int f(\Delta T_{ad}(T))\, dT \tag{6}$$

where the function f depends on the material properties and the details of the heat transfer process between the solid and the fluid.

In summary, therefore, we can see that while there are an infinite number of possible $\Delta T_{ad}(T)$ profiles for an idealized AMRR, they must all satisfy the boundary conditions of Eq. 4 and the integral work constraint of Eq. 6. Linear profiles are not required; but, they are the simplest functions that can satisfy the boundary conditions and they remain good candidates for practical devices. Moreover, in the temperature range that our group is currently interested in, from room temperature down to liquid natural gas (110 K), the best available rare-earth magnetic materials tend to have magnetocaloric effects that scale with absolute temperature anyway.[16] This has motivated us to design around a nominally linear $\Delta T_{ad}(T)$ profile, even though it is not strictly required by minimum entropy AMRR requirements.[15]

T(r) PROFILES

Real AMRR's have to be concerned with the spatial distribution of the magnetocaloric effect, $\Delta T_{ad}(r)$, in order to locate the solid material at the correct absolute temperature. This is particulary important for devices that require more than one material to span a large temperature range between the load and the environment.[9,16] Layered materials with different Curie temperatures help to maximize the magnetocaloric effect across the entire regenerator, thereby maximizing the magnetic work input per unit mass. Figure 6 presents one example of this, showing layered materials spanning the range from 100 to 250 K.. However, the placement of this material presupposes knowledge of the T(r) profile since the magnetocaloric effect is a function of temperature, not spatial position.

It is not too difficult to argue that T(r) will be close to a linear function in an AMRR. Certainly, passive regenerators have been shown to possess nearly linear T(r) profiles at steady-state operating conditions, with only minor deviations due to end effects.[17] This linearity reflects that fact that the heat transfer between the solid and the fluid is the same at all points within a uniformly porous regenerator. Given that this will also be true in an active magnetic regenerator, the question becomes to what extent can the magnetic work input in an AMRR cause non-linearity in the T(r) profile?

A precise, quantitative answer to this question is not yet available. The full magneto-thermodynamic problem can only be solved with numerical methods, and there have been relatively few studies to date. Recent work by Spearing[18] suggests that magnetic work input does not significantly alter the linearity of the T(r) profile in a rotary AMRR model; however, definitive conclusions await further work.

We can also offer two other plausibility arguments in favour of the conclusion that magnetic work inputs to an AMRR will not significantly change the nominally linear T(r) profile. First, the magnetocaloric effect is small compared to the thermal heat content of the regenerator. This scaling goes like $\Delta T_{ad}/T \sim 15/300 = .05$ for typical materials, suggesting that the magnetic work input changes the energy of the system by only a few percent. Second, this ratio will be approximately constant for a layered bed of materials spanning a large temperature range,[16] suggesting that the magnetic work will be evenly distributed. Evenly distributed magnetic work should have the same effect as evenly distributed heat transfer, thereby promoting a linear T(r) profile.

Despite the seeming inability of the magnetic work to significantly change the T(r) profile that results from passive regeneration, it is an interesting exercise to consider what T(r) profile will minimize the entropy production of a non-ideal AMRR. Although we have not yet solved this problem in a completely rigorous fashion, some insights can be obtained by considering an approximate analysis of the entropy production due to solid-fluid heat flow. This heat transfer per unit volume, Q, occurs across a small temperature difference ΔT_{sf} where,

$$\Delta T_{sf} = T_s - T_f \tag{7}$$

and T_s and T_f are the solid and fluid temperatures respectively. The volumetric entropy generation due to this process is given by

$$\dot{S}_{gen} = \frac{\dot{Q}\,|\Delta T_{sf}|}{T^2} \tag{8}$$

where T is the absolute temperature at which the heat transfer occurs.

Now, Q is itself proportional to ΔT_{sf} because of the physics of convective heat transfer; i.e.,

$$\dot{Q} = hA'''\Delta T_{sf} \tag{9}$$

where h is the heat transfer coefficient, and A''' is the surface area per unit volume. Moreover, standard regenerator theory[17] shows that

$$\Delta T_{sf} \approx \frac{dT}{dr} + time\ dependent\ terms \tag{10}$$

The time dependent terms can be neglected in the long blow limit (large thermal mass of the solid regenerator). Combination of Eqs. 7 through 10 with this assumption leads to

$$\dot{S}_{gen} \approx \frac{1}{T^2}\left(\frac{dT}{dr}\right)^2 \tag{11}$$

Integration of this term across the regenerator bed gives

$$\dot{S}_{tot} = \int \dot{S}_{gen} dV \approx \int \frac{1}{T^2}\left(\frac{dT}{dr}\right)^2 dr \tag{12}$$

If we now postulate a general temperature profile of the form $T(r)=ar^n$, where a and n are constants, then Eq. 12 becomes

This indicates that entropy production will be minimized when the exponent n is small.

$$\dot{S}_{tot} = \int \frac{(nar^{n-1})^2 dr}{a^2 r^{2n}} = \int \frac{n^2}{r^2} dr \approx \frac{n^2}{r} \tag{13}$$

Therefore, convex T(r) profiles where $n < 1$ will result in minimized entropy production under these assumptions.

Another valuable insight into the design of AMRR's can be obtained from an analysis of the interplay between the magnetic work input and heat flows from the solid to the fluid. It is important to remember that a magnetic material will only perform net work if the solid is cooled by the fluid while it is magnetized. This is equivalent to a counter-clockwise cycle in the T-s plane. Therefore, it is required that,

$$\Delta T_{sf} \sim \frac{dT}{dr} > 0 \tag{14}$$

in the magnetized region. Physically, this means that the fluid temperature must always lag the solid temperature as fluid flows from the cold to the hot side of the magnetized regenerator. Note that Eq. 14 refers to the temperature profile after magnetization; therefore, this constitutes an implicit restriction on allowable $\Delta T_{ad}(r)$ profiles. This restriction can be derived as follows.

Let T(r) be the temperature profile before magnetization, and $T^*(r)$ be the temperature profile after magnetization. Therefore,

$$T^*(r) = T(r) + \Delta T_{ad}(r) \tag{15}$$

and,

$$\frac{dT^*}{dr} = \frac{dT}{dr} + \frac{d\Delta T_{ad}}{dr} \tag{16}$$

Using the chain rule on the last term gives,

$$\frac{d\Delta T_{ad}}{dr} = \left(\frac{d\Delta T_{ad}}{dT}\right)\left(\frac{dT}{dr}\right) \tag{17}$$

Substituting Eq. 17 into Eq. 16 yields,

$$\frac{dT^*}{dr} = \frac{dT}{dr}\left[1 + \frac{d\Delta T_{ad}}{dT}\right] \tag{18}$$

Therefore, we see that the inequality of Eq. 14 can only be satisfied if $dT/dr > 0$ and

$$\frac{d\Delta T_{ad}}{dT} > -1 \qquad (19)$$

In other words, the $\Delta T_{ad}(T)$ profile cannot decrease by more than one degree per degree in the regenerator.

It is not immediately obvious that Eq. 19 will be satisfied by magnetic materials given that the typical caret-shaped profile necessarily contains negative slope regions (Fig. 2). However, a survey of rare-earth (lanthanide) magnetic refrigerants does reveal that the negative slope sections of their $\Delta T_{ad}(T)$ do satisfy the one degree per degree constraint. We have therefore proceeded to use these materials in the design of the active magnetic natural gas liquefier prototype, selecting materials that give us the maximum magneto-caloric effect possible across the entire temperature span of the device.[15] Note the additional requirement that the interfaces between materials must be properly positioned to avoid steps in the composite $\Delta T_{ad}(T)$ profile. Otherwise, these steps may violate the one degree per degree constraint.

CONCLUSIONS

It has been argued that the active magnetic regenerative refrigerator (AMRR) constitutes a unique thermodynamic process that cannot be equated to a cascade of thermodynamic cycles of each infinitesimal active element. This postulate means that the existing theory requiring a linear $\Delta T_{ad}(T)$ profile in the magnetic regenerator is incorrect. Instead, it is argued that there is no unique $\Delta T_{ad}(T)$ profile for idealized AMRR's, although there are boundary conditions that must be satisfied, as well as an integral constraint on the total magnetic work input to the system. Non-ideal AMRR's will minimize entropy production if the spatial $\Delta T_{ad}(r)$ is convex; however, the magnitude of the magnetocaloric effect appears to be too small to effect significant changes from the expected linear profile resulting from passive regeneration. Finally, a derivation has been presented to show that acceptable magnetic materials, and the interfaces between layered materials, in a regenerator must satisfy the constraint $d(\Delta T_{ad}) / dT \geq -1$.

ACKNOWLEDGEMENTS

This work is sponsored by the Natural Sciences and Engineering Research Council of Canada, and Centra Gas B.C., Inc.

REFERENCES

1. Benford, S. M. and Brown, G. V. (1981). "T-s diagram for gadolinium near the Curie temperature", *J. Appl. Phys*, Vol. 52, No. 3, pp 2110-2112.
2. Hashimoto, T., Numasawa, T., Shino, M. and Okada, T. (1981). "Magnetic refrigeration in the temperature range from 10K to room temperature: the ferromagnetic refrigerants", *Cryogenics*, Vol. 21, No. 11, pp 647-653.
3. Barclay, J. A. (1982). "The Theory Of An Active Magnetic Regenerative Refrigerator", Los Alamos Report LA-UR-82-1792. Also NASA-CP-2287 (1983).
4. Taussig, C. P., Gallagher, G. R., Smith, J. L. and Iwasa, Y. (1986). "Magnetic Refrigeration Based On Magnetically Active Regeneration", 4th International Cryocooler Conference, Easton, Maryland, September 1986.
5. Cross, C. R., Barclay, J. A., De Gregoria, A. J., Jaeger, S. R. and Johnson, J. W. (1987). "Optimal Temperature-Entropy Curves For Magnetic Refrigeration", *Advances in Cryogenic Engineering*, Vol. 33, pp 767-776.

6. Smith, J. L. and Iwasa, Y. (1990). "Material And Cycle Considerations For Regenerative Magnetic Refrigeration", *Advances in Cryogenic Engineering*, Vol. 35, pp 1157-1164.
7. Green, G., Chafe, J., Stevens, J. and Humphrey, J. (1990). "A Gadolinium-Terbium Active Regenerator", *Advances in Cryogenic Engineering*, Vol. 35, pp 1165-1174.
8. Hashimoto, T. (1991). "Recent Progress In Magnetic Regenerator Materials And Their Applications", Proceedings of the 18th International Conference of Refrigeration, Montreal, Quebec, August 1991.
9. Schuricht, S. R., De Gregoria, A. J., and Zimm, C. B. (1992). "The Effects of a Layered Bed on Active Magnetic Regenerator Performance", 7th International Cryocooler Conference, Santa Fe, New Mexico, November 1992.
10. Zimm, C. B., Ludeman, E. M., Severson, M. C. and Henning, T. A. (1992). "Materials For Regenerative Magnetic Cooling Spanning 20K to 80K", *Adv. Cryo. Eng.*, Vol. 37, pp 893-890.
11. De Gregoria, A. J. (1992). "Modeling the Active Magnetic Regenerator", *Advances in Cryogenic Engineering*, Vol. 37, pp 867-873.
12. Wang, A. A., Johnson, J. W., Niemi, R. W., Sternberg, A. A. and Zimm, C. B. (1994). "Experimental Results of an Efficient Active Magnetic Regenerator Refrigerator", 8th International Cryocooler Conference, Vail, Colorado, June 1994.
13. Barclay, J. A. (1984). "A Comparison Of The Efficiency Of Gas And Magnetic Refrigerators", 22nd National Heat Transfer Conference, Niagara Falls, New York, August 1984.
14. Waynert, J. A., DeGregoria, A. J., Foster, R. W. and Barclay, J. A. (1989). "Evaluation of Industrial Magnetic Heat Pump/Refrigerator Concepts That Utilize Superconducting Magnets", ANL-89/23.
15. Reid, C. E. (1995). "Development of Magnetic Refrigerants For an Active Magnetic Regenerative Refrigerator", Master's thesis, University of Victoria, Victoria, B.C., Canada.
16. Reid, C. E., Barclay, J. A., Hall, J. L. and Sarangi, S. (1994). "Selection of magnetic materials for an active magnetic regenerative refrigerator", *Journal of Alloys and Compounds*, Vol. 207/208, pp 366-371.
17. Hausen, H. (1983). *Heat Transfer in Counterflow, Parallel Flow and Cross Flow*, (translated by M. S. Sayer), McGraw-Hill, Inc. Chapters 11 and 13.
18. Spearing, I. G. (1995). "A Numerical Model For a Rotary Active Magnetic Regenerative Refrigerator", Master's thesis, University of Victoria, Victoria, B.C., Canada.

DESIGN OF AN EXPERIMENTAL APPARATUS FOR INVESTIGATION OF A STIRLING/MAGNETIC CYCLE

G. F. Nellis and J. L. Smith, Jr.

Massachusetts Institute of Technology
Cryogenic Engineering Laboratory
Cambridge, MA 02139, U.S.A.

ABSTRACT

Recently, a cryocooler cycle (subsequently referred to as a Stirling/Magnetic Cycle) has been proposed in which refrigeration is produced by thermodynamically cycling both helium and a paramagnetic substance. The two substances alternately regenerate one another, thereby overcoming the low temperature thermal saturation associated with a Stirling Cycle while producing more refrigeration than a simple Magnetic Cryocooler. This paper presents the design of an experimental apparatus which will be used to implement a Stirling/Magnetic Cycle. A computational model of the thermodynamic cycle is used to specify the optimum helium mass and pressure ratio, subject to constraints associated with an existing superconducting solenoid. The remaining components (e.g. heat exchangers, displacer, etc.) are designed around these requirements.

INTRODUCTION

Conventional cryocoolers employ single phase helium as the refrigerant in a regenerative refrigeration cycle (i.e. a Stirling Cycle). These refrigerators typically cannot operate effectively below 12 K due to the dramatic decrease in the specific heat capacity of regenerator materials[1].

The strong magneto-caloric effect exhibited by some paramagnetic materials make them an excellent alternative to helium as a low temperature refrigerant. Several cryocoolers have been developed in which isobaric helium has served as a fluid regenerator for a paramagnetic substance in a matrix configuration (i.e. a Magnetic Cycle)[2-4].

Although Magnetic Cryocoolers have successfully overcome some of the limitations associated with a Stirling cycle, they do not exploit the fact that two thermodynamically active substances with similar refrigeration potentials are present. It has been shown computationally that a cryocooler in which both pressure and applied magnetic field are controlled (subsequently referred to as a Stirling/Magnetic Cryocooler) can achieve up to 50% more refrigeration than either a Stirling or a Magnetic Cryocooler[5].

This paper presents the design of an experimental apparatus capable of implementing a Stirling/Magnetic Cycle. The prototype should be capable of producing 1.4 W of refrigeration at 4.2 K while rejecting heat at 12 K. The apparatus will operate with a frequency of 83 mHz, an applied field which varies between 1 and 4 Tesla, and a pressure which varies between 3 and 6 atmospheres.

SYSTEM OVERVIEW

In order to investigate the Stirling/Magnetic Cycle an experimental apparatus must be constructed. The prototype cryocooler described here implements a Stirling/Magnetic Cycle using a packed bed of Gadolinium Gallium Garnet (GGG) particles and helium-4. The warm reservoir temperature (T_{warm}) is 12 K and the refrigeration temperature (T_{cold}) is 4.2 K. The experimental apparatus is illustrated schematically in Figure 1.

The helium pressure is controlled using a compressor in conjunction with a system of valves, accumulators, and pressure regulators. In order to return the high pressure helium to the magnetic regenerator bed at the warm reservoir temperature, an upper stage regenerator/heat exchanger is required. This device uses a lead regenerator matrix to transfer energy between the low pressure, cold helium (i.e. 'cold-blow' helium) and the high pressure, warm helium (i.e. 'hot-blow' helium). An additional flow of helium gas is required to overcome regenerator ineffectiveness in the low temperature region. This 'coolant' flow is obtained from the liquid helium bath used to cool the superconducting solenoid.

The GGG is crushed and packed into a cylindrical bed within a superconducting solenoid. The large specific area associated with the packed bed matrix creates good thermal contact between the helium working fluid and the GGG. The applied field within the GGG is controlled by varying the current within the superconducting solenoid. The

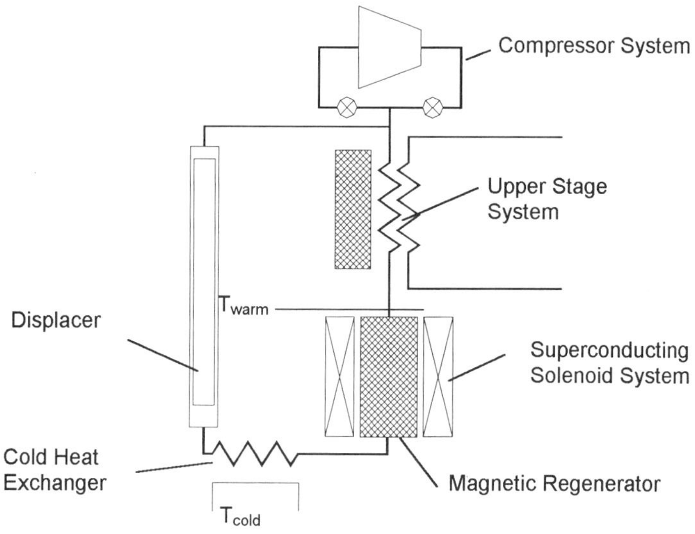

Figure 1. Schematic of proposed experimental apparatus.

refrigeration load is generated with a resistive heater placed within the cold heat exchanger. This heat exchanger is affixed to the cold end of the magnetic regenerator bed. The helium flow is controlled with a displacer.

SYSTEM DESIGN

The main design tool is a computational model of the Stirling/Magnetic Cycle[6]. This model accounts for the major sources of regenerator cycle degradation (i.e. regenerator void volume, imperfect heat transfer, and pressure gradients) while using experimentally determined properties of helium[7] and GGG[8].

The design constraints originate from the characteristics of an existing superconducting solenoid. These constraints dictate the dimensions of the magnetic regenerator bed. The characteristics of the regenerator bed, coupled with the desired refrigeration and warm reservoir temperatures, are used in conjunction with the computational model to determine the optimal helium mass and pressure variation. These results are subsequently used to design the compressor, cold heat exchanger, displacer, and upper stage.

Superconducting Solenoid

A superconducting solenoid has been constructed for a previous experiment[9] and will be used in this apparatus. This solenoid consists of a primary winding and two compensating coils.

Magnetic Regenerator Bed

The dimensions of the magnetic regenerator bed are chosen so that: (1) the uniform applied field section of the bed is completely full of GGG and (2) there is room for a vacuum space between the bore of the solenoid and the regenerator bed.

The major source of cycle degradation identified by the computational model was associated with the compression of the entrained helium within the regenerator matrix. Therefore, the GGG will be crushed and packed into the regenerator bore in order to reduce the matrix porosity. A series of experiments have been performed using glass particles in order to determine the packing force which causes a reduction in the regenerator porosity without resulting in the occlusion of the matrix.

Computational Model Results

The characteristics of the magnetic regenerator bed, together with a warm reservoir temperature (T_{warm} = 12 K) and a cold reservoir temperature (T_{cold} = 4.2 K), are used as input parameters to the computational model of a Stirling/Magnetic Cycle. The pressure variation and helium mass flow are optimized under these conditions.

The most important characteristic of the cycle is the refrigeration (q) which may be nondimensionalized with respect to the average entropy (s) change undergone by the GGG over the temperature range (T) as the applied field ($\mu_o H$) is varied. M_{GGG} and τ represent the total mass of GGG and the half-cycle time, respectively.

$$\bar{q} = \frac{q \cdot \tau \cdot \left[T_{warm} - T_{cold}\right]}{M_{GGG} \cdot \int_{T_{cold}}^{T_{warm}} T \cdot \left[s_{GGG}(\mu_o H_{low}, T) - s_{GGG}(\mu_o H_{high}, T)\right] \cdot dT} \quad (1)$$

The helium mass (M_{he}) and pressure variation are nondimensionalized as a primary mass ratio (mr_1) and a pressure ratio (pr), as defined in Equations 2 and 3.

$$mr_1 = \frac{M_{He}}{M_{GGG}} \quad (2)$$

$$pr = \frac{P_{high}}{P_{low}} \quad (3)$$

Figure 2 illustrates the nondimensional refrigeration as a function of primary mass ratio at several different pressure ratios. The selected design point is also indicated (pr = 2 and mr_1 = 0.0124). This corresponds to a helium mass of 14.26 g and a pressure which varies between 3 and 6 atm. The resulting refrigeration power is predicted to be 1.4 W.

Cold Heat Exchanger

The cold heat exchanger must transfer heat from a resistive heater to the working fluid. Furthermore, the heat exchanger must be placed directly beneath the magnetic regenerator bed (inside the solenoid) in order to minimize losses associated with void volume. Therefore, the heat exchanger will be exposed to a time varying magnetic field and eddy current heating must be avoided.

In order to satisfy these conditions, the cold heat exchanger is fashioned from a flexible heater spirally wrapped with a plastic screen. If the heater and screen are wound tightly, this geometry is characterized by a large surface area, a small amount of void volume, and no closed conduction paths. The cold heat exchanger is vacuum insulated to ensure that the refrigeration load is transferred only to the working fluid.

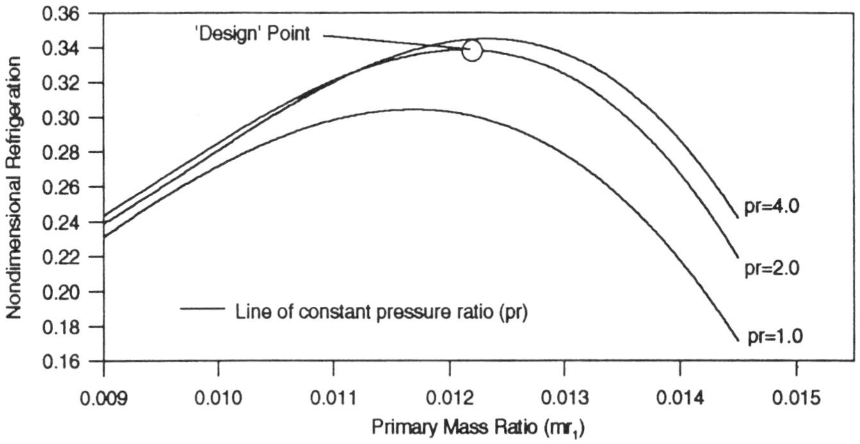

Figure 2. Nondimensional refrigeration produced by the magnetic regenerator bed as primary mass ratio and pressure ratio are varied.

Table 1. Upper stage model notation

Description	Symbol	Description	Symbol
Mass flow rate	mf	Temperature	T
Area	A	Heat transfer coefficient	h
Time	t	Axial position (from cold end)	x
Mass	M	Length	L
Matrix heat capacity	c_m	Heat capacity of working fluid & coolant	c_f
Inlet coolant temperature	T_{ci}	'Cold blow' inlet working fluid temp.	T_{wci}
'Hot blow' inlet working fluid temp.	T_{whi}	Working fluid (subscript)	w
Matrix (subscript)	m	Coolant (subscript)	c

Displacer

The displacer consists of a vacuum insulated stainless steel tube in which a segmented, phenolic cylinder is driven.

Compressor System

The compression process is accomplished with a 4-stage air compressor (modified for use with helium) and a system of pressure regulators, accumulators, and valves.

Upper Stage System

The upper stage system must function as both a regenerator and a heat exchanger. The geometry is that of a single pass shell and tube heat exchanger. The coolant stream, which flows through the shell, is obtained from the liquid helium pool used to cool the superconducting solenoid. A series of baffles forces the coolant to flow perpendicular to the tubes. The working fluid flows from the compressor to the magnetic stage during the 'hot blow' portions of the cycle (i.e. pressurization and inlet) and from the magnetic stage back to the compressor during the 'cold blow' portions of the cycle (i.e. expansion and outlet). The tubes are filled with lead shot that functions as the regenerator material.

A constant property model was used to design the upper stage system. This model is illustrated in Figure 3 and the notation is described in Table 1. The overbar symbol indicates time average. The governing equations are shown below:

$$M_m \cdot c_m \cdot \frac{\partial T_m}{\partial t} = A_{mw} \cdot h_{mw} \cdot (T_w - T_m) \tag{4}$$

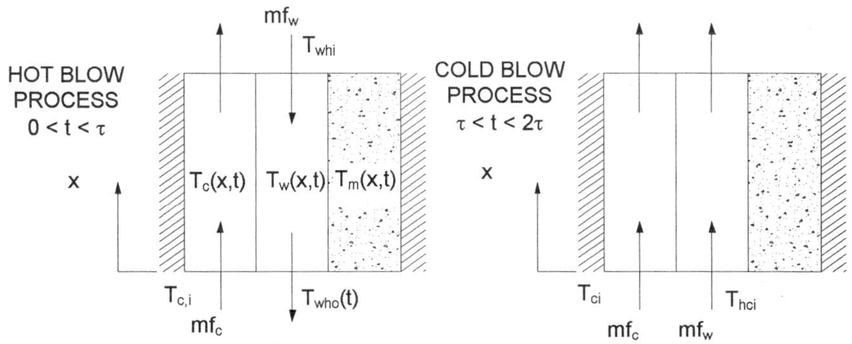

Figure 3. Model used to analyze the upper stage regenerator/heat exchanger.

$$L \cdot mf_w \cdot c_f \cdot \frac{\partial T_w}{\partial x} = A_{mw} \cdot h_{mw} \cdot (T_m - T_w) + A_{wc} \cdot h_{wc} \cdot (T_c - T_w) \quad \textbf{(cold blow)} \tag{5}$$

$$L \cdot mf_w \cdot c_f \cdot \frac{\partial T_w}{\partial x} = A_{mw} \cdot h_{mw} \cdot (T_w - T_m) + A_{wc} \cdot h_{wc} \cdot (T_w - T_c) \quad \textbf{(hot blow)} \tag{6}$$

$$mf_c \cdot c_f \cdot \frac{\partial T_c}{\partial x} = P_{wc} \cdot h_{wc} \cdot (T_w - T_c) \tag{7}$$

The problem may be nondimensionalized by defining the parameters listed below:

$$\text{reduced time} : \eta \equiv \frac{t \cdot A_{mw} \cdot h_{mw}}{M_m \cdot c_m} \tag{8}$$

$$\text{reduced period} : \Pi \equiv \frac{\tau \cdot A_{mw} \cdot h_{mw}}{M_m \cdot c_m} \tag{9}$$

$$\text{reduced axial position} : \varepsilon \equiv \frac{x \cdot P_{mw} \cdot h_{mw}}{mf_w \cdot c_f} \tag{10}$$

$$\text{reduced length} : \Lambda \equiv \frac{A_{mw} \cdot h_{mw}}{mf_w \cdot c_f} \tag{11}$$

$$\text{heat transfer ratio} : \lambda \equiv \frac{A_{wc} \cdot h_{wc}}{A_{mw} \cdot h_{mw}} \tag{12}$$

$$\text{mass ratio} : mr \equiv \frac{mf_w}{mf_c} \tag{13}$$

$$\text{nondimensional temperature} : \Theta \equiv \frac{T - T_{ci}}{T_{whi} - T_{ci}} \tag{14}$$

$$\text{nondimensional working fluid inlet temperature} : \Theta_{wci} \equiv \frac{T_{wci} - T_{ci}}{T_{whi} - T_{ci}} \tag{15}$$

The nondimensional governing equations are:

$$\frac{\partial \Theta_m}{\partial \eta} = \Theta_w - \Theta_m \tag{16}$$

$$\frac{\partial \Theta_w}{\partial \varepsilon} = \Theta_m - \Theta_w + \lambda \cdot (\Theta_c - \Theta_w) \quad \textbf{(cold blow)} \tag{17}$$

$$\frac{\partial \Theta_w}{\partial \varepsilon} = \Theta_w - \Theta_m + \lambda \cdot (\Theta_w - \Theta_c) \quad \textbf{(hot blow)} \tag{18}$$

$$\frac{\partial \Theta_c}{\partial \varepsilon} = \lambda \cdot mr \cdot (\Theta_w - \Theta_c) \tag{19}$$

The boundary conditions are:

$$\Theta_c(0,\eta) = 0 \tag{20}$$

$$\Theta_w(\Lambda,\eta) = 1 \text{ for } 0 \le \eta \le \Pi \tag{21}$$

$$\Theta_w(0,\eta) = \Theta_{wci} \text{ for } \Pi \le \eta \le 2\cdot\Pi \tag{22}$$

$$\Theta_m(\varepsilon,0) = \Theta_m(\varepsilon,2\cdot\Pi) \tag{23}$$

These equations were solved using an open finite difference technique. Band diagonal 'cold blow' and 'hot blow' stiffness matrices were generated, decomposed, and used to step forward in time until a steady state cycle was reached. The model successfully reproduces the regenerator efficiency curves for a balanced, counter-flow regenerator[10] when the heat transfer ratio is zero. Also, the model reproduces the parallel- and counter-flow heat exchanger efficiency curves[11] when the reduced period is infinite.

The results of the model are characterized by the time-average dimensionless outlet temperature of the working fluid during the 'hot blow' process. This temperature is equivalent to the warm reservoir temperature of the magnetic stage.

$$\overline{\Theta_{who}} \equiv \frac{\overline{T_{who}} - T_{ci}}{T_{whi} - T_{ci}} = \Im(\Pi,\Lambda,\lambda,\Theta_{wci},mr) \tag{24}$$

The desired warm reservoir temperature can be used to specify the required dimensionless outlet temperature of the working fluid during the 'hot blow' process.

$$\overline{\Theta_{who}} = \frac{12-4.2}{300-4.2} \tag{25}$$

The computational model indicates that the helium will leave the magnetic stage at a temperature approximately 2 K greater than the warm reservoir temperature. This can be used to further specify the problem.

$$\Theta_{wci} = \frac{14-4.2}{300-4.2} \tag{26}$$

The final constraint may be obtained from the desired cryogen boil-off rate (Q_c). A reasonable rate of liquid helium boil-off is 4.5 liters per hour. This quantity, coupled with earlier design results, can be used to specify the mass ratio (ρ_{LHe} is liquid helium density).

$$mr = \frac{M_{He}}{\tau\cdot Q_c\cdot\rho_{LHe}} = 15.23 \tag{27}$$

With these constraints, the functional form of the problem is simplified considerably.

$$\Lambda = \Im(\Pi,\lambda) \tag{28}$$

The resulting curves are shown in Figure 4. At very low reduced periods, the upper stage assembly has a very large thermal mass and behaves like a pure regenerator. Because a regenerator cannot physically achieve the constraints put forth above, the reduced length required asymptotically approaches infinity. At very high reduced periods, the upper stage assembly has a very low thermal mass and behaves like a heat exchanger. Consequently, the reduced length required asymptotically approaches a constant which is independent of reduced period.

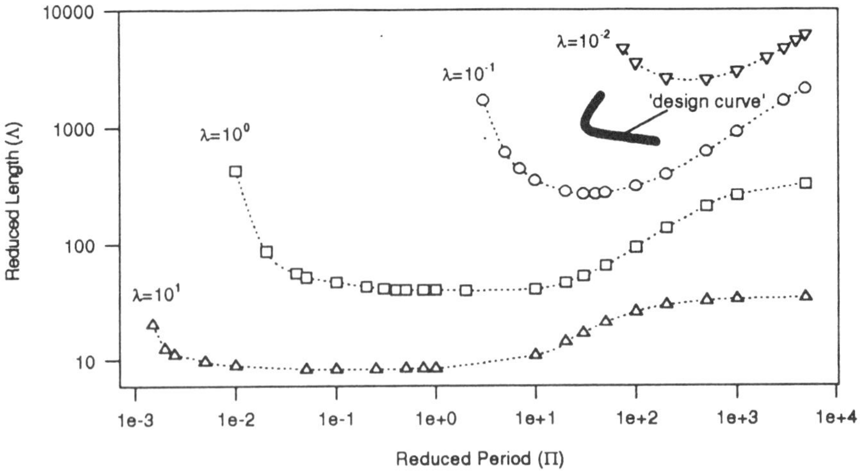

Figure 4. Solution of constrained upper stage regenerator/heat exchanger equations and the locus of the nondimensional characteristics of the upper stage design.

The constant temperature model, combined with correlations for the Nusselt Number for a packed bed[12] and a bank of tubes[13], has been used to analyze the performance of various upper stage geometries.

The reduced period and reduced length associated with the design geometry varies with temperature. The locus of this variation is also illustrated in Figure 4. The heat transfer ratio remains nearly constant with temperature ($\lambda \cong 0.05$).

ACKNOWLEDGEMENT

This work has been sponsored by the Naval Surface Warfare Center, Carderock Division. Some calculations presented in this paper were performed on the MIT CRAY X-MP.

REFERENCES

1. J. L. Smith, Jr, Cryogenic refrigeration for space exploration - a challenge for the future, *Proceedings of the XVII'th International Congress of Refrigeration*, **1**:2, (1991).
2. C. P. Taussig, "Magnetically Active Regeneration", M. I. T. (Ph.D. Thesis), Cambridge (1986).
3. F. J. Cogswell, "Cycle Control of a Regenerative Magnetic Refrigerator Operating From 4.2 K to 15 K", M. I. T. (Ph.D. Thesis), Cambridge (1989).
4. S. Jeong, "Development of the Regenerative Magnetic Refrigerator Operating Between 4.2 K and 1.8 K", M. I. T. (Ph.D. Thesis), Cambridge, (1992).
5. G. F. Nellis, "Magnetically Augmented Cryogenic Refrigeration", M. I. T. (M.S. Thesis), Cambridge, (1995).
6. G. F. Nellis, "Investigation of Magnetically Augmented Cryogenic Refrigeration", *Proceedings of the Eighth International Cryocooler Conference*, 1:647, (1994).
7. R. D. McCarty, "Thermophysical properties of Helium-4 from 2° - 1500° K with pressures to 1000 atmospheres", *NBS-TN-631*, (1972).
8. G. R. Ghallagher, "Analysis of a Magnetically Active Regenerator", M. I. T. (M.S. Thesis), Cambridge, (1986).

9. W. W. Chin, "Construction and Testing of an AC Superconducting Magnet for a Magnetic Refrigerator", M. I. T. (B.S. Thesis), Cambridge, (1985).
10. F. W. Schmidt and A. J. Willmott, "Thermal Energy Storage and Regeneration", McGraw-Hill, New York, (1981)
11. F. P. Incropera and D. P. De Witt, "Fundamentals of Heat and Mass Transfer", John Wiley and Sons, Inc, New York (1990).
12. W. M. Rohsenow, J. P. Hartnett, and E. N. Ganic, "Handbook of Heat Transfer Applications, Second Edition", McGraw-Hill, New York, (1973).
13. W. M. Kays and A. L. London, "Compact Heat Exchangers, Third Edition", McGraw-Hill, New York, (1984).

TEST RESULTS ON A 50K MAGNETIC REFRIGERATOR

C.B. Zimm[1], J.W. Johnson[1], and R.W. Murphy[2]

[1]Astronautics Technology Center
Madison, WI, 53716, USA
[2]Oak Ridge National Laboratory
Oak Ridge, TN, 37831, USA

ABSTRACT

The cost and scaling properties of superconducting magnets used in magnetic cryocoolers suggest that magnetic refrigeration is best employed in the medium power range (cooling entropy flows of about 1 W/K) rather than in the small sizes demonstrated to date. A medium scale magnetic refrigerator has been built and tested as a result of a cooperative project between the Astronautics Corporation of America and Oak Ridge National Laboratory. The device uses two active magnetic regenerative beds of $GdNi_2$ reciprocated at up to 0.5 Hz into the bore of a fixed superconducting solenoid. The beds are linked to fixed LN_2 hot heat exchangers and a fixed electrically heated cold heat exchanger by flexible metal hoses. The device developed a 35 K temperature span at no load, and a 30 watt cooling power at a reduced temperature span. Testing showed that the dominant losses in this type of design occur from the effects of pressure drop and dead volume in the flexible hoses. Mechanisms of minimizing these losses will be discussed.

INTRODUCTION

The reversibility of the magnetocaloric effect and the efficient heat transfer in packed particle beds suggest that magnetic refrigerators can, in theory, achieve high efficiency[1]. An Astronautics laboratory magnetic refrigerator has achieved high efficiency at a scale smaller than that of efficient gas cryocoolers[2]. The cost and scaling properties of superconducting magnets (the ratio of useful field volume to winding volume increases with the size of a solenoid, and the cost of power supplies and leads varies little with the size of a solenoid) suggest that magnetic refrigeration would be more cost effective at a larger scales. Hence Astronautics and Oak Ridge National Laboratory have jointly constructed and tested a single stage magnetic refrigerator that is an order of magnitude larger than the previous Astronautics device. This device uses the same hot heat sink temperature (77 K) and single-stage Active Magnetic Regenerator (AMR) concept as the previous device.

The new device is still a laboratory device not intended for direct application. Future prototypes of the device might be used to cool large high temperature superconducting magnets, motors or generators. Multiple stage AMR devices with a LN_2 heat sink might be used to liquefy hydrogen[3] or cool low temperature superconducting devices. Another future application at even larger scale might, if extremely high efficiency can be achieved, be food storage refrigeration or large building air conditioning.

Advances in Cryogenic Engineering, Vol. 41
Edited by P. Kittel, Plenum Press, New York, 1996

DESIGN CONCEPT

The Active Magnetic Regenerative (AMR) concept[4] is used for this device instead of the more conventional magnetic Carnot cycle because the change in temperature due to the magnetocaloric effect (ΔT_s) from available materials upon the application of practical magnetic fields (8 T or less) is not large enough to span the temperature region of interest. The regeneration of the AMR cycle allows a temperature span of more than 5 times the ΔT_s of the magnetic refrigerant without having to transfer large amounts of heat to an external regenerator. Figure 1 shows the overall design configuration. The magnetic field needed to perform the AMR cycle is provided by a stationary NbTi solenoid operated in persistent mode. The magnet is immersed in liquid helium contained in an annular dewar. Two packed beds of $GdNi_2$ magnetocaloric material are rigidly mounted to a G-10 (glass epoxy composite) carrier assembly. $GdNi_2$ is a ferromagnetic material that heats upon application of a magnetic field, and cools upon its removal[5]. The bed assembly is reciprocated by a room temperature drive such that the beds are alternately inserted in the bore of the magnet. Use of two beds provides partial magnetic drive force cancellation, reduces the fluctuation in magnet bore permeability, and provides a higher cooling duty cycle.

A bellows displacer causes circulation of helium gas. This gas leaves one end of the displacer, passes through the hot heat exchanger (HHEX), is cooled as it flows through the demagnetized bed, extracts heat from the cold heat exchanger (CHEX), extracts heat from the magnetized bed, and gives up heat to boiling liquid nitrogen (LN_2) in the HHEX, finally

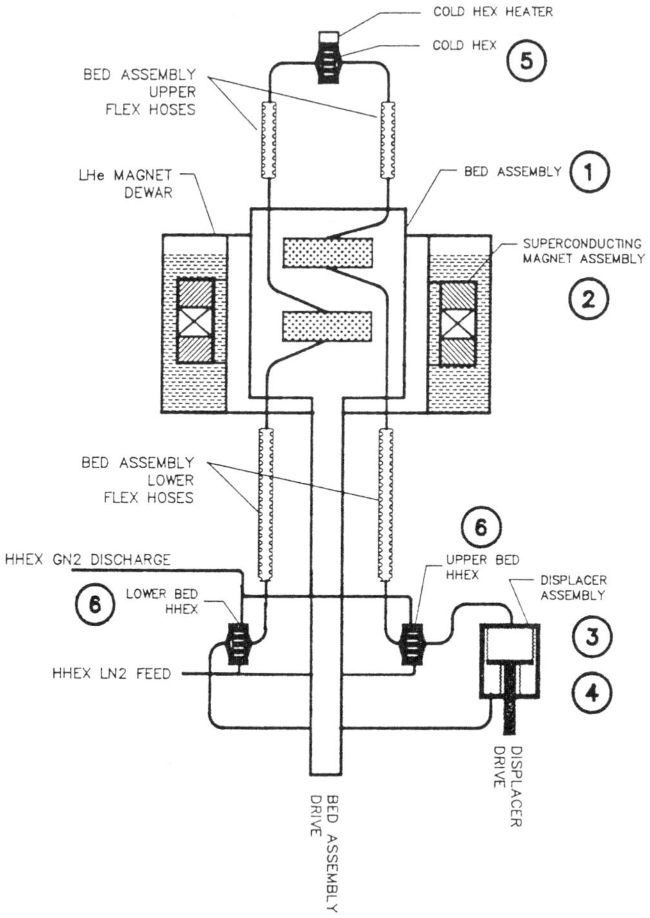

Figure 1. The helium flow circuit of the refrigerator.

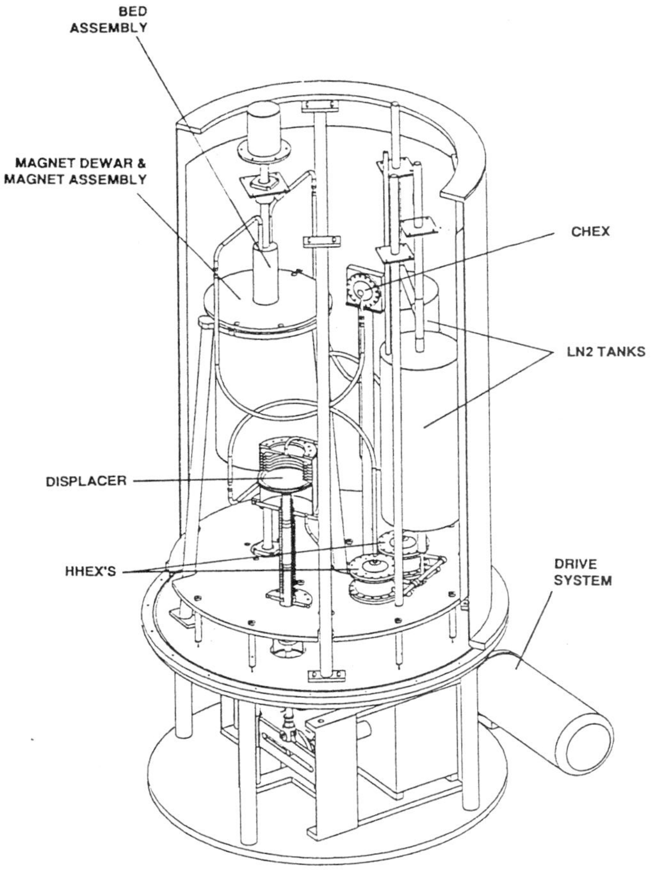

Figure 2. View of the device.

Figure 3. Photograph of the device with the dewar and heat shield removed.

returning to the other end of the displacer (see Figure 1). The displacer is moved by a room temperature drive.

The gas flow between the moving beds and the fixed heat exchangers is routed through flexible metal hoses. The cryocooler assembly is surrounded by a thermal shield attached to the HHEX's; the shield is surrounded in turn by a vacuum vessel.

The performance of an AMR refrigerator is critically dependent on the properties of the magnetocaloric material. An AMR cycle with balanced flow of helium gas with volumetric heat capacity much less than that of the magnetocaloric material obtains optimum performance when the magnetic refrigerant has a ΔT_s proportional to absolute temperature. For a ferromagnetic material such as $GdNi_2$, such dependence is approximately obtained when the hot heat exchange temperature, T_h, is at or below the Curie temperature of the material (75 K for $GdNi_2$).

COMPONENTS OF THE DEVICE

Figure 2 is a drawing of the device, and Figure 3 is a photograph of the assembled device without the front halves of the heat shields and vacuum vessel.

Figure 4 shows the design of the bed assembly. The 2 kg of 0.15 mm $GdNi_2$ particles in each magnetocaloric bed are contained in two cylinders that are bonded into a rigid cylindrical assembly. Helium flow is brought to the beds by two passages, each of which connects to one bed, and bypasses the other bed. All components of the bed assembly except the magnetic material and the end connections are made of G-10 glass-epoxy composite in order to minimize eddy currents. The initial design of the bed assembly (Figure 4) carried the helium flow to the beds through holes bored in stacked cylindrical blocks of G-10. The G-10 pieces were epoxied together using butt joints. Structural strength was augmented by a fiberglass overwrap. Although this design was strong and leak-tight at room temperature, it failed upon cooling. This was because with the low thermal diffusivity of G-10, the butt joints peeled apart at their inner radius in response to thermal contraction induced by circulating cold helium gas before the outer portions had time to cool. The problem was cured by adding sleeves at each flow passage; pins were also inserted in the adhesive joints as an additional precaution (Figure 4). Cooling of the bed assembly was still fairly slow; a possible future revision with reduced thermal mass is shown at the bottom of Figure 4.

The distance between two beds is determined from force cancellation effects and by the amount of stray field that can be tolerated at the demagnetized bed. The superconducting solenoid uses iron flux return pieces to ensure a rapid dropoff in magnetic field outside the bore. The 7 T magnet was designed by ORNL and described in a USDOE report[6].

The bed assembly is driven by a scotch yoke mechanism and indexing drive located below the vacuum vessel. A bellows serves as a seal where the drive rod penetrates the vacuum vessel. The bed assembly is centered in the bore of the magnet by two sleeve bearings operating on the drive rod below the bed assembly and a follower rod above the bed assembly. The bearings are mounted to the bottom and top plates of the radiation shield and use fiberglass-backed teflon fabric sleeves.

The bellows displacer design is similar to one previously described[7] except it has a smaller swept volume of 840 cm3. It is driven in the same manner as described above for the bed assembly.

The heat transfer surfaces exposed to the helium gas in each HEX are twelve 17 mm porous copper disks, two each in six parallel flow passages. For the HHEX's, heat is conducted to boiling LN_2 in an annular chamber around the six helium flow passages. LN_2 was fed by gravity from two tanks inside the main dewar; the vapor was vented to atmosphere by a pipe leading from the top of each HHEX. For the CHEX, the annular space is absent, and heat is supplied electrically by power resistors mounted on the outer wall of the CHEX. The porous copper disks are easier to mount than stacks of screens, but are more easily fouled if condensible impurities are present in the helium.

Helium heat transfer gas is carried from the HEX's to the magnetocaloric beds by four 1.2 m long flexible pressure hoses similar to those used to carry gas between the compressor and the cold head of G-M cryocoolers. The inner wall of each hose consists of a 12 mm OD 304 stainless steel bellows. The selection of these hoses was a difficult compromise between dead volume, pressure drop and life considerations. Maximum life is obtained with a long hose (large bend radius) of small diameter. Minimum pressure drop is obtained with a short hose of large diameter. Minimum dead volume is obtained with a short hose of small

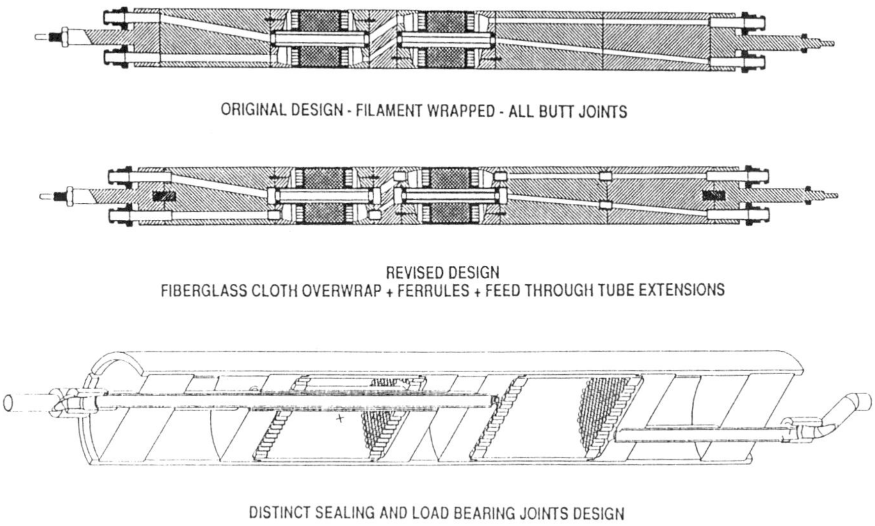

Figure 4. The magnetocaloric bed assembly. $GdNi_2$ particles are in the dark shaded chambers. The middle design was used in the tests.

diameter. The hoses chosen had a design life of at least 5,000,000 cycles (1.3 x 10^7 flexes were obtained with 20 atm internal pressure in a room temperature test). The dead volume of each hose on the hot side is 34 % of the swept volume; on the cold side it is approximately 58 % of the swept volume, allowing for a higher helium density on the cold side. The pressure drop for each hose was not separately measured, but was expected to be much higher than that for a smooth wall tube because of eddies created by the bellows corrugations. One way of improving the dead volume/pressure drop tradeoff is to insert a smooth teflon liner in the hose, which was found to significantly enhance performance.

PERFORMANCE

The performance of the device was initially measured with the nitrogen vapor leaving the HHEX bath at atmospheric pressure. Under such conditions, a maximum cooling power of 25 watts was obtained at a 2 s period with the CHEX body at 76 K. The HHEX outlet

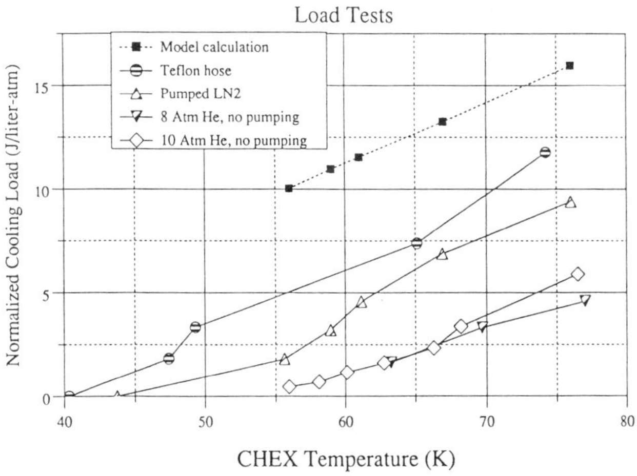

Figure 5. Cooling capability of the device. Reducing the dead volume (teflon hose) increased the cooling capability. The model calculation does not account for losses outside the magnetocaloric beds.

was at 81 K but the hot end of the magnetocaloric beds was at 88 K. The CHEX under a small (2 W) load dropped to 56 K. The high temperature at the beds was presumably due to dead volume in the flex hoses. Operation of an AMR above the Curie point of the magnetocaloric material (75 K for $GdNi_2$) is expected to reduce efficiency and temperature span. Reducing the temperature at the hot end of the beds to 82 K by pumping on the nitrogen exhaust line reduced the no-load temperature at the CHEX body to 44 K.

The cooling power of a magnetic refrigerator is initially proportional to the cycle frequency and the mass of heat transfer gas blown through the beds (at large mass flow rates, the cooling power levels off due to heat transfer and flow losses). Hence runs at different frequencies and mass flows can be compared by normalizing the cooling energy per cycle to the amount of helium blown through the beds. Such normalized cooling loads, in units of joules per liter-atmosphere, are plotted in Figure 5. The expected scaling is verified by the superimposing of the cooling load curves at 8 and 10 atmosphere helium heat transfer gas pressure. The effect of pumping on the HHEX exhaust is seen to be a large increase in cooling load for a given CHEX temperature, or a much colder CHEX temperature for a given load.

Also shown in Figure 5 is the cooling power calculated from the Astronautics AMR model[4] for temperature spans corresponding to the experimental runs. The model predicts higher cooling power for a given temperature span than that achieved in the experiment, but caution must be used in comparing model and experimental results. The following discussion of the model clarifies its limitations.

The model isolates each AMR bed. A specified mass flow rate of helium enters the cold end of the bed at a specified temperature during the second quarter of the cycle, and enters the hot end of the bed at another specified temperature during the fourth quarter of the cycle. During the first and third quarters of the cycle, the magnetic field is increased to a specified value, and then removed, respectively. Bed temperature profiles are calculated using standard correlations for the heat transfer between the helium gas and the bed particles, and experimentally measured values for bed porosity, heat capacity and magnetocaloric effect.

The cooling power and heat rejected are computing by assuming that the gas emerging from the bed is returned by the heat exchanger to the temperature at which it entered, i.e., that there is no dead volume between the heat exchangers and the beds. Axial dispersion and thermal conduction losses that occur within the bed are approximated by correcting the calculated Qhot and Qcold; pressure drop work is assumed to add entirely to Qhot. No allowance is made for pressure drop work or heat leaks that occur outside the magnetocaloric bed.

The large discrepancy between the bed model and the experimental performance suggests that dead volume effects and pressure drop work outside the beds are significantly reducing the cooling load. In order to separate the effects of dead volume from pressure drop, 8 mm ID teflon tubing was inserted into each of the flexible hoses linking the beds to the HEX's. The dead volume in the hoses was thereby reduced by a factor of three, but the pressure drop was slightly increased (the large reduction in cross sectional area available for flow was partially compensated by the reduced friction due to the smoother wall).

The insertion of teflon hose liners approximately doubled the cooling capability (Figure 5), confirming that the effects of dead volume are very important. The cooling load still falls short of that predicted by the bed model, presumably due to the effects of pressure drop. The pressure drop at the cold end of the beds occurs in the two flexible hoses, the inlet and outlet plenums of the CHEX, and the porous copper disks in the CHEX. Calculating the pressure drop precisely is not possible because of the irregular nature of the flow in the corrugated interior of the flex hoses and in the porous copper disks. An estimate of 4×10^4 Pa for the cold-end pressure drop was obtained for a 3 second period at 0.84 l displacer swept volume at 10 atm helium pressure, which is consistent with the measured total pressure drop across the whole system of about 10^5 Pa. The pressure drop work at the cold end would than be 40 Joule per cycle period, or 5 J/l-atm in the units of Figure 5. This is comparable to the difference between the model and experiment.

CONCLUSIONS

This device has demonstrated that it is feasible to build reciprocating AMR-type magnetic refrigerators in a scale ten times larger than that previously attempted. Some issues that had been a concern at the start of the project have been shown to be easily overcome, while others have been shown to be critical issues in magnetic refrigerator design. A list of these issues follows:

1. Magnet quenching due to changing bore permeability or cyclic magnetic forces: no such quenches were observed.

2. Field decay due to changing bore permeability: no detectable (<1 %) field decay observable after 12 hours operation at 0.5 Hz.

3. Failure of magnetocaloric bed assembly due to magnetic forces: so such failure occurred at full (7 T) field.

4. Failure of the bed assembly due to thermal stresses: this is a very important concern. The evolution of the design, together with a suggested future design, are given in Figure 4.

5. Flex hose failure: no failure has been observed.

6. Adverse effects of dead volume and pressure drop in a moving bed design: these were shown to be the critical issues in this temperature regime, and probably reduced cooling power by a large margin.

Previous designs of AMR devices operating with a LN_2 heat sink[2,3] have avoided the dead volume/pressure drop tradeoff by employing a moving magnet and fixed beds. If a moving magnet is considered impractical, several approaches should be considered to improve the dead volume/pressure drop tradeoff. One is to increase the swept volume, possibly with corresponding reductions in operating pressure and frequency. A second is to insert smooth liners in the flexible hoses. A third approach if boiling LN_2 is used for heat exchange is to have a moving HHEX. The flex hoses would then remove exhaust gas and carry liquid from the LN_2 tank, in which case the volume of the flex hoses would no longer matter.

The issues of dead volume and pressure drop are less severe in the 4 K temperature region because helium has a higher volumetric density at moderate pressures. Reduced helium flow circuit pressure allows the use of short straight bellows tubing, thereby reducing both pressure drop and dead volume[7]. Operation near room temperature allows the use of a liquid heat transfer agent, thereby reducing flow velocities and pressure drop. The availability of satisfactory sliding seals is allowing the investigation at Astronautics of novel approaches to minimizing dead volume.

ACKNOWLEDGEMENTS

A number of individuals have contributed to the design and construction of this device. Among them are J.J. Auringer, A.J. DeGregoria, L. Dresner, A. Jastrab, L.J. Komorowski, R. Lubasz, and R. Syphard.

REFERENCES

1. C.B. Zimm and A.J. Degregoria, "Magnetic Refrigeration, Application and Enabler for High Temperature Superconducting Magnets", Superconductivity and its Applications, AIP Conf. Proc. 273, American Institute of Physics, New York (1992) pp. 471-480
2. A.A. Wang, J.W. Johnson, R.W. Niemi, A.A. Sternberg, and C.B. Zimm, "Experimental Results of an Efficient Active Magnetic Refrigerator", Proceedings of the 8th International Cryocooler Conference, R.G. Ross, ed., Plenum Press, New York (1995), p. 665
3. D. Janda, A.J. DeGregoria, J.W. Johnson, S.F. Kral, G. Kinard, "Design of an Active Magnetic Regenerative Hydrogen Liquefier", *Adv. Cryo. Eng.*, 37B:891 (1991)
4. A.J. DeGregoria, "Modeling the Active Magnetic Regenerator", *Adv. Cryo. Eng.*, 37B:867 (1991)
5. C.B. Zimm, E.M. Ludeman, M.C. Severson, T.A. Henning, "Materials for Regenerative Magnetic Cooling Spanning 20 K to 80 K", *Adv. Cryo. Eng.*, 37B:883 (1991)5.
6. L. Dresner, "Design and Construction of the Astronautics Refrigerator Magnet," (Report ORNL/TM-12790), Oak Ridge National Laboratory, Oak Ridge, TN (1994)
7. C.B. Zimm, A.G. Jastrab, and J.W. Johnson, "Design of Active Magnetic Regenerative Stage Interfacing to a G-M Cryocooler", Proceedings of the 8th International Cryocooler Conference, R.G. Ross, ed., Plenum Press, New York (1995), p. 657

1995 CEC/ICMC Conference: Three of the winners of the 1995 Russell B. Scott Memorial Award for Cryogenic Engineering Research:(from left to right) V.K. Pecharsky, K.A. Gschneider and C.B. Zimm.

ROUTINE CALIBRATION OF CRYOGENIC THERMOMETERS IN THE RANGE 1.5-350K WITH AN ACCURACY UP TO THE MILLIKELVIN LEVEL, AND MEASUREMENT OF FIXED POINTS IN SEALED CELLS WITH A FULLY AUTOMATED AND SELF-CONTAINED MODULAR APPARATUS *

F.Pavese[1], V.M.Malyshev[2†], P.P.M. Steur[1], D.Ferri[1], D.Giraudi[1]

[1]Istituto di Metrologia "G.Colonnetti" (IMGC)
Strada delle Cacce 73, 10135 Torino, Italy

[2]VNIIFTRI
141570 Mendeleevo, Moskow Region, Russia

ABSTRACT

An apparatus realized at VNIIFTRI is presented, combining the possibility of thermometer calibration and fixed point measurements, which has a number of additional unique features generally not available at all in any other previous calibration apparatus: it is self-contained, i.e., it does not require any further apparatus or instrument outside a normal storage dewar (for liquid nitrogen or helium) with a Ø50 mm neck, where the cryogenic module is inserted; it requires only initial roughing vacuum pumping, without any permanent pumping during operation which can last up to several weeks continuously over the 24 hours; it has a built-in ^{4}He evaporation refrigerator for one-shot operation below 4.2 K, each shot lasting one working day; it is fully automated through an interactive software user interface, can be run on any inexpensive PC and is fully compatible, for data reduction, with most popular commercial packages; it has an intrinsic accuracy up to ± 1 mK in the full temperature range. Measurement results performed at IMGC are reported, on the calibration of precision and middle-quality thermometers, and on the measurement of ITS-90 fixed points in self-contained permanent IMGC sealed cells.

INTRODUCTION

Equipment necessary in laboratories of research centers and of industries generally involves a cryostat for routine calibrations of temperature sensors, most often home-made.

However, when the calibration precision must be ± 10 mK or better, the design and fabrication of such equipment become very specialized jobs, and its characterization a very time-consuming exercize. On the other hand, there exists commercially a very limited choice, and nearly none is available for a precision approaching ± 1–2 mK, especially when the temperature range must span the range from below 4.2 K up to room temperature.

---

* Partially funded by Contract CERN N° CA 1051884.

† Present address: Termis Ltd., Moskow, Russia.

It is not easy to design an apparatus convenient to use, particularly for less specialized users, and that does not require a full set of cryogenic ancillary equipment, i.e., a well-equipped cryogenic laboratory as an environment.

The main constraints to avoid user-related difficulties are:

- cryogenic environment as transparent as possible to the user;
- limited or no ancillary equipment required (self-contained);
- limited and unattended running time;
- fully automatic operation.

There are mainly two design approaches. One is based on closed-cycle refrigerators, which avoids the use, manipulation and transfer of cryogenic liquids. However, no cryostat specific to thermometer calibration is available, though few commercial complete systems are. This solution becomes very expensive when one needs to reach temperatures below ≈15 K. In addition it is nearly impossible to reach temperatures below 4 K, even with the most expensive systems. The second design makes use of cryostats based on cryogenic liquids, with provisions for easy interchange of the experimental setup (here the block with the thermometers) and for modularity. However, no specific cryostat is available for this use, and only a few are able to control the thermal environment better than ± 10 mK. The apparatus described in the paper was made according to the second approach. It comprises a thermal apparatus, an instrumentation box and a software package designed for the thermal control and data acquisition.

It was designed and used not only for calibration by comparison of thermometers, but also for calibration of thermometers at fixed points made with the sealed-cell technique developed at IMGC. [1] The basic equipment was developed at VNIIFTRI and it was upgraded to specific requirements and tested at IMGC.

THE CALIBRATION THERMAL APPARATUS

The thermal apparatus is based on the following principles (Fig.1):

- It is made as a *self-contained* stem 127 cm long and with an overall outside diameter of 44 mm, which can *directly and safely be immersed* in a standard liquid-storage vessel (for helium or nitrogen) with a Ø50 mm (2") and ≤ 500 mm length neck. A sliding collar matching the vessel flange prevents air from condensing onto the vessel neck, as required especially when working with liquid helium. The working spacé at the bottom of the cryostat has a usable diameter of 26 mm and a length of 160 mm. The head of the cryostat carries the operating valves and the gas and electrical connections.

- The cryogenic vacuum seal of the working space does not make use of an O-ring and screws, as usual, but uses a special silicon paste compound, which is vacuum tight even in superfuid helium. To open and close the working space of the cryostat *one only needs to slide* its outer jacket out or in. The vacuum space is closed by a valve located on the cryostat head.

- The thermal module *does not require vacuum pumping* during operation, due to a built-in charcoal adsorption pump. Only rough pumping is necessary before cooling down. Under these vacuum conditions, the apparatus can be fully cooled down to 4.2 K in ≈ 48 h. If a few hundreds pascal of hydrogen or neon are left inside the vacuum space, it can be cooled down to 4.2 K overnight.

- In order to reach temperatures below 4.2 K, a second *built-in adsorption pump* is placed above the working space of the cryostat, containing a charcoal getter. The adsorber can pump up to the equivalent of about 5 L of normal liquid helium and is used for pumping on the helium vapour in the evaporation chamber. This chamber is placed on the top of the shield and is connected to the adsorber via a tube. Between the adsorber and the chamber this tube is thermally anchored.

- The cryostat is cooled down to 4.2 K by a flow of helium pushed through a capillary extending down to the bottom of the dewar vessel. The cooling gas passes into the condenser, then into the adsorption pump (kept at a sufficiently high temperature) and eventually it vents

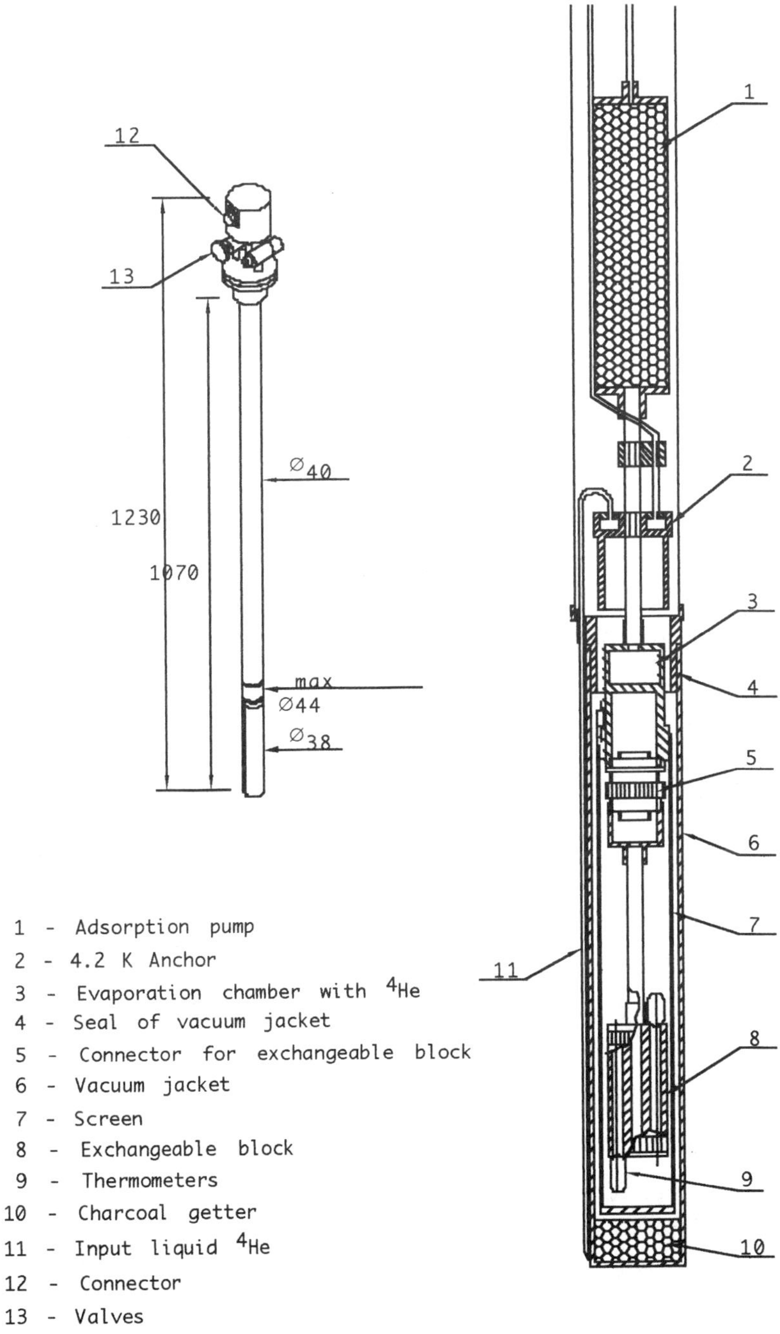

Figure 1. General design of the thermal module. (not to scale)

to the atmosphere through a valve. For temperatures below 4.2 K, the shield is cooled down to ≈ 1.3 K by pumping on the helium vapour in the evaporation chamber by means of the adsorption pump. The chamber is filled from the storage dewar via the capillary tube. When all the liquid is pumped out from the chamber, the adsorber can be regenerated by heating it up to 60 K. The liquid helium consumption during operation of the module is ≈ 2.5L/d.

• *The user has no access to any sensitive part of the cryostat.*

Measurements by comparison in a copper block

Specifically for this function the main features are the following:

• The block for the thermometers is placed inside the shield, supported by a stainless steel tube ending with a 32-pin electrical connector for the thermometers the block heater and thermocouple. Up to 9 thermometers (standard Tinsley or Leed&Northrup style) can be mounted, whose wires are thermally anchored to the block. Different blocks (or different devices, see later)) can easily be interchanged, using the same type of matching connector.

• The block works under vacuum close to adiabatic conditions. The shield temperature (absolute) and the block temperature (relative to the shield) are controlled automatically.

Measurement of a sealed cell realizing a temperature fixed point

An IMGC sealed cell is a self-contained device which realizes the triple point* of a substance which is gaseous at room temperature in a metal container permanently sealed, in which up to three standard thermometers can be fitted for calibration.[2-4] It is a cylinder Ø22x120 mm that can be mounted in the working space essentially in the same way of a block. Only the measuring procedure is different.

The thermal module is certified to provide a calibration uncertainty in the range 1.5 K to 350 K of ± 1 mK to ± 3 mK when comparing standard thermometers with 25-100 Ω at 0 °C, and an uncertainty of ± 1 mK to ± 2 mK (noise limit) when measuring a fixed point realized with IMGC sealed cells.

INSTRUMENTATION BOX

The instrumentation box comprises *all* the electronics necessary for running the thermal module and the measurements. It is self-contained (Fig.2).

• It controls the measuring modules and communications with the external computer via RS232C interface.

• It measures DC voltage and controls of external switchboards. The basic ranges of DC voltage are three: ±2.0 V, ±0.2 V and ±0.02 V (input resistance >100 MΩ, bias current –per channel– < 10 nA, source capacity < 0.015 μF). The voltage measurement error due to nonlinearity is < 0.002 % and the relative error is < 0.01 %. The time for one measurement is 80 ms. The overall voltage accuracy of the instrument in the range ±10 mV is ≈ ±0.1 μV, basically determined by random errors due by noise.

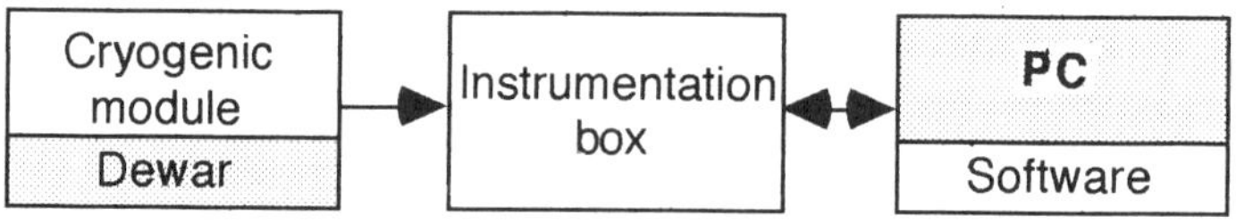

Figure 2. Layout of the system. (grayed parts: user provided)

---

* More than one fixed point is available in a single cell when also solid-to-solid phase transitions occur, as with oxygen, nitrogen and methane.

• It performs low-noise switching of DC voltages on 16 two-wire lines and switching of service lines. The residual thermal e.m.f. on a short-circuited channel is < 500 nV with an insulation resistance of an open channel > 300 MΩ. Additional channels are available for voltage switching of reference resistors used for the measurement of thermometer current.

The box includes a microprocessor-based digital unit for driving the functions and for connection with the external computer.

THE COMPUTER-BASED MANAGEMENT

The whole system can be run even on a PC/XT computer with EGA or VGA display. The requirements are very loose: even a 286 microprocessor (clock rate 12 MHz of higher) is able to run the system at normal speed. The program cannot be used with a resident program (TSR) that requires interrupt and may keep the CPU busy for more than 200 ms (this excludes Windows). MS DOS version 3.3 or later can be used. The source programs are all written in PASCAL 7.0. Only files related to calibration planning are accessible to the user. The connection to the PC is via RS232 interface.

The program has been specifically developed for calibrating thermometers by comparison with a reference thermometer. The program sets the temperature of the calibration block, controls the temperature of the block, measures the resistance of the reference thermometer and the thermometer to be calibrated, and writes a report file on the output device. The program can be used to display and control the temperature and to check and test the equipment. It can also *simulate* a full experiment, provided that the correct relevant physical data of the thermal system are supplied (which are, in any case, necessary for correctly running the real experiment). The program is interactive through control panels on the video screen. Using the control panels of the pseudo-instruments and the corresponding input boxes of the display screen the operator can set the values of the variables, the mode of measurements, and correct the process of measurements while the program is running.

At start-up a self-calibration program is run automatically. A first panel is used for setting the values of the parameters in the boxes. The thermometer calibration is then started from the main panel. The program automatically runs in sequence all temperature ranges and values, as previously set in corresponding files. For every temperature in the array the program executes the procedure that determines the block temperature and measures the resistance of thermometers to be calibrated. The temperature control before each measurement is made in three steps, each setting the temperature more precisely. The transition to each subsequent step takes place when several conditions are satisfied. The temperature controller works via timer interrupt with intervals of 2 sec. The procedure of resistance measurement at each value of temperature is executed after step number three, when stability at the millikelvin level has been obtained. After the data acquisiton has finished, the program outputs to a measurement report, containing all the relevant information, which can directly be processed with standard programs for PC, such as Excel. Then, it gets the next temperature value and repeats all over. A control panel displays continuously the measured value of the temperatures and their differences in real time or the data versus time in strip-chart mode.

The thermal apparatus can also be partially driven by external instruments connected to the PC via the RS232 interface. For example, when an accuracy higher than that provided by the built-in electronics is needed, an external high-precision bridge can be driven together with an external scanner for the thermometers, using a modified version of the program.

The system can run unattended for weeks and through the 24 hours. The system is modular. The same instrumentation box and, partially, the same software can be used also with the modules for heat capacity and thermal conductivity measurements.[5]

The system can be used for calibration of any type of resistance thermometer, from the standard type to the semiconductor ones. Work is in progress to extend its use to diode thermometers.

EXPERIMENTAL RESULTS AND DISCUSSION

In the frame of a contract with CERN the system was tested by executing comparison measurements with rhodium-iron precision thermometers over the full temperature range, and measurements at the temperature fixed points of argon and neon.

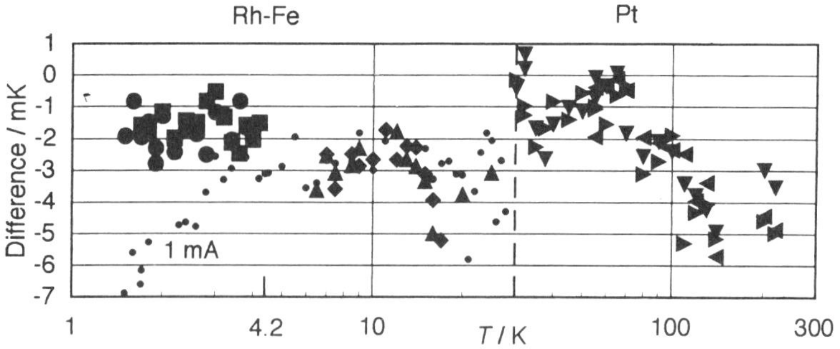

Figure 3a. Results of the calibration of rhodium-iron (Rh-Fe) and platinum (Pt) resistance thermometers in the range 1.5-300 K. Comparison of thermometers (0.5 mA): the deviations from zero are the differences between the previous calibrations; dispersion of data is within ± 1 mK. NOTE: 1 mA current causes an excessive overheating in one of the thermometers.

Comparison measurements

Eight standard rhodium-iron thermometers were mounted in the comparison block. Runs were performed in the full range, from 1.5 K to ≈ 300 K. Figure 3 shows the scatter of the results with respect to a polynomial fitting with a 1 mA current. It is possible to notice an increasing divergence below about 2.5 K, the condensation temperature of the ^{4}He filling the thermometers. When the current was lowered to 0.5 mA, increasing the noise somewhat, this effect disappeared, indicating that it was entirely due to a large increase of the overheating in the thermometers. The resulting overall calibration uncertainty is within the specifications. The full calibration run in the 1.5-300 K range (100 points at least) takes less than two full days (over the 24 hours).

Measurements of the fixed points sealed cells

The sealed cell is mounted in the experimental chamber instead of the copper comparison block, as shown in Fig.4.

The program runs a procedure which is essentially the same as that used for heat capacity measurements with the adiabatic steady-state method.

Measurements of the argon triple point (T_{90} = 83.8058 K) were performed with liquid nitrogen and of the neon triple point (T_{90} = 24.5561 K) with liquid helium. Results are shown in Fig.5.

During the heating periods some overheating is evident, then temperature recovers to equilibrium. The insert plots these equilibrium temperatures versus the inverse of the melted fraction F: from the plot the triple point temperature (tp) of the cell can be obtained by extrapolation to full melting (1/F=1, liquidus point).

At the argon triple point , where an external high precision bridge has been used, the measurement noise is less than 0.1 mK, while at the neon triple point, where the built-in electronics was used, the peak-to peak noise amounts to ≈ 1 mK ($\sigma \approx$ 0.25 mK). Fewer points

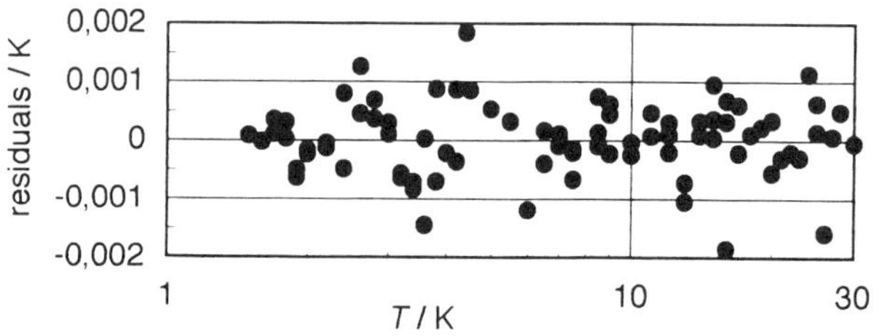

Figure 3b. Results of the calibration of rhodium-iron resistance thermometers in the range 1.5-30 K. Typical residuals from fit, showing type B (random) calibration errors, which are within ± 1 mK. For platinum resistance thermometers in the range 30-300 K, see Fig.3a.

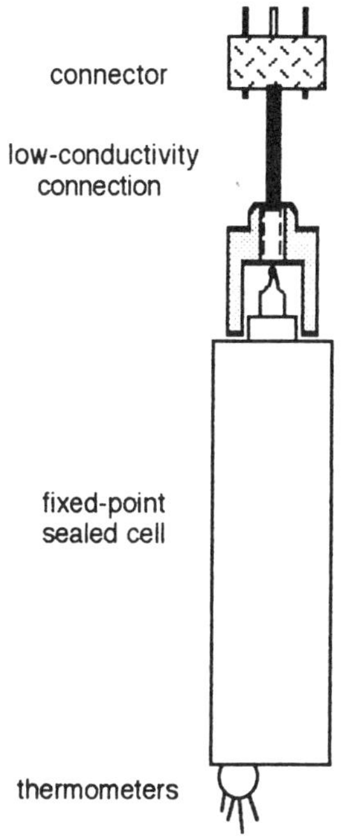

Figure 4. Mounting of the sealed cell in the thermal module for fixed-point calibration. (not to scale).

have been done for the neon triple point, to check the minimum time required to get the correct calibration: the last point is too close (≈ 90%) to the melting completion, a region where the melting plateau starts to bend upward, due to an increased effect, at helium temperatures, of the stray heat exchange of the cell. The influence of the cryostat quality becomes more critical, resulting in an increased melting range and the resulting precision of the determination of T_{tp} is limited to ≈ ± 0.5 mK. The temperature values indicated are those available from a previous calibration of the thermometer, which results to be too high by ≈ 1 mK at 83.8 K and ≈ 2 mK at 24.6 K However, these discrepancies are within the overall accuracy of the apparatus.

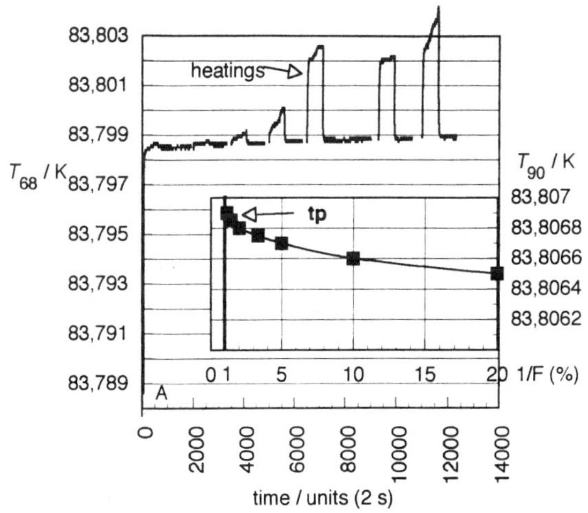

Figure 5a. Results on measurements on fixed points.: argon triple point; point A is the start of the last heating step before melting. F= melted fraction.

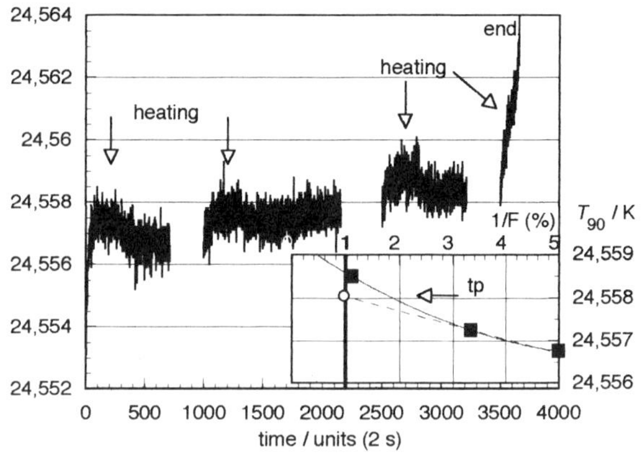

Figure 5b. Results on measurements on fixed points.: neon triple point. F= melted fraction

CONCLUSIONS

An apparatus has been described for calibration of thermometers in the range 1.5-350 K, which allows routine work to be performed at a millikelvin level, reliably and automatically 24-hours a day, without requiring ancillary equipment except a storage dewar with a 50 mm (2") neck diameter and a roughing pump for the *initial* vacuum. It has been used for calibration by comparison in a copper block and by measurements at fixed points using the sealed-cell technique. Either the block and the cell can be replaced by any other experimental equipment, in the working space and adiabatic-type measurements could be performed as well. These characteristics represent an advanced solution with respect to existing apparatuses.

REFERENCES

1. F.Pavese, D.Ferri, D.Giraudi, P.P.M.Steur, Long-term stability of permanent realizations of the triple point of gases in metal sealed cells, *Temperature, Its Measur. and Control in Science and Industry*, J.F.Schooley ed., Amer.Inst. of Phys., New York 6: 251 (1992).
2. F.Pavese, G.Cagna, D.Ferri, Miniature sealed cells as an easy–to–use temperature calibration device and a precision thermostat for cryogenic temperatures, *Proc. 6th Int.Cryo. Engin.Conf. (ICEC6)*, IPC Science and Tech. Press, Guildford (1976) p.205.
3. F.Pavese, D.Ferri, Ten years of research on sealed cells for phase transition studies of gases at IMGC, *Temperature, Its Measurement and Control in Science and Industry*, J.F.Schooley ed., Amer. Inst.of Phys., New York 5: 217 (1983).
4. F.Pavese, On the use of first-generation sealed cells in an international inter-comparison of triple point temperature of gases, *Temperature, Its Measurement and Control in Science and Industry*, J.F.Schooley ed., Amer.Inst. of Phys., New York 5: 209 (1983).
5. F.Pavese, V.M.Malyshev, Routine measurements of specific heat capacity and thermal conductivity of high-T_c superconducting materials in the range 4-300 K, *Adv.Cryo.Engin.* 40: 119 (1994).
6. F.Pavese, D.Ferri: A simplified cryostat for direct use in liquid storage dewars, *Cryogenics Suppl.* 30: 463 (1990).

SENSOR TEMPERATURE STABILITY PERFORMANCE OF THE ADVANCED LIQUID HELIUM DEWAR

J.H. Eraker

Ball Aerospace Systems Division
Boulder, Colorado 80306-1062, U.S.A.

ABSTRACT

The measured sensor temperature stability in the sensor volume of the Advanced Liquid Helium Dewar (ALHD) designed, built, and tested by Ball Aerospace for the Naval Surface Warfare Center Coastal System Station at Panama City, Florida, is described. ALHD is a horizontal, quartz epoxy composite, liquid helium dewar designed to support in-motion testing of a SQUID superconducting magnetic gradiometer. The sensor is cantilevered from the end of the neckplug and resides in an isothermal sensor volume filled with a constant volume helium exchange gas. The sensor temperature stability tests were performed at exchange gas pressures of 30-400 torr such that the peak-to-peak temperature variations were greater than the system temperature measurement noise. The sensor temperature was measured and the rms temperature stability was computed for several cases, namely: 1) stationary and in-motion, 2) stationary as a function of liquid helium level, and 3) stationary for several helium exchange gas pressures. The results are: 1) the in-motion and stationary rms temperature stabilities are similar in magnitude; for the non-optimal, 250 torr exchange gas pressure, the stability is $\sim 1 \times 10^{-6}$ K rms, 2) the stability is essentially independent of fill level, and 3) the extrapolated stability at 5 torr exchange gas pressure in the absence of self-heating is 3×10^{-8} K rms.

INTRODUCTION

The Advanced Liquid Helium Dewar (ALHD) program is part of a technology development effort for the Naval Surface Warfare Center Coastal Systems Station (CSS) in Panama City, Florida to increase the detection range of buried magnetic mines in shallow water through an improved superconducting magnetic gradiometer and dewar system. The program was designed to develop improved performance in motion over the Magnetic and Acoustic Detection of Mines (MADOM) detection system.[1] This paper describes ALHD and discusses the initial temperature stability measurements that were performed in ALHD.

A cut-away figure of ALHD is shown in Figure 1. The requirements for ALHD were to provide liquid helium cooling while *in motion* of any SQUID superconducting magnetic gradiometer with a sensor temperature stability of better than 100×10^{-6} K rms over a

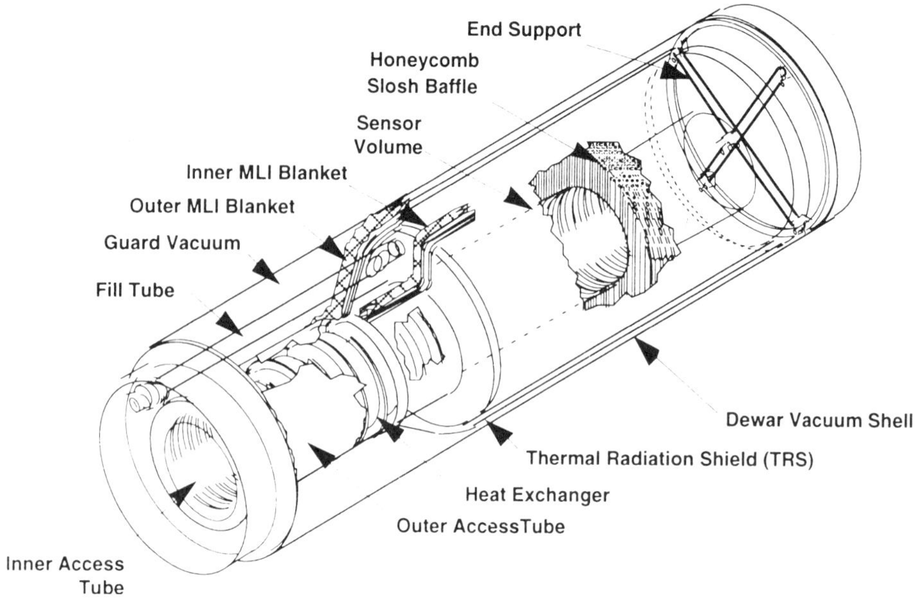

Figure 1. The Advanced Liquid Helium Dewar (ALHD) cutaway view.

bandwidth of 0.1-10 Hz. The sensor volume is cooled by the liquid helium that resides in the honeycomb slosh baffle. ALHD performance requirements in motion are such that the dewar does not limit the system performance for at least 12 hours.

The ALHD must be nonmagnetic in motion. This magnetic requirement limited the types and amounts of materials: ALHD is 90% quartz epoxy composite, 9% miscellaneous, and 1% high purity aluminum wire segments by weight. The design quandary for ALHD was how to provide high thermal stability in a dewar that is fabricated largely from thermal insulators. Several dewar thermal design features were improved or developed in novel ways to meet the program requirements.

ALHD operates similarly to other necked, helium dewars. The sensor and neckplug form a modular unit which is inserted via the dewar neck. The sensor volume and a thin annular region around the neckplug are filled with helium exchange gas. The minimum exchange gas operation pressure is ~5 torr; at lower pressures heat removal from the neckplug will be reduced. The dewar may be operated in either a horizontal or vertical orientation although the slosh baffles work only in an horizontal orientation. ALHD has a holdtime of 4.1 days horizontal and 5.2 days vertical. The dewar is operated on a Tavco pressure relief

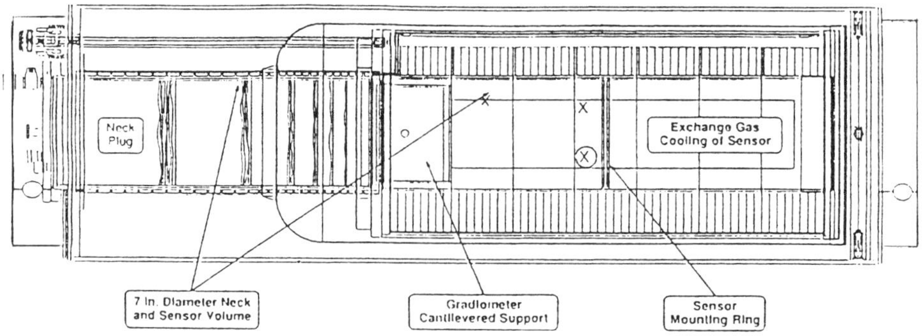

Figure 2. ALHD temperature stability measurement configuration. See text for description of sensors marked by an X.

valve with a cracking pressure of ~765 torr. The measured hysteresis in the pressure to open and close the pressure relief valve is 0.1 torr in 10 seconds, implying a temperature stability at the helium free surface of ~2×10^{-4} K at 0.1 Hz.

The dewar was designed, built, and tested by Ball Aerospace from September 1989 to September 1992. The thermal stability results reported here were obtained during the first cooldown of the dewar in July 1992. The goals of this acceptance testing were to characterize ALHD overall thermal performance and to obtain data to allow the sensor temperature stability to be maximized in operation.

TEMPERATURE MEASUREMENT APPARATUS

To measure the sensor temperature, two germanium resistance thermometers (GRTs) (model GR-200B-2500) and a silicon diode (model DT-470) from Lakeshore Cryotronics and the associated wiring were attached with GE 7031 varnish to a 2-inch diameter G-10 cylinder. The cylinder was cantilevered from the end of the neckplug as shown in Figure 2. The GRTs used in the test were located near the middle of the sensor volume with the "top" GRT located at 1:30 clock angle and the "bottom" GRT located at 7:30 clock angle since it was anticipated that the thermal environment of the sensor volume would be different between the two hemispheres. The use of a G-10 cylinder provides a conservative estimate of the temperature stability for SQUID magnetic gradiometers.

The GRTs had a nominal resistance of 2240-2250 ohms at 4.2K. An eighth order Chebychev fit derived from the GRT resistance calibration data was used to determine the temperature in the temperature range from 1.4 to 6.21K. The slope from this calibration at the measured temperature was used to convert the ac coupled GRT readings into temperature. The GRTs were packaged in a copper canister with a calculated high frequency cutoff of 0.3 Hz at 4K. The Johnson voltage noise from the GRTs is equivalent to a temperature noise of 2.8×10^{-7} K rms in the measurement bandwidth.

The GRTs were individually read out at 244 and 300 Hz with an ac, excited, active Wheatstone bridge. Two bridge circuits were used to provide simultaneous measurements with two GRTs. Each bridge circuit used a 10×10^{-6} ampere amplitude current source and synchronously demodulated the detection voltage. The joule heating from the GRTs was 0.22×10^{-6} W. This bridge was modified from the bridge developed by Ball Aerospace for the NASA Lambda-point Experiment. The advantages of this bridge, compared to a conventional ac Wheatstone bridge, are that it is relatively insensitive to: 1) dc lead resistance, 2) small phase variation in the current, and 3) variation of the parasitic capacitance in the range of several hundred picofarads. The use of low delta capacitance Triax cables with this bridge circuit allowed high precision temperature stability measurements to be made in motion.

The input to the bridge is dc coupled; the output of the bridge could be ac or dc coupled. The bridge was configured with an internal gain of 2,000. In the dc coupled mode the output was read into Macintosh IIfx computer. The output from the bridge was sampled at 10 Hz by the Labview 2.2 program which calculated and plotted temperature versus time, temperature power spectral density, and rms temperature in selected frequency intervals. Since the temperature stability of the sensor volume proved to be quite high, most measurements reported here were performed in an ac coupled mode. In that case, a preamplifier with a gain of 1000 was used to provide a system gain of 2,000,000. The measurement bandwidth was 0.03 to 0.3 Hz.

The temperature calibration of the system was checked by determining the equivalent temperature of input resistance for 2200 ohms in the dc coupled mode and 0.02 ohm variation for the ac coupled mode. The temperature measurements discussed below were taken after the temperature calibrations were verified.

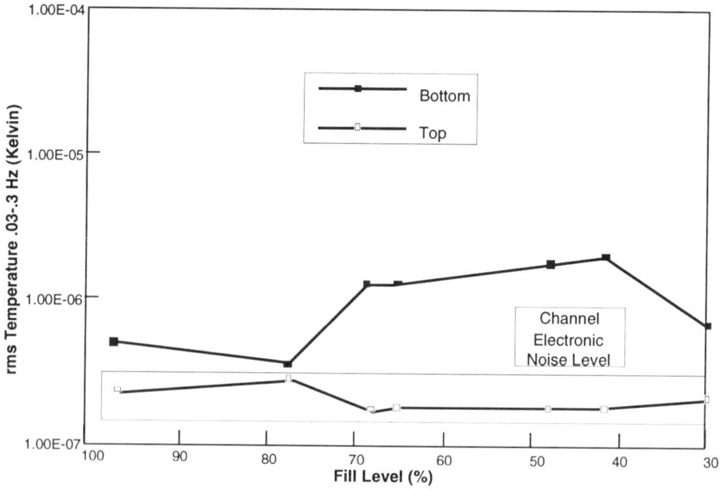

Figure 3. ALHD stationary rms temperature stability versus fill level for 250 torr exchange gas pressure.

TEMPERATURE MEASUREMENTS

The silicon diode and two GRT temperatures on the G-10 mock sensor were 3.98-4.00K after fill of ALHD corresponding to an atmospheric pressure of approximately 720 torr; with the Tavco regulator, the temperature was 4.24-4.26K corresponding to a mean regulation pressure of approximately 770 torr. The time for the temperature in the sensor volume to change after the Tavco pressure regulator was installed was ~22 hours.

We initially attempted to measure the sensor temperature stability with the silicon diode. The silicon diode did not exhibit any variation in voltage above the system noise

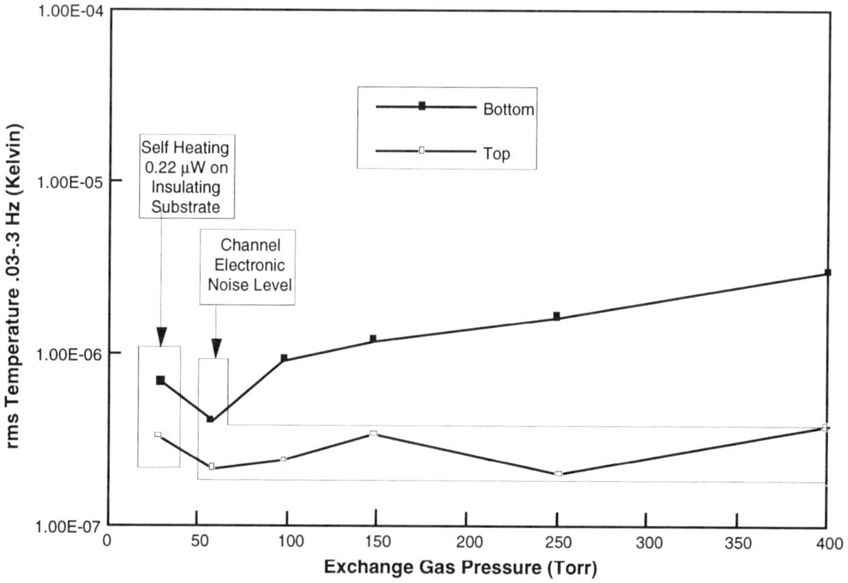

Figure 4. ALHD stationary rms temperature stability versus helium exchange gas pressure.

level of 100 microvolts corresponding to a lower limit temperature stability of 0.003K. Measurement of the temperature stability with the GRT was limited in the dc coupled mode to a similar level by a system bit noise of ~17 nV/bit and the A/D electronic drift.

To measure the sensor temperature stability, the bridge output was ac coupled. The peak-to-peak temperature variation of the bottom and top sensor was typically 40 and $20x10^{-6}$K, respectively, at an exchange gas pressure of 250 torr. This exchange gas pressure was chosen to keep the sensor temperature stability measurable. The temperature variations were incoherent leading to an overall reduction of a factor of ~10 in the rms temperature determined from a Fast Fourier power spectral analysis. The calculated power spectrum typically yielded power spectral densities of less than 10^{-11} K^2/Hz at 0.1 Hz.

To demonstrate the ALHD system performance, the stationary rms temperature stability versus percentage fill was measured for an exchange gas pressure of 250 torr. The data from this test are displayed in Figure 3. The rms temperature of the top GRT was typically at the system level (Johnson) noise for that channel of $3x10^{-7}$K. The rms temperature stability of the bottom GRT was less than $1x10^{-6}$K rms from 100% to ~80%; from 80% to 40% the rms temperature was ~$2x10^{-6}$K rms and below 40% the rms temperature was less than $1x10^{-6}$K rms. The rms temperature stability variation with fill level reflects variations in the temperature gradient of the sensor volume produced by gas and conductive cooling of the helium in the honeycomb. For a reduced exchange gas pressure, the variation of the rms temperature stability versus fill level would be smaller than measured in this case, as demonstrated by the next set of measurements.

The rms temperature stability was measured for various exchange gas pressures. Helium gas was added or removed from the exchange gas volume to change the pressure. The equilibrium rms temperature stability data versus pressure are displayed in Figure 4. The rms temperature stability at the top sensor was at the Johnson noise level of $3x10^{-7}$K for pressures greater than 50 torr. At the lowest measured pressure of 30 torr the top sensor showed temperature "bursts" probably corresponding to intermittent convection from the $0.22x10^{-6}$ W generated in the GRTs. This raised the rms temperature to $8x10^{-7}$K rms. The temperature measured by the bottom GRT as a function of pressure exhibited quasi-periodic fluctuations produced by Rayleigh-Bérnard cells for exchange gas pressures greater than 100 torr; at a pressure of 30 torr the bottom sensor exhibited the same intermittent temperature bursts observed at the top sensor.

At 54 torr, the measured temperature stabilities of the GRTs were at their lowest measured levels. The power spectral density data for this measurement are displayed in Figure 5. The specification for ALHD in these units is $33x10^{-6}$ K rms $Hz^{-1/2}$ in the bandwidth 0.1 to 10 Hz. Data only in the bandwidth from 0.03 Hz to 0.3 Hz are displayed. Note that these data fit smoothly with the outline of the electronic noise data at higher and lower frequencies. These data are consistent with a Johnson noise level temperature stability of $3x10^{-7}$ K rms in the bandwidth 0.03 to 0.3 Hz and do not measure the temperature environment in the sensor volume.

It is interesting to determine the rms temperature stability dependence on pressure from the data in Figure 6 and extrapolate that dependence to lower pressures. The data from the bottom GRT are replotted in Figure 6 versus exchange gas pressure. The data point at 30 torr was eliminated because of the self-heating effects. The rms temperature (T) in Kelvin fits a power law dependence on exchange gas pressure (P) in torr of the form

$$T = 4.5 \times 10^{-9} P^{1.18} .$$

The calculated rms temperature at the lowest operating pressure of 5 torr is ~$3x10^{-8}$ K rms. This extrapolation assumes no self-heating at the point of measurement and moreover assumes that temperature fluctuation observed in turbulent convection at several hundred torr exchange gas pressure is characteristic of the temperature fluctuation throughout this range of gas pressures. If the temperature fluctuation is reduced at low exchange gas pressure,

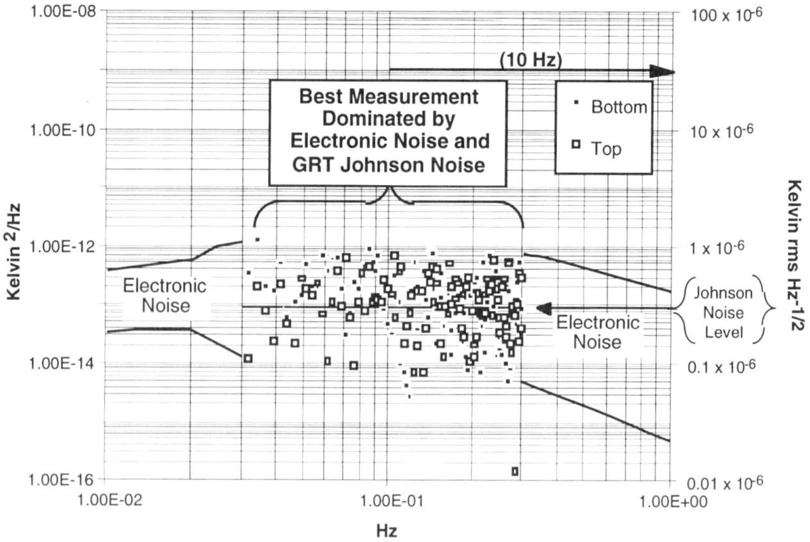

Figure 5. Best measurement of ALHD sensor temperature stability.

or if conduction dominates the heat removal at low exchange gas pressures, then the temperature stability at 5 torr may *exceed* this extrapolation.

The ALHD temperature stability in-motion was measured and is compared to measurements from the Advanced Magnetic Gradiometer (AMG) and Superconducting Gradiometer/Magnetometer Sensor (SGMS) dewars in Figure 7. All three dewars are horizontal, composite dewars that use the same Tavco mechanical pressure relief valve. Therefore all units should exhibit temperature fluctuations of ~2×10^{-4} K at 0.1 Hz at the liquid helium free surface providing a common basis for comparison. The SGMS dewar used vent gas to cool the sensor while the AMG dewar had the sensor isolated in a sensor volume and cooled with a 350 torr helium exchange gas. ALHD used a non-optimal, 250 torr helium exchange gas during the motion test. The AMG and SGMS dewars were tested at the Sperry Superconducting Electronic Technology Center circa 1986 and the best rms temperature stabilities when the dewars were full are displayed in Figure 7. For the SGMS and AMG tests, the temperature was read with silicon diodes and the rms temperature variations were ~2×10^{-3} K rms and ~0.4×10^{-3} K rms, respectively.

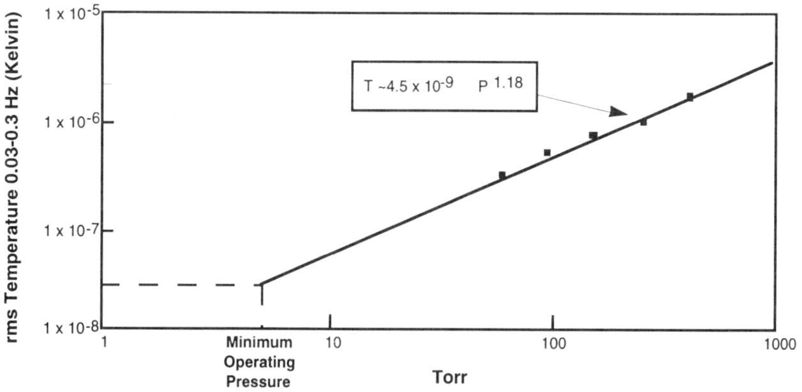

Figure 6. ALHD extrapolated rms temperature stability in the absence of self-heating.

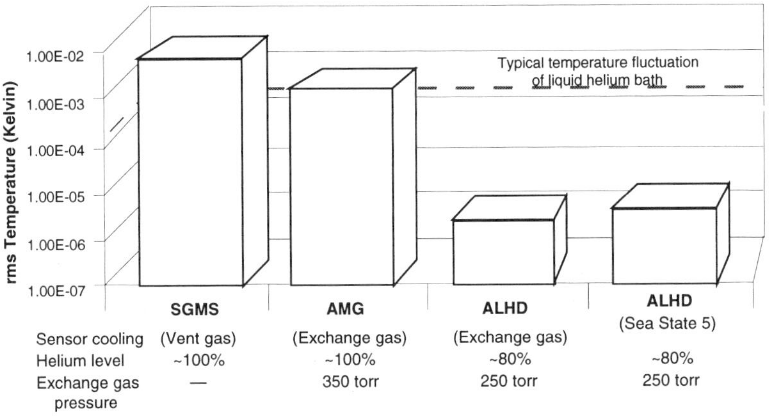

Figure 7. In-motion sensor rms temperature stability comparison.

To test the ALHD performance in motion, the temperature stability of a stationary ALHD, when the dewar was approximately 80% full, was measured and compared to the case where ALHD is stationary and the external cables between the neckplug and the bridge circuit were moved up to 5 cm from their equilibrium position. The temperature measurements for these two cases were identical indicating that this cable motion did not add any detectable noise to the measurement. To place ALHD in motion, the back end of dewar was rotated in a 12.5-cm diameter circle; a 25-cm diameter circle was also used and is described by the sea state 5 designation. Only the bottom GRT displayed an increase in the rms temperature in these tests; the measured rms temperature in motion displayed in Figure 7 was 8×10^{-7} K rms and 2×10^{-6} K rms for the two tests, respectively. Note that the measured temperature stability in the sensor volume is several orders of magnitude better than the temperature stability of the helium free surface.

DISCUSSION AND SUMMARY

During the early course of ALHD testing it became obvious that the measured temperature stability performance in the ~16,000 cm^3 sensor volume exceeded the requirements; we then improved the temperature stability measurement resolution and degraded the temperature stability performance by operating at relatively high exchange gas pressures to accumulate the data presented here. These measurements characterize the ALHD temperature stability performance required for acceptance testing which are summarized below. These measurements provide only a limited characterization of the physical conditions in the sensor volume which are discussed first.

The ALHD sensor volume may be described as a horizontal cavity with adiabatic walls in which free convection of a gas provides thermal transport. The ends of the volume are differentially heated. The parameters of interest for this model are the vertical and horizontal temperature gradients in the volume. These temperature gradients may be roughly estimated from the behavior of the top sensor. If we assume that the thermal transport from the top sensor was due to conduction we find that the vertical temperature gradient over half of the sensor volume height must be less than $\sim 1 \times 10^{-5}$ K at 54 torr for an assumed critical Rayleigh number of 200,000. The primary mode of heat transport in the sensor volume is due to the horizontal temperature gradient. If we assume that the intermittent convection at the GRTs occurs when the horizontal heat transport is less than the 0.22×10^{-6} W, the calculated horizontal temperature gradient across (an unoccupied) sensor volume is less than $\sim 1 \times 10^{-4}$ K.[2]

Table 1. ALHD thermal performance summary

Characteristics	Requirements	Performance
Versatile	All requirements are met for 12 hours with any gradiometer	Achieved for ≥70 hours* (Holdtime of dewar**)
Sensor temperature stability in-motion	100×10^{-6} K rms over entire sensor in bandwidth of 0.1 to 10Hz	8×10^{-7} K rms 0.03 to 0.3 Hz* (~30×10^{-9} K rms**)

* Measured ** Predicted

The ALHD temperature stability requirements and performance are summarized in **Table 1**. ALHD was designed as a versatile laboratory dewar which can accommodate physically large gradiometers and yet meet the temperature stability requirement in motion of 100×10^{-6} K rms in a bandwidth of 0.1 Hz to 10 Hz for a minimum 12-hour period over the entire sensor. We measured the stability on a G10 mock sensor to provide a lower limit for the temperature stability; frequencies above 1 Hz were not measured because the gas time constants of interest in the system are low frequency and the maximum range of the system is determined by the temperature stability at lower frequencies.

Figures 4 and 7 taken together revealed that the ALHD stationary and in-motion rms sensor temperature stabilities are comparable. This relative independence on motion is due to effective slosh baffling of the liquid helium and the constancy of the thermal gradients in the sensor volume. Therefore we believe that the in-motion temperature stability performance of 8×10^{-7} K rms from 0.03 Hz to 0.3 Hz displayed in Figure 7 will be characteristic of the performance in motion to fill levels of 30% (or 70 hours) as displayed in Figure 4. High sensor temperature stability should be achieved throughout the holdtime of ALHD.

The ALHD temperature stability performance depended significantly on exchange gas pressure only, as demonstrated in Figure 5. The extrapolated rms temperature stability in the sensor volume displayed in Figure 6 was 3×10^{-8} K rms over a bandwidth of 0.03 Hz to 0.3 Hz. This extrapolation assumes that similar magnitude temperature fluctuations are produced throughout the gas pressure range from 400 to 5 torr and that self-heating effects at the sensor do not initiate convection. This high temperature stability cooling performance in motion of large, insulating objects may be useful in other high performance applications.

ACKNOWLEDGEMENTS

ALHD was developed under sponsorship of Office of Naval Research and technical direction of the Coastal System Station, Dahlgren Division, Contract N61331-89-C-0072. The author wishes to thank Bill Wildhaber, Kathy Russell, and Howard Snyder for their work on the ALHD thermal design. Randy Abbott, Bill Whitehead, Bill Kiehl, and Ernie Hicks made an accurate measurement of the ALHD performance possible by developing the temperature measurement apparatus to the level demonstrated here in only six weeks.

REFERENCES

1. Ted R. Clem, "Nitrogen cooled superconducting gradiometers for mine reconnaissance from small underwater vehicles", Autonomous Vehicles in Mine Countermeasures, April 4-7, 1995, Naval Postgraduate School.
2. Warren M. Rohsenow, James P. Harnett, Ejup N. Ganic eds, "Handbook of Heat Transfer Fundamentals", McGraw Hill, 1985, pp 6-47 to 6-52.

THERMAL RESISTANCES OF MOUNTED CRYOGENIC TEMPERATURE SENSORS

D. Scott Holmes and S. Scott Courts

Lake Shore Cryotronics, Inc.
Westerville, Ohio 43081, USA

ABSTRACT

Operation of most temperature sensors requires the dissipation of power in the sensor. The flow of the heat generated by the measurement creates a temperature difference and a potential temperature measurement error. The self-heating temperature difference is directly proportional to the thermal resistance. A procedure for measuring the thermal resistance of a mounted temperature sensor is described.

Thermal resistances were measured at cryogenic temperatures (50 mK to 10 K) on several commercially available temperature sensors. The sensors were mounted to a copper block in either a vacuum, helium gas or helium liquid environment. The thermal resistance was found to depend on temperature, thermal environment and details of sensor mounting and packaging.

Minimization of the temperature measurement uncertainty requires a balance between the uncertainties due to self-heating and measurement of the output signal. Equations are provided for calculating the operating point for minimum combined temperature measurement uncertainty.

INTRODUCTION

Any difference between the temperature of the sensor and the environment the sensor is intended to measure produces a temperature measurement error or uncertainty. Power dissipation in the temperature sensor will cause its temperature to rise above that of the surrounding environment, but is necessary to make a measurement. Minimization of the temperature measurement uncertainty thus requires balancing the uncertainties due to self-heating and output signal measurement.

The self-heating can be affected by changes in the thermal resistance between the sensor and its surroundings due to remounting, temperature dependent properties, or environment. Sensor mounting techniques are reviewed by Sparks[1]. An example of how sensor mounting can affect performance is given by Kobel, et al.[2]

Calculation of an effective thermal resistance for a mounted temperature sensor is extremely difficult. Details of construction unknown to a user (e.g. thickness and composition of a ruthenium oxide thick-film, or thickness of epoxy used to mount a sensor element within a canister) can strongly affect the thermal performance. The thermal properties of many commonly used materials are frequently unknown at cryogenic temperatures. The Kapitza resistance becomes important below about 4 K and is difficult to estimate.

Following are two experimental approaches to dealing with the problem of self-heating:

1) Choose an excitation that allows acceptable instrumentation measurement uncertainty and check to make sure self-heating is negligible at one or two points where it is likely to be most significant. An easy way to check for self-heating is to increase the power dissipation and check for an indicated temperature rise. Unfortunately, this procedure will not work with diodes. An indication of the self-heating error can be made by reading the diode temperature in both a liquid bath and in a vacuum at the same temperature, as measured by a second thermometer not dissipating enough power to self-heat significantly.

2) Measure the thermal resistance in the temperature range of interest and then use the thermal resistance to select the appropriate sensor excitation. The sensor excitation can be better optimized using this approach.

Designers of cryogenic instrumentation and experiments usually must predict performance before the thermal resistance can be measured. In this case, thermal resistances measured under similar circumstances can be used to estimate self-heating effects and overall measurement system performance.

THEORETICAL DEVELOPMENT

The self-heating depends on the excitation power according to the equation

$$\Delta T_{sh} = P_s R_t = I^2 R_e R_t = V^2 R_t / R_e \tag{1}$$

where ΔT_{sh} is the temperature rise due to self-heating, P_s is the power dissipated in the sensor, I is the excitation current, R_e is the electrical resistance, and R_t is the thermal resistance between the sensor and its environment. The thermal resistance is usually temperature dependent, but a linear relationship between power dissipation and temperature rise usually holds for small values of ΔT.

The effective thermal resistance can be measured experimentally using the following procedure:

1) mount the sensor as it will be used on a temperature controlled block or directly in liquid,
2) record the output voltage as a function of excitation current (I-V curve) until significant self-heating is observed (when R = V/I is no longer constant),
3) replot the data as sensor temperature rise versus power dissipated (ΔT vs. P),
4) fit the data with a linear equation of the form $\Delta T = T - T_o = R_t P_s$ to find the effective thermal resistance, R_t.

The experimentally determined quantity is termed the effective thermal resistance because it relates the measured power dissipation to the temperature rise. In some sensor designs (e.g. germanium or carbon-glass resistors) the actual power dissipated in the sensor is significantly higher than that calculated by multiplying the voltage output by the current through the sensor because the voltage taps only extend across a portion of the resistive element. This does not affect the usefulness of the calculated effective thermal resistance.

Temperature measurement uncertainties u_T arise from many causes. As an example, equation (1) only applies to self-heating caused by power dissipated within the sensor, but heat conducted along the sensor leads or thermal radiation can produce similar effects. The combined uncertainty u_c resulting from several independent measurement uncertainties is usually estimated by assuming a statistical distribution of uncertainties, in which case the uncertainties are summed in quadrature using the formula[3]

$$u_c = \sqrt{u_1^2 + u_2^2 + \cdots + u_i^2 + \cdots + u_n^2} \qquad (2)$$

Information helpful to calculating uncertainties of cryogenic temperature measurements is given elsewhere.[4,5] Note that attempting to correct for self-heating errors by calculation or extrapolation is not considered good practice. An estimate of the self-heating error should be included in the total uncertainty calculation instead.

A temperature sensor produces some indirect temperature-dependent output such as a voltage, resistance or capacitance. Uncertainties in the measured quantity are not in units of temperature and must be converted before use in Equation (2) . In the case of an output voltage V, the temperature uncertainty u_T is related to the voltage uncertainty u_V by the dimensionless formula

$$\frac{u_T}{T} = \frac{u_V/V}{\left(\frac{T}{R}\right)\left(\frac{dR}{dT}\right)} \equiv \frac{u_V/V}{S_d} \qquad (3)$$

The quantity in the denominator, $S_d = (T/R)(dR/dT)$ is defined as the dimensionless temperature sensitivity, where R is the sensor resistance at absolute temperature T. The dimensionless sensitivity is also equal to $\partial(\ln R)/\partial(\ln T)$, the slope of the voltage versus temperature on a log-log plot. Note that Equation (3) can be made to apply to thermometers based on other temperature-dependent properties (capacitance, voltage or pressure) by replacing R with C, V, or P as appropriate.

Examination of Equation (3) leads to the conclusion that an increase in the sensor output voltage will result in a decreasing temperature uncertainty, so long as the voltage uncertainty remains constant. This is possible with an ohmic sensor by increasing the excitation current. Unfortunately, a larger excitation will dissipate more power in the temperature sensor, raising its temperature above the surroundings.

The optimum operating point minimizes the combined temperature measurement uncertainty due to instrumentation and self-heating. The optimum is determined by a) writing the instrument uncertainty (Equation 3) and the self-heating (Equation 1) in terms of sensor power, b) including these two terms on the right hand side of Equation (2), c) differentiating with respect to sensor power, d) setting the result equal to zero, and e) solving for the sensor power. The resulting expression for the optimum sensor power is

$$P_{s\,opt} = I^2 R_e = \frac{V^2}{R_e} = \frac{T(u_V/V)}{S_d R_t \sqrt{2}} + \left[\frac{u_V T}{S_d R_t \sqrt{2R_e}}\right]^{2/3} \qquad (4)$$

where the first term on the right is for instrument uncertainties expressed as a fraction of sensor output signal and the second term on the right is for instrument uncertainties expressed in volts. Note that other experimental considerations might impose more stringent limitations on the power which can be dissipated in the temperature sensor.

EXPERIMENTAL RESULTS

Several commercially available cryogenic temperature sensors were mounted in a standard Lake Shore calibration probe. In some cases, two sensors of the same model were tested, thus providing a check on sensor packaging and mounting variability. The mounting block and calibration procedure are described elsewhere[4]. Apiezon N-grease was used to improve thermal contact between the sensor and the copper calibration block. The sensors were not remounted between tests. Self-heat-

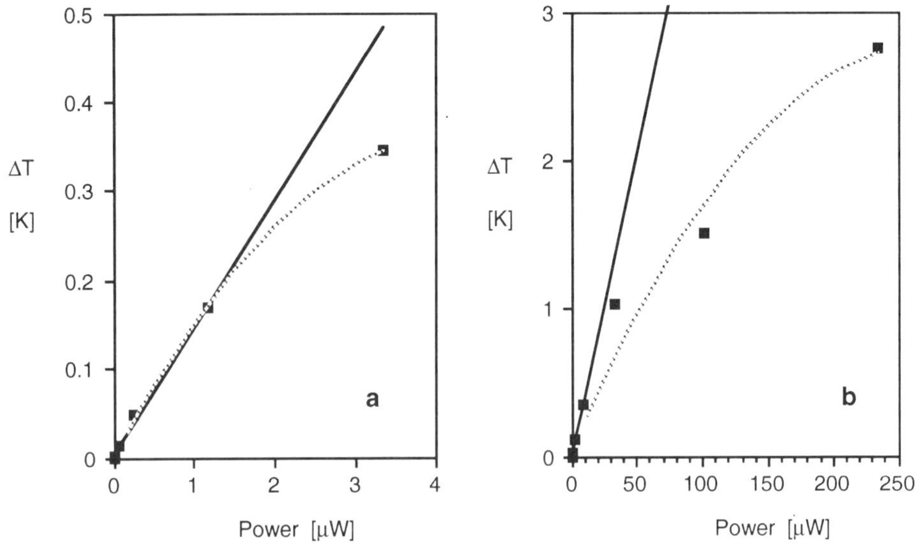

Figure 1. Self-heating temperature rise as a function of power for two sensors. Both first and second order polynomial fits are shown. **a**. Cernox™ sensor Model CX-1070-AA, serial No. X01450 at 1.6 K in vacuum, **b**. Carbon-glass sensor Model CGR-1-1000, serial No. C15352 at 4.2 K in liquid helium.

ing measurements were made at five temperatures in one of three environmental conditions: either a) vacuum, b) helium gas, or c) helium liquid surrounding the mounting block. The excitation was stepped geometrically until significant self-heating was detected. Between 2 and 13 data points were produced for each sensor and temperature.

An example of the data taken and the fit to the data is given in Figure 1. As the self-heating temperature rise ΔT approaches the temperature of the environment, T, the relationship between the temperature rise and the power is seen to become non-linear. Linear fits were performed to data points with $\Delta T/T \leq 0.2$. Second order curve fits of the form $\Delta T = a + bP + cP^2$ were also used with no restriction on the relative temperature rise. The coefficient b in the second order fit is equal to the slope of the curve at zero power, and thus equal to the thermal resistance at low powers. We consider thermal resistance values determined from second order curve fits to be more reliable.

Thermal resistance values determined from second order curve fits are given in Table 1. The thermal resistance is found to depend on temperature and the details of sensor mounting. The difference between the thermal resistance measured at 4.2 K with helium gas and with liquid helium surrounding the mounting block is one indication of the sensitivity to changes in mounting conditions.Thermal resistances typically differ by less than a factor of two for sensors of the same model. Similar germanium (GR) and carbon-glass (CGR) resistors show more variability than the Cernox™ (CX), possibly because the difference between 2-wire and 4-wire resistances are greater, which would vary the actual power dissipated.

The relative differences in the thermal resistance values calculated using first- or second-order curve fits are given in Table 2. The largest difference (-104%) was for the case shown in Figure 1b.

The thermal resistance is shown as a function of temperature for the germanium resistors in Figure 2. The 200A models are mounted in standard gold-plated copper canisters of 3 mm diameter and the 200B models are in smaller, 2.36 mm diameter canisters. The two 200B models tested appear to be much more strongly affected by the environment around the mounting block.

Table 1. Thermal resistances calculated using second order polynomial fits to ΔT versus power data

		thermal resistance, Rt [K/W]					
sensor Model No.	serial No.	1.6 K vacuum	2.0 K vacuum	3.0 K vacuum	4.2 K He gas	4.2 K LHe	10.2 K vacuum
CGR-1-500	C15303	1.714E+5	1.380E+5	9.605E+4	6.406E+4	5.953E+4	1.402E+4
CGR-1-1000	C15351	1.176E+5	9.454E+4	6.461E+4	3.123E+4	2.093E+4	7.314E+3
CGR-1-1000	C15352	1.451E+5	1.166E+5	8.287E+4	5.529E+4	2.031E+4	1.155E+4
CX-1030-BC	X01460	8.691E+4	3.935E+4	1.700E+4	9.082E+3	3.836E+3	2.444E+3
CX-1050-AA	X01770	7.878E+4	3.550E+4	9.674E+3	3.835E+3	3.737E+3	7.794E+2
CX-1050-AA	X01776	7.093E+4	2.472E+4	7.707E+3	2.846E+3	2.800E+3	4.612E+2
CX-1050-SD	X01647	1.456E+5	4.556E+4	1.554E+4	5.529E+3	4.219E+3	1.028E+3
CX-1050-SD	X01749	1.526E+5	4.661E+4	1.588E+4	5.554E+3	4.234E+3	1.070E+3
CX-1070-AA	X01450	1.680E+5	7.522E+4	1.629E+4	5.718E+3	4.282E+3	6.566E+2
CX-1070-BC	X01466	1.781E+6	3.406E+5	1.989E+5	2.772E+4	1.783E+4	4.434E+3
CX-1070-SD	#10647	1.053E+6	3.988E+5	8.481E+4	2.785E+4	2.068E+4	2.867E+3
CX-1070-SD	X01648	9.335E+5	3.542E+5	7.714E+4	2.523E+4	1.911E+4	2.637E+3
GR-200A-500	#26352	8.212E+3	5.574E+3	2.748E+3	1.508E+3	8.432E+2	4.235E+2
GR-200A-500	#26353	1.726E+4	1.182E+4	4.705E+3	2.513E+3	2.413E+3	6.390E+2
GR-200A-1000	#26392	2.300E+4	1.407E+4	6.220E+3	3.154E+3	2.378E+3	7.190E+2
GR-200A-1000	#26395	1.948E+4	1.161E+4	5.077E+3	2.558E+3	2.144E+3	6.094E+2
GR-200B-500	#25893	1.965E+4	7.898E+4	2.841E+4	4.288E+3	2.413E+3	8.627E+3
GR-200B-1000	#25891	1.299E+5	8.117E+4	4.075E+4	4.333E+3	2.629E+3	7.503E+3

Table 2. Relative differences between thermal resistances calculated using second order and first order polynomial fits to ΔT versus power data. Rt_i is notation for a fit of order i

		(Rt_2 – Rt_1) / Rt_2 [%]					
sensor Model No.	serial No.	1.6 K vacuum	2.0 K vacuum	3.0 K vacuum	4.2 K He gas	4.2 K LHe	10.2 K vacuum
CGR-1-500	C15303	6.7	2.4	3.5	-8.5	-17.8	11.9
CGR-1-1000	C15351	3.2	2.7	3.2	-12.0	-56.0	4.8
CGR-1-1000	C15352	4.8	3.2	3.9	-7.8	-104.1	4.1
CX-1030-BC	X01460	-19.6	-32.8	-41.1	9.4	-86.1	-7.5
CX-1050-AA	X01770	11.0	13.0	9.9	10.4	10.4	4.8
CX-1050-AA	X01776	9.6	-11.0	8.9	9.3	9.4	5.2
CX-1050-SD	X01647	11.0	-14.4	10.4	11.7	-14.2	9.9
CX-1050-SD	X01749	11.4	-15.8	10.5	11.8	-14.3	10.2
CX-1070-AA	X01450	14.4	35.1	31.0	20.1	9.2	5.5
CX-1070-BC	X01466	-36.6	-61.9	47.4	16.8	-19.8	12.8
CX-1070-SD	#10647	16.8	19.3	12.7	12.6	-18.7	18.0
CX-1070-SD	X01648	16.8	18.9	12.3	12.0	-17.2	15.0
GR-200A-500	#26352	7.3	-3.4	7.9	9.8	-15.2	3.4
GR-200A-500	#26353	-4.0	7.0	-2.5	5.3	-3.7	7.7
GR-200A-1000	#26392	9.7	8.8	8.4	6.7	-3.6	9.0
GR-200A-1000	#26395	10.6	-2.9	8.8	6.9	-4.1	9.5
GR-200B-500	#25893	1.6	6.2	-1.5	3.8	5.8	1.9
GR-200B-1000	#25891	6.0	6.9	7.5	6.1	6.2	-1.5

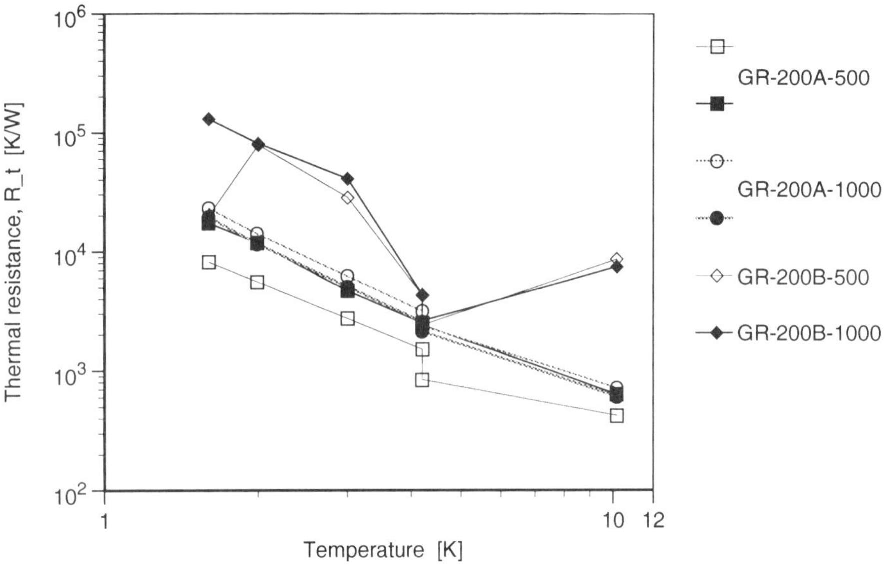

Figure 2. Thermal resistance as a function of temperature for several germanium sensors.

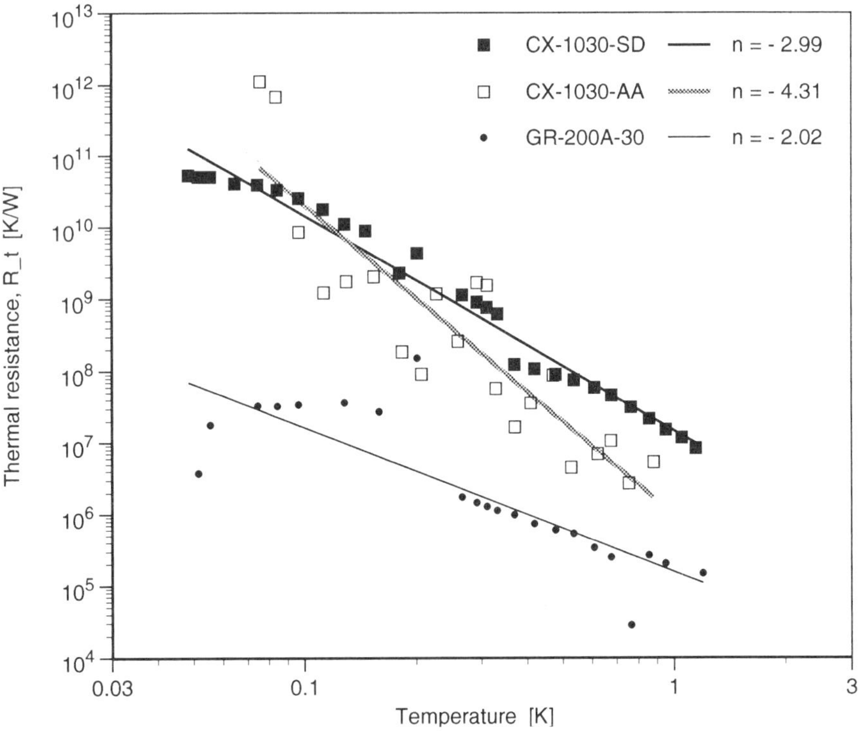

Figure 3. Thermal resistance as a function of temperature for three sensors mounted in a dilution refrigerator calibration block. The n-values are exponents for fits of the form $R_t = a\,T^n$.

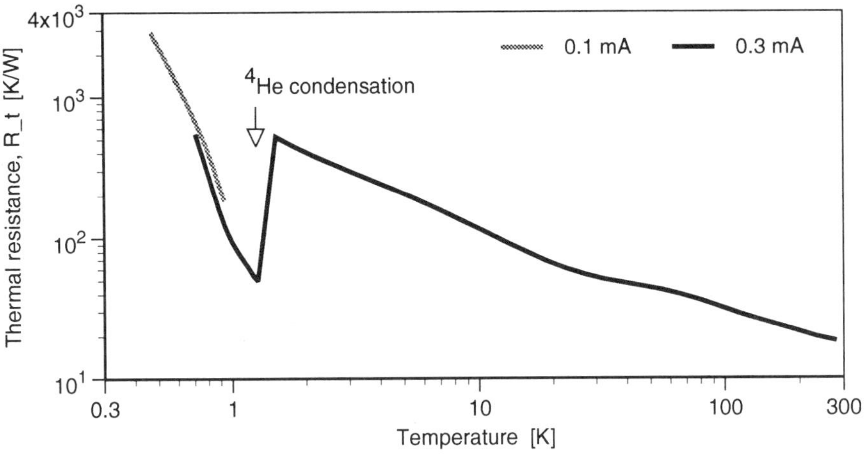

Figure 4. Thermal resistance as a function of temperature for a ^{4}He-filled, type 5187W rhodium-iron resistance thermometer (5 mm diameter, 50 mm long, ~100 Ω at 273 K).

Thermal resistance experiments were also performed on sensors mounted to a temperature controlled block in a dilution refrigerator. The mounting block was operated in a vacuum environment. The thermal resistances of three of the sensors are shown as a function of temperature in Figure 3. The thermal resistances are significantly larger at the lower temperatures and rise rapidly with decreasing temperature. The log-log slopes of the curves range from -2 to -4.3, indicating that different heat transport mechanisms may be limiting performance for these different sensors and packages.

Little data exists for comparison with the thermal resistances we measured. One rare exception is Rusby's[6] measurement of self-heating of a rhodium-iron temperature sensor between 0.4 and 280 K. The excitation currents and sensor resistance were also reported, allowing calculation of the thermal resistances using equation (1). The results are shown in Figure 4. The overlap of the thermal resistances calculated using the 0.1 and 0.3 mA data indicates that the linear relationship between power dissipation and temperature rise assumed by equation (1) is a good approximation in this range of sensor excitation. The continued decrease in thermal resistance with increasing temperature is also important to note.

CONCLUSIONS

Measurement of the thermal resistance is easy and should be standard procedure when making cryogenic temperature measurements. More data is needed on variations in thermal resistances for other sensors, other mounting configurations, and on remounting of sensors. The thermal resistance can be used to improve the accuracy of temperature measurements by allowing calculations of excitation conditions necessary to minimize the overall measurement uncertainty.

EXAMPLE CALCULATION

As an example of a thermal resistance calculation and excitation optimization, consider a Cernox model CX-1050-SD attached to a copper block operating at 1.6 K and surrounded by vacuum. The voltmeter to be used to make the measurement is a Keithley model 2000 with a dc voltage accuracy specification of ±(50 ppm of reading + 3.5 μV) and a resolution of ±100 nV on the 100 mV scale. Measurement using current reversal should eliminate the offset voltage to within the instrument resolution, yielding a voltage measurement uncertainty of ±(50 ppm + 100 nV). The sensor electrical resistance R_e = 60 000 Ω and dimensionless sensitivity S_d = 3.3 come from plots in reference[5]. The thermal resistance R_t = 150 000 K/W is found in Table 1. From the voltmeter accuracy specification, u_V/V = 50 ppm = 5e-5 with an additional voltage uncertainty u_V = 100 nV. The following values can be calculated:

$$P_{s\,opt} = \frac{(1.6\text{ K})(5\text{x}10^{-5})}{(3.3)(1.5\text{x}10^{5}\text{ K/W})\sqrt{2}} + \left[\frac{(1\text{x}10^{-7}\text{ V})(1.6\text{ K})}{(3.3)(1.5\text{x}10^{5}\text{ K/W})\sqrt{2\,(6\text{x}10^{4}\ \Omega)}}\right]^{2/3}$$

$$= 2.1\text{x}10^{-10}\text{ W} = 210\text{ pW}$$

$$I = \sqrt{\frac{P_{s\,opt}}{R_e}} = \sqrt{\frac{2.1\text{x}10^{-10}\text{ W}}{60\,000\ \Omega}} = 5.9\text{x}10^{-8}\text{ A} = 59\text{ nA}$$

$$V = IR_e = (5.9\text{x}10^{-8}\text{ A})(60\,000\ \Omega) = 3.54\text{ mV}$$

$$u_{Tsh} = \Delta T_{sh} = P_s R_t = (2.1\text{x}10^{-10}\text{ W})(1.5\text{x}10^{5}\text{ K/W}) = 31.5\ \mu\text{K}$$

$$u_{Ti} = \left(\frac{T}{S_d}\right)\left(\frac{u_{Vi}}{V}\right) = \left(\frac{1.6\text{ K}}{3.3}\right)\left(5\text{x}10^{-5} + \frac{1\text{x}10^{-7}\text{ V}}{3.54\text{x}10^{-3}\text{ V}}\right) = 37.9\ \mu\text{K}$$

$$u_{Tc} = \sqrt{u_{Tsh}^2 + u_{Ti}^2} = \sqrt{(31.5\ \mu\text{K})^2 + (37.9\ \mu\text{K})^2} = 49.3\ \mu\text{K}$$

$$\frac{u_{Tc}}{T} = \frac{49.3\ \mu\text{K}}{1.6\text{ K}} = 3.08\text{x}10^{-5} = 30.8\text{ ppm}$$

Note that the sensor excitation of 3.54 mV is only slightly above the 1-3 mV range ordinarily used in Lake Shore resistor calibrations. This is an indication that the model CX-1050-SD is nearing the lower limit of its application range.

REFERENCES

1. L.L. Sparks, Temperature, strain, and magnetic field measurements, Materials at Low Temperatures, R.P. Reed and A.F. Clark, eds., American Society for Metals, Metals Park, Ohio (1983) pp. 515-521.
2. T.A. Kobel, M.A. Kozyrczak, S.W. Schwenterly and W.M. Bell, Effects of mounting methods on temperature sensor accuracy below 10 K, Supercollider 4, J. Nonte, ed., Plenum Press, New York (1992) 619-626.
3. Guide to the Expression of Uncertainty in Measurement, ISO/TAG 4/WG 3: June 1992, International Standards Organization, Genève, Switzerland.
4. D.S. Holmes and S.S. Courts, Resolution and accuracy of cryogenic temperature measurements, Temperature: Its Measurement and Control in Science and Industry, Vol. 6, part 2, J.F. Schooley, ed., American Inst. Phys., New York (1992) 1225-1230.
5. Temperature measurement accuracy, in: "Temperature Measurement and Control," part 1, Lake Shore Cryotronics catalog (1995) A-50.
6. R.L. Rusby, The rhodium-iron resistance thermometer: Ten years on, Temperature: Its Measurement and Control in Science and Industry, Vol. 5, part 2, J.F. Schooley, ed., American Inst. Phys., New York (1982) 829-833, Fig. 3.

EFFECTS OF CRYOGENIC IRRADIATION ON TEMPERATURE SENSORS

S. Scott Courts and D. Scott Holmes

Lake Shore Cryotronics, Inc.
Westerville, OH, 43081, USA

ABSTRACT

Several types of commercially available cryogenic temperature sensors were calibrated, irradiated at 4.2 K by a gamma or neutron source, and recalibrated in-situ to determine their suitability for thermometry in radiation environments. Comparisons were made between pre- and post-irradiation calibrations with the equivalent temperature shift calculated for each sensor at various temperature in the 4.2 K to 330 K range. Four post-irradiation calibrations were performed with annealing steps performed at 20 K, 80 K, and 330 K. Temperature sensors which were gamma irradiated were given a total dose of 10,000 Gy. Temperature sensors which were neutron irradiated were irradiated to a total fluence of 2×10^{12} n/cm^2. In general, for gamma radiation environments, diodes are unsuitable for use. Both carbon glass and germanium resistance sensors performed well at lower temperature, while platinum resistance sensors performed best above 30 K. Thin-film rhodium and Cernox™ resistance sensors both performed well over the 4.2 K to 330 K range. Only thin-film rhodium and Cernox™ resistance temperature sensors were neutron irradiated and they both performed well over the 4.2 K to 330 K range.

INTRODUCTION

High energy physics experiments frequently operate in environments combining high vacuum, high radiation fluences, and cryogenic fluids in close proximity with higher temperature regions. For these applications, sensitive, stable, accurate, and easily used thermometers are needed for the temperature range of 4 K to room temperature. Research on the effects of radiation on electronic devices has been extensive,[1, 2, 3, 4] but much of the work is classified and thus not available to the public. Research has also concentrated on devices near room temperature rather than at cryogenic temperatures. Lake Shore Cryotronics, Inc. (LSCI) has performed some room temperature irradiation experiments on selected cryogenic temperature sensors, and calibration shift data is available over the 1.2 K to 330 K range for them.[5,6]

For temperature sensors, radiation damage is measured by performing temperature calibrations of the thermometers pre- and post-irradiation. In many applications involving radiation, it is unrealistic to recalibrate a temperature sensor on a timely basis if at all. The sensor must operate at cryogenic temperatures for extended periods. The user typically knows that some damage has occurred resulting in a shift in the calibration of the sensor, but he rarely has any way of quantifying that shift. The sensor can be warmed to room temperature, transferred to a calibration station and recalibrated, but this may yield misleading results. Upon warming to room temperature, the sensor may undergo thermal annealing. In this way, some of the damage caused by the radiation may be "repaired". Comparison of the new calibration to the old calibration then leads the user to underestimate the damage that has taken place and the error introduced into the experiment. To circumvent these problems, a calibration facility with radiation appropriate cryostats and probes was assembled at the site of the gamma and neutron radiation sources. Cryogenic temperature sensors were irradiated and calibrated in-situ to determine their suitability for cryogenic thermometry in high radiation environments.

EXPERIMENTAL EQUIPMENT

Performing in-situ calibrations immediately after irradiation required construction of an on-site temperature calibration system at The Ohio State University Nuclear Reactor Laboratory (OSU NRL) where the irradiations were performed. The electronics of the calibration system consisted of an HP 3456A DVM, Guildline model 9330 resistance standards, a Keithley model 224 current source, and a Keithley model 706 scanner with model 7067 4-wire scanner cards.

It was desirable to a single thermometer with high stability and high resistance to ionizing radiation. Wire-wound rhodium-iron resistance thermometers (RIRTs) met these criteria. An LSCI model RF-800-4 RIRT was used a the primary calibration thermometer. A thin-film RIRT (LSCI model RF-100) was used a backup calibration thermometer. For the wire wound RIRT, the worst case accuracy of LSCI calibrations relative to ITS-90 ranges from ± 20 mK at 4.2 K to ± 35 mK in the 20 K to 330 K temperature range. Estimates on short term stability range from about ± 2.0 mK at 4.2 K to about 3.5 mK in the 20 K to 330 K range. Estimates of resolution with the given calibration equipment range from about 0.5 mK at 4.2 K to better than 2.5 mK across the 20 K to 330 K range.

Rhodium-iron resistance sensors are fabricated from pure metals. Thus, it was expected that they would be reasonably immune to calibration shifts resulting from radiation damage. Preliminary irradiation studies in both gamma and neutron radiation environments were performed on both wire-wound RIRTs (LSCI model RF-800-4) and thin-film RIRTs (LSCI model RF-100) and they performed exceedingly well. These preliminary experiments were performed by irradiating the devices at room temperature, transferring the devices to a calibration probe at Lake Shore's calibration facility, and performing a recalibration. The results of the preliminary gamma radiation testing showed the wire-wound RIRT shifts to be less than ± 5 mK below 10 K and less than ± 30 mK from 10 K to 330 K for a 10,000 gray gamma dose (at a 30 gray/hr dose rate) from a cesium-137 gamma source. RIRTs irradiated to a total fluence of 10^{12} n/cm^2 using a nuclear pool reactor showed shifts to be less than ± 12 mK across the 1.2 K to 330 K temperature range.

The calibration probes were fabricated to be compatible with the high gamma or neutron radiation environment. The gamma irradiation cryostat and calibration probe were manufactured to LSCI's specifications by Janis Research (Wilmington, MA) while the neutron irradiation cryostat and calibration probe were manufactured to LSCI's specifications by Precision Cryogenics (Indianapolis, IN). Both calibration probes employ standard cryogenic designs. Close-up interaction with the neutron irradiation cryostat and probe was necessary during the sensor loading/unloading process, cryogen transfer and vacuum pumping periods, and it was desired to minimize the radiation exposure to project personnel during these periods. To that end, great care was taken in choosing construction materials for the cryostat and dewar that have either low radiation activation or very short half-lives (on the order of hours).

For the gamma irradiation experiment, the unknown thermometers to be calibrated were mounted together with standard wire-wound and thin-film rhodium-iron thermometers in an OFHC thermometer mounting block. For the neutron experiment, the unknown thermometers to be calibrated were mounted together with a thin-film rhodium-iron thermometer in chip form on an 85-pin pin grid array (PGA). This PGA mounted into a zero-insertion force (ZIF) socket mounted to the terminal block. The middle section of the ZIF socket was milled out to allow an aluminum finger mounted to the terminal block to pass through and make direct contact with the

Table 1. Typical timeline for gamma and neutron + gamma irradiation experiments

PROCEDURE	PROCEDURE TIME (hours)	TOTAL TIME (hours)
Initial calibration 4.2 K to 330 K	100	100
irradiation at 4.2 K	Varied (2.3 hr to 23 hr)	102-123
8 hr anneal at 4.2 K	8	108
4.2 K to 20 K calibration	6	114
8 hr anneal at 20 K	8	122
4.2 K to 80 K calibration	20	142
8 hr anneal at 80 K	8	150
4.2 K to 330 K calibration	100	250
8 hr anneal at 330 K	8	258
4.2 K to 330 K calibration	100	358
	Total time per irradiation run:	360 - 380 hours

back of the PGA. Proper design of both probes allows for the transferring of temperature scales on the millikelvin level for T less than 30 K, increasing to perhaps 10 mK at 300 K.

The radiation sources used for these experiments are operated by The Ohio State University Nuclear Reactor Laboratory (Columbus, Ohio). The gamma source used in these experiments consists of 13 cobalt-60 pins roughly 8 mm diameter by 30 cm length. Cobalt-60. The irradiation system consisted of a dry tube sitting in a 4.3 m deep pool. The cobalt pins are inserted into a pin holder on the exterior of the dry tube while the cryostat is lowered into the interior of the dry tube. The pin holder was designed to accommodate up to fifteen pins with positions at 24° intervals. The OSU NRL pool reactor produces a full spectrum of neutrons ranging from roughly 10^{-10} MeV to 10 MeV. The reactor is capable of operating at any power level in the 1 W to 500 kW range. A 9.5 inch ID, T96061-T6 aluminum dry tube was fabricated for the neutron irradiation studies. The dry tube was placed next to the reactor core with the calibration probe inserted during irradiation and moved to the opposite side of the reactor pool after irradiation..

See reference 6 for a complete description of the irradiation cryostats and probes.

EXPERIMENTAL PROCEDURES

During calibration, each data point consisted of forward and reverse voltage readings across the two standard thermometers, across the standard resistor, across an unknown thermometer, and again across the two standard thermometers. For diodes, only the forward voltage was measured. The unknown thermometer resistance was stored with two temperature values: first, the time-weighted average of the two wire-wound RIRT readings, and second, the time-weighted average of the two thin-film RIRT readings. For resistor calibrations, the current was determined by measuring the voltage across the standard resistor placed in series with the thermometer. This ratiometric technique allows the voltmeter to be used as a transfer device. In this mode, the voltmeter's short term stability is the relevant parameter. The DVM's short term stability is roughly a factor of ten better than its long-term stability.

At 4.2K and above, operation of the calibration system is totally under computer control, with preset parameters governing temperature stability and the times at which data are recorded. For these calibrations, the maximum acceptable temperature drift per minute is 5 mK/min. The drift must be less than this value for a minimum of 3 minutes before data is recorded.

A typical irradiation run took over two weeks to perform. The testing was designed around parameters important to the Superconducting Super Collider (SSC) facility. At the SSC, various maintenance would require warming portions of the cryogenic sections to either 20 K, 77 K or 295 K, while normal operation would occur at 4.2 K. Since irradiation would occur at 4.2 K, it was desirable to see if annealing would occur when the irradiated sensors were warmed to higher temperatures to perform maintenance. To this end, our irradiation experiments have included isochronal annealing steps. The calibration / irradiation / annealing steps are outlined in Table 1. The typical time for each step and accumulated time are also listed.

RESULTS

Devices tested are all cryogenic thermometers available from Lake Shore Cryotronics, Inc., located in Westerville, OH. Devices tested include silicon diodes, gallium / aluminum / arsenide diodes, carbon glass RTDs, germanium RTDs, Cernox™ RTDs, platinum RTDs, and thin-film rhodium-iron RTDs. See reference 7 for information regarding their resistance / forward voltage versus temperature, temperature sensitivity, etc.

The data obtained from a temperature calibration is in the form of resistance and temperature at several temperature points spanning the 4.2 K to 330 K temperature range. The data is smoothed by first fitting it with a Chebychev polynomial over typically three ranges. The coefficients are determined by a least squares fit to the raw data. At the end of this fitting routine, the resistance versus temperature calibration is "smoothed" by replacing the measured temperatures with the temperatures calculated from the polynomial fit. An interpolation table giving resistance as a function of temperature is created using a cubic spline routine on the smooth data. Calibration shifts are calculated as the resistance in the second calibration minus the resistance in the first calibration, divided by the temperature sensitivity. This yields a calibration shift in kelvin units. Note that Chebychev polynomial fits can give erratic results near endpoints. Large calibration shifts occurring at the extreme temperature points can be due to curve fitting and --not actual shifts. The data presented are in the form of temperature shift as a function of temperature.

Gamma Radiation Experiments

The gamma irradiation experiment was performed using 14 cobalt-60 pins producing a dose rate of roughly 3000 gray/hour. Sensors were irradiated for 3.3 hours to a total dose of 10 kGy. Figures 1-7 respectively show the temperature shift as a function of temperature after cold irradiation for a model DT-470-SD-12A silicon diode, TG-120P gallium / aluminum / arsenide diode, CGR-1-1000 carbon glass resistance thermometer device (RTD), GR-200A-000 germanium RTD, CX-1050-SD Cernox™ RTD, PT-103 platinum RTD, and RF-100-AA thin-film rhodium-iron RTD. The wire-wound RIRT standards used for thermometry showed shifts less than ± 20 mK at low temperatures to about ± 40 mK at higher temperatures when recalibrated in LSCI's Temperature Calibration Facility.

Neutron Radiation Experiments

In the neutron irradiation experiment, sensors were irradiated at 4.2 K to a total fluence of $2x10^{12}$ n/cm^2 with the reactor operating at 50 W. Total irradiation time at operating power was 450 minutes. The additional fluence from startup power to operating power was $2.2x10^9$ n/cm^2 (0.11% of the total fluence) with a background gamma dose of about 135 gray. This fluence was based on the expected and 20 year lifetime neutron fluence for sensors used at the Superconducting Super Collider Laboratory. Figures 8 and 9 show the temperature offsets induced by cold neutron irradiation for a model CX-1050-SD Cernox™ RTD and a model RF-100-AA thin-film rhodium-iron RTD respectively. The wire-wound RIRT standard was repeatable to better than ± 35 mK over the 4.2 K to 330 K range when recalibrated at LSCI's Temperature Calibration Facility. Overall, this data was noisier, which may reflect the fact that the neutron calibration probe was less suited for performing temperature calibrations. This is believable, considering the limitations on the choices of materials and the desired design features.

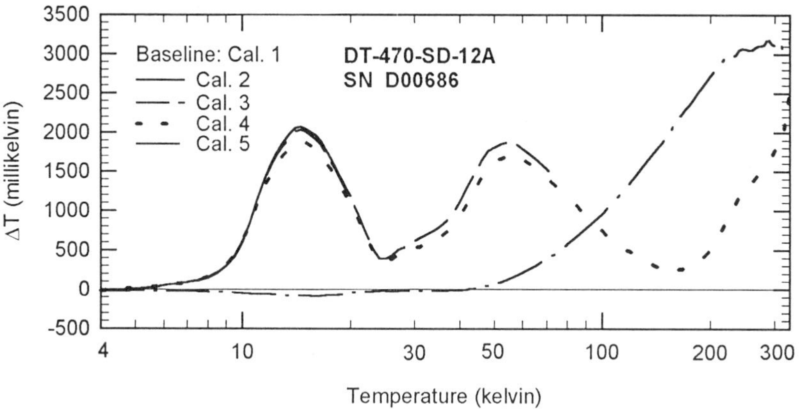

Figure 1. Calibration shift after irradiation for a DT-470-SD-12A silicon diode temperature sensor, SN D00686, irradiated at 4.2 K at dose rate of 3000 Gy/hr to a total dose of 10 kGy by a Co-60 gamma source.

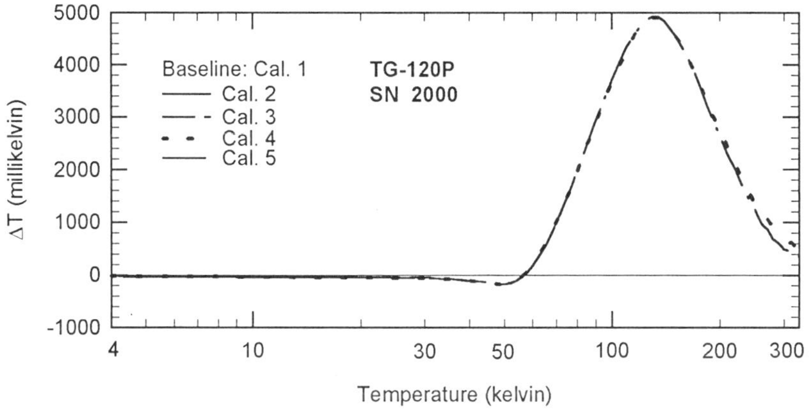

Figure 2. Calibration shift after irradiation for a TG-120P gallium/aluminum/arsenide diode sensor, SN 2000, irradiated at 4.2 K at dose rate of 3000 Gy/hr to a total dose of 10 kGy by a Co-60 gamma source.

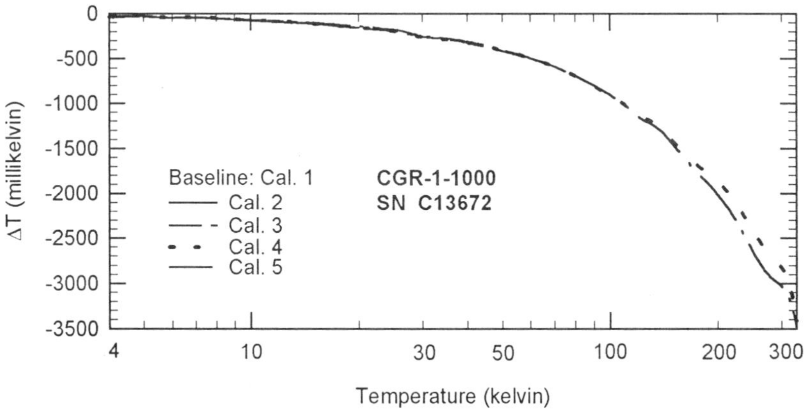

Figure 3. Calibration shift after irradiation for a CGR-1-1000 carbon glass RTD, SN C13672, irradiated at 4.2 K at dose rate of 3000 Gy/hr to a total dose of 10 kGy by a Co-60 gamma source.

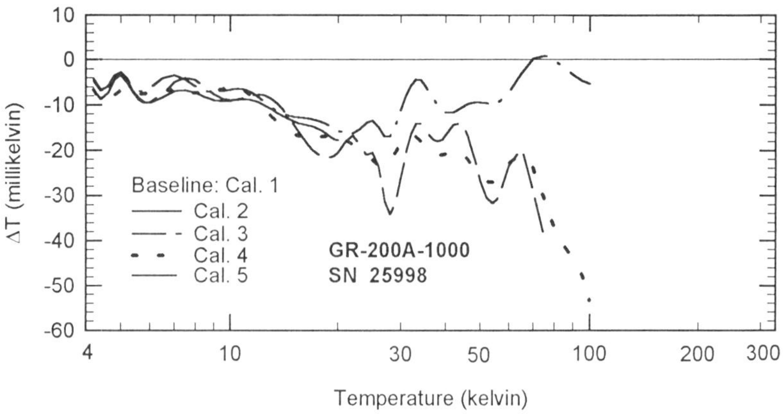

Figure 4. Calibration shift after irradiation for a GR-200A-1000 platinum RTD, SN 25998, irradiated at 4.2 K at dose rate of 3000 Gy/hr to a total dose of 10 kGy by a Co-60 gamma source.

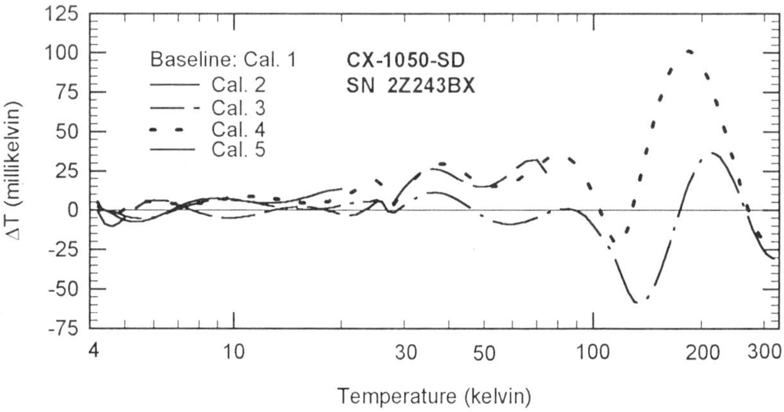

Figure 5. Calibration shift after irradiation for a CX-1050-SD Cernox™ RTD, SN 2Z243BX, irradiated at 4.2 K at dose rate of 3000 Gy/hr to a total dose of 10 kGy by a Co-60 gamma source.

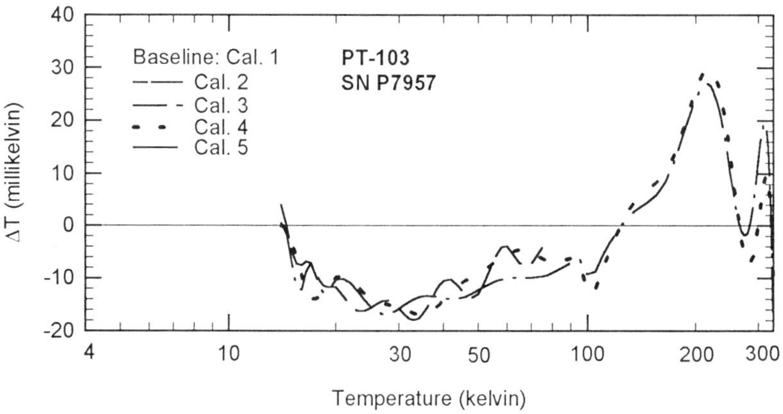

Figure 6. Calibration shift after irradiation for a PT-103 platinum RTD, SN P7957, irradiated at 4.2 K at dose rate of 3000 Gy/hr to a total dose of 10 kGy by a Co-60 gamma source.

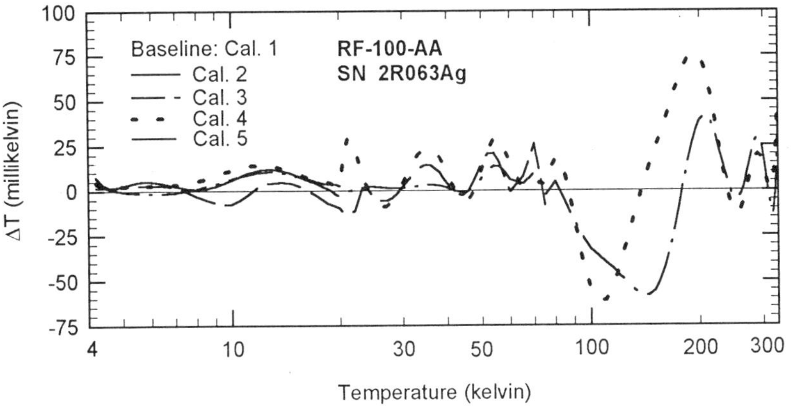

Figure 7. Calibration shift after irradiation for an RF-100-BG thin-film rhodium-iron RTD, SN 2R063Ag, irradiated at 4.2 K at dose rate of 3000 Gy/hr to a total dose of 10 kGy by a Co-60 gamma source.

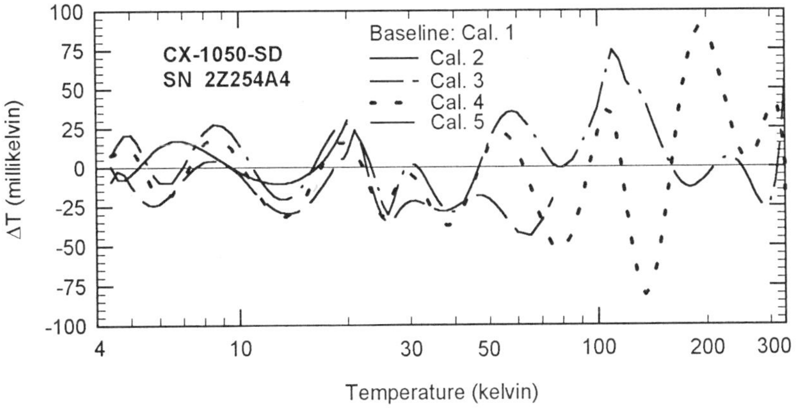

Figure 8. Calibration shift after irradiation for CX-1050-SD Cernox™ RTD, SN 2Z254A4, irradiated at 4.2 K at a flux of 7.41×10^7 n/cm2/s to a total fluence of 2×10^{12} n/cm^2 in a nuclear pool reactor.

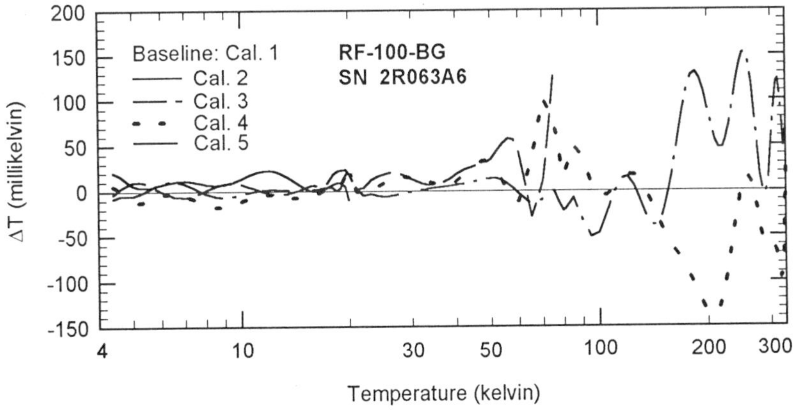

Figure 9. Calibration shift after irradiation for a RF-100-BG thin film rhodium iron RTD, SN 2R063A6, irradiated at 4.2 K at flux of 7.41×10^7 n/cm^2/s to a total fluence of 2×10^{12} n/cm^2 in a nuclear pool reactor.

CONCLUSIONS

Based on data available, thermometry is certainly possible for cryogenic, high radiation environments with reasonably good accuracy..

For gamma radiation, up to a total dose of 10,000 gray, the Cernox™ RTDs can be used with errors less than ± 25 mK below 50 K and ± 100 K in the 50 K to 330 K range. Thin-film RIRTs can also be used with virtually the same results as the Cernox™ RTDs. Platinum RTDs are also very good (± 30 mK) over their useful low end range of about 14 K to 330 K. Over the 4 K to 100 K temperature range, germanium RTDs are also good in gamma radiation showing shifts less than -55 mK. Carbon glass RTDs were worse than the above sensors showing -100 mK shifts below 10 K and over -3000 mK at room temperatures. The diode devices behaved poorly overall. The silicon diode sensor shifted over 2 K at temperatures over 15 K after irradiation. It also showed a large annealing effect after heating to 330 K for 8 hours. None of the other tested sensors exhibited an annealing effect on this order. After irradiation, the gallium / aluminum / arsenide diode showed shifts on the order of 5 K above 100 K, but only shifts of ± 50 mK below 30 K.

In the neutron radiation environment, to a total fluence of 2x1012 n/cm2, the Cernox™ RTDs can again be used with errors less than ± 25 mK below 50 K and ± 100 K in the 50 K to 330 K range. Thin-film RIRTs can also be used with errors less than ± 25 mK below 90 K and ± 75 mK in the 90 K to 330 K range.

ACKNOWLEDGMENT

We would like to thank the personnel at The Ohio State University Nuclear Reactor Laboratory for their tremendous assistance with this project. This work was supported by Department of Energy Small Business Innovative Research grants numbers DE-AC02-89ER80721 and DE-FG02-90ER81074.

REFERENCES

1. "Radiation Effects At The SSC," M.G.D. Gilchriese, ed., Superconducting Super Collider Central Design Group, SSC-SR-1035 (1988).
2. G.C. Messenger and M.S. Ash. "The Effects of Radiation on Electronic Systems," Van Nostrand Reinhold, New York (1986).
3. J.R. Srour, D.M. Long, D.G. Millward, R.L. Fitzwilson and W.L. Chadsey. "Radiation Effects on and Dose Enhancement of Electronic Materials", Noyes Publications, Park Ridge, NJ.
4. V.A.J. van Lint, T.M. Flanagan, R.E. Leadon, J.A. Naber and V.C. Rogers. "Mechanisms of Radiation Effects in Electronic Materials, Vol. 1," John Wiley and Sons, New York (1980).
5. S.S. Courts, D.S. Holmes, and P.R. Swinehart, Neutron and gamma radiation effects on cryogenic temperature sensors, in: "Temperature, Its Measurement and Control in Science and Industry," J. F. Schooley, ed., American Institute of Physics, New York (1992) p. 1237.
6. S.S. Courts, D.S. Holmes, and P.R. Swinehart, "Radiation Resistant Cryogenic Temperature Sensor for the 4 K to 80 K Range," final report for Department of Energy contract number DE-FG02-90ER81074 by Lake Shore Cryotronics, Inc., Westerville, OH (1994).
7. Lake Shore Cryotronics, Inc., "Temperature Measurement and Control, Product Catalog and Reference Guide, Part 1 of 2", published by Lake Shore Cryotronics, Inc., Westerville, OH (1995).

INDUSTRIAL-TYPE CRYOGENIC THERMOMETER WITH BUILT-IN HEAT INTERCEPTION

C. Balle and J. Casas

Accelerator Technology Division, CERN
CH-1211 Geneva 23, Switzerland

ABSTRACT

CERN is a major user of cryogenic machinery and its ancillary instrumentation. Many of these apparatus are assembled by subcontracting firms that do not have necessarily a very good knowledge of cryogenic techniques. This can result in cryogenic sensors malfunctioning with a loss of information that can be crucial during the commissioning or operation of cryogenic equipment. In particular cryogenic temperatures measured under vacuum conditions are often unreliable because of a poor thermal anchoring. In order to improve the performance of such temperature measurements a new thermometric block has been developed. It is fabricated by Prototype Circuit Boards (PCB) techniques and is thus well adapted to large-series production, it includes the thermalization of the leads and a cavity to insert a commercial temperature sensor. The main design goal was to manufacture a thermometer that can be installed by non-qualified personnel. For validating the design, prototype thermometers were mounted on an isothermal substrate as the temperature of the current and sensing leads was increased and they were exposed to thermal radiation. A temperature of 1.8 K was measured to within 50 mK when using copper leads of 2 mm diameter and about 100 mm long heated to 170 K on one end. This demonstrates the good performance of our thermometer blocks under extreme environmental conditions.

INTRODUCTION

Large-scale cryogenic machines require a huge number of cryogenic probes (temperature, level, pressure, etc.), a trouble-free operation requires these sensors to be reliable, maintain the specifications over the lifetime of the equipment and under its environmental conditions. One such machine is CERN's Large Hadron Collider (LHC) project[1] that will include a large number of superconducting magnets immersed in pressurised superfluid helium at 1.8 K. Temperature is an extremely important parameter for the LHC and furthermore its measurement is reputed to be difficult for a number of reasons like electrical leads thermalization, sensor long-term stability, signal recovery and in the case of a particle accelerator, ionising radiation hardness. The superconducting magnets for the LHC will be operated at a temperature between 1.8 K to 1.9 K, their absolute temperature should be known to within better than 0.02 K and with a 0.001 K resolution.

From a cryogenic point of view all thermal sensors should be placed inside the cryogen of which the temperature is being measured. On the other hand this complicates the cryostat design because vacuum-tight electrical connections are necessary when a cryogen-vacuum passage has to be realised. In the scope of the LHC project which will

Table 1. Parameters used for estimating the measured temperature

ρ_{cu}, [W m^{-1} K^{-1}]	ρ_{isol} [W m^{-1} K^{-1}]	$\sigma_{stycast}$ [W m^{-1} K^{-1}]	c [μm]	λ [μm]	δ [μm]
7000	0.004	0.01	140	100	6

employ about ten thousand cryogenic thermometers, it will then be desirable to make, as much as possible, the temperature measurements on the vacuum side of the coldmass. The main advantages are simpler cryostats and immunity from pressure waves produced when a superconducting magnet quenches. On the other hand thermal contact, thermal impedance across cryostat walls (usually stainless, a poor thermal conductor) through which heat losses are absorbed by the cooling cryogen, and temperature sensor self-heating are of concern. For instance the 10 meter LHC prototype dipoles employ 50 mm thick stainless steel end plates. With the type of superinsulation employed[2], a degraded vacuum of 10^{-4} mbar yields a thermal flux of the order of 200 mW m^{-2} resulting in a 0.125 K temperature increase on the surface, very far from the accuracy requirements stated in the precedent paragraph. This means that temperature measurements on the vacuum side have to be performed on interconnecting tubes that have typically 2 mm thick walls. Before deciding to place the temperature probes on the vacuum side of the cold-mass, a cryogenic thermometer has to be designed, tested under realistic LHC operating conditions, it has to provide a good thermalization of the temperature sensor and be easy to mount by non-qualified personnel.

A cryogenic thermometer has then been developed[3], the main design parameters of which are: (a) industrial robustness, (b) the thermal coupling of the sensor is done through its own wires, (c) the temperature measurement does not depend critically on the installation work and (d) it is compatible with large series fabrication techniques.

Frequently much care is given to produce a very good thermal contact between the sensor and the body under investigation. This contact is very difficult to obtain in view of

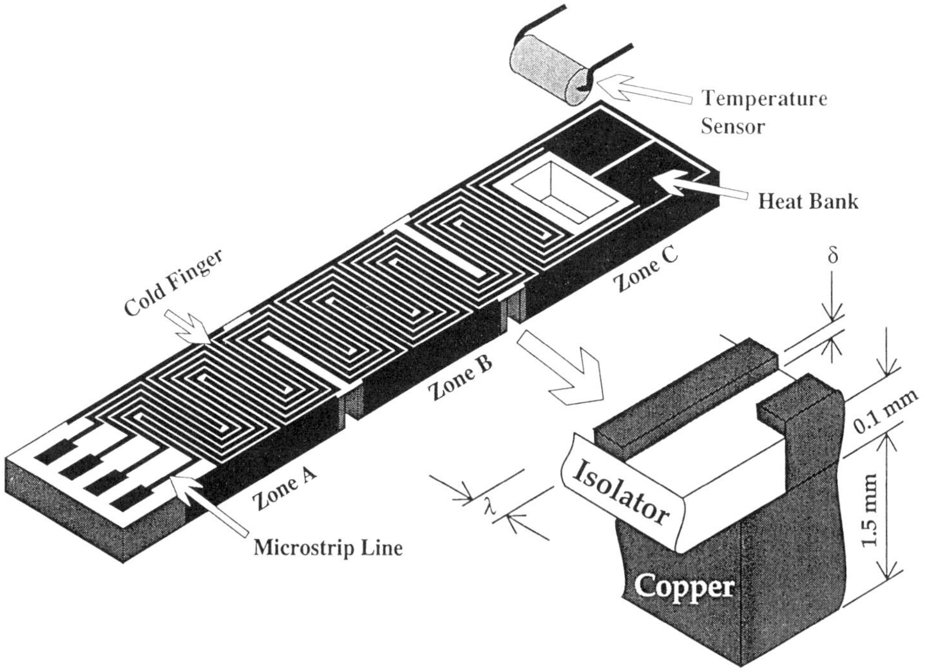

Figure 1. Thermometer layout showing the meandering microstrip lines of width λ and thickness δ, the solid copper plate thickness corresponds to that of a standard Allen Bradley carbon sensor. The thermometer size is 100 mm x 10 mm x about 2 mm.

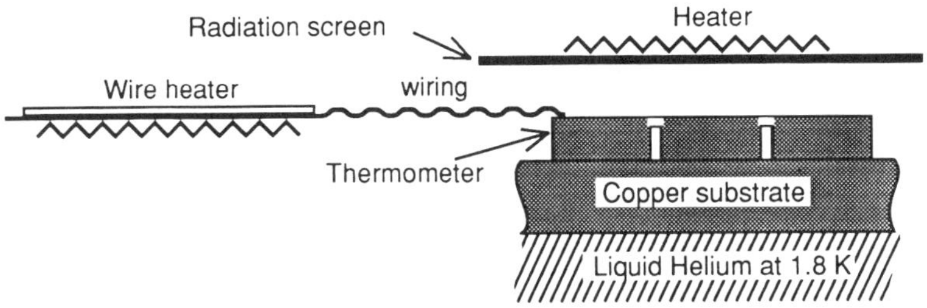

Figure 2. Schematical principle of the thermometer validation system. The thermometers are bolted to an isothermal substrate. The vacuum, the temperature of the electrical leads and the radiation screen can be regulated.

the size of many modern thin-film temperature sensors, their fragility and also because some sensors are inside a housing under vacuum conditions. In an industrial environment such a delicate and careful assembly of these sensors would be impossible to obtain at a reasonable cost.

THERMOMETER DESIGN

Cryogenic techniques require sensing leads thermalization for making proper temperature measurements. This dictates the use of long and thin wires of poor thermal conductance, with the last run in close thermal contact with the body under investigation. PCB techniques were chosen for fabricating the thermometer because it is very easy to obtain leads with small cross sections, typically the equivalent of a copper wire of a diameter of 44 μm or less, in the actual design it is the equivalent of a 670 mm long x 32 μm diameter copper wire. Furthermore PCB techniques permit large series production. Figure 1 shows schematically the thermometer design: the electrical microstrip lines follow a meandering pattern to maximise its length, this meander intercalates "cold fingers" anchored to the bottom plate, the temperature sensor is housed in a cavity and it is soldered to a "heat bank" that sinks sensor self heating thermal flow and photon radiation hitting the soldering pads. The lower part is made of solid copper of at least the same thickness as that of the sensor and it is divided in three parts in order to break longitudinal thermal conduction. In Figure 1 are not shown the screen that protects the sensor and microstrip wires from thermal radiation and the holes used for attaching the thermometers by using M4 size screws.

By neglecting transverse temperature gradients and taking into account the proper boundary conditions (i.e. heat in-leak from the wiring and from Joule effect produced by the sensor) the following two-dimensional equation permits to calculate the measured temperature assuming that it is the same as that of the microstrip wires:

$$\frac{\partial}{\partial x}\left[\rho_{cu}\frac{\partial T(x)}{\partial x}\right] = \rho_{isol}\frac{1}{e\,\delta}(T(x) - T') - \rho_{el}\left(\frac{i}{\lambda\,\delta}\right)^2 \qquad (1)$$

where ρ_{cu}, and ρ_{isol} are the thermal conductance of copper and the insulating substrate respectively, ρ_{el} the electrical resistivity, λ and δ are respectively the width and thickness

Table 2. Bottom copper plate temperature versus connecting wire terminal temperature. The measured temperature error in the absence of self-heating effects is also shown depending on the bottom plate design. The actual temperature that should be measured is 1.8 K.

T_{wire} [K]	$T_{zone\ A}$ [K]	$T_{zone\ B}$ [K]	$T_{zone\ C}$ [K]	Error [K]	$T_{monoblock}$ [K]	Error [K]
77	3.170	1.874	1.804	0.018	2.523	0.856
300	7.231	2.093	1.816	0.071	4.668	3.396

of the copper microstrip wire, e is the insulating substrate thickness, T(x) is the microstrip temperature along the longitudinal position x, T' is the bottom copper plate temperature supposed constant and i is the electrical current flowing through the microstrip. In the actual design the Joule effect along the electrical microstrip lines can be neglected.

The exact solution of Eq. (1) has to be obtained numerically because the thermal and electrical conductances depend on the temperature, a worst case analytical solution was calculated by assuming the above variables constant, the values used for the calculation are listed in Table 1. Eq. (1) predicts that the microstrip temperature decreases exponentially along the longitudinal position towards T', the exponential factor being:

$$m = \sqrt{\frac{1}{e\,\delta}\frac{\rho_{isol}}{\rho_{cu}}} \qquad (2)$$

Optimum temperature measurements require to maximise m. Its value depends strongly on the best resolution that can be obtained with PCB techniques for the microstrip and insulation thickness and also on the choice of materials through their thermal conduction.

It is very important to avoid longitudinal conduction along the bottom solid copper plate when a poor thermal contact between the thermometer and the body under investigation is present. In this case all the intercepted heat flows through a thermal resistance effectively increasing the value of T'. If the bottom plate is separated in three parts, the part that houses the sensor (zone C in Figure 1) is much cooler than the one were the connecting wires are soldered. For simulating this situation the poor thermal resistance is assumed to be that of a solid plate of 0.1 mm thick stycast whose thermal conductance is given in Table 1. Table 2 shows clearly that under poor thermal contact conditions a monoblock design for the copper plate gives very poor temperature measurements.

The effect of photon radiation can be estimated by using the Stefan-Boltzmann law[4] and assuming that the sensor is sensitive mainly to radiation hitting the window defined by the solder pads and the area without copper film shield. The measured temperature error is then given by the following formula:

$$\Delta T = \frac{e}{\rho_{isol}}\frac{\Sigma}{S}\sigma\frac{\varepsilon_1\varepsilon_2}{\varepsilon_1+\varepsilon_2-\varepsilon_1\varepsilon_2}\left(T_{shield}^4 - T_{bank}^4\right) \approx 2\ 10^{-11}\left(T_{shield}^4 - T_{bank}^4\right) \qquad (3)$$

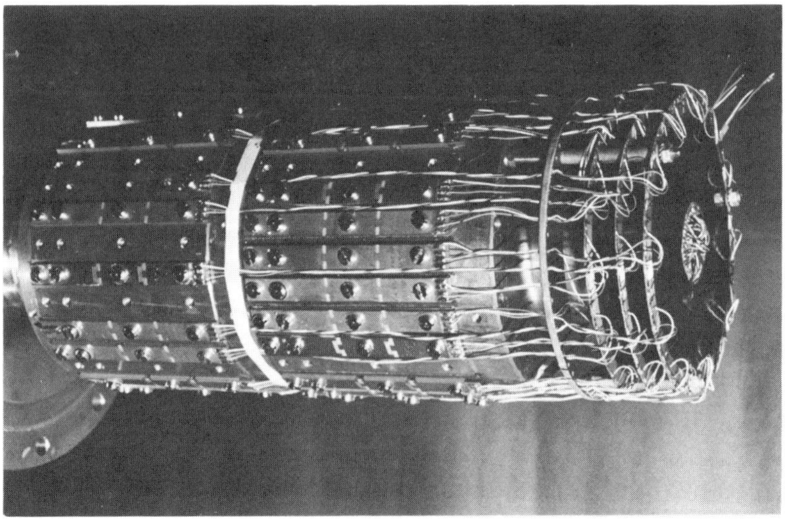

Figure 3. Photograph of the validation insert. The size of the thermometers is 100 mm x 10 mm x 2 mm and about a dozen thermometers can be seen bolted to the isothermal copper substrate. The wire terminal heaters are on the right-side. The heat radiation shield is removed. Liquid helium is present inside a cavity enclosed by the copper substrate.

where $\Sigma \approx 7$ mm^2 is the exposed window area, $S \approx 90$ mm^2 is the heat banks area, σ is the Stefan-Boltzmann constant (5.67 10^{-8} W m^{-2} K^{-4}), ε_1 and ε_2 are the emissivity factors of the exposed area and a radiation screen made of copper (assumed to be 1 and 0.2 respectively); and T_{bank} and T_{shield} are respectively the heat bank and shield surface temperatures (see Figure 1 and 2). The emissivity depends strongly on surface condition and may vary from sample to sample.

VALIDATION SYSTEM

An experimental set-up has been designed for measuring the performance of the fabricated thermometers, the measurement principle is schematically shown in Figure 2. A photograph of the actual cyogenic insert is shown in Figure 3. The thermometers are fixed by four M4 size screws to an isothermal substrate of known temperature, the temperature of the electrical leads and the radiation screen as well as the pressure of the residual vacuum can be regulated and measured. Up to 60 thermometers can be mounted in this apparatus that is used to simulate diverse environmental conditions. The isothermal substrate is a hollow cylinder machined in OFHC copper with the thermometers on the outer side and liquid helium on the inner side (Figure 3); its temperature is the same as that of the substrate. The wire heater is made of a planar heater bonded to a PCB with the sensing wires engraved on it, each copper microstrip wire is about 500 mm long, 200 µm wide and 23 µm thick and the sensing wires temperature is estimated by a platinum thermometer glued on the wires heater board. The radiation screen is made of 0.5 mm thick copper, the distance between this screen and the thermometer is of the order of 5 mm, its temperature is increased by using a planar heater and is monitored by two platinum thermometers. The thermometer chamber can be operated with or without liquid helium and when working in vacuum conditions the chamber is pumped by using a turbomolecular pump. The best residual vacuum obtained is of the order of 2 10^{-5} mbar, this relatively high value is the consequence of a long pumping line (1.5 meter x 25 mm diameter) crowded by wires and of outgassing from the many threaded holes and large area bonded heaters.

The thermometers are cabled in a true four wire configuration and the temperature sensors resistance can be measured at various excitation currents to within better than 0.05%. The thermometers resistance is measured sequentially by using a 40 x 4-wire channel scanner, a programmable d.c. current source and a digital voltmeter. During the measuring cycle the liquid helium bath temperature can be maintained to within better than 1 mK when operating in superfluid helium at about 1.8 K.

The temperature reference[5] is given by the pressure against the liquid helium bath and can be cross checked by commercially calibrated sensors as well as by using some sensors calibrated in house. When working in superfluid helium a maximum dispersion of 5 mK between the pressure and the resistive thermometric references is observed. For temperatures above the lambda point the helium gas pressure reference is awkward to use because of the relatively long stabilisation time for the liquid helium bath temperature that is in the order of half an hour, this might be due to the relatively large liquid reservoir capacity that ranges from 40 to 100 litres of liquid helium.

RESULTS

The thermometer block has been operated under various environmental conditions, and no contact grease is used to improve the thermal contact of the thermometers with the isothermal substrate. The residual vacuum is better than 3.5 10^{-5} mbar and the electrical connections between the wire heater and the thermometer (see Figure 2 and 3) is done with copper wire of 0.25 mm diameter x 120 mm length. Ten thermometers have been investigated and their measurement errors are shown in Figure 4, the error data show some dispersion that is probably due to a spread in the fabrication parameters. The microstrip wire thickness and width are near to the lowest values that can be obtained at a reasonable cost when using PCB techniques and they are thus the most likely parameters that will have a large spread in their actual values. These dimensions are impossible to measure directly because of the radiation shield that is bonded by using multi-layer PCB techniques, according to resistance data the microstrip wire thickness varies between 6 µm and 10 µm. Also the insulating substrate thickness can be expected to vary by a factor of ±20 %. Table

Table 3. Expected measurement error when varying the insulating substrate and microstrip wire thickness. The wire terminal temperature is 140K.

T_{wire} [K]	e [μm]	δ [μm]	Error [K]
140	140	6	0.012
140	140	8	0.041
140	140	10	0.096
140	140	12	0.181
140	110	6	0.003
140	170	6	0.028

3 shows the calculated measurement error depending on the most critical fabrication parameters with a wire terminal temperature of 140 K. Table 3 explains partially the dispersion of the data observed in Figure 4.

The effect of radiation heat transfer to the thermometric blocks is shown in Figure 5, the measurement error saturates below about 1 mK because of sensor stability problems and measurement accuracy. Two thermometers with unscreened temperature sensors are also shown for reference, as expected they result in much more important errors when comparing with our standard thermometers. Many of the measured errors are greater than what is expected from Eq. (2), this can be due to an increase of the PCB upper shield temperature that is relatively thin (about 30 μm) to evacuate efficiently the radiative heat to the bottom plate or to a bad choice of emissivity factors. Nevertheless temperature measurements to within 10 mK can be performed in presence of radiation shields at liquid nitrogen temperature level.

CONCLUSION

A cryogenic thermometer has been designed with the main goal being to perform reliable temperature measurements on the vacuum side of a cryostat. The thermometer has been tested and under adverse environmental conditions adequate temperature measurements can be performed.

More experimental data is necessary for assessing the effect of insulating vacuum, of mounting procedures and thermal impedance through walls separating the vacuum and the cryogen of which the temperature is being measured. Also the use of a different material for fabricating the microstrip wires of the thermometric block should be investigated, the

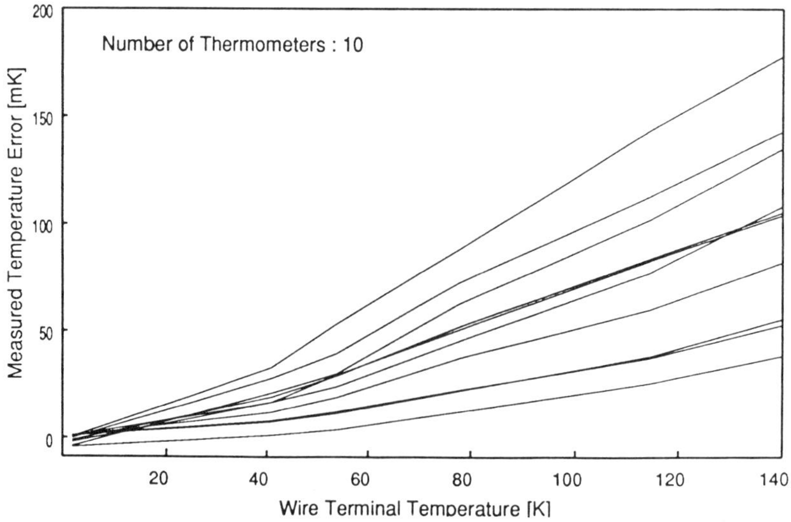

Figure 4. Measured temperature error versus wire terminal temperature. Residual vacuum is better than 3.5 10^{-5} mbar. The sensing leads are copper wires 120 mm long x 0.25 mm diameter.

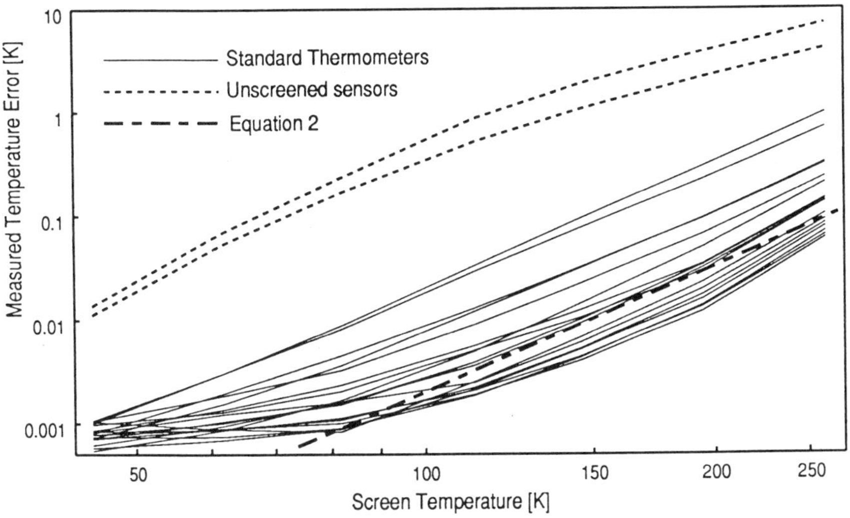

Figure 5. Measured temperature deviation when exposing the thermometers to photon radiation. The total number of thermometers investigated is 22. Two unprotected sensors are also shown and as expected they are much more sensitive to thermal radiation.

goal being to choose a material with a lower thermal conduction than copper. Copper was chosen because it is the most common material used by PCB manufacturers and then they are very familiar with the techniques necessary to obtain very small dimensions with an acceptable yield.

The type of thermometer described in this paper is being installed in different equipment at CERN and in many cases comparison thermometers are located inside the cryogen. The results obtained on the field will permit us to asses under which conditions we can use the thermometers for future installation on the LHC ring.

ACKNOWLEDGEMENT

The authors would like to thank D. Berthet and A. Monfort for fabricating the first prototypes of the thermometric block and subsequent definition of the large series production steps, and K. Henrichsen for many fruitful discussions. This project was initiated by Ph. Lebrun.

REFERENCES

1. L. R. Evans, The Large Hadron Collider, invited paper at the Particle Accelerator Conference, Dallas (1990)
2. Ph. Lebrun, L. Mazzone, V. Sergo and B. Vullierme, Investigation and qualification of thermal insulation systems between 80 K and 4.2 K, CERN LHC Note 189 (1992)
3. J. Casas, "Development of an industrial cryogenic thermometer", CERN Report, in preparation
4. S. W. Van Sciver, "Helium Cryogenics", Plenum Press, New York (1986)
5. H. Preston-Thomas, The International Temperature Scale of 1990 (ITS-90), *Metrologia* 27:3 (1990)

DEVELOPMENT OF A HIGH-RESOLUTION PENETRATION DEPTH THERMOMETER

M. J. DiPirro, J. G. Tuttle, and P. J. Shirron

Code 713, NASA/Goddard Space Flight Center, Greenbelt, MD 20771

ABSTRACT

We report on the development of a high resolution Penetration Depth Thermometer (PDT). The PDT is based on the fact that the magnetic field penetration is temperature dependent in a superconductor. The sensitivity and temperature range is broadened by the use of a thin film rather than a bulk superconductor. A dc current excitation is used with a dc SQUID for readout. Tests of a prototype PDT using an aluminum film demonstrated a maximum sensitivity of 0.8 nK/√Hz at $0.95T_c$ using limited excitation current. A recent theory by Pippard provides a good fit to the magnetic field penetration versus temperature data. Temperature resolution of better than 100 pK/√Hz is predicted for a modest sized PDT. A new version of the PDT will be fabricated according to an optimized design which overcomes several limitations. Design parameters and test results of this device are presented.

INTRODUCTION

In many low temperature experiments the key measurement parameter is the temperature. For instance, in critical phenomena, obtaining precise thermometry extremely close to the transition temperature is crucial. Many experiments have used resistance thermometry to obtain resolutions of 10^{-7} to 10^{-8} K. Recently, high resolution thermometry based on the Curie-Weiss behavior of a paramagnetic salt has been used to obtain a resolution of 10^{-10} K or better near the lambda transition in helium[1]. It has even been shown with one version of this thermometer that the resolution is limited by the fundamental statistical thermal fluctuations in the device itself[1]. Such resolution is required for helium superfluid transition experiments in space flights, because the gravitational force, which previously limited studies to further than about 10^{-7} K from the transition, has been removed. An important drawback of these thermometers, however, is that the thermal time constant between the thermometer and the liquid helium sample is a large fraction of a second. For heat capacity measurements, this would not normally be an issue, since integration times can be long. However, in space the large background of energetic particles which are absorbed by the thermometer make thermal spikes at rates of a few to hundreds per second. These spikes prove to limit the resolution to about 10^{-9} K.[2]

We are developing a thermometer based on the temperature dependence of the magnetic penetration in a superconductor. A penetration depth thermometer (PDT) can be made with a very low mass, and therefore, low cross-section for energetic particles. This thermometer also has a simple geometry, and can be easily integrated with an experiment. For instance, it could be in intimate contact with the fluid in a liquid helium experiment. The thermal time constant for such a set-up would be about one microsecond. Such a fast thermal relaxation would effectively remove all spikes mentioned above. At the same time the intimate thermal contact between helium and PDT effectively creates one large thermal reservoir reducing the statistical thermal fluctuations which can limit the resolution of the paramagnetic salt thermometers.

THEORY

Screening currents limit the degree to which a magnetic field can penetrate a superconductor. In a bulk superconductor, the field decays exponentially with distance x inside the material as $B(x) = B_o\exp(-x/\lambda)$, where λ is the magnetic penetration depth and B_o is a uniform field parallel to and just outside the superconductor. The penetration depth has a temperature dependence, $\lambda = \lambda_o(1-t^4)^{-1/2}$, where t is the reduced temperature, $t=T/T_c$, and T_c is the superconducting transition temperature. The constant λ_o (the penetration depth at T=0) is typically 10 to 100 nm and is 15.7 nm for aluminum. One can see that even very close to T_c the field drops effectively to zero inside the material within a short distance of the surface.

For a thin film the situation is somewhat more complicated. If the thickness of the film, d, is on the order of λ or less, then the field will be incompletely screened. Currents will flow in the film tending to exclude the field, but their magnitude will be temperature dependent. Pippard[3] recently derived an expression taking into account the spatial variation of these currents and the resulting field outside a planar film. His result for the mutual inductance between two coils on opposite sides (a and b) of the film is given by:

$$M_{ab} = (4\mu_o\alpha\lambda^2/d) \int h_a(k)h_b^*(k)d^2k \qquad (1)$$

The wavevector k describes the spatial variation of the fields h_a and h_b generated by unit currents within the superconducting film in response to the field applied by the primary coil. α is a factor close to 1 for $d < \lambda$, and d is the film thickness. This result is an approximation not valid close to the transition temperature where the temperature dependent term, λ, must remain within the integral. Since the mutual inductance may be measured when the film is normal, and so completely transparent to magnetic fields, we may normalize the integral

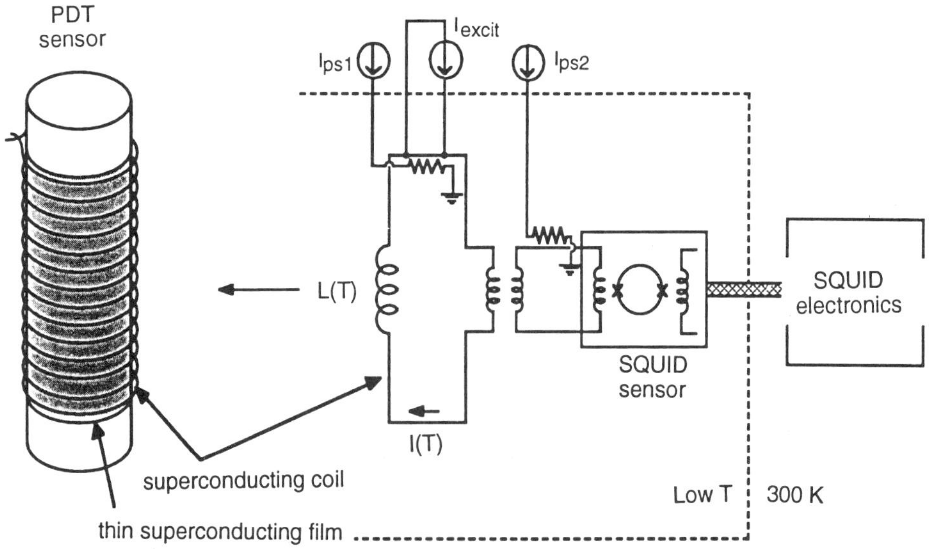

Figure 1. Schematic of the PDT and SQUID readout. The figure shows a self inductance measurement scheme, but a primary coil could be added outside the pick-up coil shown for a mutual-inductance-like measurement.

(following from Pippard) in (1). Adding a factor M_o to account for the residual inductance due to the coupling of the coils "around the film" which is not temperature dependent, (1) is simplified to:

$$M_r = M_o + M_n \frac{2\alpha\lambda^2}{Dd} \tag{2}$$

where $M_o + M_n$ is the mutual inductance between the coils when the film is not superconducting, and D is a constant resulting from the normalization of the integral. D is effectively the length scale over which the applied field varies. Note that (2) is not valid too close to the transition temperature since λ goes to infinity.

The best sensitivity of the PDT is obtained using a SQUID. The secondary coil is connected to the input coil of the SQUID. (See Figure 1.) A persistent current is stored in the superconducting primary coil. A change in temperature then produces a change in current in the superconducting secondary/SQUID input loop. Similar to (2) the total inductance of the secondary/SQUID input loop at any given temperature is:

$$L_t = L_o + L_s + L_n \left(\frac{2\alpha\lambda^2}{Dd}\right) \tag{3}$$

where L_o is the temperature independent part of the secondary coil inductance, L_s is the inductance of the SQUID input coil, and $L_o + L_n$ is the self inductance of the secondary coil when the film is above the transition temperature. The temperature dependent part of the inductance is denoted L(T):

$$L(T) = L_n \left(\frac{2\alpha\lambda^2}{Dd}\right) \tag{4}$$

The following equation describes the current in this loop:

$$I_s = kI_p \frac{L(T)-L(T_i)}{L_o+L_s+L(T)} \tag{5}$$

where I_s is the current induced in the secondary (also the SQUID input current), I_p is the current in the primary coil, k is the coupling factor between the two coils when the film is in the normal state (in the case of ideal solenoid coils, k is the turns density ratio of the secondary to the primary), and T_i is the temperature at which the current is stored in the coils. The current sensitivity to temperature is the derivative of (5):

$$\frac{\partial I_s}{\partial T} = kI_pL(T) \frac{8t^3}{T_c(1-t^4)} \frac{L_o+L_s+L(T_i)}{(L_o+L_s+L(T))^2} \tag{6}$$

To maximize the temperature sensitivity, the largest current in the primary coil should be used. This will typically be limited by the critical current density of the film, which varies with temperature as[4]:

$$J_c = J_{co}(1-t)^{1.5} \tag{7}$$

where J_{co} is the maximum current density at T=0. For aluminum films $J_{co} = 1\text{x}10^{11}$ Amp/m^2. For other type I superconductors, J_{co} is roughly proportional to the transition temperature, T_c, so one would obtain the same absolute temperature sensitivity, equation (6), for any superconducting film. The critical current in the film is $I_c = J_c wd$, where w is the width of the film. Since the maximum current decreases with increasing temperature, the optimum temperature at which to store the current will be the highest operating temperature. In this case $T_i \sim T$ and (6) simplifies to

$$\frac{\partial I_s}{\partial T} = kI_p \frac{8t^3}{T_c(1-t^4)} \frac{L(T)}{(L_o+L_s+L(T))}. \tag{8}$$

Better sensitivity is then achieved by making L_o as small as possible. It also appears from (4) and (6) that sensitivity is improved by making the film thickness very small. This is nearly balanced, however, by the critical current in the film which decreases proportionally with thickness. Since the current in the film per applied current will depend on the coil arrangement, but will usually be proportional to $\sqrt{L(T)}$, (8) will be maximized, for negligible L_o, by choosing L_n such that $L(T) \sim L_s$. For a given size device, this implies an optimum number of turns in the secondary coil.

PDT FABRICATION

The PDT consists of a thin superconducting film deposited on a substrate with two coils in close proximity acting as a primary and secondary. Alternatively, a single coil can be used with some reduction in operational flexibility. Each coil is part of a superconducting loop which is interrupted by a persistent switch. The persistent switch contains a heater wrapped on a section of the superconductor. When the heater is activated, the superconductor goes normal allowing the flux trapped within the loop to change.

The geometry of the film may be one of any number of shapes with corresponding coils arranged to produce and detect fields either parallel or perpendicular to the film. A toroidal coil around a similarly shaped film would have the salutary property of being essentially immune to outside fields. However, this shape presents some problems for fabrication of the substrate and deposition of the film. For the present work, cylindrical and planar geometries were used for mutual inductance measurements. For the SQUID results quoted, all data was from a cylindrical 7 nm aluminum film.

Sapphire substrates were chosen to provide dimensional stability and relatively high thermal conductivity to the surroundings. Metal substrates would have a higher heat capacity and would make the thermometer subject to eddy current losses when currents are changed rapidly. 0.076 mm diameter Nb-Ti wire was used for the primary and secondary coils.

Aluminum films were used. For the temperature range around the lambda transition in liquid helium, indium would seem to provide a better choice, but indium tends to puddle when deposited. To form a continuous film, thicknesses 250 nm or more were required - well over the thickness desired for good temperature sensitivity. However, thin aluminum films could easily be deposited. The transition temperature for aluminum increases with decreasing thickness from a bulk value of 1.197 K to over 4 K for the thinnest films ever made[5]. The thinnest film reported here was 7.0 nm with a transition temperature of 1.693 K, but work is progressing on thinner films which should have transition temperatures above 2.2K.

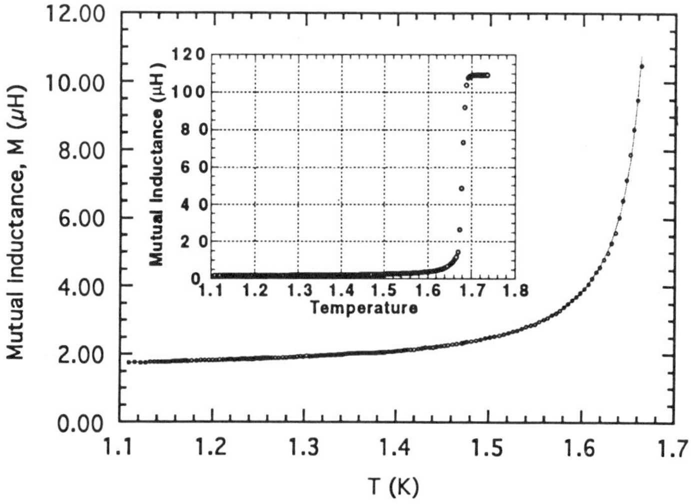

Figure 2. Mutual inductance versus temperature for two coils wrapped around a cylindrical 7 nm thick aluminum film. Inset shows the total mutual inductance change when the film is warmed to above the transition temperature.

Table 1. Apparent Critical Current in 7.0 nm Aluminum Film

Coil Arrangement	Inferred I_m (mA)	$T_c(I)$	I_m/I_c
No close coil	2.65	1.6902	1.5
No close coil	5.3	1.6883	1.4
No close coil	21.2	1.6803	1.2
Close coil not used	2.65	1.6874	0.52
Close coil not used	5.3	1.685	0.61
Close coil used as secondary	4.42	1.6782	0.20
Close coil used as secondary	21.2	1.6496	0.19
Close coil used as primary	31.8	1.595	0.086
Close coil self inductance	6.65	1.6607	0.094

MUTUAL INDUCTANCE RESULTS

The PDT was immersed in a liquid helium bucket dewar which was pumped to low temperatures. The primary coil was connected to a current source which provided a sine wave output at various frequencies and currents. The secondary was connected to a lock-in amplifier on which amplitude and phase were read. Temperature was determined by a calibrated germanium resistance thermometer.

Figure 2 shows a typical curve of mutual inductance versus temperature for the 7.0 nm film. We were concerned that the edges of the film would be non-uniform and contribute to a broadening of the transition. We see however, that the transition of even the thinnest (7.0nm) film is only a few millikelvin wide (inset, Figure 2). The transition temperatures of all the cylindrical and flat films as a function of thickness were also in good agreement with previously published data[5]. To limit the non-temperature-dependent part of the inductance, later data were taken with coils wrapped directly on the film. This is the case in Figure 2.

Excellent fits to the data were obtained using Eq. (2) up to a certain temperature limit. This limit varied for the different films and different excitation currents used. Current dependent deviations from Eq. (2) were noted at temperatures that were far enough away so that the approximation involved in Eq. (2) was still valid. This was taken to be the temperature at which the critical current in the film was exceeded, $T_c(I)$. To calculate this current, as a first approximation, the current in the coil was translated into a current in the film assuming a uniform current density. That is, the primary was assumed to be an ideal solenoid and the film was assumed to have an evenly distributed current that would completely cancel the field within the film cylinder. Table 1 shows the ratio of the measured current (inferred I_m) and calculated critical current, I_c (from Eq. (7) for the 7.0 nm cylindrical film. Since a sinusoidal excitation was applied, the values in the table are the peak currents. This may cause the measured current to be too high, explaining the excedance of the theoretical critical current in some cases.

The best agreement between these results was obtained for the primary and secondary coils not too close to the film. Once the close coil was wound onto the film to minimize L_o, the apparent critical current was lower, even if this close coil was not used as primary or secondary! It was lower still when that close wrapped coil was used for either the primary or secondary. At the same time, however, when the close wrapped coil was used the effective D value was lower, indicating better sensitivity. The explanation may be that the closeness of the coil produces both a very large local enhancement and and a large gradient of the applied field which can lower the critical current and decrease the effective length, D, respectively.

SQUID RESULTS

A schematic of the PDT connected to a SQUID is shown in Figure 1. The primary coil (not shown in the figure) surrounds the cylindrical film and secondary coil, providing a nearly uniform field in the axial direction. The primary coil is charged while the secondary coil has its current zeroed. The film then has currents shielding the inside of the cylinder, decreasing the effective inductance of the secondary coil. The primary and secondary coils are then allowed to persist. As the current in the film changes due to temperature changes, the secondary current changes to compensate. The secondary current is connected to the SQUID input. For the data shown the secondary coil is located as close as possible to the solenoid to limit the stray inductance, L_o.

A typical run was made by storing the current as above at temperatures well below the superconducting transition and letting the PDT slowly warm through T_c. (See Figure 3.) Depending on the amount of current initially stored in the primary, at a certain temperature the sensitivity of the PDT would rapidly increase. If the temperature was then lowered, a hysteresis in the output was noted. Upon once again raising the temperature, the sensitivity would be the new lower value until the highest temperature previously reached was exceeded. At this point the sensitivity would suddenly increase to the slope that the output versus temperature had before lowering the temperature.

These observations are consistent with a critical current being exceeded in the film. Once the critical current was exceeded, flux would become trapped within the cylinder lowering the current in the film back to the critical current. This process would result in a comparatively larger change in the SQUID output. For decreasing temperature, the critical current density of the film increases (Eq. 7), so the film's critical current would not be exceeded and the sensitivity would be reproducible. However, whenever some of the initial current in the film was lost through exceeding the critical current, the sensitivity would be permanently lower. While this accounts qualitatively for the observed behavior, the calculated current in the film where this hysteresis appears is substantially below the calculated critical current for a uniform current density in the film. One possible explanation for this is that the current carried by the secondary produces a large local field next to each wire. This local field then suppresses the current carrying ability of the film in that area. As evidence of this we noted that smaller currents had to be used when making mutual inductance measurements using the close coil than with a primary coil whose wires were 2 mm or more away.

The validity of equation (6) was tested by the fit shown in Figure 4. The data are derivatives of the non-hysteretic data of a run similar to that shown in Figure 3. Also, for one set of data the secondary current was zeroed at T= 1.64 K and the sensitivity remeasured without warming past 1.64 K. That is, the new T_i of equation (6) was changed from the initial current installation temperature of 1.3 K (+ in Figure 4) to 1.64 K (x in Figure 4). Improved sensitivity was seen as predicted by equation (6).

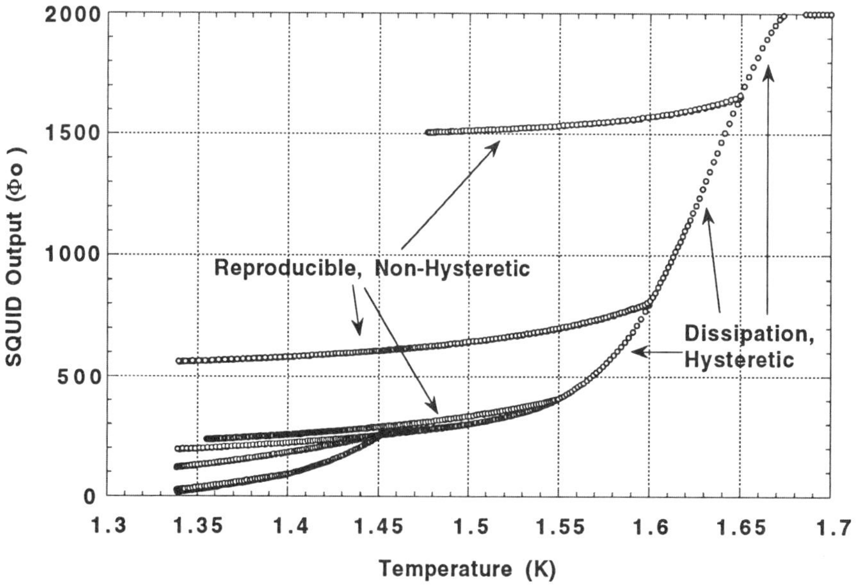

Figure 3. PDT SQUID output for 7 nm thick aluminum film. Data was taken by increasing then decreasing the temperature, advancing to a higher maximum temperature each time. Hysteresis is indicated when the decreasing temperature data had a lower slope (lower sensitivity) then the data taken while increasing the temperature.

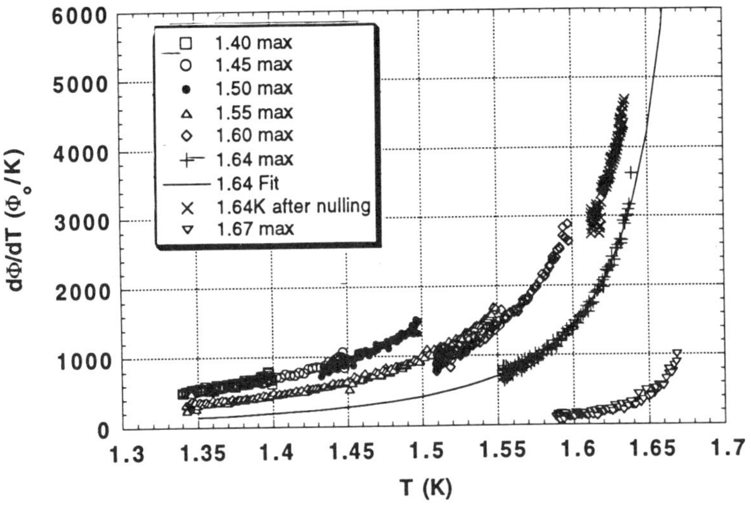

Figure 4. Temperature sensitivity of the PDT as read out by a SQUID. These data are derivatives of the non-hysteretic data like that in Figure 3. For comparison, the SQUID background noise is $4\text{x}10^{-6}\ \phi_0$. Fit uses equation (6) with $T_i = 1.35$ K.

The background SQUID noise was measured after storing a current in the primary circuit and recording the SQUID output of the secondary while the PDT was above the aluminum film transition temperature and, hence, temperature insensitive. The results matched the manufacturer's numbers for the intrinsic SQUID noise of $4\text{x}10^{-6}\ \phi_0\ \text{Hz}^{-1/2}$.

Extensive tests were made on the 7 nm film solenoid. The temperature sensitivity at various maximum temperatures was measured. Given the intrinsic SQUID noise above, a temperature resolution of down to 0.9 nK can be inferred from Figure 4.

The PDT was located in an isolated liquid helium tank within a standard bucket dewar. Unfortunately a leak in the vacuum space between the isolated tank and the bucket dewar provided a strong thermal link between the two. This in turn led to a relatively high thermal noise of 15 nK $\text{Hz}^{-1/2}$. This thermal noise was the limiting factor in demonstrating the above sensitivity of the PDT.

FUTURE WORK

A new test apparatus is being constructed with multiple temperature controlled stages to allow testing of the real temperature noise of the PDT.

It appears that high fields may exist at the film due to the discreteness of the coil windings. A possible improvement to this would involve depositing a flat coil directly onto an insulator in direct contact with the film. This will provide the close spacing needed to limit L_0 and maintain small D, but would have smaller fringing fields. This would allow much higher currents to be used. Another area for future improvement lies in the SQUID readout. Recent advances in two-stage SQUIDs[6] would allow an improvement of about a factor of three in current, and hence, temperature, resolution. The thermometer could also be increased in size, with sensitivity increasing as the cross-sectional area of the film.

With a T_c of 3.4 K, indium would allow operation at higher temperatures. We have had difficulty making thin, continuous indium films, however. Further work with different substrates and coatings will be done to try to remedy this problem.

It is envisioned that the PDT in somewhat altered form would have other uses. Because of its potential compactness and low heat capacity, a version could be made for bolometry. AC techniques could be used here which would allow operation in the hysteretic region, improving the temperature sensitivity by at least an order of magnitude. Inductance modulation at the input of a SQUID to reduce the 1/f SQUID noise is also a potential use[7].

SUMMARY

A PDT has been constructed using a thin aluminum film which has an ultimate resolution of better than 10^{-9} K/√Hz. The sensitivity of this device versus temperature is well explained using a recent theory by Pippard. The primary limitation on the temperature sensitivity of the device appears to be critical currents in the film. This limitation is being explored along with improvements in the SQUID readout and test apparatus with a goal of temperature resolution of 10^{-10} K/√Hz near the lambda transition in helium.

ACKNOWLEDGMENT

The authors thank T. Hait for expert assistance in the fabrication and wiring of the PDT. This work was supported by NASA's Microgravity Science and Applications Division through the Advanced Technology Development program.

REFERENCES

1. T.C.P. Chui, D.R. Swanson, M.J. Adriaans, J.A. Nissen, and J.A. Lipa, Temperature fluctuations in the canonical ensemble, Phys. Rev. Lett. 69:3005 (1992).
2. J.A. Lipa, D.R. Swanson, J.A. Nissen, and T.C.P. Chui, Lambda point experiment in microgravity, Cryogenics 34:341 (1994).
3. A.B. Pippard, Magnetic field penetration through a superconducting film, Superconducting Science and Technology, p.696 (1994).
4. see, for example, J. Romijn, T.M. Klapwijk, M.J. Renne, and J.E. Mooij, Critical pair-breaking current in superconducting aluminum strips far below Tc, Phys. Rev. B 26:3648 (1982).
5. see, for instance, M. Strongin and O.F. Kammerer, Superconductive phenomena in ultrathin films, J. Appl. Phys. 39:2509 (1968); and O.A.E. Cherney and J. Shewchun, Enhancement of superconductivity in thin aluminum films, Can. J. Phys. 47:1101 (1969).
6. R.P. Welty and J.M. Martinis, Two-stage integrated SQUID amplifier with series array output, IEEE Trans. Appl. Superconductivity, 3:2605 (1993).
7. B. Thorsen, private communication.

A SINGLE–BLOCK TRL TEST FIXTURE FOR THE CRYOGENIC CHARACTERIZATION OF PLANAR MICROWAVE COMPONENTS

M. Mejía [1*], A. S. Creason[1**], S. S. Toncich[2], B. T. Ebihara[1], and F. A. Miranda[1]

[1] National Aeronautics and Space Administration
Lewis Research Center, Cleveland, OH 44135, USA
[2] Bird Electronics
Solon, OH 44139, USA

ABSTRACT

The High–Temperature–Superconductivity (HTS) group of the RF Technology Branch, Space Electronics Division, is actively involved in the fabrication and cryogenic characterization of planar microwave components for space applications. This process requires fast, reliable, and accurate measurement techniques not readily available. A new calibration standard/test fixture that enhances the integrity and reliability of the component characterization process has been developed. The fixture consists of 50 Ω thru, reflect, delay, and device under test gold lines etched onto a 254 μm (0.010") thick alumina substrate. The Thru–Reflect–Line (TRL) fixture was tested at room temperature using a 30 Ω, 7.62 mm (300 mil) long, gold line as a known standard. Good agreement between the experimental data and the data modelled using Sonnet's ***em***© software was obtained for both the return (S_{11}) and insertion (S_{21}) losses. A gold two-pole bandpass filter with a 7.3 GHz center frequency was used as our device under test (DUT), and the results compared with those obtained using a Short–Open–Load–Thru (SOLT) calibration technique. Minimum insertion losses of 3.85 dB and 5.41 dB were measured with the TRL and SOLT calibration techniques, respectively. Room temperature return losses of 25.56 dB and 16.86 dB were measured using the TRL and SOLT calibration techniques, respectively. When used to perform calibrations and testing at cryogenic temperatures, the TRL fixture exhibited insertion and return loss data superior to its SOLT counterpart. A detailed description of the cryogenic components of the TRL fixture as well as results of their cryogenic characterization are presented.

---

[*] FJFP/NASA student intern; University of Pennsylvania, Philadelphia PA 19104 USA.
[**] OAI/NASA student intern; Ohio Northern University, Ada OH 45810 USA.

INTRODUCTION

Microwave systems for space applications, such as front end receivers, demand low noise, low loss components with stringent frequency response properties. Therefore, the individual microwave components that make up a particular system should be thoroughly characterized, so that their intrinsic performance (i.e., excluding the effects of the matching network) can be evaluated. It has been shown that the performance of semiconductor–based microwave components improves at cryogenic temperatures[1]. Also, it has been demonstrated that at cryogenic temperatures, High–Temperature–Superconductors (HTS) have greatly reduced conductor losses in microwave components with respect to those fabricated using normal metal conductors such as gold and copper[2]. Therefore, the overall operation of microwave systems can be improved by operating at cryogenic temperatures (e.g., 77 K), and by optimizing individual component performance at those temperatures. In this paper, we describe the design, fabrication, and testing process of a new Thru–Reflect–Line (TRL) calibration method that will allow for the fast and accurate characterization of planar microwave components at cryogenic temperatures. Results on the performance of this fixture at room and cryogenic temperatures will be presented.

EXPERIMENTAL APPROACH

An Automatic Network Analyzer (ANA) is commonly used to characterize microwave components. The ANA, however, must be calibrated to establish reference planes before any measurements can be taken. These reference planes indicate how far along the system the calibration will take into account measuring network and test–fixture losses (e.g., losses due to coaxial cables, coaxial–to–microstrip transitions, wire bonding connections, etc.), so that the losses of the device–under–test (DUT) can be determined more accurately. Often used in the characterization of microwave components, the TRL calibration establishes the reference planes beyond the electrical launchers, accounting for losses due to the coaxial–to–microstrip transition. The TRL calibration technique is typically implemented in the "split–block" configuration shown in Figure 1. While this calibration works well at a room temperature, its split–block configuration proves cumbersome for measurements at cryogenic temperatures and under vacuum conditions. For example, each calibration step would require a different block to be wire–bonded, placed under vacuum, cooled to the desired cryogenic temperatures, recorded for measurements, and returned to room temperature and pressure conditions[3]. Due to the cycling, multiple wiring, and possible drifting of the ANA reference levels, such a process would be time consuming and more prone to error. To improve the calibration process at cryogenic temperatures, the following experimental approach was initiated:

- Develop a TRL calibration standard and test fixture that would:
  1. Eliminate repeated temperature cycling and vacuum disruption (speed).
  2. Allow for electromagnetic coupling repeatability (accuracy).
  3. Have a simple and inexpensive implementation (affordability).
- Optimize TRL calibration at room temperature.
- Optimize TRL calibration at cryogenic temperatures.

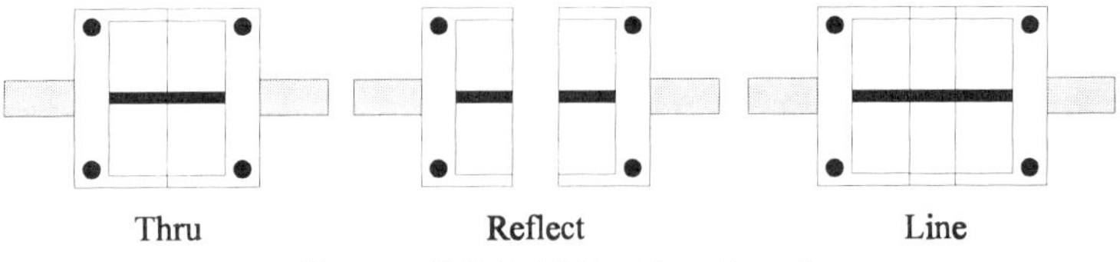

Figure 1. TRL "split-block" configuration.

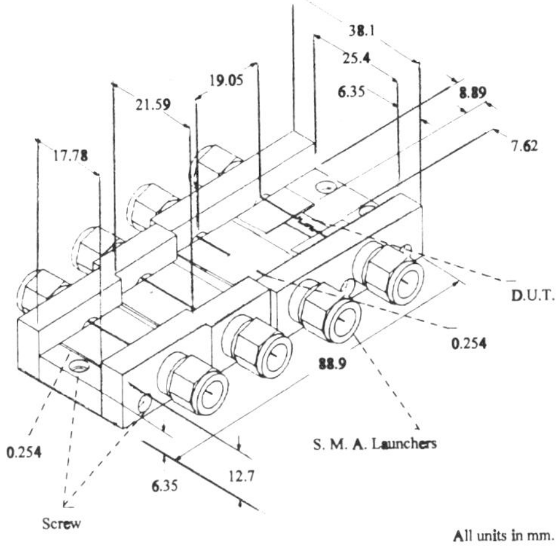

Figure 2. "Single-block" TRL schematic.

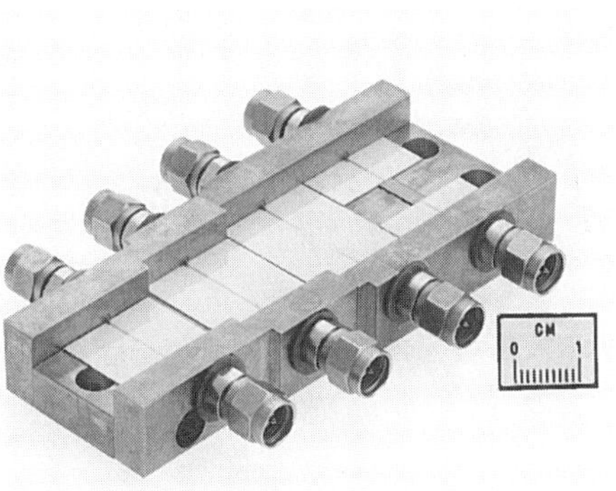

Figure 3. TRL test fixture.

The above approach led to a single–block TRL design. A schematic representation and the actual fixture are shown in Figures 2 and 3, respectively. The fixture consists of 254 μm (0.010") wide thru, reflect, delay and DUT gold lines. The 50 Ω input/output lines were patterned on a 254 μm (0.010") thick alumina substrate (88.9 x 38.1 mm (3.5" x 1.0")). A brass block supports the SMA launchers used for the coaxial–to–microstrip transition. Each microstrip line was split and mounted at 19.05 mm (0.75") intervals onto the brass piece using silver–based conductive epoxy. For the measurements, modifications corresponding to our selected standards were introduced into the TRL calibration routine of the HP-8510C ANA[4].

For the cryogenic testing, a custom made vacuum chamber was fabricated. The aluminum chamber was designed so that it can be mounted on the second stage of a closed–cycle helium gas refrigerator. It also features feed–throughs to accommodate the coaxial cables necessary to couple the microwave signal from the ANA with the TRL test fixture. Figures 4–6 show the interior and exterior of the chamber and the experimental set up for cryogenic testing. The 3.5 mm coaxial cables used to couple the TRL test fixture with the microwave signal originating in the ANA were custom made and consisted of a silver plated beryllium copper inner conductor, a Teflon (PTFE) dielectric, and a stainless steel outer conductor as shown in Figure 7. Fifty-six cables were fabricated to meet the electrical specifications of a return loss (S_{11}) better than 18.25 dB and an insertion loss (S_{21}) less than 0.7 dB in the 0.0–10.0 GHz frequency range. The test fixture was mounted inside the vacuum chamber,

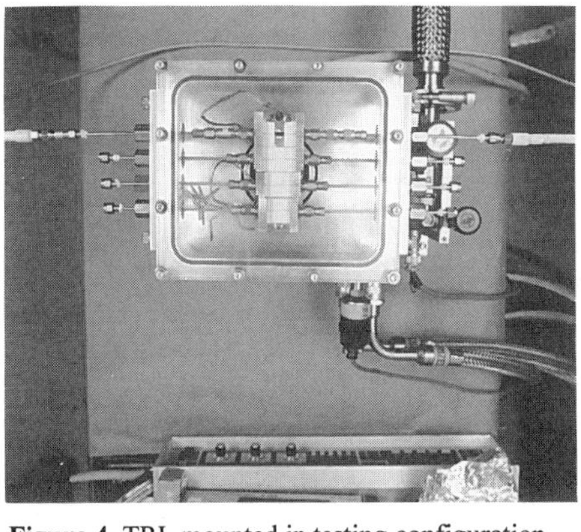

Figure 4. TRL mounted in testing configuration.

Figure 5. Vacuum chamber and refrigerator.

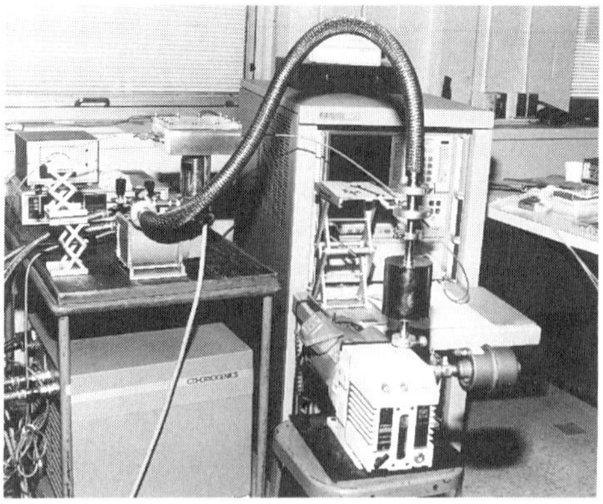

Figure 6. TRL testing set up.

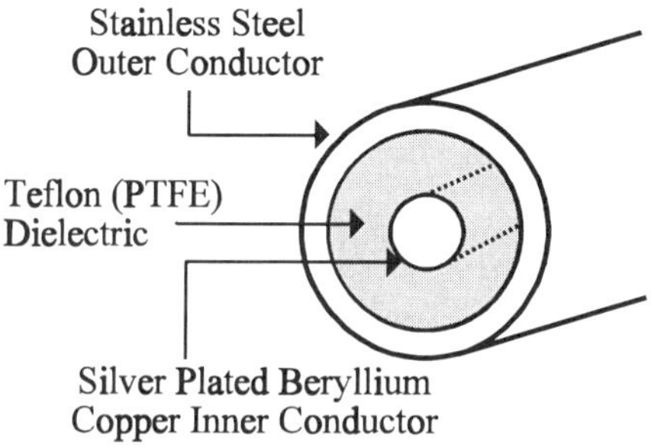

Figure 7. Custom made coaxial cable.

which was then tested to see if it could maintain vacuum conditions both at room and at cryogenic temperatures. A vacuum quality epoxy (Torr Seal ®) was applied to the joint of the coaxial cable and SMA connectors to protect against possible vacuum leakage.

RESULTS

The single–block TRL test fixture was tested at room temperature by using a 30 Ω, 7.62 mm (300 mil) long, gold line as the DUT. S_{11} and S_{21} for this line were measured, and the experimental data compared with the modelled data for the line using Sonnet's ***em***© software[5]. The experimental and the

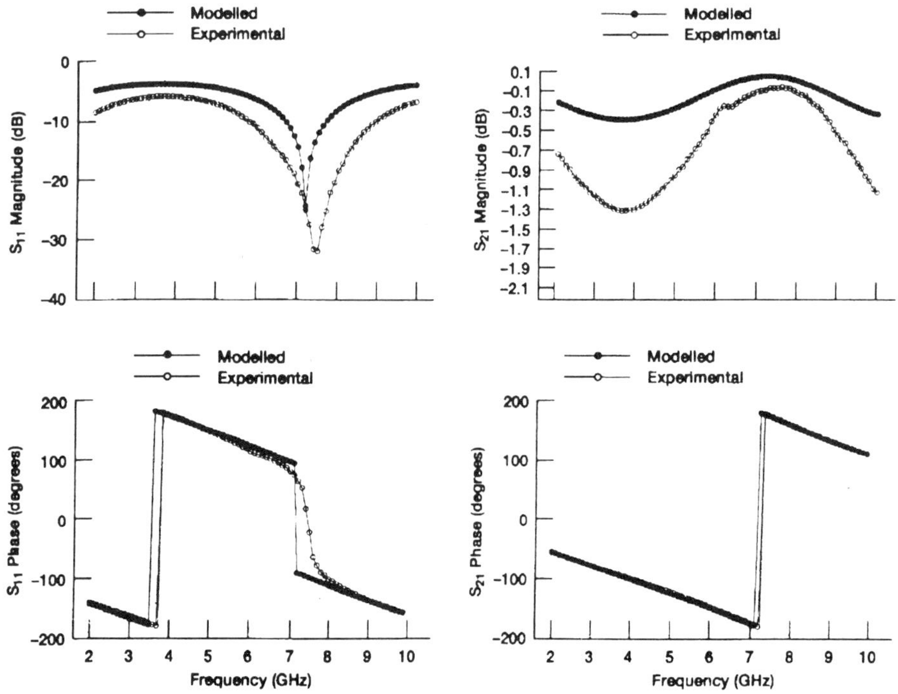

Figure 8. Insertion loss (S_{21}) and return loss (S_{11}) vs. frequency for a 30 Ω, 7.62 mm (300 mil) long, gold line on 254 μm (10 mil) thick alumina at 298 K.

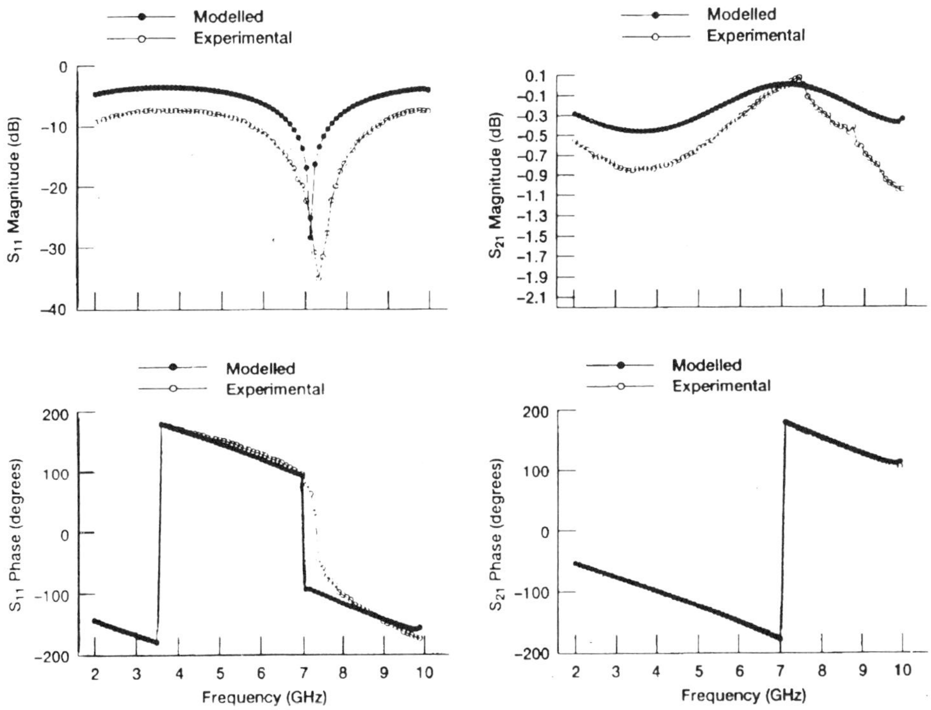

Figure 9. Insertion loss (S_{21}) and return loss (S_{11}) vs. frequency for a 30 Ω, 7.62 mm (300 mil) long, gold line on 508 μm (20 mil) thick alumina at 298 K.

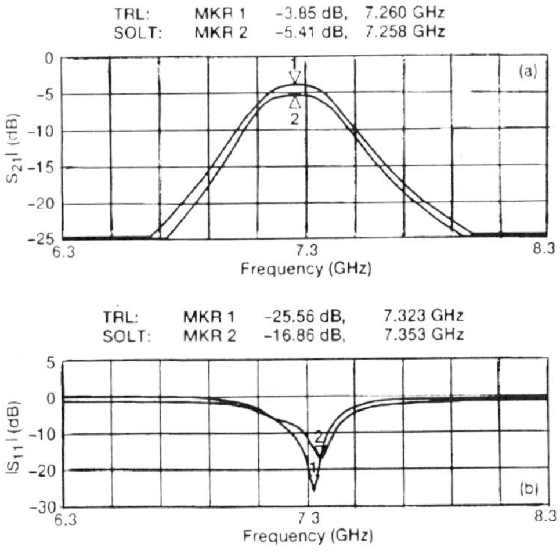

Figure 10. Insertion (S_{21}) and return (S_{11}) losses vs. frequency for a gold two-pole bandpass filter with a 7.3 GHz center frequency. Both calibrations were performed at room temperature and outside the vacuum chamber.

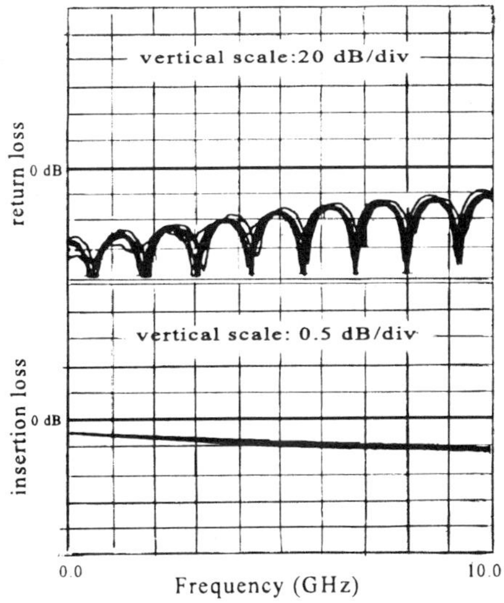

Figure 11. Insertion (S_{21}) and return (S_{11}) for coaxial cables.

Table 1. Pressure and temperature data for experimental set up

Time	Sensor A *	Sensor B**	Pressure
t=0	298 K	296 K	ATM
t=40 minutes ***	298 K	296 K	4.0 Pa (30 millitorrs)
t=3 hours	58 K	32 K	<0.133 Pa (<1 millitorr)

* mounted on test fixture
** mounted underneath of refrigerator cold finger
*** refrigerator turned on

modelled data for the magnitude and phase of S_{11} and S_{21} are shown in Figures 8 and 9. Observe that good agreement exists between the experimental and modelled data, not only for the calibration circuit on the 254 μm (10 mil) thick alumina substrate, but also for a similar circuit on a 508 μm (20 mil) thick alumina substrate. The slight discrepancies between the experimental and modelled data could be due to uncertainties in the line's dimensions and unaccounted for radiation losses.

The repeatability of the TRL calibration standard was tested by reassembling the thru standard immediately after calibration. This resulted in a return loss better than 39 dB and an insertion loss of less than 0.02 dB from 0.0 to 10.0 GHz.

The next step in the verification of the TRL fixture was to compare its performance at room temperature with that of a standard Short–Open–Load–Thru (SOLT) calibration, a method that is frequently used in the characterization of microwave components. For this test, a gold two–pole bandpass filter, designed to exhibit a 7.3 GHz center frequency, was used as a DUT[2]. The test fixture was calibrated outside the vacuum chamber using both the SOLT and TRL methods and the filter was tested in each case. As shown in Figure 10, the quality of data measured using the TRL calibration surpasses that of its SOLT counterpart, showing improvement on insertion losses near 1.5 dB and on return losses near 10 dB at the center frequency. The observed deviations between the designed and experimental center frequencies are most probably due to filter fabrication issues[2].

In preparation for testing the fixture at cryogenic temperatures, each coaxial cable was tested to determine its return and insertion losses. Typical results for a set of eight cables are shown in Figure 11. Note that the cable–to–cable signal propagation properties are very similar and satisfy the aforementioned specifications for return and insertion losses. Afterwards, the cables were assembled in the vacuum chamber (see Figures 4 and 5) and the TRL fixture was mounted as described earlier. The whole system was then tested to determine the quality of the vacuum and cooling properties. Complete, the TRL chamber held satisfactory vacuum conditions and reached acceptable cryogenic temperatures (see Table 1).

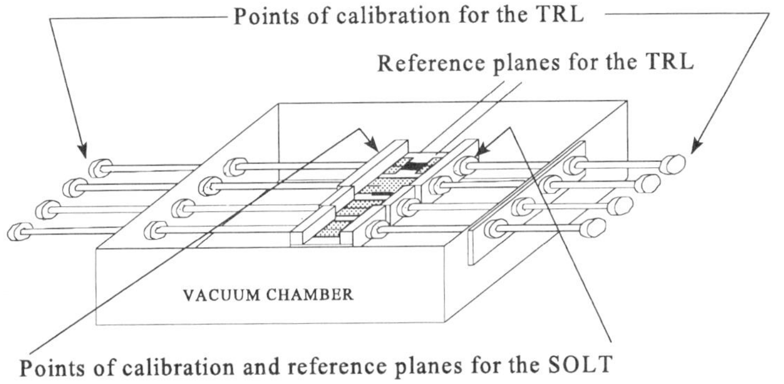

Figure 12. Points of calibration and established reference planes for the TRL and SOLT calibration techniques.

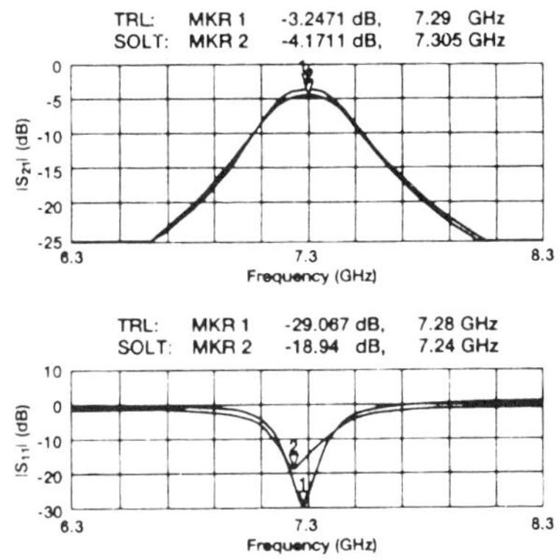

Figure 13. Insertion (S_{21}) and return (S_{11}) losses vs. frequency for a gold two-pole bandpass filter with a 7.3 GHz center frequency. Both calibrations were performed at room temperature as shown in Figure 12.

Table 2. Insertion (S_{21}) and return (S_{11}) losses for the TRL and SOLT at various temperatures

Calibration		298 K	100 K	77 K
TRL	Insertion Loss (S_{21})	-3.2471 dB	-2.5173 dB	-1.876 dB
SOLT		-4.1711 dB	-3.1511 dB	-2.0143 dB
TRL	Return Loss (S_{11})	-29.067 dB	-27.588 dB	-18.957 dB
SOLT		-18.94 dB	-15.543 dB	-13.868 dB

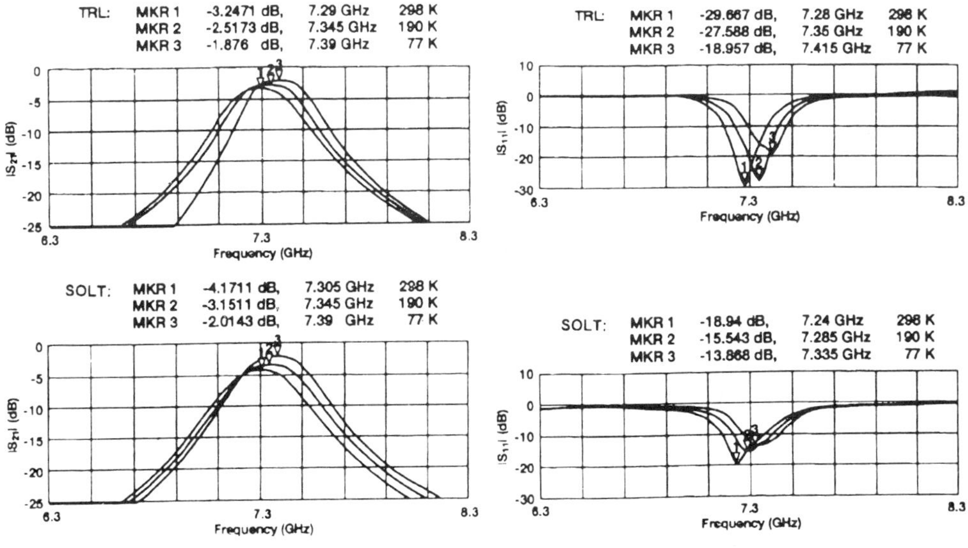

Figure 14. Insertion (S_{21}) and return losses (S_{11}) vs. frequency for a gold two-pole bandpass filter with a 7.3 GHz center frequency as a function of temperature.

To begin the initial verification of the TRL calibration at cryogenic temperatures, the test fixture was calibrated at room temperature using both the SOLT and the TRL techniques. The SOLT calibration was performed inside the vacuum can so that the reference planes were established at the plug (male) end of the SMA launchers. The TRL calibration was performed outside the vacuum chamber outside the vacuum chamber at the end of the coaxial cables (See Figure 12). For this test, a different filter of the same aforementioned type was used. Figure 13 shows room temperature results for the filter using the TRL and the SOLT calibrations. As for the previous case, the quality of the insertion and return losses measured using the TRL fixture is better than that obtained using the SOLT calibration. This remains true when comparing the data measured after calibrating at several cryogenic temperatures using the TRL technique with data obtained at the same cryogenic temperatures using the room temperature SOLT calibration (see Figure 14 and Table 2). These results demonstrate that the single-block TRL calibration technique can be effectively used for the fast and reliable characterization of microwave components at room temperature, as well as at cryogenic temperatures. Further improvements of this fixture, such as modifications to allow for the characterization of DUTs of different sizes, and the integration of solid state switches outside the vacuum chamber that would allow for the electronic switching of the standards without altering coaxial connections between the ANA and the TRL setup, are under way. This should further enhance the calibration and measurement speed, as well as the electromagnetic coupling repeatability.

CONCLUSIONS

A single-block TRL test fixture for the cryogenic characterization of planar microwave components has been designed, fabricated, and tested. The performance for this fixture was superior to that of a commonly used SOLT calibration, both at room and at cryogenic temperatures. In addition, this novel TRL fixture eliminates the repeated temperature cycling and vacuum disruption unavoidable in the performance of calibrations at cryogenic temperatures using other techniques such as the "split-block" TRL configuration. This results in a faster, more accurate, and more affordable calibration and measurement process for the evaluation of the intrinsic properties of individual microwave components. Modifications to allow for the characterization of devices of different sizes and the performance of electronic switching are being considered.

REFERENCES

1. M.W. Pospieszalski, S. Weinreb, R.D. Norrod, and R. Harris, FET's and HEMT's at cryogenic temperatures–their properties and use in low–noise amplifiers, *IEEE Trans. Microwave Theory Tech.* 36: 552–560 (1988).
2. F.A. Miranda, S.S. Toncich, and K.B. Bhasin, Performance of two–pole bandpass filters photodefined on double–sided Y-Ba-Cu-O and Tl-Ba-Ca-Cu-O thin films, *Microwave and Optical Tech. Lett.* 6: 752–755 (1993).
3. J.W. Smuk, M.G. Stubbs, J.S. Wight, S-Parameter characterization and modeling of three-terminal semiconductive devices at cryogenic temperatures, *IEEE Microwave and Guided Wave Letters* 2:111-113 (1992).
4. Product Note HP-8510-5A, Network Analysis: Specifying Calibration Standards for the HP 8510 network analyzer (1991).
5. ***em***©, Release 2.4. Sonnet Software, Inc. Liverpool NY (1993).

AUTOMATION OF DATA ANALYSIS PROGRAMS USED IN THE CRYOGENIC CHARACTERIZATION OF SUPERCONDUCTING MICROWAVE RESONATORS

A. S. Creason* and F. A. Miranda

National Aeronautics and Space Administration
Lewis Research Center
Cleveland, OH 44135, USA

ABSTRACT

Knowledge of the microwave properties at cryogenic temperatures of components fabricated using High-Temperature-Superconductors (HTS) is useful in the design of HTS-based microwave circuits. Therefore, fast and reliable characterization techniques have been developed to study the aforementioned properties. In this paper, we discuss computer analysis techniques employed in the cryogenic characterization of HTS-based resonators. The revised data analysis process requires minimal user input, and organizes the data in a form that is easily accessible by the user for further examination. These programs retrieve data generated during the cryogenic characterization at microwave frequencies of HTS based resonators and use it to calculate parameters such as the loaded and unloaded quality factors (Q and Q_0, respectively), the resonant frequency (f_0), and the coupling coefficient (κ), which are important quantities in the evaluation of HTS resonators. While the data are also stored for further use, the programs allow the user to obtain a graphical representation of any of the measured parameters as a function of temperature soon after the completion of the cryogenic measurement cycle. Although these programs were developed to study planar HTS-based resonators operating in the reflection mode, they could also be used in the cryogenic characterization of two ports (i.e., reflection/transmission) resonators.

INTRODUCTION

The discovery of High-Temperature-Superconductors (HTS) in 1986 has resulted in passive planar microwave devices such as resonators and filters with higher quality factors

---

* NASA/OAI summer intern, Ohio Northern University, Ada, OH 45810, USA.

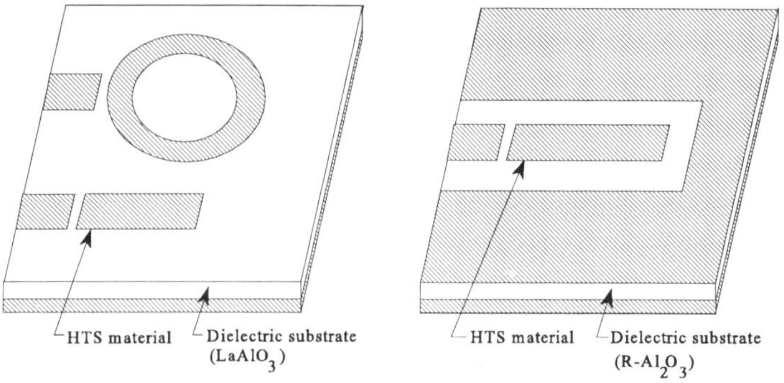

Figure 1. YBCO microstrip linear, ring, and conductor-backed coplanar waveguide resonators deposited on 1x1 cm, 508 μm (20 mil) thick, dielectric substrates.

and lower insertion loss than those fabricated using normal conductors [1-3]. Significant attention has been given to the processing and characterization of HTS-based planar microwave components as efforts continue towards the inclusion of HTS-based circuits in working systems for space- and ground-based communications. Therefore, fast and accurate characterization and more advanced data analysis capabilities become extremely important in this endeavor. This paper deals with the data analysis programs used in our laboratory in the study of HTS-based microwave resonators. Extensive modifications of preexisting programs (written in-house for data analysis purposes) were performed in order to expedite the characterization of these resonators without sacrificing accuracy. The structure of these programs will be described, and data taken using $YBa_2Cu_3O_{7-\delta}$ (YBCO) resonators on lanthanum aluminate ($LaAlO_3$) and R-plane sapphire (R-Al_2O_3) substrates will be presented.

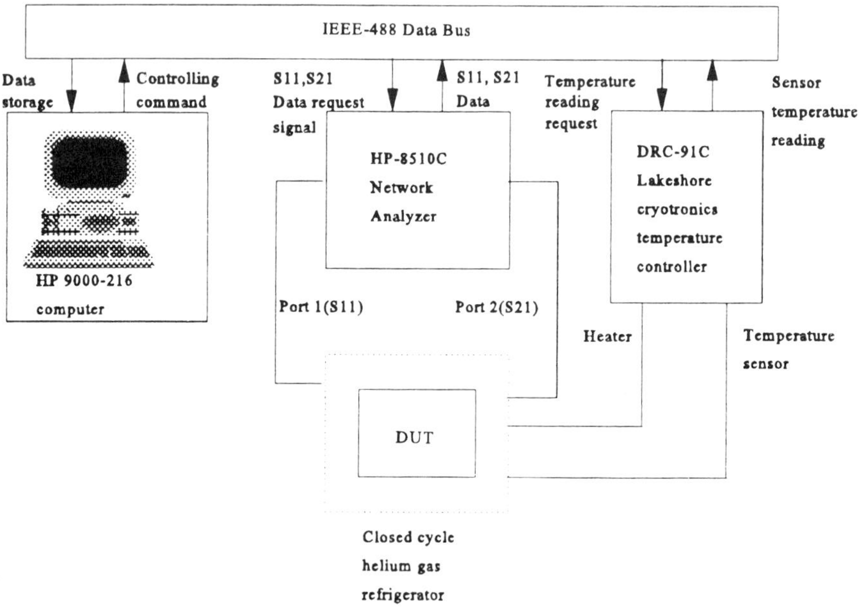

Figure 2. Experimental setup used in cryogenic testing of HTS-based resonators.

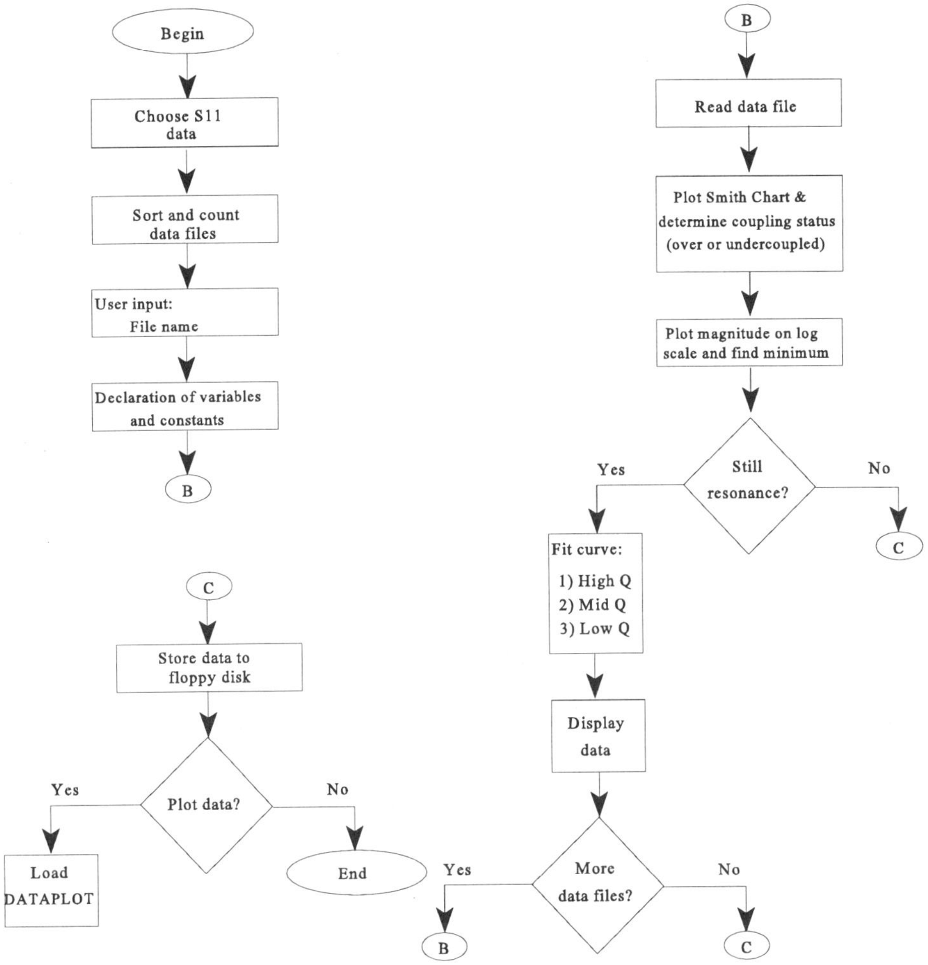

Figure 3. Flowchart for data analysis program 'QANALYSIS'.

EXPERIMENTAL

Microstrip linear and ring resonators, as well as conductor-backed coplanar waveguide (CBCPW) resonators were fabricated on YBCO thin films deposited by laser ablation on $LaAlO_3$ substrates and on R-Al_2O_3 substrates (with CeO_2 buffer layers), respectively, using standard photolithography and chemical etching techniques. A schematic of these resonators is shown in Figure 1. These resonators were tested by measuring the reflection scattering parameter (S_{11}), using the experimental configuration shown in Fig 2 [4]. Briefly, the setup consists of an HP-8510C network analyzer, an HP 9000-216 computer, and a DRC-91C Lake Shore cryotronics temperature controller all coupled through the use of an IEEE-488 data bus, along with a closed-cycle-helium-gas refrigerator containing the device under test (DUT), which allows for cryogenic measurements in the temperature range from 300K to 20K. The data measured through the use of this system were then used by the data analysis programs to determine the loaded and unloaded quality factors (Q and Q_o, respectively), the resonant frequency (f_0), and the coupling coefficient (κ) of the resonator.

Figure 3 shows a flowchart for the data analysis program we call 'QANALYSIS', after modification and automation. When automating this program, the first step was to have the computer choose the files to be analyzed from the disk. In this way, the user would not have to be familiar with the format used in naming the files in order to choose those which are important in analyzing the data. Data files are named in the following way. The first four characters are always 'DATA'. The next 2 or 3 characters are numerals which show the temperature at which the data was taken. The next character is either a 'W' or a 'C', showing whether the data were taken during the warming or cooling cycle. The final two characters are either '11' or '21', indicating whether the data are S_{11} (reflection) or S_{21} (transmission). Thus, 'QANALYSIS' can be used for the analysis of one port (reflection mode) as well as two ports (reflection/transmission modes) resonators. Since in this paper we are dealing with one port HTS-based resonators, 'QANALYSIS' catalogs the disk looking for files with the first four characters 'DATA' and the last two '11'. It counts all of the files which fit that requirement, and sets up a counting loop which will analyze only those particular files. When all of the appropriate files have been analyzed, the program ends. The program may terminate execution before all of these files have been analyzed, if certain conditions exist in the data, as will be described later in this paper.

After selecting the data files corresponding to the particular temperatures of interest, the data for the first file in the array is displayed as a Smith Chart plot on the computer's CRT screen. The Smith Chart[5] facilitates the graphical solution of transmission line problems and consists of a plot of the normalized impedance, given by

$$Z/Z_o = r + jx \tag{1}$$

with the magnitude and phase of a generalized complex reflection coefficient in a unity circle. The chart is useful for the analysis of lossless and lossy lines. It is used in this case to determine the coupling status of the resonator (i.e., overcoupled ($\kappa>1$), undercoupled ($\kappa<1$), or critically coupled($\kappa=1$)) where κ, the coupling coefficient, is defined as

$$\kappa = (1\pm|\Gamma_r|) / (1\mp|\Gamma_r|) \tag{2}$$

with the sign depending on the coupling status of the circuit. Γ_r is the reflection coefficient of the circuit at resonance[6]. For $\kappa=1$, $\Gamma_r=0$ indicating that the resonator is perfectly matched to the source. 'QANALYSIS' calculates the radius of the circle corresponding to the experimental data, then compares that value with that of the circle on the Smith Chart corresponding to critical coupling, which has a radius of 0.5. If the radius of the plot of the experimental data is less than 0.5, the circuit is undercoupled, and if it is over 0.5, the circuit is overcoupled. Next, the program plots the magnitude of the data on a logarithmic scale as a function of frequency. Depending on the quality of the resonator, this plot should show a minimum at or very near the resonant frequency for which the resonator was designed. 'QANALYSIS' then goes on to determine f_o. This step is implemented such that the correct resonance is selected; i.e., after finding the frequency at which a minimum point occurs, the program searches the entire frequency band under study for any other resonance. This accounts for cases in which a split resonance may occur, as for example, when testing microstrip ring resonators, where secondary or spurious resonances can appear due to variations in thickness, width, etc., of the line[7]. These secondary resonances could be larger than the primary resonance when the circuit is overcoupled. Note that with minor modifications, this searching capability could be applied in the analysis of multipole resonant circuits. The program also verifies that resonance still exits in the data; if the minimum point is not more than 3 dB below the value of the data where it is flat (i.e., at a frequency far from f_o), then the program indicates that a resonance no longer exists and the analysis is completed.

After the determination of f_o is complete, it is then used to calculate the loaded quality factor of the resonator, since

```
Q ANALYSIS FOR FILE:DATA77W11  76.7  KELVIN
RESONANT FREQUENCY IS:          7.96629619676 GHz
UNLOADED 'Q' IS:       1460.35560716
LOADED 'Q' IS:         449.515607637
COUPLING COEFFICIENT IS:        2.29360265696
COUPLING LOSS IS;      .0007639596078
FIT:       HIGH
HALF POWER SPAN IS:  .00539133883643      .0177219568385
```

Figure 4: Sample printout of data from 'QANALYSIS'.

$$Q = (f_o / \Delta f) \tag{3}$$

where Δf is the half power frequency span.

The final step in analyzing each data file is to fit a curve to the resonance. This assumes that the graph of the resonance resembles a parabola, and uses a second degree polynomial for the curve fitting. Depending on the shape of the curve, e.g. sharp, moderate, or shallow, the program will classify the fit as either High Q, Mid Q, or Low Q

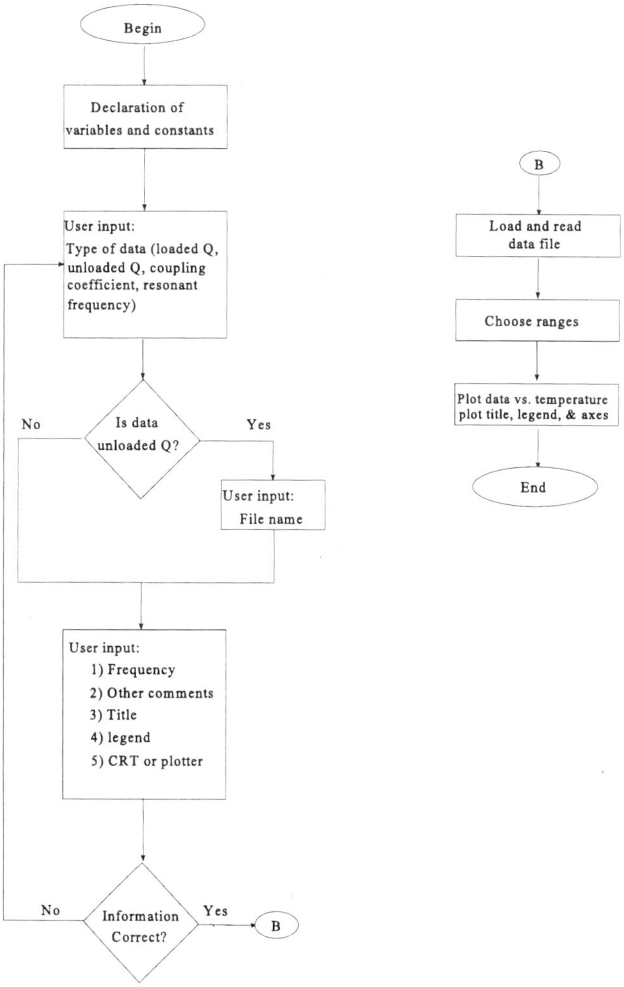

Figure 5. Flowchart for plotting program 'DATAPLOT'.

respectively, which will be used to aid in the determination of the unloaded quality factor. In calculating Q_o, the coupling effects must always be considered, regardless of the coupling state of the resonator. The coupling loss (σ), which modifies the location of the half power frequency points, can be found by

$$\sigma = (1-|\Gamma_\infty|) / (1+|\Gamma_\infty|) \tag{4}$$

where Γ_∞ is the reflection coefficient of the circuit far from f_o. In this program, Q_o is calculated in a similar manner to Q, except that for the calculation of Q_o, the half power frequency span is that obtained after correcting for coupling losses in the circuit. The program also calculates several other parameters, then prints a list which includes the file's name, the temperature, Q, Q_o, f_o, κ, σ, the data fit type, and the half power span (3 dB points) used for determining Q_o and Q, respectively, as seen in Figure 4. These parameters can be printed to the CRT or an external printer for further use. If there are more data files remaining, the process is repeated; if not, arrays of Q, Q_o, f_o, and κ versus temperature are stored to a floppy disk. The user is then given the option of plotting any of the stored parameters versus temperature or terminating the program. If the user wants to plot any of the stored quantities, then the program 'DATAPLOT' can be called to do so.

Figure 5 is a flowchart of 'DATAPLOT'. The user enters which of the parameters stored by ' QANALYSIS' is to be plotted (all are plotted as a function of temperature), the graph's title and legend, and whether the data is to be plotted to the CRT or to a plotter. Once the program has verified with the user that the information entered is correct, it loads the data from the chosen file and reads it. It determines the maxima and minima of the data set, then chooses a graphing range which will center the data on the graph, as well as making the graph easy to read. Finally, it plots the data and labels the graph with a title, legend, and any other user-specified comments.

Figure 6 shows an actual plot from 'DATAPLOT' of Q_o for both a linear and ring resonator of YBCO on $LaAlO_3$. Figure 7 shows plots of experimental data for a YBCO CBCPW resonator on $R\text{-}Al_2O_3$. These plots give an idea of the format and scope of the plots produced by 'DATAPLOT'. The reproducibility of these data was determined by performing multiple measurement cycles. The reproducibility of Q_o was found to be within two percent at 77 K, and within 1.5 percent at 20 K. The reproducibility from run-to-run for f_o was below 0.02 percent at the aforementioned temperatures.

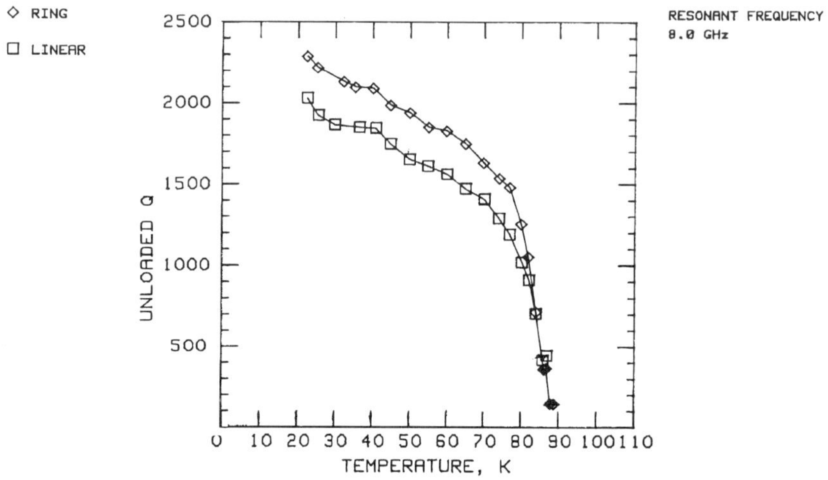

Figure 6. Sample data plot for the unloaded Q corresponding to YBCO linear and ring resonators.

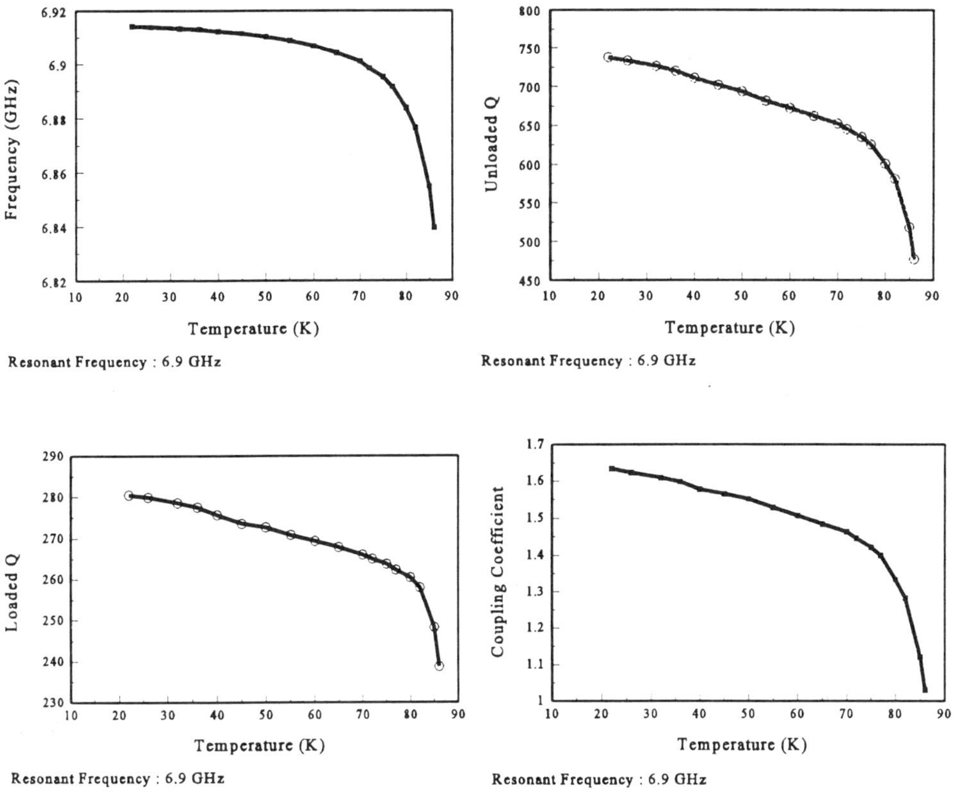

Figure 7. Plots of experimental data for a YBCO conductor backed coplanar waveguide resonator (CBCPW).

CONCLUSIONS

In the course of this work, automated data analysis programs which increase efficiency during the characterization of superconducting microwave resonators were developed. These programs significantly reduce the time commitment of the user in the analysis process, and provide quick access to the results of the cryogenic characterization of HTS-based resonators. Although these programs were originally developed for the study of one port (i.e., reflection mode) HTS-based resonators, they could also be used for the cryogenic characterization of two port (i.e., reflection/transmission) resonators.

ACKNOWLEDGMENTS

The authors wish to thank Ms. Donna Bohman for patterning the HTS circuits used in this study. The authors are grateful to the referees for their constructive comments and suggestions.

REFERENCES

1. F.A. Miranda, K.B. Bhasin, K.S. Kong, T. Itoh, and M.A. Stan, Conductor backed coplanar waveguide resonators of $YBa_2Cu_3O_{7-\delta}$ on $LaAlO_3$, *IEEE Microwave and Guided Wave Letters* 2:287 (1992).

2. S.H. Talisa et al., Low- and high-temperature superconducting microwave filters, *IEEE Trans. Microwave The. Tech.* 39:1448 (1991).
3. F.A. Miranda, Microwave properties of Tl-Ba-Ca-Cu-O thin films, *Supercond. Sci. Technol.* 6:605 (1993).
4. M. Zayas, Automated measurement system for the cryogenic characterization of planar microwave components, *Apogee-Proceedings of the Third Annual NASA/OAI Summer Intern Research Symposium* (1993).
5. T. Edwards, "Foundations for Microstrip Circuit Design," Wiley and Sons, New York (1992).
6. R.R. Romanofsky, Analytical and experimental procedures for determining propagation characteristics of millimeter-wave gallium arsenide microstrip lines, *NASA TP-2899*(1989).
7. C.M. Chorey, K.S. Kong, K. B. Bhasin, J. D. Warner, and T. Itoh, YBCO ring resona tors at millimeter-wave frequencies, *IEEE Trans. Microwave The. Tech.* 39:1480 (1991).

CHARACTERIZATION OF HYBRID FERROELECTRIC/HTS THIN FILMS FOR TUNABLE MICROWAVE COMPONENTS

M.D. Winters[1†], C.H. Mueller[2], K.B. Bhasin[1], and F.A. Miranda[1]

[1]National Aeronautics and Space Administration
Lewis Research Center
Cleveland, OH, 44135, U.S.A.
[2]Superconducting Core Technologies, Inc.
Golden, CO, 80401, U.S.A.

ABSTRACT

Since the discovery of High-Temperature-Superconductors (HTS) in 1986, a diversity of HTS-based microwave components has been demonstrated. Because of their low conductor losses, HTS-based components are very attractive for integration into microwave circuits for space communication systems. Recent advancements have made deposition of ferroelectric thin films onto HTS thin films possible. Due to the sensitivity of the ferroelectric's dielectric constant (ϵ_r) to an externally applied electric field (**E**), ferroelectric/superconducting structures could be used in the fabrication of low loss, tunable microwave components. In this paper, we report on our study of $Ba_{0.5}Sr_{0.5}TiO_3/YBa_2Cu_3O_{7-\delta}$ and $Ba_{0.08}Sr_{0.92}TiO_3/YBa_2Cu_3O_{7-\delta}$ ferroelectric/superconducting thin films on lanthanum aluminate ($LaAlO_3$) substrates. For the (Ba:Sr, 0.50:0.50) epitaxial sample, a ϵ_r of 425 and a loss tangent (tanδ) of 0.040 were measured at 298 K, 1.0 MHz, and zero applied **E**. For the same sample, a ϵ_r of 360 and tanδ of 0.036 were obtained at 77 K, 1.0 MHz, and zero applied **E**. Variations in ϵ_r from 180 to 360 were observed over an applied **E** range of $0V/cm \leq \mathbf{E} \leq 5.62x10^4 V/cm$ with little change in tanδ. However, the range of ϵ_r variation for the polycrystalline (Ba:Sr, 0.08:0.92) sample over $0V/cm \leq \mathbf{E} \leq 4.00x10^4 V/cm$ was only 3.6 percent while tanδ increased markedly. These results indicate that a lack of epitaxy between the ferroelectric and superconducting layers decreases tuning and increases microwave losses.

[†]FJFP/NASA summer intern; Department of Electrical Engineering, Cornell University, Ithaca, NY, 14853, U.S.A.

Advances in Cryogenic Engineering, Vol. 41
Edited by P. Kittel, Plenum Press, New York, 1996

INTRODUCTION

Since the discovery of High-Temperature-Superconductors (HTS) in 1986, it has been possible to fabricate HTS-based planar microwave components such as resonators and filters[1]. The use of HTS films in place of normal conductors (e.g. gold and copper) has greatly reduced conductor losses and consequently insertion losses[2]. Therefore, significant attention has been given to the quality and processing of HTS films as efforts continue towards the inclusion of HTS-based circuits in working systems. One of the areas that has received considerable attention for the insertion of HTS-based components is that of tunable microwave circuits[3]. This requires the development of hybrid superconductor/ferroelectric structures. Ferroelectric materials have a dielectric constant (ϵ_r) that can be altered by an externally applied electric field (**E**). Therefore, optimization of these structures' material properties and careful characterization of their electrical properties should result in high quality, low loss tunable microwave components, such as capacitors, filters, and phase shifters. In this paper, we report on our study of $Ba_{0.5}Sr_{0.5}TiO_3/YBa_2Cu_3O_{7-\delta}$ and $Ba_{0.08}Sr_{0.92}TiO_3/YBa_2Cu_3O_{7-\delta}$ ferroelectric/superconducting thin films on lanthanum aluminate ($LaAlO_3$) substrates.

EXPERIMENTAL

The $Ba_{0.5}Sr_{0.5}TiO_3/YBa_2Cu_3O_{7-\delta}/LaAlO_3$ (BST/YBCO/$LaAlO_3$) structure which was investigated consisted of an 800 nm thick $Ba_{0.5}Sr_{0.5}TiO_3$ film, known hereinafter as BST, deposited on a 300 nm thick $YBa_2Cu_3O_{7-\delta}$ film which coated a 5.0 mm×4.0 mm×0.50 mm $LaAlO_3$ substrate. Both layers were deposited "in-situ" using laser ablation. To elucidate the effects of film microstructure on tuning and loss, metalorganic deposited $Ba_{0.08}Sr_{0.92}TiO_3$ films, known hereinafter as MOD BST films, were prepared at room temperature and post annealed at 700°C to produce polycrystalline films[4]. After deposition of the multilayer structures,

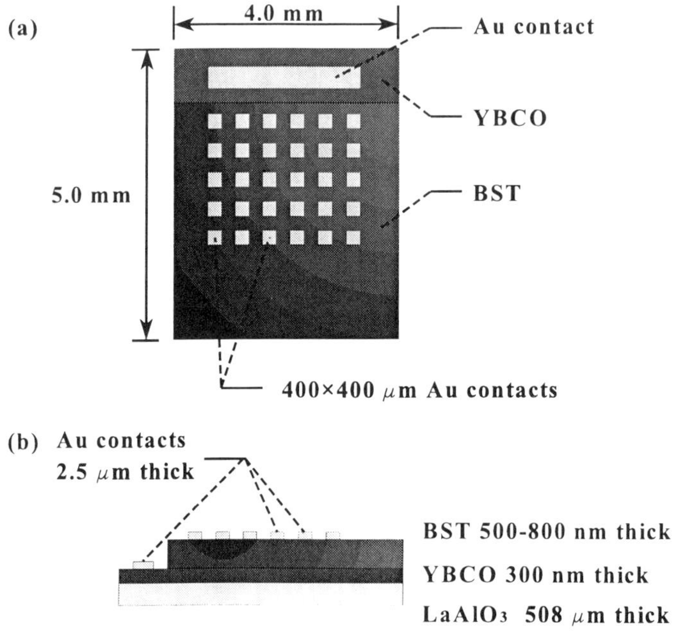

Figure 1. A schematic representation of the $Ba_xSr_{1-x}TiO_3$ (x=0.50,0.08)/$YBa_2Cu_3O_{7-\delta}/LaAlO_3$ structure. (a) Top view, (b) Side view.

(a)

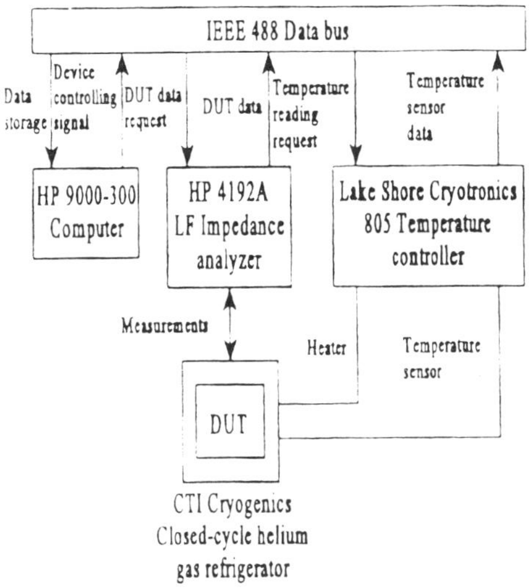

(b)

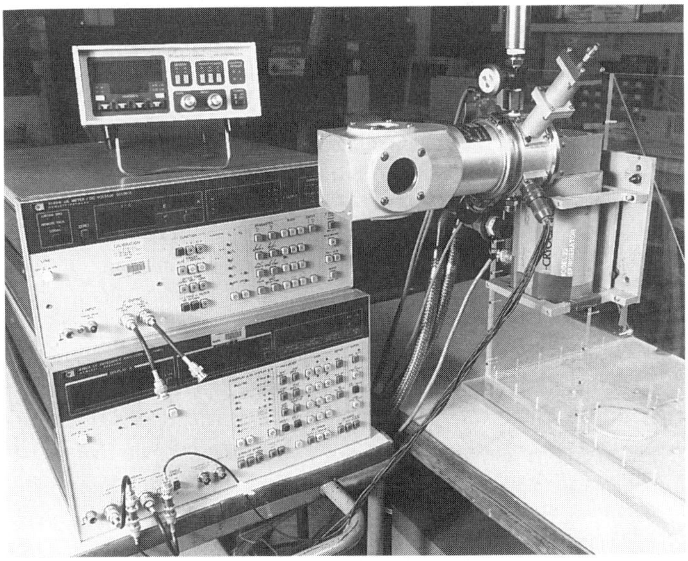

Figure 2. Experimental configuration. (a) Schematic representation, (b) Actual setup.

transition temperatures (T_c) above 89 K were measured for the YBCO films using a four point probe technique indicating that no degradation of the these films occurred as a result of the ferroelectric film deposition. Thirty $400\mu m \times 400\mu m \times 2.5\mu m$ gold (Au) contacts were fabricated on the ferroelectric layer of the structures using standard photolithography and wet etching techniques. In addition, a Au electrode was deposited onto a portion of the structures' YBCO layer, exposed by etching the ferroelectric layer away with a 7.0 percent solution of hydrofluoric (HF) acid. Figure 1 shows a schematic representation of the aforementioned structures. This configuration was used to measure the capacitance (C) at 1.0 MHz from which the ϵ_r of the ferroelectric layer was thereby calculated using Eq. (1),

$$\epsilon_r = \frac{Cd}{\epsilon_o A} \qquad (1)$$

Figure 3. Sample mounted inside vacuum chamber showing electrical feedthroughs.

where A represents the area of the gold contact, d represents the thickness of the ferroelectric layer, and $\epsilon_0 = 8.854\times10^{-12}$ F/m represents the permitivity of free space.

Values of ϵ_r for the ferroelectric layers were obtained as a function of temperature and dc bias. An HP-4192A LF impedance analyzer, a CTI-Cryogenics closed-cycle, helium gas refrigerator, and a Lake Shore Cryotronics 805 temperature controller all linked through an IEEE-488 data bus to an HP 9000-300 computer served as the main components of the experimental configuration, shown in Figure 2, which performed these measurements. As seen in Figure 3, the sample, known as the device-under-test (DUT), was mounted on a sample holder having electrical feedthroughs for dc bias. The temperature sensor, a silicon diode, was bolted directly to the sample holder. The sample holder was bolted to the cold head of the refrigerator to allow for measurements at cryogenic temperatures. An HP-Basic computer program was written to automate the measurement system allowing the user to take data as a function of temperature at a constant dc voltage and as a function of dc voltage at a constant temperature. All measurements were taken at "user-input" pre-selected conditions. Pre-selected dc voltage conditions were directly outputted to the impedance analyzer and

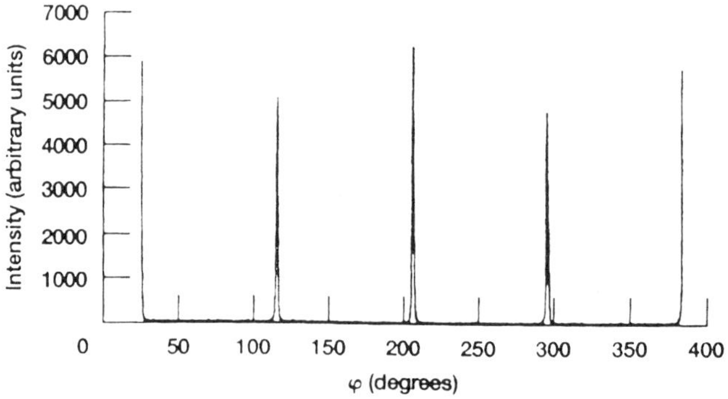

Figure 4. Phi scan about the BST (420) peak. Four-fold symmetry shows good in-plane alignment of the BST film on the YBCO film.

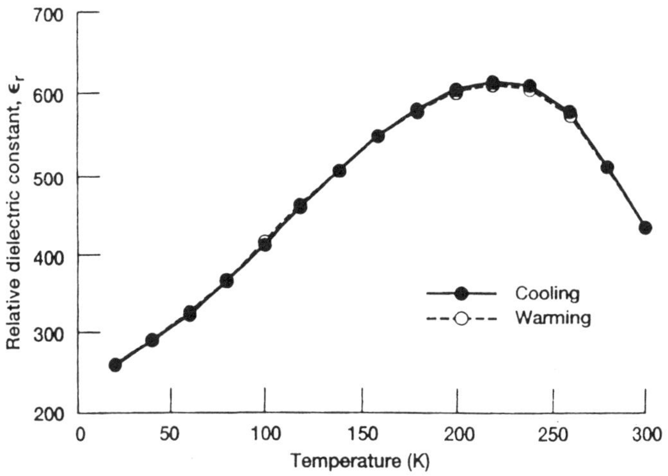

Figure 5. Dielectric constant versus temperature for the BST (Ba:Sr, 0.50:0.50) layer in the BST/YBCO/$LaAlO_3$ structure at 1.0 MHz and zero **E**.

temperatures conditions were outputted as "set-points" to the temperature controller which controlled and monitored the temperature of the DUT. When specified conditions were met, data on the DUT's capacitance and loss tangent (tanδ) were measured by the impedance analyzer and stored along with the DUT's ϵ_r, calculated by the computer program using Eq. 1, to a 3.5 inch disk for further analysis.

RESULTS

The ferroelectric layers were analyzed using x-ray diffraction (XRD). As seen in Figure 4, Phi scans about the (420) peak of the BST film show in-plane epitaxy with the underlying YBCO film. However, diffraction peaks for the MOD BST film were very weak at high angles and there was no evidence of in-plane epitaxy.

For the electrical characterization of the samples, the first test undertaken focused on measuring ϵ_r as a function of temperature at a frequency of 1.0 MHz and at zero E. Measurements were taken at intervals of 20 K during cooling from 300 to 20 K and during

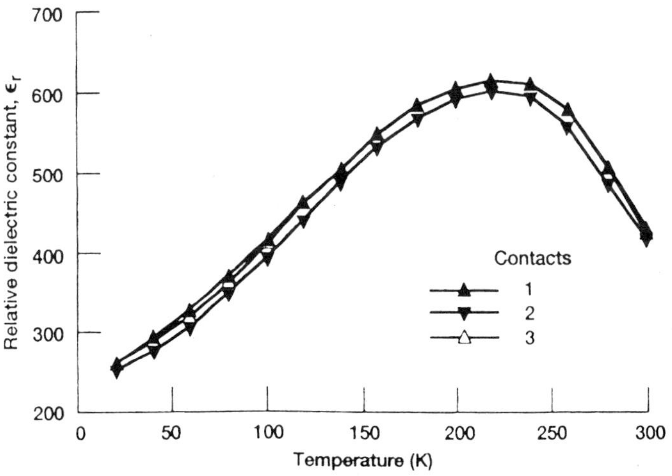

Figure 6. Dielectric constant values of different contacts versus temperature for the BST (Ba:Sr, 0.50:0.50) layer in the BST/YBCO/$LaAlO_3$ structure at 1.0 MHz and zero **E**.

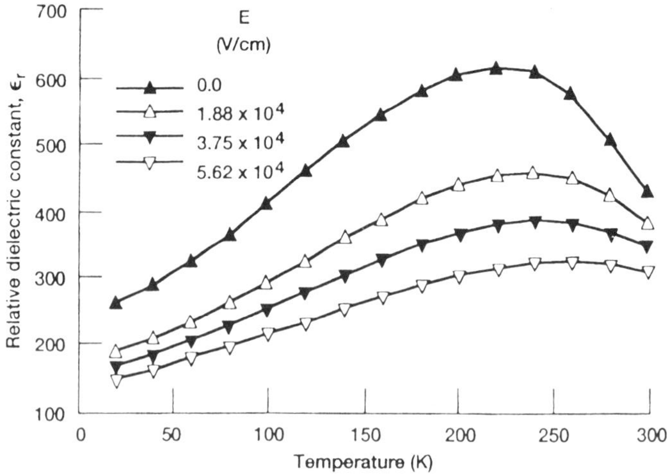

Figure 7. Dielectric constant versus temperature for the BST (Ba:Sr, 0.50:0.50) layer in the BST/YBCO/$LaAlO_3$ structure at 1.0 MHz and several values of **E**.

warming from 20 to 300 K to determine if the samples were sensitive to the temperature cycling. At room temperature, values of ϵ_r and tanδ for the BST sample of 425 and 0.040, respectively, were obtained. Neither the ϵ_r data, seen in Figure 5, nor the tanδ data showed any signs of temperature hysteresis as a result of the temperature cycling. Figure 6 shows ϵ_r data from different contacts across the sample's surface taken under the same temperature and bias conditions. This test was performed to determine the degree of uniformity of the BST film. Note that the obtained results are quantitatively and qualitatively very similar. This suggests excellent composition and structural (i.e., thickness) uniformity throughout the BST layer. The behavior of ϵ_r and tanδ as a function of **E** was also investigated. The results are shown in Figure 7 and Figure 8. In Figure 7, each curve represents the ϵ_r of the BST layer as a function of temperature at a particular **E** value. Note that there is a decrease in ϵ_r as the magnitude of the applied field is increased. Maximum variation in ϵ_r, from ϵ_r=610 to ϵ_r=300, occurred at a temperature near 220 K. For practical microwave components, such as tunable

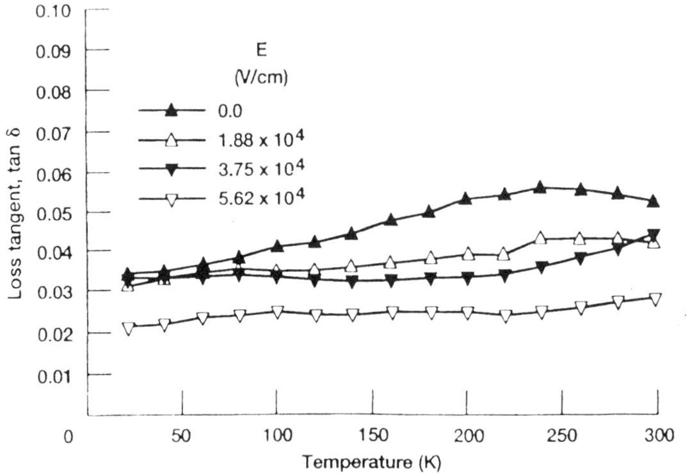

Figure 8. Loss tangent versus temperature for the BST (Ba:Sr, 0.50:0.50) layer in the BST/YBCO/$LaAlO_3$ structure at 1.0 MHz and several values of **E**.

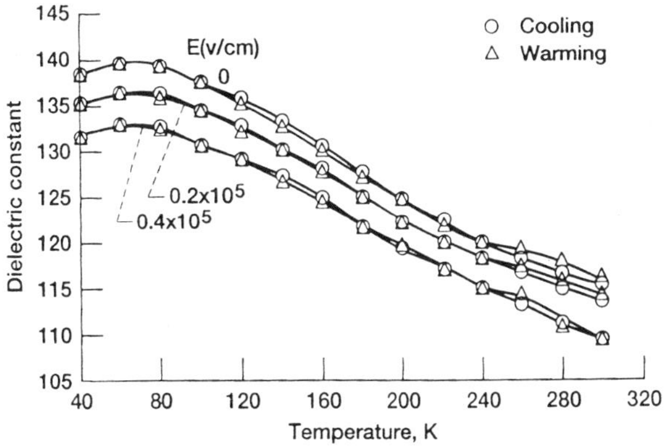

Figure 9. Dielectric constant versus temperature for a polycrystalline, 500 nm thick, metalorganic deposited MOD BST (Ba:Sr, 0.08:0.92) layer in the BST/YBCO/$LaAlO_3$ structure at 1.0 MHz and several values of **E**.

capacitors, filters, and phase shifters, a change in ϵ_r from 300 to 610 will create a frequency range within which the component can be tuned[5-7]. At 77 K, a temperature at which YBCO is in the superconducting state, values of ϵ_r changed from 180 to 360 in the range of $0V/cm \leq \mathbf{E} \leq 5.62x10^4 V/cm$. At this temperature, both the HTS film's low conductor losses as well as the ferroelectric's tunability could be fully exploited.

Figure 8 shows results of $\tan\delta$ as a function of temperature and **E**. It can be seen that the values for $\tan\delta$ vary slightly as a function of **E** in the aforementioned field range. This result is very important for microwave applications since it shows that the ϵ_r can be altered without dramatically enhancing the ferroelectric's microwave losses. The range of tunability for the polycrystalline, 500 nm thick, MOD BST film over $0V/cm \leq \mathbf{E} \leq 4.0x10^4 V/cm$ at 60 K was only 3.6 percent, see Figure 9. In addition, $\tan\delta$ was very sensitive to **E** as shown in Figure 10. We dismiss the possibility that the lower tuning in the MOD BST film is due to the composition differences since measurements on epitaxial STO and $Ba_{0.10}Sr_{0.90}TiO_3$ films deposited on YBCO displayed ϵ_r values that could be decreased by 30 and 10 percent respectively under the same electric field[8]. Therefore, the fact that the tuning was lower for the MOD BST film suggests that the lack of epitaxy decreases tuning and also increases microwave losses.

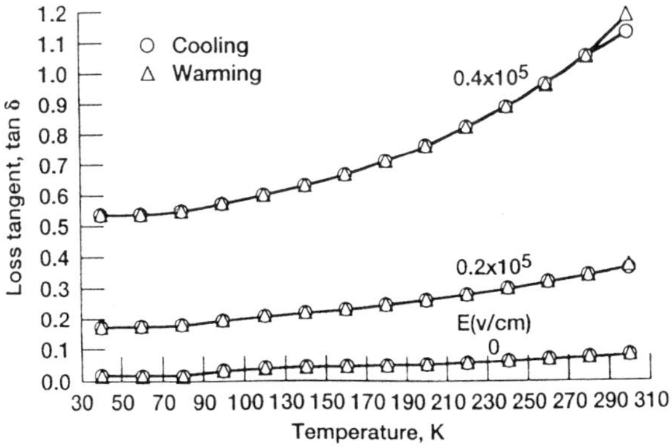

Figure 10. Loss tangent versus temperature for a polycrystalline, 500 nm thick, metalorganic deposited MOD BST (Ba:Sr, 0.08:0.92) layer in the BST/YBCO/$LaAlO_3$ structure at 1.0 MHz and several values of **E**.

CONCLUSION

A ϵ_r of 425 and a tanδ of 0.040 were measured at 298 K, 1.0 MHz, and zero **E** for an epitaxial BST ferroelectric layer. At 77 K, 1.0 MHz, and zero **E**, a ϵ_r of 360 and a tanδ of 0.036 were measured. Maximum change in ϵ_r, from 300 to 610, was observed at 220 K in the **E** range of $0 V/cm \leq \mathbf{E} \leq 5.62 \times 10^4 V/cm$ with minimal change in tanδ. The range of tunability for a polycrystalline, 500 nm thick, MOD BST film was very small, 3.6 percent, and tanδ increased markedly with **E** over $0 V/cm \leq \mathbf{E} \leq 4.0 \times 10^4 V/cm$. This suggests that the lack of epitaxy could adversely affect the tuning range and the microwave losses of ferroelectric/HTS based microwave components.

ACKNOWLEDGMENTS

The authors wish to thank Dr. R. Singh from the University of Florida at Gainesville for the fabrication of the laser ablated BST/YBCO/$LaAlO_3$ structure and NIST-Boulder for the use of their x-ray diffraction equipment. The authors are grateful to Mr. Nick Varaljay and Ms. Donna Bohman for patterning the BST and MOD BST samples and Mr. Joseph Meola for assisting with the electrical measurements. The authors are grateful to the referees for their constructive comments and suggestions.

REFERENCES

1. F.A. Miranda, S.S. Toncich, and K.B. Bhasin, Performance of two-pole bandpass filters photodefined on double-sided Y-Ba-Cu-O and Tl-Ba-Ca-Cu-O thin films, *Microwave and Optical Tech. Lett.* 6:752-755 (1993).
2. F.A. Miranda, Microwave properties of Tl-Ba-Ca-Cu-O thin films, *Supercond. Sci. Technol.* 6:605-612 (1993).
3. A.M. Hermann, A. Naziripour, A. Outzourhit, and C.H. Mueller, Tunable microwave resonators based on sputtered Tl-Ba-Ca-Cu-O and Ba-Sr-Ti-O thin films, in: "Low Temperature Electronics and High Temperature Superconductivity," C.L. Claeys, ed., The Electrochemical Society, Inc., Pennington (1995), p. 35.
4. H.D. Wu, Z. Zheng, F. Barnes, C.H. Jackson, A. Kain, and J.D. Cuchiaro, Voltage tunable capacitors using high temperature superconductors and ferroelectrics, *IEEE Trans. Appl. Superconductivity* 4:156-160 (1994).
5. O.G.Vendik, L.T.Ter-Martirosyan, A.I.Dedyk, S.F.Karmanenko, and R.A.Chakalov, High-T_c Superconductivity: New Applications of Ferroelectrics at Microwave Frequencies, *Ferroelectrics* 144:33-43 (1993).
6. S.S.Gevorgian, D.I.Kaparkov, and O.G.Vendik, Electrically Controlled HTSC/Ferroelectric Coplanar Waveguide, *IEE Proc. Microwave Antennas Propagation* 141:501-503 (1994).
7. O.G.Vendik, I.Mironenko, and L.T.Ter-Martirosyan, Superconductors Spur Applications of Ferroelectric Films, *Microwaves and RF*, 67-70 (July, 1994).
8. F.A. Miranda, C.H. Mueller, G.A. Koepf, and R.M. Yandrofski, Electrical response of ferroelectric/superconducting/dielectric $Ba_xSr_{1-x}TiO_3/YBa_2Cu_3O_{7-\delta}/LaAlO_3$ thin film multilayer structures, *Supercond. Sci. Technol.* 8:755-763 (1995).

YBCO HIGH-TEMPERATURE SUPERCONDUCTING FILTERS ON M-PLANE SAPPHIRE SUBSTRATES

J. C. Sabataitis,* C. H. Mueller,** F. A. Miranda, J. Warner, and K. B. Bhasin

National Aeronautics and Space Administration
Lewis Research Center
Cleveland, Ohio, 44135, USA

ABSTRACT

Since the discovery of High Temperature Superconductors (HTS) in 1986, microwave circuits have been demonstrated using HTS films on various substrates. These HTS-based circuits have proven to operate with less power loss than their metallic film counterparts at 77 K. This translates into smaller and lighter microwave circuits for space communication systems such as multiplexer filter banks. High quality HTS films have conventionally been deposited on lanthanum aluminate ($LaAlO_3$) substrates. However, $LaAlO_3$ has a relative dielectric constant (ϵ_r) of 24. With a ϵ_r~9.4-11.6, sapphire (Al_2O_3) would be a preferable substrate for the fabrication of HTS-based components since the lower dielectric constant would permit wider microstrip lines to be used in filter design, since the lower dielectric constant would permit wider microstrip lines to be used for a given characteristic impedance (Z_0), thus lowering the insertion losses and increasing the power handling capabilities of the devices. We report on the fabrication and characterization of $YBa_2Cu_3O_{7-\delta}$ (YBCO) on M-plane sapphire bandpass filters at 4.0 GHz. For a YBCO "hairpin" filter, a minimum insertion loss of 0.5 dB was measured at 77 K as compared with 1.4 dB for its gold counterpart. In an "edge-coupled" configuration, the insertion loss went down from 0.9 dB for the gold film to 0.8 dB for the YBCO film at the same temperature.

INTRODUCTION

Multiplexer filter banks used for channel separation in communication satellites require components that are as small and lightweight as possible. The filters used in multiplexers have

---

*Student, Electrical Engineering and Applied Physics Department, Case Western Reserve University, Cleveland, Ohio; work performed as NASA/OAI Summer Intern

**Now at Superconducting Core Technologies, Inc., Golden, Colorado

Advances in Cryogenic Engineering, Vol. 41
Edited by P. Kittel, Plenum Press, New York, 1996

been successfully made using waveguide or coaxial techniques, yet are large and heavy. Microstrip filters using conventional conductors, which potentially can be used in filter banks, are small, but are often too lossy for multiplexer applications, while conventional superconductors are less lossy, but have to be cooled with liquid helium to 4-5 K. There would be many advantages in developing circuits that could operate at 77 K, since these could be cooled at less cost with liquid nitrogen (boiling point of 77 K). The discovery of High Temperature Superconducting (HTS) materials has brought a solution to these problems. Microstrip filters using HTS films are just as small and lightweight, but can display even less power loss than their waveguide or coaxial counterparts[1-3]. Considering that filter banks are composed of many filters, the gains achieved by using a lower-loss filter are often multiplied.

HTS thin films used in microstrip filters have been deposited on substrates such as Lanthanum Aluminate ($LaAlO_3$), Magnesium Oxide (MgO), Neodymium Gallate ($NdGaO_3$), and sapphire (Al_2O_3). In general, the best performance of HTS-based microwave components has been obtained from films deposited on $LaAlO_3$. This material, however, has a relatively high dielectric constant, $\epsilon_r \sim 24$[4]. Sapphire (Al_2O_3) has a much lower dielectric constant of 9.4-11.6, depending on crystal orientations, and would be a better choice for a substrate. The lower dielectric constant would permit wider microstrip lines to be used for a given characteristic impedance (Z_0), thus lowering the insertion losses and increasing the power handling capabilities of the devices. Although power handling capabilities of microstrip filters using high ϵ_r substrates (such as $LaAlO_3$) have been improved by using low Z_0 resonance elements, the input power is limited by current crowding at junction between the feed line and first resonator[c]. In addition, sapphire has a comparable loss tangent ($tan\delta \sim 4 \times 10^{-8}$) to that of $LaAlO_3$ ($tan\delta \sim 0.76 \times 10^{-5}$) at 77 K and 10 GHz[5]. This combination of a lower dielectric constant and a low loss tangent results in an overall lower power loss[3].

One slight drawback to the use of sapphire as a substrate, however, is the fact that YBCO film cannot be deposited successfully directly on the substrate due to the reactivity of the two materials. At the temperature required for deposition, sapphire reacts with barium oxide (BaO) to form barium aluminate ($BaAl_2O_4$). Therefore a buffer layer such as $SrTiO_3$ or MgO, which are both unreactive with the two materials, must be deposited on sapphire before the YBCO film can be deposited. Unfortunately, this creates some loss due to the resultant lattice mismatch. This loss, however, should be minimal enough that the gains from using an HTS film should more than compensate for the loss due to crystal lattice mismatch[6].

HTS films have been successfully deposited on both ($1\bar{1}00$)- and ($1\bar{1}02$)- oriented sapphire (M- and R-plane, respectively). The optic axis in M-plane sapphire is aligned parallel to the substrate surface, whereas in R-plane sapphire it is oriented 32.4 degrees from the substrate surface. Circuit design is complicated by the dielectric anisotropy of the substrate, and becomes most challenging for substrate orientations for which the optic axis is neither perpendicular or parallel to the substrate plane (such as R-plane sapphire). Recent work[7] has shown that the unloaded Q values of linear, one-half wavelength resonators patterned on M-plane sapphire are considerably higher than on R-plane (2100 versus 600). Since the low Q values on the R-plane sapphire could not be attributed to conductor or dielectric losses, the only possibility is that the radiation losses in R-plane sapphire were higher than on M-plane sapphire. The difference in Q values was attributed to radiation losses associated with the fact that the optic axis in R-plane sapphire is neither parallel nor perpendicular to the optic axis, while the optic axis for M-plane sapphire is in the plane of the substrate. Thus, M-plane sapphire is a better candidate for microwave circuits.

In the following sections, we will present design and test results of microwave 3-pole bandpass filters. Four filters were made using two basic designs. Each design was fabricated on M-Plane sapphire substrates using gold or YBCO films. (The YBCO films were obtained from a commercial vendor.) All four were tested at 77 K, and display low-loss performance.

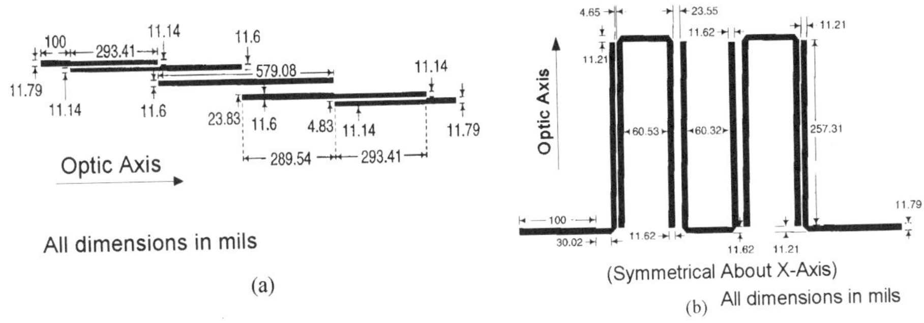

Figure 1. Dimensions of the YBCO/Al_2O_3 edge-coupled (a) and hairpin (b) filters.

FILTER DESIGN, FABRICATION, AND TESTING

Two types of bandpass filters were designed using Eagleware© software. The filters were designed to resonate at 4.0 GHz on M-plane sapphire with a dielectric constant of 9.4. One was a hairpin shape and the other was an edge-coupled design (figure 1). Both were two-port, three-pole Chebyshev configurations with 50-ohm lines and 4 percent bandwidths. Four filters were fabricated in a clean room from photo-reduced masks using standard photo-lithography and chemical etching techniques: two using gold film and ground planes (both 2.5 microns thick), and two using single-sided YBCO film and gold ground planes. The gold filters were fabricated on 17-mil thick sapphire substrates. Since the YBCO films were deposited on 12 mil thick substrates, the linewidths and gaps were scaled to yield identical center frequencies and bandwidths. The sapphire used for the YBCO filter had a buffer layer of MgO and had a 3000Å thick YBCO film deposited by magnetron sputtering[8]. A 4000Å thick patch of gold 80 microns narrower than the width of the feedline was deposited on the edges of the YBCO feedlines to improve coupling from the coaxial pin to the filter. None of the YBCO filters were post-annealed, since YBCO on M-plane sapphire films have previously responded unfavorably to this process, becoming substantially more susceptible to thermally-induced microcracks[7]. The edge-coupled filters were fabricated with feedlines parallel to the optic axis, and the hairpin filters were made with feedlines orthogonal to the optic axis, so that the coupling between resonant elements would be perpendicular to the optic axis, and hence the dielectric constant seen by the field lines in the coupling direction would be 9.4.

The filters were tested using a Hewlett-Packard 8510C Automated Network Analyzer (ANA). Both were mounted on brass test fixtures with coaxial pin launchers. The test fixture was mounted inside the vacuum can of a closed-cycle helium gas refrigerator and fed a signal by an HP8517B S-parameter test set. The temperature was controlled using a Lake Shore DRC 91C temperature controller. Both the temperature and ANA were controlled by an HP-BASIC program to record data. The ANA cables and connectors to the launchers on the test fixture were calibrated at room temperature (298 K) using a short-open-load-through calibration technique. An HP 9000-216 computer recorded measurements under vacuum (below 1 millitorr) and in the 20 K to 100 K temperature range. Data at 77 K was particularly used for insertion loss and unloaded Q analysis. Two s-parameters were recorded as a function of frequency, S_{11} (reflection), and S_{21} (transmission). The insertion losses at the maximum of the transmission parameters and at the midband points were recorded. The unloaded Q was calculated using the following equation:

$$Q = \frac{8.686 C_N \omega_1'}{w(\Delta L_A)_0} \tag{1}$$

Table 1. Insertion losses and unloaded Q values

	Insertion Loss (dB)	Unloaded Q
Gold Hairpin Filter	1.4	421
YBCO Hairpin Filter	0.5	-
Gold Edge-coupled Filter	0.9	368
YBCO Edge-coupled Filter	0.8	-

where ω_1' is the low pass prototype cutoff frequency (normally =1), w is the percent bandwidth, $(\Delta L_A)_0$ is the minimum insertion loss in dB (assuming a symmetrical and flat passband), and C_n is the constant dependent on the Chebyschev ripple and filter order[9]. The constant 8.686 is a conversion from nepers to dB.

RESULTS

The insertion losses at 77 K for edge-coupled and hairpin filters are shown in Table 1. The insertion loss of the hairpin filter changed more drastically when the material was switched to YBCO than did the loss of the edge-coupled filters. At 77 K, the insertion loss of the gold hairpin filter was 1.4 dB. When the YBCO hairpin filter was tested at 77 K, the insertion loss went down to 0.5 dB. The YBCO edge-coupled filter performed very well at lower temperatures. We observed an insertion loss of less than 0.3 dB at 35 K. However, the loss went up to 0.8 dB at 77 K, which was not significantly different from the gold edge coupled filter with a loss of 0.9 dB (also at 77 K).

The unloaded Q values of the gold filters were around 400 (table 1). The unloaded Q values were not calculated for the YBCO filters. These filters had very asymmetrical transmission parameters, unlike the gold versions (figures 2 and 3). All four filters had center frequencies around 4 GHz, but for the HTS filters, the point of minimum insertion loss was shifted a bit higher than the center frequency, making calculation of the unloaded Q using equation (1) difficult. The assymetrical shape and poor performance of the filters were attributed to radiation losses associated with the substrate's anisotropy.

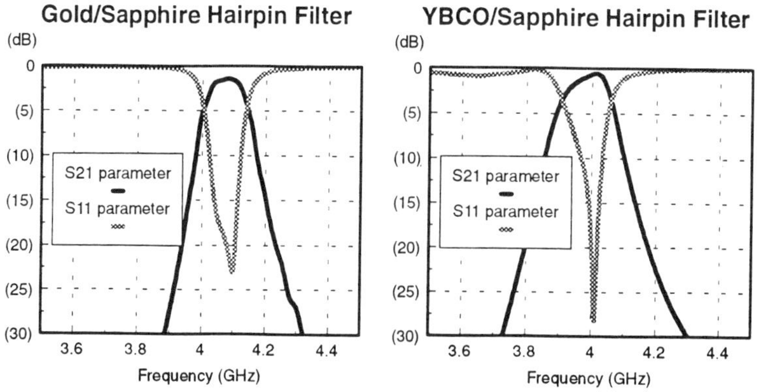

Figure 2. Reflection and transmission parameters of the gold and YBCO hairpin filters.

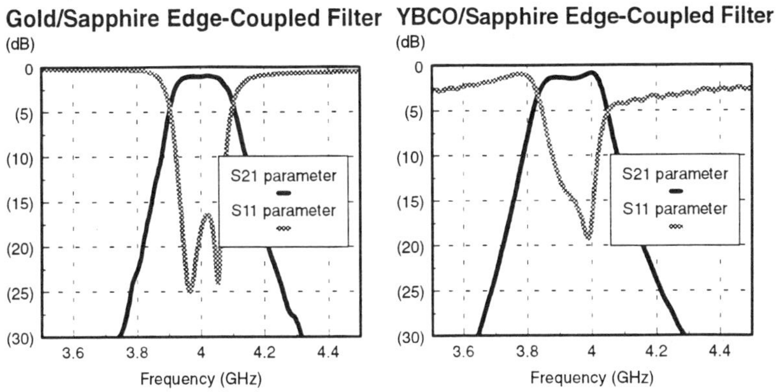

Figure 3. Reflection and transmission parameters of gold and YBCO edge-coupled filters.

CONCLUSIONS

HTS YBCO filters have been fabricated on sapphire substrates because of its lower dielectric constant and loss tangent. We attribute the poor performance of the filters to radiation losses resulting from the dielectric anisotropy of the substrates. Despite our efforts to minimize the effects of dielectric anisotropy by aligning the resonator elements parallel to the optic axis, the performance for the YBCO filters is only slightly better than their gold counterparts. Because of this, only a small improvement was obtained by using low-loss YBCO films as the conductors. Since the Q value of a single, one-half wavelength long resonator was measured to be 21000[7], we conclude that most of the microwave energy in the three-pole filters was lost in coupling between elements.

ACKNOWLEDGEMENT

We would like to thank the NASA Lewis Technical Support Staff for their support in fabrication. In particular, we would like to thank Donna Bohman for deposition and etching of the films. We would also like to thank George Readus Jr. for making the test fixtures. Finally, we are grateful to Linda D'Evelyn for helpful suggestions pertaining to the Q calculations. The authors are grateful to the referees for their constructive comments and suggestions.

REFERENCES

1. S. H. Talisa, M. A. Janocko, C. Moskowitz, J. Talvacchio, J. F. Billing, R. Brown, D. C. Buck, C. K. Jones, B. R. McAvoy, G. R. Wagner, and D. H. Watt, Low- and high-temperature superconducting microwave filters, *IEEE Trans. Microwave Theory Tech.*, 39:1448-1453 (1991)
2. F. A. Miranda, S. S. Tonich, and K. B. Bhasin, Performance of two-pole bandpass filters photodefined on double-sided Y-Ba-Cu-O and Tl-Ba-Ca-Cu-O thin films, *Microwave Opt. Technol. Lett.*, 6:752-755 (1993)
3. G. C. Liang, R. S. Withers, B. F. Cole, S. M. Garrison, M. E. Johansson, W. Ruby, and W. G. Lyons, High-temperature superconducting delay lines and filters on sapphire and thinned $LaAlO_3$ substrates, *IEEE Trans. Appl. Supercond.*, 3:3037-3041(1993)
4. T. Konaka, M. Sato, H. Asano, and S. Kubo, Relative permittivity and dielectric loss tangent of substrate materials for high-T_c superconducting film, *J. Supercond.*, 4:287, (1991)

5. N. Newman and W. G. Lyons, High-temperature superconducting microwave devices: fundamental issues in materials, physics, and engineering, *J. Supercond.*, 6:119-158 (1993)
6. J. Ceremuga, M. Barton, and F. Miranda, Analysis of influence of buffer layers on microwave propagation through high-temperature superconducting thin films, *Supercond. Sci. Technol.* 7, (1994)
7. C. H. Mueller, F. A. Miranda, S. S. Tonich, and K. B. Bhasin, YBCO X-band microstrip linear resonators on $(1\bar{1}02)$- and $(1\bar{1}00)$- oriented sapphire substrates, *IEEE Trans. Appl. Supercond.*, 5:2559-2562 (1995)
8. B. F. Cole, G. C. Liang, N. Newman, K. Char, G. Zaharchuk, and J. S. Martens, Large-area YBCO films on sapphire for microwave applications, *Appl. Phys. Lett.*, 61:1727-1729 (1992)
9. G. L. Matthaei, L. Young, and E. M. T. Jones. "Microwave Filters, Impedance-Matching Networks, and Coupling Structures," Artech House Books, Dedham (1980)

THERMAL TRANSPORT AND ELECTRICAL DISSIPATION IN ULTRA-LOW THERMAL LOAD, MULTI-GIGAHERTZ I/O CABLES FOR SUPERCONDUCTING MICROELECTRONICS

T.J. Hendricks, M.W. Bruns, and E.L. Hershberg

Systems Integration Laboratory
Technology Group - Video & Networking Division
Tektronix, Inc.
P.O. Box 500, M/S 50-470
Beaverton, OR 97077-0001

ABSTRACT

Minimizing the size and power requirements of cryosystems for operating low-temperature superconducting digital microelectronics requires minimizing the thermal load imposed by the input/output (I/O) transmission lines. Signal line/ground plane material selection, the number of transmission lines and their dimensions, and boundary temperatures dictated by cryocooler stage temperatures are critical parameters that control the design tradeoffs. The cryocooler performance plays an integral role in the transmission cable design. Achieving transmission lines with ultra-low thermal load performance necessarily creates significant design tradeoffs between thermal conduction down the cable and electrical dissipation due to signal attenuation in the cable. When minimizing the total thermal load, Q_t, to a cryocooler stage, signal dissipation in the I/O system is not generally considered, but becomes very important in the design of these high performance systems. This work describes the critical thermal load mechanisms in ultra-low thermal load transmission cables designed for high I/O bandwidths and establishes a mathematical model for analyzing and designing such cables. Temperature-dependent results from SINDA-85 thermal network models that predict total heat loads and temperature distributions for various transmission cable designs are presented. Temperature-dependent results for signal attenuation at cryogenic temperatures are also presented, and serve to quantify the I/O transmission cable electrical tradeoffs. Important design tradeoffs involving cable section lengths and signal line thickness are demonstrated.

NOMENCLATURE

f Frequency [hertz]
g_s Signal line geometry factor
g_g Ground plane geometry factor
h Dielectric thickness [m]

I_s Signal current [amperes]
J_{s1} Top surface current [amperes/m]
J_{s2} Bottom surface current [amperes/m]
l Length [m]
P Signal line or ground plane perimeter [m]
$\dot{q}$ Electrical power dissipation per unit volume [watts/m^3]
Q_t Total heat load in a cable section [watts]
t Signal line or ground plane thickness [m]
T Temperature [K]
w Signal line or ground plane width [m]
x Linear dimension in cable direction [m]
y Linear dimension normal to cable surface [m]
α_d Attenuation coefficient in dielectric [m^{-1}]
α_g Attenuation coefficient in ground plane [m^{-1}]
α_s Attenuation coefficient in signal line [m^{-1}]
δ Electrical skin depth or penetration depth [m]
κ Thermal conductivity [watts/m-K]
μ_o Free space permeability [henry/m]
v velocity of light in dielectric [m/second]
ρ Electrical resistivity [ohm-m]

INTRODUCTION

The signal input/output (I/O) cable provides the major source of heat load to state-of-the-art cryogenic cooling systems in superconducting microelectronics applications. It is critically important to minimize the I/O cable heat load at the low-temperature end (i.e., 4.2 K) in order to achieve required superconducting temperatures while minimizing cryocooler power input. Ellison et al.[1] discuss the thermal design of I/O cables in which electrical dissipation in the I/O cable was negligible. However, ultra-low thermal load cables necessarily have small signal line and ground plane thicknesses which cause electrical dissipation and signal attenuation effects to become increasingly important. This paper discusses the unique electrical and thermal design issues and tradeoffs associated with I/O cables having ultra-low thermal load characteristics.

Intermediate cryogenic stage cooling effectively reduces the heat load at the 4.2 K stage, but also complicates the cable design due to multiple cable lengths, with the problem of choosing the optimum intermediate-stage location and cable section lengths. The multi-section cable design becomes increasingly complex when electrical dissipation and signal attenuation in the I/O cable must be included. This creates a critical electrical and thermal design tradeoff in the cryosystem because the cable electrical dissipation necessarily shows up as a component of the total cryocooler thermal load.

We assumed in this study superconducting electronics supporting 256 I/O channels requiring 250 signal lines at each of four sides of a microelectronic multi-chip module. The 1000 transmission lines transported high-frequency signals between room temperature (300 K) semiconductors and the cryogenic (4.2 K) superconducting electronics. In this case, each I/O line consisting of a signal line, associated ground plane, and intervening dielectric, had to transmit 2.5 x 10^9 bits/second. Design of microwave microstrip transmission lines requires metallic signal lines separated from a metallic ground plane by a dielectric material of precise, uniform thickness. The transmission lines require a certain width and spacing for a given dielectric constant and material thickness for a controlled characteristic impedance. Furthermore, the metallic conductors must have a minimum thickness dictated by the required signal rise time. A major objective here was to determine the optimum thickness which minimizes total thermal load and satisfies electrical attenuation requirements.

Figure 1 illustrates the typical geometry of this controlled impedance structure with cable dimensions appropriate for this particular problem. A two-section I/O cable with one intermediate,

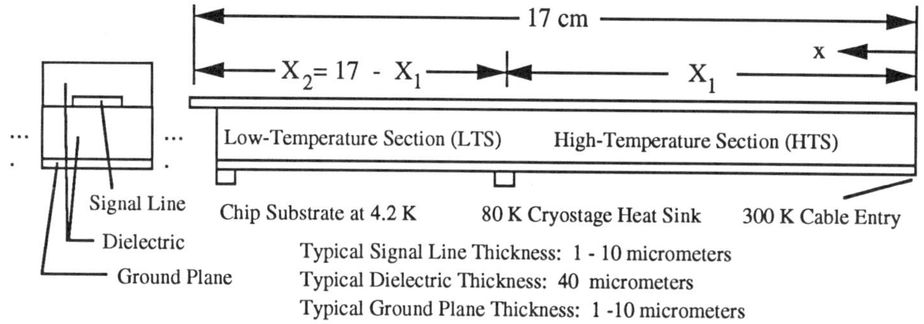

Figure 1. Typical Geometry of Signal Line/Dielectric/Ground Plane Design.

fixed-temperature heat sink was assumed. The distance between the 300 K cable end and the 80 K intermediate cryogenic stage location was allowed to vary in this investigation, while the total cable length was held constant at 17 cm. Selecting the optimum heat sink location was a major objective in this work. The cryocooler was assumed capable of maintaining constant temperatures at the 80 K and 4.2 K cold stations, regardless of the cold station heat load. Considering the cooling capacity of modern cryocoolers, the cold station heat loads were the primary performance parameters comparing various cable designs, with the added constraint of achieving minimum required bandwidth.

There were two specific goals of this work: (1) Demonstrate a design methodology for determining optimum designs, and (2) Demonstrate the design tradeoff between thermal conduction and electrical dissipation in ultra-low thermal load I/O cables associated with state-of-the-art superconducting electronics operating in cryogenic environments.

ULTRA-LOW THERMAL LOAD CABLE EFFECTS

The energy equation for the signal line and ground plane dictates the thermal response in the cable and the total heat load from the cable signal line and ground plane to the various cryocooler stages. We have simplified the analysis by performing a one-dimensional, steady-state analysis on a signal line and adjacent ground plane, with dielectric conduction normal to each line included as a loss term. The energy equation for each, neglecting thermal radiation to the cable, then becomes:

$$\frac{\partial}{\partial x}(\kappa(T)\frac{\partial T}{\partial x})+\dot{q}(T)-(\kappa(T)\cdot(\frac{P}{t\cdot w})\cdot\frac{\partial T}{\partial y})_{diel}=0 \qquad (1)$$

In this design problem, the volumetric energy dissipation is the cable electrical dissipation and is a function of temperature. Due to the temperature gradient along the cable length, it is inherently a function of position in the cable signal line and ground plane. The last term in Eq. (1) represents thermal conduction out of the signal line or ground plane across the dielectric material to the adjacent component, and strongly couples the signal line and ground plane thermal response.

The attenuation constant for electrical signals in a transmission line is described by[2]:

$$\alpha=\frac{1}{2\cdot P'}\cdot\frac{\partial P'}{\partial x} \qquad (2)$$

where the partial derivative is the power dissipated per unit length in the line.

In thin conductors (t << w), power dissipation results from surface currents on the top and bottom surface of the conductor. The "skin depth" of each surface current layer is:

$$\delta(T)=\sqrt{\frac{\rho(T)}{\pi\cdot\mu_o\cdot f}} \qquad (3)$$

In typical transmission line designs, t >> δ, and power dissipation is roughly independent of thickness. However, for ultra-low thermal load cables, the conductor thickness can approach skin depth by design to minimize thermal conduction. In this case, the current distributions of the two surfaces are no longer independent. Following the approach of Horton, et al.[3], an expression can be derived for α_S that is valid when t ~ δ:

$$\alpha_s(T) = \frac{\rho(T)}{2\cdot\delta\cdot I^2 Z_o}\mathrm{Re}\{(1+j)\int_w[(J_{s1}{}^2 + J_{s2}{}^2)\coth(\gamma t) + J_{s1}J_{s2}\,\mathrm{csc}\,h(\gamma t)]dw\} \tag{4}$$

where ρ(T) / δ is the surface resistance of a layer δ thick with resistivity ρ and Z_O is line resistance.

In the signal conductor, Eq. (4) may be simplified to yield:

$$\alpha_s(T) = \frac{g_s\cdot\rho(T)}{\delta\cdot w\cdot Z_o}\mathrm{Re}\{(1+j)[\coth(\gamma t) + \mathrm{csc}\,h(\gamma t) - [\frac{\int_w(|J_{s1}| - |J_{s2}|)^2 dw}{\int_w(J_{s1}{}^2 + J_{s2}{}^2)dw}]\mathrm{csc}\,h(\gamma t)]\} \tag{5}$$

and γ = (1+j) / δ . The dimensionless geometry factor, g_S $(0.5 \le g_s \le 1)$ is weakly dependent on w/h and t/h. In the ground conductor, the surface current on the bottom surface is zero, and:

$$\alpha_g(T) = \frac{\rho(T)\cdot g_g}{\delta\cdot w\cdot Z_o}\mathrm{Re}\{(1+j)\coth(\gamma t)\} \tag{6}$$

Note that in the limit t >> δ, the right most factor in Eqs. (5) and (6) approach unity and the left factor is the classical expression for attenuation. There is a rich body of literature describing methods and formulas to calculate g_S and g_g (e.g., see Schneider[4]).

The attenuation coefficient from the dielectric losses in imbedded microstrip is[5]:

$$\alpha_d = \frac{\pi\cdot f\cdot\tan\delta}{v} \tag{7}$$

where v = velocity of light in dielectric.

The total cable power dissipation is calculated from Eq. (2) using the total attenuation:

$$\alpha_T = \alpha_s + \alpha_g + \alpha_d \tag{8}$$

The volumetric power dissipation in Eq. (1) is then given by:

$$\dot{q}(T) = \frac{\alpha_T(T)\cdot I(x)^2\cdot Z_o}{t\cdot w} \tag{9}$$

Substituting Eq. (9) into the energy equation (Eq. (1)) then yields:

$$\frac{\partial}{\partial x}(\kappa(T)\frac{\partial T}{\partial x}) + \frac{\alpha_T(T)\cdot I(x)^2\cdot Z_o}{t\cdot w} - (\kappa(T)\cdot(\frac{P}{t\cdot w})\frac{\partial T}{\partial y})_{diel} = 0 \tag{10}$$

Cable thermal response and thermal load is then a function of thermal conductivity, κ(T), electrical resistivity, ρ(T), cable section lengths, signal line / ground plane thickness and width, and signal current, I(x) for a given signal frequency.

The cable design problem becomes a tradeoff between temperature-dependent material properties, signal current, cable cross-section geometry and cable length. Eq. (10) demonstrates four very important design issues in designing ultra-low thermal load cables. First, it shows how cable geometry (thickness, t, and width, w) influences the thermal response and total heat load. Second, it shows that signal frequency impacts the thermal response through δ in Eq. (5) for α_S. Third, it clearly defines how temperature-dependent material properties affect thermal response and total thermal loading. Finally, cable length is important because of integrating Eq. (10). All four design issues impact cable design and define design tradeoffs to consider in ultra-low thermal load cables.

FINITE DIFFERENCE MODELING

We solved this design problem by developing a high-fidelity finite-difference model of Eq. (10) using SINDA-85 thermal analysis software, an industry-standard thermal analysis package used extensively for complex thermal system analyses in the aerospace and defense industry. This software allowed easy problem set-up and easily accounted for temperature-dependent properties and temperature-dependent electrical dissipation in the cable.

The SINDA-85 finite-difference model of the cable geometry contained 720 nodes; 6 nodes across the cable cross-section and 120 length increments down the total cable length. Appropriate nodes were tied to 80 K and 4.2 K boundary temperatures to simulate the cryocooler heat sinking.

The temperature-dependent electrical dissipation per unit volume in Eq. (10) was calculated at each node along the cable. Current, I(x), was appropriately attenuated at each signal line node by:

$$I_{n+1} = I_n \exp(-\alpha_T(T) \cdot \Delta l)$$

where Δl is the transmission line length differential between nodes.

Specific transmission line parameters were chosen in the first part of this study (i.e., cable length optimization) to target total cable attenuation to be $20\log_{10}(I_{120} / I_1) \geq -3.0dB$ for frequency, f = 2.5 gigahertz (GHz). This condition represents a real electronic system requirement to maximize the amplitude of the transmitted signal. Table 1 gives the chosen transmission line parameters.

Table 1 - Specific Chosen Transmission Line Parameters (Z_O = 32 ohms)

	w	h	t	δ
High-Temperature Section (HTS)	175 μm	42 μm	2 μm	1.3 μm @ 300 K
Low-Temperature Section (LTS)	100 μm	25 μm	1 μm	0.5 μm @ 80 K

We considered only oxygen-free, high conductivity copper in examining the thermal response and thermal loads of various I/O cable designs. Temperature-dependent thermal conductivity, κ(T), and electrical resistivity, ρ(T), for the copper were taken from Childs et al.[6] and Simon et al.[7]. Figure 2 illustrates κ(T) and ρ(T) for the OFHC copper in this investigation. The thermal conductivity of the microstrip dielectric is also temperature-dependent and Figure 3 shows κ(T) for the dielectric.

Solution to the finite-difference form of Eq. (10) for the cable temperature distribution was determined with a Gauss-Seidel successive iteration scheme. Aitken convergence acceleration was applied to intermediate temperature calculations at each iteration to speed solution convergence. The temperature convergence criteria was typically 0.0003 - 0.0005 K for all analysis cases investigated. Cable heat loads at the two cryocooler stages (i.e., 80 K and 4.2 K) were calculated from the cable temperature distribution using a first-order, finite-difference expression for the net thermal conduction into the 80 K and 4.2 K boundary nodes.

RESULTS AND DISCUSSION

The SINDA-85 model can be used for numerous cryosystem and I/O cable design analyses involving the many design parameters in Eqs. (9) and (10) which impact the cable electrical and thermal response, and ultimately control the cable design. These include signal frequency, f, initial signal current, I_S, signal line thickness, width and length, and material properties, κ(T) and ρ(T). In order to demonstrate specific cable design tradeoffs, we set signal frequency at 2.5 gigabits/second (Gb/s) and initial signal current at 32 milliamperes at the 300 K entry point into the cable.

One critical design tradeoff to achieve minimal heat loads on both cryocooler stages (80 K and 4.2 K) involves properly positioning the 80 K heat sink along the cable length. The character of this tradeoff changes in ultra-low thermal load cables where line electrical dissipation becomes an important factor. The heat sink positioning then involves a critical balancing between pure cable thermal conduction and electrical dissipation in the signal lines, dielectric, and ground plane.

Figure 4 illustrates typical results for total thermal load, consisting of signal line thermal conduction and electrical dissipation, on each cryocooler stage for an OFHC copper cable of cross-section shown in Figure 1. Total thermal load, Q_t, on each cryocooler stage is plotted versus heat sink position characterized by high-temperature section (HTS) length, X_1 (See Figure 1), for two different signal line duty cycles; a 10% duty cycle (i.e., quiet line with minimal electrical dissipation) and a 50% duty cycle line (i.e., active line with very significant electrical dissipation). Duty cycle is defined as the percentage of time the signal line carries a signal current, I_S.

The total thermal load is thermal-conduction-dominated for the 10% duty cycle and the cable section lengths considered. In this case, as X_1 increases (i.e., longer HTS lengths), the 80 K-stage cryocooler load reduces substantially while the 4.2 K-stage cryocooler load increases only a little. This particular tradeoff strongly indicates that longer HTS lengths, and shorter low-temperature

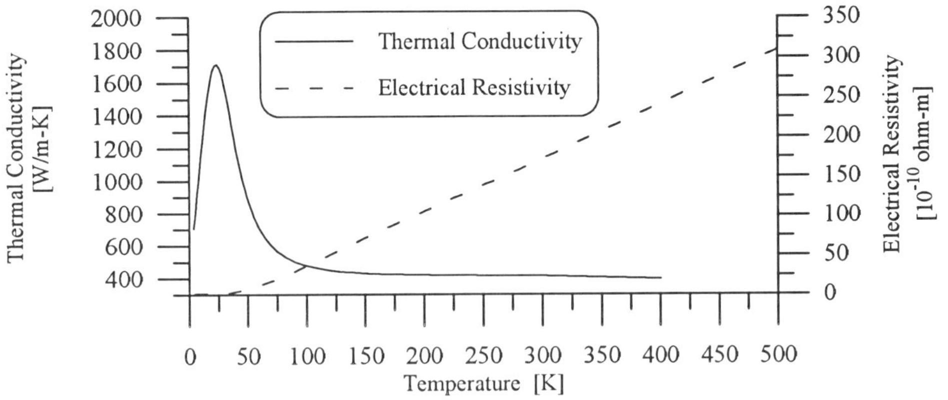

Figure 2. Thermal conductivity and Electrical Resistivity of OFHC copper vs. temperature.

section (LTS) lengths, are highly desirable. However, increasing X_1 too much will inevitably lead to unacceptable 4.2 K-stage loads due to increasing pure thermal conduction in the LTS. The best heat sink positioning would depend on cryocooler performance.

The 50% duty cycle results illustrate a completely different character to the design tradeoff for heat sink positioning. First, one must notice the dramatically higher total heat loads to the cryocooler due to the significant electrical dissipation component. Also, in this case as X_1 increases, the 80 K-stage total heat load again decreases significantly. However, as X_1 increases, the 4.2 K-stage total heat load exhibits a clear minimum value at an optimum heat sink position, $X_{1,opt}$, along the cable length. This occurs because at high HTS lengths (i.e., shorter LTS lengths) pure thermal conduction down the LTS dominates the total thermal load and becomes very large. At low HTS lengths (i.e. higher LTS lengths) signal line electrical dissipation dominates the total thermal load and becomes very large. Thus, an optimum heat sink position develops in which total thermal load on the 4.2 K-stage minimizes with neither pure thermal conduction nor electrical dissipation components having an overwhelming effect on the total thermal load.

Figure 5 illustrates the optimum cable temperature profile when the 80 K heat sink position is chosen to minimize total heat load on the 4.2 K stage. Also shown in Figure 5 is a non-optimum cable temperature profile which can occur when the heat sink position is not optimally located and the LTS length becomes too large. If the LTS length is too long (i.e., the HTS length too short), then too much electrical dissipation in the LTS increases the temperatures and temperature gradients in the LTS and total thermal load on the 4.2 K cryocooler stage. The non-optimum temperature gradient at the 80 K heat sink point is also a contributing factor in increasing the 80 K-stage total heat load

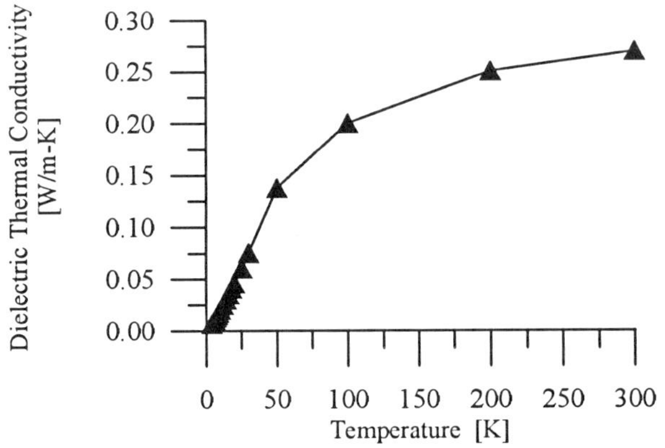

Figure 3.
Thermal conductivity of dielectric vs. temperature.

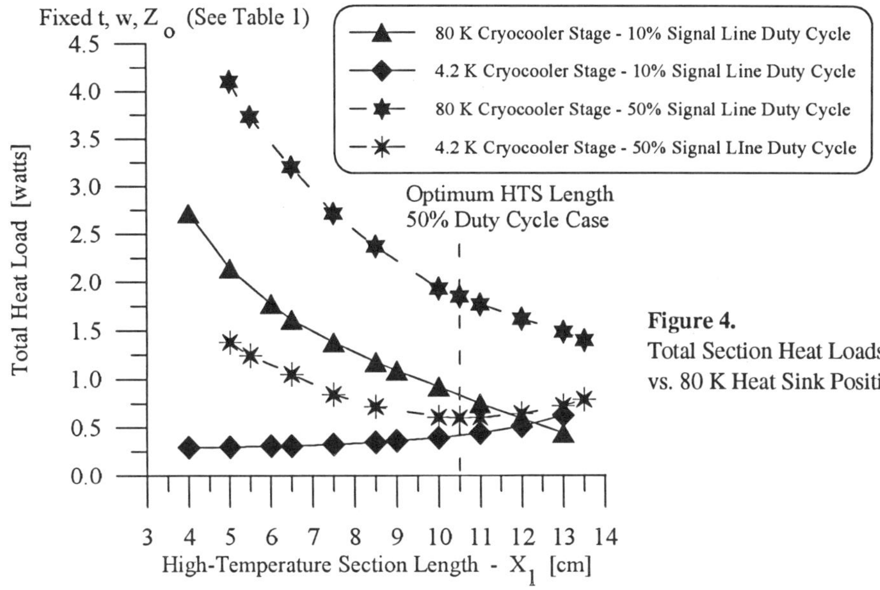

Figure 4.
Total Section Heat Loads vs. 80 K Heat Sink Position.

because of back thermal conduction up the cable from the LTS. If the HTS is too long and the LTS too short, then the LTS temperature gradient is again too high and increases total thermal load on the 4.2 K-stage above the optimum level.

Figure 5 also illustrates the difference in cable line temperature profiles for a 10% signal line duty cycle compared to the 50% signal line duty cycle case with optimum heat sink placement along the cable length. The 10% duty cycle case used the same HTS and LTS lengths as the 50% duty cycle case with optimum heat sink placement to make a consistent temperature profile comparison. Figure 5 shows the higher electrical dissipation component for the 50% duty cycle case significantly increases cable temperatures and cable temperature gradients at the 80 K and 4.2 K heat sink points. Of course, higher cable temperature gradients create higher total heat loads on the cryocooler stages.

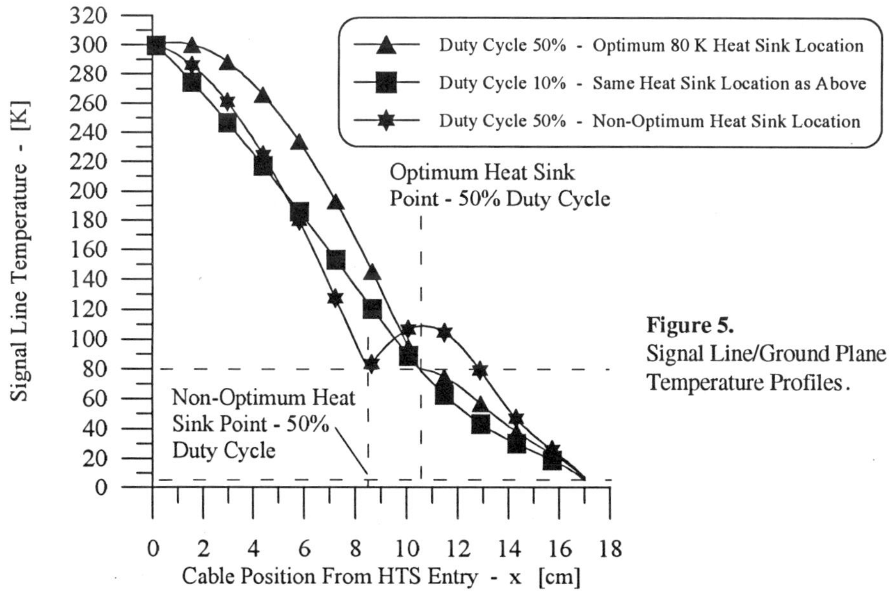

Figure 5.
Signal Line/Ground Plane Temperature Profiles.

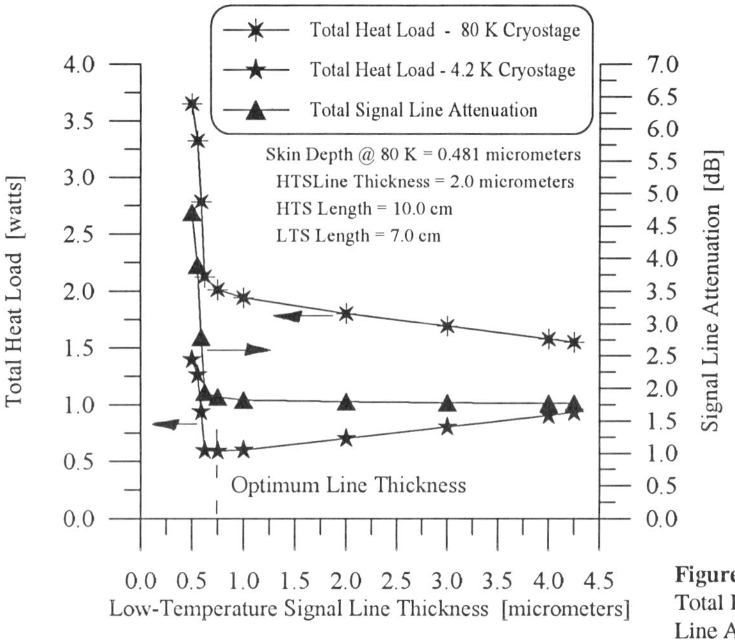

Figure 6.
Total Heat Loads / Signal Line Attenuation vs. LTS Signal Line Thickness.

Figure 6 shows total thermal load on both the 80 K and 4.2 K stages and signal attenuation versus LTS signal line thickness. It illustrates a very important design tradeoff associated with signal line thickness in ultra-low thermal load I/O cables. When signal attenuation and dissipation are not considered, one generally desires to make the signal line and ground plane as thin as possible to minimize pure thermal conduction down the cable. However, Figure 6 clearly illustrates there exists an optimum LTS signal line thickness below which there is increasing signal line attenuation and total thermal load on both the 80 K-stage and 4.2 K-stage. This is due to excessive signal attenuation in the LTS when the signal line is too thin. It is generally unacceptable from both a thermal design and signal attenuation viewpoint. This situation generally leads to the same LTS temperature profile as shown for the non-optimum heat sink location in Figure 5. When the signal line is too thick, then pure thermal conduction down the LTS also increases total thermal load on the 4.2 K-stage.

REFERENCES

1. Ellison, G.N., Hershberg, E.L., and Patelzick, D.L., "The Cryocooler Implications of Flexible, Multi-Channel I/O Cables for Low Temperature Superconducting Microelectronics", *Advances in Cryogenic Engineering*, Vol. 39B, Plenum Press, New York, (1994), pp. 1169-1176.

2. Ramo, S., Whinnery, J.R., and Van Duzer, T., *Fields and Waves in Communications Electronics,* 2nd Ed. New York, Wiley, (1984), p. 241.

3. Horton, R., Easter, B., and Gopinath, A., "Variation of Microstrip Losses with Thickness of Strip," *Electronics Letters,* Vol. 7, (1971), pp. 490-491.

4. Schneider, M.V., "Microstrip Lines for Microwave Integrated Circuits," *Bell System Technical Journal,* Vol. 48, (1969), pp. 1422-1444.

5. Gupta, K.C., Garg, R., and Bahl, I.J., *Microstrip Lines and Slotlines*, Dedham, Mass., Artech, (1979), pp. 68-69.

6. Childs, G.E., Ericks, L.J., and Powell, R.L., *Thermal Conductivity of Solids at Room Temperature and Below*, NBS Monograph 131, National Bureau of Standards, Institute for Basic Standards, Cryogenics Division, (1973), p. 97.

7. Simon, N.J., Drexler, E.S., and Reed, R.P., *Properties of Copper and Copper Alloys at Cryogenic Temperatures*, NIST Monograph 177, National Institute of Standards and Technology, Materials Science and Engineering Laboratory, Materials Reliability Division, (1992), pp. 7.1-8.58, 14.1-15.57.

A CRYOGENIC HYDROGEN MASER OPERATING AT 10K

J.J. Krupczak, Jr.[1], D.R. McAllaster[2], S.B. Crampton[2], A.L. Cole[2], and A.J. Kerman[2]

[1]Department of Physics and Engineering
Hope College, Holland, MI 49423, USA

[2]Department of Physics
Williams College, Williamstown, MA 01267-2565, USA

ABSTRACT

A cryogenic hydrogen maser has been designed and constructed which operates at temperatures near 10 K with a surface of frozen neon in the atom storage region. Cryogenic operation is expected to improve masers performance by orders of magnitude over room-temperature masers in frequency standard applications. The maser has an atom storage region and microwave cavity made from three pieces of single crystal sapphire. The neon surface is expected to be isothermal and stable to within 1 mK during operation. The sapphire is contained within a copper enclosure which is surrounded by magnetic shields and cooled by conduction to a helium-cooled copper plate. The helium cooling to the plate is in the form of cold helium gas flowing through a tube attached to the plate. The internal vacuum of the maser is maintained by a helium-cooled cryopump located between the hydrogen atom source and the transition. Atomic hydrogen is supplied by an effusive discharge operated at liquid nitrogen temperatures. The maser has operated as a self-excited oscillator and most of the required cryogenic characteristics have been experimentally verified. At temperatures near 10 K a stability of 1 mK was achieved for periods exceeding ninety minutes.

INTRODUCTION

The hydrogen maser is a device that amplifies microwaves by the process of stimulated emission. The basic underlying principle of operation is similar to that of a laser but the output is at microwave frequencies. Hydrogen masers are commonly operated as self-running oscillators producing a continuous signal at 1420 MHz, of low power but high frequency stability. Hydrogen masers operated at room temperature are the most stable frequency source for measuring times ranging from a few seconds to several hours[1]. Because the output frequency depends on the properties of the hydrogen atom, hydrogen masers are considered as one type of atomic clock. The high stability of the hydrogen masers results in their use as stable time and frequency standards with applications which include: experimental atomic physics, radio astronomy, space vehicle tracking, tests of general relativity, geodesy, and navigation.

Currently hydrogen maser frequency standards are operated at room temperature. However if a hydrogen maser could be operated at cryogenic temperatures, frequency stability might improve by as much as three orders of magnitude[2]. After the successful storage of atomic hydrogen on surfaces of frozen molecular hydrogen at liquid helium temperatures[3], the operation of a hydrogen maser at cryogenic temperatures became a possibility[4]. In a cryogenic hydrogen maser, the Teflon coating on the inside of the atom storage region is replaced by a surface of condensed inert gas of either frozen neon or helium in the superfluid state. The microwave cavity and hydrogen beam are also cooled to temperatures in the liquid helium range. One of the advantages of a cryogenic hydrogen maser is the elimination of copolymer Teflon coating from the walls of the storage region. The need to contain the hydrogen atoms in the storage bottle results in a shift in the maser output frequency caused by collisions between atoms and the bottle walls. This shift, called the wall shift, may be more repeatable by coating the bottle walls with a material to which hydrogen is only weakly adsorbed such as a surface of condensed inert gas. The frequency stability of hydrogen masers is limited by the ratio of the maser power to the thermally produced electrical noise in the maser microwave cavity and receiver electronics. The maser power is limited by spin-exchange collisions between hydrogen atoms[5]. At cryogenic temperatures the spin-exchange cross sections are significantly lower than at room temperature[6]. Operation of a hydrogen maser at cryogenic temperatures offers the possibility of greatly increased maser power and greater frequency stability due to a larger signal-to-noise ratio. The frequency stability of masers relies on spin-exchange tuning[7]. The theory of spin-exchange collision as extended by Verhaar, Koelman, Stoof, Luiten, and Crampton (VKSLC)[8] shows that at cryogenic temperatures, additional shifts exist which are not compensated by usual spin exchange tuning methods. These shifts threaten the stability improvements expected in cryogenic hydrogen masers. Using the VKSLC theory, Koelman et al. [6] have suggested ways to reduce these new sources of frequency instability. The first step in achieving the expected performance of cryogenic masers is verification of the spin-exchange frequency shifts predicted by the VKSLC spin-exchange theory in a working maser at cryogenic temperatures.

To date several groups have pursued the development of a cryogenic hydrogen maser[9-12]. While all of the groups have been able to produce operating cryogenic masers, an investigation of the capabilities of cryogenic masers in detail has not been possible. These efforts have been limited by: problems in cryogenic techniques, unexplained frequency shifts, difficulty in controlling the surface temperature, and non-uniformity in surface coating.

We have design and built a neon-surface hydrogen maser for the purposes of measuring hyperfine-induced frequency shifts in hydrogen at cryogenic temperatures and evaluating the potential of low-temperature hydrogen masers. This work was based on our earlier neon-surface hydrogen maser [12]. Here we describe the cryogenic design aspects of the maser and report our initial operating results. Other aspects of maser performance have been reported elsewhere[13].

DESIGN REQUIREMENTS

While the technology of room temperature hydrogen masers is well-established, the development of a hydrogen maser suitable for low-temperature operation presents a number of cryogenic engineering challenges. The design requirements are sumarized in Table 1. During maser operation, the temperature of the neon surface must be cooled to a temperature in the range of 10 to 11.5 K. Within this range the precise operating temperature will vary with surface conditions. The neon surface must be isothermal to within ±0.001 K and the surface temperature must remain stable to ±0.001 K for periods of several hours. These conditions must be maintained with the possibility of a heat flux to the surface due to the recombination of hydrogen atoms on the order 0.5 mW. During experiments the period of continuous operation at low temperatures is expected to be 8 to 24 hours. To prepare the neon surface, the maser must have the capability of condensing a frozen neon surface 20 microns thick on the inside of the atom storage region. Surfaces are to be frozen slowly over periods of several hours at 24.6 K near the triple point of neon. The frozen surface must then be cooled slowly, over a period of hours, from 24.6 K to the operating temperature at 10 K. During cooling the surface temperature must remain uniform so that the surface does not crack during cooling.

Table 1. Neon-surface cryogenic hydrogen maser design requirements

Category	Requirement	Specification
General H Maser	Avoid H beam attenuation	H mfp >> L
	Homogeneous magnetic field	< 0.050 mGauss
	Microwave cavity	1420.40 MHz
	Coarse and fine tuning	1 MHz, 50 kHz
	Coupling and waveguide output	4 loops
Neon-Surface H Maser	Operating Temperature	10-11.5 K
	Surface temperature uniformity	±0.001K
	Temperature stability	±0.001K
	Duration of continuous operation	24 hours
	State-selected beam intensity	10^{14} atoms/s
	Surface preparation temperature	25 K
	Duration of surface freezing period	several hours
	Neon surface thickness	20 microns
	Duration of cooldown from 10 to 11.5 K	several hours
General	Heat generation	Q < Cooling
	Frequency shift due to H re-entry	<< 10^{-13}

Besides the requirements unique to the neon surface, the maser must fulfill general hydrogen maser requirements[5]. Requirements include the need for a microwave cavity with a frequency of 1420.405 MHz and sufficiently narrow resonance. Hydrogen beam attenuation is minimized if vacuum conditions are such that the mean-free-path between collisions of the hydrogen atoms is long compared to the distance between the source and the storage region. In the maser microwave cavity, a homogeneous magnetic field is required with gradients of the magnetic field small enough not to cause atomic relaxations. The microwave cavity must be equipped with some capability for external tuning while cold, and suitable coupling and waveguides must be also provided to carry the maser output signal to the external electronics. A hydrogen beam intensity of 10^{14} state-selected atoms/sec is desired.

During steady-state operation, hydrogen atoms which exit the microwave cavity must have negligible chance of returning to the cavity in the atomic state. Hydrogen atoms which exit and reenter the cavity lead to undesired frequency shifts. In practice, this requirement implies that the neon must not be allowed to freeze anywhere in the maser except on the surface of the atom storage region.

DESCRIPTION OF MASER

Overview

The maser is shown in Figure 1. The design is based on an atom storage region and microwave cavity made of three pieces of single crystal sapphire. The high thermal conductivity of sapphire at liquid helium temperatures promotes an isothermal substrate for the frozen neon. The sapphire is cooled by conduction to a copper plate which is in turn cooled by cold helium gas flowing through a heat exchange tube. The sapphire microwave cavity and storage region, along with the magnetic shields and the copper plate, are contained within a stainless steel vacuum enclosure. A low thermal conductivity transition region connects the sapphire cavity to the region above, and maintains the required temperature profile outside of the storage region both during preparation of the neon surface and steady-state operation. The internal vacuum of the maser is maintained by a helium-cooled cryopump located between the hydrogen atom source and the transition. Atomic hydrogen is supplied by an effusive discharge operated at liquid nitrogen temperatures.

The maser is surrounded by a bath of liquid nitrogen contained in a superinsulated storage dewar (not shown). Figure 2 shows an alternate view of the contents in the lower maser vacuum can.

This maser contains a number of features included to help achieve the design objectives. The use of continuous flow cooling allows operation over the 10 to 25 K temperature range, and facilitates neon surface formation by insuring that cooling is localized to the atom storage region. The internal cryopump produces high vacuum with low contamination and can be both turned off and regenerated without warming the rest of the maser. The transition section has also been optimized for the requirements of the neon-surface hydrogen maser.

Cryogenic Hydrogen Source

Hydrogen gas enters the source from the inlet line at the top. The hydrogen inlet is a 0.8 mm outer diameter stainless-steel capillary tube which connects to the hydrogen gas handling system. The capillary is soldered through a series of adapter pieces to the stainless-steel end piece of the source tube. The source tube is Pyrex glass of length 127 mm. The wall

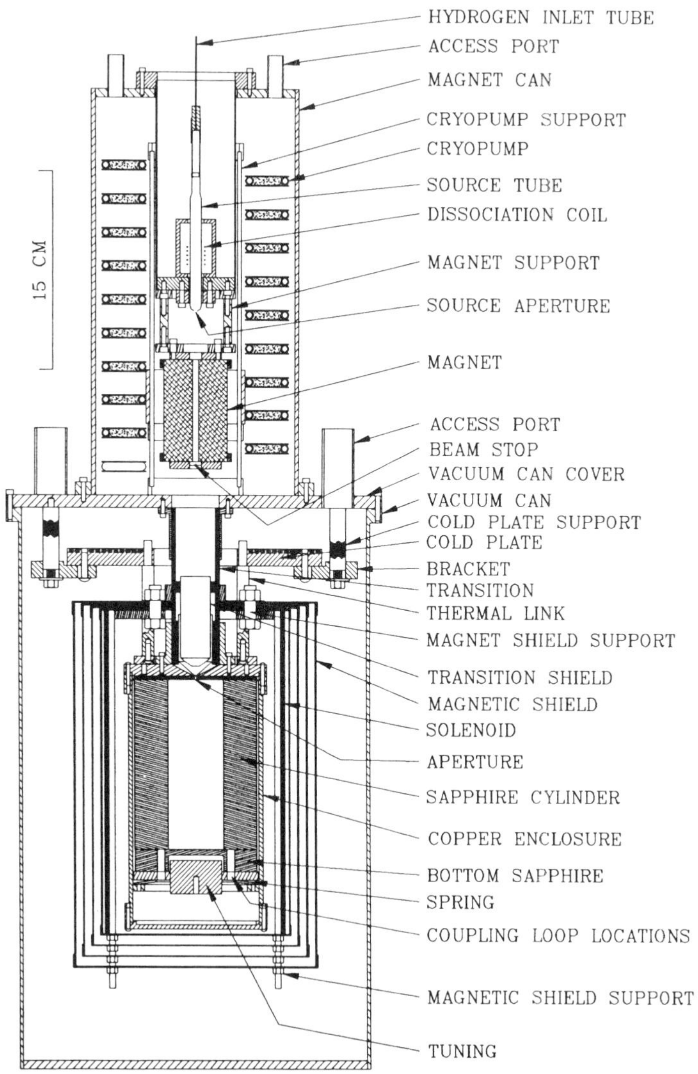

Figure 1. Cryogenic Hydrogen Maser.

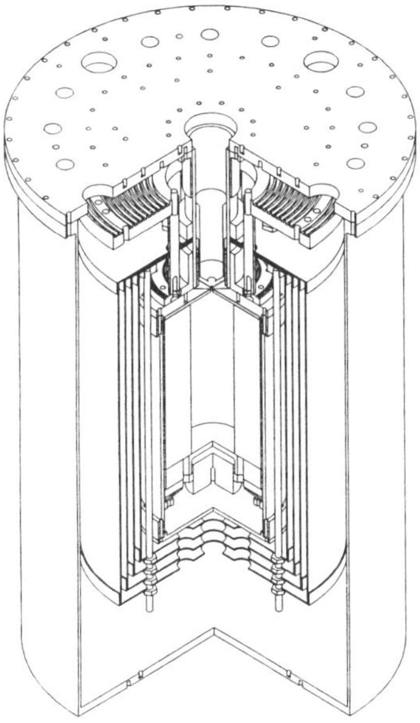

Figure 2. View of cryogenic hydrogen maser lower vacuum can.

thickness is 0.8 mm except near the aperture where the wall thins to 0.4 mm. The source aperture diameter is 2 mm. The Pyrex source tube is bonded with indium to a stainless steel flange. This bond forms a cryogenic vacuum seal between the Pyrex tube and the flange. The dissociation coil and associated resonant cavity are located around the Pyrex source tube. The coil and cavity have a resonance at 200 MHz. A transmission line of semi-rigid coaxial cable (not shown) is used to provide RF power to the discharge cavity. The remainder to the hydrogen source consists of the state-selection magnet, beam stop, and the necessary mounting hardware. The magnet and beam stop were provided by R. Vessot of the Harvard-Smithsonian Center for Astrophysics. The magnet support consists of a set of four aluminum posts and a pair of mounting rings. The rings attach to the magnet and the bottom flange of the source mounting structure, the posts determine the position of the magnet with respect to the source aperture. To prevent non-state selected atoms or contaminants from reaching the sapphire cylinder, a beam stop of a 2 mm diameter glass bead formed over a length of tungsten wire. An enclosure called the magnet can surrounds the hydrogen source.

Cryopump

A cryopump surrounds the hydrogen source and the state-selecting magnet. During operation, the vacuum inside the maser is maintained by the cryopump. The cryopump is fabricated from size 14 mesh charcoal granules sandwiched between an 0.8 mm thick sheet of OFHC copper and stainless steel screen. The copper tubing through which the helium coolant flows is soldered to the edges of the copper sheet. Soft copper refrigeration tubing with an outer diameter of 6 mm is used. The cryopump is wound in a continuous 9-turn spiral arrangement. A separate coaxial helium line (not shown) supplies coolant to the cryopump. A structure made of G-10 rings and phenolic rods supports the cryopump. The charcoal is regenerated using heaters (not shown) located on the bottom side of the copper sheet. The heaters are stainless steel conductors sealed in Kapton film. The cryopump operates at 6 K and is designed to pump hydrogen at a rate of 200 standard liters per minute for 24 hours.

Cooling System

The cooling system and the other portions of the maser which operate at temperatures in the range of 10 to 25 K are located in the lower stainless steel enclosure called the vacuum can. This can has a diameter of 298 mm in and a height of 425 mm. The 10 mm thick can cover is sealed with an indium gasket.

The maser cooling system is continuous flow cooling using cold helium gas. This method was chosen over a closed-cycle cryocooler based on considerations of temperature stability, mechanical vibration, and the ability to integrate the cooling system with the other maser components. Continuous flow cooling control over was favored over bath immersion in liquid helium, because of the greater degree of localization of cooling possible with a continuous flow system.

The scheme used resembles that described by Klipping et al.[14]. Cold helium vapor at approximately 4.5 K travels from the storage dewar to the coldplate through a vacuum-insulated transfer line. The inlet in the storage dewar is made of a coaxial pipe packed with copper wool. The coldplate consists of an 203 mm diameter and 12.7 mm thick OFHC copper plate into which a coil of 3.2 outer diameter and 1.6 mm inner diameter copper tube has been indium soldered. The tubing forms a continuous spiral with a total length of 4 meters. The cold helium gas is always confined inside the transfer lines or the copper tubing. A pressure regulating valve an vacuum pump in the exhaust line are used to maintain a constant exhaust pressure. This system provides approximate control of the temperature by matching the flow rate of gas to the heat input to the coldplate. The cold plate is fixed in place by a set of copper brackets held by G-11 posts attached to the vacuum can cover.

Transition Section

A thermal transition lies between the state-selecting magnet and the atom storage region. This component is called the transition because a temperature change from 77 K to the maser operating temperature must occur across this section. The transition section thermally isolates the atom storage region from the rest of the maser. In addition to establishing the desired temperature distribution, the transition provides for vacuum isolation, beam conductance, magnetic field homogeneity, structural alignment ,and support.

The transition consists of a G-11 fiberglass tube with an aluminum flange at the magnet end, and a copper flange at the microwave cavity end. The total length is 120 mm and the outer diameter in the narrowest section is 38 mm. The G-11 is bonded to the copper and aluminum flanges with epoxy. Bolted connections with indium seals secure the flanges to the maser components at either end. The flanged G-11 tube provides the basis for the thermal isolation, physical strength, magnetic field homogeneity, and flow conductance requirements of the transition.

Within the main G-11 tube is a component called the transition shield. The shield is a thin tube made of copper. The lower end tapers into a truncated cone made of 0.025 mm thick OFHC copper. The shield is epoxied to a support made of G-10. To hold the shield in position, the G-10 shield support is epoxied onto the interior of the G-11 tube. The function of the transition shield is to insure that hydrogen atoms leaving the storage region through the aperture have a negligible chance of reentering the aperture. This is accomplished by creating conditions such that the hydrogen atoms effusing from the cavity recombine into molecules. In steady state maser operation near 10 K, the shield serves as a warm surface of 30 K or greater on which the hydrogen atoms leaving the storage region can recombine into molecules. The geometry and dimensions of the shield were determined through thermal analysis using the ANSYS finite element analysis package (Swanson Analysis, Inc. Hudson, PA, 15342, USA.)

The lower portion of the transition tube contains a section called the insert. The insert is a thin copper tube with an open cone attached to the lower end. During surface formation, the copper insert is maintained at a temperature above the freezing temperature of neon. This insures that the neon will freeze inside the sapphire rather than in the transition region. The lower conical portion of the insert is held in place and insulated from the cavity cover by cast epoxy. The insert was designed such that the surface area colder than the freezing temperature of neon is confined to the 0.8 mm of the epoxy below the conical portion of the insert. The thermal transition to temperatures below the freezing temperature of neon takes place across this small portion of the transition section just above the microwave cavity.

Microwave Region

The microwave region of the hydrogen maser is the area where the output signal is generated. A primary requirement for this component is a surface of the atom storage region that is isothermal and stable to within ±0.001 K The atom storage region and microwave cavity consists of three pieces of sapphire and a cylindrical copper enclosure. By providing a hollow interior, the sapphire is both the storage region and the microwave cavity. The frozen neon surface is formed on the inside surface of the sapphire.

The sapphire cavity consists of three separate cylindrical pieces of single-crystal sapphire (Union Carbide Corporation, Danbury CT 06817, USA.), and is similar to that used by Walsworth et al.[9]. The 2.5 mm thick top piece has a 1.93 mm diameter central aperture to allow about a 1 second storage time for hydrogen atoms in the cavity. The main sapphire cylinder has an outer diameter of 98.1 mm, an inner diameter of 44.4 mm, and a length of 128.3 mm. A bottom sapphire piece 17.1 mm in length closes the storage region and contains a large central channel and 4 holes for the tuning and coupling loops. An OFHC copper cylinder mounted on the cavity axis provides 4 MHz of coarse tuning. Fine tuning of about 10 kHz is accomplished through a varactor diode coupled into the cavity. Two coupling loops (not shown) are also installed in the bottom piece. On the outer diameter of all sapphire pieces there is a 0.5 to 1.0 mm thick silver coating. By coating the outer surfaces with a conducting material, and properly selecting the cylinder dimensions, a microwave cavity at the desired frequency can be established in the sapphire itself. The high thermal conductivity of the sapphire promotes temperature uniformity, and the high dielectric constant results in small physical dimensions.

An OFHC copper enclosure contains the sapphire pieces, and provides a thermal shield and vacuum enclosure for the sapphire. A phosphorous bronze belleville spring (Key Bellvilles, Inc. Leechburg, PA 15656, USA), provides a 0.7 kN force and holds the sapphire pieces together. The spring also compensates for the different amounts of thermal contraction between the copper enclosure and the sapphire. Heat flow to and from the sapphire is through the top. The sapphire does not touch the sides of the copper cylinder. To minimize the thermal contact resistance, the contacting surfaces of the sapphire top piece and the cover of the enclosure are coated with about 0.5 micron of gold. To establish a homogeneous magnetic field, the maser cavity is surrounded by a solenoid and a set of 4 cryoperm 1 mm thick low-temperature magnetic shields.

The temperature of the neon surface covering the sapphire is controlled by controlling the temperature of the copper enclosure. Temperatures measured by four germanium resistance sensors (Lakeshore Cryotronics, Westerville, OH, USA, series GR-200) mounted on the exterior of the copper cavity enclosure (not shown). The sensors are repeatable to 1 mK and, as read by the ohmmeter in the Lakeshore DRC-93C controller, and have a resolution of 0.2 mK The temperature of the enclosure is regulated by a bilfilar heater of phosphorous bronze wire operating in a feedback loop with a sensor and the temperature controller.

The maser requires auxiliary systems for hydrogen and neon gas handling and vacuum pumping. These systems are located external to the maser. A 65 Liter/s turbopump is used to initially pumpdown the apparatus and regenerate the cryopump.

RESULTS AND CONCLUSIONS

Preliminary experiments have been carried out with this cryogenic hydrogen maser operating under self-sustained oscillation. The initial measurements for the drift removed intrinsic stability show frequency residuals comparable to our room temperature masers. A problem has been a persistent frequency drift. The drift may be due to contamination of the neon surface by impurities or from a thinning in the neon coating.

Cryogenic performance has been found to generally as expected. The temperature distribution in transition region has been measured and found to be consistent with the ANSYS analysis. The cryopump maintained a suitably low vacuum pressure of less 10^{-5} Torr for eight hours under maximum discharge from the hydrogen source. The temperature uniformity of the copper enclosure has been worse than expected possibly due to greater than anticipated heat conduction through the coaxial waveguides. At both 10 K and 25 K operation, temperature stability in the microwave region of better than ±0.001 K was observed for periods exceeding 90 minutes. The duration of the stability was limited only by the observation time.

Presently, experimental effort is focused on improving the methods used to create the neon surface. The possibility of adding a quartz liner to the storage region is under consideration. This is expected to reduce the effects of paramagnetic impurities in the sapphire. After implementation of these changes it should be possible to use this maser to measure spin-exchange induced frequency shifts at 10 K and investigate some of the characteristics expected from cryogenic hydrogen masers.

ACKNOWLEDGMENTS

This research was carried out at the Williams College Hydrogen Maser Laboratory with support from the National Science Foundation under contracts: PHY-8807983, and PHY-9313811, and from the Air Force Office of Scientific Research under contract: AFOSR-91-0312. The authors would like to thank: Bryce Babcock, Andy Beveridge, Ernie Bombardier, Peter Bronk, Kevin Forkey, Heidi Hazlett, Lisa Kable, Barb Madden, Emile Oulette, Craig Pedercini, Howard Rhett, Ed Shearer, Carter Smith, Katrina Spraggins, Dillion Tracy, R.F.C. Vessot, Phat Vu, and George Walter, for their many contributions made during the course of this work.

REFERENCES

1. Robert F. C. Vessot, "State of the art and future directions for the atomic hydrogen maser," Proc. of the 22nd Precise Time and Time Interval (PTTI) Applications and Planning Meeting, 1990.
2. W. N. Hardy and M. Morrow, *J. de. Physique*, 42:C-171 (1981).
3. S. B. Crampton, T. J. Greytak, D. Kleppner, W. D. Phillips, D.A. Smith, and A. Weinrib, *Phys. Rev. Lett.* 42:1039 (1979).
4. S. B. Crampton, W.D. Phillips, and D. Kleppner, *Bull. Am. Phys. Soc.* 23:86 (1978).
5. D. Kleppner, H. C. Berg, S. B. Crampton, N. F. Ramsey, R. F. C. Vessot, H. E. Peters, and J. Vanier, *Phys. Rev.*, 138:A972, (1965).
6 J. M. V. A. Koelman, S. B. Crampton, H. T. C. Stoof, O. J. Luiten, and B. J. Verhaar. *Phys. Rev.* A38:3535 (1988).
7. S. B. Crampton, *Phys. Rev.* 158:57 (1967).
8. B. J. Verhaar, J. M. V. A. Koelman, H. T. C. Stoof, O. J. Luiten, and S. B. Crampton, *Phys. Rev.* A35:3825 (1987).
9. R. L. Walsworth , I.F. Silvera, H. P. Godfried, C. C. Agosta, R. F. C. Vessot, and E. M. Mattision. *Phys. Rev. A.* 34:2550 (1986).
10. H.F. Hess, G. P. Kolchanski, J. M. Doyle, T. J. Greytak, and D. Kleppner. *Phys. Rev. A.* 34:1602 (1986).
11. M. D. Hurlimann, W. N. Hardy, A. J. Berlinsky, and R.W. Cline, *Phys. Rev.* A34:1605 (1986).
12. S. B. Crampton, K.M. Jones, G. Nunes, Jr., and S.P. Souza, Proc. 16th Precise Time and Time Interval (PTTI) Applications and Planning Meeting, Greenbelt MD 1984, p.339.
13. D.R. McAllaster, J.J. Krupczak, Jr., A.L. Cole, A.J. Kerman, and S.B. Crampton, Proc. of the 1994 IEEE International Frequency Control Symposium, (1994) p. 677.
14. G. Klipping, W.D. Schonherr, and W. Schulze, *Cryogenics*, 10:501(1970).

NOVEL METHOD FOR MEASUREMENT OF THE PARA-CONTENT OF A HYDROGEN-STREAM

J. Clausen[1], A. Hofmann[2], M. Wanner[2]

[1] LOTEPRO Corporation, Valhalla, NY 10595, USA
[2] LINDE AG, Process Engineering and Contracting Division
D-82049 Höllriegelskreuth, Germany

ABSTRACT

Hydrogen at ambient temperature is a mixture of 75% ortho- and 25% para-hydrogen whilst it is 100% para-hydrogen at boiling temperature. Hydrogen liquefaction processes use catalysts to stepwise convert normal-hydrogen to para-hydrogen during cooldown to cryogenic temperatures. This conversion at different temperature levels is necessary to avoid release of the heat of conversion in the liquid state which would result in losses due to enhanced evaporation. Therefore customers can expect to receive a liquid hydrogen quality with a para-content of almost 100% from their gas supplier.

A novel technique was developed[1] to precisely measure the para content during cool-down of the hydrogen as well as of the liquid product . The measurement principle is based on the caloric differences of ortho- and para-hydrogen. A direct determination of the para-concentration is accomplished by using a simple measurement of a temperature difference and is therefore superior to previous techniques which are based on the thermal conductivity of hydrogen. The new method was experimentally verified. It has shown a very close agreement with theory.

This paper describes the novel method along with the experimental apparatus and the test results. Based on the experimental setup an industrialized instrument was developed and is now available. It features a direct readout of the para-content of the measured hydrogen stream.

INTRODUCTION

Hydrogen has a great potential for expanded use as an energy carrier in the future. This is due to its unlimited availability as a component of water, its low-polluting nature after combustion, its storage capability and transportability in the liquid phase. The hydrogen molecule, H_2, exists in two modifications, as ortho-hydrogen and as para-hydrogen, which are distinguished by the spin of their atomic nuclei. The two different orientations of the nuclear spin result in different magnetic, optical and thermal properties.

The equilibrium composition between ortho- and para-hydrogen is dependent on temperature and changes from 25% para concentration at ambient temperature to almost 100% at 20 K, the boiling point of hydrogen. The conversion of ortho- to para-hydrogen is exothermic and automatically takes place over a period of several days. It can be accelerated with the help of catalysts. Since the heat of transition at the boiling point is about 1.2 times as great as the heat of evaporation it is advantageous that the heat of transition is removed by catalytic converters at different temperatures before liquefaction. A purchaser of liquid hydrogen who would like to store the hydrogen for long periods would require a para content of at least 98%.

Up to now, the determination of the para content has been based on the different thermal conductivities of ortho- and para-hydrogen[2]. The conductivity difference between equilibrium and para-hydrogen amounts to only 2% at ambient temperature and is in addition dependent on the temperature. Therefore the measurement has to compare the thermal conductivity of the gas being measured against a para-hydrogen reference at exactly the same temperature. This reqires very accurate measurements of the thermal conductivity.

PRINCIPLE OF MEASUREMENT

The novel method introduced makes use of the different enthalpies of ortho- and para-hydrogen. Para-hydrogen has the lower enthalpy which makes it the predominant component at low temperatures. The enthalpy of the hydrogen mixture increases proportionally with the ortho content and the difference in enthalpies shows up as the heat of conversion.

If for instance a para-hydrogen stream is heated to an elevated temperature (as e.g. to liquid nitrogen temperature) and is then brought to its equilibrium composition in an adiabatic catalytic converter, the gas will cool by about 20 K. This occurs because the heat of conversion has to be extracted from the gas mixture.

In a similar way a flow of normal-hydrogen (which describes the equilibrium mixture at ambient temperature with 25% para and 75% ortho content) will warm by about 12 K if it is first cooled to 77K and then passed through the same reactor.

This effect is shown in Figure 1 where the equilibrium para-concentration is shown as a function of temperature.

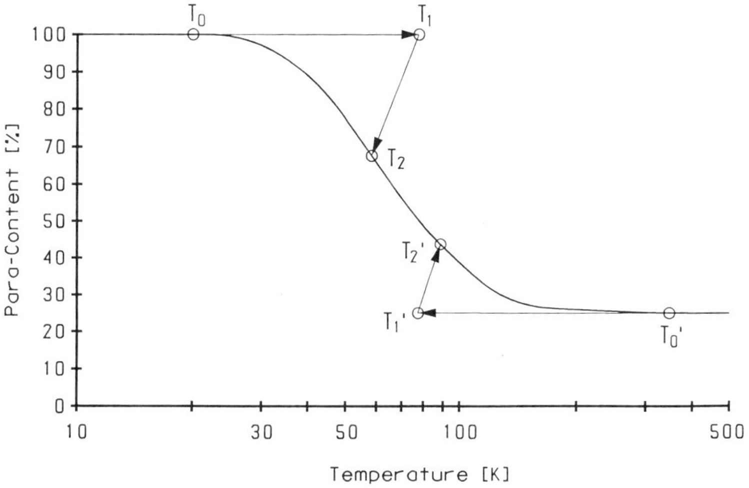

Figure 1. Equilibrium para-content as a function of the temperature. The arrows point out the two situations described in the text.

Let x_1 be the initial para-content of the hydrogen and T_1 and P_1 the temperature and pressure at the inlet of the adiabatic reactor. After the stream passes through the catalyst an equilibrium is achieved where temperature and pressure have changed to P_2, T_2 and the para content x_2 is in equilibrium with the temperature T_2.

If h(x, T, P) denotes the enthalpy of hydrogen[3], the para-content x_1 can be calculated from

$$h_1(x_1, T_1, P_1) = h_2(x_2, T_2, P_2)$$

since in thermal equilibrium x_2 is only a function of the temperature T_2. The dependence of the enthalpy on pressure is negligible, therefore the measurement of the para content can be simply reduced to a measurement of the temperature T_1 and the difference $T_2 - T_1$. Because the thermodynamic properties of hydrogen are well known this technique can be used to determine x_1.

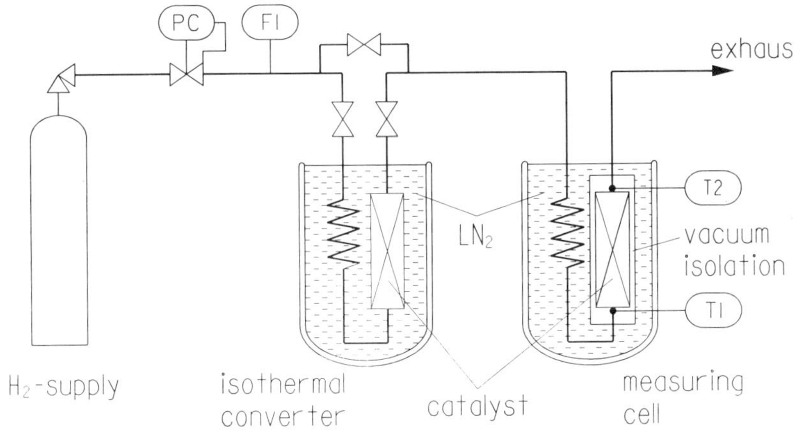

Figure 2. Flow schematic of the experimental setup.

The maximum temperature effect $T_2 - T_1$ would be at an inlet temperature T_1 into the reactor of appr. 120 K. It is however advantageous to use the constant temperature of 77 K of a boiling liquid nitrogen bath to define the inlet temperature. In that case the sensitivity of the instrument would be appr. 0.4 K per percent change of the para content over the full measuring range from 25 - 100 % para-hydrogen.

DESIGN OF THE INSTRUMENT

A prototype instrument was designed, built and tested to verify the measuring method. For that purpose an experimental setup was established according to Figure 2. Hydrogen with a para content of 25% is taken from a gas bottle at a constant flow rate and throttled to slightly above ambient pressure.

For reference purposes this gas is either fed directly to the measuring cell, which acts as an adiabatic reactor or first passed through an additional isothermal converter consisting of a heat exchanger and a fixed bed reactor filled with catalyst.

When placed in a liquid nitrogen bath this isothermal converter can achieve a para content of 50.5% as a reference point.

Another reference point is obtained by using almost pure para-hydrogen from a large liquid hydrogen storage tank which is connected to a hydrogen liquefier. This superinsulated tank will stay at a temperature of 21.6 K for very long periods. For that reason one can assume with good confidence an equilibrium concentration of 99.5 %.

Upon leaving the isothermal converter the gas is warmed-up to ambient temperature and is fed to the measuring cell. There the measuring gas stream is brought into thermal equilibrium in a heat exchanger which is placed in a liquid nitrogen bath. After controlling the inlet temperature T_1 the gas enters the adiabatic reactor section which is thermally insulated from the nitrogen bath. Once the temperature difference $T_2 - T_1$ is measured, the gas is vented.

Several aspects were considered during the design of the instrument:

- To prevent the change of the para-content of the hydrogen stream occuring during flow through the feed-line to the instrument, nonmagnetic materials such as high grade steel or copper pipes are used for the feed line.

- To maintain adiabatic conditions in the converter all potential sources of heat input were minimized. This holds for the thermal insulation of the reactor where high reflective multilayers were combined with a high vacuum. Longitudinal heat conduction along the gas exit pipe as well as through the electric connections to the thermometers is minimized by sufficient thermal length and thermal anchoring of the cables to the liquid nitrogen bath.

- The reactor vessel is thin walled in order to keep longitudinal heat conduction effects along the catalyst bed to below 0.2% of the temperature difference.

- The catalyst bed is designed to achieve at least 99.5% of the equilibrium conversion. Fluidization of the catalyst bed is avoided by passing the hydrogen flow from top to bottom. On the other hand channelling effects and wall dispersion have to be avoided by proper determination of the gas flow rate.

- To achieve a short time constant of the instrument the mass of the reactor was kept low resulting in a nominal hydrogen flow of $0.18 m^3/h$.

The design of the measuring cell is shown in Figure 3. The dimensions of the instrument are about 100 mm diameter and a length of about 250 mm. A helical wound heat exchanger pipe surrounds the vacuum can which insulates the fixed bed reactor. Hydrous Ferric Oxide ($Fe OH_3$) produced by IONEX [4] was used as catalyst and was baked to 120 °C according to the manufacturer's instructions. Thermal insulation is performed by a permanent high vacuum which is maintained by a special getter material sized for the lifetime of the instrument. In addition a copper radiation shield, which is thermally connected to the exit pipe, reduces the heat input to the reactor bed.

The temperature difference is measured by thermocouples of the NiCr-Ni-type using a differential wiring. The reference temperature T_1 is given by the liquid nitrogen bath. The sensitivity of the temperature measurement at liquid nitrogen temperature resulted in a temperature resolution of 0.2K.

The measurement of the thermocouple voltage is performed by a digital voltmeter. The instrument uses a microprocessor controller to directly convert the voltage signal into the corresponding para content.

The entire measuring cell is immersed in dewar filled with liquid nitrogen.

The isothermal converter is used to establish a reference flow with a constant and well defined para content. The converter consists of a pipe type heat exchanger to precool the gas and a heat-exchanger filled with catalyst. In order to enhance the heat exchange between the converter and the surrounding liquid nitrogen bath the catalyst bed has the shape of a cylinder ring. The dimensions of the heat-exchanger and the converter are designed to minimize deviations of the para content well below 1% of the equilibrium value.

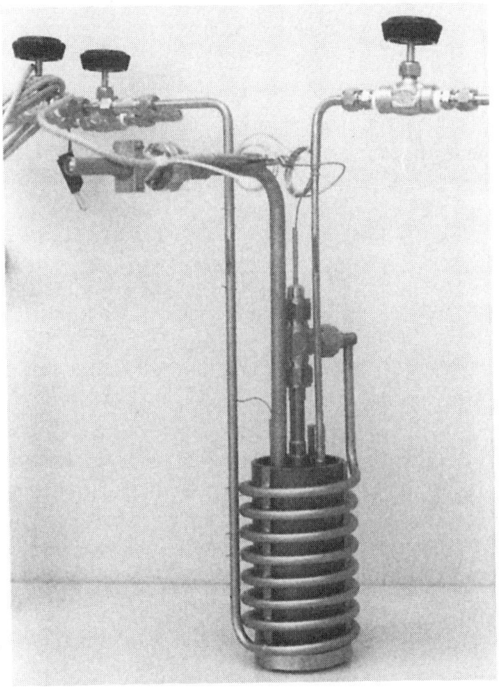

Figure 3. Photograph of the measuring cell.

EXPERIMENTAL RESULTS

As a first step, the zero-reading of the measuring cell was checked. For that purpose hydrogen is first sent through the isothermal converter placed in a nitrogen bath and then to the measuring cell. Since the gas is already converted to 50.5% para-hydrogen in the

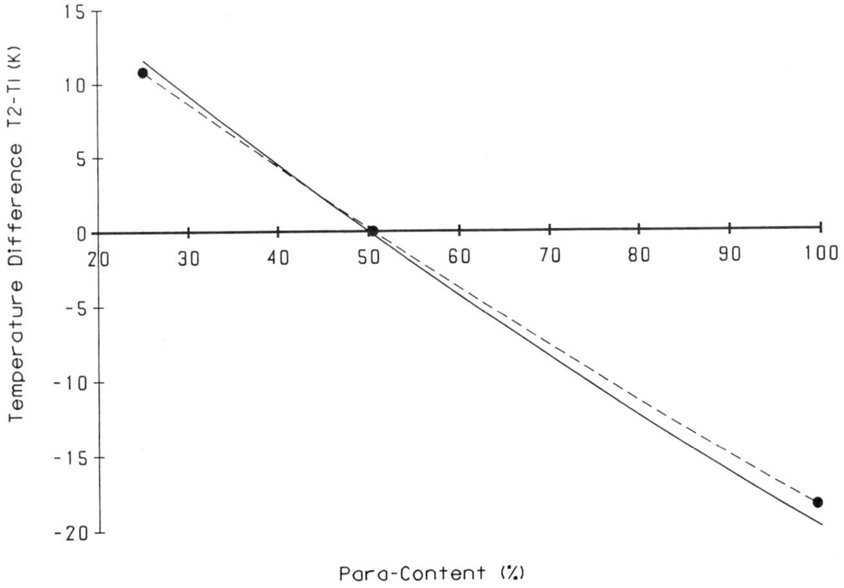

Figure 4. Comparison of the theoretical characteristic with the experimental data (dots).

isothermal reactor, no further conversion will take place in the measuring cell. Since both reactors work at the same temperature the resulting temperature difference will be zero.

In a next step the isothermal reactor is bypassed and the gas from the hydrogen supply bottle, which now has a para content of 25%, is sent directly to the instrument. The conversion to the new equilibrium concentration causes the reactor bed to warm. Due to the heat capacity of the catalyst and the pipe material it takes approximately 10 minutes until the hydrogen flow has stabilized to it's new equilibrium temperature. A temperature difference of 10.8 K was measured in this case.

Additional measuring points can be taken by placing the isothermal reactor into other cryogenic liquids such as oxygen, neon or even hydrogen. For simplicity it was easier to use a gas sample from a liquid hydrogen storage tank. The pressure in this tank is kept at 1.4 bar(a), corresponding to a bath temperature of 21.6 K and a para content of 99.5%. It is assumed that no conversion takes place in the stainless steel pipe connecting the storage tank with the instrument. The observed temperature difference was 18.4 K.

Figure 4 shows the measurements (dots) taken by the instrument together with the theoretical curve. The graph, which shows the temperature difference as a function of the para content, confirms that there is a very close agreement with the predictions from the theory.

It is observed that the temperature differences are be slightly less than the theoretical predictions from the ideal change of the thermodynamic state. However this deviation corresponds only to 3% of the para-content at 100% para-hydrogen and 2% at 25% para-hydrogen. This deviation is somewhat larger than the estimated contributions from longitudinal heat conduction within the catalyst bed and the reactor housing the wiring as well as the heat leak from the surrounding. The remaining discrepancy may be attributed to uncertainties in the thermal conductivity data of the catalyst bed as well as in the caloric data of ortho- and para-hydrogen. Moreover the contribution of the longitudinal heat conduction has to consider the nonlinear temperature change within the catalyst which complicates the analysis. Uncertainties from the temperature measurement are within 0.5% of the para-concentration.

The instrument has an almost straight characteristic. Therefore it can be represented with sufficient accuracy by two linear sections between the three measurement points at 25, 50 and 100% para content

Changes in the barometric pressure in the range between 900 and 1000 mbar which affect the boiling temperature of the liquid nitrogen bath, add only less than 1% change to the measurement and may be electronically compensated.

CONCLUSIONS

The novel method relates the para content of a hydrogen stream to a temperature difference. The method was verified experimentally and allows a precise measurement of the ortho-para composition of a hydrogen stream. The temperature signal is processed electronically to give a readout over the full range between 25 and 100% para content The instrument has an accuracy of better than 1%.

REFERENCES

1. M.Wanner, Process and Device for Determination the Para Content of a Hydrogen Gas Stream, Patent applications in Germany, US and Japan pending
2. V.J.Johnson, Properties of Materials at Low Temperature (Phase1) A Compendium, Pergamon Press (1961)
3. Hydrogen Technology Survey - Thermophysical Properties NASA-SP 3089 (1975)
4. Hydrous Ferric Oxide catalyst made by IONEX Research Corporation, Broomfield CO (USA)

SENSITIVE HELIUM LEAK DETECTION IN CRYOGENIC VACUUM SYSTEMS

M. G. Rao

Continuous Electron Beam Accelerator Facility
12000 Jefferson Avenue
Newport News, Virginia, 23606, USA

ABSTRACT

Conventional leak detectors are not useful for cold leak detection of LHe cooled cryogenic vacuum systems since He is adsorbed onto the cold walls. Even though the He desorption leak detection system[1] developed at the Continuous Electron Beam Accelerator Facility (CEBAF) is very sensitive, the cryogenic system needs to be warmed up to at least 10.5 K for leak detection purposes, and it still would be impossible to identify the location of the leak. Recently hot cathode extractor gauges have been found to be reliable, very useful and convenient for very low pressure measurement (10^{-14} torr) at LHe temperatures. This paper discusses the operation of extractor gauges at LHe temperatures and their in situ use for fast and convenient cold leak detection of cryogenic vacuum systems. Further, the location and identification of the leaks based on the principle of the pressure wave travel of He and time-of-flight measurements of the He signal are also discussed.

INTRODUCTION

Pressure measurement in cryogenic vacuum systems is generally carried out with room temperature vacuum gauges. In practical systems such measurements do not reflect the true pressure, not only because of the outgassing of the gauge electrodes as well as the room temperature parts of the system, but also because of the large errors (up to 30%) in the estimation of the thermomolecular pressure correction for obtaining the real pressure. Such low pressure measurement difficulties can now be eliminated by operating the extractor gauges at LHe temperatures.[2] These cold gauges are extremely sensitive to the presence of He leaks. Since all the other gas species are cryopumped, they can also be used for the in situ location of leaks with the help of the pressure wave travel of He. This paper presents the details of the operation of the extractor gauges at LHe temperatures and their use in cold leak detection. We will also discuss the identification and location of leaks in situ with time-of-flight measurements. Further, the needed parameters (DRK constants) for technical surfaces (stainless steel, niobium, niobium oxide and aluminum oxide) are also included in this paper based on our measurements.

OPERATION OF EXTRACTOR GAUGES AT LHe TEMPERATURES

The vacuum integrity of CEBAF linacs is ensured by the integral cold leak checking of the cavity pairs.[3] A Faraday cup residual gas analyzer (RGA) system was used for these tests at room temperature since a vacuum gauge does not have the required sensitivity for small He partial pressure (pp) in the presence of large hydrogen and water background pressures. However, if the vacuum gauge could be operated cold, one could detect He leaks since the rest of the gas species are cryopumped onto cold surfaces at LHe temperatures. With this in mind the operation of extractor gauges and inverted magnetrons was tested in LHe. The latter gauges did not work satisfactorily at LHe temperatures due to the inconsistent discharge at the low pressures $< 1 \times 10^{-10}$ torr. The extractor gauges (Leybold-Inficon) have relatively low power dissipation (~ 3 W) compared to the conventional Bayard-Alpert ion gauges. These gauges also have a low x-ray limit of $\sim 1 \times 10^{-12}$ torr at room temperature. In the following the calibration details of the extractor gauges at LHe temperatures are presented.

Calibration of Extractor Gauge

The cold extractor gauge calibration setup is shown in Figure 1. It incorporates an ion pump, an RGA, a calibrated He leak and an extractor gauge (Leybold Inficon Model IE514) mounted at the low temperature part of the setup. For the baseline test, the setup was baked to 475 K with a blank flange in place of the CEBAF higher order mode load (a ceramic absorber mounted on a stainless steel flange). After cooldown to room temperature, the cryostat was filled with LHe and maintained at 4.2 K for a period of 5 hours. The extractor gauge reading was less than 1×10^{-12} torr at 4.2 K and did not increase with the warmup of the system to 20 K indicating that there was no He leak in the system. The RGA did not show any trace of He either.

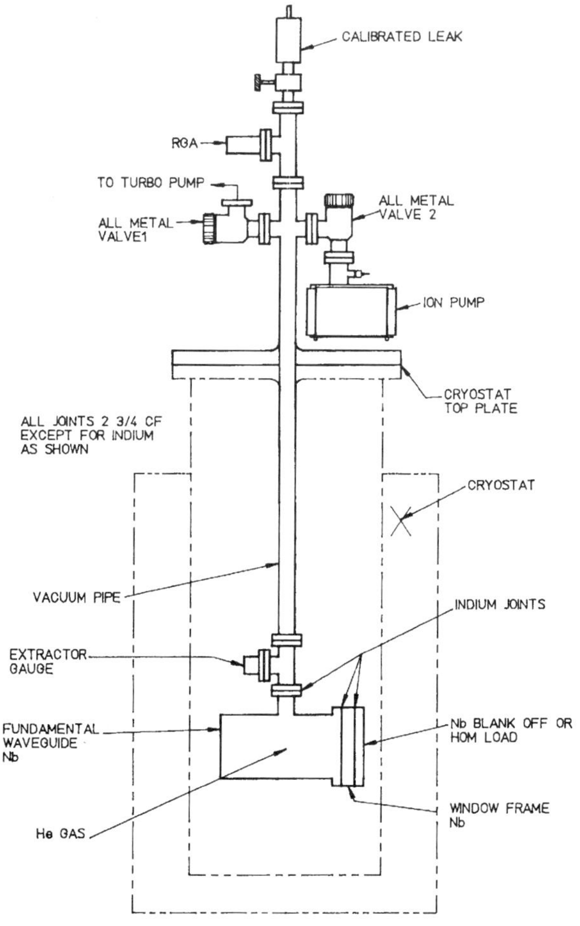

Figure 1. Cold extractor gauge calibration setup.

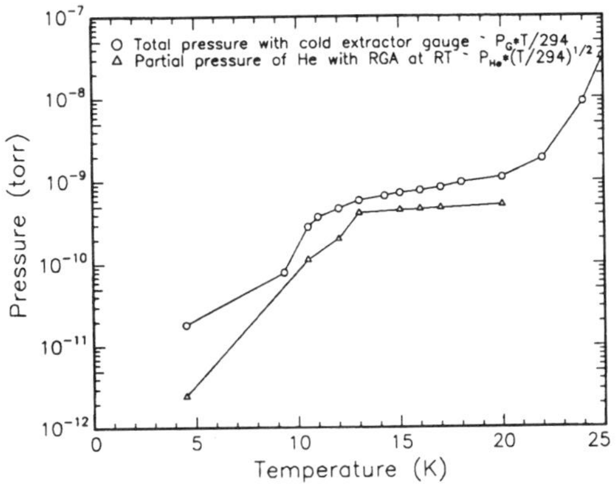

Figure 2. The corrected cold extractor gauge and RGA readings as a function of temperature.

In a subsequent test, the blank-off flange was replaced with a higher order mode load. The cryostat was filled with LHe after an overnight bake out of the system (limited to 355 K due to the possible creeping of indium seals at higher temperature). The pressure in the system at 4.2 K was < 10^{-12} torr. The calibrated He leak (9.65×10^{-10} atm cc s^{-1} at 294 K) was opened to the system for 65 minutes. The pressure as indicated by the cold extractor gauge reached a value of 2×10^{-11} torr towards the end of the He leak-in period, whereas the RGA at the ambient part of the system did not indicate any change in the pp of He (the Faraday cup RGA has a sensitivity limit in the low 10^{-10} torr range). Subsequently He was boiled off and the setup was warmed-up with a 1 K/min temperature rise. During this warm up period the extractor gauge as well as the RGA pp of He readings were recorded as a function of temperature.

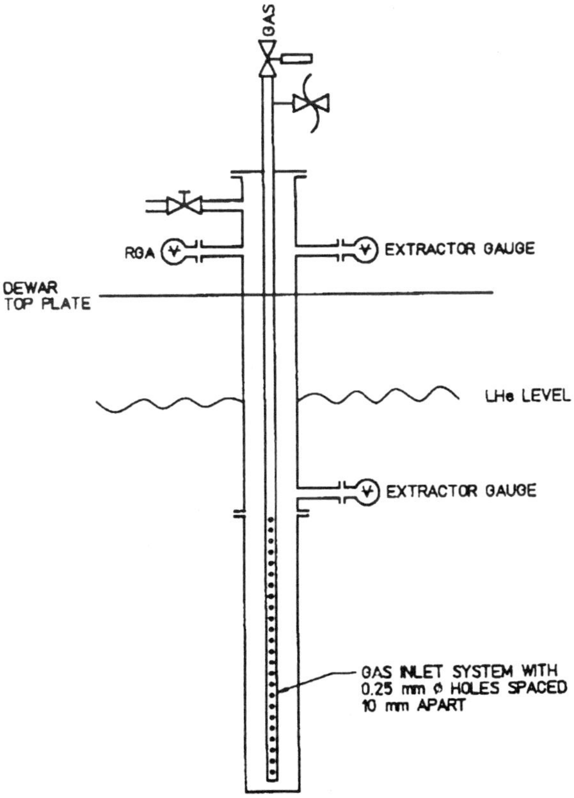

Figure 3. Adsorption isotherm measurement setup.

The RGA readings were corrected for thermomolecular effect (multiplied by the square root of T/294). The cold extractor gauge readings were also corrected for the operating temperature of the gauge (multiplied by T/294) since the gauge controller assumes that the gauge is operating at 294 K.[4] The corrected pressure readings are shown in Figure 2 as a function of temperature.

The cold extractor gauge reading starts increasing rapidly at ~ 9K and reaches a near plateau at ~ 11 K. Further small steady increase in pressure with temperature is due to the desorption of hydrogen. The pp of He as indicated by the RGA at ambient temperature starts increasing at ~ 10 K and reaches a constant value of 4×10^{-10} torr at 12 K. There is a time lag between the cold extractor gauge and the RGA readings which may be attributable to the non-equilibrium situation between the cold and the ambient pressures due to the continuous heating for warming-up purposes. The RGA reading is in good agreement with the near plateau value of the cold extractor gauge reading.

The quantity of He introduced into the system during the 65 minute period was 3.7×10^{-6} atm cc. The amount of He in the system (1.5 and 2 liters are the estimated cold and warm volumes) calculated from the extractor gauge and RGA readings is 3.5×10^{-6} atm cc. These two values are within 10% of each other and hence confirm that the cold extractor gauge works well at LHe temperatures. The following section presents additional details on the usefulness of extractor gauges for accurate measurement of low pressures at LHe temperatures.

Hydrogen Saturated Vapor Pressure Measurement with Cold Extractor Gauges

The correct operation of the cold extractor gauges is further checked by measuring the saturated vapor pressure of hydrogen at 4.25 K. The all-stainless-steel setup as shown in Figure 3 is used for these measurements. The system includes two extractor gauges, one operating at room temperature (rt) and the other at LHe temperature. Additionally an RGA is also used in this setup. Matheson's research grade (99.999%) hydrogen was introduced into the system through the capillary with gas distribution holes of 0.25 mm diameter, spaced 10 mm apart. The flow of hydrogen was stopped after the gauges indicated stable equilibrium pressures as shown in Figure 4.

The rt gauge indicates a very high background pressure (three orders of magnitude higher) in comparison to the cold extractor gauge. The rt gauge starts responding only after introducing a large quantity of hydrogen into the system (~ 5×10^{15} molecules cm^{-2}). The cold extractor gauge reading was low in comparison to the rt gauge during the flow of hydrogen into the system as seen in Figure 4. During these measurements the vapor pressure of LHe in the cryostat was noted to be 812 torr. The corresponding temperature of the LHe bath is 4.25 K. The saturated vapor pressure of hydrogen at 4.25 K is calculated to be 1.39×10^{-6} torr.[5] In the absence of hydrogen flow and at equilibrium the saturated vapor pressure of hydrogen as indicated by the corrected cold and ambient

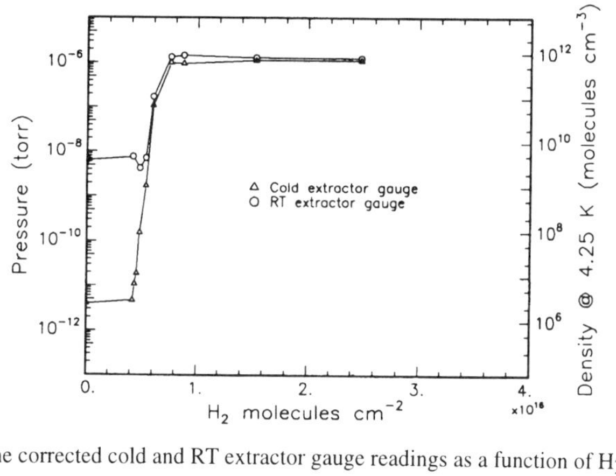

Figure 4. The corrected cold and RT extractor gauge readings as a function of H_2 molecules.

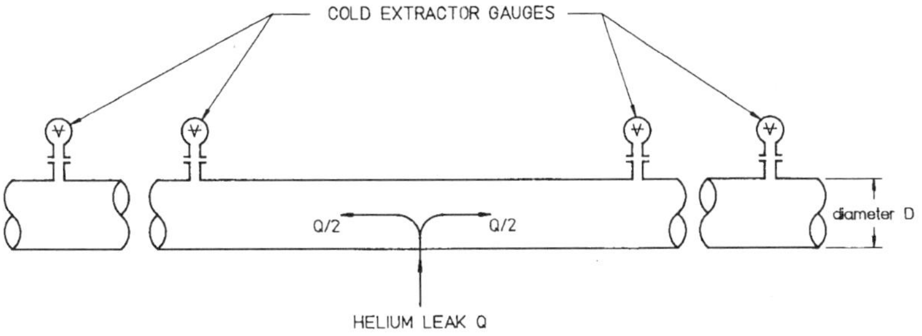

Figure 5. Schematic of pressure wave cold leak detection with cold extractor gauges.

Table 1. DRK parameters of technical surfaces from measured He adsorption isotherms

The local equilibrium pressure is determined by the DRK equation: [9]

$\sigma = \sigma_m - B\,(RT \ln P/P_o)^2$ where:

σ number of molecules adsorbed per unit area
σ_m monolayer capacity of an adsorbed layer
P_o saturation pressure (torr)
P equilibrium pressure and (torr)
B constant $(cal/mol.)^{-2}$
R gas constant (per mole)
T temperature (K)

Surface	B $(cal/mol.)^{-2}$	σ_m molecules cm^{-2}
Stainless Steel	7.6×10^{-5}	5.9×10^{15}
Nb	2×10^{-5}	$5.6\ 10^{14}$
Nb_2O_5	5×10^{-5}	1.3×10^{15}
Al_2O_3	7.9×10^{-5}	5.6×10^{19}

extractor gauge readings is 1.48×10^{-6} and 1.32×10^{-6} torr respectively. They are with in $\pm$ 6.5% of the theoretical value. Such a pressure measurement accuracy is satisfactory for all practical purposes. However, the cold extractor gauge can be expected to do much better in the low pressure range since the above measurements are almost at the higher end of the gauge's operating density range.

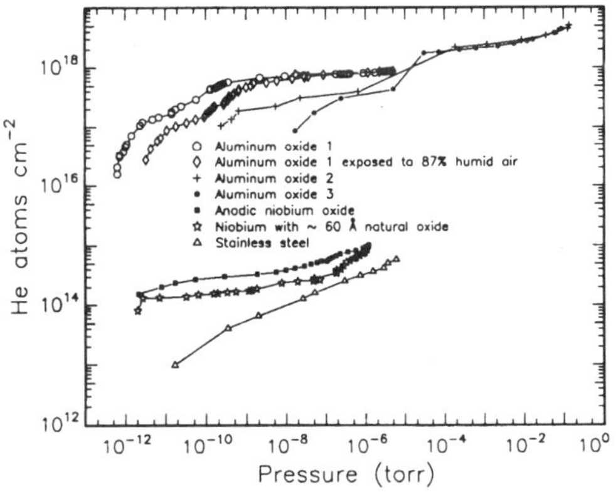

Figure 6. The adsorption isotherms of He at 4.3 K on various technical surfaces.

PRESSURE WAVE TRAVEL OF HELIUM FROM LEAKS

It is widely known that due to adsorption of He onto the cold walls of LHe-cooled cryogenic vacuum systems, cold leaks cannot be easily detected with conventional leak detector systems. Further He reaches an equilibrium pressure locally and after saturation travels as a pressure wave. The local equilibrium pressure follows a well known Dubinin, Raduskavich and Kaganer (DRK) adsorption isotherm. Recently Hobson et al. have discussed the time-dependent helium pressure profiles in a long, cryogenically cooled tube pumped at periodic intervals.[6] The cold extractor gauges can be used for in situ cold leak checking of such cryogenic vacuum systems as shown schematically in Figure 5.

Cold Leak Detection With Time of Flight Measurement

A helium pressure wave first reaches the nearest cold extractor gauge. Knowing the time taken by the pressure wave reaching the gauge from the time the system was cooled down, one could estimate the location of the leak from the equations presented in Ref. 5. Similarly another estimate of the location of the leak also can be made from the time the second-nearest gauge records the arrival of He pressure wave. The DRK parameters that are needed for accurate identification of the location of the leaks are presented in Table 1 for technical surfaces (viz.stainless steel, niobium, niobium oxide and aluminum oxides) and are derived from the measured adsorption isotherms shown in Figure 6.[7] A detailed analysis of the low pressure adsorption isotherms of hydrogen and helium at 4.3 K will be presented elsewhere.[8]

CONCLUSIONS

The extractor gauges are useful for accurate low pressure measurement in cryogenic vacuum systems at LHe temperatures. In situ cold leak detection of cryogenic vacuum systems is possible with the pressure wave travel principle using the cold extractor gauges. The needed DRK parameters for the estimation of the location of the cold leaks are presented from the measured adsorption isotherms of He on technical surfaces (stainless steel, niobium, niobium oxide and aluminum oxide).

ACKNOWLEDGMENTS

The author is grateful to many of his CEBAF colleagues, including P. Kneisel, J. Susta, J. Brock, D. Machie and G. Sundeen, for their support of this work. This work was carried out under D.O.E Contract No. DE-AC05-84ER40150.

REFERENCES

1. M.G. Rao, Helium Desorption Leak Detection Method, *J. Vac. Sci. Technol.* A 11, 1598 (1993).
2. M.G. Rao and P. Kneisel, The Use of Extractor Gauge at LHe Temperatures, CEBAF Tech Note #93-057, available from the Librarian, CEBAF, 12000 Jefferson Avenue, Newport News, VA 23606 USA.
3. M.G. Rao, Recent Advances in UHV Techniques for Particle Accelerators, CEBAF Tech Note #95-011, available from the Librarian, CEBAF, 12000 Jefferson Avenue, Newport News, VA 23606 USA.
4. J.G. Hengevoss, Gas Sorption by Surfaces Cooled to Low Temperatures, Trans. 3rd Int. Vac. Cong., Vol. 1, 51 (1965).
5. T.J. Lee, The condensation of H_2 and D_2: Astrophysics and Vacuum Technology, *J. Vac. Sci. Technol.* 9, 257 (1971).
6. J.P. Hobson and K. M. Welch, Time-Dependent Helium and Hydrogen Pressure Profiles in a Long, Cryogenically Cooled Tube, Pumped at Periodic Intervals, *J. Vac. Sci.Technol.* A 11, 1566 (1993).
7. M.G. Rao, P. Kneisel, and J. Susta, Cryosorption Pumping of H_2 and He With Metals and Metal Oxides at 4.3 K, Cryogenics ICEC Supplement 34, 377 (1994).
8. M.G. Rao and W.C. Turner, to be published in *J. Vac. Sci. Technol.*
9. M.G. Kaganer, Proc. Acad. Sci. USSR Phys. Chem. Sect. 116, 603 (1957).

LHe FLOWMETERS : STATE OF THE ART AND FUTURE DEVELOPMENTS

A.Rivetti, G.Martini, and G.Birello

IMGC - Istituto di Metrologia "G.Colonnetti"
Strada delle Cacce 73, I-10135 Torino (Italy)

ABSTRACT

In general flowmeters for use with cryogenic helium are not commercially available; when the measurement of flowrate is essential in a cryogenic facility, the designer is not likely to achieve the most appropriate transducer for his application, also because the literature existing on the topic is rather scarce. In addition, owing to the lack of primary calibration facilities, the majority of the transducers used in this field have not been calibrated against SI standards, so that measurement values, corrected through theoretical calculations, can hardly guarantee accuracies better than 10%.

The characteristics of the different families of transducers tested in cryogenic helium during recent years are reported and discussed. Directions are given to facilitate the designer in choosing the most suitable type of transducer, according to the thermodynamic conditions of helium (saturated, sub-cooled, supercritical, superfluid) and depending on particular requirements of specific applications (e.g. long-term reliability, withstanding vibrations and shocks).

INTRODUCTION

Flowrate measurements of traditional fluids at room temperature are commonly made in a large number of industrial and commercial applications; however, despite many studies started at the end of the last century and the development of a variety of transducers based upon many different principles, flowrate is rarely measured to an accuracy better than 1%: only with special (and expensive) installations, for example when a twin-turbine transducer set or a Coriolis-based mass-flow transducer are employed, measurement uncertainty can reach values as low as 0.2%. In any case, it must be noted that, when the required accuracy is better than a few percent, most flowmeters should be individually calibrated (against a higher-level transducer or a primary calibration facility) under the same conditions as those of actual operation.

In the low-temperature domain, particularly in almost all the applications where cryogenic helium is used, the importance of accurate flowrate measurements was largely undervalued till not long ago: whenever flowrate determination was considered necessary, transducers generally not designed for cryogenic service were used, essentially Venturi and turbine flowmeters. In the absence of calibrations made at cryogenic temperature, for determining the actual LHe flowrate, the transducer output readings were corrected to take account of the different values of helium density and/or viscosity, and, in some cases, also of the thermal contraction of materials. Moreover, in the case of Venturi meters, the pressure-difference (ΔP) transducer was usually placed at room-temperature outside the LHe environment, and connected to the Venturi body through long, small-size tubes, with obvious problems of signal accuracy and stability.

The precise knowledge of cryogenic helium flowrate begins to be considered important and useful especially in large-size facilities. Unfortunately, reliable flowmeters expressly designed for use with cryogenic helium are not commercially available to date. Moreover, at present cryogenic helium is seldom used in normal state (saturated fluid at room pressure): more and more frequently today's applications require helium under different thermodynamic conditions, for example, supercritical helium (characterised by an absolute pressure exceeding 2.25 bar), sub-cooled helium (generally at room pressure but at temperatures between 2.2 and 4K); nevertheless, in most cases, superfluid helium ($T < 2.17K$), either saturated or sub-cooled, is the fluid preferred[1], mainly because of its very high thermal conductivity which makes this unique liquid most valuable especially when used as coolant in superconducting applications.

METHODS AND TRANSDUCERS FOR CRYOGENIC SERVICE

Transducers to be used at low temperatures are usually derived from the classic room-temperature flowmeters, generally modified in some way to withstand hostile environments. There are three main categories of transducers mostly used in cryogenic conditions: *turbine flowmeters* , *Venturi meters* and *thermal-pulse anemometers*.

In addition to these families, which will be analysed in detail in the next sections, a number of flowmeters working on the basis of different principles have been proposed and experimented at low temperatures. Among them:

- *thermal mass-flowmeters*[2], in the room-temperature version are extensively adopted, but only for measuring gas flowrates: the mass-flowrate is evaluated by measuring the temperature increase of the fluid downstream an electric heater and by taking the specific heat of the fluid into account. At a low temperature the method is suitable only for single-phase helium I, essentially supercritical or sub-cooled. It cannot be used with saturated LHe I nor with LHe II;
- *ultrasonic anemometers*[3,4], where the flow speed is evaluated by measuring the difference between the transit times of sound waves travelling in the flow direction and against it. Special ultrasonic transducers have been recently developed[5] in order to comply with the low acoustic impedance of helium. This technique has the advantage of being non-intrusive but, on the other hand, having an uncertainty hardly below 2%, it requires a complex and expensive instrumentation. It may be used also with superfluid helium;
- *second-sound anemometers*[6] which, working only in superfluid helium, exploit the high thermal conductivity of LHe II: their operation principle is similar to that of ultrasonic anemometers except as regards the flying wave: thermal-pulses instead of trains of pressure oscillations;
- *fluidic flowmeters*[1] based on frequency measurement of pressure (or temperature) oscillations due to hydrodynamic instabilities. Compared with Venturi meters, these transducers cover a wider flowrate range because of their linear relationship between volumetric flowrate and output signal; on the other hand they produce a large unrecovered pressure drop. They are hardly applicable to LHe II.

TURBINE FLOWMETERS

The operation of transducers of this kind is theoretically independent of the fluid temperature, only weakly dependent on fluid density and slightly more dependent on its viscosity[7]. This means that, in principle, there are no reasons against the use of a commercial turbine flowmeter designed for room temperature service in a cryogenic environment. In practice, some contrivance is necessary for ensuring reliable operation of the turbine when working at low temperature: first of all, it is vital that both the body and the rotor are made of the same material, otherwise, when the meter is cooled down, the differential thermal contraction could change the clearance between the moving parts and too large a clearance would result in an irregular movement of the rotor, while clearance reduction could block the rotor itself or even produce the rupture of the bearings. Rupture of the bearings, which are generally of the journal type and made of hard and brittle material (widia or jewels), may also result from thermal shock during cool-down, especially in the case of pre-existent internal stresses; for this reason the turbine meters having journal bearings should be qualified, before use at low temperature, through repeated thermal cycling in LHe or, at least, in LN_2.

In principle, the angular speed of a turbine rotor is directly related to the mean speed of the flow passing through the turbine, the relationship depending on the cross sectional area and on the angle between the blades and the pipe axis. In practice, the actual angular speed is always lower than the theoretical: in fact, the rotor is slowed down by the viscous drag exerted by the fluid on the blades, and by the counter-torque due to the mechanical friction on the bearings supporting the rotor. If both the torque and friction values remain constant, the turbine has excellent repeatability (and accuracy, if calibrated) down to (0.1-0.2)%. While viscous drag can be considered dependent only on the physical characteristics of the fluid (essentially viscosity and, partially, density), the value of the mechanical friction on the bearings may, instead, change in an unpredictable way because of the natural wear and tear of the surfaces in contact, or even as a result of accidents occurred to the turbine in operation or during cool-down (over-speeding, impact of solid particles).

The existence of the bearing friction appears then as the real weak point of transducers of this kind. In fact, because of the slight forces exerted by the fluid on the rotor, a small increase of the friction torque results in notable slowing down of angular speed, without any apparent failure; measurement errors as high as 5-10 % are usual in such a case, and only a continuous and automatic cross-check with a second flowmeter can reveal them. It can be concluded that, although very repeatable when under control, turbine flowmeters show rather poor long-term reliability, which advises against their use in important applications where instrumentation maintenance is impossible, as in space missions, or economically disadvantageous, as in large scale facilities.

In order to overcome the bearing friction and to take advantage of the LHe low temperature, a turbine flowmeter with the rotor magnetically levitated by Meissner effect was developed at IMGC[8] in 1987. The meter, which has niobium caps on the two ends of the rotor and Nb-Ti superconducting windings for generating the sustaining magnetic fields, was designed only for LHe service and was intended as a laboratory transfer standard. The main advantage of this flowmeter is the virtual absence of mechanical friction and, consequently, a good long-term repeatability; however, because of the limited value of the repulsive forces acting by Meissner effect, both the radial and axial stiffness of the suspension are rather weak and unable to withstand vibrations, accelerations and shocks typical of practical applications.

Turbine flowmeters, either traditional or levitated, are theoretically suitable for use with helium under all practical thermodynamic conditions, including the superfluid. Nevertheless, Daney[9], during the calibration of a traditional turbine flowmeter in superfluid helium, observed a jump of the calibration factor, occurring above a certain value of LHe II flowrate. This phenomenon, which was attributed to cavitation of LHe II, could be suppressed only by adding an orifice at the exit of the meter, in order to produce a suitable back-pressure.

VENTURI METERS

Based on Bernoulli's law, they are the oldest flowrate sensors that have been used extensively with ordinary fluids, either liquids or gases and, since the beginning, also at cryogenic temperatures. Owing to the absence of moving parts, these flowmeters are intrinsically reliable and especially suitable for applications where the possibility of maintenance is reduced. Flowrate is evaluated across Venturi meters by measuring the difference of pressure between the upstream and throat sections, according to the following equation:

$$G = C_d \cdot A_t \cdot \sqrt{\frac{2 \cdot \rho \cdot \Delta P}{\left(1-\beta^4\right)}} \qquad (1)$$

where:
G = mass flowrate [kg/s]
A_t = throat section area [m^2]
ΔP = pressure difference [Pa]
C_d = discharge coefficient [dimensionless]
ρ = fluid density [kg/m^3]
β = throat-to-upstream diameter ratio [dimensionless]

Today the availability of accurate and reliable ΔP meters, suitable also for low temperature service, allows compact and rugged Venturi flowmeters to be constructed, which, when individually calibrated under operation conditions, can ensure acceptable medium-term accuracies. Two different types of pressure sensors have been checked at low temperatures by several authors, with tests based on the displacement measurement of an elastic diaphragm: in the Siemens KPY type, the measurement is made through the piezo-resistive technique, whereas Validyne DP10 sensor, equipped with a stainless steel magnetic diaphragm, applies the variable-reluctance technique.

Siemens KPY transducers, designed for direct connection to a printed circuit, were specially tested at LHe temperature by Walstrom and Maddocks[10]. They found them rather delicate, of limited over-pressure capability and having a fairly high failure rate upon cool-down to LHe temperature (4 sensors out of a total of 20); in some of the survived specimens, effects of hysteresis and apparent non-repeatability were also observed.

Validyne AP10 and DP10 sensors, in which the diaphragm is welded to the meter body, have been extensively investigated by Kashani et al.[11] in order to qualify pressure transducers for the SHOOT experiment, undertaken by NASA to demonstrate the possibility of transferring superfluid helium in space: of the 17 transducers which were put through the flight qualification tests, 16 completed them successfully showing linear and repeatable responses within 1% of the full scales. It was found that all the transducers (and especially the differential-pressure type, DP10) needed multiple thermal cycling to LHe temperature for their response to remain repeatable.

The outcome of our experience concerning Venturi meters designed and built at the IMGC and having Validyne DP10 transducers, is in good agreement with the above results; in addition, a notable and non-repeatable zero-shift, probably due to thermal settling of the diaphragm, was found to occur each time the meters were cooled down, forcing us to adjust the instrument zero at the beginning of low temperature operations. According to the IMGC experience[12], some directions should be observed during the construction and installation of Venturi transducers, in particular as regards:

- *mechanical construction:* the body of a Venturi meter may be designed in different ways, provided that some main rules are observed, essentially those indicated by the ISO Standard 5167; although concerning room temperature conditions, and meter dimensions larger than those for low-temperature applications, these specifications proved to be successfully applicable also to cryogenic Venturi flowmeters. In particular, the following dimensioning ratios should be respected:
  - the length-to-diameter ratio of the upstream (≥ 1) and of the throat ($=1$) sections,
  - the value of the throat-to-upstream diameter ratio ($0.4 \leq \beta \geq 0.75$), and
  - the semi-angle aperture of the converging $(10.5 \pm 0.5)°$ and diverging $(4 \pm 0.5)°$ sections.

  Great care must be taken also in adjusting the inner surface of the meter, which must be well polished, while section changes must be accurately radiused;
- *choice of the maximum ΔP:* when designing the inner profile of the meter, one of the most important parameters to be established is the value of the maximum ΔP obtainable between the upstream and throat sections. In the determination of this value, it must be considered that while a high *ΔP* value means a high meter sensitivity, it can also cause the cavitation phenomenon in the throat section if the pressure in that point becomes lower than the fluid vapour pressure. In the IMGC Venturi meters, a value of 15 mbar is generally adopted as the maximum ΔP at flowrate full scale;
- *positioning:* in principle, a Validyne meter should be installed with the diaphragm in the vertical position, so that its displacement would not be affected by its own weight. However, if the pressure difference acts always along the same direction, as in the case of a Venturi meter, the horizontal position with the low pressure chamber below it could be more favourable, since in this way the diaphragm would always be stressed in the same direction (thus reducing the risk of zero drift). It is obvious that, in both cases, the meter must be calibrated in the same conditions as those of the final installation;
- *hydraulic and electric connections:* hydraulic connections between the ΔP meter and the Venturi body must be leak-proof, especially for use with superfluid helium: we successfully experimented 1/8" stainless steel piping and Swagelok® connectors, welded at the Venturi side and tightened into the Validyne ports by replacing the original plastic gaskets with suitably-built copper washers. According to manufacturer's specifications, it is recommended to connect the Validyne ΔP meter to the carrier/demodulator through a shielded cable; tests

made at IMGC demonstrated that a single meter can work quite well even with an ordinary flat cable, whereas, if two or more ΔP meters are working close to each other, significant reading errors can occur, owing to the coupling of the single supply frequencies. In a low-temperature environment, shielded cables expressly designed for cryogenic service (PTFE-insulated) should be used;

- *magnetic shielding:* owing to the meter operation principle which implies the use of a magnetic diaphragm, care must be taken as regards the possible presence of magnetic fields in the close vicinity of the meter. In such a case, some kind of magnetic shielding must be provided to avoid measurement errors due to the interaction between the external magnetic field and the diaphragm: it must be pointed out that even moderate magnetic fields may produce permanent damage to the meter by changing the magnetic remanence of the diaphragm and affecting the flowmeter calibration curve. Suitable shielding design should be based on the characteristics of the actual magnetic field; in any case, since the sensor works at cryogenic temperature, the traditional shielding materials (high permeability alloys, e.g. Permalloy® or Mumetall®) may be combined with superconducting materials, either traditional (like niobium), or high-T_c (like YBCO or BSCCO). Because of the low value of the first critical field (H_{c1}) of these superconductors, they should work close to the sensor and surrounded by traditional high-permeability materials, which act as primary shielding. In a recent installation of a Venturi flowmeter made by IMGC at CERN in Geneva[13], a double-effect magnetic shielding demonstrated its capability of shielding magnetic fields as high as 0.07 T.

It is believed that a Venturi meter expressly designed for cryogenic service, skilfully made and equipped with a suitable ΔP meter, if individually calibrated under the same conditions as those of the actual operation, can provide measurements repeatable and accurate within (1.5-3)% of cryogenic helium flowrates in almost all practical thermodynamic conditions. Only with saturated superfluid helium, as reported by Daney[9] and confirmed by IMGC experience[14], may Venturi meters (like turbines) undergo the cavitation phenomenon at high flowrates. According to Daney, this problem may be overcome in the same way as that used with turbines, that is with the generation of a suitable back-pressure by installing a small-diameter orifice downstream the meter.

THERMAL - PULSE ANEMOMETERS

This method is based on the measurement of the flight-time of a short heat pulse injected into the stream and passing in front of two thermal sensors placed at a known distance apart. The heat pulse is assumed to be transferred to the small volume of liquid which, at that moment, is surrounding the heater, and this heated volume to continue travelling inside the stream at its mean speed. The two thermal sensors will then detect a sudden temperature increase as soon as the heated volume passes in front of each of them; the distance between the two thermal sensors, divided by the time interval between the two temperature peaks, provides the mean speed of the stream.

This technique, first tested in liquid and in supercritical helium at IMGC[15], can give acceptable results (repeatability within 3%) provided that some conditions[16] are observed, particularly as regards:

- *the shape of the thermal pulses injected inside the stream:* the pulses should have short duration (less than 50 ms, typically 20 ms), and energy level dependent upon the physical state of helium (below 1 J for supercritical, between 5 J and 10 J for LHe I);
- *the characteristics of the thermal sensor:* they should provide a short response time (small size) and a high signal-to-noise ratio (high sensitivity); on the contrary, "good" characteristics in the metrological sense (accuracy, repeatability, linearity) are not required. As a result of an IMGC investigation on a number of commercial semiconductors, the Philips diode BAS21 in package SOT23 was selected as the best candidate, showing (in at least 50% of the specimens tested) a sensitivity around -500 mV/K near 4.2 K, with a signal-to-noise ratio of about 1:50 and a response time less than 5 ms/K.
- *the evaluation of the time interval between temperature peaks:* it is difficult to determine the exact time interval between the temperature peaks on the two thermometers, since the shape of these pulses is rather large with respect to the distance between them. The best ac-

curacy can be achieved by using the cross-correlation technique, which is based on the following equation:

$$R(t) = \int_0^T X(t) \cdot Y(t - \tau) \cdot dt \tag{2}$$

where $X(t)$ and $Y(t)$ are, respectively, the amplitudes of the output signals from the first and the second thermal sensor at time t. The maximum of $R(t)$ is reached with a value of τ which, in principle, equals the actual transit time of the thermal-pulse in front of the two thermometers.

Thermal-pulse anemometry proved to be specially effective in single-phase helium, both supercritical and sub-cooled, in which the heat pulses are transferred to the surrounding volume without changing the physical state of the fluid. Although possible, the method is less suitable for working in saturated LHe, because the heat pulse in this case produces unavoidably one or more bubbles or even vapour fog: owing to the very different density of liquid and vapour, the buoyancy effect would accelerate or retard the vertical motion of the heated vapour with respect to the liquid flow, thus affecting measurements especially at the lowest flowrates. In the case of horizontal motion, the vapour tends to move toward the upper part of the tube, with higher probability of missing the thermal sensors. Both the above situations have been experimented at IMGC. The possibility of using this method also in superfluid helium was considered and experimented by several authors[17,18], but the actual performances appear generally poor and inadequate to practical applications.

CHOICE CRITERIA

From the above discussion one can conclude that there exists at least one type of flowrate transducer which is characterised by reliability and repeatability values suitable for the majority of practical cryogenic applications. Although the long-term reliability of Validyne transducers at low temperature is still under investigation and some characteristics (sensitivity to thermal cycling, zero stability during cool-down) should be improved, their use in LHe flow-metering has to be considered positively. Venturi meters equipped with ΔP transducers of this type could be commercially produced and cover the 90% of the demand in the cryogenic field; the remaining 10% concerns special requirements, such as:

- *operation in presence of magnetic fields above 0.1 T:* in this environment a Venturi meter could work, provided that the Validyne ΔP transducer is replaced by a non-magnetic sensor (i.e. the Siemens KPY type);
- *applications with high-level vibrations, centrifugal fields, mechanical shocks:* in these conditions, which can affect the correct operation of the diaphragm of the ΔP transducer, a possible alternative is the thermal-pulse anemometer, whose components are fixed and light;
- *operation in saturated LHe II:* though with the already mentioned limitation of possible cavitation at the highest flowrates, Venturi meters can work also with superfluid helium; however, other possibilities are offered by techniques based on the unique characteristics of LHe II, such as the second sound anemometry.

Thermal-pulse anemometers, expressly developed to monitor the flow of supercritical helium in cryogenic rotating machinery, have been designed to withstand vibrations, notable magnetic fields and high centrifugal forces. Owing to the inability of these transducers to provide continuous flowrate measurements, to their limited performances (especially as regards the achievable accuracy) and to the relative complexity of the instrumentation needed to evaluate the speed of the stream, they appear unpractical for many applications (apart from those mentioned above). However, the small-size and low-cost of their in-situ components make them particularly suitable for redundant or emergency installations: for instance, in a large facility, a number of thermal-pulse sets could be installed together with turbine and/or Venturi flowmeters and, with a single switchable electronic instrument, be used to confirm the validity of the output of the main transducers, or to replace them in case of failure.

Turbine flowmeters, like Venturi meters, are not able to work correctly in the presence of magnetic fields, centrifugal forces or vibrations, and their use with saturated superfluid helium should also be avoided. In addition, one must remember that turbines are delicate instruments, requiring care and periodical maintenance: they should not be used in space missions, nor in large-size facilities, except in case where high accuracy is much appreciated (but not vital), or if the number of flowmeters installed is redundant.

POSSIBLE FUTURE DEVELOPMENTS

It has already been said that the turbine flowmeter with the rotor magnetically levitated by Meissner effect shows rather weak centering and sustaining forces, unable to operate safely in industrial applications. The recent development of high T_c superconducting materials, especially of textured YBCO, offers the opportunity of designing a *turbine flowmeter* with the rotor sustained by much stronger forces, owing to the property of these materials of trapping the magnetic flux during the superconducting transition. Whereas by the Meissner effect the magnetic field is expelled from the superconductor with consequent generation of repulsive forces, by the trapping phenomenon the magnetic field is "frozen" inside the superconductor and the forces consequently generated in all directions strongly counteract any change in the magnetic distribution inside the superconductor. As a result of this opposition, a superconductor immersed in an axially-symmetric magnetic field can only rotate around that axis, whereas all the other types of motion, like the axial, the radial and the tilting are prevented. At IMGC a project is at present being studied for the design and the construction of a turbine prototype based on this property and intended as a flowrate transducer for different cryogenic fluids (nitrogen, oxygen, hydrogen, helium). The aim is the repeatability typical of a turbine (± 0.2%) and a reliability much higher than that based on the Meissner effect, thus suitable for practical use as an alternative to Venturi meters.

Another possibility, which, to our knowledge has never been thoroughly investigated as regards its ability to measure cryogenic helium flowrate, is the application of the *vortex-shedding technique*, based on the alternate generation of two trains of vortices at the two sides of an obstacle placed in the middle of a flow stream. The flowrate value is obtained by measuring the frequency of these vortices, which proved to be proportional to the mean speed of the flow. Although under certain conditions this measurement may be made through a couple of sensitive and fast-response thermometers, in most cases a ΔP transducer is preferred. With traditional fluids at room temperature, the best commercial transducers can perform accurate measurements, generally with an uncertainty below 1%, only limited by very low values of the Reynolds number ($R_e<60$). Apparently, there are no reasons why such transducers should not work also in cryogenic helium, since its low viscosity guarantees a Reynolds number always higher than 10^3 in any condition, and thanks to the availability of ΔP transducers working at low temperatures. The only doubt as to the actual applicability of this technique with cryogenic helium arises from the possibility of cavitation on the two sharp edges of the obstacle, especially with LHe II, which has proved to be particularly liable to this phenomenon. In case of favourable results of the experimental tests, a meter of this kind would appear attractive, as it combines the main advantages of Venturi meters (no moving parts), and of turbine flowmeters (frequency output).

Finally, *the Coriolis-type flowmeter*, characterised by excellent room-temperature performances[19] (repeatability within 0.1%), could be considered for possible low-temperature service, although with remarkable hardware modifications. This meter is based on the Coriolis effect by which, when a fluid flows inside a bent pipe excited to its resonance frequency, Coriolis inertia forces are produced, which act on the tube itself, producing elastic deformations, which are proportional to the mass flowrate passing through the meter. The ability to measure the mass flowrate (instead of the volumetric flowrate) is certainly an advantage with respect to turbine or Venturi meters, whereas the main obstacles to its development in view of its use with liquid helium are the relatively large dimensions and the notable amount of power dissipated in the vibrating motion.

CALIBRATING LHe FLOWMETERS

In the room-temperature range, there are few flowmeter types which can work with measurement uncertainties down to 2% without requiring individual calibration. Such are, for example, Venturi meters in a limited range of dimensions and Reynolds numbers, for fluids whose rheological properties (density and viscosity) are well known, and provided that the meters are built, installed and used following exactly the standard specifications. The possibility of dispensing with calibration is the result of a large number of experimental tests carried out world-wide during the whole present century. A similar experience is not possible at low temperatures, so that not even a highly repeatable transducer could provide, under cryogenic conditions, flowrate measurements accurate within acceptable limits unless directly or indirectly calibrated against SI standards.

If an accuracy better than 10% is required, flowmeters for low-temperature service must then be calibrated, preferably under the same conditions as those of actual operation. For this purpose, apparatus designed to generate and measure stable flowrates of cryogenic helium in different thermodynamic conditions and able to accommodate one or more flowmeters should be employed.

Although hydraulic test-circuits of different kinds were built by individual authors to evaluate the performances of their flowmeters, as far as we know only three flowrate calibrators are described in the literature[20,9,21] which were purposely designed for use with cryogenic helium, and metrologically characterised. The three calibrators, all of the volumetric type, are based on the axial displacement of a bellows immersed in a LHe bath and operated from outside the cryostat. Usually, since the volume generated by a bellows is not exactly proportional to the axial displacement and a theoretical calculation on the basis of bellows dimensions would be affected by unacceptable errors, a direct volumetric calibration is essential. The bellows of the most recent calibrator made at IMGC[21] was calibrated at room temperature with water and with application of the gravimetric method: the resulting calibration curve, corrected for the thermal contraction of the bellows when working at low temperature, allows flowmeters to be calibrated in cryogenic helium to an estimated accuracy of ±0.5%.

CONCLUSIONS

Accurate flowrate measurements begin to be required also in cryogenic applications, especially in large-size facilities. Different transducers may be used, depending on the thermodynamic state of helium and on special application requirements. In most cases, the best candidate is still the Venturi flowmeter, owing to the absence of moving parts and to the availability of ΔP transducers suitable for cryogenic service. In any case, if an accuracy better than 10% is required, flowmeters must be individually calibrated under the same conditions as those of actual operation.

REFERENCES

1. S.W. Van Sciver, D.S. Holmes, X. Huang and J.G. Weisend, *Cryogenics*, 31:75 (1991)
2. R. Schweiger, H.W. Lorenzen and J. Schaller, in "Adv. Cryo. Eng.", 37:147 (1992)
3. A. Hofmann and B. Vogeley, in "Proc. ICEC 10", Butterworts publ., Guildford (1984), pp.448-451
4. J.H. Sondericker, in "Adv. Cryo. Eng.", Vol.29 (1984)
5. P. Lavocat, in "Proc. ICEC 11", Butterworts publ., Guildford (1986), pp.577-581
6. D.S. Holmes and S.W. Van Sciver, in "Adv. Cryo. Eng.", 35:247 (1990)
7. L.A. Salami, *Trans. Inst. Mc*, 7:183 (1985)
8. A. Rivetti, G. Martini, R. Goria and S. Lorefice, *Cryogenics*, 27:8 (1987)
9. D.E. Daney, in "Adv. Cryo. Eng.", 33:1071 (1988)
10. P.L. Walstrom and J.R. Maddocks, *Cryogenics*, 27:439 (1987)
11. A. Kashani, R.A. Wilcox, A.L. Spivak, D.E. Daney and C.E. Woodhouse, *Cryogenics*, 30:286 (1990)
12. A. Rivetti, G. Martini and G. Birello, *Cryogenics ICEC Suppl.*, 34:4449 (1994)
13. A. Rivetti, G. Martini and G. Birello, IMGC Internal Report P208 (1994)
14. A. Rivetti, G. Martini and G. Birello, in "Adv. Cryo. Eng.", 39:1051 (1994)
15. A. Rivetti, R. Goria, G. Martini and S. Lorefice, in "Adv. Cryo. Eng.", 29:895 (1984)
16. R. Goria, G. Martini, A. Rivetti and S. Lorefice, *Sensors and Actuators*, 6:215 (1984)
17. L. Dresner, in "Adv. Cryo. Eng.", 29:323 (1984)
18. A. Kashani and S.W. Van Sciver, in "Adv. Cryo. Eng.", 33:417 (1988)
19. F. Cascetta, G. Cignolo, R. Goria, G. Martini, A. Rivetti and P. Vigo, *Trans. Inst. Mc*, 14:99 (1992)
20. A. Rivetti, G. Martini and R. Goria, in "Adv. Cryo. Eng.", 29:903 (1984)
21. A. Rivetti, G. Martini and G. Birello, in "Adv. Cryo. Eng.", 39:1043 (1994)

WEIR FLOW METER FOR SATURATED LIQUID HELIUM

R. C. Sanders, R. J. Rabehl, and M. W. McGee

Fermi National Laboratory
Batavia, Ill., 60510, USA

ABSTRACT

This paper presents the design and test results of a weir used to measure the flow rate of saturated liquid helium in a two phase helium flow stream. The flow meter requires the presence of saturated vapor to measure the liquid flow rate but does not measure the vapor flow rate. The weir, located at a high point in the flow path, acts as a dam and creates a reservoir of liquid behind it. Liquid helium drains through a notch in the weir. The liquid helium level behind the weir is measured to determine the flow rate. The level increases with flow rate. The unmeasured vapor passes over the top of the weir and mixes with the liquid downstream of the weir. A weir was installed on the outlet of a string of superconducting magnets.

INTRODUCTION

Weirs have been used for over two thousand years in the form of dams and structures to raise and divert the flow of water in streams and channels. In the last two hundred fifty years, people have formally studied the use of weirs as flow meters, primarily for water in open channels[1]. Fermilab has recently built and tested a liquid helium flow meter that uses a weir. The liquid helium must be saturated and mixed with vapor for the flow meter to work. The flow meter monitors helium discharge from a string of superconducting magnets. These magnets are not ready for operation until the outlet has a minimum liquid helium flow rate.

The primary goal of being able to detect changes in the liquid flow rate has been accomplished, but the secondary goal of calibrating the flow meter remains to be done. Fermilab will install the same design at four, possibly six, other locations. The crowded conditions at these locations caused by existing equipment dictates a very compact flow meter design. The small size compromised to some extent the performance of the flow meter.

PHYSICAL DESCRIPTION

Figure 1 shows the assembly of the weir flow meter. It is made of AISI 304 stainless steel and designed to have a pressure rating of 11.0 MPa gauge pressure, with a full vacuum in the vacuum space. The outer rectangular vacuum jacket is 18.3 cm high with a 10.2 cm by 9.4 cm cross section. Inside the jacket is the 13.7 cm high, 6.1 cm outer diameter cylindrical inner vessel. Multilayer insulation (MLI) covers the inner vessel. Two phase helium flow enters and exits through the 1.9 cm diameter tubes on the side and bottom respectively.

The weir, shown in Figure 2, is a stainless steel plate; the sides and bottom are seal welded to the inside of the inner vessel. The weir has a 4° angle V-notch cut into it by an electrodischarge machining (EDM) using a 0.25 mm diameter wire. The plate had a slot milled to produce a thin edge for the notch. The upstream surface was ground flat to make the edge sharp.

The American Magnetics Inc. (AMI) liquid helium level sensor is clamped rigidly to the weir to ensure repeatable level measurements. This sensor is 10.2 cm long and 6.4 mm in diameter. Feed-throughs in the vacuum jacket and inner vessel provide electrical connections to it. The sensor can only make level measurements over its 8.9 cm long active length which extends above the top of the weir and below the bottom of the V-notch. An AMI chassis model 134 Liquid Helium Level Monitor reads the sensor and sends a 4-20 mA signal to a Moore APACS controller that calculates the flow rate[2].

FLOW DESCRIPTION

The flow meter is installed at a locally high spot in the transfer line downstream from a string of superconducting magnets. The normal operating pressure is about 0.14 MPa, and the flow rates are estimated to be about 4 g/s of liquid helium and 16 g/s of helium vapor.

The inlet tube (shown in Figure 3) discharges two phase helium into the flow meter above the height of the weir. Gravity separates the liquid helium from the vapor; the liquid falls into a small reservoir behind the weir and the vapor passes over the top of the weir. A baffle on the inlet tube stops liquid from splashing over the top of the weir. Liquid drains out of the reservoir through the V-notch in the weir. The vapor mixes with the liquid and together they exit through a tube from the bottom of the flow meter. Liquid helium should be less likely to cling to the weir than water, since its surface tension is well over one hundred times that of water. Figure 3 shows the liquid stream striking the bottom of the vessel. However, this is strictly conjecture. The liquid stream may instead impinge on the vessel wall. The AMI level sensor measures the liquid level in the reservoir to determine the flow rate.

To improve the repeatability of the flow measurement, the nappe (the stream of liquid flowing through the notch) must discharge freely and not cling to the side of the weir. This requires ample

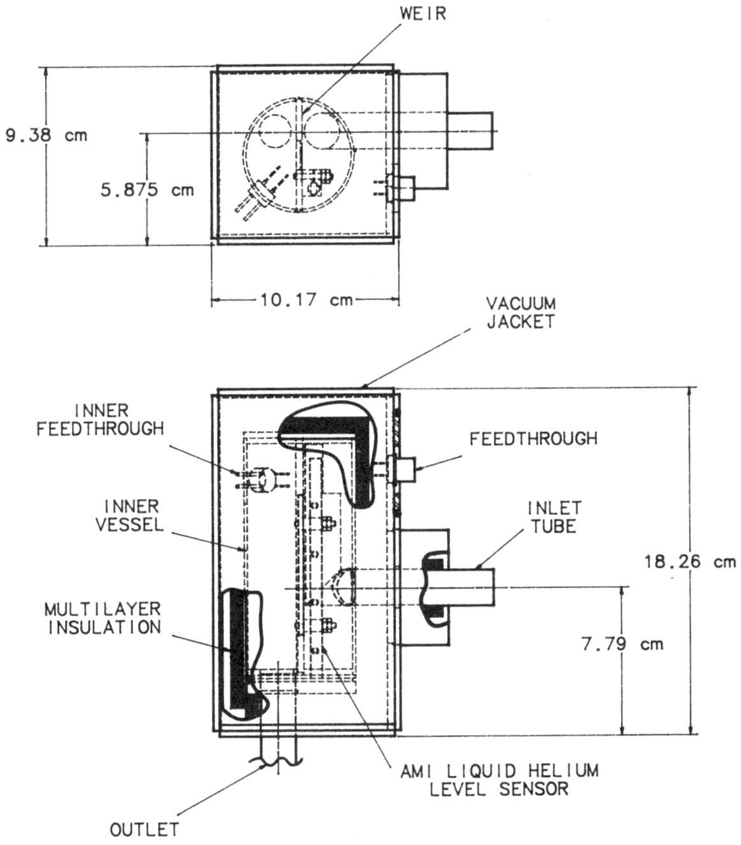

Figure 1. Assembly drawing of weir flow meter.

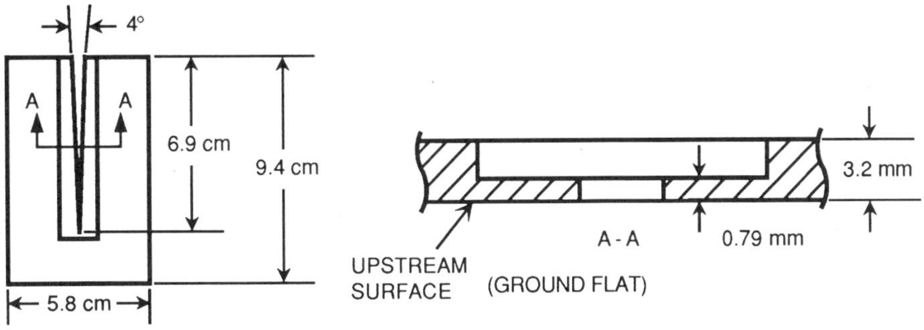

Figure 2. Stainless steel weir.

space for ventilation under the nappe to equalize pressures around it, and the notch must have sharp thin edges. The notch must also be sufficiently high so that the accumulation of liquid on the downstream side of the weir does not reach the bottom of the notch.

The theoretical equation for the flow rate through the V-notch weir is:

$$Q = \frac{8}{15} C_d \sqrt{1 - \left(\rho_V / \rho_L \right)} Tan\left(\phi / 2 \right) \sqrt{2g}\, h^{5/2} \tag{1}$$

where h is the height of the liquid surface above the bottom of the V-notch. It is derived by applying Bernoulli's equation to the liquid and vapor flow paths. The assumptions are that no head loss occurs in the liquid flow, the liquid velocity in the reservoir is negligible, and the liquid nappe has the same cross sectional shape as the V-notch as it leaves the crest of the weir. Also, the vapor flow must have no frictional pressure drop and the vapor velocity must be negligible. None of these assumptions are strictly true, and the coefficient C_d corrects for these effects. From experimental data, Lenz developed a nondimensional correlation that predicts a C_d of about 0.6 for

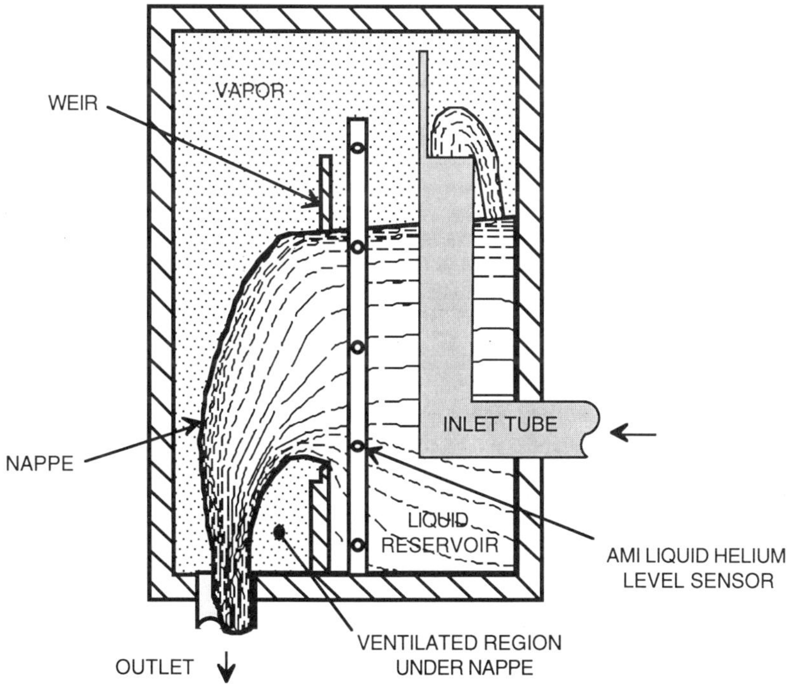

Figure 3. Cross section of weir flow meter through middle of V-notch.

the Fermilab flow meter[3]. However, the unusual fluid and construction of the Fermilab flow meter warrant the experimental determination of its C_d.

Substituting the approximation $\rho_v = 0$ into Eq. (1) generates another simpler and more commonly used expression. However, for two phase helium, this approximation should not be used since helium vapor density is a significant fraction of liquid helium density.

An undesirable consequence of the flow meter's small size is the constricted vapor flow path causes higher vapor velocities and friction losses that affect vapor pressure. As a result, increased vapor pressure over the reservoir surface with respect to the nappe will push additional liquid through the V-notch and change the level. These effects are not accounted for in Eq. (1) and degrade flow meter accuracy. We estimate that at normal operating conditions, such vapor pressure changes may alter the liquid level 1.5 to 3.5 mm.

The flow meter relies upon gravity to separate the vapor from liquid helium. Baffles and screens are not used for the separation process because they would impart additional pressure drop on the vapor flow. Calculations determined that, by comparison to experimental data with helium on the Baker diagram, stratified two phase flow would exist in the flow meter reservoir[4].

A change of the flow rate into the flow meter causes a time delayed change in the level of the reservoir. The larger the cross sectional area of the reservoir, the longer it takes the reservoir level to fully respond to flow rate changes. A full reservoir will drain empty in about 1 minute if the flow into it is suddenly stopped. This flow meter response time is sufficient for its intended use. However, this quick response time may have caused problems with the attempt to calibrate the flow meter as described below.

THEORETICAL ACCURACY

The flow meter accuracy depends upon the accuracy of the liquid level measurement. Assume that Δh is the error in the level measurement and ΔQ is the resulting error in the calculated flow rate. Then:

$$Q + \Delta Q = \frac{8}{15} C_d \sqrt{1 - \left(\rho_V / \rho_L\right)} Tan\left(\phi / 2\right) \sqrt{2g}\, (h + \Delta h)^{5/2} \tag{2}$$

is derived from Eq. (1). If the flow coefficient C_d is constant over the range of flow rates being measured, then the accuracy expressed as percentage of flow rate is found by combining Eq. (1) and Eq. (2).

$$\frac{\Delta Q}{Q} = 100\left(\left(1 + \frac{\Delta h}{h}\right)^{5/2} - 1\right) \% \tag{3}$$

For the weir flow meter described in this paper, Eq. (1) and Eq. (3) are plotted in Figure 4a. Figure 4b shows the plot of accuracy versus flow rate. A flow coefficient of 0.6 is assumed. The possible error of the AMI level measurement error Δh is 0.05 cm and fairly uniform over the length of the sensor[5].

Figures 4a and 4b demonstrate the theoretical accuracy for the weir flow meter and not the measured performance. The coefficient C_d is unknown at this time and may vary with the liquid

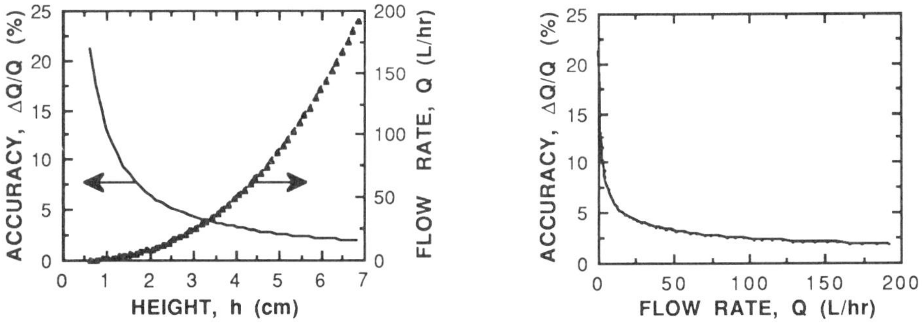

Figure 4. Theoretical Accuracy, height and flow rate of V-notch weir flow meter, a) Theoretical flow rate and accuracy versus height, b) Theoretical accuracy versus flow rate.

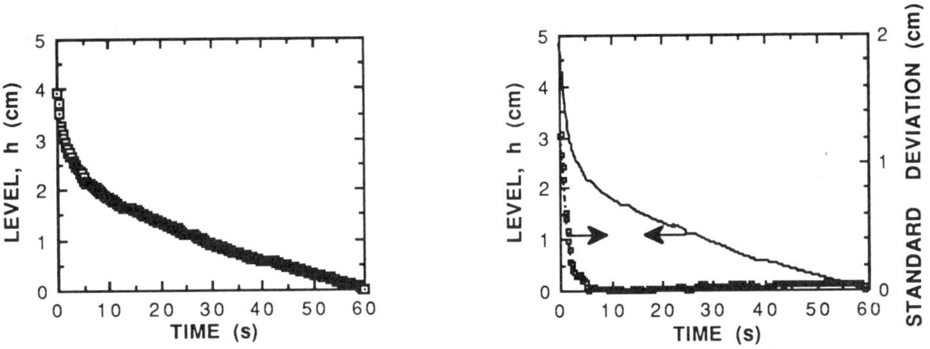

Figure 5. Results from calibration tests, a) Level data versus time from a typical test, b) Average level and standard deviation versus time for 27 test runs after eliminating outlying data points.

level. For this application, a flow meter that is sensitive over a wide range of flows is more important than high accuracy. As shown in Figure 4a, this flow meter could theoretically measure flow rates from 3.47 L/hr to 198 L/hr with a 10% error. The ratio of these two flow rates, called the turndown ratio, is 57.1. The V-notch weir was selected for this application because of its high turndown ratio for the desired accuracy.

In open channel hydraulics, V-notch, square notch, trapezoidal notch and other shaped notches have been used in weirs[6]. The V-notch is not necessarily the best for all applications. It is interesting and worthwhile to compare the accuracy and range of different notch shapes for a given application. However, for almost any application, increasing the depth of the reservoir and the working length of the liquid helium sensor will increase a flow meter's accuracy.

TESTING

The weir flow meter was used in April and March of 1995 during operation of the cryogenic plant at Fermilab's Proton East Beam Line service building (PS4). The flow meter responded in a believable manner to changes in valve positions, magnet string cooldowns, warmups, and quenches. It was a reliable and quick indication of system problems. The normal operating level of the weir flow meter was between 5.6 cm and 6.4 cm, about 90% of the maximum height and 77% of the estimated maximum flow rate.

The cryogenic control system (Moore APACS) automatically performed 32 calibration tests with the weir flow meter. Before a calibration test, the reservoir was filled as high as possible. The controller closes the inlet and outlet valves to the flow meter. Flow is suddenly stopped into the flow meter; since the inlet tube into the flow meter discharges above the height of the weir (see Figure 2), all of the liquid in the reservoir drains out through the notch. After waiting to allow for flow in the transfer line to stop, the controller serves as a datalogger storing and saving 320 consecutive level measurements, one every 200 ms over a period of 64 seconds. The data was sent via Dynamic Data Exchange (DDE) to an Excel spread sheet on a PC for analysis.

In five of the tests, slugs of liquid apparently sloshed into the reservoir from the upstream piping, causing sudden increases in the level. Of the remaining 27 tests, a statistical study was made on the data. For each data set, the time was offset to compensate for variations in the initial level. In effect the plot of each data set was shifted along the time axis so they would all overlay each other. As a typical example, in Figure 4a is the level data from one of the tests plotted versus time. Figure 4b shows a plot of the average of the data versus time from the 27 tests with extraneous outlying points (outside of the range of 95% probability of occurrence) ignored following a statistical method assuming a Gaussian distribution of the data[7].

It is difficult to draw conclusions from Figures 5a and 5b. In theory, the volumetric flow rate at a given reservoir level can be calculated by multiplying the cross sectional area of the reservoir times the slope of the level curve in Figure 5b. These calculations have been made and do not agree well with Eq. (1). Space does not permit a detailed discussion of the test data, and further testing and study need to be performed. Perhaps the liquid helium is "clinging" to the weir and Eq. (1) is not valid below a certain level. Also, the time constant of the level instrumentation is unknown and the level measurements may not have changed as quickly as the actual level, perhaps causing considerable error in the data.

Other errors exist in the data. The total estimated loss of liquid helium due to heating from the AMI liquid helium sensor and external heat load is 0.22 L/hr. When taking data the current source to the AMI sensor was not pulsed. The pressure varied in the flow meter during testing inducing some error in the data. In the APACS controller, the exact time during a scan period that a level measurement is stored is not known.

CONCLUSIONS

A weir flow meter that indicates the liquid helium flow rate of a two phase mixture has been built and tested. Although uncalibrated, the flow meter has proven useful in the operation of a string of superconducting magnets. The meter responds in a plausible manner to changes in the cryogenic system. Further study and more data are required to calibrate the flow meter and fully understand its behavior. The upcoming Fixed Target run at Fermilab will provide considerable opportunity for gathering data on the flow meter.

ACKNOWLEDGMENTS

Work supported by the U.S. Department of Energy under contract No. DE-AC02-76CHO3000. The efforts of Margaret Stone, James Peifer, James Seeman, William Gatfield and Donald Friend contributed greatly to this project.

NOMENCLATURE

C_d	Coefficient of discharge, dimensionless
Δh	Error in level measurement, m
ΔQ	Error in volumetric flow rate measurement, L/hr
g	Acceleration due to gravity, m/s^2
h	Reservoir level measured from bottom of V-notch, m
ϕ	V-notch angle in weir, °
Q	Volumetric flow rate, m^3/s
ρ_L	Density of liquid helium, Kg/m^3
ρ_v	Density of vapor helium, Kg/m^3

REFERENCES

1. N.S. Lakshmana Rao, Theory of Weirs, in: "Advances in Hydroscience," Vol. 10, Academic Press, New York (1975).
2. American Magnetics Inc., "Installation, Operation And Maintenance Instruction for The AMI Model 134 Liquid Helium Level Monitor," Oak Ridge, Tenn. (1992).
3. A.T. Lenz, Viscosity and Surface Tension Effects on V-Notch Weir Coefficients, in : "American Society of Civil Engineers," New York (1943).
4. J. C. Theilacker and C. H. Rode, An Investigation Into Flow Regimes for Two-Phase Helium Flow, in: "Advances in Cryogenic Engineering," Vol. 33, Plenum Press, New York, (1988), p.393.
5. E.R. Efferson and S. Smith, private communication.
6. P. Ackers, W.R. White, J.A. Perkins and A.J.M. Harrison, "Weir and Flumes for Flow Measurement, John Wiley and Sons, New York (1978).
7. R.S. Figliola and D.E. Beasley, "Theory and Design for Mechanical Measurements," Wiley and Sons, New York (1995), p.158.

COMPRESSION MASS GAUGE TESTING IN A LIQUID HYDROGEN DEWAR

J.M. Jurns [1] and A.C. Rogers[2]

[1]NYMA Inc., NASA LeRC Group
Brook Park, OH 44142, U.S.A.
[2]Southwest Research Institute
San Antonio, TX 78228, U.S.A.

ABSTRACT

This paper describes testing that was conducted using a mass gauge in a liquid hydrogen environment. The mass gauge, referred to herein as the "compressibility gauge," is being developed as a means to accurately determine the mass of liquid contained in a tank in a low-gravity environment. The concept, described by Mord et al.,[1] is based on the thermodynamic principle that the pressure of gas or vapor changes when its volume changes. Previous work has been conducted by Southwest Research Institute in collaboration with NASA Lewis Research Center. This consisted of testing the concept with water and other cryogenic simulant fluids. The purpose of conducting liquid hydrogen tests is to test the concept in actual cryogenic conditions, and address hardware issues that arise in fabricating a test article for use in liquid hydrogen.

NOMENCLATURE

α tank stiffness factor

C boundary layer coefficient

f volume change frequency

P tank total pressure

ΔP tank pressure change with volume change

V gas or vapor volume

ΔV tank volume change

γ polytropic constant, specific heat ratio of gas or vapor for adiabatic process

BACKGROUND

The ability to accurately measure quantities of cryogenic fluids in a low gravity environment is a critical requirement for future space exploration missions. In low gravity, the position of the liquid in the container may be markedly different than it is in a one-g gravity field environment. Instead of settling in the bottom of the container, the fluid may become a mixture of gas bubbles interspersed within the liquid, which may be located randomly throughout the container. Consequently, the familiar gauging methods used on earth are not generally applicable in space. The need to develop a gauging system that will work in low gravity is evident. Many "low-g quantity gauges" have been investigated in concept or by laboratory testing over the past 30 years. These systems have been based on the use of a variety of physical principles such as radio frequency microwaves, gas bubble resonant frequency, liquid heat capacity, optical absorbency, ultrasonics, acoustics, gamma ray densitometry, and flow meters for monitoring liquids leaving and entering the tank (mass balancing). To this point, however, they have all proved to have significant limitations in gauging accuracy, complexity, or weight. Of all the potential technologies, only compressibility gauging offers the benefits of simplicity of design, minimum intrusion into the tank, high accuracy, and functionality independent of liquid orientation.

COMPRESSIBILITY GAUGE CONCEPT

The concept of compressing the ullage gas bubble as a means of gauging liquid volume has been developed previously for ground applications at the Southwest Research Institute[3] and recently for space applications.[4] The physical basis of such a gauge is the relation between gas volume V and gas pressure P, when a small adiabatic volume change ΔV is used to produce a corresponding small change in pressure ΔP:

$$V = -\gamma P \frac{\Delta V}{\Delta P} \tag{1}$$

In principle, a compressibility gauge does not require knowledge of the way gas is distributed in the tank. In practice, however, several considerations may limit the applicability of Eq. (1) and cause a need for more elaborate gauging controls and data analyses:

(1) Liquid compressibility
(2) Tank elasticity
(3) Heat, mass, and momentum effects
(4) Thermodynamic phase change
(5) Multiple ullage bubble resonances
(6) Density stratification
(7) Noncondensible pressurant gas
(8) Liquid motion

Each of these factors can be addressed by modifications to Eq. (1), or by modifying the compressibility gauge operating procedure. A more detailed explanation of the modification for each of these factors has been covered in previous works.[2] Taking into account these effects, the resulting equation for V is:

$$V = -\gamma P\left(\frac{\Delta V}{\Delta P} - \alpha\right)\left(1 + \frac{C}{2\sqrt{f}}\right)^{-1} \tag{2}$$

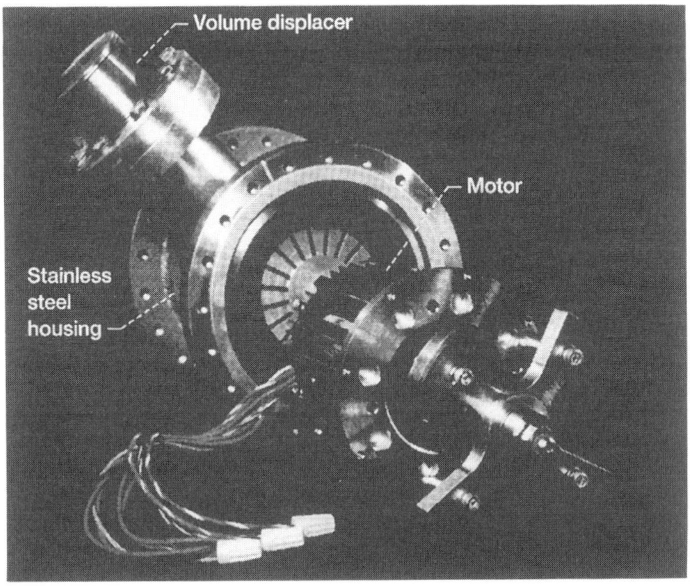

Figure 1.—Compressibility gauge-exploded view.

By algebraically equating V for two values of frequency f, the parameter C can be eliminated from Eq. (2) to yield Eq. (3):

$$V = \frac{\gamma P\left[\left(\Delta V/\Delta P_1 - \alpha\right)\sqrt{f_1 f_2} - \left(\Delta V/\Delta P_2 - \alpha\right)\right]}{\sqrt{f_1 f_2} - 1} \tag{3}$$

PREVIOUS TESTS

To properly assess the viability of the compressibility gauge in a cryogenic tank in a microgravity environment, the different effects that influence its accuracy must be examined. Some of these effects have been investigated in previous tests conducted with water and a cryogenic simulant fluid (either Freon R-11, or HCFC-123).

Initial tests to validate the gauging concept were conducted by Ball Aerospace.[1] These tests used water in a commercial 210 liter stainless steel process drum modified to include a compression driver and pressure sensing head. The tests confirmed the accuracy of the gauging system, showed dependence of boundary layer coefficient "*C*" on fill level, investigated the effect of cooperative bubble resonance, and gave some insight into pressure transducer location and orientation.

Further tests conducted by Southwest Research Institute,[3,4] used R-11 refrigerant as a cryogenic simulant fluid. These tests confirmed the accuracy of the gauging system with a condensible fluid. They also investigated tank total pressure effect, pulser frequency and fluid quantity effects, and unsteady heat transfer effects. They also reported progress in pressure sensing by using and evaluating a "spark plug" type of piezoelectric transducer that mounted directly to the tank wall in contact with the tank contents.

The ability of the compressibility gauge to function in microgravity was also examined by NASA LeRC. A compressibility gauge using R-11 refrigerant was flown on a Lear jet, and subjected to low-gravity trajectories. Results from these tests were inconclusive because of the noise and vibration in the Lear jet. However, these tests did reveal some interesting results that will aid in developing better pressure signal processing procedures and gauge sensing head design.

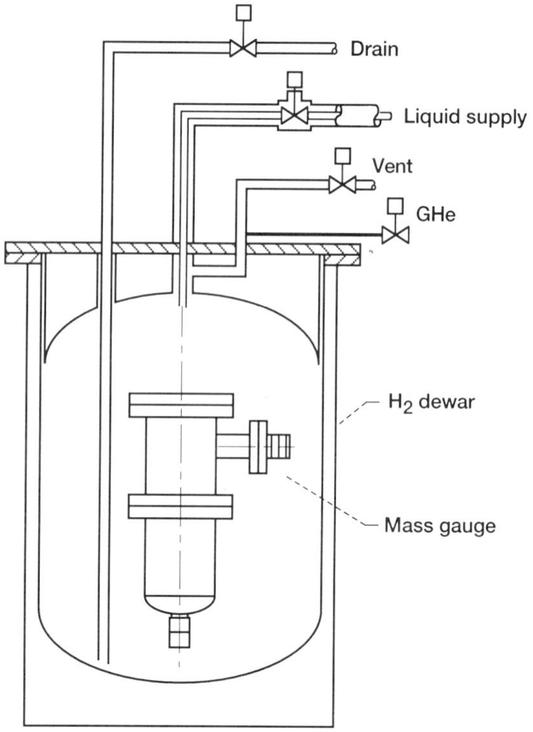

Figure 2.—Piping schematic diagram.

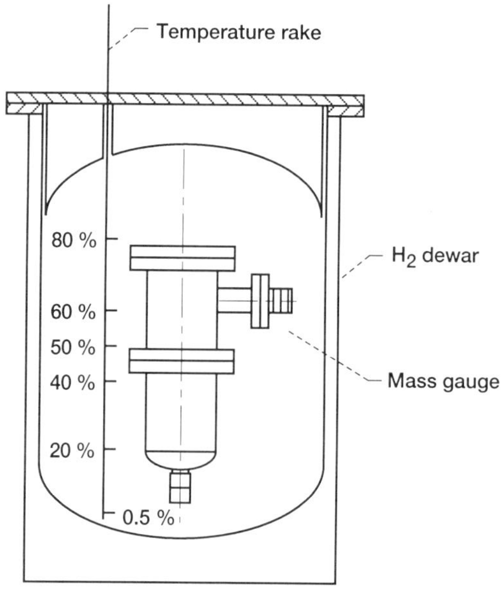

Figure 3.—Silicon diode locations.

LIQUID HYDROGEN TESTING

The applicability of the compressibility gauge in cryogenic fluids requires the development of components that will operate in a cryogenic environment. Previous tests, both water and cryogenic simulant, were conducted at ambient temperatures and pressures. The tests described in this report were conducted with a compressibility gauge which operated in a liquid hydrogen dewar. The gauge was designed and built by Southwest Research Institute,[5] and is shown in Fig. (1). It was tested in a liquid hydrogen dewar at the NASA LeRC Cryogenic Components Lab cell 7 research facility.

Cryogenic Compressibility Gauge

The compressibility gauge shown in Fig. (1) consisted of a 186.4 W (1/4 hp) non-synchronous 3-phase induction motor which was enclosed in a stainless steel canister, and coupled by means of an eccentric drive shaft to a stainless steel bellows assembly which was also attached to the canister. The rotation of the motor drive shaft was converted to a linear motion which pulsed the bellows assembly. The eccentric for the drive shaft was built such that the linear displacement of the bellows could be varied. For these tests, the stroke was set at 1.63 cm, which resulted in a total ΔV of 38.1 cm^3. The ratio of displacer ΔV to test tank volume V was on the order of 10^{-4}, which is typical for compressibility gauging devices.[1] The motor was an off the shelf 186.4 W motor which was modified by cleaning out the lubrication, and replacing bearings with greaseless bearings suitable for cryogenic service. The motor speed was controlled by using a variable frequency drive controller, and could vary the pulser frequency from a fraction of a hertz up to about 20 Hz.

A piezoelectric differential pressure transducer suitable for liquid hydrogen service was attached to the cryogenic compressibility gauge to measure the differential pressure in the dewar resulting from the action of the bellows. The transducer had a range of ±34.5 kPa. Motor speed was monitored by a speed sensor attached to the motor drive shaft.

Liquid Hydrogen Test Facility

The test facility where the cryogenic compressibility gauge was tested is the NASA LeRC Cryogenic Components Laboratory Cell 7. This multipurpose test facility is used for testing of cryogenic components. A schematic of the test dewar with compressibility gauge is shown in Fig. (2). A supply dewar (not shown) with an internal volume of approximately 305.8 L supplies liquid hydrogen to a test dewar which is 164.2 L internal volume. The cryogenic compressibility gauge was installed in the test dewar. Electrical power wiring was connected to the motor, instrument wiring was connected to the piezoelectric transducer, and a helium line connected to provide positive pressure to the gauge.

Both the supply and test dewars have flat flanges for lids. The lids have a short cylindrical section with an inverted dome bottom. The space between the flange and cylindrical section is evacuated and insulated with MLI to minimize heat transmission through the lid. With this arrangement, heat transfer into the test dewar due to the design of the dewar and piping penetrations is approximately 12.6 W/m^2.

Liquid hydrogen is transferred between the supply and test dewars through vacuum jacketed piping. The dewars can be pressurized with either helium or hydrogen gas, and can also be evacuated by means of a series of air ejectors to approximately 13.8 kPa absolute. Maximum working pressure of the dewars is 377.1 kPa absolute.

Instrumentation and Data Acquisition

Temperature sensors are positioned throughout the rig and on the tank walls, selected fluid lines, and components. Temperatures are measured with type T (copper-constantan) thermocouples and silicon diodes. Tank wall sensors are located in the annular vacuum space of the supply and test dewars, and are mounted to the inner tank wall. Within each dewar is an instrument tree with silicon diodes attached at various heights. These diodes can be used both to measure temperature of the contents to within ±0.1 K, and also as point level sensors to determine liquid level to within ±0.5 percent (±3 mm).[6] See Fig. (3)for diode locations. A capacitance type level probe is also available to monitor liquid level. Transducers provide continuous pressure measurement throughout the system with an estimated accuracy of ±0.5 percent.

Data acquisition is controlled by a PC computer with software written in the C programming language. Nominally, 240 channels of data are taken on the test rig. Data is displayed on the CRT screen of the PC. Toggle switches, selector knobs, and continuously adjustable dials mounted on the control panel are used to control the system valves and fluid routing.

The facility data acquisition system has a nominal recording rate of one sweep every two seconds. As the compressibility gauge pulser had an nominal operating frequency of 1 to 20 Hz, an additional data acquisition system had to be used to record the differential pressure signal generated by the gauge. This data acquisition was done by a portable notebook computer connected to a high speed data acquisition board. It was set up to record at a rate of 250 points/sec. At a maximum pulsing rate of 20 Hz, this would provide approximately 12.5 data points per pressure pulse. An electrical voltage signal was used as a flag on both data systems to synchronize the data records.

Test Procedure

After the compressibility gauge was installed in the test dewar, the gauge was pressurized with helium. A transducer measured the internal pressure in the gauge, and pressure was typically maintained several psi above dewar pressure. The dewar was evacuated and purged with hydrogen, and then filled with liquid hydrogen to the 80 percent full level (as measured by the silicon diode tree). The gauge was run at five frequencies varying from 3 to 15 Hz. The tank pressure, fluid temperature, gauge internal pressure and temperature, and dewar pressure were all recorded. The tank was drained to the next level, and the test repeated. Measurements were made at the 80, 60, 50, 40, 20, and 0.5 percent full tank levels.

These tests were repeated varying several parameters. For example, the dewar pressure was maintained at 103.4, 137.9, and 172.4 kPa absolute for three different series of tests. Also, for three of the tests, the ullage volume was pressurized with helium gas instead of hydrogen gas. (The test matrix is shown in Table 1.)

TEST RESULTS

Test results were used as input to Eq. (3) which is referred to as a *Dual Frequency Model*. This model assumes that the Boundary Layer Coefficient C is not zero, and thus requires test data at two different frequencies to calculate the volume. Plots are presented in Figs. 4 and 5 for gauging accuracy for 103.4 kPa tests and for 137.9 and 172.4 kPa tests, as functions of fill level in the tank.

Ambient Pressure Tests

These tests were conducted with the dewar at approximately 103.4 kPa pressure. Several preliminary tests were conducted to gain experience with the system, and to become familiar with what outputs could be expected from the instrumentation. The piezoelectric differential pressure transducer originally gave spurious results. It was discovered that the transducer needed to be electrically isolated from the body of the gauge to function correctly. After the transducer was electrically isolated, it functioned properly. The tests discussed here are four series of tests which were run after test conditions were fairly well understood and controlled.

Table 1. Compressibility Gauge Test Matrix

Test	Frequency, Hz	Dewar press, kPa	Ullage gas
1	3 to 15	103.4	Hydrogen
2	3 to 15	103.4	Helium
3	3 to 15	103.4	Hydrogen
4	3 to 15	103.4	Hydrogen
5	3 to 15	172.4	Hydrogen
6	3 to 15	172.4	Helium
7	3 to 15	137.9	Helium
8	3 to 15	137.9	Hydrogen

Figure 4 shows gauging accuracy as a function of vapor volume in the dewar. The gauging error was typically within ±5 percent of actual level as determined by the point level sensors.

The gauge functioned well in all tests. The pulser frequency was varied from 3 to 15 Hz. At each setting, the pulser frequency remained constant (that is, no apparent fluctuations in frequency). During one test, the gauge internal pressure was intentionally increased to determine the effect of differential pressure between the gauge and the dewar. At a differential pressure of 34.5 kPa, the pulser did exhibit fluctuations in frequency. This was expected, as the motor for the gauge was designed to push only against the minimum pressure induced by the pulser.[5]

A small amount of helium has was introduced into the ullage during test 2 (also tests 6 and 7 of higher pressure tests). This was done to try to suppress vaporization by raising liquid pressure above the thermodynamic equilibrium pressure at the liquid temperature.

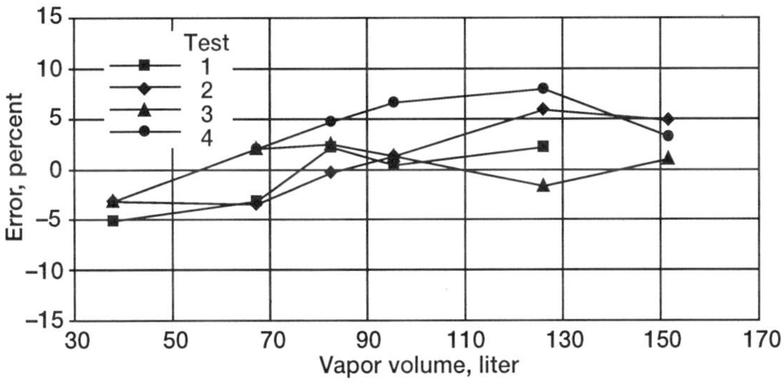

Figure 4.—Gauge error, LH_2 test, 103.4 KPa test pressure.

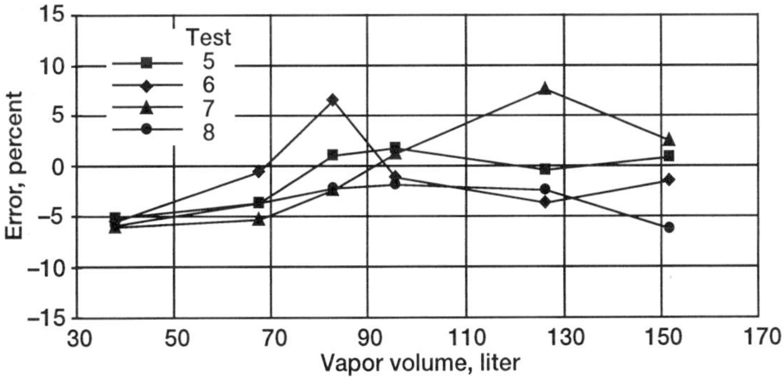

Figure 5.—Gauge error, LH_2 test, 137.9 and 172.4 KPa test pressure.

Higher Pressure Tests

These tests were similar to the ambient pressure tests, except that the dewar pressure was increased to either 137.9 or 172.4 kPa absolute, depending on the test. Figure 5 shows the results of these tests. The pressure inside the gauge was maintained at approximately the same as the dewar pressure to minimize the additional load on the motor. Results were similar to those obtained in the ambient pressure tests. For the lowest gas volumes, however, there appeared to be a negative bias to the error.

Possible Sources of Error

Heat flux.—The test dewar, although vacuum jacketed, had a parasitic heat leak into the system of about 12.6 W/m^2. This heat leak caused the pressure in the dewar to rise immediately upon sealing the dewar and the operating the gauge. So, although the pressure in the dewar was nominally 103.4 kPa, it actually rose while the gauge was operating. The gauge only operated for several seconds at a time, but even for that short time, conditions were not actually steady state. As a reference for comparison, a typical space based storage tank has a heat leak of approximately 0.31 W/m^2, which is 40 times less than the test bed used here. The adverse effects of unsteady heat transfer are consistent with poor results reported in earlier simulated cryogen tests.[2]

Vapor stratification.—The heat leaked into the test dewar mostly through its lid. This resulted in significant stratification of the vapor temperature in the ullage. Refer to Fig. 6. Temperatures typically ranged from 22 K at the liquid vapor interface to 78 K at the top of the vapor ullage. The algorithm used to calculate vapor volume assumes a constant specific heat ratio. A specific heat ratio for the averaged ullage vapor temperature and composition was used in calculations.

CONCLUSIONS

Based on the results of this test program, it has been shown that a compressibility gauge can be successfully built and operated in a cryogenic liquid hydrogen environment. The gauge built for this test was a breadboard unit. That is, the purpose of building this gauge was to verify the function of each of the critical components, not to optimize the overall design. Specifically, the following was shown:

(1) An off the shelf 186.4 W nonsynchronous 3-phase induction motor which has been modified for hydrogen service by cleaning out the hydrocarbon based lubrication and replacing the factory bearings with greaseless bearings suitable for cryogenic service functions suc-

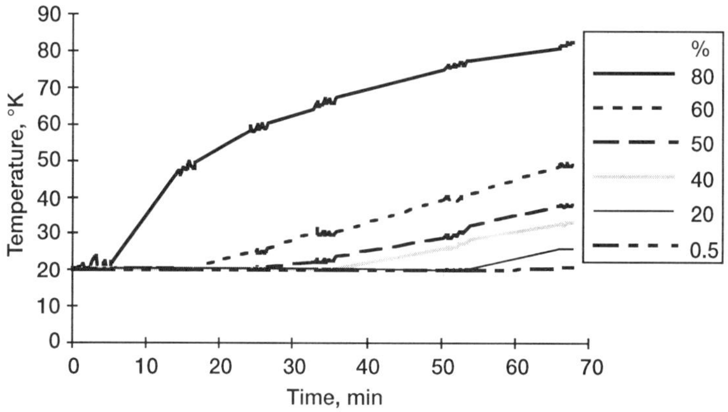

Figure 6.—Test dewar vapor temperature profile, showing stratification.

cessfully at liquid hydrogen temperatures. This motor provides sufficient torque to operate the gauge bellows without any significant variation in the motor speed.

(2) The piezoelectric differential transducer designed specifically for liquid hydrogen service functioned properly once it was electrically isolated from the compressibility gauge.

(3) The entire compressibility gauge could be successfully immersed in liquid hydrogen, and still function properly. The gauge with the integrated motor provides minimum design restriction. That is, it could be placed anywhere in the test dewar. The thermal impact to the dewar is minimum. A few instrument wires, the motor power wires, and a small (3.2 mm diameter tube) pressurization line were the only items protruding through the dewar. In future designs, the pressurization line may also be eliminated.

(4) The use of helium in the ullage did not improve accuracy for these tests. Calculation of vapor volume requires knowledge of the specific heat ratio γ, and our inability to accurately meter the amount of helium used was a likely source of error in determining γ and thus vapor volume.

(5) The compressibility gauge produced results that were typically within ±5 percent of actual volume. This was not as good as results from previous tests.[2,3] However, this could be attributed to heat flux and vapor stratification previously mentioned. Further tests in a dewar with a lower heat leak would be desirable.

REFERENCES

1 A.J. Mord, H.A. Snyder, K.A. Kilpatrick, L.A. Hermanson, R.A. Hopkins, and D.A. Vangundy, "Fluid Quantity Gaging," Ball Aerospace Systems Report DRD MA183T, Contract NAS9 17616, 1988.

2. A.C. Rogers, F.T. Dodge, and K.A. Behring II, "Feasibility Development of a Cryo Fluid Gauging System for Space Vehicle Applications," Southwest Research Institute Report AIAA 93–1801, presented at AIAA 29th Joint Propulsion Conference June 28–30, 1993.

3. A.C. Rogers, F.T. Dodge, and K.A. Behring II, "Feasibility Development of a Cryo Gauging System for Space Vehicle Applications," Southwest Research Institute Final Report, SwRI Project No. 04–9664, April, 1993.

4. R. Monti, University of Naples, Italy, W. Berry, ESTEC, The Netherlands, "Liquid Gauging in Space: The G-22 Experiment", ESA Journal 1994, Vol 18.

5. A.C. Rogers, "Design Review—Cryo Hydrogen Quantity Gauge Test Hardware Design and Fabrication," Southwest Research Institute Project 04–2453–330, NASA Contract NAS3–25365, May 25, 1993.

6. Paula J. Dempsey and Richard H. Fabik, "Using Silicon Diodes for Detecting the Liquid-Vapor Interface in Hydrogen," NASA TM–105541, presented at ISA 38th International Instrumentation Symposium, April 26–30, 1992.

USING Bi-2223 Ag CLAD TAPE AS A LIQUID NITROGEN LEVEL SENSOR

J. X. Jin,[1] H. K. Liu,[1] C. Grantham,[2] and S. X. Dou[1]

[1]Center for Superconducting and Electronic materials
University of Wollongong, Wollongong, NSW 2522, Australia
[2]Dept. of Electric Power
University of NSW, Sydney 2052, Australia

ABSTRACT

The dramatic resistance change in the superconducting-normal transition temperature range enables a high T_c superconductor to be considered for use as a liquid nitrogen level sensor. A $(Bi,Pb)_2Sr_2Ca_2Cu_3O_{10+x}$ Ag clad superconducting wire is tested as a continuous liquid nitrogen level sensor to investigate the possibility for this application. The $(Bi,Pb)_2Sr_2Ca_2Cu_3O_{10+x}$ Ag clad superconducting wire has approximately 110 K critical temperature, with more flexible and stable properties compared with bulk shape ceramic high T_c superconductors. The voltage drops across the sensor are tested with different immersion length in liquid nitrogen. The voltage signal is sensitive to liquid nitrogen level change, and this signal can be optimised by controlling the transport current. The problems of the Ag clad superconductor are that the Ag sheath thermal conductivity is very high, and the sensor normal resistance is low. These are the main disadvantages for using such a wire as a continuous level sensor. However a satisfactory accuracy can be achieved by control of the transport current. A different configuration of the wire sensor is also designed to avoid this thermal influence.

INTRODUCTION

Liquid nitrogen is widely used in superconductivity research work and superconducting devices. In many cases the liquid nitrogen level is necessary to be monitored and controlled in liquid nitrogen containers. High T_c superconductors (HTS) enable a new simple solution to be a sensor to monitor and control the amount of liquid nitrogen in the container, and this has been considered by using a $Y_1Ba_2Cu_3O_{7-x}$ sensor produced by a technique of flame and plasma spraying.[1]

The well developed $(Bi,Pb)_2Sr_2Ca_2Cu_3O_{10+x}$ Ag clad superconducting wire has about 110 K critical temperature, with more flexible and stable properties compared with previous bulk shape HTS. This kind of wire is easy to achieve in a long length and high critical current density compared with the bulk shape HTS. For the application of a continuous liquid nitrogen level sensor, the critical current, length and mechanical flexibility of this kind of wire is satisfied.[2,3]

A $(Bi,Pb)_2Sr_2Ca_2Cu_3O_{10+x}$ Ag clad HTS wire is selected, and tested to achieve the experimental identification of this wire as a suitable material for the liquid nitrogen level sensor.

PRINCIPLE OF OPERATION

Transition resistance signal between superconducting state and normal state of a HTS is used for this liquid nitrogen level sensor. Four-point I-V measurement is a simple method to detect small resistance change precisely, and is therefore employed in this investigation. The sensor is linked with a sensor controller which consists of a program controlled current source and voltage detector.

The Ag sheath of this HTS wire has a relative high resistivity $\rho_{100K} = 0.4 \times 10^{-8}\ \Omega m$, compared to the HTS core in liquid nitrogen, therefore when a piece of long HTS wire is partly immersed in liquid nitrogen, the voltage drop across the total sensor is proportional to the Ag sheath resistance of the level sensor, which appears above the liquid nitrogen surface. The Ag sheath is a very good thermal conductor, so it will cool the superconducting core of the sensor above the liquid nitrogen surface to be in superconducting state. However, this problem can be improved by controlling the transport current within the critical current value of the sensor, in order to reduce the length of the superconducting part above the liquid nitrogen surface.

Figure 1 shows the configuration of the HTS nitrogen level sensors and the controller. Normal four-point I-V measurement connection is used to detect the sensor resistance which changes with different liquid nitrogen immersing level. Sensor S-1 is designed for a continue liquid nitrogen level controller sensor, and sensor S-2 is for multi-point local nitrogen level detection.

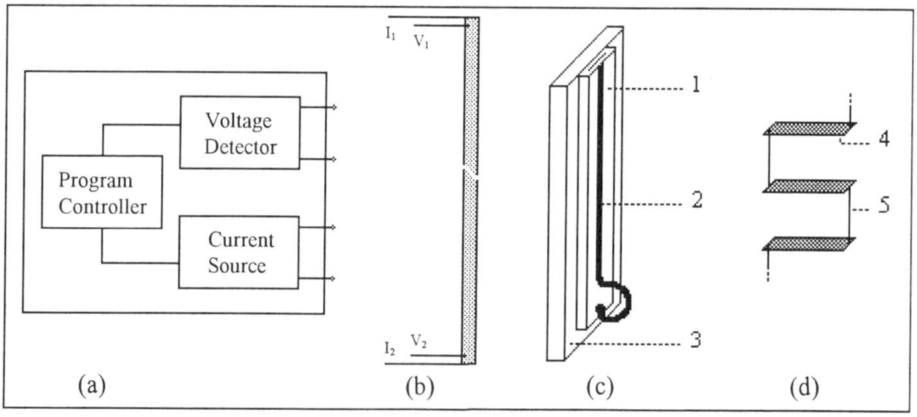

Figure 1. Schematic of the liquid nitrogen level sensor. (a) Liquid nitrogen level detector controller. (b) Continuous HTS liquid nitrogen level sensor (S-1). (c) Configuration of S-1. 1-Bi-2223 Ag clad tape; 2-Thin Cu current lead; 3-Thermal insulating support. (d) Local liquid nitrogen level detector. 4-Bi-2223 Ag clad short tape; 5-Thin Cu wire.

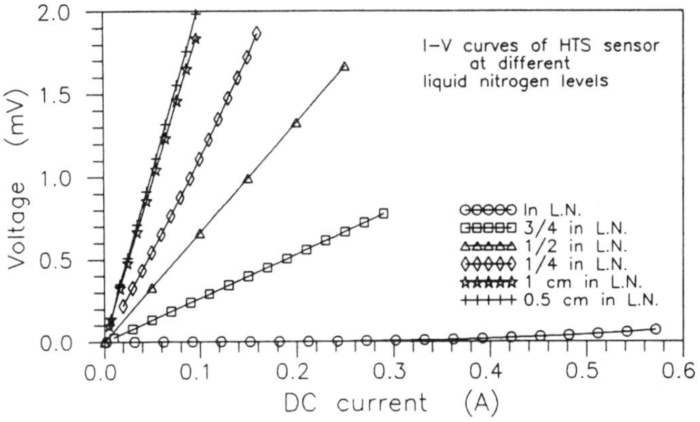

Figure 2. I-V curves of the continuous HTS sensor S-1 with different liquid nitrogen levels.

EXPERIMENTAL INVESTIGATION

For testing purposes, a $(Bi,Pb)_2Sr_2Ca_2Cu_3O_{10+x}$ Ag clad superconducting wire is selected which was made by a powder-in-tube method in 1991, with initial molar composition of Bi:Pb:Sr:Ca:Cu = 1.8:0.35:2:2.05:3. The wire was made through 3 rolling-sinter procedure with 66 - 80 h and 829 - 832 °C at each step. The long wire was cut to a few pieces of about 7 m long to wind small coils (ϕ_{in} = 28 mm), and the long wire having sectional critical current I_c about 1 A along the wire. The HTS sensor tapes used were cut from the small coils and strengthen.

The $(Bi,Pb)_2Sr_2Ca_2Cu_3O_{10+x}$, wire, used as a liquid nitrogen level sensor, has 0.17 mm × 3.2 mm total cross section and 0.0397 mm^2 HTS core cross section. For testing purpose, it was cut 470 mm for the sensor S-1, and cut to a number of 12 mm short tapes for the level sensor S-2.

The configurations of the sensors are shown in Figure 1. The HTS continuous liquid nitrogen level sensor S-1 is the 470 mm long wire. The HTS wire is fixed on a piece of wood. A thin Cu wire is selected to be the current lead and it is attached on the HTS tape to warm up the part of the sensor above the liquid phase.

For local or position detection, S-2 is designed with a number of short HTS tapes in series by very thin Cu wires for multi-point liquid nitrogen level detection, and every HTS tape is in parallel with the liquid nitrogen surface to detect the nitrogen level in a certain range. The S-2 is fixed in a notched piece of wood.

I-V curves of the S-1 were then tested with different immersing levels in a open Dewar bottle, as shown in Figure 2. The total voltage drops across the entire sensor were recorded according to different transport currents and different immersing levels. These I-V curves show a clear voltage signal change of the sensor with different immersing levels.

Figure 3 is the sensor S-1 signal with different liquid nitrogen levels. The L in the figure is the distance from the top of the sensor S-1 to the liquid nitrogen surface. A 0.15 A transport current was used in this figure. The higher transport current used, within its critical current, the higher voltage signal will be generated to achieve high accuracy. However, this will generate more thermal energy in the resistive part. Consequently the optimum current is related to accuracy and thermal loss.

The S-2 is designed for local level detection. The HTS sensor tapes are secured to a piece of wood. Figure 4 shows one tape voltage signal change with different distance L from the sensor to the liquid nitrogen surface. A 0.4 A transport current was used for this test. The result shows that the wooden supporter reduces the sensitivity by thermal transfer to cool the HTS sensor tapes in a certain length above liquid nitrogen surface.

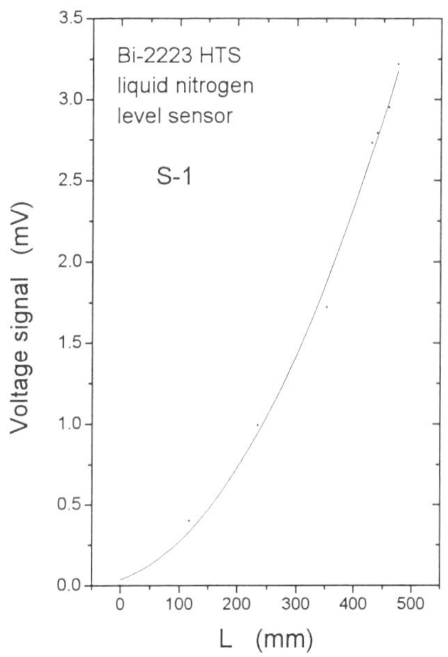

Figure 3. The signal of the continuous HTS sensor S-1 with different liquid nitrogen level.

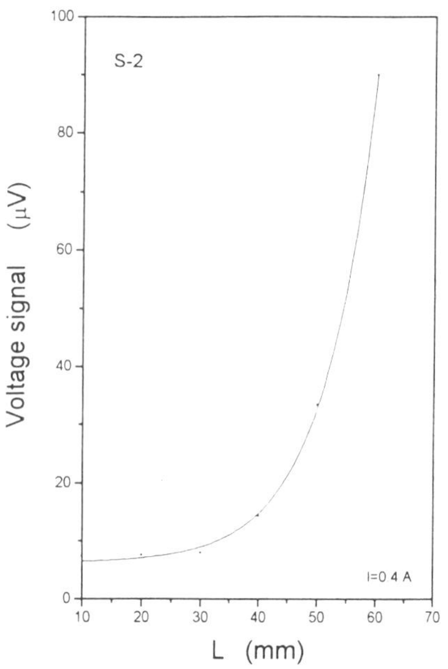

Figure 4. Liquid nitrogen level change and signals of local sensor S-2.

CONCLUSIONS

The $(Bi,Pb)_2Sr_2Ca_2Cu_3O_{10+x}$ Ag clad wire is identified to be used as a practical liquid nitrogen level sensor, since its resistance is sensitive to the liquid nitrogen level change. Consequently, a simple liquid nitrogen level monitor and controller can be built with a designed length of the HTS wire sensor, and with programmed current source and voltage detector.

The history of the selected HTS wire, i.e., long time since production, deformations due to winding, cutting to smaller lengths and the many cycles tests in liquid nitrogen, shows that this wire has a better mechanical flexibility, and stability compared with bulk shape HTS. It is also fairly easily produced with higher critical current density, therefore a smaller sensor cross-section can be achieved. By reducing the cross-section, the sensitivity and thermal balance can be improved.

Ag has very good thermal conductivity, it will transfer the temperature of the liquid nitrogen along the sensor above the liquid nitrogen surface. However, this effect could be balanced or reduced by increasing transport currents, therefore the sensor can be optimised.

Bi-based HTS $(Bi,Pb)_2Sr_2Ca_1Cu_2O_{8+x}$ has about 85 K critical temperature, and therefore can achieve better accuracy by reducing the superconducting part above the liquid nitrogen surface. The $(Bi,Pb)_2Sr_2Ca_1Cu_2O_{8+x}$ Ag clad wire is suggested as being ideal for use in this application.

REFERENCES

1. St. Siegmann, T. Frey, J.P. Ramseyer, and H.-J. Güntherodt, Continuous level sensor for cryogenic liquids using high-T_c superconductors, *Rev. Sci. Instrum.*, **61** 1946 (1990).
2. H.K. Liu, J. Yau, Y.C. Guo, Q.Y. Hu, J.X. Jin, N. Savvides, and S.X. Dou, Long Ag-clad Bi-based superconducting tapes and coils, *Physica B*, **194-196** 2213 (1994).
3. S.X. Dou, Y.C. Guo, J. Yau, J.X. Jin, and H.K. Liu, Mechanical properties of Ag-clad Bi-based superconducting composite tapes, *Physica B*, **194-196** 1577 (1994).

HIGH SENSITIVITY SQUID DAMAGE DETECTOR IN A HIGH-Tc SUPERCONDUCTING SHIELD

K.Shintomi,[1] S.Nagaya,[2] N.Hirano[2]
A.Yahara,[1] M.Sakano,[1] K.Imai,[1] H.Matsuba[1]

[1]Yokohama R&D Laboratories, Furukawa Electric Co.,Ltd.
Yokohama, Kanagawa, 220, Japan

[2]Electric Power Research & Development Center
Chubu Electric Power Co.,Inc.
Nagoya, Aichi, 459, Japan

ABSTRACT

A non-destructive evaluation (NDE) system for material damages of metals has been developed. The system enables very low magnetic field measurements by using a SQUID magnetometry system in a high-Tc superconducting shield. The shield has large internal dimensions of D0.87 m x L2.4 m, so that large specimen of up to 50 cm x 10 cm can be inspected. Size and location of magnetization from various material damages were contact less estimated with a Nb-based multi-channel SQUID magnetometer system which was placed directly in the shield.

Magnetization of austenitic stainless steels induced by plastic strain was successfully detected. The magnetization of austenitic stainless steels increased with plastic strain. The magnetization of type 304 stainless steels varied more with plastic strain than that of types 310 and 316 stainless steels. X-ray diffraction measurements showed that the magnetization increases were induced by a ferromagnetic martensitic transformation from the austenitic phase. These measurements suggest that it could be possible to detect in this way anticipatory signs of damage in the materials.

INTRODUCTION

Metals such as stainless steel are widely used in machinery which expose the metals to a high temperature and/or a high-stress and/or corrosive environments. The machines base their reliability on the integrity of the metal properties, therefore, the detection of metal damages with non-destructive methods is an important issue. The examination of the damage has been carried out with various kinds of technologies, e.g. electrical, electrochemical, ultrasonic, magnetic, or electromagnetic.[1,2,3] These conventional approaches were quite successful in detecting damages and have been widely used. However, the development of a method using simpler procedure and with a better ability in damage detection is required, to overcome many disadvantages of the conventional methods.

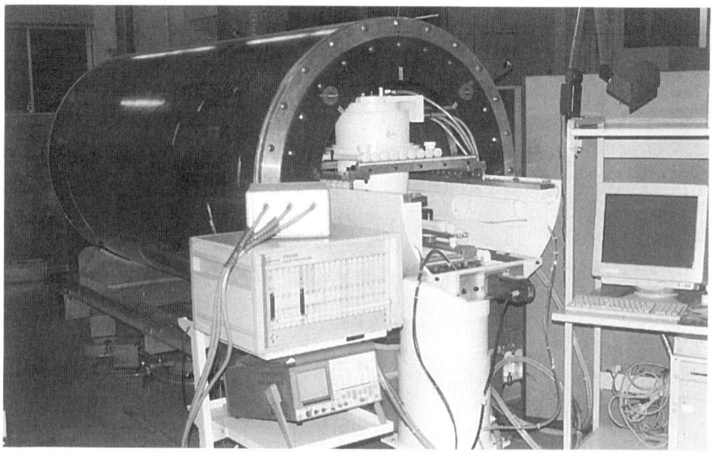

Photograph 1. System for magnetization measurements using SQUID magnetometers in a high-Tc superconducting shield.

As an alternative to conventional methods, non-destructive evaluation (NDE) on metals by means of SQUIDs has been investigated, since the method is expected to be able to detect a damage located deeply in a metal without contact and to detect magnetic anomalies in the DC to low-frequency range.[4,5,6] The fundamental principle is that, scanning with a SQUID pick-up coil the surface of a test specimen, damages can be estimated on the basis of the detection of magnetic anomalies in specific locations. It remains difficult to test specimens with complex shapes or where it is difficult to position the pick-up coil close to the surface.

We have developed a NDE system which consists of a 3-axes vector magnetometer and of a high-Tc superconducting shield which have a value of the shielding factor never attained before with conventional shields. The system is expected to offer the ability to determine the size and location of damage from a remote position by vector sensing of the magnetic field in a very low noise environment produced by the high-Tc superconducting shield, and also the ability to find anticipatory signs of damages thanks to its very high sensitivity to magnetic field anomalies. We prepared specimens of non-ferromagnetic austenitic stainless steels which were subjected to a plastic deformation. In this paper we show the resulting changes of magnetization in the specimens.

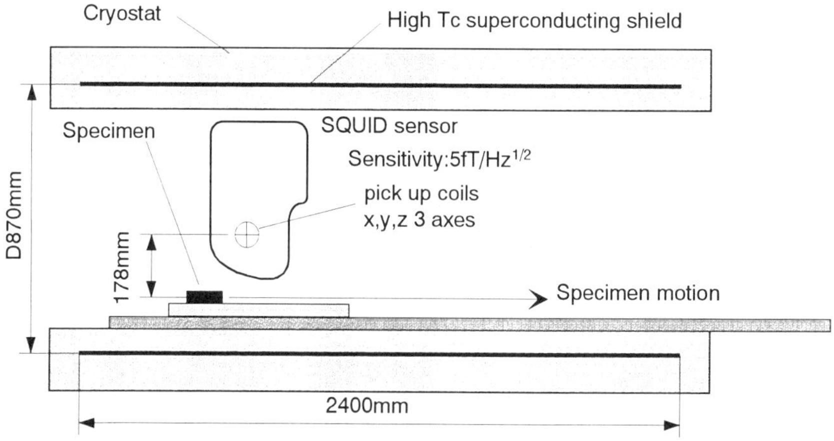

Figure 1. Schematic drawing of the system for magnetization measurements s using SQUID magnetometers in a high-Tc superconducting shield.

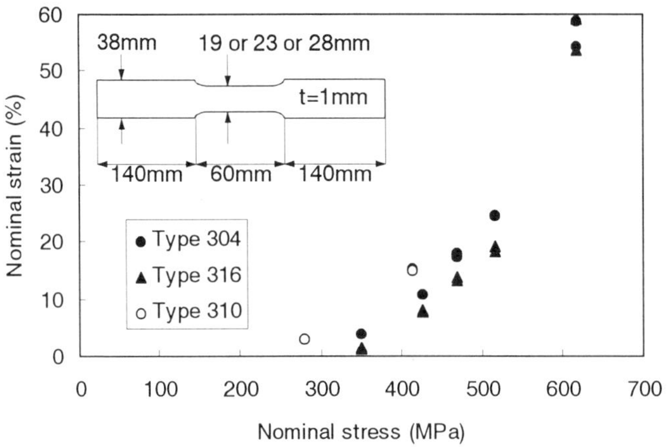

Figure 2. Nominal stress and nominal strain curves for types 304, 316 and 310 stainless steels due to tensile tests. Dimensions of the specimens used for the tensile tests are shown.

EXPERIMENTAL ARRANGEMENT

A SQUID system was set in a high-Tc superconducting shield with dimensions D0.87 m x L2.4 m. The shield attenuated the environmental field by 120 dB, so that the external noise fell in the shield bellow the sensitivity of SQUID system. The SQUID system was equipped with a three pick-up coil arrangement for vectorial magnetic detection; the Nb-based DC-SQUID magnetometers have a magnetic sensibility of 5 $fT/Hz^{1/2}$. The x-, y- and z- components of the magnetic signal were detected by the three pick-up coils, while a specimen travels in the z direction of the magnetic shield, with a minimum distance of 178 mm between the pick-up coils and the specimen. The magnetic fields were measured at a 5 Hz sampling rate, with a 1.5 Hz high cut off frequency and while the specimen was traveling at a speed of 5 cm/s. Figure 1 shows the test arrangement.

Figure 2 shows nominal stress and nominal strain curves of types 304, 316 and 310 stainless steels, and the dimensions of the specimens used for the tensile test. Table 1 shows the chemical composition of the specimens. Specimens having dimensions of 16 mm x 240 mm for the magnetic tests were cut out from the stressed specimens by means of the electric discharge cutting technique to avoid any further plastic deformation.

EXPERIMENTAL RESULTS

Figure 3 shows a typical SQUID x-, y- and z- output for a type 304 stainless steel specimen with 10.8 % nominal strain.

We assumed that the magnetization can be represented by a single dipole in the specimen and calculated the magnitude and the direction of the dipole from the magnetic measured when the specimen was located just below the three pick-up coils. Zero magnetic field was taken as the magnetic field measured when the specimen was far from the SQUIDs.

Figure 4 shows the magnetic moment for types 304, 316 and 310 stainless steels as a function of the nominal strain. The volume of the specimens ranged from 3.56 cm^3 to 3.84 cm^3.

Table 1. Chemical composition of the austenitic stainless steels (wt%)

Specimen	%C	%Si	%Mn	%P	%S	%Ni	%Cr	%Mo
Type 304	0.07	0.49	1.21	0.03	0.028	8.18	18.39	
Type 316	0.06	0.53	1.65	0.033	0.03	10.19	16.28	2.06
Type 310	0.02	0.42	1.17	0.024	0.001	19.39	24.31	

Balance : Fe

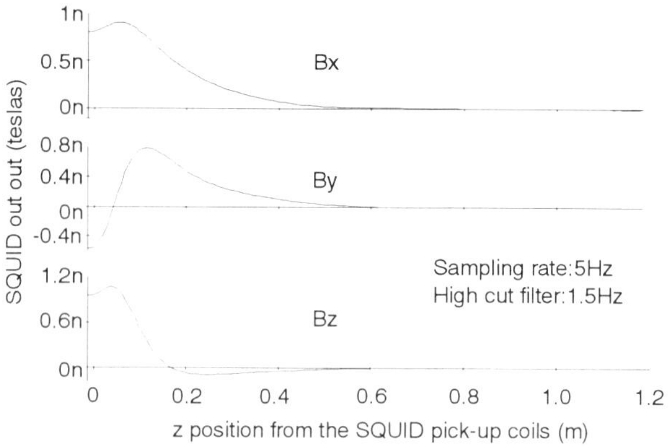

Figure 3. SQUID output signals for type 304 stainless steel having suffered a 10.8% plastic strain.

Thus, quantitative material damages can be evaluated by magnetic moments which are independent of the distance between the specimen and the pick-up coils.

The magnetic moment of type 304 stainless steels increased with increasing plastic strain, while types 316 and 310 stainless steels exhibited only little changes of the magnetic moment with increasing plastic strain.

It is well-known that the non-ferromagnetic austenitic phase is transformed into a ferromagnetic martensitic phase by applying stress on austenitic stainless steels.[7] It seems that the magnetization of the martensitic phase is determined by the environmental field (e.g. earth's field) at the moment when the martensitic phase is generated. We did not control the environmental field at the tensile test site, therefore, the magnetic moment values might show some inconsistencies depending upon the nominal strain, especially for types 310 or 316 stainless steels.

We then magnetized the specimens in a DC magnetic field. Figure 5 shows the dependence of the magnetization of specimens on the DC magnetic field. Magnetization approaches its maximum when the DC magnetic field exceeds 85 mT. Therefore we used a magnetic field of 255 mT.

Figure 6 shows the dependence of the magnetic moment on nominal strain for magnetized specimens of types 304, 316 and 310 stainless steels. Comparing these results with those of figure 4, we observe a consistent increase of the magnetic moment with the plastic strain for all types 304, 316 and 310 stainless steels. The magnetic moments after the magnetization are larger and their consistency is better than in figure 4.

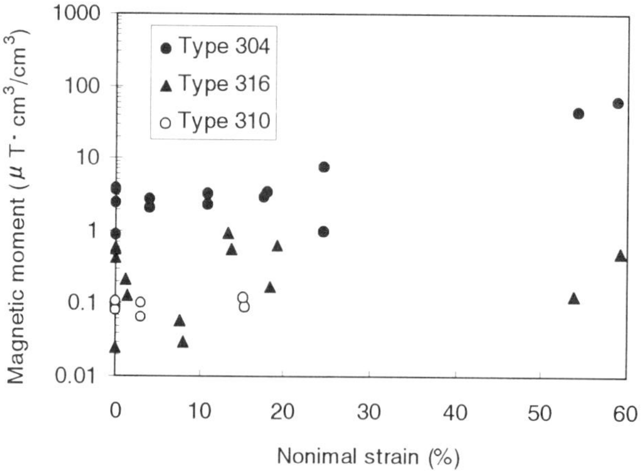

Figure 4. Magnetic moment versus nominal strain for types 304, 316 and 310 stainless steels.

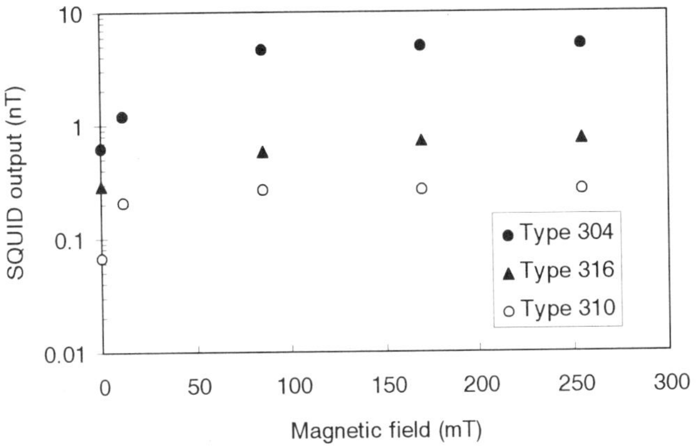

Figure 5. Dependence of specimen magnetization on DC magnetic field strength.

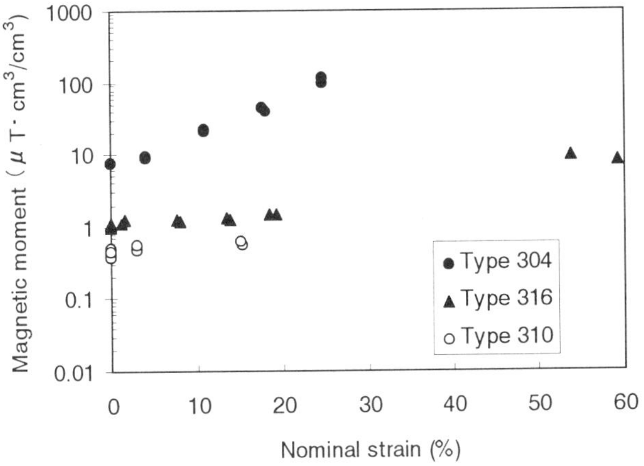

Figure 6. Magnetic moment versus nominal strain for specimens of types 304, 316 and 310 stainless steels magnetized in a DC magnetic field of 255 mT.

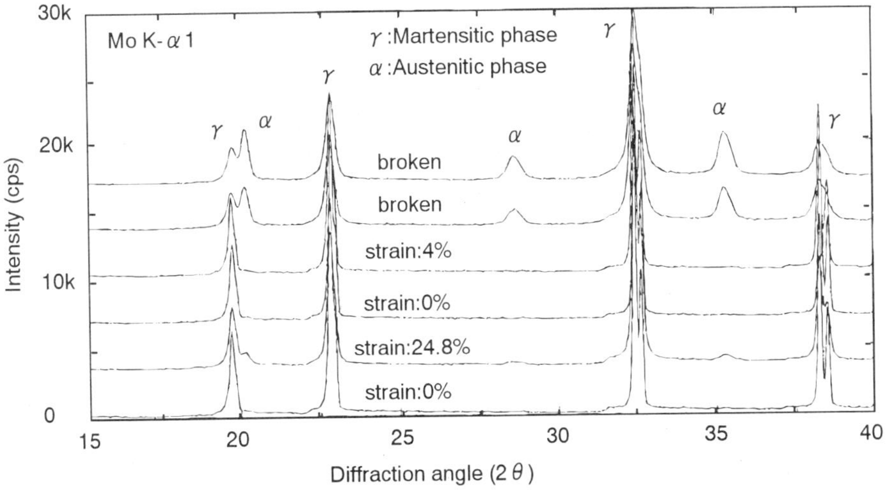

Figure 7. X-ray diffraction patterns for type 304 stainless steel.

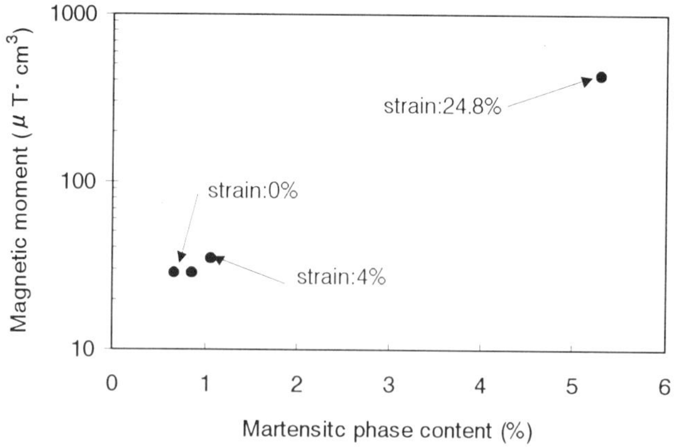

Figure 8. Magnetic moment versus martensitic phase content for type 304 stainless steel.

DISCUSSION

Figure 7 shows the X-ray diffraction patterns for type 304 stainless steel specimens at various nominal strain. The diffraction peaks of the martensitic phase are clearly observed for the specimen with 24.8 % nominal strain, while the peaks are not observed for the specimen with 4 % nominal strain. The maximum peak of the martensitic phase is observed for the broken specimen, which suffered the maximum nominal strain. The results clearly indicate that the martensitic phase was induced from the austenitic phase by the deformation.

The content of the martensitic phase was estimated from the relative amount of the austenitic phase in figure 7. Figure 8 shows the magnetization as a function of the content of martensitic phase and suggests that the magnetization increases with the content of the martensitic phase.

Therefore, the magnetization of type 304 stainless steel is due to the ferromagnetic martensitic phase which is induced from the austenitic phase by plastic deformation, and the ferromagnetic martensitic phase is magnetized by the environmental field (i.e. earth's field). If the earth field is not controlled at the site where ferromagnetic martensitic phase is produced, the magnetization of the ferromagnetic martensitic phase will show inconsistent values.

We observed that also strain-free specimens showed some magnetic moment. This result may be explained by the fact that some martensitic phase may be induced by plastic strain during the steel manufacture process, and/or during the (cold) rolling process.

According to Schaeffler's phase diagram,[8] the chemical composition of type 304 stainless steel gives the equilibrium state for a mixture of austenitic phase with martensitic phase and with ferritic phase; the chemical composition of type 316 stainless steel gives the equilibrium state at the boundary between pure austenitic phase and a mixture of austenitic phase with ferritic phase; and the chemical composition of type 310 stainless steel gives the equilibrium state only for pure austenitic phase. The magnetization of types 310, 316 and 304 stainless steels may reflect these equilibrium conditions.

CONCLUSIONS

We have developed a magnetization measurement system consisting of a high-Tc superconducting shield and of a three-axis SQUID magnetometer which could represent a new way to detect very small magnetization signals from a specimen placed in a remote location. We found that the magnetization of types 310, 316 and 304 austenitic stainless steels increases with plastic strain, and that the increase of the magnetization is more clearly detected with additional magnetization of the specimens in an external DC field. The results suggest that it could be possible to detect anticipatory signs of material damage by this technique.

REFERENCES

1. Y.Yokono, Y.Minami, S.Mukae, M.Katoh, N.Nishio, The Japanese Society for Non-Destructive Inspection, 39-11 (1990), p. 937
2. S.Yamaoka, F.Imataki, O.Ishikawa, The Japanese Society for Non-Destructive Inspection, 42-2 (1993), p. 86
3. A.Yoshinaga, J.Abe, M.Nakao, K.Endo, C.Takizawa, The Japanese Society for Non-Destructive Inspection, 37-9 (1988), p. 730
4. J.B.Bellingham, M.L.A.MacVicar, M.Niesenoff; IEEE Transaction on Magnetics, Vol. MAG-23, No. 2, March 1987
5. J.C.Moulder, T.E.Capobianco; Journal of Research of the National Bureau of Standards, Vol. 92, Number 1, Jan-Feb. (1987)
6. A.Cochran,J. C. Macfarlane, L. N.C.Morgan, J. Kuznik, R.Weston, L. Hao, R.M.Bowman, and G.B.Donaldson; IEEE Transactions on applied superconductivity, Vol. 4, No. 3, Sep. 1994
7. For example, J. F. Breedis; Acta Met., 13 (1965) 239
8. A.L.Schaeffler, Metal Prog., 56 (1949) 680

PREPARING A $(Bi,Pb)_2Sr_2Ca_2Cu_3O_{10+X}$ Ag CLAD HTS WIRE FOR A FAULT CURRENT LIMITER PROTOTYPE

J. X. Jin,[1] S. X. Dou,[1] C. Grantham,[2] R. Bhasale,[1] and H. K. Liu[1]

[1]Center for Superconducting and Electronic Materials
University of Wollongong, Wollongong, NSW 2522, Australia
[2]School of Electrical Engineering
University of NSW, Sydney, NSW 2052, Australia

ABSTRACT

A pair of rated 3.5 kVA magnetic cores are connected as a magnetic saturable reactor, and tested as a fault current limiting device, in which a high current ampere-turns dc coil is required to saturate the magnetic core. The reactance of the device ac limiting winding is tested with different dc bias currents. The magnetic field and power loss generated by the normal dc coil are also measured. These experimental results provide information of the required dc bias ampere-turns and generated magnetic fields. A 27-filament $(Bi,Pb)_2Sr_2Ca_2Cu_3O_{10+x}$ Ag clad wire, is selected and prepared under react and wind procedure to be a high T_c superconducting coil. A sample coil is formed from the superconducting wire and has achieved about 1500 AT at 77 K. The dc coil of the designed small prototype fault current limiter is possible to be implemented by using several such high T_c superconducting coils in series, with suitable critical current ampere-turns at 77 K.

INTRODUCTION

Developing a fault current limiting device by using a high T_c superconducting (HTS) coil has been studied with a small laboratory prototype.[1] Currently a pair of 3.5 kVA rated magnetic cores with conventional copper windings are used to build a magnetic core saturable type fault current limiting device, and a $(Bi,Pb)_2Sr_2Ca_2Cu_3O_{10+x}$ Ag clad HTS ribbon is prepared to wind a HTS coil for use in the device as a dc bias coil.

The Ag clad $(Bi,Pb)_2Sr_2Ca_2Cu_3O_{10+x}$ HTS wire has achieved high critical transport currents and long lengths.[2,3] The flexible mechanical property of this HTS wire enables it to be used to make a HTS coil under a react and wind procedure.

The normal resistance of this HTS wire is decided by the Ag sheath, so it is not ideal to develop a resistive fault current limiter by using this Ag clad HTS wire. Therefore, to develop a fault current limiter with this HTS wire, the use of a coil form of this HTS wire is

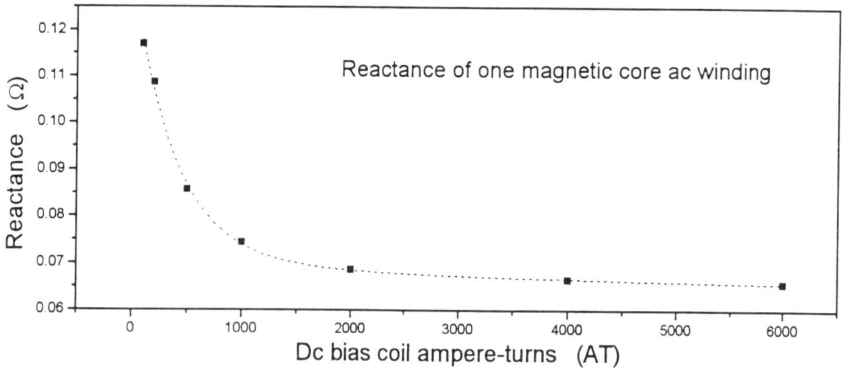

Figure 1. Fault current limiter normal reactance with dc bias coil ampere-turns.

considered. In the above mentioned fault current limiter the HTS wire is required to produce a HTS coil which generates a high dc mmf.

The fabrication of a HTS coil from a 27-filament $(Bi,Pb)_2Sr_2Ca_2Cu_3O_{10+x}$ Ag clad HTS wire has been studied in this work and a prototype fault current limiting device was tested at a low current and voltage level to evaluate the use of this HTS wire in this electric power application.

PROTOTYPE DEVICE

The principle of a magnetic core saturable reactor type fault current limiter has been introduced,[4] and the HTS coil has been investigated for this application.[1]

For the prototype device, the cross-section of the selected magnetic core is 38.1 mm × 38.1 mm with a 254 mm × 101.6 mm window. Windings consist of ac windings in one branch of the magnetic core and a thermal insulted container prepared to operate the HTS dc coil with liquid nitrogen on the other branch. There are also magnetic flux search coils linked with the magnetic core.

From a previous small prototype, the transient behaviour demonstrates that the device is able to limit the fault current in its first cycle.[1] When a simple operation is considered where there is no screen for the dc coil, the required dc ampere-turns can be estimated based on the magnetic core B-H curve by:

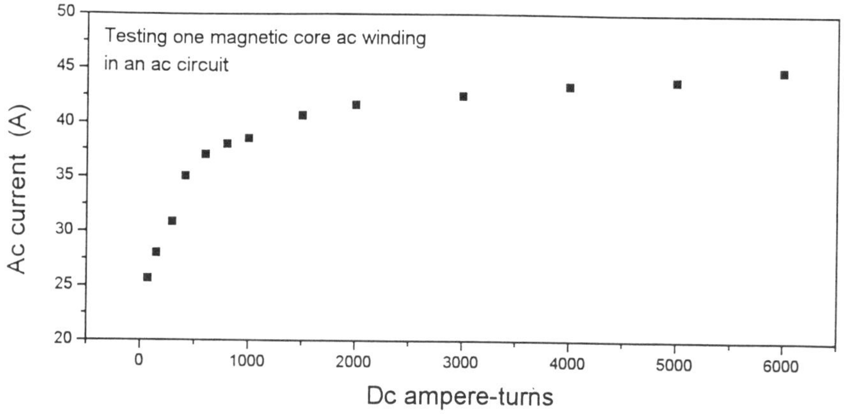

Figure 2. Ac circuit current with dc bias ampere-turns.

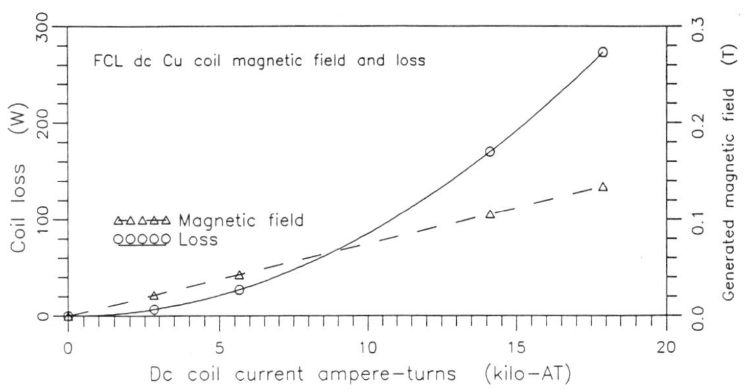

Figure 3. Magnetic field generated by dc coil and normal Cu dc coil loss.

$$I_{dc}n_{dc} = I_{limit}n_{ac} + H_sMPL \quad (1)$$

where the I_{dc} is dc bias current, n_{dc} is the number of turns in the dc winding, I_{limit} is the ac fault current peak value at which the device starts to limit the increased ac current, n_{ac} is the ac winding turns, H_s is the saturable field of the magnetic core, and MPL is the magnetic path length.

The dc bias winding required is related to the rated ampere-turns of the ac winding. A 20 turn ac winding was selected for testing its ac reactance with increasing dc bias ampere-turns, and the magnetic core was temporally energized with a 1000 turn Cu winding. Figure 1 shows one set of the magnetic core ac winding reactance changing with different dc bias coil ampere-turns.

Figure 2 is the test result of a steady state low current situation of this device. One set of the magnetic core with a 20 turn ac winding is connected to a 50 Hz - 5.1 V ac power supply without additional resistors in series, and a 1000 turn dc coil on the magnetic core is supplied from a dc power supply. The rms ac circuit current is then measured with increasing dc current. The result shows that the ac circuit reactance is reduced by increasing the dc bias ampere-turns. The fairly flat part of the curve is expected where the magnetic core is saturated.

The dc coil magnetic field and its power loss are tested with different dc current ampere-turns. The Cu dc coil is made from a 2 mm diameter enamel wire with a total of 2824 turns on a 46 mm × 47 mm × 252 mm former. Figure 3 shows the dc coil magnetic field and power loss. From the loss curve, it is clear that the coil is warmed by increased transport currents, and this is why a normal Cu wire is prohibited in a practical fault current

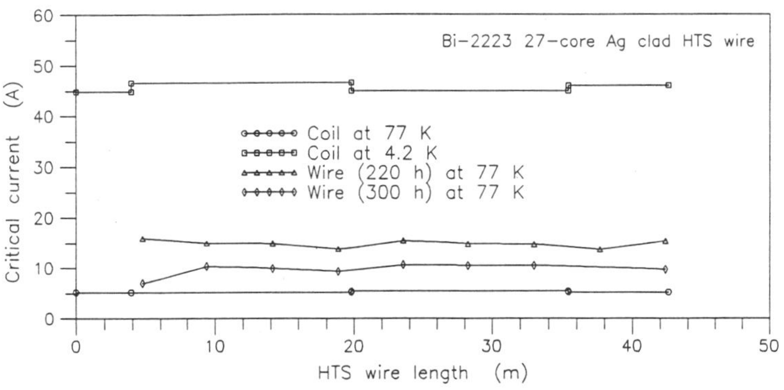

Figure 4. Producing a HTS coil by the HTS wire under react and wind procedure.

limiting device. The measured magnetic field at the coil centre is shown in the figure to be about 0.15 T at 20 kilo-AT.

PREPARING A HTS COIL

A Ag-clad ribbon with 27-filaments of the $(Bi,Pb)_2Sr_2Ca_2Cu_3O_{10+x}$ HTS cores has been produced through the normal powder-in-tube techniques with the wire final cross-section of $0.18 \times 4\ mm^2$, a superconductor-core/CS ~ 1:6, and a short tape critical current of $I_c = 21$ A (1 μV, 77 K) at $H_a = 0$ and 10 A at $H_a = 0.15$ T, where H_a is the applied magnetic field.

The sintered long wire has been fabricated into a solenoid coil using a react and wind procedure. A sample coil was formed with 292 turns with a 30 mm id, a 55 mm od, and a length of 65 mm. Its sectional critical currents were measured as shown in Figure 4. The critical currents were also measured at some points along the long HTS wire during the wire processing. The results show a better critical current is achieved after a total of 220 hours of sintering. The critical current after 300 hours of sintering is reduced by about 1/3 comparing with the value at 220 hours. The winding property of this HTS wire during the react and wind procedure is illustrated in Figure 4 by comparing the critical currents at different stage of the wire processing.

CONCLUSIONS

The prototype fault current limiting device requires a dc bias current ampere-turns related to the rated ac current. Higher dc ampere-turns are required for increased rated ac current ampere-turns. The dc coil magnetic field of this device is about 0.15 T with a 20 kilo-ampere-turns dc coil. The testing results on the HTS show a short piece of this HTS wire in a 0.15 T applied magnetic field can maintain 50% of the 77 K zero applied field critical current in liquid nitrogen. This HTS wire can be used at 77 K in a low field application.

The results also show that a multifilament $(Bi,Pb)_2Sr_2Ca_2Cu_3O_{10+x}$ HTS Ag clad wire with a fairly long length, mechanical flexibility and with uniform critical currents can be achieved. The sample coil produced under the react and wind procedure has achieved 1518 AT at 77 K and 13140 AT at 4.2 K. Several such coils in series would be required in the prototype fault current limiter.

REFERENCES

1. J.X. Jin, C. Grantham, H.K. Liu, and S.X. Dou, Prototype fault current-limiter with a high T_c superconducting coil, *J. Electr. & Electronics, Australia*, **15** 117 (1995).
2. S.X. Dou, Y.C. Guo, J. Yau, J.X. Jin, and H.K. Liu, Mechanical properties of Ag-clad Bi-based superconducting composite tapes, *Physics B*, **194-196** 1577 (1994).
3. H.K. Liu, J. Yau, Y.C. Guo, Q.Y. Hu, J.X. Jin, N. Savvides, and S.X. Dou, Long Ag-clad Bi-based superconducting tapes and coils, *Physics B*, **194-196** 2213 (1994).
4. B.P. Raju, K.C. Parton, and T.C. Bartram, Fault current limiting reactor with superconducting DC bias winding, CIGRE 1982, **23** 03.

SUPERCONDUCTING ANGULAR ACCELEROMETERS FOR THE SUPERCONDUCTING GRAVITY GRADIOMETER EXPERIMENT

P. J. Shirron[1], M. J. DiPirro[1], M. V. Moody[2], E. R. Canavan[2], and H. J. Paik[2]

[1]Code 713, NASA/Goddard Space Flight Center
Greenbelt, Maryland 20771 USA

[2]University of Maryland
Department of Physics and Center for Superconductivity Research
College Park, Maryland 20742 USA

ABSTRACT

A test of the University of Maryland's Superconducting Gravity Gradiometer[1] (SGG) on the Space Shuttle is currently being studied. The SGG can measure the spatial gradient of Earth's gravity to very high precision, having a sensitivity of 10^{-3} E Hz$^{-1/2}$ (1 E$\equiv 10^{-9}$ m/s^2 per m) in each axis. For successful application of this instrument in an orbital platform, a number of difficulties must be overcome. One is that the measured gravity signal is strongly dependent on the orientation of the instrument with respect to Earth. Another is that centrifugal acceleration and gravity gradients are indistinguishable in a rotating reference frame. Also, angular accelerations can couple to the gradient output through imperfections in the instrument. It is therefore necessary to measure the angular position, velocity and acceleration of the gradiometer in order to recover the true gravity signal. Superconducting angular accelerometers are described which can meet the sensitivity requirements for all three quantities. The acceleration signal will be integrated once to obtain angular velocity and then again for position. Angular position measurement is most critical, and sets a requirement of measuring accelerations to 10^{-10} rad/s^2 Hz$^{-1/2}$. The sensors must be compatible with a cryogenic environment since they will be mounted onto the SGG. Two designs are being pursued, differing significantly only in how the proof mass is confined to rotations about a single axis. Both use niobium proof masses, superconducting position sensing circuits, and SQUIDs for signal readout as does the SGG. Design parameters and anticipated performance are described.

INTRODUCTION

In recent years, increasing attention has been given to studies of planet Earth. Of particular importance to many of these studies is Earth's gravitational field, which can yield information about the structure and composition of the Earth, stresses along plate boundaries which precipitate earthquakes, and dynamic processes associated with global climate. An instrument has been developed at the University of Maryland which can measure gravity with unprecedented precision from an orbital platform. The Superconducting Gravity Gradiometer (SGG) is a cryogenic instrument which measures the relative motion of paired proof masses which are constrained by cantilever springs to move along a common axis. The sensing circuit provides a signal which is proportional to the relative motion of the paired masses, thereby measuring the difference in the gravitational field over a fixed baseline. For objects more than a few baselines distant, this closely approximates the gravity gradient. Three mass pairs simultaneously measure the gravity gradient in three orthogonal axes. By taking advantage of properties of superconductors and the dimensional stability of materials at low temperature, the device can measure differential accelerations with a sensitivity of 10^{-13} g_E $Hz^{-1/2}$ in the presence of 10^7 times larger common-mode (platform) accelerations. This corresponds to a gradient sensitivity of 10^{-3} E $Hz^{-1/2}$, where E, the Eötvös unit, is defined as 10^{-9} m/s^2 per m. The gravity gradient at the Earth's surface is 3000 E. The frequency band of the SGG is 2×10^{-4} to 10^{-1} Hz, allowing the instrument to measure gravity features from orbit as small as 100 km.

The gravity gradient signal is affected by the instrument's angular position, velocity, and acceleration through several mechanisms. The Earth's gradient is a tensor quantity. The gravity gradient along each of the instrument's axes is therefore a function of orientation. Also, if the instrument rotates, centrifugal acceleration will generate motion of the proof masses relative to each other indistinguishable from the gravity gradient, just as linear and gravitational accelerations are indistinguishable Finally, if the axes of the component accelerometers are not perfectly colinear, angular accelerations will cause relative motion of the masses. Thus, to remove the residual sensitivity to angular acceleration and determine the gravity gradient in an inertial coordinate system, one must have an independent measure of the angular position, velocity, and acceleration of the SGG. The SGG is weakly coupled to the dewar and orbital platform through a set of springs to avoid down-conversion of high frequency vibration through nonlinear processes. Hence, one cannot simply mount angular sensors to the spacecraft. Instead, angular motion of the instrument must be measured directly and the sensors used must operate at liquid helium temperature.

Superconducting angular accelerometers are being designed for use with the SGG. Angular rate and position will be obtained by integrating the angular acceleration. Several other methods for measuring angular motion were investigated, including hemispherical resonator gyros and fiber optic gyros. Hemispherical resonator gyros are incompatible with the cryogenic environment. While fiber optic gyros will theoretically operate well at low temperature, the projected cost for this option, including development, was prohibitive. The superconducting angular accelerometer will be relatively simple to implement since it operates on the same principles as the SGG and uses the same excitation and readout electronics. In fact, the sensing circuits of the two devices are identical; the devices differ in that the mechanical element of the angular accelerometer is a bar pivoted about its center of mass, so that it responds only to accelerations about the pivot axis.

PERFORMANCE REQUIREMENTS

Since the angular accelerometers will also be used to provide angular velocity and position, determining the performance requirements is relatively complicated and depends strongly on the characteristics of the spacecraft. On the Shuttle, knowledge of angular position imposes the most severe requirements on the instrument. The signal measured by the SGG is a quadratic function of angle, θ, with respect to vertical. In its "quiet" gravity stabilized attitude, periodic deflections of ~0.03 rad are common. Because of the double integration, the angular position noise at frequency f is degraded by $1/f^2$. In order to obtain sufficient resolution of θ, angular acceleration must be measured to 10^{-9} rad/s^2 $Hz^{-1/2}$ at low frequency, ~0.001 Hz. The device is expected to be limited by SQUID noise. SQUIDs have a typical noise spectra which is flat down to 0.1 Hz, then increases as $(0.1\ Hz/f)^{1/2}$ at lower frequency. The accelerometer will therefore have resolution of 10^{-10} rad/s^2 $Hz^{-1/2}$ at frequencies above 0.1 Hz, far better than is actually required at those frequencies.

The instrument must also be insensitive to linear accelerations. Data obtained with SAMS (Shuttle Acceleration Measurement System) indicates that during "quiet" periods the linear acceleration noise is 10^{-5} m/s^2 $Hz^{-1/2}$. To achieve the desired resolution at low frequency in the presence of this linear acceleration noise background, the center of gravity, c.g., must be either controlled or known to $(10^{-4}\ m^{-1}) \times r_g^2$, where r_g is the radius of gyration of the bar. Although precise positioning of the c.g. would eliminate this noise source, knowledge of the c.g. is sufficient because the SGG measures linear acceleration.

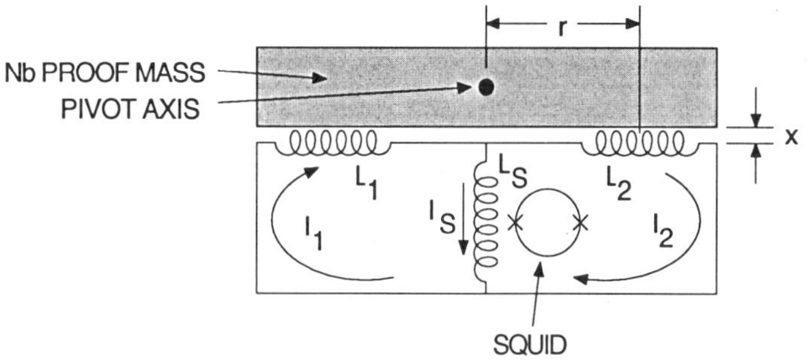

Figure 1. Schematic of sensing circuit for the angular accelerometer. The transformer used to couple the signal to the SQUID is not shown.

On a dedicated free-flier satellite, angular position errors will be two to three orders of magnitude smaller and correcting for centrifugal acceleration will dominate. However, the requirements are not significantly eased. In addition to sensitivity requirements, the direction of the sensitive axis of each angular accelerometer should be aligned to within 10^{-4} rad of the sensitive axis of the corresponding gradiometer to simplify data reduction.

PRINCIPLES OF OPERATION

The basic angular accelerometer consists of a proof mass attached to a housing by a torsional spring which limits motion to a single rotational degree of freedom. The housing is rigidly attached to the gradiometer assembly. The instantaneous torque, τ, on the proof mass is equal to both

$$\tau = \kappa\theta = I\alpha, \tag{1}$$

where κ and I are the torsional spring constant and the moment of inertia about the sensitive axis, θ is the angular displacement relative to the housing, and α is the angular acceleration of the proof mass about its sensitive axis. At frequencies well below the resonant frequency of the proof mass, ω_{res}, given by

$$\omega_{res} = \left(\frac{\kappa}{I}\right)^{1/2}, \tag{2}$$

any difference between α and the acceleration of the housing is negligible, and θ correctly gives the angular acceleration of the SGG.

The angle θ is measured using the same displacement sensing scheme as for the gradiometer. As depicted in Figure 1, the proof mass will consist of a niobium bar which rotates about a principal axis. At operating temperature (<4.2 K) the niobium is a superconductor. Spiral coils of niobium wire are located very close to the vertical faces of the bar. Because of the Meissner effect, the inductances of the coils are a function of the coil-bar spacing, and are approximately linear with the separation. The two coils are connected in parallel with the input of a SQUID amplifier through an impedance matching transformer. We assume that initially $L_1=L_2=L$. Excitation currents I_1 and I_2 are established in the two halves of the circuit. If initially we set $I_1=I_2\equiv I_{EXCIT}$, no current will flow through the inductor coupled to the SQUID, L_S. As the bar is rotated, for example CCW, L_1 will decrease and L_2 increase. Because the flux enclosed by the two loops must remain constant, the currents I_1 and I_2 will increase and decrease, respectively. The SQUID measures the difference I_1-I_2, which is proportional to the angular displacement.

In the limit of small gaps, x, between the coil and proof mass, each coil's inductance varies linearly with x. As the proof mass rotates by a small angle $\delta\theta$, the fractional change in inductance is $(r\delta\theta/x)$, where r is the distance of the coil center from the pivot. Solving the circuit equations and optimizing the parameters, we find that the change in current through the SQUID is related to $\delta\theta$ by

$$\delta I_S = I_{EXCIT} \sqrt{\frac{2L}{L_S}} \frac{r\delta\theta}{x} \tag{3}$$

The expression has been somewhat simplified by assuming the coupling constant of the transformer is 1. For excitation currents on the order of 5 A and using single layer, close wound excitation coils of 2 cm diameter, a SQUID resolution of 10^{-12} A $Hz^{-1/2}$ will allow the coil-bar gap to be resolved to 2×10^{-16} m $Hz^{-1/2}$. For a sensor having a 7 cm long proof mass, this corresponds to an angular displacement sensitivity of 8×10^{-15} rad $Hz^{-1/2}$. Combining Eqs. (1) and (2), we see the response of the proof mass to accelerations is inversely proportional to ω_{res}^2. In order to obtain the required sensitivity of 10^{-10} rad/s^2 $Hz^{-1/2}$, the resonant frequency must therefore be less than about 18 Hz.

While reducing the resonant frequency improves the sensitivity of the device, it also reduces the effective bandwidth. To remove centrifugal acceleration, which appears as a nonlinear component of the gradient signal, ω_{res} must be greater than the resonant frequency of the SGG. For the test units, the pivot is designed to give a 10 Hz resonant frequency.

To measure angular acceleration in all three degrees of freedom, three angular accelerometers will be mounted into pockets in the faces of the central cube which is 7 cm on a side, as shown in Figure 2. This limits the size of each angular accelerometer to approximately 2 × 5 × 7 cm. The SGG is formed by mounting a linear accelerometer on each face of the cube. Each linear accelerometer, which itself is a 7 cm cube, is mounted with its sensitive axis perpendicular to the face of the cube.

ANGULAR ACCELEROMETER CONCEPTS

Two spring concepts are currently being considered: a machined flexure and commercial flexural pivots. In both cases, the spring must display a high degree of linearity, low hysteresis, and rigidity in the other degrees of freedom. The main design and fabrication issues are the mass balance and relative alignment of the axes for three orthogonal sensors. The sensors must have extremely small c.g. offsets to limit sensitivity to linear accelerations. The structural design of the sensor must also take into account high accelerations levels present during launch. This will have an impact on both the flexure design and caging of the proof mass.

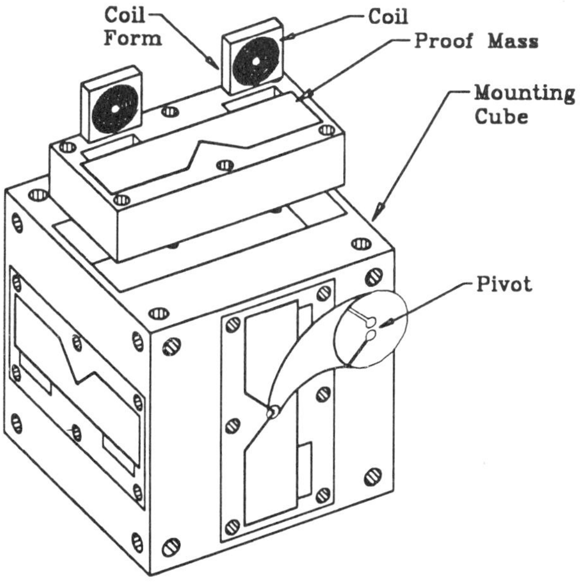

Figure 2. The angular accelerometers installed in the faces of the SGG mounting cube.

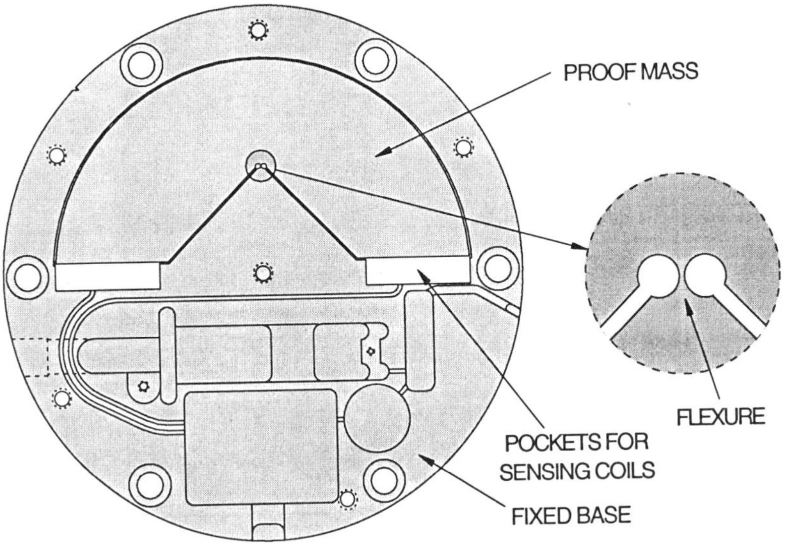

Figure 3. Accelerometer concept in which the proof mass (upper half-circular mass) and rotational axis are defined by a wire EDM cut. The millwork defines pockets for the SQUID sensor, persistent switches and transformer. Expanded region shows detail of flexure.

Machined Flexure Design

In the first concept, the proof mass, the spring and part of the housing are machined from a single block of niobium[2]. Figure 3 shows a prototype unit, which is cut from a 10.2 cm diameter disk 2 cm thick. The nearly semicircular proof mass is defined by a wire EDM (electric discharge machine) cut originating in one of the two holes (shown enlarged), tracing around the outside, and terminating in the other hole. Macor coil forms with spiral niobium coils slide into the two rectangular slots just below the proof mass. The displacement sensing circuits, including the SQUID, fit into the channels and pockets in the lower half.

The proof mass is supported by and can rotate about a flexure whose width is set by the closest approach of the two holes. The torsional spring constant of the flexure is proportional to the cube of its width[3]. The orthogonality of the sensitive axis with respect to the face of the instrument housing is determined by the orthogonality of the wire EDM cut with respect to the niobium block, which is 10^{-4} rad. In other degrees of freedom, the flexure is very rigid. Structural analysis has shown that by limiting the coil-proof mass gap, thereby limiting the maximum pivot deflection, the stresses in the pivot can be kept below the normal (0.2%) yield stress of niobium. However, because the requirement on mass balance demands exceptional dimensional stability, microyielding of the pivot is a concern. Avoiding microyielding in a niobium pivot would require a gap of less than 10 μm, which would greatly increase the fabrication cost. To circumvent this, a titanium alloy pivot, with a much higher yield strength, is used. This allows the gap between the coils and the proof mass to be 80 μm. At this spacing, the proof mass need not be caged during launch. The pivot has sufficient strength to withstand launch loads, and the energy inputs to the proof mass about the compliant axis are within acceptable limits.

To form the pivot, a 0.3 cm diameter hole, centered on the flexure point, is drilled through the niobium block before the wire EDM cutting and filled with a pin of Ti-6Al-4V alloy. The pin will be machined 7.5 μm larger than the hole and pressed into position. The unit will then be heated to diffusion bond the pin into place. Measurements on test pieces show the bond area to be stronger than either parent material.

The prototype sensor has been designed to be tested with an existing SGG which is 40% larger than the flight version. The flexure is defined by 508 μm holes and is 132 μm wide, and the proof mass's moment of inertia is 1.97×10^{-4} kg m^2, yielding a resonant frequency of 10 Hz which varies as the cube of the flexure width. The radius of gyration, r_g is 23.6 mm, so the c.g. must be known to within $(10^{-4}\ m^{-1} \times r_g^2) = 0.06$ μm of the center of bending of the flexure. The quoted tolerance for wire EDM machining is ±5 μm, which results in relatively large mass moments. By using capacitor plates in place of the niobium sensing coils the c.g. can be determined at room temperature prior to final assembly. As mentioned above, knowledge of the c.g. should be sufficient, though precise balancing of the proof mass will simplify data reduction. Several methods for mass balancing are being considered. These include balancing screws, and mass removal using laser ablation or punch EDM. The entire proof mass (388 g) offset by 0.06 μm is equivalent to an extra 0.5 mg of niobium at the outer edge of the proof mass. Controlling the overall balance to this level should be simple matter.

To date, two test pieces have been fabricated and tested at room temperature. Since the goal was only to resolve the c.g. to sufficient precision, these test pieces were made from a single piece of Ti-6Al-4V alloy, rather than the composite material. The c.g. of the proof mass was measured by applying a sinusoidal tilt of known amplitude, ϕ, to the sensors in three orthogonal orientations. Plotting the capacitance signal amplitude verses the frequency squared generates a straight line. When the tilt axis is parallel to the pivot axis, the slope of the line gives the sensor scale factor, H. The intercept is equal to $Hm\delta\boldsymbol{r}\bullet(g_E\times\phi)$, where δr is the c.g. offset and m is mass. For these test pieces, mistakes in machining produced a δr of 38 μm, and a torsional spring constant approximately 25 times too high. In spite of the high spring constant which lowers H, δr was measured with a one sigma error of 0.4 μm. With the proper torsional spring constant, achieving a resolution of 0.01 μm is straightforward.

Flexural Pivot Design

The second design shown schematically in Figure 4 is functionally equivalent to the first, but replaces the machined flexure with a commercial flexural pivot[4] which consist of two adjacent, cylindrical collars connected by thin, flat flexures. The two halves rotate relative to each other about a single, concentric axis. The flexural pivot is very rigid in the axes perpendicular to the compliant axis. Off-the-shelf items are made of stainless steel, but they can be custom made with other materials. A variety of sizes and torsional spring constants are available. Recent improvements in design yield flex pivots which are highly linear and have very low frictional losses and hysteresis.

The sensor consists of a fixed niobium base containing the readout coils or capacitor plates and a separate niobium proof mass which are connected by the flex pivot. The flex pivot is non-magnetic Inconel. The flex pivot is anchored in the base using a split clamp, which allows it both to be removed and rotated to correct misalignments. Because the proof mass and rotating portion of the flex pivot must have a well controlled center of mass, they will not be clamped together. Instead, the proof mass will have an interference fit hole drilled near its center to receive the flex pivot. At least 12 μ interference is needed to ensure sufficient friction remains after cooling to liquid helium temperature to prevent slippage during launch. The two are assembled by cooling the pivot and heating the proof mass to generate a clearance fit. A temperature difference of 400 K has been shown to be sufficient. There was also enough residual compressive force to prevent slippage between the proof mass and flex pivot under 20 g_E launch accelerations.

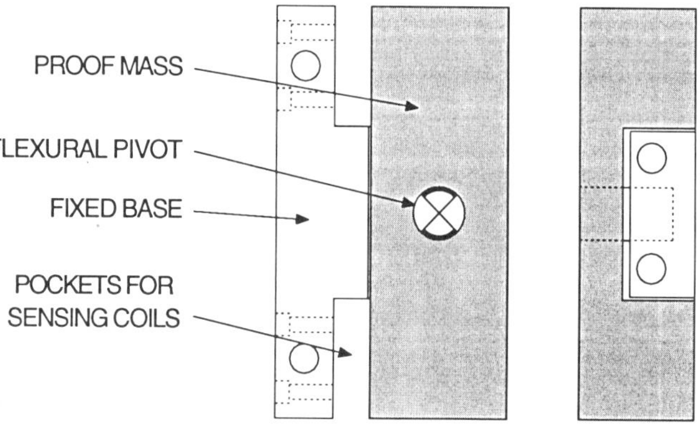

Figure 4. Schematic of angular accelerometer using commercial flexural pivots.

For the prototype sensor, a 3/8 inch flex pivot having a spring rate of 0.31 N•m/rad will be used. The proof mass will have basic dimensions of 7 × 2 × 2.5 cm, positioned so that the c.g. of the rotating mass is located in the geometric center of the flex pivot. The moment of inertia is 1.22×10^{-4} kg•m^2, giving a resonant frequency of 8 Hz. The radius of gyration, r_g is 23.1 mm, so the c.g. must be known to within $(10^{-4}\ m^{-1} \times r_g^2) = 0.05\ \mu m$.

The advantages of this design are standardization of the spring, modularity of the proof mass, and less reliance on precision machining. The disadvantages are a less well defined center of mass and rotation axis (centrode) which in fact displaces as the pivot is rotated, and the possibility of hysteresis. The centrode shift, which varies as θ^2, and hysteresis are negligible for angular excursions less than about 10 degrees.

At present, the design has been completed and the first unit is being fabricated. Because the rotating portion of the flex pivot is not symmetric, it must be offset from the geometric center of the proof mass. The uncertainty in locating the c.g. is several times the machining tolerances and could result in a c.g. offset 2 orders of magnitude larger than acceptable. Therefore the unit will have to be dynamically mass balanced. Initial testing will determine the relative perpendicularity of the rotation axis with respect to the proof mass surfaces and the response to angular accelerations in the off axes.

SUMMARY

The superconducting angular accelerometer described in this paper has been designed for use with the University of Maryland's Superconducting Gravity Gradiometer to remove errors related to angular motion. The mechanical element of the device is simply a bar suspended by a pivot that permits rotation about only a single axis. To make the instrument insensitive to linear acceleration, the center of mass of the bar must be accurately located along the center of bending of the pivot. Given the expected level of linear accelerations, the center of mass needs to be located to within 0.06 μm to achieve the desired resolution. Although balancing the mass to this accuracy is desirable and appears feasible, residual error in the location of the center of mass can be compensated for with the SGG linear accelerometers if the offset is known. A prototype angular accelerometer, using a monolithic mechanical element cut from a single block of niobium, is being fabricated. The technique to accurately determine the center of mass has been demonstrated on two tests cuts for this device. A second design, using commercial pivots, is being investigated.

ACKNOWLEDGMENTS

This work was supported by NASA/Goddard Space Flight Center's Director's Discretionary Fund and Air Force Phillip's Laboratory under contract number F19628-93-C-0150. The authors wish to thank Russ Ekblad, Mike Mooney and David Goldstein for their design and drafting assistance.

REFERENCES

1. H. A. Chan, M. V. Moody, and H. J. Paik, Superconducting gravity gradiometer for sensitive gravity measurements, *Phys. Rev. D* 35:3551 (1987).
2. F.J. van Kann, M.J. Buckingham, M.H. Dransfield, C. Edwards, A.G. Mann, R. Mathews, R.D. Penny, and P.J. Turner, "Laboratory Tests of a Mobile Superconducting Gravity Gradiometer," *Physica B* 165:93 (1990).
3. J. M. Paros and L. Weisbord, "How to design flexure pivots", Machine Design November 25, 1965, 151-156.
4. Flexural Pivots are commercially available from Lucas Aerospace, 211 Seward Ave, PO Box 457, Utica, NY 13503.

A NOVEL SILICON MICROMACHINED CRYOGENIC CAPACITIVE PRESSURE TRANSDUCER

P.M. Echternach, Inseob Hahn, and U.E. Israelsson

Jet Propulsion Laboratory, California Institute of Technology
Pasadena, California 91109

ABSTRACT

A novel cryogenic capacitive pressure transducer was developed using Silicon micromachining techniques. It employs a flexible Silicon membrane which is sealed to a helium sample chamber with an Indium seal. A thin film capacitor plate is attached to the back side of the membrane. A second thin film capacitor plate is located a short distance away on a fixed Silicon wafer. Pressure variations in the sample cell will flex the membrane thereby adjusting the relative capacitance of the circuit. The membrane can be easily fabricated to precise dimensions, allowing fine tuning of the spacing between the capacitor plates. This allows easy optimization of the sensitivity and dynamic range of the transducer.

INTRODUCTION

Capacitive pressure gauges are extremely useful devices in low temperature science, being conceptually simple, well analyzed and yielding a resolution of up to $\delta P/P=10^{-8}$.[1] However, current designs are somewhat complicated to build and not very reproducible. Once the transducer is assembled, then its characteristics can be measured, and if the characteristics are not desirable, one has to essentially rebuild the transducer. Another problem with the conventional design is the degree of parallelism between the capacitor plates and the distance between them, both of which will affect the sensitivity of the device. We developed a new type of capacitive pressure sensor in which the gap between the electrodes can be set precisely to one's specification, allowing tailoring of the shorting pressure and dynamic range, while maintaining a high degree of parallelism between electrodes. This new design borrows techniques from integrated circuit fabrication technology to build mechanical structures using a single crystal silicon wafer as the starting material.

DESIGN AND CONSTRUCTION

Figure 1 shows a schematic drawing of the transducer. A silicon membrane seals a helium sample chamber with an Indium O-ring. The back side of the membrane has a well etched on it. A capacitor plate is defined inside the well by photolithography and metal deposition. A second identical wafer is placed against the membrane so that the gap between the capacitor plates is given by the depth of the wells. By measuring the capacitance as the membrane displaces due to changes in pressure (relative to ambient pressure, which is close to zero in a cold vacuum can), one can then measure the change in pressure in the chamber. An identical set of wafers can be placed under this first set to act as a reference capacitance, in case a cold reference capacitance is needed. In a preliminary version, a block of Silicon replaced the reference capacitor. The gap between the capacitor plates can be adjusted by timing the etch of the silicon wafers as will be described below. By varying this gap we can tailor the shorting pressure to be close to the pressure to be measured, yielding greater resolution. The trade-off is the dynamic range, which becomes smaller as the pressure to be measured gets closer to the shorting pressure.

The fabrication of the wafers is as follows. One inch wafers polished on both sides are furnished by the vendor. The wafers are then oxidized in a furnace at 1050 °C in a flow of oxygen saturated with water vapor. The oxide acts as a mask during the etch step. The well pattern is defined by standard photolithography. The result is a pattern where the oxide is exposed in the regions to be etched and protected by the photoresist in the regions that will not be etched. Reactive Ion Etching with CF_4 is than performed on the wafer, removing the oxide from the regions left exposed by the photoresist mask. After removing the leftover resist in acetone the wafer is placed in the standard anisotropic etch bath (KOH at 95 °C). The KOH will etch preferentially the regions without the protective oxide. The etch rate is typically 1 μm per minute. For short etch times, the bottom of the well remains optically flat. The remaining protective oxide is removed in Buffered Oxide Etch and the whole wafer is oxidized again in a furnace at 1050 °C under a flow of oxygen saturated with water vapor. The capacitor plate

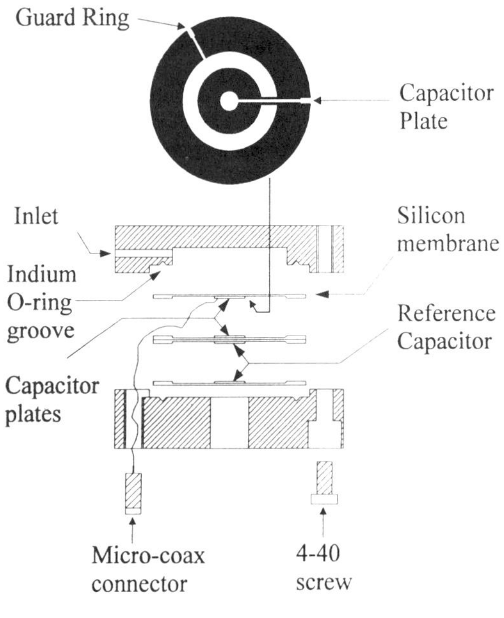

Figure 1. Schematic drawing of the silicon micromachined capacitive pressure transducer.

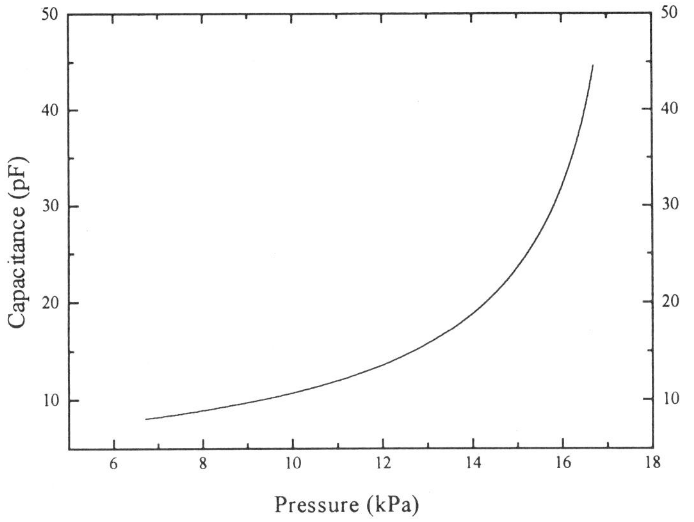

Figure 2. Pressure Transducer Calibration.

patterns are defined by standard photolithography, metal evaporation and lift-off in acetone. The capacitor plates are composed of a bottom layer of Titanium (300 Å), an intermediate layer of Platinum (300 Å) and a top layer of Gold (3000 Å). Notches are cut on the wafers so external contact can be made to the coaxial connectors.

A typical depth of the well is 12 μm, yielding a separation between capacitor plates of 24 μm. Such small separation yields reasonably large capacitances (a few to tens of pico-Farads) even for small (1.5 mm radius) capacitor plates. Moreover, small capacitance plates ensure that the plates displace with respect to one another while still remaining flat, therefore minimizing the effect of bowing of the plates, which degrades the performance of the transducer. The distance between plates can be controlled to within 1 μm. The design also includes a guard ring around the capacitor plates. The guard ring avoids errors in capacitance measurements due to stray capacitances, and they should be always grounded.

The membrane and bottom wafer plates can be replaced if one desires different characteristics, with thicker wafers yielding higher shorting pressures. The fabrication time for a set of wafers is about two days.

TESTS AND RESULTS

In order to test the ultimate resolution of the transducer, a preliminary version was built that had a shorting pressure of about 17 kPa, which is equivalent to the vapor pressure of helium at a temperature of about 2.8 K. This allowed the use of the known vapor pressure versus temperature relation of helium[2] for calibrating the performance of the transducer. An existing cryostat with a high resolution thermometer was used to accurately regulate the temperature of the pressure transducer which was filled with helium in equilibrium with its vapor. Figure 2 shows the calibration of this version of the transducer. The calibration was done by slowly cooling the partially filled transducer and passively measuring the temperature in the cell and the transducer capacitance. The pressure is obtained from the vapor pressure of helium at a given temperature.[2] We have used a low temperature valve[3] to isolate the sample chamber from pressure fluctuations due to the temperature gradient along the fill line.

In order to measure the resolution of the transducer, the temperature in the cell was stabilized at 2.75 K (vapor pressure = 16.1 kPa) to within 2 nK rms using a High Resolution Thermometer (HRT)[4]. The temperature was determined by measuring the temperature dependent magnetization of a paramagnetic salt ($GdCl_3$) placed in a constant magnetic field.

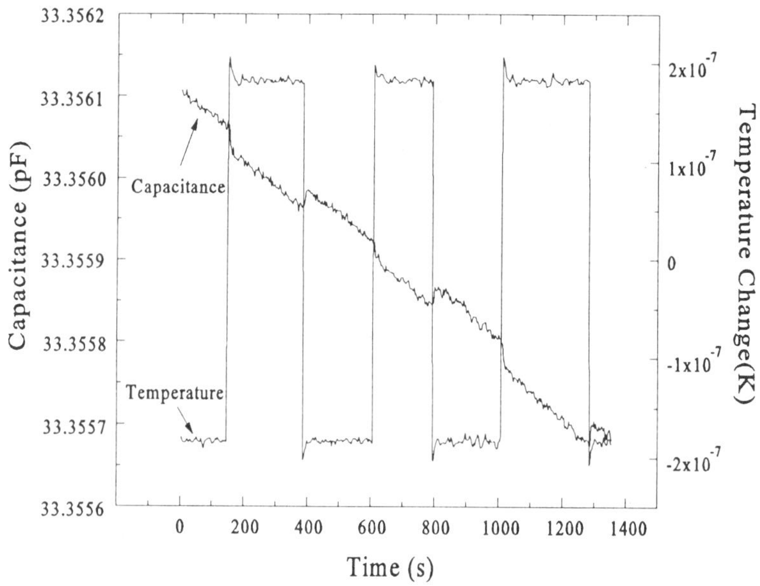

Figure 3. Temperature in sample cell and capacitance versus time.

Changes in magnetization induce current changes in a superconducting pick-up loop coupled to an RF SQUID. The HRT was calibrated against a germanium resistor. To stabilize the temperature the output of the RF SQUID is used as the input signal for a Linear Research LR130 Temperature controller.[5] While measuring the capacitance, the temperature was changed by 365 nK and stabilized to within 2 nK rms. After roughly five minutes the temperature was changed backed to the original point and again stabilized to within 2 nK rms. This was done several times and the corresponding change in capacitance was recorded. Figure 3 shows typical data. The averaging time for the capacitance measurements was 0.5 seconds.

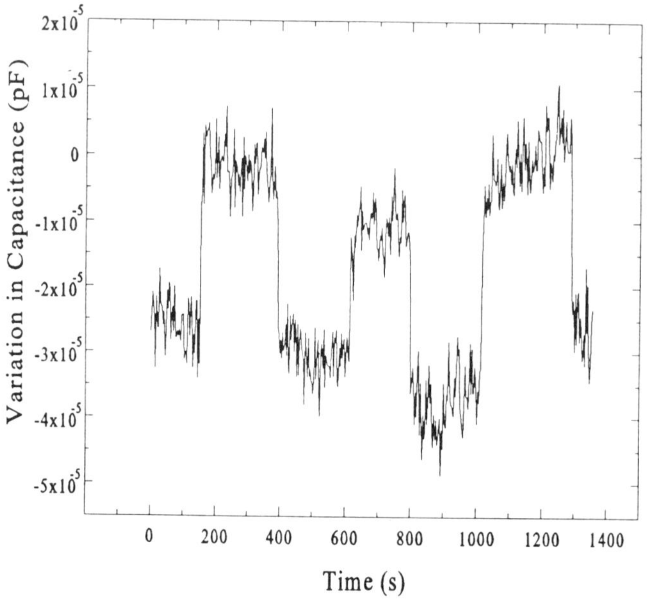

Figure 4 . Variation in Capacitance as a function of time without the constant drift.

The capacitance measurements were performed with an Andeen-Hagerling model 2500A self balancing capacitance bridge. The rms variation of the temperature at each plateau was 2 nK. The capacitance had a constant drift of -3.25×10^{-7}pF per second. In order to obtain the rms value of the fluctuation in capacitance, this drift was subtracted from the data as shown in Figure 4. The rms fluctuation of the capacitance at each plateau was $\delta C=3.74\times10^{-6}$ pF. Based on the change in capacitance with the change in temperature, we estimate dC/dT=79 pF/K. At this temperature (2.75 K) the slope of the helium vapor pressure line is dP/dT=26.82 kPa/K.[2] The fluctuation of pressure at the plateaus is then $\delta P=1.22\times10^{-3}$ Pa. This gives a pressure resolution of $\delta P/P=7.6\times10^{-8}$. Note that the fluctuation in pressure due to temperature fluctuations is only 5.4×10^{-5} Pa, more than one order of magnitude smaller. This indicates that the source of fluctuations in capacitance is intrinsic to the transducer. At this time, the cause is still unknown, and further investigation is necessary. One possible source is the measuring setup since, even though the bridge used is self balancing (i.e., it is never run off-balance), the ultimate resolution can still be affected by the cable length. In order to correct that, a cold reference capacitor has to be included in the setup.

The next version of the transducer will have this cold reference. The other problem to be solved is the drift in capacitance as a function of time. One possibility that was considered to explain this behavior is the ductility of the Indium O-ring. It is possible that the O-ring relaxes with time, therefore affecting the pressure that presses the two silicon wafers together and thus the distance between the capacitor plates. To correct this, different O-rings will be tested. The two possibilities considered are gold and kapton. It is possible that a cold reference will also minimize this problem, since the pressurethat holds the measurement capacitor together will be exactly the same as the pressure that holds the reference capacitor, therefore canceling the effect.

CONCLUSION

We have designed and built a novel pressure transducer based on silicon micromachining techniques. Preliminary tests show a resolution $\delta P/P=7.6\times10^{-8}$, roughly one order of magnitude worse than the state-of-the-art and two orders of magnitude worse than our target. An additional problem is a constant drift of the capacitance with time. New versions of this transducer will have a cold reference built in, and possibly a different kind of O-ring for sealing the chamber. We believe this will bring us close to the state-of-the-art. In order to achieve a resolution of $\delta P/P$ of 10^{-9}, the length of the leads may have to be reduced, implying the potential necessity of a cold amplifier.[6] Once these problems are solved, we believe this instrument will be extremely useful and versatile, providing the opportunity of tailoring the desired resolution and dynamic range with great ease.

ACKNOWLEDGMENT

This work was carried out by the Science and Technology Section, Jet Propulsion Laboratory, California Institute of Technology under a contract with the Microgravity Science and Applications Division of NASA. We thank the Micromachining Group of the Microdevices Laboratory at JPL for allowing us to use their facilities.

REFERENCES

1. E. Dwight Adams, High Resolution Capacitive Pressure Gauges, *Rev. Sci. Instrum.* 64, 601 (1993).
2. F. Pobell. " Matter and Methods at Low Temperatures", Springer-Verlag, Berlin (1992).
3. I. Hahn, M. Barmatz, and A. Clark, A new superfluid leak tight low temperature valve using a magnetostrictive actuator, *Mat. Res. Soc. Symp. Proc.* 360, 253 (1995).

4. T.C.P. Chui, P. Day, I. Hahn, A.E. Nash, D.R. Swanson, J.A. Nissen, P.R. Williamson, and J.A. Lipa, High resolution thermometers for ground and space applications, *Cryogenics* 34, ICEC Supplement, p. 417 (1994).
5. Linear Research Inc., 5231 Cushman Place, Suite 21, San Diego, CA 92110.
6. J. Imai and R. Flores, Low temperature metal-oxide semiconductor field-effect transistor pre-amplifier, *Rev. Sci. Inst.* 64, 3024 (1993).

CRYOGENIC DIELECTRIC TESTING PERFORMED FOR THE SMES-ETM PROGRAM

J. L. Wu

Westinghouse Science & Technology Center
Pittsburgh, Pennsylvania, 15235, USA

ABSTRACT

The operation of superconducting magnetic energy storage (SMES) systems such as the SMES-ETM requires charge and discharge of electrical energy to and from the SMES coil. These charge and discharge operations produce voltage stresses between the coil windings, as well as between the coil and the grounded cryostat and its accessories. Electrical insulation design for withstanding these electrical stresses typically have employed solid insulators, such as G-10 and Kapton, and the cryogen as liquid dielectrics. For a large scale SMES such as SMES-ETM, the insulation design requires consideration of the presence of mechanical stress, the large surface area and volume under electrical stress, the possibility of contamination in the cryogen, and the requirement of a long device lifetime. Experimental database of cryogenic dielectric breakdowns in all of these areas are either not sufficient for reliable design or are non-existing. A comprehensive testing program was thus undertaken in the SMES-ETM program to generate the required database. The tests produced data of dielectric breakdown of liquid helium in the temperature range covering the λ-transition region and data which include the effects on dielectric breakdown associated with large surface area/volume under electrical stress, mechanical stress, cryogen contamination and electrical stress aging of solid insulators. The testing and the analysis of test results are presented.

INTRODUCTION

The SMES-ETM (Superconducting Magnetic Energy Storage - Engineering Test Model) Program was intended to develop a SMES system capable of storing up to 20 MWH of electromagnetic energy in a superconducting coil. It is a sub-scaled model for the full-size 5,500 MWH SMES envisioned for utility load leveling and other applications.

The SMES-ETM coil design pursued by the Ebasco/Westinghouse/UW/CBI team is to be cooled by pool-boiling superfluid helium at 1.8 K and is to carry a nominal rated current of 60 kA. The coil system consist of four layers, 88 turns per layer, of Nb-Ti superconductor with stabilizer wound in a solenoid configuration. The radius of the innermost layer is 66.722 m, and that of the outermost layer, 67.278 m; the height of the coil stacks is approximately 4.2 m. The coil turns are wound over structure aluminum blocks and are electrically insulated turn-to-turn by G-10CR sheets. The aluminum blocks are constrained axially by preloaded insulated stainless steel clamping studs with G-10CR clamping plates. In the radial direction, mechanical supports are provided by G-10CR finger plates between the dewar walls and the adjacent layers, and by G-10CR loading bars between layers. Both the finger plates and the loading bars also serve as electrical insulation members. Insulation design of the coil was described in more detail in a previous publication.[1]

Each time energy is charged to or withdrawn from the coil, or when the coil quenches, a voltage develops in the coil system. The associated electrical stresses have to be withstood by the electrical insulation of the coil system in order to avoid dielectric breakdown and arcing damages. Due to the large amount of energy stored, damages caused by a dielectric breakdown in the coil could be catastrophic. The magnitude of voltage stress developed on the coil system depends on many factors. Among them are the charging and discharging power level, the connection configurations of coil modules and layer/turn and, during quench protection sequence, the quench characteristics of the superconducting coil and the presence or absence of protection arrangement such as voltage limiting shorting switches. For the baseline coil design and operating at a charging/discharging power of approximately 325 MW, the maximum local voltage stresses for the charge/discharge operations were found to be approximately 13 kV/cm on conductor surface and 6.7 kV/cm on insulator surface.[1] In cases of a coil quench, with stored energy dissipated within the cryostat and shorting switches employed, the maximum voltage stress was estimated at 0.17 kV/cm.

To generate data for the insulation design and thus ensuring the insulation integrity and reliability of the SMES system, we have undertaken comprehensive testing programs during Phase I of the SMES-ETM Program and in the follow-on SMES Risk Reduction Program. The testing and the data obtained are summarized and discussed below.

TESTING REQUIREMENTS

Discussion of insulation design and testing requirements for the SMES-ETM and the full-scaled SMES needs to consider the unique design and operation characteristics of these systems as compared to other superconducting devices currently in use. These characteristics are:

- Large System Size
  Due to the large size of the device, the effects of large conductor surface and large volume of insulating medium subjected to voltage stress need to be evaluated. As found in transformers, a large electrically stressed surface area/volume tends to decrease the dielectric breakdown voltages. In the SMES-ETM, the area subjected to high voltage stress is estimated as 1.78×10^5 cm^2, and for the full-scale SMES, 4.46×10^6 cm^2 . These numbers are two to four orders of magnitude higher than the maximum electrode surface areas used in data available in the literature. Testing to extend the data range is clearly required.

- Superfluid Helium (HeII) as the Insulating Medium
  There is only very limited amount of dielectric breakdown data of HeII available in the literature and virtually no insulator aging data for insulators immersed in HeII. Testing to establish database in these areas is certainly warranted.

- Short Duration of Voltage Stress
  In SMES, the high voltage stress appears only when the coil is performing charging/discharging operations or a quench protection energy dump sequence is activated. Therefore, the total accumulated time in which voltage stresses appear is relatively short, estimated to be in equivalent to weeks over a SMES lifetime of, e.g., fifty years. The aspect of material aging is thus much less severe as compared to, e.g., power transmission lines in utility systems. The short accumulated duration also allows accelerated life testing to be completed in a laboratory in an affordable fashion.

- Electrical Stress and Mechanical Stress Are Out of Phase
  In SMES operations with a constant power input or output, the maximum electrical stress occurs at the time at which the coil current is at its minimum, i.e., the times at which the electromagnetic force is at the minimum. Therefore, the electrical stress and the mechanical stress are out of phase. This minimizes the interaction of these two stresses with respect to material aging effects and permits the testing of these two effects to be conducted separately.

In addition to the above considerations, we need also to look into the effects of contamination of either the liquid helium or the components of the coil upon the voltage withstand capability of the insulation system. Even though strict quality control can be implemented during fabrication and assembly of the coil and its components, the possibility of contamination can not be precluded over the lifetime of the SMES system.

TESTS PERFORMED AND TEST FIXTURES

Table 1 summarized the dielectric tests performed for the SMES-ETM Program, the major test parameters and the objectives for performing these tests. The majority of the testing used liquid helium as the dielectric medium, although tests were also performed using gaseous helium at room temperature and liquid nitrogen. For tests using liquid helium as the dielectric medium, efforts were also spent in obtaining data closer to the operating regime of the SMES-ETM system, i.e., with helium at the superfluid state of 1.8 K and 1 atm. These tests were designed to produce (a) basic dielectric breakdown data which are not only important for the designing of the SMES-ETM insulation system but are also useful for the designing of other cryogenic electrical equipment, (b) to assess the effects of design and operation parameters such as large surface area, electrical and mechanical stress aging and cryogen contamination which may be more specifically related to SMES-ETM and (c) to check the insulation system design of SMES-ETM through testing of mock-up model of coil section.

Several test fixtures have been fabricated and some existing test apparatus were employed for performing these tests. Fig. 1 is the photo showing the external view of the basic test fixture used for testing with liquid helium. Fig. 2 is a sketch of this fixture showing the internal components and dimensions. This fixture employs a "double dewar" feature which enables the outer dewar be pumped down to very low pressures, thus cooling down the liquid helium within the inner dewar test electrode region to below the λ-transition point while maintaining its pressure at near the atmospheric level. The test electrode region, shown with a plane-plane uniform field electrode pair in a photo as Fig. 3, can easily be modified to allow mounting of different electrode arrangements/configurations for performing other tests. Some of these electrode arrangements/ configurations are shown: electrode configuration for non-uniform field flashover test (Fig. 4), large surface area coaxial cylindrical electrodes (Fig. 5), and for liquid helium contamination test (Fig. 6).

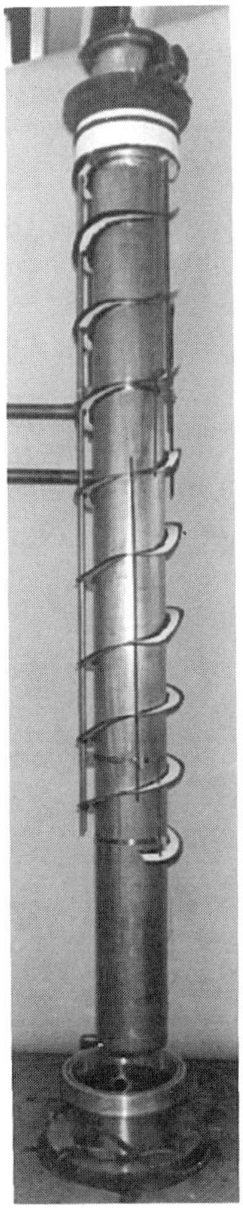

Figure 1. Liquid helium test fixture - external view.

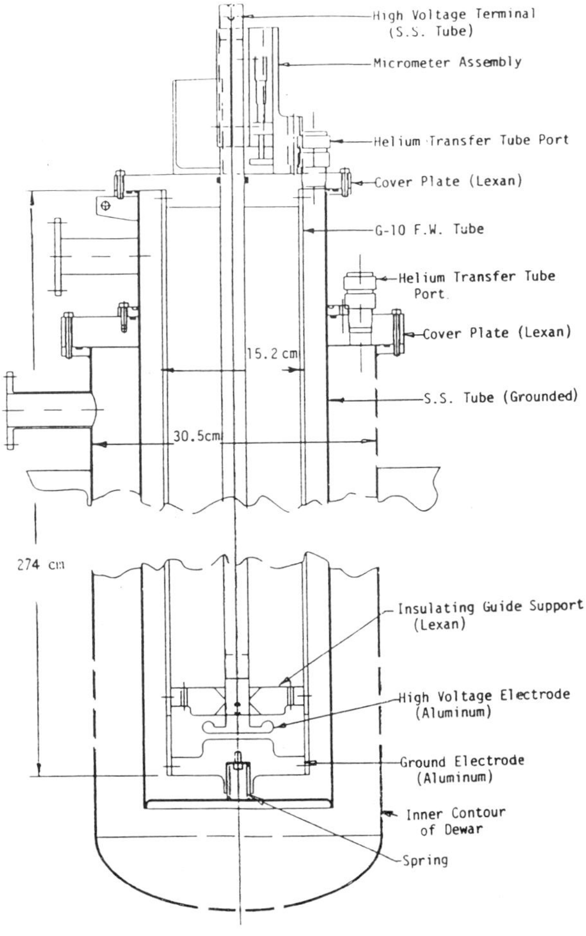

Figure 2. Simplified drawing showing the liquid helium test fixture.

Table 1. Dielectric Tests Performed for SMES-ETM Program

Dielectric Medium	Tests Performed	Major Test Parameters	Testing Objectives
Liquid helium	Uniform field gap breakdowns in LHe	T: 1.93 - 4.18K P: ~ 1 atm Gap: 0.25 - 1.27 mm	Basic breakdown data
	Uniform & non-uniform field flashovers on G-10CR surfaces	T: 187 - 4.24 K P: ~ 1 atm Pin-plane geometry for non-uniform field tests Sample thickness: 1 mm	Basic breakdown data
	Electrical stress aging effects on G-10CR sample	T: 2.14 & 4.2K P: ~ 1 atm 60,000 cycles at 8kV/mm	Repetitive stress on reduction of dielectric strength of G-10CR
	Breakdowns with edge crack or internal void in G-10CR samples	T: 2.14 & 4,2 K P: ~ 1 atm. Uniform field electrodes	Reduction of dielectric strength due to cracks or internal voids
	LHe contamination effects	T: 2.15 & 4.2 K P: ~ 1 atm Particles: G-10CR, Al & solder	Reduction of dielectric strength due to the presence of contaminating particles
	Breakdowns of uniform field gap with large electrode surface area	T: 1.8 & 4.2 K P: ~ 1 atm Gap: 2.54 mm Area: 685 cm^2	Reduction of dielectric strength due to large surface areas
Gaseous Helium	Near-uniform field gap breakdowns	T: 20°C P: ~ 1-3 atm Gap: 2.54-12.7 cm	Basic breakdown data
	Surface flashover on G-10CR samples	T: 24°C P: 1 - 3 atm Spacing: 2.54-12.7 cm	Basic breakdown data
	Dielectric tests on mock-up of coil section	T: ~ 20°C P: 0.25-1.0 atm	Check insulation design
Liquid Nitrogen	Mechanical loading effects on mock-up of coil section	T: ~ 78 K P: 1 atm 36,000 mech. cycles	Check insulation design

Figure 3. Test electrode region.

Figure 4. Electrodes for non-uniform field flashover tests.

Figure 5. Coaxial cylindrical electrodes for large electrode surface tests.

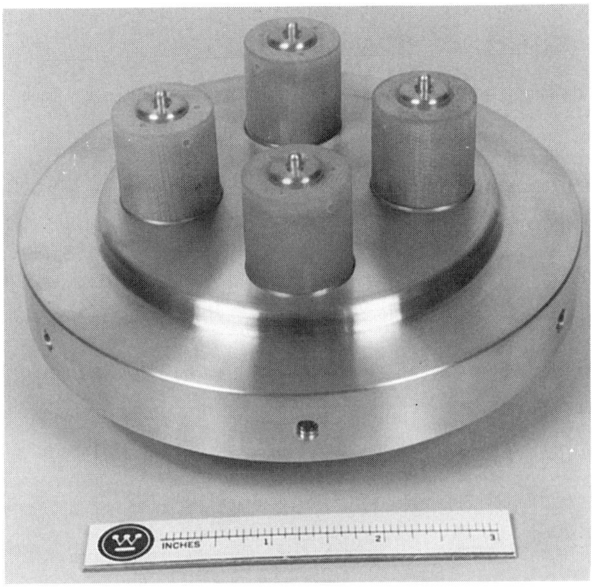

Figure 6. Electrode modules for LHe contamination tests.

SUMMARY OF LIQUID HELIUM TEST RESULTS AND DISCUSSIONS

Due to space limitation, only test results with liquid helium as dielectric medium will be summarized and briefly discussed:

Uniform Field Gap Breakdowns in Liquid Helium

Fig. 7 shows the dc breakdown voltages of liquid helium at temperature of 1.93 ± 0.02K and pressure of 684 to 708 Torr, plotted with respect to electrode spacing. As expected, the data shows a general trend of increasing breakdown voltage with increasing electrode spacing. The breakdown voltage increases slightly less than linear with the electrode spacing. Substantial variations from test to test for a given spacing are also observed.

Dielectric breakdown data between electrodes in liquid helium at different temperatures are shown as Fig. 8. These data cover tests for electrode spacing of 0.254 to 1.27 mm and temperature of between 1.93 to 4.18K. Substantial fluctuations in data are clearly evident. Increase of breakdown voltage with increasing electrode spacing are generally evident at all temperatures. However, the trend for voltage dependence on temperature is less clear. Generally speaking, the breakdown voltage seems to increase for tests with 0.254 mm electrode spacing but decrease for larger electrode spacing as the temperature decreases through the λ-transition of around 2.16K. These trends seem to be consistent with the results of Blank and Edwards[2] for the larger spacing tests and with Hwang[3] for the smaller spacing tests.

Uniform and Non-Uniform Field Flashovers on G-10CR Sample Surfaces

Under uniform electrical field, tests of dc flashover of 1 mm thick G-10CR sample show a dielectric breakdown strength of between 9 to 13.5 kV/mm, indicating that the presence of the insulator has a negligible effect as compared to gap breakdown in liquid helium. There is, however, substantial variation from test to test.

For non-uniform field flashover on 1 mm thick G-10CR sample surface in a pin-plane electrode geometry (Fig. 4), the results clearly show a local field dominated phenomena. The results generally indicate a lower breakdown voltage for 1.87K than for 4.2K. Furthermore, the data show a slight increase in breakdown voltage with the increase of horizontal creepage distance from 0 to about 4 mm, but seems to have no further improvement with additional increase in creepage distance. This indicates, as expected, a breakdown process dominated by local field. More detailed discussion of these data was given in Ref. 4.

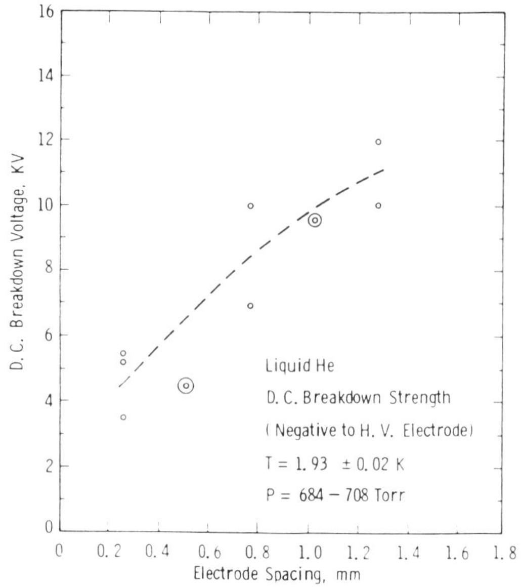

Figure 7. Uniform field dc dielectric breakdown of liquid helium at T=1.93K.

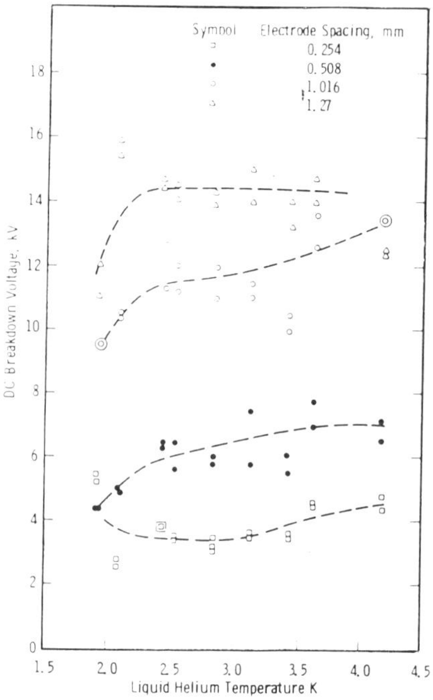

Figure 8. Uniform field dc dielectri breakdown voltage vs. temperature.

Electrical Stress Aging Effects on G-10CR Sample

G-10CR insulator sample, 2.54 cm in diameter and 1 mm thick, was installed between a plane-plane uniform field electrode set and subjected to repetitive dc voltage of 8 kV in liquid helium maintained, most of the time, at 4.2K. Dielectric breakdown tests were performed after completing pre-determined number of stress cycles. Fig. 9 shows the results of the dielectric breakdown tests before the electrical stress cycling started and the results after completing 60,000 cycles. The data show no significant degradation in breakdown field strength through the testing cycles for either at approximated 4.2 K or at approximately 2.14 K. The variation in breakdown field is in effect smaller than the experimental data variation margin at a given cycling point.

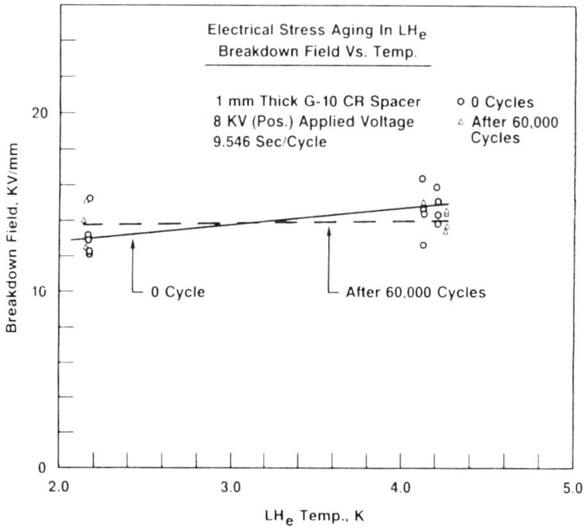

Figure 9. Electrical stress aging effects on G-10CR sample.

Breakdowns with Edge Cracks or Internal Voids in G-10CR Samples

As expected, edge crack in insulator samples which does not appreciably distort the electrical fields within the samples shows little degradation effects in breakdown strength either in HeI or in HeII. The average breakdown field strength values, with or without cracks, are between 14.4 to 18.4 kV/mm at 4.16 to 4.19 K and 12.5 to 13.8 kV/mm at 2.12 to 2.13 K. It is to be noted that even though no degradation of dielectric breakdown field strength was detected with the presence of cracks in G-10CR samples, the long term effects of these cracks subjected to electrical stresses, which could be a region for initiation of partial discharges, have not been tested. Additional testing in this area is required in the future.

For testing of effects of internal void in insulators, one sample was fabricated with three layers of 3.81 cm diameter G-10CR sheets and epoxy glued together. The mid-layer, 0.2 mm thick, has a machined hole of 0.397 mm in diameter at its center. Other samples, without void, were also tested for comparison. The tests were performed with uniform field electrode configuration. Table 2 summarizes the test results. There is a reduction in the breakdown strength for the sample with void. However, it is not clear that this reduction is caused by the void or by the increase of total sample thickness. Additional tests are required to clarify these data and also to assess the effects of possible partial discharges initiation in the void.

Table 2. Summary of Test Results on Effects of Void in G-10CR Sample

Sample No.	Description	Dielectric Breakdown Strength, kV/mm T = 4.2 K	T = 2.14 K
1	Single layer, 0.391 mm thick; no void	30.5*	
2	Two layers, 0.845 mm thick; no void	18.5	15.02
3	Three layers, 1.140 mm thick; void of 0.397 mm dia. and 0.2 mm deep at the mid-layer	15.7	11.66

*Sample punctures after 3 breakdowns.

Liquid Helium Contamination Effects

Three types of prospective liquid helium contamination particles (G-10CR, Al-6061 and 60/40 Sn-Pb solder) were tested. The particle sizes distribution are from under 50 μm to 200 μm. These particles were released into the uniform field test electrode region (1 mm electrode spacing) within the electrode modules by pushing down the module plunger with a rotatable arm which is part of the high voltage bus. This action also connects the selected electrode module to the high voltage test circuit.

Table 3 summarizes the test results, which also include the referenced data of one electrode module containing no contamination particles. As expected, G-10CR particles, being non-metallic, produce no degradation effects. Both the aluminum and the solder particles show an appreciable degradation effect, with the reduction in breakdown strengths of 20 to 50%.

Table 3. Summary of LHe Contamination Effects Tests

Contamination Particles	Dielectric Breakdown Strength, kV/mm T = 2.15 K	T = 4.2 K
None	13.67	15.09
G-10CR	13.91	14.74
Al-6061	11.08	7.78
Sn-Pb Solder (60/40)	8.42	7.63

Breakdown of Uniform Field Gap with Large Electrode Surface Area

Testing of the coaxial cylindrical large surface area (684 cm^2) electrodes allows an assessment of the electrode surface area effects by comparing the breakdown data to those obtained with a plane-plane electrode set with a much smaller surface area (31.65 cm^2). By increasing the electrically stressed areas

21.6 times, the large surface area electrode configuration shows a reduction of 32 to 39% in breakdown electrical field, from 9.84 to 6.02 kV/cm for T=4.2K and 9.06 to 6.19 kV/cm for T=1.8 to 1.9K. The factor of reduction in dielectric strength follows the same trends shown in the data published by Girdinio, et al.[5] However, it is to be noted that part of this reduction in breakdown electrical field is a result of increase in electrode spacing.

CONCLUSION

As part of the SMES-ETM Program, we have undertaken a comprehensive test program to obtain dielectric breakdown data in liquid helium, gaseous helium and liquid nitrogen. These data not only provide an important database for the designing of the insulation system of SMES-ETM but also can be very valuable for the designs of other cryogenic power equipment.

ACKNOWLEDGMENT

The work reported here was funded, in most part, by the Defense Nuclear Agency under contract no. DNA001-88-C-0027 and DNA001-92-C-0064, and, in smaller part, by Westinghouse Electric Corporation.

REFERENCES

1. J. L. Wu, J. F. Roach, D. C. Johnson, J. T. Dederer, S. K. Singh and D. T. Hackworth, "Design and Analysis of SMES-ETM Electrical Insulation," Adv. Cryo. Eng., 39A:813, Plenum Press, New York (1994).
2. C. Blank and M. H. Edwards, "Dielectric Breakdown of Liquid Helium," Phys. Rev. 119:50 (1960).
3. K. F. Hwang, "Dielectric Strength of Helium Vapor and Liquid at Temperature Between 1.4 and 4.2K," Adv. Cryo. Eng. 23:110 (1978).
4. J. L. Wu and J. F. Roach, "DC Dielectric Breakdown Tests of Liquid Helium at Temperatures 1.8 to 4.2K," Supercollider 4, Plenum Press, New York (1992), p. 731.
5. P. Girdinio, G. Libreti, P. Molfino and G. Molinari, "A Statistical Investigation on the Brekakdown Strength of Liquid Helium with IEC and Cylindrical Electrodes," IEEE International Symposium on Electrical Insulation, p. 272, (1982).

A PORTABLE INSTRUMENT FOR MEASURING EMISSIVITIES

G. Perinić[1], W. Scherber[2], K. Schulz[2]

[1] Institute for Technical Physics, Forschungszentrum Karlsruhe GmbH
PO Box 3640, D-76021 Karlsruhe, Germany

[2] Dornier GmbH
D-88039 Immenstaad, Germany

ABSTRACT

A portable instrument for measuring the mean emissivity in the spectral range 0.6μm to 40μm has been developed for quality control purposes. The paper presents the working principle and the design of the instrument as well as results of calibration measurements. The instrument has proved to measure emissivities with an accuracy better than ±0.03.

INTRODUCTION

One factor that has to be considered in the design of cryogenic equipment is the surface emissivity of the components. In order to increase efficiencies and to decrease losses, the surface emissivities can be adjusted by the means of surface treatment or surface coating techniques. During the manufacturing process it must be ensured that the surface emissivities comply with the specification. It is therefore necessary to establish quality control means, especially when non standard coating techniques are used.

Surface emissivity values can be obtained from theory[1] or by calorimetric[2] or optical[3] measurements. Of the measurement methods the calorimetric method requires a large experimental set-up. Hence, for quality control measurements where mobility is necessary optical measurement methods are favourable.

The paper describes a portable instrument applying a pyrometric measurement method. It has been designed for the on site quality control of cryopump components and other equipment for fusion reactors. The instrument is based on the working principle of a table top instrument developed by Dornier. In the following the theory of the measurement, the principle of the instrument and calibration experiments to check the accuracy are described.

THEORY

The emissivity of the sample S is determined by measuring the radiation emitted from a surface element of the size A that is exposed to the irradiation of a black body radiator R. According to Kirchhoff's law the amount of radiation leaving the surface element is:

$$Q_S = \varepsilon_S \, \sigma \, A \, T_S^4 + (1 - \varepsilon_S) \, \sigma \, A \, T_R^4 \tag{1}$$

Where ε_S is the emissivity of the sample, 1-ε_S is the reflectivity and σ is Stefan-Boltzmann's constant.

In order to eliminate the temperatures of the sample T_S and the radiator T_R calibration measurements are carried out with two reference samples A and B of known emissivity and of the same temperature T_S as the sample S:

$$Q_A = \varepsilon_A \, \sigma \, A \, T_S^4 + (1 - \varepsilon_A) \, \sigma \, A \, T_R^4 \tag{2}$$

$$Q_B = \varepsilon_B \, \sigma \, A \, T_S^4 + (1 - \varepsilon_B) \, \sigma \, A \, T_R^4 \tag{3}$$

Resolving the expressions (1) to (3) yields:

$$\frac{Q_S - Q_A}{Q_B - Q_A} = \frac{\varepsilon_S - \varepsilon_A}{\varepsilon_B - \varepsilon_A} \tag{4}$$

which can also be written as

$$\varepsilon_S = \varepsilon_A + (\varepsilon_B - \varepsilon_A) \frac{Q_S - Q_A}{Q_B - Q_A} \tag{5}$$

Thus, in equation 5 the emissivity ε_S is only a function of the known emissivities of the reference samples A and B and of the radiation emitted by the samples S and the reference samples A and B.

DESCRIPTION OF THE INSTRUMENT

The emissivity measuring instrument consists of two portable components - the measuring unit and the control and power supply unit (see figure 1). The instrument can be used stationary to measure emissivities of small samples or it can be used hand held to measure surface emissivities of larger components.

The measuring unit houses an infrared radiation thermometer as well as a black radiator. The black radiator emits radiation by a blackened concave opening in order to approximate a black body. In front of the opening there are two small slide-rails to keep a defined distance in between the sample and the instrument. The radiation emitted and reflected by the sample is measured by the infrared radiation thermometer through a hole in the radiator at an angle of twelve degrees off the normal. The measured spot on the sample has a diameter of 10-15mm.

The infrared radiation thermometer is a type KT15.99 by Heimann GmbH, Wiesbaden. It covers a wavelength range from 0.6µm to 40µm and yields an output signal that is proportional to the mean radiation flux. The adjustment time of the sensor is one second. The output voltage is displayed on the measuring unit.

Figure 1. The emissivity measuring instrument consisting of measuring unit (right) and control unit (left).

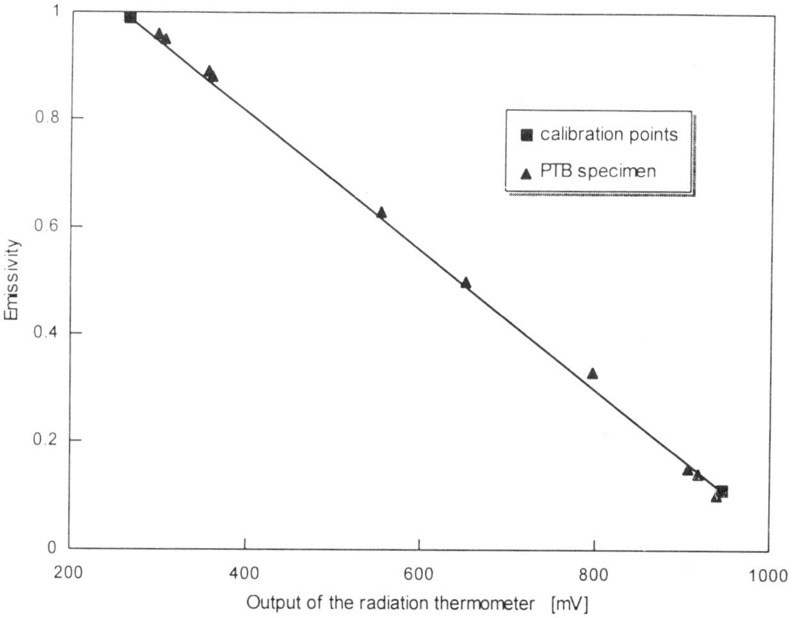

Figure 2. The calibration curve of the emissivity measuring instrument using two PTB reference samples and emissivities of PTB standards plotted against the output of the radiation thermometer.

The radiator is an aluminium block heated by four 30W rod heaters (Ø3.1x31.75). The temperature is kept at 100°C by the means of a two point temperature controller in the control and power supply unit and a PT100 platinum temperature sensor installed near the centre of the block. For the blackening of the concave opening a temperature resistant black varnish (ε=0.95) was used.

THE MEASUREMENT PRINCIPLE

Before the measurement can start, a calibration of the instrument must be carried out. The calibration is performed with two reference samples of known emissivity - one sample with high and one sample with low emissivity. The reference samples and the samples and surfaces to be measured must be flat and the size of the area to be measured must be at least 30mm in diameter. It is prerequisite for the calibration and for the measurement that all samples have the same temperature which is ensured by placing instrument and samples or components into the same room at least one hour prior to the measurement.

The calibration is accomplished by noting the output corresponding to the two reference samples. In the course of the measurement this is repeated several times in order to detect any drift of the output. For the measurement the instrument output corresponding to the samples is noted and the emissivity is calculated using equation 5.

The main interference effect which needs to be observed during the measurement is a possible temperature rise of the sample surface due to the radiation from the black radiator. The temperature rise can be observed in particular with thin or low thermal conductivity samples of high emissivities. It can be limited by short measurement times.

MEASUREMENTS

In order to determine the accuracy of the instrument a series of measurements has been carried out with test specimen of the Physikalisch-Technische Bundesanstalt (PTB) in Braunschweig. The PTB has measured the normal emissivities of the specimen - sandblasted stainless steel, black paint coated surfaces and coated glass samples - at room temperature in vacuum with an absolute accuracy better than ±0.005.

Figure 2 shows the calibration curve of the emissivity measuring instrument using two PTB samples - a polished stainless steel plate (ε=0.11) and a black painted aluminium plate (ε=0.99) - as reference samples. It also shows the emissivities of PTB standards plotted against the output of the infrared radiation thermometer. The standard deviation of the emissivities measured by PTB compared with the calibration curve is 0.014. The largest difference in between calibration curve and PTB results being 0.027. A straight line fitted to all twelve reference points (not shown) yields a standard deviation of 0.011 in between fitted line and reference points. This shows that the linearity of the emissivity measuring instrument is very good. The fitted line can also be regarded as a twelve point calibration curve. However, compared with the two point calibration the improvement in standard deviation is less than 0.5%.

CONCLUSIONS

A portable instrument for measuring emissivities was designed and constructed for the use in the manufacturing of cryopump components and other fusion reactor components. Measurements on emissivity standards have shown that the linearity of the instrument is very good across the whole emissivity range and that the measurement error is less than ±0.03. Therefore, the described instrument can well be used for the quality control of surface emissivities.

REFERENCES

1. W.M. Toscano and E.G. Cravalho, Thermal radiative properties of the noble metals at cryogenic temperatures, *J. of Heat Trans., Trans. ASME, Ser. C,* **98** (3) (1976) 438.
2. T. Amano and A. Ohara, An experimental study of thermal radiation at cryogenic temperatures, *Heat Trans.-Jap. Res.,* **20** (4) (1991) 307.
3. R.R. Willey et al. Total reflectance properties of certain black coatings, *Proc. SPIE - Int. Soc. Opt. Eng.,* **384** (1983) 19.

CRYOGENIC SYSTEM FOR A HIGH TEMPERATURE SUPERCONDUCTOR NMR PROBE

Vincent Kotsubo and Robert Nast

Conductus, Inc.
Sunnyvale, CA 94086

ABSTRACT

We have developed a cryogenic NMR probe using HTS detection coils for high resolution chemical spectroscopy and have demonstrated a signal-to-noise improvement by more than a factor of four over conventional technology. The probe is cooled by a remotely located closed-cycle refrigeration system and coil temperatures below 20 K have been attained. The probe is designed to be inserted into a 37 mm diameter bore of a high homogeneity superconducting magnet with the coils located less than 2 mm from a room temperature NMR sample. The samples can be loaded using the standard pneumatic loading system, and sample temperatures can be regulated using the standard gas flow method. Standard sample spinning can also be performed. The closed-cycle refrigeration system consists of a closed loop of circulating helium gas cooled by a heat exchange network on a Gifford-McMahon refrigerator. A vacuum insulated transfer line is used to flow the helium from the cooling system to the probe. The probe has been optimized for minimal heat leak while maintaining RF performance for the NMR experiments.

INTRODUCTION

Within the last few years, a wide variety of commercial applications for High-Temperature Superconducting (HTS) thin-film devices have been explored. Much of the effort has been on microwave or RF devices, where the low electrical loss allow performance superior to normal metal devices. One promising near term application is NMR detection coils. There is a large commercial market in high-resolution spectroscopy for liquid sample chemical analysis, where narrow intrinsic linewidths allow resolution of the small chemical shifts of the NMR frequencies (few ppb), which is then used to determine molecular structure. Here, demands for signal-to-noise (SNR) are so great that a 20% increase using conventional technology is a significant achievement. Recent work on HTS coils for NMR imaging microscopy has demonstrated a factor of 10 SNR improvement over conventional technology[1], and the present work for NMR spectroscopy has demonstrated an SNR improvement by a factor of more than 4, so it is anticipated that HTS coils will have a significant impact on the commercial NMR industry.

To penetrate this market, there must be a commercially viable cryogenic system for cooling the coils. This system must be closed-cycle, relatively easy to use and maintain by technician level personnel, and have an existing infrastructure for manufacturing and servicing. Cost is an issue, but since typical NMR spectrometer systems costs several hundred thousand dollars to over a million dollars, it is not overly constraining.

Conductus, Inc. and Varian Associates have been jointly developing a cryogenic HTS NMR probe for high resolution spectroscopy. The refrigeration system uses a Gifford-McMahon (GM) refrigerator to cool a helium gas stream, which is then used to cool HTS detection coils. The complete functionality of the system has been demonstrated, and excellent NMR performance has been achieved. This paper describes the cryomechanical system; details of the HTS coils and NMR performance are presented elsewhere.[2,3]

NMR SYSTEM DESCRIPTION/TECHNICAL REQUIREMENTS

Modern commercial NMR spectrometer systems consists of a room temperature bore superconducting magnet (7-17 Tesla), a computer controlled RF console for control of the experiments and data acquisition, and a probe which is inserted into the bore of the magnet. The probe contains the NMR excitation and detection circuitry, which are RF coils in a tank circuit tuned to the NMR frequency of the nuclei of the sample. High signal to noise is obtained from large Q of the tank circuit. Chemical spectroscopy probes contain two or more detection circuits such that more than one nuclear species can be detected. Typical coils are on the order of 1 - 2 cm in length and width, with the samples contained in NMR compatible glass or quartz tubes that are a few cm in length and several mm in diameter. The entire NMR experimental region of the probe occupies a volume of a few cubic centimeters in the center of the main superconducting magnet.

Modern high-resolution chemical spectroscopy NMR experiments have reached such a high level of sophistication that probe design and construction, particularly in the sample region, is very demanding. The most stringent requirement is a uniformity of the dc magnetic field of better than one part in 10^9, since non-uniformities will artificially broaden the spectral lines. At this level, the magnetic susceptibility of materials usually considered non-magnetic is problematic. Uniformity is achieved by carefully controlling the materials and material quality near the sample, designing the sample region for a high degree of symmetry, and the application of adjustable low level gradient magnetic fields to compensate any local field distortions.

The RF performance requirements are also demanding. The high Q's of the tank circuit ($\sim10^4$) must be maintained, so the amount of electrically lossy materials near the coils must be minimized. High excitation fields are applied to the sample, so the coils must be able to handle high currents. Organic materials such as epoxies in this region must also be minimized, since they can result in background NMR signals.

Probes also have severe space constraints. The coils must be placed as close to the sample as possible to maximize signal to noise. Components for tuning the tank circuit frequency, and an adjustment for an impedance matching network between the coil and preamplifier must also fit within the probe.

Finally, high stability is required for the more sophisticated experiments. Fluctuations in the tank circuit resonant frequency, Q, or coupling to the preamplifier can modulate the signals and lead to spurious peaks in the spectra. Phase errors in complicated excitation pulse sequences can also degrade the spectra. Vibration is a major source of these fluctuations, so internal probe structural resonances must be damped, and the highest performance systems usually require isolation from room vibrations.

Commercial systems also require user interface features. Because NMR is performed on several nuclear species with a variety of experimental techniques, several probes with different RF characteristics are typically used with one spectrometer system. Probes must be easily interchanged, so the superconducting magnets are supported off of the floor and the probes are inserted from the bottom. Samples are loaded from the top, and an axial bore down the probe allows the sample tubes to be placed within the RF field of the coils. Because of the high cost of spectrometer systems, rapid sample turnaround is necessary to maximize spectrometer usage. Typically, a probe will remain in the magnet while samples are sequentially loaded and analyzed, with sample loading and unloading accomplished with a pneumatic system. This system also spins the samples, which spatially averages the magnetic field to reduce spectral line broadening due to field inhomogeneities. Sample temperature control is also required, and is accomplished by flowing temperature regulated gas past the sample. The gas manifold, temperature sensor, and heater all interface to the bottom of the probe.

SUPERCONDUCTING PROBE AND CRYOGENICS

The HTS prototype probe developed in this collaboration targeted 400 MHz proton spectroscopy. The operation of the complete probe/closed-cycle cooling system has been demonstrated in spectrometer measurements, and a signal to noise improvement in excess of a factor of four over conventional probes has been achieved. In other functions such as magnetic field homogeneity and excitation field strength, it has successfully achieved performance comparable to conventional probes.

Cryogenic System

The cryogenic system is based on the highly reliable G-M refrigerator.[4] These refrigerators easily reach the 20 K operating temperatures necessary to achieve the required coil performance in high magnetic fields[1,2]. In addition, the cost effectiveness in low volumes, and the existence of a manufacturing and service infrastructure are significant for commercial reasons.

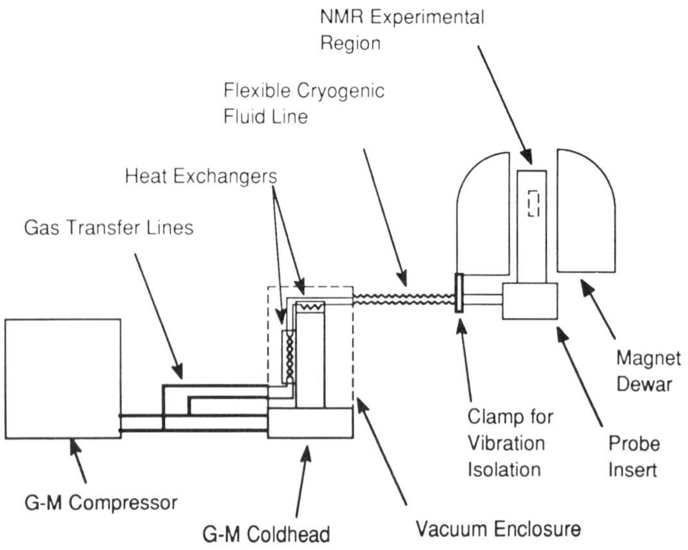

Figure 1. Schematic of the probe and closed-cycle cooling system.

The system, shown schematically in Figure 1, consists of the G-M refrigerator, probe, and a secondary helium gas flow loop which cools the coils within the probe. The gas loop taps off a small quantity of gas from the G-M compressor, and circulates it through a heat exchange system on the G-M coldfinger to cool the gas to the required operating temperature. A flexible vacuum insulated cryogenic transfer line containing both supply and return lines in parallel connects the gas loop to the probe, with detachability provided by a bayonet-style connector. The flexibility of the transfer line, aided by a hard mount of the transfer line to the magnet dewar, vibration isolates the probe from the cryocooler. The G-M coldfinger is mounted upright, and is about 1.5 meters off the axis of the magnet. At this distance the stray magnetic field strength is less than about 100 gauss so there are no detrimental effects, either due to the magnetic field affecting motor performance or lifetime, or the AC magnetic fields from the motor affecting the NMR spectra.[5]

Figure 2a. HTS NMR coil pattern.

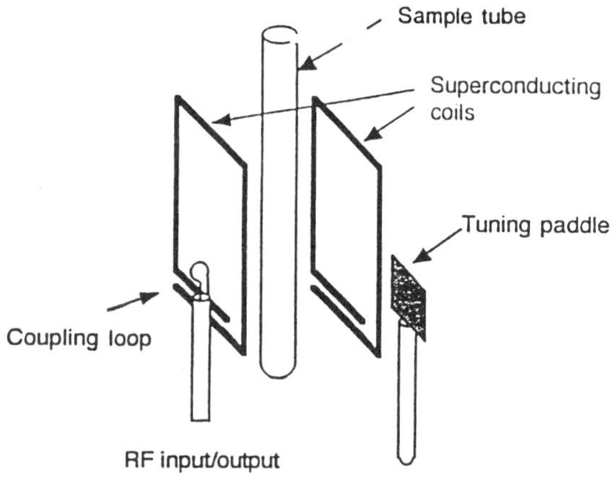

Figure 2b. Layout of the experimental region.

Because of commercial drivers toward rapid turnaround, turnkey systems, one objective is to facilitate rapid probe installation by having fast cooldowns and eliminate the need to evacuate the system prior to cooldown. While commercial G-M refrigerators have the capacity to cool to temperatures where appreciable cryopumping can occur even with some residual gas in the insulating vacuum space, some level of residual gas will increase the thermal conduction load during cooldown and increase the cooldown time. Therefore, minimizing contamination due to outgassing is an objective, so no superinsulation was used. Radiative thermal loads were reduced by minimizing surface areas and gold plating or electropolishing as required. Thermal mass within the system was also minimized.

Probe/Coil description

Figure 2a shows the layout of a typical NMR coil.[2] The coils are fabricated in $YBa_2C_3O_{7-\delta}$ on standard substrates using photolithographic techniques. The tank circuit consists of the inductance of the single loop and the interdigitated capacitor formed by the parallel strips of superconductor. Within the probe, fine tuning of the resonant frequency is performed by adjusting the position of a small metal plate within the RF field pattern of the coil. The resulting change in the inductive loading shifts the frequency. Coupling to the 50 ohm transmission line to the preamplifier is also inductive. Coupling and tuning are adjusted by mechanical linkages inserted through the bottom of the probe. Both the tuning plate and transmission line were heat sunk to the coldhead to reduce their Johnson noise contributions. The arrangement of the experimental region of the probe is shown in Figure 2b.

To minimize distortions of the applied dc magnetic field, materials used in the probe were carefully selected. The probe body was aluminum, and in the experimental region, only quartz, high-purity copper, brass, teflon, and non-filled epoxies were used. For gold plating, diffusion barriers or adhesion layers of magnetic materials were avoided. Because this probe was designed for proton detection, the regions near the coils were kept free of epoxies or organic contaminants as these would cause a background NMR signal.

The probe inserts into a 37 mm bore, 9 T magnet. It has an axial room temperature sample access tube, which allows samples to be placed to within 2 mm of the HTS coils. The top of the probe interfaces this tube to the sample pneumatic manipulation system, while the bottom of the probe interfaces to the sample temperature control gas flow unit.

The coils are temperature regulated using a Conductus commercial temperature controller (LTC10), with a ruthenium oxide temperature sensor located near the coils and a heater on the input gas stream. Temperature stability requirements are set by the RF phase stability requirements of the NMR experiments, since the Q and the resonant frequency of the HTS tank circuit are both temperature dependent.[2] Near 25 K the temperature dependence is weak such that the required stability is less than 80 mK, and is achieved quite readily.

The total probe thermal load has been determined by measuring the flow rate and temperature rise of the helium gas flowing through the probe. This was measured to be about 2.2 watts, and losses includes radiation and conduction through internal supports, coaxial signal cables, and linkages.

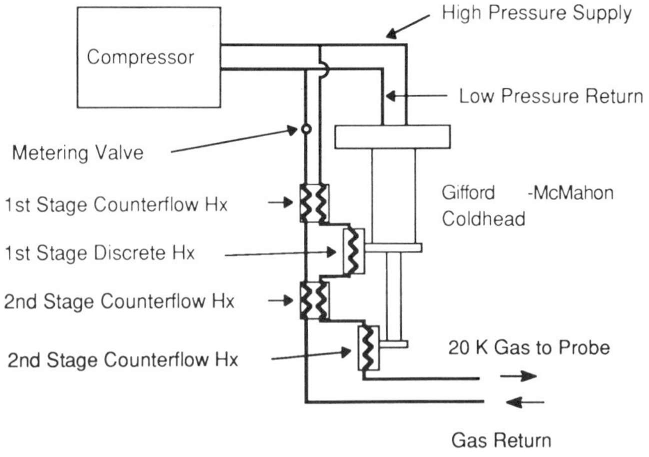

Figure 3. Schematic of the secondary gas flow loop.

Secondary Gas Flow Loop

Figure 3 shows the secondary gas flow loop heat exchange system, consisting of two discrete heat exchangers tied to the first and second stage of the G-M coldfinger, and two counterflow heat exchangers. Using the G-M compressor as the gas source conveniently eliminates the need for a second circulating pump. In addition, the high gas pressure and large differential pressure between the supply and return lines simplified the design of the heat exchangers and flow loop in that pressure drop was not a consideration.

The thermal design of the counterflow heat exchangers was also not demanding since the load on the cooler is dominated by the probe, the transfer line, and radiation load on the coldtip rather than the load from heat exchanger inefficiency. In addition, typical commercial GM machines have large first stage cooling capacities that can easily absorb the load from counterflow heat exchanger inefficiency and cool the incoming gas to 50 - 70 K, reducing the demand on the more inefficient 2nd stage counterflow exchanger. Designs were driven primarily by ease of manufacturing at low volumes.[6]

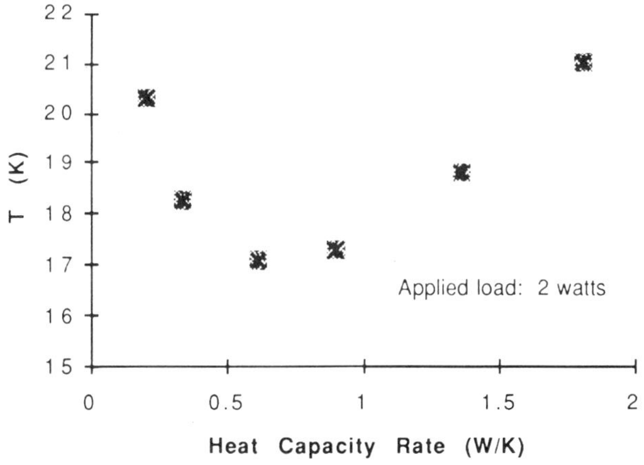

Figure 4. Gas temperature after the applied load as a function of flow rate.

Flow rates were up to 2 grams/sec and was controlled by a metering valve on the return side of the loop. Since commercial GM compressors have excess capacity over coldhead demands, the load of the gas loop did not degrade coldhead performance.

Figure 4 shows a typical performance curve for the secondary gas flow loop as a function of gas flow rate for a fixed applied heat load. The NMR probe was not inserted for this test; the load was applied by a heater in the flow loop. By measuring the temperature rise of the gas before and after the heater, the heat capacity flow rate was determined. The figure plots the gas temperature after the heater as a function of flow rate for a two watt load. This behavior is expected; at low flow rates there is a large temperature increase because of the low heat capacity flow rate; at high flow rates, the counterflow heat exchangers become increasingly ineffective, and the additional load on the refrigerator causes the coldtip temperature to rise.

Cooldown time tests of the complete probe-flow loop system were also performed. The time for the NMR coils to cool to an operating temperature of 20 K was under 60 minutes, in comparison to about 35 minutes for the bare refrigerator.

SUMMARY

We have demonstrated a prototype probe and cryogenic system for an HTS NMR coil for high resolution chemical spectroscopy. Cooling is achieved with a closed-cycle helium gas flow cooling loop based on a Gifford-McMahon refrigerator, and coil operating temperatures below 20 K has been achieved. The system was designed for commercial customers, and is convenient, reliable, and cost effective. The probe interfaces with the standard user interface features such as the pneumatic systems for sample loading, spinning, and temperature control.

The probe itself showed a substantial signal-to-noise improvement over conventional technology by more than a factor of four. Performance in other aspects are comparable to conventional technology. We anticipate that this system will be introduced commercially within the near future.

ACKNOWLEDGMENTS

Howard Hill and Weston Anderson of Varian Associates provided expertise in NMR probe design, and supported testing in spectrometers at the Varian NMR laboratory. William Brey, Felipe Fuks, Richard Withers, and Wai-Ha Wong developed the RF section of the probe. The superconducting coils were fabricated by Brady Cole, Marie Johannson, and Michael Krivoruchko. Kevin Delin provided expertise on the properties of HTS materials in high magnetic fields. Fred Holy was responsible for the cryomechanical fabrication of the prototype probe and secondary gas flow loop. The basic concept for the manufacturable counterflow heat exchangers was provided by Ray Sarwinski.

This work was sponsored in part by the National Institutes of Health.

REFERENCES

1. Black, R. D. et al., A High Temperature Superconducting Receiver for Nuclear Magnetic Resonance Microscopy, *Science*, 259:793, 1993.
2. Wong. W-H. et al., this proceedings.
3. Anderson, W. A. et al., High Sensitivity NMR Spectroscopy Probe Using Superconductive Coils, to be presented at the *12th Conference of the International Society of Magnetic Resonance ISMAR* '95.
4. Longsworth, R. C., Experience with Displex GM Type-Refrigerators on Magnetic Resonance Imaging Cryostats, *16th Energy Sources Technology Conference, Houston*, TX Jan. 31, 1993.
5. Longsworth, R. C., The Use of Small Cryogenics Refrigerators Near High Homogeneity Magnets," *Cryogenics*, 24:243, 1984.
6. To be published.

BEHAVIOUR OF IGBT MODULES IN THE TEMPERATURE RANGE FROM 5 TO 300 K

F. Rosenbauer, H. W. Lorenzen

Lehrstuhl für Elektrische Maschinen und Geräte
Technische Universität München
80290 München, Germany

ABSTRACT

The behaviour of insulated gate bipolar transistors (IGBT) in cryogenic environment was investigated. The test program covered IGBT modules rated for 300 – 500 A in single and parallel operation. The data recorded were the static forward characteristic the dynamic switching, the blocking voltage and the characteristic of the integrated antiparallel diode. The investigated modules show very different temperature dependency. Some do not work below a threshold temperature.

INTRODUCTION

Currently there is a 1 MJ superconducting magnetic energy storage (SMES) prototype under construction in our laboratory.[1 2] It consists of six coils assembled as a torus. Especially because of the low copper-superconductor-ratio of our superconductor, each coil must be protected from thermal destruction in the case of a quench. Then the current has to be removed from that coil as fast as possible.

The fastest way to perform this task is by bridging the "sick" coil. The SMES can then be de-energized to the network by demand. Diodes used for bypassing coils of accelerator magnets are not suitable because of the high reactive voltage across the coil during loading and unloding. Therefore we propose switchable semiconductors. To avoid losses in (five) additional current leads, these semiconductors have to be placed within the cryogenic area (Figure 1).

Previous work showed that bipolar semiconductors like silicon controlled rectifiers (SCR) are — because of their thermal charge carrier generation — not suitable for this purpose.[3] That is why we investigated the hybrid IGBT. This device combines the advantages of the low on-resistance of a bipolar transistor with the low driving power of a MOSFET because the switching is performed by the field effect. IGBTs are manufactured as punch-through (PT) and non-punch-through (NPT) types which differ in their layer arrangement and doping.[4 5]

TEST SETUP

The device under test (DUT) is mounted on the controlled temperature block of a LHe vaporization cryostat. This allows the adjustment of the DUT's temperature in the range from 300 down to 5 K which is measured with a Si-diode thermometer placed within the controlled temperature block.

Advances in Cryogenic Engineering, Vol. 41
Edited by P. Kittel, Plenum Press, New York, 1996

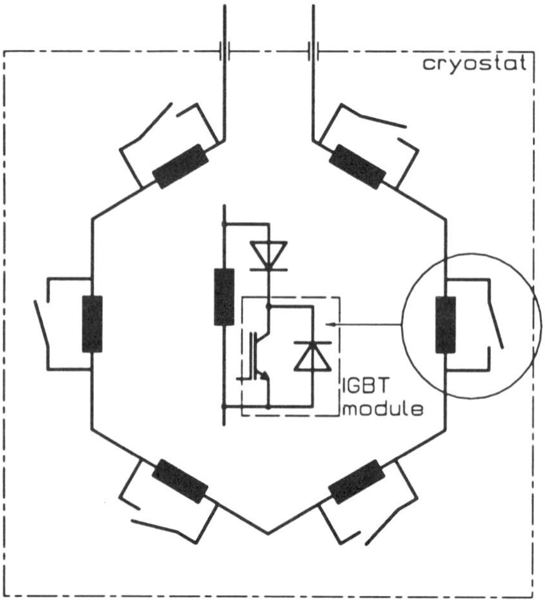

Fig. 1. Principle of the switchable bypass. The switch is realized with an IGBT module and a diode which compensates the antiparallel diode of the IGBT module.

The block is large enough to accomodate IGBT modules with maximum dimensions up to 108 × 74 mm and therefore with a maximum rated current of 500 A. Our test program covers PT and NPT IGBTs from different manufacturers and with different ratings (Table 1). Using a special holding device made of copper it is possible to cool down and test a pair of IGBT modules switched in parallel.

The output characteristics and the measurements of the maximum blocking voltage were recorded with a high power curve tracer. Due to the limited output power of this device, the maximum current for static measurements is 250 – 300 A. For higher currents and the investigation of the dynamic behaviour we use a large capacitor (10 mF) which can be discharged via the DUT (Figure 2). U_{CE} and I_{C} (using a wide-band current probe) were monitored with an oscilloscope.

For the dynamic measurements we also need a gate driver. It is placed outside the croystat so it does not need to work at cryogenic temperatures. Our layout (Figure 3) permits to vary the gate-emitter-voltage U_{GE} in a range from 12 to 24 V. Although there is a long line between driver and gate and the input capacity of the IGBT is relatively large (≈ 100 nF), this driver can perform a fast turn-on. It is already suitable for use in the SMES.

RESULTS

Output characteristics

Fundamental behaviour. As a first step we investigated whether IGBTs are switchable at cryogenic temperatures at all. The experiments showed that all PT

Table 1. Summary of examined IGBTs

Identifier	Type	$I_{\mathrm{C,\,r}}$ / A	$U_{\mathrm{CE,\,max,\,r}}$ / V	$T_{\mathrm{sw,\,min}}$ / K
1MBI 300 L-120	PT	300	1200	125
FZ 300 R12 KF	PT	300	1200	50
FZ 400 R12 KF	PT	400	1200	30
FZ 500 R12 KF	PT	500	1200	125
FZ 360 R17 KF	NPT	360	1700	5.9*
BSM 300 GA 120 D	NPT	300	1200	5.6*

*Lowest temperature reached in experiments.

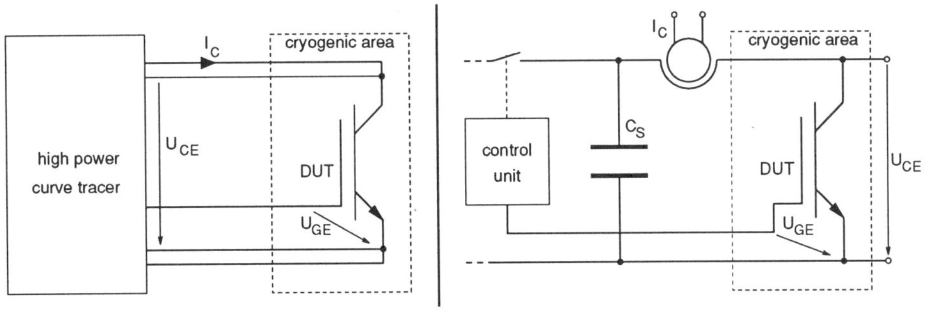

Fig. 2. Test setups used for static (left) and dynamic (right) measurements.

IGBTs loose their switchability below a certain temperature $T_{\mathrm{sw,\,min}}$ (Table 1). If the temperature rises above this threshold, switchabilty returns. Only the NPT IGBTs remain switchable even at the lowest reached temperature.

For a better comparison of the temperature behaviour of the different modules, the collector-emitter-voltage U_{CE} with constant collector current I_{C} is plotted as a function of temperature (Figure 4). This temperature characteristic has a positive temperature coefficient for MOSFETs. Here the charge carrier mobility decreases with rising temperature as there are more collisions with the lattice of the semiconductor. This increases the on-resistance of the device. At bipolar transistors the number of

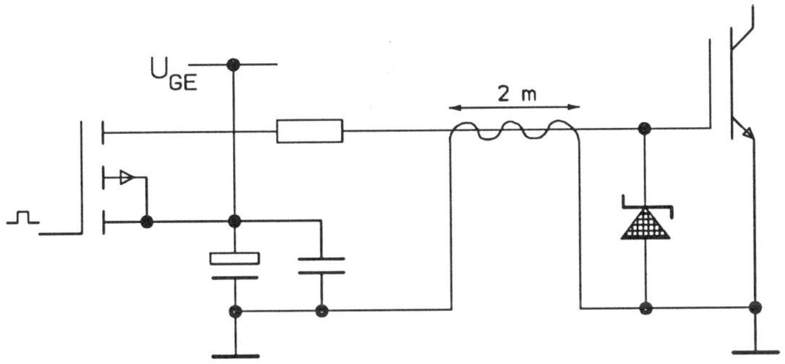

Fig. 3. IGBT driver with variable U_{GE} for fast turn-on. Used for dynamic measurements.

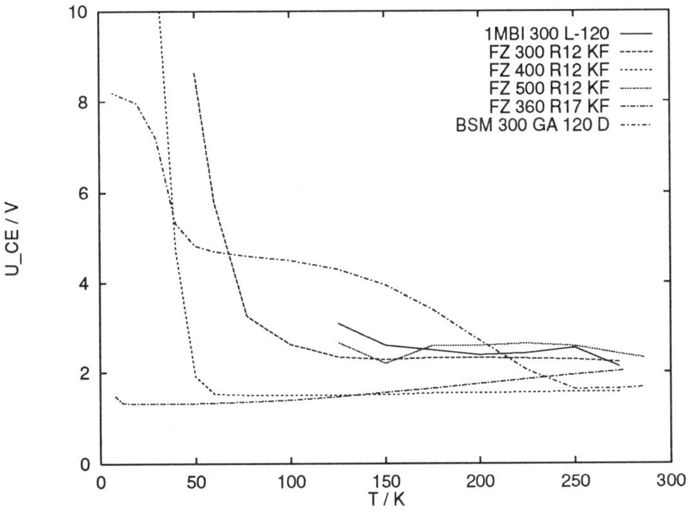

Fig. 4. Comparison of U_{CE} of the different modules as a function of temperature ($I_{\mathrm{C}} = 100$ A, $U_{\mathrm{GE}} = 20$ V).

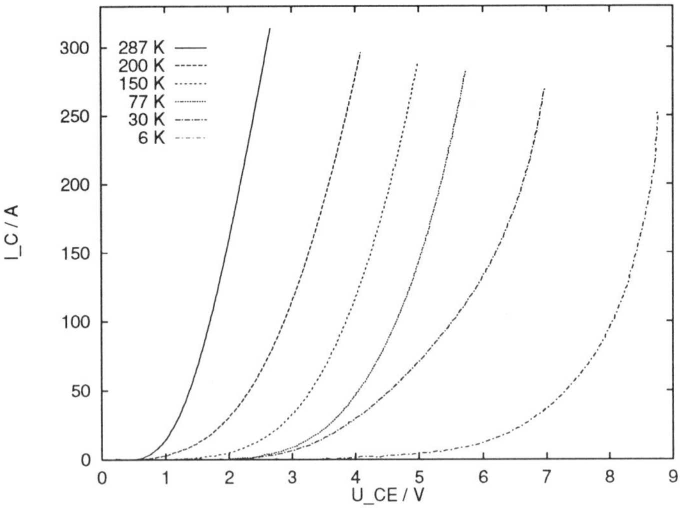

Fig. 5. Output characteristic of BSM 300 GA 120 D as function of temperature ($U_{\mathrm{GE}} = 20$ V). Note the limited output power of the curve tracer.

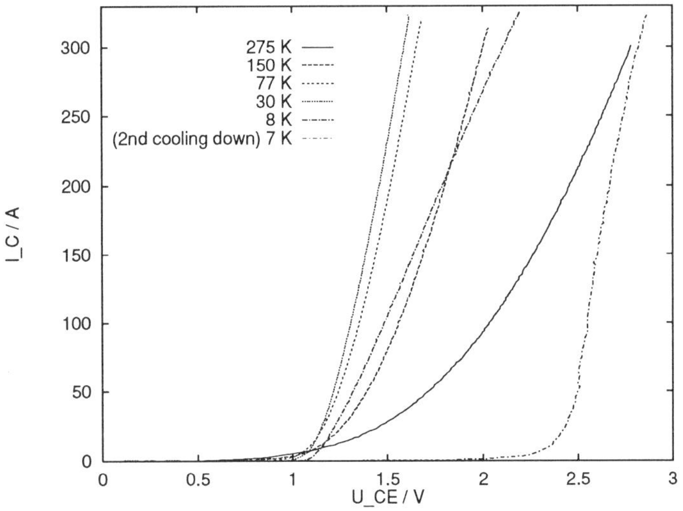

Fig. 6. Output characteristic of FZ 360 R17 KF as function of temperature ($U_{\mathrm{GE}} = 20$ V).

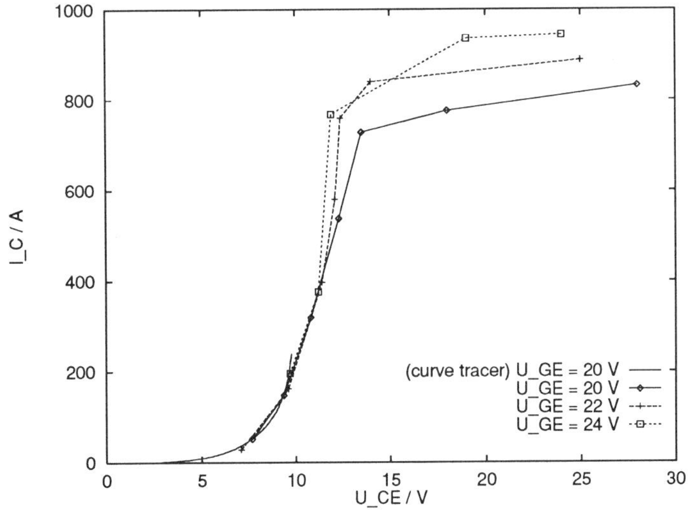

Fig. 7. Output characteristic of BSM 300 GA 120 D as function of U_{GE} ($T = 6$ K; data taken from dynamic measurements). Note warming up of the device when curve tracer is used.

free charge carriers rises with increasing temperature. So this device has a negative temperature coefficient. The hybrid IGBT is a combination of both devices. Depending on the actual micro-structure and the temperature one of them becomes dominant. If the temperature is too low, charge carrier freezeout may occur.

The behaviour of the PT IGBTs corresponds with that of power MOSFETs.[6 7] This applies especially to the FZ 400 R12 KF (slight decrease with falling temperature and then steep increase at 50 K). An explanation could be that both PT IGBT and power MOSFET incorporate a n^+-buffer layer, but the NPT IGBT does not. This does not explain the behaviour of the FZ 360 R17 KF. Although this is a NPT device, it also has a positive temperature characteristic. The steep increase begins here probably at 10 K. In contrast to that is the BSM 300 GA 120 D: It has a globally negative temperature coefficient which can be split into two sections of different gradients. Both effects are subject to further investigation.

From the output characteristics at different temperatures it can be seen that the equivalent resistance of the BSM 300 GA 120 D remains nearly constant, but the threshold voltage increases with falling temperature (Figure 5). In the case of the FZ 360 R17 KF the threshold voltage and the equivalent resistance decrease. Only at temperatures below 30 K do both values begin rising again (Figure 6).

Static losses. Due to the repetitive measuring done by the curve tracer, the high U_{CE}, the low specific thermal capacities and the low thermal conductivity, the BSM 300 GA 120 D quickly warms up at temperatures below 10 K. This results in a vertical output characteristic at currents over 200 A (Figure 5). This effect was noticable neither during the dynamic measurements (Figure 7) nor with the FZ 360 R17 KF. For the latter, the static losses at cryogenic temperature are only 25 % compared with the BSM 300 GA 120 D.

Variation of U_{GE}. The on-state or saturation area of an IGBT is limited by the saturation voltage $U_{CE,\,sat}$. If U_{CE} exeeds this voltage the device enters the active region and the current becomes constant. The maximum current $I_{C,\,sat}$ is controlled by the gate-emitter-voltage U_{GE} and is also temperature dependent. For higher values of U_{GE} the limits of the on-state area can be reached only with the capacitor discharge circuit (Figure 7). For the BSM 300 GA 120 D, this current has a distinct minimum at about 50 K.

Paralleling. For the intended use in the SMES, the current that will flow through the bypass is higher than the maximum rated current of a single module. The IGBTs have to work in parallel. Since current sharing is not equal, a module working in parallel can carry only 75 – 90 % of the current it would carry in single operation. This results from measurements with two paralleled BSM 300 GA 120 D over the whole temperature range. Due to its negative temperature coefficient of U_{CE}, this is not the ideal device for parallel operation. The FZ 360 R17 KF is more suitable because with its positive temperature coefficient, a sharing of the current between the two devices with rising temperature is done automatically. Measurements have not been performed yet.

Aging. The results achieved with the BSM 300 GA 120 D can be reproduced with modules from different series. Even repeated coolings do not seem to change the output characteristic significantly. This is different for the FZ 360 R17 KF. At the second cooling down, the threshold voltage at temperatures below 30 K increased from 1 to 2.5 V (Figure 6).

Switching

The turn-on losses of the devices behave like static losses (Figure 8). It is interesting to note, that the turn-on losses of the FZ 360 R17 KF at 6 K are much smaller than those of the BSM 300 GA 120 D at room temperature. When reaching the active region, the FZ 360 R17 KF shows oscillations of U_{CE}, but the BSM 300 GA 120 D does not. The striking differences of the turn-off losses are caused by the different types of diodes used (see Diodes).

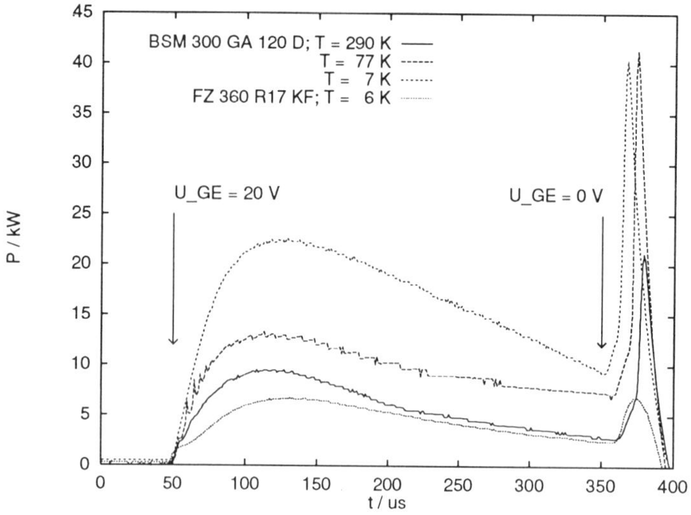

Fig. 8. Switching losses $P = U_{CE} \cdot I_C$ ($I_{C,\ max} \approx 800$ A)

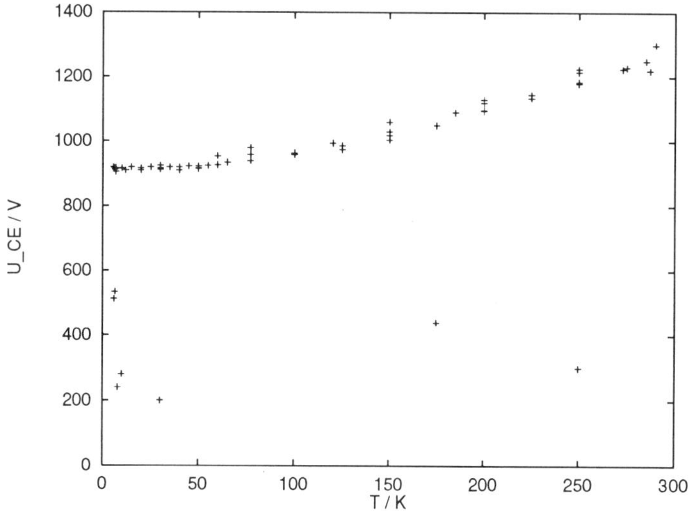

Fig. 9. Maximum blocking voltage of BSM 300 GA 120 D as a function of temperature. The outliers are caused by partial discharges.

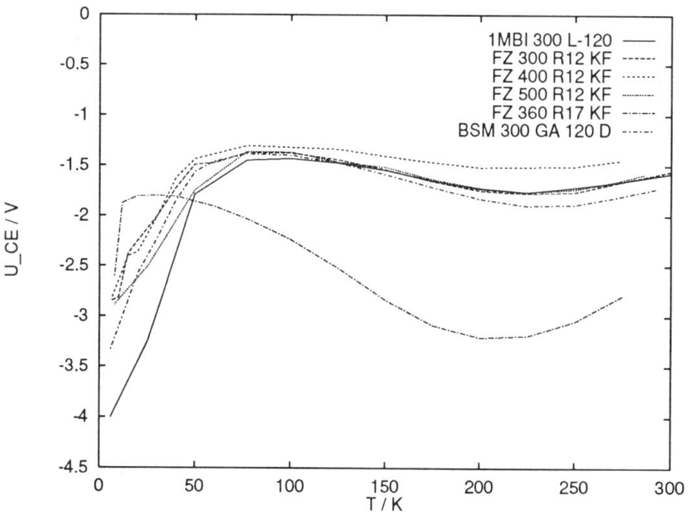

Fig. 10. Behaviour of the antiparallel diodes ($I_C = -100$ A)

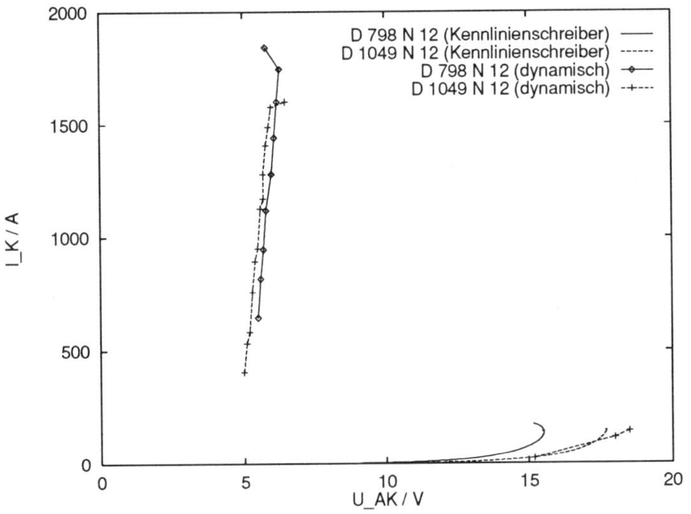

Fig. 11. Cathode current I_K versus anode-cathode voltage U_{AK} of power diodes. In consequence of the repetitive acquisition of the curve tracer the diodes warm up. ($T = 6$ K)

Blocking voltage

As important as the switchability of the semiconductor is its ability to block voltages when it is switched off. The maximum blocking voltage $U_{CE,\,max}$ of the BSM 300 GA 120 D decreases from 1300 V at room temperature to approx. 900 V at 5 K (Figure 9). This positive temperature coefficient again is caused by the decreasing charge carrier mobility with higher temperatures. Therefore they need a higher electric field to reach the kinetic energy necessary to initiate an electron avalanche and finally a breakdown.

In addition to this effect within the semiconductor, $U_{CE,\,max}$ is reduced by partial discharges. The cases and terminals of the IGBT modules are not suited for use in vacuum, so the electric field around the sharp-edged terminals has to be controlled. During the experiments this was done with insulating tape. It can be easily and quickly applied, but the glue has a short lifetime in the cryogenic environment. As soon as there is no adherence any more, the partial discharges reappear and destroy the tape.

Because of different case layouts, insulation is more difficult with the FZ 360 R17 KF. Here an $U_{CE,\,max}$ of only 350 V at 5 K can be reached. This is sufficient for our application as the maximum voltage an IGBT must block will be 120 V.

Diodes

The standard application of IGBT modules are power converters. For this purpose they contain an antiparallel diode as a freewheeling and protection element (Figure 1). These diodes were also included into our test program. All have the same temperature behaviour (Figure 10). In the first section, down to a temperature of 77 K, the voltage drop across the diode remains nearly constant. At lower temperatures this value increases evenly. The only exception is the FZ 360 R17 KF which has a totally different temperature behaviour. This module contains a new diode with reduced reverse recovery charge for soft and fast switching and has therefore a higher on-resistance.

The integrated diodes are inconvenient for our use and must be compensated with power diodes (Figure 1). That is why some press-pack diodes were also included into our tests (Table 2). Their temperature behaviour is similar to that of the IGBT diodes

Table 2. Summary of examined power diodes

Identifier	$I_{K,\,r}$ / A	$U_{AK,\,max,\,r}$ / V	$U_{AK,\,max}$ / V $T = 300$ K	$U_{AK,\,max}$ / V $T = 6$ K
D 798 N 12	1780	1200	1400	1140
D 1049 N 12	2800	1200	1800	530

and their reverse voltage decreases with falling temperature as well as the blocking voltage of the IGBT modules does.

At temperatures below 77 K the diodes start to behave like diacs or SCRs: If the anode-cathode voltage rises over a breakover voltage the diodes "switch on" and change to a different characteristic (Figure 11). This breakover voltage is about 30 V for the power diodes and 8 V for the IGBT diodes (T = 6 K). This effect has already been reported from signal diodes.[8] Even the there mentioned oscillations in the forward characteristics can be observed with the IGBT diodes.

CONCLUSIONS

The FZ 360 R17 KF seems to be the ideal device for working in a cryogenic environment. It is switchable even at the lowest temperatures achieved where its on-state and switching losses are smaller than at room temperature. Its blocking capability is heavily reduced, but is still sufficient for our application. Further experimental work must cover long term aging tests and tests in paralleled operation. Additionally warming up at continous current must be investigated.

ACKNOWLEDGMENT

The authors thank Deutsche Forschungsgemeinschaft for the provision of the funds and eupec, Fuji Electric and Siemens, Bereich HL for supplying test devices and useful discussion.

REFERENCES

1. U. Brammer, H.G. Herzog, J.F. Kärner, H.W. Lorenzen, J. Schaller, and R.M. Schöttler, A fast-acting 1 MJ superconducting magnetic energy storage, in: *Proc. ICEM '94* p. 373
2. J.F. Kärner, H.W. Lorenzen, F. Rosenbauer, J. Schaller, and R.M. Schöttler, A protection system for small high power SMES with power semiconductors working at cryogenic temperature, presented on ASC '94, Boston, in: *IEEE Trans. Applied Superconductivity* [in press]
3. J.F. Kärner, H.W. Lorenzen, and W. Rehm, Semiconductors at low temperature, in: *Proc. EPE '91, Firenze* p. 2-500.
4. N. Mohan, T.M. Undeland, and W.P. Robbins. "Power Electronics: Converters, Applications and Design", John Wiley & Sons, New York (1989)
5. H. Yilmaz, J.L. Benjamin, R.F. Dyer, L.S. Chen, W.R. van Dell, and G.C. Pifer, Comparison of the punch-through and non-punch-through IGT structures, in: *IEEE Trans. Ind. App.* IA-22, p. 466 (1986).
6. O. Mueller, On-resistance, thermal resistance and reverse recovery time of power MOSFETs at 77 K, in: *Cryogenics* p. 1006 (1989).
7. R. Karunanithi, A.K. Raychaudhuri, Z. Szücz, and G.V. Shivashankar, Behaviour of power MOSFETs at cryogenic temperatures, in: *Cryogenics* p. 1065 (1991).
8. A. Szmyrka-Grzebyk and L. Lipiński, Low temperature current voltage characteristics of silicon diodes used as thermometers, in: *Cryogenics* p. 222 (1993).

COLD SEMICONDUCTING CONVERTER, A COMPATIBILITY DEVICE FOR A SUPERCONDUCTING MAGNET

O. A. Shevchenko, H. J. G. Krooshoop, M. A. Fedorovsky[c], and H. H. J. ten Kate

Faculty of Applied Physics, University of Twente
P. O. Box 217, 7500 AE Enschede, The Netherlands
[c]Institute for Electrodynamics, National Academy of Sciences
Peremoga Prospect, 56, Kiev 57, 252680 Ukraine

ABSTRACT

A method to increase the efficiency of a power system for a superconducting (SC) magnet is presented. The system links a magnet and the mains (which serves as a power source and/or a load). It contains both room and low temperature electrical components.

In order to reduce heat leaks, it is proposed to use a semiconducting converter unit located in the same cryostat and which is connected via high current leads to a magnet and via low current leads to the mains. In the standby mode the temperature of the unit is close to that of a magnet. While operating, the temperature of the unit will rise up to a level normally allowed for such components (~400 K for the switches). The efficiency of the proposed scheme is close to the conventional one which is based on a semiconducting power supply placed at room temperature in combination with retractable current leads.

To demonstrate the approach, we developed such a converter. The performance of the converter unit is demonstrated while ramping a current of a small SC magnet operating at 4 K. Characteristic in- and output currents are 2 A and 300 A respectively. The unit standby temperature is measured to be ~40 K, while the operating temperature is about 300 K. Reasonable energy efficiency is demonstrated even at a low powering voltage.

INTRODUCTION

Superconducting magnets have some unique requirements that call for novel design configurations of their powering systems[1]. In general, a magnet can be connected to the mains as shown in Figure 1 (here and below only single phase schemes are discussed). The magnet current I_M, the source current I_P and the load current I_L can differ. Depending on the case, various design solutions are possible[2,3]. In this paper we focus on the case $I_M \geq (I_P, I_L)$. When I_M is of the same order as (I_P, I_L) it can be essential to have both transformers $TR_{1,2}$.

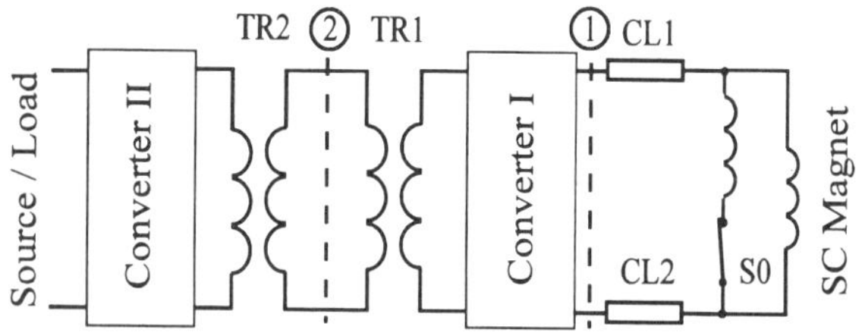

Figure 1. Basic power scheme to link a magnet and a primary source (or a load): S_0 is a persistent mode switch; TR 1 and 2 are transformers; CL 1 and 2 are current leads; the dashed lines indicate alternative locations of the 300 K border. The protection circuit of the magnet is not shown.

A number of different semiconducting and superconducting devices can be used as components to build such a powering system for a magnet[4,5]. The scheme with the interface at position 1 in Figure 1 is of common use. Semiconducting power supply is placed at room temperature and connected via current leads to the magnet located in a cryostat. The border between room and the cryogenic environment crosses the leads CL permanently (as it marked in Figure 1 by the dashed line 1), so that the warm end of the leads are at room temperature.

The leads introduce a heat load to the cryogenic system both in the standby and normal operation modes. To reduce the thermal load, other components can be included in the scheme[4]: like a persistent mode switch S_0, 'cold' mechanical contacts or a SC loss make-up rectifier.

A possibility to apply a superconducting power supply (interface 2 in Figure 1) is presented elsewhere[6]. Such a rectifier would operate at the same temperature as the magnet, which explains the position 2 of 300 K interface in Figure 1. In this case the heat load from the leads is negligible. But, there is an extra heat load to the cryogenic system due to the energy conversion at low temperatures.

Various systems based on the schemes 1 and 2 have been classified and a comparison has been made[4]. The efficiency of the scheme depends on the specific circumstances set by the application and depends on the magnet inductance L, current I_M, storage time t_0 and powering voltage U_p. If the product $L \times I_M / t_0$ is below ~0.1 V scheme 1 must include 'cold'

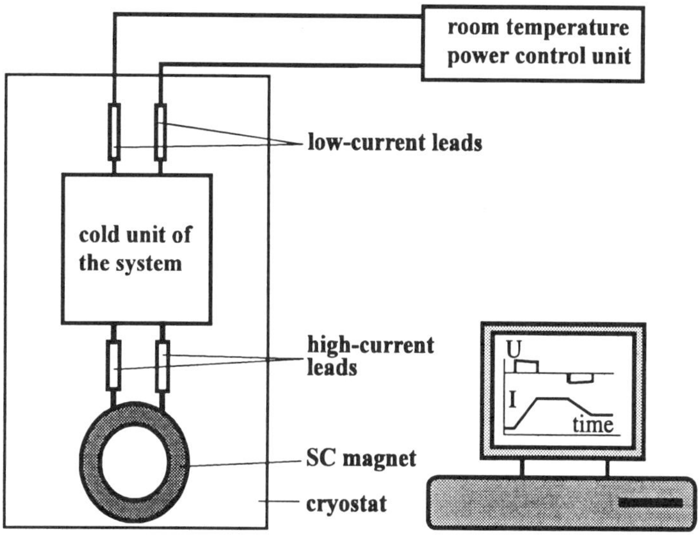

Figure 2. Basic scheme of a cold converter system to ramp the magnet current.

mechanical contacts and the persistent mode switch in order to compete with scheme 2 with regard to the efficiency.

The presence of the 'cold' contacts in a high current circuit is not always acceptable. To avoid them, one can close the entire high current loop within a cryostat. The 300 K interface will be moved from position 1 to position 2. This step requires that both the converter I and the transformer TR 1 are included in the cold unit and are able to operate at low temperatures.

In several papers the operation of semiconducting devices at low temperatures has been considered[7,8]. The energy loss due to the forward voltage drop across a semiconducting device must be removed by the cooling system. This can bound certain applications of the power devices at low temperatures, especially at high current and low voltage.

PRINCIPLE OF OPERATION

A sketch of the proposed powering system is presented in Figure 2. It consists of a low temperature (cold) unit and a room temperature power and control unit. The cold unit is electrically connected via high current leads to the magnet and via low current leads to the room temperature unit.

The cold unit of the system is placed inside the cryostat. It includes a transformer and the switches, protection and measuring circuits, current leads and a temperature control system. In general, it provides both AC/DC and DC/AC conversion.

The room temperature unit handles the connection between the cold unit and the mains. It supplies control pulses to the switches and protects the low current circuit. The logic part is PC-based (in the case) and contains a control and measuring system, a code and a user shell. It controls the operation of the various units and supports the user interface.

Further on we will focus on the development of the cold part of the system, a key part which enables one to close the high current loop within a cryostat and to build an efficient powering system for superconducting magnets.

Both the converter and the transformer in the cryostat have a temperature close to the magnet temperature in the standby mode. The loss introduced by the high current leads is reduced substantially. During the operational mode, the temperature of the converter or at least of the switches, may be close to ~400 K, which is within their standard specifications. The energy dissipated in the switches can be removed in a simple and efficient way.

The temperature of semiconducting switches will be in between of the magnet temperature and the acceptable 400 K. The position of the dashed line in Figure 1 will not be fixed. In the standby mode it crosses the scheme as it is marked by the line 2, and during the normal operation it remains close to the position 1. This step requires certain changes in the powering system.

Electrical Scheme

The electrical scheme of the device is presented in Figure 3. For the sake of simplicity, only the single phase scheme will be is considered here. The same approach can be expanded to other single and multiple phase schemes.

The transformer. In general low T_c, high T_c superconductors or normal metals can be used for the transformer coils. Copper coils can be superior, especially at higher frequency. The transformer can have either an air- or ferromagnetic core. The particular design depends on the specific application requirements.

The repetitive switches. The switches provide both a rectification and an inversion of the current. Diodes, thyristors, and MOSFETs[7,8] can operate at low temperature.

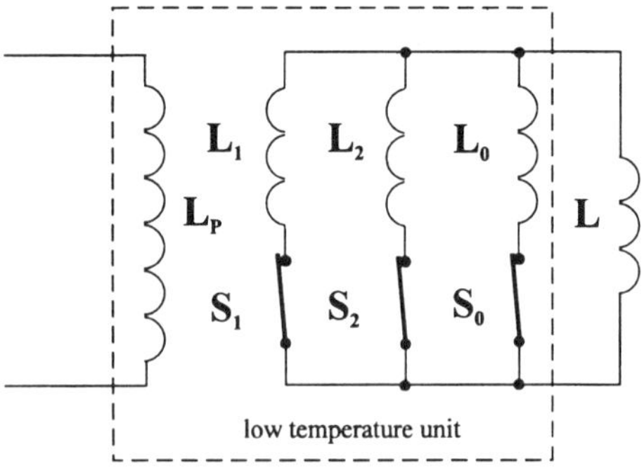

Figure 3. Equivalent electrical scheme of a full-wave, single-phase low temperature converter: L_p, L_1, L_2, are self inductances of the centre tapped power transformer (the mutual inductances are not shown); S_1 and S_2 are repetitive switches; S_0 is a persistent mode switch (the inductance of the branch is also shown); L is a superconducting magnet. Current leads connecting the switches, the transformer and the magnet are not shown.

The persistent mode switch belongs to the magnet or to the converter itself. It can be made superconducting (of low or high temperature superconductor) or based on mechanical contact.

This mode in the power circuit eliminates the ripple of the magnet current under DC conditions. The design of a power supply can be much simpler, as the critical requirement to the output filtering is reduced. This reduces the price of the supply remarkably, as it is shown in the Appendix.

Connectors. The superconducting connectors of high Tc materials are commercially available now and can be used in the device as high current leads between the converter and the magnet. If it is applicable, they also can connect the switches and the transformer. Otherwise, connectors of copper or low Tc superconductor are the best.

Theory of Operation

V-I characteristic of a semiconductor is non-linear and temperature dependent. Precise calculation of the ramping process requires, therefore, extended calculations. To enable an approximate analytical approach, one can consider the forward voltage across the switch constant (or a step like function), independent on the current and depending only on the

Table 1. Input data of the converter*

Parameter	Value
Operating frequency, [Hz]	50
Transformation coefficient	145.6
Primary voltage amplitude, [V]	360
Primary commut. inductance, [mH]	46
Magnet inductance, [mH]	37.9
Persistent mode switch (S_0):	
Min. propagation current, [A]	0.03
Recovery time, [s]	10
Heater resistance, [Ohm]	80
Control voltage amplitude, [V]	6
Control pulse width, [ms]	5

* the repetitive switches are described in the text.

operating temperature. The temperature can be also considered constant, as temperature control of the switches is present in the device.

Under these assumptions the theory developed for superconducting converters[9] can be modified to describe transient processes in the converter while ramping the magnet current up or down in the inductive commutation mode. The input data required for the theory are summarised in Table 1.

For the precise calculations, we developed a numerical model of the converter using the PSPICE circuit simulator.

Control of the Device

The same control unit, PC and the control shell as described earlier[6] was used to provide a two-quadrant operation for the device. The only difference is that a thyristor switch is a normally opened device, while a superconducting repetitive switch is normally closed when the control signal is zero. To start an inductive commutation, it is enough to supply a control signal to a proper thyristor. Where applicable, a diode based rectifier can be used and no special control (except for the switch S_0) is required.

Temperature control of the switches is provided by the thermometer and the heater both placed close to the switches and the controller at the room temperature.

DESIGN AND CONSTRUCTION

The specifications of the converter are:
-input voltage amplitude of 360 V at 50-60 Hz;
-input current below 2 A;
-output current of 300 A;
-energy efficiency better than 99 %;
-operating temperature in the range of 4 to 350 K;
-cold volume less than 3 litres;

The main goal of the design is to demonstrate the possibility of the high efficiency operation under close to real conditions, rather than to create a device for a particular application. To demonstrate the approach, we developed a small power converter able to operate inside the helium cryostat. The device is thyristor based and provides a two-quadrant operation.

The Repetitive Switches

A pair of silicon based phase control thyristors of type T 210 N from the EUPEC[10] is used as the converter switches. Each of the thyristors has RMS current of 330 A and forward voltage drop of 0.8 V at the operating temperature of about 410 K. The reverse voltage under the same condition is 400 V. The switches are fixed on the insulator plate. As the connectors, 3 pieces of the flat braided copper cable are used with the cross-section of about 30 mm^2, each about 1.5 m long; 6 pieces of the same type of cable having somewhat larger cross-section, each 0.2 m long, are mounted under the switches as radiators. A thermometer and a heater to control the temperature are mounted on the power connector close to the switches.

Assembled switch unit of the converter has a volume of less than 0.5 litre, and weight less than 1 kg. It is placed in the foam about half way between the liquid helium level and upper flange of the cryostat. Channels for the helium gas are made in the foam in order to provide proper cooling for the switches and the connectors.

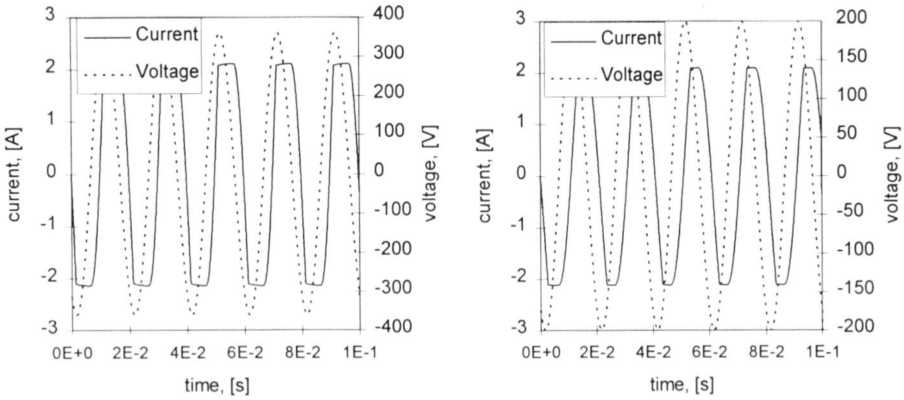

Figure 4. Voltage and current on the high-voltage side of the converter while ramping up (left) and down (right) the magnet current (measured).

The Transformer

At 300 A and zero average output voltage the converter dissipates about 250 W in the thyristors. High electrical efficiency of the device can be achieved at output voltages much higher, than 1 V. This requires a rather powerful transformer. The same superconducting transformer as described earlier[11], operating at 4 K, is used to power the demonstration converter.

RESULTS AND DISCUSSION

In the standby mode the temperature of the converter switches lowers to about 40 K. At temperatures less than ~50 K the semiconductor-insulator transition takes place and control of the thyristor in a conventional way becomes difficult.

The reverse voltage of the thyristors was checked separately in the opened state. No decrease of the voltage is found due to the low temperature up to 360 V.

Heating of the switches to the initial operating temperature (about 60 K) is provided by the external heater within less than 10 s. Heat dissipation in the switch gate provides further heating of the switch. The operating temperature of the switch depends on the magnet current, cooling conditions and is measured less than 350 K for 300 A.

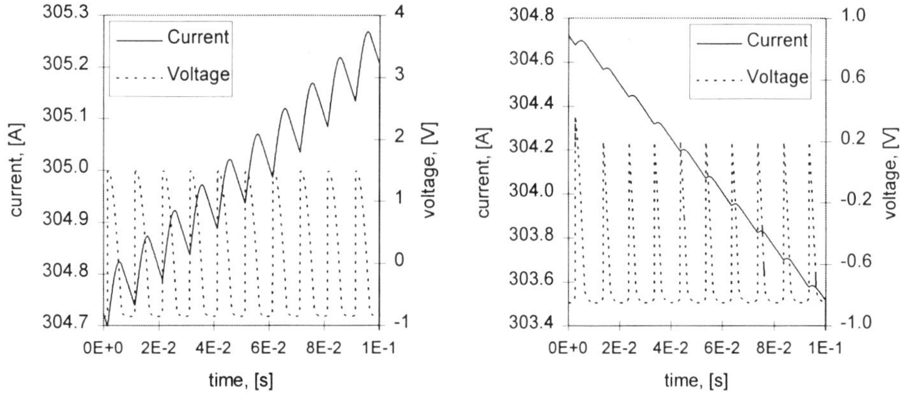

Figure 5. Voltage and current of the high current side of the converter while ramping up (left) and down (right) the magnet current (measured).

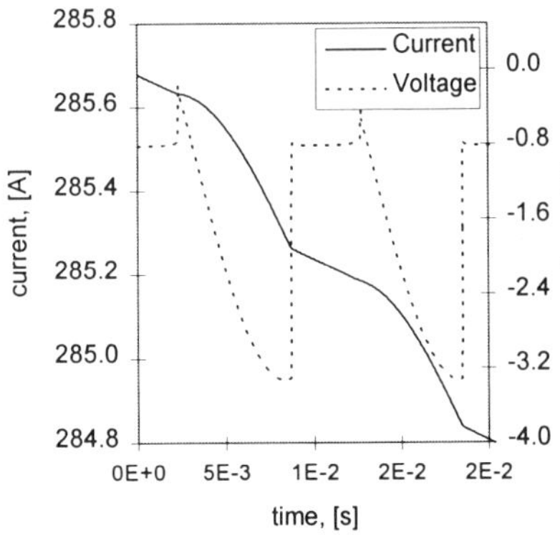

Figure 6. Voltage and current on the high current side of the converter while ramping down the magnet current (simulated).

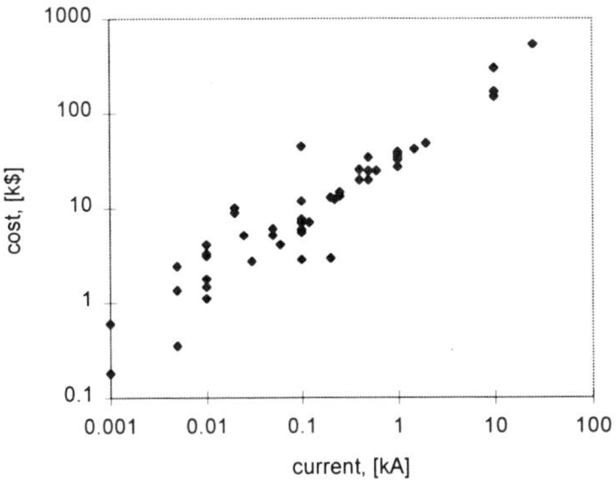

Figure 7. Cost of a room temperature DC power supply as a function of the output current: points-various data for the supplies commercially available in Europe.

The V-I characteristics of the separate thyristor and some silicon diodes were measured in liquid nitrogen and compared with the curve at room temperature. We did not conduct detailed comparison with the theoretical behaviour yet. The first impression is, that between 77 and 400 K the V-I characteristic follows temperature dependence predicted by the known theoretical models.

Operation

The operation of the converter in the inductive commutation mode is illustrated in Figures 4 and 5. Primary voltage amplitude is varied by means of the autotransformer connected between the mains and the high voltage side of the converter. For the cases presented in Figure 4, it is constant and equal to 360 V and 200 V respectively. The inductive commutation mode is provided automatically. The primary current amplitude is proportional to the magnet current. Both ramping up and down of the magnet current are available during the device operation.

Slow ramping down of the magnet current is possible in the rectifier mode by dissipating the energy of the magnet in the switches. For the particular device considered here the sign of the magnet current change is controlled by the amplitude of the voltage applied across the primary coil. At the primary voltage of 280 V and magnet current of 300 A the average output voltage generated by the rectifier is zero (while output voltage ripple is about 1 V). Recovery of the switch S_0 is possible under these conditions. This enables the persistent mode of the magnet and reduction of the current ripple from the level indicated in Figure 5 to zero. During ramping at higher than 280 V primary voltage, the magnet current rises. Lower primary voltage ramps down the magnet current.

Faster ramping down is provided by controlling the switches in the same way as it is done for the superconducting converter[9]. Few cycles of the process are presented in Figure 6 for the primary voltage of 360 V. This option complicates the control system of the converter.

Qualitatively the same results are obtained up to the operating frequency of 1000 Hz. In this case maximum current is limited by the commutation inductance of the transformer.

Efficiency

Excellent energy efficiency of the system was found during the test. Total loss in helium of the converter (including the transformer) is measured calorimetrically while ramping the magnet current from 0 to 300 A at 50 Hz. It is found less than 17 J at the average voltage across the magnet of 0.94 V. This corresponds to the efficiency of 99 % for the energy stored in the magnet of 1705 J. Higher operating voltage will provide even better efficiency of the converter. In a standby mode the loss of the powering system is measured as less than 0.1 W. The efficiency can be further improved by applying HTS current leads.

CONCLUSIONS

A new semiconducting converter able to operate at low temperature is developed and tested. The repetitive switches of the converter are connected via current leads to the magnet and to the transformer.

High energy efficiency of the thyristor based converter system ≥99 % even at low operating voltage of ~1 V was demonstrated and measured experimentally. In a standby mode the heat leak introduced by the system to the helium was less than 0.1 W.

APPENDIX

Cost C, [k$] of the low voltage and high current room temperature power supply versus the output current I, [kA] is plotted in Fig. 7. An average trend can be described by the power law:

$$C=38\times I^{0.8}$$

The spread of the points is due to the difference in type, precision requirements (as ripple), year of production, producer, etc. Bipolar, high precision power supply is always more expensive, than an unipolar and rough device of similar design. Switchers[5] are available only for current below 250 A. At higher currents mainly thyristor based supplies are in use.

REFERENCES

1. J. McCarthy, and C. C. Smith, "Power supply design for superconducting magnet applications", Superconductor Industry, pp. 22, (Spring 1992).
2. W. E. Buckels, M. A. Daugherty, and E. L. Kostecki, "The SSD: a commercial application of magnetic energy storage", IEEE Trans. on Applied Superconductivity, 30:N1:328 (March 1993).
3. Y. M. Eyssa, X. Huang, "DC-DC converter for discharging energy storage magnets", IEEE Trans. on Magn., 30:N4:2665 (July 1993).
4. O. A. Shevchenko, H. J. G. Krooshoop, H. H. J. ten Kate, and M. A. Fedorovsky, "Development of 50 Hz superconducting converters", *Applied Superconductivity,* DGM INFORM.GESSELSCHAFT VERLAG Germany vol. 2 (1993) p. 1537.
5. "3000 Watts DC power supplies", DELTA ELEKTRONIKA B.V. Zierikzee, The Netherlands, (June 1991).
6. O. A. Shevchenko, H. H. J. ten Kate, H. J. G. Krooshoop, and N. V. Markovsky, "New 50 Hz superconducting power supply for a 2 kA DC magnet", *IEEE Trans. on Magn.* 30:N 4:1831 (July 1994).
7. T. Ise and Y. Murakami, "Control of a superconducting coil by a MOSFET power converter operating at near liquid nitrogen temperature", IEEE Trans. on Magn. 27: N2:2020 (March 1991).
8. V. S. Vysotsky, et. al. Proc. of the P. N. Ljebedev Physical Institute, vol. 205 (1991).
9. O. A. Shevchenko, H. H. J. ten Kate, H. J. G. Krooshoop, and N. V. Markovsky, Operation modes of a 50-60 Hz superconducting converter, Adv. in Cryog. Engineering, 39: part A:933 N. Y.-London, Plenum Press, (1994).
10. Power Semiconductors, EUPEC (AEG/Siemens) Catalogue, (1992).
11. O. A. Shevchenko, H. J. G. Krooshoop, H. H. J. ten Kate, and M. A. Fedorovsky, "Transformer and switch characteristics of a 50 Hz superconducting rectifier", Cryogenics, ICEC Supplement, 34:745 (1994).

WHAT HAPPENED TO CRYOGENIC AND SUPERCONDUCTING EQUIPMENT IN THE GREAT HANSHIN EARTHQUAKE

T. Haruyama, K. Fujioka, H. Hirabayashi, T. Ishigohka, A. Ishiyama, Y. Kawate, S. Nishijima, T. Noguchi, O. Ogino, H. Ogiwara, O. Okazaki, and A. Sato

Special Committee for Investigation of Cryogenic Equipment Damage
Cryogenic Association of Japan
Palais Dor's Hongo 302, 6-12-8 Hongo, Bunkyo-ku, Tokyo 113, Japan

ABSTRACT

At 5:46 a.m. on 17 January 1995, a big earthquake with a magnitude of 7.2 on the Richter scale struck the Kansai area in western Japan. It was a shallow dislocation earthquake and resulted in huge damages in this area. Kobe, with a population of one million, was greatly affected by this disastrous earthquake. There are many hospitals, laboratories and universities in this area, and various kinds of cryogenic and superconducting equipment were about to go into operation. The Cryogenic Association of Japan immediately organized a special committee for damage investigation on this equipment purely from the technical point of view. This paper is presented as a quick report on what happened to MRI magnets, cold evaporators and other cryogenic and superconducting equipment in the universities affected by this earthquake.

INTRODUCTION

Although many people do not necessarily recognize it, cryogenic and superconducting equipment is penetrating deeply into our daily lives. We can easily find liquid oxygen or liquid nitrogen cold evaporators in factories and hospitals. Someone has already been diagnosed one's brain by using a full-body size MRI (magnetic resonance imaging) body scanner equipped by a large superconducting magnet. One can find many vacuum insulated tankers carrying liquid nitrogen, liquid oxygen and other cryogenic liquids on the road. High speed magnetic levitating trains are now under development in Japan, the green house effect due to carbon dioxide is pushing the development of hydrogen fuel vehicles, and realistic applications of high Tc superconductors will push up the use of cryogenic liquid even in a home. In these situations, what can we do or what do we have to do when a big earthquake hits a city? It is a natural disaster and no one can predict an earthquake even by using most advanced technology.

Early this year, a disastrous earthquake hit the Kansai area in western Japan with a magnitude of 7.2 on the Richter scale killing over 5,500 and injuring 415,000. Awaji Island, the epicenter, and Kobe city recorded a 7th degree earthquake (maximum) on the Japanese seismic intensity scale ("Sin-do"). There are a million people in Kobe city, and much cryogenic and superconducting equipment is located in the area.

Table 1. Data on the Great Hanshin earthquake as announced by the Meteorological Agency

Date and time	17 January, 1995, 5:46 a.m.
Epicenter	Awaji Island (Hyogo prefecture)
Depth	20 km
Magnitude	7.2

Table 2. Area assigned as "Sin-do" 5 to 7 on the Japanese seismic intensity

Japanese seismic intensity "Sin-do"	Area
7	Certain areas in Kobe and Awaji Island (Hyogo)
6	Kobe, Sumoto (Hyogo)
5	Kyoto, Hikone (Shiga), Toyooka (Hyogo)

A task force of the Cryogenic Association of Japan was organized for obtaining information in order to find out exactly what happened to this equipment as a result of the earthquake and to make suggestions on what we have to do for safe operation. The committee made inquiries at hospitals concerning MRI magnet damages and also collected damage reports on cold evaporators and cryogenic equipment in the universities. In May of this year, the committee organized a special session at the Spring Meeting on Cryogenic and Superconductivity in Tokyo, and reported these results as a preliminary report.

THE GREAT HANSHIN EARTHQUAKE

Detailed data of the Great Hanshin earthquake as formally announced by the Meteorological Agency on 10 May 1995 were listed in Table 1. It must be noted that the earthquake struck early Tuesday morning following a bank holiday. This situation is one of the big factors when we consider the effects of the earthquake. Presumably, there were no patients in full-body MRI scanners, only a few workers were in factories and few experiments were going on in laboratories. The Agency also released magnitude data by area using the Japanese seismic intensity scale "Sin-do". Table 2 lists the "Sin-do" 5 to 7 areas. We have no experience of "Sin-do" 7 until this earthquake. "Sin-do" 7 corresponds to the 9th - 12th degree in Mercalli seismic intensity scale (MMI).

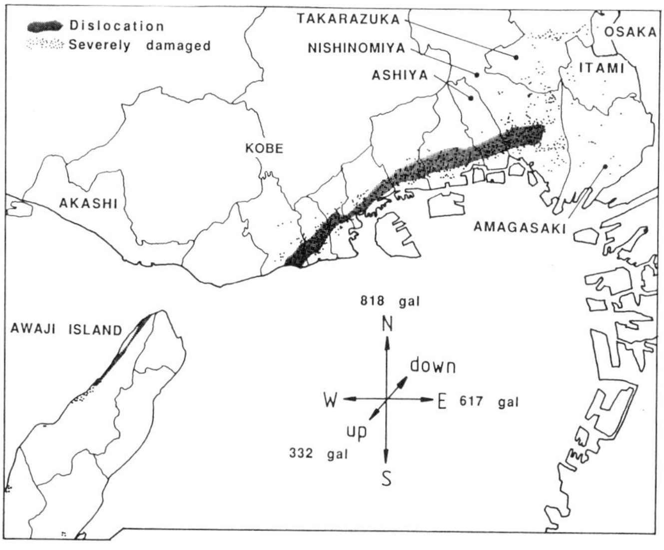

Figure 1. Dislocation belt and the severely damaged area. Acceleration levels are also shown.

Figure 2. Special session for preliminary reports at the Spring Meeting on Cryogenic and Superconductivity.

The dislocation belt and the severely damaged area are shown in Figure 1. The acceleration levels in each direction which were measured at the Kobe Marine Meteorological Observatory are also shown in the figure.

DAMAGE INVESTIGATION COMMITTEE

The Cryogenic Association of Japan immediately organized a special committee for damage investigation on cryogenic and superconducting equipment in the earthquake area.

The primary concern of this ad hoc committee is to investigate what things happen to equipment such as MRI magnets, cold evaporators, etc. when a unusually big earthquake strikes. For this purpose, the committee made inquiries at hospitals in Hyogo and Osaka prefectures asking about the situation on MRI magnets. This inquiries were made under the auspices of Japan Society of Magnet Resonance in Medicine (JMRM). The investigation committee also collected information on cold evaporators in order to find out what happened to them. Damage reports on cryogenic equipment in the universities were offered by two universities in the area.

A secondary concern is to look at the operation manuals of this equipment, specially for emergency case, and to make proposals for improving the contents of those manuals,if necessary. This activity is dedicated to establishing a safety guide for using cryogenic and superconducting equipment in daily lives. We feel that the results of this investigation will be important for future applications of this specialized equipment.

The Damage Investigation Committee is chaired by Prof. H. Hirabayashi of NIFS, the National Institute for Fusion Science (also, Emeritus Professor of KEK, the National Laboratory for High Energy Physics). The name and affiliation of committee members are listed below:

K. Fujioka	Daido-Hoxan Co., Ltd.	
T. Haruyama	KEK(Natl. Lab. High Energy Physics)	
H. Hirabayashi	NIFS(Natl. Inst. Fusion Science)	(Chairman)
T. Ishigohka	Seikei University	
A. Ishiyama	Waseda University	
Y. Kawate	Kobe Steel, Ltd.	
S. Nishijima	Osaka University	
T. Noguchi	Ikegami Technology	(Secretariat)
O. Ogino	Mitsubishi Electric Corp.	
H. Ogiwara	Shonan Institute of Technology	(Secretariat)
O. Okazaki	Daido-Hoxan Co., Ltd.	
A. Sato	NRIM(Natl. Research Inst. Metals).	

Table 3. Results of inquiries on MRI magnet answered by hospitals

Fields (T)	Number of MRI	Damaged Total	Position shift	Others
0.5	11	2 (18%)	1	1
1.0	10	4 (40%)	3	1
1.5	22	11 (50%)	4	7
Permanent	2	0	0	0
TOTAL	45	17 (38%)	8	9

The Kansai Branch, Refrigeration Commission and Applied superconductivity Research Group of the Cryogenic Association of Japan also cooperates with the committee.

PRELIMINARY REPORTS ON DAMAGE INVESTIGATION AT THE SPRING MEETING ON CRYOGENIC AND SUPERCONDUCTIVITY IN JAPAN

The Spring Meeting on Cryogenic and Superconductivity was held in Tokyo during 17-19 May, 1995. The committee organized a special session for preliminary reports at this conference. The session continued for 2 and half hours including 4 reports and discussions. About one hundred of delegates participated in this special session. The meeting atmosphere is shown in Figure 2.

MRI body scanner magnet

Results of the inquiries on MRI magnets in hospitals were reported by T. Noguchi (Committee secretariat) , and Dr. T. Kamei, Chairman of the Technical Committee for the Safety of Equipment for MRI of JMRM. The total number of MRI magnets in Hyogo and Osaka prefecture is 240 (185 superconducting magnets, 52 permanent magnets and 3 resistive magnets) as of September 1994. The investigation committee received 44 replies from hospitals covering 45 MRI magnets (43 for superconducting and 2 for permanent).

The details of the results are listed in Table 3. No magnet quenches were reported. Nearly half of the reported damages were classified as position shifts. The degree of shifts varied from <1 cm to several tens of cm, depending mainly upon the places where the MRI magnets were installed and methods of fixing the magnets to the floor. It was reported that, in general, damage was not severe for MRI magnets installed in basements. Anchor bolts were not used in many cases. In one case (obtained by direct inquiry to a person in a hospital), an MRI magnet shifted about 80 cm, but the helium recovery pipe was not torn off because a flexible pipe was used.

Damages classified in 'others' include cases of a broken refrigerator, a position shift of a computer console, etc.

Because the earthquake struck early in the morning, there were no patients in MRI full body scanners. The situation could be changed drastically if diagnoses were going on.

Many users in hospitals pointed out that the safety manuals for emergency cases should be re-examined and arranged for easier usage. They also intend to change the water-cooled refrigerators to air-cooled ones to eliminate torn-off cooling pipes, and they are fixing MRI magnets to the floor by using anchor bolts .

NMR magnet

Inquiries on NMR magnets were not made at this time. However, several reports regarding damages on NMR magnet indicate that the situation is much severe than that of

Table 4. Number of irregularly sunk cold evaporators (1398 in total; largest sinking ratio was 21.3%)

irregularly sinking ratio (%)	number of cold evaporators
0.5-1.0	25
> 1.0	41

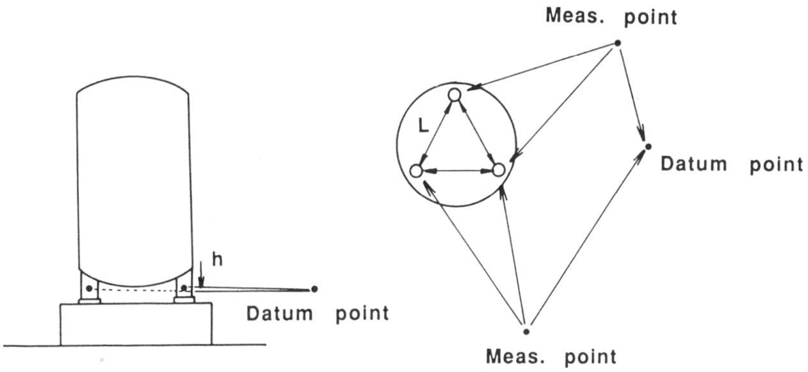

Figure 3. An irregular sinking ratio of a cold evaporator is defined as h/L X 100 (%).

Table 5. Damages on cryogenic and superconducting equipment in the universities

University	Kobe Univ. of Mercantile Marine	Kobe Univ.
Ground condition	Reclaimed, seashore (Liquidization)	Hill side
Helium liquifier	OK	OK
Purifier	--	fallen, broken
LN_2 CE	3% declined	--
LHe Dewar(100L)	OK	--
LHe Dewar(glass)	broken	broken
Gas manifold	fallen down	fallen down
Cryostat	fallen down	fallen down
Dilution refrigerator	--	OK
NMR	--	OK
Superconducting magnet	OK	--
Superconducting ship	OK	--

Figure 4. Severe crack near the Cryogenic Center of the Kobe University of Mercantile Marine.

MRI magnets. The committee is now making inquiries on these NMR magnets in the universities, companies and hospitals.

Cold evaporator

Damage reports on cold evaporators was presented by O. Okazaki (committee member). As of the end of March 1995, 1398 cold evaporators in Hyogo, Osaka and a part of Kyoto prefectures were investigated. The LNG and LPG tanks are not included in this report. No cold evaporator fell over. Also, there are no reports on cryogenic liquid floods. Slight irregular sinking took place on 66 cold evaporators out of 1398. The definition of the irregular sinking ratio, h/Lx100 (%), is shown in Figure 3. A number of irregularly sunk cold evaporators and their sinking ratios are listed in Table 4.

Generally speaking, irregular sinking could be attributed to ground subsidence. The structures of cold evaporators had enough strength even under the trembling of "Shin-do" 7. It was fortunate that many valves were shut off because the day when the earthquake struck the city was after a 3 day-long holiday.

Cryogenic equipment in the universities

Prof. K. Nishigaki of Kobe University of Mercantile Marine (KUMM) reported damages on cryogenic and superconducting equipment at universities in Kobe city. The seismic intensity scale near the universities is estimated as the 6th degree on "Sin-do". The KUMM and Kobe University had various kinds of equipment in their cryogenic laboratories. Detailed damage reports on experimental equipment are listed in Table 5. There is a big difference in the ground conditions between the two universities. The KUMM was built on reclaimed land, and liquidization was caused by the earthquake. No liquidization was observed at Kobe University which is located at hillside area. Figure 4 shows a severe crack near the cryogenic center of KUMM showing a trace of liquidization and the intensity of the earthquake. Figure 5 shows an inside view of laboratory devastated by the earthquake.

One heavy superconducting magnet was set in a cryostat with casters and it was not damaged after moving on the floor many times, perhaps because the center of gravity was sufficiently low. However, it is hard to say that this method of using casters could be applied in every cases.

Figure 5. Inside view of a laboratory in the Cryogenic Center devastated by the earthquake.

Helium transportation

T. Noguchi presented one report regarding liquid helium transportation in Kobe. Helium gas is not produced in Japan and mostly it is imported from USA by ship with a liquid helium tank. Kobe is the only port of entry for these tankers in the Kansai area. When the earthquake hit the city, one ship with 11,000 gallons of liquid helium tank was near Kobe bay about to unload. As the functionality of Kobe port was mostly destroyed, the ship moved to Osaka port, the nearest to Kobe. However, the ship could not clear the custom due to an administrative inexperience. Liquid helium continued to vaporatize and had to be blown off several times. This ship finally moved to the Tokyo port and cleared customs as a case of emergency. Liquid helium was unloaded and transported by land.

CONCLUDING REMARKS (PRELIMINARY) AND FUTURE WORKS

A preliminary result of damage investigation on cryogenic and superconducting equipment in the Great Hanshin Earthquake was compiled and reported by the special committee of the Cryogenic Association of Japan as follows:

-MRI magnets
- 45 out of 240 MRI systems were investigated.
- No magnet quenches were reported.
- 50% of MRI magnets moved on a floor.
- Fixing anchor should be used.
- Manual for safety, installation, emergency should be re-examined and arranged.

-Cold evaporator
- 1398 CE were investigated and no hardware damage was reported.
- 66 CE sunk irregularly mostly at levels of a few % (Max. 21.3%).
- Irregulary sinking occurred due to ground subsidence.
- No cryogenic liquid flooding out.

-In the universities
- Damages greatly depended upon the ground condition.
- Education and exercises regarding earthquake must be emphasized.
- Fixing methods for various kinds of equipment should be studied.

Following these preliminary results, the committee is continueing to collect and compile more detailed information on damages of cryogenic and superconductivity equipment. A final report will be published in autumn this year.

ACKNOWLEDGMENTS

The members of the damage investigation committee wish to thank Dr. T. Kamei, Chairman of the Technical Committee for the Safety of Equipment for MRI of Japan Society of Magnetic Resonance in Medicine for his great contribution to this investigation. They wish to thank Prof. K. Nishigaki of Kobe University of Mercantile Marine for offering much precious information. They wish to thank Mr. T. Kurino, Executive Director of the Society of Non-traditional Technology for helpful financial support.

REFERENCES

General information on the Great Hanshin Earthquake is available in articles of the following journals.

•*Science*, Vol. 267, 27 Jan. 1995, pp. 444-446
•*New Scientist*, No. 1962, 28 Jan. 1995, pp. 4-6
•*Kagaku Asahi* special issue (in Japanese), Feb. 1995, Asahi Newspaper Publishing Co.
•*Newton* special issue (in Japanese), March 1995, Kyoikusha

1995 CEC/ICMC Conference: Klaus Timmerhaus (right) presenting Ronald Fast (left) the 1995 Samuel C. Collins Award

ELECTRICAL SAFETY OF A THIN SUPERCONDUCTING SOLENOID IN AN IRON YOKE*

R. P. Smith, K. Krempetz, A. Kristalinski, A. Lee, D. Markley, S. Orr, R. Rucinski, R. L. Schmitt, B. Squires, R. P. Stanek, A. M. Stefanik, R. Wands, R. Yamada

Fermi National Accelerator Laboratory
Batavia, IL, 60544, USA

ABSTRACT

Fabrication of a thin superconducting solenoid[1,2] for the Upgrade Tracking System[3] for the DØ Detector[4] at the Fermilab Proton-Antiproton Collider has begun. The 2.0 T magnet is 2.75 m long, 1.2 m in diameter and stores 5.6 MJ magnetic energy at full excitation. The magnet is novel in that no thin superconducting solenoid magnet for a particle detector has yet been fabricated which operates at this field level.

The magnet is to be installed in the existing DØ detector which has a thick magnetized steel muon absorber which surrounds the superconducting solenoid. In the event of an unexpected electrical short in the magnet is is desireable that the resulting asymmetric forces generated between the magnet and the muon steel not cause collateral damage to the detector.

Although the magnet is designed to sustain a quench without a protection resistor such a resistor is provided to extract a portion of the stored energy from the magnet during a quench to permit faster recool after the quench. This resistor cannot be used for routine discharging of the magnet as its use at full current would in fact cause a quench. To enable timely routine discharge it can be switched into the circuit at some lower current to speed the discharge without causing a quench. It is necessary to estimate the current at which the protection resistor can be used to safely speed discharge.

INTRODUCTION

An unexpected electrical short in a solenoid magnet which is near steel can cause the appearance of unanticipated forces between the coil and the steel. The probability

*Work performed under US DOE Contract No. DE-AC02-76CH03000.

of such a short occurring in a superconducting detector magnet is evidently low as no such accident has been reported to date. This is especially noteworthy as the typical thin superconducting magnet has been one-of-a-kind and has pioneered new design concepts. The only known failure of this type in a detector magnet occurred with a conventional (i.e. water cooled) coil[5] .

The occurrence of a short from the conductor to the winding bobbin in a superconducting magnet can be more or less benign depending on how the bobbin is grounded to the magnet cryostat and how the magnet energization circuit is designed. If there is only one shorted path to ground one can in principle ensure that no current can flow via this path by properly isolating the magnet cryostat from the power supply ground.

The occurrence of a turn-to-turn short is vastly less benign. Provided the short has sufficently small resistance current flow can be induced through it every time the magnet is charged or discharged. This current flow can be destructive in the event of a too-rapid charge or discharge; a quench is likely to generate just such a discharge. In what follows we concern ourselves with turn-to-turn shorts.

ANALYSIS OF A SHORTED TURN

The equivalent circuit for a turn-to-turn short in a coil is:

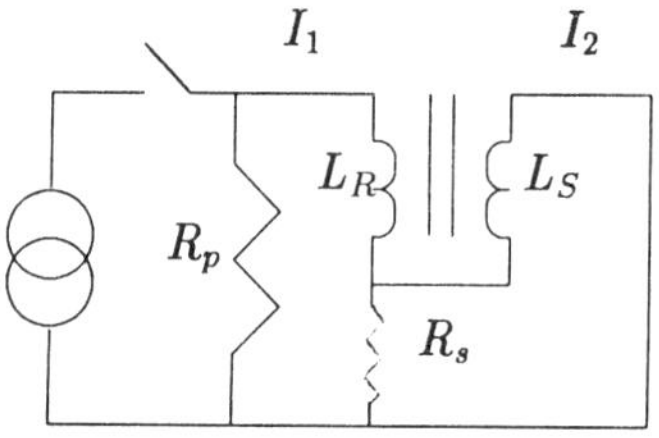

Here R_P is the protection resistor of the magnet and R_S is the resistance of the interturn short. The turns in the shorted section of the coil are taken to be perfectly coupled to those in the remainder of the coil.

For the circuit shown:

$$L_R \frac{dI_1}{dt} + M \frac{dI_2}{dt} + (R_P + R_S)\, I_1 \; = \; 0 \tag{1}$$

$$L_S \frac{dI_2}{dt} + M \frac{dI_1}{dt} + R_S\, I_2 \; = \; 0 \tag{2}$$

where $M = \sqrt{L_R L_S}$ (i.e. perfect coupling) and L_R is the portion of the original inductance L_o not contained in L_S. It is conservative to assume perfect coupling between the two sections of the coil since imperfect coupling would reduce the current I_2. It is also conservative to ignore the presence of the outer support cylinder of the DØ magnet which acts as another "shorted secondary" because the support cylinder also reduces the current I_2 induced by rapid current change in the coil. In any case the current induced in the support cylinder is uniformly distributed axially and does not generate decentering forces on the magnet.

Because we seek the conditions which maximize I_2, we solve the equations for the discharge of the primary circuit (which is much faster than charge of the primary circuit) to yield:

$$I_1 = I_o e^{-\lambda t} \tag{3}$$

$$I_2 = I_0 \left[\frac{M\lambda}{L_S(\lambda_S - \lambda)} exp(-\lambda t) + \left\{ 1 - \frac{M\lambda}{L_S(\lambda_S - \lambda)} \right\} exp(-\lambda_S t) \right], \tag{4}$$

where

$$\lambda = \frac{\lambda_R \lambda_S}{\lambda_R + \lambda_S}. \tag{5}$$

In fact a short in a superconducting magnet[6,7] can generate ohmic heating at the short, which can in turn cause the magnet to quench (or even lead to the destruction of the coil), depending on the resistance of the short, its location in the coil, the rapidity of the charge or discharge, the total stored energy of the coil and the current density in the windings. We are not concerned with the consequences of such heating.

For simplicity we assume R_P and R_S are constant during the discharge, although they would in fact increase somewhat during a quench. Quench studies of the DØ coil do not show a great dependence of the time constant of the decay on these effects.

Now $L_i \sim n_i^2$, so that for n turns in the entire coil and m turns in the shorted section, $L_S = m^2 L_0/n^2,\ and\ L_R = (n-m)^2 L_0/n^2$. Thus the current in the secondary can be written

$$I_2 = I_0 \frac{n-m}{m} \frac{\lambda}{(\lambda_S - \lambda)} [exp(-\lambda t) - exp(-\lambda_S t)] + I_0 exp(-\lambda_S t). \tag{6}$$

For a single-layer magnet a turn-to-turn short generally links only one turn of the coil. The DØ magnet requires two winding layers to achieve 2 T so a layer-to-layer short can cause any fraction of the magnet to be included in the secondary circuit. For any choice m of the number of turns in the short we can calculate the maximum current I_2 in the shorted section during a discharge using this result. Care must be taken to avoid numerical imprecision when λ_S and λ are nearly equal.

For $\lambda_R \ll \lambda_S$, we have

$$I_2 = I_0 \frac{m}{n-m} \frac{R_P + R_S}{R_S} [exp(-\lambda_R t) - exp(-\lambda_S t)] + I_0 exp(-\lambda_S t). \tag{7}$$

This current rises rapidly (governed by λ_S) from I_0 at t = 0 to some intermediate larger value then it decays less rapidly (governed by λ_R). In the limit of very small R_S this current has the simpler form

$$I_2 = I_0 \frac{m}{n-m} \frac{R_P + R_S}{R_S} [exp(-\lambda_R t)] \leq I_0 \frac{m}{n-m} \frac{R_P + R_S}{R_S}. \tag{8}$$

Because $\lambda_R \ll \lambda_S$,

$$\frac{R_P + R_S}{R_S} \ll \frac{(n-m)^2}{m^2}. \tag{9}$$

Thus

$$I_2 \ll I_0 \frac{n-m}{m}. \tag{10}$$

In fact we can locate the maximum value of I_2 as a function of time either analytically or numerically so that this pessimistic limit ("the transformer limit") can be avoided by use of the actual maximum.

Similar inspection shows that for $\lambda_R \simeq \lambda_S$, and for $\lambda_R \gg \lambda_S$, the current $I_2 \leq I_0$.

Decentering Forces

For any particular m the resulting shorted segment of the coil is not centered in the steel of the DØ muon system and so will experience a net axial force. By assuming the image currents in the iron are instantaneous (i.e. by neglecting the time constant of the iron) and by assuming the iron is far from saturation, an upper limit to the decentering force is obtained when the current in the shorted segment reaches its peak value during the magnet discharge.

By considerations of the interactions between the coil segment and its images mirrored in the steel one can show that the net force on the off-centered segment is

$$F = \frac{3\mu_0 M^2}{32\pi}\left[\frac{1}{(D-\Delta)^4} - \frac{1}{(D+\Delta)^4}\right], \tag{11}$$

where D is the distance from the center of the coil to the steel and Δ is the distance the shorted segment is from the center of the system. M is the dipole moment of the shorted segment, $M = \pi R^2 I_2$.

Numerical Results

Using the values of the magnet parameters pertaining to the DØ solenoid, these formulas can be evaluated to yield numerical results. In Figure 1 is shown the force on a shorted segment when the resistance of the short $R_S = 0.05$ Ohms, the same as R_P, the protection resistor of the DØ magnet. It is seen that the worst case force is not severe. The cold mass support system has been designed to support this force.

In Figure 2 is shown the force for the case of $R_S = 0.5 \times 10^{-4}$ Ohms. This case reflects a short typical of contact resistance between two metal surfaces[6] . The decentering force remains readily manageable.

As the short resistance is reduced to increasingly smaller values the peak in the force for small shorted fractions increases (as does the numerical sensitivity of the calculation to small values of the difference between λ and λ_S) toward the "transformer limit". But even for the unphysically small value for R_S of 0.5×10^{-10} Ohms the peak force remains less than 1900 N (with one turn in the shorted segment) when the exact expression is evaluated.

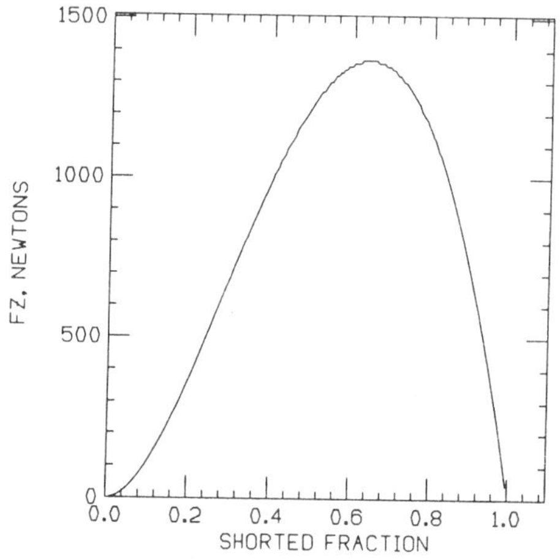

Fig. 1. Force on shorted segment with short resistance 0.050 Ohms.

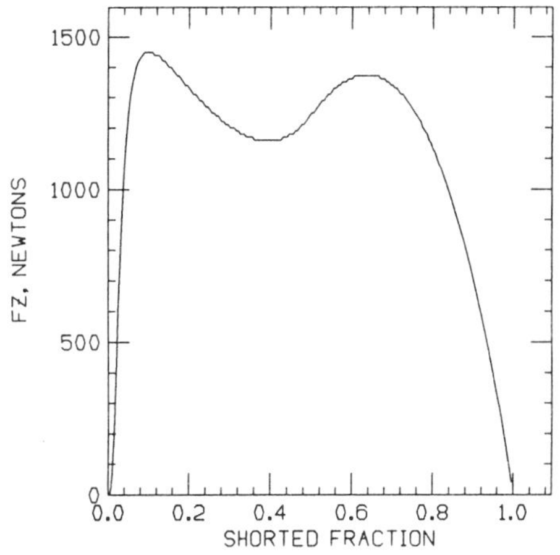

Fig. 2. Force on shorted segment with short resistance 0.0005 Ohms.

FACILITATING THE ROUTINE DISCHARGE

As noted previously the magnet would quench if it were discharged from full current into the protection resistor. This happens because the rapid discharge generates sufficient heating in the support cylinder to overwhelm the cooling system and drive the coil temperature above the superconductor transition temperture. To ensure that the coil can be routinely discharged without quenching, the resistance of the discharge circuit must be limited at the outset and the resulting time constant of this decay (the "slow discharge") for the DØ magnet is more than 5 minutes.

Coupled equations similar to Eq. (1) and Eq. (2) (where the second circuit is that of the support cylinder instead of a shorted turn) can be solved for the current induced in the support cylinder by the discharge of the primary circuit (the coil itself):

$$I_2 = \frac{I_0 M \lambda}{L_2(\lambda_2 - \lambda)} \left[exp(-\lambda t) - exp(-\lambda_2 t)\right]. \qquad (12)$$

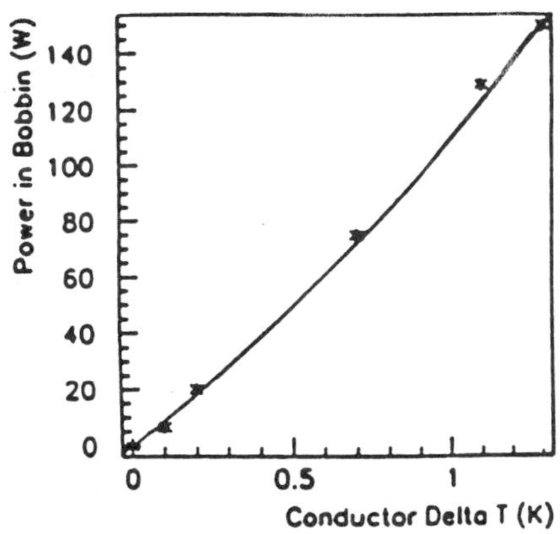

Fig. 3. Heating power in the coil support cylinder vs. temperature elevation of the conductor .

Here $M = \sqrt{L_1 L_2}$, $\lambda_i = R_i/L_i$, and the power in the support cylinder is just $P(t) = I_2^2 R_2$. The time at which this power is maximum can be found by differentiating Eq. (12), equating the result to zero and solving for the time. This time can be used to find the peak power generation, and the result is(where N is the number of turns in the magnet):

$$P(t_{peak}) = \frac{I_0^2 R_1^2}{R_2 N^2}. \tag{13}$$

To this discharge power is added eddy current losses in the conductor stabilizer, and magnetization and AC coupling losses in the superconductor.

Since the discharge heating rate peaks at the beginning of the discharge and decreases as the current decays and since the superconductor stability margin increases as the current and field decay, at some current less than maximum it is possible to switch in the protection resistor and speed the discharge substantially without quenching the magnet.

A finite-element heat conduction calculation was made[2] of the temperature elevation of the magnet conductor above the helium coolant in the cooling tubing on the support cylinder as a function of the heating power in the support cylinder and the cooling power of the helium in the cooling tubes. This curve is shown in Figure 3 and the conductor operating margin at full current is expected to be about 1.5 K with proper cryogenic operating conditions.

For the slow discharge described above for the DØ magnet the heating power in the coil is about 15 W at the beginning of the discharge. The conductor does not approach the current sharing region.

The power corresponding to a fast discharge from only about 500 amperes however is about 180 W, sufficient to elevate the conductor temperature about 1.5 K. Because the conductor current sharing margin is greater than 1.5 K at such a low current this is a pessimistic result. Be integrating the discharge power $P(t)$ throughout the time of discharge and equating this to the enthalpy of the support cylinder and one layer of conductor, it is found that for initial currents below about 1800 amperes the coil does not rise above the current sharing temperature of the conductor.

CONCLUSIONS

Although the DØ magnet itself would in all probability be damaged internally as a result of a turn-to-turn short, from a consideration of the upset forces that might appear due to such a short it is seen that it is practical to design it so that it might not damage other components of the detector in such an event.

From a consideration of the heating induced in the coil support cylinder by a discharge we conclude that for routine discharge of the magnet the protection resistor must be omitted from the circuit at the outset of the discharge. Once the current has decayed to about one third its initial value the protection resistor can be switched into the circuit and the discharge speeded substantially without quenching the magnet.

REFERENCES

1. B. Squires *et al.* Design of the 2 Tesla Superconducting Solenoid For the Fermilab DØ Detector Upgrade, Advances in Cryogenic Engineering 39:301, Plenum, New York (1993).
2. J. Brzezniak *et al.* Conceptual Design of a 2 Tesla Superconducting Solenoid for the Fermilab DØ Detector Upgrade, FERMILAB-TM-1886, Fermilab, Batavia (1994).

3. E823(DØ Upgrade) Magnetic Tracking, FERMILAB-DØ Upgrade E823, Fermilab, Batavia (1993).
4. S. Abachi, *et al.*, The DØ Detector, Nuclear Instruments and Methods in Physics Research A338:185, (1994).
5. Private communications from various individuals cognizant of the magnet failure. The incident involved an electrical short from the inner layer of the conventional water cooled coils to ground via a plywood thermal shield that had become water soaked due to leaks from the coils. In this case collateral damage to detector elements followed since the coil package moved violently in response to the unanticipated decentering forces generated by the suddenly asymmetrized currents in the coils.
6. S. H. Kim, *et al.*,Effect of Electrical Shorts on Cryostatic Stable Superconducting Magnets, Advances in Cryogenic Engineering 23:214, Plenum, New York (1978).
7. L. R. Turner, Consequences of Shorted Turns in a Superconducting Magnet, Advances in Cryogenic Engineering 29:255, Plenum, New York (1984).

THE PRESSURE RISE SIMULATION WHEN HELIUM PIPES ARE BROKEN IN THE ITER CRYOSTAT

K. Nishida,[1] T. Honda,[1] K. Hamada,[1] K. Matsui,[1] T. Kato,[1] H. Tsuji[1]
N. Mitchell,[2] and K. Yoshida[2]

[1]Japan Atomic Energy Research Institute
[2]Joint Central Team, ITER Naka Joint Work Site
801, Mukouyama, Naka-Machi, Naka-gun, Ibaraki, 311-01, Japan

ABSTRACT

The superconducting coil has a potential risk of its cryostat pressure rising as a result of cold helium leaked from coolant pipes. If the cryostat pressure rapidly rises until all inventory helium expands to room temperature in a narrow cryostat space. All components inside of the cryostat must be designed for the saturated pressure. The cryostat pressure rise caused by helium leakage may be slower than ideal heat input to inventory helium. Thus, it is necessary to estimate the case when the pipes are broken in the ITER cryostat as the worst fault for safety aspects. A computer simulation code has been developed to calculate the pressure and temperature rise for the above fault conditions so that safety measures can be adopted.

INTRODUCTION

The Engineering Design Activities (EDA) work on the International Thermonuclear Experimental Reactor (ITER) project started in 1992, as a collaboration between the government of Japan, the European Atomic Energy Community, the government of the United States of America and the government of the Russian Federation. The ITER superconducting coils consist of 7 poloidal field coils, 20 toroidal field coils and a central solenoid coil. The total cold mass of the coil system is about 21,000 tons and the total helium inventory in the coils is 28.3 tons at 0.6 MPa, 4.5K. Several hundreds of cooling pipes are required to supply large amount helium to the forced-cooling coils.

The reason for pipes failing is the thermal stress of expansion and contraction during cool-down and warm-up, and cyclic stress of the pulsed plasma operation. However, these pipes can be designed to withstand the above stresses. The accidental failure is only considered in this study. For example, the pipe line of a coil is cut when some heavy structure falls down or when a coil moves beyond its allowable displacement. However for complete protection of the coils and the cryostat, a guillotine break of several cooling pipes must be simulated to investigate countermeasures to eliminate the possible cryostat failure and loss of the coil isolation.

WORST FAULT SCENARIO

If the cooling pipes are broken , helium at 4.5 K and 0.6 MPa will escape into the cryostat vacuum. The pressure of the helium gas rises with the heat input from the cryostat wall within the constant volume of the cryostat. The escaped helium can not be pumped out because the mass flow rate of the vacuum pump is negligible smaller than the mass flow rate of escaped helium gas. The pressure rise of helium inside the cryostat accelerates the temperature rise of helium inside a coil. The thermal conduction and AC loss during the coil current-dump contribute to the temperature rise of the coils. The coil quench will happen after the coil temperature exceeds the critical temperature of its superconductor. All the coil current will be dumped after the coil quench is detected.

SOLUTION METHOD

Analysis Object

A cross-section view of the ITER is shown in Figure 1. The coils, mechanical supports and cooling pipes are located inside of the cryostat. A thermal shield surrounds the coils. The coils, the mechanical supports and the thermal shield are cooled by supercritical helium. The TF coils contain the largest helium inventory. If cooling pipes connected to the TF coils are broken, the largest pressure rise will happen. The TF coils, the cryostat and the thermal shield are considered in this simulation. The TF coils consist of winding and steel case as shown in Figure 2. All of cold mass of coils and structure are considered in this simulation.

Modeling

Geometry and connections. The simulation is modeled with a concentric column model of the TF coils and the cryostat as shown in Figure 3. One dimensional thermal

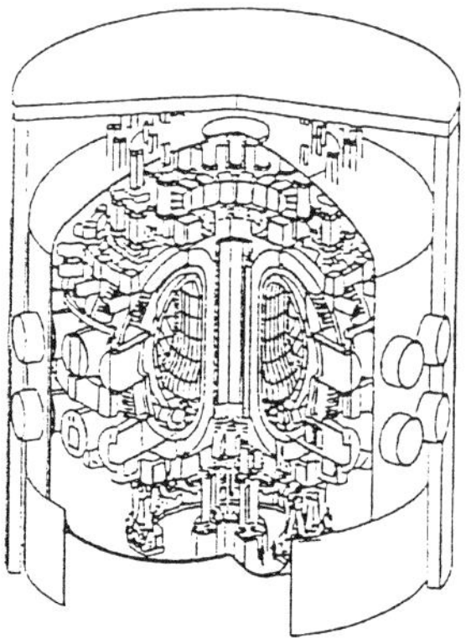

Figure 1. Cross section view of the ITER .

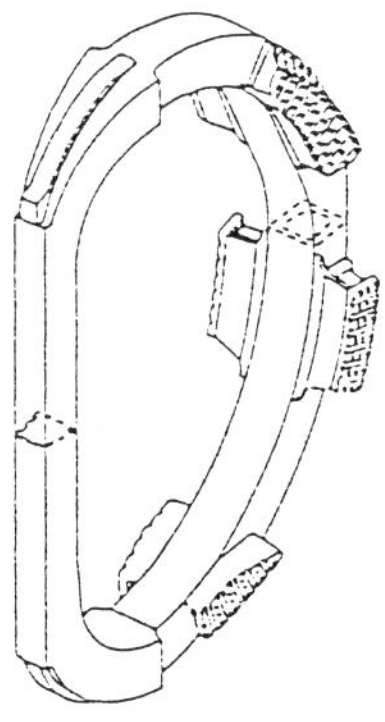

Figure 2. 3-D view of a TF coil .

Table 1. Nusselt number for natural convection of a space between two cylinder

$Ra<2\times10^3$	$2\times10^3 \le Ra<2\times10^5$	$2\times10^5 \le Ra<1.1\times10^7$	$Ra \ge 1.1\times10^7$
$Nu=1$	$Nu=0.197Ra^{\frac{1}{4}}(H/L)^{\frac{-1}{9}}$	$Nu=0.073Ra^{\frac{1}{3}}(H/L)^{\frac{-1}{9}}$	$Nu=0.021Ra^{\frac{2}{5}}$

The H means height of convection space.

conduction equation is used for the heat flow. The heat flows and the temperature are shown in Figure 4. The symbols T_{He}, T_{SS}, T_{G10}, T_{VI}, T_{TS}, T_{VO} and T_W are the temperatures of helium in the coil, stainless steel (case), fiber reinforced plastic (insulator), helium in the inner vacuum, the thermal shield, helium in the outer vacuum and the cryostat wall, respectively. The symbols Q1, Q2, Q3, Q4, Q5, Q6, Q7, Q8 and Q9 are the helium gas heat transfer in the outer vacuum, helium gas heat capacity in the outer vacuum, helium gas heat capacity in the inner vacuum, helium gas heat transfer in the inner vacuum, heat conduction in the insulator, heat capacity of insulator, heat conduction in the case, heat capacity of the case, heat capacity of the helium in the coil, respectively.

Heat transfer. The heat transfer through helium gas in the vacuum chamber is determined by the mean free path of the helium gas. It is assumed that the heat transfer process is free molecular heat conduction in the case that the mean free path is shorter than 1% of distance between boundaries. If the mean free path is longer than 1% of the distance between the heat transfer process is vertical closed space free convection.

The natural convection heat transfer of a space between two cylinders[3] is shown as Eq. (1) and Nusselt number (Nu) is listed in Table 1. If Rayleigh number (Ra) is larger than 1.1×10^7, then the vertical plate model is adopted. The geometric effect of the vertical closed space free convection is negligible.

$$Q = \frac{Nu \cdot \lambda}{L} A(T_h - T_l) \tag{1}$$

where Nu is Nusselt number , λ is thermal conductivity of helium gas, A is surface area , L is distance, T_h is high temperature, T_l is low temperature.

Mass flow rate. Helium flow for the guillotine break is adopted as convergent nozzle model. The helium expands at a nozzle and extends into the vacuum. From the energy equation of isentropic flow, Eq. (2) called the St. Venant formula is introduced.

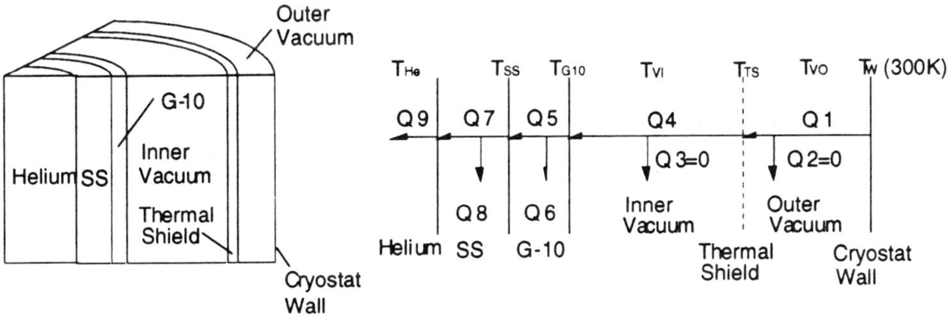

Figure 3. Calculation model (left) and heat flow (right) for the cooling pipes break simulation
The inner and the outer vacuum is as same as pressure.

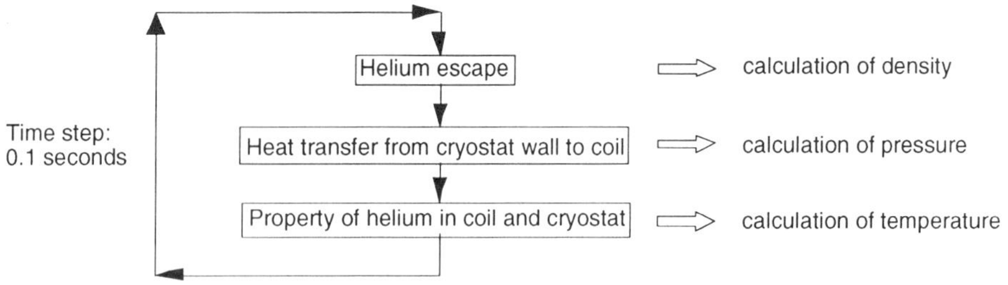

Figure 4. Calculation flow .

$$m = A\sqrt{\frac{2\kappa}{\kappa - 1} p_0 \rho_0 \left\{ \left(\frac{p_1}{p_0} \right)^{\frac{2}{\kappa}} - \left(\frac{p_1}{p_0} \right)^{\frac{\kappa+1}{\kappa}} \right\}} \tag{2}$$

where κ, p, ρ, A and m mean specific heat ratio, pressure, density, nozzle cross section and mass flow rate, respectively. The subscripts 0, 1 mean in the coil and at the nozzle, respectively.

When the helium in the coils is two phase, the helium flow can be regarded as the mixture of the liquid and vapor helium. The average density and the specific heat ratio κ are obtained from liquid and vapor of the helium. By using Eq. (2), the mass flow rate can be calculated from the above averages.

Solution

Numerical method. The one dimensional heat conduction equations are solved with the forward difference numerical method as shown in Figure 4. The mass flow rate and

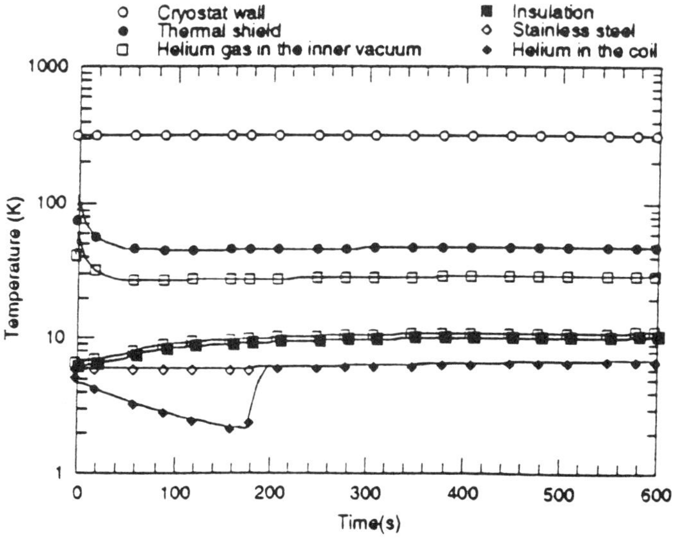

Figure 5. Temperature profiles in the cryostat on Case A .

the heat transfer of helium are simultaneously solved under the condition of the mass balance of helium and the energy balance of all material. The helium pressure in the cryostat are calculated from temperature and density of helium. The time step of differential equation is automatically subdivided when the helium properties are at the helium phase transition.

Assumption and boundary conditions. The following assumptions and the boundary conditions are considered in the numerical analyses;

1) The cooling paths within a coil and the cryogenic system are neglected. The coil is assumed to consist of one volume.
2) The cryostat temperature is constant at 300 K.
3) The weight of the thermal shield is ignored.
4) Pressure of discharged helium gas in the inner vacuum is as same as that in the outer.
5) The heat capacity of discharged helium gas is assumed to be negligible.
6) Temperature of the helium gas in the inner vacuum is the average of the insulation temperature and the thermal shield temperature.
7) Temperature of the helium gas in the outer vacuum is the average of the thermal shield temperature and the cryostat temperature.

INPUT DATA

A TF coil has 28 cooling pipes which consist of 14 inlet pipes and 14 outlet pipes. Many pipes are installed in the cryostat. Pipes for each coil should be grouped for simple piping location. It is assumed that the guillotine break of the cooling pipes occurs at the inlet or the outlet grouped pipes. Other data about the TF coils and the cryostat are obtained from the reference design[1]. The following cases are calculated for the parametric survey;

Case A: 14 broken pipes of one TF coil, reference volume of cryostat vacuum,
Case B: 28 broken pipes,
Case C: half of vacuum volume,
Case D: helium inventory of 20 TF coils,
Case E: current dump.

The input values for this simulation are listed in Table 2.

ANALYSIS RESULTS

Temperature Rise

Results of Case A, B, C, D. The temperature profiles on the Case A, B, C and D are very similar. The temperature profiles on Case A are shown in the Figure 5. The thermal shield temperature (T_{TS}) falls during helium leakage because the escaped helium gas cools the thermal shield. Helium gas temperature in the vacuum (T_{VI}) rapidly decreases because T_{VI} is average between T_{G10} and T_{TS}. The temperature of the coil case (T_{SS}) slowly rises with large heat input from the insulator because of its huge mass. Helium temperature in the coil (T_{He}) decreases during helium leakage because helium adiabatically expands into the vacuum. T_{He} gradually saturates to T_{SS} after 200 s when

Table 2 . The input data for calculation

	Reference	Number of broken pipes	Vacuum volume	Helium inventory	Fast discharge
Case	A	B	C	D	E
Helium inventory(m3)	4.26	←	←	85.2	4.26
Mass of TF coils (tons)	663	←	←	13260	663
Surface of TF coils (m2)	200	←	←	4000	200
Thickness of TFcoils (cm)	5	←	←	←	←
Mass of insulation(tons)	7	←	←	140	7
Surface of insulation(m2)	200	←	←	←	←
Thickness of insulation(cm)	1	←	←	←	←
Distance of outer vacuum (m)	0.1	←	←	←	←
Surface of thermal shield (m2)	2500	←	←	←	←
Height of thermal shield (m)	10	←	←	←	←
Distance of inner vacuum (m)	2	←	←	←	←
Volume of vacuum (m3)	10000	←	5000	10000	←
Cooling inner diameter (mm)	10.6	←	←	←	←
Number of broken pipes	14	28	14	←	←
Fast discharge	none	←	←	←	including

helium mass flow rate fades out. The moderate temperature rise after 200 s indicates that the helium heat capacity is much smaller than the coil heat capacity. The insides of the cryostat warm up slowly by the heat input from the cryostat wall.

Result of Case E. The temperature profiles for Case E is shown in Figure 6. The temperature of the coil heated by Joule heat on current dump reached around 35 K. The temperature of the insulation does not rise, because most of the heat is conducted to the helium in the coils and not to the insulation, have low thermal conductivity. The helium in the inner vacuum and the thermal shield temperatures profiles are similar to the other case.

Phenomena. The convection heat transfer is larger than the conduction by helium gas. So the thermal shield is used for preventing helium gas convection within the cryostat near the wall. Even if the cooling pipes break, the thermal shield can maintain the low temperature of helium gas around the coils .

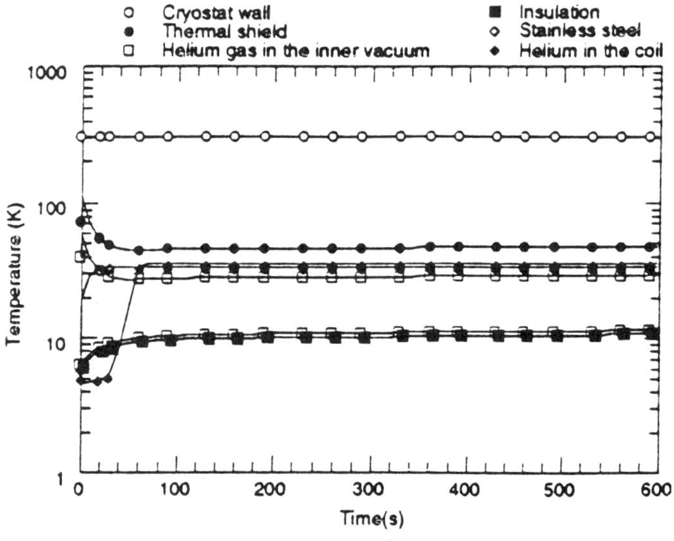

Figure 6. Temperature profiles in the cryostat on Case E .

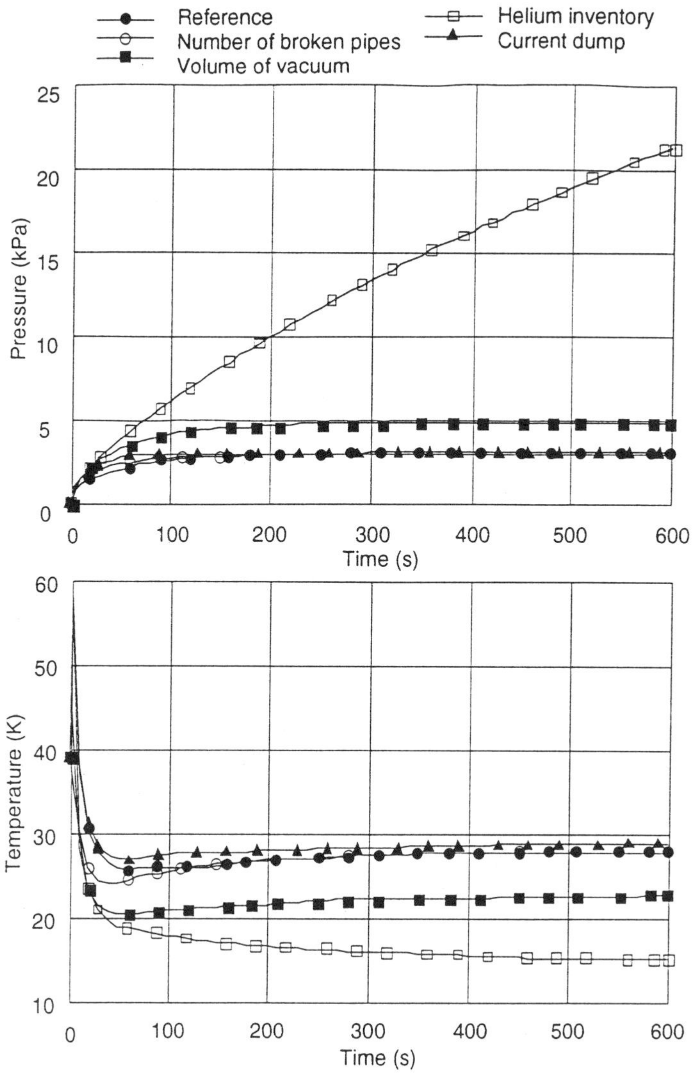

Figure 7. The pressure and temperature profiles of the helium gas in the inner vacuum The upper and the lower figures show the pressure and temperature profiles.

Effect to superconducting coil. At operating condition, current shearing temperature of the TF coil is 6.75 K. Case A, B, C and D indicate that the TF coil may quench after around 600 seconds . The discharge time constant of the TF coil is 15 seconds and therefore, the current of the TF coil may be dropped without quench. So it is expected to avoid the quench and insulation break on the guillotine break of the cooling pipes.

Pressure Rise

The pressure and temperature profiles of the helium gas in the inner vacuum are shown in Figure 7.

Effects of the number of broken pipes (Case B). Doubling the number of broken pipes increases the mass flow rate of the helium from the coil to the vacuum. The temperature profile of Case B is similar to Case A because amount of leaked helium gas into the vacuum is the same. The speed of pressure rise of Case B is faster than that of Case A. However, pressure profile after 100 s for both cases is almost same because the helium mass flow fades out.

Effects of volume of vacuum (Case C). The pressure rising profile of Case C is faster and higher in the case of a half volume of the cryostat vacuum. The mass flow rates in these cases are constant. The convection heat transfer is increased with higher density in the cryostat vacuum. The pressure at 600 s in Case C is 81% of ideal gas expansion because the thermal shield prevents thermal conduction from the cryostat wall.

Helium inventory effects (Case D). All helium inventory of 20 TF coils escapes through the broken pipes of one TF coil in Case D condition. The speed of pressure rise is the most rapidly of all the cases. The pressure at 600 s is the highest of all the cases. The supercritical helium becomes two phase helium as the pressure in the coil falls down to 0.11 MPa. It takes two or three days to warm-up the whole magnet system. This saturated pressure is 0.70 MPa which is 20 times one TF coil inventory. The interlock system will be installed for closing valves of inlet and outlet on each TF coil in order to avoid Case D accident.

Current dump phoneme (Case E). During pipe failure and coil quench, the heat of 1.5 GW from AC loss and Joule heating inputs to coil case and helium. Total energy of 13.4 GJ is consumed in 20 TF coils. Calculation results show that pressure rising profile is the most rapidly during current dump. The pressure at cryostat is saturated to the value of Case C after 100s. Thus, current dump of TF coil does not influence pressure rise of the cryostat.

During the current dump, the temperature of the helium in the coil rises rapidly and reaches around 35 K. It is the same as the coil temperature. The helium pressure rises up to around 0.2 MPa from 0.1 MPa by Joule heating. After the heating, the temperature is kept around 35 K and the pressure decreases slowly.

CONCLUSION

A simplified representation of the ITER TF coils and cryostat has been used to allow numerical simulation of the escape of helium from the coils into the cryostat. The model simulates fluid flow effects as the helium expands into the vacuum as well as heat transfer from the 300K wall by conduction and convection. The model can be used to predict the onset of coil quench as a result of heat flow (assuming that the fast discharge system does not operate) and the worst case pressure rise when many cooling pipes fail simultaneously. It is shown that the TF coils may not quench for 300 seconds or so with a small fault and that the pressure in the cryostat with a large fault is still under 0.5 atmospheres after 600 seconds. This preliminary assessment suggests the helium leaks are not a major safety issue for ITER.

ACKNOWLEDGMENT

Authors would like to thank Dr. Y. Tanaka and Dr. Shimamoto for their encourage and support during this work. Authors would like to thank Joint Central Team NAKA work site staff.

REFERENCE

1. K. Yoshida et al., Design of the outer poloidal field coil system for ITER, MT-14
2. R. J. Corruccini. "Advances in Cryogenic Engineering **13**" (1967) 353
3. J.P. Holman. "Heat Transfer", McGraw-Hill Book Company, "USA (1976)

PORTABLE NEON PURIFICATION SYSTEM*

R. A. Richardson and R. L. Schmitt

Fermi National Accelerator Laboratory
Batavia, Illinois, 60510-0500, United States

ABSTRACT

This paper describes the principle design features of a portable neon purification system and the results of the system performance testing. Neon gas replaces air in the Ring Imaging Cherenkov (RICH) detector without using vacuum, in experiment E781(SELEX) at Fermilab. The portable neon purification system purifies neon gas by, first purging air with CO_2, freezing the CO_2, then cryo-adsorbing the remaining contaminants. The freezer removes carbon dioxide from a neon gas mixture down to a maximum concentration of 500 parts-per-million (ppm). The charcoal bed adsorber removes nitrogen from neon gas down to a maximum concentration of 100 ppm. The original RICH vessel was designed to hold vacuum but its photo-multiplier tube plates were not.

INTRODUCTION

The neon purification system was initiated as a cost effective means of removing air and replacing it with purified neon without using vacuum, in the phototube Ring Imaging Cherenkov (RICH) detector, for experiment E781 at Fermilab. The RICH vessel is unable to withstand vacuum due to the use of thin window plates containing approximately 2850 quartz windows. Consequently, a small positive pressure will be maintained in the RICH vessel. Experimental data on purifying neon gas by adsorption using activated coconut shell carbon as an adsorbent with our conditions, was not available.

The physics requirement for experiment E781 is to provide charged particle identification during the next fixed target run. Their goals includes high efficiency detection of the generated photons and good ring radius resolution. [1] The presence of oxygen and water vapor contaminants adsorb photons needed for particle identification. Therefore, a pure neon gas was selected because its refractive index is acceptable for particle identification.

---

* Work supported by the U.S. Department of Energy under contract No. DE-AC02-76CHO3000

SYSTEM DESCRIPTION

The neon purification system purifies neon gas to remove O_2 and water vapor contaminants. Initially, the RICH is filled with air. Cylinders of CO_2, with a densities greater than that of air, are introduced into the bottom of the vessel, to purge air out the top of the vessel and discharged to atmosphere. Neon gas is introduced at the top of the RICH and purges CO_2 out the bottom. The freezer is then used to remove CO_2 from the neon, followed by the adsorber which is used to remove all remaining contaminants. Care is taken not to waste the expensive neon gas. The system design objective is to reach a maximum contamination of 500 ppm CO_2 and 100 ppm O_2 within 24 hours.

Components

The major purification system components and their associated piping are housed on a 0.914 m x 1.12 m x 1.80 m portable skid. The system's components consist of (1.) a freezer, used to remove carbon dioxide and water vapor (2.) a heat exchanger, to achieve required temperatures, (3.) an adsorber, to remove oxygen from the neon gas mixture and (4.) the RICH, a 43.6m^3 (1540 ft^3) carbon steel vessel and associated instrumentation. Several 160 liter LN_2 dewars supply the adsorber jacket at 0.25 MPa and 84 K. A 15 hp Corken two stage gas compressor, which compresses and flows neon gas into the adsorber, via a heat exchanger, to pressures of approximately 0.96 MPa.

Instrumentation and Controls

A Fischer & Porter (F&P) controller via a custom valve positioner box, controls the actuation of two Fermi-built control valves. One valve controls the flow of liquid nitrogen

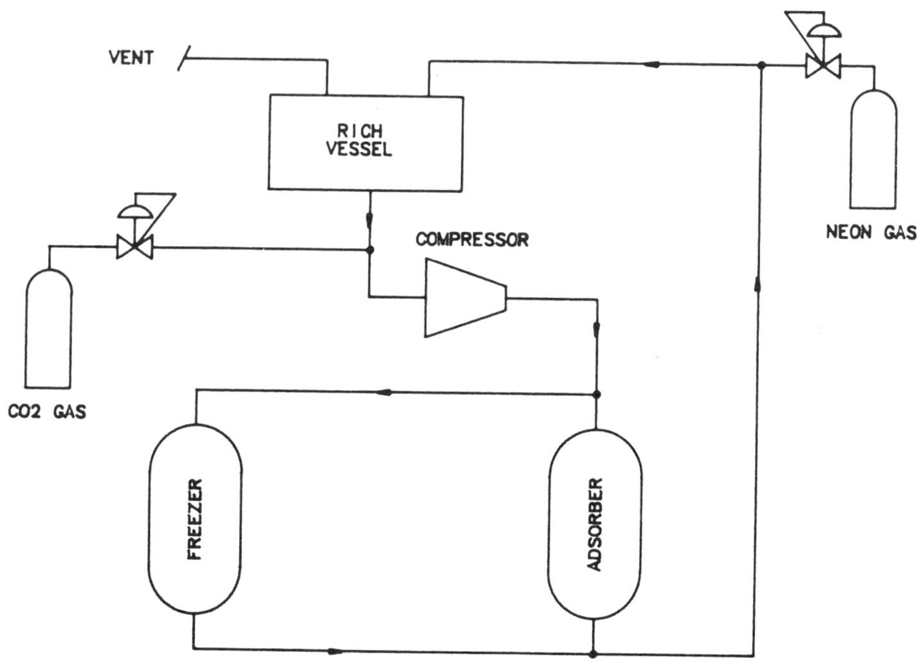

Figure 1. Neon Purification System Flow Schematic.

Table 1. Freezer Design Data

	CO_2 Inlet	LN_2 Inlet
Pressure	0.10 MPa	0.27 MPa
Temperature	294 K	84 K
Fluid Density	1.83 kg/m^3	7.30 kg/m^3
Mass Flow Rate	0.047 kg/s	0.09 kg/s

to the freezer, to achieve and maintain temperatures. The F&P controller monitors and controls the temperature of the freezing process, by the signal it receives from an resistance temperature device (RTD). The second actuator valve controls the level in the adsorber liquid nitrogen jacket. The F&P controller also monitors the temperature of the adsorption process from the signal it receives, from an RTD, located inside the charcoal bed. The controls instrumentation, as well as the Rosemont CO_2 and Teledyne O_2 gas analyzers, are located on an instrumentation rack, near the purification skid.

THEORETICAL DESIGN RESULTS

Freezer

The density of CO_2 is greater than that of air so, initially CO_2 will be introduced from the bottom of the vessel to purge air from the top of the RICH, to atmosphere. Neon gas is introduced into the top of the RICH to purge CO_2 out the bottom. CO_2 and neon enters the compressor, then the freezer, where CO_2 and water vapor are frozen out. The freezer was designed as a counter flow heat exchanger using the number of transfer units (NTU) and log mean temperature difference methods (LMTD). It is a 0.305 m OD pressure vessel constructed of 304 stainless steel. It's volume is sized at 0.141 m^3 (5 ft^3), contains 0.203 m OD and 0.254 m OD helical wound copper cooling coils (125 turns each) and a 0.152 m OD excluder which directs the flow of gas. The freezer is insulated with 0.076 m (3 in) of Armaflex foam insulation. This unit was designed to remove carbon dioxide, of a minimum equal to one-half the vessel volume 21.8 m^3 (770 ft^3) from the neon gas mixture to a maximum concentration of 500 ppm and to operate at pressures and temperatures below 1.34 MPa and 193K, respectively. Table 1 contains parameters used in the design of the freezer.

Adsorber

The gas flowing during the process enters downward, to minimize adsorbent movement and pressure-drop through the bed, into the charcoal bed under isobaric and isothermal conditions. The bed design is critical to the adsorption process. In this fixed bed adsorber design, the most critical factors to size were the required volume of adsorbent and gas velocity, which typically result in height to diameter ratio of about 2:1 and velocities between 0.15 m/s and 0.50 m/s based on an empty vessel. In this design the pressure-drop of the gas flowing through the bed was calculated using the following generalized pressure-drop correlation given by

$$\frac{\Delta P}{L} = \frac{f C_t^2 G^2}{\rho D_p} \tag{1}$$

Table 2. Adsorber Design Data

Volumetric Flow Rate	1529 slpm
Gas Flow Velocity	849 slpm
Operating Pressure	0.96 MPa
Operating Temperature	77 K
Neon Gas Density	31.1 kg/m^3
Initial Contamination Level	3% N_2
On-Stream Time	10 hours
Bed Pressure Drop	0.12 MPa
Bed Height to Diameter Ratio	2
Required Desiccant Area	0.167 m^2
Required Volume of Adsorbent	0.51m^3

where C_t is the pressure drop coefficient, D_p is the effective particle diameter, f is the friction factor, G is superficial mass velocity, L is the distance from the bed entrance, ΔP is pressure drop, ρ is fluid density and $\Delta P/L$ is the pressure drop per unit length of the bed. [2] The adsorber design data is listed in Table 2 .

The resulting adsorber is a 0.458 m O.D. pressure vessel, constructed of 304 stainless steel. It's volume is sized at 0.153 m^3 (5.4 ft^3), it is filled with 64 kg of type PCB (4X8 mesh) activated coconut-shell carbon, contains four 600 watt heaters for adsorber regeneration and is equipped with an outside liquid nitrogen jacket. The adsorber is insulated with 0.076 m (3") Armaflex foam insulation. This component was designed to remove a minimum of 1.31 m^3 (46.2 scf) of nitrogen (3% of RICH volume) from a neon gas mixture down to a maximum concentration of 100 ppm, operate at a temperature of 77K, a pressure of 0.96 MPa (139.7 psi) and a gas flowrate of 1529 slpm (54 scfm). Several 160 liter LN_2 dewars supply the adsorber jacket at 2.5 atm and 84 K. A 15 hp Corken two stage gas compressor compresses and flow neon gas into the adsorber, via a heat exchanger, to pressures of approximately 1.06 MPa.

EXPERIMENTAL PERFORMANCE RESULTS

Freezer

Performance testing was conducted to determine the performance characteristics of the neon purification system freezer. The objective of this test was to determine the amount of pure carbon dioxide the freezer could freeze. The first test was conducted using a small 0.934 m^3 (33 scf) vessel. The test results indicated that the neon purification system freezer's capacity is 29.7 m^3 (1048 scf) of pure carbon dioxide, at an operating pressure of 0.96 MPa (140.7 psi), temperatures below 193 K and a gas flowrate of 339 slpm (12 scfm).

A final performance testing was performed whereby, the actual RICH vessel, filled with air, was purged with carbon dioxide gas. Neon gas was then introduced to purge one-half the carbon dioxide volume out the vessel, before starting the compressor. To begin the freezing process, liquid nitrogen was allowed to flow through the cooling coils, activating the freezer. The RICH was maintained at pressures up to 0.114 MPa (14.9 psi).
The compressor compressed and circulated the gas mixture to an average system pressure of 9.6 atm, which stabilized the system at an average flowrate of 425 slpm (15 scfm) and freezer temperatures below 123 K (-238 ^oF). The freezer removed about 22.54 scm (800 scf) of CO_2, from the Ne-CO_2 mixture, down to 300 ppm. The performance data is presented in Figure 2.

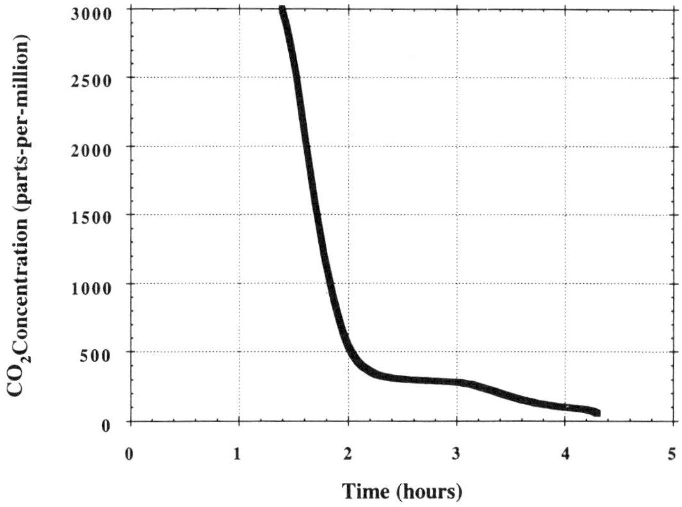

Figure 2. Freezer Purification Rate.

Adsorber

Initially, the adsorber capacity was determined via testing. The RICH was maintained at pressures between 1.03 and 1.14 atm. Several 160 liter LN_2 dewars were used to supply the adsorber jacket at 0.25 MPa (36.7 psi) and 84 K. Initially, the jacket was filled to the 75% level. The gas compressor was used to compress and flow neon gas into the adsorber, via a heat exchanger, to pressures of approximately 1.06 MPa. Due to the unavailability of an O_2 analyzer, nitrogen gas analyzer was used. N_2 is more difficult to remove, from neon gas, than O_2. The adsorption process was monitored and the system operated until the adsorber capacity, the breakthrough point, was reached. The breakthrough point was the point at which the adsorber became saturated, which occurred at 131 ppm N_2 (by volume) in less than 3 hours, after 16.1 m^3 (570 scf)of N_2 was added.

The Teledyne O_2 analyzer was used during the final adsorber performance test to monitor the O_2 content in the neon gas. The adsorber removed oxygen from neon gas in the RICH to a contamination level of less than 2 ppm. The average system pressure was

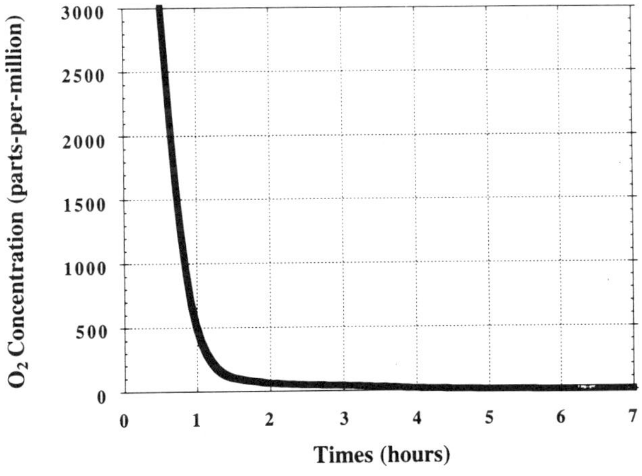

Figure 3. Adsorber Purification Rate .

0.99 MPa (144.7 psig) and flowrate of 510 slpm (18 scfm). Figure 3 represents the rate at which the adsorber removed O_2 from neon.

COMPARISON OF RESULTS

The freezer was designed to freezer 21.7 m^3 (770 ft^3) of CO_2 from neon gas, at a pressure less than 1.34 MPa (194.7 psig) and temperatures lower than 193 K, to a maximum contamination level of 500 ppm. The initial performance test results indicate that the freezer is capable of freezing (1048 ft^3) of pure CO_2. The final test results, show that the freezer is capable of freezing out CO_2 from neon gas at an average pressure of 0.97 MPa (141.7 psi), at temperature below 123 K and at an average flowrate of 425 slpm (15 scfm) to a concentration level of 300 ppm.

The adsorber was designed to remove a minimum of 131 m^3 (46.2 scf) of nitrogen from a neon gas mixture down to a maximum concentration of 100 ppm, operate at a temperature of 77 K, a pressure of 1.01 MPa and a gas flowrate of 1529 slpm (54 scfm). The adsorber capacity was determined to be 570 scf of N_2. As a result of the final performance testing, the adsorber removed oxygen from neon gas to a contamination level of less than 2 ppm at an average system pressure was 0.99 MPa (144.7 psi) and flowrate of 510 slpm (18 scfm). The portable purification system was successful at freezing out CO_2 to less than 500 ppm CO_2, followed by cryo-adsorbing the remaining contaminants to a concentration level less than 2 ppm O_2.

ACKNOWLEDGMENT

The authors acknowledge the advice of James Kilmer, as well as the technical efforts of James Loskot, Greg Johnson, Byron Rodewalt and Gerry Szabo.
The work was performed at Fermi National Accelerator Laboratory, which is operated by Universities Research Association, Inc. under contract No. DE-AC02-76CHO3000 with the U.S. Department of Energy.

REFERENCES

1. Pommot Maia, M., Cooper, P. S., Stutte, L., Solyanik, V., Filimov, I. and, Nemitkin, A., *A Phototube RICH Detector*, Fermi National Accelerator Laboratory, FERMILAB-PUB-92-155, (June 1992).
2. Kohl, A. L. and Riesenfeld, F. C., "Gas Purification", Gulf Publishing Company, Houston (1985).
3. Kotchenda, L.M. and Schmitt, R.L., *Neon Filling Collection System for RICH - Cherenkov Detector*, Petersburg Nuclear Physics Institute, Preprint 1888 EP-19, (1993).

A CONTINUOUSLY COOLED MECHANICAL SUPPORT FOR CRYOGENIC EQUIPMENT

K. Iyengar and S.W. Van Sciver

FAMU/FSU College of Engineering
2525 Pottsdamer St.
Tallahassee, FL 32310, U.S.A.

National High Magnetic Field Laboratory
1800 E. Paul Dirac Dr.
Tallahassee, FL 32310, U.S.A.

ABSTRACT

A scheme is presented for interception of the heat flowing into a cryogenic environment, through its mechanical support. In the present method, the mechanical support is cooled continuously by means of a coolant-carrying tube, wound in a helix around the support. A numerical simulation of the present support cooling method has been conducted. An experimental study of this support-cooling method is proposed.

INTRODUCTION

Cryogenic systems operated at low temperatures are supported by structural members, which transmit the exerted mechanical loads to the ground or to other load-bearing environments, which are at room temperature. The difference between the temperatures of the cryogenic equipment and the load-bearing environment imposes a steep temperature gradient across the mechanical support, resulting in a large quantity of undesirable heat being conducted from the ground into the cryogenic equipment, via the mechanical support. The heat leaking into the cryogenic equipment necessitates removal by the refrigeration system, thus significantly increasing the required refrigeration capacity.

Increasing the thermal resistance of the mechanical support suggests itself as a means of reducing the heat leak into cryogenic apparatus. This implies supports with large length, small cross-sectional area and low thermal conductivity. However, these parameters may not be adjusted beyond certain limits, imposed by considerations such as rigidity, buckling, vibration, mechanical strength, fatigue resistance, corrosion resistance, and cost as applied to the mechanical support.

Since the coefficient of performance (COP) of refrigerators decreases with decreasing temperature of the refrigerated space, removal of the heat leak in its entirety at the low temperature of the cryogenic apparatus is extremely inefficient as compared to interception of the heat leak at some intermediate temperature along the mechanical support. However, intermediate heat interception causes steeper temperature gradients at the warm end of the support, resulting in increased heat inleaks into the warm end, and thus increasing the total amount of heat to be removed. It is therefore necessary to optimize the intermediate heat interception process, so as to minimize the total refrigeration work.

VARIOUS SUPPORT COOLING METHODS

Some of the practical support cooling methods studied are illustrated in Fig. (1) and Fig. (2). The hot and cold ends of the support are, respectively, at temperatures of T_H and T_L. The heat inleak at the hot end is $\dot{Q}_H$, and the heat current conducted through the support is $\dot{Q}$. The heat removed at the cold end of the support is $\dot{Q}_L$. Fig. (1a) shows a support cooled by a single cooling station located at the cold end. Fig. (1b) shows a support cooled by multiple discrete cooling stations, placed along the length of the support, at locations indicated by L_1, L_2, etc., and maintained, respectively, at temperatures of T_1, T_2, etc. The heat removed by each cooling station is indicated by $\dot{Q}_1$, $\dot{Q}_2$, etc. The continuous methods of cooling the support are shown in Fig. (2). In Fig. (2a), the support is cooled by means of a surrounding envelope of flowing coolant, such as helium gas. The coolant is injected at the cold end of the support, at a mass flow rate of $\dot{m}$, and flows counter to the direction of the heat current in the support. The support is cooled continuously by means of convective heat transfer to the coolant. In Fig. (2b), the support is again cooled by means of an envelope of counter-flowing coolant, with the envelope now consisting of two or more streams of coolant in series, with mass flow-rates of $\dot{m}_1$, $\dot{m}_2$, etc. The first of the multiple streams is injected at the cold end of the support, while the remaining streams are injected at suitable locations along the strut.

Various studies have been conducted to optimize the cooling of the mechanical supports, so as to minimize the refrigeration work. Bejan and Smith[1] analytically calculated the optimal heat current distribution along the length of the support. They suggested cooling by means of a continuous distribution of infinitesimal refrigerators, placed along the length of the support, or a counter-flow coolant envelope, with a continuously variable mass-flow rate, as possible methods to theoretically achieve this optimum heat current distribution in the support. The discrete cooling station method is optimized[2,3] in terms of the temperatures and locations of the stations, for a specified number of cooling stations. The counter-flow coolant envelope method is optimized in terms of the mass-flow rate of the coolant, while the multiple-stream coolant envelope method is optimized in terms of the mass-flow rates, and the location and temperatures of injection of the coolant streams.

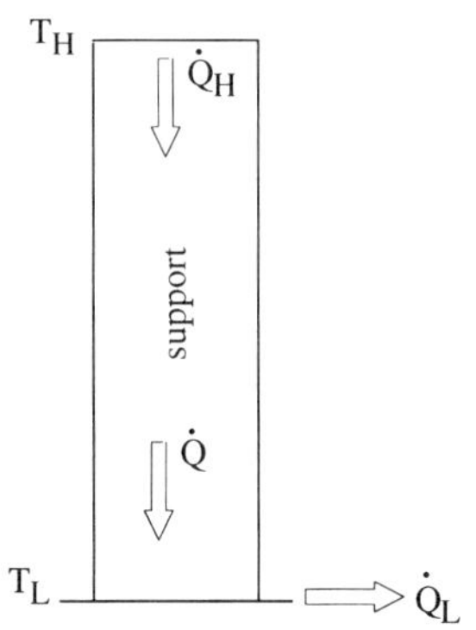

(a) Single cooling station at cold end

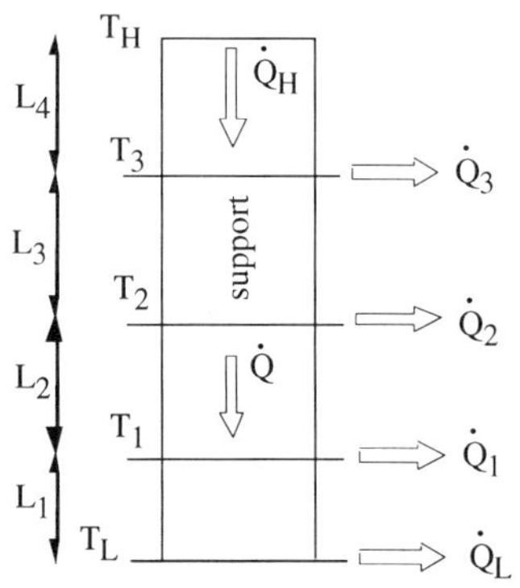

(b) Multiple discrete cooling stations

Figure 1. Discretely cooled mechanical supports.

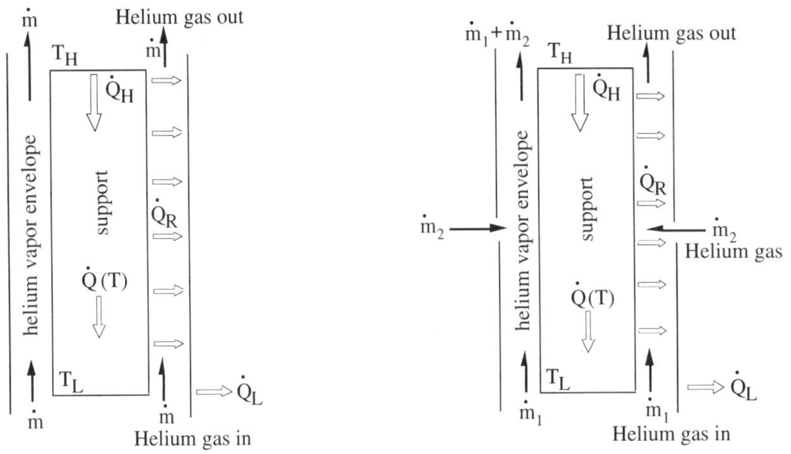

(a) Envelope of single counter flow stream of coolant.

(b) Envelope of multiple counter- flow streams of coolant in series.

Figure 2. Continuously cooled mechanical supports.

The continuously distributed cooling method, with the optimum heat current distribution in the support, involves the least refrigeration work, and represents the best method of support cooling. However, this method is not practically achievable. The other cooling methods are less efficient, but easier to implement.

In the present method, it is proposed that the mechanical support be cooled continuously by means of a coolant-carrying tube, wound in a helix around the support, as shown in Fig. (3). The helium gas coolant, which is injected into the helical tube near the cold end of the support, at a mass flow rate of m, flows in the tube, and exits the tube near the hot end of the support. The support is cooled over its length by convective heat transfer to the coolant. The envelope cooling method can be thought of as a special case of the helical coil cooling method, with the pitch of the helix being equal to zero for the envelope cooling method.

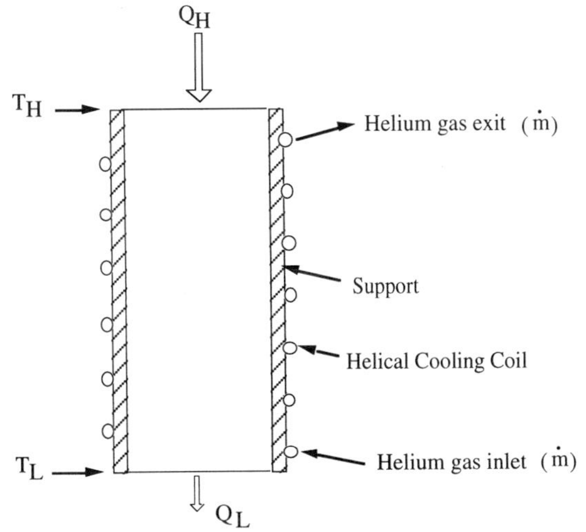

Figure 3. Proposed method for cooling of mechanical support, by means of a coolant-carrying tube wound in a helix around the support. The mass-flow rate of the helium gas coolant is $\dot{m}$.

COMPUTATIONAL STUDY

The cooling of the mechanical support by means of the helically coiled coolant-carrying tube has been computationally studied. A FORTRAN code has been written to solve for the steady-state temperature distributions in the strut and the coolant stream, as well as the other state properties of the coolant fluid. The contact resistance between the outer wall of the cooling tube and the surface of the helical groove on the strut is neglected, as is the thermal resistance across the thickness of the wall of the cooling tube. Thus, the contact between the coil and the strut is assumed to be perfect.

The time-dependent, three-dimensional heat equation in cylindrical co-ordinates governs the temperature distribution in the interior of the strut.

$$\rho C_p \frac{\partial T}{\partial t} = \frac{1}{r}\frac{\partial}{\partial r}\left(kr\frac{\partial T}{\partial r}\right) + \frac{1}{r^2}\frac{\partial}{\partial \phi}\left(k\frac{\partial T}{\partial \phi}\right) + \frac{\partial}{\partial z}\left(\frac{\partial T}{\partial z}\right) \tag{1}$$

The steady-state properties of the helium vapor (coolant) are governed by the following equations.[4]

$$\frac{dp}{ds} = \frac{\dfrac{1}{\rho D}\left(-2fG^2 - \dfrac{4qG\beta}{\left(C_p + u^2\beta\right)}\right)}{1 - \dfrac{G^2(\kappa - \beta\phi)}{\rho}} \tag{2}$$

$$\frac{dT_f}{ds} = \frac{\left(\dfrac{4q\left(1 - \kappa G^2/\rho\right)}{GD\left(C_p + u^2\beta\right)} - \dfrac{2fG^2\phi}{\rho D}\right)}{1 - \dfrac{G^2(\kappa - \beta\phi)}{\rho}} \tag{3}$$

$$with\ \phi = \frac{\mu_j C_p + u^2\kappa}{C_p + u^2\beta} \tag{4}$$

$$q = h\ (T_s - T_f) \tag{5}$$

where : p is the fluid pressure, T is the temperature at a point in the interior of the strut, T_s is the temperature of the strut at a point of contact with the cooling coil, T_f is the corresponding fluid temperature at this point of contact, u is fluid velocity, ρ is fluid density, s is the distance along the flow-path, D is diameter of flow channel, $G = \rho u$, f is the Fanning friction factor, μ_j is the Joule-Thomson co-efficient, C_p, κ and β are, respectively, the specific heat at constant pressure, the bulk compressibility, and the bulk expansivity of the helium, and q is the convection heat flux.

The boundary conditions governing the temperature distribution in the strut are the insulated boundary on the inner cylindrical surface of the strut, the prescribed temperatures at the top and bottom faces of the strut, and the combination of prescribed heat flux and convection heat transfer on the outer cylindrical surface of the strut. The flux on the outer cylindrical surface is everywhere prescribed to be zero, except at those portions of the surface, which are in contact with the helically-wound cooling tube. The portions of the outer cylindrical surface in contact with the coolant tube are cooled by convective heat transfer between the coolant and the strut, and the appropriate boundary condition here is a heat flux whose magnitude is given by Eq. (5). The initial condition for Eq. (1) is a prescribed temperature distribution. Eq. (2) and Eq. (3) are a pair of coupled ordinary differential equations. The boundary conditions for Eq. (2) and Eq. (3) are, respectively, the prescribed pressure and temperature of the helium gas at the inlet of the flow. Eq. (1) is coupled to the system of Eq. (2) and Eq. (3) via the convection boundary condition Eq. (5).

NUMERICAL SOLUTION PROCEDURE

A finite difference method is employed in conjunction with a Runge-Kutta method to solve, respectively, the governing partial and ordinary differential equations. The cylindrical domain of the strut is divided discretely in the radial, azimuthal and axial directions, while the fluid domain is divided discretely in the direction of the flow-path. As a starting point, a suitable temperature distribution, consistent with the prescribed end temperature boundary conditions, is assumed for the strut. Similarly, the temperature, pressure and other property distributions are assumed for the fluid in the direction of the flow-path, consistent with the prescribed inlet fluid properties.

a) The fluid equations Eq. (2) and Eq. (3) are solved simultaneously, using a fourth-order Runge-Kutta algorithm to integrate the system of ordinary differential equations along the flow-path, with the latest known strut temperature distribution being used to generate the convection boundary condition Eq. (5).

b) Next, the time-dependent, 3-dimensional heat equation Eq. (1) is solved in cylindrical co-ordinates, for one time step, using a 3-dimensional alternating-direction-implicit (ADI) finite difference method,[5] with the convection boundary conditions, Eq. (5), being generated by the most recently known fluid condition.

c) The solutions are advanced in time by successively repeating steps (a) and (b) until the solution for a particular time-step converges, within a specified tolerance, to that from the previous time-step. At this stage, the steady-state temperature distributions in the strut and the fluid have been obtained.

RESULTS

The refrigeration work for various support cooling methods was computed. These results are displayed in Fig. (4). All of the cases studied pertain to the configuration of the proposed experiment, which will consist of a brass strut, with 50.8 mm outer diameter, 38.1 mm inner diameter, and 240.0 mm height. The low temperature end of the strut is at 4.5 K, while the high temperature end is maintained at 77 K. Helium vapor with an inlet temperature of 4.5 K is used as the coolant. The pitch of the helical cooling coil is 13.33 mm or 5.55 % of the total strut height. The results are reported in terms of non-dimensional refrigerator power, $\dot{s}\frac{L}{Ak_h}$, versus the non-dimensional mass flow rate of the helium gas coolant $\dot{m}\frac{LC_p}{Ak_h}$, where L is the length of the support, A is the cross-sectional area of the support, k_h is the thermal conductivity of the support material at the temperature of the hot end, C_p is the specific heat of the helium gas at the temperature of the hot end of the support, $\dot{m}$ is the mass flow rate of the helium coolant, and $\dot{s}$ is the rate of entropy generation in the support. Perfect heat transfer between the coolant and the support has been assumed for the envelope cooling method, whereas the Dittus-Boelter correlation is used to compute the local convection heat transfer coefficient in the case of the helical coil cooling method. For the given conditions of the support, and for a given helix pitch, there exists an optimum coolant mass flow rate for the helical cooling coil method, as in the case of the coolant-envelope cooling method, which minimizes the refrigeration work. The results for the helical cooling coil method are comparable to those obtained using multiple discretely cooled stations.

PROPOSED EXPERIMENTAL STUDY

An experimental study has been proposed, in which a brass strut is to be cooled by a helically-wound brass cooling tube, containing flowing helium gas as the coolant. Brass was selected as the material for the strut, in order to maintain the mass flow rate of the helium gas coolant within limits that would be experimentally convenient. In this proposed experimental study, the cooling coil is soldered into a helically-running groove on the surface of the brass strut. The strut assembly is placed inside a vacuum can. The top end of

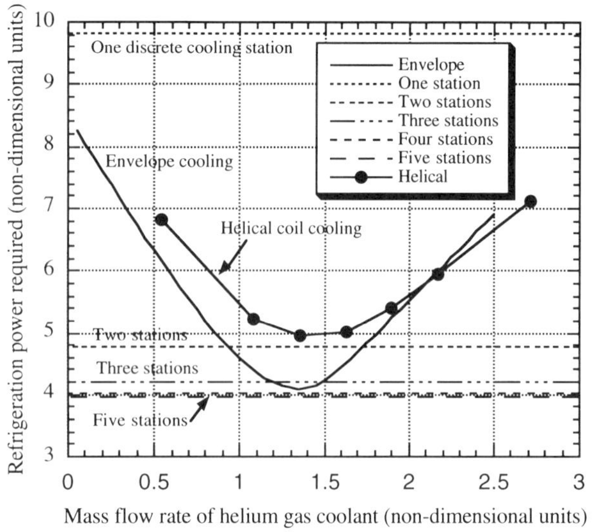

Figure 4. Non-Dimensional refrigeration power for various methods of cooling the support.

the strut will be maintained at an elevated temperature (T > 4.2 K) by means of a heater and a temperature controller, while the bottom end will be maintained at 4.2 K by means of intimate thermal contact with the liquid helium reservoir. The mass-flow rate of the helium gas coolant will be varied by means of a heater, attached to the liquid-helium reservoir. The parameters to be studied include the heat fluxes at the ends of the strut, the mass-flow rate, pressure drop and temperature rise of the coolant.

CONCLUSIONS

Cooling of the mechanical support of a cryogenic equipment by means of a coolant-carrying tube, wound helically around the support, has been numerically studied. For each helix pitch, there exists an optimum mass flow rate of coolant in the cooling tube, which minimizes refrigerator work. The helical coil cooling method seems to be a viable alternative to the method of discretely cooled stations. An experimental study is planned to confirm the numerical results obtained.

ACKNOWLEDGMENTS

Research supported by the State of Florida and the National Science Foundation through NSF Cooperative Grant No. DMR9016241. The support provided by the National High Magnetic Field Laboratory, Tallahassee is gratefully acknowledged.

REFERENCES

1. A. Bejan and J.L. Smith, Jr., " Thermodynamic optimization of mechanical supports for cryogenic apparatus," *Cryogenics (March 1974) p. 158*
2. M.A. Hilal and R.W. Boom, " Optimization of mechanical supports for large superconductive magnets," *Adv. Cryog. Eng. (1977)* ***22****, p. 224*
3. S.Y. Seol, M.C. Chyu, T.M. Kuzay and J.R. Hull, " Minimization of refrigeration power requirements for mechanical supports in cryogenic systems," *Cryogenics (1992) Vol. 32, No. 4, p. 409.*
4. Steven W. Van Sciver, "Helium Cryogenics," Plenum Press, New York, (1986).
5. Dale A. Anderson, John C. Tannehill and Richard H. Pletcher." Computational Fluid Mechanics and Heat Transfer," Hemisphere Publishing Corporation, (1984).

CRYOGENIC SKIRT SUPPORT POST

R. C. Niemann,[1] W. E. Buckles,[2] and B. R. Weber[2]

[1]Consultant
Downers Grove, IL 60516 USA
[2]Superconductivity, Inc.
Madison, WI 53705 USA

ABSTRACT

The cold masses of cryostats having vertical axes, like vertical pressure vessels, can be effectively supported by means of a cylindrical skirt that wraps concentrically around the cold mass. The skirt is a cryogenic support post connected at its upper end to the cold mass and at its lower end to the cryostat vacuum vessel. A heat intercept connection to an intermediate temperature refrigeration source can be employed to control heat leak. The support post consists of a composite; e.g., epoxy fiberglass, cylinder with bolted or thermal interference fit end connections. The support post, being a single element support, simplifies cryostat assembly and alignment. The composite cylinder, with a relatively large diameter, lends itself to structural soundness and stability under both static and dynamic loading conditions. Its relatively long length and intermediate temperature heat intercept allows low heat leak to the cold mass. The details of the design of a cryogenic skirt support post as applied to superconducting magnetic energy storage cryostats are presented. Included are support post fabrication, cryostat assembly, and predicted structural and thermal performance. Fabrication of and operational experiences with a prototype support post assembly are discussed.

The cold masses of cryostats having vertical axes, like vertical pressure vessels,[1] can be effectively supported by means of a cylindrical skirt that wraps concentrically around the cold mass.

The skirt is a cryogenic support post[2,3] connected at its upper end to the cold mass and its lower end to the cryostat vacuum vessel. The advantages of a cylindrical post type support include 1) load carrying versatility, a cylindrical section results in a support that can carry tension, compression, bending, and torsional loads; 2) low heat leak, the use of composite materials with effective heat intercepts results in predictably low heat leaks; and 3) installation ease, the post, involving only a single support member at a suspension point, simplifies installation and alignment.

The cryogenic support post was the major suspension system element of the Superconducting Super Collider magnet cryostats. Extensive experience exists in the construction and installation[4] and performance[5] of such support posts.

DESIGN

The support post consists of a composite cylinder connected at its upper end to the cold mass and at its lower end to the cryostat vacuum vessel. A heat intercept connection to an intermediate temperature refrigeration source can be employed to control heat leak. The goal of the support post design is to select a geometry and a material that results in an assembly that satisfies both the specified structural and thermal design constraints.

Structural

Structural loads include lateral and vertical loads that are applied by the cold mass. Load sources can include static weight; shipping and handling; seismic; and quench, in the case of a cold mass that incorporates a superconducting magnet.

The height and diameter of the support post and the ratios of the thermal path lengths are determined in large part by the cryostat geometry and the conductive heat load constraints. The design optimization of the complete assembly essentially consists of determining the composite tube materials and wall thicknesses. The predicted maximum tensile and shear stresses and an elastic stability constraint determine the optimum value for the wall thickness.

Thermal

In addition to determining the structural performance of the support post, it is critical that an accurate prediction of the conductive heat load be made to each thermal station. Given the support post geometry and the thermal conductivity integral for the composite tube, the steady state heat loads to the top of the post and to intermediate temperature heat intercepts can be predicted.

Connections

Connections between composite tubes and metallic structural and heat intercept fittings can be made using some form of mechanical fastener, chemical bond, or by thermal interference fitting[3] the composite tube between an inner metal disk and an outer metal ring.

Material Selection

Suitable tube materials are structural materials consisting of glass reinforced composites in an epoxy matrix such as NEMA grades G-10, G-10CR, G-11, and G-11CR. These materials are readily available, have well characterized structural and thermal properties, are relatively strong, have low thermal conductivity, and are inexpensive. Recent developments in new fibers for use in composites offer advantages in terms of strength, some in terms of thermal conductivity, some in both. For example, graphite composites can offer greater strength and stiffness than their glass counterparts.

SUPERCONDUCTING MAGNETIC ENERGY STORAGE APPLICATIONS

Superconductivity, Inc. produces superconducting magnetic energy storage (SMES) systems in which electrical energy is stored in a superconducting magnet.[6] The economic justification for SMES utilization is based on the value of avoiding an interruption of the power supply to a critical load. Electric utilities have adopted low resistance delivery systems which have virtually eliminated long-term power outages. Momentary voltage disturbances are increasing, however, in part because utilities are replacing fuses with automatic reclosing circuit breakers which automatically reenergize a line after a fault has cleared. At the same time, companies are using more sophisticated and highly sensitive electronics to increase efficiency and control in critical processes. This equipment is especially vulnerable to momentary voltage disturbances. For these reasons sags and momentary outages in many processes result in serious economic environmental and/or safety consequences.

To ensure that the electrical energy required to avoid these sags and outages is always available in the superconducting magnet, the cryostat must be of a robust design which is capable of operating unattended for periods of many months.[7] The heat load through the cryostat and into the liquid helium must be minimized to reduce the load on the refrigeration system. Since some industrial or utility customers prefer to have the ability to move the system from site to site, the systems are installed in semi-trailers. The cryostat must therefore fit within the trailer. The cryostat support structure must also support the energy storage magnet against dynamic transportation loading.

Design Criteria

Representative suspension system design criteria for the initial cryogenic skirt support post application is as given by Table 1.

The SMES magnet cryostat configuration is as shown by Figure 1.

Table 1. Suspension system design criteria

Design Aspect	Design Value	
Geometry		
• Cold mass diameter	0.76 m	(30 in.)
• Cold mass height	2.17 m	(85.6 in.)
• Cold mass center of gravity height	0.99 m	(39 in.)
Structural		
• Static		
Cold mass weight	829 kg	(1828 lb)
• Dynamic		
- Vertical (3 g handling shock)	2488 kg	(5484 lb)
- Lateral (2g handling shock)	1658 kg	(3656 lb)
Thermal		
• Heat intercept temperature	60 K	
• Required heat load		
- To liquid helium (4 K)	2 W maximum	
- To heat intercept (60 K)	5 W maximum	

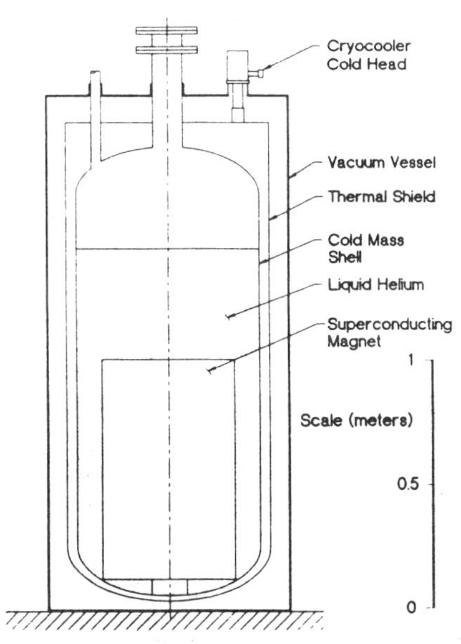

Figure 1. SMES magnet cryostat configuration.

Design Assumptions

A single section skirt supports the cold mass relative to the vacuum vessel.

The skirt connection to the cold mass is at the elevation corresponding to the overall center of gravity of the cold mass; i.e., 0.99 m (39 in.) to the outer bottom of the vacuum vessel. This was done to control the bending moment at the helium vessel-skirt structural connection.

Material is an epoxy fiberglass composite with properties similar to G-10.

A skirt nominal diameter of 0.86 m (34 in.) provides 0.05 m (2 in.) space to the helium vessel and 0.08 m (3 in.) space to the vacuum vessel to accommodate multilayer insulation, thermal shield, and assembly clearances.

Analysis

The design methodology is to 1) determine minimum thickness required for structural purposes, 2) estimate heat loads to helium and the heat intercept with skirt thickness and shield intercept location as variables, and 3) develop a design that satisfies performance requirements, is reliable, is simple in nature and lends itself to a straightforward cryostat assembly.

The minimum skirt thickness required is determined by the axial and bending loads at the base of the skirt. Compressive stressing of the skirt governs. A maximum allowable axial compressive tube stress of 138 MPa (20,000 psi) was employed. The resulting minimum thickness is 0.0013 m (0.052 in.).

Since an 0.86 m (34 in.) diameter x 0.0013 m (0.052 in.) thick cylinder would present fabrication and possible stability problems, thicker walls were considered. The analysis results for a 0.86 m (34 in.) diameter x 0.0032 m (0.125 in.) thick x 0.99 m (39 in.) long skirt are as follow:

- Maximum axial compressive stress 57.4 MPa (8320 psi)
- Axial compression buckling margin (critical vs. applied stress) 84:1
- Bending buckling margin (critical vs. applied stress) 8:1
- External pressure buckling margin 0.5:1. Thus adequate pressure equalization must be incorporated into the design to prevent possible collapse of the tube due to differential pressures.
- The heat loads vs. 60 K intercept height, h, are shown in Figure 2. It should be noted that considerable latitude exists in the location of the 60 K intercept on the skirt; i.e., 0.26 m (10.1 in.) $\leq h \leq$ 0.96 m (37.8 in.), while satisfying the heat load criteria:

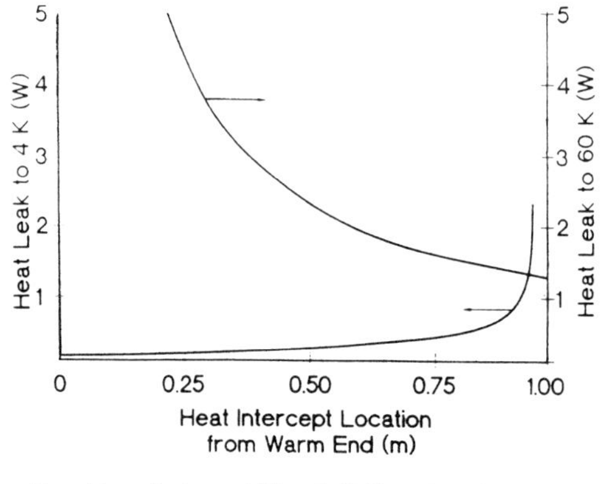

Figure 2. Support skirt predicted heat leaks to 4 K and 60 K vs. heat intercept location from 300 K end of skirt tube.

- For $q_{4\,K} \leq 2.0$ W, $h \leq 0.96$ m (37.8 in.)
- For $q_{60\,K} \leq 5.0$ W, $h \geq 0.26$ m (10.1 in.)

An h of 0.29 m (11.3 in.) was selected as the design value.

- Operation without the heat intercept; i.e., intercept refrigeration system failure, was considered. The heat load to helium is 1.4 W which is less than the 2.0 W requirement.

Connections

The end connections to the composite tube were by means of bolting to structural support rings made of rolled 304 stainless steel angle. The support rings were then connected to the cold mass and vacuum vessel by bolting.

The skirt-heat intercept interface was by means of clamping, by sheet metal forming, of the skirt tube between the upper and lower sections of the cryostat heat shield. The shield is constructed of 1100 aluminum. The connection was configured to reduce thermal resistance along the heat intercept path.

The details of the skirt and its connections are shown by Figures 3, 4, and 5.

Construction

The skirt was a filament wound epoxy fiberglass tube fabricated by a commercial manufacturer. The properties of the tube material were specified to be similar to G-10. The composite consisted of E-glass filaments at a 65° wind angle in an epoxy matrix. A thickness of 0.0064 m (0.25 in.) was required for tube fabrication reasons.

The skirt assembly was then integrated into the cryostat assembly. The assembly sequence was 1) mount skirt to vacuum vessel plate, 2) insulate lower shield, 3) mount lower shield to skirt, 4) insulate cold mass, 5) mount cold mass to skirt, 6) mount upper shield to skirt, 7) heat intercept cold mass piping penetrations to upper shield, 8) insulate upper shield and skirt, 9) mount vacuum vessel shell to vacuum vessel bottom plate, 10) mount vacuum vessel top plate to vacuum vessel shell, and 11) weld cold mass piping penetrations to vacuum vessel top plate bellows assemblies.

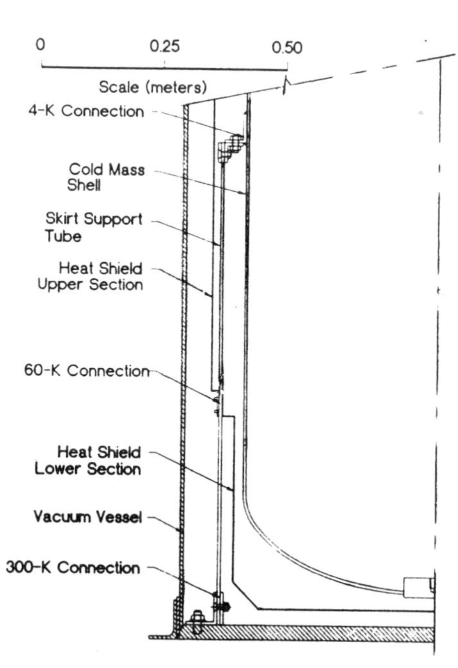

Figure 3. SMES magnet support skirt and connection details.

Figure 4. Support skirt assembly.

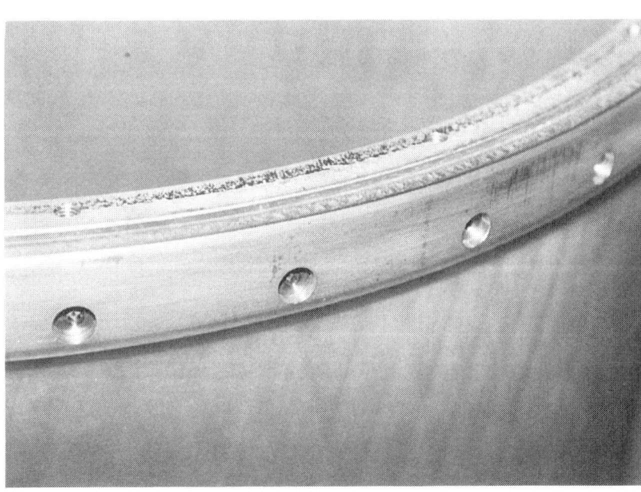

Figure 5. Support skirt 4.2 K connection detail.

Prototype Unit Operating Experience

The prototype cryostat incorporating the skirt support post was extensively instrumented for the measurement of thermal and electrical performance.[7]

The results of the steady state skirt and heat shield intercept temperature measurements are as follow:

- The thermal coupling of the tube to the shield is effective. A ΔT of 9 K corresponds to the difference of the average inside and outside tube temperature (62 K) and the average above and below shield temperature (53 K) at the intercept connection point.
- The thermal coupling of the upper to lower shield sections through the tube is effective. A ΔT of 5 K corresponds to the difference of the shield temperature above the intercept (50 K) and the shield temperature below the intercept (55 K).
- The measured and predicted conduction heat leaks along the tube are in good agreement. The measured heat leaks were analyzed as one-dimensional conduction. Inputs to the analysis were tube geometry, sensor location, sensor temperature, and G-10 warp direction thermal conductivity. The predicted heat leaks were determined from Figure 2. Inputs were heat intercept location and a tube thickness adjustment of two. The measured and predicted heat leaks to the 60 K intercept are 4.2 W and 6.2 W respectively. The measured and predicted heat leaks to the 4 K cold mass are 0.4 W and 0.2 W respectively.

Field Unit Operating Experience

Since 1992, seven cryostats incorporating the original configuration skirt support post have been constructed and operated. The skirt support has proven to be cost effective; straightforward to fabricate and install; has successfully withstood the structural loadings associated with cryostat construction, handling and shipping; and has performed flawlessly with no apparent degradation in performance.

Four cryostats are being constructed that accommodate a larger cold mass. The skirt support post is an epoxy fiberglass cylinder with a 1.05 m (41.5 in.) nominal diameter, a 0.0051 m (0.2 in.) thickness, and a 0.94 m (37.0 in.) length.

COST

The cost breakdown for a cryostat incorporating the original configuration skirt support post are as given by Table 2. The skirt support post cost is 11% of the total cryostat cost.

CONCLUSIONS

Cryogenic skirt support posts provide a straightforward, functional, reliable, and cost effective means of supporting cold masses of cryostats having vertical axes.

The skirt support posts can be readily fabricated with commercially available composite materials and with bolted structural and heat intercept connections.

Table 2. Cryostat cost breakdown

Subassembly	Cost
• Helium vessel	$14 K
• Skirt support post and hardware	5 K
• Thermal shield and utility column	15 K
• Vacuum vessel	13 K
Total	$47 K

The predicted and measured thermal performance are in good agreement.

A 829 kg (1828 lb) cold mass, subjected to both static and dynamic structural loading, can be supported with a single element skirt support measured heat leaks of 0.4 W to 4 K and 4.2 W to 60 K.

Seven SMES system cryostats employing the skirt support post have been constructed and successfully operated in the field while four similar units are under construction.

The skirt support post and associated hardware cost is ≈11% of the total cryostat cost.

ACKNOWLEDGMENTS

The authors gratefully acknowledge the engineering contributions of M. A. Daugherty and the technical support of M. Bordson.

REFERENCES

1. M.H. Jawad and J.R. Farr. "Structural Analysis and Design of Process Equipment," John Wiley & Sons, New York (1984).
2. T.H. Nicol, R.C. Niemann, et al. Design, construction and performance of a post type cryogenic support, in "Adv. Cryo. Engr.," Plenum Press, New York (1985) Vol. 31 pp. 73-80.
3. T.H. Nicol, R.C. Niemann, and J.D. Gonczy. Design and analysis of the SSC dipole magnet suspension system, in "Supercollider 1," M. McAshan, ed., Plenum Press, New York (1989), pp. 637-649.
4. R.C. Niemann. Model SSC dipole magnet cryostat assembly at Fermilab, in "Supercollider 1," M. McAshan, ed., Plenum Press, New York (1989), pp. 747-757.
5. J.D. Gonczy, R.C. Niemann, et al. Cryogenic support thermal performance measurements, in "Adv. Cryo. Engr.," Plenum Press, New York (1987), Vol. 33, pp. 243-250.
6. W.E. Buckles, M.A. Daugherty, B.R. Weber, and E.L. Kostecki. The SSD: a commercial application of magnetic energy storage, *IEEE Trans. Applied Superconductivity* 3: 328-331 (1993).
7. M.A. Daugherty, W.E. Buckles, and R.C. Niemann. Design and operation of a robust cryostat for commercial superconducting energy storage systems, in "Adv. Cryo. Engr.," Plenum Press, New York (1994), Vol. 39, pp. 855-862.

BASIC EXPERIMENTS ON A CRYOGENIC WIND TUNNEL OF FLASHING-BLOWDOWN TYPE

K. Maeno,[1] S. Ozaki,[2] H. Kitajima,[2] and Y. Hanaoka[3]

[1]Department of Mechanical Engineering
Faculty of Engineering, Chiba University
1-33, Yayoi, Inage-ku, Chiba City, Chiba #263 JAPAN

[2]Graduated Course of Department of Mechanical System Engineering
Muroran Institute of Technology

[3]Department. of Mechanical System Engineering
Muroran Institute of Technology
27-1, Mizumoto, Muroran, Hokkaido #050 JAPAN

ABSTRACT

This paper deals with the experimental approach to a new concept of cryogenic wind tunnel of short duration. In the last two decades various closed-cycle and continuous operation (CCCO) cryogenic wind tunnels have been developed and they have shown many advantages for high Reynolds number experiments. These cryogenic wind tunnels, however, suffer from problems of purging and cooldown required before their operations and warming up too. And it is also rather difficult to perform mixing jet experiments by different gases. Therefore, the development of intermittent cryogenic wind tunnels seems to be necessary in the same way that many short duration wind tunnels have been developed for atmospheric temperatures. In this paper we report a flashing blowdown cryogenic wind tunnel developed at the Muroran Institute of Technology [MH(Maeno-Hanaoka)-Cryo.Flash.W.Tunnel]. This cryogenic wind tunnel utilize the rapid flashing phenomenon of pressurized liquid nitrogen under rapid depressurization by a quick opening valve. The operating principal and details of the insturmentation for the pressure and temperature measurement are given. Test results and flow visualization results are presented. The usefulness of this small-scale pilot wind tunnel for large scale apparatus is discussed.

INTRODUCTION

In accordance with the growing demand on improved aerospace transprotation technique, high Reynolds number testing has become an effective and important method for the realization of new vehicle concepts. Among the advanced high Reynolds number testing facilities, cryogenic wind tunnels indicate their specially high potentialitis. There have been variously developed cryogenic wind tunnels of closed-cycle and continuous operation (CCCO) in cooled down modes in these decades[1,2] . Actually they have much advantage in high Reynolds number experiments, and many fruitful results have been obtained. And CCCO cryogenic wind tunnels are still developing to their optimized configuration. These cryogenic wind tunnels, however, have been suffering from the problems of long duration purging and cooldown before their operations, and of heating up and opening to atmospheric circumstances. The consumption rate of nitrogen is inevitably high, and their operations become expensive ones. Furthermore, it is sometimes difficult to perform the experiments on jet mixing problems of different gases with high Reynolds number[3,4] maintaining the constant gas compositions.

There exists a special cryogenic blowdown wind tunnel (Lutwig tube) in Germany, DLR Goettingen. By the cryogenic Lutwig tube of DLR the well-controlled flow can be obtained. The scale is, however, so large for enough heat shielding, and fast actuating cryogenic valve of large scale has to be developed. So the development of intermittent cryogenic wind tunnels is necessary as many short duration tubes and tunnels have been developed in atmospheric or high temperature range.

This paper deals with the experimental approach to a new concept of cryogenic wind tunnel of short dudration, which utilizes the flashing and blowdown phenomena of pressurized liquid nitrogen[5] . A type of single quick piston valve named MO (Maeno-Oguchi) Valve series is employed to develop our Cryogenic Flashing-blowdown Wind Tunnel (CFWT). We named this tunnel as MH (Maeno-Hanaoka) CFWT. The concept and detail of MH-CFWT will be described in the following sections.

FLASHING AND ITS APPLICAT ION TO WIND TUNNEL

Flashing is a well known phenomenon in high pressure fluid technology as nuclear plants. It is the rapid transient bulk boiling from the initially pressurized high temperature equilibrium liquid to the new depressurized equilibrium conditions. Depending on the amount of initial liquid and thermodynamical conditions and the rate of depressurization, the liquid indicates the fast pressure undershoot less than the local pressure equilibrated to the teperature process, then it boils violently to recover almost to the initial pressure, and afterward the pressure shows gradual decay with much slower rate than the gas phase does. During these processes, a large amount of gas can be released. The merit of employing liquid nitrogen flashing-blowdown to cryogenic wind tunnel can be indicated in the following example. Consider the volume of low temperature gas nitrogen of 1m diameter and 60m length at 0.1MPa. It corresponds to about 1m diameter and 0.1m volume of liquid nitrogen (it is about 1/600 in atmospheric pressure). Even at 1MPa it is about 1m length of liquid nitrogen, i.e., even in the range of higher order pressure operation, the apparatus scale can be 1 order smaller than the gas phase. This makes us realize easier cooling and isolating system.

The intermittent operation of cryogenic wind tunnel has no problem for jet mixing experiments, and wide variety of gases can be tested. The problems, however, may be considered in this new flashing wind tunnel scheme, which are for the time dependency of flow parameters and the quality of tested flow. Pressure variation at test section, inhomogeneity in density and temperature should be clarified carefully. The disturbance and turbulent characteristics must be studied. And the temperature boundary layers on the surfaces of model and walls present the strong effects in the experiments.

EXPERIMENTAL APPARATUS

The schematic diagram of our cryogenic flashing-blowdown wind tunnel (MH-CFWT) settled in Muroran Institute of Technology is shown in Fig.1. The dimension of whole apparatus is 3m x 4m with dump tank of 2.5 m^3 and the test section has 0.1m x 0.1m square cross section. The tunnel is made of stainless steel. This wind tunnel utilizes a rapid flashing-blowdown phenomenon of initially pressurized liquid nitrogen under rapid depressurization by the quick opening single piston valve named MO(Maeno-Oguchi)-Valve developed by us. The reservoir of 200mm inner diameter and about 1.2m length is filled by liquid nitrogen, and it is kept silently. Heat intrusion into the reservoir makes it pressurized. The time variations of pressure and temperatures inside the reservoir are shown in Fig.2, from which it is seen that at about 130s we obtain the pressure up to 800kPa. Pressurization range has not been very wide, which ranges from 400kPa to 700kPa, as it is originally handmade, and from the safety

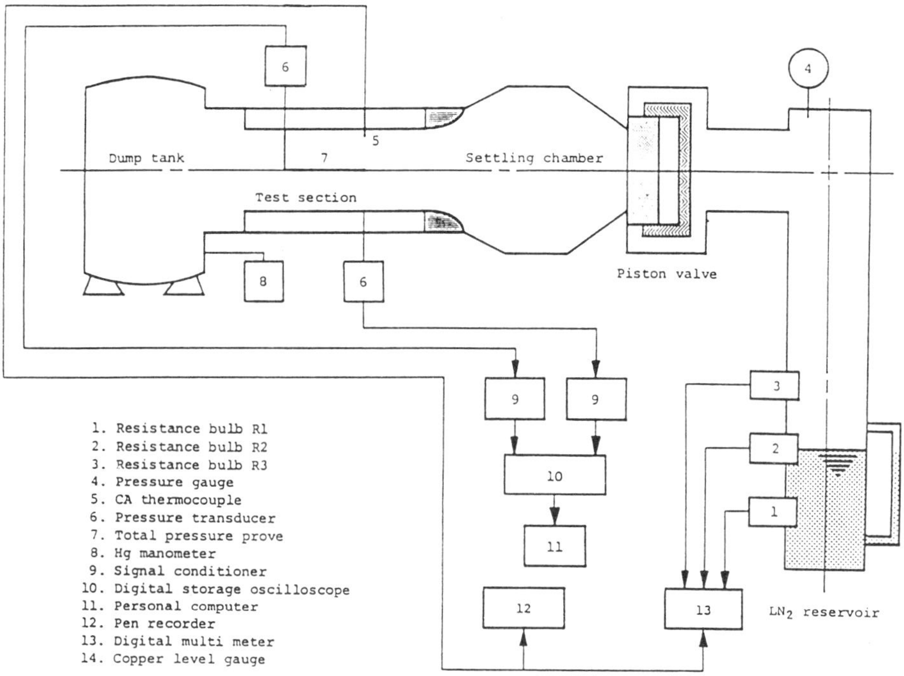

Fig.1 Schematic diagram of flashing-blowdown cryogenic wind tunnel (MH-CFWT).

problems in university the pressurization in reservoir has been carefully improved (designed maximum operation is 2MPa). The reservoir pressure is measured by Bourdon gauge, and the reservoir temperature distribution is monitored by three resistance bulb thermometers.

The total temperature in the test section is measured by CA thermocouple, and the total and static pressures are measured at 40mm from the entrance of test section. We calculate Mach number and Reynolds number based on the length of 0.1m in the test section from these measured values, by assuming the isentropic flow for perfect gas. The viscosity is evaluated by using correlations for nitrogen from the data by Jacobsen

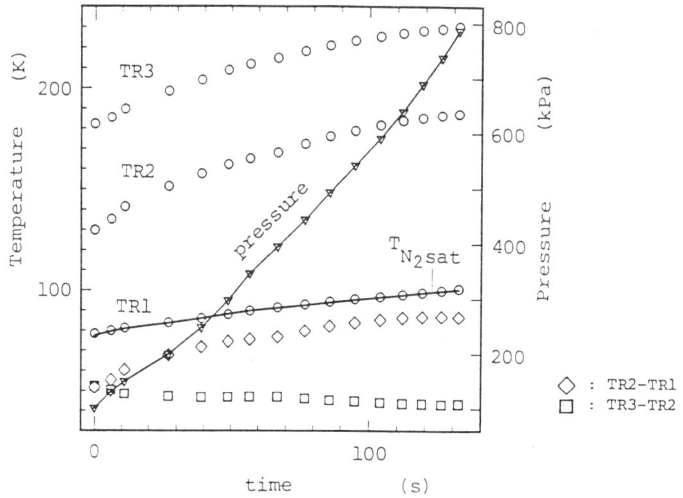

Fig.2 Temperatures and pressure time histories in the reservoir of liquid nitrogen .

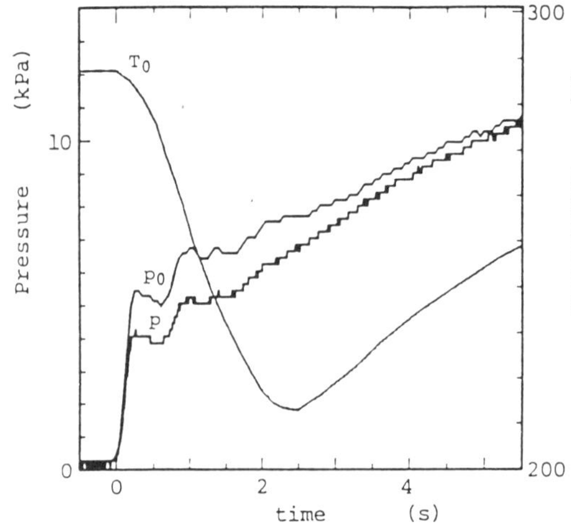

Fig.3 Time variations of total and static pressures and total temperature in the test section.

et al[6] . Flow visualization by color/monochrome schlieren method is also performed with semi-2-dimensional models and CCD camera system.

RESULTS AND DISCUSSION

Among our experimental results obtained, the typical time variations of total pressure, static pressure, and total temperature are indicated in Fig.3. The data show the minimum of temperature time variation, and the pressure vibrations in time. The calculated Mach number and Reynolds number variations are presented in Fig.4. In these figures the reservoir pressure measured is 493kPa. It is remarked that the Reynolds number of 1 x 10^5 has been attained for Mach number range of 0.4-0.6 with our low reservoir conditions. And the flow varies with time, and it continues several seconds. The total pressure in the test section, however, cannot exceed 10kPa in this case.

If we increase the mass of liquid nitrogen and reservoir pressure, the flow conditions in test section become more satisfactory. Figures 5 and 6 indicate such effects, where in Fig.5 the minimum total temperature in the test section is plotted with respect

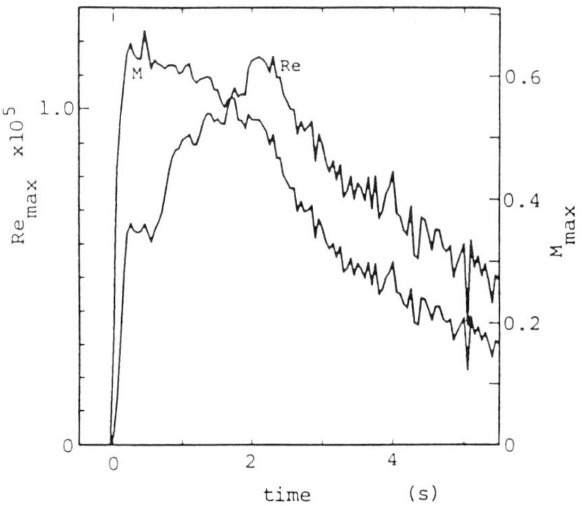

Fig.4 Time variations of Mach number and Reynolds number estimated in test section.

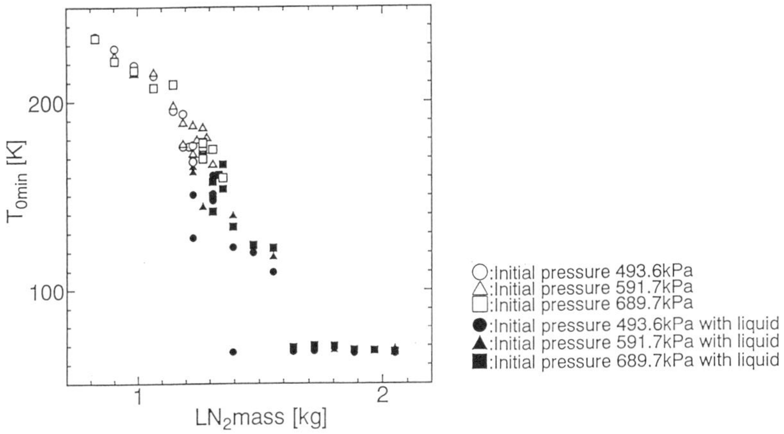

Fig.5 Minimum total temperature in the test section with liquid nitrogen mass in reservoir.

to the nitrogen amount in the inital reservoir. With increasing the reservoir liquid nitrogen, the total temperature in the test section decreases, while the mist flow with liquid droplets begins to be measured, and condensation effects become more dominant in higher range of liquid nitrogen mass. In Fig.6 the total pressure in test section at the minimum temperature is plotted with repect to minimum total temperature. The saturation curve of nitrogen is also indicated in the figure. From the figure it is seen that with increasing the liquid nitrogen amount the flow in test section approaches its saturation conditions. So we find the necessity of heating up of the cold nitrogen flow before entering the test section or of some kind of heat/mass addition into the process.

One of the results of improved scheme can be seen in Fig.7, where the total pressure at minimum total temperature is plotted with liquid nitrogen mass. The data are taken under the normal temperature gas injection conditions into the liquid nitrogen in reservoir. The total pressure indicates the increasing tendency as liquid nitrogen mass is increased, where no mist flow is observed. Figure 8 indicates the maximum Reynolds number evaluated from experimental data with respect to the liquid nitrogen mass in reservoir. The Reynolds number ranges over 2 x 10^5 with about 700kPa reservoir pressure of about 1.3kg liquid nitrogen.

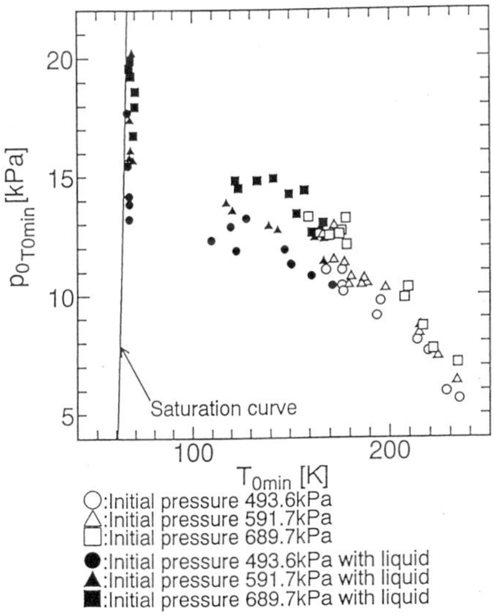

Fig.6 Total pressure at minimum total temperature vs. minimum total temperature.

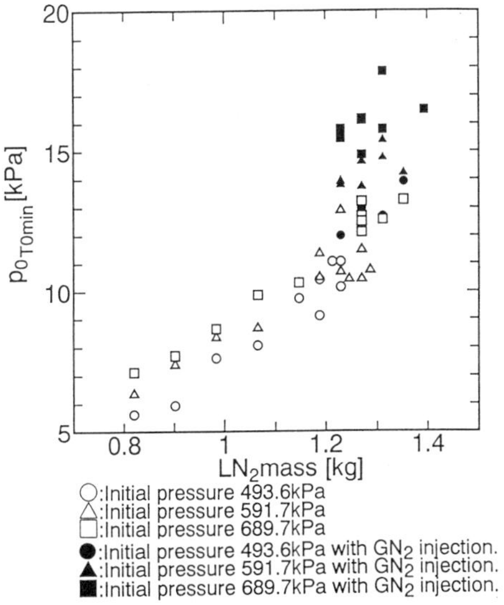

Fig.7 Total pressure in test section at minimum total temperature with nitrogen mass.

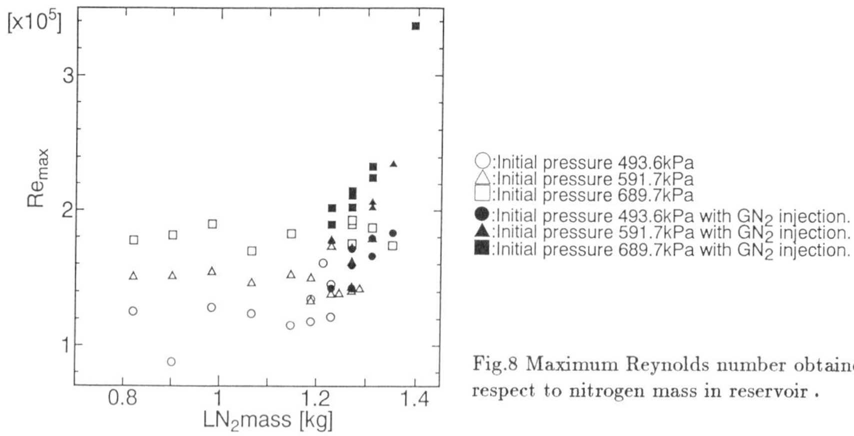

Fig.8 Maximum Reynolds number obtained with respect to nitrogen mass in reservoir .

We also performed the flow visualization by schlieren technique. The schematic diagram of visualization is shown in Fig.9. The dimensions of the model of semi-2-dimensional type are indicated Fig.10a. The shclieren photograph obtained is shown in Fig.10b. The outer flow observed is quite homogeneous in density from this photograph, it varies, however, with time. The flow separations from upper and lower ends of wedge surface can be observed clearly and the development of boundary layer on the surfaces is also seen. There exists the strong temperature boundary layer on the normal temperature surface in the cryogenic flow, which makes us possible to obtain such a clear schlieren photograph in low pressure and small model conditions. This temperature boundary layer, however, produces a different flow field from expected conditions, so precooling of the model surface may be necessary.

CONCLUSIONS

We have developed a new type of short duration cryogenic wind tunnel by flashing-blowdown of pressurized liquid nitrogen. Total and static pressures and total tempera-

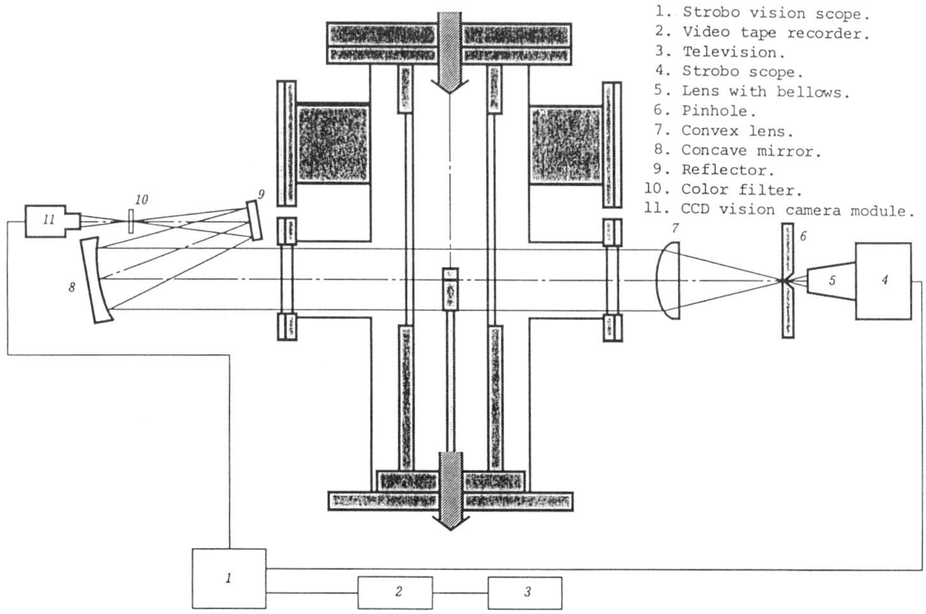

Fig.9 Schematic diagram of flow visualization by schlieren method.

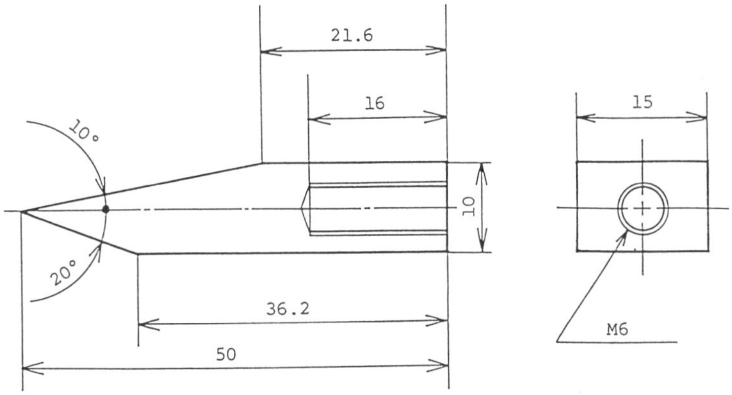

(a) Semi-2D sharp wedge model

(b) Schlieren photograph (originally colored)

Fig.10 Model tested and schlieren photograph obtained.

ture of the flow in test section are measured to estimate flow Mach number and Reynolds number. Flow visualization by color/monochrome schlieren technique is also performed. Liquid nitrogen of 1-2kg up to 700kPa can produce the Mach number 0.1-0.8 flow of 10-20kPa in our test section. Maximum Reynolds number ranges over 2 x 10^5 for L = 0.1m and duration ranges 1 to 4 seconds in our configuration. Measured total and static pressures and temperature are varying with time. This is the typical characteristics in short duration wind tunnels. This, however, is also mainly originated from our low amount of superheated liquid nitrogen in flashing process. With increasing the reservoir liquid amount and pressure, mist flow and condensation are observed. It is necessary to have the combination of heat source or additional gas nitrogen injection at suitable location in the process of flashing blowdown.

Though our apparatus is a small scale pilot tunnel, it can produce the fruitful results on cryogenic intermittent wind tunnel. Extrapolating our experimental results, with 0.1m x 0.1m test section, about 20kg of 5-10MPa liquid nitrogen can produce the flow of 1-2 seconds with Reynolds number of 3 x 10^7/m under suitable combination of heat exchanging system.

REFERENCES

1. R.A. Kilgore, M.J. Goodyer, J.B. Adcock, and E.E. Davenport, The cryogenic wind-tunnel concept for high Reynolds number testing, *NASA TN D-7762* (1974).
2. C.L. Ladson and E.J. Ray, Evolution, calibration, and operational characteristics of the two-dimensional test section of the Langley 0.3-meter transonic cryogenic tunnel, *NASA TP-2749* (1987).
3. K. Asai, Hot-jet simulation in cryogenic wind tunnels, *NASA RP-1220* (1989).
4. K. Asai and T. Aoki, Experiments to evaluate hot-jet simulation capabilities in cryogenic wind-tunnel testing, *AIAA Paper, No.92-0567* (1992).
5. T. Fukuda, Y. Hanaoka, and K. Maeno, Trial manufacture of intermittent cryogenic wind tunnel, *CRYO Newsletter, No.8* (1992) p.3.
6. R.T. Jacobsen et al., Thermophysical properties of nitrogen from the fusion line to 3500R for pressures to 150,000psia, *NBS TN-648* (1973).

ENGINEERING ASPECTS OF THE EUROPEAN TRANSONIC WINDTUNNEL

Patrick Wagner

European Transonic Windtunnel GmbH
Ernst-Mach-Str.
51147 Cologne, Germany

ABSTRACT

This paper will review several aspects of the ETW design which are unique and have been specially developed for this application.

The design of the high speed leg components (Fixed Contraction, Supersonic Nozzle, Test Section, Model Carts, Second Throat, High Speed Diffuser) was based on very stringent specifications regarding deflection of flow liner under thermal and pressure induced loads. ETW specifications are of a very high standard in terms of flow quality and call for a very stiff structural arrangement, however ETW has as well been designed for a high productivity and flexibility. The need to perform cooldowns from ambient temperature to less than 100 K in three hours and to perform short range temperature changes (± 40 K) in a few seconds without affecting the dimensional stability of the whole high speed leg has led to very specific design features. The tight control requirements of remotely actuated walls and model support had also to be integrated in the design. This paper gives an overview of the different subsystems involved.

The concept selected for housing the drives and actuators is to have all actuation systems in thermally conditioned boxes (about 50 actuators altogether). A number of design aspects had to be looked at such as the thermal break between warm and cold environment, and various heating and cooling concepts have been selected depending on the local constraints.

INTRODUCTION

A windtunnel can be divided in several packages. One of the most critical to meet the tunnel ultimate performances is the one formed by the high speed leg components.

When we try to list all the requirements called for by a windtunnel with very stringent specifications in terms of flow quality, efficiency, flexibility and economic criteria we realize immediately that the type of design necessary for the high speed leg components to meet all these requirements has to be a trade off between conflicting specifications.

While a stiff and play free type of design is required on one hand to keep the tunnel geometry stable under all types of loads (thermal, aerodynamic), a rapid thermal stabilization time, minimum cold mass and aerodynamic considerations requires on the other hand light structures with thin flow plates and stress relief design details. The level of accuracy to which

Advances in Cryogenic Engineering, Vol. 41
Edited by P. Kittel, Plenum Press, New York, 1996

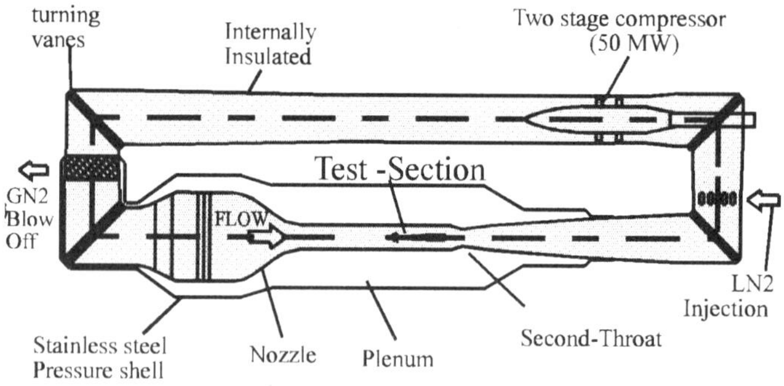

Fig. 1. ETW Aerodynamic Circuit.

flow surfaces shall be positioned requires the machining of flow surfaces over large dimensions which might conflict with the type of material selected to meet the strength, fatigue and material stability requirements at cryogenic conditions. The necessity to protect actuators from the cold environment conflicts with the criteria for controlling the flow plate positions with high accuracy and fast reaction time.

The high productivity requires the model to be easily accessible which means remotely removable from the tunnel. Still a very high position accuracy of the model relative to the flow and the flow plates relative to each other are necessary to compare with the standards of measurements in already existing conventional tunnels.

This paper will give some indications on how some of these design aspects have been looked at and in which way the initial years of operation at ETW validate the selections made.

SHORT DESCRIPTION OF THE ETW FACILITY

A number of papers have already described the facility, its aerodynamic specifications, its model handling system, its control system and some calibration results. This is well documented in existing reports. We can briefly recall that ETW has a closed, pressurized, internally insulated aerodynamic circuit (fig. 1).

The high speed leg is the section where the gas is accelerated to its nominal Mach number value. The model is located in the Test Section supported by a structure which can be moved out of the tunnel: the Model Cart. Upstream of the Test Section the flow is accelerated via a Fixed Contraction and a Nozzle where the contour can be adjusted from a fully convergent

test section dimensions	**2.4 m x 2.0 m**
Mach number range	**0.15 - 1.3**
operating pressure range	**1.25 - 4.5 bar**
operating temperature range	**90 - 313 K**
maximum Reynolds number	**50 million**
productivity	**3 to 6 runs / day**

Fig. 2. ETW specifications.

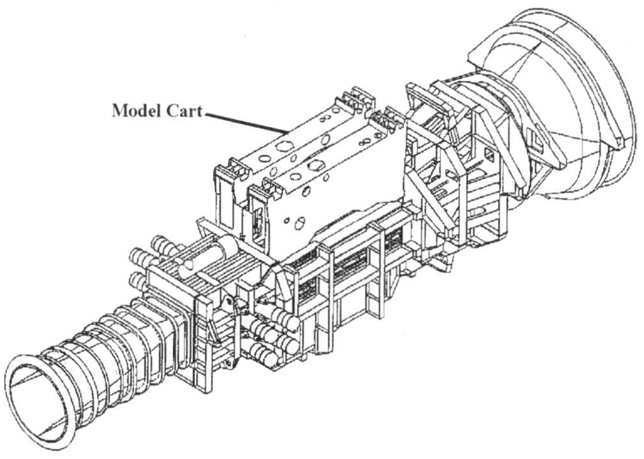

Fig. 3.
High speed leg with a Model Cart.

contour to supersonic contours. Downstream of the Test Section we find the Second Throat and the Diffuser. Fig. 2 indicates the ETW operation conditions.

This paper will concentrate mainly on the high speed leg components which forms a 35 meter long structure, weighing 750 t of stainless steel. It comprises about 50 actuators in thermal conditioned boxes. The movable structure (the Model Cart) weighs 200 t and is remotely moved in and out of the Test Section with a 0.1 mm accuracy. Fig. 3 and 4 show an isometric view of the high speed leg with and without a Model Cart.

DESIGN DEVELOPMENT

Design Difficulties

A major difficulty which ETW had to face from the beginning was the limited experience amongst suppliers being capable of handling major contracts in the field of cryogenic engineering. Many of the potential contractors were very reluctant to invest in costly research and development with limited future applications for other projects and customers. The company NFM (Neyrpic Framatome Mecanique) located in Lyon (France) was considered as having the design and manufacturing capabilities and was awarded the contract for the high

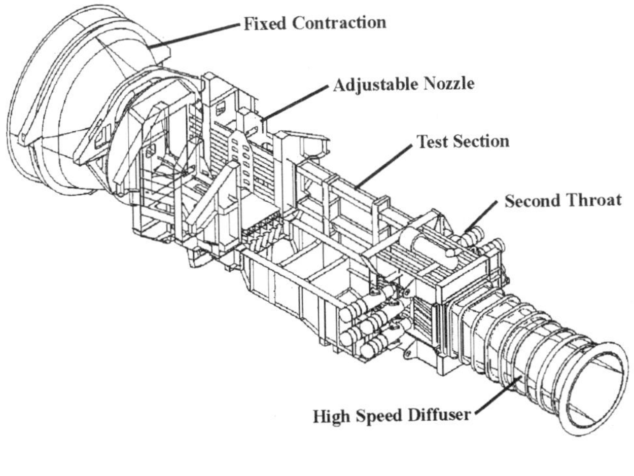

Fig. 4.
High speed leg view inside flow path.

speed leg internals. The functional specifications written by ETW were used as a basis for the design requirements.

Already at the stage of the very initial conceptual design it was recognized that a number of components would be at the limit of the technical know-how: spindle/nut arrangement working in cold conditions (size, lubrication, backlash free arrangement, temperature gradient effects); rollers for the main model support (size, loads, temperature gradient effects); ball-joints and sliding surfaces coated with teflon; temperature effects not very well known on some grades of steel (fatigue curve, stability, weldability).

Developments have been undertaken either to adapt existing products or to design new concepts. For instance the model pitching movement is done through the displacement of an arc of a sector. This is a massive piece of steel (16 t) rolling on roller tracks and moved by an actuator via a spindle and a nut. A major difficulty was the lubrication of the spindle under cold conditions to prevent seizing with the nut. A number of tests in cryogenic conditions have shown that with a specific surface treatment and a permanent lubrication with teflon pads of the spindle during its movement the difficulty could be overcome. The main rollers of the sector were as well specially developed to work in cold conditions under full load at maximum speed. This concept was selected against sliding pads. The technological risk was higher with rollers but the risk of having stick-slip effect on a sliding pad was considered as detrimental to the model support system function and for this reason it was not selected.

However, apart from these considerations on design details the initial analysis concentrated on a global modelization of the facility and primarily the high speed leg structures, to try to get an understanding of the deformation of the structures induced by a cooldown or a warm-up of the flow.

Tolerances - Dimensional Stability

The flow liner plates in the high speed leg had to be positioned inside the tunnel within 0.1 to 0.2 mm of design position. It was aimed as well to try to keep the deflections within the same range of accuracy under aerodynamic and thermal loads. The thermal loads are induced not so much by the temperature absolute level but by the gradients within the components as discussed later and by components of different thermal expansion coefficients.

To achieve the tolerance requirements the flow surfaces were decoupled from the backup structures as much as practicable, only for the fixed contraction and the diffuser this concept could not be applied.

Typical examples are shown on fig. 5, 6 and 7. The plates are pressed against a sliding pad with an accurately calculated spring washer arrangement to keep a permanent pre-stress whatever the tunnel conditions might be. Ball joint connections have been used to allow the thermal growth to occur in all directions and to allow movements of moving surfaces. On the

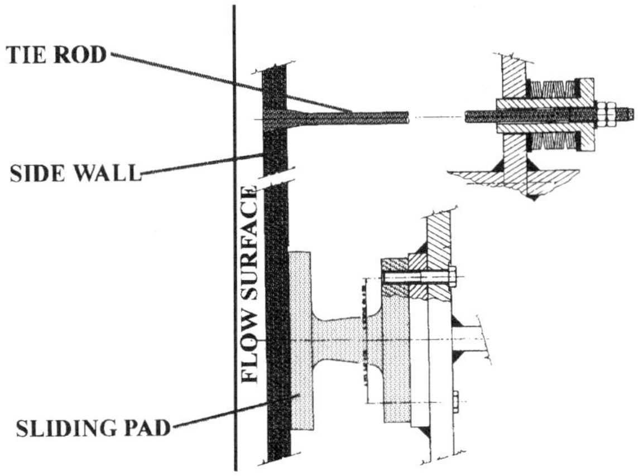

Fig. 5. Typical flow plate attachment.

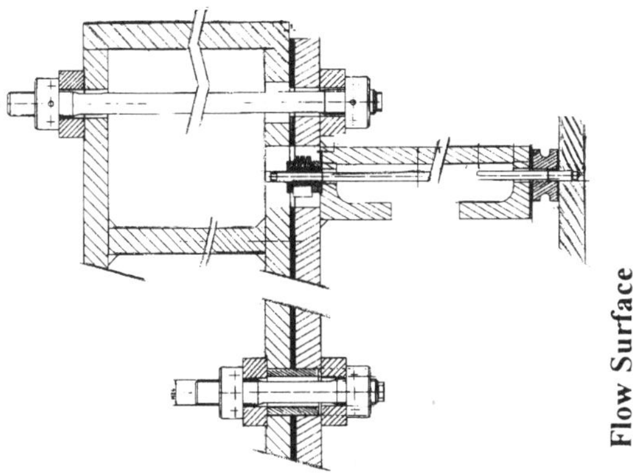

Fig. 6. Attachment between structure components.

whole complete assembly only one fixed point was selected, the flow plates being free to expand relative to this point.

The various concepts have proven until now their validity. The unknown is how the sliding surfaces will wear after a number of cycles and if the spring washers will keep their initial characteristics. Regular checks during maintenance will provide the necessary survey.

The important conclusion is that all flow surfaces could be initially very accurately positioned and this accuracy is kept at all pressure and temperature conditions.

Temperature Conditioning of the Tunnel. Flexibility of Tunnel Operation

ETW has been designed for very rapid temperature changes. A typical scenario is presented in fig. 8 and 9. This requirement has been expressed by potential clients at a very early stage of the project. The test programs of the first campaigns, in which comparisons are being made with low Reynolds number windtunnels, show that cooldowns and warm-ups are requested almost daily. Although variations of flow temperature will be less frequent in normal production testing the behavior of the tunnel during rapid temperature changes will still be an important factor in productivity. The possibility to perform temperature steps of ±40 K around a nominal value within a few seconds without affecting the flow quality provides for a very flexible means of exploring Reynolds number effects, thermal boundary layer effects and flow visualization experiments.

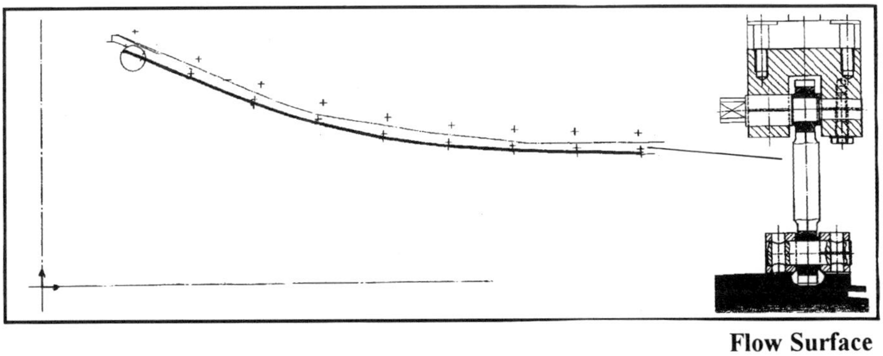

Fig. 7. Attachment of supersonic nozzle flexible plate.

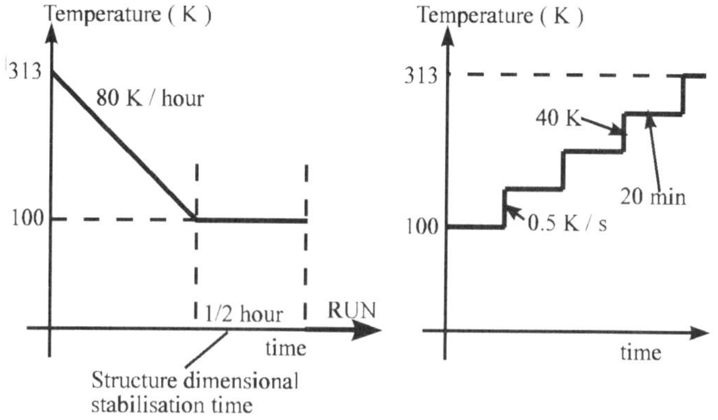

Fig. 8. Scenario for tunnel temperature conditioning.

A complete cooldown of the facility can be achieved in 3 hours plus ½ hour for stabilization of the structures.

As already shown in fig. 2 and 3 the structures have been designed as open frame structures having the advantage of a large exposed surface, very stiff construction, minimum weight to reduce energy consumption. However, the only way to achieve the rapid temperature conditioning was to force gas from the main stream to flow through the structures. This is achieved via remotely controlled pneumatic actuators fixed in the upstream bulkhead of the plenum. By opening these valves part of the main flow (around 10%) is forced into the plenum and reenters the main flow again in the test section. The objective is to maintain the gas temperature of the plenum identical to the main flow temperature. The temperature difference between the structures and the surrounding gas is then kept as high as possible to drive the cooldowns and warm-ups in an efficient way. The speed with which the gas flows through the plenum increases as well as the rate of heat exchange. Calculations have shown that without the bulkhead valves the tunnel temperature conditioning requirements of ETW would be impossible to meet and that the thermally induced stresses would be higher than allowable.

As a consequence, the type of structure selected has a much more complex shape than conventional beam structures. The available space within the shell was as well a difficult constraint on the sizing of the components.

As mentioned before about 50 actuators are integrated in the structure, the very high loading on the flow plates combined with the sometimes very limited space available within the framework has created major design problems. One of the most difficult areas was the "re-entry" located at the downstream end of the test section where the top wall of the test section finishes, the "finger flaps" of the re-entry start and the model support structure penetrates inside

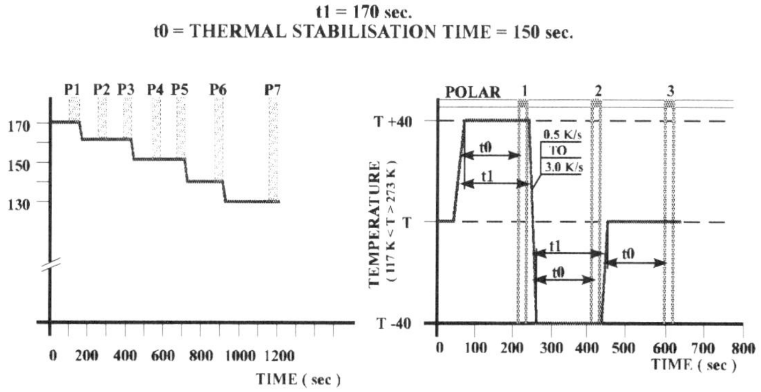

Fig. 9. Short term gas flow temperature changes.

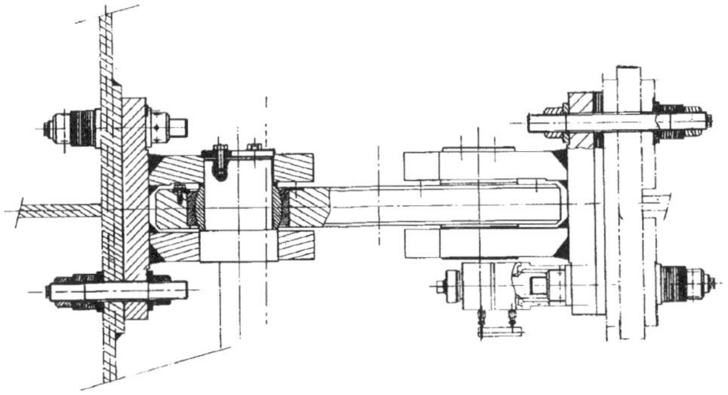

Fig. 10.
Typical bearing attachment.

the wall and the flaps at high incidence of the sector. In this section the design had to cope with 6 different movable components, all remotely actuated and in a very sensitive area for the flow.

To cope with thermal gradients within the structure induced by the large time constant difference amongst the components (different thickness, different heat exchange coefficients) many design details had to be incorporated. Fig. 10, for instance, shows how ball bearings had to be used for all attachments, the behavior of these bearings and the wear of the coating is one of the aspects to be carefully looked at in the future.

Fig. 11 gives a typical example of a bolting arrangement between structures, the pre-stress on the washers is regularly checked. Fig. 12 gives a typical sealing arrangement between the top and side walls of the flow channel.

Example of a Cooldown at ETW

Cooldowns are performed on a regular basis at ETW. The structures of the high speed leg are equipped with a number of thermocouples and the temperatures are monitored throughout the cooldown.

Fig. 13 indicates the status at the start of a cooldown with a flow temperature at 290 K. The internals are initially in equilibrium with the gas. Fig. 14 shows how the various components have changed their temperature after 2 hours of cooldown (at a Mach number maintained between 0,2 and 0,25) and 0,5 hour of temperature stabilization. The flow temperature has reached 180 K. During the cooldown the bulkhead valves are permanently open.

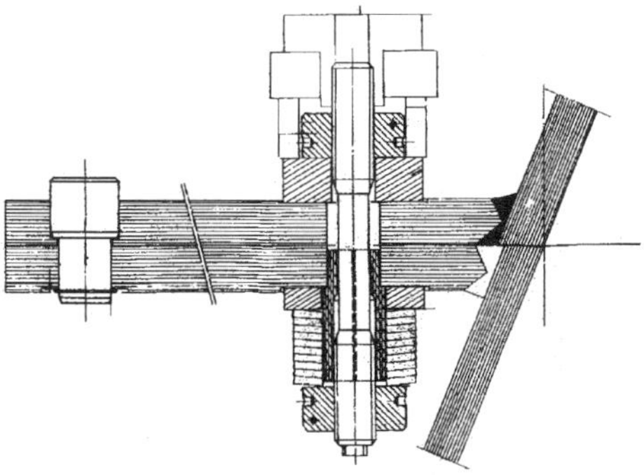

Fig. 11. Framework bolting concept.

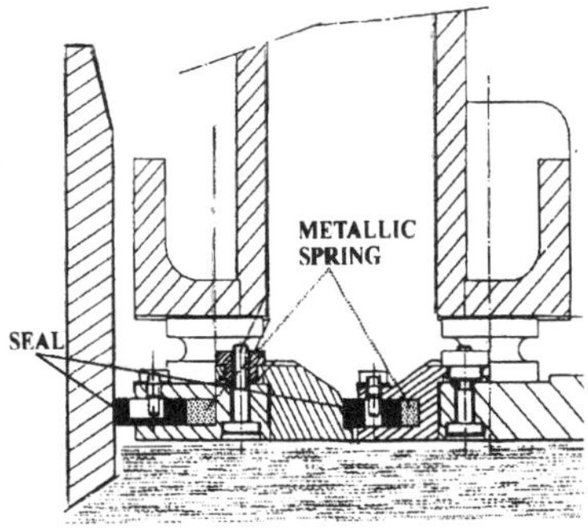

Fig. 12. Concept for sealing.

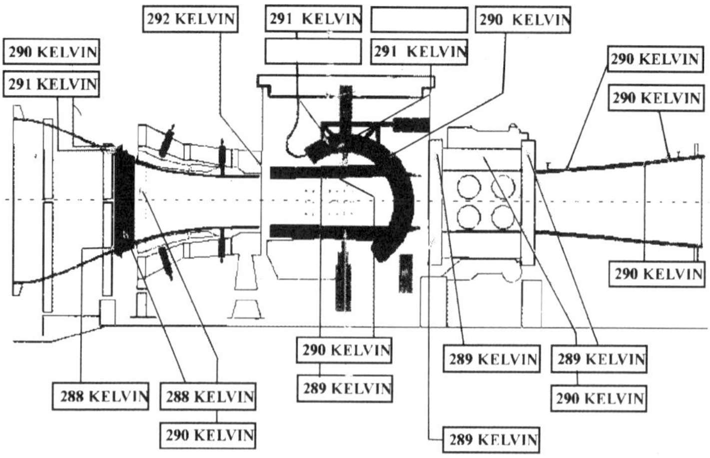

Fig. 13. Start of a cooldown, initial gas temperature 290 K.

If we look at the temperature of the structural elements of the internals, they are within 20 K of the gas temperature, this is the case for the complete Fixed Contraction, the nozzle structure and the test section walls. The second throat and the diffuser have also followed the cooldown despite the fact they are far away from the bulkhead valves. The second throat structures are however slightly lagging behind the average temperature.

We can notice some temperatures higher than the average but they are measured on thick holding pins and not on the main framework. The sector itself which is a thick component has followed the cooldown (189 K).

These results are very encouraging to prove the validity of the analysis made on the global behavior of the internals. It has been noticed as well that by opening the bulkhead valves the flow mixing in the plenum is strong enough to prevent temperature stratification.

DESIGN OF TEMPERATURE CONDITIONED ENCLOSURES

Design Philosophy

The ETW internals have a large number of actuators (about 50). All of them are electrically driven, hydraulic actuators have been totally excluded from the design to prevent

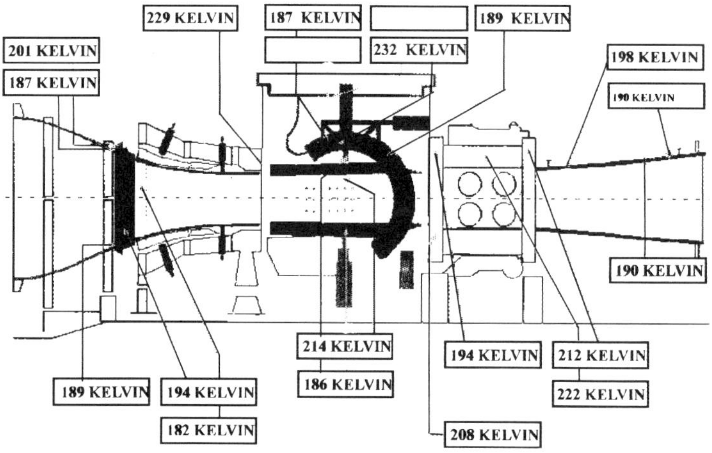

Fig. 14. End of a cooldown with a gas temperature at 180 K.

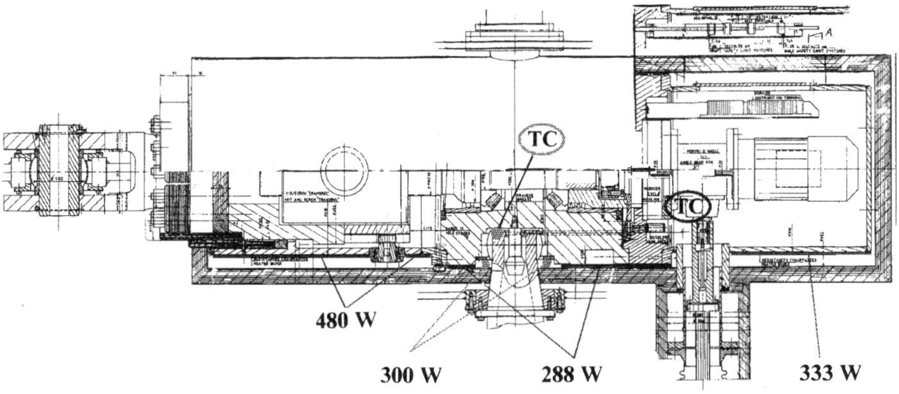

Fig. 15. Actuator heating concept.

oil contamination of the flow. It was decided to select standard equipment normally used for ambient conditions and to protect it from the cryogenic environment. All actuators have connections to the cold environment through their main shaft and they are all trunnion mounted (see fig. 15). The control requirements part of the ETW specifications for the position accuracy of movable components (plates of the Adjustable Nozzle, walls of the test section, second throat, model support, etc.) call for a backlash free design. The thermal gradients, which will inevitably occur between the inside warm actuator and its outside connections, shall then be extremely well under control to avoid seizing which can occur easily on a tight fit arrangement. If the ball bearings external cage supporting the actuator deforms in some sort of oval shape, the actuator drive will not be powerful enough to move the actuator. It is essential for the operation of the facility to size the heaters correctly, determine the most appropriate locations inside the heated boxes, determine where to measure the temperatures controlling the heaters. This problem was one of the most difficult of the design.

The difficulties the design shall consider are:

- Prevent large temperature differences (positive and negative) between the metallic parts of the actuators and the gas inside the box. Heaters shall not either stay on or off permanently for a long period of time.

- The temperature variations in the actuators induced by the heating might create fatigue problems. Heaters shall not be switched on and off at short intervals.

- The actuator box shall not be overheated at non-cryogenic temperature when the tunnel is run at ambient conditions.

- Temperature gradients inside the actuator box shall be prevented (hot on top and cold in the bottom).

- In this environment with components made of different material, different thicknesses, different time constant, it is not obvious where and how to set the temperature alarms (over or under heating).

Design Concepts (Fig. 15)

1. Heaters are organized in clusters so that different heaters have different tasks.

- a "normal heating" which is controlled by an air temperature measurement in the vicinity of metallic masses at a point which is supposed to be warm and will be driven by the inertia of the components around. Its set point (between +10°C and +25°C) shall be tuned after a few cooldowns.

- a "high density heating" with heaters at the point of cold entry into the box, the temperature measurement is made in depth in the steel structure with a low set point (-20°C to 0°C) and at a point supposed to be directly exposed to cold conditions.

- an alarm directly on the skin to protect the heating foil.

2. For a better heat transfer, contact foil heaters have been used, directly glued on the components to be kept warm. The problems of applying the foils on the surface has revealed to be more difficult to solve than originally anticipated. When foil heaters could not be used standard grid heaters have been installed. In some instances standard heaters with a circulation fan have been used. This technology is a lot easier to apply than the one with foils but the temperature control inside the actuator box is a lot more difficult. The foil technique has the advantage of introducing heat directly where it is requested on the actuator.

3. The points of temperature measurements have been selected from calculations of temperature distributions (see fig. 16).

4. The heat density (W/m^2) has been strictly limited to avoid any hot spot on the heating foil itself, specially if the foil is glued on surfaces with bad heat transfer coefficients (epoxy, etc.).

5. The boxes have been internally insulated with soft polyamide which is a non flammable material.

6. Thermal breaks have been used at the shaft entry into the heated boxes, a minimum gap has been left between shaft and insulation blocks, just enough to avoid seizing.

7. Purge lines have been connected to all actuator boxes, purge can be done with GN2 or dry air to remove humidity or to cool down the boxes if necessary when the tunnel is operated warm for an extended period.

The operation of ETW for the last 2 years has shown that the design selected was correct and the actuators have been cycled and operated at cryogenic conditions a number of times. A few problems occurred and some foils have been destroyed by overheating, mainly in boxes where there was hardly any room for a proper insulation. It has been noticed that the foils start burning at determined hot spots which propagate after the phenomenon has been initiated. An analysis involving the foil manufacturer has led to a redesign of some foils including a modification of the foil geometry to avoid gluing them on components having bad heat transfer coefficients, a further limitation of the heat density, a reduction of the area of each individual foil.

Stress Analysis

An extensive stress analysis has been performed on the high speed leg components. The complete structure has been checked against excessive deformation due to primary mechanical stresses. This criteria is anyhow not the sizing criteria of the structure, the specifications related to the maximum deflections allowable have led to a very stiff structure.

More critical was the progressive deformation analysis including the effect of combined primary and secondary thermal stresses. The thermal loading has been computed using the temperature conditioning scenario shown in fig. 8 and fig. 9. The criteria for the range of primary plus secondary stresses to be less than twice the yield stress is met everywhere, the most critical area being the stiffeners of the Fixed Contraction in the downstream part. This is easily understandable, the flow surface is <u>not</u> decoupled from the stiffeners and the flow speed (i.e. the heat exchange coefficient) is high due to the section contraction. A special arrangement has been made at the downstream end of the Fixed Contraction where during the cooldowns

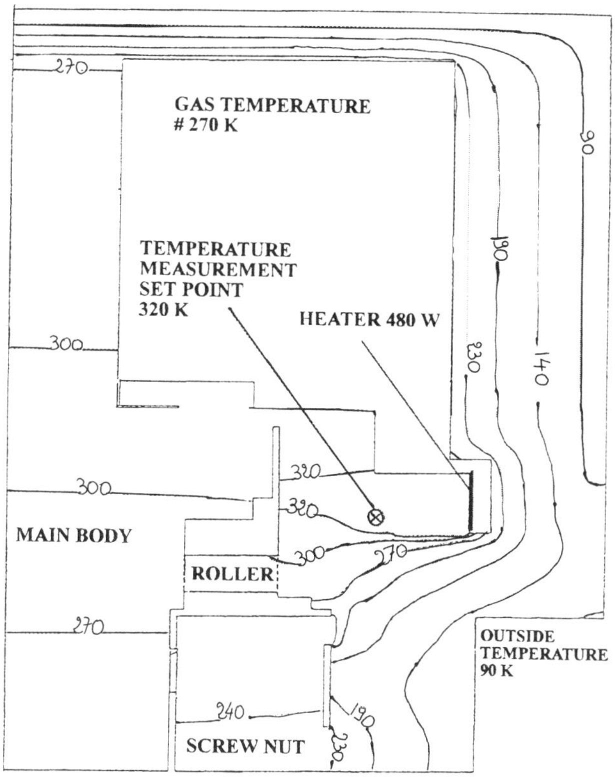

Fig. 16. Modelization of the temperature distribution in an actuator box

and warm-ups flow is diverted from the main stream and is forced to flow inside a box-type structure to limit the temperature gradients between stiffeners and flow plate in a similar way to the bulkhead valves effect already discussed.

Finally a fatigue life analysis was performed using Finite Element Model and stress concentration factors. Generally all components have a calculated fatigue life far in excess of 50 years. The two most critical elements being the Fixed Contraction downstream end again and the movable plates of the adjustable nozzle where a permanent pre-stressing is always present.

A modal analysis was as well done in the model cart, test section arrangement. As far as practicable the first modes of the model support structure were kept outside a range 10 to 50 Hz, with the large masses of the components involved this criteria was not too difficult to meet.

CONCLUSION

The design of the high speed leg of ETW was faced with a number of major challenges. Solutions, sometimes innovative, have been applied to meet all the specifications. The experience today after 2 years of operation has not revealed any major problems, some tuning had to be done as anticipated but all components had a satisfactory behaviour. The flow quality is as expected, the cooldown rate is as foreseen and the model support system has the expected performance.

The initial goal was to build a cryogenic tunnel having the quality of the existing conventional tunnels and however still keep a high flexibility in the operation of the tunnel. The first test campaigns have demonstrated that this goal has been achieved.

ETW has not yet been tested in its full operating envelope (temperatures below 150 K, proper supersonic calibration, routine cold model transfers) and some time will still be necessary to explore all the possibilities of ETW but the tunnel can already provide a powerful testing tool to all clients. The behaviour and wear of all the various components will have to be watched over carefully in the coming years to validate the long term validity of the design concepts selected.

PHYSICAL ENGINEERING PROBLEMS OF HYPERCONDUCTING TRANSFORMER CONSTRUCTION

S.E.Demyanov[1], V.N.Khazov[2], L.A.Mastrukov[2],
Yu.V.Tkach[3], and S.P.Zakatov[2]

[1] Institute of Physics of Solids and Semiconductors
Minsk, 220072, Belarus
[2] LANTEP Co. Ltd.
Moscow, 119146, Russia
[3] Institute of Electromagnetic Researches
Kharkov, 310164, Ukraine

ABSTRACT

The problem of construction of a space vehicle-borne 1 MVA cryogenic converter transformer, which provides serviceability of 12-pulse rectifying circuit has been studied. As a winding conductor pure aluminum foil is used, which is cooled by liquid hydrogen, which makes it possible to decrease the mass-to-power ratio several fold. Physical - technical problems, arising due to high current density in windings, for solution of which original approaches of windings and magnetic core design were suggested, have been described. Calculations of main parameters, demonstrating that hyperconducting transformers of medium power are promising devices, were made.

INTRODUCTION

Interest in the problem of cryogenic transformer construction, which appeared in the middle of 70s, was connected with the necessity of their power increase to 1 - 3 GVA[1]. Conventional facilities for constructing industrial transformers for these power values were considerably exhausted.

The principal distinction of deep cooling is that it allows a decrease in load loss in the transformer (short-circuit loss and cooling loss) without additional expenditures of active materials, or it allows an increase in transformer power by means of current density increase, without the increase of transformer loss.

Preliminary evaluations of cryogenic cooling efficiency were made on the basis of the most simple considerations of proportional change of the transformer dimensions with the rise of its power. It was shown that, with a decrease in

Table 1. The influence of windings resistance decrease by factor of n at cryogenic cooling on the main transformer parameters[2]

No. of variant	Power	Winding weight	Weight of magnetic core	Loss in winding	Loss in steel	Short circuit power	Short circuit voltage	Current density
1	c*	$n^{-3/5}$	$n^{-3/8}$	$n^{-3/8}$	$n^{-3/8}$	$n^{3/8}$	$n^{3/8}$	$n^{1/2}$
2	c	$n^{-3/5}$	$n^{-3/5}$	c	$n^{-3/5}$	$n^{3/5}$	$n^{3/5}$	$n^{4/5}$
3	$n^{1/2}$	c	c	c	c	n	$n^{1/2}$	$n^{1/2}$
4	$n^{4/5}$	c	c	$n^{8/5}$	c	$n^{8/5}$	$n^{4/5}$	$n^{4/5}$

* - remains invariable

winding resistance by factor of n, by the increase of current density it is possible to change the main parameters of transformer, as it is shown in Table 1.

Variants No. 1 and 2 show the change of transformer characteristics at the stable power, and variants No. 3 and 4 show characteristics change at the stable load loss.

These evaluations do not take acount of economical factors, the main of these being high cost of cryogenic cooling. From this point of view the advantageous solution of the task should be the following. When dimensions of magnetic core and windings are specified, short circuit loss should be decreased, and then, during the increase of current density in windings, the economically optimal regime of cryogenic transformer power should be found.

Though this method of transformer design is very promising, only a few examples of construction and use of such transformers are known at present. These are: the low power transformer with aluminum winding, cooling by gaseous helium at T = 14 K[1] and the air impulse transformer, cooling by liquid nitrogen[3,4].

In the present paper we demonstrate the possibility of construction of a medium power conversion cryogenic transformer, we describe the concept of design of hyperconducting windings and magnetic core, which offers considerable reduction of dimensions and weight of the transformer.

THE CHOICE OF INITIAL CRITERIA FOR THE CRYOGENIC TRANSFORMER DESCRIPTION

The use of hyperconductor as winding material is most effective in transformers, which are installed in transport systems, for example in a space power station. In that case the main criteria for the transformer, as a space vehicle equipment - its low weight should be satisfied.

We will consider the problem of cryogenic transformer construction on the concrete example of space vehicle-borne conversion transformer, which is able to operate in the following electric circuit: generator - transformer - rectifier - load. The following specifications were chosen for the transformer:

Rate power - up to 1000 kVA;
Frequency - 400 - 600 Hz;
Efficiency - not less than 0.98;
Mass to power ratio - no more than 0.07 kG/kW;
Linear primary voltage - 2000 V;
Rectified voltage output - 100 V;
Cooled with liquid hydrogen - 20.4 K;

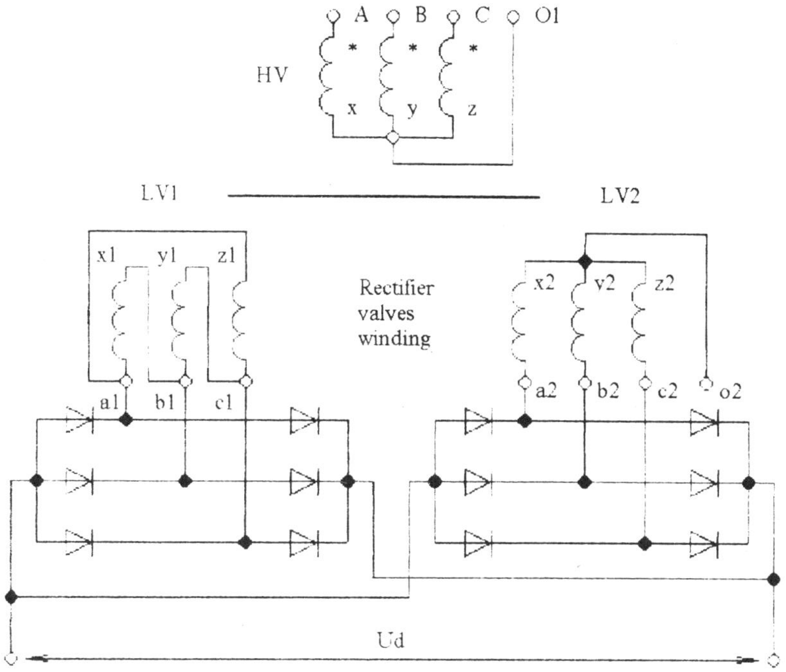

Figure1. 12-pulse rectifier circuit with parallel bridges connection.

In order to ensure 12-pulse rectifying circuit, 2 sets of secondary windings should be applied: one set - with a star connection (low voltage (LV1)) and the second set - with a delta connection (LV2). Along with that, vectors of voltages in secondary winding are shifted by the angle of 30 electrical degrees, and are used to energize two bridges with uncontrolled rectifier valves, the outputs of which are connected in parallel and feed a common load (see Fig.1).

The absence of reactors at bridge outputs forces the need for relatively high values of transformer short circuit voltage, because only observing the above condition, it is possible to limit the variable component of the current in the loop, formed by parallel connection of the windings via rectifying bridges.

THE CHOICE OF CRYOGENIC TRANSFORMER DESIGN

Some Problems of the Transformer Construction

The rise of alternating current frequency up to 400 - 600 Hz and multiple increase of current density in windings when cooling with liquid hydrogen makes it possible to increase the transformer power per mass unit by the order of two. Current density rise does not cause a proportional increase of main loss, as it occurs in typical transformers, since at low temperatures resistance of a high pure metal decreases considerably faster, than linear decrease. At the same time with cryogenic cooling a considerable rise of additional loss will take place, which is proportional to the square of frequency and induction of scattering field, and is inversely proportional to the active winding resistance. Additional loss occurance is connected with a number of effects, emerging in a conductor of alternating current flow. The principal effects of this kind are eddy currents, skin effect, size effect, and effect of current forcing out to winding boundaries. In principle, the task of additional loss reduction was considered earlier[2], and its solution was reduced to the necessity of the choice of optimal conductor size and winding design, depending on type and purpose of transformer.

The increase of current density leads to a proportional rise of the main transformer parameter - short circuit voltage, which in principle could make a cryogenic transformer inoperative. Therefore, design decisions should be found, which will afford reduction of short-circuit voltage and, correspondingly, reduction of short-circuit loss. The short- circuit loss does not depend on winding material resistance and is determined only by geometrical dimensions of conductor and winding. This leads to the necessity of reduction of radial size of winding and its height rise. These conditions can be successfully satisfied by dividing the windings into parts, by these the number of high-voltage (HV) and low-voltage (LV) windings pairs would be increased.

The Choice of Magnetic Core Type

The value of loss in the transformer produces a decisive influence on the basic approaches to transformer design, which can have two principally different versions. In the first version the transformer is developed as a transformer with a steel magnetic core, the second version envisages an air transformer.

The advantages of the transformer version with a steel core are small values of no-load running current and loss in its windings. The disadvantages of such a transformer are high labour content of its construction and relatively big power loss in its magnetic cores.

The disadvantage of an air transformer is high no-load running current, which should be taken into consideration when determining the parameters of power supply generator, and respectively, added loss in primary winding, connected with no-load running current.

The requirements to reduce the no-load running current and loss in windings of air transformers are controversial.

In order to reduce the no-load running current, it is necessary to increase the diameters of coaxially located windings, which leads to increase of the volume occupied by them and, respectively, to increase of the loss in them.

The requirements to reduce the transformer overall dimensions, hinder the increase of the windings diameter.

The calculations done for comparison of both transformer versions showed, that with respect to dimensions and loss, the preference should be given to the transformer with a steel magnetic core.

Along with that, much attention should be paid to the issue of optimum relation between the masses of the magnetic core and the windings.

The extent of magnetic core utilization is determined by the values of magnetic induction. It should be adopted within the limits from 1.2 to 1.4 T. Naturally, with the increase of induction B, the mass of the transformer reduces. Still, along with this, increase of loss in steel occurs.

In order to minimize the loss, the decision was taken to fabricate the magnetic core from textured tape, having width $1.5 * 10^{-4}$ m and laquer insulation. The tape is manufactured of cold-rolled electrotechnical steel of E3425 type. At present, positive experience of such magnetic cores functioning at low temperatures has been accumulated.

The choice of magnetic core design type was made proceeding from the necessity of minimizing its mass, convenience of disposition and mounting of current leads and its installation in a cryostat. From this point of view, we gave preference to the magnetic core built up of laminations with spatial placement of legs. Main characteristics of the magnetic core will be given below.

The choice of conductor and windings design

To minimize mass of the windings and loss in them, it is expedient to use high purity aluminum. At 20.4 K such material has residual resistivity lower by a factor of 200 - 500 than that at room temperature. Along with that, this material has considerably lower density, than that of other conducting materials.

We should point out that along with main loss there exist added loss, resulting from eddy currents, caused by longitudinal and transverse components of induction due to scattering fields. To reduce loss, it is expedient to use aluminum foil with thickness of $0.6 - 1.0 * 10^{-4}$ m.

In order to ensure the required tightness of coils, which should have cylindrical shape, foil width should not exceed 0.1 m. With this the necessity appears to reduce distances between insulation in windings, and consequently, it is expedient to use insulating film (having width not exceeding $0.5 * 10^{-4}$ m), which withstands an electric field gradient up to $2 * 10^{8}$ V/m.

Additional loss in the transformer windings could be reduced by two methods: reducing foil thickness, or suppressing the magnetic scattering field. Reduction of the transformer magnetic scattering field is achieved by division in parts of LV and HV windings, which then should be alternating in radial direction, as follows: LV - HV - LV - HV - LV - ... etc., with it the transformer windings division could be brought to one turn. This procedure is similar to the capacitor windings, when two foils, which form capacitor coverings and which are divided by insulation are winded simultaneously. "Capacitor" winding is accomplished by a package of aluminum and insulation strips, consisting of two LV strips and two HV strips, which alternate as follows: LV - Insulation - HV - HV - Insulation - LV.

When the necessary number of LV wraps are applied, the strips relating to LV are slit in the longitudinal direction and are brought outside in loop form, producing the end of the first section and the beginning of the following section. Then the process of joint application of LV and HV windings is continued; the bringing outside of LV wraps process is repeated, and so on. The winding of HV strips is made continuously, without leads taking out, and as a result, all HV sections are connected in series. LV sections, brought outside, could be connected in any necessary circuit. With this LV and HV strip alternation, ideal periodical structure of scattering field is obtained, which facilitates alignment of current distribution in LV windings.

Transformation coefficient of such a transformer can not be optional. The product of LV windings number by LV sections number (m) always equals to HV windings number: $W_{HV} = m\ W_{LV}$.

The space between LV - HV strips always affects the complete voltage of the HV winding, therefore insulation between them should possess high electric strength, such as that of polymide film. For the increase of strength, the edges of insulating film should exceed the foils edges, and LV and HV strips are shifted relative to each other, as it is made in capacitors. Moreover, winding faces should be potted in electric isolating compound. The capacitor windings pattern completely removes eddy current loss and makes it possible to use foils of large thickness as a conductor, which considerably increases the coefficient of winding cross-section filling in. The windings, fabricated by the capacitor pattern possess large initial capacity, which can cause undesirable over-excitation of the transformer. The windings capacitance could be compensated by putting a small air gap in the magnetic core leg.

Table 2. Basic electric specifications of cryogenic 1 MVA transformer

Linear primary voltage HV (V)	2000
Phase primary voltage HV (V)	1155
Phase primary current HV(A)	289
Rectified voltage LV (V)	100
Phase secondary voltage LV_1(V)	73.5
Phase secondary voltage LV_2(V)	42.5
Phase secondary current LV_1(A)	2267.5
Phase secondary current LV_2(A)	3927.5

For the windings of a conversion transformer it is suggested to place two coils of LV - HV windings on the leg. HV windings are connected in series between coils, and as a result, numbers of HV windings are summarized.

Every coil contains two low voltage windings LV_1 and LV_2 , which are connected in parallel between the coils. At the three-phase switch LV_1 windings are connected by the "delta" scheme, and LV_2 windings are connected by the "star" scheme (see Fig.1).

Every system of linear LV voltages operates on a separate three-phase rectifying bridge by the six-pulse circuit. At joint operation of the bridges twelve-pulse circuit is formed, because linear LV voltages are shifted by 30°.

TRANSFORMER CALCULATION AND ITS PROPOSED DESIGN

The main parameters of the transformer, such as rated power, linear primary voltage, rectified voltage and the diagram of its switching into the electric circuit make it possible to calculate other electric parameters of the transformer, which are given in Table 2.

In accordance with these parameters, and also in accordance with the principles of winding design, which were stated in the previous section, the transformer windings should have the following parameters:

Primary winding HV:
- number of turns per phase - 112;
- number of coils per phase - 2;
- number of turns per coil - 56;

Secondary winding LV_1:
- number of turns per phase - 7;
- number of parallel sections - 8;
- number of wires per phase - 56;

Secondary winding LV_2:
- number of turns per phase - 4;
- number of parallel sections - 14;
- number of wires per phase - 56;

Every phase has two coils, which are placed one over another on the magnetic core leg, and with this HV winding is applied continuously and simultaneously with LV_1 winding or with LV_2 winding. LV windings are applied by the following pattern: 7 turns LV_1 + 4 turns LV_2 + $4LV_2$ + $7LV_1$ + $4LV_2$ + $4LV_2$ + $7LV_1$+ $4LV_2$ + + $4LV_2$ + $7LV_1$ + $4LV_2$.

Table 3. Basic specifications of magnetic core

active cross-section of the leg(m^2)	$3.75 * 10^{-3}$
active cross-section of the yoke (m^2)	$2.15 * 10^{-3}$
mass of three legs (kg)	20.4
mass of two yokes (kg)	21.6
magnetic induction of the leg (T)	1.15
magnetic induction of the yoke (T)	1.16
total loss in magnetic core (W)	755

The windings are applied by the wire which consists of two parallel pure aluminum foils with thickness 10^{-4} m and width $7 * 10^{-2}$ m. Two foil layers (packet) are placed between polymide films with thickness $0.4 * 10^{-4}$ m, which form insulation between HV and LV turns. Thickness of such packet equals $2.8 * 10^{-4}$ m, and the total thickness of winding is $3.2 * 10^{-2}$ m. Therefore, the following values of current density will be implemented in windings:

HV - $20.6 * 10^6$ A/m^2;
LV_1 - $20.3 * 10^6$ A/m^2;
LV_2 - $20.1 * 10^6$ A/m^2;

For a resistance ratio of the aluminum foil $R_{273\,K} / R_{20.4\,K} = 400$, main loss in the windings at 20.4 K equals 115 W.

The magnetic core consists of three legs, placed on the vertexes of an equilateral triangle, which are connected between each other by two circle yokes. For maximal filling of the glass-textolyte tube inner volume, on which the coils are wound, the legs are composed of packages of steel plates having various width. The yoke cross-sections repeat the form of the legs cross-sections. Legs and yokes connection was made by their interleaving. Basic specifications of magnetic core are given in Table 3.

The total loss of the described transformer does not exceed 1000 W, without accounting for the loss on current leads. The three-phase windings connection is expedient to make on a special panel directly over the upper yoke of the magnetic core in order to provide maximal compensation of current leads-out, right up to their output from the cryostat.

CONCLUSIONS

1. Physical-technical aspects of construction of a 1 MVA space vehicle-borne cryogenic transformer with windings of pure aluminum foil, operating at liquid hydrogen cooling are considered. The problems of increase of additional loss and short-circuit loss, connected with multiple rise of current in windings are studied.
2. Design of phase windings, which provides suppression of eddy-current loss by means of HV and LV windings currents compensation is proposed. Magnetic core geometry of spatial legs placement provides minimal scattering fields occurance.
3. On the base of a proposed transformer design, calculations of electric and geometric transformer parameters are made, which prove that the transformer is a promising device of large power per mass unit.

REFERENCES

1. V.N.Elagin, V.L.Karanazuk, E.L.Trukhin, and L.A.Mastrukov, Experimental study of low-power cryogenic transformer, *Electrotechnika*, 3:17 (1973).
2. L.A.Mastrukov, Yu.V.Petrovskii, V.N.Elagin, and E.L.Trukhin, Deep cooling of windings in transformers of ultimate power, *Electrotechnika*, 7:20 (1975).
3. D.E.Johnson, J.P.Barber, and H.L.Laquer, Energy storage transformers - MAPPS test results, in: "4th Symposium on Electromagnetic Launch Technology", Austin, TX (1988).
4. D.E.Johnson and N.D.Clements, A hyperconducting 5 MJ 1MA pulse transformer, *Adv. in Cryog. Eng*. 35: 1653 (1990).

A LARGE SCALE HIGH-Tc SUPERCONDUCTING SHIELD AND ITS CRYOSTAT

K.Shintomi, A.Yahara, D.Irisawa, K.Imai, H.Matsuba

Yokohama R&D Laboratories, Furukawa Electric Co.,Ltd.
Yokohama, Kanagawa, 220, Japan

ABSTRACT

We have developed a cryostat for a large scale high-Tc superconducting shield enclosing a human body to measure biomagnetism. The cryostat has an interior space of ϕ 0.8 m x L2.8 m and cools a high-Tc superconducting shield ϕ 0.87 m x L2.4 m. The cryostat is made of non-magnetic materials to avoid internal magnetic noise. It also includes other provisions to achieve a very low magnetic noise environment. We investigated the optimal design of the radiation shields for the cryostats using liquid nitrogen and concluded that the one radiation shield has sufficient performance for the thermal insulation. We designed the liquid nitrogen reservoir to keep the superconductor at 77.3 K until the reservoir is nearly emptied.

A magnetic shielding capacity of the superconducting shield for an external magnetic fields reaching 1.8 mT was found. This sufficed for biomagnetic measurements. The critical current density distribution of the superconductor was evaluated using a magnetic field source formed by a pair of small opposing coils placed near the wall of the cryostat. The mean critical current density of the superconductor was about 450 A/cm^2. External noise was attenuated by 160 dB within the shielded space. This kept the external noises below the internal noise of the SQUID magnetometer.

INTRODUCTION

A high-Tc superconducting shield for biomagnetic measurements has been developed.[1,2] The shield has superior properties which enable its use for SQUID magnetometers, because the external magnetic field is attenuated by 160 dB in the DC to low frequency range. In this paper, we introduce a cryostat design suitable for refrigerating high-Tc superconducting shields to be used for biomagnetic measurements. The superconducting shield performance is evaluated.

Several investigations have been carried out using liquid helium cryostats. Usually a helium cryostat has three radiation shields. This design is based on the fact that the enthalpy change from 4 K to room temperature is much larger than the vaporization enthalpy and that the enthalpy change must be used to obtain an effective thermal insulation. On the contrary, liquid nitrogen enthalpy change from 77 K to room temperature is nearly equal to the vaporization enthalpy. This fact gives a different design for liquid nitrogen cryostats that for liquid helium cryostat. Thus, fewer shields are needed for a liquid nitrogen cryostat that for a liquid helium cryostat. A theoretical investigation on radiation shields was made by Paivanas et al..[3,4] However, that theory did not offer the optimum arrangement for the

Advances in Cryogenic Engineering, Vol. 41
Edited by P. Kittel, Plenum Press, New York, 1996

radiation shield nor gave the increase in the heat flux in arrangements different from the optimal.

We investigated the optimal radiation shield for a liquid nitrogen cryostat and made a design for it. In this paper, we describe the performance of the liquid nitrogen cryostat used with the shield for biomagnetic measurements, made using such a design.

Furthermore, we describe, a method for evaluating the shield performance in biomagnetic measurements of the superconducting shield, by estimating the critical current density distribution arising from the trapped magnetic field in a small region of the superconductor.

DESIGN AND PERFORMANCE OF THE CRYOSTAT

Radiation shield

Figure 1 shows a model of a cryostat having N radiation shields. The temperature of the cooling media reservoir is T_0. The radiation shield covers the p part of the reservoir surface; R_n denotes the heat resistance between radiation shields n and n-1, T_n is the temperature of radiation shield n, and G is the gas vaporization rate; Q_s is heat flux to the reservoir. Assuming that the vaporized gas absorbs heat from the radiation shields so that their temperature equals that of the gas, the following equations hold:

$$Q_n = Q_{n-1} + (T_n - T_{n-1}) \cdot G \cdot C_p \tag{1}$$

$$Q_{n-1} = \frac{T_n - T_{n-1}}{R_n} \tag{2}$$

$$G = \frac{Q_s}{L_v} = \frac{Q_0 + Q_d}{L_v} \tag{3}$$

From these equations one can derive that the heat flux to the reservoir reaches its minimum when the heat resistances between adjacent radiation shields has the same value. At the minimum of the heat flux, the equation:

$$\left[1 + \frac{1}{N+1} \cdot \frac{q_r \cdot \gamma}{p}\right]^{N+1} = 1 + \frac{q_r \cdot \gamma}{q_r - (1-p)} \tag{4}$$

holds where,

$$q_r = \frac{Q_s}{Q_t} \tag{5}$$

denotes the ratio of the minimum heat flux to the heat flux when no radiation shield is present, and

$$\gamma = \frac{(T_e - T_0) \cdot C_p}{L_v} \tag{6}$$

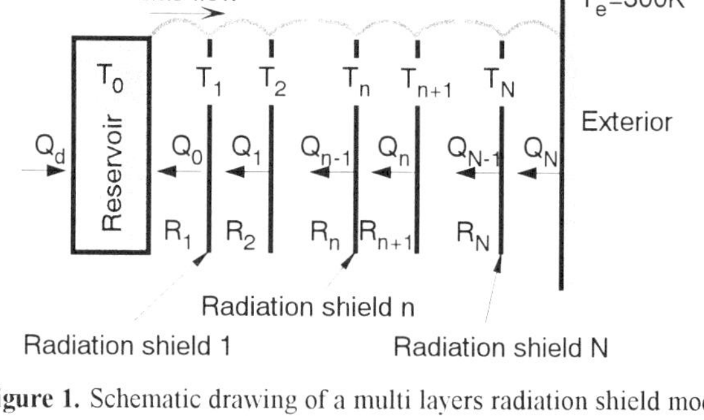

Figure 1. Schematic drawing of a multi layers radiation shield model.

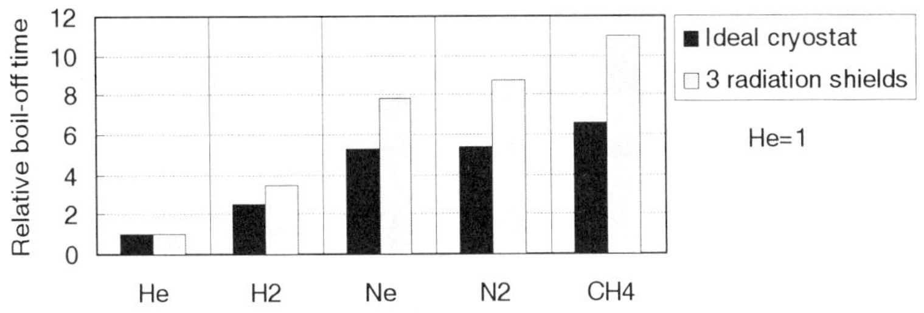

Figure 2. Relative boil-off times for various cooling media and for an optimal designed cryostat having a three radiation shields or having infinite radiation shields.

is the ratio of the enthalpy change of the gas between the vaporizing and room temperature to the heat of vaporization.

Figure 2 shows the comparison between the boil-off times for various cooling media for an optimal designed cryostat having three radiation shields and for a cryostat having infinite radiation shields (i.e. the ideal cryostat). It suggests that a boil-off time for a liquid nitrogen cryostat is as much as 8 times longer than that of a liquid helium cryostat, in the case of a three radiation shields. Since the price of liquid nitrogen is about 1/20 of that of liquid helium, the running cost of a liquid nitrogen cryostat becomes only 1/160 of that of a liquid helium cryostat.

Figure 3 shows the heat leak for a cryostat having one radiation shield as a function of its location and of the coverage area. In figure 3, the minimum heat leak is obtained at the location where heat resistances are divided identically. However, the heat leak shows only small deviation from the optimal value. The heat leak of a liquid nitrogen cryostat decreases with the coverage area and approaches its minimum at more than 50 % coverage, while the heat leak of a liquid helium cryostat decreases monotonically up to 100 %.

Figure 4 shows the heat leak for the optimal designed cryostats as a function of the number of radiation shields. In conclusion, a liquid nitrogen cryostat requires only one radiation shield coverings about 50 % of the reservoir, while liquid helium cryostats require three radiation shields covering 100 % of the reservoir.

Liquid nitrogen cryostat

The cryostat was designed on this basis of the investigation on the radiation shields. Figure 5 shows its schematic drawing with the high-Tc superconducting shield. The cryostat has dimensions ID0.8 m x OD1.3 m x L2.8 m and weights about 1000 kg. The superconducting shield is ϕ 0.87 m x L2.4 m x t0.3 mm in size and is made of BSCCO (Tc

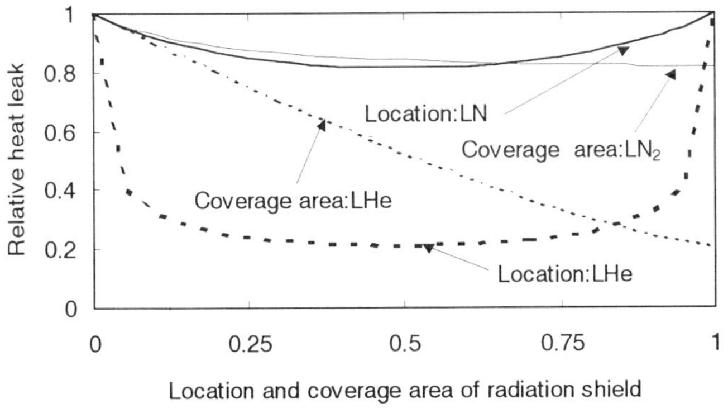

Figure 3. Heat leak of a cryostat having one radiation shield as a function of its location and of the coverage area.

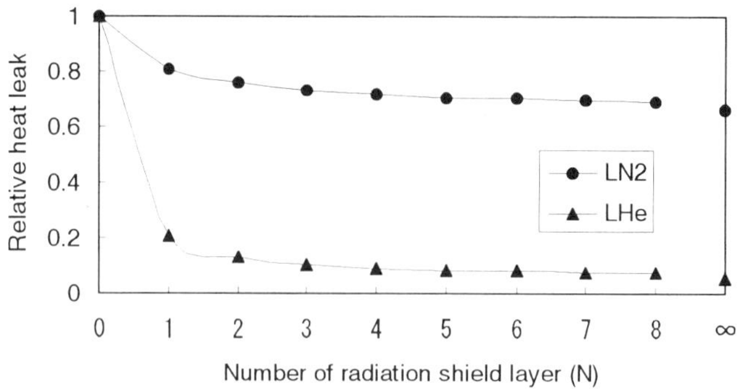

Figure 4. Heat leak of an optimal designed cryostat as a function of the number of radiation shields.

= 90 K) high-Tc superconductor. All the components of the cryostat are made from non-magnetic materials to avoid an increase of the internal magnetic noise. The inner cylinder, which must enclose a human body, is made from G-FRP, because the magnetic Johnson noise arising from a high conductivity metal like aluminum would disturb the biomagnetic measurements.[5] The cryostat is equipped with one radiation shield covering 65 % of the reservoir surface. The calculated heat leak of the cryostat is 30 W.

The superconductor is cooled by a liquid nitrogen reservoir located outside the superconducting shield. This arrangement eliminates all metal parts inside the superconductor which might be a source of magnetic noise and leaves the maximum inner space for the biomagnetic measurements. The reservoir is fabricated with copper pipes in a cage pattern. Each pipe has dams inside which contains liquid nitrogen. The dams enable the superconductor temperature to stay stable at 77.3 K until the reservoir nearly empties. The superconductor is cooled by heat conduction through flexible copper wires attached to the reservoir. The flexible wires also help in decreasing the thermal stress induced by thermal shrinkage during the cooling process.

Cryostat performance

The superconductor was cooled by supplying liquid nitrogen from an external tank. It took eight hours to cool the superconductor. The liquid nitrogen consumption and the superconductor temperature were monitored. Figure 6 shows the temperature at upper and

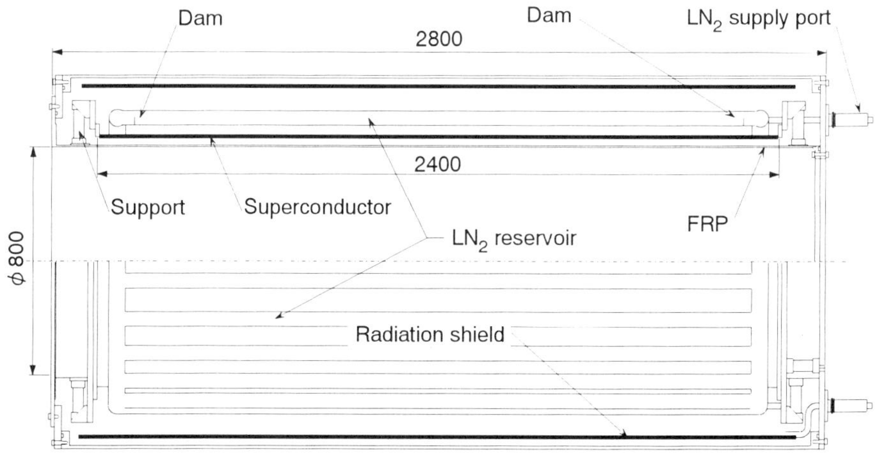

Figure 5. Schematic drawing of the cryostat for the superconducting shield.

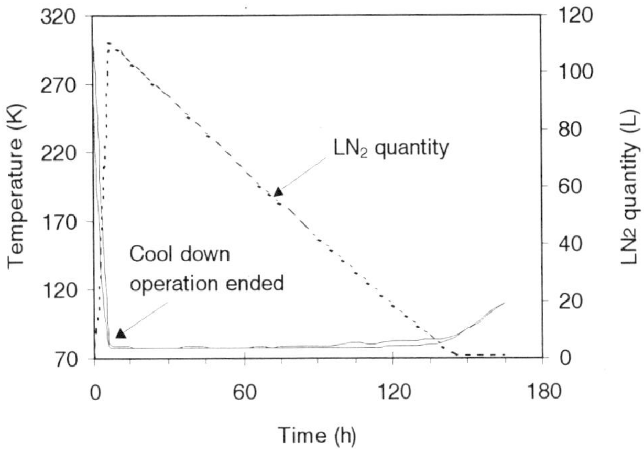

Figure 6. Temperature of the superconductor and liquid nitrogen quantity in the reservoir after the starting of the cool-down operations.

lower points of the superconducting cylinder and the liquid nitrogen quantity in the reservoir. Liquid nitrogen in the reservoir lasts six days without refilling, while the temperature of the superconductor stayed at 77.3 K. The boil-off rate of the cryostat was 19 liter/day (15 kg/day), the measured heat leak was 35 W. This performance agrees well with the design estimate.

EVALUATION OF THE SUPERCONDUCTING SHIELD

Experimental procedures

The maximum shielding capacity and the trapped magnetic field strength of the superconducting shield were evaluated. The cryostat was placed in a magnetic field axial to the superconducting cylinder, produced by a 52-turn magnetic coil D1.3 m x L2.6 m, which contained the full cryostat. A magnetic sensor was placed at the center of the superconducting cylinder. The axial magnetic field in the superconducting cylinder was measured while the external magnetic field was increased up to 8 mT and then decreased to zero. The result is shown in figure 7. The trapped field defined as the residual magnetic field after 90 minutes since the coil current is turned off, was measured on the full axis length of the superconducting cylinder. The result is shown in figure 8.

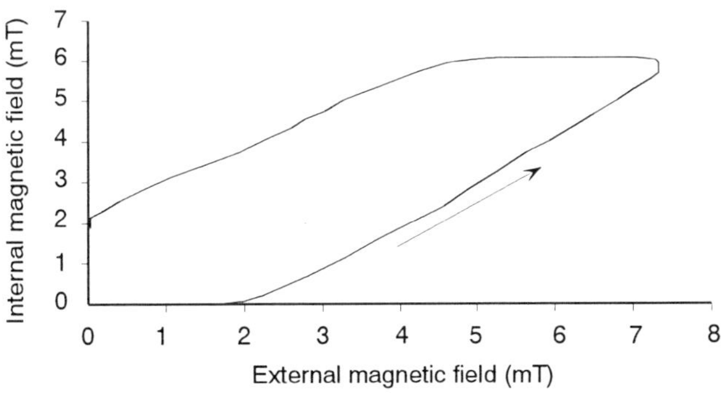

Figure 7. Magnetic field at the center of the superconducting cylinder versus external magnetic field.

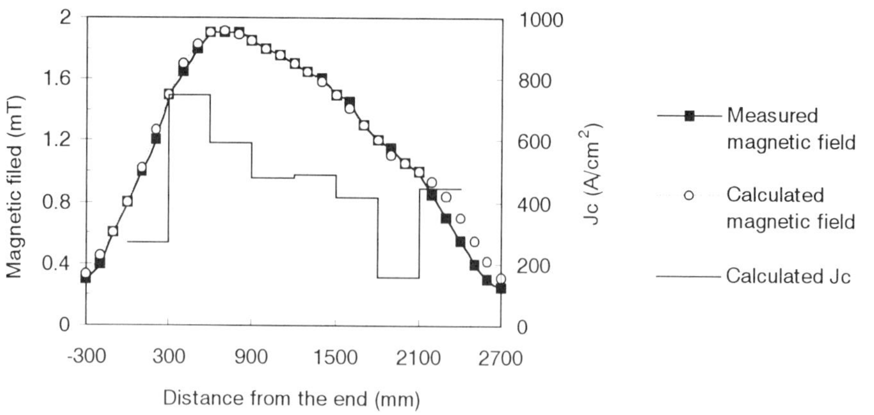

Figure 8. Trapped magnetic field distribution along the axis of the superconducting cylinder, calculated critical current densities, and derivative magnetic field distribution.

The distribution of the trapped magnetic field over the full superconductor surface was measured with a pair of coils applying a magnetic field at a small part of the superconductor. Figure 9 shows the pair of coils and the position of a Hall magnetic sensor. The coils generate opposite fields resulting in a magnetic field applied only to a small area on the superconductor, as shown in figure 9.[6] The coils were energized with a 4 A current, then the trapped magnetic field was measured 60 s after the current was turned off with the Hall device. The distance of the Hall device from the superconductor was 40 mm. The results are shown in figure 10.

The shielding factor and the magnetic noise within the shield were measured with a DC-SQUID magnetometer with intrinsic noise 5 $fT/Hz^{1/2}$. The shielding factor was calculated as the ratio of the AC magnetic field in the shield to the external AC magnetic field. The results are shown in figures 11 and 12.

Discussion

Figure 7 indicates that the external magnetic field can be as high as 1.8 mT and that also the residual magnetic field is about 1.8 mT. Therefore, 1.8 mT represents the maximum shielding field of the superconducting cylinder.

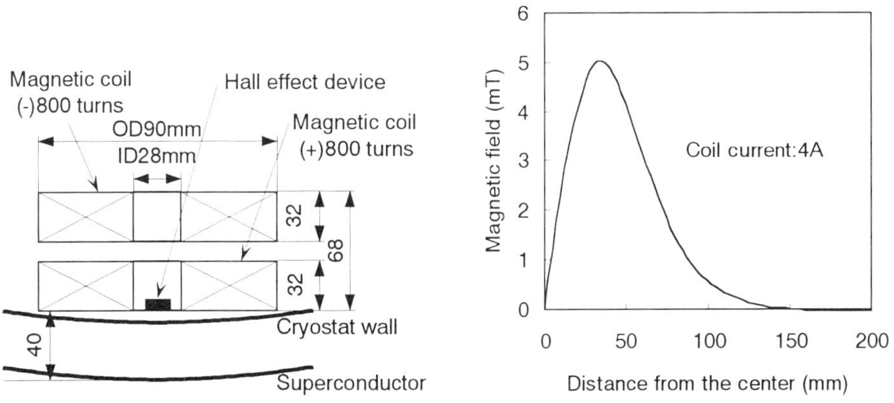

Figure 9. Schematic drawing of the small magnetic coils generating opposite fields and the magnetic field distribution at the surface of the superconductor.

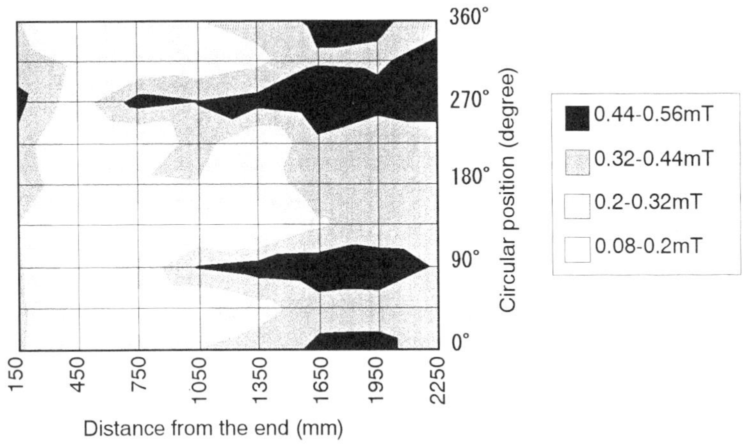

Figure 10. Map of the trapped magnetic field 40 mm above the superconductor .

According to the critical state model,[7] The maximum shielding field strength (Bc) of a infinite plate is given by

$$Bc = -\mu_0 \cdot Jc \cdot d \tag{7}$$

Using this equation, the critical current density Jc of the superconducting cylinder was calculated to be 480A/cm^2 for the maximum shielded field Bc = 1.8 mT.

We calculated a distribution of the critical current density over the axial direction, assuming that the current flows in the circular direction and that the critical current density is constant in each of the eight sections in which the cylinder was subdivided. The calculated critical current densities and the corresponding calculated magnetic fields are shown in figure 8. The critical current density of the superconducting cylinder varies from 150 A/cm^2 to 750 A/cm^2. The mean value roughly agrees with the above calculated overall value of Jc (480 A/cm^2).

The two dimensional distribution of the critical current densities of the superconducting cylinder was derived from figure 10, assuming that the permanent currents flow in an area ID16 mm x OD149 mm x t0.3 mm and that the magnetic field at 40 mm above the surface of the superconductor equals the field in figure 10. The mean values also roughly agreed with the above calculated overall value of Jc (480 A/cm^2). Thus we have used three different methods to evaluate the performances of the superconducting cylinder.

Figure 11 indicates a shielding factor of 140 - 160 dB for magnetic fields below 10 Hz.

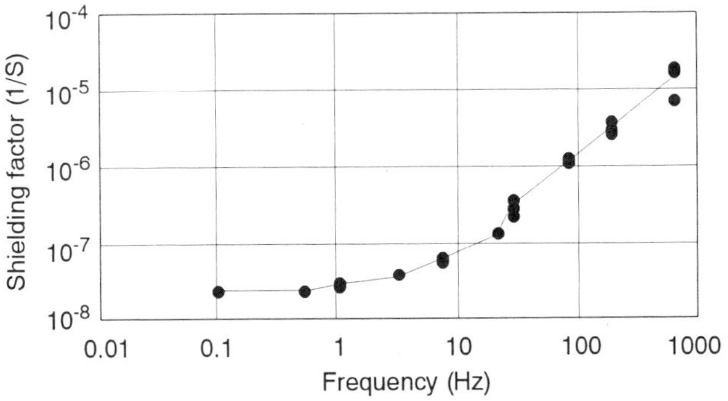

Figure 11. Shielding factor of the superconducting cylinder depending on the frequency of the applied magnetic field.

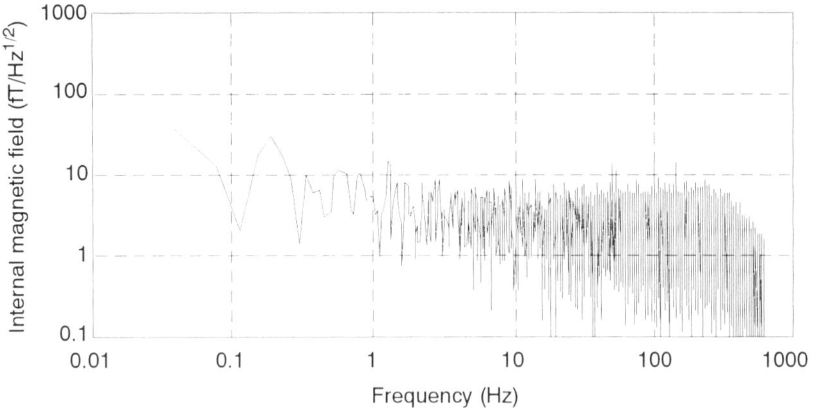

Figure 12. Environmental magnetic noise spectrum within the superconducting shield .

The low frequency behavior is very important for biomagnetic measurements, and it is the most difficult to achieve with the conventional ferromagnetic shields.

The magnetic noise spectrum within the superconducting shield is shown in figure 12. The noise in the shield fell bellow the sensitivity of the SQUID magnetometer that no other signal than the intrinsic SQUID noise was observed.

We proved that the superconducting shield can operate in an environmental magnetic field below 1.8 mT that results in compatibility with a MRI or other equipment generating a strong magnetic fields.

CONCLUSIONS

We have developed a way to obtain the optimal design for the shield of a liquid nitrogen cryostat. The optimal cryostat is designed with one radiation shield. According to this principle, we have fabricated the cryostat for a high-Tc superconducting shield to be used for biomagnetic measurements. This cryostat kept liquid nitrogen in the reservoir for 6 days without refilling, while the temperature of the superconductor stayed stable at 77.3 K. The boil-off rate of the cryostat was 19 liter/day (15 kg/day), and the heat leak 35 W.

Secondly, we have developed three different methods to evaluate a superconducting cylinder, which can estimate the full map of the critical current density of the superconductor.

Finally, we demonstrated that the external noise in the shield is bellow the sensitivity of a SQUID magnetometer so that no other signal than its intrinsic noise is observed.

REFERENCES

1. A.Yahara, K.Shintomi, K.Imai, D.Irisawa, H.Yoshida, S.Seike, H.Shimizu, H.Matsuba; 8th The Journal of Japan Biomagnetism and bioelectromagnetics society, I-1 (1993)
2. A.Yahara, K.Shintomi, K.Imai, D.Irisawa, M.Masuda, H.Yoshida, S.Seike, H.Shimizu, H.Matsuba; 9th The Journal of Japan Biomagnetism and bioelectromagnetics society, B1-13 (1994)
3. J. A. Paivanas, O. P. Roberts and D.I-J. Wang; Advances in Cryogenic Engineering vol. 10, Plenum Press New York (1965) p. 197
4. J. A. Paivanas, A. W. Francis and D. I-J. Wang; US Patent 3, 133, 422 May 19 1964
5. T. Varpula, T. Poutanen; J. Appl. Phys., (1984)55, 4015-4021
6. K.Shintomi, H.Matsuba; 2nd International Symposium on Superconductivity (ISS `89), p. 717
7. C. P. Bean; Rev. Mod. Phys., (1964) 36, 31-39

IRRADIATION OF QUENCH PROTECTION DIODES AT CRYOGENIC TEMPERATURES IN A NUCLEAR RESEARCH REACTOR

D. Hagedorn,[2] H. Gerstenberg[1], and H. Schönbacher[2]

[1]Technical University Munich, Department of Physics E21
D-85748 Garching, Germany
[2] CERN, CH-1211 Geneva 23, Switzerland

ABSTRACT

Within the framework of the Large Hadron Collider (LHC) R&D programme, CERN and the Department of Physics E21 of the Technical University Munich have established a collaboration to carry out irradiation experiments at liquid helium and liquid nitrogen temperatures on epitaxial diodes for the superconducting magnet protection.

Small diode samples of 10 mm wafer diameter from two different manufacturers were submitted to doses of up 50 kGy and neutron fluences up to 10^{15} n/cm^2 and the degradation of the electrical characteristics was measured versus dose. During irradiation the diodes were submitted to current pulse annealing and after irradiation to thermal annealing. After exposure some diodes show a degradation in forward voltage drop of up to 600 % which, however, can be reduced to about 15 % - 20 % by thermal annealing. The degradation at liquid helium temperature is very similar to the degradation at liquid nitrogen temperature.

These degradations of electrical characteristics during the short term irradiation in a nuclear reactor are compared with degradations during long term irradiation in an accelerator environment at liquid nitrogen temperature.

INTRODUCTION

Irradiation tests at liquid helium temperatures are essential for a decision as to whether these diodes can be used for the current by-pass at super fluid helium temperatures for the LHC-superconducting magnets [1].

Our original intention to irradiate full size (75 mm wafer) diodes of the epitaxial type in liquid helium in a nuclear reactor has been given up because of the very high costs for adequate cryogenic installations. Instead, small diode samples of 10 mm wafer diameter have been irradiated in the nuclear research reactor at Munich in liquid helium and also at liquid nitrogen temperature for comparison, up to gamma doses of 50 kGy and neutron fluences up to 10^{15} n/cm^2 (E > 0.1 MeV) [2].

IRRADIATION FACILITY AND DOSIMETRY

The low temperature irradiation facility (TTB) at the 4 MW swimming pool type Munich Research Reactor (FRM) offers the possibility to perform neutron and gamma irradiation experiments in the temperature range between 4.6 K and 450 K. The irradiation facility has been described in detail elsewhere[2,3].

The layout of the irradiation cryostat restricts the sample size to a maximum length of l_{max} = 120 mm and a maximum diameter of $Ø_{max}$ = 16 mm. The maximum weight of the TTB-specimens is limited by the gamma-heating in the irradiation position to be about 35-40 g. For the diode irradiations the core position TTB was used. In order to perform irradiations during reasonably long time periods the nominal power of the reactor had to be reduced from 4 MW to 40 kW. During irradiation at 4.6 K the temperature was controlled to within about ± 0.1 K and during 77 K irradiation to within about + 5 K and -15 K.

Dosimetry measurements were carried out prior to the irradiations at reduced reactor power of 50 kW. The neutron flux was determined with standard activation detector methods. The value adjusted to the reactor power was 3.3 10^{11} n/cm^{2}s (E>0.1 MeV). The gamma dose was measured by means of Alanine and RPL glass dosimeters which were calibrated in a Cobalt-60 source at 4.2 K and at 77 K[4]. The value found was 16.7 Gy/s. The accumulated gamma dose and neutron fluences are presented together with the various irradiation steps in Table 1. It shows, that during all irradiations the gamma dose is a rather important part of the radiation spectrum. This is, however, not in disagreement with the situation to be expected in the LHC-dipoles where also the integrated dose will be a more important factor compared to the neutron fluence [5].

DIODE TYPES AND SAMPLE HOLDERS

The irradiation tests described here were performed on 10 mm diameter samples of the same wafer material used for the 75 mm diameter wafers. The small diameter of only 10 mm was chosen in order to meet the dimensional and cryogenic requirements of the low temperature irradiation facility. All diodes are of the epitaxial type with an epitaxial width of about 22 µm.

For the diodes from EUPEC (Siemens) - diode number starting with the letter E - rectangular pieces were sawn out of large diameter wafers and cylindrically ground to disk-like samples of 10 mm diameter. Since the rims were not passivated the reverse blocking voltages are very poor (in the range of a few volts) and neither reverse voltage nor junction capacity measurements could be carried out. The diodes from GEC-PLESSEY - diode number starting with the letter M - were fabricated in the normal way with passivated rims. Diode samples No. E1, E2, E5, E6, and M4N - M7N were irradiated at 4.6 K and diode samples No. E3, E4, E7, E12, M1, M3, M4, and M14 at about 77 K in the nuclear reactor. The diodes E8 - E11 and M8 - M11 were exposed to target irradiation at liquid nitrogen temperature only in the SPS Accelerator at CERN. Each of the six sample holders contained four diode samples. The diodes were stacked - together with copper disk electrodes between each of the diodes - and clamped under a force of 1 kN as shown in Fig. 1. Two Carbon (Allan Bradley) resistors were mounted in the 4.6 K-sample holder and two Copper-Constantan thermocouples in the 77 K-sample holder to control the temperature during irradiation.

TEST PROCEDURES

Table 1 shows the irradiation procedure which was applied to the six sets of diodes. For the reactor irradiation at 4.6 K the procedure consisted of 5 exposure periods of increasing duration. At 77 K the irradiation was carried out in three steps up to the maximum dose of about 50 kGy for the EUPEC diodes and up to 30 kGy for the GEC diodes. After each exposure step the samples were removed from the reactor core and the electrical characteristics measured. The electrical tests have been described in detail

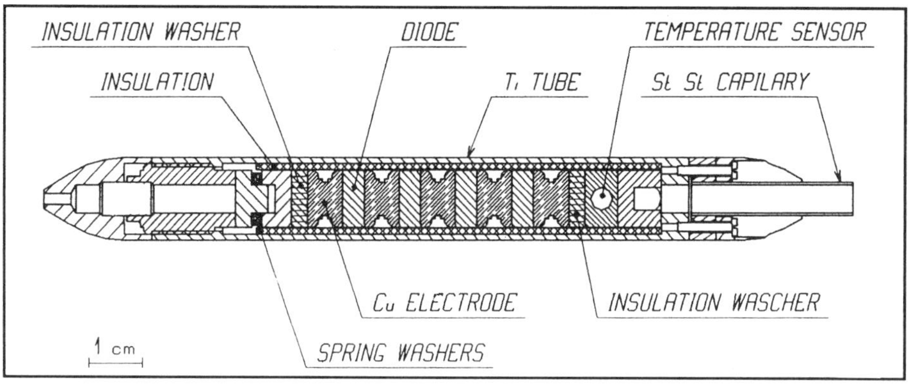

Figure 1. Sample holder with diodes and temperature sensor.

already elsewhere[3]. On the diode samples marked with an asterisk (*) current pulses up to about 110 A were applied after each exposure step and the forward voltage at peak current level (dI/dt = 0) monitored. On the other diodes high current pulse measurements were carried out before and at the end of the total irradiation procedure only, to minimise the self-annealing effects by carrier injection and temperature rise (induced by the tests themselves).

One of the four diodes of each of the six sets was submitted to intermediate current pulse annealing during irradiation. After the last irradiation step at 4.6 K all four diodes were warmed up to room temperature for about 15 min and then measured again at 4.6 K to

Table 1. Irradiation procedures in the nuclear reactor at 4.6 K and at 77 K and in the accelerator target area at 77 K

Irrad. Step No. /Temperature	Diode No.	Acc.Dose [kGy]	Acc.Neutrons 10^13[n/cm²]	Acc.Irrad. Time
Reactor Irrad.				
1 4.6 K	E1*, E2, E5*, E6	1.5	3	90 s
2 4.6 K	E1*, E2, E5*, E6	5	10	300 s
3 4.6 K	E1*, E2, E5*, E6	15	30	900 s
4 4.6 K	E1*, E2, E5*, E6	30	60	1800 s
5 4.6 K	E1*, E2, E5*, E6	50	100	3000 s
1 77 K	E3*, E4, E7*, E12	5	10	300 s
2 77 K	E3*, E4, E7*, E12	15	30	900 s
3 77 K	E3*, E4, E7*, E12	50	100	3000 s
1 4.6 K	M4N, M5N, M6N*, M7N*	1.5	3	90 s
2 4.6 K	M4N, M5N, M6N*, M7N*	5	10	300 s
3 4.6 K	M4N, M5N, M6N*, M7N*	15	30	900 s
4 4.6 K	M4N, M5N, M6N*, M7N*	30	60	1800 s
5 4.6 K	M4N, M5N, M6N*, M7N*	50	100	3000 s
1 77 K	M1*, M3, M4, M14*	5	10	300 s
2 77 K	M1*, M3, M4, M14*	15	30	900 s
3 77 K	M1*, M3, M4, M14*	30	60	1800 s
Target Irrad.				
1 77 K	E8, E9, E10*, E11*, M8, M9, M10*, M11*	6.80	10.0	505 h
2 77 K	E8, E9, E10*, E11*, M8, M9, M10*, M11*	13.90	16.2	750 h
3 77 K	E8, E9, E10*, E11*, M8, M9, M10*, M11*	20.48	19.0	879 h
4 77 K	E8, E9, E10*, E11*, M8, M9, M10*, M11*	36.98	34.6	1641 h
5 77 K	E8, E9, E10*, E11*, M8, M9, M10*, M11*	51.60	55.2	2370 h

* diodes measured versus dose

verify the effect of temperature annealing. Finally all diodes were re-measured again at 4.6 K and 77 K after staying for about two months at room temperature.

MEASUREMENT RESULTS

The forward voltage drop U_f of these diodes, which governs the thermal transient in the diodes during de-excitation of the magnets and the dimensioning of the heat sinks, is the most important characteristic affected by irradiation. If during irradiation this voltage exceeds a certain value then by excessive joule heating the diode will be damaged because the generated heat cannot be transferred sufficiently fast to the heat sink[6].

Reactor Measurements at 4.6 K

Figure 2 shows as typical examples the change of the current-voltage characteristic $I_f = f(U_f)$ of the diodes E5 and M6N before irradiation and after irradiation up to 50 kGy and thermal annealing at 300K. The deformation of the curve of diode M6N at higher doses is due to heating effects at higher current and voltage levels causing a decrease in forward voltage. The degradation - increase in forward voltage U_f - of the diode M6N is significant.

In Fig. 3 the relative increase of forward voltage $\Delta U_f/U_{fo}$ at the peak current level of 100 A is shown versus dose and fluence for the 8 diodes irradiated at 4.6 K. The maximum degradation of almost 600 % was measured on the diode M7N whereas the diodes E2 and E6 show an increase in U_f of about 180 %. The degradation in forward voltage versus dose is smaller for the diodes E1 and E5 which had intermediate annealing caused by the measurement itself. After 50 kGy and one thermal cycle of 15 minutes to room temperature the degradation in forward voltage drops to about 22 % for all of the EUPEC-diodes and to about 50 % for all of the GEC diodes.

The effect of annealing during 2 months at 300 K amounts to a further reduction of the degradation in forward voltage down to about 20% for the EUPEC diodes and to about 25 % for the GEC diodes. Intermediate current pulse annealing gave a final reduction in U_f of a few per cent only.

Reactor Measurements at about 77 K

In Fig. 4 the relative increase in forward voltage $\Delta U_f/U_{fo}$ for the EUPEC diodes E3, E4, E7, E12 and for the GEC diodes M1, M14 at about 77 K is shown. Similar to the

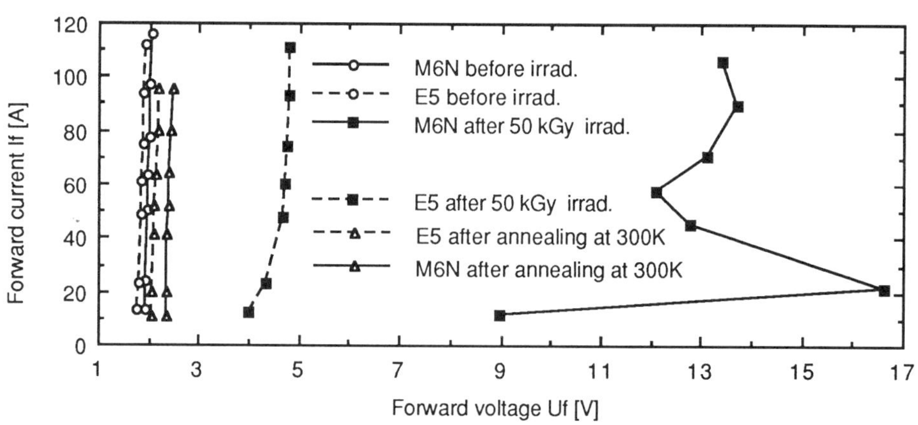

Figure 2. I_f-U_f-characteristic of diode E5 and M6N at 4.6 K before and after irradiation.

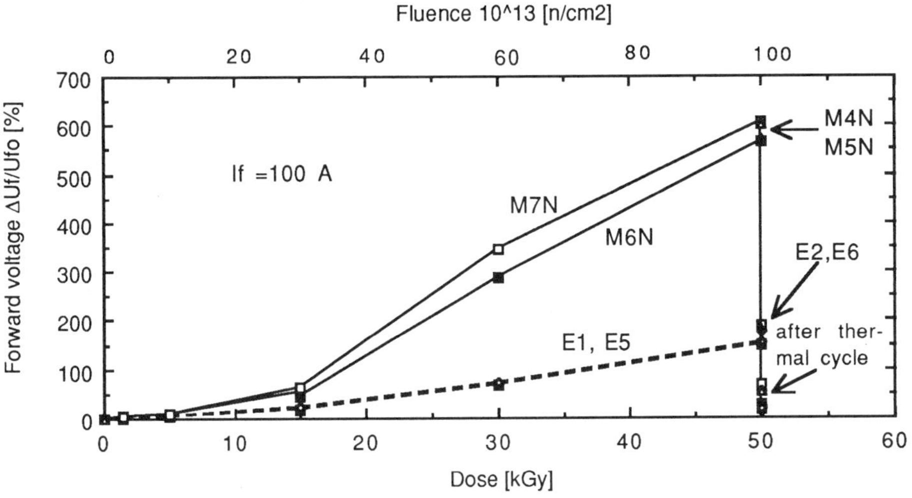

Figure 3. Relative increase of the forward voltage versus dose and fluence at 4.6 K.

results for the irradiation at 4.6 K the increase in forward voltage amounts to about 150 %for the EUPEC diodes E3 and E7 which were measured versus dose and up to about 175 % for the diodes E4 and E12 which were measured only before and after irradiation. The relative increase in U_f for the GEC diodes amounts to about 400 % after a dose of 30 kGy. After about 2 months at 300K the degradation is reduced to about 10 to 20 % for all of the diodes.

Comparison of Results at 4.6 K and 77 K

The comparison of results obtained for EUPEC diodes at 4.6 K and 77 K is shown in Fig 5. The degradation versus dose is surprisingly similar, although a larger degradation was expected at 4.6 K due to the reduced electrical conductivity. The corresponding results for the GEC diodes are given in Fig. 6 up to a dose of 30 kGy for the irradiation at 77 K. The GEC diodes irradiated at 4.6 K and 77 K gave similar results up to a dose of 30 kGy. If

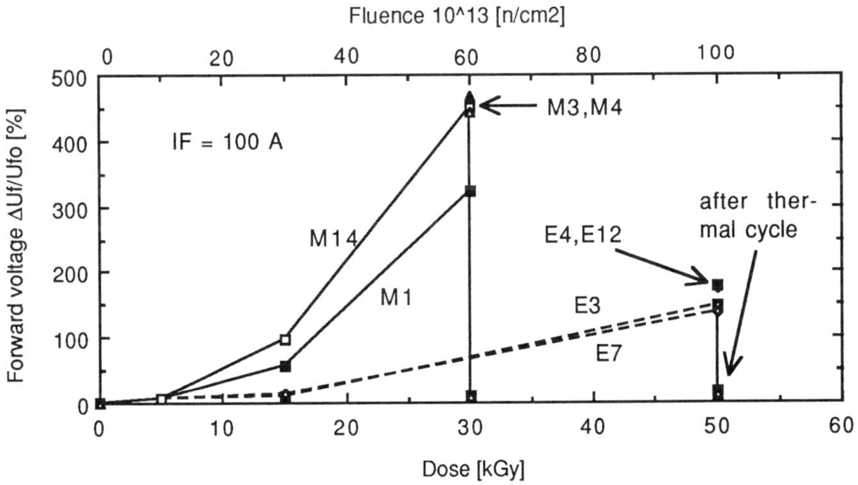

Figure 4. Relative increase of the forward voltage versus dose and fluence at 77 K.

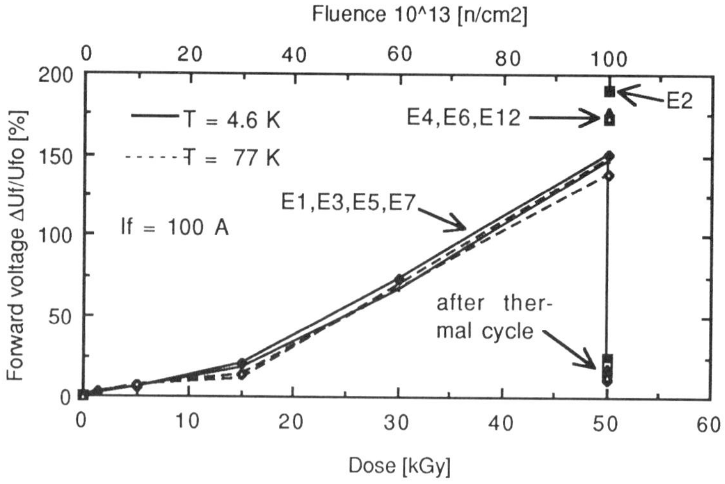

Figure 5. Relative increase of forward voltage versus dose and fluence at 4.6 K and 77 K for EUPEC diodes.

further tests on similar diodes should confirm this results then irradiation tests could be limited to tests at 77 K only.

The full reverse characteristic $I_r = f(U_r)$ was monitored on the GEC diodes after each exposure period in the I_r-range from 1 μA to about 1 mA. The reverse voltage U_r versus dose at $I_r = 1$ mA is shown in Fig. 7 during irradiation at 4.6 K up to a dose of 50 kGy and during irradiation at 77 K up to a dose of 30 kGy. During 77 K irradiation the reverse voltage increased for most of the diodes (up to a maximum of about 50 % for diode M4) without any significant decrease after annealing at 300 K. During 4.6 K irradiation the reverse voltages also increased for all diodes (up to a maximum of about 100 % for diode M4N). After annealing at 300 K however the reverse voltage decreased for most of the diodes to a value less than before irradiation. On diode M5N even a short circuit was measured.

Accelerator Target Area Irradiation Measurements

Four GEC diodes and four EUPEC diodes of 10 mm diameter were exposed to irradiations at 77 K in the SPS target area at CERN up to a dose of about 52 kGy and a

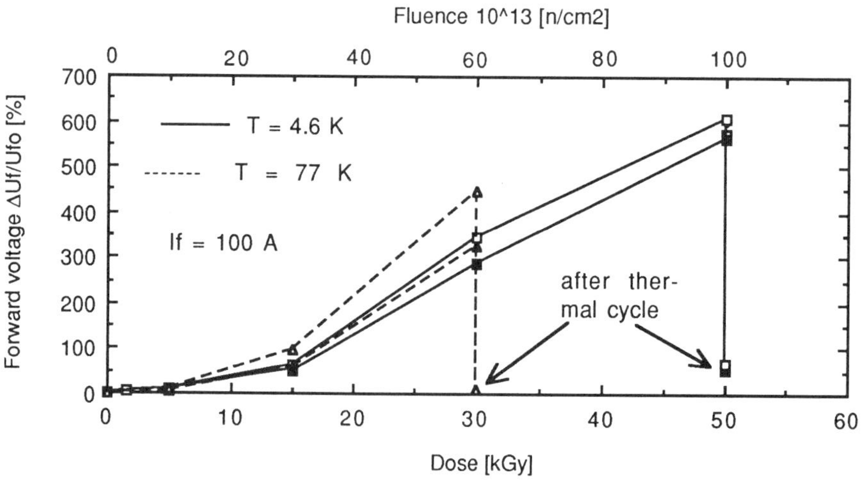

Figure 6. Relative increase of forward voltage versus dose and fluence at 4.6 K and 77 K for GEC diodes.

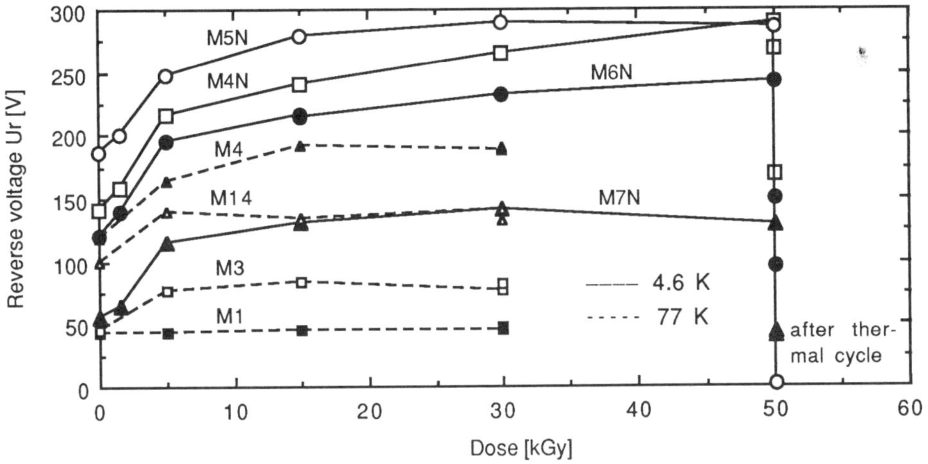

Figure 7. Reverse voltage (at Ir=1 mA) versus dose at 4.6 K and 77 K irradiation.

neutron fluence of up to about 50 10^{13} n/cm^2 which is about half of the fluence compared to the reactor irradiations. The relative increase in forward voltage $\Delta Uf/Uf_o$ at $I_f = 100$ A for these diodes is shown in Fig. 8. Intermediate current pulses annealing was applied to the GEC diode M11 and to the EUPEC diode E11 at about 37 kGy. The reduction in forward voltage especially for diode M11 is significant and mainly due to partial thermal annealing caused by the measurement itself.

Comparison of Reactor Irradiation and Accelerator Target Area Irradiation

For the comparison between reactor irradiation and accelerator target irradiation typical results are presented in Fig. 9 for the EUPEC diodes E7 and E10 and for the GEC diodes M10 and M14 where the relative increase in forward voltage is given versus dose. For the EUPEC diodes the increase in forward voltage during reactor irradiation and target irradiation are almost identical up to a dose of about 15 kGy and more than a factor of two higher during reactor irradiation at a dose of about 50 kGy. For the GEC diodes the

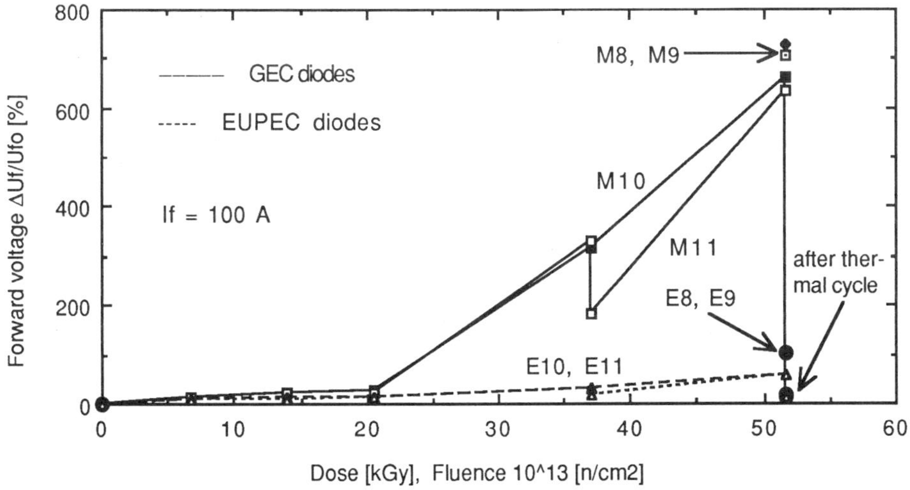

Figure 8. Relative increase of forward voltage versus dose and fluence for EUPEC and GEC diodes during target irradiation at 77 K.

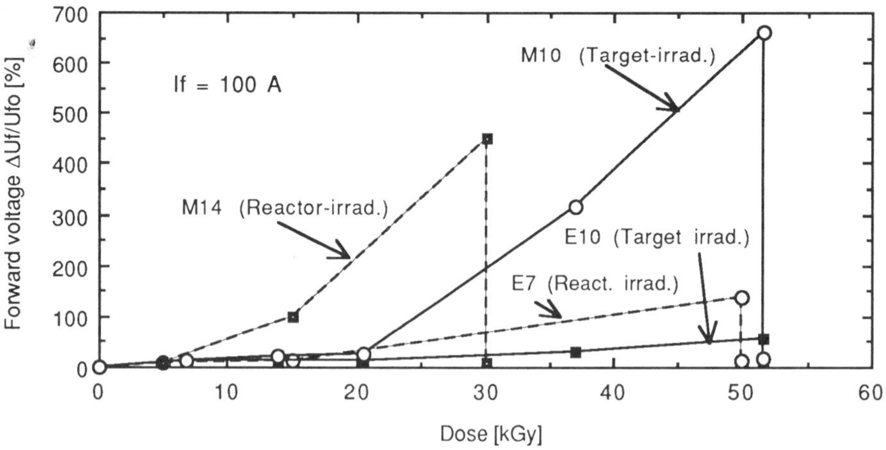

Figure 9. Forward voltage increase versus dose during reactor- and target irradiation at 77 K.

increase in forward voltage is similar up to a dose of about 5 kGy only and also more than double during reactor irradiation at a dose of 30 kGy.

CONCLUSIONS

Epitaxial diodes of the same epitaxial width (22μm) of two different manufacturers show substantial differences when irradiated, which can be attributed to different fabrication processes. Results from irradiations at 4.6 K and 77 K are similar, at least for the EUPEC diodes, and it should be possible in future to restrict irradiation tests to 77K. For the EUPEC diodes, reactor and target irradiations show almost identical degradation in forward voltage up to about 15 kGy, so that long term irradiations (over 6 month) in the target area could be replaced by short term irradiations in a nuclear reactor (a few days).

ACKNOWLEDGMENT

The authors wish to thank G. Brianti, L. Evans, and R. Perin for their active support of these tests, V. Berland and L. Coull for many helpful discussions, M. Tavlet for his assistance in dosimetry, A. Gharib for careful carrying-out of the measurements, and J.M. Fraigne for the careful design of the sample holders.

REFERENCES

1. D. Hagedorn and W. Nägele, "Quench Protection Diodes for the LHC", ICMC, Huntsville, Alabama, 1991.
2. H. Gerstenberg and W. Gläser, "Neutron Irradiation at Temperatures below 6 K at the Munich Research Reactor (FRM)", Nuclear Sience and Engineering, 110 (1992), p. 50-55.
3. H. Gerstenberg et al., "Irradiation of Quench Protection Diodes at Liquid Helium Temperatures in a Nuclear Research Reactor", CERN, AT-MA Internal Note 94-98, (1994).
4. F. Conincks et al., "Response of Alanine Dosimeters to Irradiations at Cryogenic Temperatures", CERN-TIS-CFM/ 95-09.
5. L. Burnod et al., "Expected Doses Inside and Around LHC Dipoles", AC/DI/FA/Note 93-06 (1993).
6. V. Berland et al., "Testing of High Current By-pass Diodes for the LHC Magnet Protection", Magnet Technology Conference, MT14, June 11-16 (1995), Tampere, Finland.

LOW RADIOACTIVITY CRYOSTAT SYSTEM FOR A 20mK PARTICLE DETECTOR

J. D. Taylor,[1,3] D. S. Akerib,[1,2] P. D. Barnes, Jr.,[1,2] P. A. Luft,[3] R. R. Ross,[1,2,3] B. Sadoulet,[1,2,3] R. V. Schafer,[3] S. White,[1,2] R. C. Wolgast[3]

[1]Center for Particle Astrophysics, University of California at Berkeley Berkeley California 94720
[2]Department of Physics, University of California at Berkeley, Berkeley California 94720
[3]Lawrence Berkeley National Laboratory, 1 Cyclotron Road, Berkeley California 94720

ABSTRACT

A system for cooling detectors used in particle astrophysics is described. The volume accommodating particle detectors is 30 cm diameter by 30 cm high and attained a temperature of 8.5 +/- 0.5 mK. The need for low radioactive backgrounds has driven design, selection of materials and assembly techniques. Unusual features include a detector package cooled by conduction by tubes ~1m in length connected to the port of a commercial side access dilution refrigerator, swinging support system for detectors and heat shields, fabrication using electron beam welding, assembly techniques for thermally conducting bolted joints, flexible thermal conduction paths, and copper flange on copper gasket vacuum seals at LHe temperature. A first cool down of the system resulted in temperature distributions and thermal loads consistent with thermal modeling.

INTRODUCTION

There is strong evidence that the portion of the universe which is visible is only a small portion of the total mass. The portion which is not visible is referred to as "dark matter". Many hypotheses have been advanced to provide explanations of the unseen dark matter and how they might be verified.[1,2] The equipment we are reporting on is part of an effort to detect one of the possible candidates, a new particle described as a Weakly Interacting Massive Particle (WIMP). These WIMPs would encounter the earth in a way similar to solar neutrinos. They would interact with the nuclei of the detector material, producing both ionization and heat. The detectors being developed for use with this equipment operate at temperatures around 20mK. Their design is such that they detect conventional particles emerging from natural radioactivity as well as WIMPs. They are described more completely in other publications.[3,4,5] This report describes the hardware used to house and cool these detectors, with emphasis on some unusual design problems and solutions.

Advances in Cryogenic Engineering, Vol. 41
Edited by P. Kittel, Plenum Press, New York, 1996

RADIOACTIVITY

The response of the detectors to radioactivity other than WIMPs poses a background which must be kept at low levels relative to the interaction rate of WIMPs (expected to be 1 or 2 per day for a detector mass of 1 kg). This imposes limitations in the choice of construction materials. Experience building other low radioactivity detectors has produced a short list of materials which usually pass (or fail) when monitored for radiation at very low levels.[6,7] 101 Copper (oxygen free) and many plastics usually pass while stainless steel, aluminum, and indium usually fail due to impurities which are radioactive. A sample from each lot of material used in the construction of our cryostat had to pass a radiation counting test. Since these tests required several days each, we tried to limit the total number of lots needed for the design. We chose to use brass screws to limit radioactivity. These screws were custom machined from material we supplied and tested. While not as radio-pure as 101 copper, the material we selected was acceptable if used sparingly. Comparing rates of thermal contraction for brass and copper, it is seen that using brass screws also provided an increase in tension as the system cooled. Stainless steel contracts at a lower rate than copper and would lose tension. This helps make up for the lower yield strength of brass fasteners. The main body of the cryostat was made from 101 copper. In most cases we used electron beam welding to avoid impurities from the welding process. This produced full penetration welds with good thermal conductivity.

In order to minimize the effects of radioactivity from outside the cryostat, a shielding system utilizing lead, polyethylene, and scintillator was designed to surround the cryostat. The source of cooling is a dilution refrigerator located outside this shield. The dilution refrigerator chosen has a side-access port near the bottom of its cryostat and thermal radiation shields at (approx.) 77K, 4.2K, 600mK, 50mK as well as the base temperature around 5mK with no thermal load. The base temperature was rated by the manufacturer at 100mK for a 400 micro-watt load. The design goal was to be capable of detector temperatures <20mK. The radiation shields were arranged concentrically. Our design has a set of nested "cans" surrounding the detectors and connected horizontally to the sideports of the dilution refrigerator. A path for electronic wiring to the detectors is also necessary. Figure 1 shows this design. We have minimized the diameter of the horizontal "stems" in order to minimize the opening in the radioactivity shield. To achieve this goal, multi-layer insulation was used between the 77K and room temperature cans and in the dilution refrigerator but not in the horizontal stems. The aluminized mylar used has a small, but measurable, radioactivity level presumed to be carried in the aluminum coating. This may require replacement in the future. The copper shields at room temperature, 77K, and 4K were gold plated to reduce emissivity.

DIMENSIONS

The space for detectors was specified as 30 cm in diameter and 30 cm in height with another 10 cm added at 4K for electronics. The outside dimensions of the room

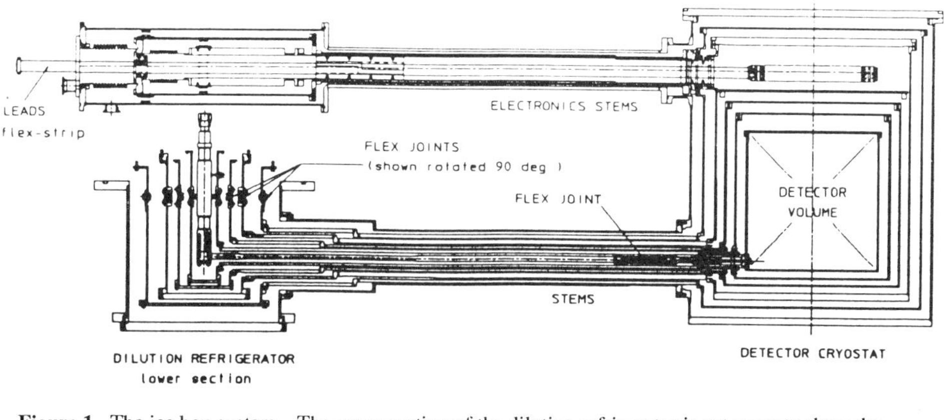

Figure 1. The ice box system. The upper section of the dilution refrigerator is cut away to show the electronics stems. The layers of the cryostst operate at (nom.) 300K, 77K, 4K, 600mK, 50mK and 10mK.

Table 1. Calculated heat loads

shield	radiation	conduction: hangers	conduction: other	total: this test	future added load
LN (nom.77K)	33 W	6.2 mW	980 mW	34 W	1.8 W
LHe (nom. 4K)	400 mW	1.7 mW	210 mW	610 mW	460 mW
Still (nom 600mK)	210 uW	13 uW	100 uW	320 uW	1.1 mW
cold plate (nom. 50mK)	24 nW	35 nW	0.76 uW	0.82 uW	1.7 uW
base	3.2 E-13 W	3.0 E-11 W	4.8 nW	4.8 nW	33 nW

temperature can are 60.4 cm in diameter at the top flange and 70 cm in height. The size and cost of the radioactivity shield made it desirable to keep these dimensions as small as practical. The room temperature stem is 12 cm in diameter where it goes through the shield. The step-down of stem diameter near the dilution refrigerator is intended to block line of sight gamma radiation down the axis of the stems. A second stem for electronics exits near the top of the cans. Its orientation is such that it clears along the side of the dilution refrigerator. This stem is designed to provide heat sinking for the wiring at room temperature, LN, and LHe, and limit conduction heat loads. All cooling for this stem is via conduction from the ice box cans. It is also designed to block line-of-sight radiation along its axis.

THERMAL CONTRACTION AND THERMAL CONDUCTION

As the dilution refrigerator is cooled, the cold stages move vertically relative to the room temperature portion by about 5 mm. In addition, the link between the detector region and the dilution refrigerator contracts about 6mm horizontally when cooled. These motions had a strong influence on our suspension system design for the cans. We selected a suspension system where the bottom of each can was hung by sets of loops from the top of the next warmer one. This system allows long thermal paths with small cross section. The hanger loops were fabricated from kevlar cord and epoxy, with copper spools at each end. Figure 2 shows details of this system. Table 1 shows the anticipated heat loads for the system based on the materials and dimensions chosen for the design. The emissivities used in the calculation were chosen to be pessimistic, and are probably higher than realized. For the 77K, 4K, and 600mK shields, the radiation heat load predominates in the detector cryostat without the detectors installed. The column for "conduction: other" is significant at the lowest temperature stages. Much of this load is from the standoff bumpers at the

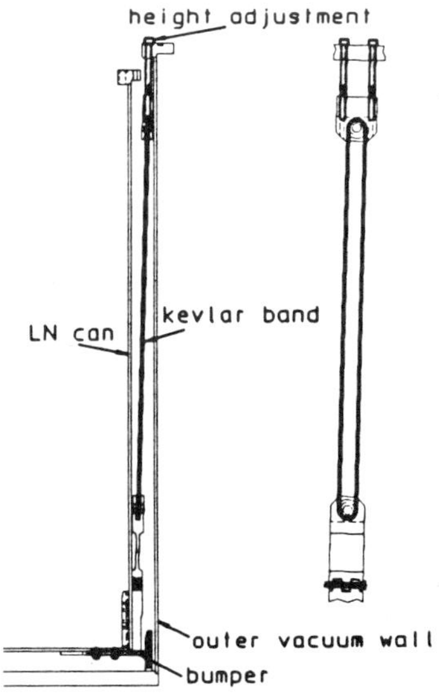

Figure 2. Suspension design. Kevlar hanger loops allow for thermal contraction.

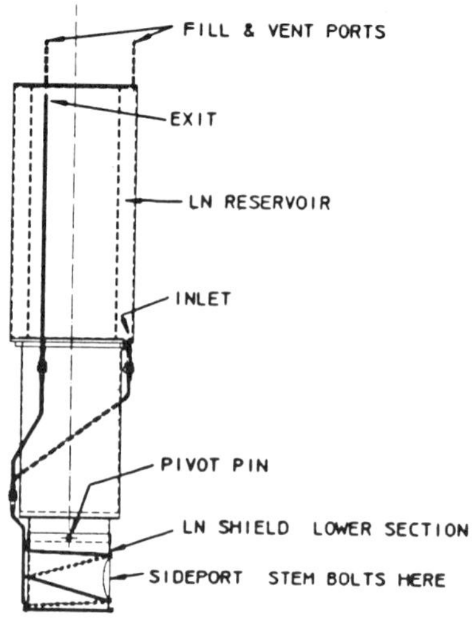

Figure 3. LN shield of the dilution refrigerator, modified with a cooling loop and pivot joint.

bottom of the cans. These position the cans at right angles to the axis of the stems. They are designed to have clearance if the can is hanging perfectly centered. For this calculation, a solid contact is modeled, which corresponds to more conductivity than realized. The column for "future added load" represents addition of 12 detectors to the system, along with their wiring and electronics. For the colder stages of the system, this added heat will clearly dominate the heat loads.

The suspension geometry is such that during cooling and warming, a flex joint is required in each cold layer of the system. These must conduct the anticipated heat loads as

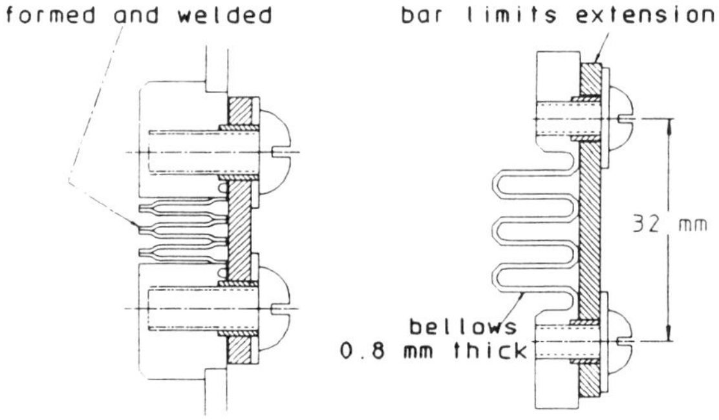

Figure 4. Bellows flex joints (cross section of cylindrical bellows). The design on the left is formed and welded. The design on the right is machined from a solid copper pipe.

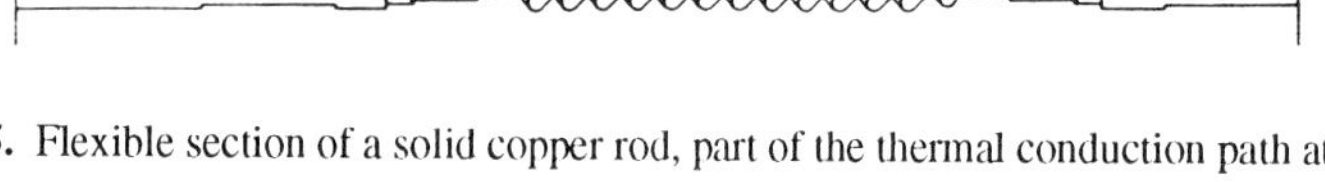

Figure 5. Flexible section of a solid copper rod, part of the thermal conduction path at 10 mK.

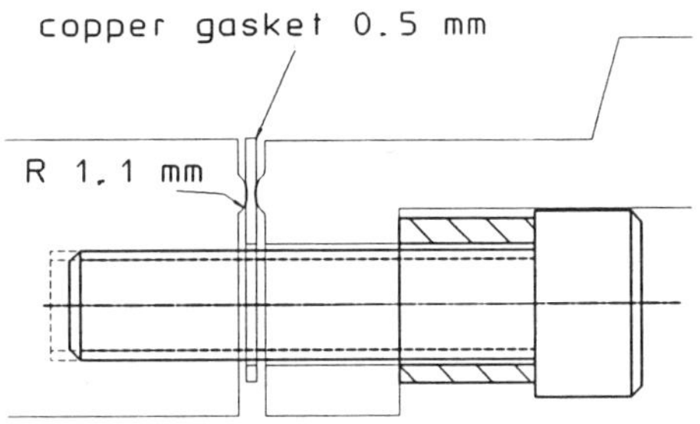

Figure 6. 4K Vacuum seals, an all-copper design.

well as flex easily. The flex joints took three different forms in our design. The need to transmit 40W was a major concern in designing the flex joint for the 77K shield. Our design uses two pivot pins near the bottom of the dilution refrigerator (see Figure 3). These pins are not expected to transmit any cooling. Cooling is provided by a tube with both ends plumbed into the LN reservoir of the dilution refrigerator and coiled around the shield below the pivots. The path of the tube is such that it allows this pivoting without requiring bellows. An additional access port through the room temperature vacuum wall of the dilution refrigerator was added to give access to two connections in this cooling loop to permit disassembly. Liquid nitrogen enters the inlet side of the tube and is directed to the bottom of the radiation shield before contacting the shield. As heat from the shield boils the LN, the bubbles rise along the upwardly spiraling section of the line to exit at a point higher than the inlet. No cryogenic fluids traverse the horizontal stems; conduction is the only cooling mechanism in the stems and cans for the 77K and 4K shields. The stems are made from heavy walled tubes of annealed copper in order to conduct the expected heat loads.

The 4K, 600mK, and 50mK shields utilize a bellows design to allow pivoting motion around the pivot axis defined by the pins in the LN shield (see Figure 4). Short copper links at this axis carry the weight of the parts and fix the vertical spacing. During cool-down of the shields which will operate below 4K, an exchange gas is used for convection

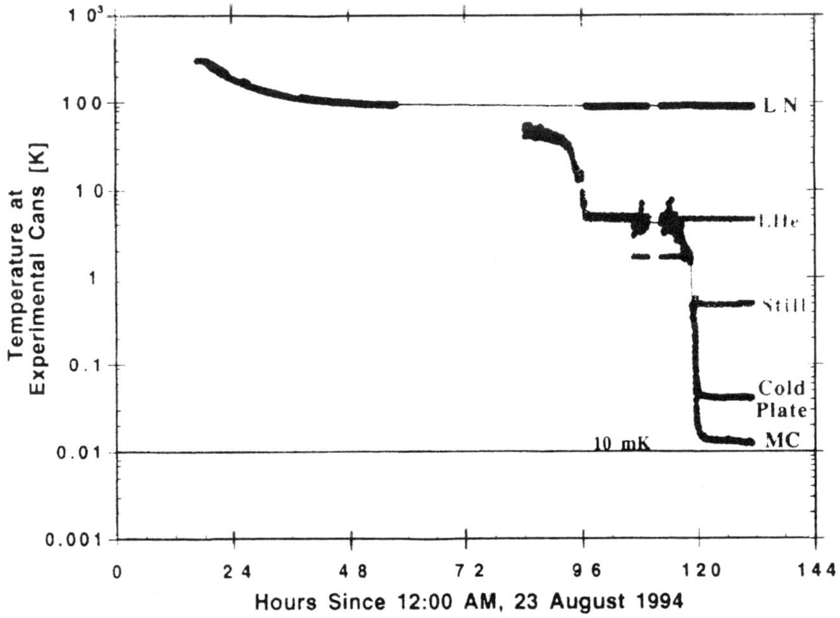

Figure 7. First cooldown of the ice box.

cooling of the colder stages. This requires the 4K shield to be leak-tight to retain the insulation vacuum surrounding it. The bellows length and thickness are chosen to provide adequate thermal conduction without undue resistance to pivoting. After fabricating a welded bellows for the 4K shield, it was decided that it would be more economical to machine subsequent bellows.

The pivot in the base temperature path is accomplished thru flexing of a 10 cm length of copper welding cable located in the horizontal stem (see Figure 5). The 2000 fine strands of the cable transmit little force to the detector region where vibration is of concern. The cable is electron beam welded to solid copper rod at both ends and provides good thermal conduction. The cable is installed slightly compressed so that there is remaining travel along its axis.

The thermal conductivities of the bellows and welding cable were checked by cooling the parts to LHe temperatures and measuring their electrical resistance. The Wiedemann-Franz law gives the correlation between electrical and thermal conduction. This was a useful technique for testing relative effectiveness of various thermal joints and comparing them to the resistance of the bulk copper. This technique was also used to test some of the bolt-together connections and methods of surface preparation prior to bolt-up. Test results showed that the limited torques achievable with brass screws would produce acceptable thermal contact. Surfaces to be bolted together were scrubbed at assembly time with Scotch-Brite wet with ethyl alcohol, then cleaned with low lint wipes and alcohol. Gold plated surfaces or fragile vacuum seals were flushed with alcohol.

VACUUM SEALS

One of the problems caused by the choice of copper for most of the metal parts was creation of vacuum seals at flanges in the 4K shield. Indium seals were retained in the dilution refrigerator, but indium did not meet the requirements for low radioactive background inside the radioactivity shield. The design used for the 4K can seals is shown in Figure 6. The two flanges to be sealed have opposing ridges machined in their faces. A thin gasket of annealed copper is clamped between the flanges using brass screws. The parts are held in alignment by two or more pins. The flanges must be thick enough to resist flexing and the tension in the screws must be sufficient to press the ridges into the replaceable gasket. In the example shown, the length of the bolt under tension has been extended by the use of a spacer washer. The gasket remains sealed when cooled to 4K and conducts heat satisfactorily.

MODIFICATIONS TO THE DILUTION REFRIGERATOR

While the dilution refrigerator cooled to a reasonable base temperature in the as-delivered condition, tests using heaters to simulate the expected additional heat loads from the detector cryostat and electronics revealed some inadequacies. To improve thermal conduction, the stainless steel bottom plate of the LHe reservoir was replaced with one made from copper. The connections between the still and the 600mK shield, and between the cold plate section of the heat exchanger and the 50mK shield were tightened and retro-fitted with highly conductive thermal shunts made from high purity copper wire annealed in a low pressure mixture of helium and oxygen.[8] In addition, the coiled counter-flow heat exchanger was wrapped with monofilament line to limit turn-to-turn thermal shorts. These modifications produced much more satisfactory temperatures under the same heater loads.

PERFORMANCE

The cryostat was assembled and first cooled at Berkeley in August 1994. A plot showing temperatures of the various nested cans while cooling is given in Figure 7. Initially, the LN and LHe reservoirs of the dilution refrigerator were filled with LN. Nitrogen exchange gas was used inside the volume of the LHe shield to cool the (eventual) colder shields. On the fourth day of the cooldown, LHe replaced LN in the LHe shield. Helium exchange gas replaced the nitrogen gas which had been pumped from the system the previous night. The He exchange gas was pumped from the system at the end of the

Table 2. Thermal test results

shield	refrigerator only temp.	with cryostat temp. at cans	heaters as a set load	heaters as a set delta temp.	heaters individually load	heaters individually delta temp.
LHe (nom. 4K)	4.25 +/- 0.2 K	5.14 +/- 0.2 K	0.60 W	0.9 K	1.2 W	0.48 K
Still (nom 600mK)	530 +/- 100 mK	510 +/- 100 mK	0.14 mW	7 mK	3.0 mW	160 mK
cold plate (nom. 50mK)	39 +/- 5 mK	45 +/- 5 mK	1.4 uW	5 mK		
base	5 +/- 0.5 mK*	8.5 +/- 0.5 mK*	0.090 uW	0.4 mK	1.0 uW	6.6 mK
* spin orientation thermometry						

day, and cooling of the lower temperature stages began. These stages were cold at the end of the 5th day.

Thermometers were located throughout the system to permit detection of poorly conducting connections and flexible sections. Resistive heaters were installed to model the calculated future heat loads. Temperatures near 77 K were measured using diodes, colder temperatures were measured using pre-calibrated RuO_2 resistors. Thermal data recording was accomplished by computer at preset intervals. Table 2 gives thermal results from this test and the previous test of the dilution refrigerator without the detector cryostat. The base temperature was also measured at the detector volume using nuclear spin orientation thermometry with a ^{60}Co source. After adjusting operation of the dilution refrigerator, a base temperature of 8.5 +/- 0.5 mK for the detector volume was achieved. Temperatures realized at the various shields are close to predicted values. Based on the temperature rises resulting from powering heaters, it is anticipated that the additional heat from electronics and mechanical connections between layers will not raise the base temperature above 10mK when detectors are added to the system. Heating individual layers with larger heat loads had little effect on the temperatures of the other layers. A temperature controller is anticipated which will apply heat to the base temperature shield and provide constant temperature operation.

The LN cooling loop worked well in earlier tests of the dilution refrigerator with heaters installed. While operating well at the beginning of this system cooldown, the cooling loop later appeared to have been blocked with ice. The LN reservoir was often open to air and this probably resulted in frost and condensation. The fill system for the LN bath is now redesigned to prevent moist air from entering.

After this successful cold test, the equipment was moved to an underground site at Stanford University. The dilution refrigerator has subsequently been cooled to comparable base temperatures and the ice box has been reassembled. At the time of this writing, it is in the beginning stages of a second cool-down.

ACKNOWLEDGMENT

This work was supported by the Center for Particle Astrophysics, a National Science Foundation Science and Technology Center operated by the University of California at Berkeley, under the Cooperative Agreement No. AST-9120005 and by the Department of Energy under Contract No. DE-AC03-76SF00098.

The preliminary design included much work by Paul Pipersky. Tom Shutt provided much useful input, particularly relating to dilution refrigerator operation. Several people assisted in preparation of the instrumentation and running the dilution refrigerator, including P.Andre, E.Aubourg, A.Cummings, A.DaSilva, R.Diaz, S.Golwala, S. Murgulies, B.Pritychenko, S.Sadoulet, D.Seitz, T.Shutt, A.Sonnenschein, and W.Stockwell. D.Osheroff assisted with advice and equipment in the annealing of high purity wire. Many fruitfull discussions were held with collaborators at Stanford Universisty and the University of California at Santa Barbara. A large part of the fabrication was done by Lawrence Berkeley National Laboratory's shops and the University of California at Berkeley's Physics Department shop .

REFERENCES

1 P.F. Smith and J.D. Lewin, Physics Reports 187:203 (1990).
2 J.R. Primack, D Seckel and B. Sadoulet, Annu Rev. Nucl. Part. Sci. 38:751 (1988).
3 PD.Barnes et al.,Workshop on Low Temperature Detectors V, Journal of Low Temp. Phys. 93:791 (1993),.
4 T.Shutt, et al., Phys. Rev. Lett., 69:3452, (1992).
5 T.Shutt et al., Phys. Rev. Lett., 69:3531, (1992).
6 R.L.Brodzinski et al., AIP Proceedings - Solar Neutrinos and Neutrino Astronomy, 50 (1984).
7 A. Smith, Lawrence Berkeley National Laboratory, private conversations.
8 F.R.Fickett, J. Phys. F:Met. Phys., 12:1753 (1982).

SIMPLE SPREADSHEET THERMAL MODELS FOR CRYOGENIC APPLICATIONS

Alfred Edward Nash III

Jet Propulsion Laboratory
California Institute of Technology
Pasadena, California 91109 USA

ABSTRACT

Self - consistent circuit analog thermal models that can be run in commercial spreadsheet programs on personal computers have been created to calculate the cooldown and steady state performance of cryogen cooled dewars. The models include temperature dependent conduction and radiation effects. The outputs of the models provide temperature distribution and dewar performance information. These models have been used to analyze the SIRTF Telescope Test Facility (STTF). The facility has been brought on line for its first user, the Infrared Telescope Technology Testbed (ITTT), for the Space Infrared Telescope Facility (SIRTF) at JPL. The model algorithm as well as a comparison between the models' predictions and actual performance of this facility will be presented.

INTRODUCTION

The current economic climate dictates that the designed performance of large scale cryogenic systems be traded off against cost to minimize the resources necessary to develop the system while still meeting the technical requirements. In keeping with NASA's desire to simplify and reduce the cost of flight operations, the Jet Propulsion Laboratory (JPL) undertook the task of a quick development, low cost facility for the optical interferometric testing of mirrors of diameter $\leq$ 1 m, f $\leq$ 6, at temperatures from 300 to 5 K. The project was constrained by an allocation of one year in which to complete the design, fabrication, and optical qualification test, and fixed funding.

In order to meet these design challenges, self - consistent circuit analog thermal models that could be run in a commercial spreadsheet program on a personal computer were developed to calculate the cooldown and steady state performance of cryogen cooled dewar designs. These models were used to identify design aspects where the resources necessary to develop the system could be minimized while still meeting the technical requirements.

MODEL DESCRIPTIONS

Steady State Model

The equation for one dimensional conductive heat flow Q through an area A of a material of conductivity λ under a temperature gradient $\partial T/\partial x$, $Q = \lambda(T)\ A\ \partial T/\partial x$, and the equation for radiative heat transfer Q between parallel surfaces of area A, and emissivities ε_1 and ε_2 and at respective temperatures T_1 and T_2, $Q = \sigma A\,(T_2^4 - T_1^4)\,(\varepsilon_1\varepsilon_2/(\varepsilon_1 + \varepsilon_2 - \varepsilon_1\varepsilon_2))$ (where σ is Stefan's constant) can be used to write equations for one dimensional heat flow between two points in terms of effective thermal resistances R and $\Delta T = (T_2 - T_1)$, $Q = R^{-1}\,\Delta T$.

The conductive resistance is

$$R_C = \left[\frac{A}{l}\,\overline{\lambda}\right]^{-1} \tag{1}$$

where l is the length of conductive segment and $\overline{\lambda}$ is the average thermal conductivity of the segment found by integrating an analytic function through published material data

$$\overline{\lambda} = \int_{T_1}^{T_2} \lambda(T)\,dT\,/\Delta T \tag{2}$$

The radiative resistance (with $\varepsilon = \varepsilon_1\varepsilon_2/(\varepsilon_1 + \varepsilon_2 - \varepsilon_1\varepsilon_2)$) is

$$R_R = \frac{\Delta T}{\varepsilon\,\sigma\,A\,(T_2^4 - T_1^4)} \tag{3}$$

A network of conductive and radiative heat transfer paths that model an entire thermal system can then be solved by treating the system as an electrical resistor network

$$Q = R^{-1}\,\Delta T \Rightarrow I = R^{-1}\,V \tag{4}$$

Although the radiative and conductive thermal resistances are highly temperature dependent, a self - consistent steady state solution to the circuit analog thermal model that includes the thermal properties of the materials can be found by executing the following steps:

- An initial distribution of temperatures at each resistor network node point $[T_1, T_2, \ldots, T_i, T_j, \ldots, T_N]$ is assumed.
- The temperatures at each resistor network node point, and the known thermal properties of the link (conductive or radiative) are used to calculate the (thermal) resistance of each link in the system network.

$$R_{ij}^{C} = \left[\frac{A}{l\,(T_j - T_i)} \int_{T_i}^{T_j} \lambda(T)\,dT\right]^{-1} \tag{5}$$

$$R_{ij}^{R} = \frac{T_j - T_i}{\varepsilon_{ij}\,\sigma\,A\,(T_j^4 - T_i^4)} \tag{6}$$

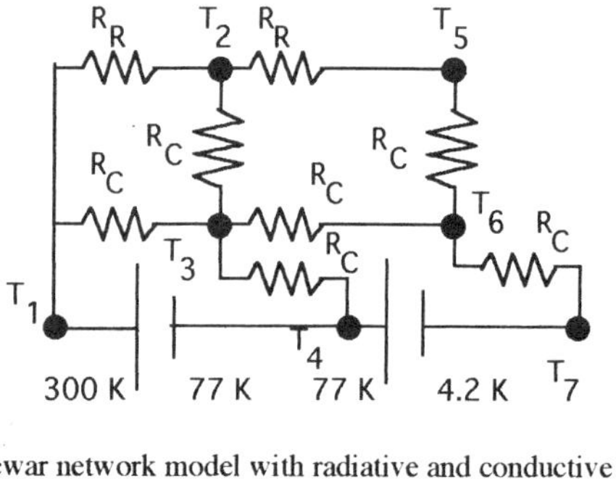

Figure 1 Dewar network model with radiative and conductive heat paths.

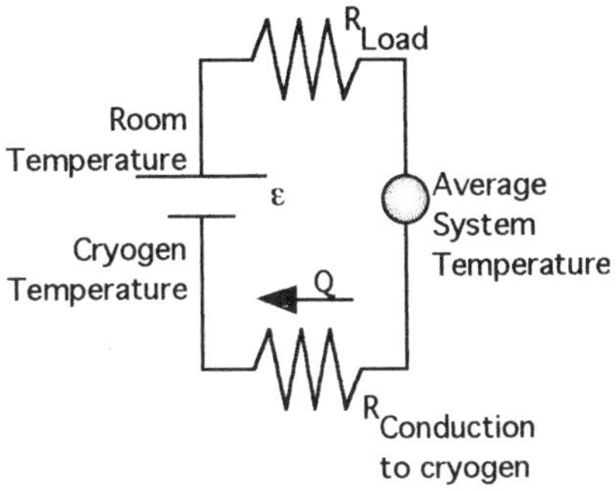

Figure 2 Equivalent resistor circuit model for the average system temperature and cryogen heat load.

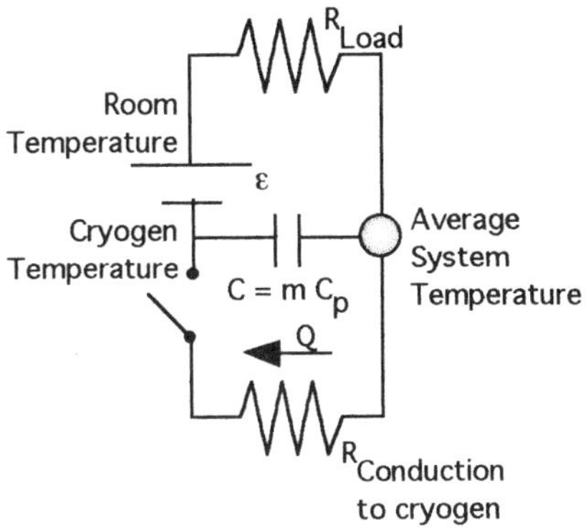

Figure 3 Equivalent RC circuit model for the average system temperature and cryogen heat load.

- Using Kirkoff's rules, the (heat) current through each resistive element is calculated (using room temperature and the cryogen boiling points as points of fixed "potential").
- The calculated heat currents and thermal resistances are used to calculate a new distribution of temperatures at the resistor network node points (using room temperature and the cryogen boiling points as fixed points).

$$T_j - T_i = R_{ij} \bullet Q_{ij} \tag{7}$$

- The steps are repeated with the system temperature distribution calculated in the previous step until the heat flow though each element calculated by the network solution and the heat flow though each element calculated from thermal properties of the individual element differ by less than 0.1%.

This series of steps can be implemented on a personal computer within a spreadsheet program that can invert matrices. To calculate the network solution to the heat flow in each resistive element, Kirkoff's laws are used to write a matrix **K** of equations multiplying the element currents. The matrix **K** can be written as three sub-matrices: $\mathbf{K}_1$, $\mathbf{K}_2$ and $\mathbf{K}_3$. The matrix $\mathbf{K}_1$ multiplied by the vector of element currents **Q** is the set of equations for the temperature drop across all thermal paths in the system, $\{R_{ij}Q_{ij}\}$, and is equal to a sub matrix $\mathbf{\Delta T}$. The matrix $\mathbf{K}_2$ multiplied by the vector of element currents is the set of equations for the total temperature drop around all closed loops within the resistor network, $\{R_{12}Q_{12} + R_{23}Q_{23} + ...\}$, and is equal to a null vector **0**. The matrix $\mathbf{K}_3$ multiplied by the vector of element currents is the set of equations for the difference between the sum of the currents going into each node and the sum of the currents out of each node. It is also equal to a null vector **0.**

$$[\mathbf{K}] \begin{bmatrix} Q_{12} \\ \cdot \\ Q_{ij} \\ \cdot \\ Q_N \end{bmatrix} = \begin{bmatrix} [\mathbf{K}_1] \\ [\mathbf{K}_2] \\ [\mathbf{K}_3] \end{bmatrix} \begin{bmatrix} Q_{12} \\ \cdot \\ Q_{ij} \\ \cdot \\ Q_N \end{bmatrix} = \begin{bmatrix} [\mathbf{\Delta T}] \\ [\mathbf{0}] \\ [\mathbf{0}] \end{bmatrix} \tag{8}$$

The network solution to the heat flow though each element can then be calculated by inverting the matrix of equations obtained from Kirkoff's laws **K**.

$$\begin{bmatrix} Q_{12} \\ \cdot \\ Q_{ij} \\ \cdot \\ Q_N \end{bmatrix} = [\mathbf{K}]^{-1} \begin{bmatrix} [\mathbf{\Delta T}] \\ [\mathbf{0}] \\ [\mathbf{0}] \end{bmatrix} \tag{9}$$

Time Dependent Cooldown Model

From the information obtained with the steady state model it is possible to create a time dependent cooldown model for the average cooling rate of the entire system by executing the following steps:

- Create an equivalent resistor circuit model for the average system temperature T and cryogen heat load Q (Fig. 2).
- The time dependent cooldown behavior can then be modeled with an RC circuit obtained by adding a capacitor in parallel to, and a switch in series with, the thermal resistance to the cryogen bath (Fig. 3).

The average system temperature is then given by

$$T_{\text{system avg.}} = T_{\text{cryogen}} + \tilde{\varepsilon}\,(1 + \rho\, \exp^{-t/\tilde{R}C}) \tag{10}$$

where

$$\tilde{\varepsilon} = \frac{\varepsilon R_C}{(R_L + R_C)} \tag{11}$$

$$\rho = \frac{R_L}{R_C} \tag{12}$$

$$\tilde{R} = \frac{R_L R_C}{(R_L + R_C)} \tag{13}$$

$$C = m\, C_p \tag{14}$$

where R_L and R_C are the thermal load and conduction to cryogen resistances, respectively, and m and C_p are the mass and specific heat of the cooled parts of the system, respectively.

The time - dependent heat load to the cryogen is given by

$$Q = \frac{\tilde{\varepsilon}}{R_C}(1 + \rho\, \exp^{-t/\tilde{R}C}) \tag{15}$$

DEWAR STRUCTURE

The STTF dewar has a vertically oriented inner cylindrical test chamber with a diameter of 1.4 m and an internal height of 2.3 m. The experiment mounting surfaces, the upper and lower surfaces of the chamber, are the exterior surfaces of two liquid helium tanks. The cylindrical wall of the test chamber is conductively cooled by the two 300 liter capacity tanks. Surrounding the test chamber is a bell shaped aluminum radiation shield. This bell is conductively cooled through a bolted contact joint with a 300 liter capacity liquid nitrogen tank directly below the helium cooled test chamber. The outer vacuum shell of the dewar is constructed of an aluminum dome that makes an O - ring seal with a stainless steel lower shell surrounding the lower cryogen tanks. Structural and thermal separation between the outer vacuum shell, the nitrogen tank, and the lower helium tank is provided by titanium struts and a stainless steel tension/compression system necessary for earthquake safe operation of the facility in southern California. The entire dewar had to be designed to withstand 1.0 g in the vertical and 0.5 g in the horizontal axis without serious hazard to human safety. Optical access is provided through a window at the bottom of the dewar, with cryogenically cooled baffles and shutters in the apertures through the lower cryogen tanks. The dewar was fabricated through a partnering arrangement with Janis Research Company, Inc.

The nitrogen cooled bell shaped radiation shield and liquid nitrogen tank were covered with multi layer insulation (MLI) blankets that were constructed of layers of aluminized mylar ($\varepsilon = .03$) with spacer material[1]. The following equation[2]

$$\frac{Q}{A} = \frac{\varepsilon \frac{\sigma}{100}\left(T_2^{4.67} - T_1^{4.67}\right)}{n} + C\,\frac{N^{2.56}}{1+n}\,\frac{\left(T_2^2 - T_1^2\right)}{2} \tag{16}$$

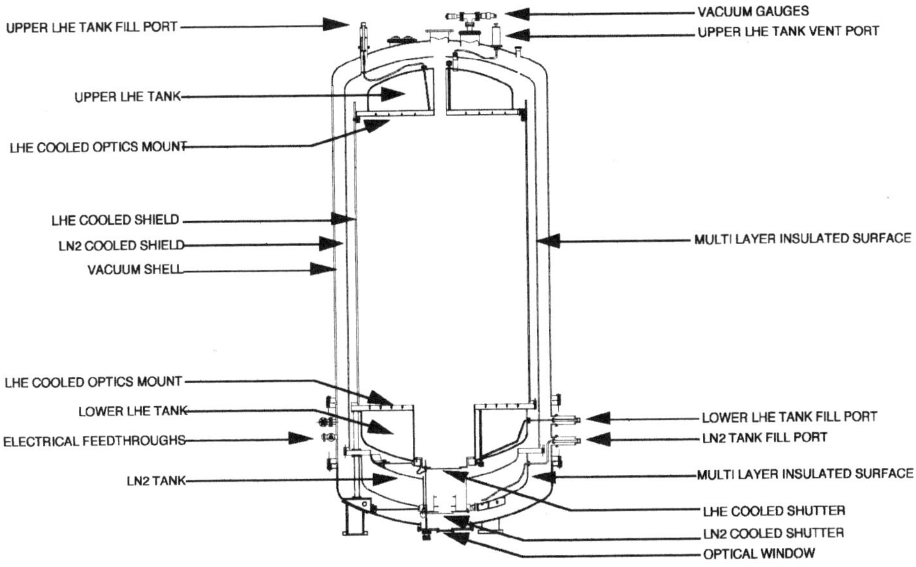

Figure 4 STTF dewar structure.

was used to model the heat transfer through the MLI in W/m^2 where C = 8.85 × 10^{-8}, N = # Layers/cm, n = # Layers, and T is in Kelvin. Ten layers was chosen to meet the design performance requirements while minimizing fabrication cost and schedule. All exterior helium cooled surfaces (tank walls and shroud) were covered with a single layer of aluminized mylar to lower the surface emissivity[3].

COMPARISON TO SYSTEM PERFORMANCE

Thirty-one node "design phase" spreadsheet models, which produced identical results to SINDA models run with the same parameters, were performed during the trade study/design phase of the project to estimate the dewar performance. Thirty-one node "as built" spreadsheet models were later run with certain input parameters adjusted to match the as built characteristics of the dewar (20% less loft in the MLI than planned, less than perfect conduction across bolted joints). Gas conduction was ignored since the dewar

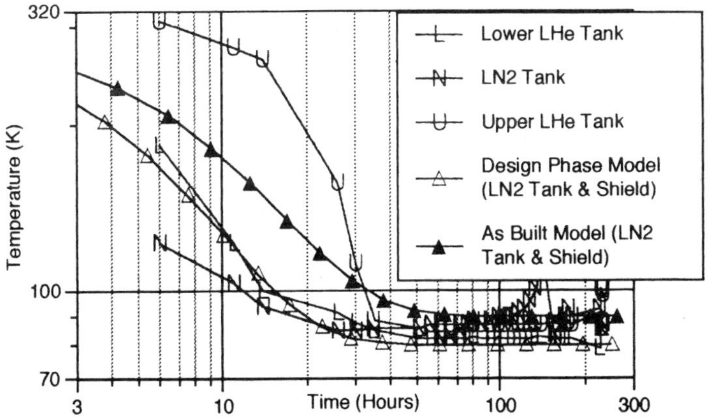

Figure 5 Temperatures within the STTF as a function time during the liquid nitrogen cooldown.

vacuum is always less than 5 × 10^{-5} Torr. Results for both sets of models are presented below.

Liquid nitrogen cooldown

The cooldown of the STTF begins with the introduction of liquid nitrogen into all three cryogen tanks simultaneously. Cooldown to steady state takes about 72 hours and requires approximately 2000 liters of liquid nitrogen when transferring at an average rate of 40 liters/hour into each tank.

The model is consistent with the approach to steady state for the nitrogen tank and radiation shield. It is beyond the scope of this simple model to predict the LN_2 pre-cool of the test chamber (helium tanks & shroud), but mass and construction similarities would indicate any difference in the total time to reach equilibrium should be no more than 20 %.

Liquid nitrogen steady state performance

At steady state, with liquid nitrogen in all three tanks, the parasitic heat load (as measured by cryogen boil off rate) to the upper helium, lower helium and nitrogen tanks were 2.4 W, 2.8 W and 99 W, respectively. This is well within an order of magnitude of the design phase estimate, and well modeled when using the as built parameters.

Liquid helium cooldown

Liquid helium cooldown is accomplished by transferring into both tanks simultaneously. Cooldown to steady state takes about 24 hours and requires approximately 1200 liters of liquid helium when transferring at an average rate of 1 liter/minute into each tank. Although the model correctly predicts the total time required for the test chamber to reach steady state, it does not correctly model the precipitous drop in temperature that occurs when liquid starts accumulating in the tanks.

Liquid helium steady state performance

At steady state, with liquid helium in both the upper and lower helium tanks, and liquid nitrogen in the nitrogen tank, the parasitic heat load (as measured by cryogen boil off

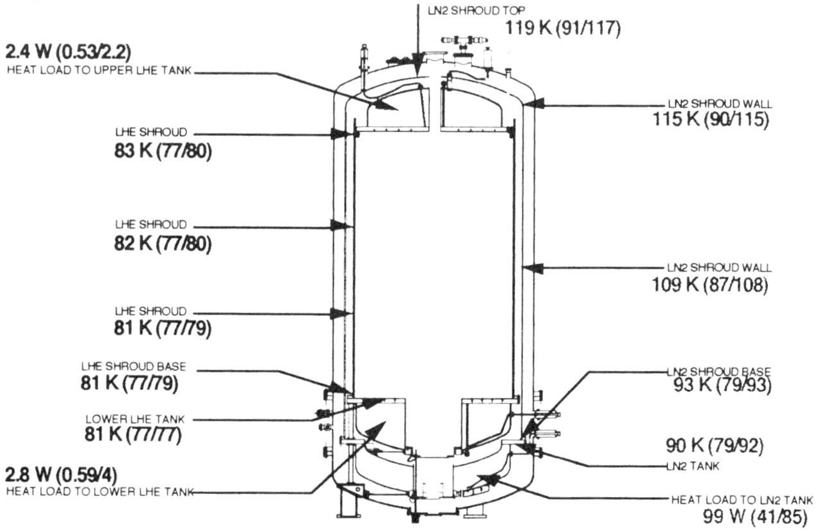

Figure 6. Steady state performance of the STTF with liquid nitrogen in all three tanks. In parenthesis are the design phase/as built model values.

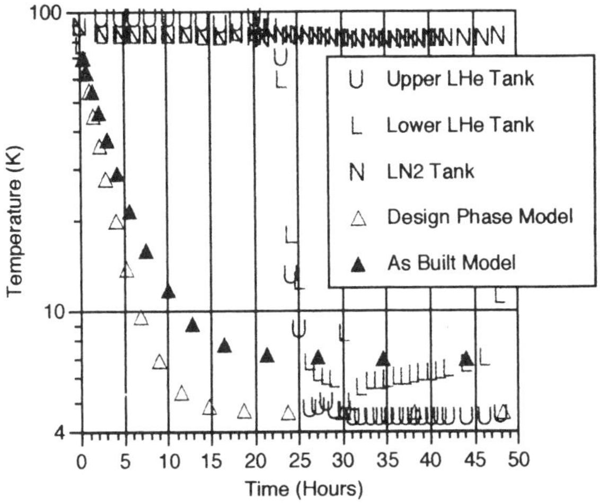

Figure 7 Temperatures within the STTF as a function time during the first liquid helium cooldown. The transfer of liquid helium began at 0 hours.

rate) to the upper helium, lower helium and nitrogen tanks were 2.2 W, 9.7 W and 57 W, respectively. This is well within an order of magnitude of the design phase estimate, and well modeled when using the as built parameters . The as built thermal model has shown that the thermal profile and the difference in the boil off rates in the upper and lower helium tanks are consistent with a factor of 4 difference in the thermal conductivity of the joints between the helium shroud and the upper and lower helium tanks respectively. A minor modification has been made to the upper tank mounting scheme that should provide over 36 hours of cryogen lifetime in both helium tanks. The majority of the heat load in the as built thermal model, which reproduces the measured thermal profile and cryogen heat loads, is through radiative links. This thermal model indicates that over 95% of the heat load to the nitrogen cooled surfaces and over 66% of the heat load to the helium cooled surfaces are radiative in nature.

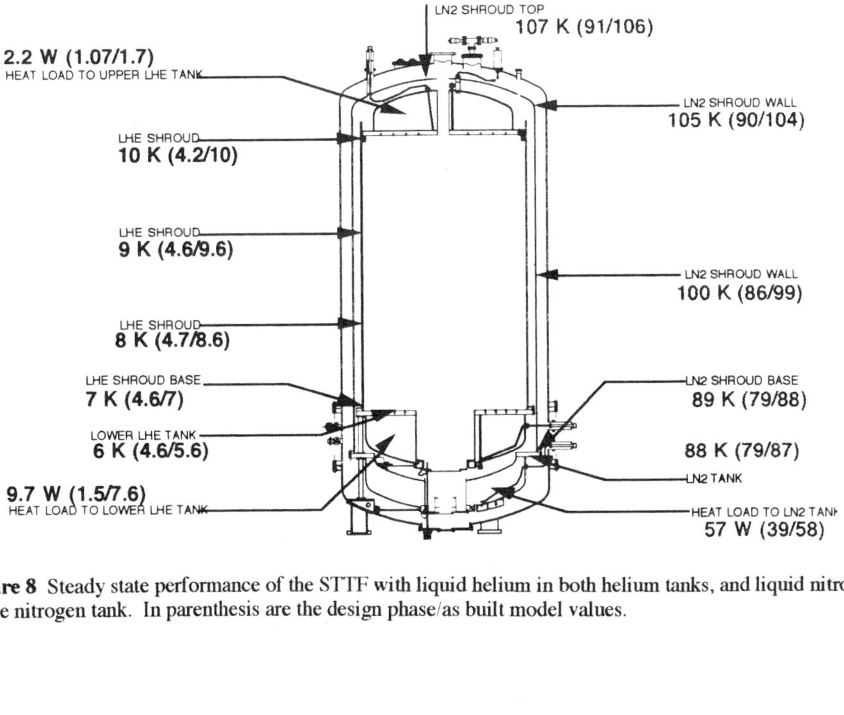

Figure 8 Steady state performance of the STTF with liquid helium in both helium tanks, and liquid nitrogen in the nitrogen tank. In parenthesis are the design phase/as built model values.

SUMMARY

Self - consistent circuit analog thermal models that can be run in commercial spreadsheet programs on personal computers have been developed to predict and analyze the SIRTF Telescope Test Facility cooldown and steady state performance. These models were used to successfully select design parameters to minimize the resources necessary to develop the system while still meeting the technical requirements. Furthermore, these models have been used to analyze the dewar operation and develop modifications to enhance the system performance.

ACKNOWLEDGMENTS

The work described in this paper was carried out at the Jet Propulsion Laboratory, California Institute of Technology, under a contract with the National Aeronautics and Space Administration.

REFERENCES

1 T. C. Nast, "A review of multilayer insulation, theory, calorimeter measurements, and applications", *Recent Advances in Cryogenic Engineering*, Vol. 267, pp. 29 - 43, 1993.

2 I. E. Spradley, T. C. Nast, and D.J. Frank, "Experimental studies of MLI systems at very low boundary temperatures", *Recent Advances in Cryogenic Engineering*, Vol. 35, pp. 477 - 485, 1990.

3 More than a single layer has been shown ineffective at low temperatures. See I. E. Spradley, T. C. Nast, and D.J. Frank, "Experimental studies of MLI systems at very low boundary temperatures", *Recent Advances in Cryogenic Engineering*, Vol. 35, pp. 477 - 485, 1990.

DEVELOPMENT OF MINIATURE HELIUM EVAPORATORS FOR CONTINUOUS FLOW CRYOSTATS

M.Chorowski,[1] M.Wilczek,[1] G.Strychalski,[2] W.Sumarokow[2], S.Trojanowski[3]

[1]Technical University of Wrocław
50-370 Wrocław, Poland

[2]Kriosystem sp. z o.o.
53-529 Wrocław, Poland

[3]Institute for Low Temperature
and Structure Research, Wrocław, Poland

ABSTRACT

Some applications /e.g. microprobing, optical microscopy/ of continuous flow helium cryostats demand an essential miniaturization of the cold finger connected thermally to helium evaporator. The miniaturization of the evaporator is limited by a necessary development of the heat exchange area of the evaporator. To compensate all heat leaks to the cold finger the cooling power of the analysed continuous flow cryostat should be no less than 0.35 W. A minimal heat exchange area of helium evaporator was found. Problem of helium evaporators miniaturization was analyzed and some technologies of evaporator production /machining, wire mesh filling, and powder metallurgy/ were taken into account.

INTRODUCTION

Continuous flow cryostats are a very convenient tool for measurements of such physical phenomena like materials thermal features (heat capacity, heat conductivity, thermal expansion), electric features (conductivity, including superconductivity, permittivity, Hall effect), magnetic features (magnetic permeability, susceptibility, magnetization), laser effect, NMR resonance, EPR resonance, Mössbauer effect, Zeeman effect, photoelectric effect, photomagnetic effect, tunnel microscopy and many others. The continuous flow cryostats are easy in use and they are characterised by the following features:
-only one cryogenic liquid is used,
-the sample can be cooled either in vacuum or in a heat exchange gas - helium,
-cryostats are flexible in shape and size and they can be adapted to use with magnets, optical systems, radiation sources, microprobe systems, detectors, microscopes etc.,

-the cryostat cool-down time is short - usually less then an hour, same for warm up period,
-an access to a measured sample is fast and easy, the sample can quickly be replaced with a new one.

In some applications (like microprobing, microscoping) the miniaturization of the sample area is a must. It means that the outer dimensions of helium vaporizer connected thermally with a cold finger are strongly limited. As it is necessary to provide a sufficient heat exchange area in a limited volume of sample area, helium evaporator should be characterized by a high value of the factor "heat exchange area"/"heat exchanger volume".

ANALYSIS OF HELIUM EVAPORATOR OPERATION

The general way of operation of a continuous flow cryostat is shown in Fig.1. Liquid helium from a separate storage vessel is drawn through a high efficient transfer tube and circulated through the heat exchanger 10 - helium evaporator. During the cool-down period the temperature of the heat exchanger falls down from about 300 K to 4.2 K.

First the supplied LHe is fully vapourized within the transfer line and the heat exchanger is cooled by helium gas or vapours. When the transfer line is cooled down to LHe temperature, liquid helium reaches the heat exchanger.In this moment the temperature difference between the boiling liquid and the heat exchange surface can be as high as tens of K.

Thanks to upper heat exchanger 4 the enthalpy of the cold helium vapours is used to cool

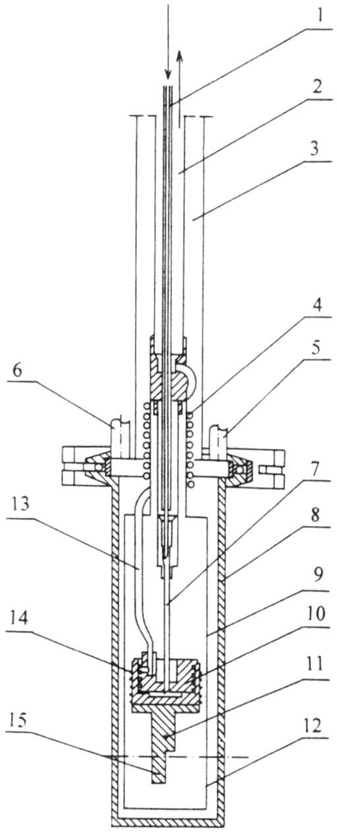

Figure 1. Continuous flow cryostat. 1- LHe inlet, 2- gaseous helium outlet , 3-vacuum insulation, 4- heat exchanger, 5- vacuum valve, 6- electrical feedthrough, 7- capillary inlet to helium evaporator, 8- cryostat outer body, 9- thermal shield, 10- helium evaporator, 11-temperature sensor, 12- optical access window, 13- outlet tube from helium evaporator, 14- heater, 15- cold finger, sample area.

Figure 2. Continuous flow cryostat, helium evaporator and cold finger visible /manufactured by Kriosystem/.

down the thermal shield 9 and compensate parasitic heat fluxes conducted from the room temperature parts of the cryostat.Temperature control is obtained using temperature sensor 11 and heater 14.

A typical continuous flow cryostat is illustrated in Fig 2.

Working parameters of any continuous flow cryostat depend mostly on the design and production technology of helium evaporator heat exchanger - Fig. 1. , pos 10. It should be effective both during the cool down period and during the steady state operation of the cryostat. The heat exchanger should provide a sufficient cooling power during a cool-down period and steady state operation as well as a temperature stability and stable temperature control conditions. In result a heat exchanger is a complicated system of a properly chosen heat exchange surface, mass and hydraulic resistance. The cooling power of the heat exchanger must be high enough to compensate all the parasitic heat fluxes from the environment to the sample area together with heat streams generated during the measurement procedure (light radiation, laser radiation, temperature sensor heat dissipation etc.)

For an average continuous flow cryostat a cooling power during a cool down period should be no less than 0.35 W. Fig. 3. showes the dependence of parasitic heat fluxes to the helium evaporator versus its temperature. The data were calculated for the following conditions:

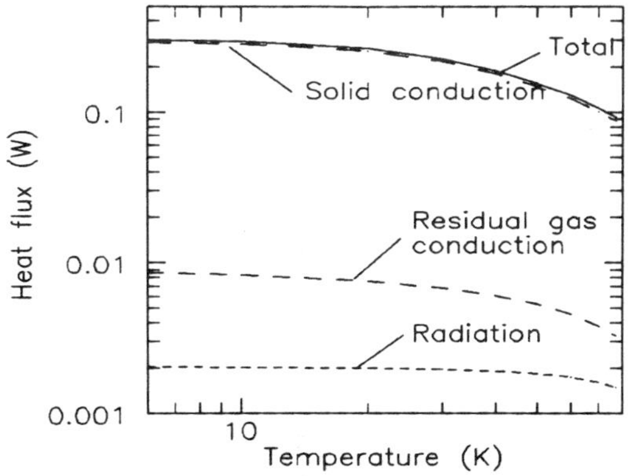

Figure 3. Heat fluxes to helium evaporator versus temperature.

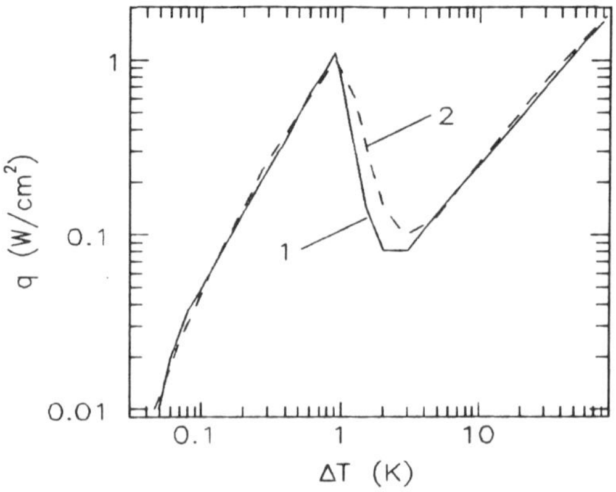

Figure 4. Heat flux versus temperature difference between liquid helium and evaporator surface, curve 1 - Ref.[1], curve 2 - Ref.[2].

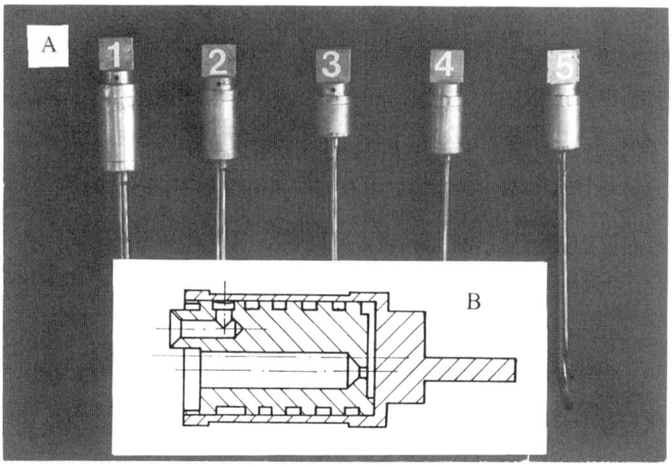

Figure 5. Miniature helium evaporators, a - general view, b - construction schema.

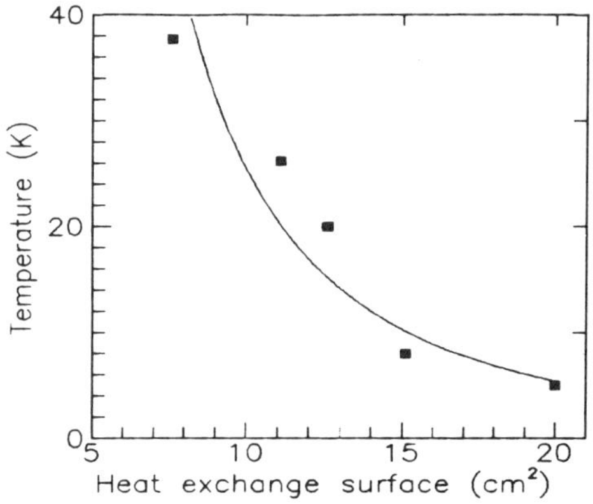

Figure 6. Minimal operation temperature of the helium evaporator versus heat exchange surface.

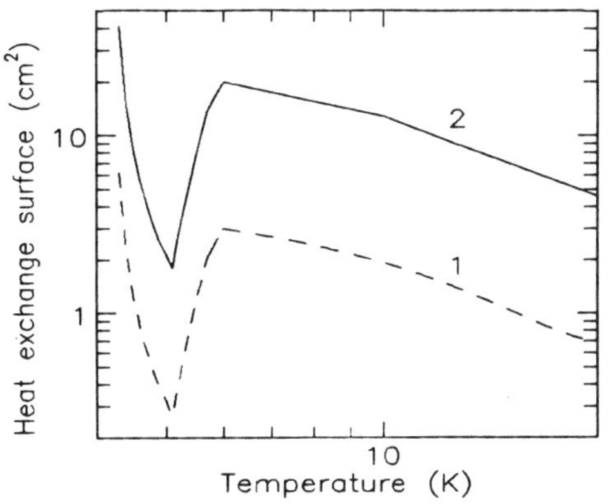

Figure 7. Helium evaporator heat exchange area versus the minimal surface temperature.

-ambient temperature 300 K,
-cryostat length 400 mm,
- vacuum level /Fig.1.pos.3/ 10^{-2} Pa,
-thermal shield /Fig.1.pos.9/ temperature 120 K,
-capillary inlet /Fig.1.pos.7/ made of stainless steel, outer diameter 2 mm, wall thickness 0.3 mm
-outlet tube /Fig.1, pos. 13/, outer diameter 3 mm, wall thickness 0.3 mm
-10 electrical connectors /wires/, copper.

The obtained value of the heat fluxes to the evaporator were increased by 100 % to provide a positive net cooling power during dynamic stages and control processess.
To achieve a cooling power equal to 0.35 W /Fig.3/ in 4.2 K temperature it is necessary to vapourize 0.49 L of liquid helium per hour. It is assumed that similar to film boiling type of heat transfer is observed in the canals of the heat exchanger. Fig. 4 presents the relation between the specific heat flux and temperature difference between the heat exchanger surface and boiling liquid helium.
A number of helium evaporators /Fig.5/ manufactured with machining technology and characterized by different helium flow canal areas were investigated and the minimum operation temperature for each of them was found. The results are presented in Fig. 6.
A temperature 4.2 K was reached only in case when the area of helium flow canals exceeded 20 cm^2 . For lower heat exchange areas the temperature of the heat exchanger remained high, independently of helium flow rate.
Both theoretical and experimental relation between the heat exchange area of the evaporator and operation temperature of the cryostat /sample temperature/ is presented in Fig.7. The figure is a combination of data from Figures 3, 4 and 6. Curve 1 is a theoretical one calculated on basis of literature data concerning film boiling[1,2], while curve 2 takes a real effectiveness of heat exchange process into account. Although both curves are similar qualitatively, quantitative differences are as high as 600 %. It means that a real heat transfer within the helium evaporator is much less effective then film boiling.
The curves shown in Fig.7. are characterized by two singular points corresponding to temperatures 6 K and 5 K. For temperatures above 6 K a film boiling is observed and the lower the temperature the higher heat exchange surface is necessary to vapourize fully liquid helium flowing through the heat exchanger. The biggest value of the heat exchange area is necessary for a temperature of the heat exchanger equal to 6 K. If the area is lower then the value corresponding 6 K, a fraction of liquid helium passes the evaporator and parasitic heat fluxes are not fully compensated. In result the temperature of the evaporator body remains higher than 6 K. For temperatures below 5 K a bubble boiling is observed.

Table 1. Comparison of helium evaporators

Technology	Factor F [cm2/cm3]	Equivalent diam. [mm]	Re at 6 K [/]	Length [mm]
Machining	24	1.9	12970	28
Mesh filling	101	5.9	4720	7
Powder met.	66	5.6	5010	9

LIMITATIONS OF HELIUM EVAPORATOR MINIATURIZATION

In some application an important problem is a miniaturization of the helium evaporator. A useful factor describing the miniaturization limits of the heat exchanger is the ratio of heat exchange /helium canal/ area to its volume, F[1/m]. The ratio was calculated for the following technologies of the evaporator fabrication /all heat exchangers were cylindrical shape, outer diameter 11 mm/:

-machining - a spiral helium canal of a cross section area 2,8 mm^2, F=24 cm^2/cm^3

-wire mesh filling- cylinder filled with a mesh, wire diameter 0,075 mm, F=101 cm^2/cm^3

-powder metallurgy - cylinder filled with a sintered powder, single ball diameter 0.5 mm, F=66 cm^2/cm^3

A comparison of the evaporators manufactured with the above technologies is given in Table 1. The evaporators were of cylindrical shape, outer diameter 11 mm, wall thickness 1.5 mm, heat transfer area 30 cm^2. Reynolds number was calculated for 6 K temperature.

Dynamics and Control

A high accuracy control of a cold finger temperature is usually demanded. The dynamic features of a cold finger depend on its the material features (ρ – density, c - heat capacity, k - heat conductivity), x - distance between a heater and temperature sensor. As material features depend strongly on temperature, dynamic characteristics change rapidly in low temperatures (below 20 K). Thanks to the heat exchanger miniaturization it is possible to reduce a distance between a heater and a temperature sensor. Let ω_o be a normalized pulsation

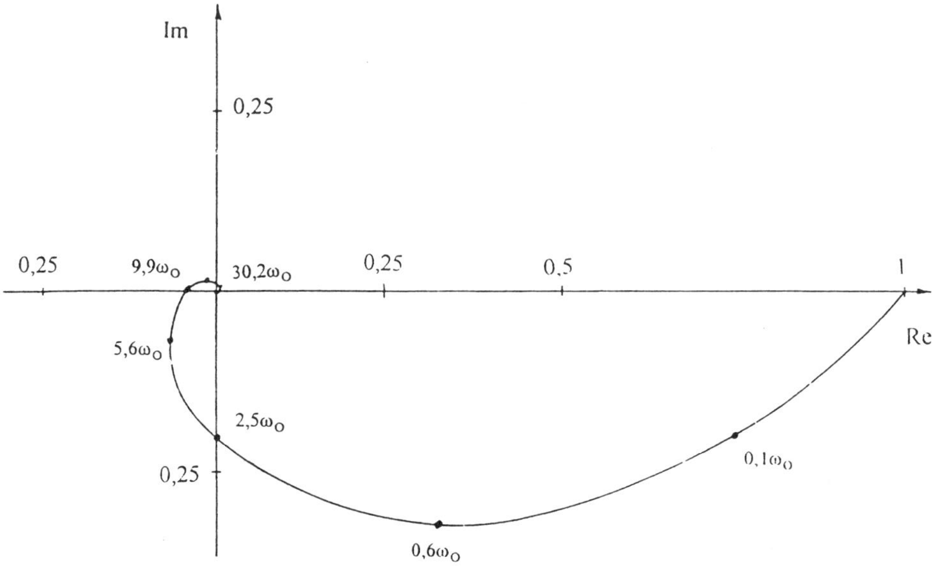

Figure 8. Cold finger amplitude-phase characteristic as a function of a normalized pulsation ω_0.

$$\omega_0 = \frac{2k}{c\rho x^2} \tag{1}$$

Figure 8 showes a cold finger amplitude -phase characteristic as a function of a normalized pulsation calculated from Eq.(1).[3]

CONCLUSIONS

A heat flow duty in a cotinuous flow cryostat helium evaporator is the lowest at temperature 6 K. It is assumed that a heat transfer of film boiling type occurs in temperatures above 6 K. It has been stated experimentally that the minimum heat transfer area for the helium evaporator being a part of a continuous flow cryostat is about 20 cm^2 . For lower value of the heat exchange area a temperature of a cold finger higher than 4.2 is measured - Fig. 7. Helium evaporators can be significantly miniaturized - in case of cylindrical shape of 11 mm outer diameter they can be as short as about 10 mm. The miniaturization makes possible to shorten the distance between a heater and temperature sensor , what improves the dynamic features of the cryostat.

REFERENCES

1. S. W.Van Sciver. "Helium Cryogenics," Plenum Press, New York (1986).
2. S. Tomasz. "Hel Wlasnosci Produkcja Zastosowanie,", Wydawnictwa Naukowo-Tcchnicznc, Warszawa (1978).
3. M.Chorowski, S.Trojanowski, W.Sumarokow, Temperature control in research cryostat below 20 K, to be presented at XIX International Congress of Refrigeration, The Hague 1995.

1995 CEC/ICMC Conference: The Program Chairs of the 1995 CEC/ICMC Conference: (from left to right) K. Theodore (Ted) Hartwig, ICMC, and J. Patrick Kelly, CEC .

DIRECTIONS IN CRYOGENIC TECHNOLOGY - 1995

J. Patrick Kelley

Babcock and Wilcox
Product Development Department
Lynchburg, VA 24505, U.S.A.

ABSTRACT

The 29th Cryogenic Engineering Conference (CEC) was held as a joint conference with the International Cryogenic Materials Conference (ICMC) in Columbus, Ohio, during July of 1995. The CEC is the largest regularly held conference in the United States dedicated to all topics in cryogenic engineering. This paper briefly examines directions in cryogenic engineering by considering the CEC's technical content as represented by submitted abstracts for the '95 and past conferences. Specific comments are made about the direction of large-scale refrigeration. It also remarks on the general health of the field in the U.S. by considering the origin of the abstracts. Finally, the demographics of the overall attendance, based on nationality, are presented.

INTRODUCTION

The CEC held the most successful conference in its 40 year history in 1993 in terms of technical content, with 325 abstracts on a broad variety of topics. Shortly after this conference, the Superconducting Supercollider (SSC) was terminated which had far reaching negative impact on the U.S. cryogenic community. Indeed, SSC related activities accounted for 10% of the '93 CEC program[1]. In November of '94, there was a change in the party controlling Congress and the federal budget. For deficit reduction, government agency budgets were slashed. Programs critical to U.S. cryogenic technology were eliminated in order to meet new spending profiles. For example, the Tokamak Physics Experiment (TPX), a program with significant activity in superconducting magnet, cryopump, and cryostat technology, was canceled due to reduced DOE fusion research funding.

Given this history, it seems reasonable to ask whether events (the '95 conference) lent credit to a perception of reduced U.S. cryogenic activity. To evaluate this, the trends in the origin of CEC abstracts, and of the general attendance are considered. Directions in cryogenic technology are also discussed using observations derived principally from the CEC abstracts as a source.

TECHNICAL CONTENT

Technical content is established from the conference abstract ('93, '95) and program ('91) booklets. While the papers presented number less than abstracts received, the abstracts received should still present a reasonable representation of the work being conducted in the field. Papers not presented tend to represent either a lack of financial resources for travel, or work that wasn't completed in time for the conference, not a decrease in topical interest. Therefore, reasonable conclusions can be drawn about trends in the cryogenic field from an analysis of the abstract booklet.

Table 1 presents the subjects addressed at the '95, '93, and '91 CECs, and the number of abstracts received for each subject. For the '95 CEC data, abstracts from the '95 ICMC organized special joint sessions are included since these session topics traditionally fall under the CEC scope.

Plenary Topics

The first 4 topics in Table 1 correspond to sessions organized around the '95 plenary topics. Selected to compliment the theme, "New Directions in Cryogenic Engineering and Materials Development," these abstracts were heavily solicited. *Power Generation & Transmission*, built around the keynote plenary *Superconducting Electric Power Applications*, saw a large increase due to the DOE Superconductivity Partnership Initiative Program. This industry-led, cost-sharing program is intended for the development of the technology base needed for industry to proceed with commercial development of high temperature superconductor (HTS) electric power equipment[2]. Industry is making progress in material fabrication, and some of the HTS components are nearing reality; however, successful application will depend on the reliability and availability of the components and the support infrastructure, and the intangible needs of a particular end user[3].

Cryoelectronics, electronics operating in the 50-150 K temperature region, has enjoyed modest growth over the period considered. Concerned principally with superconductor, semiconductor and magneto-electronic materials, economics and enhanced performance needs will ultimately lead to over-coming the barriers that each currently encounter, and drive their operating temperatures into this region. This will permit creative integration of these materials. However, cryoelectronics will not be generally accepted unless the support systems become invisible to the user. Hence, cryocooler reliability, availability, and efficiency must improve, and their cost and size must be reduced[4].

Cryobiology has, in the history of the CEC prior to '95, only provided 18 papers to the conference. Cryobiology generally draws from the medical and biological communities, whereas many of the problems faced are inherently engineering in nature. By introducing this topic to the CEC, it was hoped that synergy would develop between communities leading to future opportunities.

Despite a dedicated plenary session, *Cryofuels* did not enjoy an increase in interest. This can in part be attributed to the magnitude of the current investment in the production and distribution network for diesel and gasoline. This results in a great deal of inertia and, unless there is another eruption in the Middle East or a firm enforcement of emission reduction laws, there is little to motivate the large oil and engine manufacturing companies to change the way they do business and seriously consider the investment required to develop alternative fuels[5].

Significant Topics

Table 2 summarizes data presented in Table 1, listing the 7 topics that have consistently drawn the most abstracts, and the percentage of the CEC abstracts that these topics have drawn. *Cryocoolers*, which accounted for almost 20% of the '95 abstracts, have also seen the greatest growth for the period considered, with pulse tubes and interfacing cryocoolers accounting for most of the growth. *Cryocoolers* and the next largest category, *Superconducting Magnets*, accounted for 36.5% of the abstracts. Considering the demise of the SSC and the near commissioning of RHIC, the size of the superconducting magnet category is somewhat surprising until one considers the work that has been done with ITER, the recently canceled TPX, mine countermeasures, micro-SMES, and NMR magnets. The latter three areas help account for the growth in cryocoolers and vice versa since workers in these areas, through the use of Nb_3Sn and HTS materials, are relying on conduction cooling with cryocoolers to replace purchased liquid and to reduce system size. Additionally, by using new regenerator materials and multiple stages, cryocoolers are seeing limited use for NbTi magnets operating at 4.5 K, particularly when HTS materials are used in the current lead assembly.

Increased activity in *Current Leads,* reflected by Table 1, is a matter of note. Ignoring *Feedthroughs*, there was little change in this category between '91 and '93, but, due to the HTS

Table 1. Breakdown of the '91, '93, and '95 CEC by Number of Papers per Topic

Session Title	'91	'93	'95	Session Title	'91	'93	'95
Power Generation & Transmission	**5**	**1**	**17**	**Superconducting Magnets**	**50**	**41**	**58**
				Fusion Magnets			10
Cryoelectronics	**5**	**10**	**13**	Large Scale SC Magnets			7
				Conduction Cooled SC Magnets			5
Cryobiology	**0**	**2**	**13**	Issues Impacting Accelerator Magnets	7		11
Eng. Aspects of Cryobiology		1	13	Superconducting Magnets - General	33	40	9
Cryobiology		1		SC Magnets - Power Interface			4
				Hi Tc Superconducting Magnets			7
Cryofuels	**1**	**6**	**5**	Stability	10		
				SC Magnets for Mine Countermeasures	0	1	5
Small Refrigerators	**26**	**41**	**66**				
Pulse Tube Cryocoolers	6	7	23	**Current Leads**	**2**	**7**	**11**
Stirling Cryocoolers	7	11	7	Conventional Vapor Cooled	2	2	2
G-M Cryocoolers	3	8	13	Hi Tc Leads		1	9
J-T Cryocoolers		2	5	Feedthroughs		4	
Interfacing Cryocoolers			13				
Cryocooler Component Heat Exchange	2	5	5	**Aerospace Cryogenics**	**43**	**18**	**16**
Sorption Cryocoolers	3	6		Aerospace Cryogenics	26	18	16
Boreas Cryocoolers		1		Space Fluid Management & Transfer	7		
Misc. Cryocoolers	5	1		Space Cryogenic Systems	8		
				Microgravity Fluid Mechanics	2		
Helium	**6**	**9**	**11**				
Helium Phenomena			8	**Heat Transfer & Thermal Properties**	**14**	**28**	**18**
Helium II	6	9	3	Thermal Properties	7	10	
				Insulation & Thermal Properties			9
Large Refrigerators	**25**	**40**	**28**	Convective Heat Transfer	3	3	
Large Refrigerator Systems	10	18	16	Insulation	1	7	
Process Simulation & Control	2	4	3	Boiling Heat Transfer	3	8	9
Compressors, Pumps and Expanders	13	10	7				
Refrigerator System Design		8	2	**Miscellaneous**	**16**	**67**	**27**
				High Frequency Cryogenic Applications			4
Cryogenic Facilities & Equipment	**39**	**37**	**31**	Applications of Superconductivity		8	
Thermal Design of Cryostats			8	SC Magnetic Bearings	2		3
Cryostat Design	2	21		Cryogenic Wind Tunnels		7	3
Cryogenic Cooling Systems		12	4	High Field Magnet Laboratories		3	
Cryogenic Facilities & Equipment	23		16	Cryogenics in Other Countries		1	
Cryogenic System Performance	9			Safety		1	
Heat Exchangers	5	4	3	Superconducting Supercollider		9	
				TESLA			11
Instrumentation	**15**	**18**	**26**	Magnetic Levitation	1	2	
Flowmeters			4	Magnetic Refrigeration	12	5	6
Thermometry			6	Cryo. Techniques and Systems		31	
Misc. Instrumentation	15	18	16	History of Cryogenics	1	0	0

materials revolution, a 367% increase was experienced between '93 and '95. While the mechanical properties of these materials still prohibit their use as conductors in large magnets, the existence of these materials in tape and bulk form permits integration into hybrid leads. Hybrid leads have the potential to reduce lead losses by 50% or more[6]. This is significant since lead losses can account for 20-50% of the total magnet system's refrigeration load. Near term use of HTS hybrid leads will make NbTi and Nb_3Sn magnet systems more appealing to an end user through reductions in operational and capital costs of ancillary refrigeration equipment.

Table 2 shows an interesting decline in the number of abstracts in *Aerospace Cryogenics* (which also made up 11.6% of the '89 CEC). Additionally, the U.S. content is down significantly, from 94% of the abstracts in '93 to 69% in '95. This can be attributed in part to the fiscal problems experienced by NASA over the past few years, and NASA's focus on the shuttle and space station, instead of on a number of smaller, more diverse programs[7].

Large Refrigerators and *Cryogenic Facilities & Equipment* have experienced a decline. This is influenced by the cancellation of the SSC. SSC facilities have been disassembled for use on other government programs leaving little need to create new facilities from scratch. Considering *Large Refrigerators*, 15 of the 27 abstracts were from the U.S. (56%), down from 21 in '93 when 6 abstracts were associated with the SSC (2 of these abstracts were included in the *SSC* category under *Miscellaneous* and not in the Table 1 *Large Refrigerators* total). Most of the '95 abstracts dealt with upgrades, recommissioning or operation of old equipment, or commissioning of often discussed equipment. Only 3 were significant: one dealing with new equipment, a prototype gas-bearing turboexpander currently being tested at CCI Cryogenics; and 2 dealing with the long awaited commissioning and operation of 4 stages of cold compressors. This is a major hurdle in realizing future large programs requiring He II, such as TESLA and the Large Hadron Collider (LHC).

Non-U.S. contributions in *Large Refrigerators* were down from 21 abstracts in '93 to 12 in '95. CERN workers summarized their commissioning experience with the 6 cryoplants for LEP2, and with the LHC superfluid string test facility. Other abstracts were concerned with analytical studies or plant operations. The five notable Japanese contributions included commissioning descriptions of 2 systems for the Large Helical Device (LHD) and for the ITER CS Model Coil, respectively, with the latter also reporting on new turboexpanders developed for this facility. Another paper presented an interim report on the Super-GM component development program. At the '93 conference, the 8 Japanese contributions were even more notable. A partial list of the '93 presentations includes new warm screw compressors, 80 K cold compressors, turboexpander bearing systems, and oil removal systems. Considering the lack of developmental activities in the U.S., and that one of the major refrigerator manufacturers, CVI, Inc., was acquired by Chart Industries, Inc. in mid-1994, who then made the decision to no longer pursue custom cryogenic work[8], one could well conclude that Japan has surpassed the U.S. with respect to innovations in this field.

A recurring theme is that new systems have been commissioned and development programs have come to fruition. Large refrigerator construction depends heavily upon large scientific construction programs or the need for a gas company to develop new helium resources. The latter is generally not reported in the technical literature; however, with the recent Praxair Ulysses Plant upgrade[9], and the recently commissioned Helios facility in Algeria[10], there appears to be sufficient world wide capacity to meet current helium needs. As to the former, the refrigeration systems for current programs have been built[11,12,13,14,15]. ITER is far in the future and projected LEP2 upgrades for LHC, and the additional LHC cryoplants, are sometime at the beginning of the next century[14]. Current TESLA activities are developmental only. Recalling the Chart Industries, Inc. decision to stop pursuing custom cryogenic work, and considering that Linde AG has restructured and consolidated its Cryogenic Departments[16], one could assume that, with the exception of upgrades, there will be further near term declines in *Large Refrigerator* activities. An exception could be with systems that are developed for superconducting electric power industry applications. These systems will be mid-size (100-1,000 W) and could be accommodated by modifications to existing designs. If the power equipment is successfully applied, a significant market could develop.

DEMOGRAPHICS

Table 3 presents a breakdown of the CEC abstracts and the CEC/ICMC attendance by country of origin. This is given as the percentage of CEC abstracts, and the percentage of the total joint conference attendance, respectively. Data were unavailable for the missing years. If the number of abstracts is an indicator of the health of the cryogenic field, the decline in percentage of U.S. content between '85 and '93 does not imply a decline in U.S. activity. In '85, there were 109 abstracts from U.S. authors, in '93 there were 197. This indicates that activity in the U.S. has increased at a slower rate than elsewhere. Considering '95, there were 159 U.S. abstracts indicating a significant decline in overall cryogenic activity in the U.S. A decline of 6% in U.S. attendance also support this indication. As noted, this is primarily caused by a decline in government spending for programs supporting cryogenic activities, and the surplus of relatively new SSC equipment which puts the commercial side of the field in the position of competing with equipment they recently sold. The data also indicates that cryogenic activities in other countries are still increasing.

Japan, by a factor of 2, is the 2nd largest source of '95 CEC abstracts. Since '85, Japan has also provides the 2nd largest group of attendees for the conference. In '95, 75% of the abstracts on

Table 2. Topics that have Drawn the Largest Number of Abstracts Over the Last Three Conferences

Session Title	% of '91 CEC	% of '93 CEC	% of '95 CEC
Small Refrigerators (Cryocoolers)*	10.5	12.6	19.4
Superconducting Magnets	20.2	12.6	17.1
Cryogenic Facilities & Equip.	15.8	11.4	9.1
Large Refrigerators	10.1	12.3	8.2
Instrumentation	6.1	5.5	7.6
Heat Transfer & Thermal Properties	5.7	8.6	5.3
Aerospace Cryogenics	17.4	5.5	4.7
Total Number of Abstracts	247	325	340**

* Includes Cryocoolers for Aerospace Applications.

** Includes ICMC Organized Joint sessions addressing topics that typically are handled by the CEC.

Table 3. Percentage by Country of CEC Abstracts and of CEC/ICMC Attendance for Given Years

Countries	CEC Abstracts				CEC/ICMC Attendance					
	% '83	% '85	% '93	% '95	% '85	% '87	% '89	% '91	% '93	% '95
United States*	61.0	65.3	60.7	50.48	77.59	70.59	71.39	71.43	71.69	65.57
Japan	10.1	16.8	11.8	13.33	7.81	8.46	9.20	13.32	9.92	12.48
Germany (West)	2.5	2.4	4.3	7.94	3.23	5.70	4.34	3.05	4.10	5.31
Russia (or former USSR)	11.3	0.6	5.0	6.98	0.34	0.55	2.81	1.93	1.19	0.86
Peoples Rep. of China	0.6	1.8	3.7	4.13	0.68	1.65	1.15	1.12	1.32	1.15
Switzerland	1.3	2.4	3.4	4.13	1.36	2.21	1.28	1.77	2.65	2.73
Ukraine	N/A	N/A	0.0	2.86	0.00	0.00	0.00	0.00	0.66	1.72
France	1.9	2.4	1.2	2.54	1.87	2.57	2.04	1.61	1.98	2.15
Australia	0.0	0.0	0.0	1.27	0.00	0.00	0.13	0.00	0.26	0.72
Canada	1.9	3.0	1.9	1.27	1.02	0.74	0.64	0.32	0.79	1.72
Italy	5.0	0.0	0.9	0.95	0.68	1.29	1.40	0.48	0.66	0.86
United Kingdom	3.1	3.6	1.5	0.63	2.72	2.39	2.30	1.77	1.59	0.86
New Zealand	0.0	0.0	0.0	0.63	0.00	0.00	0.00	0.00	0.00	0.14
Netherlands	0.6	1.8	1.2	0.63	1.36	1.47	1.15	0.96	1.19	1.00
Bulgaria	0.0	0.0	0.0	0.63	0.00	0.00	0.00	0.00	0.00	0.00
South Korea	0.0	0.0	0.0	0.32	0.17	0.37	0.89	0.16	0.13	0.29
India	0.6	0.0	1.9	0.32	0.00	0.00	0.13	0.00	0.00	0.14
Belarus	N/A	N/A	0.0	0.32	0.00	0.00	0.00	0.00	0.13	0.29
Romania	0.0	0.0	0.3	0.32	0.00	0.00	0.00	0.00	0.13	0.00
Poland	0.0	0.0	0.3	0.32	0.00	0.00	0.00	0.00	0.13	0.14
Spain	0.0	0.0	0.3	0.0	0.00	0.00	0.00	0.00	0.13	0.00
Kazakhstan	N/A	N/A	0.3	0.0	0.00	0.00	0.00	0.00	0.00	0.00
Brazil	0.0	0.0	0.3	0.0	0.17	0.00	0.26	0.16	0.00	0.00
Taiwan	0.0	0.0	0.3	0.0	0.00	0.00	0.26	0.16	0.00	0.29
Austria	0.0	0.0	0.3	0.0	0.17	0.37	0.13	0.96	0.26	0.57
Israel	0.0	0.0	0.3	0.0	0.51	0.55	0.00	0.32	0.66	0.43
Slovakia					0.00	0.00	0.00	0.00	0.00	0.14
Finland					0.00	0.37	0.26	0.16	0.40	0.14
Chile					0.00	0.00	0.00	0.00	0.00	0.29
Denmark					0.00	0.00	0.00	0.16	0.00	0.00
Belgium					0.00	0.18	0.13	0.16	0.00	0.00
Saudi Arabia					0.00	0.00	0.13	0.00	0.00	0.00
Yugoslavia					0.17	0.18	0.00	0.00	0.00	0.00
Egypt					0.00	0.18	0.00	0.00	0.00	0.00
Sweden					0.17	0.18	0.00	0.00	0.00	0.00
# of CEC Abstracts	159	167	325	315						
# of CEC/ICMC Abstracts	---	355	611	609						
Total # of Attendees					589	544	783	623	756	697

*U.S. attendance may be slightly less than indicated since some participants with non-U.S. residences were using the address of the their company's U.S. subsidiary.

Helium Phenomena were from Japanese contributors, who also contributed notably in the area of pulse tube and other cryocoolers. Technical contributions from Germany and the Peoples Republic of China have increased over the years, while those from the United Kingdom have steadily declined.

CONCLUSIONS

Evaluating CEC abstract origins indicates that cryogenic activities in the U.S. had in past years increased at a slower rate than elsewhere, and has declined significantly in the past 2 years. Growth appears to have been sustained in many other locations. *Cryocoolers*, particularly pulse tubes, have seen the greatest growth in recent years. *Superconducting Magnet* activities have remained relatively constant with emphasis within the field changing due to cryocooler improvements and to applications of HTS. *Large Refrigerators* have seen a decline, both in the U.S. and elsewhere, and is expected to continue this decline unless developments in superconducting electric power applications see significant success. *Aerospace Cryogenics* has experienced a serious decline which will continue unless government spending profiles change.

Potential growth areas include *Power Generation & Transmission* equipment, which could see near term use by the power industry if drivers other than improved efficiency can be targeted and exploited. *Cryoelectronics* also shows potential for enhanced performance systems at reduced size, weight and power. Large and small refrigerators with high reliable, availability and efficiency are almost a prerequisite for implementation in both of these areas. *Cryobiology* offers opportunities both in the development and manufacture of equipment for the freezing and preservation of biological tissues, and in materials research for an understanding of tissue characteristics during freezing and thawing. *Cryofuels* have a great deal of potential but need significant development work in the production, liquefaction, and distribution areas before economics dictate widespread transition from traditional fuel sources. While none of these areas appear to offer a quick fix for the ailing U.S. cryogenic industry, they do offer the potential for near term research and development activities, both in the U.S. and internationally, with long term rewards for mankind.

REFERENCES

1. Statistics provided were developed from the '85-'95 Conference Attendance Sheets; from the '93 and '95 Abstract Books; from the '85-'91 Special CEC/ICMC Issues of *Cryogenics*; and from a CEC Board Memo from the 1985 Program Chairman, which included the data given in Table 3 for the '83- '85 conferences.
2. Superconductivity Partnership Initiative Solicitation for Cooperative Agreement Applications, #DE-PS02-10554, U.S. DOE, Golden Field Office, Golden CO, January 19, 1993.
3. R. Blaugher, Superconducting Electric Power Applications, 1995 CEC/ICMC, Columbus, OH, July 1995.
4. D.U. Gubser, Cryo-electronics: the Promise and the Challenge, 1995 CEC/ICMC, Columbus, OH, July 1995.
5. J.A. Barclay, Cryofuels: Now and in the Future, 1995 CEC/ICMC, Columbus, OH, July 1995.
6. J.R. Hull, High-Temperature Superconducting Leads, IEEE Trans. App. of Superconductivity, Vol. 3, No. 1, Mar. 1993, p. 869.
7. A.R. Urbach: Ball Electro-Optic/Cryogenics, Private Conversations.
8. Chart Industries, Inc. 1994 Annual Report.
9. Chemical Economics Handbook, Industrial Gases, 743.4000 J, SRI International, May 1995.
10. Chemical Economics Handbook, Industrial Gases, 743.4000 C, SRI International, May 1995.
11. S. Satoh et al., Construction and Commissioning Test of 10 kW Class Helium Refrigerator for the Large Helical Device, 1995 CEC/ICMC, Columbus, OH, July 1995.
12. F. Spath et al., Performance Tests of a 2 kW He Refrigerator for SC Magnet Tests Down to 3.3 K, in: Adv. Cryo. Eng., Vol. 39A, P. Kittel, ed., Plenum Press, NY (1994) p. 563.
13. T. Kato et al., Cryogenic System for ITER CS Model Coil, 1995 CEC/ICMC, Columbus, OH, July 1995.
14. M. Barranco-Luque, et al., Conclusions From Procuring, Installing and Commissioning Six Large-Scale Refrigerators at CERN, 1995 CEC/ICMC, Columbus, OH, July 1995.
15. G. Grygiel et al., Status of the TTF Cryogenic System, 1995 CEC/ICMC, Columbus, OH, July 1995.
16. J.J. Clausen: Lotepro, Inc., Valhalla, NY, USA, Private Conversations.

AUTHOR INDEX, Vol. 41

SUBJECT INDEX, Vol. 41